THE ELEMENTS

	1B	2B	3A	4A	5A	6A	7A	8A
							1 H 1.0079	2 He 4.00260
			5 B 10.81	6 C 12.011	7 N 14.0067	8 O 15.9994	9 F 18.99840	10 Ne 20.179
			13 Al 26.98154	14 Si 28.086	15 P 30.97376	16 S 32.06	17 Cl 35.453	18 Ar 39.948
28 Ni 58.70	29 Cu 63.546	30 Zn 65.38	31 Ga 69.72	32 Ge 72.59	33 As 74.9216	34 Se 78.96	35 Br 79.904	36 Kr 83.80
46 Pd 106.4	47 Ag 107.868	48 Cd 112.40	49 In 114.82	50 Sn 118.69	51 Sb 121.75	52 Te 127.60	53 I 126.9045	54 Xe 131.30
78 Pt 195.09	79 Au 196.9665	80 Hg 200.59	81 Tl 204.37	82 Pb 207.2	83 Bi 208.9804	84 Po (210)	85 At (210)	86 Rn (222)

63 Eu 151.96	64 Gd 157.25	65 Tb 158.9254	66 Dy 162.50	67 Ho 164.9304	68 Er 167.26	69 Tm 168.9342	70 Yb 173.04	71 Lu 174.97	2 8 18 32 9 2
95 Am (243)	96 Cm (247)	97 Bk (247)	98 Cf (251)	99 Es (254)	100 Fm (257)	101 Md (258)	102 No (255)	103 Lr (256)	2 8 18 32 32 9 2

ALTERNATE SI VERSION
OF
MASTERTON & SLOWINSKI

"The SI units and symbols used in the text CHEMICAL PRINCIPLES have been reviewed by the Canadian Government Specifications Board and found to be in accordance with the two National Standards of Canada; the International System of Units, CAN3-Z234.2-73, and the Metric Practice Guide, CAN3-Z234.1-76."

T. MELSKI
Standards Officer

SAUNDERS COMPLETE PACKAGE
FOR TEACHING GENERAL CHEMISTRY

Masterton & Slowinski: **Chemical Principles,** 4th edition

Rochow: **Modern Descriptive Chemistry**

Masterton & Slowinski: **Chemical Principles,** 4th edition **SI Version**

Boyington & Masterton: **Student's Guide to Masterton & Slowinski's Chemical Principles,** 3rd edition

Slowinski, Masterton & Wolsey: **Chemical Principles in the Laboratory,** 2nd edition

Shakhashiri, Schreiner & Meyer: **General Chemistry Audio-Tape Lessons**

Shakhashiri, Schreiner & Meyer: **Workbook for General Chemistry Audio-Tape Lessons**

Clouser: **Keller Plan for Self-Paced Study Using Masterton & Slowinski's Chemical Principles,** 2nd edition

Masterton & Slowinski: **Overhead Projectuals for Chemical Principles,** 4th edition

Peters: **Problem Solving for Chemistry,** 2nd edition

Masterton & Slowinski: **Elementary Mathematical Preparation for General Chemistry**

Masterton & Slowinski: **Mathematical Preparation for General Chemistry**

Slowinski & Masterton: **Qualitative Analysis and the Properties of Ions in Aqueous Solution**

Slowinski, Wolsey & Masterton: **Chemical Principles in the Laboratory with Qualitative Analysis**

FOURTH EDITION

CHEMICAL PRINCIPLES

USING THE INTERNATIONAL SYSTEM OF UNITS

WILLIAM L. MASTERTON

Professor of Chemistry
University of Connecticut, Storrs, Connecticut

EMIL J. SLOWINSKI

Professor of Chemistry
Macalester College, St. Paul, Minnesota

ILLUSTRATED BY GEORGE KELVIN

SAUNDERS GOLDEN SUNBURST SERIES

W. B. SAUNDERS COMPANY / Philadelphia / London / Toronto

W. B. Saunders Company: West Washington Square
Philadelphia, PA 19105

1 St. Anne's Road
Eastbourne, East Sussex BN21 3UN, England

1 Goldthorne Avenue
Toronto, Ontario M8Z 5T9, Canada

Library of Congress Cataloging in Publication Data

Masterton, William L

Chemical principles.

(Saunders golden sunburst series)
Includes index.

1. Chemistry. I. Slowinski, Emil J., joint author.
II. Title.

QD31.2.M383 1977 540 76-58603

ISBN 0-7216-6163-7

The front cover illustrations, *Variation within a Sphere, No. 10,* a metal sculpture by Richard Lippold, courtesy of The Metropolitan Museum of Art.

Chemical Principles—Using the International System of Units ISBN 0-7216-6163-7

Last digit is the print number: 9 8 7 6 5 4

to the students

PREFACE

While working for the past two years on the fourth edition of *Chemical Principles,* we have frequently been asked, "Why revise a successful text?" There are many answers to this question, some more honest than others. Our objective has been to improve what has always been a student-oriented book, hopefully making it clearer and more interesting than its predecessors. We have been guided primarily by reactions from large numbers of college freshmen who have used the third edition. Included among these are the wife and son of one of the authors; their candid (occasionally caustic) suggestions about arguments that needed clarification have been particularly helpful.

We have also been influenced by the changing composition of the student body in general chemistry. Very few of our students are chemistry majors—most of them are preparing for careers in engineering, the biological sciences, medicine, and allied professions. They are not interested in abstract theory; instead, they want to know how the principles of chemistry can be applied to their field of interest and, more generally, to the world around them. With this in mind, we have consistently blended fact with theory, emphasizing the application of principles rather than their derivation. The environmental flavor of the third edition has been revised to include discussions of such topics of current interest as the energy crisis (Chapter 4) and the broader problem of the depletion of natural resources (Chapter 7).

The basic framework and sequence of topics common to previous editions have been retained. As before, introductory chapters (1–5) emphasize the quantitative, experimental aspects of chemistry. Descriptive inorganic chemistry is again organized around types of reactions (precipitation reactions in Chapter 18, acid-base reactions in Chapters 19 and 20, complex-ion formation in Chapter 21, and redox reactions in Chapters 22 and 23). Three chapters are essentially new: Chapter 7 (Periodic Table), Chapter 10 (Introduction to Organic Chemistry), and Chapter 25 (Natural and Synthetic Polymers). One effect of these additions is to increase somewhat the amount of descriptive chemistry relative to theory.

Those chapters which deal mainly with chemical principles have been reviewed for clarity and simplicity of argument. The discussion of thermochemistry in Chapter 4, which caused difficulties in earlier editions, has been completely rewritten (several times!). Less extensive but significant changes have been made in the areas of electronic structure (Chapter 6), chemical bonding (Chapter 8), gaseous equilibrium (Chapter 15), and chemical kinetics (Chapter 16). Virtually all of the problems are new. The "historical perspectives," a popular feature of the third edition, have been expanded to include brief sketches of G. N. Lewis and Michael Faraday.

Several features are new in the fourth edition. The most immediately apparent is the series of color plates which appears in the center of the book. For these we are indebted to Ray Boyington and Ruven Smith, a pair of amateur but (by their own admission) talented photographers. Included for the first time in this edition, at the back of the book, is a glossary of chemical terms used throughout the text. At the end of each chapter, we have added a few particularly straightforward problems, each illustrating a single principle. We suggest that students, immediately after reading a chapter, work these problems to test their understanding of the material covered. Later, they can proceed to the other problems in the set, which are of the "matched pair" type used in the third edition.

Every author of a chemistry text nowadays has to decide how far to go in the use of the International System of Units. This system has now been adopted in most countries outside the United States. In particular, it is in use throughout secondary schools in Canada. For this reason, it seemed desirable to prepare a separate version of the fourth edition of *Chemical Principles* in which SI units are used consistently throughout. In this *SI Version* we express:

—energy in joules or *kilojoules*
—distance in metres or nanometres
—volume in cubic metres, *cubic decimetres,* or cubic centimetres
—pressure in pascals, *kilopascals,* or, in some cases, atmospheres

A regular version is also available, in which we emphasize units more familiar to students (and faculty) in the United States (e.g., calories, Angstroms, liters, millimeters of mercury).

A variety of supplementary materials are available for use with this text. Those which are new with this edition include:

Student's Guide to Chemical Principles, by Raymond Boyington and W. L. Masterton, which includes chapter summaries, self-tests, and basic skills sections.

Keller Plan for Self-Paced Study with Chemical Principles, by Joseph L. Clouser.

Workbook for General Chemistry with Audio-Tape Lessons, by B. K. Shakhashiri.

Modern Descriptive Chemistry, by Eugene Rochow, a short (250-page) paperback for those instructors who desire more material on the properties and reactions of the elements.

It is a pleasure to acknowledge the contributions to this edition by our colleagues at Macalester College and the University of Connecticut. Special recognition is due Chic Waring, whose many comments about content and organization are always appreciated and frequently followed. We are grateful to Jon Bellama of the University of Maryland, Clark Bricker of the University of Kansas, Bill Fisher of Clayton Junior College, Curt Sears of Georgia State University, Conrad Stanitski of Randolph Macon College, and Ted Williams of the College of Wooster, all of whom provided us with detailed, down-to-earth critiques of the third edition, and to Andy Ternay, University of Texas, Arlington; Ed Mellon, Florida State University; John Chandler, University of Massachusetts, Amherst; Grover Willis, California State University, Chico; Peter Berlow, Dawson College, Canada; and Carl von Frankenberg, University of Delaware. Joe Wiebush of the Naval Academy, who went over the manuscript for this edition, is our all-time favorite among reviewers: brief and flattering. For their advice on the construction of the SI Version, we are indebted to Professor E. B. Robertson, of The University of Calgary, and to Mr. T. Melski of the Canadian Government Specifications Board. Finally, we should acknowledge the many contributions of the staff of the W. B. Saunders Company and, in particular, of our editor, John Vondeling. We are indebted to John for his unlimited energy, limited patience, and modest skill as a fly fisherman.

WILLIAM L. MASTERTON

EMIL J. SLOWINSKI

CONTENTS

1

CHEMISTRY: AN EXPERIMENTAL SCIENCE

Throughout all of history, the very existence of human life has been threatened by natural disasters, among them famine and disease. The degree of success achieved in our continuing struggle against such calamities is due in no small part to contributions from chemistry. During World War I, a German chemist, Fritz Haber, developed a practical process for the conversion of atmospheric nitrogen into ammonia, a principal component of the synthetic fertilizers now used to provide food for the world's expanding population. A generation later, another chemical, DDT, was used successfully to control malaria and other infectious diseases in war-ravaged areas of Asia and Europe. Since World War II, DDT and other chemical insecticides have been applied to increase agricultural production throughout the world.

In recent years we have become aware of some of the undesirable side-effects of chemicals developed to meet the needs of society. Chemical fertilizers promote the growth of algae that clog many of our lakes and streams. Pesticides such as DDT can have adverse effects on wildlife and, in some cases, on human life as well. In a more general sense, we have come to realize that the quality of our environment is threatened by materials produced in an attempt to achieve "a better life." Now we find that efforts to clear up our air and water are restricted by another, potentially catastrophic problem: the depletion of our natural resources. The "energy crisis" of the 1970's is only one indication of the fact that we are running out of the cheap raw materials upon which our economy is based.

Chemists, along with other scientists, are deeply involved in efforts to find solutions to the problems caused by pollution and dwindling resources. The sophisticated instruments used to measure pollutants in automobile exhaust at the part per million level were developed and applied by analytical chemists. Inorganic chemists were involved in the research that produced the "catalytic converters" now used to reduce harmful emissions from automobiles. Organic chemists, along with biochemists and biologists, have synthesized a variety of products that open up new approaches to the control of harmful insects. The search for new energy sources is being carried out by physical chemists working with physicists and engineers.

The approaches used by chemists and other scientists in solving problems are varied. Many significant discoveries have come about partly by chance. A biologist studying media for growing bacteria may accidentally contaminate his culture and thereby find a new antibiotic. A chemist studying the mechanism of a particular reaction may get a clue about a good catalyst for a very different reaction. Such discoveries cannot be attributed to luck alone; they require a mental attitude that is conducive to new ideas and an experimental environment in which they can be tested rigorously.

Regardless of how a new idea is generated, the method of testing it is one that has been used with considerable success in all the sciences for two hundred years. The so-called "scientific method" starts with carefully designed experiments carried out in the laboratory under closely controlled conditions. Ordinarily, the system that a chemist works with is a relatively simple one, consisting perhaps of a single pure substance or a solution containing two or three such substances. Measurements on these systems, when properly interpreted, can lead to conclusions applicable to the more complex world that exists outside the laboratory.

In this beginning course in chemistry, we will discuss a great many different experiments. Many of these you will carry out in the laboratory; others you will have to visualize as they are described in the text or by your instructor. We will be interested in the principles or "laws" that can be developed from these experiments and in the application of these chemical principles to practical problems. In this chapter we will consider some very simple experiments that chemists carry out to identify pure substances and separate them from one another. Before doing so, it will be helpful to review the types of measurements that are fundamental to all experiments.

Chemistry deals with the properties and reactions of the materials which make up the earth and the universe

1.1 MEASUREMENTS

Most of the experiments carried out in chemistry laboratories are quantitative in nature. That is, they involve assigning numbers to such quantities as length, volume, mass, and temperature. We will now consider the instruments used to measure these quantities and the various units in which they may be expressed.

Measuring Devices

LENGTH. Most of us are familiar with a simple measuring device found in the general chemistry laboratory, the metre stick, which reproduces, as accurately as possible, the fundamental unit of length in the metric system, the **metre.** When we examine a metre stick, we see that it is divided into one hundred equal parts, each one *centimetre* in length (1 cm = 10^{-2} m). A centimetre is, in turn, subdivided into ten equal parts, each one *millimetre* long (1 mm = 10^{-3} m). A much larger unit, familiar to track and field runners, is the *kilometre* (1 km = 10^{3} m). The prefixes "kilo," "centi," and "milli" are used in the metric system to designate units obtained by multiplying by 1000, 0.01, and 0.001, respectively. (For a more complete list of prefixes, see Table 1.4.) Two other units frequently used in chemistry to express the dimensions of tiny particles such as atoms and molecules are the angstrom (1 Å = 10^{-8} cm) and the nanometre (1 nm = 10^{-9} m = 10^{-7} cm).

VOLUME. Units of volume in the metric system are simply related to those of length. A **cubic metre** (m^3) represents the volume of a cube one metre on an edge. A much smaller unit is the *cubic centimetre* (cm^3):

$$1\ cm^3 = (10^{-2}\ m)^3 = 10^{-6}\ m^3$$

A unit of intermediate size, widely used in expressing the volume of liquids, is the litre (ℓ), which was redefined in 1964 to be exactly 1000 cm^3; a millilitre (ml) has a volume exactly equal to that of one cubic centimetre.

The device most frequently used to measure volumes in the general chemistry laboratory is the graduated cylinder. When greater accuracy is required, we use a pipet or buret (Fig. 1.1). A pipet is calibrated to deliver a fixed volume of liquid (e.g., 25.00 ± 0.01 cm^3) when filled to the mark and allowed to drain normally.

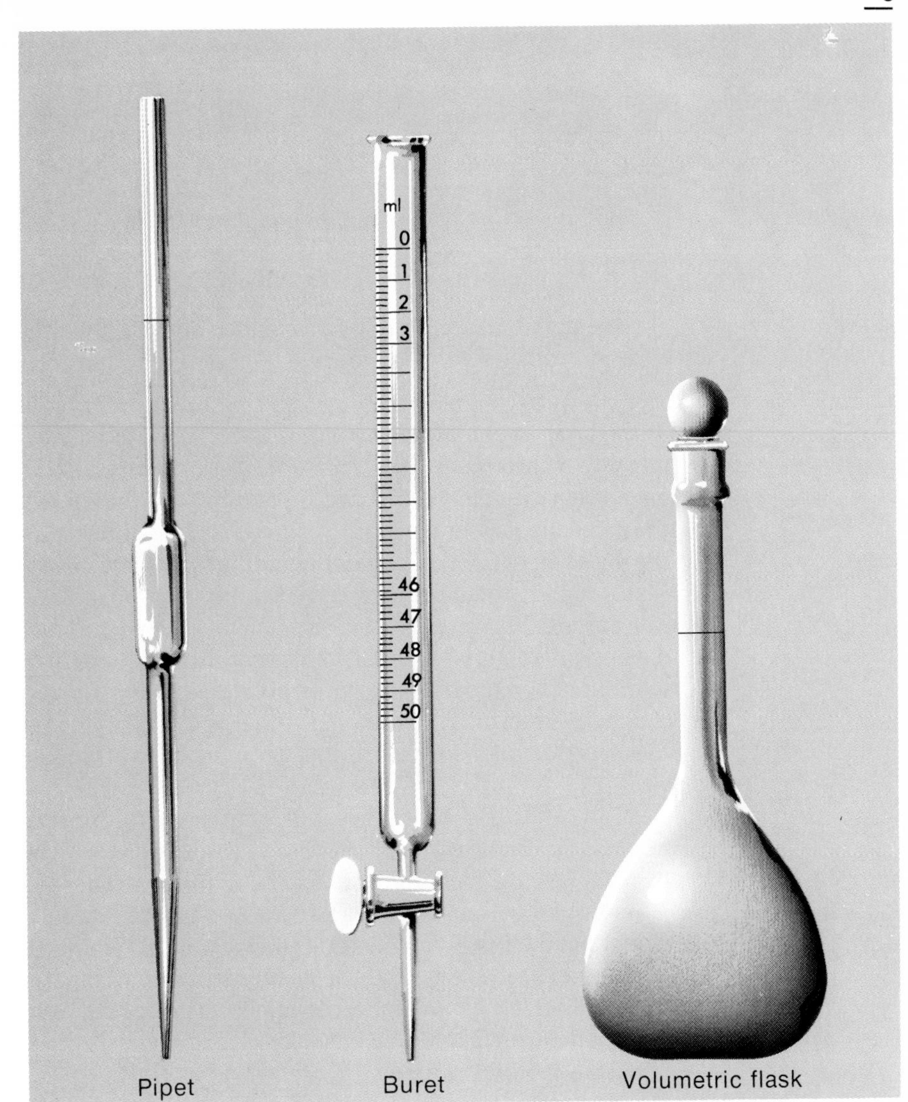

Figure 1.1 Instruments used with liquids to deliver a fixed volume (pipet), deliver a variable volume (buret), or contain a fixed volume (volumetric flask). On the buret the numerical markings from 4 to 45 ml have been omitted.

Variable volumes can be delivered with about the same accuracy from a buret. Here, the final and initial readings must be made carefully to calculate the volume of liquid withdrawn. The volumetric flask shown at the right of Figure 1.1 is designed to contain a specified volume of liquid (e.g., 50, 100, . . . 1000 cm^3) when filled to a level marked on the narrow neck.

Volumetric flasks are often used to prepare solutions to a desired concentration

MASS. The amount of matter in a sample, its *mass,* is most frequently expressed in the metric system in *grams,* kilograms (1 kg = 10^3 g), or milligrams (1 mg = 10^{-3} g). In the chemistry laboratory, mass is ordinarily measured by means of a balance (Fig. 1.2). To understand what is involved, consider the simple two pan balance shown at the left of the figure. We will assume that it has been adjusted so that, with nothing on either pan, the balance comes to rest with the two pans at the same height. To weigh an object, we place it on the left-hand pan. Metal pieces of known mass are then added to the right-hand pan to restore balance, i.e., to bring the pans to the same height. Under these conditions, the gravitational force acting on the sample, *f,* is equal to that acting on the pieces of metal.

$$f \text{ sample} = f \text{ metal}$$

But Newton's first law of motion tells us that gravitational force is directly proportional to mass

$$f = \text{k (mass)} \tag{1.1}$$

where the proportionality constant k has a fixed value at a given location. It follows that, at balance,

$$\text{k (mass sample)} = \text{k (mass metal)}$$

or

$$\text{mass sample} = \text{mass metal}$$

We see then that the double pan balance detects not only *equality of gravitational force but also equality of mass.*

Nowadays most teaching and research laboratories use single pan balances of the type shown at the right of Figure 1.2. Here a pan (A) and a set of movable masses (B) are suspended from one arm of the balance; their mass is exactly balanced by that of a fixed mass (C) attached to the other arm. Adding a sample to the pan deflects it downwards; balance is almost, but not quite, restored by turning the dials (D) to remove one or more of the movable masses. At this point, the beam is still slightly tilted from its original horizontal position. An optical system is used to translate this deflection into a small mass correction which appears as a number on a brightly lighted scale. The mass of the sample is obtained by adding this mass to that read on the dials. Fortunately, it takes less time to use this instrument than it does to describe it. Beginning students can carry out weighings to ± 0.001 g in a few seconds on a single pan balance in good working order.

Single pan balances used in general chemistry weigh to ± 0.001 g or better; they should *not* be used for crude weighings

TEMPERATURE. The concept of temperature is familiar to all of us, largely because our bodies are so sensitive to temperature differences. When we pick up a piece of ice, we feel cold because its temperature is lower than that of our hand. After drinking a cup of coffee, we may refer to it as "hot," "lukewarm," or "atrocious"; in the first two cases at least, we are describing the extent to which its temperature exceeds ours. From a slightly different point of view, temperature may be regarded as the factor that determines the direction of heat flow. Anyone brave enough to swim in a Minnesota lake in January feels cold because heat is absorbed

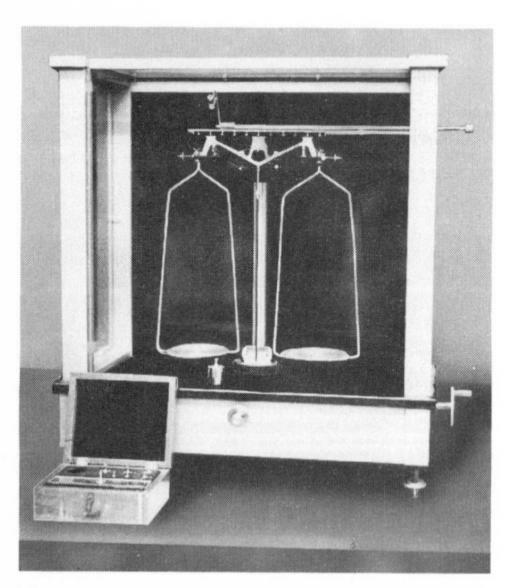

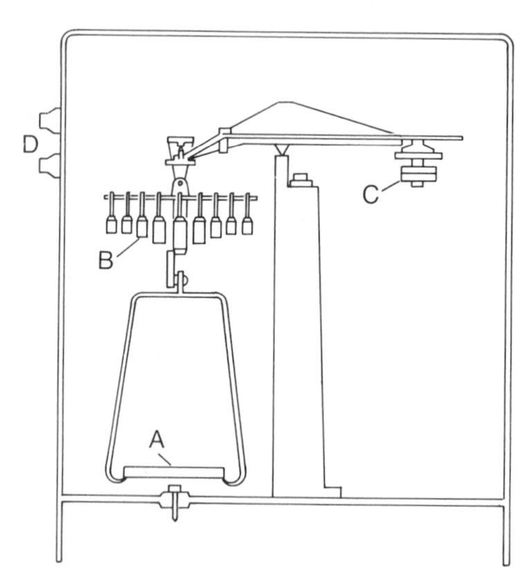

Figure 1.2 Two pan balance and schematic drawing of single pan balance. In most single pan balances, the masses are hung from the same frame that holds the balance pan. When a sample is put on the pan, masses are removed until balance is restored. With this design the balance works under a constant load.

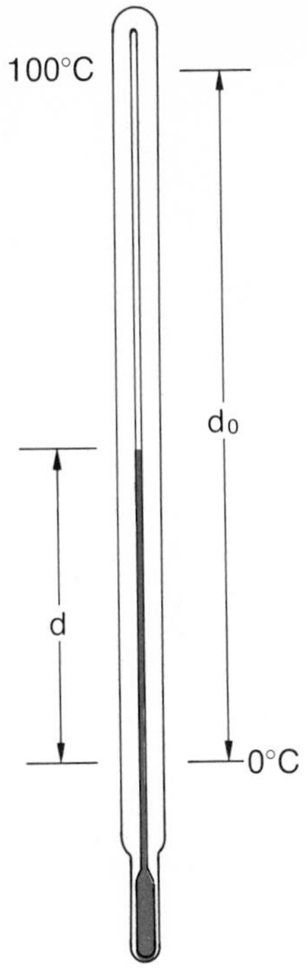

Figure 1.3 At 45°C, as read on a mercury-in-glass thermometer, d equals 0.45 d_0, where d_0 is the distance from the mercury level at 0°C to the level at 100°C.

from his body. If he takes a hot shower afterward, which he certainly will, heat flows in the reverse direction. In general, whenever two objects at different temperatures come in contact with each other, heat flows spontaneously from the higher to the lower temperature.

To measure temperature with a mercury-in-glass thermometer, we take advantage of the fact that mercury, like other substances, expands as the temperature rises. The thermometer is designed so as to make readily visible a rather small fractional change in volume (Fig. 1.3). The total volume of the thin capillary column is only about 2 to 3 per cent of that of the mercury reservoir at the base.

The Celsius scale is the common one in most countries

Thermometers used in the chemistry laboratory are marked in degrees Celsius after the Swedish astronomer, Anders Celsius (1701–1744). On this scale, the freezing point of water is taken to be 0°C and the boiling point at one atmosphere pressure to be 100°C. When we place a good quality mercury-in-glass thermometer in a beaker containing crushed ice in equilibrium with water, the mercury will come to rest exactly at the 0°C mark. In a beaker of boiling water, the mercury will rise to the 100°C mark. The distance between these two marks is divided as accurately as possible into one hundred equal parts, each of which corresponds to a degree Celsius. Thus, a temperature reading of 45°C corresponds to a mercury level 45 per cent of the way from the 0°C mark to the 100°C mark.

An older temperature scale still used in several countries, including the United States, is based on the work of Daniel Fahrenheit (1686–1736), a German instrument maker who was the first to use the mercury-in-glass thermometer. On this scale the normal freezing and boiling points of water are taken to be 32°F and 212°F, respectively. This leads to a rather simple relationship between degrees Fahrenheit and degrees Celsius. From Figure 1.4 we see that degrees Fahrenheit is a linear function of degrees Celsius. Since the equation of a straight line is $y = ax + b$, it follows that

$$°F = a°C + b$$

The constants a and b are readily evaluated; b is the intercept on the vertical axis, 32, while a is the slope of the line

$$a = \frac{\Delta y}{\Delta x} = \frac{212 - 32}{100 - 0} = \frac{180}{100} = 1.8$$

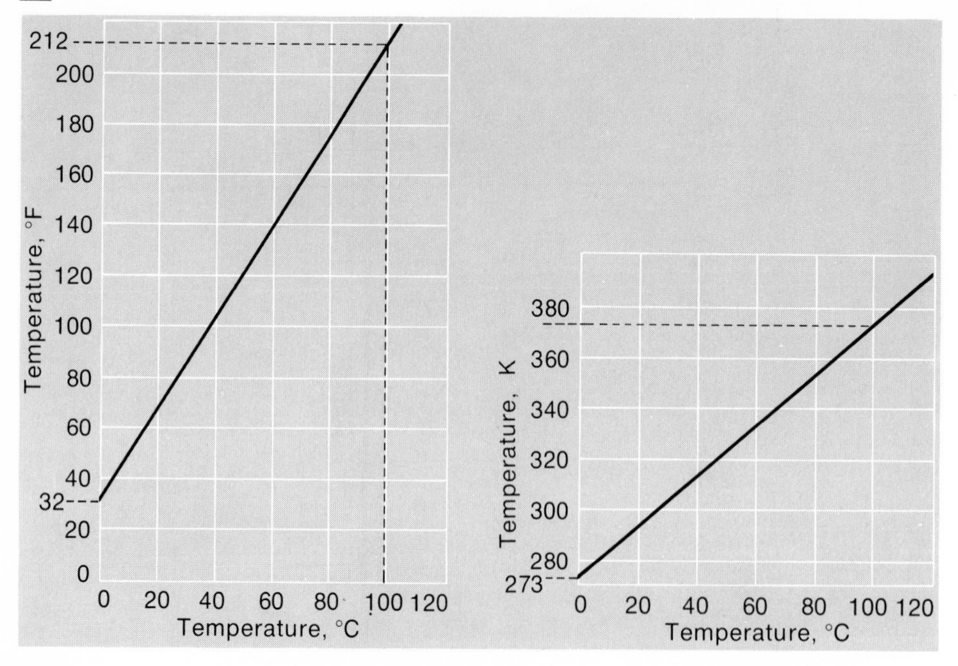

Figure 1.4 Relationship between two temperature scales. °F = 1.8°C + 32; slope is 1.8 and intercept is 32: K = °C + 273; slope is 1 and intercept is 273. A degree Celsius is equal in size to the kelvin and is 1.8 times as large as the degree Fahrenheit.

Making these substitutions we obtain

$$°F = 1.8°C + 32 \tag{1.2}$$

Another temperature scale that we will use extensively is the absolute or kelvin scale. The relationship between degrees Celsius (°C) and kelvin (K) is

$$K = °C + 273 \tag{1.3}$$

Lord Kelvin, who entered college at the age of 10, was an old man of 26 when this work was published

The scale is named after Lord Kelvin, an English physicist who demonstrated by a mathematical development based upon both theory and experiment that it is impossible to achieve a temperature lower than 0 K.

Example 1.1 Thermometers in hospitals are commonly marked in degrees Fahrenheit. Normal body temperature, as read on such a thermometer, is 98.6°F. Express this in degrees Celsius and kelvin.

Solution Substituting in Equation 1.2, we obtain

$$98.6 = 1.8(°C) + 32$$

Solving:

$$1.8\,(°C) = 98.6 - 32 = 66.6$$

$$°C = 66.6/1.8 = 37.0$$

Applying Equation 1.3:

$$K = 37.0 + 273.0 = 310.0$$

Uncertainties in Measurements: Significant Figures

Associated with every experimental measurement is a degree of uncertainty whose magnitude depends upon the nature of the measuring device and the skill with which it is used. If, for example, we attempt to measure out 8 cm^3 of liquid using a 100-cm^3 graduated cylinder, the volume delivered is likely to be in error by at least 1 cm^3. With such a crude measuring device, we will be fortunate indeed to obtain a volume closer to 8 than to 7 or 9 cm^3. To improve on the accuracy of the measurement, we might use a narrow 10-cm^3 cylinder on which the divisions are considerably farther apart. The volume we measure now may be within 0.1 cm^3 of the desired value of 8 cm^3; i.e., in the range 7.9 to 8.1 cm^3. Using a buret, we might be able to do even better; if we are very careful, the uncertainty in our measurement may be reduced to 0.01 cm^3.

The person who carries out an experiment has a responsibility to indicate the uncertainty associated with his measurements. Such information is vital to anyone who wants to repeat the experiment or to judge its validity. There are many ways to do this; we might, for example, report the three volume measurements referred to above as

8 ± 1 cm^3 (large graduated cylinder)
8.0 ± 0.1 cm^3 (small graduated cylinder)
8.00 ± 0.01 cm^3 (buret)

Throughout this text, we will drop the ± notation and simply write

8 cm^3; 8.0 cm^3; 8.00 cm^3

with the understanding that there is an *uncertainty of at least one unit in the last digit* (1 cm^3, 0.1 cm^3, 0.01 cm^3).

This method of specifying the degree of confidence in a measurement is often described in terms of **significant figures.** We say that in "8.00 cm^3" there are three significant figures; each of the three digits is experimentally meaningful. Similarly, there are two significant figures in "8.0 cm^3" and one significant figure in "8 cm^3."

Frequently we need to know the number of significant figures in a measurement reported by someone else. The manner in which this is deduced is illustrated in Example 1.2.

Example 1.2 An instructor asks his class in general chemistry to weigh a gold nugget. Among the masses reported are the following

20.03 g, 20.0 g, 0.020 03 kg, and 20 g

How many significant figures should be assumed in each case?

Solution The student who reports 20.03 g clearly believes that each of the four digits is meaningful; he is specifying 4 significant figures. Similarly, the student reporting 20.0 g indicates 3 significant figures. She has placed the zero after the decimal point to make it clear that the nugget was weighed to the nearest 0.1 g.

A moment's reflection should convince you that the third student, like the first, has shown 4 significant figures. The zero immediately to the right of the decimal point is not significant; it is there only because the mass is expressed in kilograms rather than grams. By the same token, there are 2 significant figures in the quantity "0.064 g" and only 1 in "0.007 g."

We cannot be sure of the number of significant figures in "20 g." Perhaps the student weighed the nugget to the nearest gram and wants to indicate 2 significant figures. Then again he may be trying to tell us that his balance weighs only to the nearest 10 g, in which case only the first figure in "20 g" is significant. This ambiguity could be avoided by giving the mass in exponential notation (Appendix 4) as either

2.0×10^1 g; 2 significant figures

or

2×10^1 g; 1 significant figure

The number of significant figures in a measured quantity is equal to the number of digits shown when the quantity is expressed in exponential notation

Most of the quantities that we measure in the laboratory are not end results in themselves; they are used to calculate other quantities. We might, for example, measure the mass and volume of a sample in order to determine its density (Example 1.3). Clearly, the precision of any such derived result is limited by that of the measurements upon which it is based. In particular, **when experimental quantities are multiplied or divided, the number of significant figures in the result cannot exceed that in the least precise measurement.**

Example 1.3 A student checks the purity of a water sample by determining its density (mass per unit volume) at 25°C. (The value given for pure water in a handbook is 0.9970 g/cm³.) He measures out 25 cm³ of the sample from a cylinder and determines its mass to be 25.624 g. What should he report for the density?

Solution Unfortunately the precision of the density determination is limited by that of the volume measurement, which has only 2 significant figures.

$$\text{density} = \frac{\text{mass}}{\text{volume}} = \frac{25.624\ \text{g}}{25\ \text{cm}^3} = 1.0\ \text{g/cm}^3$$

The five significant figures in the mass measurement are wasted; the student might as well have used a crude balance weighing to the nearest gram.

When experimental quantities are added or subtracted, the precision of the sum or difference is limited by the **absolute uncertainty** (rather than the number of significant figures) in the *least precise measurement.* Suppose, for example, we wish to calculate the mass of a solution prepared by adding 10.21 g of instant coffee and a "pinch" of sugar, 0.2 g, to 256 g of water. The implied uncertainties in these masses are

Note that the result is good to 3 significant figures, even though the mass of sugar is accurate to only 1

instant coffee:	10.21 ± 0.01 g
sugar:	0.2 ± 0.1 g
water:	256 ± 1 g
total mass:	266 ± 1 g

The sum of the masses cannot be more precise than that of the water (±1 g). The total mass should be reported as 266 g rather than 266.4 g or 266.41 g.

In applying the rules governing the use of significant figures, you should keep in mind that certain numbers involved in calculations are exact rather than approximate. Thus, if you were asked to express in centimetres a measured length of 6.10 mm, using the relation

$$1\ \text{cm} = 10\ \text{mm}$$

your answer should be given to 3 significant figures, the number in the measured quantity (6.10 mm). The "10" that appears in the equation just written is a defined quantity; there are exactly 10 mm in 1 cm. Similarly, in the equation relating the Fahrenheit and Celsius scales

$$°\text{F} = 1.8°\text{C} + 32$$

the numbers 1.8 and 32 are exact and so do not affect the accuracy of any calculation involving a temperature conversion (Example 1.1).

Conversion of Units

LENGTH, VOLUME, AND MASS. Frequently it is necessary to convert measurements expressed in one unit (e.g., mg) to another unit (g or kg). Conversions within the metric system, in which the units are related to each other by powers of ten, are readily carried out. The arithmetic is somewhat more complicated when we make conversions within the English system where, for example:

$$1 \text{ ft} = 12 \text{ in}; \ 1 \text{ gallon} = 4 \text{ qt}; \ 1 \text{ lb} = 16 \text{ oz}$$

Because of its simple, decimal interrelations, the metric system is universally employed by scientists. It has long been in use on the continent of Europe; within the past few years many other countries, including England and Canada, have shifted to the metric system. Conservatism, coupled with a certain amount of stubbornness, is responsible for the retention of the English system in the United States, although conversion to the metric system now seems likely within five to ten years. Maybe

Despite the worldwide trend toward the use of metric units, it seems likely that the familiar English units will continue to appear, at least in the nonscientific literature, for some time to come. Conversions from English to metric units can be carried out using conversion factors such as those given in Table 1.1.

With the aid of conversion factors, we can readily convert experimental quantities from one set of units to another. To illustrate the general approach, suppose we need to convert a length of 25.6 in to centimetres. To do this we use the conversion factor

$$1 \text{ in} = 2.54 \text{ cm}$$

Dividing both sides of this equation by 1 in gives a quotient equal to unity:

$$\frac{2.54 \text{ cm}}{1 \text{ in}} = 1$$

Multiplying 25.6 in by the quotient 2.54 cm/1 in does not change the value of the length but does accomplish the desired conversion of units:

$$25.6 \text{ in} \times \frac{2.54 \text{ cm}}{1 \text{ in}} = 65.0 \text{ cm}$$

The conversion equation, 1 in = 2.54 cm, could be used equally well to convert a length in centimetres, let us say 4.00 cm, to inches. In this case, we divide both sides of the equation by 2.54 cm to obtain

$$\frac{1 \text{ in}}{2.54 \text{ cm}}$$

Multiplying 4.00 cm by the quotient 1 in/2.54 cm converts the length from centimetres to inches:

$$4.00 \text{ cm} \times \frac{1 \text{ in}}{2.54 \text{ cm}} = 1.57 \text{ in}$$

TABLE 1.1 CONVERSION FACTORS RELATING LENGTH, VOLUME, AND MASS UNITS

	Metric		English		Metric-English	
Length						
	1 km	$= 10^3$ m	1 ft	= 12 in	1 in	= 2.54 cm
	1 cm	$= 10^{-2}$ m	1 yd	= 3 ft	1 m	= 39.37 in
	1 mm	$= 10^{-3}$ m	1 mile	= 5280 ft	1 mile	= 1.609 km
	1 nm	$= 10^{-9}$ m				
Volume						
	1 m^3	$= 10^6$ cm^3 $= 10^3 \ell$	1 gallon	= 4 qt = 8 pt	1 in^3	= 16.39 cm^3
	1 cm^3	= 1 ml $= 10^{-3} \ell$	1 qt (Can.)	= 69.35 in^3	1 ℓ	= 0.8799 qt (Can.)
			1 qt (U.S. liq.)	= 57.75 in^3	1 ℓ	= 1.057 qt (U.S. liq.)
Mass						
	1 kg	$= 10^3$ g			1 lb	= 453.6 g
	1 mg	$= 10^{-3}$ g	1 lb	= 16 oz	1 g	= 0.035 27 oz
	1 tonne	$= 10^3$ kg	1 short ton	= 2000 lb	1 tonne	= 1.102 short ton

Notice that a single conversion factor (e.g., 1 in = 2.54 cm) gives us two quotients (2.54 cm/1 in or 1 in/2.54 cm), both of which are equal to unity. In making a conversion, we choose the quotient which will enable us to cancel out the unit we wish to eliminate. Example 1.4 illustrates the application of the conversion factor approach to a slightly more complicated problem.

Example 1.4 Analysis of an air sample taken from the Holland Tunnel in New York City shows that it contains 1.9×10^{-10} lb/in³ of carbon monoxide. Express the concentration of carbon monoxide in the metric units, grams per cubic centimetre.

Solution Two conversions are required, one from pounds to grams using the factor

$$1 \text{ lb} = 453.6 \text{ g}; \text{ (Table 1.1)}$$

the other from cubic inches to cubic centimetres

$$1 \text{ in}^3 = 16.39 \text{ cm}^3; \text{ (Table 1.1)}$$

We can set up the arithmetic in a single expression

$$1.9 \times 10^{-10} \frac{\text{lb}}{\text{in}^3} \times \frac{453.6 \text{ g}}{1 \text{ lb}} \times \frac{1 \text{ in}^3}{16.39 \text{ cm}^3} = 53 \times 10^{-10} \text{ g/cm}^3$$

or, in standard exponential notation (Appendix 4)

$$= 5.3 \times 10^{-9} \text{ g/cm}^3$$

Note how the conversion factors are set up to cancel the undesired units

The conversion factor approach will be used repeatedly throughout this text. If this is your first contact with it, it may seem awkward or artificial. You will find, however, that it is the most straightforward way to solve a wide variety of problems in chemistry. It is particularly useful in dealing with unfamiliar units or carrying out a multistep conversion as in Example 1.4.

OTHER MEASURED QUANTITIES. Experiments carried out in chemical laboratories frequently involve the measurement of quantities other than length, volume, mass, and temperature. We may, for example, specify the **time** required to carry out a reaction. All of us are familiar with the units in which time is expressed (days, hours, minutes, or seconds):

$$1 \text{ d} = 24 \text{ h}; \ 1 \text{ h} = 60 \text{ min}; \ 1 \text{ min} = 60 \text{ s}$$

In experiments involving gases, it is important to specify the **pressure** of the gas as well as its volume, mass, and temperature (Chapter 5). Associated with every chemical reaction is an **energy change** whose magnitude is often specified in reporting the results of experiments (Chapter 4). Many different units are used to express pressure and energy; the more common ones are listed in Table 1.2 along with appropriate conversion factors.

Conversions involving these units are carried out in exactly the same way as with the units discussed earlier.

Example 1.5 On a certain day, the pressure read on a barometer is 742 mmHg. Express this in kilopascals.

Solution Looking at Table 1.2, it seems reasonable first to convert millimetres of mercury to atmospheres (760 mmHg = 1 atm) and then to kilopascals (1 atm = 101.3 kPa). Setting up the calculation as two consecutive conversions:

$$742 \text{ mmHg} \times \frac{1 \text{ atm}}{760 \text{ mmHg}} \times \frac{101.3 \text{ kPa}}{1 \text{ atm}} = 98.9 \text{ kPa}$$

TABLE 1.2 CONVERSION FACTORS RELATING PRESSURE AND ENERGY UNITS

Unit	Conversion
PRESSURE	
pascal (Pa)	
kilopascal (kPa)	1 kPa = 10^3 Pa
atmosphere (atm)	1 atm = 1.013×10^5 Pa = 101.3 kPa
millimetre of mercury (mmHg)*	1 atm = 760 mmHg
ENERGY	
joule (J)	
kilojoule (kJ)	1 kJ = 10^3 J
thermochemical calorie (cal)†	1 cal = 4.184 J
electron volt (eV)	1 eV = 1.602×10^{-19} J

*A "millimetre of mercury" is the pressure exerted by a column of mercury one millimetre high.
†A thermochemical calorie is the amount of heat required to raise the temperature of one gram of water one degree Celsius (more exactly, from 14.5°C to 15.5°C).

SI Units

As indicated by the large number of entries in Tables 1.1 and 1.2, many different units are often used to express a single quantity. For example, pressure may be expressed in atmospheres, millimetres of mercury, pascals, or even pounds per square inch. This proliferation of units has long been of concern to scientists. In 1960, the General Conference of Weights and Measures recommended the adoption of a self-consistent set of units based upon the metric system. In this so-called International System of Units (SI), it is recommended that a single unit be used for each measured quantity. Table 1.3 lists the SI units which are of particular interest in general chemistry and which we will use extensively throughout this text.

TABLE 1.3 SI UNITS

BASE UNITS			DERIVED UNITS		
Quantity	**Unit**	**Symbol**	**Quantity**	**Unit**	**Symbol**
length	metre	m	volume	cubic metre	m^3
mass	kilogram	kg	force	newton*	N
time	second	s	pressure	pascal†	Pa
temperature	kelvin	K	energy	joule**	J
amount of substance	mole††	mol	electric charge	coulomb††	C
electric current	ampere††	A	electric potential	volt††	V

*A newton is the force required to give a mass of one kilogram an acceleration of one metre per second squared ($N = kg \cdot m/s^2$).
†A pascal is the pressure exerted by a force of one newton acting on an area of one square metre [$Pa = N/m^2 = kg/(m \cdot s^2)$].
**A joule is the work done when a force of one newton acts through a distance of one metre ($J = N \cdot m = kg \cdot m^2/s^2$).
††Defined in later chapters.

Certain of the SI units listed in Table 1.3 have rather awkward magnitudes. For example, the cubic metre represents a very large volume; the medium-sized test tube used in the general chemistry laboratory has a volume of about 0.000 01 m^3. At the opposite extreme, the pascal represents a very small pressure; ordinary atmospheric pressure is about 100 000 Pa. An obvious way to get around this problem would be to resort to exponential notation. That is, we might express the volume of a test tube as 1×10^{-5} m^3 and the pressure of the atmosphere as 1.0×10^5 Pa. Alternatively, we can make use of the prefixes listed in Table 1.4.

TABLE 1.4 SI PREFIXES

FACTOR	PREFIX	SYMBOL	FACTOR	PREFIX	SYMBOL
10^9	giga	G	10^{-1}	deci	d
10^6	mega	M	10^{-2}	centi	c
10^3	kilo	k	10^{-3}	milli	m
10^2	hecto	h	10^{-6}	micro	μ
10^1	deca	da	10^{-9}	nano	n

Thus we might express 1.0×10^5 Pa as 100 kPa; i.e.,

$$1.0 \times 10^5 \text{ Pa} \times \frac{1 \text{ kPa}}{10^3 \text{ Pa}} = 1.0 \times 10^2 \text{ kPa} = 100 \text{ kPa}$$

or 1×10^{-5} m^3 as 10 cm^3: $1 \times 10^{-5} \text{ m}^3 \times \dfrac{1 \text{ cm}^3}{(10^{-2} \text{ m})^3} = \dfrac{1 \times 10^{-5} \text{ cm}^3}{10^{-6}} = 10 \text{ cm}^3$

Throughout the remainder of this text, we will use SI units wherever possible. We will, for example, express atomic dimensions in nanometres rather than angstroms (1 nm = 10^{-9} m = 10 Å), volumes in cubic centimetres or cubic decimetres rather than litres (1 ℓ = 1 dm^3 = 10^3 cm^3), and energy changes in joules rather than calories (1 cal = 4.184 J). From time to time, however, we will make use of a familiar pressure unit, the standard atmosphere, which is outside the SI: 1 atm = 1.013×10^5 Pa = 101.3 kPa.

1.2 KINDS OF SUBSTANCES

Chemists, by methods which we will consider shortly, have isolated many thousands of pure substances from the earth's crust, the oceans, and the atmosphere. Approximately 100 of these are unique in one important respect. No one has ever succeeded in resolving these substances, called **elements,** into two or more substances that differ in their properties from the original. All other pure substances are classified as **compounds.** A compound, by definition, is a pure substance that can be resolved into two or more elements.

Many different methods have been used to resolve compound substances into elements. The English chemist Joseph Priestley decomposed mercuric oxide into the elements mercury and oxygen by exposing it to an intense beam of sunlight focused by a powerful lens. Sir Humphry Davy showed that ordinary quicklime, long believed to be an elementary substance, was a compound by passing an electric current through melted lime and demonstrating that two different substances (calcium and oxygen) were produced at the electrodes.

In Table 1.5 are listed the percentages by mass of the 20 most abundant elements in nature. Notice that a single element, oxygen, makes up nearly half of the total. Water, our most common compound, contains 89% oxygen. Many of the most common minerals contain compounds of oxygen with such other elements as silicon, aluminum, and iron.

TABLE 1.5 ABUNDANCE OF ELEMENTS (EARTH'S CRUST, OCEANS, AND THE ATMOSPHERE)

Oxygen	O	49.5%	99.2%	Chlorine	Cl	0.19%	0.7%
Silicon	Si	25.7		Phosphorus	P	0.12	
Aluminum	Al	7.5		Manganese	Mn	0.09	
Iron	Fe	4.7		Carbon	C	0.08	
Calcium	Ca	3.4		Sulfur	S	0.06	
Sodium	Na	2.6		Barium	Ba	0.04	
Potassium	K	2.4		Chromium	Cr	0.033	
Magnesium	Mg	1.9		Nitrogen	N	0.030	
Hydrogen	H	0.87		Fluorine	F	0.027	
Titanium	Ti	0.58		Zirconium	Zr	0.023	
				All others			<0.1%

You will find it helpful to learn the symbols of the more common elements

The symbol of an element (Table 1.5) consists of one or two letters, taken most frequently from the name (O, Si, Al, Ca). Some symbols are derived from the Latin name of the element (iron, *ferrum*) or one of its compounds (sodium carbonate, *natrium*). A complete list of all known elements, arranged in alphabetical order with their symbols, appears on the inside back cover of this book.

1.3 IDENTIFICATION OF PURE SUBSTANCES

Substances may be identified on the basis of their *chemical properties*. These are properties which are observed when the substance undergoes a *chemical change* in which it is converted to one or more other substances. One might, for example, demonstrate that a certain red solid is mercuric oxide by heating it in air and observing that it decomposes to form elementary mercury and oxygen.

More commonly, elements or compounds are identified by measuring their *physical properties*. These are properties which can be measured without changing the chemical identity of a substance. They include such characteristics as melting point, boiling point, solubility, and density. Tables of such properties are available in the literature (*Handbook of Chemistry and Physics*, Chemical Rubber Publishing Co.; Lange's *Handbook of Chemistry*, McGraw-Hill). By measuring several such properties and comparing to recorded values, it is ordinarily possible to identify the substance and get some idea of its purity. For example, if you find that a certain liquid boils at 81°C, freezes at 4°C, is soluble in alcohol but not in water, and has a density of 0.882 g/cm³, the chances are you are dealing with a sample of slightly impure benzene (Table 1.6).

If all the physical properties of a substance are identical with those of penicillin, it must *be* penicillin

TABLE 1.6 PHYSICAL PROPERTIES OF A FEW COMMON SUBSTANCES*

SUBSTANCE	MELTING PT (°C)	BOILING PT (°C)	SOLUBILITY AT 25°C (g/100 g) Water	Ethyl Alcohol	DENSITY (g/cm³)
Acetic Acid	16.6	118.1	infinite	infinite	1.05
Benzene	5.5	80.1	0.07	infinite	0.879
Bromine	−7.1	58.8	3.51	infinite	3.12
Iron	1530	3000	insoluble	insoluble	7.86
Methane	−182.5	−161.5	0.0022	0.033	6.67×10^{-4}
Oxygen	−218.8	−183.0	0.0040	0.037	1.33×10^{-3}
Sodium Chloride	808	1473	36.5	0.065	2.16

*All values given at 1 atm pressure.

Behavior of Pure Substances During Melting and Boiling

A study of the melting point behavior of a substance can help us not only to identify it but also to check its purity rather closely. A pure, crystalline solid, if heated slowly, melts sharply at a characteristic temperature (*mp* benzene = 5.5°C). The temperature stays constant as long as any solid remains (Fig. 1.5*A*); only when it is completely melted does the temperature rise again.

The melting point behavior of an impure solid differs from that of a pure substance in one important respect. It is ordinarily found, for reasons discussed in Chapter 12, that the solid starts to melt at a temperature below that of the pure substance and that the temperature rises steadily during the melting process (Figure 1.5*B*). Any evidence of a deviation from the horizontal in a temperature-time plot leads us to suspect the presence of impurities.

The purity of a liquid can be tested in a manner quite similar to that described for a solid. In this case a constant temperature during the boiling process is the criterion of purity. The heating curve for a pure liquid looks very much like that shown at the left in Figure 1.5. If the liquid is impure, we ordinarily find that the temperature rises steadily during the boiling process.

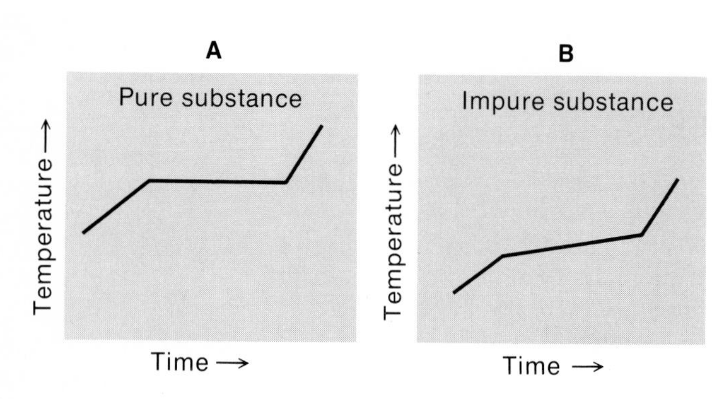

Figure 1.5 Time-temperature curves for the melting of a pure substance (left) and a mixture (right). A pure substance will melt at a constant temperature. If the substance contains an impurity, its melting point will increase as melting proceeds and will at all stages be lower than when pure.

Color: Absorption Spectrum

Some of the substances that we work with in general chemistry can be identified at least tentatively on the basis of their color. Gaseous nitrogen dioxide has a brown color; vapors of bromine and iodine are red and violet, respectively. An aqueous solution of copper sulfate is blue, a solution of potassium permanganate, purple, and so on.

The colors of gases and liquids are due to the selective absorption of certain components of visible light. Bromine, for example, absorbs in the violet and blue regions of the spectrum (Table 1.7). The subtraction of these components from visible light accounts for the red color that we see when we look through a sample of bromine vapor or liquid. The purple (blue-red) color of a potassium permanganate solution results from absorption in the green region (Plate 1).

White solids and colorless liquids do not absorb visible light

Many colorless substances absorb light in the ultraviolet or infrared regions of the spectrum. Ozone in the atmosphere absorbs harmful ultraviolet radiation from the sun. Carbon dioxide absorbs infrared radiation given off from the earth's surface, preventing excessive loss of heat to outer space. Most organic substances absorb at certain well characterized wavelengths in the infrared. By exposing a substance to infrared radiation covering a range of wavelengths and measuring absorption as a function of wavelength, a spectrum of the type shown in Figure 1.6 is obtained. Such an infrared spectrum serves as a "fingerprint" to identify the substance. Organic chemists use this technique routinely; one of its major advantages is that it can be applied to samples weighing a milligram or less.

TABLE 1.7 COLOR OF SUBSTANCES WHICH ABSORB LIGHT IN THE VISIBLE REGION

WAVELENGTH REGION	COLOR ABSORBED	COLOR TRANSMITTED
<400 nm	Ultraviolet	Colorless
400–450	Violet	Red, Orange, Yellow
450–490	Blue	
490–550	Green	Purple
550–580	Yellow	
580–650	Orange	Blue, Green
650–700	Red	
>700 nm	Infrared	Colorless

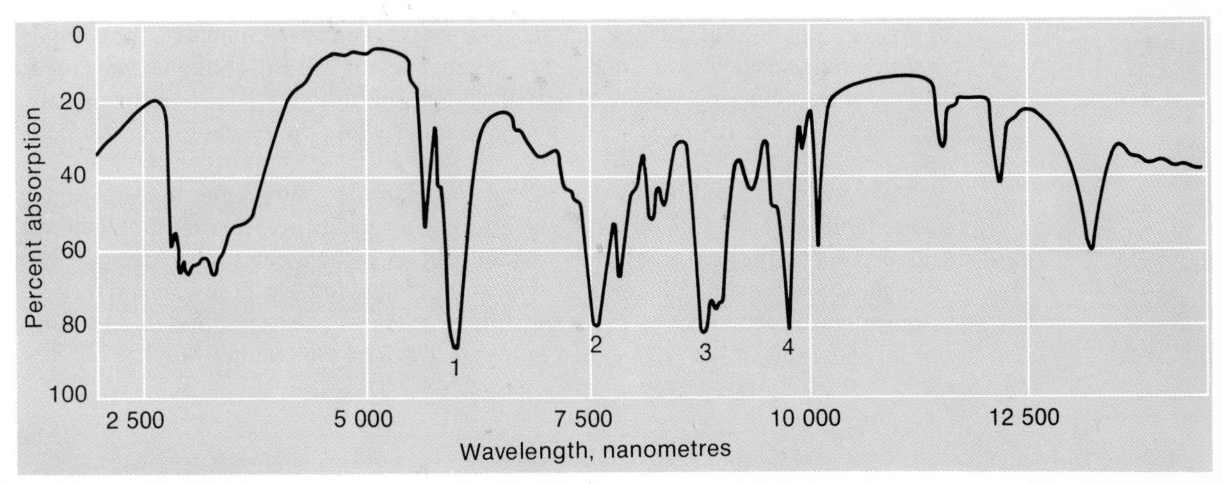

Figure 1.6 Infrared absorption spectrum of ascorbic acid, vitamin C. Note the strong absorptions at 1 (6 100 nm), 2 (7 600 nm), 3 (8 800 nm), and 4 (9 600 nm) which are characteristic of this particular compound.

1.4 SEPARATION OF MATTER INTO PURE SUBSTANCES

Very few elements and compounds occur in nature in the pure state. Most often, they are found in mixtures with other substances. Air, for example, is a *homogeneous (i.e., uniform) mixture* (a **solution**) of several different gases including the elementary substances nitrogen, oxygen, and argon and the compounds water and carbon dioxide. Seawater is a solution of various salts, including sodium chloride, in water. Most of the rocks and minerals found in the earth's crust are complex *heterogeneous (i.e., nonuniform) mixtures* of many different chemical substances.

Ordinary granite is one such mixture

A continuing task of chemists at all levels is to separate pure substances from mixtures. A great many different techniques have been worked out for this purpose. We will consider three of the more common methods:

1. *Distillation,* used to separate a solid from a liquid, and *fractional distillation,* by which two liquids can be separated from each other.
2. *Fractional crystallization,* a technique used routinely to purify solids.
3. *Chromatography,* one of the most versatile separation methods, which can be applied to solid, liquid, or gaseous mixtures.

The first two of these methods have been used for centuries. Chromatography, on the other hand, is a relatively new technique that has been in large-scale use for only about twenty years.

A mixture of two substances, only one of which is volatile, is readily separated by *distillation*. A simple distillation apparatus which can be used to separate sodium chloride from water is shown in Figure 1.7. When the solution is heated, the water boils off, leaving a residue of sodium chloride in the distilling flask. The water may be collected by passing the vapor down a condenser through which cold water is circulated. In many arid regions, particularly in the Mideast, distillation is used to convert seawater to fresh water.

If both components of a mixture are volatile, simple distillation will not achieve a complete separation. Suppose, for example, that we heat a solution of ether (*bp* = 35°C) in benzene (*bp* = 80°C). The mixture will begin to boil at a temperature somewhat above 35°C. As heating is continued, we find that the vapor entering the condenser is somewhat richer in the more volatile component, ether, than was the original mixture. This *distillate* will still, however, contain an appreciable amount of the less volatile component, benzene. Similarly, the *residue* in the distilling flask will become richer in benzene but will still contain some ether. Thus, for example, if we distilled half of a 50–50 mixture of ether and benzene, we might find that the distillate contained 70% ether while the residue was 70% benzene.

We could improve upon this partial separation by subjecting both distillate and residue to further distillation. Each time we repeat the process, the distillate will become richer in ether, the residue richer in benzene. If we are patient enough, we will eventually end up with fractions which are pure or nearly pure. A more convenient way to carry out this process of *fractional distillation* is to use an apparatus of the type shown in Figure 1.8. The tall column which separates

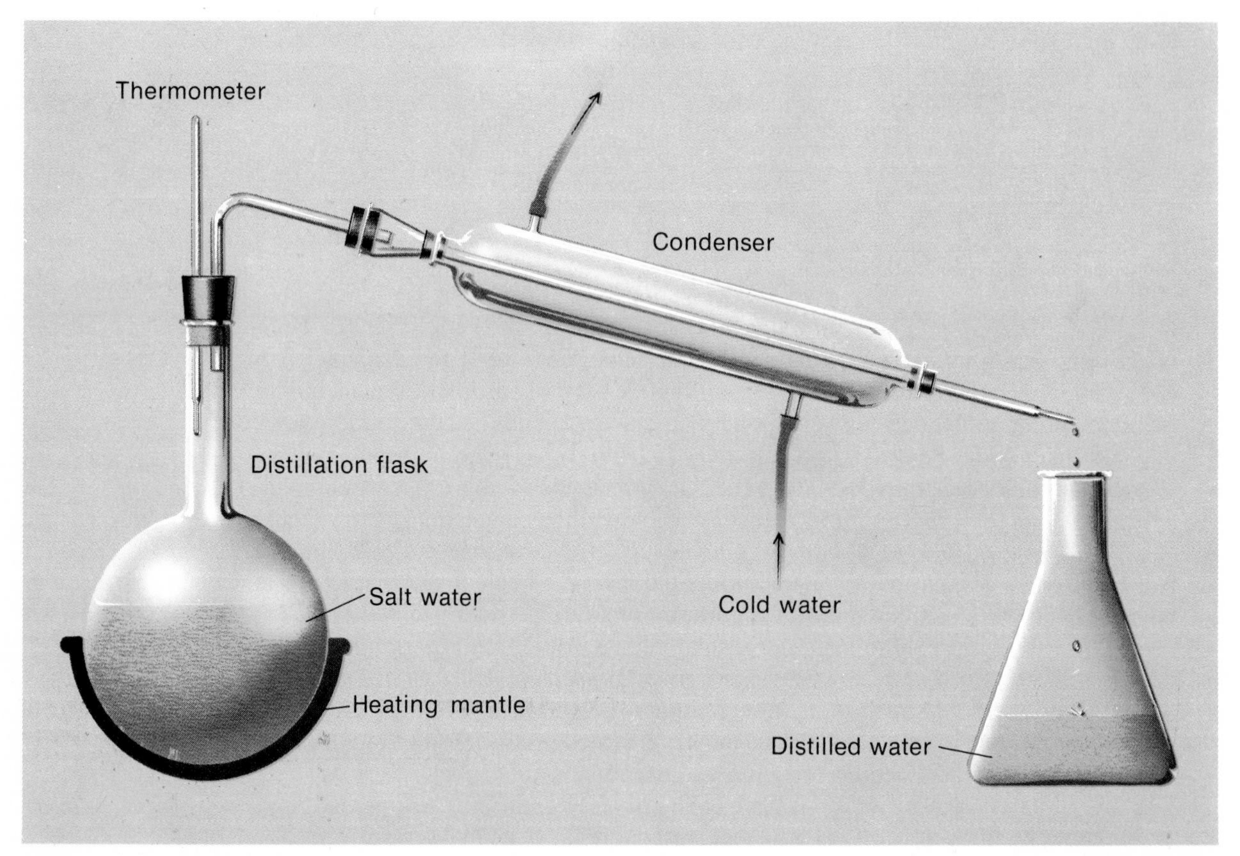

Figure 1.7 Separation of a liquid from a solid by simple distillation. Since salt is not volatile, it all remains in the distillation flask.

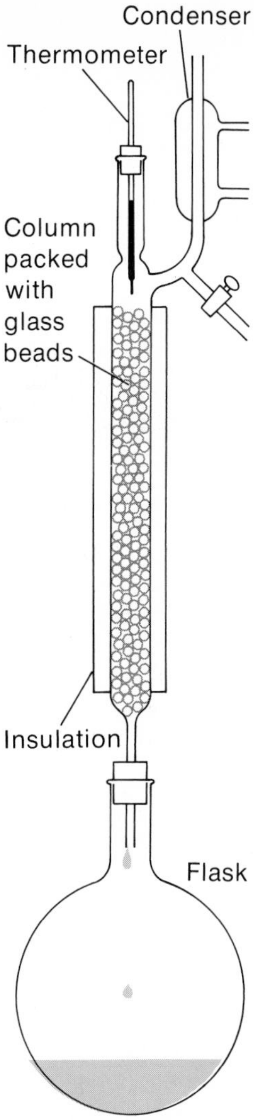

Condenser
Thermometer
Column packed with glass beads
Insulation
Flask

In a fractionating column, the upper region is kept cooler than the lower

Figure 1.8 A simple fractionating column. The temperature gradually decreases as one goes up the column, which tends to condense the less volatile species and allow only the most volatile one to reach the condenser and be recovered.

the distilling flask from the condenser is packed with a solid material, often small glass beads. This gives a large surface area upon which the less volatile component of the vapor (e.g., benzene) can condense and fall back into the distilling flask. The more volatile component (e.g., ether) more easily moves up the column and passes over into the distillate. If the rate of distillation is carefully controlled, a good separation can be achieved. Huge fractionating columns of very sophisticated design are used routinely in the petroleum industry to separate crude oil fractions such as gasoline (*bp* 40–200°C), kerosene (*bp* 175–325°C), and diesel fuel (*bp* > 275°C).

Fractional Crystallization

Most of the reagent grade solid chemicals that you will work with in the general chemistry laboratory have been purified by *fractional crystallization.* In this process a solid, A, containing a relatively small amount of an impurity, B, is dissolved in the minimum amount of hot solvent. The solution is then cooled, often to room temperature. If all goes well, the solid that separates out on cooling will be pure A, which can be filtered off and dried. The filtrate, which should contain all of the impurity B and relatively little A, is ordinarily discarded.

TABLE 1.8 SOLUBILITIES OF TARTARIC AND SUCCINIC ACIDS (g/100 g WATER)*

°C	Tartaric Acid	Succinic Acid
20	18	7
30	25	11
40	37	16
50	50	24
60	65	36
70	81	51
80	98	71

*Throughout the text discussion, it is assumed that the presence of one solid in solution has no effect on the solubility of the other. For example, at 80°C, the solubility of succinic acid is taken to be 71 g/100 g of water, regardless of how much tartaric acid is present.

To illustrate the process of fractional crystallization, suppose we wish to purify a sample of tartaric acid, an organic compound isolated from grape juice, contaminated by a small amount of another organic substance, succinic acid. Specifically, suppose the sample contains 5 g of succinic acid mixed with 95 g of tartaric acid. Referring to Table 1.8, we see that this sample should dissolve completely in 100 g of water at 80°C (100 g of water at 80°C dissolves 98 g of tartaric acid and 71 g of succinic acid). Consider now what happens when this solution is cooled to 20°C. Since the solubility of tartaric acid at 20°C is only 18 g/100 g of water, we see that

$$95\text{ g} - 18\text{ g} = 77\text{ g}$$

of tartaric acid crystallizes out of solution. All the succinic acid (5 g) stays in solution, since 100 g of water at 20°C can hold as much as 7 g of succinic acid. Hence, by dissolving this sample (95 g tartaric acid, 5 g succinic acid) in 100 g of water at 80°C and cooling to 20°C, we recover 77 g of *pure* tartaric acid, 81 per cent of the amount we started with.

This discussion implies two of the limitations of fractional crystallization.

1. The compound to be purified must be much more soluble at high than at low temperatures. Otherwise, much of it will be "lost"; i.e., it will remain in solution upon cooling. It would be futile to try to purify a sample of sodium chloride by dissolving in boiling water and cooling to room temperature, since its solubility at 20°C (36 g/100 g water) is nearly as great as at the boiling point (40 g/100 g water).

2. To obtain a pure sample by a single fractional crystallization, the amount of impurity must be relatively small. In the example discussed, consider what would have happened if the sample had contained 10 g of succinic acid instead of 5 g. Since the solubility of succinic acid at 20°C is 7 g/100 g water, it is evident that the tartaric acid isolated from the solution would have been contaminated by

$$10\text{ g} - 7\text{ g} = 3\text{ g}$$

of succinic acid. A second recrystallization would have been required to give a pure product.

Alternatively, we could carry out the crystallization at 30°C

The separation technique known as chromatography* can be applied to extremely complex mixtures; as many as 20 amino acids in a hydrolyzed protein sample have been identified by this method. Moreover, it can be used for very small samples or for components present at very low concentrations. Chromatographic techniques have been developed to analyze for air pollutants at concentrations of one part per million or less.

To illustrate the use of this separation method, let us consider a typical experiment in **paper chromatography** which is often carried out in the general chemistry laboratory (Plate 2). The sample, a liquid solution of two or more solids, is applied near the lower edge of a long strip of filter paper. The spot is allowed to dry and the paper suspended vertically in a stoppered flask. Enough solvent is added to almost, but not quite, reach the spot. As the solvent rises by capillary action, the components of the sample are carried along with it. A component which is more soluble tends to move farther up the paper. A component which is less soluble or more strongly adsorbed by the paper moves a smaller distance. In this way, a separation occurs. If the substances are colored they are easily visible, as in Plate 2. If not, some developing process must be used to detect them and make their positions clear. If desired, the separated components can be recovered by cutting the strip into sections and extracting with a suitable solvent.

The process just described is suitable for separating two or more solids from each other. Many other chromatographic techniques have been developed. One of the most widely used is **vapor phase chromatography,** in which the components are separated as vapors. The mixture is added to one end of a glass tube containing a solid packing material coated with a high-boiling, nonvolatile liquid such as silicone oil. An unreactive "carrier" gas such as helium is passed through the column. The components of the sample gradually separate as they continuously vaporize into the helium and either adsorb onto or dissolve in the packing.

*The word chromatography is derived from the Greek *chroma,* meaning color; in most of the early experiments the separated components were identified by their colors.

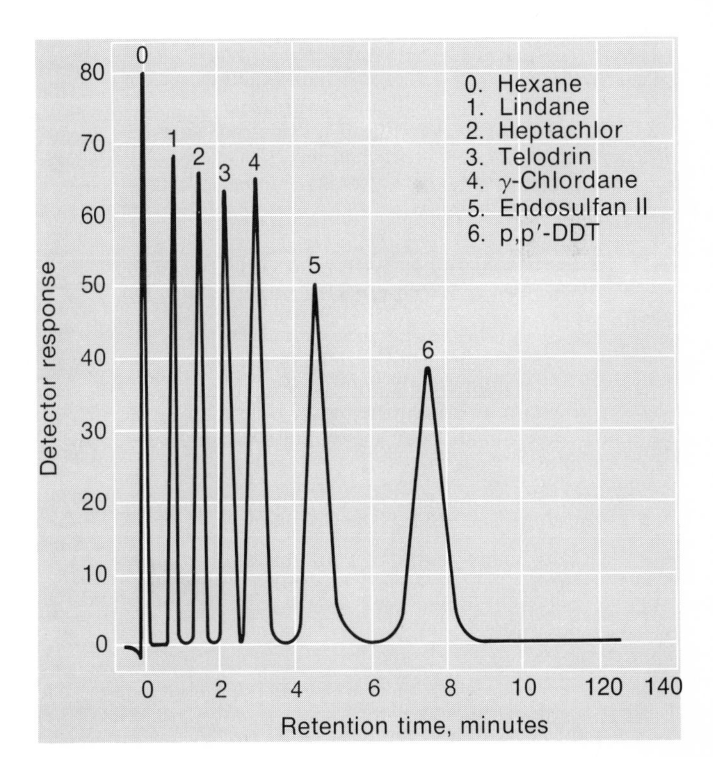

Figure 1.9 Separation of chlorinated pesticides dissolved in hexane, using VPC.

Ordinarily the more volatile components move faster and come out of the column first. As successive fractions leave the column, they activate a detector and recorder to give a plot such as that shown in Figure 1.9.

PROBLEMS*

1.1 *Significant Figures* Aluminum has a density of 2.70 g/cm^3. Calculate the volume of a piece of aluminum weighing 35 g.

1.2 *Conversion of Units* On an old road map, the distance from Ottawa to Quebec is given as 292 miles. Express this distance in kilometres; in nanometres.

1.3 *Temperature Scales* Express 209 K in degrees Celsius and degrees Fahrenheit.

1.4 Explain, in terms that would be understood by someone who has had no exposure to chemistry,

a. what a temperature of 50°C means.
b. the distinction between an element and a compound.
c. how a pure substance can be identified.

1.5 Which of the following statements are always valid? never valid?

a. A sample whose temperature remains constant during melting is a pure substance.
b. A given temperature is higher if expressed in degrees Fahrenheit rather than degrees Celsius.
c. A solid whose solubility in water is the same at all temperatures cannot be purified by fractional crystallization.
d. Substances which absorb only in the ultraviolet are red in color.

1.6 Describe experiments by which you could

a. determine whether a certain liquid is pure or contaminated with another liquid.
b. measure the density of a sample of gasoline to three significant figures.
c. demonstrate that water is a compound rather than an element.

1.7 You are asked to perform each of the following experiments. List all the equipment you would need in each case.

a. Determination of the density of a liquid.
b. Separation of salt from water.
c. Determination of the purity of a solid.
d. Determination of the volume of an Erlenmeyer flask.

1.8 An aluminum atom has a radius of 0.143 nm.

a. How many Al atoms would have to be lined up in a row to form a line 1.923 cm long?

1.18 Explain in your own words the principle behind

a. paper chromatography.
b. fractional crystallization.
c. the rules for the use of significant figures in multiplication.

1.19 Criticize each of the following statements.

a. When two quantities, each known to three significant figures, are added, the sum will contain three significant figures.
b. Fractional distillation is required to separate sugar from water.
c. The temperature of an object in degrees Celsius is directly proportional to its temperature in degrees Fahrenheit.
d. A sample of matter containing two or more elements is a compound.

1.20 Describe what you would observe in an experiment in which

a. a solution of sodium chloride in water freezes.
b. a solution saturated with succinic acid and tartaric acid at 80°C is cooled to 20°C.
c. visible light covering the range 400 to 800 nm is passed through liquid bromine.

1.21 Suggest two different ways in which you might

a. measure out a 5.00-cm^3 sample of water.
b. show that methane is a compound rather than an element.
c. separate nitrogen from air.

1.22 The radius of an iron atom is 1.26×10^{-8} cm; its mass is 55.8 u (1 u = 1.67×10^{-24} g).

a. Calculate the volume of an iron atom in cubic centimetres.

*Answers are given in Appendix 5 to all problems except those listed in the left column (1.4–1.17). The first few problems (1.1–1.3) are designed to be particularly straightforward, requiring only that you apply a single principle. In contrast, those marked with an asterisk (p. 22) ordinarily require extra effort or ability.

b. Estimate the volume of an Al atom in cubic centimetres. ($V = 4\pi r^3/3$)

b. Calculate the density of an iron atom in grams per cubic centimetre.
c. Explain why the density of iron atoms is greater than that of a bulk sample of iron (7.86 g/cm^3).

1.9 A student determines the density of a piece of metal by weighing it (mass = 12.54 g) and adding it to a flask which has a volume of 25.00 cm^3. He finds that 20.54 g of water (d = 0.9970 g/cm^3) is required to fill the flask with the metal in it. What value should he report for the density of the metal?

1.10 A sample is balanced by the following masses: one 100 g; two 20 g; three 1 g; one 500 mg; two 100 mg; one 5 mg. Assuming each mass to be accurate to the nearest milligram, what is the mass of the sample?

1.11 Using Tables 1.1 and 1.2, convert

a. 1.30 qt (Can.) to litres.
b. 1.16 atm to pascals.
c. 1.29 cal to joules.

1.12 Referring to Table 1.6, give

a. the melting point of iron in K.
b. the number of grams of bromine required to saturate 350 cm^3 of water (d = 1.00 g/cm^3).
c. the volume of a piece of iron weighing 29.0 g.

1.13 A certain substance which is a liquid at room temperature is infinitely soluble in ethyl alcohol but has a limited solubility in water. Which one(s) of the substances in Table 1.6 might it be? Suggest a further experiment to identify it.

1.14 On a certain day, the concentration of carbon monoxide in the air over Toronto reached 1.8×10^{-8} g/cm^3. How many tonnes of carbon monoxide were present, assuming a total volume of air of 5.5×10^{11} m^3?

1.15 Using the data in Table 1.5 and assuming a total mass of 3×10^{24} g, estimate the number of tonnes of phosphorus available to man.

1.16 Referring to Table 1.7, what would you expect to be the color of a gas which absorbs at 520 nm?

1.17 You are given a sample containing 75 g of tartaric acid and 12 g of succinic acid. Using the data in Table 1.8,

a. how much water should be used to dissolve the sample at 80°C?
b. how much tartaric acid will come out of solution on cooling to 20°C? how much succinic acid?
c. suggest two ways in which this procedure might be modified to avoid contamination of the final product with succinic acid.

1.23 A student calibrates a 10-cm^3 pipet by filling it with pure water (d = 0.9970 g/cm^3) and allowing the water to run into a beaker which weighs 74.242 g. In four trials he obtains masses (water + beaker) of 84.136 g, 84.151 g, 84.141 g, and 85.279 g. What value would you advise him to report for the volume of the pipet?

1.24 A reagent bottle of copper sulfate in the general chemistry lab is filled with 125.8 g of that chemical at the beginning of the period. Successive students remove 19 g, 0.68 g, 52.1 g, and 12 mg. How much remains?

1.25 Convert

a. the solubility of sodium chloride in water, 0.804 lb/kg water, to grams per hundred grams of water.
b. 1.29 kPa to atmospheres.
c. 12.4 kcal to joules.

1.26 Referring to Table 1.6, calculate

a. the melting point of sodium chloride in K.
b. the number of cubic centimetres of bromine required to saturate 1000 g of water.
c. the density of methane in kilograms per cubic metre.

1.27 You are given a flask at 80°C containing a colored vapor and told that it contains one or more of the substances listed in Table 1.6. How would you go about determining what substances are present? (There is no liquid or solid in the flask when you get it.)

1.28 Fish containing more than 0.50 part per million of mercury (e.g., 0.50 g Hg/10^6 g) cannot be sold commercially. How many grams of mercury can be present in a two-pound fish without exceeding this limit?

1.29 The atmosphere contains about 78% by mass of nitrogen. Taking the mass of nitrogen in the atmosphere to be 9.4×10^{22} g, estimate the total mass of the atmosphere in tonnes.

1.30 Estimate the wavelength of light absorbed by a solution of copper sulfate, which is blue.

1.31 Suppose you have a mixture of 50 g of succinic acid and 50 g of tartaric acid. If this mixture is dissolved in 100 g of water at 80°C and cooled to 20°C, how much succinic acid will come out of solution? How much tartaric acid?

*1.32 Oil spreads on water to form a film about 100 nm thick. How many square kilometres of ocean will be covered by the slick formed when one barrel of oil is spilled (1 barrel = 26.2 Can. gallons)?

*1.33 A certain automobile consumes fuel at the rate of 17.8 miles per gallon (Can.). Express this in litres per hundred kilometres, the units in which gasoline consumption is now to be expressed in Canada.

*1.34 In paper chromatography, we frequently refer to the R_f value of a substance, defined as the ratio: distance moved by substance/distance moved by solvent. The five substances A, B, C, D, and E are known to have R_f values of 0.10, 0.25, 0.40, 0.60, and 0.80, respectively. Assuming that the spots in Plate 2 correspond to certain of these substances and that the solvent has reached the bottom of the plate, identify them.

*1.35 Assume that essentially all of the hydrogen found in nature occurs as the compound water, which contains 11.1% by mass of hydrogen. With this assumption, calculate the fraction of the total amount of oxygen in nature (Table 1.5) which occurs as water.

2

ATOMS, MOLECULES, AND IONS

Every pure substance has its own characteristic properties, such as melting point, boiling point, and density. As we saw in Chapter 1, differences in properties between pure substances can be used to isolate them from a mixture. We can, for example, take advantage of the difference in volatility between sodium chloride and water to separate a salt solution into its two pure components by distillation.

We are naturally curious as to why substances differ so markedly from one another in their properties. How do we explain the difference in boiling point between water (100°C) and sodium chloride (1473°C)? Why does gold have a density (19.3 g/cm^3) more than seven times that of aluminum (2.70 g/cm^3)? To answer such questions, we have to consider the particle structure of matter, i.e., the tiny "building blocks" of which elements and compounds are composed. We begin our study of the microstructure of matter in this chapter by examining three such particles: the atom, the molecule, and the ion.

2.1 ATOMIC THEORY

The notion that matter consists of discrete particles is an old one. About 400 B.C. this idea appeared in the writings of Democritus, a Greek philosopher, who apparently had been introduced to it by his teacher, a man named Leucippus. The idea was rejected by Plato and Aristotle, and it was not until about 1650 A.D. that it again was suggested, this time by the Italian physicist Gassendi. Sir Isaac Newton (1642–1727) supported Gassendi's arguments with these words:

> . . . it seems probable to me that God, in the Beginning, formed Matter in solid, massy, hard, impenetrable, movable Particles, of such Sizes and Figures, and with such other Properties, and in such Proportions to Space, as most conduced to the End for which he formed them. . . .

As far as we know, Newton did no experiments to test his ideas

Prior to 1800 the concept of the particle nature of matter was based largely on speculation and intuition. Then, in 1808, an English schoolteacher, John Dalton, using scientific insight of a remarkable quality, developed an explanation of several of the then known laws of chemistry which became known as the **atomic theory.** Some of the ideas of Dalton had to be discarded as chemists learned more about the structure of matter, but the essentials of his theory have stood the test of time. Three of Dalton's main postulates, which comprise modern atomic theory, are given here with examples to illustrate the meaning of each.

Atoms are the "no-see-ems" of the chemical world

1. *An element is composed of extremely small particles called atoms. All atoms of a given element show the same chemical properties.* The element oxygen is made up of oxygen atoms. These atoms are very small indeed; they have a diameter of the order of 10^{-10} m and a mass of about 10^{-23} g. All oxygen atoms behave chemically in the same way.

2. *Atoms of different elements have different properties. In the course of an ordinary chemical reaction, no atom of one element disappears or is changed into an atom of another element.* The chemical behavior of oxygen atoms is different from that of hydrogen atoms or any other kind of atom. When the elementary substances hydrogen and oxygen combine with each other, all the hydrogen atoms and all the oxygen atoms that react are present in the water formed, and no atoms of any other element are formed in the process.

3. *Compound substances are formed when atoms of more than one element combine. In a given pure compound the relative numbers of atoms of the elements present will be definite and constant. In general, these relative numbers can be expressed as integers or simple fractions.* In the compound substance water, hydrogen atoms and oxygen atoms are combined with each other. For every oxygen atom present, there are always two hydrogen atoms. Ammonia, a gaseous compound of nitrogen and hydrogen, always contains three hydrogen atoms for every nitrogen atom.

The atomic theory offers a simple explanation for two of the basic laws of chemistry:

1. The **Law of Conservation of Mass** which, in modern form, states that *there is no detectable change in mass in an ordinary chemical reaction.* If atoms are "conserved" in a reaction (postulate 2), mass must also be conserved.

2. The **Law of Constant Composition,** which tells us that a *compound,* regardless of its origin or method of preparation, *always contains the same elements in the same proportions by mass.* Clearly, if the atom ratio of the elements in a compound is fixed (postulate 3), their proportions by mass must also be fixed.

The validity of this law became generally recognized at about the same time that Dalton's theory appeared. Prior to 1808 many people agreed with the French chemist, Berthollet, who believed that the composition of a compound could vary over wide limits, depending on how it was prepared. Joseph Proust, a French expatriate working in Madrid, refuted Berthollet by showing that the "compounds" Berthollet had cited were actually mixtures.

It is now known that in certain compounds, particularly metal oxides and sulfides, the atom ratio may vary slightly from a whole number ratio. Careful analyses of apparently homogeneous samples of nickel oxide prepared by heating nickel with oxygen at high temperatures give an atom ratio of 0.97:1.00 rather than the expected 1:1 ratio. Compounds of this type are sometimes referred to as "Berthollides" or, more frequently, *nonstoichiometric* compounds. The deviations from constant composition arise because of defects in the crystal structure (Chapter 11).

The third postulate of the atomic theory led Dalton to formulate another of the basic laws of chemistry, the **Law of Multiple Proportions,** which states that *when two elements combine to form more than one compound, the masses of one element which combine with a fixed mass of the other element are in a ratio of small whole numbers* such as 2:1. Dalton reasoned that elements A and B might form two compounds, in one of which two atoms of A were combined with one of B, while in the other, one atom of A was combined with one atom of B. If this happened, the mass of A combined with a fixed mass of B in the first compound would be twice that in the second. The validity of this law was tested and soon established, partly on the basis of experiments that Dalton himself carried out.

Example 2.1 The elements carbon and oxygen form two different compounds. The first contains 42.9% by mass of carbon and 57.1% by mass of oxygen. In the second compound, the percentages of carbon and oxygen are 27.3 and 72.7, respectively.

a. Show that these data are consistent with the Law of Multiple Proportions.
b. Suggest a possible explanation in terms of the atomic theory.

See also Problems 2.9 and 2.25

Solution

a. According to the Law of Multiple Proportions, the masses of oxygen in the two compounds that combine with a fixed mass of carbon should be in the ratio of small whole numbers. To show that this is the case here, it is convenient to base our calculations on one gram of carbon. In the first compound, the mass of oxygen per gram of carbon is

$$\frac{57.1 \text{ g O}}{42.9 \text{ g C}} = 1.33 \text{ g O/g C}$$

In the second compound, we have

$$\frac{72.7 \text{ g O}}{27.3 \text{ g C}} = 2.66 \text{ g O/g C}$$

Clearly the masses of oxygen that combine with one gram of carbon are in a simple 2:1 ratio, i.e., 2.66/1.33 = 2:1.

b. A simple explanation (which happens to be the correct one!) is that in the first compound (carbon monoxide) there is one atom of oxygen per atom of carbon while in the second (carbon dioxide) there are two oxygen atoms for every carbon atom.

2.2 COMPONENTS OF THE ATOM

Like any useful scientific theory, the atomic theory raised more questions than it answered. Even before Dalton's ideas had been generally accepted, philosophers and scientists were speculating as to whether atoms, tiny as they are, could in turn be broken down into still smaller particles. Nearly one hundred years were to pass before this question was answered on the basis of experimental evidence. Three physicists did pioneer work in this area: J. J. Thomson, an Englishman working at the Cavendish laboratory at Cambridge; Ernest Rutherford, a native of New Zealand who carried out his research at McGill University in Montreal and at Manchester and Cambridge in England; and Robert A. Millikan at the University of Chicago.

Electrons

The first convincing evidence for subatomic particles came from experiments involving the conduction of electricity through gases at low pressures. When an apparatus of the type shown in Figure 2.1 is partially evacuated and connected to a high voltage source such as a spark coil, an electric current flows through the tube. Associated with the flow are colored rays of light, which originate at the negative electrode (cathode). The properties of *cathode rays* were studied extensively during the last three decades of the nineteenth century. In particular, it was found that the rays were bent by both electric and magnetic fields. From a careful study of the nature of this deflection, J. J. Thomson demonstrated in 1897 that the rays consist of a stream of negatively charged particles, which he

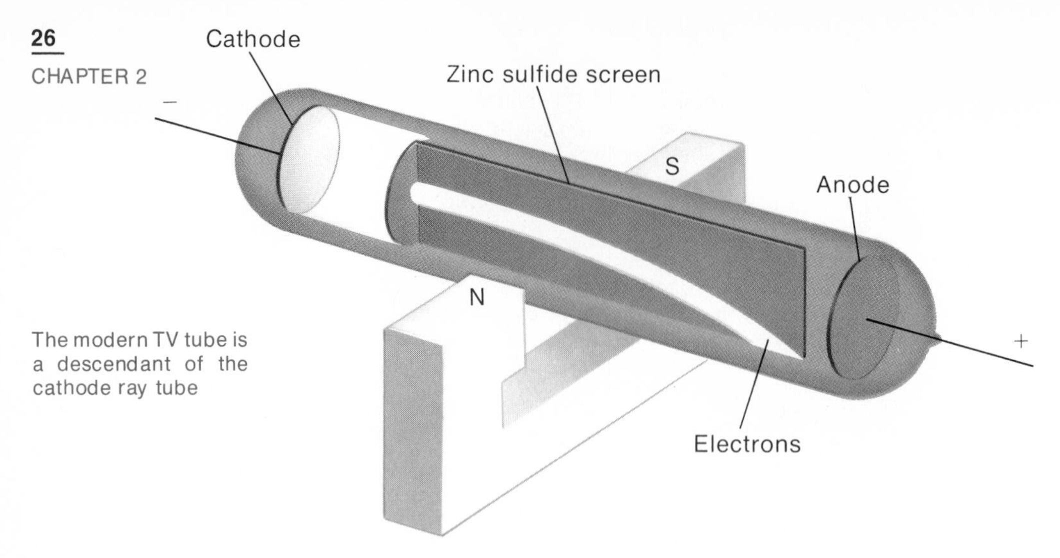

The modern TV tube is a descendant of the cathode ray tube

Figure 2.1 Cathode ray tube. The cathode ray is made up of a beam of fast-moving electrons. In an electric or magnetic field the beam is deflected in such a way as to indicate that it carries a negative charge.

called *electrons*. Thomson went on to measure the mass-to-charge ratio of the electron, in grams per coulomb, finding that

$$m/e = 5.69 \times 10^{-9} \text{ g/C}$$

The fact that this ratio is the same, regardless of what gas is in the tube, implies that the electron is a fundamental particle, common to all atoms.

In 1909 Millikan determined the charge on the electron, using the apparatus shown schematically in Figure 2.2. He measured the effect of an electrical field

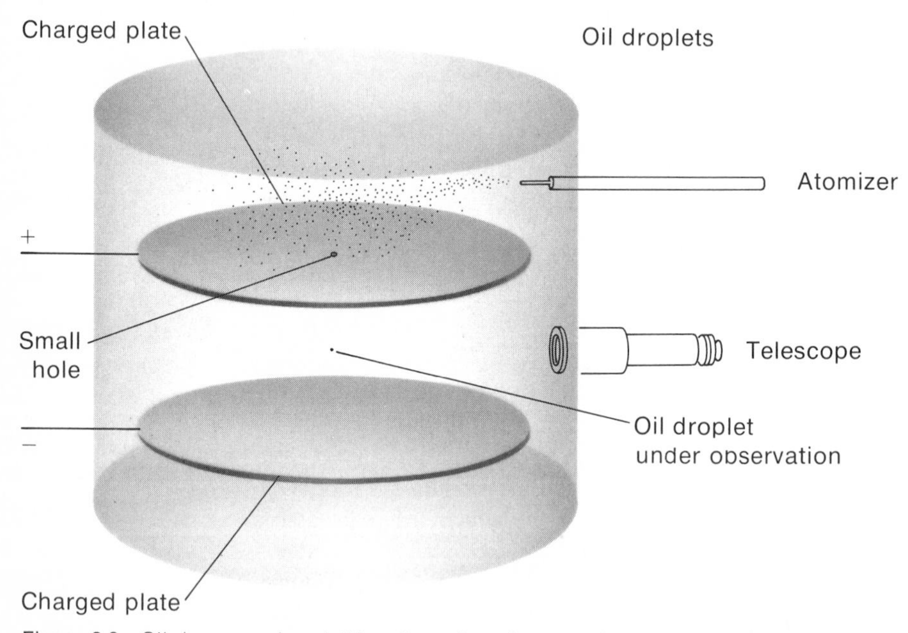

Figure 2.2 Oil drop experiment. When the voltage between the plates is increased, the negatively charged drop falls more slowly, since it is attracted to the positively charged plate. At one particular voltage, the electrical and gravitational forces on the drop are exactly balanced, and the drop remains stationary. Knowing this voltage and the mass of the drop, it is possible to calculate the charge on the drop.

on the rate at which charged oil drops fall under the influence of gravity. From his data Millikan calculated the charge on the drops, which he found always to be an integral multiple of a smallest charge. Assuming the smallest charge to be that of the electron, he arrived at a value of 1.60×10^{-19} C. Combining this number with the mass-to-charge ratio quoted previously, we obtain for the mass of the electron

$$m = (1.60 \times 10^{-19}\ \text{C}) \times (5.69 \times 10^{-9}\ \text{g/C})$$
$$= 9.11 \times 10^{-28}\ \text{g}$$

This is roughly 1/2000 of the mass of the lightest atom, that of the element hydrogen.

All atoms contain an integral number of electrons. This number, which may vary from 1 to over 100, is characteristic of an atom of a particular element. All hydrogen atoms contain one electron; all atoms of the element lawrencium contain 103 electrons. We shall have more to say in Chapter 6 about the way in which these electrons are distributed relative to one another. For the time being, it is sufficient to point out that electrons are found in the outer regions of atoms, where they comprise what amounts to a cloud of negative charge about the positively charged atomic nucleus.

The Atomic Nucleus

In 1911 Ernest Rutherford and his students carried out a series of experiments that profoundly influenced our ideas about the nature of the atom. Using a radioactive source, they bombarded a piece of thin gold foil (Figure 2.3) with α-particles (helium atoms stripped of their electrons). With a fluorescent screen they observed the extent to which the α-particles were scattered. Most of them went through the foil almost undeflected. A few, however, were reflected back from the foil at acute angles. The relative numbers of α-particles reflected at different angles were determined by counting on the screen the scintillations caused by individual particles. By a beautiful mathematical analysis of the electrostatic forces involved, Rutherford was able to show that the scattering was caused by a positively charged center within the gold atom which had a mass nearly equal to that of the atom but a diameter (about 10^{-14} m) only 1/10 000 of that of the atom. The experiment was repeated, with similar results, with foils of many different elements. In this way Rutherford established that an atom contains a tiny, positively charged, massive center, called the **atomic nucleus.**

Before this experiment was done, it was generally supposed that + and − particles were more or less uniformly distributed throughout an atom

Protons and Neutrons. Atomic Number and Mass Number

Since the time of Rutherford we have learned a great deal about the properties of atomic nuclei, although we still do not have a clear physical picture of the forces that hold the nucleus together. For our purposes, we may consider the nucleus of an atom to be made up of two different types of particles:

1. The **proton,** which has a mass nearly equal to that of the hydrogen atom and carries a unit positive charge, equal in magnitude but opposite in sign to that of the electron.

2. The **neutron,** an uncharged particle with a mass about equal to that of the proton.

All nuclei contain an integral number of protons, exactly equal to the number of electrons in the neutral atom. In the nucleus of every hydrogen atom there is one proton; the nucleus of every lawrencium atom contains 103 protons. The number of protons in the nucleus of an atom is a fundamental property of the corresponding element, known as its **atomic number.**

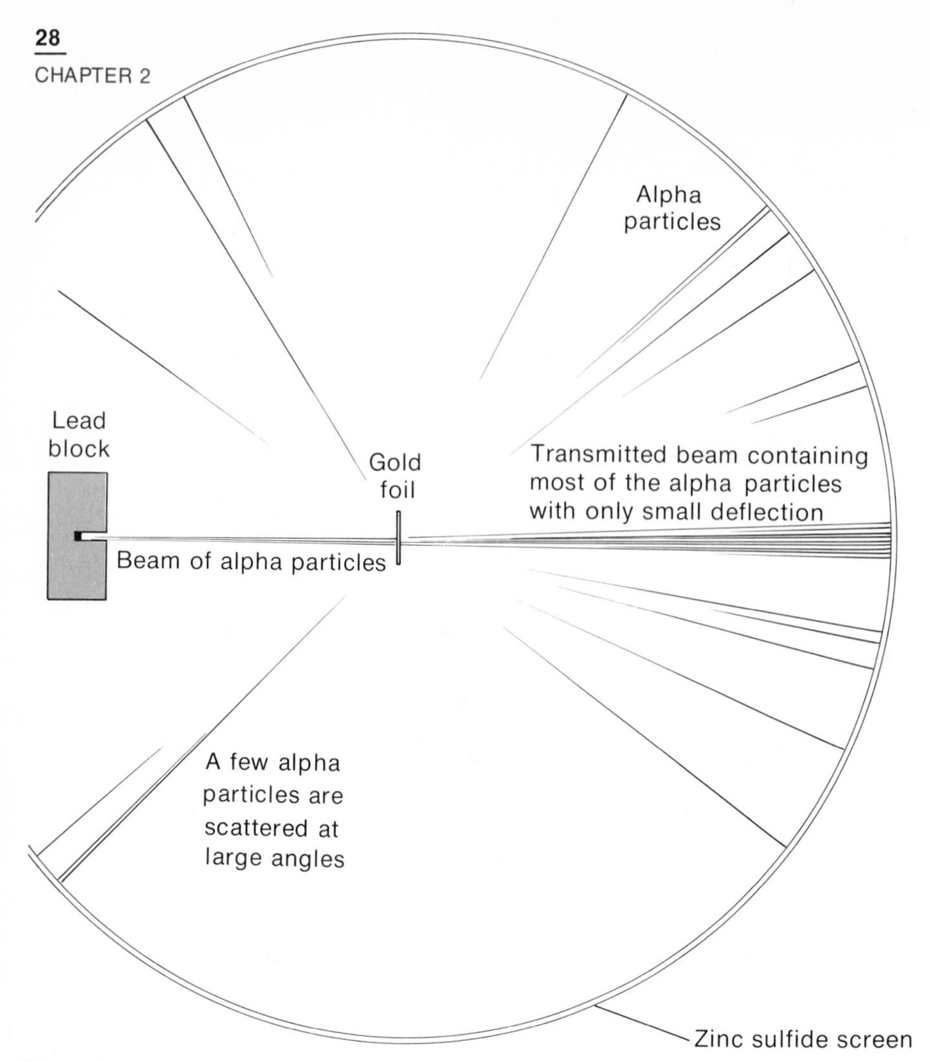

Figure 2.3 Rutherford scattering experiment. Most of the α-particles are essentially undeflected, but a few are scattered at large angles. In order to cause the large deflections, atoms must contain heavy, positively charged nuclei.

$$\text{atomic number} = \text{number of protons}$$

Thus, we say that the atomic number of the element hydrogen is 1, while that of the element lawrencium is 103.

Atoms of the same element may vary in the number of neutrons found in the nucleus. In the element hydrogen, for example, we find three different kinds of atomic nuclei containing 0, 1, and 2 neutrons, respectively. These three species are often referred to as **isotopes** of the element hydrogen. They differ in mass; the heaviest isotope of hydrogen (tritium) has a mass about three times as great as that of the lightest isotope. A deuterium atom (1 proton, 1 neutron) is about twice as heavy as a "light" hydrogen atom (1 proton, 0 neutrons). As another example, two well known isotopes of uranium, "uranium-235" and "uranium-238," both contain the same number of protons, 92, but differ in the number of neutrons, 143 vs. 146.

The **mass number** of a nucleus is found by adding the numbers of protons and neutrons.

$$\text{mass number} = \text{number of protons} + \text{number of neutrons}$$

The three isotopes of hydrogen have mass numbers of 1, 2, and 3, respectively, while the mass numbers of the two isotopes of uranium are

"light isotope": mass number = 92 + 143 = 235 (uranium-235)

"heavy isotope": mass number = 92 + 146 = 238 (uranium-238)

We often indicate the composition of a nucleus by writing the atomic number as a subscript at the lower left of the symbol of the element and the mass number as a superscript at the upper left. For the species discussed above, we would write

$$^{1}_{1}H,\ ^{2}_{1}H,\ ^{3}_{1}H;\ ^{235}_{92}U,\ ^{238}_{92}U$$

Example 2.2

a. Write nuclear symbols for three isotopes of oxygen (atomic no. = 8) in which there are 8, 9, and 10 neutrons, respectively.

b. One of the most harmful species in nuclear fallout is a radioactive isotope of strontium, $^{90}_{38}Sr$. How many protons are there in this nucleus? how many neutrons?

Solution

a. The mass numbers must be: 8 + 8 = 16; 8 + 9 = 17; 8 + 10 = 18. Thus we have:

$$^{16}_{8}O,\ ^{17}_{8}O,\ ^{18}_{8}O$$

b. The number of protons is given by the atomic number, 38. To obtain the number of neutrons we subtract the number of protons from the mass number.

$$\text{number of neutrons} = 90 - 38 = 52$$

2.3 MOLECULES AND IONS

Isolated atoms are rarely encountered in matter. Only in a very few elementary substances, the so-called noble gases (He, Ne, Ar, Kr, Xe, Rn), is the individual atom the structural unit of which the substance is composed. Most elements and compounds are made up of other types of structural units, two of the most important of which are molecules and ions.

Molecules

The basic structural unit in most volatile substances, both elementary and compound, is the molecule, which is a small group of atoms held together by relatively strong forces called chemical bonds. In contrast, the forces between molecules are relatively weak. As a result, molecules do not interact strongly with one another but behave more or less as independent particles.

Hydrogen chloride is an example of a molecular substance. The hydrogen chloride molecule is a simple one consisting of a hydrogen atom joined by a chemical bond to a chlorine atom. The structure of the hydrogen chloride molecule is often indicated as

H—Cl

where the dash represents the bond holding the two atoms together. Examples of elementary molecular substances are hydrogen and chlorine. At ordinary temperatures and pressures, the "building block" in both of these substances is a diatomic molecule and may be represented as

Most free atoms are very reactive; Cl atoms would combine instantly to form molecules

$$\mathrm{H{-}H} \qquad\qquad \mathrm{Cl{-}Cl}$$

Most of the molecular substances with which you are familiar are built up of molecules more complex than those just cited. The water molecule, for example, consists of a central oxygen atom bonded to two hydrogen atoms. In the ammonia molecule, a central nitrogen atom is bonded to three individual hydrogen atoms. Methane, the major constituent of natural gas, has as its basic structural unit a molecule in which a carbon atom is bonded to four hydrogen atoms. The structures of these molecules can be represented in two dimensions as

```
                                    H
     O           H—N—H              |
    / \            |            H—C—H
   H   H           H                |
                                    H
```

Two-dimensional "structural formulas" do not, in general, represent the true geometry of any but the simplest of molecules. Space-filling models of the type shown in Figure 2.4 give a more adequate picture of the positions of atoms within a molecule. We will have more to say about molecular geometry and other aspects of molecular structure when we discuss chemical bonding in more detail in Chapter 8.

Ions

If sufficient energy is available, it is possible to remove one or more electrons from a neutral atom, leaving a positively charged particle that is somewhat smaller than the original atom. Alternatively electrons may be added to certain atoms to form negatively charged species that are somewhat larger than the atom from which they are derived. These charged particles are referred to as **ions.** Examples of positive ions **(cations)** include the sodium ion, Na^+, formed from a sodium atom by the loss of a single electron, and the Ca^{2+} ion, derived from a calcium atom by extraction of two electrons

$$\text{Na atom } (11\ p^+,\ 11\ e^-) \rightarrow \text{Na}^+ \text{ ion } (11\ p^+,\ 10\ e^-) + e^-$$

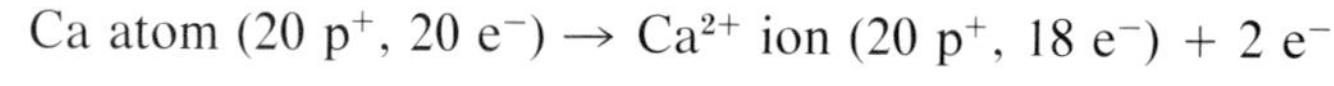

$$\text{Ca atom } (20\ p^+,\ 20\ e^-) \rightarrow \text{Ca}^{2+} \text{ ion } (20\ p^+,\ 18\ e^-) + 2\ e^-$$

Figure 2.4 Structures of some simple molecules. The atoms in molecules are held together by strong bonds. Many molecules have a considerable amount of symmetry.

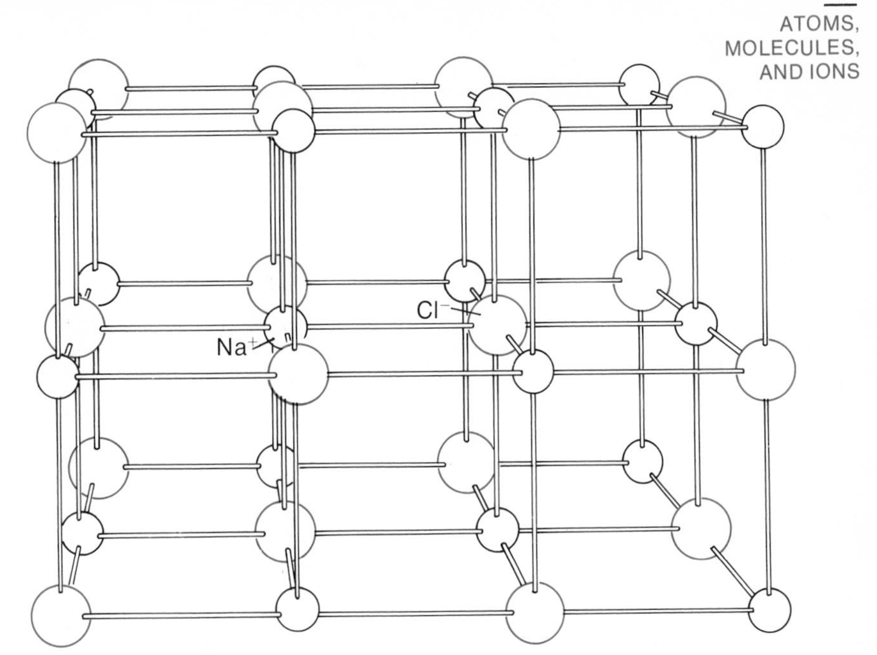

Figure 2.5 Crystal structure of NaCl, expanded for clarity. Like the Cl^- ion in the center of the cube, all chloride ions are surrounded by 6 sodium ions. Each Na^+ ion is similarly surrounded by 6 Cl^- ions. The crystal includes billions of ions in the pattern shown.

Two common negative ions **(anions)** are the chloride ion, Cl^-, and the oxide ion, O^{2-}, formed when atoms of chlorine or oxygen acquire extra electrons.

$$\text{Cl atom } (17\ p^+,\ 17\ e^-) + e^- \rightarrow \text{Cl}^- \text{ ion } (17\ p^+,\ 18\ e^-)$$

$$\text{O atom } (8\ p^+,\ 8\ e^-) + 2\ e^- \rightarrow \text{O}^{2-} \text{ ion } (8\ p^+,\ 10\ e^-)$$

The superscript on the symbol for an ion indicates its charge. The Na^+ ion has a positive charge equal in magnitude to the charge of an electron. The O^{2-} ion has a negative charge equal to that of two electrons.

Since a macroscopic sample of matter must be electrically neutral, ionic compounds always contain both cations and anions as structural units. Ordinary table salt, sodium chloride, is made up of an equal number of Na^+ and Cl^- ions. The structure of a small portion of a sodium chloride crystal is shown in Figure 2.5; that of calcium oxide (quicklime), in which there are equal numbers of Ca^{2+} and O^{2-} ions, is quite similar. In the ionic compounds sodium oxide and calcium chloride there are unequal numbers of cations and anions. In order to maintain electroneutrality there must be two Na^+ ions for every O^{2-} ion in sodium oxide; similarly two Cl^- ions are required to balance one Ca^{2+} ion in calcium chloride.

You can't buy a bottle of Na^+ ions

It is important to recognize that in all ionic compounds there is a continuous, three-dimensional network of chemical bonds holding oppositely charged ions together. Notice, from Figure 2.5, that there are no discrete "molecules" in the sodium chloride structure analogous to those in molecular substances such as hydrogen chloride or water. Ionic bonds, which arise because of the electrical attraction between cations and anions, are extremely strong. Since these bonds have to be broken for melting or vaporization to occur, ionic compounds have relatively high melting and boiling points and are typically solids at room temperature.

2.4 RELATIVE MASSES OF ATOMS: ATOMIC MASSES

In the preceding two sections of this chapter, we have examined the composition of the basic structural units of matter—atoms, molecules, and ions. Now we shift our attention to an important property of these particles—their masses. Here and in the next two sections we shall see how various atoms, molecules, and ions compare to one another in mass.

Relative Atomic Masses; The Carbon-12 Scale

On the inside back cover of this book there is a table of *relative* **atomic masses** of the various elements. The numbers in this table indicate how heavy, on the average,* an atom of one element is compared to an atom of another element. When we find that the numbers listed for hydrogen and oxygen are 1.0079 and 15.9994, respectively, we conclude that a hydrogen atom is about 1/16 as heavy as an oxygen atom. Again, since the relative atomic mass of sulfur is 32.06, an atom of sulfur must be a little more than twice as heavy as an oxygen atom. More exactly, a sulfur atom weighs 32.06/15.9994 = 2.004 times as much as an oxygen atom. In general, for two elements X and Y,

$$\frac{\text{mass of an atom of X}}{\text{mass of an atom of Y}} = \frac{\text{atomic mass of X}}{\text{atomic mass of Y}}$$

(In this equation, and henceforth, we use the simpler term "atomic mass" to represent the relative atomic mass as defined above.)

In order to set up a scale of atomic masses, it is necessary to establish a standard value for one particular species. During the 19th century, several different standards were used, including H = 1.00 and oxygen = 100. In the first decade of this century, chemists reached general agreement on a scale that set the average atomic mass of oxygen to be exactly 16. After the discovery of isotopes, physicists decided that it was more convenient to assign the most common isotope of oxygen, $^{16}_{8}O$, an atomic mass of exactly 16. Since naturally occurring oxygen contains small amounts of heavier isotopes, the two scales differed slightly from one another. In 1961, this confusing situation was resolved by agreeing to establish a single scale in which the atomic mass 12 was assigned to the most common isotope of carbon. Such a compromise satisfied the physicists, who insisted that a single isotope should be used as a standard. For chemists, it had a very small effect. The atomic mass of oxygen, for example, changed from 16.0000 to 15.9994. The atomic masses used in this text and listed on the inside of the back cover are based on the "carbon-12" scale.

Determination of Atomic Mass. The Mass Spectrometer

In Section 2.7 we will comment on some of the classical approaches that were followed in the 19th century to arrive at atomic masses. Today these methods are largely of historical interest; they have been replaced by techniques that use the modern mass spectrometer, shown in Figure 2.6. This instrument measures the mass-to-charge ratio of charged particles. It can be used to investigate a wide variety of chemical problems; indeed, an early prototype of the mass spectrometer was used by J. J. Thomson to determine the mass-to-charge ratio of

*The word "average" is used here because most elements consist of a mixture of isotopes. (See Example 2.3.)

the electron. Before discussing how the mass spectrometer can be applied to atomic mass measurements, let us consider the principle upon which it is based.

In a mass spectrometer (Fig. 2.6), a gas sample at very low pressure is bombarded by a stream of high-energy electrons. A few of the gas particles are excited enough to be converted to positive ions. These ions are focussed into a narrow beam and accelerated by a voltage of 500 to 2000 V toward a magnetic field. The field deflects the ions from their straight-line path toward the collector.

The extent to which a beam of ions is deflected in a mass spectrometer depends upon several factors. Two of these can be controlled by the operator:

1. *The strength of the magnetic field.* The stronger the field, the greater will be the extent of deflection.

2. *The magnitude of the accelerating voltage.* The greater this voltage, the more rapidly the ions will be moving and hence the less they will be bent by the field.

Two other factors are determined by the nature of the ions themselves:

Cadillacs are harder to push around than Volkswagens

3. *Their mass.* The heavier the ion, the less it will be deflected.

4. *Their charge.* The greater the charge of the ion, the stronger will be its interaction with the field and hence the more it will be bent off course.

To illustrate how the mass spectrometer can be used to determine atomic masses, consider the element helium, which consists essentially of only one isotope, $^{4}_{2}He$. To make things simple, we will assume that only +1 ions are formed. Using a fixed accelerating voltage, we carefully measure the field strength at which a $^{12}_{6}C^{+}$ ion arrives at point A. In a second experiment we might find that, with the same voltage and field strength, a lighter $^{4}_{2}He^{+}$ ion would be bent to point B. We now gradually decrease the field strength and hence the deflection of the $^{4}_{2}He^{+}$ beam until it falls exactly at A. By comparing the field strength at this stage to that required to bring $^{12}_{6}C^{+}$ ions to the same point, we can readily calculate the relative

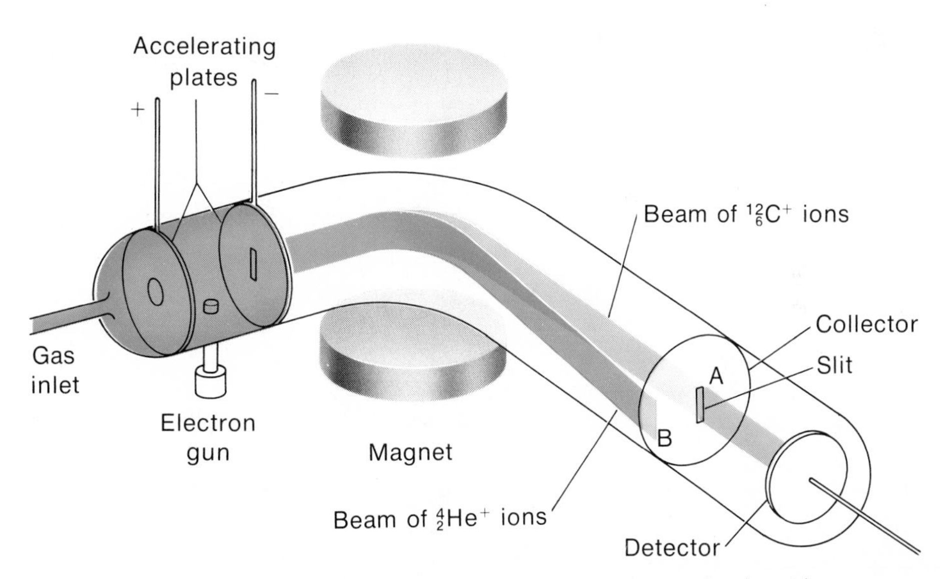

Figure 2.6 The mass spectrometer. In the mass spectrometer gas molecules at low pressure are ionized and accelerated by an electric field. The ion beam is then passed through a magnetic field; in that field the beam is resolved into components, each containing particles of equal mass. Lighter particles are deflected more strongly than heavy ones. In a beam containing $^{12}_{6}C^{+}$ and $^{4}_{2}He^{+}$ ions, the lighter $^{4}_{2}He^{+}$ ions would be deflected more than the heavier $^{12}_{6}C^{+}$ ions. The spectrometer shown is adjusted to detect the $^{12}_{6}C^{+}$ ions. By changing the magnitude of the magnetic or electric field, one could move the beam of $^{4}_{2}He^{+}$ ions striking the collector from B to A where it would be detected.

masses of the two ions.* When this is done, we find that He-4 has a mass about 1/3 (more exactly, 0.3336) of that of C-12. Hence its atomic mass must be

$$(0.3336)(12.00) = 4.003$$

But, since naturally occurring helium contains essentially 100 per cent of this isotope, the number 4.003 must also represent the atomic mass of the element helium.

The determination of the atomic mass of an element which contains more than one isotope is somewhat more difficult. In the case of neon, where there are three isotopes, a beam of +1 neon ions is split into three segments ($^{20}_{10}Ne^+$, $^{21}_{10}Ne^+$, and $^{22}_{10}Ne^+$) when it passes through the magnetic field. To obtain the average atomic mass of the element, we need to know not only the mass but also the relative abundance of each isotope. The latter can be estimated by measuring the relative heights of the peaks produced on a recorder when the three beams reach the detector (Fig. 2.7). Knowing the mass of each isotope and its per cent abundance, we can readily calculate the average atomic mass (Example 2.3).

Example 2.3 From mass spectra it can be determined that the element neon consists of three isotopes whose masses on the carbon-12 scale are 19.99, 20.99, and 21.99. The abundances of these isotopes are, respectively, 90.92 per cent, 0.25 per cent, and 8.83 per cent. Calculate an accurate value for the atomic mass of neon.

Solution We can always obtain the average atomic mass of an element by adding the contributions of each of the isotopes. That is, for an element consisting of isotopes X, Y, . . . :

$$\text{atomic mass} = (\text{mass X})(\text{fraction X}) + (\text{mass Y})(\text{fraction Y}) + \ldots$$

where the fraction of the isotope is its percentage divided by 100. Applying this relation to neon, where there are three isotopes:

$$\begin{aligned}\text{atomic mass Ne} &= (19.99)(0.9092) + (20.99)(0.0025) + (21.99)(0.0883) \\ &= 18.17 + 0.05 + 1.94 = 20.16\end{aligned}$$

*The relation is: $\frac{\text{mass He-4}}{\text{mass C-12}} = \left(\frac{H \text{ for He-4}}{H \text{ for C-12}}\right)^2$, where H is the magnetic field strength.

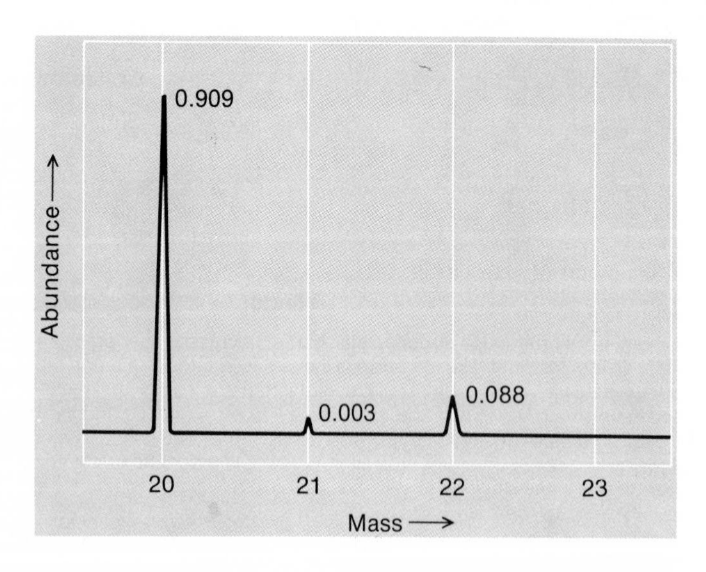

Figure 2.7 Mass spectrum of neon (+1 ions only). Neon contains three isotopes, of which neon-20 is by far the most abundant. The mass of that isotope, to five decimal places, is 19.992 44 on the carbon-12 scale.

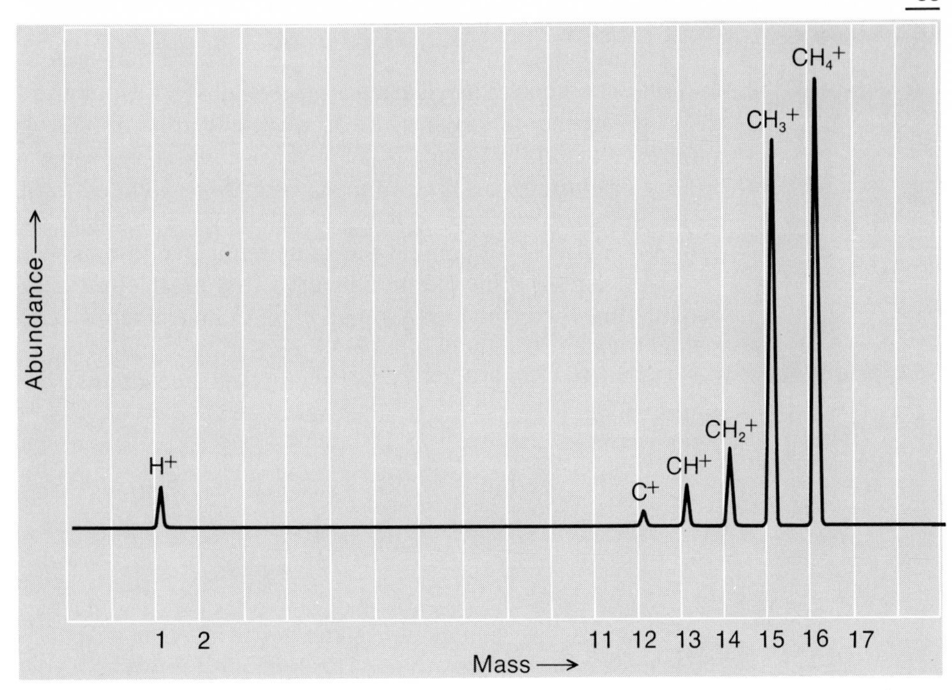

Figure 2.8 Mass spectrum of methane. The mass spectra of most organic compounds are much more complex than the spectrum shown.

For most elements, isotopic abundances remain constant regardless of where the element comes from or how it is prepared. Consequently, we are working with an average "mix" of isotopes and the average atomic mass listed in tables is the one used in all chemical calculations. Slight variations in natural abundances have been detected for a few elements (notably lead); here the accuracy of the listed atomic mass (207.2 ± 0.1) is limited by this effect.

When the substance entering the mass spectrometer is made up of polyatomic molecules rather than individual atoms, the pattern obtained is more complex. In Figure 2.8, we show the mass spectrum of the simplest gaseous hydrocarbon, methane, CH_4. The parent peak, at mass 16, is formed by the loss of an electron from the methane molecule. The other peaks are attributed to fragments of the molecule formed by breaking C—H bonds. Ethane, C_2H_6, and other hydrocarbons, in which there are C—C as well as C—H bonds, show considerably more complicated mass spectra.

At a given voltage, the mass spectrum of a substance, like its infrared spectrum, is characteristic of that species and can be used to identify it. Mass spectroscopy is widely used to identify molecular substances. It is particularly useful for detecting trace constituents in mixtures, which show up as detectable peaks even when present at concentrations as low as a few parts per million. The nature and amount of air pollutants such as the oxides of sulfur and nitrogen are often determined by using the mass spectrometer.

2.5 MASSES OF ATOMS; AVOGADRO'S NUMBER

From experiments such as those described in Section 2.4, we can establish the relative masses of atoms of different elements. Having found, for example, that the atomic masses of helium and hydrogen are 4.003 and 1.008, respectively, we conclude that a helium atom is about four times as heavy as a hydrogen atom.

The question remains as to the absolute masses of these atoms—i.e., what is the mass in grams of a helium atom? A hydrogen atom?

To answer this question, we start by realizing that since a helium atom weighs four times as much as a hydrogen atom, we must take four grams of helium to obtain the same number of atoms as there are in one gram of hydrogen. More exactly, a sample of helium weighing 4.003 g must contain the same number of atoms as 1.008 g of hydrogen. This argument is readily extended to other elements: 32.06 g of sulfur ($AM = 32.06$), 4.003 g of helium ($AM = 4.003$), and 1.008 g of hydrogen ($AM = 1.008$) all contain the same number of atoms.

The general principle just developed is most often expressed in terms of a quantity known as the *gram atomic mass* of an element (the mass in grams of that element which is numerically equal to its atomic mass):

One gram atomic mass (*GAM*) of any element contains a certain fixed number of atoms, *N*.

Thus we have:

$$1\ GAM\ \text{H} = 1.008\ \text{g H} = N \text{ atoms of hydrogen}$$

$$1\ GAM\ \text{He} = 4.003\ \text{g He} = N \text{ atoms of helium}$$

$$1\ GAM\ \text{S} = 32.06\ \text{g S} = N \text{ atoms of sulfur}$$

Clearly, if we can establish the value of N, it will be possible to calculate the masses of individual atoms. This number, known as Avogadro's number, may be determined in various ways; two different methods are illustrated by Problems 2.40 and 2.41 at the end of this chapter.

N was first calculated in 1865; Avogadro died in 1856

Avogadro's number, to four significant figures, has been found to be 6.022×10^{23} (i.e., 602 200 000 000 000 000 000 000). This is a number so large as to defy comprehension. Some idea of its magnitude may be gained when one realizes that if the entire population of the world were to be assigned to counting the number of atoms in one gram atomic mass of an element, each person counting one atom per second and working a 48-h week, the task would require more than three billion years. From another point of view, the fact that Avogadro's number is so huge means that the atom itself is almost inconceivably small. Individual atoms are far too small to be seen with the most powerful microscope or to be weighed on the most sensitive analytical balance (Example 2.4).

Example 2.4

a. Calculate the mass of a hydrogen atom.

b. Calculate the number of magnesium atoms in a sample weighing 1.0×10^{-6} g. (This is the smallest sample of magnesium that can be weighed on the most sensitive analytical balance.)

Solution

a. Knowing that one gram atomic mass of hydrogen weighs 1.008 g and contains 6.022×10^{23} atoms, we have

$$1.008\ \text{g} = 6.022 \times 10^{23} \text{ atoms H}$$

$$\text{mass of 1 atom H} = 1 \text{ atom H} \times \frac{1.008\ \text{g}}{6.022 \times 10^{23} \text{ atoms H}}$$
$$= 1.674 \times 10^{-24}\ \text{g}$$

b. $1\ GAM\ \text{Mg} = 24.30\ \text{g Mg} = 6.022 \times 10^{23} \text{ atoms Mg}$

$$\text{number of atoms Mg} = 1.0 \times 10^{-6}\ \text{g} \times \frac{6.022 \times 10^{23} \text{ atoms}}{24.30\ \text{g}}$$
$$= 2.5 \times 10^{16} \text{ atoms (25 000 000 000 000 000)}$$

2.6 MASSES OF MOLECULES. MOLECULAR MASS

In dealing with substances such as water, ammonia, or methane, in which the fundamental building block is the molecule, it is convenient to define a number known as the **molecular mass** (*MM*), which tells us how heavy a molecule is compared to an atom of $^{12}_{6}C$. When we say, for example, that the molecular mass of water is "18," we mean that a water molecule is 18/12 as heavy as a carbon-12 atom (or 18/16 as heavy as an oxygen atom, or . . .).

If we know the composition of a molecule, we can readily calculate its molecular mass. Referring back to Figure 2.4, we see that for the three molecules shown

$$\begin{aligned}
MM \text{ water} &= AM \text{ oxygen} + 2(AM \text{ hydrogen}) \\
&= 16.00 + 2(1.008) = 18.02 \\
MM \text{ ammonia} &= AM \text{ nitrogen} + 3(AM \text{ hydrogen}) \\
&= 14.01 + 3(1.008) = 17.03 \\
MM \text{ methane} &= AM \text{ carbon} + 4(AM \text{ hydrogen}) \\
&= 12.01 + 4(1.008) = 16.04
\end{aligned}$$

In general, the molecular mass is always the sum of the atomic masses of the atoms in the molecule. Experimentally, molecular masses can be determined directly in a mass spectrometer in much the same way that atomic masses are established. Two classical experimental approaches to the determination of molecular masses will be discussed later (Chapters 5 and 12).

The *gram molecular mass* (*GMM*) of an elementary or compound substance is defined as the mass in grams which contains Avogadro's number (6.022×10^{23}) of molecules. (Note the analogy to gram atomic mass defined previously.) The molecular mass of water is 18.02; its gram molecular mass is 18.02 g. Similarly, the molecular and gram molecular masses of methane are 16.04 and 16.04 g, respectively.

Knowing the gram molecular mass of a substance, we can readily calculate the mass of an individual molecule (Example 2.5).

Example 2.5 Calculate the mass in grams of a water molecule.

Solution We know that the gram molecular mass of water is 18.02 g; one gram molecular mass of any substance contains 6.022×10^{23} molecules. Therefore

$$18.02 \text{ g water} = 6.022 \times 10^{23} \text{ molecules water}$$

$$\text{mass} = 1 \text{ molecule} \times \frac{18.02 \text{ g}}{6.022 \times 10^{23} \text{ molecules}} = 2.992 \times 10^{-23} \text{ g}$$

It takes a while to adjust to the idea that atoms and molecules are so small.

A HISTORICAL PERSPECTIVE; ATOMIC MASSES IN THE 19TH CENTURY

Too often, the beginning student is left with the impression that scientific principles are developed in a beautifully logical sequence by a succession of gifted individuals, all of whom understand and appreciate the works of their predecessors. Occasionally this happens; more frequently it does not. Scientists, even great scientists, are capable of mistakes in reasoning.

Sometimes they continue to maintain a point of view in spite of an overwhelming amount of evidence to the contrary. Perhaps we can illustrate the human qualities of scientists and the tortuous path that leads to scientific principles if we follow the atomic mass concept as it evolved in the 19th century.

DALTON. In his paper introducing the atomic theory in 1808, John Dalton wrote: "It is one great object of this work to show the importance and advantage of ascertaining the relative weights of the ultimate particles. . . ." Dalton realized that the problem of determining the relative masses of atoms of different elements was intimately related to that of finding the atom ratio in which they combine. Consider, for example, the elements hydrogen and oxygen, which react to form the compound water. Analysis shows that water contains about 1 part by mass of hydrogen to 8 parts by mass of oxygen. If we assume that the atom ratio in water is 1:1, it follows that a hydrogen atom must be 1/8 as heavy as an oxygen atom. If, on the other hand, two atoms of hydrogen are combined with one atom of oxygen, then a hydrogen atom need be only 1/16 as heavy as an oxygen atom to explain the observed combining mass ratio of 1:8. Many other atom ratios are possible, each leading to a different value for the relative masses of hydrogen and oxygen atoms. In general, one cannot determine the atomic masses of elements by chemical analysis alone.

Dalton saw no simple way of determining the atom ratio in which two elements combine. Consequently, he made what seemed to him to be a reasonable assumption:

"When only one combination of two . . . (elements) . . . can be obtained, it must be presumed to be a binary one (1 atom A + 1 atom B) . . . ."

Applying this reasoning to the water molecule, Dalton assumed it to be made up of one hydrogen and one oxygen atom. This would require that hydrogen be assigned an atomic mass of 2 on a scale based on oxygen = 16. From our vantage point in time, it is obvious that Dalton was wrong. Indeed, even in 1808, there was evidence which, properly interpreted, would have led to the formula H_2O for water rather than HO.

DULONG AND PETIT, BERZELIUS. The first direct approach to the determination of atomic masses was proposed in 1819 by two French scientists, Pierre Dulong and Alexis Petit. They suggested that the amount of heat required to raise the temperature of an atom of a solid element by a given amount, let us say 1°C, should be independent of the type of atom; in their words, "the atoms of all simple bodies have the same capacity for heat." If this is true, then since one gram atomic mass of every element contains the same number of atoms, a fixed amount of heat should be required to raise the temperature of one gram atomic mass of a solid element by 1°C.

The Law of Dulong and Petit is expressed most simply in terms of the property known as specific heat, which is the amount of heat required to raise the temperature of one gram of a substance by 1°C. In modern form, the Law states that the product of the specific heat of a solid element multiplied by its gram atomic mass is approximately 25 J/°C.

$$GAM \times \text{specific heat} \approx 25\ \text{J/°C} \qquad (2.1)$$

Equation 2.1 is far from exact; for many elements the product differs from 25 J/°C by 10% or more. It does, however, yield approximate gram atomic masses for metals which can be refined using precise data available from chemical analysis. To understand what is involved, consider the element zinc. As early as 1818, the Swedish chemist Berzelius had shown that, in the oxide of zinc, 4.032 g of zinc were combined with one gram of oxygen. For reasons which we will not go into here, Berzelius was convinced that in this compound one atom of zinc was combined with two atoms of oxygen. If this were the case, the gram atomic mass of zinc would be the mass that combines with two gram atomic masses of oxygen:

$$4.032 \times 32.00 \text{ g} = 129.0 \text{ g}$$

However, Dulong and Petit in 1819 reported the specific heat of zinc to be 0.39 J/(g·°C). Applying Equation 2.1, we calculate the approximate atomic mass of zinc to be

$$\frac{25 \text{ J/°C}}{0.39 \text{ J/(g·°C)}} \approx 64 \text{ g}$$

Berzelius realized what this result implied. The atom ratio of zinc to oxygen in zinc oxide must be 1:1 rather than 1:2. Going back to his analytical data, he recalculated the gram atomic mass of zinc to be the mass that combined with one gram atomic mass of oxygen:

$$4.032 \times 16.00 \text{ g} = 64.51 \text{ g}$$

After making several revisions of this type, Berzelius published an extensive table of atomic masses in 1826, portions of which are reproduced in Table 2.1. With a few exceptions (notably silver) the values he quoted were very close to those accepted today. Considering the crude equipment he had to work with, this was a truly amazing accomplishment. Ironically, no one seems to have paid much attention to the atomic masses of Berzelius; they disappeared from the chemical literature, to be "rediscovered" 40 years later.

TABLE 2.1 ATOMIC MASSES OF BERZELIUS*

	1818	1826	Modern Value
Ag	432.51	216.26	107.87
Al	54.72	27.39	26.98
Bi	283.86	212.86	208.98
C	12.05	12.25	12.01
Ca	81.93	40.96	40.08
Fe	108.55	54.27	55.85
Mg	50.68	25.34	24.30
Pb	414.24	207.12	207.2
S	32.19	32.19	32.06
Zn	129.03	64.51	65.38

*Corrected to the modern C-12 scale.

GAY-LUSSAC, AVOGADRO, AND CANNIZZARO. The Law of Dulong and Petit, as useful as it was, did nothing to resolve the controversy about the atomic masses of gaseous elements. The key that would eventually unlock that door was provided by the Frenchman Gay-Lussac, who, in 1808, discovered the law of combining volumes of gases (Chapter 5). Gay-Lussac seems not to have grasped the theoretical implications of his work.

However, Dalton saw clearly where Gay-Lussac's Law led and he didn't like what he saw. Consider, for example, the reaction between hydrogen and oxygen, where Gay-Lussac showed that

2 volumes hydrogen + 1 volume oxygen → 2 volumes water vapor

Dalton realized that this simple, integral relationship between combining volumes implied an equally simple relationship between reacting particles

2 particles hydrogen + 1 particle oxygen → 2 particles water

At this point, Dalton, equating particles with atoms, was in trouble. One atom of oxygen could hardly yield two particles of water, both of which must contain at least one atom of oxygen (in Dalton's words, "Thou canst not split an atom"). Faced with what he took to be a direct challenge to the atomic theory, Dalton attempted to dispute Gay-Lussac's results by citing contradictory experiments of his own. His argument accomplished little except to prove that he was a better theoretician than experimenter.

In 1811 an Italian physicist with the improbable name of Lorenzo Romano Amadeo Carlo Avogadro di Quaregua e di Cerreto called attention to the error in Dalton's reasoning. He pointed out that Dalton had confused the concepts of atoms and molecules. If one assumed that the oxygen molecule were *diatomic* then two molecules of water, each containing one atom of oxygen, could be formed from one oxygen molecule. Avogadro interpreted Gay-Lussac's Law to mean that

2 H_2 molecules + 1 O_2 molecule → 2 H_2O molecules

After thinking about this for a while, Avogadro concluded that Gay-Lussac's Law really implied that *equal volumes of all gases at the same temperature and pressure contain the same number of molecules.*

Avogadro's hypothesis suggested a simple method of determining molecular mass. If it were correct, then the densities of gases at the same temperature and pressure must be in the same ratio as their molecular masses. Unfortunately Avogadro's ideas made very little impact on the scientific community; examination of the literature of that period suggests that many eminent chemists did not bother to read Avogadro's paper (this still happens from time to time). Those who were familiar with Avogadro's ideas tended to dismiss them because they refused to believe that an element could form diatomic molecules.

After all, why *should* two identical atoms combine with each other?

Nearly fifty years later Avogadro's hypothesis was revived by a fellow countryman, Stanislao Cannizzaro, professor of chemistry at Genoa. He pointed out that it could be used to determine not only molecular masses but atomic masses as well. To see how this is possible, let us consider the problem of determining the atomic mass of hydrogen, which forms a large number of gaseous compounds with other elements. One cubic decimetre of each of these compounds at, let us say, 25°C and 1 atm, must, by Avogadro's hypothesis, contain a fixed number of molecules, which we shall call x. The number of atoms of hydrogen in 1 dm^3 of a gaseous compound will be x if there is one atom of hydrogen per molecule, 2x if there are two, and so on. With this idea in mind, let us consider the experimental data in Table 2.2.

TABLE 2.2 MASSES OF HYDROGEN IN 1 dm^3 OF GASEOUS COMPOUNDS AT 25°C, 1 ATM

Compound	Mass 1 dm^3	×	%H	=	Mass H
Hydrogen chloride	1.491 g		2.76		**0.0412** g
Ethylene	1.147 g		14.37		0.1648 g
Hydrogen cyanide	1.105 g		3.73		**0.0412** g
Hydrogen sulfide	1.393 g		5.92		0.0824 g
Methane	0.655 g		25.15		0.1648 g
Ammonia	0.696 g		17.76		0.1236 g
Hydrogen fluoride	0.818 g		5.04		**0.0412** g
Phosphine	1.390 g		8.90		0.1236 g

Examination of Table 2.2 suggests that the compounds hydrogen chloride, hydrogen cyanide, and hydrogen fluoride each contain one atom of hydrogen per molecule and, hence, that x atoms of hydrogen weigh 0.0412 g. From a similar table for gaseous compounds of oxygen, we find that the smallest mass of oxygen is 0.654 g. Accordingly, 0.654 g must then be the mass of x atoms of oxygen. It follows that a hydrogen atom is

$$\frac{0.0412}{0.654} = \frac{1}{16}$$

as heavy as an oxygen atom, or that, on the modern scale, hydrogen has an atomic mass of about 1.

Cannizzaro presented his reasoning at a conference at Karlsruhe, called in 1860, to try to resolve the confusion concerning atomic masses of gaseous elements. His talk appears to have made few converts, but Cannizzaro was shrewd enough to distribute copies of his paper to all the delegates as they left the conference. Apparently, it was read widely enough to persuade the scientific community of the validity of Cannizzaro's argument. Within twenty years, all but a few diehards had been convinced and the atomic mass scale was finally established.

PROBLEMS

2.1 *Law of Multiple Proportions* Copper forms two compounds with oxygen. In compound (1), the percentages by mass of Cu and O are 79.89 and 20.11, respectively. In compound (2), these percentages are 88.82 and 11.18.

a. Calculate the masses of copper combined with one gram of oxygen in compounds (1) and (2).
b. Show how the numbers calculated in (a) illustrate the Law of Multiple Proportions.

2.2 *Composition of Atoms*

a. Consider the $^{19}_{9}F$ atom. This atom contains ____ protons, ____ electrons, and ____ neutrons.
b. Give the nuclear symbol of the isotope of tin (Sn) that contains 50 protons and 71 neutrons.

2.3 *Relative Masses of Atoms* The atomic masses of arsenic and oxygen are 74.9 and 16.0, respectively.

a. What is the ratio of the mass of an arsenic atom to that of an oxygen atom?
b. If the atomic mass of oxygen were taken to be 100, what would be the atomic mass of arsenic?

2.4 *Average Atomic Mass* The element boron consists of two isotopes of masses 10.02 and 11.01 whose abundances are 18.83 and 81.17%, respectively. Calculate the average atomic mass of boron.

2.5 *Gram Atomic Mass; Masses of Atoms* The atomic mass of Cd is 112.

a. What is the mass in grams of a Cd atom? ($N = 6.02 \times 10^{23}$)
b. How many atoms are there in 28.0 g of Cd?

2.6 *Gram Molecular Mass; Masses of Molecules* The acetylene molecule contains two carbon atoms (*AM* C = 12.01) and two hydrogen atoms (*AM* H = 1.008).

a. What is the gram molecular mass of acetylene?
b. How many molecules are there in 13.0 g of acetylene?

2.7 Describe an experiment which would

a. demonstrate the Law of Conservation of Mass.
b. demonstrate the Law of Constant Composition.
c. determine the mass of a proton.
d. show that elementary chlorine is a mixture of isotopes.

2.8 Which of the following statements is always true? never true?

a. A +1 ion is heavier than the corresponding atom.
b. The gram atomic mass is the mass in grams of an atom.
c. In two different compounds of A and B, the masses of A that combine with one gram of B are in a 2:1 ratio.

2.9 Referring to Example 2.1, show that the Law of Multiple Proportions could be demonstrated just as well by calculating the mass of carbon that combines with one gram of oxygen.

2.10 An element X forms two different compounds with Y. The percentages by mass of X in the compounds are 60.0 and 69.2. Show how these data illustrate the Law of Multiple Proportions.

2.11 Fill in the blanks below, using the table on the inside back cover where necessary.

	No. protons	No. neutrons	No. electrons	Charge
$^{45}_{21}Sc$	____	____	21	____
$^{33}_{16}S^{2-}$	____	____	____	−2
____	8	9	____	0
____	27	31	____	+2

2.12 Teflon is a polymer in which there are two fluorine atoms for every carbon atom.

a. What is the mass ratio of F to C in Teflon?
b. What are the mass percentages of F and C in Teflon?

2.23 Describe an experiment to

a. demonstrate the Law of Multiple Proportions.
b. determine the atomic mass of fluorine, which consists of a single isotope.
c. show that the ammonia molecule has one N and three H atoms.

2.24 Criticize each of the following statements.

a. Molecules are heavier than atoms.
b. There are 6.02×10^{23} molecules in 58.44 g of sodium chloride.
c. +2 ions are deflected more in the mass spectrometer than +1 ions.
d. Ionic compounds contain equal numbers of + and − ions.

2.25 Which of the following atom ratios would be consistent with the data in Example 2.1?

	Compound 1	Compound 2
a.	1 C:2 O	1 C:4 O
b.	2 C:1 O	1 C:1 O
c.	2 C:1 O	1 C:2 O

2.26 A certain metal M forms two oxides. In the first oxide, three atoms of M are combined with four of oxygen. In the second, the atom ratio is 2:3. Show that the ratio

$$\frac{\text{g M per g of O in compound 1}}{\text{g M per g of O in compound 2}}$$

must be 9/8.

2.27 Consider the isotope of plutonium used in nuclear fission, $^{239}_{94}Pu$.

a. How many protons are there in the nucleus?
b. How many neutrons are there in the nucleus?
c. How many electrons are there in a neutral atom of plutonium?
d. How many protons and electrons are there in a +3 ion of this isotope?

2.28 Petroleum, essentially a mixture of hydrocarbons, is about 13% H and 87% C by mass.

a. How many gram atomic masses of each element are there in 100 g of petroleum?
b. What is the atom ratio of H to C in petroleum?

2.13 Gallium consists of two isotopes of masses 68.95 and 70.95 with abundances of 60.16 and 39.84%, respectively. What is the average atomic mass of gallium?

2.14 Chlorine, with an average atomic mass of 35.45, consists of two isotopes of masses 34.98 and 36.98. Estimate the abundances of the isotopes.

2.15 What would be the total volume, in cubic metres, of Avogadro's number of golf balls, each of which has a volume of 20 cm^3 (two significant figures)?

2.16 Calculate

a. the number of atoms in 1.43 g of sodium (Na).
b. the mass in grams of a sodium atom.

2.17 A molecule of octane contains 8 carbon and 18 hydrogen atoms.

a. What is the molecular mass of octane?
b. What is the mass in grams of 1.20×10^{20} octane molecules?

2.18 A gold atom can be considered to be a sphere with a radius of 0.144 nm. Estimate the density of a gold atom ($V = 4\pi r^3/3$; $AM = 197$).

2.19 An investor pays $4500 for a kilogram of gold. How much is he paying per atom? per gram atomic mass?

2.20 DDT, an organic compound with a molecular mass of 354, has one of the lowest water solubilities of all known substances. Its solubility is estimated to be 1×10^{-6} g/dm^3. How many molecules of DDT are there in one cubic centimetre of saturated solution?

2.21 For use in semiconductor devices, germanium crystals are prepared in which only one atom out of every billion is an impurity. If such a crystal weighs one milligram, how many foreign atoms does it contain?

2.22 In silver oxide, the atom ratio of Ag to O is now known to be 2:1. Referring to Table 2.1, what ratio do you think Berzelius assumed in 1826? in 1818? If his 1826 value for the atomic mass of silver were correct, what would you predict for the specific heat of Ag? [The correct value is 0.23 J/(g · °C).]

2.29 The element carbon consists of two isotopes, $^{12}_{6}C$ (98.9%) and $^{13}_{6}C$ (1.1%). Taking the most common isotope to have a mass of 12.00 and assuming that of $^{13}_{6}C$ to be 13/12 as great, calculate the average atomic mass of carbon.

2.30 Silicon ($AM = 28.086$) consists mostly of an isotope of mass 27.985 (92.28%). It contains smaller amounts of two other isotopes of masses 28.99 and 29.98. Estimate their abundances.

2.31 One gram molecular mass of carbon monoxide is released in a garage where the total volume of air is 150 m^3. Assuming complete mixing, how many molecules of CO are there per cubic centimetre?

2.32 Work Problem 2.16, substituting silicon for sodium.

2.33 Arrange in order of increasing mass:

a. a copper atom
b. 1×10^{-21} *GAM* of copper
c. 1×10^{-21} *GMM* of water, H_2O
d. a Cu^{2+} ion
e. six water molecules

2.34 Taking the atomic radius of aluminum to be 0.143 nm, estimate the density of an Al atom. Can you explain why gold is about seven times as dense as aluminum?

2.35 The prices of copper, zinc, and lead are $1.50/kg, $0.88/kg, and $0.55/kg, respectively. Which metal is most expensive per gram atomic mass?

2.36 Male silkworm moths are attracted by very small amounts of an organic sex attractant with a molecular mass of 238. If 1×10^{-6} g is released in 10^3 m^3 of air, most of the male moths in the area are excited. How many molecules of the compound are there in one cubic centimetre of air?

2.37 When light strikes a photographic film covered with silver bromide, some of the Br^- ions are converted to bromine atoms which escape from the film. How many Br^- ions would have to be "lost" before one could detect a change in mass, using the most sensitive balance, which weighs to 1.0×10^{-6} g?

2.38 Berzelius found the percentages of Mg and O in magnesium oxide to be 61.30 and 38.70, respectively. Show how, assuming an atom ratio of 1 Mg:2 O, he arrived at an *AM* of 50.68 for Mg. Explain how, using a specific heat of 1.0 J/(g·°C) for Mg, he could arrive at an *AM* of 25.34 and the correct atom ratio in magnesium oxide.

*2.39 Taking the smallest mass of oxygen in 1 dm^3 of any of its gaseous compounds at 25°C and 1 atm to be 0.654 g and applying Cannizzaro's reasoning to the data given below, determine the atomic mass of nitrogen.

Compound	Mass 1 dm^3 (25°C, 1 atm)	%N	Compound	Mass 1 dm^3 (25°C, 1 atm)	%N
1	1.799 g	63.6	4	1.758 g	97.7
2	0.696 g	82.2	5	1.308 g	87.5
3	4.415 g	25.9	6	1.104 g	51.8

*2.40 One way to find Avogadro's number is to determine the number of electrons required to plate out a given mass of a particular metal. It is found that 893 C are required to form one gram of silver metal from Ag^+ ions. Using the known atomic mass of silver and the charge of the electron in coulombs given in this chapter, calculate the number of atoms in one gram atomic mass of silver.

*2.41 By x-ray diffraction, it is possible to determine the geometric pattern in which atoms are arranged in a crystal and the distances between atoms. It is found that in a crystal of copper, four atoms effectively occupy the volume of a cube 0.363 nm on an edge. Taking the density of copper to be 8.92 g/cm^3, calculate the number of atoms in one gram atomic mass of copper.

*2.42 In 1894, Lord Rayleigh found that the molecular mass of "nitrogen" extracted from air was 0.45% greater than that of chemically pure nitrogen. He showed that this effect was due to the presence of a previously unknown gas, argon, which has a molecular mass of 39.95 as opposed to 28.01 for pure nitrogen. What was the percentage of argon in the sample he extracted from air?

*2.43 In the text, we explained how the mass spectrometer could be used to determine the ratio of the masses of ${}^{12}_{6}C^+$ and ${}^{4}_{2}He^+$. A small correction has to be applied to obtain the ratio of the masses of the corresponding atoms. What is the nature of this correction? What is the percentage difference between the two ratios?

*2.44 The gram equivalent mass of an element is the mass that combines with eight grams of oxygen. Suppose the gram equivalent mass of a certain metal is 18.0 g. What is its gram atomic mass if the atom ratio of metal to oxygen is 2:1? 1:1? 2:3? 1:2? (The general relationship here is summarized by the equation $GAM = n \times GEM$, where n is a positive integer, i.e., 1, 2, 3, 4, ...)

3

CHEMICAL FORMULAS AND EQUATIONS

In Chapter 2, we examined the masses and compositions of the building blocks of matter: atoms, molecules, and ions. In chemistry, we represent the structures of substances containing these particles by chemical formulas. We start this chapter by considering two different types of formulas, the simplest and the molecular formulas. In later sections, we will discuss the use of these formulas in the balanced equations used to represent chemical reactions.

3.1 CHEMICAL FORMULAS

The **simplest** or **empirical** formula of a compound gives the simplest whole number ratio between the numbers of atoms of the different elements making up the compound. An example is the simplest formula of water, H_2O, which tells us that there are twice as many hydrogen atoms as oxygen atoms. In potassium chlorate, it has been established by experiment that the three elements potassium, chlorine, and oxygen are present in an atom ratio of 1:1:3; the simplest formula of potassium chlorate is $KClO_3$.

The **molecular** formula indicates the number of atoms in a molecule of a molecular substance. The molecular formula may be identical with the simplest formula, as is the case with water, H_2O: there are two atoms of hydrogen and one atom of oxygen per molecule of water. Alternatively the molecular formula may be an integral multiple of the simplest formula. This is the case with hydrogen peroxide, where we write the molecular formula H_2O_2 to indicate that two hydrogen atoms are combined with two oxygen atoms to form a single structural unit, the hydrogen peroxide molecule. The simplest formula of this compound would be HO.

Elements as well as compounds can be represented by molecular formulas. At ordinary temperatures and pressures, the element chlorine consists of diatomic molecules; its molecular formula is Cl_2. For the same reason, the gaseous elements oxygen, nitrogen, and fluorine are given the molecular formulas O_2, N_2, and F_2. In white phosphorus, a reactive, highly toxic solid, the basic building block is the P_4 molecule.

Chemical formulas are often used instead of names of substances

Where there is a choice between representing a substance by its simplest or molecular formula, we always use the latter because it is more informative. Thus we would write the molecular formulas H_2O_2 and Cl_2 for hydrogen peroxide and elementary chlorine, respectively. For substances which do not contain molecules, we are restricted to the simplest formula. An example is the formula NaCl written to represent the ionic compound sodium chloride. Elements in which there are no small, discrete molecules, such as iron and carbon, are represented by their symbols, i.e., Fe, C.

3.2 SIMPLEST FORMULAS FROM ANALYSIS

In order to determine the simplest formula of a compound, we must establish by chemical analysis the proportions by mass of the elements making up the compound. With this information and the known atomic masses of the elements, we can readily calculate the simplest formula. The reasoning involved is indicated in Example 3.1.

Example 3.1 Vitamin C, which may or may not be effective in preventing colds, is an organic compound containing the three elements carbon, hydrogen, and oxygen. Analysis of a carefully purified sample of vitamin C gives the following results for the mass percentages of the elements:

$$\%C = 40.9; \%H = 4.58; \%O = 54.5$$

From these data, determine the empirical formula.

Solution We need to know the relative numbers of atoms of C, H, and O. To obtain this information, let us calculate the relative numbers of gram atomic masses of C, H, and O in a given mass of vitamin C, for convenience, 100 g.

In 100 g, we have 40.9 g of C, 4.58 g of H, and 54.5 g of O. Converting to numbers of gram atomic masses, we obtain, for C, H, and O:

The result would be the same if we started with 1 g, 1000 g, or any fixed mass of vitamin C. Why?

$$40.9 \text{ g C} \times \frac{1 \; GAM \text{ C}}{12.0 \text{ g C}} = 3.41 \; GAM \text{ C}$$

$$4.58 \text{ g H} \times \frac{1 \; GAM \text{ H}}{1.01 \text{ g H}} = 4.53 \; GAM \text{ H}$$

$$54.5 \text{ g O} \times \frac{1 \; GAM \text{ O}}{16.0 \text{ g O}} = 3.41 \; GAM \text{ O}$$

These quantities represent not only the relative numbers of gram atomic masses of C, H, and O, *but also the relative numbers of atoms of these three elements,* since the conversion factor relating gram atomic masses to atoms (1 *GAM* = 6.02×10^{23} atoms) is the same for all elements. In other words, the numbers of atoms of C, H, and O in vitamin C are in the ratio 3.41:4.53:3.41.

To deduce the simplest formula, we need to know the simplest, whole number ratio between these three numbers. To obtain this, we divide each number by the smallest, 3.41:

$$\text{C: } \frac{3.41}{3.41} = 1.00; \text{ H: } \frac{4.53}{3.41} = 1.33; \text{ O: } \frac{3.41}{3.41} = 1.00$$

We see that for every C atom, there are 1.33 = 4/3 H atoms and one oxygen atom. Multiplying by three to obtain the simplest whole number ratio, we have

$$3 \text{ C:4 H:3 O; simplest formula} = C_3H_4O_3$$

Once the percentages by mass of the elements in a compound have been determined, the calculation of the simplest formula is straightforward. Finding the percentages, however, is not necessarily a simple matter. One way in which we might obtain the data used in Example 3.1 is shown schematically in Figure 3.1.

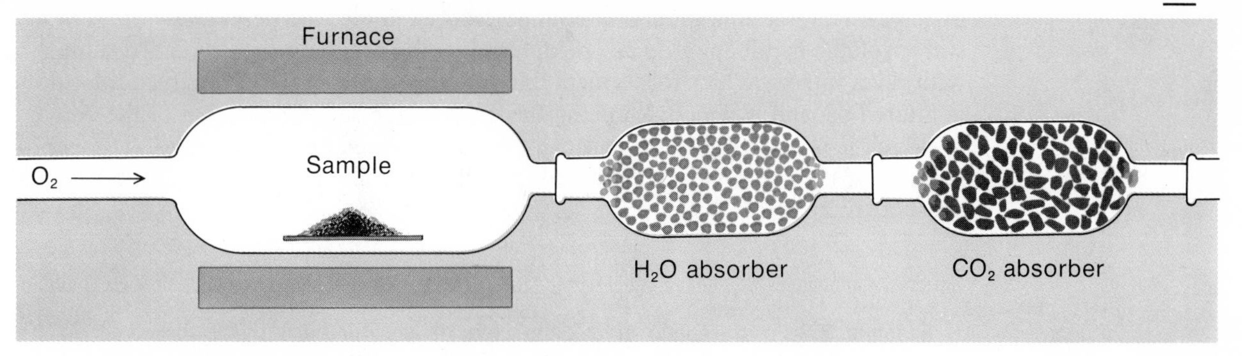

Figure 3.1 Combustion train used for carbon-hydrogen analysis. The absorbent for water is magnesium perchlorate, $Mg(ClO_4)_2$. Carbon dioxide is absorbed by finely divided sodium hydroxide supported on asbestos. Only a few milligrams of sample are needed for an analysis.

A weighed sample of the pure compound, usually only a few milligrams, is burned in oxygen to form carbon dioxide and water. The amounts of CO_2 and H_2O produced, which are directly dependent on the amounts of C and H in the sample, are determined by measuring the increases in mass of the two absorption tubes. From the masses of these two compounds, the masses of C and H in the sample and hence their percentages can be calculated. The percentage of oxygen can be determined by difference from 100.

Example 3.2 The percentages of C, H, and O in vitamin C (Example 3.1) are determined by burning a sample weighing 2.00 mg; the masses of CO_2 and H_2O formed are 3.00 mg and 0.816 mg, respectively. What are the percentages of C, H, and O?

Solution Let us first calculate the mass of carbon in 3.00 mg of CO_2. Since one gram molecular mass of CO_2, 44.0 g, contains one gram atomic mass of carbon, 12.0 g: 12.0 g C $\simeq$ 44.0 g CO_2*

$$\text{milligrams C} = 3.00 \text{ mg } CO_2 \times \frac{12.0 \text{ g C}}{44.0 \text{ g } CO_2} = 0.818 \text{ mg C}$$

Similarly, since there are 2.02 g of hydrogen in 18.0 g of H_2O:

$$\text{milligrams H} = 0.816 \text{ mg } H_2O \times \frac{2.02 \text{ g H}}{18.0 \text{ g } H_2O} = 0.0916 \text{ mg H}$$

Noting that the sample weighed 2.00 mg, we have

$$\%C = \frac{0.818}{2.00} \times 100 = 40.9\% \qquad \%H = \frac{0.0916}{2.00} \times 100 = 4.58\%$$

We obtain the percentage of oxygen by difference:

$$\%O = 100.0 - (40.9 + 4.6) = 54.5\%$$

We see from Figure 3.1 and Example 3.2 that the determination of the percentage of carbon in an organic compound depends upon converting a weighed sample to a compound of known composition, CO_2. Similarly, to find the percentage of hydrogen, we need to know the formula of the compound to which it is converted, H_2O.

*The symbol $\simeq$ means "chemically equivalent to." Mathematically, it can be treated as an equals sign.

The same principle can be applied to determine the percentages by mass of the elements in inorganic compounds. For example, the percentage of Cl in a water-soluble metal chloride can be determined by precipitating a weighed sample with silver nitrate, which forms insoluble silver chloride, AgCl. The silver chloride is filtered off and weighed; knowing the mass of the sample and that of the AgCl to which it is converted, we can calculate the percentage of chlorine. The percentage of the metal can be obtained by difference if it is the only other element present (Example 3.3).

Example 3.3 A sample of calcium chloride weighing 1.641 g is dissolved in water and treated with Ag^+. A precipitate of AgCl weighing 4.238 g forms. Determine the percentage composition and the simplest formula of calcium chloride.

Solution Let us first calculate the mass of chlorine in the AgCl. Since in AgCl one gram atomic mass of Cl, 35.45 g, is combined with one gram atomic mass of Ag, 107.87 g, to give a total mass of 35.45 g + 107.87 g = 143.32 g, we have

$$35.45 \text{ g Cl} \simeq 143.32 \text{ g AgCl}$$

Hence

$$\text{grams Cl} = 4.238 \text{ g AgCl} \times \frac{35.45 \text{ g Cl}}{143.32 \text{ g AgCl}} = 1.048 \text{ g Cl}$$

This is the mass of chlorine not only in the precipitated silver chloride but also in the original sample of calcium chloride. The percentages of chlorine and calcium in the 1.641-g sample must be:

When possible, it is best to analyze for all the elements in a compound. Why?

$$\% \text{ Cl} = \frac{\text{mass Cl}}{\text{mass sample}} \times 100 = \frac{1.048 \text{ g}}{1.641 \text{ g}} \times 100 = 63.86$$

$$\% \text{ Ca} = 100.00 - 63.86 = 36.14$$

To obtain the simplest formula, we proceed as usual. In a 100-g sample there would be

$$36.14 \text{ g Ca} \times \frac{1 \ GAM \text{ Ca}}{40.08 \text{ g Ca}} = 0.902 \ GAM \text{ Ca}$$

$$63.86 \text{ g Cl} \times \frac{1 \ GAM \text{ Cl}}{35.45 \text{ g Cl}} = 1.80 \ GAM \text{ Cl}$$

Clearly, since 1.80/0.90 = 2, there are two gram atomic masses of chlorine for one of calcium. The simplest formula is $CaCl_2$.

Incidentally, as you may or may not have noticed, it was not really necessary in Example 3.3 to calculate the mass percentages of calcium and chlorine if our only objective had been to obtain the simplest formula. We could have worked directly with the masses of Ca and Cl in the sample. That is, once we had shown that there was 1.048 g of Cl present, it followed that the mass of Ca in the sample must be 1.641 g − 1.048 g = 0.593 g. Hence, in the sample there are:

$$0.593 \text{ g Ca} \times \frac{1 \ GAM \text{ Ca}}{40.08 \text{ g Ca}} = 0.0148 \ GAM \text{ Ca}$$

$$1.048 \text{ g Cl} \times \frac{1 \ GAM \text{ Cl}}{35.45 \text{ g Cl}} = 0.0296 \ GAM \text{ Cl}$$

These two numbers, like those we worked with in the last step of Example 3.3, are in a 1:2 ratio, so we decide, as before, that the formula of calcium chloride must be $CaCl_2$.

3.3 MOLECULAR FORMULA FROM SIMPLEST FORMULA

As we pointed out in Section 3.1, the molecular formula must be a whole-number multiple of (e.g., 1, 2, 3, . . . times) the simplest formula. In order to decide what that multiple is, we need one additional piece of information—the molecular mass (Example 3.4).

Example 3.4 The empirical formula of vitamin C was determined in Example 3.1 to be $C_3H_4O_3$. From another experiment, its molecular mass is found to be about 180. Determine its molecular formula.

Solution The formula mass corresponding to the simplest formula, $C_3H_4O_3$, is

$$(3 \times 12) + (4 \times 1) + (3 \times 16) = 88$$

Clearly the approximate molecular mass, 180, is twice the formula mass. It follows that the simplest formula must be multiplied by two to obtain the molecular formula.

$$\text{Molecular formula vitamin C} = C_6H_8O_6$$

Note that an approximate value of the molecular mass is sufficient to obtain the molecular formula from the simplest formula. An experiment that establishes the molecular mass of vitamin C to be in the range from 160 to 200 would demonstrate that the simplest formula should be multiplied by 2 rather than 1, 3, or some other integer to obtain the molecular formula.

3.4 THE MOLE

In discussing mass relations in chemical reactions, we make frequent use of a quantity known as the **mole** (abbreviation: **mol**). Depending upon the context in which the word "mole" is used, it may refer to a specific number of particles or to a specific mass in grams. That is, it may represent:

1. **Avogadro's number of items**

$$\text{one mole of Fe atoms} = 6.02 \times 10^{23} \text{ Fe atoms}$$
$$\text{one mole of } CO_2 \text{ molecules} = 6.02 \times 10^{23}\ CO_2 \text{ molecules}$$
$$\text{one mole of } Cl^- \text{ ions} = 6.02 \times 10^{23}\ Cl^- \text{ ions}$$
$$\text{one mole of C—C bonds} = 6.02 \times 10^{23} \text{ C—C bonds}$$

In this sense, the mole is analogous to the dozen, a term which always refers to 12 items of a particular type.

2. **One gram formula mass of a substance**

$$\text{one mole Fe} = 1\ GFM \text{ Fe} = 55.85 \text{ g Fe}$$
$$\text{one mole } CO_2 = 1\ GFM\ CO_2 = 12.01 \text{ g} + 2(16.00 \text{ g}) = 44.01 \text{ g } CO_2$$
$$\text{one mole NaCl} = 1\ GFM \text{ NaCl} = 22.99 \text{ g} + 35.45 \text{ g} = 58.44 \text{ g NaCl}$$

Whenever the mole is used in this context it is necessary to specify the formula referred to. It would be impossible to say how much "1 mol of nitrogen" or "2.65 mol of sodium carbonate" represent unless we know their formulas.

You will note from the examples given that, for iron, the mole is identical to the gram atomic mass as defined in Chapter 2. That is,

$$1\ GAM\ \text{Fe} = 6.02 \times 10^{23} \text{ atoms Fe} = 55.85 \text{ g Fe} = 1 \text{ mol Fe}$$

For carbon dioxide, the mole is identical with the gram molecular mass.

$$1\ GMM\ CO_2 = 6.02 \times 10^{23} \text{ molecules } CO_2 = 44.01 \text{ g } CO_2 = 1 \text{ mol } CO_2$$

More generally, we can say that:

When a substance is built up of individual atoms (e.g., Fe, C, . . .), the mole and the gram atomic mass weigh the same and represent the same number of particles (6.02×10^{23}).

For molecular substances (e.g., CO_2, H_2O, . . .), the mole and the gram molecular mass weigh the same and represent the same number of particles (6.02×10^{23}).

How many ions are there in a mole of NaCl?

From the other examples cited, it should be clear that the mole is a more general term than gram atomic or gram molecular mass. For sodium chloride, where the basic structural units are the Na^+ and Cl^- ions, it would be misleading to refer to a "gram atomic mass" or a "gram molecular mass" of the substance. We can, however, use the mole to refer to a specific mass of NaCl, 58.44 g. Indeed, the mole can be used in this sense for any substance, regardless of its structure, provided we know its formula. Consider, for example, the oxide of tin, SnO_2. Without worrying about whether this compound is made up of molecules, ions, or some other unspecified building block, we can say with complete confidence that

$$1 \text{ mol } SnO_2 = 1\ GFM\ SnO_2 = 118.69 \text{ g} + 2(16.00 \text{ g}) = 150.69 \text{ g } SnO_2$$

Conversions between moles and grams can be carried out by the same conversion factor approach followed in Chapters 1 and 2 (Example 3.5).

Example 3.5

a. Determine the mass in grams of 2.60 mol of baking soda, $NaHCO_3$.
b. How many moles of penicillin, $C_{16}H_{18}O_4N_2S$, are there in 218 g?

Solution

a. One mole of $NaHCO_3$ weighs

$$23.0 \text{ g} + 1.0 \text{ g} + 12.0 \text{ g} + (3 \times 16.0 \text{ g}) = 84.0 \text{ g}$$

$$\text{grams } NaHCO_3 = 2.60 \text{ mol } NaHCO_3 \times \frac{84.0 \text{ g } NaHCO_3}{1 \text{ mol } NaHCO_3}$$

$$= 218 \text{ g } NaHCO_3$$

b. Mass of one mole of penicillin:

$$16(12.0 \text{ g}) + 18(1.0 \text{ g}) + 4(16.0 \text{ g}) + 2(14.0 \text{ g}) + 32.0 \text{ g} = 334 \text{ g}$$

$$\text{moles penicillin} = 218 \text{ g penicillin} \times \frac{1 \text{ mol penicillin}}{334 \text{ g penicillin}}$$

$$= 0.653 \text{ mol penicillin}$$

In this and succeeding chapters, we shall make extensive use of the mole concept. Example 3.6 illustrates how it is used in a problem dealing with the percentage by mass of an element in a compound of known formula.

Example 3.6 One of the ores of copper is malachite, a bright green mineral which has the simplest formula $Cu_2CO_5H_2$.

a. What is the percentage of copper in malachite?
b. How much copper can be obtained from 340 g of malachite?

Solution

a. In one mole of malachite there are

$$\begin{aligned} 2\ GAM\ \text{Cu} &= 2(63.55\ \text{g}) = 127.10\ \text{g} \\ 1\ GAM\ \text{C} &= 12.01\ \text{g} \\ 5\ GAM\ \text{O} &= 5(16.00\ \text{g}) = 80.00\ \text{g} \\ 2\ GAM\ \text{H} &= 2(1.008\ \text{g}) = \underline{2.02\ \text{g}} \\ & \quad 221.13\ \text{g} \end{aligned}$$

$$\%\ \text{Cu} = \frac{127.10\ \text{g}}{221.13\ \text{g}} \times 100 = 57.48\%$$

b. mass Cu = 340 g × 0.5748 = 195 g

3.5 CHEMICAL EQUATIONS

So far in this chapter, we have focussed attention on the formulas used to represent substances. We now consider how such formulas are involved in writing a **balanced equation** to describe a chemical reaction. Writing a chemical equation is by no means a simple, mechanical process, as beginning students are sometimes led to believe. One seemingly obvious point is often underemphasized; *you cannot write an equation to describe a reaction unless you know what happens.*

To illustrate how one arrives at a balanced equation, consider the reaction that takes place in a rocket motor which uses hydrazine as a fuel and dinitrogen tetroxide as an oxidizer. It has been established experimentally that the molecular formulas of these two species, both of which are liquids, are N_2H_4 and N_2O_4, respectively. Analysis of the mixture coming out of the rocket exhaust shows that it consists of gaseous elementary nitrogen, N_2, and liquid water. To write a balanced equation for this reaction, we proceed as follows:

To write an equation, we must know the formulas of all reactants and products

1. We first write an unbalanced equation in which the formulas of the reactants appear on the left and those of the products on the right. In this case, we have

$$N_2H_4 + N_2O_4 \rightarrow N_2 + H_2O$$

2. We balance the equation by taking into account the Law of Conservation of Mass, which requires that there be the same number of atoms of each element on the two sides of the equation. To accomplish this, we might begin by writing a coefficient of 4 for the H_2O, thereby obtaining 4 oxygen atoms on both sides.

$$N_2H_4 + N_2O_4 \rightarrow N_2 + 4\ H_2O$$

Looking now at the hydrogen atoms, we notice that we have $4 \times 2 = 8$ H atoms on the right. To obtain 8 H atoms on the left, we write a coefficient of 2 for N_2H_4.

$$2\ N_2H_4 + N_2O_4 \rightarrow N_2 + 4\ H_2O$$

Finally, we consider the number of atoms of nitrogen; there are a total of $(2 \times 2) + 2 = 6$ nitrogen atoms on the left. We balance the nitrogen by writing a coefficient of 3 for N_2.

$$2\ N_2H_4 + N_2O_4 \rightarrow 3\ N_2 + 4\ H_2O \quad (3.1)$$

3. Throughout this text, the physical states of reactants and products will be indicated in the equation. We will use the letters (g), (l), and (s) to represent gases, *pure* liquids, and solids, respectively. For the reaction between hydrazine and dinitrogen tetroxide, we write

Using the letters (s), (l), (g) and (aq) helps give physical meaning to an equation

$$2\ N_2H_4(l) + N_2O_4(l) \rightarrow 3\ N_2(g) + 4\ H_2O(l) \quad (3.2)$$

When we deal with reactions taking place in water solutions, we will use the symbol (aq) to designate a dissolved species (ion or molecule). Thus, the equation for the reaction that takes place when a water solution containing Ag^+ ions is mixed with one containing Cl^- ions is written as

$$Ag^+(aq) + Cl^-(aq) \rightarrow AgCl(s) \quad (3.3)$$

Again, the equation

$$Zn(s) + 2\ H^+(aq) \rightarrow Zn^{2+}(aq) + H_2(g) \quad (3.4)$$

is written to represent the reaction observed when solid zinc is added to an acidic water solution containing H^+ ions; the products are Zn^{2+} ions in aqueous solution and molecules of H_2 in the gas state.

The approach that we have followed in this section is often referred to as "balancing an equation by inspection." It will be sufficient for our purposes throughout most of this text. However, there are some equations, involving many different reactants and products, where it is difficult to proceed in this way. When we meet such equations for the first time, in Chapter 22, we will develop special procedures for balancing them.

3.6 MASS RELATIONS IN REACTIONS

Our main interest in balanced chemical equations lies in the information they give us about the relative masses of reactants and products in a reaction. We can use a balanced equation to decide how many grams of one substance will be needed to react with a given mass of another, or how much product we can expect to obtain from a specified amount of reactant. Calculations of this sort are readily made if we realize that **the coefficients of a balanced equation represent the relative numbers of moles of reactants and products.**

To show that the statement just made is valid, consider the reaction between N_2H_4 and N_2O_4, where we arrived at the balanced equation

$$2\ N_2H_4(l) + N_2O_4(l) \rightarrow 3\ N_2(g) + 4\ H_2O(l)$$

Perhaps the simplest way to interpret the coefficients of this equation is to say that they represent numbers of molecules. That is,

$$2 \text{ molecules } N_2H_4 + 1 \text{ molecule } N_2O_4 \rightarrow 3 \text{ molecules } N_2 + 4 \text{ molecules } H_2O$$

However, a balanced chemical equation, like an algebraic equation, remains valid if each of the coefficients is multiplied by the same number. That number might be 1/2, 100, . . . , *or, in particular, Avogadro's number, N.* That is:

$$2\,N \text{ molecules } N_2H_4 + N \text{ molecules } N_2O_4 \rightarrow 3\,N \text{ molecules } N_2 + 4\,N \text{ molecules } H_2O$$

But, as we saw in Section 3.4, a mole represents Avogadro's number of items. So, we can write

$$2 \text{ mol } N_2H_4 + 1 \text{ mol } N_2O_4 \rightarrow 3 \text{ mol } N_2 + 4 \text{ mol } H_2O$$

which is the relationship we set out to demonstrate. Looking at this relation from a slightly different viewpoint, we might say that, in this particular reaction,

$$2 \text{ mol } N_2H_4 \simeq 1 \text{ mol } N_2O_4 \simeq 3 \text{ mol } N_2 \simeq 4 \text{ mol } H_2O$$

where the symbol $\simeq$ means "is chemically equivalent to."

To obtain a relationship between masses of reactants and products in grams, we need only convert moles of each species in the balanced equation to grams. The gram formula masses of N_2H_4, N_2O_4, N_2, and H_2O are readily shown to be 32.0 g, 92.0 g, 28.0 g, and 18.0 g, respectively. Since a mole represents a gram formula mass, we can write

A real advantage of a balanced chemical equation is that it leads readily to information of this sort

$$2(32.0 \text{ g } N_2H_4) + 1(92.0 \text{ g } N_2O_4) \rightarrow 3(28.0 \text{ g } N_2) + 4(18.0 \text{ g } H_2O)$$

$$64.0 \text{ g } N_2H_4 + 92.0 \text{ g } N_2O_4 \rightarrow 84.0 \text{ g } N_2 + 72.0 \text{ g } H_2O$$

Or, we might say that

$$64.0 \text{ g } N_2H_4 \simeq 92.0 \text{ g } N_2O_4 \simeq 84.0 \text{ g } N_2 \simeq 72.0 \text{ g } H_2O$$

The procedure that we have followed for the N_2H_4–N_2O_4 reaction can be applied to any reaction. Thus for the reaction between aluminum and oxygen, where the balanced equation is

$$4 \text{ Al(s)} + 3 \text{ O}_2\text{(g)} \rightarrow 2 \text{ Al}_2\text{O}_3\text{(s)} \qquad (3.5)$$

we can write

$$4 \text{ mol Al} \simeq 3 \text{ mol O}_2 \simeq 2 \text{ mol Al}_2\text{O}_3$$

Or, since the gram formula masses of Al, O_2, and Al_2O_3 are 27.0 g, 32.0 g, and 102.0 g, respectively

$$4(27.0 \text{ g Al}) \simeq 3(32.0 \text{ g O}_2) \simeq 2(102.0 \text{ g Al}_2\text{O}_3)$$

$$108.0 \text{ g Al} \simeq 96.0 \text{ g O}_2 \simeq 204.0 \text{ g Al}_2\text{O}_3$$

The use of these relationships is illustrated in Example 3.7. Here again we use the conversion factor approach.

Example 3.7 For the reaction

$$2\ N_2H_4(l) + N_2O_4(l) \rightarrow 3\ N_2(g) + 4\ H_2O(l)$$

determine
a. the number of moles of N_2O_4 required to react with 2.72 mol of N_2H_4.
b. the number of grams of N_2 produced when 2.72 mol of N_2H_4 are consumed.
c. the mass in grams of H_2O formed when 1.00 g of N_2O_4 reacts.

Solution In each case, we use the coefficients of the balanced equation to obtain the conversion factor required.

a. Here the conversion factor follows directly from the coefficients

$$2 \text{ mol } N_2H_4 \simeq 1 \text{ mol } N_2O_4$$

$$\text{moles } N_2O_4 = 2.72 \text{ mol } N_2H_4 \times \frac{1 \text{ mol } N_2O_4}{2 \text{ mol } N_2H_4} = 1.36 \text{ mol } N_2O_4$$

b. In this case, we need a conversion factor relating moles of N_2H_4 to grams of N_2. From the coefficients of the balanced equation

$$2 \text{ mol } N_2H_4 \simeq 3 \text{ mol } N_2$$

To obtain the desired conversion factor, we need only translate 3 mol of N_2 into grams. Since 1 mol of N_2 weighs 28.0 g we have

$$2 \text{ mol } N_2H_4 \simeq 3\ (28.0 \text{ g } N_2) = 84.0 \text{ g } N_2$$

Conversion factors can help to keep units straight in calculations like these

$$\text{Hence, grams } N_2 \text{ formed} = 2.72 \text{ mol } N_2H_4 \times \frac{84.0 \text{ g } N_2}{2 \text{ mol } N_2H_4} = 114 \text{ g } N_2$$

c. Again, we start with the mole relationship

$$1 \text{ mol } N_2O_4 \simeq 4 \text{ mol } H_2O$$

This time we translate both quantities into grams (1 mol N_2O_4 = 92.0 g; 1 mol H_2O = 18.0 g)

$$92.0 \text{ g } N_2O_4 \simeq 4\ (18.0 \text{ g})\ H_2O = 72.0 \text{ g } H_2O$$

$$\text{grams } H_2O \text{ formed} = 1.00 \text{ g } N_2O_4 \times \frac{72.0 \text{ g } H_2O}{92.0 \text{ g } N_2O_4} = 0.783 \text{ g } H_2O$$

3.7 LIMITING REAGENT. THEORETICAL AND ACTUAL YIELDS

When we carry out a reaction between two substances, we rarely use exactly equivalent amounts of reactants. Consider, for example, the reaction between benzene, C_6H_6, and nitric acid, HNO_3, to form nitrobenzene, $C_6H_5NO_2$:

$$C_6H_6(l) + HNO_3(l) \rightarrow C_6H_5NO_2(l) + H_2O(l) \tag{3.6}$$

Suppose you want to convert 1 mol of benzene to 1 mol of nitrobenzene. In principle, you could do this by adding precisely 1 mol (63.0 g) of nitric acid. In practice, if you want to get as much nitrobenzene as possible, it is desirable to use an excess of the cheaper reagent, nitric acid. Thus for every mole of benzene you might use 3/2, 2, or even 5 mol of HNO_3.

In a situation such as the one just described, it is important to distinguish between the reagent which is in excess (e.g., HNO_3) and the **limiting reagent** (e.g.,

C_6H_6). When the amounts are given in moles, it is relatively easy to make this distinction. For example, if we are told that in carrying out Reaction 3.6 we should use

$$1.20 \text{ mol } C_6H_6 + 3.56 \text{ mol } HNO_3$$

it is clear that nitric acid is in excess and benzene is the limiting reagent. From the coefficients of the balanced equation we see that the required mole ratio of C_6H_6 to HNO_3 is 1:1; only 1.20 mol of HNO_3 are required to react with 1.20 mol of C_6H_6. Hence we have an excess of 3.56 − 1.20 = 2.36 mol of HNO_3.

If you want to make martinis, starting with 1 bottle of vermouth and 1 bottle of gin, which is the "limiting reagent"?

When amounts of reactants are given in grams rather than moles, it is not as easy to spot the limiting reagent. Suppose, for example, we were asked to make this decision for a reaction mixture consisting of 100 g of C_6H_6 and 100 g of HNO_3. Here it is perhaps simplest to start by converting grams to moles. Noting that the gram formula masses of C_6H_6 and HNO_3 are 78.0 and 63.0 g, respectively, we have

$$\text{no. moles } C_6H_6 = 100 \text{ g } C_6H_6 \times \frac{1 \text{ mol } C_6H_6}{78.0 \text{ g } C_6H_6} = 1.28 \text{ mol } C_6H_6$$

$$\text{no. moles } HNO_3 = 100 \text{ g } HNO_3 \times \frac{1 \text{ mol } HNO_3}{63.0 \text{ g } HNO_3} = 1.59 \text{ mol } HNO_3$$

Having made this conversion, it should be clear that benzene is again the limiting reagent. From one point of view, it would require 1.59 mol of C_6H_6 to react with all the HNO_3; since we have less than that (1.28 mol), benzene must be the limiting reagent.

To calculate the amount of product to be expected from a reaction, we must use the limiting reagent as our base. In the case just cited, we could not hope to obtain more than 1.28 mol of nitrobenzene via Reaction 3.6, since we had only 1.28 mol of benzene to start with. Noting that the molecular mass of $C_6H_5NO_2$ is 123, we might expect to get

$$1.28 \times 123 \text{ g} = 157 \text{ g } C_6H_5NO_2$$

The amount just calculated is often referred to as the *theoretical yield* of nitrobenzene in the reaction. In general, **the theoretical yield is the amount of product that will be obtained if the limiting reagent is completely consumed in forming it.**

Example 3.8 Aspirin is made by adding acetic anhydride to a water solution of salicylic acid. The equation for the reaction is

$$\underset{\text{salicylic acid}}{2\ C_7H_6O_3(aq)} + \underset{\text{acetic anhydride}}{C_4H_6O_3(l)} \rightarrow \underset{\text{aspirin}}{2\ C_9H_8O_4(aq)} + H_2O$$

If 1.00 kg of salicylic acid is used with 2.00 kg of acetic anhydride, determine

a. the limiting reagent.
b. the theoretical yield of aspirin in grams.

Solution

a. We first calculate the numbers of moles of salicylic acid (MM = 138) and acetic anhydride (MM = 102).

$$\text{no. moles } C_7H_6O_3 = \frac{1000 \text{ g}}{138 \text{ g/mol}} = 7.25 \text{ mol salicylic acid}$$

$$\text{no. moles } C_4H_6O_3 = \frac{2000 \text{ g}}{102 \text{ g/mol}} = 19.6 \text{ mol acetic anhydride}$$

We see from the coefficients of the balanced equation that 2 mol of salicylic acid are required to react with 1 mol of acetic anhydride. Consequently, we would need 2(19.6) = 39.2 mol of salicylic acid to react with all the acetic anhydride. Obviously, we have much less than that; salicylic acid, $C_7H_6O_3$, must then be the limiting reagent.

b. The theoretical yield must be based on the limiting reagent, salicylic acid. From the balanced equation, we see that one mole of aspirin is formed for every mole of salicylic acid consumed. Thus, we can obtain 7.25 mol of aspirin (*MM* = 180).

$$\text{Theor. yield} = 7.25 \text{ mol aspirin} \times \frac{180 \text{ g aspirin}}{1 \text{ mol aspirin}} = 1.30 \times 10^3 \text{ g aspirin}$$
$$= 1.30 \text{ kg}$$

As you might guess, the actual yield of product in a reaction is almost always less than the theoretical yield. There are many reasons for this. In the first place, the reaction may not go to completion; significant amounts of the limiting reagent may remain unreacted. Again, side reactions of various types may lower the yield of desired product by consuming part of the limiting reagent. Finally, some of the product is likely to be lost in the separation and purification steps that follow the reaction.

All of these factors tend to reduce the **percentage yield,** defined by the relation

$$\%\text{ yield} = \frac{\text{actual yield}}{\text{theoretical yield}} \times 100 \qquad (3.7)$$

If the calculated % yield is > 100, something is wrong

In the preparation of aspirin described in Example 3.8, most organic chemists would consider themselves fortunate to obtain as much as 1.12 kg of product, which would correspond to a yield of slightly over 86%.

$$\frac{1.12 \text{ kg}}{1.30 \text{ kg}} \times 100 = 86.2\%$$

PROBLEMS

3.1 *Simplest Formula from % Composition* Analysis of the hydrocarbon ethylene shows that it contains 85.6% by mass of C and 14.4% H. What is its simplest formula? (*AM* C = 12.0, H = 1.01)

3.2 *Molecular Formula from Simplest Formula* The molecular mass of ethylene is about 30. Using the simplest formula obtained in Problem 3.1, determine the molecular formula.

3.3 *The Mole*

a. How many moles of CaO are there in 19.6 g? (*AM* Ca = 40.1, O = 16.0)
b. What is the mass in grams of 2.19 mol of CaO?

3.4 *% Composition from Formula* What are the percentages by mass of the elements in aspirin, $C_9H_8O_4$? (*AM* C = 12.0, H = 1.0, O = 16.0)

3.5 *Balancing Equations* When phosphine, a poisonous gas with molecular formula PH_3 is burned in air, the products are liquid water and a solid of molecular formula P_4O_{10}. Write a balanced equation for this reaction.

3.6 *Mass Relations in Reactions* Referring to the reaction in Problem 3.5: (*AM* P = 31.0)

a. How many moles of PH_3 are required to form 1.16 mol of P_4O_{10}?
b. How many grams of water are formed from 0.198 mol of O_2?
c. How many grams of O_2 are required to react with 12.0 g of PH_3?

3.7 *Limiting Reagent: Theoretical vs. Actual Yield* Referring again to the reaction in Problem 3.5, suppose 6.80 g of PH_3 is reacted with 6.40 g of O_2.

a. Which is the limiting reagent?
b. What is the theoretical yield in grams of P_4O_{10}?
c. If the actual yield of P_4O_{10} is 5.20 g, what is the % yield?

3.8 The simplest formula of mica is $NaAl_3Si_3H_2O_{12}$.

a. What are the percentages by mass of the elements in mica?
b. How many grams of aluminum could be extracted from one kilogram of mica?

3.9 A sample of toothpaste contains stannous fluoride, SnF_2, as an additive. Analysis of a 1.340 g sample gives 1.20×10^{-3} g of F.

a. How much SnF_2 is there in the sample?
b. What is the percentage of SnF_2 in the sample?

3.10 What are the simplest formulas of compounds with the following compositions?

a. 65.2% Sc, 34.8% O
b. 12.6% Li, 29.2% S, 58.2% O
c. 24.8% Co, 29.8% Cl, 40.3% O, 5.1% H

3.11 A certain hydrate of nickel chloride has the formula $NiCl_2(H_2O)_x$. To determine the value of x, a student heats a sample of the hydrate until all the water is driven off. 1.650 g of hydrate gives off 0.590 g of H_2O. What is x?

3.12 Glycerin contains the three elements C, H, and O. When a sample weighing 0.673 mg is burned in oxygen, 0.965 mg of CO_2 and 0.527 mg of H_2O are formed. What is the simplest formula of glycerin?

3.13 In a certain hydrocarbon there are equal numbers of C and H atoms. The molecular mass is about 65. What is the molecular formula?

3.14 A sample of lanthanum chloride weighing 0.862 g is treated with excess Ag^+; 1.511 g of AgCl is formed. What is the simplest formula of lanthanum chloride?

3.15 An organic compound containing C, H, and O is burned in a stream of Cl_2 gas. The hydrogen is converted to HCl, the carbon to CCl_4. A sample weighing 1.254 g gives 9.977 g of CCl_4 and 4.730 g of HCl. What is its simplest formula?

3.26 The formula of rust may be represented most simply as $Fe(OH)_3$.

a. What are the percentages by mass of the elements in rust?
b. How many grams of rust are formed when one kilogram of iron corrodes?

3.27 A certain drug has aspirin, $C_9H_8O_4$, as its major ingredient. None of the other substances in the drug contain carbon. Analysis shows the percentage of carbon in the drug to be 48.0. What is the percentage of aspirin?

3.28 What are the simplest formulas of the following compounds?

a. 45.9% K, 16.5% N, 37.6% O
b. 74.0% C, 8.7% H, 17.3% N
c. 24.8% Cr, 50.8% Cl, 20.1% N, 4.3% H

3.29 In analyzing a different hydrate of nickel chloride, a student finds that 1.390 g of the hydrate gives, after heating, 0.758 g of $NiCl_2$. How many moles of H_2O are there per mole of $NiCl_2$ in the hydrate?

3.30 The insecticide lindane contains C, H, and Cl. When a sample weighing 1.642 mg is burned, 1.491 mg of CO_2 and 0.305 mg of H_2O are formed. What is the simplest formula of lindane?

3.31 A certain organic compound has the simplest formula CH_2O. Which of the following molecular masses are possible?

a. 15
b. 30
c. 45
d. 60
e. 150

3.32 A sample of a fluoride of uranium weighing 1.760 g is treated with excess Ca^{2+}; 1.170 g of CaF_2 is formed. What is the simplest formula of the uranium compound?

3.33 A sample of an oxide of tin weighing 3.014 g is heated at a low temperature; part of the oxygen is driven off to form another oxide weighing 2.694 g. Further heating with H_2 removes the rest of the oxygen to give 2.374 g of Sn. What are the simplest formulas of the oxides?

3.16 Caffeine has the molecular formula $C_8H_{10}O_2N_4$. Calculate

a. the mass in grams of 1.20×10^{-2} mol.
b. the number of moles in 62.0 g.
c. the number of molecules in (a) and (b).

3.17 How many moles are there in

a. 1.00 kg of quartz, SiO_2?
b. 1.56×10^{-4} g of $K_2Cr_2O_7$?
c. a dozen electrons?

3.18 Write balanced equations to represent

a. the combustion of butane gas, C_4H_{10}, to give carbon dioxide and water vapor.
b. the precipitation reaction that occurs when water solutions containing Sb^{3+} and S^{2-} are mixed.
c. the reaction in the blast furnace, where iron ore, Fe_2O_3, is heated with carbon monoxide to form iron and carbon dioxide.

3.19 For the reaction

$$N_2H_4(l) + O_2(g) \rightarrow N_2(g) + 2\ H_2O(l)$$

complete the following:

a. 1.60 mol $N_2H_4 \rightarrow$ _____ mol H_2O
b. _____ mol $O_2 \rightarrow$ 14.1 g H_2O
c. _____ g $N_2H_4 \rightarrow$ 3.80 mol N_2
d. 12.6 g $N_2H_4 \rightarrow$ _____ g N_2

3.20 Hydrogen sulfide, given off by decaying organic matter, is converted to sulfur dioxide in the atmosphere by the reaction

$$2\ H_2S(g) + 3\ O_2(g) \rightarrow 2\ SO_2(g) + 2\ H_2O(l)$$

a. How many moles of H_2S are required to form 8.19 mol of SO_2?
b. How many grams of O_2 are required to react with one mole of H_2S?
c. How many grams of H_2S are required to give an average SO_2 concentration of 6.0×10^{-8} mol/dm^3 in a column of air one kilometre high over a city 50 km^2 in area?

3.21 Sulfide minerals associated with coal deposits are responsible for "acid mine drainage," contamination of streams in coal mining areas by sulfuric acid. A typical reaction of such minerals is

$$4\ FeS(s) + 9\ O_2(g) + 4\ H_2O(l) \rightarrow 2\ Fe_2O_3(s) + 4\ H_2SO_4(l)$$

How many grams of FeS are required to form 100 dm^3 of a solution containing 2.0×10^{-4} mol/dm^3 of H_2SO_4?

3.34 The simplest formula of ammonium phosphate is $(NH_4)_3PO_4$. Determine

a. the number of moles in one gram.
b. the mass in grams of 0.204 mol.
c. the number of N atoms in (a) and (b).

3.35 Estimate the mass in grams of a mole of ice cubes, each 1.82 cm on an edge, taking the density of ice to be 0.91 g/cm^3.

3.36 Write balanced equations to represent

a. the combustion of the rocket fuel diborane, $B_2H_6(l)$, which burns to form $B_2O_3(s)$ and liquid water.
b. the precipitation reaction that occurs when Ag^+ and CO_3^{2-} ions are mixed in water solution.
c. the detonation of nitroglycerin, for which the unbalanced equation is

$$C_3H_5N_3O_9(l) \rightarrow CO_2(g) + N_2(g) + H_2O(l) + O_2(g)$$

3.37 For the reaction

$$2\ SO_2(g) + O_2(g) \rightarrow 2\ SO_3(g)$$

complete the following table.

	mol SO_2	g O_2	mol SO_3	g SO_3
a.	1.68	_____	_____	_____
b.	_____	12.0	_____	_____
c.	_____	_____	3.49	_____

3.38 In a rocket propellant composed of hydrazine, N_2H_4, and hydrogen peroxide, H_2O_2, the reaction is

$$N_2H_4(l) + 2\ H_2O_2(l) \rightarrow N_2(g) + 4\ H_2O(l)$$

a. What is the total number of moles of product (N_2 and H_2O) formed when 8.11 mol of N_2H_4 react?
b. How many moles of water are formed from 6.19 g of H_2O_2?
c. What should be the mass ratio of H_2O_2 to N_2H_4 to achieve complete reaction?
d. What should be the volume ratio of H_2O_2 to N_2H_4 (density H_2O_2 = 1.44 g/cm^3, N_2H_4 = 1.01 g/cm^3)?

3.39 At least in principle, SO_2 can be removed from stack gases by treatment with slaked lime. The reaction is

$$2\ Ca(OH)_2(s) + 2\ SO_2(g) + O_2(g) \rightarrow 2\ H_2O(l) + 2\ CaSO_4(s)$$

What volume of $Ca(OH)_2$ solution, containing 0.10 mol/dm^3, would be required to remove one kilogram of SO_2?

3.22 One way to remove carbon dioxide from the air in a spacecraft is to react it with lithium hydroxide:

$$CO_2(g) + 2\ LiOH(s) \rightarrow Li_2CO_3(s) + H_2O(l)$$

In one day, a person exhales about one kilogram of CO_2. How many grams of LiOH are required to remove the CO_2 formed in a six-day lunar expedition with three astronauts?

3.23 The fermentation of glucose may be represented by the equation

$$\underset{\text{glucose}}{C_6H_{12}O_6(aq)} \rightarrow 2\ \underset{\text{alcohol}}{C_2H_5OH(aq)} + 2\ CO_2(g)$$

Fermentation of one gram of glucose will give _____g of alcohol and _____g of CO_2. If the density of the $CO_2(g)$ is 1.80 g/dm³, the volume of CO_2 produced is _____cm³.

3.24 Consider the reaction described in Example 3.8. If one starts with the following amounts of salicylic acid (SA) and acetic anhydride (AA), which will be the limiting reagent and what will be the theoretical yield of aspirin in grams?

a. 1.50 mol SA + 1.26 mol AA
b. 10.0 g SA + 10.0 g AA
c. 10.0 g SA + 3.00 g AA

3.25 A student prepares HBr by reacting sodium bromide with phosphoric acid:

$$3\ NaBr(s) + H_3PO_4(l) \rightarrow 3\ HBr(g) + Na_3PO_4(s)$$

She needs 50.0 g of HBr. If NaBr is the limiting reagent and a 50% excess of H_3PO_4 is to be used, how much of each reagent should she weigh out, assuming a yield of

a. 100%
b. 80%

3.40 Another way to remove carbon dioxide from a spacecraft is to use quicklime, CaO, which is converted to calcium carbonate, $CaCO_3$.

a. Write a balanced equation for the reaction.
b. How many grams of CaO would be required for the same amount of CO_2 as in Problem 3.22?

3.41 When cider ferments, the product contains about 12% alcohol by mass. Assuming the reaction is that given in Problem 3.23,

a. how many grams of glucose are required per kilogram of cider?
b. what percentage of the cider by mass comes from glucose?
c. how many moles of CO_2 are produced when one kilogram of cider is formed?

3.42 A student in the organic chem lab prepares ethyl bromide, C_2H_5Br, by reacting ethyl alcohol with phosphorus tribromide:

$$3\ C_2H_5OH(l) + PBr_3(l) \rightarrow 3\ C_2H_5Br(l) + H_3PO_3(s)$$

He is told to react 34.0 g of ethyl alcohol with 59.0 g of PBr_3.

a. Which is the limiting reagent?
b. What is the theoretical yield of C_2H_5Br?
c. If the actual yield is 26.0 g, what is the percentage yield?

3.43 A student in the inorganic laboratory wants to make 20 g of the compound $[Co(NH_3)_5SCN]Cl_2$ from $[Co(NH_3)_5Cl]Cl_2$.

The reaction is:

$$[Co(NH_3)_5Cl]Cl_2(s) + KSCN(s) \rightarrow KCl(s) + [Co(NH_3)_5SCN]Cl_2(s)$$

He is told to use a 60% excess of KSCN and that he can expect to get a 55% yield in the reaction. How many grams of each reagent should he use?

*3.44 A sample of LSD (D-lysergic acid diethylamide, $C_{24}H_{30}N_3O$) is diluted with sugar, $C_{12}H_{22}O_{11}$. When a 1.00-mg sample is burned, 2.00 mg of CO_2 is formed. What is the percentage of LSD in the sample?

*3.45 In a blast furnace, iron ore, Fe_2O_3, is reduced to the metal by heating with carbon monoxide; carbon dioxide is formed as a by-product. The carbon monoxide in turn is formed by burning carbon in a limited supply of pure oxygen.

a. Write balanced equations for the two reactions involved.
b. How many grams of oxygen are required to produce a gram of iron?

*3.46 The active ingredients of a certain biscuit mix are cream of tartar, $KHC_4H_4O_6$, and baking soda, $NaHCO_3$. The reaction that occurs when the biscuits "rise" is

$$KHC_4H_4O_6(s) + NaHCO_3(s) \rightarrow KNaC_4H_4O_6(s) + CO_2(g) + H_2O(l)$$

According to the label on the package, the mix contains 1.40% by mass of cream of tartar and 0.60% by mass of baking soda. What volume of CO_2 in cubic centimetres is produced per gram of mix? Assume that the volume of one mole of $CO_2(g)$ is about 30 dm³.

*3.47 When a sample of $CoCO_3$ weighing 1.000 g is heated in a vacuum, it decomposes to form an oxide of cobalt weighing 0.630 g. When this oxide is exposed to air it gains mass, forming a second oxide which weighs 0.675 g.

a. What are the formulas of the two oxides?
b. Write balanced equations for the two reactions involved.

4

THERMO-CHEMISTRY

Chemical reactions, such as those considered in Chapter 3, are accompanied by energy changes which may take any of several different forms. Consider, for example, the reactions

$$CH_4(g) + 2\ O_2(g) \rightarrow CO_2(g) + 2\ H_2O(l)$$

$$6\ CO_2(g) + 6\ H_2O(l) \rightarrow C_6H_{12}O_6(s) + 6\ O_2(g)$$

The first reaction, which takes place when a mixture of natural gas (methane) and air is ignited, supplies the heat required to cook a steak on a gas range or boil water with a Bunsen burner. In contrast, the formation of glucose, $C_6H_{12}O_6$, by the process of photosynthesis requires the absorption of light energy from the sun.

Throughout most of this chapter, we will be concerned with a particular type of energy change: the heat flow associated with a chemical reaction. Most of the reactions with which we are familiar are ones that evolve heat **(exothermic);** an example is the combustion of methane referred to above. In an exothermic reaction, heat flows from the reaction mixture into the surroundings. Usually the effect of this heat flow is to raise the temperature of the surroundings, which might be a beaker of water over a Bunsen burner or a steak on a gas range. Reactions which absorb heat **(endothermic)** are perhaps less common; one familiar example is the phase change that occurs when ice melts. In an endothermic reaction, heat flows into the reaction mixture from the surroundings. The usual effect, observed when an ice cube melts in a glass of tea, is to lower the temperature of the surroundings, in this case the tea.

4.1 THE ENTHALPY CHANGE: ΔH

In a general way, we can relate the energy change observed when a reaction takes place to the difference in energy between products and reactants. If the products have a higher energy than the reactants, we must supply energy to make the reaction go. Conversely, if the products are in a lower energy state than the reactants, we should be able to get energy out of the reaction.

To make this argument quantitative, let us consider reactions taking place at a constant pressure. Most reactions that we study in the general chemistry laboratory meet this condition; the "constant pressure" is that of the atmosphere. Such is the case, for example, when a fuel burns in an open flame or when any reaction takes place in a beaker, test tube, or crucible. The heat flow associated with such reactions is directly related to an important property of the substances involved, known as their "heat content," or **enthalpy**, and given the symbol H. **For any reaction carried out directly at a constant pressure, the heat flow is exactly equal to the difference between the enthalpy of the products and that of the reactants.** Using Q_p to represent the heat flow for a constant pressure process and ΔH to denote the enthalpy change, we have

$$Q_p = H_{products} - H_{reactants} = \Delta H_{reaction} \tag{4.1}$$

To understand what Equation 4.1 means, consider what happens when a lighted splint is inserted into a test tube filled with a mixture of hydrogen and oxygen. A vigorous, exothermic reaction occurs, which may be represented by the equation

$$H_2(g) + \frac{1}{2} O_2(g) \rightarrow H_2O(l) \tag{4.2}$$

Both products and reactants are at 25°C

We find that when this reaction takes place at 25°C and one atmosphere pressure, 285.8 kJ of heat is evolved to the surroundings per mole of liquid water formed. This means that the enthalpy of one mole of $H_2O(l)$ under these conditions is 285.8 kJ less than that of the reactants, one mole of $H_2(g)$ and one half mole of $O_2(g)$:

$$Q_p = \Delta H = H \text{ of 1 mol } H_2O(l) - (H \text{ of 1 mol } H_2 + H \text{ of } \tfrac{1}{2} \text{ mol } O_2)$$

$$= -285.8 \text{ kJ (at 25°C and 1 atm)}$$

Exothermic reactions, such as 4.2, are always associated with a *decrease* in enthalpy ($\Delta H < 0$). The enthalpy "lost" by the reaction mixture is the source of the heat that flows into the surroundings. In contrast, an *endothermic* reaction is accompanied by an *increase* in enthalpy ($\Delta H > 0$). Heat flowing into the reaction system "raises" its enthalpy. An example is the endothermic reaction that occurs when mercuric oxide is heated in a test tube:

$$HgO(s) \rightarrow Hg(l) + \frac{1}{2} O_2(g) \tag{4.3}$$

We find that 90.7 kJ of heat must be absorbed by the reaction mixture to decompose one mole of mercuric oxide. This means that the enthalpy of the products,

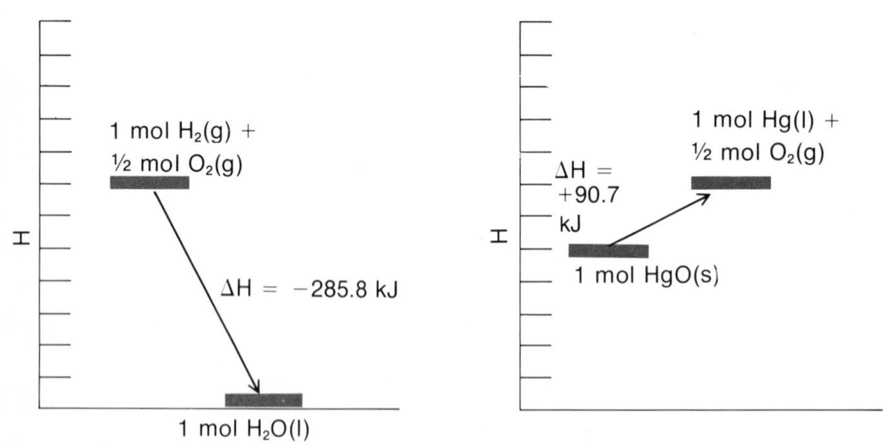

Figure 4.1 For an exothermic process, such as $H_2(g) + \frac{1}{2} O_2(g) \rightarrow H_2O(l)$, $H_{products} < H_{reactants}$, and $\Delta H < 0$. For an endothermic reaction, such as $HgO(s) \rightarrow Hg(l) + \frac{1}{2} O_2(g)$, $H_{products} > H_{reactants}$, and $\Delta H > 0$.

one mole of Hg(l) and one half mole of $O_2(g)$, must be 90.7 kJ greater than that of one mole of the reactant, HgO(s):

$$Q_p = \Delta H = (H \text{ of 1 mol Hg} + H \text{ of } \tfrac{1}{2} \text{ mol } O_2) - H \text{ of 1 mol HgO}$$

$$= +90.7 \text{ kJ (at 25°C and 1 atm)}$$

Nature of Enthalpy

Measurements of heat flow in reactions lead only to values of ΔH, the difference in enthalpy between products and reactants. We cannot experimentally determine the enthalpy of an individual substance. Nevertheless, we have every reason to believe that the enthalpy (''heat content'') of a fixed amount of a pure substance, at a given temperature and pressure, is a characteristic property of that substance. In this sense, enthalpy, H, resembles volume, V. Just as one gram of liquid water at 25°C and 1 atm has a definite volume, it also has a fixed, definite enthalpy under these conditions.

Quantities such as H and V whose values are fixed when one specifies the temperature and pressure (i.e., the ''state'' of the substance) are often referred to as *state properties*. Their magnitude depends only upon the state of the substance and not upon its history. Thus the enthalpy of one gram of $H_2O(l)$ at 25°C and 1 atm is the same regardless of how the water was formed, whether by melting ice, by condensing steam, by Reaction 4.2, or by some other means.

The *cost* of a gram of water would not be a state property

The enthalpy of a substance changes at least slightly when its temperature changes. One gram of liquid water has a somewhat larger enthalpy at 100°C than it does at 0°C. The fact that we have to add 418 J of heat to raise the temperature of one gram of liquid water from 0°C to 100°C implies that its enthalpy has increased by that amount.

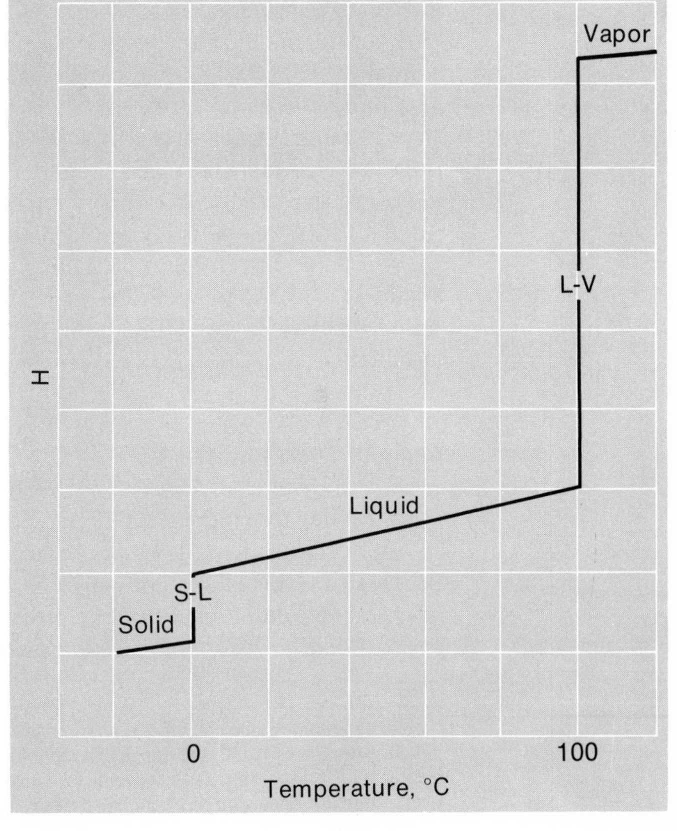

Figure 4.2 Variation with temperature of the enthalpy of one gram of water at 1 atm pressure. Enthalpy of a substance increases with increasing temperature. Note the large increases where phase changes occur; at 0°C, ΔH_{fusion} = 333 J/g and at 100°C, $\Delta H_{vaporization}$ = 2257 J/g.

$$H \text{ of } 1 \text{ g } H_2O(l) \text{ at } 100°C - H \text{ of } 1 \text{ g } H_2O(l) \text{ at } 0°C = 418 \text{ J}$$

When a substance undergoes a phase change, its enthalpy changes. Since we must add 333 J of heat to melt a gram of ice at 0°C and 2257 J to boil a gram of liquid water at 100°C, we conclude that:

$$H \text{ of } 1 \text{ g } H_2O(l) \text{ at } 0°C - H \text{ of } 1 \text{ g } H_2O(s) \text{ at } 0°C = 333 \text{ J}$$

$$H \text{ of } 1 \text{ g } H_2O(g) \text{ at } 100°C - H \text{ of } 1 \text{ g } H_2O(l) \text{ at } 100°C = 2257 \text{ J}$$

Enthalpy, like volume, is directly proportional to mass. The enthalpy of one mole (18.0 g) of $H_2O(l)$ at a given temperature is 18 times as great as that of one gram of liquid water; two moles of water have an enthalpy twice that of one mole, and so on.

4.2 THERMOCHEMICAL EQUATIONS

A thermochemical equation is simply a balanced equation of the type discussed in Chapter 3 with one addition: the heat flow accompanying the reaction is specified. We will ordinarily do this by listing, at the right of the equation, the value of ΔH in kilojoules. Thus for the two reactions discussed in Section 4.1, we write the thermochemical equations:

$$H_2(g) + \tfrac{1}{2} O_2(g) \rightarrow H_2O(l); \Delta H = -285.8 \text{ kJ} \qquad (4.2)$$

$$HgO(s) \rightarrow Hg(l) + \tfrac{1}{2} O_2(g); \Delta H = +90.7 \text{ kJ} \qquad (4.3)$$

Several conventions are used in writing and interpreting equations such as these. In particular:

1. The coefficients always refer to number of moles. Thus, −285.8 kJ is the enthalpy change when one mole of $H_2O(l)$ is formed from one mole of $H_2(g)$ and one half mole of $O_2(g)$.

Unless (s), (l), or (g) is indicated, an equation like 4.2 is ambiguous

2. Since the enthalpy of a substance depends upon whether it is a liquid, solid, or gas, this should be specified in writing the equation. If, in Reaction 4.2, $H_2O(g)$ were produced rather than $H_2O(l)$, less heat would be evolved because some of it (about 44.0 kJ) would be absorbed in vaporizing the water.

3. Since the enthalpy of a substance depends upon temperature, we should really specify the temperature at which the reaction is carried out.* Unless indicated otherwise, we will take the temperature to be 25°C for both reactants and products.

Laws of Thermochemistry

In order to make effective use of thermochemical equations, it is necessary to apply certain basic laws of thermochemistry. These laws are most simply expressed in terms of the enthalpy change, ΔH.

1. Since enthalpy is directly proportional to mass (Section 4.1), *ΔH is directly proportional to the amount of substance that reacts or is produced in a reaction.*

*The enthalpy change for a reaction, ΔH, is much less sensitive to temperature than the individual enthalpies of the substances involved. If t increases, the enthalpies of both products and reactants increase. These effects tend to cancel, often leaving ΔH nearly unchanged over a rather wide temperature range.

Example 4.1 Hydrogen peroxide, a common bleaching agent, decomposes as follows:

$$H_2O_2(l) \rightarrow H_2O(l) + \frac{1}{2} O_2(g);\ \Delta H = -98.2 \text{ kJ}$$

How much heat is released when one gram of H_2O_2, stored in an open bottle, decomposes?

Solution From the equation, ΔH for the decomposition of one mole of H_2O_2 is -98.2 kJ. Therefore

$$1 \text{ mol } H_2O_2 = 34.0 \text{ g } H_2O_2 \simeq -98.2 \text{ kJ}$$

Using the conversion factor approach:

$$1.00 \text{ g } H_2O_2 \times \frac{-98.2 \text{ kJ}}{34.0 \text{ g } H_2O_2} = -2.89 \text{ kJ}$$

We deduce that the decomposition of one gram of hydrogen peroxide gives off 2.89 kJ (2890 J) of heat.

2. *ΔH for a reaction is equal in magnitude but opposite in sign to ΔH for the reverse reaction.*

The validity of this law should be evident from Figure 4.1. We see from the diagram on the right that the enthalpy of HgO(s) lies 90.7 kJ below the sum of the enthalpies of Hg(l) and ½ O_2(g). Consequently, 90.7 kJ of heat must be absorbed to decompose one mole of HgO to the elements.

$$HgO(s) \rightarrow Hg(l) + \frac{1}{2} O_2(g);\ \Delta H = +90.7 \text{ kJ}$$

If we were to go in the opposite direction, forming one mole of HgO from the elements, 90.7 kJ of heat would be evolved.

$$Hg(l) + \frac{1}{2} O_2(g) \rightarrow HgO(s);\ \Delta H = -90.7 \text{ kJ}$$

3. Since H is a state property, ΔH must be independent of the path followed in going from an initial to a final state. In particular, ΔH must be the same regardless of the number of steps involved in the path. This means that *if a reaction can be regarded as the sum of two or more other reactions, ΔH for the overall reaction must be the sum of the enthalpy changes for the other reactions.*

Consider the following reactions (Fig. 4.3):

$$Sn(s) + Cl_2(g) \rightarrow SnCl_2(s);\ \Delta H_1 = -349.8 \text{ kJ} \quad (4.4a)$$

$$SnCl_2(s) + Cl_2(g) \rightarrow SnCl_4(l);\ \Delta H_2 = -195.4 \text{ kJ} \quad (4.4b)$$

If we add these two equations together, we obtain the equation for the formation of $SnCl_4$ from the elements:

$$Sn(s) + 2\ Cl_2(g) \rightarrow SnCl_4(l);\ \Delta H = \Delta H_1 + \Delta H_2 = -545.2 \text{ kJ} \quad (4.4)$$

The enthalpy change for Reaction 4.4 is the sum of the ΔH's of Reactions 4.4a and 4.4b.

This relationship between enthalpy changes in related reactions is called **Hess's Law.** It is very useful for determining the heat flow of a reaction which is difficult or impossible to carry out directly in a single step.

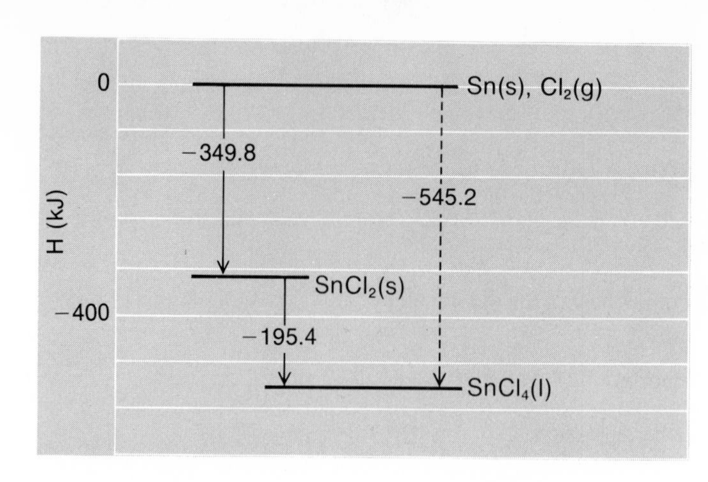

Figure 4.3 ΔH for the reaction $Sn(s) + 2\ Cl_2(g) \rightarrow SnCl_4(l)$ is the same whether it is carried out in one step (dotted arrow) or two steps (solid arrow).

In all the equations we have written, ΔH is expressed in joules or kilojoules. We will continue to do this throughout this chapter, since the joule is the preferred unit of energy in the International System of Units (SI). You should realize, however, that energy changes are readily expressed in other units by using appropriate conversion factors. To convert from joules to calories, an energy unit still in common use in many countries, or from kilojoules to kilocalories, we need only divide by 4.184, since

$$1 \text{ cal} = 4.184 \text{ J}; \ 1 \text{ kcal} = 4.184 \text{ kJ}$$

Thus for Reaction 4.4 we calculate

$$\Delta H = -545.2 \text{ kJ} \times \frac{1 \text{ kcal}}{4.184 \text{ kJ}} = -130.3 \text{ kcal}$$

so the thermochemical equation could be written

$$Sn(s) + 2\ Cl_2(g) \rightarrow SnCl_4(l); \ \Delta H = -130.3 \text{ kcal}$$

4.3 HEATS OF FORMATION

We have now written several thermochemical equations, in each case citing the corresponding enthalpy change, ΔH. Literally thousands of such equations would have to be written to list ΔH values for all the reactions that have been studied. Clearly we need some more concise way of recording thermochemical data in a form which could easily be used to calculate ΔH for any reaction in which we are interested. It turns out that this can be accomplished by listing quantities known as heats of formation. Let us consider first how these quantities are defined and then how they can be used to calculate ΔH for reactions.

The molar heat of formation of a compound, ΔH_f, is equal to the enthalpy change, ΔH, when one mole of the compound is formed from the elements in their stable forms at 25°C and 1 atm. Thus, from the equations

$$Ag(s) + \tfrac{1}{2}\ Cl_2(g) \rightarrow AgCl(s); \ \Delta H = -127.0 \text{ kJ} \quad (4.5)$$

$$\tfrac{1}{2}\ N_2(g) + O_2(g) \rightarrow NO_2(g); \ \Delta H = +33.9 \text{ kJ} \quad (4.6)$$

we conclude that the heats of formation of silver chloride and nitrogen dioxide are −127.0 kJ/mol and +33.9 kJ/mol, respectively.

Heats of formation are listed for a variety of compounds in Table 4.1. Notice that, with a few exceptions such as NO, NO_2, and acetylene, C_2H_2, heats of formation are usually negative quantities. This reflects the fact that the formation of a compound from the elements is ordinarily exothermic. Conversely, the decomposition of a compound into the elements usually requires the absorption of heat.

Compounds for which $\Delta H_f > 0$ are usually unstable

To show how heats of formation can be used to calculate enthalpy changes for reactions, let us suppose that we wish to determine ΔH for

$$SnO_2(s) + 2\ H_2(g) \rightarrow Sn(s) + 2\ H_2O(l);\ \Delta H = ? \qquad (4.7)$$

We can imagine that this reaction takes place in two steps, the first of which is the decomposition of SnO_2 to the elements:

$$SnO_2(s) \rightarrow Sn(s) + O_2(g);\ \Delta H_a = -\Delta H_f\ SnO_2(s) \qquad (4.7a)$$

The oxygen produced then reacts with hydrogen to form two moles of water:

$$2\ H_2(g) + O_2(g) \rightarrow 2\ H_2O(l);\ \Delta H_b = 2\ \Delta H_f\ H_2O(l) \qquad (4.7b)$$

Since Equations 4.7a and 4.7b add to give 4.7, Hess's Law tells us that the enthalpy change for Reaction 4.7 must be

$$\Delta H = \Delta H_a + \Delta H_b = 2\ \Delta H_f\ H_2O(l) - \Delta H_f\ SnO_2(s)$$

We could go through this sort of analysis to relate ΔH for any reaction to the heats of formation of the compounds involved. In practice, it is unnecessary to do this. We need only apply the general rule illustrated by the above example: **ΔH for any reaction is equal to the sum of the heats of formation of the product compounds** (e.g., two moles of H_2O in Reaction 4.7) **minus the sum of the heats of formation of the reactant compounds** (e.g., one mole of SnO_2 in Reaction 4.7).

TABLE 4.1 HEATS OF FORMATION (kJ/MOL) AT 25°C AND 1 ATM

Compound	ΔH_f	Compound	ΔH_f	Compound	ΔH_f	Compound	ΔH_f
AgBr(s)	−99.5	C_2H_2(g)	+226.7	H_2O_2(l)	−187.6	NH_3(g)	−46.2
AgCl(s)	−127.0	C_2H_4(g)	+52.3	H_2S(g)	−20.1	NH_4Cl(s)	−315.4
AgI(s)	−62.4	C_2H_6(g)	−84.7	H_2SO_4(l)	−811.3	NH_4NO_3(s)	−365.1
Ag_2O(s)	−30.6	C_3H_8(g)	−103.8	HgO(s)	−90.7	NO(g)	+90.4
Ag_2S(s)	−31.8	n-C_4H_{10}(g)	−124.7	HgS(s)	−58.2	NO_2(g)	+33.9
Al_2O_3(s)	−1669.8	n-C_5H_{12}(l)	−173.1	KBr(s)	−392.2	NiO(s)	−244.3
$BaCl_2$(s)	−860.1	C_2H_5OH(l)	−277.6	KCl(s)	−435.9	$PbBr_2$(s)	−277.0
$BaCO_3$(s)	−1218.8	CoO(s)	−239.3	$KClO_3$(s)	−391.4	$PbCl_2$(s)	−359.2
BaO(s)	−558.1	Cr_2O_3(s)	−1128.4	KF(s)	−562.6	PbO(s)	−217.9
$Ba(OH)_2$(s)	−946.4	CuO(s)	−155.2	KOH(s)	−425.8	PbO_2(s)	−276.6
$BaSO_4$(s)	−1465.2	Cu_2O(s)	−166.7	$MgCl_2$(s)	−641.8	Pb_3O_4(s)	−734.7
$CaCl_2$(s)	−795.0	CuS(s)	−48.5	$MgCO_3$(s)	−1113	PCl_3(g)	−306.4
$CaCO_3$(s)	−1207.0	$CuSO_4$(s)	−769.9	MgO(s)	−601.8	PCl_5(g)	−398.9
CaO(s)	−635.5	Fe_2O_3(s)	−822.2	$Mg(OH)_2$(s)	−924.7	SiO_2(s)	−859.4
$Ca(OH)_2$(s)	−986.6	Fe_3O_4(s)	−1120.9	$MgSO_4$(s)	−1278.2	$SnCl_2$(s)	−349.8
$CaSO_4$(s)	−1432.7	HBr(g)	−36.2	MnO(s)	−384.9	$SnCl_4$(l)	−545.2
CCl_4(l)	−139.5	HCl(g)	−92.3	MnO_2(s)	−519.7	SnO(s)	−286.2
CH_4(g)	−74.8	HF(g)	−268.6	NaBr(s)	−359.9	SnO_2(s)	−580.7
$CHCl_3$(l)	−131.8	HI(g)	+25.9	NaCl(s)	−411.0	SO_2(g)	−296.1
CH_3OH(l)	−238.6	HNO_3(l)	−173.2	NaF(s)	−569.0	SO_3(g)	−395.2
CO(g)	−110.5	H_2O(g)	−241.8	NaI(s)	−288.0	ZnO(s)	−348.0
CO_2(g)	−393.5	H_2O(l)	−285.8	NaOH(s)	−426.7	ZnS(s)	−202.9

In mathematical form, using the symbol $\sum$ to represent "sum," we have

A very useful equation

$$\Delta H = \sum \Delta H_f \text{ products} - \sum \Delta H_f \text{ reactants} \quad (4.8)$$

Any elementary substance in its stable form taking part in the reaction (e.g., Sn(s) and H_2(g) in Reaction 4.7) is omitted in taking this sum, since its heat of formation is **zero** as a consequence of the way in which ΔH_f is defined.

The use of Equation 4.8 to calculate ΔH of a reaction from heats of formation is illustrated by Example 4.2.

Example 4.2 Using Table 4.1, calculate ΔH for the reaction

$$8\ Al(s) + 3\ Fe_3O_4(s) \rightarrow 4\ Al_2O_3(s) + 9\ Fe(s)$$

Solution Following Equation 4.8, and omitting terms for elementary substances

$$\Delta H = 4\ \Delta H_f\ Al_2O_3(s) - 3\ \Delta H_f\ Fe_3O_4(s)$$

From Table 4.1

$$\Delta H = 4(-1669.8\text{ kJ}) - 3(-1120.9\text{ kJ}) = -3316.5\text{ kJ}$$

Equation 4.8 can also be applied to obtain heats of formation from experimentally determined heats of reaction (Example 4.3).

Example 4.3 The heat of combustion of benzene is −3268.5 kJ/mol. That is,

$$C_6H_6(l) + \frac{15}{2}\ O_2(g) \rightarrow 6\ CO_2(g) + 3\ H_2O(l);\ \Delta H = -3268.5\text{ kJ}$$

Using Table 4.1, calculate the heat of formation of benzene.

Solution From Equation 4.8

$$\Delta H = 6\ \Delta H_f\ CO_2(g) + 3\ \Delta H_f\ H_2O(l) - \Delta H_f\ C_6H_6(l)$$

$$-3268.5\text{ kJ} = 6(-393.5\text{ kJ}) + 3(-285.8\text{ kJ}) - \Delta H_f\ C_6H_6(l)$$

Solving:

$$\Delta H_f\ C_6H_6(l) = -2361.0\text{ kJ} - 857.4\text{ kJ} + 3268.5\text{ kJ}$$

$$= +50.1\text{ kJ/mol of benzene}$$

This approach is perhaps more widely used than any other to determine heats of formation. It is particularly useful for organic compounds, which are difficult if not impossible to form directly from the elements.

4.4 BOND ENERGIES (ENTHALPIES)

We have seen how a table of heats of formation can be used to calculate ΔH for a variety of reactions. In some cases, ΔH turns out to be a large negative number, indicating that the reaction is strongly exothermic. In other reactions, ΔH is positive; heat must be absorbed for reaction to occur. At this stage, you might well wonder why ΔH should vary so widely from one reaction to another. More to the point, is there some fundamental property of reactant and product molecules which determines the size and magnitude of ΔH?

These questions can be answered on a molecular level in terms of a quantity known as bond energy or, more properly but less commonly, bond enthalpy. The **bond energy is defined as ΔH when one mole of bonds is broken in the gas state.** From the equations

$$H_2(g) \rightarrow 2\ H(g);\ \Delta H = +436\ \text{kJ} \quad (4.9)$$

$$2\ Cl(g) \rightarrow Cl_2(g);\ \Delta H = -243\ \text{kJ} \quad (4.10)$$

we conclude that the bond energies for the H—H bond and the Cl—Cl bond must be +436 and +243 kJ/mol, respectively (Fig. 4.4). In Reaction 4.9, one mole of H—H bonds in H_2 molecules is broken; the bond energy is equal to the quoted ΔH. In Reaction 4.10, one mole of Cl—Cl bonds is formed; the bond energy is the enthalpy change for the reverse reaction, +243 kJ. Notice that in both cases we are dealing with gaseous atoms and molecules; *the bond energy can be equated to ΔH only when all reactants and products are gases.*

Bond energies for a variety of single bonds are listed in Table 4.2. You will note that the bond energy is always a positive quantity; energy is always absorbed by a molecule when one of its chemical bonds is broken. Conversely, heat is given off when bonds are formed from gaseous atoms. Thus we have

$$HCl(g) \rightarrow H(g) + Cl(g);\ \Delta H = B.E.\ \text{H—Cl} = +431\ \text{kJ}$$

but

$$H(g) + F(g) \rightarrow HF(g);\ \Delta H = -B.E.\ \text{H—F} = -565\ \text{kJ}$$

where the abbreviation *B.E.* is used to represent bond energy (kJ/mol of bonds).

The concept of bond energy helps us to understand, at the molecular level, why some reactions are exothermic and others endothermic. If the bonds in the product molecules are stronger than those in the reactants, the reaction will be exothermic.

$$\text{``weak'' bonds} \rightarrow \text{``strong'' bonds};\ \Delta H < 0$$

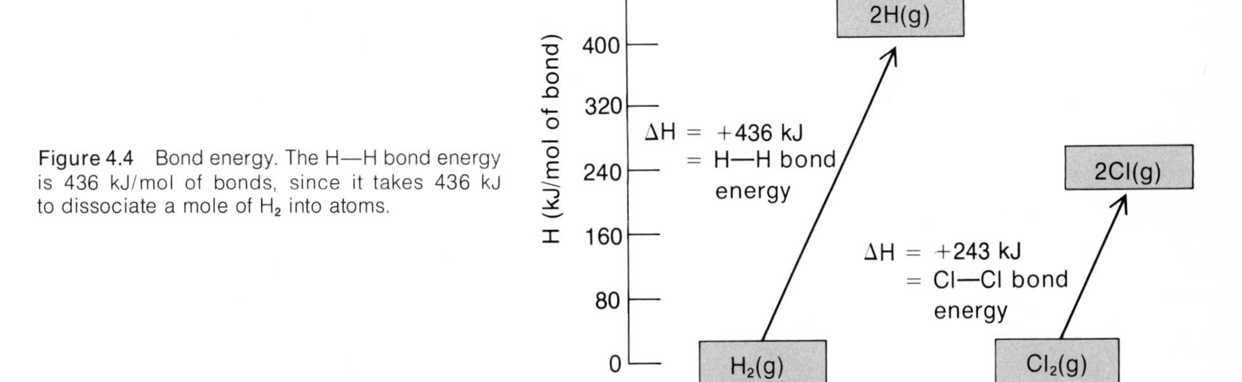

Figure 4.4 Bond energy. The H—H bond energy is 436 kJ/mol of bonds, since it takes 436 kJ to dissociate a mole of H_2 into atoms.

TABLE 4.2 SINGLE BOND ENERGIES (kJ/MOL) AT 25°C

	H	C	N	O	S	F	Cl	Br	I
H	436	414	389	464	339	565	431	368	297
C		347	293	351	259	485	331	276	238
N			159	222	—	272	201	243	—
O				138	—	184	205	201	201
S					226	285	255	213	—
F						153	255	255	—
Cl							243	218	209
Br								193	180
I									151

If the reverse is true, heat will have to be absorbed to bring about reaction

$$\text{"strong" bonds} \rightarrow \text{"weak" bonds; } \Delta H > 0$$

To cite a specific example, consider the reaction between hydrogen and fluorine:

$$H_2(g) + F_2(g) \rightarrow 2\ HF(g);\ \Delta H = -541\ kJ$$

The fact that this reaction evolves heat implies that the bonds in HF are "stronger" than those in the elementary substances H_2 and F_2. More precisely, the amount of heat evolved when two moles of H—F bonds are formed is 541 kJ greater than that absorbed in breaking one mole of H—H and one mole of F—F bonds.

The thermal stability of molecules is directly related to the strengths of the bonds holding them together. The fluorine molecule, F_2, which contains the relatively weak F—F bond, undergoes appreciable decomposition at 1000°C:

What is ΔH_f for F(g)?

$$F_2(g) \rightarrow 2\ F(g);\ \Delta H = +153\ kJ$$

In contrast, the hydrogen fluoride molecule, HF, is one of the most stable known. The bond between hydrogen and fluorine is so strong that the HF molecule is stable at temperatures as high as 5000°C.

$$HF(g) \rightarrow H(g) + F(g);\ \Delta H = +565\ kJ$$

This explains, at least in part, the interest in fluorine compounds as components of rocket propellants. The fuel-oxidizer combination is designed so that hydrogen fluoride will be a major product in the rocket exhaust.

In principle, a table of bond energies can be used to calculate ΔH for chemical reactions, applying Hess's Law in a straightforward manner (Example 4.4).

Example 4.4 Using Table 4.2, estimate ΔH for the reaction

$$H_2(g) + Cl_2(g) \rightarrow 2\ HCl(g)$$

Solution Let us imagine this reaction as taking place in two steps:
(1) The reactant molecules, H_2 and Cl_2, break down to free atoms

$$H_2(g) + Cl_2(g) \rightarrow 2\ H(g) + 2\ Cl(g)$$

(2) The atoms formed in (1) combine to form HCl molecules

$$2\ H(g) + 2\ Cl(g) \rightarrow 2\ HCl(g)$$

In the first step, we *break* a mole of H—H bonds and a mole of Cl—Cl bonds

$$\Delta H_1 = B.E.\ \text{H—H} + B.E.\ \text{Cl—Cl} = +436\ \text{kJ} + 243\ \text{kJ} = +679\ \text{kJ}$$

In the second step, *two* moles of H—Cl bonds are *formed:*

$$\Delta H_2 = -2\ B.E.\ \text{H—Cl} = -2(431\ \text{kJ}) = -862\ \text{kJ}$$

By Hess's Law: $\Delta H = \Delta H_1 + \Delta H_2 = +679\ \text{kJ} - 862\ \text{kJ} = -183\ \text{kJ}$

It may have occurred to you that the calculation called for in Example 4.4 could have been carried out very simply using heats of formation. That is, for the reaction

$$H_2(g) + Cl_2(g) \rightarrow 2\ HCl(g)$$

$\Delta H = 2\ \Delta H_f\ \text{HCl} = 2(-92.3\ \text{kJ}) = -184.6\ \text{kJ}$. Clearly, in this case it was unnecessary to work through bond energies. Consider, however, the reaction

$$CH_4(g) + OH(g) \rightarrow CH_3(g) + H_2O(g) \qquad (4.11)$$

which is believed to be one of the steps involved in the combustion of methane. We cannot use Table 4.1 to calculate ΔH since heats of formation are not given for the unstable species OH and CH_3. However, if we look at the reaction from the standpoint of breaking and forming bonds, we see that what is really involved is

the breaking of a C—H bond:

```
    H              H
    |              |
H — C — H   →   H— C  + H      (4.11a)
    |              |
    H              H
```

followed by

the formation of an O—H bond: H + O—H → H—O—H (4.11b)

Hence ΔH for Reaction 4.11 must be the sum of the enthalpy changes for 4.11a and 4.11b, both of which can be related to bond energies:

$$\Delta H = \Delta H_a + \Delta H_b = B.E.\ \text{C—H} - B.E.\ \text{O—H} = 414\ \text{kJ} - 464\ \text{kJ} = -50\ \text{kJ}$$

In practice, we seldom calculate ΔH from bond energies if heat of formation data are available for all of the compounds taking part in the reaction. The reason is that in most cases it is not possible to assign an exact value to a bond energy; it varies to some extent with the species in which the bond is found. For example, the enthalpy changes for the two reactions

$$\text{H—O—H}(g) \rightarrow H(g) + \text{O—H}(g);\ \Delta H = +502\ \text{kJ} \qquad (4.12)$$

$$\text{O—H}(g) \rightarrow H(g) + O(g);\ \Delta H = +426\ \text{kJ} \qquad (4.13)$$

are not exactly the same, even though both involve the breaking of an O—H bond. The bond energy of the O—H bond given in Table 4.2, +464 kJ/mol, is an average value calculated from enthalpy changes for a variety of reactions such as 4.12 and 4.13, in which an O—H bond is broken.

4.5 MEASUREMENT OF HEAT FLOW. CALORIMETRY

The amount of heat evolved or absorbed in a process is measured in an apparatus known as a calorimeter. A simple "coffee cup" calorimeter commonly used in the general chemistry laboratory is shown in Figure 4.5. It consists essentially of a styrofoam cup partially filled with water. A thermometer is lowered through the cover of the calorimeter until its bulb is well below the surface of the water.

To illustrate the principles involved in measuring heat flow, let us consider an experiment designed to determine the specific heat of a metal. The general procedure is to measure the heat flow from a hot sample of the metal to water contained in the calorimeter. We start by weighing into the calorimeter a known amount of water, let us say 50 g. The temperature of the water is read to be 25.0°C. Now in a separate container (perhaps a beaker of boiling water) we bring a sample of metal weighing 20 g to a known temperature, 100.0°C. The hot metal is dropped into the calorimeter and the cup swirled to bring its contents to a constant temperature. Let us suppose that this temperature is 31.0°C.

To analyze the heat flow here, we start by noting that since styrofoam is a good insulator there should be no heat transfer through the walls of the calorimeter. Consequently, the amount of heat given off by the metal in cooling from 100°C to 31°C should be numerically equal to that absorbed by the water as it warms from 25°C to 31°C. Stated another way, *the heat flow for the metal is equal in magnitude but opposite in direction and hence in sign to that for the water.* Using Q to represent heat flow,

If the calorimeter isn't covered, Eq. 4.14 won't apply exactly. Why?

$$Q_{metal} = -Q_{water} \quad (4.14)$$

The magnitude of the heat flow, in either direction, can be calculated by using the general relation

$$Q = (\text{specific heat}) \times m \times \Delta t \quad (4.15)$$

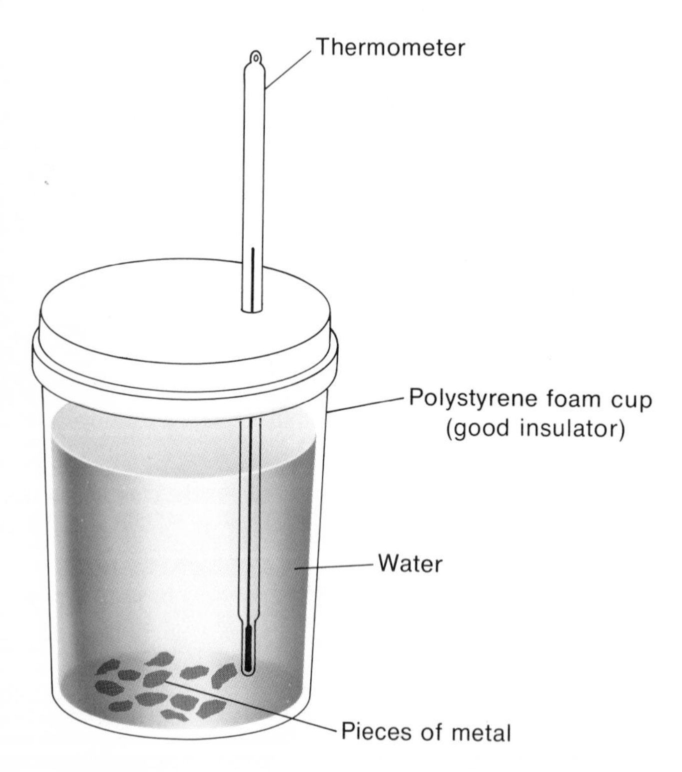

Figure 4.5 Coffee cup calorimeter. Since polystyrene foam is a good heat insulator, the heat flow is limited to that which passes from the metal chunks into the water. This type of calorimeter can also be used to measure ΔH for reactions between species in solution or to measure heats of solution.

where Q is the heat flow, m is the mass of the substance in grams, and Δt is the difference between the final and initial temperatures in degrees Celsius, i.e.,

$$\Delta t = t_{final} - t_{initial}$$

The specific heat, defined by this equation, can be regarded as the amount of heat required to raise the temperature of one gram of a substance by one degree Celsius; with Q in joules, it has the units of joule per gram degree Celsius.

To obtain Q_{water}, we start by noting that since it takes approximately 1.00 cal or 4.18 J to raise the temperature of one gram of water one degree Celsius, the specific heat of water is 4.18 J/(g · °C). The mass of water in this experiment is 50 g, its final temperature is 31.0°C, and its initial temperature is 25.0°C.

$$Q_{water} = 4.18 \frac{\text{J}}{\text{g°C}} \times 50 \text{ g} \times (31.0°\text{C} - 25.0°\text{C}) = +1250 \text{ J}$$

It follows from Equation 4.14 that the heat flow for the metal, Q_{metal}, must be −1250 J.

It may be well to pause briefly here to comment upon the signs of Q_{water} and Q_{metal}. When a substance absorbs heat, as the water did in this experiment, its temperature rises, Δt is positive, and Q as calculated from Equation 4.15 must be positive. Conversely, when a substance (the metal in this case) evolves heat, its temperature drops, Δt is a negative quantity, and hence, Q must be negative.

At this stage, you should be able to calculate the specific heat of the metal involved in this experiment.

Example 4.5 Complete the above discussion by obtaining the specific heat of the metal.

Solution We need only apply Equation 4.15 to the metal. We have seen that Q_{metal} is −1250 J; the metal has a mass of 20 g. Its final temperature must be the same as that of the water, 31.0°C; its initial temperature was given as 100.0°C. Hence:

$$-1250 \text{ J} = (S.H.) \times 20 \text{ g} \times (31.0°\text{C} - 100.0°\text{C})$$

Solving:

$$S.H. \text{ metal} = \frac{-1250 \text{ J}}{20 \text{ g } (-69°\text{C})} = 0.91 \text{ J/(g} \cdot °\text{C)}$$

Notice that the minus sign took care of itself!

A coffee cup calorimeter is quite adequate for measuring heat flows in simple processes such as the one just described. It can also be used to study many chemical reactions taking place in water solution. However, it would hardly be suitable for reactions involving gases or ones in which the products reach high temperatures. The *bomb calorimeter,* shown in Figure 4.6, is a more versatile instrument; indeed, for precise heat flow measurements a device such as this is required.

To use a bomb calorimeter, we start by adding a weighed sample of the reactant(s) to the heavy-walled steel container, called a "bomb," which is then sealed and lowered into a metal container which fits snugly inside the insulating walls of the calorimeter. A weighed amount of water sufficient to cover the bomb is added to the container and the entire apparatus closed. After the initial temperature is measured precisely, the reaction is started, perhaps by electrical ignition. If all goes well, reaction occurs instantly and goes to completion. Assuming we are dealing with an exothermic reaction, the hot reaction products give off heat to

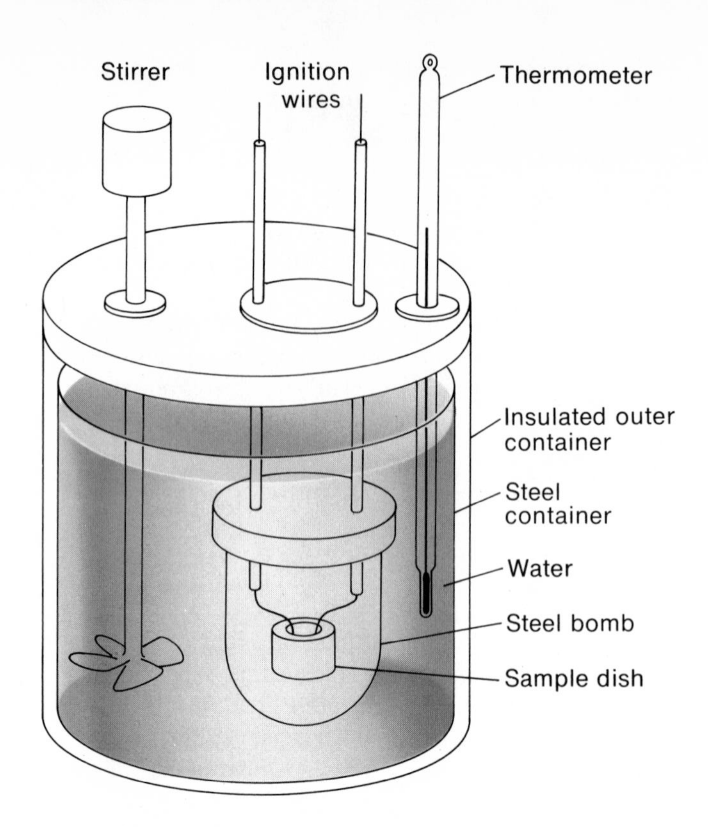

Figure 4.6 Bomb calorimeter, used to determine heat flow for combustion reactions. The heat liberated by the reaction is absorbed by the bomb and the surrounding water.

the walls of the bomb, the water, and the metal container. The final temperature is taken to be the highest value read on the thermometer.

Analysis of the heat flow is similar but slightly more involved than that described earlier in connection with Figure 4.5. Specifically, we cannot ignore the heat flow into the metal parts of the calorimeter. The amount of heat *evolved* in the reaction is numerically equal to that *absorbed* by the water plus that *absorbed* by the bomb. That is,

$$Q = -(Q_{water} + Q_{bomb}) \tag{4.16}$$

where Q, Q_{water}, and Q_{bomb} are the heat flows associated with the reaction, the water, and the bomb, respectively. Q_{water} is evaluated as before:

$$Q_{water} = \frac{4.18 \text{ J}}{\text{g}°\text{C}} \times m_{water} \times \Delta t$$

The value of Q_{bomb} is obtained in a slightly different way. The total "heat capacity," C, of the bomb assembly (the amount of heat required to raise its temperature by 1°C) is first determined in a preliminary experiment (see Problems 4.17 and 4.33, p. 87). Having established that it takes C joules to raise the temperature of the bomb 1°C, we have:

$$Q_{bomb} = C \times \Delta t$$

The use of these relationships to determine the heat flow for a chemical reaction is illustrated in Example 4.6.

Example 4.6 A 1.000-g sample of the rocket fuel hydrazine, N_2H_4, is burned in a bomb calorimeter containing 1200 g of water. The temperature rises from 24.62 to 28.16°C. Taking the heat capacity of the bomb to be 840 J/°C, calculate

a. Q for the combustion of the one-gram sample.
b. Q for the combustion of one mole of hydrazine in the bomb calorimeter.

Solution

a. Using Equation 4.16:

$$Q = -(Q_{water} + Q_{bomb})$$

$$= -(4.18 \frac{J}{g°C} \times m_{water} \times \Delta t + C \times \Delta t)$$

$$= -(4.18 \frac{J}{g°C} \times m_{water} + C)\ \Delta t$$

Note that the mass of the water is 1200 g, C = 840 J/°C, Δt = 28.16°C − 24.62°C = 3.54°C. Hence:

$$Q = -(4.18 \frac{J}{g°C} \times 1200\ g + 840 \frac{J}{°C})(3.54°C) = -20\ 700\ J \text{ or } -20.7\ kJ$$

b. From (a), we see that 20.7 kJ of heat is evolved per gram of hydrazine burned. One mole of hydrazine, N_2H_4, weighs 32.0 g. Hence, for the combustion of one mole of hydrazine:

$$Q = 32.0\ g \times -20.7 \frac{kJ}{g} = -662\ kJ$$

4.6 THE FIRST LAW OF THERMODYNAMICS

Up to this point in our discussion of thermochemistry, we have been concerned exclusively with one type of energy change, heat flow. Yet, as pointed out at the beginning of this chapter, many other types of energy can be absorbed or evolved in chemical reactions. For example, the combustion of gasoline in an automobile engine generates mechanical energy as well as heat. To start your car in the morning or operate its headlights at night, you use electrical energy produced by a chemical reaction taking place in the car battery.

To develop the principles governing the total energy change in chemical reactions, we turn to the area of thermodynamics, which considers all kinds of energy effects in all kinds of processes. Here, we will present some of the basic ideas of thermodynamics, stressing their application to the relatively simple kinds of processes described in Sections 4.1 through 4.5. Those of you who go on to more advanced courses in the physical sciences or engineering will be able to explore these ideas in greater depth.

In any thermodynamic treatment we must carefully define the sample of matter to be studied; this is called the *system*. The system may be a reaction mixture whose enthalpy change we are studying, a piece of metal whose specific heat we are asked to determine, or a sample of gas undergoing an expansion. In general terms, the thermodynamic system is simply that portion of the universe upon which we focus our attention. The system is separated from its *surroundings* by a boundary, which may be real or imaginary. For practical purposes, the surroundings may be a beaker of water or the air in a laboratory, but in principle they include the rest of the universe.

Thermodynamicists are very careful people

$$\text{System} + \text{Surroundings} = \text{Universe} \qquad (4.17)$$

Thermodynamics arbitrarily divides energy effects into two categories, heat (Q) and work (W). As you might imagine, a great many different kinds of work are possible. A chemical system might do electrical work on the surroundings, as is the case when the battery in an automobile is used to operate the headlights. Again, a chemical system might have electrical work done upon it, as when a sample of water is electrolyzed to produce hydrogen and oxygen. Later in this text we will have more to say about electrical work. Here, we will restrict our attention to a quite different type of work, that associated with *expansion or compression of a system.*

In Section 4.5, we pointed out that Q has direction as well as magnitude. Heat can flow into a system, raising its temperature, or it can flow out of the system, in which case its temperature drops. The same situation holds for W. A system such as a confined gas can do work on the surroundings by expanding. Alternatively, work may be done on the system, as is the case when the gas is compressed by an external force. In thermodynamics, the direction in which energy flows in the form of heat or work is indicated by specifying its sign. The usual conventions are:

Q is + when the system *absorbs* energy in the form of heat
W is + when the system *evolves* energy by doing work
Q is − when the system evolves energy in the form of heat
W is − when the system absorbs energy by having work done upon it

Following these conventions, the total energy change for a system is given by the difference between Q and W. If a gas absorbs 400 J of heat ($Q = +400$ J) and expands in such a way as to do 150 J of work ($W = +150$ J), its total energy increases by:

$$400 \text{ J} - 150 \text{ J} = 250 \text{ J}$$

The calculation we have just made illustrates a very general relationship called the First Law of Thermodynamics, which tells us that: **In any process, the total change in energy of the system, ΔE, is numerically equal to the heat, Q, absorbed by the system, minus the work, W, done by the system.** In other words:

$$\Delta E = Q - W \tag{4.18}$$

The quantity $\boldsymbol{E}$ in this equation is often referred to as the **internal energy.**

Properties of E and ΔE

If X is a state property, then $\Delta X = X_2 - X_1$, and is independent of path

The internal energy of a system (like its enthalpy, Section 4.1) is a *state property*. That is, its value is fixed when we specify the state of the system. Since this is true, it follows that *ΔE for any change is determined by the final and initial states of the system* and cannot depend upon the "path," i.e., the way in which the change is carried out. Thus, ΔE for the heating of one gram of water from 0°C to 100°C

$$1 \text{ g } H_2O(l, 0°C, 1 \text{ atm}) \rightarrow 1 \text{ g } H_2O(l, 100°C, 1 \text{ atm})$$

is the same regardless of how the process is carried out.

Values of ΔE are readily obtained from Equation 4.18 if W and Q are known (Example 4.7).

Example 4.7 When a system moves from state 1 to state 2 by a certain path, Q and W are found to be +120 J and +70 J, respectively. Calculate ΔE.

Solution Substituting in Equation 4.18:

$$E_2 - E_1 = \Delta E = Q - W = 120\text{ J} - 70\text{ J} = +50\text{ J}$$

A moment's reflection should convince you that there are many different ways in which the system referred to in Example 4.7 could gain 50 J of energy, in addition to the one specified. For example, we could go from state 1 to state 2 by paths in which the system:

absorbs 50 J of heat ($Q = +50$ J) and does no work at all ($W = 0$)
absorbs 80 J of heat ($Q = +80$ J) and does 30 J of work ($W = +30$ J)
absorbs no heat ($Q = 0$) and has 50 J of work done upon it ($W = -50$ J)

All of these paths produce the same ΔE, +50 J. In contrast, Q and W individually can have any values whatsoever. In other words, Q and W, unlike ΔE are *not* state properties. Their values depend upon the path by which the change in state occurs.

Application of the First Law to Chemical Reactions

For a reaction system, ΔE represents the difference in internal energy between products and reactants

$$\Delta E_{reaction} = E_{products} - E_{reactants} = Q - W \qquad (4.19)$$

while Q and W refer to the heat and work effects accompanying the reaction. Although the First Law can be applied to reactions occurring under any conditions, we will restrict our discussion to two particular cases.

1. *Constant Volume, with* $W = 0$. This condition applies when a reaction is carried out in a bomb calorimeter. Since volume is constant within the sealed bomb where reaction occurs, there is no expansion or contraction of the system so no work is done, i.e., $W = 0$. The First Law equation becomes simply

$$\Delta E = Q_v \qquad (4.20)$$

where Q_v represents the heat flow at constant volume. In other words, the heat flow under these conditions is exactly equal to the difference in internal energy between products and reactants. Stated another way, *the heat flow measured in a bomb calorimeter gives directly* ΔE *for the reaction under study.*

2. *Constant Pressure, with* $W = P\Delta V$. This condition applies when, as is ordinarily the case in the laboratory, a reaction is carried out in an open container. The "constant pressure," P, is simply that exerted by the atmosphere on the system, the reaction mixture. Ordinarily, there will be at least a small change in volume as the reaction occurs, so work of the expansion type is done. By the argument indicated in Figure 4.7, p. 78, the value of W is given by the product of the pressure, P, and the volume change, ΔV. That is,

$$W = P\Delta V$$

where $\Delta V = V_{products} - V_{reactants}$. Substituting in the First Law equation with $W = P\Delta V$:

$$\Delta E = Q_p - W = Q_p - P\Delta V \qquad (4.21)$$

where Q_p refers to the heat flow under constant pressure conditions.

If we solve Equation 4.21 for Q_p

$$Q_p = \Delta E + P\Delta V \tag{4.22}$$

we see that for a reaction occurring in an open container against atmospheric pressure, the heat flow is *not* exactly equal to the change in internal energy, ΔE. The two quantities differ by the term $P\Delta V$, which may be either positive or negative, depending upon whether the reaction results in a volume increase ($V_{products} > V_{reactants}$) or decrease ($V_{products} < V_{reactants}$).

ΔH vs. ΔE for Chemical Reactions

For any reaction, $\Delta H = Q_p$; $\Delta E = Q_v$

To complete our discussion of the First Law, it will be helpful to relate the quantity that we have emphasized in this section, the change in internal energy (ΔE), to that discussed at length earlier in the chapter, the enthalpy change (ΔH). You will recall that in Section 4.1 we identified ΔH with the heat flow observed, Q_p, when the reaction is carried out directly at constant pressure. Substituting ΔH for Q_p in Equation 4.22, we obtain the simple relation

$$\Delta H = \Delta E + P\Delta V \qquad (4.23)^*$$

where, as pointed out before, P will ordinarily be the atmospheric pressure and ΔV the volume change when the reaction is carried out at that pressure. In other words, the two quantities ΔH and ΔE differ only by the work term, $P\Delta V$, observed when the reaction occurs against the constant pressure of the atmosphere.

*This equation can be derived in a more rigorous way by using the basic thermodynamic definition of enthalpy

$$H = E + PV$$

where P is the pressure of a system and V its volume. For a change occurring at constant pressure, we can write

$$\Delta H = \Delta E + P\Delta V$$

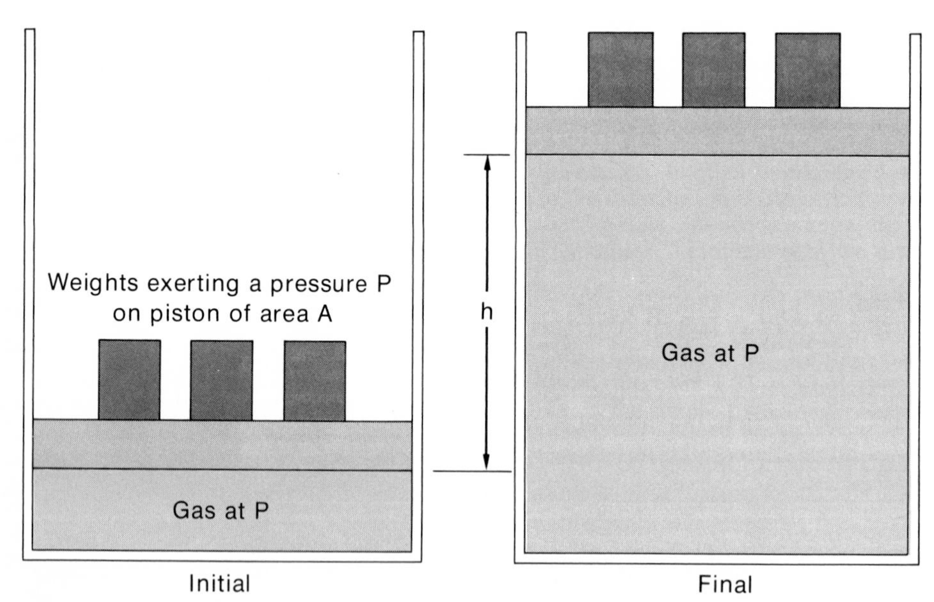

Figure 4.7 Work of expansion. When the gas expands against a constant pressure P, the general expression for the work done is $W = f \times h$, where f is the opposing force and h the distance the piston moves. But since pressure is force per unit area, $P = f/A$, or $W = P \times A \times h$. Since the product ($A \times h$) is simply the volume change, ΔV, the work of expansion is given by the equation: $W = P\Delta V$.

For reactions involving only liquids and/or solids, the volume change is so small that the $P\Delta V$ term in Equation 4.23 is completely negligible. For example, for the melting of ice at 0°C and 1 atm

$$H_2O(s) \rightarrow H_2O(l)$$

one can show (Problem 4.40) that $P\Delta V$ amounts to only about 0.003% of ΔH. Clearly, for reactions of this type the enthalpy change is, to all intents and purposes, equal to the change in internal energy.

For reactions involving gases, the $P\Delta V$ term in Equation 4.23 can be expected to be somewhat larger. By the use of the Ideal Gas Law (Chapter 5), it is possible to show that in this case

$$P\Delta V \text{ (in kJ)} = (0.0083)\Delta n_g T$$

where Δn_g is the change in the number of moles of gas when the reaction takes place

$$\Delta n_g = \text{no. moles gaseous products} - \text{no. moles gaseous reactants}$$

and T is the temperature in K. At 25°C, i.e., 298 K

$$P\Delta V = (0.0083)(298)\Delta n_g = 2.5\Delta n_g$$

Substituting in Equation 4.23, we obtain finally

$$\Delta H \text{ (in kJ)} = \Delta E \text{ (in kJ)} + 2.5\Delta n_g \text{ (at 25°C)} \qquad (4.24)$$

Equation 4.24 is perhaps most useful in calculating ΔH for a chemical reaction from ΔE as determined directly in a bomb calorimeter.

Example 4.8 For the reaction referred to in Example 4.6

$$N_2H_4(g) + O_2(g) \rightarrow N_2(g) + 2\ H_2O(l)$$

ΔE, as determined from the heat flow in a bomb calorimeter, is −662 kJ. Calculate ΔH for this reaction, using Equation 4.24.

Solution There is only one mole of *gaseous* product (N_2), and a total of two moles of gaseous reactants (one mole N_2H_4 + one mole O_2). Hence:

$$\Delta n_g = 1 - 2 = -1$$

$$\Delta H = \Delta E + 2.5(-1) = -662 \text{ kJ} - 2.5 \text{ kJ} = -664 \text{ kJ}$$

For most reactions, ΔH and ΔE have about equal values

You will note that for the reaction considered in Example 4.8 the $P\Delta V$ term makes a relatively small contribution to ΔH, changing it by less than 0.5%. This situation is typical; only rarely do we find that the $P\Delta V$ term makes a large contribution.

4.7 SOURCES OF ENERGY

Our major source of energy today is a particular type of exothermic reaction: the combustion of the fossil fuels coal, natural gas, and petroleum. Reactions such as

$$CH_4(g) + 2\ O_2(g) \rightarrow CO_2(g) + 2\ H_2O(l);\ \Delta H = -891 \text{ kJ}$$

and

$$C_8H_{18}(l) + \frac{25}{2} O_2(g) \rightarrow 8\ CO_2(g) + 9\ H_2O(l);\ \Delta H = -5440\ kJ$$

are used directly to heat our homes and places of business. Even more important, the development of such devices as the gas turbine and the internal combustion engine has made it possible to use the heat evolved in the combustion of fuels to generate electrical and mechanical energy. In a very real sense, modern civilization is dependent upon this process of energy conversion.

Today, with our resources of petroleum and natural gas dwindling, we face an "energy crisis" whose magnitude we are just beginning to grasp. If we are to maintain our standard of living in the face of an expanding population, we must develop new sources of energy. Some of the possibilities in this area will be discussed shortly. First, let us see where we stand today with respect to the supply and demand of the fossil fuels that furnish more than 95% of the world's energy.

Consumption of Fuels

Until about a century ago a single fuel, wood, supplied virtually the entire energy requirement of the United States. At about the time of the Civil War, coal began to replace wood as a fuel for use in industry and transportation. The production of coal in the United States reached 5×10^{11} kg in 1910 and has fluctuated around that figure ever since. The increasing demand for energy during the past sixty years has been met almost entirely by increased consumption of the hydrocarbon fuels, petroleum and natural gas.

From the standpoint of conservation of resources, there is a disturbing trend evident from Table 4.3. Over the past century the major energy source has shifted from wood, which takes a few years to grow, first to coal and then to oil and natural gas, fossil fuels formed millions of years ago. This trend is typical of most countries; in Canada, for example, about 80% of the total energy requirement is now being met by petroleum and natural gas, with the remainder about equally divided between coal and water power. Clearly, we are no longer living on our income from stored solar energy; instead we are dipping into our principal at an ever-increasing rate.

The world's dependence upon petroleum as a major source of energy was dramatized by the Arab oil embargo of 1974 and the price increase that followed.

Which of these percentages do you think will be higher in 1980? Lower?

TABLE 4.3 CONSUMPTION OF ENERGY IN THE UNITED STATES

	Per Cent of Total						
	Wood	Coal	Petroleum	Natural Gas	Water Power	Nuclear	Total (kJ)
1800	99.3	0.5	0.0	0.0	0.2	0.0	0.063×10^{16}
1850	91.1	8.6	0.0	0.2	0.1	0.0	0.31
1900	28.3	66.8	2.4	2.4	0.1	0.0	1.1
1920	10.9	71.8	13.1	3.8	0.4	0.0	2.3
1940	7.6	50.6	31.3	9.8	0.7	0.0	2.7
1950	5.6	39.7	36.2	17.5	1.0	0.0	3.9
1960		25.0	43.8	30.0	1.2	0.0	5.0
1970		22.8	41.4	33.0	2.6	0.2	7.5

Within a single year, the cost of oil imported from OPEC countries went up by a factor of four. Among industrialized countries, the economic effect has been particularly severe in Japan and Western Europe, where domestic oil supplies are virtually nonexistent (Table 4.4). The United States, which now imports more than 40% of its oil, is becoming increasingly vulnerable to the inflationary effect of price increases. Even in Canada, the future is hardly promising; in 1975, for the first time since the development of the western oil fields in the 1950's, Canada became a net importer of oil.

TABLE 4.4 ESTIMATED POTENTIAL RESERVES OF FOSSIL FUELS (ENERGY EQUIVALENT)

	Petroleum, Natural Gas*	Coal
Middle East	520×10^{16} kJ	—
U.S.S.R.	430	$11\,400 \times 10^{16}$ kJ
Africa	220	300
Latin America	190	100
United States	170	3 900
Oceania	160	200
Canada	80	600
Europe	20	1 000
Far East	10	1 800
	$1\,800 \times 10^{16}$ kJ	$19\,300 \times 10^{16}$ kJ

*Proved reserves are considerably lower (about 25% of estimated potential reserves) and distributed quite differently (63% in the Middle East).

Regardless of how the world's supply of petroleum is distributed in years to come, we must realize that our supply of petroleum and natural gas is by no means inexhaustible. The energy equivalent of worldwide reserves (Table 4.4) is estimated to be $1\,800 \times 10^{16}$ kJ. Each year, about 22×10^{16} kJ are consumed (about one third in the United States alone). Simple division, with no allowance for increase in population or per capita consumption, gives

$$\frac{1\,800 \times 10^{16}}{22 \times 10^{16}} \approx \text{eighty years}$$

as a time limit for the exhaustion of reserves of these fuels. Adding in an annual growth rate of 2% would reduce this number to fifty years; at the current 4% growth rate, the world will run out of petroleum in about thirty-five years.

It should be evident from the above discussion that the "energy crisis" of the 1970's is in no sense a temporary phenomenon. Indeed, it seems inevitable that the price of petroleum and natural gas will rise steadily in the future as supplies dwindle. To minimize the economic effects of this factor, it will be necessary to take one or more of the following steps.

1. *Reduce the consumption of fuel.* The principal target areas here are space heating and transportation, which together account for nearly half of the total energy requirement of most industrialized countries. Three of the more practical suggestions for cutting back in these areas are:

—better insulation of homes and places of business. We could save about one third of the energy used for space heating (and save money at the same time) through proper insulation of existing buildings and design changes in new construction (e.g., less plate glass). This would also reduce the need for air conditioning, which accounts for a small but rapidly increasing fraction of our energy budget.

—improving the fuel economy of automobiles, which consume more than half of the gasoline used for transportation. In 1974, cars built in the United States

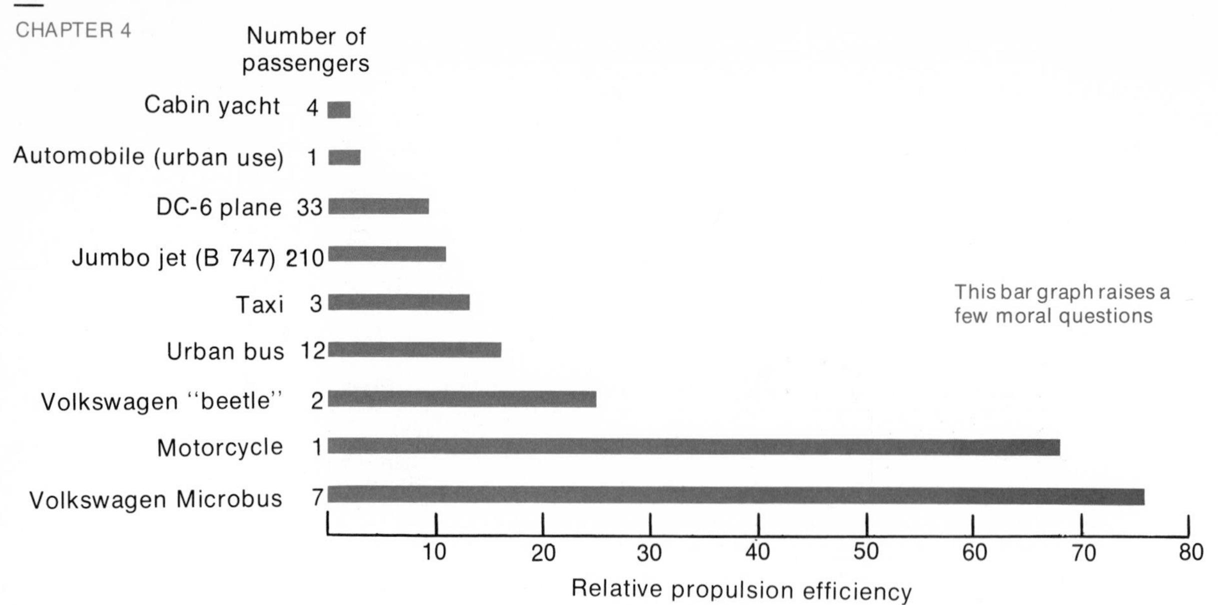

This bar graph raises a few moral questions

Figure 4.8 Relative efficiencies of various vehicles in terms of passenger kilometres per litre of gasoline. Note the dramatic increase in efficiency as more passengers are accommodated. (Adapted from *Technology Review,* Jan., 1972, p. 34.)

averaged 5 km/ℓ; it is projected that by 1980 this figure can be raised to at least 9 km/ℓ, primarily by reducing body mass.

—using more efficient means of transportation. Figure 4.8 suggests some interesting possibilities here.

2. *Use more coal.* From Table 4.4, it appears that reserves of coal in the United States and Canada have an energy equivalent about 20 times that of petroleum and natural gas; on a worldwide basis, the multiple is about 10. By shifting to a coal-based economy, we could satisfy our energy requirements for at least a century, probably longer. In some cases, this would be relatively easy to do. As late as 1950, virtually all electrical energy was generated by the combustion of coal; in principle, there is no reason why oil and natural gas need be consumed for this purpose. On the other hand, for most of our transportation needs, converting to a solid fuel would be essentially impossible. The use of coal for this purpose requires that processes be developed to convert it to gaseous or liquid fuels at a reasonable cost (Chapter 10).

At least for automobiles

3. *Develop new sources of energy.* Two such sources are discussed below.

Alternatives to Fossil Fuels

The "energy crisis" has promoted the development of energy sources other than the heat available from the combustion of fossil fuels. One such source, *water power,* has been used for centuries, first to drive water wheels in small mills and more recently to propel electric generators. Although the contribution from this source has been increasing steadily since 1900 (Table 4.3), its future potential is less promising. Engineers estimate that even if it were possible to harness the water power of all the major rivers of the world, this would yield at most 2.5×10^{16} kJ each year, about 10% of present energy requirements.

Another possibility is *geothermal energy,* the heat energy associated with deposits of superheated steam located several thousand feet below the surface of the earth. A power plant in Larderello, Italy, has been generating electrical energy from one such deposit since 1904. More recently the Geysers area of northern California has been developed to provide electricity at a cost lower than

M. C. Escher's fine lithograph offers a possible solution to the energy problem. Just put a water wheel under the falls, attach it to an electric generator, and we're in business. One potential disadvantage of the system, however, is that it appears to violate the First Law; does it?

Waterfall, by Maurits C. Escher. (Reproduced by permission from the collection of C. V. S. Roosevelt, Washington, D.C.)

that derived from the combustion of fossil fuels. Here again, however, the potential is strictly limited; full development of all known sources of geothermal steam would at most supply 5 to 10% of the world's energy requirements.

NUCLEAR ENERGY. The only energy source, known to be economically feasible in the present and for the near future, with the potential to replace fossil fuels, is *nuclear fission* (Chapter 24). Here, a heavy atom splits under neutron bombardment into smaller fragments, accompanied by the evolution of energy. In nuclear power plants now in operation, the fuel used is the relatively rare isotope of uranium, U-235. A typical fission reaction is

$$^{235}_{92}U + ^{1}_{0}n \rightarrow ^{90}_{38}Sr + ^{144}_{54}Xe + 2\,^{1}_{0}n; \quad \Delta H = -17 \times 10^{9}\ kJ \qquad (4.25)$$

To get an idea of the enormous amount of energy available from fission reactions, compare the heat given off per gram of uranium

$$\frac{17 \times 10^{9}\ kJ}{235\ g} = 7 \times 10^{7}\ \text{(i.e., 70 million) kJ/g}$$

to that for fossil fuels, which range from about 20 kJ/g for wood to 45 kJ/g for petroleum and natural gas.

Balanced against this obvious advantage of nuclear fuels is the problem of storing and ultimately disposing of the radioactive products of fission, such as strontium-90, which are extremely dangerous pollutants.

Moreover, as nuclear power plants increase in number, the possibility of a nuclear accident which could release radioactive material to the surroundings cannot be entirely discounted. For these reasons, among others, the development of nuclear energy has been much slower than was predicted some years ago. Never-

For a more extensive discussion, see Chapter 24

theless, it is projected that by 1980 there will be over 100 fission reactors operating in the United States and Canada, supplying about 5% of energy needs. Beyond that, prospects are clouded by the likelihood that we will effectively run out of U-235 by 1990. For nuclear fission to make a long-range contribution, it will be necessary to use *breeder reactors* where the fuel is the more abundant U-238 isotope (99.3% of natural uranium).

In many ways, the ideal energy source of the future would be *nuclear fusion* (Chapter 24) in which light nuclei such as deuterium, ${}^{2}_{1}H$, and tritium, ${}^{3}_{1}H$, combine to form heavier, stable nuclei. A fusion reaction now under intensive study is

$$ {}^{2}_{1}H + {}^{3}_{1}H \rightarrow {}^{4}_{2}He + {}^{1}_{0}n; \; \Delta H = -17 \times 10^{8} \text{ kJ} \qquad (4.26) $$

This reaction, per gram of fuel, evolves about four times as much energy as fission.

$$ \frac{17 \times 10^{8} \text{ kJ}}{5 \text{ g}} = 3 \times 10^{8} \text{ kJ/g} $$

The products of the fusion process are not radioactive, so the safety hazards associated with fission reactors are greatly reduced. Most important, the light isotopes required for fusion are sufficiently common to supply all of our energy needs for hundreds of years.

Unfortunately, no one up to now has been able to use Reaction 4.26 to generate a sustained flow of energy. In order to overcome the electrostatic repulsion between two small, positively charged H nuclei, they must be accelerated to enormous velocities, corresponding to temperatures of millions of degrees. Maintaining such temperatures has required the development of a whole new technology. The year 2000 appears to be a reasonable target date for production of energy on a commercial scale from fusion reactors. Optimists suggest this point could be reached by 1990; pessimists are convinced it will never happen.

SOLAR ENERGY. The problem of trapping and using solar energy is one that has intrigued scientists for generations (and federal granting agencies since 1974). Each year the earth receives from the sun the enormous total of 2×10^{21} kJ of energy. A tiny fraction of this, through the evaporation and condensation of water, supplies us with hydroelectric power. If this fraction could be increased to one hundredth of one per cent (0.0001), the world's annual energy requirement (2×10^{17} kJ) could be met from this source alone.

The most promising application of solar energy is in the field of space heating. The roof collector shown in Figure 4.9 consists of a shallow metal tank, the base of which is painted black to absorb as much sunlight as possible. The glass cover exerts a "greenhouse" effect, allowing sunlight to penetrate but preventing radiant heat from escaping. Water, circulating repeatedly through the collector, is heated to a temperature close to its boiling point. Hot water from the storage tank is used to heat the house, in much the same way as with an ordinary hot water system.

This system has one serious drawback. To provide heat during an extended period of cold, cloudy weather it would be necessary to have a huge storage tank. In areas where such climatic conditions are common during the winter months, which includes most of the United States, an auxiliary heating system of the conventional type must be available. This factor in the past has discouraged home owners from installing solar heating units, which have an initial cost considerably greater than that of ordinary heating systems. However, recent increases in the prices of heating oil and natural gas make solar heating look more attractive. It is now economically feasible to install solar collectors which would supply at least

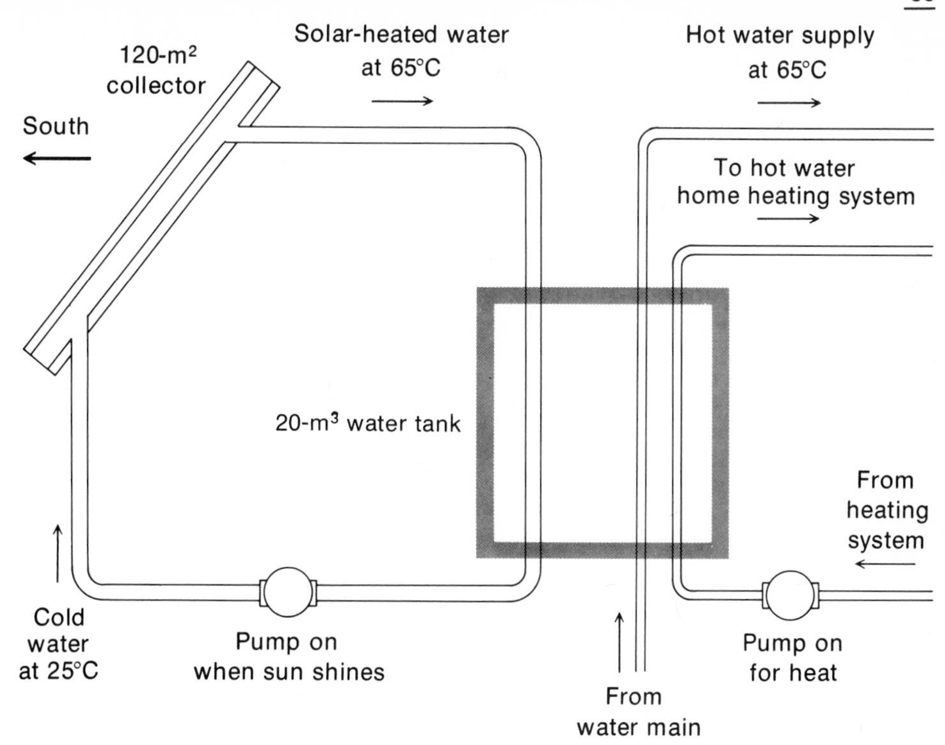

Figure 4.9 A simple solar heating system. In effect, the large, insulated storage tank (shaded box) acts as a "heat reservoir" to transfer heat from the roof collector to hot water faucets and radiators in the house.

half of the space heating requirements of the United States, thereby reducing the consumption of fossil fuels by about 10%.

The prospects for solar energy in areas other than space heating are less encouraging. In principle, sunlight can be converted directly to electrical energy by means of a semiconductor device known as a solar battery (Chapter 11). However, to be competitive with other sources of electricity, the prices of these units must be reduced by a factor of at least 50. Alternatively, solar energy might be used to produce steam or hot gases to run an ordinary electric generator. In order to do this efficiently, it is necessary to reach temperatures well above 100°C. One proposal involves focussing the sunlight falling on a large area of desert (at least five square kilometres) with a huge number (about half a million!) of small plastic lenses. The feasibility of such an ambitious project remains to be demonstrated.

PROBLEMS

4.1 *Laws of Thermochemistry* Given: $2\ Mg(s) + O_2(g) \rightarrow 2\ MgO(s)$; $\Delta H = -1204$ kJ Calculate

a. ΔH per gram of MgO formed. (*AM* Mg = 24.3, O = 16.0)
b. the number of grams of Mg which must be burned to evolve one kilojoule of heat.*

4.2 *Heats of Formation* Using Table 4.1, calculate ΔH for the reaction

$$2\ C_2H_2(g) + 5\ O_2(g) \rightarrow 4\ CO_2(g) + 2\ H_2O(l)$$

---

*Unless otherwise specified, all problems refer to reactions at constant pressure, where $\Delta H = Q_p$.

4.3 *Bond Energies* Using Table 4.2, estimate ΔH for the reaction

$$\mathrm{H{-}\underset{\underset{\displaystyle H}{|}}{N}{-}H(g) + Cl{-}Cl(g) \rightarrow H{-}\underset{\underset{\displaystyle H}{|}}{N}{-}Cl(g) + H{-}Cl(g)}$$

4.4 *Specific Heat* When 10.0 g of a certain metal at 90.0°C is added to 30.0 g of water at 20.0°C, the final temperature is 24.0°C. Taking the specific heat of water to be 4.18 J/(g · °C), determine

a. the amount of heat absorbed by the water.
b. the specific heat of the metal.

4.5 *Bomb Calorimeter* When one gram of ethane, C_2H_6, is burned in a bomb calorimeter, the temperature rises from 25.00°C to 33.84°C. The bomb (heat capacity = 837 J/°C) contains 1200 g of water [specific heat = 4.18 J/(g · °C)]. Calculate

a. the amount of heat absorbed by the bomb and water.
b. Q if one mole of ethane (MM = 30.0) were burned under the same conditions.

4.6 *First Law* In a certain process, a system evolves 1.62 kJ of heat and does 0.36 kJ of work. Calculate ΔE.

4.7 *Relation between ΔH and ΔE* When the reaction: $C_3H_8(g) + 5\ O_2(g) \rightarrow 3\ CO_2(g) + 4\ H_2O(l)$ is carried out in an open flame at 25°C. $Q_p = -2220$ kJ.

a. What is ΔH for this reaction?
b. What is Δn_g?
c. Calculate ΔE using Equation 4.24.

4.8 For the reaction $H_2(g) + \frac{1}{2}\ O_2(g) \rightarrow H_2O(l)$, $\Delta H = -286$ kJ; calculate

a. ΔH when one gram of H_2 burns.
b. ΔH when one gram of $H_2O(l)$ is formed.
c. the number of grams of H_2 that must be burned to evolve 100 kJ.

4.9 When 2.50 g of nitroglycerin, $C_3H_5(NO_3)_3$, decomposes to form $N_2(g)$, $O_2(g)$, $CO_2(g)$, and $H_2O(l)$, 19.9 kJ of heat is evolved.

a. Write a balanced equation for the reaction.
b. Calculate ΔH for the decomposition of one mole of nitroglycerin.
c. How much heat is evolved per mole of gas formed in the decomposition?

4.10 When one gram of sugar (MM = 342) is metabolized by the body, 16.7 kJ of heat is given off.

a. How much heat is produced when you eat a candy bar containing 50 g of sugar?
b. How much sugar would you have to take in to supply 8.79×10^3 kJ of heat?

4.11 Given:

$$Cu_2O(s) + \tfrac{1}{2}\ O_2(g) \rightarrow 2\ CuO(s);\ \Delta H = -143.7\ \text{kJ}$$

$$CuO(s) + Cu(s) \rightarrow Cu_2O(s);\ \Delta H = -11.5\ \text{kJ}$$

Calculate the heat of formation of CuO.

4.24 For the reaction $HgO(s) \rightarrow Hg(l) + \frac{1}{2}\ O_2(g)$, $\Delta H = 90.7$ kJ; calculate

a. ΔH to form one gram of Hg(l).
b. the number of grams of HgO that can be formed when one kilojoule is absorbed.

4.25 Recall Equation 3.2 for the reaction of hydrazine, N_2H_4, with dinitrogen tetroxide, N_2O_4. ΔH for that reaction is -1220 kJ.

a. How much heat is evolved per gram of hydrazine?
b. How much heat is evolved per gram of gas formed?

4.26 The reaction that occurs when a typical fat, glyceryl trioleate, is metabolized by the body is $C_{57}H_{104}O_6(s) + 80\ O_2(g) \rightarrow 57\ CO_2(g) + 52\ H_2O(l)$; $\Delta H = -3.35 \times 10^4$ kJ.

a. How much heat is evolved when one gram of this fat is metabolized?
b. How much energy in the form of heat would have to be disposed of to get rid of one kilogram of this fat?

4.27 Given:

$$MnO_2(s) \rightarrow MnO(s) + \tfrac{1}{2}\ O_2(g);\ \Delta H = +134.8\ \text{kJ}$$

$$MnO_2(s) + Mn(s) \rightarrow 2\ MnO(s);\ \Delta H = -250.1\ \text{kJ}$$

Calculate ΔH_f of MnO_2.

4.12 Using Table 4.1, calculate ΔH for

a. the combustion of one mole of ethylene, C_2H_4, to form $CO_2(g)$ and $H_2O(l)$.
b. the combustion of one gram of ethylene to form $CO_2(g)$ and $H_2O(g)$.

4.13 For the reaction $2\ C_8H_{18}(l) + 25\ O_2(g) \rightarrow 16\ CO_2(g) + 18\ H_2O(l)$, $\Delta H = -11\ 003$ kJ. Using the data in Table 4.1, calculate the heat of formation of octane, C_8H_{18}.

4.14 Estimate ΔH for each of the following, using Table 4.2:

a. $H_2(g) + S(g) \rightarrow H{-}S{-}H(g)$

b. $H{-}\underset{H}{\overset{H}{\overset{|}{\underset{|}{C}}}}{-}O{-}H(g) + H{-}Br(g) \rightarrow$

$H{-}O{-}H(g) + H{-}\underset{H}{\overset{H}{\overset{|}{\underset{|}{C}}}}{-}Br(g)$

c. Explain why your answer to (a) is not equal to the heat of formation of $H_2S(g)$.

4.15 When a silver spoon weighing 45.0 g, originally at 25.00°C, is dropped into a cup containing 80.0 g of boiling water (100.00°C) the final temperature is 97.70°C. What is the specific heat of silver?

4.16 When one mole of $CH_4(g)$ burns in a bomb calorimeter (heat capacity = 5.02 kJ/°C) containing 21.00 kg of water, the temperature rises 9.53°C. Calculate

a. Q for the combustion of one mole of $CH_4(g)$.
b. Q per gram of methane burned.

4.17 To determine the heat capacity of a bomb calorimeter, a student adds 150 g of water at 50.0°C to the bomb, which is initially at 25.0°C. The final temperature is 33.0°C. What is the heat capacity of the bomb in joules per degree Celsius?

4.18 Which of the following are state properties?

a. internal energy
b. heat flow
c. density
d. airline distance from New York to Chicago ("as the crow flies")
e. population of the U.S. in 1970

4.19 Calculate ΔE for each of the following:

a. $Q = +50$ J, $W = +10$ J
b. $Q = +20$ J, $W = -30$ J
c. 25 J of heat is evolved, 15 J of work is done

4.28 Using Table 4.1, calculate ΔH for

a. the decomposition of 3.50 g of $NH_3(g)$ to the elements.
b. $2\ Ca(OH)_2(s) + 2\ SO_2(g) + O_2(g) \rightarrow 2\ CaSO_4(s) + 2\ H_2O(l)$

4.29 Using the information given in Problem 4.26, calculate the heat of formation of glyceryl trioleate.

4.30 Estimate ΔH for each of the following, using Table 4.2.

a. $H_2O(g) + Cl_2(g) \rightarrow H{-}O{-}Cl(g) + HCl(g)$
b. the first step in the thermal decomposition of hydrazine, $H{-}\underset{H}{\underset{|}{N}}{-}\underset{H}{\underset{|}{N}}{-}H$, assuming two NH_2 radicals are formed.
c. $Cl_2(g) \rightarrow 2\ Cl(g)$. What is the heat of formation of Cl(g)?

4.31 Using the specific heat of Ag calculated in Problem 4.15, calculate the final temperature to be expected when the spoon at 100.0°C is added to 50.0 g of water at 20.0°C.

4.32 When 1.50 g of the rocket fuel dimethylhydrazine, $(CH_3)_2N_2H_2$, is burned in a bomb calorimeter (heat capacity = 1840 J/°C) containing 5.00 kg of water, the temperature rises from 22.05 to 24.13°C. Calculate the amount of heat that would be evolved in the combustion of one mole of dimethylhydrazine.

4.33 When one gram of benzoic acid is burned in a bomb calorimeter containing 2.95 kg of water, the temperature rises from 24.33 to 26.25°C. The heat of combustion of benzoic acid under these conditions is 26.42 kJ/g. What is the heat capacity of the bomb in joules per degree Celsius?

4.34 Which of the following are state properties?

a. work
b. enthalpy
c. volume
d. highway distance from New York to Chicago
e. population of the U.S. in 2000.

4.35 For a certain change in state: $E_{final} - E_{initial} = -56$ J. Calculate W if

a. $Q = 90$ J
b. $Q = -27$ J
c. 54 J of heat is absorbed

4.20 For the reaction $N_2(g) + 3\ H_2(g) \rightarrow 2\ NH_3(g)$, 87.2 kJ of heat is evolved in a bomb calorimeter at constant volume. Calculate

a. ΔE b. Δn_g c. ΔH at 25°C

4.21 Using the data in Table 4.3, determine

a. the number of kilojoules of energy obtained by burning coal in 1900; in 1970.
b. the number of kilograms of coal burned in 1900 and in 1970, assuming that 29 kJ are obtained per gram of coal.
c. the per capita consumption of coal (kilograms) in 1900 and in 1970 (cf. Problem 4.37).

4.22 Assume that 30% of the petroleum used in the U.S. is consumed in automobiles. By what percentage could our petroleum requirements be reduced if the average kilometres per litre figure were increased from 5.0 to 9.0?

4.23 Which of the following energy sources do you think will make the biggest contribution in 1980?

a. solar energy
b. nuclear fission
c. nuclear fusion
d. petroleum

Explain your reasoning and indicate how your answer might change if the date were 2000.

4.36 For the reaction $2\ KClO_3(s) \rightarrow 2\ KCl(s) + 3\ O_2(g)$, 89.5 kJ of heat is evolved upon heating in an open test tube. Calculate

a. ΔH b. Δn_g c. ΔE at 25°C

4.37 The population of the United States in 1900 was 7.6×10^7; in 1970, it was 20.4×10^7. Using the data in Table 4.3, find

a. the per capita consumption of energy in 1900; in 1970.
b. what the total consumption would have been in 1970 if the per capita value had remained at its 1900 value.
c. what the total consumption would have been in 1970 if the population had remained constant, assuming the per capita figure for 1970.

4.38 Referring to Figure 4.8,

a. how many kilometres per litre does one get with a Volkswagen Microbus?
b. where would the Microbus fall on the chart if there were two people in it rather than seven?

4.39 Discuss the advantages and disadvantages of concentrating research efforts on the development of the following sources of energy as alternatives to petroleum

a. geothermal energy
b. hydroelectric power
c. breeder reactors
d. nuclear fusion

*4.40 For the melting of ice, $H_2O(s) \rightarrow H_2O(l)$, ΔH is 333 J/g at 0°C and 1 atm. The densities of ice and liquid water are 0.917 and 1.000 g/cm³, respectively. Calculate

a. ΔV in cubic centimetres for the melting of one gram of ice.
b. $P\Delta V$ in joules, using the conversion factor $1\ cm^3 \cdot atm = 0.101$ J.
c. ΔE in joules.
d. the percentage of ΔH represented by the $P\Delta V$ term.

*4.41 Calculate the heating values of the following foods (kilojoules evolved per gram), taking the heating values of fats, proteins, and carbohydrates to be 38 kJ/g, 17 kJ/g, and 17 kJ/g, respectively.

	% fat	% protein	% carbohydrate	% water
Apples	0.5	0.4	14.2	84.9
Soy beans	17.2	37.0	28.0	17.8
Coconut	57.4	6.3	31.5	4.8
Chocolate	34.9	8.0	51.1	7.0

How many "calories" would be given off when 100 g of each of these foods is metabolized in the body? (The calorie referred to in nutrition is really a kilocalorie.)

*4.42 When methane burns in a Bunsen burner, it combines with the oxygen of the air to give carbon dioxide and water vapor. Suppose that the burner is adjusted so that there is a five-fold excess of air; i.e., the amount of air admitted is five times that required for the combustion of the methane. Calculate the maximum temperature in °C that can be reached with the burner under these conditions. In your calculations, make the following assumptions: (1) the air and methane enter the burner at 20°C; (2) air consists of N_2 and O_2 in a 4:1 mol ratio; (3) the heat evolved in the combustion of the methane is absorbed in raising the temperature of the "products," CO_2, $H_2O(g)$, and the N_2 and O_2 from the excess air; (4) the specific heats of the products in J/(g · °C) are: CO_2, 0.845; H_2O, 1.866; O_2, 0.874; N_2, 1.042.

5
THE PHYSICAL BEHAVIOR OF GASES

Depending on temperature and pressure, pure substances exist as solids, liquids, or gases. At low temperatures all substances become solids. At some intermediate temperature range they behave as liquids, and at high temperatures they are gaseous. What is "low" or "high" temperature for one substance is not necessarily "low" or "high" for another, so that while oxygen, ammonia, and carbon dioxide are gases at room temperature, iron, silicon carbide, and tungsten become gases only at temperatures well in excess of 2000°C.

Gases have been involved, in one way or another, with many of the fundamental developments of chemistry. Because of their invisibility, they contributed to the confusion which reigned for centuries with regard to the nature of chemical substances and the chemical changes associated with combustion. Avogadro's hypothesis made it possible for Cannizzaro (Chapter 2) to use gas analyses to establish for the first time a rational method for finding atomic masses. More recently, the spectra of gases have allowed theorists to show the validity of some of the relationships obtained from quantum theory. Although at present we know far more about the properties of gases than we do about liquids and solids, the possibilities for research using gases are far from exhausted, with a current area of research being the elucidation of the details of simple chemical reactions between gas molecules.

5.1 SOME GENERAL PROPERTIES OF GASES

When a liquid is heated sufficiently, it begins to evaporate or boil. In this process the substance is said to make a transition from the liquid to the gaseous phase. During the change in phase the particles in the liquid become free from one another and pass into space as molecules of gas. In general this process is accompanied by a great change in volume. If half a cupful of water is evaporated, the resulting water vapor (water in the gas phase) at one atmosphere and 100°C occupies a volume roughly equal to that of a 200-ℓ oil drum. Since molecules in the gas phase are the same size as they are in the liquid phase, it follows that the distances between them in the gas are much greater than they are in the liquid.

In the air you breathe, the molecules are roughly five diameters apart

In view of the rather large distances between gas molecules, we might expect that it would be fairly easy to compress a gas, at least as compared to a liquid. Experimentally we find that this is so, and that in general the volume of a gas varies inversely with the applied pressure; that is, doubling the pressure reduces the volume to about half its previous value. Similarly, if we double the amount (mass) of gas in a container we find that the pressure approximately doubles. Increasing the temperature of a gas in a closed container will increase the pressure of the gas.

Gases can be expanded indefinitely and will always tend to occupy their containers completely and uniformly. If one cubic centimetre of hydrogen gas at one atmosphere pressure is let into an evacuated 10-m^3 container, the hydrogen almost instantly diffuses to give a constant density and pressure throughout the container. The pressure of the gas would be very low under such conditions, about 10^{-7} atm, but could be measured easily. You might wonder whether the situation would be the same if there had already been another gas at one atmosphere in the container. Under such circumstances the amount of hydrogen in any part of the container would ultimately be the same as if the container were initially evacuated, but the time required to attain a homogeneous mixture would be of the order of hours rather than a fraction of a second; the diffusion of the hydrogen molecules would be hindered greatly by collisions with the other gas molecules.

The H_2 molecule would encounter the same difficulty as a commuter trying to get on a subway train at rush hour

All gases mix readily with one another to form completely homogeneous solutions. Ordinary air is such a solution. No one has ever been able to prepare a mixture of gases which tended to settle out into two or more regions of different composition. This situation is very different from that with mixtures of liquids and solids, in which the solubility of one substance in another is usually limited.

5.2 ATMOSPHERIC PRESSURE AND THE BAROMETER

The most common gas we encounter, and the only one known until about 1750, is the air about us. This gas lies over the earth in a blanket about 80 km thick. Like all earthly matter, the air is subject to the gravitational pull of the earth. The air near the earth is compressed by the mass of the air above it. The pressure of the air at the earth's surface is by no means negligible, amounting to about 101 kPa, nearly 15 lb/in^2; this means that our bodies are at all times subject to a rather gigantic force. The pressure of the atmosphere varies with the height above sea level and weather conditions. In Denver, Colorado (altitude, 1.6 km above sea level), the atmospheric pressure averages about 93 kPa. During a hurricane the pressure may become that low at sea level. Above 3 km, breathing becomes uncomfortable for people not accustomed to such altitudes, and for that reason modern aircraft have pressurized cabins. Spacecraft, which operate at much higher altitudes, are also pressurized; on landing and takeoff, however, astronauts wear pressurized suits and oxygen masks. (Why?) At a height of 20 km the atmospheric pressure is only about 10 per cent of that at sea level. Uncomfortable is not the word to describe the sensations of an unpressurized human being at such an altitude.

Although the facts of atmospheric pressure are really very simple, they were not clearly understood until about 1650. Men had learned earlier in a very practical way that they could not lift water more than about 10 m (33 ft) with a suction pump; the reason for this was unknown, and the explanation given by philosophers was that "nature abhors a vacuum." Admittedly, that wasn't much of an answer; in

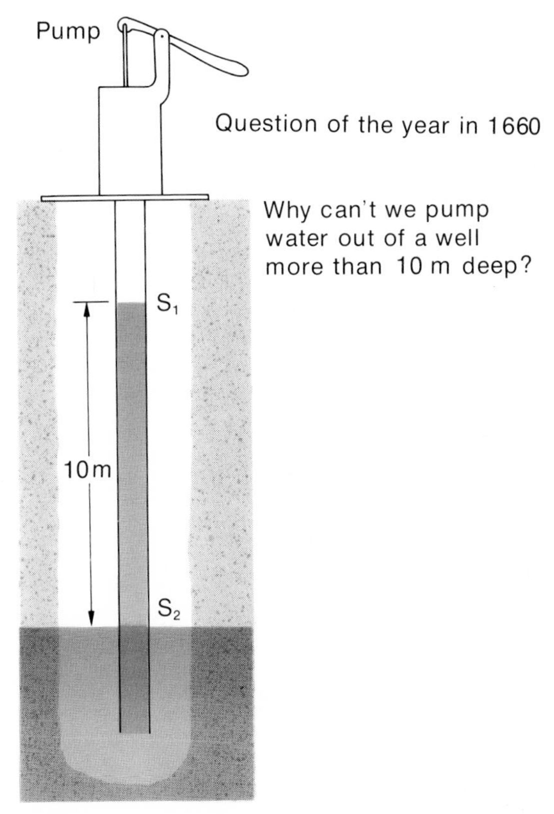

Figure 5.1 The suction pump. Would you expect that the maximum height would remain constant for a well like this, or would it vary from day to day?

fact, it wasn't an answer at all. The solution to this ancient problem was given by Torricelli, an Italian scientist. He applied the natural law that in a liquid equal pressures exist at equal heights. With this idea in mind let us consider Figure 5.1.

At the surface S_1 of the water in the pipe the pressure is essentially zero, since the surface there is in contact with a vacuum. The pressure on the lower surface S_2 is that of the atmosphere, and by the law just cited must be equal to the pressure on the water at the same level *inside* the pipe. This pressure, however, must also be equal to the pressure exerted by the water column, 10 m high, above that level. Once the pump lowers the gas pressure at the top of the pipe to zero, the water in the pipe will stop rising, no matter how fast the pump works.

Torricelli recognized that the maximum height of the water column was essentially a measure of the atmospheric pressure (can you see why this must be so?), and made a device for making the measurement more conveniently. His instrument, called a barometer, is shown in Figure 5.2; it consists of a closed glass tube filled with mercury and inverted over a pool of mercury. Provided the tube is sufficiently long, liquid mercury flows into the reservoir when the tube is first inverted, leaving a nearly perfect vacuum above the liquid; the height of the liquid column remaining in the tube is then a measure of the atmospheric pressure. Since the density of mercury is high, the height of the column is much less than with water; it ranges from 740 to 760 mm, varying, as we have mentioned, with both altitude and weather conditions. The unit of pressure called the **standard atmosphere,** or simply **atmosphere,** is defined to be equal to the pressure exerted by a column of mercury 760 mm high with the mercury at 0°C:

$$1 \text{ atmosphere} = 760 \text{ mmHg} \qquad (\text{Hg at } 0°\text{C})$$

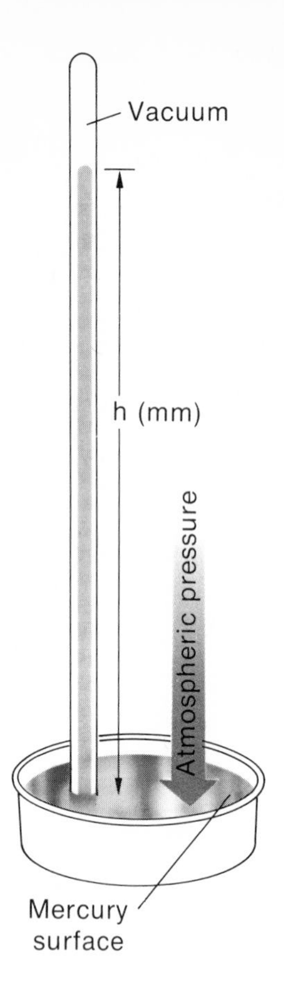

Figure 5.2 The barometer. At the level of the lower mercury surface, the pressure both inside and outside the tube must be that of the atmosphere. Inside the tube the pressure is exerted by the mercury column h mm high. Hence, the atmospheric pressure must equal h mm Hg.

The Manometer and the Measurement of Gas Pressure

The measurement of the pressure exerted by a gas in a closed container is ordinarily accomplished by a device called a manometer (Fig. 5.3), which operates on the same principle as a barometer.

A manometer consists of a glass U-tube partially filled with mercury. One side of the U-tube is connected to the container in which the pressure is to be measured, and the other side is connected to a region of known pressure. The gas in the container will exert a force on the mercury column and will tend to make it go down. This force will be opposed by that created by the gas over the other surface. The difference between the mercury levels at equilibrium is directly proportional to the difference between the two gas pressures.

Noting Figure 5.3, we can say that at the level h_l the pressure on the mercury must have the same value in the left arm, where the mercury is under the unknown pressure, as in the right arm, where it is under the sum of the known pressure and the pressure due to the mercury column of height Δh. Mathematically,

$$\text{Pressure}_{\text{unknown}} = \text{Pressure}_{\text{known}} + \text{Pressure due to } \Delta h \text{ mmHg}$$

Often the known pressure is expressed in terms of the height of a column of mercury to which it is equivalent; usually the known pressure is either that of the air in the laboratory or that of a vacuum. In the former case, the known pressure would be barometric pressure; in the latter case, it would be zero.

Throughout the remainder of this chapter, we will express pressures either in standard atmospheres (atm), pascals (Pa), or **kilopascals (kPa):**

$$1 \text{ atm} = 1.013 \times 10^5 \text{ Pa} = 101.3 \text{ kPa} \quad (5.1)$$

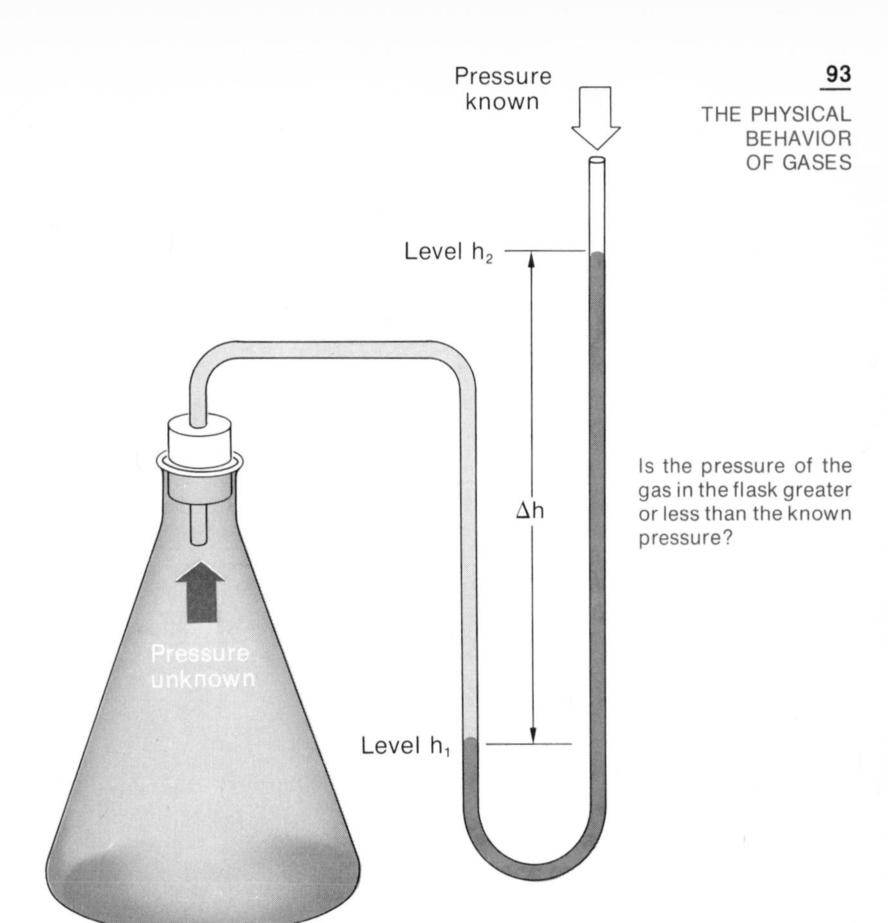

Figure 5.3 The manometer. At the level h_1 the total pressure on the mercury in the left arm must equal the total pressure on the mercury in the right arm.

Is the pressure of the gas in the flask greater or less than the known pressure?

As Equation 5.1 implies, the pascal is a very small unit; it has been estimated that a dollar bill lying flat on a table exerts a pressure of about one pascal! For this reason, we will most frequently express pressures in kilopascals. A kilopascal is approximately 1/100 of normal atmospheric pressure; from another point of view, it corresponds to the pressure exerted by a column of mercury about 7.5 mm high.

5.3 THE IDEAL GAS LAW

When we compress a sample of gas at constant temperature, we find that the volume change which occurs does not depend on the kind of gas used, but is determined only by the initial volume and the ratio of the initial and final pressures. Similarly, heating a gas sample at constant pressure results in an increase in volume which is not dependent on the nature of the gas in the sample. Under ordinary conditions, with pressures of the order of an atmosphere or so, gases can be considered to exhibit volume-pressure-temperature behavior that can be described in terms of general relations, which are as valid for methane as for helium, for air as for fluorine. The most important of these relations, and the one with which this chapter is mainly concerned, is called the Ideal Gas Law.

The Ideal Gas Law is an equation which describes the relationship among four of the fundamental properties of a gas. The Law is usually stated in the following form:

$$P V = n R T \tag{5.2}$$

where P is the pressure, V is the volume and n is the number of moles of gas. In the equation T is the absolute temperature of the gas in kelvins and is related to the Celsius temperature t in the following way:

$$T = t + 273 \tag{5.3}$$

The quantity R is called the gas constant; it has the same value for all gases and, using SI units for pressure, volume and temperature, is equal to 8.31 Pa · m³/(mol · K).

As you can see, the Ideal Gas Law is an expression in four variables, P, V, n, and T. It tells us how these properties of a gas depend on each other and is really a rather remarkable equation in that it is one of the very few natural laws that involves four variables. Usually scientists perform experiments in which, hopefully, there are only two variables, since then it is easiest to recognize how one depends on the other. Actually the Ideal Gas Law is based on experiments in which only two quantities are allowed to vary; it took its final form when it was realized that it was possible to summarize the results from such experiments, several of which are described below, in a single general equation.

Boyle's Law

Probably the simplest experiment that can be done on a gas is to trap a sample of air and measure its volume at several different pressures, holding the temperature constant. This experiment will determine the relationship between the two variables, pressure and volume, under conditions in which the amount and temperature are fixed. When we perform this experiment we find that as the pressure on the sample is increased, the volume decreases in such a way that the product of pressure and volume remains essentially constant. In Table 5.1 we have listed some experimental data for the compression of air at 25°C. Pressures are listed in kilopascals; volumes are given in *cubic decimetres*.

$$1\ \text{dm}^3 = 1\ \ell = 10^3\ \text{cm}^3 = 10^{-3}\ \text{m}^3$$

As you can see, although the pressure and volume change a great deal, their product remains constant within the accuracy of the experiment. Mathematically,

$$P\,V = k_1 \qquad \text{or } V = k_1/P \qquad \text{at constant } n \text{ and } T \tag{5.4}$$

This was one of the first physical laws to be discovered

where k_1 is a constant; in the experiment cited, k_1 would equal 4.0 kPa · dm³.

In Figure 5.4 we have plotted the data in Table 5.1 and connected the data points with a smooth curve on which we would expect any other data in this experiment to fall.

Equation 5.4 is called Boyle's Law, since it was Robert Boyle who first dis-

TABLE 5.1 COMPRESSION OF A SAMPLE OF AIR AT 25°C

State	Pressure kPa	Volume dm³	Pressure × Volume (kPa · dm³)
I	40	0.100	4.0
II	67	0.060	4.0
III	80	0.050	4.0
IV	100	0.040	4.0
V	120	0.033	4.0
VI	160	0.025	4.0

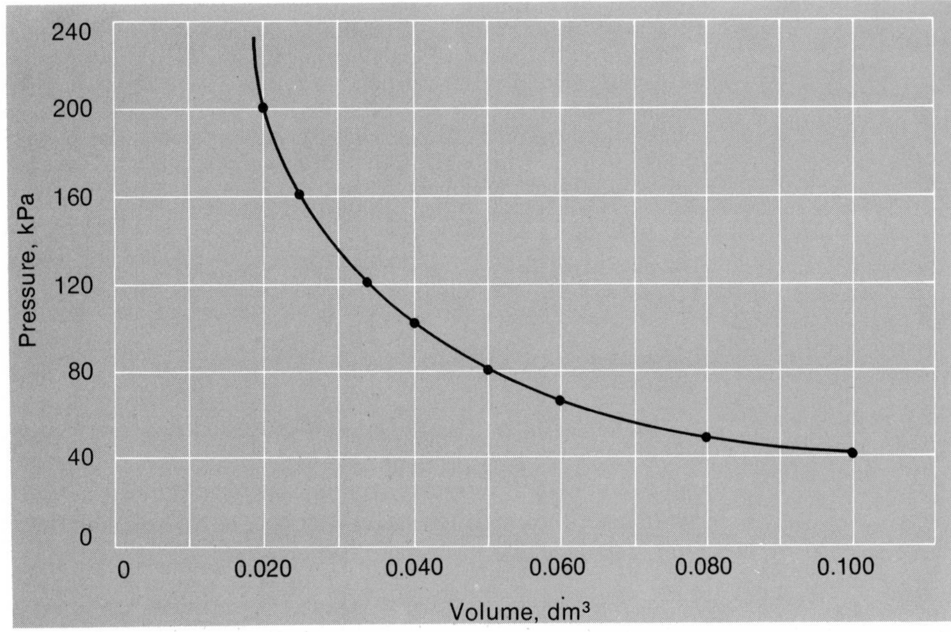

Figure 5.4 Boyle's Law. When a sample of gas is compressed at constant temperature, the product of the pressure and the volume remains constant.

covered that a gas behaves this way when compressed at constant temperature. Boyle, a British natural philosopher, did his experiments on air in 1660, in a glass U-tube like that shown in Figure 5.5, using mercury to compress the entrapped gas. By assuming that the diameter of the tube was constant, Boyle was able to obtain all the data he needed to establish his law with just a metre stick and a barometer. Any schoolboy these days could perform Boyle's experiment in a few hours, but even now he might have a bit of trouble interpreting his data properly. (See Problem 5.6.)

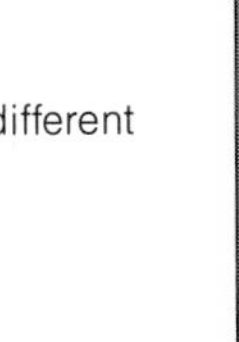

Figure 5.5 Boyle's experiment. Boyle measured the length d of the trapped air column at different pressure heads Δh of mercury; he also measured the barometric pressure.

Robert Boyle was one of the first experimental scientists. In addition to his discovery of the law which bears his name, he made many contributions to scientific thought. One of the most important of these was a book entitled *The Sceptical Chemist* in which he challenged, on the basis of his experiments, the then prevailing notion that salt, sulfur and mercury were the true principles of nature. In this regard he proposed that matter was ultimately composed of particles of various sorts which could arrange themselves into groups, and that groups of one kind constituted a chemical substance. Boyle thus used concepts of atomic and molecular theory similar to those we have today, and by some has been called one of the fathers of modern chemistry. Since his ideas on chemistry had of necessity to be extremely primitive and were not in any sense quantitative, we would prefer to consider the first modern chemists to be men like Dalton, Lavoisier, and Priestley, who actually performed experiments to establish fundamental chemical relationships.

Charles and Gay-Lussac's Law

We can readily extend the approach just described to another pair of variables. This time let us look at how the volume of a gas sample changes when it is heated at constant pressure. In this case V and T will be the variables and n and P will be kept fixed. If we study a sample of hydrogen under such conditions, we obtain data like that in Table 5.2.

TABLE 5.2 HEATING A SAMPLE OF HYDROGEN AT 1 ATM PRESSURE

STATE	VOLUME, dm^3	T, K	t, °C	V/T, dm^3/K
I	0.075	100	−173	7.5×10^{-4}
II	0.150	200	−73	7.5×10^{-4}
III	0.225	300	27	7.5×10^{-4}
IV	0.300	400	127	7.5×10^{-4}
V	0.375	500	227	7.5×10^{-4}

In this kind of experiment we find that the volume of the gas is directly proportional to the absolute temperature, doubling every time the temperature is doubled. The mathematical equation to express this dependence is simply

$$V = k_2T \qquad \text{or } V/T = k_2 \qquad \text{at constant } n \text{ and } P \qquad (5.5)$$

where k_2 is constant through the experiment; for the data in Table 5.2, k_2 is equal to 7.5×10^{-4} dm^3/K. In Figure 5.6 we have plotted the data in Table 5.2, again connecting the data points, this time with a straight line.

The simple dependence of gas volume on absolute temperature which we observe in Figure 5.6 is not fortuitous. It is partly the result of the experimental properties of gases and partly due to the way the absolute temperature scale was set up. The relevant experimental observations were made by Gay-Lussac and Charles, two French scientists, around 1800. They studied the expansion of gases with temperature, and found that the rate of expansion with Celsius temperature was constant and had the same value for all gases. In modern terms we would say that any two gases, when each was heated at constant pressure from one given temperature to another, would have the same percentage increase in volume. If the two temperatures involved are 0° C and 100° C, any gas will increase in volume by about 37 per cent. [Actually, Charles and Gay-Lussac were not so interested in the fundamental laws of gases as they were in the use of hot, or light, gases in lighter-than-air balloons. Charles was on board the second balloon ever to lift man off the face of the earth. This happened on December 1st, 1783 (1783 12 1), with a balloon filled with hydrogen gas; in the first ascent, ten days earlier, a hot-air balloon was used. In 1804 Gay-Lussac made a solo flight in a hydrogen balloon to 7 km, which set an altitude record that lasted fifty years.]

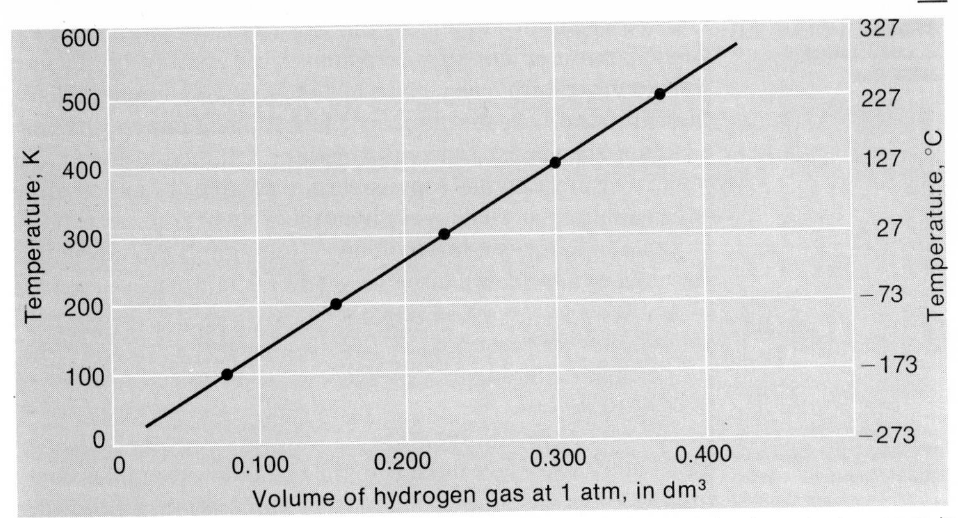

Figure 5.6 The hydrogen gas thermometer. The volume of a gas sample at constant pressure is directly proportional to its absolute temperature, and so can be used to establish its own temperature or that of a bath in which it is immersed.

It was nearly a century after Charles and Gay-Lussac did their experiments on gas expansion that scientists first realized that gas volume could be used to set up the absolute temperature scale shown in Figure 5.6. Essentially what is done is to define absolute temperature in terms of the volume of a gas heated at constant pressure; that is, Equation 5.5 is *defined* to be true. In order to retain a simple relationship between the absolute scale and the Celsius scale of temperature, the size of the degree is kept the same, so that 100°C separate the freezing and boiling points of water on both scales. The actual numerical relation between volume and temperature on the two scales is determined by the ratio of the volume of a gas at 100°C to that of the same sample at 0°C and the same, low, pressure; that ratio is found in accurate experiments to be 1.3661. Rewriting Equation 5.5 to obtain volume ratios, we have

$$\frac{V_{100}}{V_0} = 1.3661 = \frac{T_{100}}{T_0} = \frac{T_0 + 100}{T_0} = 1 + \frac{100}{T_0}$$

Solving this equation, we see that T_0, the temperature on the absolute scale corresponding to 0°C, must satisfy the equation

$$T_0 = \frac{100}{0.3661} = 273.15$$

In general,

$$T = t + 273.15$$

In light of the above line of reasoning, it should be clear that a gas system such as the hydrogen in Figure 5.6 could be used, after being calibrated, to determine both absolute and Celsius temperatures and so act as a standard reference thermometer. Indeed, the best thermometers are themselves calibrated against a gas volume thermometer such as this. (Somewhat fortuitously, the Celsius scale on a mercury-in-glass thermometer made by simply dividing the distance between mercury levels at 0° and 100°C into 100 equal parts agrees quite well with the scale as found by the gas thermometer.)

Both the Celsius and absolute scales are based ultimately on the gas volume thermometer

Equation 5.5 is, as you may have suspected, called Charles' and Gay-Lussac's Law, although perhaps it should have been named Kelvin's Law, since he had far more than they to do with its derivation. It is, indeed, a rather remarkable relation, and far less simple than it looks. One thing that might bother you, if you worry about matters of this sort, is the extrapolation of the volume of the gas to zero in Figure 5.6. Obviously this cannot go all the way, since at some finite temperature every gas will condense to a liquid. This is not a serious objection, however, to the concept of absolute temperature, since the fixing of the absolute zero is established from data obtained at 0 and 100°C, where the intermolecular attractions responsible for condensation are of negligible importance. It is true, though, that a very tight theoretical argument, borne out by experiment, can be made for the impossibility of attaining the absolute zero of temperature. The best we have been able to do so far is to get to about 0.0001 K.

Hydrogen boils at about 20 K

Avogadro's Law

Thus far we have dealt with the behavior of samples containing a fixed amount of gas. However, one of the most useful features of the Ideal Gas Law is that it allows us to relate the amount of gas, in moles, to its pressure, volume and temperature. There are two relations involving the amount of gas and its dependence on other properties that we should mention.

The first of these is concerned with the relation between the volume of a gas and the amount, as measured under conditions of constant temperature and pressure. As you might expect, we find that the volume is directly proportional to amount, doubling each time the amount is doubled. The equation relating volume and amount is

$$V = k_3 n \qquad \text{at constant } P \text{ and } T \tag{5.6}$$

Here k_3 is the proportionality constant between the volume and the amount in moles of different samples of the same gas, all measured at the same pressure and temperature.

The other relation concerning amount of gas was first stated in 1813 by Avogadro. In modern terminology we would state his Law as follows:

Equal volumes of different gases under the same conditions of temperature and pressure contain the same number of moles.

The mathematical equivalent of the above statement is:

$$n_A = n_B, \text{ for two gases A and B having the same values of } V,\ T, \text{ and } P \tag{5.7}$$

The Combined Gas Law

The Ideal Gas Law incorporates in one equation all of the relationships in Boyle's Law, Charles' Law, Avogadro's Law, and Equation 5.6. To see that this is the case, consider the form of the Ideal Gas Law:

$$P\,V = n\,R\,T \tag{5.2}$$

If we hold n and T constant, as Boyle did, the right side of the equation will maintain a constant value (remember that R is constant), which we might call k_1; under such conditions Equation 5.2 takes the form:

$$P\,V = k_1 \text{ (at } n,\ T \text{ constant), which is Boyle's Law}$$

Similarly, if we hold n and P constant, we can rewrite Equation 5.2 as

$$V/T = nR/P = k_2 \text{ (at } n, P \text{ constant), which is Charles' Law}$$

Thirdly, holding P and T constant, and again rewriting Equation 5.2, we get

$$V/n = RT/P = k_3 \text{ (at } P, T \text{ constant), which is Equation 5.6}$$

Lastly, considering samples of gases A and B, each at the same volume, temperature, and pressure, we obtain for these samples,

$$P\,V = n_A\,R\,T \qquad \text{and} \qquad P\,V = n_B\,R\,T$$

Since P, V, and T have the same values in the two equations, it follows that

$$n_A = n_B, \text{ which is Avogadro's Law}$$

Note that R must have the same value for gas A as it does for gas B if Avogadro's Law is to be obeyed. This means that R, the gas constant, has the same value for all gases. Since the Ideal Gas Law reduces to the four simpler laws when the various conditions are imposed, it indeed properly expresses the relationship between P, V, n, and T for any gas.

Evaluation of R, The Gas Constant

To establish the magnitude of R one merely needs to determine under known conditions the volume of a mole of gas of known molecular mass. For example, at 0°C and one atmosphere pressure (101.3 kPa), one litre (one cubic decimetre) of oxygen weighs 1.429 g. Since one mole of oxygen weighs 32.00 g, its molar volume under these conditions is

$$V_{molar} = 32.00 \text{ g } O_2 \times \frac{1.00 \text{ dm}^3}{1.429 \text{ g } O_2} = 22.4 \text{ dm}^3$$

R follows directly by substitution into the Ideal Gas Law:

$$PV = nRT \text{ or } R = \frac{PV}{nT}$$

where $P = 101.3$ kPa, $V = 22.4$ dm^3, $n = 1.00$ mol, $T = 0 + 273 = 273$ K

$$R = \frac{101.3 \text{ kPa} \times 22.4 \text{ dm}^3}{1.00 \text{ mol} \times 273 \text{ K}} = \mathbf{8.31\ kPa \cdot dm^3/(mol \cdot K)}$$

In most of our work in this chapter, we will use the value of R just calculated. However, it is possible to express R in other units. Since 1 kPa = 10^3 Pa, and 1 dm^3 = 10^{-3} m^3, we could write R as 8.31 Pa · m^3/(mol · K). Furthermore, since a pascal is the pressure exerted by one newton acting on one square metre (N/m^2), and a joule is the work done when one newton acts through a distance of one metre (N · m), the product Pa · m^3 is equivalent to the joule. That is,

$$\text{Pa} \cdot \text{m}^3 = \frac{\text{N}}{\text{m}^2} \times \text{m}^3 = \text{N} \cdot \text{m} = \text{J}$$

so that R can be expressed equally well as **8.31 J/(mol · K).** Alternatively, since the joule has the dimensions kg · m^2/s^2, we can express R in SI base units as **8.31 kg · m^2/(s^2 · mol · K).**

5.4 USING THE IDEAL GAS LAW

The Ideal Gas Law can be used to solve a wide variety of problems dealing with the experimental behavior of gases. It can, for example, be used to determine the effect of a change in conditions upon a variable such as volume (Example 5.1) or temperature (Example 5.2).

Example 5.1 In a McLeod gauge a gas at a low pressure, to be determined, is compressed at room temperature to a much smaller volume, where the pressure is increased sufficiently to be measured directly on a manometer. In a certain experiment a sample of helium in a vacuum system was compressed at 25°C from a volume of 200 cm^3 to a volume of 0.240 cm^3, where its pressure was found to be 3.00 kPa. What was the original pressure of the helium?

Solution For the gas in both states, $PV = nRT$

Initially: $P_1 = ?,\ V_1 = 200\ cm^3,\ n_1 = n,\ T_1 = T$

$$P_1V_1 = nRT$$

Finally: $P_2 = 3.00\ kPa,\ V_2 = 0.240\ cm^3,\ n_2 = n,\ T_2 = T$

$$P_2V_2 = nRT$$

Since both the temperature and amount of gas are held constant in this problem, the right sides of the two final equations are equal, and so

$$P_1V_1 = P_2V_2 \qquad P_1 = \frac{P_2V_2}{V_1}$$

$$P_1 = \frac{3.00\ kPa \times 0.240\ cm^3}{200\ cm^3} = 3.60 \times 10^{-3}\ kPa$$

Including the units helps you to see that they cancel

This problem is typical of many encountered with gases. The state of the gas is changed, with one or more of its variables remaining fixed; in this case, at constant n and T, we have essentially a Boyle's Law problem. By writing the Gas Law for the two states, one can usually quickly recognize which terms in the two equations are equal and use that information and the given data to find the unknown quantity. Note that in the problem here the units of volume (cubic centimetres) must be consistent if they are to cancel properly. The pressure is obtained in kilopascals, but can be readily converted to pascals or atmospheres if necessary:

$$3.60 \times 10^{-3}\ kPa \times \frac{10^3\ Pa}{1\ kPa} = 3.60\ Pa$$

$$3.60 \times 10^{-3}\ kPa \times \frac{1\ atm}{101.3\ kPa} = 3.55 \times 10^{-5}\ atm$$

Example 5.2 A hydrogen gas volume thermometer has a volume of 100.0 cm^3 when immersed in an ice-water bath at 0°C. When immersed in boiling liquid chlorine, the volume of the hydrogen at the same pressure is 87.2 cm^3. Find the temperature of the boiling point of chlorine in K and °C.

Solution For the hydrogen in the two states, $PV = nRT$

Initially: $P_1 = P \qquad V_1 = 100.0\ cm^3 \qquad n_1 = n \qquad T_1 = 0 + 273 = 273\ K$

$$PV_1 = nRT_1$$

Finally: $P_2 = P \quad V_2 = 87.2\ \text{cm}^3 \quad n_2 = n \quad T_2 = ?$

$$PV_2 = nRT_2$$

In this problem, P, n, and R are the same in the two states; we collect those variables on the left side of the two equations, obtaining

$$\frac{P}{nR} = \frac{T_1}{V_1} = \frac{T_2}{V_2}$$

Therefore

$$T_2 = \frac{V_2 T_1}{V_1}$$

$$T_2 = \frac{87.2\ \text{cm}^3 \times 273\ \text{K}}{100.0\ \text{cm}^3} = 238\ \text{K} \qquad t_2 = T_2 - 273 = -35°\text{C}$$

In the previous examples the value of the gas constant R was not needed, since R cancelled from the calculations. Cases in which explicit evaluation or use of the amount of gas is involved require the full use of the Ideal Gas Law, including the value of R. The following problems are illustrative and show the large amount of information that the Law allows one to obtain about gases.

Example 5.3 2.50 g of XeF_4 gas is introduced at 80°C into an evacuated container with a volume of 3.00 dm^3. Find the pressure in kilopascals.

Solution In this problem only one state is involved, and for it, $PV = nRT$.

$$P = ?;\ V = 3.00\ \text{dm}^3;\ n = 2.50\ \text{g XeF}_4 \times \frac{1\ \text{mol}}{207.3\ \text{g XeF}_4} = 0.0121\ \text{mol}$$

$$R = 8.31\ \text{kPa} \cdot \text{dm}^3/(\text{mol} \cdot \text{K});\ T = 273 + 80 = 353\ \text{K}$$

Substituting:

$$P = \frac{nRT}{V} = \frac{0.0121\ \text{mol} \times 8.31\ \frac{\text{kPa} \cdot \text{dm}^3}{\text{mol} \cdot \text{K}} \times 353\ \text{K}}{3.00\ \text{dm}^3} = 11.8\ \text{kPa}$$

Here all the elements in the Ideal Gas Law enter the calculation directly; if we use 8.31 kPa · dm^3/(mol · K) for R, the units of all the terms are of necessity those that appear in R, and any quantities which do not have these units must be converted before substituting in the Gas Law. Here, for example, we converted the number of grams of XeF_4 to the number of moles by dividing by the gram molecular mass, 207.3 g.

Example 5.4 A lighter-than-air balloon is designed to rise to a height of 10 km, at which point it will be fully inflated. At that altitude the atmospheric pressure is 27.9 kPa and the temperature is −40°C. If the full volume of the balloon is 100 m^3, how many kilograms of helium will be needed to inflate the balloon?

Solution To calculate the number of kilograms of helium required, we shall first use the Ideal Gas Law to obtain the number of moles, n, and then convert from moles to grams and finally to kilograms.

Since balloons have flexible walls, the pressure in the balloon is substantially equal to the outside atmospheric pressure. Hence, for the helium at a height of 10 km, $P = 27.9$ kPa. Expressing V in dm^3:

$$V = 1.00 \times 10^2\ \text{m}^3 \times \frac{10^3\ \text{dm}^3}{1\ \text{m}^3} = 1.00 \times 10^5\ \text{dm}^3$$

$$n = ?;\ R = 8.31\ \text{kPa} \cdot \text{dm}^3/(\text{mol} \cdot \text{K});\ T = 273 - 40 = 233\ \text{K}$$

$$n = \frac{PV}{RT} = \frac{27.9\ \text{kPa} \times 1.00 \times 10^5\ \text{dm}^3}{8.31\ \frac{\text{kPa} \cdot \text{dm}^3}{\text{mol} \cdot \text{K}} \times 233\ \text{K}} = 1440\ \text{mol}$$

Since 1 mol of helium weighs 4.00 g, the mass of helium is

$$1440\ \text{mol} \times 4.00\ \text{g} = 5760\ \text{g} \times \frac{1\ \text{kg}}{10^3\ \text{g}} = 5.76\ \text{kg}$$

There are some problems in which the use of the Ideal Gas Law is not quite so direct as in the previous examples. Two important cases involve the determination of the density of a gas under given conditions and the evaluation of the molecular mass from experimental data. These problems are most readily treated by rewriting the Ideal Gas Law so that it contains explicitly the number of grams of gas, g. Recalling that

$$\text{number of moles} = \frac{\text{number of grams}}{\text{gram molecular mass}} \qquad n = \frac{g}{(GMM)}$$

we can write:

$$PV = nRT = \frac{gRT}{(GMM)} \tag{5.8}$$

where GMM is the gram molecular mass.

Example 5.5 Uranium hexafluoride, UF_6, is perhaps the heaviest of all gases. What is the density in g/dm³ of UF_6 at 100°C and 100 kPa?

Solution The problem essentially is to find the number of grams per cubic decimetre of UF_6 present in a container at this temperature and pressure. We solve Equation 5.8 for the density, d, of the gas:

$$\text{density} = \frac{\text{mass}}{\text{volume}};\ d = \frac{g}{V} = \frac{(GMM)P}{RT}$$

$$GMM\ UF_6 = 238 + 6(19) = 352\ \text{g/mol};\ P = 100\ \text{kPa}$$

$$R = 8.31\ \text{kPa} \cdot \text{dm}^3/(\text{mol} \cdot \text{K});\ T = 273 + 100 = 373\ \text{K}$$

$$d = \frac{352\ \text{g/mol} \times 100\ \text{kPa}}{8.31\ \frac{\text{kPa} \cdot \text{dm}^3}{\text{mol} \cdot \text{K}} \times 373\ \text{K}} = 11.4\ \text{g/dm}^3$$

Example 5.6 A sample of chloroform weighing 0.495 g is collected as a vapor (gas) in a flask having a volume of 127 cm³. At 98°C the pressure of the vapor in the flask is 0.992 atm. Calculate the molecular mass of chloroform.

Solution Rewriting Equation 5.8, this time to obtain an explicit expression for the gram molecular mass,

$$GMM = \frac{gRT}{PV}$$

We need merely to express the variables in the equation in the proper units and solve for the gram molecular mass by substitution.

$$g = 0.495\ \text{g};\ R = 8.31\ \text{kPa} \cdot \text{dm}^3/(\text{mol} \cdot \text{K});\ T = 98 + 273 = 371\ \text{K}$$

$$P = 0.992\ \text{atm} \times \frac{101.3\ \text{kPa}}{1\ \text{atm}} = 100.5\ \text{kPa};\ V = 127\ \text{cm}^3 \times \frac{1\ \text{dm}^3}{10^3\ \text{cm}^3} = 0.127\ \text{dm}^3$$

$$GMM = \frac{0.495\ \text{g} \times 8.31\ \frac{\text{kPa} \cdot \text{dm}^3}{\text{mol} \cdot \text{K}} \times 371\ \text{K}}{100.5\ \text{kPa} \times 0.127\ \text{dm}^3} = 119\ \text{g/mol}$$

The calculated molecular mass of chloroform is, therefore, 119.

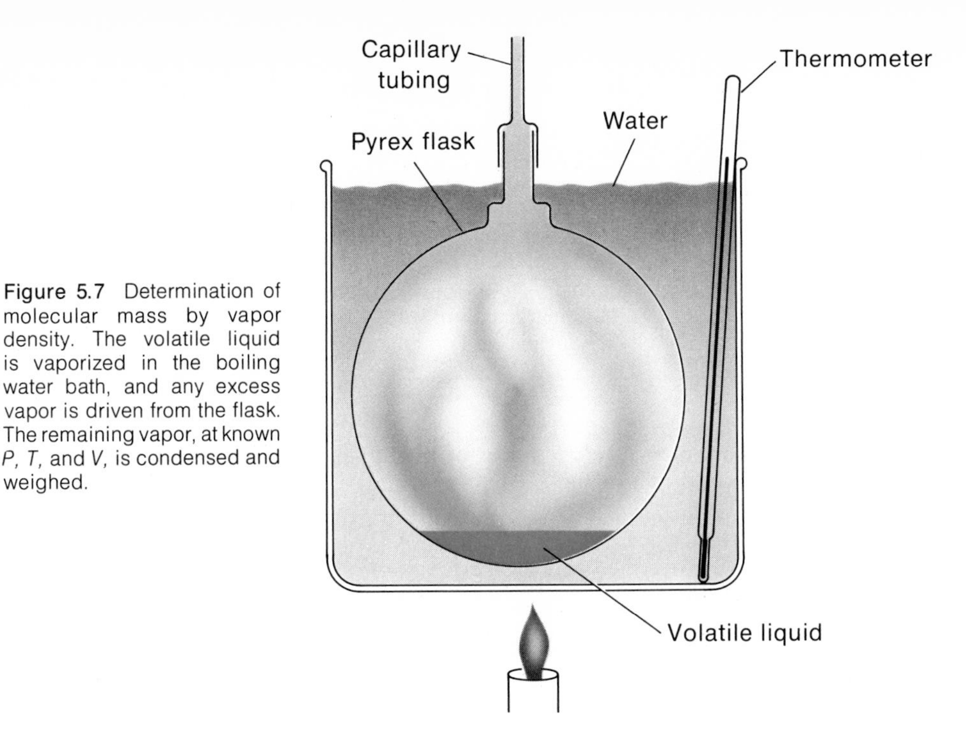

Figure 5.7 Determination of molecular mass by vapor density. The volatile liquid is vaporized in the boiling water bath, and any excess vapor is driven from the flask. The remaining vapor, at known *P*, *T*, and *V*, is condensed and weighed.

The data in Example 5.6 are typical of those obtained in one of the simplest experimental methods for the determination of molecular masses. In the experiment a small amount of a volatile liquid is placed in a flask fitted with a stopper in which there is a fine orifice (Fig. 5.7). The flask is then heated in a water bath to a temperature somewhat above the boiling point of the liquid. The liquid evaporates, and its vapor replaces the air in the flask. After all the liquid has evaporated and the flask is filled with vapor, the flask is removed from the bath and allowed to cool. The vapor condenses and air re-enters the flask. The mass of the vapor is taken to be the difference between the mass of the flask containing the condensed vapor and its mass when it contains just air. The method is an approximate one for several reasons, but when used properly it gives results accurate to within a few per cent.

How, if at all, would you modify this experiment to measure the *MM* of water?

Another application of the gas laws arises in connection with our interpretation of chemical reactions. We have seen that, given any chemical equation, it is possible to calculate the relative masses and numbers of moles of reactants and products. Where gases are present it is possible to extend the interpretation of chemical equations to include the volumes of the gases that would be involved under given conditions. The procedure is shown in the following example.

Example 5.7 How many cubic decimetres of pure oxygen, measured at 0.974 atm and 24°C, would be required to burn 1.00 g of benzene, $C_6H_6(l)$, to carbon dioxide and water?

Solution We must first write the balanced equation for the reaction, using the approach described in Chapter 3:

$$C_6H_6(l) + \frac{15}{2}\,O_2(g) \rightarrow 6\,CO_2(g) + 3\,H_2O(l)$$

Recognizing that for this reaction 1 mol $C_6H_6 \simeq 15/2$ mol O_2, we can find the number of moles O_2 required:

$$1.00 \text{ g } C_6H_6 \times \frac{1 \text{ mol } C_6H_6}{78.1 \text{ g } C_6H_6} \times \frac{15/2 \text{ mol } O_2}{1 \text{ mol } C_6H_6} = 0.0960 \text{ mol } O_2$$

Once the number of moles of O_2 has been determined, its volume is found by substitution into the Gas Law:

$$P = 0.974 \text{ atm} \times \frac{101.3 \text{ kPa}}{1 \text{ atm}} = 98.7 \text{ kPa}; \ T = 273 + 24 = 297 \text{ K}$$

$$V = \frac{nRT}{P} = \frac{0.0960 \text{ mol} \times [8.31 \text{ kPa} \cdot \text{dm}^3/(\text{mol} \cdot \text{K})] \times 297 \text{ K}}{98.7 \text{ kPa}} = 2.40 \text{ dm}^3 \text{ (i.e., } 2.40 \ \ell)$$

Since this reaction is very analogous to the one which occurs in an automobile engine, it should be apparent that a rather large volume of air (which is only 21% O_2 by volume) must pass through the engine during the combustion of gasoline.

A rather interesting interpretation of chemical reactions can be made when the substances involved are all gases whose volumes are measured under the same conditions of temperature and pressure. Consider the following chemical reaction:

$$4\ NH_3(g) + 5\ O_2(g) \rightarrow 4\ NO(g) + 6\ H_2O(g)$$

According to the usual interpretation, we would say

$$4 \text{ mol } NH_3 + 5 \text{ mol } O_2 \rightarrow 4 \text{ mol } NO + 6 \text{ mol } H_2O$$

If all these gases are measured at the same temperature and pressure, their molar volumes will all be equal to some fixed value, V_m cubic decimetres. Under such conditions it would be true that

$$4\ V_m \text{ dm}^3\ NH_3 + 5\ V_m \text{ dm}^3\ O_2 \rightarrow 4\ V_m \text{ dm}^3\ NO + 6\ V_m \text{ dm}^3\ H_2O$$

or, dividing through by V_m, simply

$$4 \text{ dm}^3\ NH_3 + 5 \text{ dm}^3\ O_2 \rightarrow 4 \text{ dm}^3\ NO + 6 \text{ dm}^3\ H_2O$$

Volumes of gases, then, when measured under the same conditions, have the same simple numerical relationships that exist between moles of substances in chemical reactions. The remarkable fact is that, whereas the relation between moles is deduced from theory, that between volumes can be found experimentally.

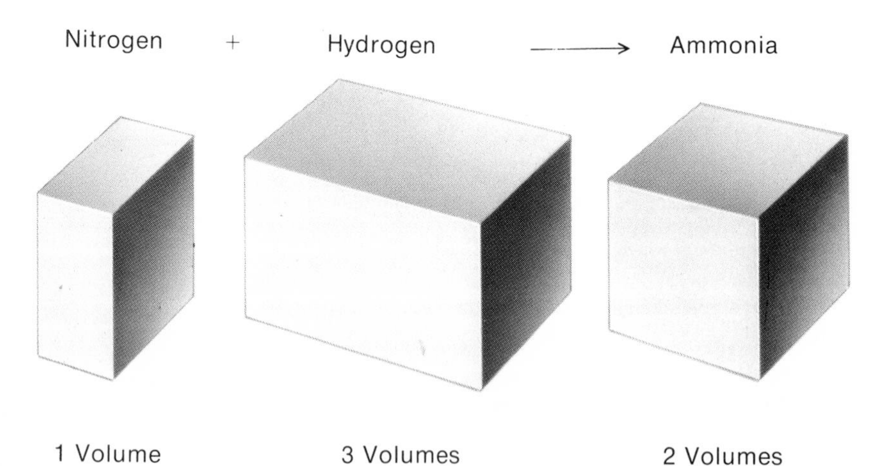

Figure 5.8 Gay-Lussac's Law of Combining Volumes as applied to the reaction: $N_2(g) + 3\ H_2(g) \rightarrow 2\ NH_3(g)$. When measured under the same conditions, the volumes of reacting gases have the same ratios as their coefficients in the equation for the reaction.

Indeed, in 1808 Gay-Lussac discovered the relationship and stated his Law of Combining Volumes: **In any chemical reaction involving gaseous substances the volumes of the various gases reacting or produced are in the ratios of small whole numbers.** (The gases are measured at the same temperature and pressure.)

5.5 MIXTURES OF GASES: DALTON'S LAW OF PARTIAL PRESSURES

So far we have considered the behavior of gaseous systems in which only one component is present. If several substances are present in a gaseous solution, we can still use the gas laws, but must take proper account of the presence of the different substances. Suppose we have a gaseous mixture of A, B, and C, confined in a container of volume V. The Ideal Gas Law will apply to the mixture, provided we let n equal the total number of moles present. That is,

$$PV = nRT \qquad \text{or} \qquad P = \frac{nRT}{V}$$

in which P is the total pressure in the container and n equals $n_A + n_B + n_C$.

The three substances, A, B and C, each contribute to the total pressure in the system. Their individual contributions, P_A, P_B and P_C, may be obtained by the Gas Law. We can say that

$$P_A = \frac{n_ART}{V} \qquad P_B = \frac{n_BRT}{V} \qquad P_C = \frac{n_CRT}{V}$$

P_A, P_B, and P_C are called the partial pressures of A, B, and C in the container. The partial pressure of a gas is the pressure the gas would exert if it alone were present in the container at the same temperature as the mixture. It is clear that since

$$P = \frac{nRT}{V} = (n_A + n_B + n_C)\frac{RT}{V} = \frac{n_ART}{V} + \frac{n_BRT}{V} + \frac{n_CRT}{V}$$

it follows that

$$P = P_A + P_B + P_C \tag{5.9}$$

This equation is a mathematical statement of Dalton's Law of Partial Pressures, first proposed by John Dalton in 1807. In words the law is: *The total pressure in a container is equal to the sum of the partial pressures of the component gases.*

Dalton's Law makes it possible to handle problems in which gaseous solutions are involved. The following example illustrates a typical application.

Example 5.8 Air from the prairies of Manitoba in winter contains essentially only nitrogen, oxygen, and argon. A sample of air collected at Winnipeg at −22°C and 97 kPa was analyzed; the mole fractions of N_2, O_2, and Ar were 0.78, 0.21, and 0.01, respectively. (The mole fraction of a substance X is the number of moles of X in the sample divided by the total number of moles in the sample.) Find the partial pressures of each of the gases in the sample when it was collected.

Solution The analysis tells us that 78 per cent of the molecules, or moles, of gas in the air are nitrogen, 21 per cent are oxygen, and 1 per cent are argon. For the whole sample of air,

$$P = \frac{nRT}{V}$$

where
$$n = n_{N_2} + n_{O_2} + n_{Ar}$$

For the nitrogen in the sample,

$$P_{N_2} = \frac{n_{N_2}RT}{V}$$

Dividing P_{N_2} by P, we obtain

$$\frac{P_{N_2}}{P} = \frac{n_{N_2}}{n} = 0.78$$

Therefore,
$$P_{N_2} = 0.78P = 0.78 \times 97 \text{ kPa} = 76 \text{ kPa}$$

By similar reasoning,

$$P_{O_2} = 0.21 \times 97 \text{ kPa} = 20 \text{ kPa}$$
$$P_{Ar} = 0.01 \times 97 \text{ kPa} = 1 \text{ kPa}$$

Example 5.8 illustrates the general principle that the partial pressure of a gas, P_A, in a gas mixture may be found by multiplying its mole fraction, X_A, by the total pressure:

$$P_A = X_A P_{tot}$$

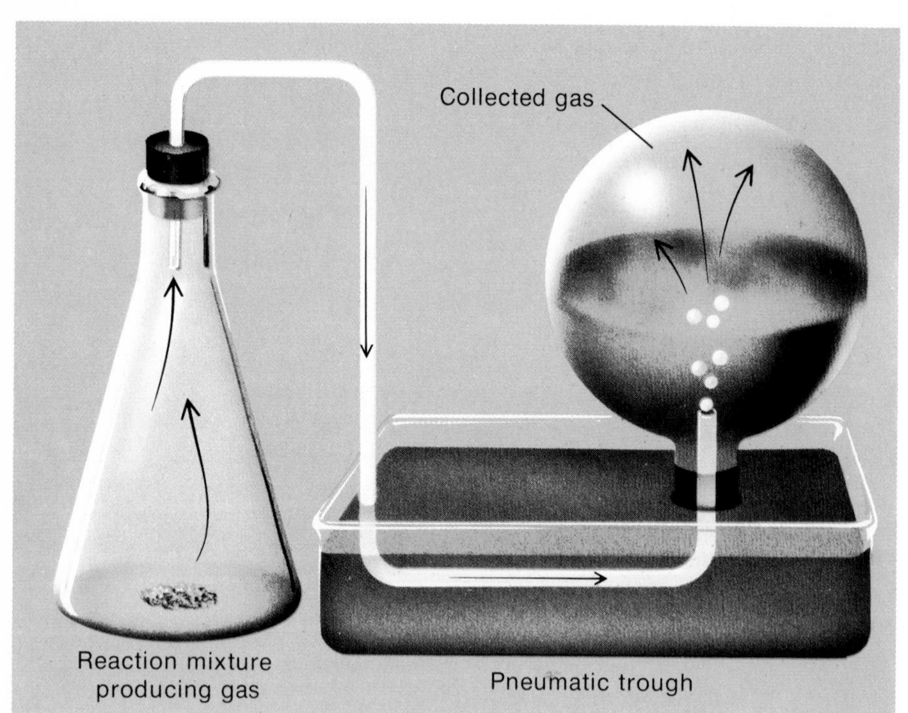

Gases collected this way cannot be appreciably soluble in water

Figure 5.9 The collection of gases over water. With this apparatus the collected gas will contain water vapor at a partial pressure equal to the vapor pressure of water at the temperature of the system.

Dalton's Law often finds practical use in experiments involving gases. In order to measure the amount of a gas produced in a chemical reaction we must collect the gas under known conditions. Probably the easiest way of doing this is to let the gas displace water in a system such as that shown in Figure 5.9. In this way we can measure the volume of gas at atmospheric pressure and known temperature. If the gas were pure, we could immediately use the Gas Law to calculate the number of moles produced by the reaction. However, under the conditions of the experiment the gas collected contains water vapor in addition to the gas of interest. The true pressure of the gas produced is, therefore, by Dalton's Law, equal to the total pressure minus the partial pressure of the water vapor. It is found experimentally that the pressure of water vapor in the presence of liquid water is a constant at a given temperature; its value can be obtained from a table of vapor pressures (Appendix 1). The following example is illustrative.

Example 5.9 In a laboratory experiment, concentrated hydrochloric acid was reacted with aluminum. Hydrogen gas was evolved and collected over water at 25°C; it had a volume of 0.355 dm³ at a total pressure of 100.0 kPa. The vapor pressure of water at 25°C is known to be about 3.2 kPa.

a. What was the partial pressure of the hydrogen in the sample?
b. How many moles of hydrogen were collected?

Solution

a. The collected gas was a mixture of hydrogen and water vapor. The partial pressure of water in the sample would equal its vapor pressure, 3.2 kPa. By Dalton's Law: $P_{total} = P_{H_2} + P_{H_2O}$. Therefore, $P_{H_2} = P_{total} - P_{H_2O}$.

$$P_{H_2} = (100.0 - 3.2)\text{kPa} = 96.8 \text{ kPa}$$

b. By the Gas Law:

$$P_{H_2} = \frac{n_{H_2}RT}{V}; \qquad n_{H_2} = \frac{P_{H_2}V}{RT} = \frac{96.8 \text{ kPa} \times 0.355 \text{ dm}^3}{8.31 \frac{\text{kPa}\cdot\text{dm}^3}{\text{mol}\cdot\text{K}} \times 298 \text{ K}} = 0.0139 \text{ mol}$$

5.6 REAL GASES

In this chapter we have applied the Ideal Gas Law to all gases, tacitly assuming that the Law was obeyed exactly. Under ordinary conditions (and in nearly all problems involving gases in this text) the assumption is a very good one. Actually, however, all gases deviate to some degree from the ideal laws to an extent that depends on the pressure, the temperature, and the nature of the gas, particularly the ease with which it could be condensed to a liquid. For gases in the vicinity of room temperature and 1 atm pressure the deviation is small, at most a few per cent. Gases like oxygen and hydrogen, which boil at very low temperatures (−183 and −253°C respectively), have molar volumes at 25°C and 1 atm that are within 0.1 per cent of the ideal value. On the other hand, sulfur dioxide and chlorine, which boil at 10° and −35°C, are, at 25°C and 1 atm, not nearly so ideal and have molar volumes that are 2.4 and 1.6 per cent lower, respectively, than the value predicted by the Ideal Gas Law.

We can illustrate the actual behavior of real gases graphically by plotting the ratio of the observed volume, V_{obs}, to the ideal volume, V_{ideal}, for a gas as a function of pressure. Figure 5.10 is such a graph for several gases at 25°C. For an ideal gas, V_{obs}/V_{ideal} would be constant and equal to one at all pressures. As you can see from the figure, V_{obs}/V_{ideal} at relatively low pressures (less than 100 atm) is less than one in the vicinity of 25°C for all the gases except hydrogen and helium, and approaches one as the pressure approaches zero. Generalizing, we can say that the observed volume of most gases at moderate pressure and temperature is slightly lower than would be predicted by the Gas Law. This behavior is caused, we believe, by the weak attractions that exist between molecules even in the gas phase (see Chapter 9). These forces make the gas easier to compress, and so tend to make the value of V_{obs}/V_{ideal} less than one at any given pressure.

As the temperature of the gas is lowered, the energy of motion of the molecules is lowered, and the relative effect of the attractive forces becomes more important until, at the boiling point of the substance, these forces become large enough to cause the gas to condense to a liquid at any pressure above 1 atm. As you might expect, intermolecular attractions are more important at temperatures near the boiling point of the substance, which explains in a qualitative way why SO_2 and Cl_2 are less nearly ideal in their behavior at room temperature than are oxygen and hydrogen. The behavior of the other substances in Figure 5.10 bears out this line of reasoning (see Table 5.3).

The behavior of gases at very high pressures (500 atm and above) is shown at the right of Figure 5.10. For all gases, as the pressure is increased, V_{obs}/V_{ideal} ultimately begins to increase, finally becoming larger, even much larger, than the ideal value. Such an effect is best understood by consideration of the volumes of gas molecules. As we noted, in general the volume occupied by a gas is much larger than the volume of its constituent molecules. However, the volume of the molecules is not zero (being roughly equal to the volume of the liquid obtained

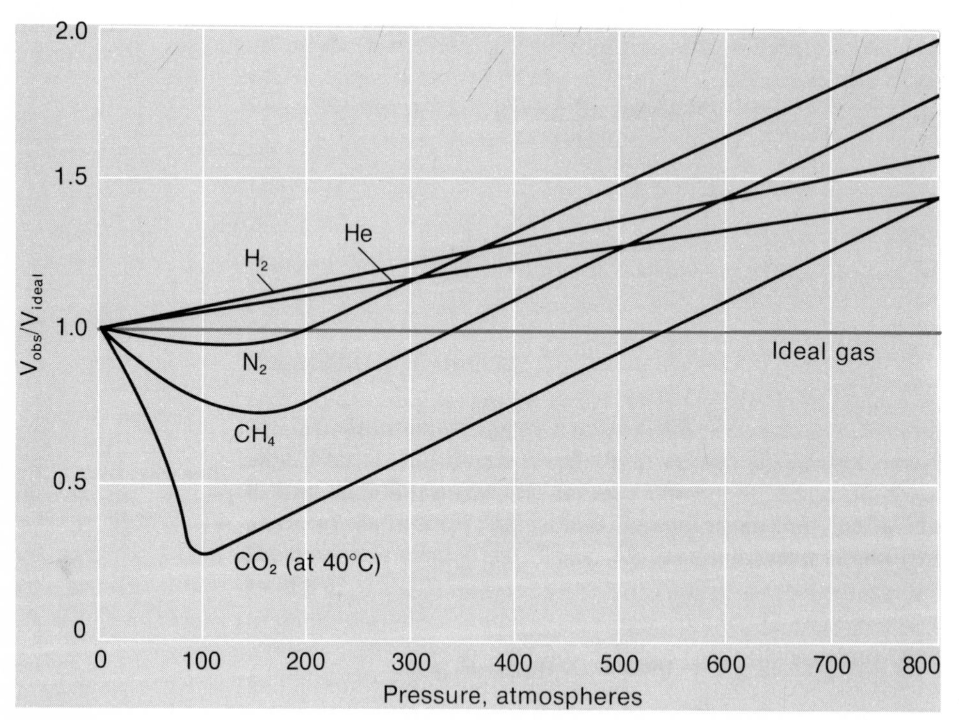

Figure 5.10 The behavior of real gases at 25°C. At very low pressures all gases become ideal. At moderate pressures most gases have volumes which are lower than the ideal value; the deviation from the ideal increases as the boiling point of the substance increases. At very high pressures all gases have volumes that are larger than the ideal value.

when the gas is condensed); as the gas is compressed, the volume of the molecules becomes a larger and larger fraction of the gas volume. Since the observed volume at a given pressure would tend to be increased with respect to the ideal volume by essentially the volume occupied by the molecules, the net effect is to make V_{obs}/V_{ideal} larger than 1 at very high pressures.

These two causes of nonideal gas behavior, attraction between molecules and finite molecular volumes, act in opposite directions, with the former making a gas easier to compress and the latter making compression more difficult. At relatively low pressures, the attractive forces between molecules dominate deviations from the Ideal Gas Law, while at very high pressures molecular volumes are the more important. For CH_4 at about 350 atm and 25°C the two effects just about cancel each other, making CH_4 appear to behave ideally at that point (Fig. 5.10).

It is possible to write equations of state for gases which take into account both intermolecular attractions and finite molecular volumes. Perhaps the best known of these relations is the van der Waals equation, which for one mole of gas takes the form

$$P = \frac{RT}{V - b} - \frac{a}{V^2} \tag{5.10}$$

where a and b are constants selected for each gas to give the best possible agreement between the equation and actual experimental behavior. The term a/V^2 is associated with the attractive forces between the molecules. The van der Waals b corrects for the effect of molecular volume.

In Table 5.3 we have listed values of van der Waals a and b for a few gases, along with their boiling points and molar volumes of the liquid. The van der Waals b is about equal to the molar volume of the liquid, as it should be if it represents the effective volume occupied by the molecules of the gas. The van der Waals a is not so readily accounted for, but its increase with increasing boiling point seems very reasonable, since high boiling point would imply strong intermolecular attractive forces. If you used the van der Waals equation to calculate V_{obs}/V_{ideal} for the gases in Figure 5.10, you would find that the experimental curves in the Figure agree qualitatively, but not exactly, with the curves obtained from the equation. The van der Waals equation is often used in dealing with problems in which the nonideality of the gas is important and a better analytical relation between P, V, and T is required than is obtained from the Ideal Gas Law.

TABLE 5.3 VAN DER WAALS CONSTANTS FOR SOME TYPICAL GASES

SUBSTANCE	a IN $\frac{\text{kPa} \cdot \text{dm}^6}{\text{mol}^2}$	b IN $\frac{\text{dm}^3}{\text{mol}}$	bp (°C) AT 1 ATM	V_{molar} (DM^3) OF LIQUID
H_2	24.7	0.027	−253	0.022
O_2	137.8	0.032	−183	0.028
N_2	140.8	0.039	−196	0.035
CH_4	228.2	0.043	−161	0.039
CO_2	363.9	0.043	−78 (subl)	0.040
SO_2	680.1	0.056	−10	0.044
Cl_2	657.7	0.056	−35	0.054
H_2O	553.5	0.030	100	0.018

5.7 THE KINETIC THEORY OF GASES

The fact that the Ideal Gas Law can be used to describe reasonably accurately the physical behavior of all gases, whatever their degree of molecular complexity, is a clear indication that the gaseous state of matter is a relatively simple one to treat from a theoretical point of view. There must be certain properties common to all gases which cause them to follow the same natural law and exhibit so many generally similar characteristics.

Between about 1850 and 1880, Maxwell, Boltzmann, Clausius, and others, using the notion that the properties of gases are the result of molecular motions, developed the kinetic theory of gases. The theory was highly successful in explaining the known laws of gas behavior as discovered by Boyle, Gay-Lussac, Charles, and Graham, and was consistent with the ideas on heat, energy, and temperature as developed by Mayer, Joule, and Kelvin. Since the time of its creators, who, incidentally, rank among the very best theoreticians the world has known, the theory has had to be modified to only a small extent to make it consistent with quantum theoretical principles. In its present form it is one of the most successful of scientific theories, ranking in stature with the Copernican theory for the motion of the planets and the atomic theory of the nature of matter.

Postulates of the Kinetic Theory of Gases

1. **Gases consist of molecules in continuous, random motion.** The molecules undergo collisions with one another and with the container walls. The pressure of a gas arises from the forces associated with wall collisions.

2. **Molecular collisions are elastic.** During collisions there are no frictional losses which result in loss of energy of motion. The temperature of a gas insulated from its surroundings does not change.

The energy of a rocket is mostly translational; that of a spinning top is rotational

3. **The average energy of translational motion of a gas molecule is proportional only to the absolute temperature.** The energy associated with the motion of a molecule from one place to another depends on the temperature, but not on the pressure or the nature of the molecule. Mathematically, we can state this postulate in the following way:

$$\epsilon_T = (\tfrac{1}{2})\, mu^2 = \mathrm{c}T \qquad (5.11)$$

where ϵ_T is the average molecular energy of translation, m is the mass of the molecule, u is its average speed, and c is a proportionality constant which has the same value for all gases.

In addition to these postulates, it is often assumed that the volumes of the molecules are negligible as compared to container volume and that molecules do not exert forces on each other except by collisions.

The postulates of the kinetic theory are easily stated. Their implications, however, are by no means obvious. The difficulty arises essentially from the fact that gas molecules move about in a completely random manner, with velocities that are constantly changing both in magnitude and direction because of molecule-molecule and molecule-wall collisions. To take account of this kind of motion in anything resembling a rigorous way requires, as you can well imagine, mathematics of a very high level of sophistication. In this text we shall not attempt even a simplified mathematical development using the kinetic theory, but shall proceed directly to some of the relationships which the theory has produced.

One of the important relations which can be derived from the kinetic theory and the laws of mechanics relates the pressure of a gas to its volume and the number, mass, and speed of its molecules (see Fig. 5.11). The final equation which one obtains is really quite simple:

$$PV = (1/3)(KMM)nu^2 = nRT \tag{5.12}$$

where KMM is the molar mass in kilograms (e.g., $KMM\ O_2 = 0.0320$ kg/mol), n is the number of moles in the volume V, and u is the average molecular speed. The first equality in Eq. 5.12 is obtained from the analysis of the forces exerted by the molecules on the container walls. The last equality is introduced to make the theory consistent with the Ideal Gas Law, and allows evaluation of the constant c in the third postulate (Eq. 5.11). Much of the rest of this chapter will be devoted to exploring some of the implications of Equation 5.12.

MOLECULAR SPEEDS AND ENERGIES. One of the most interesting results of the kinetic theory is that it allows us, through Eq. 5.12, to calculate average speeds of molecules in a gas. A simple manipulation of that equation furnishes an explicit expression for the average molecular speed u:

$$u^2 = \frac{3RT}{(KMM)} \quad \text{or} \quad u = \left(\frac{3RT}{KMM}\right)^{1/2} \tag{5.13}$$

where R is the gas constant and KMM is the kilogram molecular mass. It is clear from Equation 5.13 that **molecular speeds increase with increasing temperature, decrease with increasing molecular mass, and,** rather surprisingly, **are not directly influenced by gas pressure.**

If we wish to actually calculate an average molecular speed, we can do so by substitution into Equation 5.13. Here, we must be careful to use the proper units for each of the quantities that enter into that equation. As usual, we use the absolute temperature in kelvins (K). We take R to be 8.31 kg · m^2/(s^2 · mol · K). As pointed out above, KMM is the molar mass in kilograms and hence has the units

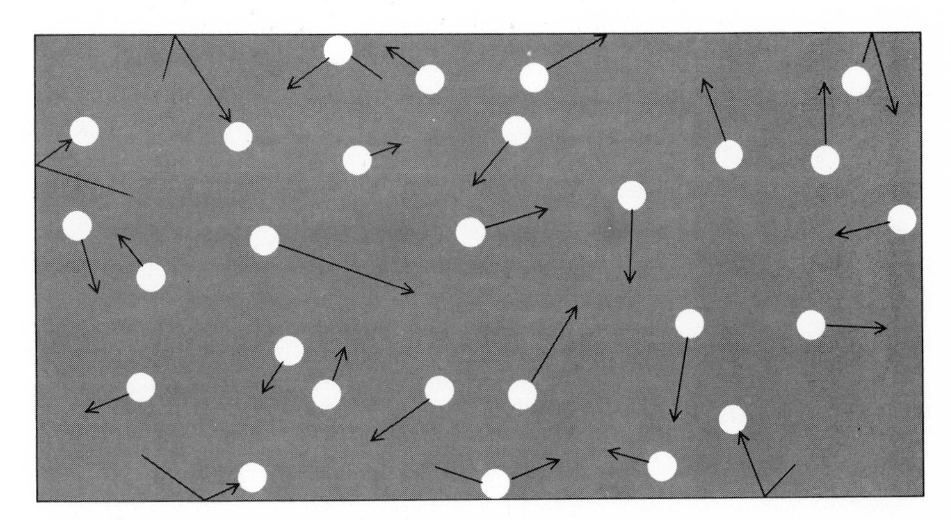

Figure 5.11 The pressure of a gas is the result of the forces on the wall caused by molecule-wall collisions. The pressure is proportional to the product of the force per collision ($\sim mu$) and the number of collisions per second on unit wall area ($\propto nu/V$), so $P \propto nmu^2/V$ (see Equation 5.12).

kilograms per mole (kg/mol). Substituting in Equation 5.13, we see that the units of u will be

$$\left(\frac{\frac{\text{kg}\cdot\text{m}^2}{\text{s}^2\cdot\text{mol}\cdot\text{K}}\times\text{K}}{\text{kg/mol}}\right)^{1/2} = \left(\frac{\text{m}^2}{\text{s}^2}\right)^{1/2} = \frac{\text{m}}{\text{s}};$$

i.e., the average molecular speed will be calculated in metres per second.

Example 5.10 Find the average speed of an oxygen molecule ($MM = 32.0$) in air at room temperature (25°C).

Solution Substituting $R = 8.31\ \text{kg}\cdot\text{m}^2/(\text{s}^2\cdot\text{mol}\cdot\text{K})$, $T = 298$ K, and $KMM = 0.0320$ kg/mol, we have

$$u = \left(\frac{3RT}{KMM}\right)^{1/2} = \left(\frac{3\times 8.31\times 298}{0.0320}\right)^{1/2}\frac{\text{m}}{\text{s}} = 482\ \text{m/s}$$

According to the kinetic theory, molecular speeds are, on the average, very high by ordinary standards. It has been possible to test this prediction by several direct experiments, and the quantitative data agree very well with the theoretical values. Qualitatively, the very high average speed of molecules appears reasonable when one considers the speed of sound in air. Since sound is propagated by molecular motion, one would expect that the speed of sound and that of the molecules in the gas through which it passes would be roughly equal. The speed of sound in air is about 360 m/s, which is reasonably close to the average speed of an oxygen or nitrogen molecule at 25°C.

It is possible to use Equation 5.12 in another manner to determine molecular energy instead of speed. A single molecule moving through space will have a translational energy given by Equation 5.11:

$$\epsilon_T = (1/2)\, mu^2$$

A mole of such molecules will have an energy of motion E_{trans} which is equal to $N\epsilon_T$, where N is Avogadro's number:

$$E_{trans} = N\epsilon_T = (1/2)\, mNu^2 = (1/2)\, (KMM)u^2$$

where KMM is the kilogram molecular mass. By substitution into Equation 5.12 we obtain, on solving for the energy,

$$E_{trans} = (1/2)\, (KMM)u^2 = (3/2)\, RT \tag{5.14}$$

Consistent with a postulate of the theory, the *energy* of translation of a mole of gas is directly proportional to temperature and *has the same value for all gases.* At 25°C, using R with units of heat energy, 8.31 J/(mol · K), we find the molar energy,

At 100 K, a gas has an energy of translation 1/3 that at 300 K

$$E_{trans} = (3/2)\, RT = (3/2)\times 8.31\ \text{J/(mol}\cdot\text{K)}\times(25+273)\ \text{K} = 3710\ \text{J/mol}$$

This amount of energy, about 4 kJ/mol, is rather small by comparison with the energy changes typically observed in chemical reactions. E_{trans}, however, is by no means of negligible importance. The fact that E_{trans} turns out to have the same value at any given temperature for every gas, from He to UF_6, is, more than any other single factor, responsible for the common physical behavior we observe for gases.

GRAHAM'S LAW. Several years before the kinetic theory of gases was developed, Thomas Graham performed some experiments which relate to molecular speeds. Graham studied the rates at which different gases flowed through a small orifice from a container into a vacuum. This phenomenon is called *effusion* and, as one might expect, it occurs at a rate that is proportional to the average speed of the effusing molecules. If the rates of effusion of two gases, A and B, were measured in the same container under the same initial conditions, we would predict that

$$\frac{\text{rate of effusion of A}}{\text{rate of effusion of B}} = \frac{\text{average speed of molecule A}}{\text{average speed of molecule B}} = \frac{u_A}{u_B} \qquad (5.15)$$

By Equation 5.12, if gases A and B were studied at the same temperature,

$$(1/3)\,(KMM)_A u_A^2 = RT = (1/3)\,(KMM)_B u_B^2$$

or

$$\frac{u_A^2}{u_B^2} = \frac{(KMM)_B}{(KMM)_A} = \frac{(MM)_B}{(MM)_A} \qquad (5.16)$$

If Equation 5.16 is solved for the ratio of the average molecular speeds, and this is substituted into Equation 5.15, we find that

$$\frac{\text{rate of effusion of A}}{\text{rate of effusion of B}} = \left(\frac{MM_B}{MM_A}\right)^{1/2} \qquad (5.17)$$

This relation, that the rate of effusion of a gas varies inversely as the square root of its molecular mass, was discovered experimentally by Graham in 1828. (At that time molecular masses were not thoroughly understood, and Graham stated the relation in terms of the relative densities of the gases he studied. Since we now know that the ratio of the densities of two gases measured under the same conditions is equal to the ratio of their molecular masses, Graham's Law is often stated in the form given in Equation 5.17.)

Graham's Law gives us an alternate method for measuring the molecular masses of gases, and one of its main uses is in this area. One needs merely to compare the rate of effusion of the unknown gas or vapor to that of a known reference gas. What is usually done is to measure the time required for the pressure of a known gas A, confined in a container which is connected to a vacuum pump via a fine capillary tube, to drop by a predetermined amount. The experiment is repeated with an unknown gas B under the same conditions of temperature and pressure, so that the same number of moles of gas effuse in the two runs. The measured times will vary inversely with the rates of effusion, which makes the pertinent ratios relate to one another as follows:

If you move twice as fast as I do, you'll get to your destination in half the time

$$\frac{\text{time}_A}{\text{time}_B} = \frac{\text{rate}_B}{\text{rate}_A} = \frac{u_B}{u_A} = \left(\frac{MM_A}{MM_B}\right)^{1/2} \qquad (5.18)$$

Example 5.11 In an effusion experiment it required 45 s for a certain number of moles of an unknown gas to pass through a small orifice into a vacuum. Under the same conditions it required 18 s for the same number of moles of oxygen to effuse. Find the molecular mass of the unknown gas.

Solution The first and last terms of Equation 5.18 relate time and molecular mass, so we can substitute directly:

$$\frac{\text{time}_{O_2}}{\text{time}_X} = \frac{18}{45} = \left(\frac{MM_{O_2}}{MM_X}\right)^{1/2} = \left(\frac{32}{MM_X}\right)^{1/2}$$

Squaring both sides, $\frac{32}{MM_X} = \left(\frac{18}{45}\right)^2 = 0.16; MM_X = \frac{32}{0.16} = 200$

During World War II Graham's Law had a rather unexpected application in connection with a very complicated chemical problem. It had been found that the isotope of uranium having a mass of 235, ^{235}U, has a nucleus unstable to collisions with neutrons. Such collisions result in a splitting of the uranium nucleus into lighter fragments (fission) and the liberation of large amounts of energy in the form of heat and γ-rays (see Chapter 4). It became necessary to separate ^{235}U from the much more plentiful (but not fissionable) isotope, ^{238}U. Because of the great chemical similarity of the isotopes of an element, the chemical resolution of uranium into its isotopes was not feasible, and some physical method was sought. Since the rate of effusion of a gas varies with its molecular mass, the composition of a gas mixture coming through an orifice will not be quite the same as that in the original sample and, hence, the resolution of a gas mixture by successive effusions is possible, at least in principle. Preliminary effusion experiments with uranium hexafluoride, UF_6, a volatile uranium compound, indicated that $^{235}UF_6$ could indeed be separated from $^{238}UF_6$ by effusion, and an enormous plant was built for the purpose in Oak Ridge, Tennessee. In the process, UF_6 effuses many thousands of times through porous barriers, with the lighter fractions moving on to the next stage and the heavier fractions being recycled through earlier stages.

bp UF_6 = 56°C

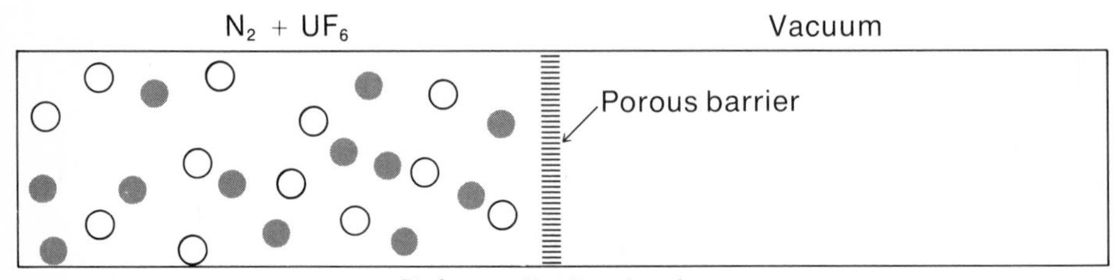

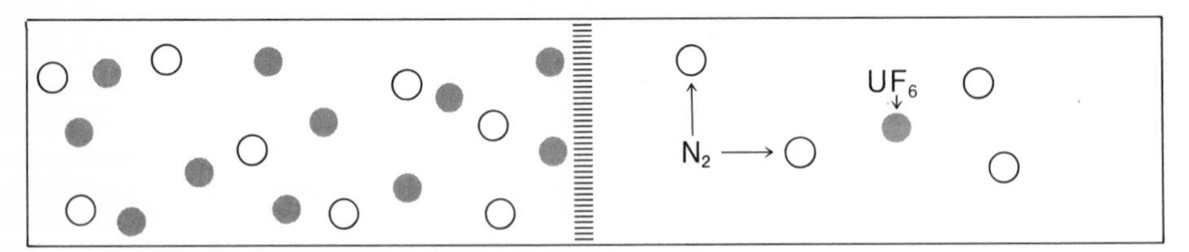

Figure 5.12 Effusion of gases. Since N_2 molecules are much lighter than UF_6 molecules, they will move faster and will effuse through the porous barrier much more quickly than will UF_6. The same sort of effect would be observed if $^{238}UF_6$ and $^{235}UF_6$ were allowed to effuse through a barrier, but the separation of those species would be much less efficient since their masses are so nearly equal.

DISTRIBUTION OF MOLECULAR SPEEDS AND ENERGIES. Although it is useful to be able to calculate an average speed or energy for a molecule, one must remember that not all the molecules will have that speed and energy. The motion of molecules in a gas is utterly chaotic. In the course of a second a molecule will typically undergo millions of collisions with other molecules, with each collision resulting in a change of the molecule's speed and direction of motion. In view of this situation you might well wonder whether one could hope to determine anything more than average molecular properties; in fact, the calculation of average speeds and molar energies seems a rather remarkable achievement.

Surprisingly, we do know considerably more about molecular motion in a gas than would first appear. In 1860 James Clerk Maxwell showed that different possible speeds were distributed among molecules in a definite way, and by a careful analysis of the gaseous system he derived a mathematical expression for this distribution. Figure 5.13 illustrates graphically the relation he obtained for oxygen gas at 25°C and 1000°C. On the graph the relative number of oxygen molecules having the speed u is plotted as a function of the speed. We see that there are very few molecules having speeds near zero, and that the number having a given speed increases rapidly with speed, up to a maximum at about 400 m/s at 25°C. This is the most probable speed of an oxygen molecule at 25°C, with the majority of the molecules having speeds between 200 and 600 m/s. Above about 400 m/s, the number of molecules moving at any particular speed decreases, so that the likelihood of finding a molecule moving at about 800 m/s is only about one fifth as large as that of finding a molecule moving at about 400 m/s. For speeds in excess of about 1200 m/s the fraction of molecules drops off rapidly.

As the temperature of the gas increases, the speeds of the molecules increase, and the distribution curve for molecular speeds is displaced to the right and becomes broader. From the curves at 25 and 1000°C it is clear that the probability of a molecule having a given high speed is larger at high temperatures than it is at low.

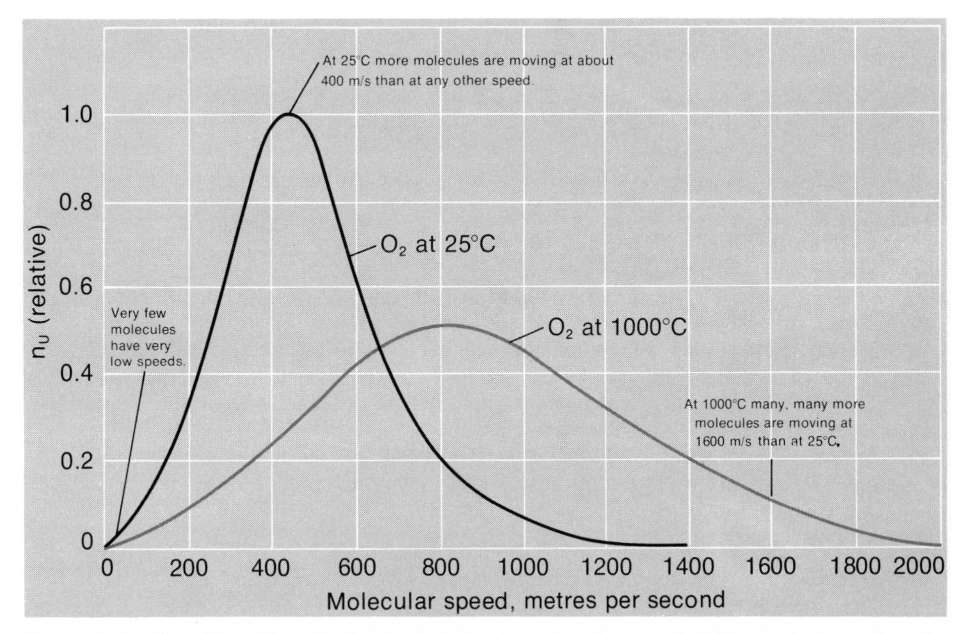

Figure 5.13 The Maxwellian distribution function for molecular speeds. This graph shows the relative numbers of O_2 molecules having any given speed at 25°C and at 1000°C. At 25°C most O_2 molecules have speeds between 200 and 600 m/s.

Since translational energy for a molecule is simply related to the speed u, it is true that one can also speak of a distribution of energy among molecules which is analogous to that for speed among molecules. At high molecular energies we find very few molecules; the fraction of molecules having a given energy E or greater per mole drops off exponentially with increasing energy and to a good degree of approximation is given by the equation

$$f = e^{-E/RT} \tag{5.19}$$

where f is the fraction of molecules having molar energy E or greater in a gas at temperature T, and e is the base of natural logarithms. Since molecules having high energies are the ones which tend to participate in chemical reactions, Equation 5.19 is important in studies of reaction rates (Chapter 16).

The mathematical relation obtained by Maxwell for the distribution of molecular speeds is one of the most formidable in all of physical science. It is

$$n_u\,(\text{rel}) = 4\pi u^2 \left(\frac{KMM}{2\pi RT}\right)^{3/2} e^{-(KMM)u^2/2RT}$$

The fact that Maxwell was able to develop this equation from first principles is indicative of the intellectual ability of this remarkable man.

Maxwell was born in 1831 in Scotland. After education at the University of Edinburgh and at Cambridge he became a professor of natural philosophy, first in Scotland and later at Cambridge. His mathematical abilities became apparent at an early age and were applied in many areas. In addition to his accomplishments with gases, which laid the foundation of the science now known as statistical mechanics, Maxwell worked extensively in thermodynamics, developing several fundamental equations which bear his name. His greatest successes, however, were in connection with the theory of light and electricity, where he discovered and formulated the general equations of the electromagnetic field. He was first to recognize that light is a form of electromagnetic radiation and anticipated the development of what we now call radio waves. For several years during his life Maxwell was forced to be inactive because of illness: he was 48 when he died, having completed much of his work by the time he was 30.

The number of truly outstanding theoreticians the world has known is very small, certainly numbering less than one hundred. James Clerk Maxwell belongs among the elite of this group. He was truly an intellectual giant, to be ranked with Newton, Einstein, and J. Willard Gibbs (Chapter 14).

PROBLEMS

5.1 *Uses of the Ideal Gas Law*

a. If 600 cm^3 of H_2 at 25°C and 1.00 kPa is compressed to a volume of 120 cm^3, holding temperature constant, what does the pressure become?
b. A sample of CO_2 occupying a volume of 200 cm^3 at 10°C and 93.3 kPa pressure is heated to 100°C at that pressure. What is its volume under those conditions?
c. How many moles of O_2 would there be in a 24-dm^3 tank at 20°C if the pressure of the O_2 in the tank were 1.5×10^3 kPa?
d. What is the density in grams per cubic decimetre of CCl_4(g) at 100°C and 56.6 kPa?

5.2 *Dalton's Law* What would be the partial pressure of N_2 in a container at 50°C in which there is 0.20 mol of N_2 and 0.10 mol of CO_2 at a total pressure of 1.00 atm?

5.3 *Kinetic Theory*

a. How could you double the average speed of the O_2 molecule in a gas at 25°C? How could you double its energy of motion?
b. Arrange the following components of air at 25°C in order of increasing average speed: N_2, O_2, Ar, H_2O, CO_2.

5.4 Describe what would happen if you put the barometer in Figure 5.2 in a closed container and gradually pumped out the air.

5.5 In a McLeod gauge, 240 cm^3 of gas from a low-pressure system was compressed at room temperature to a volume of 0.039 cm^3, where it had a pressure of 2.48 kPa. What was the pressure in the system?

5.6 In Boyle's original experiment he measured the length d (directly proportional to volume) of an air column as a function of the pressure head Δh of mercury (Fig. 5.5). The following data were obtained by his procedure:

d (cm)	50.0	45.0	40.0	35.0	30.0
Δh (cmHg)	0.0	8.3	18.8	32.1	50.0

The barometric pressure was 75.0 cmHg and the temperature was 25°C. Show that these data support the validity of Boyle's Law.

5.7 The volume of a hydrogen gas thermometer is 150 cm^3 at 25°C and 1.00 atm pressure. When immersed in boiling ammonia its volume at 1.00 atm falls to 121 cm^3. Find the boiling point of NH_3 in kelvins and degrees Celsius.

5.8 A sample of N_2 is heated at constant pressure from −25°C to +75°C with the following results:

t (°C)	−25	0	25	50	75
V (cm^3)	125	138	150	163	175

Show that these data are consistent with Charles' Law.

5.9 Calculate the number of moles of gas in a high-pressure cylinder of volume 56.4 dm^3 if the pressure is 2.15×10^4 kPa at a temperature of 25°C.

5.10 How much will the steam (H_2O) filling a 970-litre oil drum at 100°C and 1.00 atm weigh? What would be the volume of the liquid water required to produce that amount of steam? ($d_{H_2O(l)}$ = 1.00 g/cm^3).

5.11 Given that the mole fractions of N_2 and O_2 in air are about 0.79 and 0.21, respectively, calculate the average "molecular mass" of air. What is the density of air in grams per cubic decimetre at 25°C and 100 kPa?

5.12 A sample of a volatile organic liquid is vaporized in an apparatus like that shown in Figure 5.7. The flask has a volume of 218 cm^3 and the experiment is carried out at 99°C and 0.967 atm pressure. The vapor that condenses on cooling the flask weighs 1.063 g. Find the molecular mass of the liquid.

5.23 A suction pump in Storrs, Connecticut was able to lift water 10.2 m, 10.0 m, and 10.1 m on successive days. Explain these observations.

5.24 What volume of Ne at one atmosphere and 25°C would have to be added to a neon sign having a volume of 250 cm^3 to create a pressure of 1.00×10^{-3} atm at that temperature?

5.25 In measuring the pressure of a sample of Ar in an apparatus like that in Figure 5.3, it was found that the mercury level in contact with Ar was 48.0 cm lower than the level in contact with air, when the atmospheric pressure was 746 mmHg. What was the pressure of the Ar in millimetres of mercury? in kilopascals?

5.26 The natural gas in a storage tank in Winnipeg has a volume at atmospheric pressure of 2.00×10^6 m^3 at 20°C. The next day the temperature falls to −20°C, but the pressure remains constant. What does the volume of the gas become?

5.27 The gauge pressure at 23°C in an automobile tire was 165 kPa in Indianapolis on May 31. Gauge pressure is the difference between total pressure and atmospheric pressure, which is about 101 kPa. After driving on the speedway for an hour the gauge pressure in the tire became 207 kPa. If the volume of the tire didn't change, what was the temperature of the air in the tire?

5.28 What pressure will be exerted by 15.0 g of CO_2 in a 20.0-dm^3 container at 0°C?

5.29 Standard temperature and pressure, *STP*, is taken to be 0°C and 1.00 atm. Find the volume occupied by one mole of an ideal gas at *STP*. Find the molar volume of the gas at 25°C and one atmosphere.

5.30 The lightest gas is hydrogen, H_2, and one of the heaviest is UF_6. What are the densities of these two gases at 100°C and 98.6 kPa?

5.31 An organic liquid contains 92.3% carbon and 7.7% hydrogen. What is its simplest formula? Using the apparatus in Figure 5.7 it was found that 226 cm^3 of the vapor at 100°C and 100.6 kPa weigh 0.573 g. What is the molecular mass of the liquid? What is its molecular formula?

5.13 A 0.675-dm^3 sample of wet CO, saturated with water vapor at 22°C, exerts a total pressure of 101.0 kPa. At 22°C, vp_{H_2O} = 2.7 kPa.

a. What is the partial pressure of CO in the sample?
b. How many grams of CO does the sample contain? How many molecules?

5.14 Hydrogen chloride will burn in air at 25°C according to the reaction

$$4\ HCl(g) + O_2(g) \rightarrow 2\ Cl_2(g) + 2\ H_2O(l)$$

If all gases are measured under the same conditions, how many cubic decimetres of chlorine would be produced by combustion of 7.0 dm^3 of HCl?

5.15 Octane, C_8H_{18}, is one of the major constituents of gasoline.

a. Write a balanced equation for the reaction of octane with oxygen (the products are carbon dioxide and water vapor).
b. What volume of pure O_2, at 1.00 atm and 20°C, is required to burn one gram of octane?

5.16 A 0.136-g sample of an Al-Zn alloy when treated with HCl evolves 129 cm^3 of H_2, measured dry at 27°C and 1.00 atm. Calculate the % Al in the alloy (1 mol Al → 3/2 mol H_2; 1 mol Zn → 1 mol H_2).

5.17 A mixture of O_2, N_2, and CO_2 exerts a pressure of 93.3 kPa in a 10.0-dm^3 container at 30°C. Analysis shows that P_{O_2} is 26.7 kPa and that 5.00 g of CO_2 are present.

a. Find P_{CO_2} in the container.
b. Find P_{N_2} in the container.
c. Find the mole fraction of O_2.

5.18 The molar volume of O_2(l) is about 26 cm^3. Calculate the pressure required to compress one mole of O_2(g) to twice that volume at 25°C using

a. the Ideal Gas Law.
b. the van der Waals equation.

Would you expect V_{obs}/V_{ideal} to be greater or less than 1 under these conditions?

5.19 A gas mixture at 25°C and a total pressure of 1 atm contains 1.00 g of each of the following: H_2, He, N_2, CO_2. Arrange these gases in order of

a. increasing partial pressure.
b. increasing molecular speed.
c. increasing molecular energy.

5.32 Hydrogen gas is often produced by reaction of zinc with hydrochloric acid:

$$Zn(s) + 2\ H^+(aq) \rightarrow Zn^{2+}(aq) + H_2(g)$$

If the hydrogen is collected over water at 22°C and a total pressure of 98.6 kPa, how many grams of Zn would have to react to produce 2.50 dm^3 of wet H_2?

5.33 In the electrolysis of water to give H_2 and O_2, what would be the relative volumes produced of these two gases if both were collected at 24°C and 100 kPa? How might information of this sort have led Avogadro to propose his Law?

5.34 In a Volvo engine the maximum cylinder volume is about 0.500 dm^3. If air enters the cylinder at 50°C and one atmosphere, how many grams of octane, C_8H_{18}, should the fuel injection system send into the cylinder if it is to burn completely in that air when the spark plug fires at the end of the compression stroke? Assume air is 21 mol % O_2.

5.35 A 0.326-g sample of XH_2 reacts with water according to the equation

$$XH_2(s) + 2\ H_2O(l) \rightarrow X(OH)_2(s) + 2\ H_2(g)$$

The hydrogen evolved is found to have a volume when dry of 0.375 dm^3 at 21°C and one atmosphere. What is the atomic mass of X?

5.36 0.200 dm^3 of N_2 at 25°C and 100 kPa and 0.300 dm^3 of He at 125°C and 160 kPa are put in a 1.00-dm^3 container at 25°C. Calculate

a. the numbers of moles of N_2 and He.
b. the partial pressures of N_2 and He.
c. the total pressure in the container.

5.37 Using the van der Waals equation, find the molar volume of H_2O(g) at 100°C and 1.00 atm. (You'll have to find V by trial and error; use V_{ideal} as your first guess). Assuming the VDW equation is exact at this point, calculate the per cent deviation from ideal behavior of H_2O(g) under these conditions.

5.38 In an air sample at 25°C having the composition given in Problem 5.11, compare

a. the speeds of O_2 and N_2 molecules.
b. the molar energies of motion of O_2 and N_2.
c. the partial pressures of O_2 and N_2.

5.20 At what temperature will H_2 molecules have the same average speed as UF_6 molecules at 100°C? At what temperature will they have the same average energy of motion?

5.21 It took 68 s for a sample of N_2 to effuse down a capillary and 85 s for a gas that might have been either C_3H_8 or C_4H_{10} to effuse under the same conditions of T and P. Which gas was it?

5.22 For a mole of ideal gas, sketch graphs of:

a. P vs. V at constant T.
b. P vs. T at constant V.
c. u vs. T.
d. E_{trans} vs. P at constant T.

5.39 At what temperature will H_2 molecules have an average speed of 10.0 km/s? Estimate the speed of sound in hydrogen at this temperature.

5.40 If it takes 100 s for a sample of O_2 to effuse down a capillary, how long would it take for the same number of moles of SO_2 to effuse down the tube?

5.41 For an ideal gas, sketch graphs of

a. V vs. T at constant P, n.
b. P vs. n at constant V, T.
c. E_{trans} vs. T at constant P, n.
d. n vs. T at constant P, V.

*5.42 In order for a balloon to rise, the mass of the air it displaces must be equal to the sum of the masses of the balloon, the gas it contains, and the balloonist. If the balloon and balloonist together weigh 170 kg, what would the volume of a helium-filled balloon have to be if the rig were to go up at 25°C and one atmosphere? If the balloon were filled with hot air at 250°C, how big would it have to be? (Take the average molecular mass of air to be 28.8.)

*5.43 A thermometer on the dark side of a satellite on the way to the moon records a temperature of −270°C. The molecules in the vicinity of the thermometer have speeds corresponding to a temperature of over 1000°C, as calculated from Equation 5.13. Explain these observations.

*5.44 It takes 12 J to raise the temperature of one mole of any monatomic gas by 1°C when the gas is heated at constant volume. Can you suggest how this experimental observation could be used to test the validity of Equation 5.14?

*5.45 The escape velocity v_e from the surface of the moon is about 2.4×10^3 m/s. Unless $u \leq 0.20\, v_e$, all of a gas will escape in a billion years or less. If the temperature on the surface of the moon is about 25°C when the moon is in sunlight, would you expect to find any H_2 in the lunar atmosphere? O_2? Given that the atmospheric pressure on the moon is essentially zero, what can you say about the history of the moon over geologic time?

*5.46 In a classic experiment the density of carbon dioxide was carefully measured at 0°C at several low pressures, with the following results:

Pressure (atm)	*Density (g/dm³)*
1.000 00	1.976 76
0.666 67	1.314 85
0.333 33	0.655 96

Use the Ideal Gas Law to calculate the apparent molecular mass of carbon dioxide at each pressure. Use a hand calculator for the calculations, since the accuracy of the data exceeds that possible with a slide rule. $R = 8.3141$ kPa · dm³/(mol · K) and 0°C = 273.15 K. The differences between values of MM obtained are due to deviations of the gas from the ideal. Make a graph of $MM_{apparent}$ vs. pressure, and extrapolate the (straight) line to zero pressure, where the gas behaves ideally. The limiting value of $MM_{apparent}$ at zero pressure is the best value for the molecular mass of CO_2. Assuming the molecular formula for carbon dioxide and the atomic mass of oxygen, calculate the atomic mass of carbon from the molecular mass of CO_2 you obtained.

The approach employed in this problem is called the limiting density method for molecular mass determination. It was used extensively in the early part of this century to find molecular and atomic masses.

6

THE ELECTRONIC STRUCTURE OF ATOMS

The atomic theory as put forth by Dalton furnished the base that made possible the rapid development of chemistry in the eighteenth century. We have seen how this theory facilitates our understanding of chemical formulas and mass and energy relationships in chemical reactions. The kinetic theory is based on the atomic theory and allows us to relate the behavior of gases to the properties of their constituent molecules. We are now ready to extend the atomic theory in yet another direction, this time with a view toward explaining such properties as molecular geometry, atom combining ratios in compounds, and chemical reactivity. To deal with these areas, we must first attain some understanding of the electronic structure of atoms.

In Chapter 2 we presented some experimental and theoretical information about the general nature of atoms. You will recall that we believe each atom consists of a central, positively charged nucleus, surrounded by a group of negatively charged electrons. The nucleus is very small, about 10^{-5} nm in diameter, and includes just about all the mass of the atom. According to our model, the nucleus is made up of protons and neutrons in sufficient number to account for its mass and charge. The number of protons in the nucleus of an atom is characteristic of the element to which that atom belongs and is called the atomic number of that element.

Since electrons have charges equal in magnitude but opposite in sign to those of protons, a neutral atom of atomic number Z will have Z electrons outside its nucleus. These electrons move about in a spherical region roughly 0.2 to 0.3 nm in diameter. Atoms may gain or lose one or two electrons relatively easily, becoming charged particles called ions with chemical properties which differ greatly from those of the atoms from which they were derived.

6.1 SOME PROPERTIES OF ELECTRONS IN ATOMS AND MOLECULES

The behavior of electrons in atoms and molecules has been the subject of extensive research, both experimental and theoretical, during all this century. It must, however, be admitted that at present our knowledge of the detailed electronic structure of all but the simplest atoms is still incomplete. Much progress has been made on this problem, but much more work remains to be accomplished. In this book we shall present some features of the current model for electron arrangements in atoms and molecules.

The main obstacle to our understanding of the properties of chemical substances in terms of the electrons and nuclei of which they are composed is that small particles, such as atoms, molecules, nuclei, and particularly electrons, appear to obey different laws regarding energy and motion than do larger objects such as billiard balls and rotating bicycle wheels. Systems with which we are ordinarily concerned, with masses many, many times those of atoms and molecules, follow the laws of motion first formulated by Isaac Newton. These laws constitute that part of physics called classical mechanics. Small particles obey the laws of a somewhat different kind of mechanics, called quantum mechanics. (Actually it turns out that classical mechanics is, in a very real sense, a special case of quantum mechanics and is valid for all but those particles which have exceedingly small masses.)

A dust particle would obey classical mechanics

Quantum mechanics is part of a general theory, called the quantum theory, which had its beginnings early in this century. Like the atomic theory, the quantum theory has evolved considerably during its development; some of its original postulates have been retained, while others have been modified or discarded. It is at present the fundamental theory used to explain the behavior of electrons and other small particles. Some experiments which led to the quantum theory will be considered in the next section after we have stated and discussed three of the underlying principles of the theory that are of chemical interest.

Postulates of the Quantum Theory

1. Atoms and molecules can only exist in certain states, characterized by definite amounts of energy. When an atom or molecule changes its state, it must absorb or emit an amount of energy just sufficient to bring it to another state.

Atoms and molecules can, as we have seen, possess various kinds of energy. One form of energy of particular importance when considering atomic structure arises from the motion of electrons about the atomic nucleus and from the charge interactions among the electrons and between the electrons and the nucleus. This kind of energy is called *electronic energy*. Only certain values of electronic energy are allowed to an atom. When an atom goes from one allowed electronic state to another, it must absorb or emit just enough energy to bring its own energy to that of the final state.

Analogous considerations apply to the other forms of energy possessed by atoms, molecules, and other small particles. Translational energy of motion, rotational and vibrational energy, in addition to electronic energy, are subject to the limitations of the quantum theory.

The energy of systems that can exist only in discrete states is said to be *quantized*. A change in the energy level of such a system involves the absorption or emission of a definite amount ("quantum") of energy.

2. When atoms or molecules absorb or emit light in the process of changing their energies, the wavelength λ of the light is related to the magnitude of the energy change $|\Delta E|$ by the equation

$$|\Delta E| = hc/\lambda \tag{6.1}$$

where h is a physical constant, called Planck's constant, and c is the speed of light.

A ray of light can be considered to consist of *photons,* which appear to have some properties characteristic of particles. In particular, each photon of wavelength λ has an associated energy equal to hc/λ. The energy of a photon in joules can be easily found, given the values of Planck's constant, 6.626×10^{-34} J · s, the speed of light, 2.998×10^{8} m/s, and the wavelength of the photon in metres (see Example 6.1).

The energy lost by an atom or molecule in going from a higher to a lower energy state will equal the energy of the photon which is emitted during the transition. Similarly, to absorb a photon, an atom or molecule must, during the absorption, be able to make a jump between two energy levels separated by an amount equal to the energy of the absorbed photon. Total energy, therefore, is conserved in these transitions.

3. The allowed energy states of atoms and molecules can be described by sets of numbers called quantum numbers.

The mathematical solutions to problems regarding the energies of atoms and molecules as obtained by quantum mechanics usually result in sets of integral numbers, which serve to denote the allowed states and permit calculation of their energies. These numbers are called *quantum numbers*.

Associated with each electronic state of an atom is a group of quantum numbers that identify the state (see Section 6.5). In the usual model, the quantum numbers are associated with the individual electrons in the atom. Each electron is assigned a set of quantum numbers according to a set of rules. A statement of all the quantum numbers of all the electrons in an atom is used to designate the energy level, or quantum state, of the atom.

6.2 EXPERIMENTAL BASIS OF THE QUANTUM THEORY

Classical mechanics could not explain these phenomena

Probably the best way to become acquainted with the quantum theory is to examine some of the experiments and theoretical relations on which it is based. There are actually a great many experiments which are best explained by the theory; among them should be mentioned black body radiation, the photoelectric effect, atomic spectra, and the several kinds of molecular spectra. Though these are all important phenomena, we shall restrict our attention here to atomic spectra, postponing a discussion of the other experiments to more advanced courses of study in physics and chemistry. Atomic spectra bear directly on the problem of atomic structure and nicely illustrate an area of experiment in relation to the general quantum theory.

Atomic Spectra

When atoms are exposed to high energy from nearly any source, they tend to become excited and to give off energy in the form of radiation. The nature of the radiation emitted depends on the excitation which is used.

If we heat a metal in a furnace or in a flame, it will, depending on its temperature, give off visible light; at 1000°C it will look red, at 1500°C it will appear white. If the emitted light is examined with a spectroscope, a device which breaks up light into its component colors, the light is found to contain essentially all colors. More precisely, we would say that, over the region in which the metal radiates, its *spectrum is continuous,* containing light at all wavelengths.

Not all emitters of light radiate at all wavelengths. If we observe in a spectroscope the light emitted by sodium chloride when it is placed in a flame, say from a Bunsen burner, we see only a few bright lines, which indicate the few wavelengths at which sodium atoms, excited by the flame, are emitting light. In this case we are seeing the *atomic spectrum* of sodium, which, since it contains light at only specific wavelengths, is said to be *discrete*.

Atomic spectra are emitted when atoms are mildly excited; this can be accomplished by a flame, by a spark, or by electrons which have been accelerated by falling through a few volts of potential. Our common fluorescent lights and

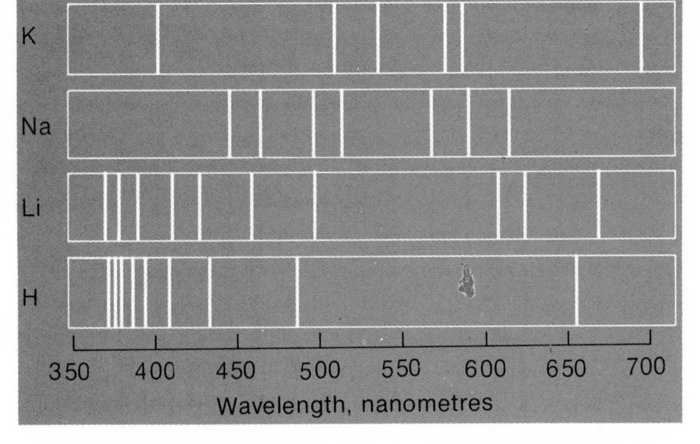

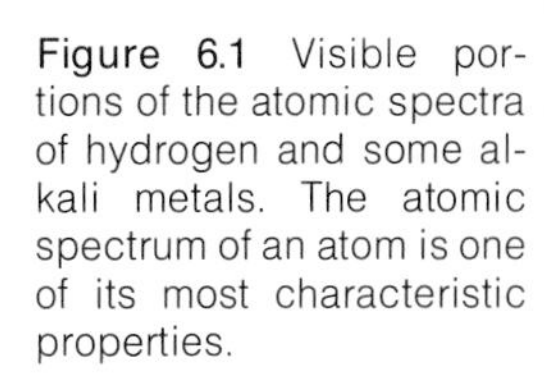
Figure 6.1 Visible portions of the atomic spectra of hydrogen and some alkali metals. The atomic spectrum of an atom is one of its most characteristic properties.

mercury vapor highway lights give off light of this sort, with components (or lines) which can be resolved with a spectroscope (Fig. 6.1).

As the figure suggests, every atom has its *own characteristic spectrum*. Kirchoff was first to recognize this fact, and Bunsen, his coworker in Germany, used it in his discovery of two elements, rubidium (1861) and cesium (1860). Spectral wavelengths can be measured very accurately ($\pm 1 \times 10^{-12}$ m) and detection of lines due to trace components of a sample is possible. These factors have made spectral analysis one of the most powerful tools of the analytical chemist.

The existence and character of atomic spectra are readily treated by the quantum theory. When a gas containing sodium atoms is heated to a high temperature or is exposed to fast-moving electrons, some of the sodium atoms will, through collisions, absorb energy and become excited, moving in the process to higher allowed electronic energy levels. Once excited, the atoms will be unstable and will tend to return to lower electronic energy levels, ultimately reaching the lowest energy level, the ground state of the atom. Under excitation the sodium gas will quickly reach a state of dynamic equilibrium in which some atoms are releasing energy at the same rate that it is being absorbed by others.

One of the ways in which electronically excited sodium atoms can lose energy is by radiating it as light. In a process in which an excited atom makes a transition to a lower energy level, a photon of light may be emitted. The wavelength λ of the photon is related to the magnitude of the change in energy $|\Delta E|$ of the atom by Postulate 2 of the quantum theory:

$$|\Delta E| = \frac{hc}{\lambda} = E_{photon} \qquad (6.2)$$

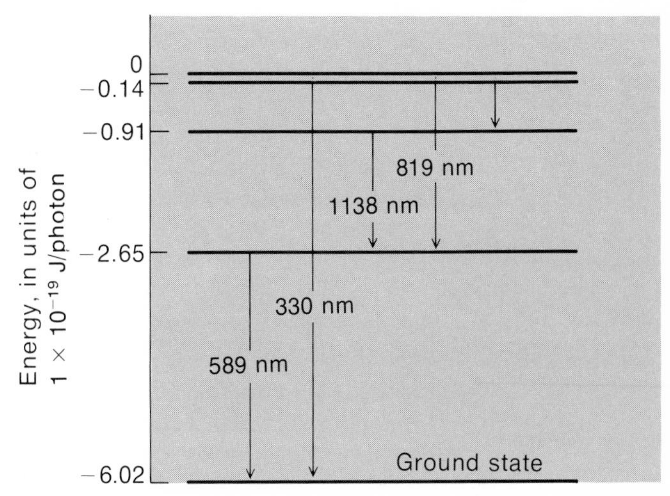

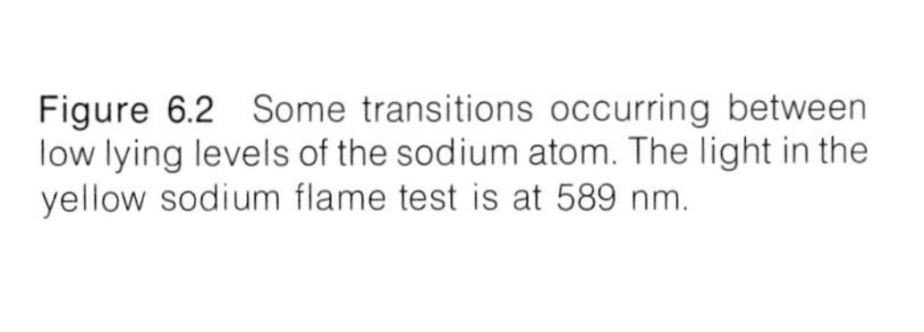
Figure 6.2 Some transitions occurring between low lying levels of the sodium atom. The light in the yellow sodium flame test is at 589 nm.

This equation was first proposed by Einstein in 1905, in the very early days of the quantum theory.

Atomic spectra, then, are believed to consist of the radiation resulting from transitions made by electronically excited atoms from higher to lower allowed energy levels. In Figure 6.2 we have arranged, in somewhat simplified form, some of the lower electronic energy levels of the sodium atom, along with the transitions that are observed between those levels. Each transition gives rise to a line in the atomic spectrum of sodium, with a wavelength given by Equation 6.2. One of the lines in the sodium spectrum in Figure 6.1 is the result of a transition shown in Figure 6.2.

Example 6.1 Excited sodium atoms may emit radiation having a wavelength of 589.0 nm.

a. What is the energy in joules of the photons in this radiation?
b. What is the energy of a mole of these photons in kilojoules?

Solution

a. The energy of a photon is given by Equation 6.2:

$$|\Delta E_{atom}| = E_{photon} = \frac{hc}{\lambda}$$

Given that Planck's constant, h, equals 6.626×10^{-34} J · s, and the speed of light, c, is 2.998×10^8 m/s, the wavelength, λ, must be expressed in metres to obtain the photon energy in joules. Since 1 nm = 1×10^{-9} m,

$$\lambda = 589.0 \text{ nm} \times \frac{1 \times 10^{-9} \text{ m}}{1 \text{ nm}} = 5.890 \times 10^{-7} \text{ m}$$

You can always relate the energy of a photon to its wavelength by the approach used here

$$E_{photon} = \frac{6.626 \times 10^{-34} \text{ J} \cdot \text{s} \times 2.998 \times 10^8 \text{ m/s}}{5.890 \times 10^{-7} \text{ m}} = 3.37 \times 10^{-19} \text{ J}$$

b. Recalling that 1 mol = 6.02×10^{23} particles,

$$E_{mole\ of\ photons} = \frac{3.37 \times 10^{-19} \text{ J}}{1 \text{ photon}} \times \frac{6.02 \times 10^{23} \text{ photons}}{1 \text{ mol}}$$

$$= 2.03 \times 10^5 \text{ J/mol}$$

And, since 1 kJ = 10^3 J

$$E_{photons} = 2.03 \times 10^5 \frac{\text{J}}{\text{mol}} \times \frac{1 \text{ kJ}}{10^3 \text{ J}} = 203 \text{ kJ/mol}$$

In ordinary terms, the energy of a photon is very small indeed. Although it is occasionally helpful to work with the energy of a single photon, we will ordinarily express energy in systems of this sort in kilojoules per mole of particles, since then the energies correlate more easily with those observed in thermochemistry. The following conversion factor may be used in relating energy per particle to energy per mole:

$$1 \text{ J/particle} = 6.02 \times 10^{20} \text{ kJ/mol} \tag{6.3}$$

It is of interest to compare the energy changes which occur when atoms emit their characteristic spectra to the energy changes involved in chemical reactions. In Example 6.1 a mole of sodium atoms undergoing the transition would lose about 203 kJ of energy. This is about half the amount of energy that would be evolved if a mole of sodium metal reacted with chlorine to form NaCl (see

Table 4.1). *Energy changes involved in the emission of atomic spectra are, mole for mole, of the same order of magnitude as energy changes observed in chemical reactions.*

REGULARITIES IN ATOMIC SPECTRA. Beginning about 1880 it was recognized that there was a certain amount of order in atomic spectra. It was found possible in some cases to sort out the lines in a spectrum into series, with lines being assigned to a given series on the basis of wavelength, intensity, and breadth. In the atomic spectrum of sodium we still speak of the principal (intense), sharp, and diffuse series.

Certainly the simplest of all atomic spectra is that of hydrogen (Fig. 6.1). Even a casual examination of its visible spectrum reveals a regular progression of lines. The wavelengths of the first five of these lines are given in Table 6.1.

TABLE 6.1 WAVELENGTHS OF THE VISIBLE LINES IN THE ATOMIC SPECTRUM OF HYDROGEN

Line	1	2	3	4	5
Wavelength (in nm)	656.279	486.133	434.048	410.175	397.009

Balmer, in 1885, suspected that there was a mathematical relationship between the wavelengths of these lines. By a graphical procedure (see Fig. 6.3), he obtained the following expression for the wavelengths of all of the lines listed in Table 6.1:

Johann Balmer was a Swiss high school teacher and a devotee of numerology

$$\lambda = 364.600\ n^2/(n^2 - 4) \qquad (6.4)$$

in which λ is the wavelength in nm and n is an integer which has the values 3, 4, 5, for the first, second, third, . . . lines (i.e., $n = q + 2$, where q is the number of the line).

Following Balmer's discovery, it was found that there were several other lines in the series that fit his equation; all the wavelengths which fall on the line

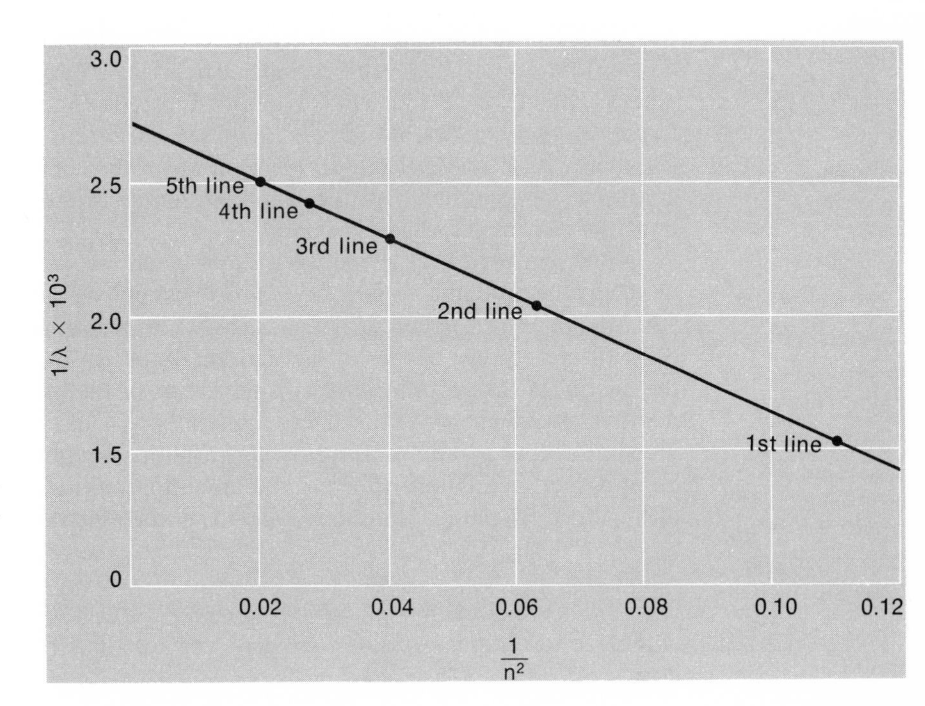

Figure 6.3 Balmer discovered, probably by trial and error, that for the lines in the visible spectrum of H, a plot of $1/\lambda$ vs. $1/n^2$ is linear. The equation of the straight line may be written:

$$\frac{1}{\lambda} = \frac{1}{364.6}\left(1 - \frac{4}{n^2}\right) = \frac{1}{364.6}\left(\frac{n^2 - 4}{n^2}\right)$$

Taking reciprocals, we obtain Equation 6.4 above.

There are at least a dozen lines in the Balmer series

in Figure 6.3 comprise what is now called the **Balmer series** for hydrogen. Those lines follow Equation 6.4 so well that, if we calculate their wavelengths by that equation, we find that they agree with observed values to about one part in one million, making Equation 6.4 one of the most exact in all of physical science.

Example 6.2 Find the wavelength of the line in the Balmer series for which n = 4 in Equation 6.4 (second line listed in Table 6.1).

Solution

$$\lambda = 364.600\left(\frac{4^2}{4^2 - 4}\right) = 486.133 \text{ nm (correct to at least one part in 500 000)}$$

As you might expect, Balmer's discovery touched off a great deal of research aimed at finding other relationships between lines in spectral series for other atoms. In spite of a tremendous amount of work, to this day no equations have been found for atoms other than hydrogen that approach the Balmer equation in exactness.

At the time Balmer made his discovery scientists had no idea as to why the hydrogen atomic spectrum should have the properties it did. The answer to that question came about thirty years later.

6.3 THE BOHR THEORY OF THE HYDROGEN ATOM

By the beginning of this century knowledge of atomic structure had advanced to the point where scientists could begin to speculate on the way in which positive and negative charges were arranged in atoms. Part of the problem was solved when Rutherford demonstrated the existence of atomic nuclei (Chapter 2). It was only two years later, in 1913, that Niels Bohr, a young Danish physicist, presented a theory for the structure of the hydrogen atom that added greatly to our ideas regarding the behavior of electrons in atoms.

Bohr based his approach on Rutherford's nuclear atom and on Planck's suggestion that atoms and other small particles can possess only certain definite amounts of energy. He also had available Einstein's equation relating wavelength to photon energy and, through the Balmer series, accurate information relating to the energy levels in the H atom. His goal was to formulate a model for the hydrogen atom, compatible with the facts of atomic structure, which explained the equation that Balmer had discovered.

Bohr assumed that a hydrogen atom consisted of a central proton, around which an electron moved in a circular orbit, much as the earth moves around the sun. By relating the force of attraction of the proton for the electron to the centrifugal force due to the circular motion of the electron, he was able to express the energy of the atom in terms of the radius of the electron's orbit; this was a purely classical analysis, based on Coulomb's Law of electrostatic attraction and Newton's Laws of motion. Bohr then incorporated quantum theory into his model by imposing, boldly and arbitrarily, a condition on the angular momentum, mvr, of the electron, namely that it was given by the equation

$$mvr = \mathrm{n}h/2\pi$$

where m is the electron's mass, v its speed, r the radius of the orbit, n a quantum number which can take on any positive integral value (i.e., 1, 2, 3, . . .) and h is Planck's constant.

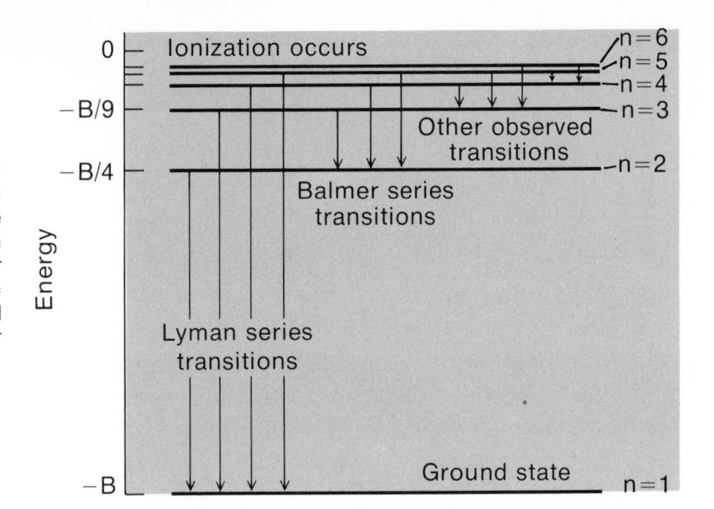

Figure 6.4 Some energy levels and transitions in the hydrogen atom. Lines in the Balmer series arise from transitions from upper levels (n > 2) to the n = 2 level. In the so-called Lyman series, the lower level is n = 1.

Bohr found that his quantum condition restricted the allowed energies of the hydrogen atom to those values given by the equation

$$E = -B/\mathrm{n}^2 \tag{6.5}$$

where n was the quantum number and B was obtainable directly from the theory, from which it was found to have the value 2.179×10^{-18} J.

In Figure 6.4 we have shown the energy levels of the hydrogen atom as predicted by Bohr's theory. In setting up his model, Bohr took the zero of energy to be at the point where the proton and electron were completely separated—that is, in the state where the atom was ionized. As the electron approaches the nucleus, the atom becomes more stable, so *its energy lies below zero and is negative in all its allowed states.* The lowest possible energy is in the state where n equals one and the energy, by Equation 6.5, is $-B$. This level is called the **ground state** of the atom, and is the state in which the atom is ordinarily found.

When n equals two, the energy, by Equation 6.5, is $-B/4$, considerably higher (i.e., less negative) than that of the ground state. An H atom in this state is said to be excited, and will tend to quickly make a transition to the ground state. As the quantum number n increases further, the energy of the atom increases (becomes less negative), approaching zero as n becomes very large.

The triumph of Bohr's theory was its ability to deal with the spectral properties of the H atom. By the quantum theory, the lines in the hydrogen spectrum arise from transitions between energy levels of the atom, as shown in Figure 6.4. The wavelengths of these lines can be obtained from Einstein's Equation 6.2. Bohr recognized that the lines in the Balmer series all arose from transitions to the n = 2 level and made his calculations for that series accordingly.

Example 6.3 Find the wavelength in nm of the line in the Balmer series that is associated with the n = 4 → n = 2 transition (second line in the series).

Solution By Bohr's theory, $E_4 = -B/16$ and $E_2 = -B/4$

In joules, $E_4 = -(2.179 \times 10^{-18}\ \text{J})/16 = -1.362 \times 10^{-19}$ J
$E_2 = -(2.179 \times 10^{-18}\ \text{J})/4 = -5.448 \times 10^{-19}$ J

The energy of the photon equals the change in energy $\Delta E = E_4 - E_2$:

$$E_{photon} = (-1.362 + 5.448) \times 10^{-19}\ \text{J} = 4.086 \times 10^{-19}\ \text{J}$$

By Einstein's equation, $E_{photon} = hc/\lambda$, so $\lambda = hc/E_{photon}$

$$\lambda = \frac{6.626 \times 10^{-34}\ \text{J}\cdot\text{s} \times 2.998 \times 10^{8}\ \text{m/s}}{4.086 \times 10^{-19}\ \text{J}} = 4.861 \times 10^{-7}\ \text{m}$$

$$= 4.861 \times 10^{-7}\ \text{m} \times \frac{1 \times 10^{9}\ \text{nm}}{1\ \text{m}} = 486.1\ \text{nm}$$

(Compare this value with that of the second line in Table 6.1.)

The calculated values of the wavelengths of *all* of the lines in the Balmer series agree very well with those that are observed. Indeed, *if you use Bohr's theory to derive an expression for these wavelengths as a function of n, you essentially obtain Equation 6.4, which Balmer found empirically*. Bohr went on to predict that there would be other spectral series for the H atom, in particular one for which the lower state for all transitions was the n = 1 level. This series was subsequently discovered by Lyman, and was found to contain wavelengths (all in the ultraviolet region) which agree almost perfectly with those predicted by Bohr. As you might expect, correlations of this sort between theory and experiment lent great support to Bohr's approach.

A corollary prediction made possible by the theory is the ionization energy of the H atom. To ionize the normal atom one must give it enough energy to raise it from the ground state (n = 1, $E = -B$) to the zero of energy, where the electron is completely removed from the proton. This means that the atom must receive an amount of energy equal to B, or about 2.179×10^{-18} J. Using Equation 6.3 we can convert this value into energy per mole of H atoms:

$$E_{ionization} = \frac{2.179 \times 10^{-18}\ \text{J}}{1\ \text{atom}} \times \frac{6.02 \times 10^{20}\ \text{kJ/mol}}{1\ \text{J/atom}} = 1310\ \text{kJ/mol}$$

The observed value is about 1318 kJ/mol.

There are two other results from Bohr theory that might be mentioned. Both the radius of the electronic orbit and the speed of the electron are predicted to be quantized and can be expressed in terms of the quantum number n. The equations that are obtained from the theory are:

$$r\ (\text{in nm}) = 0.0529\ \text{n}^2 \qquad v = \frac{2.18 \times 10^{7}}{\text{n}}\ \text{m/s}$$

The diameter of an H atom in its ground state, n = 1, is predicted to be about 0.106 nm, and while this value cannot be confirmed very well experimentally it does seem reasonable in light of internuclear distances as obtained from, say, x-ray diffraction analysis. The theory predicts that the size of the atom increases rapidly with increasing n. The speed of the electron in the orbit in the ground state turns out to be enormous, about 7 per cent of the speed of light; this value cannot be tested experimentally.

6.4 WAVES AND PARTICLES

Bohr's theory for the structure of the hydrogen atom (and all other one-electron species such as He^{+} and Li^{2+}) was indeed highly successful, and scientists of the day must have thought that they were on the verge of being able to predict the allowed energy levels of all the atoms. However, it soon became obvious that the extension of Bohr's ideas to atoms with two or more electrons gave, at best, only qualitative agreement with experiment. For example, if you apply Bohr's theory to the helium atom the errors in calculated energy and wavelength are of the order of 5 per cent instead of the less than 0.1 per cent error obtained with hydrogen. There appeared to be no way that the theory could be modified to make it work well with helium or other atoms. It also failed to predict properly the energy levels of some other simple systems, such as vibrating or rotating diatomic molecules. These limitations of the Bohr theory ultimately led to an extensive search, beginning about 1920, for other approaches to the problem of the energies of electrons in atoms and molecules.

In 1924 a young French physicist, Louis de Broglie, reasoning as physicists do, sometimes with strikingly successful results, suggested that if light rays have particle properties, then perhaps particles may, under some circumstances, exhibit wave properties. By an argument which is beyond the level of this text, de Broglie predicted the wavelength associated with a particle of mass m moving at velocity v to be given by the relation

$$\lambda = h/mv \tag{6.6}$$

where h is, as you might now expect, Planck's constant.

Within a few years Davisson and Germer, working at the Bell Telephone Laboratories, tested de Broglie's prediction by diffracting electrons of known energy from crystals. They established that a beam of electrons did indeed have wave properties and that the *wavelength to be associated with electrons of known velocity was exactly that predicted by the de Broglie equation.*

An electron which falls through a potential of 20 V has a wavelength of about 3×10^{-10} m

Some Wave Properties of Particles

The de Broglie relation is a key to the wave properties of small particles, much as the Einstein equation is a key to the particle properties of light. We will now consider one problem involving de Broglie-type waves, so as to show you the kinds of waves that appear to be associated with small particles, how quantum numbers arise in the wave treatment, and how the allowed energy levels of a system can be found. We shall then apply the solution obtained to a few examples to illustrate how the observed properties of small particles are correlated with their masses and locations.

The system we will work with consists simply of a particle confined to move in one dimension between two impenetrable walls. We assume that the particle has only kinetic energy and moves across the space allowed to it at a constant speed, v, bouncing elastically every time it hits a wall (Fig. 6.5). By the classical relations of mechanics, the energy ϵ and the momentum are given by the relations

$$\epsilon = \tfrac{1}{2} mv^2 \qquad \text{momentum} = mv \tag{6.7}$$

The de Broglie relation states that there must be a wave associated with the particle. Our theory tells us that the wave must just fit into the space allowed to

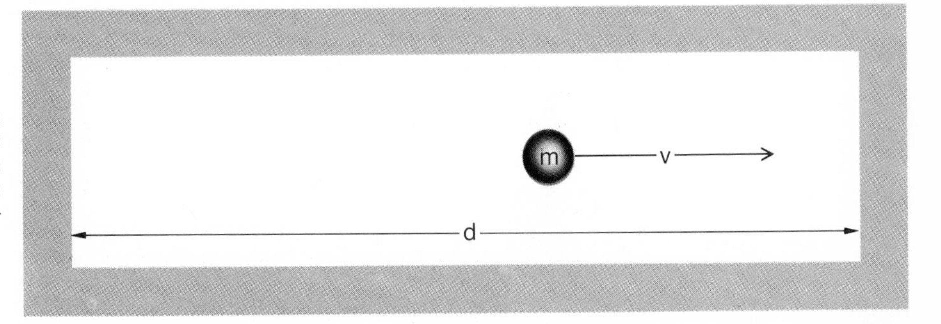

Figure 6.5 Particle in a one-dimensional box. The particle is confined to move in a straight line between the walls, maintaining constant energy.

the particle. The wave can have finite amplitudes only inside the system; the amplitude at the walls must be zero. Examples of some waves which meet these conditions are shown in Figure 6.6.

Mathematical expressions called wave functions can be written which give the amplitude of each wave as a function of position; the waves of Figure 6.6 result from plotting the amplitude of the wave function against position in the container. Wave functions are often given the symbol ψ, and, as you can see in Figure 6.6, they are sinusoidal for the particle in the system we are examining.

According to current theory, the square of the value of the wave function at any point is equal to the relative probability of finding the particle at that point. If we could examine experimentally a system which had a wave function like that in Figure 6.6a, we would be most likely to find the particle in the middle of the box and would essentially never be able to detect it at the wall! A strange situation indeed, but one which, insofar as it can be tested, is consistent with the behavior of such systems.

When we treat this problem by classical methods we find that we can, if we wish, predict the path a single particle would follow, given constant speed and perfectly reflecting walls. When we use the wave approach, all we can hope to do in this regard is determine the waves to be associated with a particle, and so predict the likelihood of finding it at any given point in space. Bohr's theory was classical in the sense that it allowed one to predict the path, or orbit, of the electron as well as its speed. Apparently this is not possible for small particles; their behavior seems somewhat less defined than that of classical systems, and such properties as position and speed have associated probabilities, or uncertainties, that a theory like Bohr's could not handle.

Although the paths of particles are not obtainable by the wave approach,

Note that for n = 1, $\lambda/2 = d$; for n = 3, $3\lambda/2 = d$; for n = 6, $3\lambda = d$. In general, $n\lambda/2 = d$ (Equation 6.8)

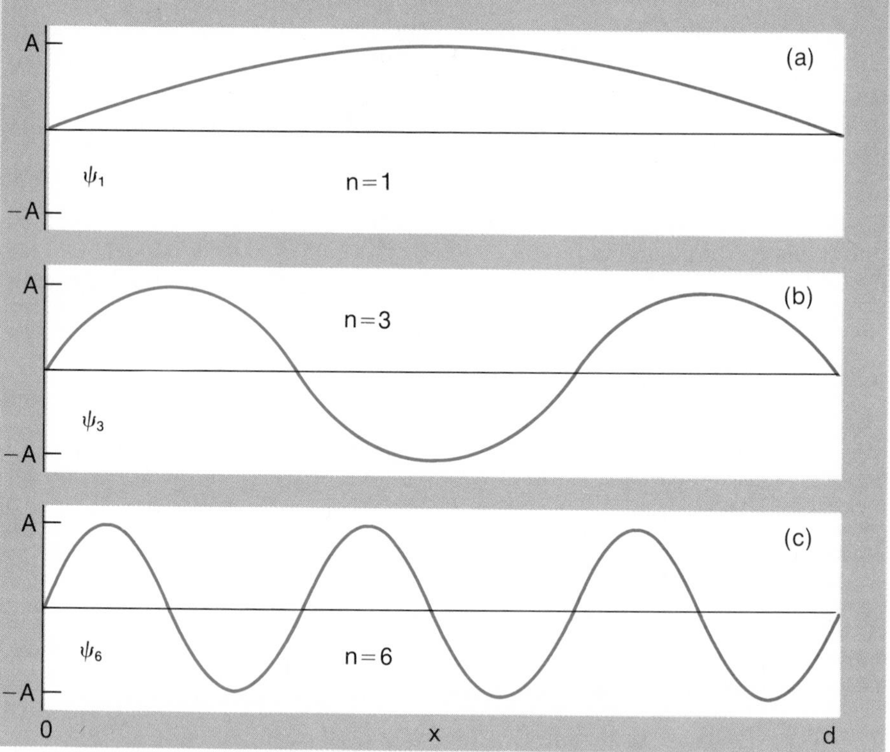

Figure 6.6 Some wave functions of a particle in a one-dimensional box. According to the theory, the probability of finding the particle at a point x is proportional to the square of the amplitude of the wave at that point. (The amplitude is the vertical distance from the base line.)

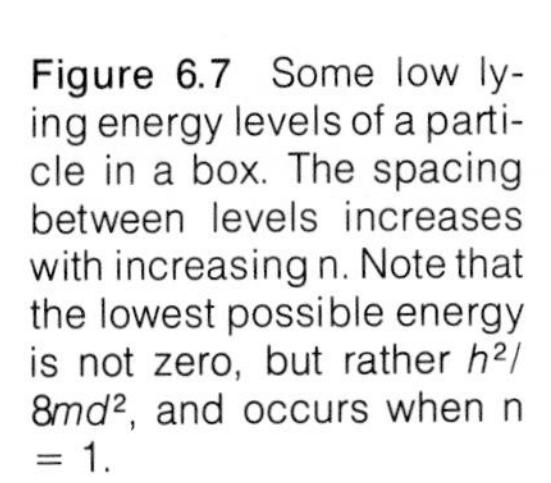

Figure 6.7 Some low lying energy levels of a particle in a box. The spacing between levels increases with increasing n. Note that the lowest possible energy is not zero, but rather $h^2/8md^2$, and occurs when n = 1.

their energies are. From the properties of the waves in Figure 6.6, we can see that for this system there are many waves that satisfy the condition that the amplitude be zero at the walls. The wavelengths, λ, of allowed waves all can be seen to satisfy the relation

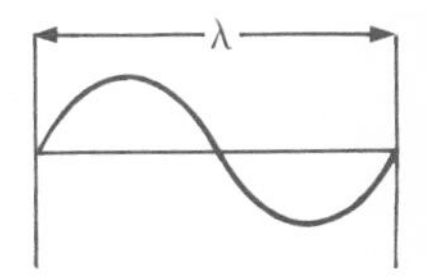

$$\lambda = \frac{2d}{\text{n}} \tag{6.8}$$

where d is the distance between the walls and n is an integer equal to 1, 2, 3, ... The number n is the quantum number for this system and arises here naturally as a result of the condition that the waves must fit properly in the space allowed to the particle.

By combining de Broglie's Equation 6.6 with the values of λ in Equation 6.8, we obtain

$$\lambda = \frac{h}{mv} = \frac{2d}{\text{n}} \quad \text{or} \quad mv = \frac{\text{n}h}{2d} \tag{6.9}$$

Equation 6.9 effectively imposes a quantum condition on the momentum mv of the particle, which limits its energy ϵ to certain allowed values:

$$\epsilon = (\tfrac{1}{2})mv^2 = \frac{1}{2m}(mv)^2 = \frac{1}{2m}\left(\frac{\text{n}h}{2d}\right)^2 = \frac{\text{n}^2h^2}{8md^2} \tag{6.10}$$

This equation tells us that a particle, of any kind, confined to move in a region of any length, will have only certain energies allowed to it. Those energies depend, as indicated by Equation 6.10, on the mass m of the particle, the interval d through which it can move, and the quantum number n. The energy levels available to such a particle are shown in Figure 6.7. For each allowed energy there will be an associated value of the quantum number, n, and a wave describing a wave function, ψ_n, with n loops, analogous to ψ_3 in Figure 6.6. The implications of Equation 6.10 with regard to the behavior of electrons, atoms, molecules, and matter in general are surprisingly far-reaching and we shall briefly explore a few of them.

Implications of the de Broglie Relation

Using Equation 6.10 one can calculate the energies available to essentially any particle confined to a region of any length. There are several features of the equation that should be recognized. The first is that a confined particle cannot have zero energy, since n cannot be zero if a wave is to be present. Since n can equal 1, 2, 3 . . . , the particle may take on many energies; of these, the lowest value will be for n = 1.

Secondly, the lowest energy available to a particle depends markedly on its mass and on the size of the space to which it is confined. This energy varies inversely with the mass; every time the mass is decreased by a factor of two, the energy is doubled. The minimum energy also varies inversely with the square of the length through which the particle can move; cutting this length in half will always quadruple the lowest energy the particle can have. This means that particles of low mass such as electrons confined to small regions, like atoms or molecules, have very much higher minimum energies than relatively heavy particles such as gas molecules kept in a rather large space like a room. It turns out that these two particular systems are important to the argument, so let us pursue their properties further.

We first apply Equation 6.10 to obtain the minimum energy (n = 1) of an electron ($m = 9.1 \times 10^{-31}$ kg) in a hydrogen atom. As a very approximate model, we can assume that an electron in a hydrogen atom behaves as a particle confined to a box of the order of 0.1 nm long ($d = 1 \times 10^{-10}$ m). Substituting into Equation 6.10:

$$\epsilon_{min} = \frac{n^2h^2}{8md^2} = \frac{1^2(6.626 \times 10^{-34})^2}{8(9.1 \times 10^{-31})(1 \times 10^{-10})^2} = 6.0 \times 10^{-18}\ \text{J} \qquad (6.11)$$

For a nitrogen molecule moving about in a room, the situation is quite different. In the first place, the N_2 molecule ($m = 4.7 \times 10^{-26}$ kg) is about 50 000 times as heavy as an electron. Even more important, taking a typical room to be about three metres across (3×10^9 nm), we see that the space available to the N_2 molecule in a room is 3×10^{10} as great as that for an electron restricted to a hydrogen atom. Since both m and d appear in the denominator of Equation 6.10, it follows that the minimum energy of an N_2 molecule wandering about in a room must be much smaller than that of an electron in a hydrogen atom. Calculation confirms this conclusion:

$$\epsilon_{min} = \frac{n^2h^2}{8md^2} = \frac{1^2(6.626 \times 10^{-34})^2}{8(4.7 \times 10^{-26})(3)^2} = 1.3 \times 10^{-43}\ \text{J} \qquad (6.12)$$

Clearly, an N_2 molecule in the translational ground state has an energy which is only a tiny fraction of that for an electron in its lowest energy state in a hydrogen atom or, indeed, in any atom or simple molecule.

In Figure 6.8 the minimum energies of several systems, including the two just discussed, are indicated. Shown as a heavy horizontal arrow in the figure is a very important quantity, the *average kinetic energy* of a gas molecule at 25°C.

Notice from the figure that the *minimum* translational energy of an N_2 molecule is much *smaller,* by a factor of about *10^{23}*, than that of an *average* N_2 molecule. This means that a nitrogen molecule in its ground state would of necessity gain energy, and increase its quantum number, by colliding with other molecules. As a result of such collisions, the N_2 molecules in a gas arrive at an equilibrium state in which the energy of the gas is shared among the molecules according to the Maxwellian distribution described in Chapter 5. The molecules populate energy states consistent with that distribution and, since most of the molecules have energies fairly near the average, they are in states with very large quantum numbers. In general, we find that when the particles in a system have a minimum energy which is much lower than that of an average gas molecule, those particles,

A typical N_2 molecule in the gas will have a quantum number of about 10^{11}!

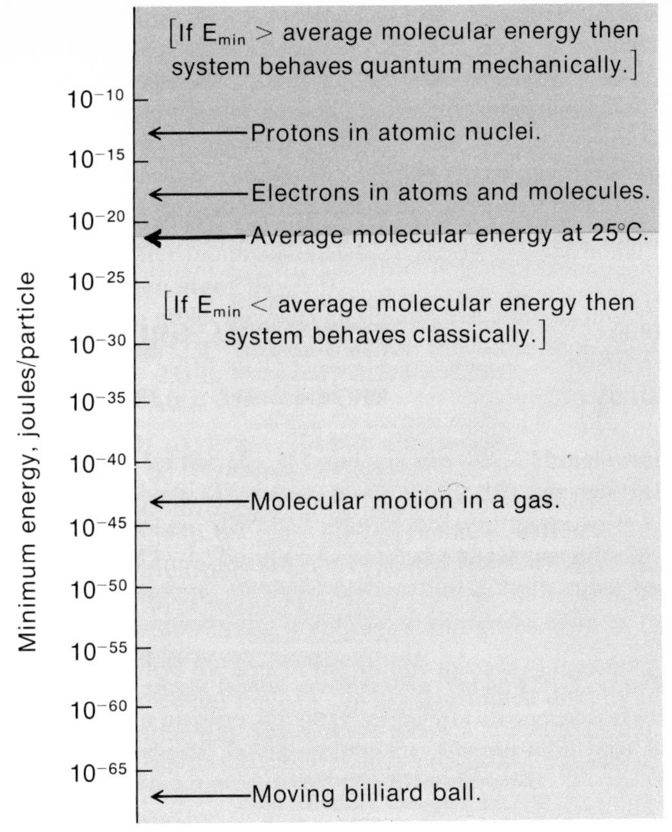

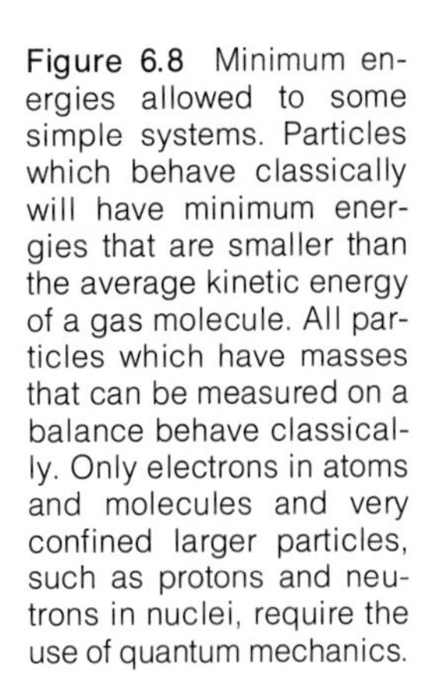
Figure 6.8 Minimum energies allowed to some simple systems. Particles which behave classically will have minimum energies that are smaller than the average kinetic energy of a gas molecule. All particles which have masses that can be measured on a balance behave classically. Only electrons in atoms and molecules and very confined larger particles, such as protons and neutrons in nuclei, require the use of quantum mechanics.

The average energy of a gas molecule at 25°C is 6.2×10^{-21} J

like N_2 molecules, will exist in states with high quantum numbers. It is an experimental fact that the particles in such systems behave classically, obeying Newton's laws of motion.

In contrast, we see in Figure 6.8 that the *minimum* energy of an electron in a hydrogen atom is about *1000 times larger* than the energy of an *average* gas molecule. This means that the electrons in H atoms, or H_2 molecules, in a gas could not gain energy as a result of molecular collisions, since the energies involved in such collisions are simply much too small to excite electrons to higher energy states. The net result of this situation is that electrons in atoms and molecules at ordinary temperatures are found in their *lowest possible energy states.* There is essentially no distribution of electrons among a large number of energy states. Systems of this sort, where the particles are of necessity in their lowest possible energy states, are found to obey the laws of quantum mechanics.

As we can deduce from Figure 6.8, protons in atomic nuclei resemble electrons in their behavior, while billiard balls, or professors meandering about in their laboratories, follow the same laws of motion as nitrogen molecules. In general, relatively heavy particles that have a bit of room to move about in can be treated by classical mechanics. Quantum mechanics exists because it is needed if one is to predict the behavior of light particles confined to tiny, submicroscopic regions.

The de Broglie relation can be applied rather easily to small systems in which the energy of the particle can be described, as in the case of the particle in a box, in terms only of its speed. There are many systems, including those involving electrons in atoms and molecules, in which the energy of a particle is partly due to its speed and partly due to its position. For such systems it is not possible to apply the de Broglie equation except as an approximation, and we are forced to resort to a more sophisticated approach if we are to obtain exact solutions.

Even before the experimental proof of de Broglie's theory was established, Erwin Schrödinger had applied de Broglie's ideas in a somewhat different way than we have done. He developed in 1926 another equation, now called the *Schrödinger wave equation,* which allowed him to determine the energy levels and wave properties of the hydrogen atom, the particle in the box, and a few other relatively simple systems. Unlike the Bohr theory, the wave equation appears to be applicable both to atoms other than hydrogen and to molecules. Unfortunately the extension of the wave equation to these other systems has proved to be much more difficult in fact than in principle, since the form of the equation for such systems is, in general, so complicated mathematically as to make it insoluble. However, in the relatively few cases in which a satisfactory solution has been possible, the results have agreed essentially perfectly with experiment. This has led scientists to believe that the approach to atomic and molecular properties through the Schrödinger wave equation is a correct although complex one.

So far we have not been able to solve the wave equation exactly for any 3 electron system; it begins to look like we never will

Although we shall not deal with any mathematical problems using the Schrödinger wave equation, we write it down so that you may have some idea of its nature. For a one-particle system, the equation takes the form

$$\frac{\partial^2\psi}{\partial x^2} + \frac{\partial^2\psi}{\partial y^2} + \frac{\partial^2\psi}{\partial z^2} + \frac{8\pi^2 m}{h^2}(E - V)\psi = 0 \tag{6.13}$$

where m is the mass of the particle, E and V are its total and potential energies, respectively, h is Planck's constant, and the first three terms are partial derivatives with respect to x, y and z, of ψ, the wave function to be associated with the particle.

The wave equation is a formidable mathematical relation, as you will probably agree. The quantum mechanical problem is to solve the equation for the wave functions, ψ, allowed to the system. In general, for a given problem there are a great many wave functions that satisfy Equation 6.13. In Figure 6.6 we sketched some of these functions for the one-dimensional particle in a box problem. Those functions were obtained by the rigorous solution of the wave equation for that problem. Similar wave functions are found in the exact solution to the H atom problem and to the other problems which have been treated exactly.

As we noted, the interpretation of the wave function in a given state is that ψ^2 at any point in space is proportional to the probability of finding the particle

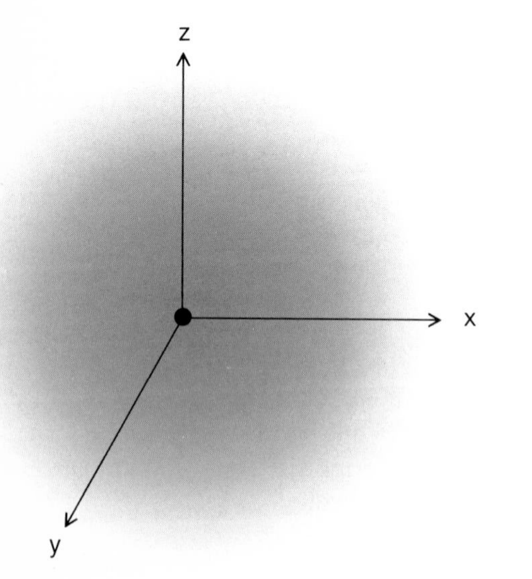

Figure 6.9 Electron cloud surrounding an atomic nucleus. The charge density drops off rapidly but smoothly as distance from the nucleus increases.

at that point. In problems involving electrons, we can interpret the value of ψ^2 as being proportional to the electric charge density at the point. "Electron cloud" diagrams showing charge density in atoms and molecules are frequently drawn on this basis. In Figure 6.9 we have drawn schematically the "electron cloud" surrounding the nucleus of a typical atom.

6.5 ELECTRON ARRANGEMENTS IN ATOMS

Quantum Numbers of Electrons

Schrödinger found that the electron in the H atom could be described by three quantum numbers, which are now called **n**, ℓ, and m_ℓ. There are three such numbers because the electron requires three coordinates to determine its motion. The quantum numbers of the electron establish its state, fixing both its energy and wave function. These quantum numbers are also used in describing other atoms, where they furnish similar but less quantitative information about the electrons. We shall now discuss the quantum numbers of electrons as they are used with atoms in general.

FIRST QUANTUM NUMBER, **n**. The first or *principal* quantum number, given the symbol **n**, is of primary importance in determining the energy of an electron. For the H atom, the energy is completely fixed by the value of **n**:

$$E = -\frac{B}{\mathbf{n}^2}$$

In other atoms, in which there are more electrons, the energy of each electron is dependent mainly (but not completely) on the value of the principal quantum number. As the quantum number increases, the energy of the electron increases, and the average distance of the electron cloud from the nucleus also increases. The principal quantum number is always integral and can take on the values

$$\mathbf{n} = 1, 2, 3, 4, 5, \ldots, \text{ but not } 0$$

In an atom, electrons having the same value of **n** move about in roughly the same region and are said to be in the same principal *level* or *shell.*

SECOND QUANTUM NUMBER, ℓ. Each level of electrons in an atom includes one or more *sublevels* or subshells. The sublevels are denoted by the second quantum number, ℓ. The general geometric shape of the electron cloud associated with the electron is determined by ℓ. The quantum number ℓ for an electron is related to the quantum number **n**; for a given value of **n**, ℓ is limited to the values:

$$\ell = 0, 1, 2, \ldots (\mathbf{n} - 1)$$

An electron with a quantum number **n** equal to 1 will have of necessity an ℓ quantum number equal to 0; this means that in the **n** = 1 level, there will be only one sublevel, in which ℓ equals 0. Electrons in the second level, where **n** equals 2, can have an ℓ quantum number of either 0 or 1, so that there are two sublevels in the **n** = 2 level, denoted by $\ell = 0$ and $\ell = 1$. Similarly, in the **n** = 3 level, electrons can have ℓ values of 0, 1, or 2. In general, in the **n**th level, there will be **n** sublevels, with ℓ quantum numbers equal to 0, 1, 2, . . . , **n** − 1.

We find from experience that electrons in sublevels within a given principal level will have energies which increase slightly with increasing ℓ value. In the $\mathbf{n} =$ 3 level, an electron with ℓ equals 0 will have less energy than one with an ℓ value of 1. The most energetic electrons in the third level will be those in the sublevel in which ℓ equals 2.

THIRD QUANTUM NUMBER, $\mathbf{m}_\ell$. Each sublevel contains one or more *orbitals,* each of which is designated by the third quantum number, $\mathbf{m}_\ell$. This quantum number is associated with the orientation of the electron cloud with respect to a given direction, usually one which is imposed on the atom by a strong magnetic field. For a given value of ℓ, $\mathbf{m}_\ell$ can have any integral value between ℓ and $-\ell$. That is:

$$\mathbf{m}_\ell = \ell, \ \ell - 1, \ldots 0, -1, \ldots, -\ell$$

Each electron in an atom will be in an orbital within a sublevel within a principal level

By our rule, all of the electrons in an ℓ equals 0 sublevel must have quantum number $\mathbf{m}_\ell$ equal to 0. Since there is one orbital for each different value of $\mathbf{m}_\ell$, there is only one orbital in the ℓ equals 0 sublevel. If ℓ equals 1, $\mathbf{m}_\ell$ may take on the values 1, 0, and −1; this means that an electron in a sublevel in which ℓ equals 1 may be in any one of three orbitals, denoted by $\mathbf{m}_\ell = 1$, 0, and −1. In general, within a given sublevel of quantum number ℓ, there will be $2\ell + 1$ orbitals. Each of these $2\ell + 1$ orbitals will have essentially the same energy.

FOURTH QUANTUM NUMBER, $\mathbf{m}_s$. In order to completely describe an electron in an atom, we must specify, in addition to its $\mathbf{n}$, ℓ, and $\mathbf{m}_\ell$ quantum numbers, a fourth quantum number $\mathbf{m}_s$, called the spin quantum number. The spin quantum number can be loosely associated with the spin of the electron about its own axis. It was introduced into the theory to make the properties of atoms consistent with experiment, so it is somewhat different from the first three quantum numbers, which come from the solution of the hydrogen atom problem. The quantum number $\mathbf{m}_s$ is not related to the values of $\mathbf{n}$, ℓ, or $\mathbf{m}_\ell$ for an electron and can have one of two possible values, +1/2 or −1/2, depending on the direction of rotation of the electron about its axis ($\mathbf{m}_s = +1/2$ or $-1/2$). Two electrons in the same orbital having $\mathbf{m}_s$ values of +1/2 and −1/2 are said to be *paired;* both of these electrons would have the same amount of energy.

PAULI EXCLUSION PRINCIPLE. According to the rules just cited, we could have an electron in a sodium atom with the quantum numbers 2, 0, 0, +1/2. The question we might then ask is: How many electrons in the atom can have this particular set of quantum numbers? The answer to this question is furnished by the Pauli exclusion principle, which tells us: *No two electrons in an atom can have the same set of four quantum numbers.* This condition, which has important implications with regard to our theory of atomic structure, was first stated by Pauli in 1925, again to make the theory consistent with the observed properties of atoms.

By the Pauli exclusion principle we conclude that there is only *one* electron in a sodium atom with the quantum numbers 2, 0, 0, +1/2. Similarly, there is only one electron with quantum numbers 2, 0, 0, −1/2. These two electrons would both belong to the same *orbital* of the atom, namely the 2, 0, 0 orbital. If we try to write another set of four quantum numbers for a third electron in that same orbital we find that it is not possible, since the spin quantum number $\mathbf{m}_s$ can only equal +1/2 or −1/2. Since, by the exclusion principle, each electron in an atom, or an orbital in that atom, must have a different set of four quantum numbers, we conclude that only two electrons can fit in the 2, 0, 0 orbital of the sodium atom. The same argument could be made for any orbital, with a given set of values of $\mathbf{n}$, ℓ, and $\mathbf{m}_\ell$, in any atom, so we can state as a general rule that *no more than two electrons can fit into any atomic orbital.*

TABLE 6.2 ALLOWED SETS OF QUANTUM NUMBERS FOR ELECTRONS IN ATOMS

Level **n**	1	2				3								
Sublevel ℓ	0	0	1			0	1			2				
Orbital m_ℓ	0	0	1	0	−1	0	1	0	−1	2	1	0	−1	−2
Spin m_s ↑ = +1/2, ↓ = −1/2	⇅	⇅	⇅	⇅	⇅	⇅	⇅	⇅	⇅	⇅	⇅	⇅	⇅	⇅

The rules that we have given for assigning quantum numbers fix the capacities of orbitals, sublevels, and principal levels. To summarize:

—each principal level of quantum number n contains a total of n sublevels
—each sublevel of quantum number ℓ contains a total of $2\ell + 1$ orbitals
—each orbital can hold two electrons

These rules are illustrated in Table 6.2 for the principal energy levels **n** = 1 through **n** = 3. Their application is indicated in Example 6.4.

Example 6.4

a. How many electrons can fit into the principal level for which **n** = 2?

b. the sublevel for which **n** = 3, ℓ = 1?

Solution

a. In the **n** = 2 level there are two sublevels (ℓ = 0 and ℓ = 1). The first of these sublevels contains one orbital (m_ℓ = 0), which can hold two electrons with quantum numbers +1/2 (↑) and −1/2 (↓). The second sublevel contains three orbitals (m_ℓ = 1, 0, and −1), each of which can contain two paired electrons (⇅). The capacity of the **n** = 2 level is equal to the sum of that of its two sublevels and is therefore equal to 2 + (3 × 2), or 8 electrons.

b. The 3,1 sublevel includes three orbitals (m_ℓ = 1, 0, and −1), each of which can hold two electrons. The capacity of the sublevel is six electrons.

Using the approach in Example 6.4 one can readily obtain the total electron capacities of any level or sublevel in an atom. In Table 6.3 we have listed the capacities of the various levels and sublevels of an atom up to **n** = 4. It is interesting to note that the capacity of a level is always equal to $2n^2$, where **n** is the quantum number fixing the level.

An ℓ = 1 sublevel has a capacity of 6 electrons, regardless of the principal level to which it belongs

TABLE 6.3 CAPACITIES OF ELECTRONIC LEVELS AND SUBLEVELS IN ATOMS

Level n	Total Number of Electrons in Level	Number of Electrons in Sublevels ℓ = 0	1	2	3
1	2	2	–	–	–
2	8	2	6	–	–
3	18	2	6	10	–
4	32	2	6	10	14

You will recall that the energy of an electron is dependent almost completely on the values of its **n** and **ℓ** quantum numbers, increasing sharply with increasing **n** value and only slightly as the **ℓ** value goes up. The $\mathbf{m}_\ell$ and $\mathbf{m}_s$ quantum numbers of an electron have only a very small effect on its energy. It is, then, the electronic sublevels, characterized by different **n, ℓ** values, which have appreciably different energies in atoms. The electron configuration of an atom is a statement giving the populations of each of these electronic sublevels.

To arrive at the electron configurations of atoms, we must know the order in which the various sublevels become populated. In general, electrons enter the available sublevels in order of their increasing energy, filling each sublevel to capacity before entering the next one of higher energy. Although the energies of the different sublevels in atoms cannot be calculated exactly, their relative magnitudes can be obtained from experiments and are qualitatively quite similar in different atoms. In Figure 6.10 we have shown the relative energies of some of the lower-lying sublevels for electrons in atoms.

From Figure 6.10, we see that the lowest sublevel is the one for which **n** = 1, **ℓ** = 0. The single electron associated with the hydrogen atom will ordinarily be found in that sublevel, since it would seek the lowest possible energy. The electron configuration of the H atom might be written as 1, 0, indicating that there is one electron in the **n** = 1, **ℓ** = 0 sublevel.

For some reason, probably traditional, we do not express electron configurations in quite this way. The notation for the principal quantum number, **n**, is as we have written it, but a new symbol is used for the second quantum number, **ℓ**. Electrons for which **ℓ** = 0 are called *s* electrons; if **ℓ** = 1 they are called *p* electrons; for **ℓ** = 2 they are called *d* electrons; if their **ℓ** value is 3, they are called *f* electrons. The symbols *s, p, d,* and *f* come from the sharp, principal, diffuse, and fundamental series observed in atomic spectra early in this century. Spectroscopists were among the first to describe atoms in this way, and their notation has persisted. On this basis the electron in the ordinary H atom is a 1s electron, and the electron configuration of the H atom is simply 1s.

In the helium atom, with two electrons, we have an electron in each of the two states in the lowest lying level, the quantum number descriptions are:

$$1, 0, 0, +1/2: \quad 1, 0, 0, -1/2$$

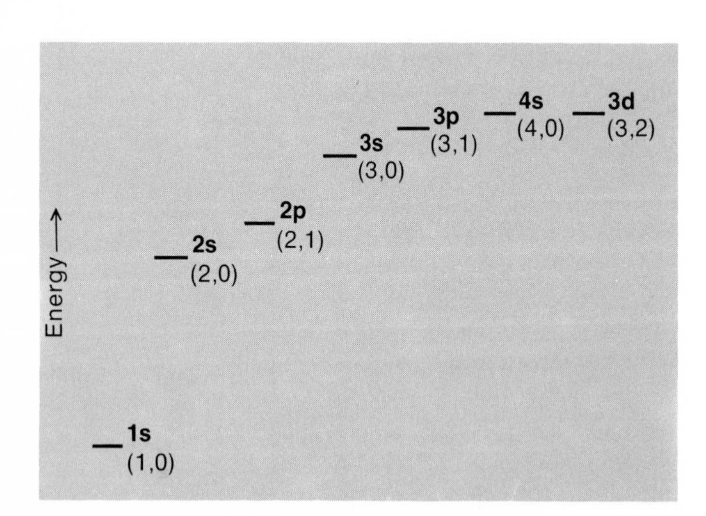

Figure 6.10 Relative energies of electronic sublevels. Energy depends on both **n** and **ℓ**. The 4s sublevel is slightly lower in energy than the 3d sublevel, at least for atoms where these levels are filling.

To indicate that the two electrons are in the same sublevel, the electron configuration of helium is written $1s^2$, pronounced "wun ess too," meaning there are two electrons with **n**, ℓ values of 1, 0, respectively.

The three electrons in the lithium atom cannot all be in the 1s sublevel, since as we have seen (Table 6.3) the **n** = 1 level has a capacity of only two electrons. We see from Figure 6.10 that the third electron would be assigned to the 2s sublevel, since that is the one having the next higher energy. With the first two electrons in the 1s level, we have an electron configuration for the Li atom of $1s^22s$.

Since the 2s sublevel contains only one orbital ($\mathbf{m}_\ell = 0$), it can hold only two electrons, and so will be filled at beryllium, atomic number 4, with electron configuration $1s^22s^2$.

The next sublevel to be filled is 2p. The fifth electron in the boron atom will be in this sublevel, giving boron an electron configuration of $1s^22s^22p$. Since there are three orbitals in the 2p sublevel, it can hold up to six electrons (Tables 6.2 and 6.3). That sublevel will gradually take on electrons as we go on to carbon (at. no. = 6) and up to neon (at. no. = 10), at which point it will be filled; the electron configuration of neon is $1s^22s^22p^6$. Here, as always, the superscripts (2,2,6) indicate the number of electrons in each sublevel.

We can use this general approach to find the electron configuration of any atom. The method is sometimes called the Aufbau, or building up, principle, adding electrons one by one as the atomic number increases. *The electrons are added to sublevels in order of increasing energy, in general filling each sublevel to capacity before beginning the next.* The order of sublevels is that in Figure 6.10; the complete list in proper order is as follows:

What would be the atomic number of the first element to fill all these sublevels?

1s, 2s, 2p, 3s, 3p, 4s, 3d, 4p, 5s, 4d, 5p, 6s, 4f, 5d, 6p, 7s, 5f, 6d

As noted in Table 6.3, the capacities of the sublevels depend on their ℓ values: for an s sublevel ($\ell = 0$), two electrons; a p sublevel ($\ell = 1$), six electrons; a d sublevel ($\ell = 2$), ten electrons, and an f sublevel ($\ell = 3$), fourteen electrons.

You will note several cases where there is an "overlap" between principal energy levels. For example, the 4s sublevel (**n** = 4, $\ell = 0$) is filled before the 3d sublevel (**n** = 3, $\ell = 2$). This means that in potassium and calcium, electrons enter the 4s sublevel in preference to the 3d. The electron configuration of potassium is $1s^22s^22p^63s^23p^64s$ rather than $1s^22s^22p^63s^23p^63d$.

Example 6.5 Find the electron configuration of the sulfur atom and the nickel atom.

Solution S atom: at. no. 16; 16 electrons. We fill the sublevels to capacity in order of increasing energy. There are two 1s electrons, two 2s, six 2p, two 3s, and four 3p electrons, making a total of sixteen. The configuration of the S atom is therefore $1s^22s^22p^63s^23p^4$.

Ni atom: at. no. 28; 28 electrons. Proceeding as before, this time until we have added 28 electrons, we obtain as the electron configuration of the Ni atom: $1s^22s^22p^63s^23p^64s^23d^8$. (Note that the 4s-sublevel fills before the 3d.)

In Table 6.4 (p. 140) we have listed the electron configurations of the elements of atomic numbers 1 through 101. The configurations generally follow the rules we have given. The major exceptions occur for the elements of medium to high atomic number in which a sublevel is close to being filled or half-filled. It turns out that having a full or half-full sublevel confers a measure of stability on the electronic structure of an atom. Where 3d sublevels are involved, with energies close to

TABLE 6.4 THE ELECTRON CONFIGURATIONS OF THE ATOMS OF THE ELEMENTS

Element	Atomic Number	Populations of Subshells 1s	2s	2p	3s	3p	3d	4s	4p	4d	4f	5s
H	1	1										
He	2	2										
Li	3	2	1									
Be	4	2	2									
B	5	2	2	1								
C	6	2	2	2								
N	7	2	2	3								
O	8	2	2	4								
F	9	2	2	5								
Ne	10	2	2	6								
Na	11				1							
Mg	12				2							
Al	13		Neon		2	1						
Si	14		core		2	2						
P	15				2	3						
S	16				2	4						
Cl	17				2	5						
Ar	18	2	2	6	2	6						
K	19							1				
Ca	20							2				
Sc	21						1	2				
Ti	22						2	2				
V	23						3	2				
Cr	24						5	1				
Mn	25			Argon			5	2				
Fe	26			core			6	2				
Co	27						7	2				
Ni	28						8	2				
Cu	29						10	1				
Zn	30						10	2				
Ga	31						10	2	1			
Ge	32						10	2	2			
As	33						10	2	3			
Se	34						10	2	4			
Br	35						10	2	5			
Kr	36	2	2	6	2	6	10	2	6			
Rb	37											1
Sr	38											2
Y	39									1		2
Zr	40									2		2
Nb	41					Krypton				4		1
Mo	42					core				5		1
Tc	43									6		1
Ru	44									7		1
Rh	45									8		1
Pd	46									10		
Ag	47									10		1
Cd	48									10		2

TABLE 6.4 THE ELECTRON CONFIGURATIONS OF THE ATOMS OF THE ELEMENTS *(Continued)*

Element	Atomic Number		Populations of Subshells									
			4d	4f	5s	5p	5d	5f	6s	6p	6d	7s
In	49		10		2	1						
Sn	50		10		2	2						
Sb	51		10		2	3						
Te	52		10		2	4						
I	53		10		2	5						
Xe	54		10		2	6						
Cs	55		10		2	6			1			
Ba	56		10		2	6			2			
La	57		10		2	6	1		2			
Ce	58		10	2	2	6			2			
Pr	59		10	3	2	6			2			
Nd	60		10	4	2	6			2			
Pm	61		10	5	2	6			2			
Sm	62		10	6	2	6			2			
Eu	63		10	7	2	6			2			
Gd	64		10	7	2	6	1		2			
Tb	65		10	9	2	6			2			
Dy	66		10	10	2	6			2			
Ho	67		10	11	2	6			2			
Er	68		10	12	2	6			2			
Tm	69		10	13	2	6			2			
Yb	70		10	14	2	6			2			
Lu	71		10	14	2	6	1		2			
Hf	72		10	14	2	6	2		2			
Ta	73		10	14	2	6	3		2			
W	74		10	14	2	6	4		2			
Re	75	Krypton	10	14	2	6	5		2			
Os	76	core	10	14	2	6	6		2			
Ir	77		10	14	2	6	9					
Pt	78		10	14	2	6	9		1			
Au	79		10	14	2	6	10		1			
Hg	80		10	14	2	6	10		2			
Tl	81		10	14	2	6	10		2	1		
Pb	82		10	14	2	6	10		2	2		
Bi	83		10	14	2	6	10		2	3		
Po	84		10	14	2	6	10		2	4		
At	85		10	14	2	6	10		2	5		
Rn	86		10	14	2	6	10		2	6		
Fr	87		10	14	2	6	10		2	6		1
Ra	88		10	14	2	6	10		2	6		2
Ac	89		10	14	2	6	10		2	6	1	2
Th	90		10	14	2	6	10		2	6	2	2
Pa	91		10	14	2	6	10	2	2	6	1	2
U	92		10	14	2	6	10	3	2	6	1	2
Np	93		10	14	2	6	10	5	2	6		2
Pu	94		10	14	2	6	10	6	2	6		2
Am	95		10	14	2	6	10	7	2	6		2
Cm	96		10	14	2	6	10	7	2	6	1	2
Bk	97		10	14	2	6	10	9	2	6		2
Cf	98		10	14	2	6	10	10	2	6		2
Es	99		10	14	2	6	10	11	2	6		2
Fm	100		10	14	2	6	10	12	2	6		2
Md	101		10	14	2	6	10	13	2	6		2

those of the 4s sublevel, the effect is large enough in chromium to allow "promotion" of a 4s electron to the 3d sublevel, giving chromium five 3d electrons. In the copper atom a 4s electron is also promoted to the 3d sublevel, thereby filling that sublevel.

Orbital Diagrams—Hund's Rule

Ordinarily electron configurations such as we have been discussing are adequate for describing the electron arrangements in atoms. In some cases, however, more detailed descriptions are helpful and furnish more information than do the electron configurations. There are various ways of stating such descriptions; the most detailed would be a listing of the four quantum numbers for each of the electrons in the atom. More commonly, we use what is known as an *orbital diagram,* which indicates the number of electrons in each orbital and their relative spins. To illustrate how such a diagram is constructed, consider the boron atom (at. no. 5), which as we have seen has the electron configuration $1s^22s^22p$. We write its orbital diagram as follows, using parentheses to indicate an orbital.

$$\begin{array}{ccc} 1s & 2s & 2p \\ (\uparrow\downarrow) & (\uparrow\downarrow) & (\uparrow\ )(\ \ )(\ \ ) \end{array}$$

The two arrows in the parentheses indicate the two electrons in the filled orbitals; the fact that they point in opposite directions means that the electrons have opposite spins and so are paired. The fifth electron could be in any of the three 2p orbitals and could have spin either down or up, since all of these states have the same energy and are equally likely; the diagram includes only one of the possibilities, usually the one shown.

Proceeding now to the orbital diagram for the carbon atom (at. no. 6), electron configuration $1s^22s^22p^2$, we find that there is a question as to how the sixth electron should be added to the diagram we have given for boron. That electron could go into the same orbital as the other 2p electron, or it could go into one of the other two available orbitals. In the first case the two p electrons would have to have opposite spins and so would be paired; in the latter the two electrons might be paired or they might be unpaired, with parallel spins (both spin quantum numbers having the same sign). The experimental properties of the carbon atom indicate that there is an energy difference between the different possible arrangements, and that the most stable configuration is the one in which the two electrons are unpaired, with parallel spins, and so are in different orbitals. The orbital diagram for the carbon atom would therefore take the form

$$\begin{array}{ccc} 1s & 2s & 2p \\ (\uparrow\downarrow) & (\uparrow\downarrow) & (\uparrow\ )(\uparrow\ )(\ \ ) \end{array}$$

in which the two 2p electrons are in different orbitals and have the same spin.

Hund's rule is based on experiment, not theory

The condition we observe on the electron arrangement in the carbon atom can be generalized to include other atoms for which similar situations exist. There is a general principle governing all such cases, Hund's rule, which states that

In an atom in which orbitals of equal energy are to be filled by electrons, the order of filling is such that as many electrons remain unpaired as possible.

The orbital diagrams for atoms beyond carbon which contain 2p electrons in the outermost sublevel follow immediately by the approach we used for carbon. In Figure 6.11 we have shown the orbital diagrams for the atoms from boron to neon. You can see that the neon atom has no unpaired electrons, boron and fluorine both have one, carbon and oxygen two, and in the nitrogen atom the three

2p electrons can all remain unpaired. As we shall see in succeeding chapters, the orbital diagram notation for electronic arrangements in atoms is useful in discussions of chemical bonding in many substances. Hund's rule applies to electronic structures in any atoms, ions, or molecules in which there are several orbitals with equal, or nearly equal, energies. Those arrangements will be energetically preferred where the maximum number of electrons are unpaired.

Example 6.6 Construct the orbital diagrams of the sulfur and nickel atoms.

Solution We first need to know the electron configurations of these atoms. In Example 6.5 we found them to be: S atom, $1s^22s^22p^63s^23p^4$; Ni atom, $1s^22s^22p^63s^23p^6 4s^23d^8$. In the orbital diagrams we list all orbitals, pairing all electrons in orbitals in filled sublevels. With unfilled sublevels, electrons are put one by one in the available orbitals, keeping spins parallel, unpaired, as much as possible.

S atom:	(⇅)	(⇅)	(⇅)(⇅)(⇅)	(⇅)	(⇅)(↑)(↑)		
	1s	2s	2p	3s	3p		
Ni atom:	(⇅)	(⇅)	(⇅)(⇅)(⇅)	(⇅)	(⇅)(⇅)(⇅)	(⇅)	(⇅)(⇅)(⇅)(↑)(↑)
	1s	2s	2p	3s	3p	4s	3d

We would expect that there would be two electrons left unpaired in the S atom. In Ni there would also be two unpaired electrons, in the 3d sublevel.

As we have noted, although each electron in an atom has four quantum numbers, the descriptions of the electrons in atoms are usually somewhat abbreviated, furnishing either the first two quantum numbers, as in electron configurations, or populations and spin arrangements in orbitals, as in orbital diagrams. One can, however, within limits, state full sets of quantum numbers for electrons, as is illustrated in Example 6.7.

Atom	Orbital diagram 1s	2s	2p	Electron configuration
B	(↑↓)	(↑↓)	(↑)( )( )	$1s^22s^22p$
C	(↑↓)	(↑↓)	(↑)(↑)( )	$1s^22s^22p^2$
N	(↑↓)	(↑↓)	(↑)(↑)(↑)	$1s^22s^22p^3$
O	(↑↓)	(↑↓)	(↑↓)(↑)(↑)	$1s^22s^22p^4$
F	(↑↓)	(↑↓)	(↑↓)(↑↓)(↑)	$1s^22s^22p^5$
Ne	(↑↓)	(↑↓)	(↑↓)(↑↓)(↑↓)	$1s^22s^22p^6$

Figure 6.11 Orbital diagrams showing electron arrangements for atoms with 5 to 10 electrons. Electrons entering orbitals of equal energy remain unpaired as long as possible.

Example 6.7 Write a complete set of quantum numbers for the four 3p electrons in the sulfur atom.

Solution For all of the 3p electrons, $\mathbf{n} = 3$, and $\ell = 1$. There are three 3p orbitals, corresponding to $\mathbf{m}_\ell = +1$, 0, and -1. In each orbital the spin quantum numbers can be $+1/2$ or $-1/2$. A 3p electron can, therefore, be described by any one of the six sets of four quantum numbers listed below:

(I) 3, 1, 1, +1/2	(III) 3, 1, 0, +1/2	(V) 3, 1, −1, +1/2
(II) 3, 1, 1, −1/2	(IV) 3, 1, 0, −1/2	(VI) 3, 1, −1, −1/2
$\mathbf{m}_\ell = 1$	$\mathbf{m}_\ell = 0$	$\mathbf{m}_\ell = -1$

It would not be possible to assign unequivocally four of the above sets of quantum numbers to the four 3p electrons in the sulfur atom. The best we could do is say that, by Example 6.6, two of the electrons would be paired and would fill one of the 3p orbitals. The other two electrons would have the same spin quantum number (would have parallel spins), with one electron in each of the two remaining p orbitals. One possible complete set of quantum numbers for the 3p electrons, then, might be I, II, III, and V, but there are many others (see Problem 6.21).

Geometric Representations of Electron Charge Clouds

You will recall that the value at any point of the square of the wave function, ψ, associated with an electron is proportional to the probability of finding the electron at that point, and hence to the charge density at that point. The electron cloud sketched in Figure 6.9 (p. 134) indicates that the charge density in the atom is largest near the nucleus and drops off rapidly as the distance from the nucleus is increased.

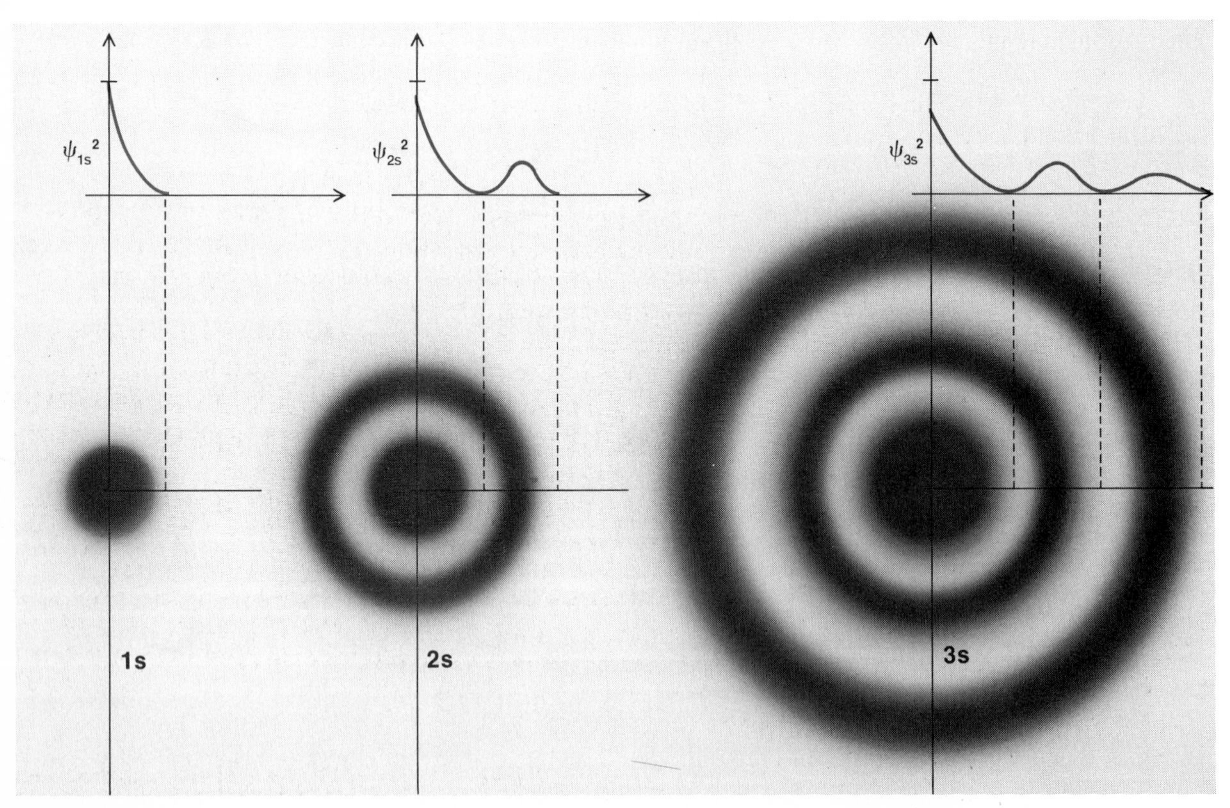

Figure 6.12 Charge clouds associated with s orbitals. Sketches show cross sections in plane of atomic nucleus.

The electron charge cloud associated with a single electron is dependent, as you might expect, on the values of $\mathbf{n}$, ℓ, and $\mathbf{m}_\ell$ for the state under consideration. The dependence on $\mathbf{n}$ can be expressed in a function that includes only r, the distance of the electron from the nucleus, and so is always spherically symmetric. As $\mathbf{n}$ increases, the charge cloud moves, on the average, farther from the nucleus and takes on a very roughly shell-like structure, with maxima and minima that depend in their position and magnitude on the value of $\mathbf{n}$. The charge clouds for an electron with different $\mathbf{n}$ values and a value of $\ell = 0$ (an s electron) are shown in Figure 6.12. The charge cloud for any electron in an s orbital is spherically symmetric.

For states in which ℓ is not equal to zero the associated electron clouds are more complicated in their geometric structure. If we are working with p electrons, where $\ell = 1$, the value of $\mathbf{m}_\ell$ is no longer restricted to zero, but may take on the values $+1$, 0, and -1. We would anticipate that there would be three different electron clouds possible for $\ell = 1$ states, one for each possible value of $\mathbf{m}_\ell$. This is indeed the case. The three clouds are identical in their overall structure but differ in their orientation with respect to a given set of axes, whose position might be established, for instance, by a strong magnetic field. The three clouds may be considered to be concentrated along the x, y, and z axes in the manner indicated in Figure 6.13. The three possible p orbitals are often represented by these charge clouds called p_x, p_y, and p_z orbitals. (The actual structure of the charge cloud for a p orbital will depend on the value of $\mathbf{n}$ as well as that of $\mathbf{m}_\ell$ and in Figure 6.13 $\mathbf{n}$ has been taken to be 2.)

The total charge in each cloud in Figures 6.12 and 6.13 is that of one electron

It is possible to draw charge clouds for orbitals associated with a d electron. In this case there are five different orbitals, corresponding to the five possible values for $\mathbf{m}_\ell$. We shall consider d orbitals in some detail in Chapter 21.

SOME COMMENTS ON THE ELECTRON CONFIGURATION THEORY. The theory of electron configurations associates the electrons in an atom with quantum states appropriate to their energies and space properties. The electron configuration and orbital diagram of an atom describe, as best we can at present, the properties of an atom in terms of the properties of its electrons. Associated with each electron quantum state is its set of quantum numbers and a cloud of negative charge, whose density at any point is proportional to the probability of finding the electron at that point in space. An atom, according to the model, consists of a positive nucleus surrounded by a cloud of negative charge resulting from the superposition of the charge clouds associated with each of its electrons.

Since the charge clouds for different sets of quantum numbers, or different quantum states, may differ in their properties, particularly in the distance from the atomic nucleus at which they have appreciable densities, our model of the electronic structure of the atom is not quite so nebulous as it might first appear. Theory tells us that electrons with high principal quantum number $\mathbf{n}$ have high energy and, on the average, are much more likely to be found in regions relatively distant from the nucleus than are electrons with low $\mathbf{n}$ values.

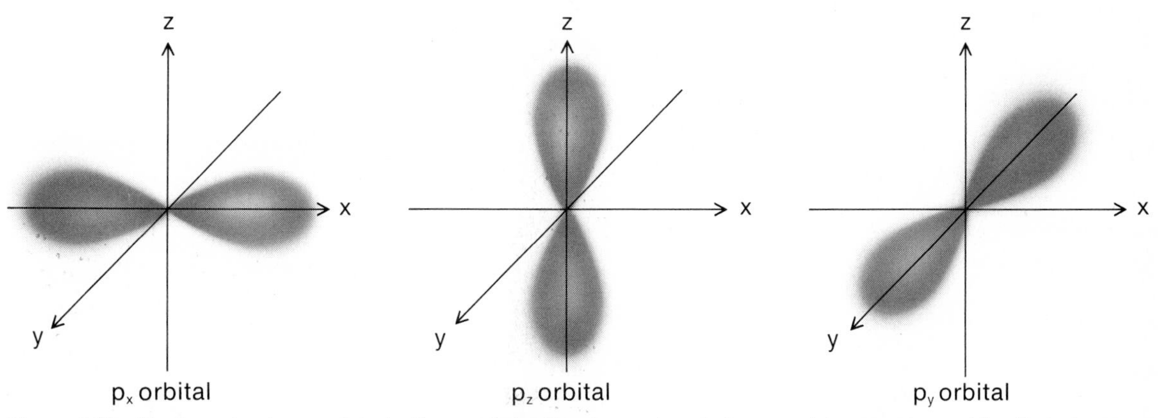

p_x orbital p_z orbital p_y orbital

Figure 6.13 Electron clouds associated with p orbitals. The charge density is symmetric around one of the Cartesian axes.

On this basis we can assign electrons to regions, which are usually called "shells" or "levels," one shell or level for each value of **n**, with shells associated with low **n** values near the nucleus and those with higher **n** values more distant. To discriminate between electrons in the same level (same **n** value) but different values of the quantum number ℓ, we use the term "same sublevel" or "same subshell" to denote those electrons with the same values of both **n** and ℓ.

There is one difficulty with the electron configuration theory that should be mentioned. The theory is based on the assumption that the electrons in an atom can be described by assigning them quantum numbers. This means that in the theory we consider the electrons to have properties as individuals. (Indeed the quantum numbers we use and the interpretations we give them are based on the wave mechanical solution of the one-electron problem and, hence, involve only the electron-nucleus interactions.) In such an atomic model the electronic properties of the atom will simply be equal to the sum of the properties of all the electrons in the atom, with individual electron properties determined by solutions to one-electron problems. Such a model is clearly incorrect, since in an atom there are important electron-electron charge interactions and electron-electron spin interactions, as well as the electron-nucleus charge interaction covered by the model. So far, the theory of atomic structure has been unable to properly treat electron-electron interactions in atoms. Such interactions are usually considered to be "averaged in" when one interprets electron configurations. The qualitative structure of atoms, involving the existence of shells and their populations, does not seem to be appreciably altered by such an averaging; this explains the usefulness of electron configurations. Clearly, however, the necessity for such an averaging renders impossible the use of electron configurations in quantitative calculations of atomic energy levels and atomic dimensions.

Sometimes models can be very useful even if they aren't quite right

6.6 EXPERIMENTAL SUPPORT FOR ELECTRON CONFIGURATIONS

In the next and succeeding chapters we will make extensive use of electron configurations in correlating and explaining many chemical and physical properties of pure substances. Before proceeding to such topics, however, we should present at least some experimental evidence in support of these configurations, with a view toward adding a physical aspect to their so far highly theoretical nature. Let us examine ionization energies in this regard, since they correlate well with electron configurations and are relatively easy to understand.

Ionization Energy

The first ionization energy, ΔE_I, of an atom is taken to be the energy change for the formation of a +1 ion from the gaseous atom:

$$X(g) \rightarrow X^+(g) + e^- \qquad \Delta E_I = \text{ionization energy}$$

For the above reaction ΔE_I is always positive; it takes energy to remove an electron from an atom. The magnitude of ΔE_I reflects the ease with which the electron is removed; the smaller the ionization energy, the more readily the atom will give up an electron. A given element will have several ionization energies, one for each electron in its atom. For the sodium atom for example,

$$Na(g) \rightarrow Na^+(g) + e^- \qquad \Delta E_1 = \text{first ionization energy}$$
$$Na^+(g) \rightarrow Na^{2+}(g) + e^- \qquad \Delta E_2 = \text{second ionization energy}$$
$$Na^{2+}(g) \rightarrow Na^{3+}(g) + e^- \qquad \Delta E_3 = \text{third ionization energy}$$

Successive ionization energies are always related in the order first < second < third . . . , since the charge on the species becomes more and more positive and the electrons being removed are closer to the nucleus.

For the sodium atom, electron configuration $1s^22s^22p^63s$, the first ionization energy is that required to remove the outermost, 3s, electron from the atom; the second ionization energy measures the energy required to remove a 2p electron from Na^+; the third, the energy needed to remove a second 2p electron from Na^{2+}, and so on. In Figure 6.14 we have plotted the square roots of the ionization energies of sodium as a function of the electron being removed.

By the electron configuration theory we would expect that the first electron to be removed from sodium, being a 3s electron and relatively far away from the rest of the atom, would come off fairly easily. Then we would predict that there would be a group of eight electrons, in the 2s and 2p sublevels, each of which would be somewhat less easily removed than the one before it, but which would not be grossly different in behavior as far as energy of removal is concerned. The last two electrons, in the n = 1 level, should be very tightly bound and very resistant to removal from the nucleus.

Figure 6.14 clearly supports the theory. The ionization energies are grouped as expected, with one very low value, eight of gradually increasing magnitude, and two very large values. The eight electrons for which n = 2 fall in a single group, with no sharp break between the six 2p and the two 2s electrons.

Using a somewhat different approach it is possible to show the resemblances between species having what our theory would predict to be the same electronic configuration. Let us consider, for example, the series of species

$$Li \quad Be^+ \quad B^{2+} \quad C^{3+} \quad N^{4+} \quad O^{5+} \quad F^{6+}$$

These seven species all belong to different elements, but all have three electrons; since electron configurations in general depend only on number of electrons, these species should all have the same electron configuration, $1s^22s$. The electronic structures of these species should all be similar to one another, differing only in the charge on the central nucleus, which would vary from +3 for Li to +9 for F^{6+}. This charge increase should make it harder to remove the 2s electron from F^{6+} than from Li, but the change in ionization energy for that electron should, if the theory is valid, be gradual and smooth as one proceeds from Li to F^{6+}.

In Figure 6.15 we have plotted the square roots of the ionization energies versus nuclear charge for each of the species in the series mentioned. As you can see, the line obtained is essentially straight, clearly indicating that those species

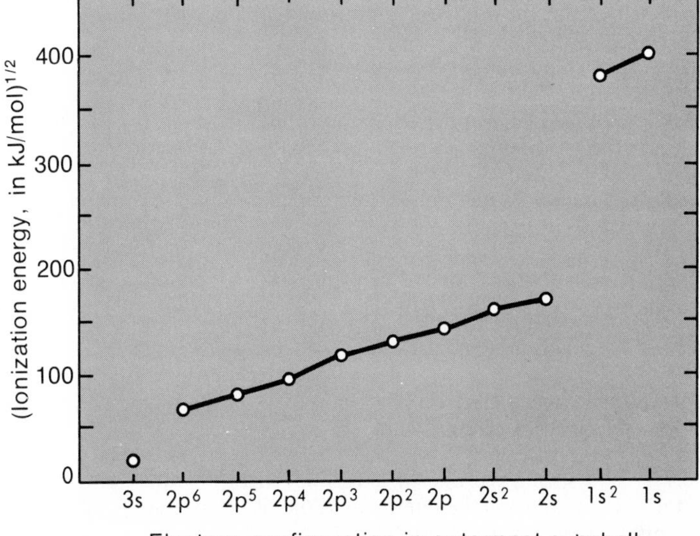

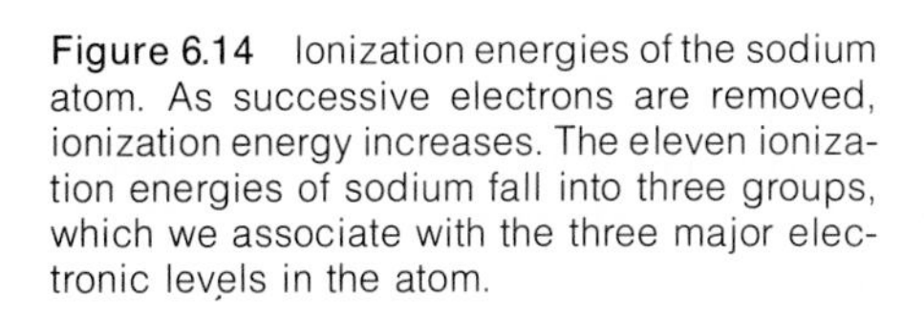
Figure 6.14 Ionization energies of the sodium atom. As successive electrons are removed, ionization energy increases. The eleven ionization energies of sodium fall into three groups, which we associate with the three major electronic levels in the atom.

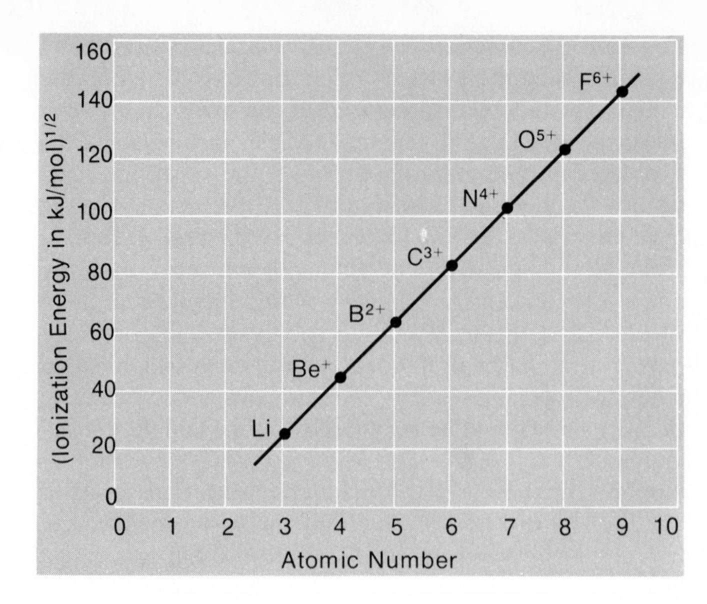

Figure 6.15 Ionization energies of an isoelectronic series. Each of the species has the same electron configuration, $1s^22s$. The energies shown are for removal of the 2s electron. The regular increase of energy with atomic number reflects the very similar electronic structures of these species.

are closely related as far as electronic structure is concerned. Species such as these, with the same number of electrons, are said to form an **isoelectronic series,** and, as here, typically show resemblances that can be attributed to their having the same electron configurations.

PROBLEMS

6.1 *Atomic Spectra*

a. How does quantum theory explain the fact that excited potassium atoms emit light at only a relatively few wavelengths?
b. One of the principal lines in the spectrum of potassium has a wavelength of 401.4 nm. What is the energy in joules of the photon in this radiation?

6.2 *Bohr Theory* Using Bohr's equation for hydrogen, find the difference in energy between the $\mathbf{n} = 2$ and $\mathbf{n} = 1$ levels in the hydrogen atom in joules per atom; kilojoules per mole.

6.3 *Electron Configurations* What is the electron configuration of the N atom (at. no. = 7) in the ground state?

6.4 *Orbital Diagrams* Draw the orbital diagram for the N atom. How many unpaired electrons would you expect the atom to have?

6.5 *Quantum Numbers* Write a complete set of four quantum numbers for each of the electrons in the nitrogen atom.

6.6 Explain what is meant by each of the following terms:

a. quantum number
b. photon
c. unpaired electron
d. electron configuration

6.7 The strong red line in the Li spectrum occurs at 670.8 nm. What is the energy in kilojoules per mole of photons of this wavelength?

6.22 What were the contributions to atomic theory of each of the following scientists?

a. de Broglie
b. Balmer
c. Einstein
d. Bohr

6.23 A certain electron transition in the mercury atom involves an energy change of 274 kJ/mol. What wavelength is associated with this transition?

6.8 Predict the wavelength in nanometres of the tenth line ($n = 12$) in the Balmer series (Table 6.1).

6.9 Use Bohr's equation to calculate the wavelength of the first line in the Lyman series for hydrogen ($n = 2 \rightarrow n = 1$).

6.10 The energy of any one-electron species, like He^+ or Li^{2+}, in its nth state is found by the Bohr theory to be $-Z^2B/n^2$, where Z is the charge on the nucleus and B is 2.179×10^{-18} J. On this basis, find the ionization energy of the He^+ ion, first in joules per atom, and then in kilojoules per mole.

6.11 What would be the minimum energy in joules of an electron confined to a 0.20-nm box? Why would the electron obey only quantum mechanical laws?

6.12 Calculate the minimum energy of an electron confined to an atomic nucleus of diameter equal to 1×10^{-14} m. Compare this energy with that available for binding nuclear particles, which is about 1.3×10^{-12} J. Suggest why nuclear theoreticians do not feel that there are electrons in atomic nuclei.

6.13 If the **n** quantum number of an electron equals 2, what values could its ℓ quantum number have? its $\mathbf{m}_\ell$ quantum number? its $\mathbf{m}_s$ quantum number?

6.14 The electron configuration of the B atom is $1s^22s^22p$. What does this tell you about the electrons in that atom?

6.15 What are the electron configurations of the following atoms?

a. Si
b. Zn
c. Zr

6.16 When atoms are excited, it is usually the outermost electrons that are moved to higher levels. Which of the following configurations would correspond to atoms in excited states? to ground state configurations? Which are incorrect?

a. $1s^22s$
b. $1s^22s^22d$
c. $1s^22s2p^2$
d. $1s^22s^22p^43s$
e. $1s^22s^42p^2$
f. $1s^22s^22p^63d$

6.17 Draw orbital diagrams for atoms with the following electron configurations:

a. $1s^22s^22p^2$
b. $1s^22s^22p^63s^23p^5$
c. $1s^22s^22p^6$

How many unpaired electrons would each atom have?

6.24 One of the lines in the Balmer series has a wavelength equal to 379.791 nm. Use Equation 6.4 to find n for that line. What is the order number, q, of the line in the series?

6.25 Use Bohr's Equation 6.5 to derive Equation 6.4 for the Balmer series (note that $n = 2$ for the lower energy state).

6.26 Referring to Problem 6.10 for information, find the energy in joules of the Li^{2+} ion in the ground state. What would be the ionization energy of this ion in kilojoules per mole?

6.27 Calculate the minimum energy allowed to a 100-kg man confined to a closet one metre long. What would be his quantum number if he moved at a speed of 1×10^{-4} m/s (Equation 6.10)? Would his movements in the closet obey classical laws (Fig. 6.8)?

6.28 Calculate the minimum energy of a proton confined to an atomic nucleus. Compare this energy with that available for binding nuclear particles. Would this binding energy be enough to hold a proton? a neutron? an alpha particle? See Problem 6.12 for data.

6.29 How many electrons can fit into an $\mathbf{n} = 1$ level? Explain how you arrived at your answer.

6.30 Explain how you would derive the electron configuration of the atom X. What would the result tell you about X?

6.31 Noting that ions have configurations that, at least for the lighter elements, are determined by the number of electrons they have, find the electron configurations of the following species:

K^+ F^- Al Fe

6.32 Which of the following electron configurations would be those of atoms in ground states? in excited states? Which are just plain wrong?

a. $1s^22s^32p$
b. $1s^22p^2$
c. $1s^22s^2$
d. $1s^22s^22p^63s3d$
e. $1s^22s^22p^54f$
f. $1s^22s2p$

6.33 Draw orbital diagrams for the following atoms: Ti, Si, Mn. How many unpaired electrons would each atom have?

6.18 An atom has the following orbital diagram. What is its electron configuration? Name the element to which the atom belongs.

1s	2s	2p	3s	3p
(⇅)	(⇅)	(⇅)(⇅)(⇅)	(⇅)	(↑)(↑)( )

6.19 Write a complete set of quantum numbers for the electrons in the B atom. Explain why there is more than one correct answer to this problem.

6.20 Given electrons with the following sets of quantum numbers, arrange them in order of increasing energy. If they have essentially the same energy put them together.

a. 3, 2, 1, 1/2
b. 2, 1, 1, −1/2
c. 2, 1, 0, 1/2
d. 3, 1, −1, −1/2
e. 2, 0, 0, −1/2
f. 3, 1, 0, 1/2

6.21 Referring to Example 6.7, which of the following would represent a correct set of quantum numbers for the four 3p electrons of sulfur?

a. I, II, III, IV
b. I, II, IV, VI
c. I, III, V, VI
d. I, II, IV, V
e. I, III, IV, V
f. I, III, IV, VI

6.34 An ion with charge +2 has the following orbital diagram. Name the ion and state its electron configuration.

1s	2s	2p	3s	3p
(⇅)	(⇅)	(⇅)(⇅)(⇅)	(⇅)	(⇅)(↑)(↑)

6.35 Write a complete set of quantum numbers for each of the electrons in an atom whose configuration is $1s^22s^22p^5$.

6.36 Given the following sets of quantum numbers for electrons, indicate those which could not occur, and state why they are wrong.

a. 3, 2, 2, 1/2
b. 3, 0, −1, 1/2
c. 2, 2, 2, 2
d. 1, 0, 0, 0
e. 2, −1, 0, 1/2
f. 2, 0, −2, 1/2

6.37 Referring to Example 6.7, list all the possible correct sets of quantum numbers for the four 3p electrons of sulfur.

*6.38 If the electronic charge were only one tenth as large as it is, atoms would be about 100 times as large as they are now. Under such conditions, would the electrons in atoms behave quantum mechanically? Explain your reasoning.

*6.39 Sodium atoms can be made to emit yellow light at a wavelength of 589.0 nm if they are heated sufficiently in a Bunsen flame. This light is the basis for the test for sodium in qualitative analysis. In the flame the atoms are excited by fast-moving molecules having an average energy per mole of $3RT/2$. Account for the fact that a sodium-containing substance will give off yellow light when in a flame at 1500°C.

*6.40 Suppose the rules for assigning quantum numbers were as follows:

$$\mathbf{n} = 1, 2, 3, \ldots$$
$$\ell = 0, 1, 2, \ldots, \mathbf{n}$$
$$\mathbf{m}_\ell = 0, 1, 2, \ldots, \ell$$
$$\mathbf{m}_s = +1/2 \text{ or } -1/2$$

Prepare a table similar to Table 6.2, based on these rules, for **n** = 1 and **n** = 2. Give the electron configuration for an atom with eight electrons.

*6.41 Suppose that the spin quantum number could have the values 1/2, 0, and −1/2. Assuming that the rules governing the values of the other quantum numbers and the order of filling sublevels were unchanged,

a. what would be the electron capacity of an s sublevel? a p sublevel?
b. how many electrons could fit in the **n** = 2 level?
c. what would be the electron configuration of the element with atomic number 8? 17?

7

THE PERIODIC TABLE AND THE PROPERTIES OF ELEMENTS

In Chapter 6, we pointed out the correlation between the electron configurations of atoms and certain of their physical properties, in particular emission spectra and ionization potentials. However, for our purposes in chemistry, we are primarily interested in the relationship between electron configuration and chemical properties of elements. This relationship is most readily expressed in terms of the Periodic Table of the elements, which was developed over a century ago (see Section 7.6) at a time when electrons were unknown and the very existence of atoms and molecules was still a matter of dispute. Proposed as a means of correlating the chemical properties of known elements and predicting the properties of new ones, the Periodic Table remains today the single most valuable predictive device in all of chemistry.

The modern form of the Table, printed on the inside cover of this book, arranges the elements in order of increasing atomic number in horizontal periods of such length that elements with similar chemical properties fall directly beneath one another in the same vertical group or family. To understand what this means, let us consider how the Periodic Table is constructed.

7.1 STRUCTURE OF THE PERIODIC TABLE; VERTICAL RELATIONSHIPS

Referring to the Periodic Table on the inside cover, we see that it consists of seven horizontal periods (Table 7.1). Each period, except the last, is of such length that it ends with a noble gas in Group 8A ($_{2}He$, $_{10}Ne$, $_{18}Ar$, $_{36}Kr$, $_{54}Xe$, $_{86}Rn$). This requires that we start off with a very short period of only 2 elements, followed by two relatively short periods of 8 elements each. The next two periods, each with 18 elements, are relatively long. They are followed by a very long period of 32 elements; to save space, 14 of these (at. nos. 58 through 71) are taken out and placed in a horizontal row below the main body of the Table. The last period is incomplete; in principle at least it should contain 32 elements and end with another noble gas of atomic number 118.

To illustrate similarities within a group in the Periodic Table, consider the elements in Group 1A at the far left. These so-called *alkali metals* include three quite common elements (lithium, sodium, and potassium) and two relatively rare ones (rubidium and cesium). All of these elements are soft, white, extremely

TABLE 7.1 STRUCTURE OF THE PERIODIC TABLE

PERIOD	NUMBER OF ELEMENTS	BEGINS WITH:	ENDS WITH:
1	2	$_1H$	$_2He$
2	8	$_3Li$	$_{10}Ne$
3	8	$_{11}Na$	$_{18}Ar$
4	18	$_{19}K$	$_{36}Kr$
5	18	$_{37}Rb$	$_{54}Xe$
6	32	$_{55}Cs$	$_{86}Rn$
7	—	$_{87}Fr$	—

active metals; they are commonly stored under oil to prevent reaction with oxygen of the air. Moreover, all of the alkali metals react vigorously with:

1. Chlorine, to form salts (chlorides) of general formula MCl.

The heaviest members of 1A and 7A, Fr and At, do not occur in nature

$$2\ M(s) + Cl_2(g) \rightarrow 2\ MCl(s);\ M = Li, Na, K, Rb, Cs \qquad (7.1)$$

2. Sulfur, to form salts (sulfides) of general formula M_2S.

$$2\ M(s) + S(s) \rightarrow M_2S(s) \qquad (7.2)$$

3. Cold water, to form hydrogen gas and a metal hydroxide, MOH.

$$2\ M(s) + 2\ H_2O(l) \rightarrow 2\ MOH(s) + H_2(g) \qquad (7.3)$$

Reaction 7.3 is strongly exothermic; the hydrogen produced often catches fire or explodes.

At the right of the Periodic Table is another well-known family of elements, the *halogens* in Group 7A. Two members of this group, fluorine and chlorine, are gases at room temperature and atmospheric pressure; bromine is a liquid and iodine a solid under these conditions. All of the halogens are nonmetals; all of them form stable diatomic molecules (F_2, Cl_2, Br_2, I_2). Moreover, they all react with:

1. Sodium, to form salts of general formula NaX.

$$2\ Na(s) + X_2 \rightarrow 2\ NaX(s);\ X = F, Cl, Br, I \qquad (7.4)$$

2. Calcium, to form salts of general formula CaX_2.

$$Ca(s) + X_2 \rightarrow CaX_2(s) \qquad (7.5)$$

3. Hydrogen, to form gaseous, molecular compounds of general formula HX.

$$H_2(g) + X_2 \rightarrow 2\ HX(g) \qquad (7.6)$$

The first element listed in Group 7A, hydrogen, is in a class by itself. It shows some of the chemical properties of the halogens. In particular, hydrogen:

forms a diatomic molecule, H_2, analogous to F_2, Cl_2, Br_2, and I_2.

reacts with sodium and calcium to form salts (hydrides) which have the formulas NaH and CaH_2 (compare Reactions 7.4 and 7.5).

On the other hand, hydrogen resembles, at least superficially, the alkali metals in forming a chloride of formula HCl and a sulfide of formula H_2S (compare Reactions 7.1 and 7.2). For this reason, it is sometimes listed in Group 1A. The unusual properties of hydrogen are related to its electron configuration, $1s^1$. This structure is unique in the sense that hydrogen has only one outer s electron, like the alkali metals, but also is only one electron shy of a noble gas configuration, like the halogens.

TABLE 7.2 PROPERTIES OF THE HALOGENS

	MP (°C)	*BP* (°C)	COLOR	REACTIVITY TOWARD H_2
F_2	−223	−188	light yellow	explosive, even at −180°C
Cl_2	−101	−34	yellow-green	vigorous at 25°C
Br_2	−7	+59	red-brown	slow at 25°C
I_2	+114	+187	deep violet	slow below 200°C

Although elements in a given group show many similarities in chemical properties, they also differ from one another in a systematic way. Ordinarily we find that both physical and chemical properties follow a smooth trend as we move down a given group. Among the halogens, for example, melting point and boiling point increase steadily from fluorine to iodine; the color of the elementary substance deepens (Plate 3) and its chemical reactivity decreases (Table 7.2).

7.2 CORRELATION WITH ELECTRON CONFIGURATION; HORIZONTAL RELATIONSHIPS

Even before the Periodic Table had been generally accepted, chemists began to speculate as to the basis for its existence. Why, in particular, should the members of a group behave chemically in the same way? This question and others related to it were answered by the modern theory of atomic structure discussed in Chapter 6. We now know that similarity in chemical properties reflects similarity in electronic structures. Specifically, **all of the elements in a group have the same electron configuration in their outermost principal energy level.**

To see what is meant by the statement just made, consider the electron configurations of the 1A elements. From Table 7.3 we see that in atoms of each of these metals there is a single s electron in the outermost principal energy level. We describe the general outer configuration of this group as "ns^1," where n refers to the period in which the element is located (2 for Li, 3 for Na, . . .).

The chemical similarities between the 1A elements can be explained quite simply in terms of their electron configurations. When atoms of these elements take part in chemical reactions, they lose the outer s electron to form +1 ions (Li^+, Na^+, K^+, Rb^+, Cs^+). A similar situation applies with other families of elements. As we will see in Chapter 8, when atoms react to form ions or molecules, it is ordinarily the electrons in outer energy levels which are involved. Hence, it is hardly surprising that elements with the same outer electron configuration have similar chemical properties.

Inner electrons are too stable to be affected by chemical reactions

To explore further the correlation between the position of an element in the Periodic Table and its electron configuration, it is helpful to divide the groups into three categories, discussed separately on the following page.

TABLE 7.3 ELECTRON CONFIGURATIONS OF THE ALKALI METALS

Li	$1s^22s^1$	[He] $2s^1$
Na	$1s^22s^22p^63s^1$	[Ne] $3s^1$
K	$1s^22s^22p^63s^23p^64s^1$	[Ar] $4s^1$
Rb	$1s^22s^22p^63s^23p^64s^23d^{10}4p^65s^1$	[Kr] $5s^1$
Cs	$1s^22s^22p^63s^23p^64s^23d^{10}4p^65s^24d^{10}5p^66s^1$	[Xe] $6s^1$

Representative Elements (A Subgroups)

In the A groups, the group number is equal to the number of electrons in the outermost principal energy level

In Groups 1A through 8A of the Periodic Table, which include both metals and nonmetals, electrons are being added to outer s and p sublevels. In the first period, two elements, hydrogen and helium, are required to fill the 1s sublevel. In all succeeding periods, there are 8 so-called "representative elements"; the first two elements in each period (Groups 1A and 2A) add electrons to an s sublevel while the last six members (Groups 3A through 8A) fill the corresponding p sublevel. The outer electron configurations of these elements are:

1A	2A	3A	4A	5A	6A	7A	8A
ns^1	ns^2	ns^2np^1	ns^2np^2	ns^2np^3	ns^2np^4	ns^2np^5	ns^2np^6

where n = 1 for the 1st period, 2 for the 2nd period, and so on.

Example 7.1 Using the Periodic Table, give the outer electron configuration of atoms of

a. Ba
b. As

Solution We locate barium in the 6th period in Group 2A. It must then have two electrons in the 6s sublevel; its outer electron configuration is $6s^2$. Arsenic, in Group 5A, is in the 4th period. Its outer electron configuration is $4s^24p^3$.

Transition Metals (B Subgroups)

In the 10 "extra elements" ($_{21}Sc \rightarrow {}_{30}Zn$) that appear in the 4th period, electrons are being added to an inner d sublevel (3d). Directly below, in the 5th and 6th periods, are two other series of 10 elements each in which the 4d and 5d sublevels are filling. These elements, which comprise the B subgroups of the Periodic Table, are properly classified as metals; they are known collectively as *transition metals*. Their properties differ in many ways from those of the 1A and 2A metals that precede them in each period. In particular, the transition metals as a class are:

1. *Higher melting*. All of the metals in the first transition series save one (Zn, *mp* = 420°C) melt at temperatures well above 1000°C. In contrast, their neighbors to the left, potassium and calcium, melt at 64 and 851°C, respectively.

2. *Less reactive*. Both potassium and calcium react vigorously with water to form hydrogen:

$$K(s) + H_2O(l) \rightarrow KOH(s) + \tfrac{1}{2}\, H_2(g) \qquad (7.7)$$

$$Ca(s) + 2\, H_2O(l) \rightarrow Ca(OH)_2(s) + H_2(g) \qquad (7.8)$$

Of the 10 transition metals that follow K and Ca in the 4th period, only the first one, scandium, reacts with water at room temperature:

$$Sc(s) + 3\, H_2O(l) \rightarrow Sc(OH)_3(s) + \tfrac{3}{2}\, H_2(g)$$

The other members react only at high temperatures (Ti, Mn, Fe, Co, Ni, and Zn) or not at all (V, Cr, Cu).

3. *More likely to form colored compounds*. All of the metals in the first transition series except scandium and zinc form a variety of brilliantly colored compounds (Plate 4). In contrast, solutions of potassium and calcium salts are ordinarily colorless.

The two sets of 14 elements each, located at the bottom of the Periodic Table, are adding electrons to f sublevels. The *lanthanides,* or *rare earths* (at. nos. 58 through 71), fill the 4f sublevel. Adding inner f electrons appears to have little effect on chemical properties; all of these elements closely resemble lanthanum in their chemistry. All the lanthanides are relatively active metals, reacting readily with chlorine to form salts of general formula MCl_3:

$$2\ M(s) + 3\ Cl_2(g) \rightarrow 2\ MCl_3(s);\ M = \text{La, rare earths} \qquad (7.10)$$

and with water to evolve hydrogen and form hydroxides of general formula $M(OH)_3$:

$$M(s) + 3\ H_2O(l) \rightarrow M(OH)_3(s) + 3/2\ H_2(g) \qquad (7.11)$$

Because of similarities in their properties, compounds of the rare earths are very difficult to separate from one another by classical methods such as fractional crystallization (Chapter 1). Until quite recently, samples of pure compounds of these elements were not commercially available except for those of cerium, the most abundant member of the series. Within the past few years, chromatographic processes have been adapted to separate salts of these elements and they have become much more accessible. The availability of rare earth compounds in highly pure form has led to several commercial applications. The brilliant red phosphor used in many late-model color TV receivers contains a small amount of europium oxide, Eu_2O_3, added to a base of yttrium oxide, Y_2O_3, or gadolinium oxide, Gd_2O_3. Another rare earth oxide, Nd_2O_3, is being used as part of a liquid laser system.

Summarizing: 1A and 2A metals fill s sublevels, 3A to 8A elements fill p sublevels, transition metals fill d sublevels, inner transition metals fill f sublevels

The *actinide* metals (at. nos. 90 through 103), all of which are radioactive, fill the 5f sublevel. Of these 14 elements, only two, uranium and thorium, are found in appreciable amounts in nature. The other elements were all first observed in the products of controlled nuclear reactions, and in most cases have been produced in only very small amounts. Uranium and plutonium are used as the fuel elements in nuclear reactors and nuclear weapons. These applications have been primarily responsible for the interest in this series, which includes the heaviest elements in the Periodic Table.

7.3 TRENDS IN ATOMIC PROPERTIES

Many of the characteristic properties of atoms can be correlated with their positions in the Periodic Table. In this section we will consider how four of these properties (atomic radius, ionization energy, electronegativity, and metallic character) change as we move across or down in the Table.

Atomic Radius

The "size" of an atom is a rather nebulous concept, since the electron cloud surrounding the nucleus does not have a sharp boundary. It is possible, however, to define and measure a quantity known as the atomic radius. To do this, it is assumed that those atoms which are closest to each other in an elementary substance are actually touching. On this basis, the atomic radius of a solid element such as copper is taken to be one half of the distance between the nuclei of adja-

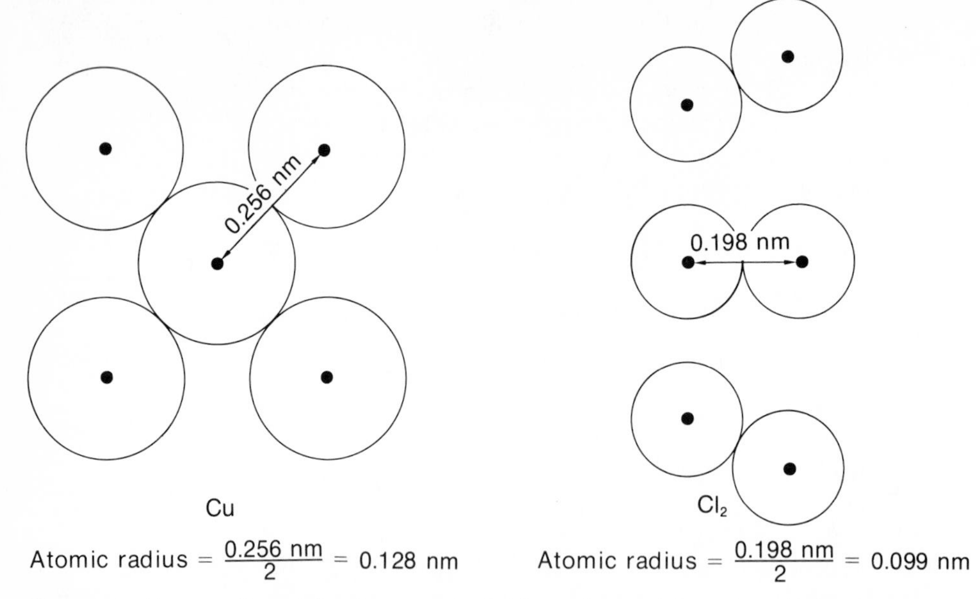

Figure 7.1 Atomic radii are defined by assuming that those atoms which are closest in the element are touching.

cent atoms (Fig. 7.1). For gaseous elements which form molecules, such as Cl_2, the atomic radius is taken to be one half of the distance between centers of atoms in the molecule.

The atomic radii of the A group elements are shown in Figure 7.2 (numerical values for all elements are given in Appendix 2, p. A.8). Notice that in general **atomic radius decreases as one moves across the Periodic Table from left to right and increases as one moves down within a given group.**

It is possible to rationalize the trends observed in Figure 7.2 in terms of the electron configurations of the corresponding atoms. Consider first the increase in radius observed as we move down the Table, let us say among the alkali metals (1A). All of these elements have a single s electron outside a filled level or filled p sublevel. Electrons in these inner levels are much closer to the nucleus than the outer s electron and hence effectively shield it from the positive charge of the nucleus. To a first approximation, each inner electron cancels the charge of one proton in the nucleus, so the outer s electron is attracted by a net positive charge of +1. In this sense, it has the properties of an electron in the hydrogen atom. Since the average distance of the electron from the hydrogen nucleus increases with the principal quantum number, we would expect the radius to increase as we move from Li (2s electron) to Na (3s electron), and so on down the group.

The decrease in atomic radius observed as we move across the Periodic Table can be explained in a similar manner. Consider, for example, the 3rd period, where electrons are being added to the 3rd principal energy level. The added electrons should be relatively poor shields for each other, since they are all at nearly the same distance from the nucleus. Only the 10 electrons in inner, filled levels ($n = 1$, $n = 2$) should be effective in shielding the outer electrons from the nucleus. This means that the "effective nuclear charge" should increase steadily as we move across the period; it would be about +1 for Na (at. no. = 11), +2 for Mg (at. no. = 12), +3 for Al (at. no. = 13), and so on. As the effective nuclear charge increases, outer electrons are pulled in more tightly and atomic radius decreases.

To be a good shield, an electron must be between the nucleus and the electron being shielded

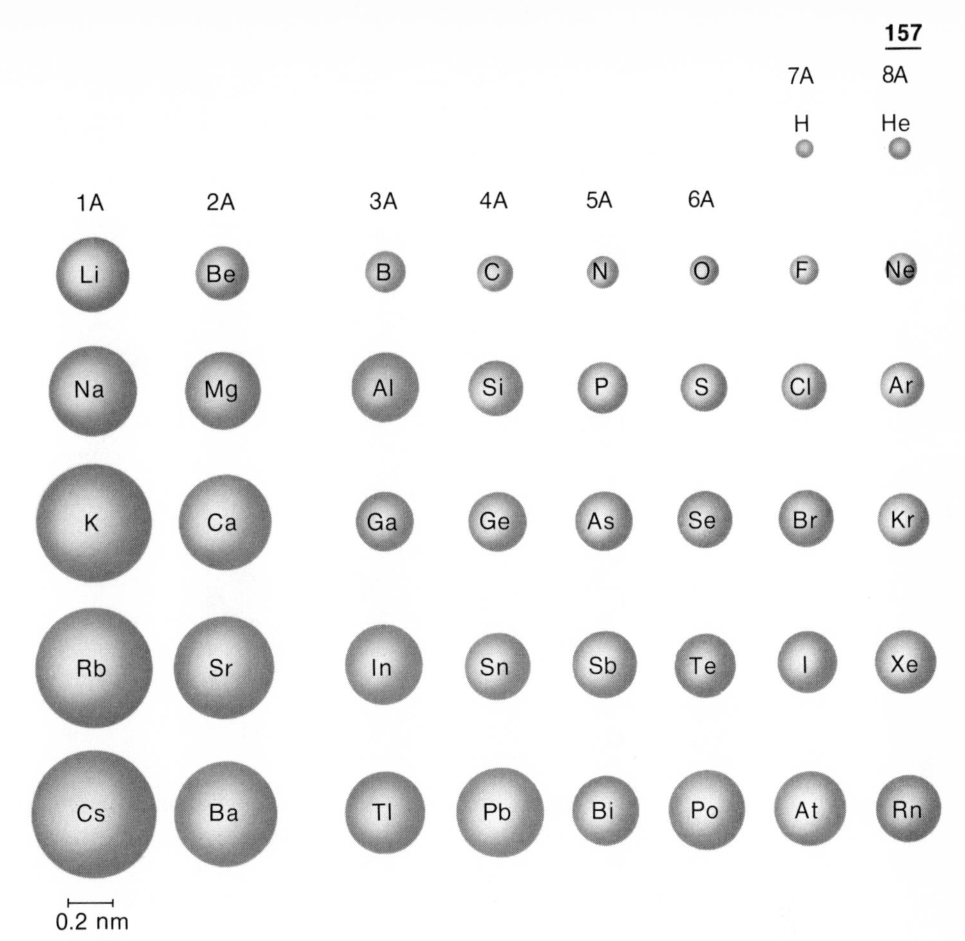

Figure 7.2 Atomic radii of the A group elements. Atomic radii increase as one goes down a group and in general decrease in going across a row in the Periodic Table. Hydrogen has the smallest atom, cesium the largest.

First Ionization Energy

As pointed out in Chapter 6, the 1st ionization energy of an element is the energy change, ΔE, for the removal of the outermost electron from a gaseous atom to form a +1 ion:

$$M(g) \rightarrow M^+(g) + e^- \qquad (7.12)$$

Ionization energy is always a positive quantity; energy must be absorbed to remove an electron. The magnitude of ΔE measures the ease with which an atom gives up an electron; the smaller the ionization energy, the more readily the electron is removed.

Ionization energies of the A group elements are listed in Figure 7.3. Notice that there is a general tendency for **ionization energy to increase as one moves across the Periodic Table and decrease as one moves down.** Comparing Figure 7.3 with Figure 7.2, we see that there is an inverse correlation between ionization energy and atomic radius. The smaller the atom, the more tightly its electrons are held to the nucleus and the more difficult they are to remove. Large atoms such as those of the alkali metals have relatively low ionization energies because the outer electrons are far away from the positive charge of the nucleus.

1A	2A	3A	4A	5A	6A	7A	8A
						H 1318	He 2377
Li 527	Be 904	B 808	C 1092	N 1410	O 1322	F 1686	Ne 2088
Na 502	Mg 745	Al 586	Si 791	P 1021	S 1004	Cl 1264	Ar 1527
K 427	Ca 594	Ga 586	Ge 770	As 954	Se 946	Br 1146	Kr 1356
Rb 410	Sr 556	In 565	Sn 715	Sb 841	Te 879	I 1017	Xe 1176
Cs 377	Ba 510	Tl 594	Pb 724	Bi 711	Po 820	At	Rn 1042

Figure 7.3 First ionization energies of the A group elements (kJ/mol). Ionization energies vary inversely with atomic radius, decreasing as one goes down a group and increasing across a row. Cesium has the lowest ionization energy, helium the highest.

Electronegativity

Ionization energy is, as we have seen, a quantitative measure of the ease with which an atom gives up electrons. It is considerably more difficult to find a useful measure of the opposite process: the tendency of an atom to attract electrons. The approach we will follow here was first suggested by Linus Pauling. On the basis of bond energy calculations (Chapter 8), he assigned to each atom an electronegativity value which is proportional in a qualitative way to its attraction for electrons. On this scale, atoms are assigned numbers ranging from 4.0 for the most electronegative atom, fluorine, to 0.7 for the element having the least attraction for electrons, cesium. Figure 7.4 lists electronegativity values for all the A group elements for which information is available.

Oxygen is second only to fluorine in electronegativity

H 2.1						
Li 1.0	Be 1.5	B 2.0	C 2.5	N 3.0	O 3.5	F 4.0
Na 0.9	Mg 1.2	Al 1.5	Si 1.8	P 2.1	S 2.5	Cl 3.0
K 0.8	Ca 1.0	Ga 1.6	Ge 1.8	As 2.0	Se 2.4	Br 2.8
Rb 0.8	Sr 1.0	In 1.7	Sn 1.8	Sb 1.9	Te 2.1	I 2.5
Cs 0.7	Ba 0.9	Tl 1.8	Pb 1.8	Bi 1.9	Po 2.0	At 2.2

Figure 7.4 Electronegativity values (Pauling) for the A group elements. Electronegativity varies in the same way as ionization energy, decreasing down a group in the Periodic Table and increasing across a row. Fluorine atoms have the highest electronegativity.

It is readily apparent from Figure 7.4 that **electronegativity increases moving across the Periodic Table and decreases moving down the Table.** Comparing this figure with Figure 7.2, we see again an inverse correlation with atomic radius. The smallest atoms, located toward the upper right corner of the Table, have the strongest attraction for electrons. As we move down a given group, the atoms get larger and electronegativity decreases.

Metallic Character

Of the 106 known elements, approximately 81 can be classified as metals. All metals possess to varying degrees the following physical properties:

1. *High electrical conductivity.* Metals as a group have electrical conductivities several orders of magnitude greater than those of typical nonmetals. Copper, one of the best metallic conductors, is used extensively in electrical wiring. Mercury is actually one of the poorest metallic conductors, but is used in many electrical devices where a liquid conductor is necessary.

2. *High thermal conductivity.* Among solids, metals are by far the best conductors of heat.

3. *Luster.* The pleasing appearance of polished metal surfaces is due to their ability to reflect light. Most metals have a silvery white color, indicating that light of all wavelengths is reflected. Gold and copper absorb some light in the blue region of the spectrum and hence appear yellow and red, respectively.

4. *Ductility, malleability.* Most metals are ductile (capable of being drawn out into wire) and malleable (capable of being hammered into thin sheets).

Perhaps the single most distinguishing characteristic of **metals** is their tendency to **lose electrons readily.** From a chemical standpoint, this means that they tend to form positive ions (e.g., Na^+, Mg^{2+}, Al^{3+}) when they react with nonmetals. Physically, we can associate metallic character with *low ionization energy.* Referring back to Figure 7.3, we would predict that **metallic character should decrease as we move across the Periodic Table and increase as we move down.** Figure 7.5

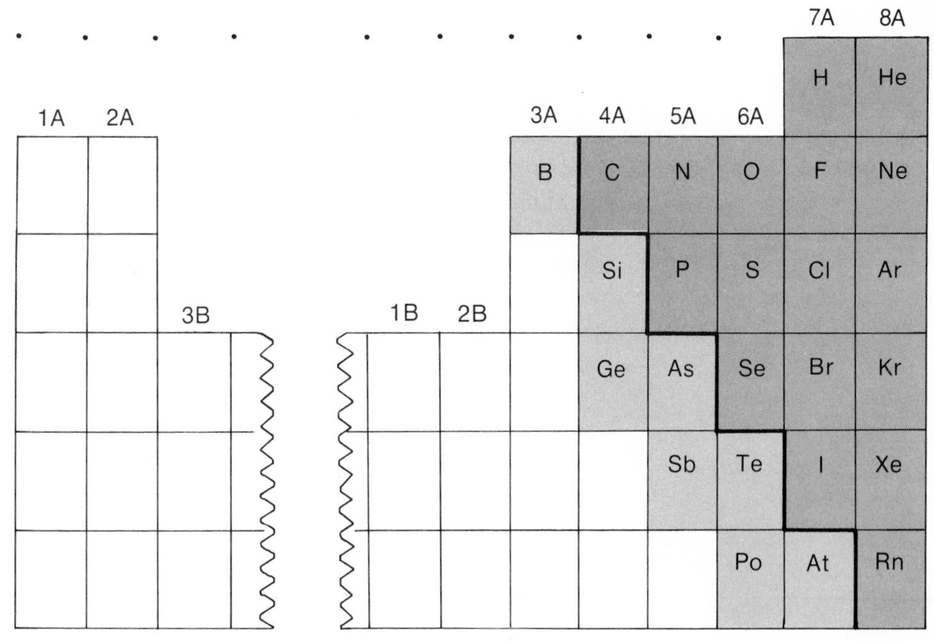

Figure 7.5 Metallic character and the Periodic Table. Most of the elements are metals (colorless). There are 17 nonmetals (dark color) and about eight metalloids (medium color).

confirms these general trends. The elements we commonly consider to be metals include:

—All of the elements in Groups 1A and 2A.
—The heavier elements in 3A (Al, Ga, In, Tl), 4A (Sn, Pb), and 5A (Bi).
—All of the transition elements (B subgroups).
—All of the lanthanides and actinides.

NONMETALS. Clustered toward the upper right corner of the Periodic Table are 17 elements ordinarily classified as nonmetals. They have few if any of the characteristic properties of metals. Except for selenium and the graphite form of carbon, they are nonconductors of electricity. With a few exceptions, notably diamond, crystals of nonmetals have a dull rather than shiny appearance. All solid nonmetals shatter if drawn out or hammered.

From a chemical point of view, the most important characteristic of **nonmetals** is their tendency to **gain electrons in chemical reactions.** Thus we find that elements such as oxygen and fluorine acquire electrons to form negative ions (O^{2-}, F^-) when they react with metals. The tendency of nonmetal atoms to gain electrons readily is also reflected in their high electronegativities (Fig. 7.4), which range from 2.1 to 4.0. In contrast, all metals have electronegativities less than 2.0.

Of the elements listed as nonmetals in Figure 7.5, the so-called "noble gases" in Group 8A are unique in two important respects. Atoms of these elements show:

1. *No tendency to combine with one another.* All of these gases (He, Ne, Ar, Kr, Xe, Rn) are monatomic. In contrast, most of the other nonmetals form polyatomic molecules in the gaseous state (e.g., N_2, P_4, O_2, S_8, F_2, Cl_2, Br_2, I_2).

The most reactive element, F, is next to a completely inert one, Ne, in the Periodic Table. Can you suggest an explanation?

2. *Little tendency to react with atoms of other elements.* As late as 1962, no stable compounds were known for any of the noble gases. Since that time, a few compounds of the heavier elements in the group (Kr, Xe, Rn) have been prepared. To this day, no stable compounds are known for the first three members of the family (He, Ne, Ar).

Both of these characteristics are related to the unusual stability of the electron configurations of the noble gases (ns^2np^6). We will have more to say on this topic when we discuss chemical bonding in Chapter 8.

METALLOIDS. Located along a diagonal line in the Periodic Table (Figure 7.5) are several elements (B in 3A, Si and Ge in 4A, As and Sb in 5A, Te in 6A) that have properties intermediate between those of metals and nonmetals. These elements are referred to as *metalloids.* All of them show metallic luster, yet none of them form positive ions in chemical reactions. They have ionization energies clustered around 800 kJ/mol (Fig. 7.3) and electronegativities close to 2.0 (Fig. 7.4).

Typically, metalloids are semiconductors, although antimony and arsenic actually have electrical conductivities which approach those of metals. Silicon and germanium are the principal elements used in commercial semiconductor devices such as transistors (Chapter 11). Perhaps the single most distinctive characteristic of metalloids is the temperature dependence of their electrical conductivity; in contrast to metals, they become better conductors when the temperature is raised.

The two very heavy elements, polonium and astatine, neither of which occurs naturally, are ordinarily classed as metalloids. Actually, since very little is known about their chemical and physical properties, any such classification must be rather arbitrary.

7.4 PREDICTIONS BASED ON THE PERIODIC TABLE

Given an understanding of the Periodic Table, one can, on the basis of relatively little information, predict the properties of many substances, including both elements and compounds. Here we will consider how such predictions can be made in three general areas.

1. *Physical properties of elements, given those of neighboring elements in the Table.* The principle here is very simple: the properties of elements ordinarily vary smoothly as we move down a group. Consequently, an "unknown" element should have physical properties approximately midway between those of its neighbors above and below it. The same principle can be used, with somewhat less reliability, to estimate the properties of an element from those of its neighbors within the same period (i.e., to the left and right of the element in question).

Example 7.2 Predict the melting point and density of strontium, given those of Ca (851°C, 1.5 g/cm³) and Ba (710°C, 3.5 g/cm³).

Solution Taking a simple average of the properties of the elements above and below strontium (Ca and Ba) in Group 2A, we obtain

melting point $$\frac{851°\text{C} + 710°\text{C}}{2} = 780°\text{C}$$

density $$\frac{1.5\ \text{g/cm}^3 + 3.5\ \text{g/cm}^3}{2} = 2.5\ \text{g/cm}^3$$

The observed values are 770°C and 2.6 g/cm^3.

It is only fair to point out that the excellent agreement obtained in Example 7.2 is rather unusual. Sometimes this approach predicts values for physical properties that vary widely from those measured in the laboratory. A more dependable (but more tedious) approach is to plot the desired physical property vs. atomic number, using values for all known members of a group or period rather than only two. The property of the unknown element can then be estimated reliably by interpolation from a smooth curve drawn through the known points.

2. *Relative values of properties of elements in comparison to those of other elements.* We can use the principles discussed in Section 7.3 to compare different elements with respect to such properties as metallic character, ionization energy, electronegativity, or atomic radius.

Example 7.3 Referring to the Periodic Table, arrange the following elements in order of decreasing electronegativity: Be, B, Mg, Al.

Solution Recall from the discussion in Section 7.3 that electronegativity increases as we move to the right and decreases as we move down in the Table. The four elements listed form a block in the Table:

Be	B
Mg	Al

Since boron is at the upper right corner of the block, it should have the highest electronegativity. Beryllium, which is to the left of boron, and aluminum, which is below boron, should have lower electronegativities, about equal to each other. The least electronegative element of the four should be magnesium, since it is located at the lower left corner. Thus our prediction is:

$$\text{B} > \text{Be} \approx \text{Al} > \text{Mg}$$

The actual values, taken from Figure 7.4, are B = 2.0, Be = Al = 1.5, Mg = 1.2.

3. *Formulas of compounds, given that of a compound containing elements from the same groups of the Periodic Table.* We might expect that elements in the same groups would form compounds with the same type of formula. Given that the formula of magnesium chloride is $MgCl_2$, we would predict correctly that the formulas of other chlorides of the 2A elements would be $CaCl_2$, $SrCl_2$, and $BaCl_2$, or that the formulas of the other magnesium halides would be MgF_2, $MgBr_2$ and MgI_2. The same approach can be applied to predict the formulas of compounds containing three elements (Example 7.4).

Example 7.4 Given the formulas for sodium chlorate, $NaClO_3$, barium chromate, $BaCrO_4$, and sodium phosphate, Na_3PO_4, predict the formulas of:

a. potassium arsenate
b. strontium tungstate
c. rubidium bromate

Solution

a. Potassium is in the same group as sodium; arsenic is in the same group as phosphorus. Hence, we predict K_3AsO_4, analogous to Na_3PO_4.
b. $SrWO_4$, analogous to $BaCrO_4$.
c. $RbBrO_3$, analogous to $NaClO_3$.

Like any general rule, the one just cited has many exceptions. In some cases it leads us to predict formulas for compounds that are unknown. To illustrate, suppose we were to take carbon monoxide, CO, as a reference. Our approach would predict the existence of a sulfide of carbon with the formula CS. The compound, if it exists at all, appears to be highly unstable. Exceptions of this type occur most frequently when one of the elements involved is the first member of a group of nonmetals, 4A through 7A (e.g., C, N, O, F). Referring back to Example 7.4, we might point out that:

—No fluorine-containing compound analogous to $NaClO_3$ exists.

—Although sodium nitrate is a stable compound, its formula, $NaNO_3$, is quite different from that of sodium phosphate, Na_3PO_4.

7.5 SOURCES OF THE ELEMENTS

Of the 90 elements found in nature, six (N_2, O_2, and four noble gases, Ne, Ar, Kr, Xe) occur in elementary form in the atmosphere. They are separated from one another by the fractional distillation of liquid air (Chapter 17). Four others (Na, Mg, Cl_2, and Br_2) can be extracted from the oceans or salt brines, where they are present as monatomic ions (Na^+, Mg^{2+}, Cl^-, Br^-). Chemical reactions, often carried out in electrical cells, are required to convert these ions to the elementary substances (Chapters 22, 23).

Nearly all of the remaining elements are obtained from mineral deposits on or beneath the surface of the earth. Most such deposits are of little value as sources of elements, either because they are too impure or because it is too difficult to obtain the element from them. The relatively few minerals from which elements can be extracted profitably are referred to as *ores*. The principal ores of the various elements are listed in Figure 7.6; other, less important ores are available for many of these elements.

Figure 7.6 reveals a correlation between the position of an element in the Periodic Table and the chemical composition of its principal ore. In particular:

1. Metals which are sufficiently unreactive to occur uncombined (i.e., in elementary form) are clustered in Groups 8B and 1B of the 2nd and 3rd transition series (e.g., platinum, gold; free silver is also found in nature).

2. Metalloids (e.g., Ge, As, Sb) and neighboring metals, all of which have relatively large ionization energies, tend to occur as sulfides.

3. The more strongly metallic elements that form positive ions readily are usually found as oxides (transition metals), carbonates (2A metals), or chlorides (1A metals).

Very few ores occur as chemically pure substances. Ordinarily, they are heterogeneous mixtures in which the desired mineral is contaminated by large amounts of impurities (*gangue*). Many different separation methods, both physical and chemical, are used to get rid of this waste material. To cite one example, many sulfide ores, notably Cu_2S, are concentrated by a process known as flotation (Fig. 7.7). The finely divided ore is mixed with oil and agitated with soapy water in a large tank. When compressed air is blown through the mixture, the oil-coated particles of metal sulfide are carried to the top of the tank, where they form a froth which can be skimmed off. The gangue, which may comprise as much as 99% of the ore, settles to the bottom.

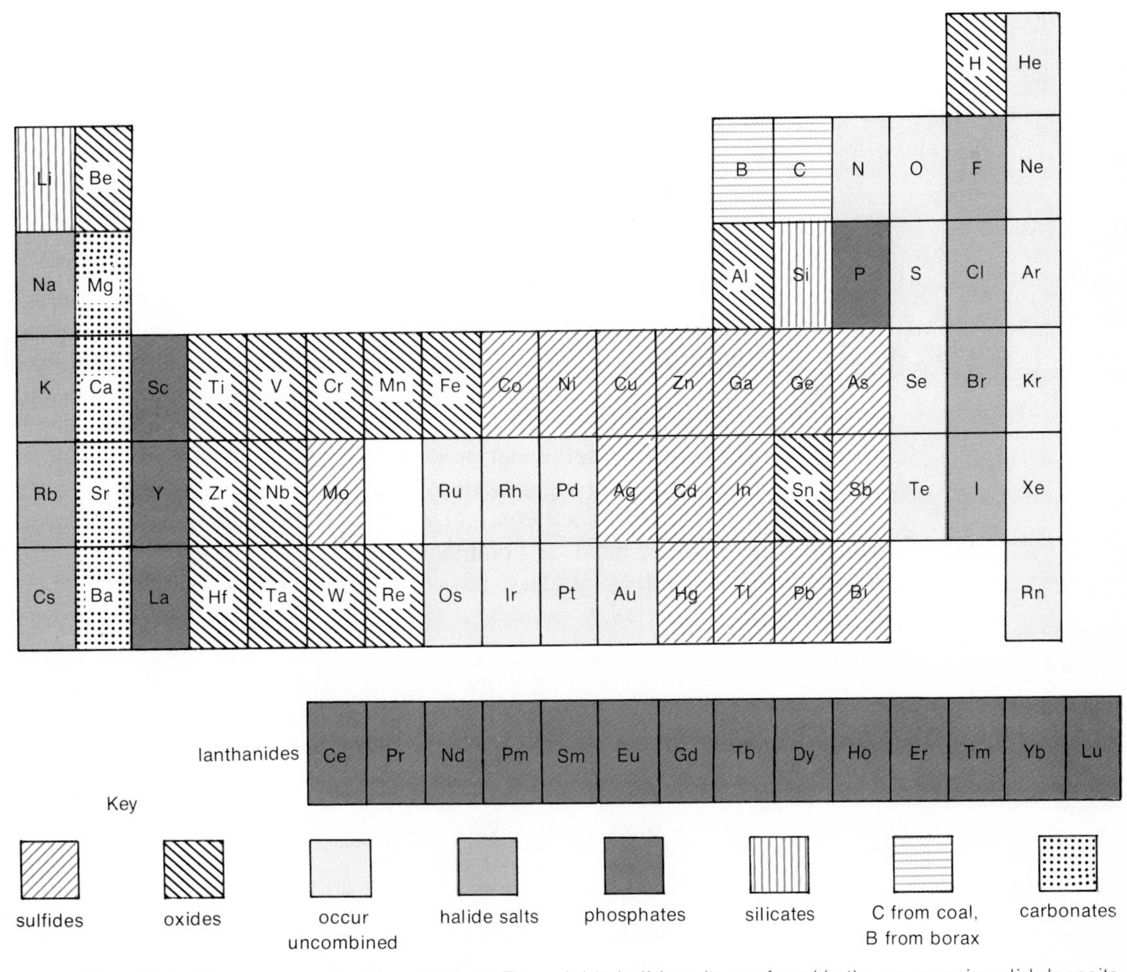

Figure 7.6 Natural sources of the elements. The soluble halide salts are found in the ocean or in solid deposits. Most of the noble gases are obtained from air.

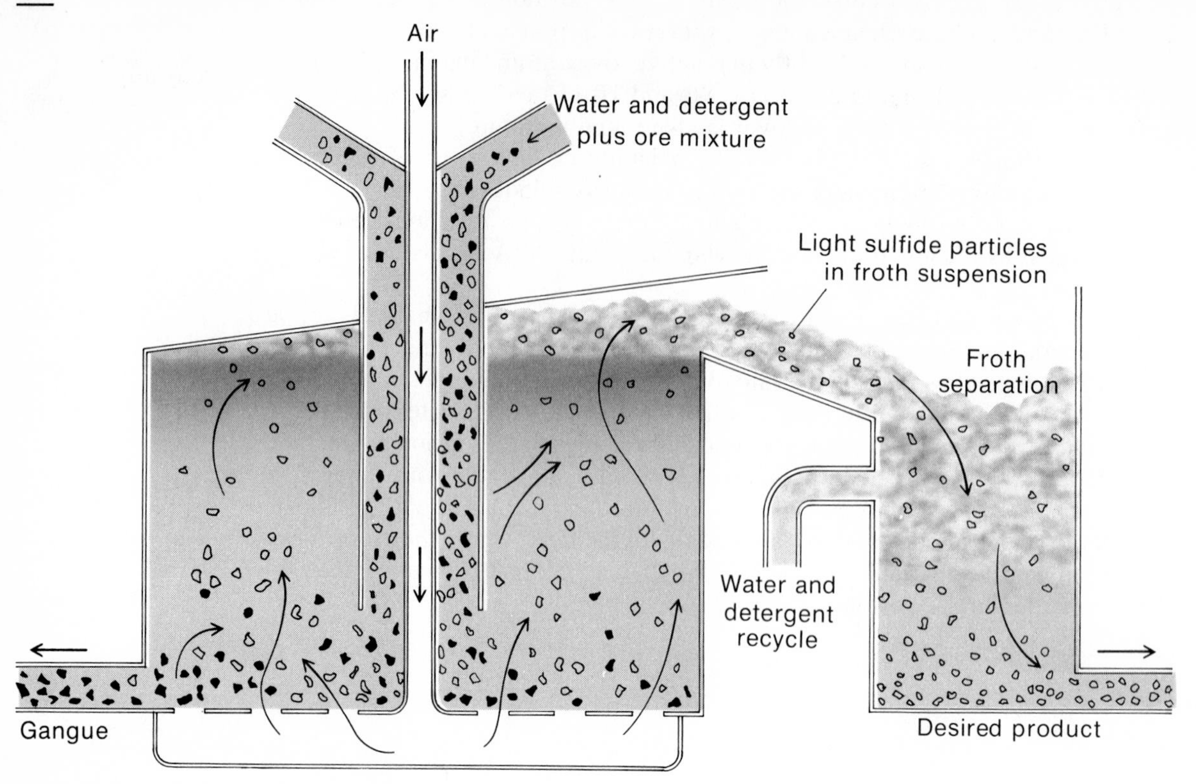

Figure 7.7 Sulfide ores are often enriched by flotation. The relatively light sulfide particles are entrapped in soapy bubbles and separated from the heavier gangue.

Formation of Elementary Substances from Their Ores

Gold metal can be separated by dissolving it in mercury

Metals which occur in the elementary state, such as gold and platinum, can be obtained in nearly pure form by simple physical separations. The only solid nonmetal in this category, sulfur, occurs in large underground deposits of 99% purity. It is mined by a process worked out in 1890 by a chemical engineer, Herman Frasch. Superheated water under pressure is used to melt the sulfur (*mp* = 119°C), which is then brought to the surface with compressed air.

More commonly, ores are compound rather than elementary substances. In this case, the element must be obtained chemically. With a few sulfides of the less active metals such as mercury, this can be accomplished simply by heating in air (*roasting*):

$$HgS(s) + O_2(g) \rightarrow Hg(g) + SO_2(g) \quad (7.13)$$

With more active metals such as zinc, roasting produces the oxide rather than the free metal

$$2\ ZnS(s) + 3\ O_2(g) \rightarrow 2\ ZnO(s) + 2\ SO_2(g) \quad (7.14)$$

and further treatment of the oxide is necessary.

Most metal oxides can be converted to the metal by heating with an element which has a strong affinity for oxygen. Carbon in the form of coke is most frequently used. The reactions that occur are rather complex; the effective oxygen-removing agent appears to be carbon monoxide, formed from the coke by a two-step, exothermic process:

$$C(s) + O_2(g) \rightarrow CO_2(g)$$

$$C(s) + CO_2(g) \rightarrow 2\ CO(g)$$

$$2\ C(s) + O_2(g) \rightarrow 2\ CO(g);\ \Delta H = -221\ \text{kJ} \qquad (7.15)$$

The CO produced then reacts with the metal oxide, forming the metal and carbon dioxide. In the case of zinc oxide, the reaction is

$$ZnO(s) + CO(g) \rightarrow Zn(g) + CO_2(g) \qquad (7.16)$$

The process forms zinc vapor at 1200°C. Finely divided zinc dust forms on cooling.

The most important metallurgical process involving carbon is the reduction of hematite ore, essentially Fe_2O_3 with 10 to 30% SiO_2. More than four hundred million tonnes of crude "pig iron" are produced annually in the world, using blast furnaces of the type shown in Figure 7.8. A mixture of iron ore, coke, and limestone, $CaCO_3$, is charged at the top of the furnace, which may be as tall as

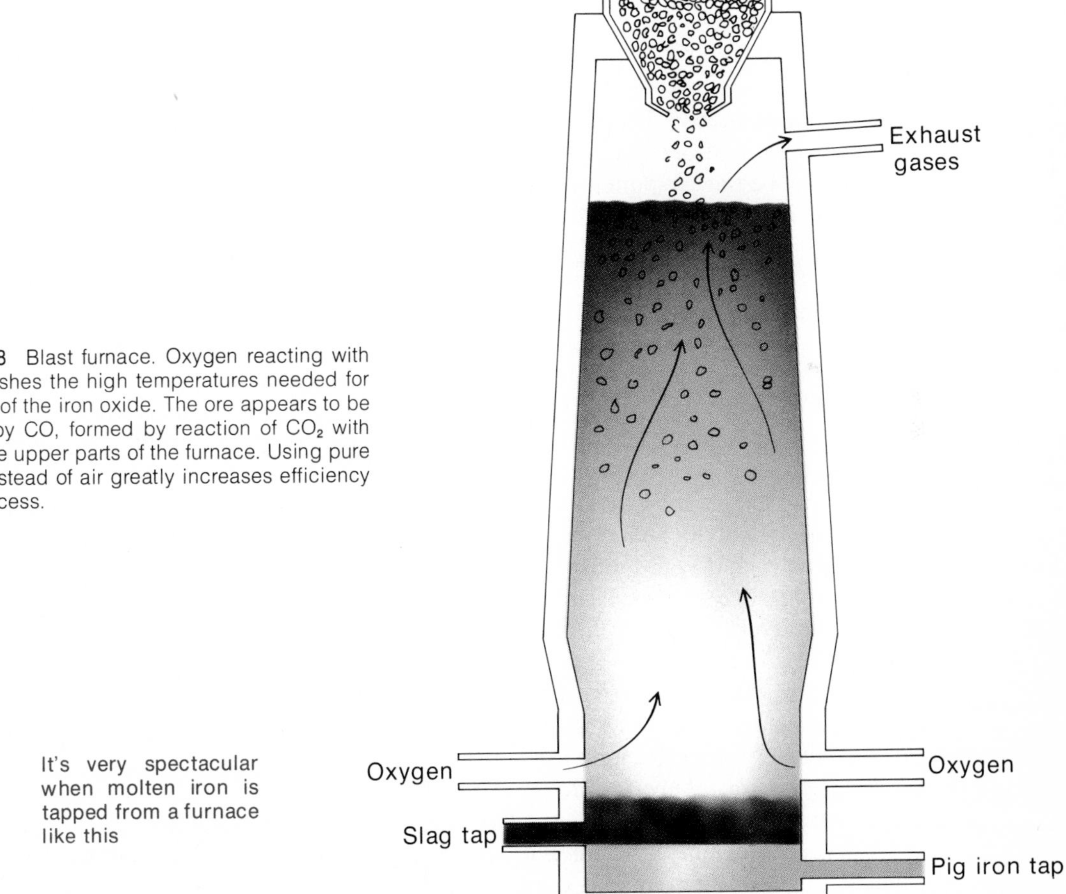

Figure 7.8 Blast furnace. Oxygen reacting with coke furnishes the high temperatures needed for reduction of the iron oxide. The ore appears to be reduced by CO, formed by reaction of CO_2 with coke in the upper parts of the furnace. Using pure oxygen instead of air greatly increases efficiency of the process.

It's very spectacular when molten iron is tapped from a furnace like this a ten-story building. Pure oxygen or compressed air admitted near the bottom burns the coke, liberating a great deal of heat and producing some carbon monoxide which reacts with the Fe_2O_3:

$$Fe_2O_3(s) + 3\ CO(g) \rightarrow 2\ Fe(l) + 3\ CO_2(g) \qquad (7.17)$$

Molten iron (mp = 1530°C) collects at the base of the furnace, where the temperature may be as high as 2000°C. A liquid "slag" with a density about one third that of iron is formed by a reaction which may be represented most simply as

$$CaCO_3(s) + SiO_2(s) \rightarrow CaSiO_3(l) + CO_2(g) \qquad (7.18)$$

Oxides of very active metals cannot be reduced by heating with coke. The metals may often be obtained, however, by electrolysis (Chapter 22) or by heating with a metal which has an even stronger affinity for oxygen, such as aluminum. Chromium metal is produced in this way:

$$Cr_2O_3(s) + 2\ Al(s) \rightarrow 2\ Cr(s) + Al_2O_3(s);\ \Delta H = -542\ kJ \qquad (7.19)$$

This type of reaction, known as the *thermite* process, is highly exothermic. Indeed, with a mixture of Fe_2O_3 and powdered aluminum, a temperature of 3000°C can be reached. The molten iron formed was used at one time to weld steel rails and repair broken machinery.

Prospects for the Future

In Table 7.4, projections are made as to how many years we can reasonably expect to extract the common metals from ores now in commercial use. The "static index" listed in column 3 is obtained by simply dividing known global reserves of ores by the annual consumption (in 1970). A more realistic estimate may be the "exponential index" in column 5, which takes into account the rate at which consumption of metal products is growing. On this basis, we seem likely to run out of many of the heavy metals by the end of this century.

There are at least three ways in which the picture presented in Table 7.4 might change. We could:

1. *Substitute* renewable or more abundant resources such as wood, glass, or granite for metals. This would appear to be feasible with at least a few metals such as iron and aluminum which are used extensively in construction.

2. *Recycle* metal products. Despite all the attention that recycling projects have received in recent years, they make a relatively minor contribution to our supplies of most metals. Only about one-tenth of the ten billion aluminum and tin-plated steel cans thrown away each year are recycled. One difficulty is that waste metal products are often contaminated with materials which are difficult to remove and have an adverse effect on the properties of the base metal. The relatively small amount of copper used in electrical wiring in automobiles is essentially impossible to remove from molten iron or steel and forms a low strength alloy with it. One solution here would be to replace copper with aluminum, which can be separated easily from steel. This would have the added benefit of conserving copper, a metal in short supply (Table 7.4).

This is one reason why there are 15 million abandoned cars in the U.S.

3. *Use lower-grade ores.* The reserves listed in Table 7.4 include only those ores from which metals can now be commercially extracted. In most cases, they represent only a small fraction of the total amount of the metal in the earth's crust. As an extreme example, bauxite ore, (Al_2O_3), from which aluminum is obtained, accounts for less than one billionth (10^{-9}) of the total supply of that metal potentially available. Most of the aluminum is tied up in silicate minerals (e.g.,

TABLE 7.4 RESOURCES OF METALS (1970)*

1	2	3	4	5	6
METAL	KNOWN RESERVES (KG)	STATIC INDEX** (YEARS)	ANNUAL % GROWTH†	EXPONENTIAL INDEX††	LOCATION OF RESERVES (% OF TOTAL)
Gold	1.10×10^{7}	11	4.1	9	Rep. So. Africa (40)
Mercury	1.15×10^{8}	13	2.6	13	Spain (30); Italy (21)
Silver	1.71×10^{8}	16	2.7	13	Communist countries (36); U.S. (24)
Tin	4.4×10^{9}	17	1.1	15	Thailand (33); Malaysia (14)
Zinc	1.23×10^{11}	23	2.9	18	U.S. (27); Canada (20)
Lead	9.1×10^{10}	26	2.0	22	U.S. (39)
Copper	3.08×10^{11}	36	4.6	22	U.S. (28); Chile (19)
Tungsten	1.3×10^{9}	40	2.5	28	China (73)
Molybdenum	4.90×10^{9}	79	4.5	34	U.S. (58); U.S.S.R. (20)
Manganese	8×10^{11}	97	2.9	46	Rep. So. Africa (38); U.S.S.R. (25)
Aluminum	1.17×10^{12}	100	6.4	31	Australia (33); Guinea (20)
Cobalt	2.2×10^{9}	110	1.5	60	Rep. Congo (31); Zambia (20)
Pt metals	1.33×10^{7}	130	3.8	47	Rep. So. Africa (47); U.S.S.R. (47)
Nickel	6.67×10^{10}	190	3.4	53	Cuba (25); New Caledonia (22); Canada (14)
Iron	1×10^{14}	240	1.8	93	U.S.S.R. (33); Canada (14)
Chromium	7.75×10^{11}	420	2.6	95	Rep. So. Africa (75)

*From Meadows, D. H., et al., *Limits of Growth.* Universe Books: New York, 1972.
**Column 3: number of years reserves will last at 1970 rate of consumption.
†Column 4: average rate at which consumption is growing.
††Column 5: number of years reserves will last if growth continues at rate given in (4).

clay, feldspar, granite) which cannot be treated by present metallurgical methods. In principle, there is no reason why the aluminum in these minerals could not be extracted if we had a cheap and essentially limitless source of energy. Here, as in so many other areas, the answer to the depletion of resources caused by population growth and economic development lies in finding new sources of energy, the ultimate resource.

A HISTORICAL PERSPECTIVE ON THE PERIODIC TABLE: NEWLANDS, MENDELEEV, AND MEYER

In 1867, Dmitri Mendeleev, professor of chemistry at the University of St. Petersburg, started writing a textbook of general chemistry. One problem he faced was that of finding a logical way to organize the descriptive chemistry of the more than 60 elements known at that time. After several false starts, he hit upon the idea of devoting separate chapters to families of elements with similar properties. Early in 1869, he was working with the chapters on the halogens, alkali metals, and alkaline earth metals. Listing successive members of these families one after the other in order of increasing atomic mass,

F = 19	Cl = 35.5	Br = 80	I = 127
Na = 23	K = 39	Rb = 85	Cs = 133
Mg = 24	Ca = 40	Sr = 88	Ba = 137

Mendeleev was struck by the nearly constant increase in atomic mass between corresponding members of the three families. Further speculation along these lines led him to the discovery of the Periodic Table, which first appeared in a paper presented before a meeting of the Russian Chemical Society on 1869 3 6. The acclaim that greeted this and successive papers lifted Mendeleev from obscurity to fame. One result was that he had no trouble finding a publisher for his textbook, *Principles of Chemistry*. It appeared in 1870, went through eight editions, and was translated into English, French, and German.

Although Mendeleev was unaware of it at the time, he was not the first to come up with the idea of a Periodic Table. That distinction belongs to J. A. R. Newlands, professor of chemistry at the City of London College. In 1864 he arranged the elements in order of increasing atomic mass in rows of seven and found that such a sequence tended to put elements with similar properties into the same group. Unfortunately, Newlands' Periodic Table, based on what he called the "Law of Octaves," met with only ridicule and scorn from members of the English Chemical Society. Indeed, they refused to publish his paper.

Looking at Figure 7.9, you can readily see why these two versions of the Periodic Table met such different receptions. Newlands' arrangement worked very well for the elements through calcium, but beyond that he was in trouble. One can make a logical case for locating chromium below aluminum (they both form oxides of formula R_2O_3), but this requires that chromium (AM = 52) precede titanium (AM = 48). Things get even worse as we move across this period. Clearly, iron should not be in the same group with oxygen. Neither cobalt nor nickel (let alone both of them!) belong with the halogens.

Mendeleev, facing the same problem that Newlands "solved" so unsatisfactorily, got around it in a novel way. He was shrewd enough to realize that the elements beyond calcium would fall in the proper sequence only if he left gaps in certain places in the Table. This he justified by arguing that the elements which should fill these slots had not yet been discovered. Mendeleev went out on a very long limb by predicting in detail the chemical and physical properties of three such elements: "ekaboron" (scandium), "ekaaluminum" (gallium), and "ekasilicon" (germanium). By 1886, all of these elements had been isolated and shown to have properties virtually identical with those predicted (Table 7.5). The spectacular agreement between prediction and experiment removed any doubts about the validity and value of Mendeleev's Periodic Table.

This story would not be complete without mentioning the contribution of a man whom history seems almost to have forgotten, the German chemist Lothar Meyer. In July of 1868, in the process of revising his highly successful text *Modern Theories of Chemistry,* Meyer compiled a Periodic Table containing 56 elements. He also prepared extensive graphs showing that

Newlands (1864)

						H
Li	Be	B	C	N	O	F
Na	Mg	Al	Si	P	S	Cl
K	Ca	Cr	Ti	Mn	Fe	Co, Ni
Cu	Zn	Y	In	As	Se	Br
Rb	Sr	La.Ce	Zr	Nb, Mo	Ru, Rh	Pd
Ag	Cd	U	Sn	Sb	Te	I
Cs	Ba, V					

Mendeleev (as revised, 1871)

I	II	III	IV	V	VI	VII	VIII
R_2O	RO	R_2O_3	RO_2	R_2O_5	RO_3	R_2O_7	RO_4
H							
Li	Be	B	C	N	O	F	
Na	Mg	Al	Si	P	S	Cl	
K	Ca	—	Ti	V	Cr	Mn	Fe,Co, Ni
Cu	Zn	—	—	As	Se	Br	Ru,Rh, Pd
Ag	Cd	In	Sn	Sb	Te	I	
Cs	Ba						

Figure 7.9 Two early versions of the Periodic Table. Both are in the "condensed" form used by early chemists, with transition metals placed with A group elements. The elements in color were out of place in Newlands' table.

such properties as molar volume were a periodic function of atomic mass. Unaware of Mendeleev's work, Meyer published his results in 1870. In 1882, the two men jointly were awarded the Davy Medal, the highest honor of the Royal Society. Five years later, the Society belatedly awarded the same medal to its own member, J. A. R. Newlands.

TABLE 7.5 PREDICTED AND OBSERVED PROPERTIES OF GERMANIUM

	PREDICTED BY MENDELEEV (1871)	OBSERVED (1886)
Atomic mass	72	72.3
Density	5.5 g/cm^3	5.47 g/cm^3
Specific heat	0.31 J/(g · °C)	0.32 J/(g · °C)
Melting point	very high	960°C
Formula of oxide	RO_2	GeO_2
Formula of chloride	RCl_4	$GeCl_4$
Density oxide	4.7 g/cm^3	4.70 g/cm^3
Boiling point of chloride	100°C	86°C

PROBLEMS

7.1 *Electron Configurations* Give the outer electron configuration of polonium (at. no. = 84).

7.2 *Prediction of Properties* The melting points of chlorine, iodine, selenium, and krypton are −101°C, 113°C, 218°C, and −157°C, respectively. Use this information to obtain two different estimates for the melting point of bromine and compare to the observed value, −7°C.

7.3 *Trends in the Periodic Table* Compare the two elements phosphorus (at. no. = 15) and germanium (at. no. = 32) as to metallic character, ionization energy, electronegativity, and atomic radius.

7.4 *Formulas of Compounds* The formula of aluminum sulfate is $Al_2(SO_4)_3$. What is the formula of indium selenate?

7.5 Using information given in this chapter, write balanced equations for the reactions, if any, of the following metals with cold water:

a. Rb
b. Fe
c. Nd
d. Ca

7.6 Give the symbol of the element which

a. has the outer electron configuration $6s^2$.
b. is in Group 8A but has no p electrons.
c. has a single electron in the 4p level.
d. starts to fill the 4d sublevel.

7.7 Give the outer electron configurations of:

a. Sr
b. Cs
c. Te

7.8 Consider the element technetium (at. no. = 43). How would you expect it to compare to Rb in

a. atomic radius?
b. number of colored compounds?
c. reactivity toward nonmetals?
d. melting point?

7.20 Write balanced equations for the reactions of the following metals with water and with chlorine gas:

a. Cs
b. Ba
c. La

7.21 Give the symbols of all elements which

a. have the outer configuration ns^2np^3.
b. react with water to give $H_2(g)$ and a metal hydroxide of formula MOH.
c. have at least one but no more than four 4p electrons.

7.22 What outer electron configuration would you expect for the element of atomic number 114?

7.23 Consider the element Eu (at. no. = 63). How would you expect it to compare with gold (Au) in

a. atomic radius?
b. chemical reactivity?
c. electronegativity?
d. density?

7.9 Compare the three elements Si, Ge, and As as to

a. metallic character.
b. ionization energy.
c. electronegativity.
d. atomic radius.

Use only the Periodic Table inside the front cover.

7.10 Note from Figure 7.3 that the 1st ionization energies of the 3A elements are slightly lower than those of the 2A elements. Suggest an explanation in terms of electron configurations.

7.11 Explain, in terms of electron configuration, why

a. the atomic radius of Ca is greater than that of Mg.
b. As resembles Sb in its chemical properties.
c. the ionization energies of Sb and Po are about the same.

7.12 The densities of Ne(g) and Kr(g) at 25°C and 1 atm are 0.82 and 3.42 g/dm^3.

a. Estimate, by averaging, the density of Ar(g) at 25°C and 1 atm.
b. The observed density of Ar is 1.63 g/dm^3. Explain the discrepancy, using the Ideal Gas Law.

7.13 Using the data in Appendix 2, p. A.8, prepare a plot of atomic radius vs. atomic number for elements 39 through 45 and use it to estimate the atomic radius of Tc.

7.14 Given the following formulas
silver sulfate, Ag_2SO_4
calcium phosphate, $Ca_3(PO_4)_2$
cesium chloride, CsCl
rubidium chromate, Rb_2CrO_4
predict the formulas of magnesium arsenate, lithium fluoride, sodium molybdate, and copper sulfate.

7.15 Given the following known substances, which of the possible predicted substances is more likely to exist. Give your reasoning.

Known Substance	Possible Substances
SF_6	OF_6 or TeF_6
PCl_5	NCl_5 or $AsCl_5$
XeF_4	RnF_4 or HeF_4
Na_2SiF_6	Na_2SiCl_6 or Na_2GeF_6

7.16 Referring to Figure 7.6, state the number of

a. elements which occur uncombined.
b. metals which occur as sulfides.
c. elements which occur as phosphates.

7.24 Follow the directions of Problem 7.9 for the three elements S, As, and Se.

7.25 It appears from Figure 7.3 that oxygen has a slightly lower ionization energy than nitrogen. Write out the electron configuration for these two atoms and suggest an explanation for this reversal in trend.

7.26 Explain, in terms of electron configuration, why

a. the atomic radii of metals are larger than those of nonmetals in the same period.
b. H shows properties similar to both Li and F.
c. the decrease in atomic radius from Ca to Ga is greater than that from Mg to Al (Figure 7.2).

7.27 Use appropriate data from Figure 7.4 to estimate, by averaging, the electronegativity of Ge.

7.28 Plot the melting points of the 2A metals (Be, 1283°C; Mg, 650°C; Ca, 851°C; Sr, 770°C; Ba, 710°C) vs. atomic number and draw a smooth curve through your data. Which element(s) lie below the curve? above?

7.29 Given the information in Problem 7.14, predict the formulas of substances containing the elements

a. Mg, Bi, and O.
b. Fr and Br.
c. Au, S, and O.
d. K, W, and S.

7.30 The oxides of N and P have the formulas N_2O_5, N_2O_3, NO_2, NO, N_2O, P_2O_5, P_2O_3. Suggest likely formulas for two oxides of Sb.

7.31 Explain why

a. aluminum is obtained from its oxide rather than silicate minerals.
b. copper is ordinarily obtained from sulfide ores, while gold occurs free in nature.
c. HgS gives Hg on roasting while ZnS gives ZnO.

7.17 Write balanced equations for the

a. roasting of Ag_2S (behaves like HgS).
b. roasting of Bi_2S_3 (behaves like ZnS).
c. reduction of SnO_2 with CO.

7.18 Using data in Table 7.4, estimate the number of kilograms of gold, copper, and iron that are used each year.

7.19 The earth's crust weighs about 3×10^{21} kg. The percentage of Cr in the crust is 0.033%. Compare the total amount of Cr available to that given in Table 7.4 and explain the discrepancy.

7.32 Write balanced equations for the

a. formation of Ni from NiS (two steps are required).
b. formation of Ba from $BaCO_3$ (two steps; the last involves heating with Al).

7.33 Explain, using the data in Table 7.4, why the "exponential indices" for Pb and Cu are the same but the "static index" is much higher for copper.

7.34 The concentration of gold in seawater is about 4×10^{-9} g/dm^3. Taking the volume of the oceans to be 1.2×10^{21} dm^3, estimate the total amount of gold available and compare it to the figure given in Table 7.4.

*7.35 Using information given in this chapter, predict as many physical and chemical properties as you can for the elements francium and astatine.

*7.36 A certain solid element is a semiconductor with an electronegativity less than that of any element in Group 6A. It forms a chloride which contains 83.5% by mass of chlorine and has a density of 5.55 g/dm^3 in the gaseous state at 100°C and one atmosphere. Identify the element and give the formula of its chloride.

*7.37 Consider the data given in Table 7.4 for copper. Show by calculation that at an annual increase in consumption of 4.6%, the known reserves would be exhausted in twenty-two years. Suppose new discoveries raise the amount of copper ore available by a factor of two. What would the "exponential index" become then?

8 CHEMICAL BONDING

In Chapter 6, we described the electron configurations of isolated, gaseous atoms. Ordinarily, such atoms are not chemically stable. When they approach each other closely, a reaction occurs in which their configurations change through formation of a chemical bond. In this chapter, we will be concerned with two bonding processes:

1. Electrons are *transferred* from one atom to another to form positive and negative ions. When sodium and chlorine atoms are brought into contact with each other, they react vigorously to form Na^+ and Cl^- ions. These ions are the basic building blocks in ordinary table salt, NaCl. They are held together by strong electrostatic forces called *ionic bonds*.

2. Electrons are *shared* by atoms to form a molecule. An example is the formation of the Cl_2 molecule, the building block of elementary chlorine, from two isolated Cl atoms. The strong forces that hold the atoms together in a molecule are referred to as *covalent bonds*.

The electrons involved in forming ionic or covalent bonds are ordinarily those in outer, incomplete energy levels. Compare, for example, the configurations of Na and Cl atoms to those of the corresponding ions, Na^+ and Cl^-.

$$_{11}Na(1s^22s^22p^63s^1) \rightarrow {}_{11}Na^+(1s^22s^22p^6) + e^-$$

$$_{17}Cl(1s^22s^22p^63s^23p^5) + e^- \rightarrow {}_{17}Cl^-(1s^22s^22p^63s^23p^6)$$

In both cases, only those electrons in the third principal energy level are involved in the electron transfer. Those in inner levels ($n = 2$ or $n = 1$) are unaffected. This explains the observation made in Chapter 7; it is the *outer* electron configuration that determines the chemical properties of an element.

When atoms interact with one another to form ions or molecules, there is a tendency for them to acquire particularly stable configurations. The most important structure of this sort is that of the noble gases (ns^2np^6). Notice that both the Na^+ ion and the Cl^- ion have this configuration, often referred to as an *octet* structure.

8.1 IONIC BONDING

Electron transfer to form oppositely charged ions is favored when the atoms involved differ greatly in electronegativity. Typically, ionic bonds are found in compounds formed by the reaction between a metal of low ionization energy (Groups 1A, 2A, or transition metal) and a highly electronegative nonmetal

(Groups 6A, 7A). An example of such a reaction is that which occurs when sodium metal is exposed to chlorine gas:

$$Na(s) + \frac{1}{2}\,Cl_2(g) \rightarrow NaCl(s) \qquad (8.1)$$

The product, sodium chloride, has the characteristic physical properties of an ionic compound. It is a high-melting solid (*mp* = 800°C) and is a good conductor of electricity either in the molten state or in water solution, where the Na^+ and Cl^- ions are free to move.

The formation of an ionic solid from the elements is always an exothermic process. In Reaction 8.1, for example, 411 kJ of heat is evolved per mole of NaCl formed, which accounts in large measure for the stability of sodium chloride. To understand where this energy comes from, let us imagine a possible three-step mechanism for the reaction:

(1) *The elementary substances in their stable states at 25°C and 1 atm are converted to gaseous atoms.*

$$Na(s) + \frac{1}{2}\,Cl_2(g) \rightarrow Na(g) + Cl(g);\ \Delta H_1 = +230\ \text{kJ}$$

As you might expect, this step is endothermic. It requires that enough energy be absorbed to break the bonds holding sodium atoms together in sodium metal and those joining chlorine atoms in Cl_2 molecules.

(2) *Electron transfer occurs between the gaseous atoms to form ions.*

$$Na(g) + Cl(g) \rightarrow Na^+(g) + Cl^-(g);\ \Delta H_2 = +130\ \text{kJ}$$

Interestingly enough, this step is also endothermic. The energy which must be absorbed to remove the electrons from a mole of sodium atoms ($\Delta H = +494$ kJ/mol) exceeds that which is evolved when these electrons are acquired by chlorine atoms ($\Delta H = -364$ kJ/mol).

(3) *The gaseous ions combine to form an ionic solid.*

$$Na^+(g) + Cl^-(g) \rightarrow NaCl(s);\ \Delta H_3 = -771\ \text{kJ}$$

Ionic bonds are strong because this reaction is so exothermic

This is a highly exothermic process because of the strong electrostatic attraction between oppositely charged ions.

The value of ΔH for Reaction 8.1 must be, by Hess's Law, the sum of the enthalpy changes for the individual steps:

$$\Delta H = \Delta H_1 + \Delta H_2 + \Delta H_3$$

$$= +230\ \text{kJ} + 130\ \text{kJ} - 77\ \text{kJ} = -411\ \text{kJ}$$

From this analysis we see that the evolution of energy in the overall reaction is due to the exothermic nature of the final step. This enthalpy change, in which the crystal lattice is formed from gaseous ions, is referred to as the **lattice energy.**

It is possible to go through an analysis of the type just described for the formation of any ionic solid. The magnitudes of the several enthalpy changes vary with the nature of the atoms and ions involved, but the signs are always the same. Steps 1 and 2 are endothermic. The lattice energy, ΔH_3, is always a large enough negative quantity to make the overall reaction exothermic and the ionic compound stable.

TABLE 8.1 IONS WITH NOBLE GAS CONFIGURATIONS

1A	2A	3B	3A	4A	5A	6A	7A
							H^-
Li^+	Be^{2+}				N^{3-}	O^{2-}	F^-
Na^+	Mg^{2+}		Al^{3+}			S^{2-}	Cl^-
K^+	Ca^{2+}	Sc^{3+}				Se^{2-}	Br^-
Rb^+	Sr^{2+}	Y^{3+}				Te^{2-}	I^-
Cs^+	Ba^{2+}	La^{3+}					

Monatomic Ions: Charges and Electron Configurations

The fact that so many positive ions (cations) and negative ions (anions) have a noble gas configuration reflects the high stability of this electronic structure. As shown in Table 8.1, ions with such structures are generally formed by:

1. *Metals whose atoms have one, two, or three electrons beyond a noble gas configuration.* Atoms of 1A and 2A metals acquire noble gas structures by losing one and two electrons respectively, forming +1 and +2 ions. The 3B metals and aluminum in 3A, all of which have three more electrons than the preceding noble gas, form +3 ions. (Boron, the first member of group 3A, does not form ions.)

Metals below Al in 3A are far removed from the preceding noble gas

2. *Nonmetals whose atoms have one or two electrons less than the next noble gas.* Nonmetals in Group 6A (outer configuration ns^2np^4) gain two electrons to form −2 ions. Those in 7A (ns^2np^5) gain a single electron to form −1 ions. Nitrogen, the most electronegative element in 5A, forms −3 ions with a few very active metals such as sodium and magnesium.

The metals located to the right of Group 3B in the Periodic Table cannot form monatomic cations with noble gas configurations. To do so, they would have to lose four or more electrons, for which the energy requirement is prohibitively large. These metals do, however, form cations with charges ranging from +1 to +3. Table 8.2 lists some of the more common ions formed by the *transition metals* in Groups 6B through 2B and the *post-transition metals* (Groups 3A through 5A). Two features of this table are particularly interesting:

1. When positive ions are formed from transition metal atoms, *it is always the outer s electrons that are lost first.* Consider, for example, the formation of the Mn^{2+} ion from the Mn atom:

$$_{25}Mn\ (Ar\ 4s^23d^5) \rightarrow {}_{25}Mn^{2+}\ (Ar\ 3d^5) + 2e^-$$

TABLE 8.2 OUTER ELECTRON CONFIGURATIONS OF SOME TRANSITION AND POST-TRANSITION METAL CATIONS*

	6B	7B	8B				1B		2B	3A	4A	5A
First series	Cr^{3+}	Mn^{2+}	Fe^{3+}	Fe^{2+}	Co^{2+}	Ni^{2+}	Cu^{2+}	**Cu^+**	**Zn^{2+}**	**Ga^{3+}**		
	$3d^3$	$3d^5$	$3d^5$	$3d^6$	$3d^7$	$3d^8$	$3d^9$	$3d^{10}$	$3d^{10}$	$3d^{10}$		
Second series	Mo^{2+}					Pd^{2+}		**Ag^+**	**Cd^{2+}**	**In^{3+}**	**Sn^{2+}**	
	$4d^4$					$4d^8$		$4d^{10}$	$4d^{10}$	$4d^{10}$	$5s^2$	
Third series						Pt^{2+}		**Au^+**	**Hg^{2+}**	**Tl^+**	**Pb^{2+}**	**Bi^{3+}**
						$5d^8$		$5d^{10}$	$5d^{10}$	$6s^2$	$6s^2$	$6s^2$

*Ions with completed sublevels are shown in bold type.

From a slightly different point of view, we can say that **in transition metal ions, the inner d sublevels are lower in energy than outer s sublevels** (3d < 4s; 4d < 5s; 5d < 6s).

2. Most of the ions formed by metals in Groups 1B through 5A have completed sublevels (d^{10} or s^2). Metals which can acquire such a structure by losing one, two, or three electrons commonly do so. It would appear that these structures, like that of the noble gases, are particularly stable.

Sizes of Monatomic Ions

The fact that it is impossible to obtain a sample of matter made up of only one kind of ion makes it difficult to determine ionic radii. From x-ray diffraction studies (Chapter 11), we obtain the distances between the centers of touching ions. For most ionic crystals, where the contact is between positive and negative ions, this distance is the sum of the two ionic radii, i.e.,

$$d = \text{radius cation} + \text{radius anion}$$

Without further information, it is impossible to decide how the internuclear distance should be split up among the two ions.

One way out of this dilemma is to make x-ray measurements on crystals in which anions are in contact with each other. One such crystal is lithium iodide, where the small Li^+ ions fit into holes in the lattice between adjacent I^- ions. In this case the measured internuclear distance (Fig. 8.1) is twice the radius of the anion, i.e.,

This is the simplest way to establish ionic radii

$$\text{radius } I^- = 0.432 \text{ nm}/2 = 0.216 \text{ nm}$$

Once the radius of the I^- ion is known, it is possible to choose cationic radii which fit the observed internuclear distances in such compounds as RbI and CsI, where the cations are large enough to prevent anion-anion contact. From the radii of Rb^+ and Cs^+, we can establish radii for F^-, Cl^-, Br^-, and so on. The ionic radii shown in Figure 8.2 were obtained by an approach similar to that just described.

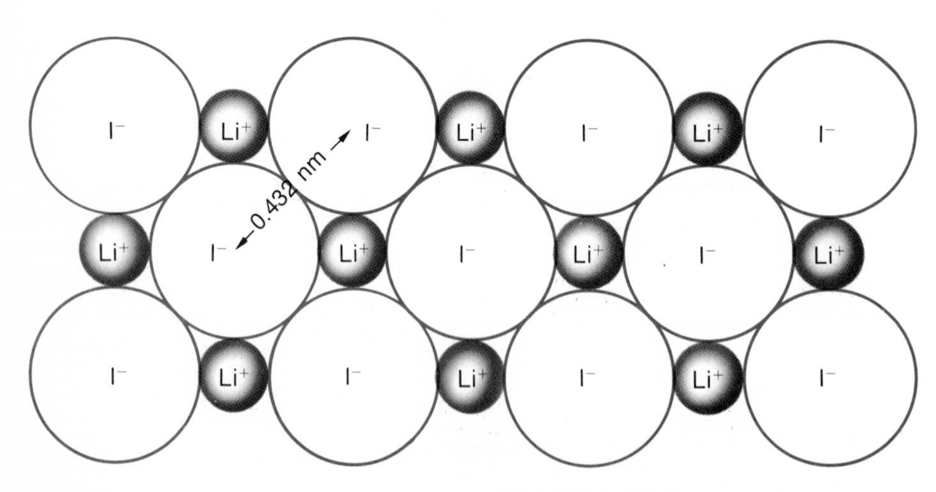

Figure 8.1 In LiI, the anions are touching. The radius of the I^- ion can be taken to be equal to half the distance between centers of adjacent anions. The Li^+ ions are so small they fit into "holes" between I^- ions.

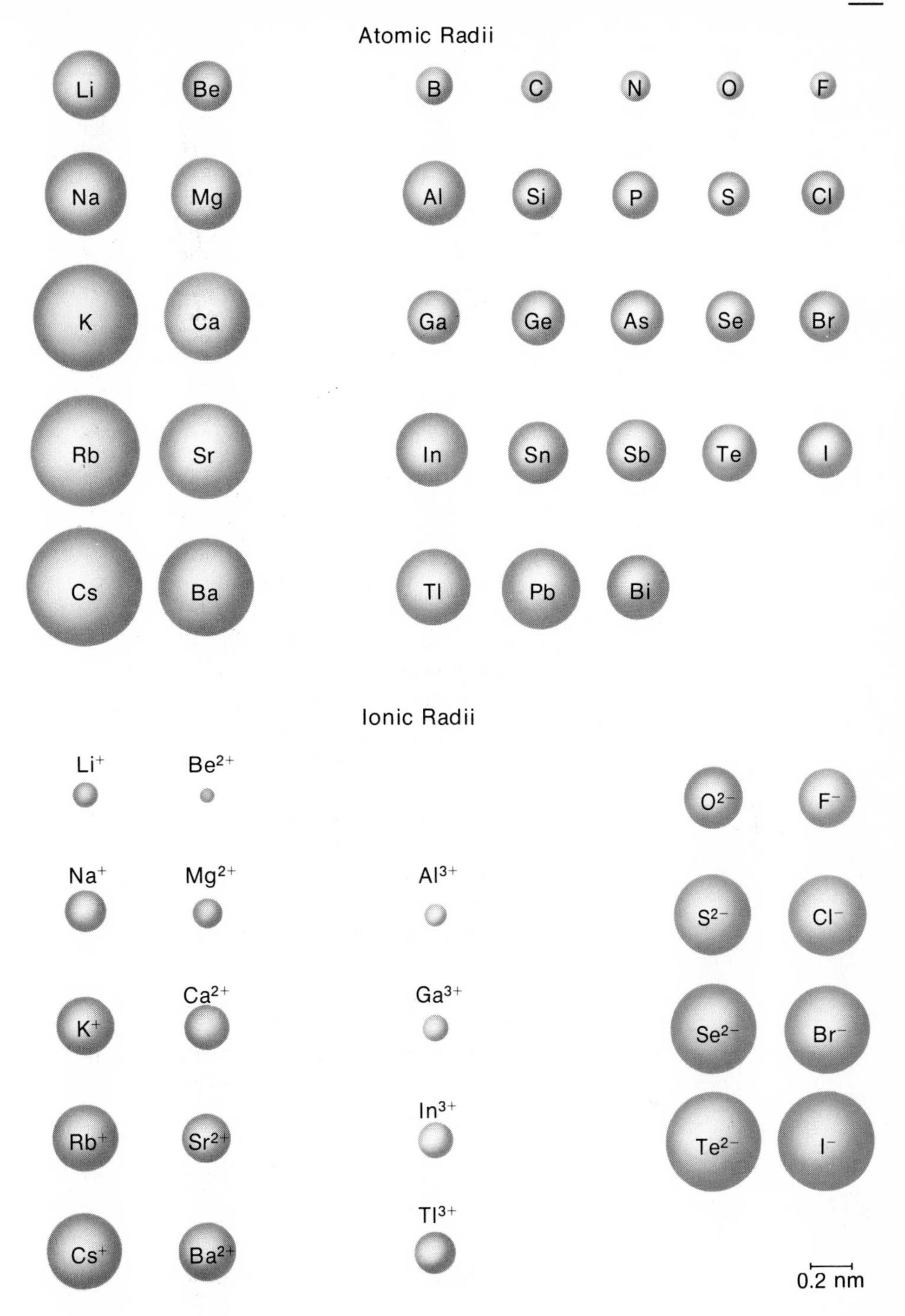

Figure 8.2 Sizes of atoms and ions of the A group elements. Negative ions are always larger than the atoms from which they are derived, whereas positive ions are smaller.

Comparing ionic radii to those of the corresponding atoms (Fig. 8.2), we see that positive ions are smaller than the metal atoms from which they are formed, while negative ions are always larger than the corresponding nonmetal atoms. As a result, anions are generally larger than cations. Compare, for example, the Cl^- ion ($r = 0.181$ nm) to the Na^+ ion ($r = 0.095$ nm). In contrast, the chlorine atom ($r = 0.099$ nm) is smaller than the sodium atom ($r = 0.186$ nm).

TABLE 8.3 SOME COMMON POLYATOMIC IONS

+1	−1	−2	−3
NH_4^+ (ammonium)	OH^- (hydroxide)	CO_3^{2-} (carbonate)	PO_4^{3-} (phosphate)
	NO_3^- (nitrate)	SO_4^{2-} (sulfate)	
	ClO_3^- (chlorate)	CrO_4^{2-} (chromate)	
	ClO_4^- (perchlorate)		
	MnO_4^- (permanganate)		

Polyatomic Ions

Many of the ions most frequently encountered in general chemistry contain more than one atom. The electronic structures of such *polyatomic* ions are discussed in Section 8.4. The names and charges of a few of the more common ions of this type are given in Table 8.3.

Formulas and Names of Ionic Compounds

All pure substances carry a net charge of zero

The simplest formula of an ionic compound is readily derived by using the principle of electroneutrality: the total charge of the positive ions must just balance that of the negative ions. Using this principle, we conclude that compounds containing Na^+ cations and Br^-, S^{2-}, or PO_4^{3-} anions must have the formulas

$NaBr$ (1 Na^+ ion, 1 Br^- ion)

Na_2S (2 Na^+ ions, 1 S^{2-} ion)

Na_3PO_4 (3 Na^+ ions, 1 PO_4^{3-} ion)

Note that in giving the formula of an ionic compound, the positive ion (Na^+) always precedes the negative ion (Br^-, S^{2-}, PO_4^{3-}).

The name of an ionic compound consists of two words: the name of the positive ion followed by that of the negative ion. Thus, for the three compounds referred to above, we write

$NaBr$: sodium bromide

Na_2S: sodium sulfide

Na_3PO_4: sodium phosphate

To assign names to individual ions, note that:

1. Monatomic positive ions (e.g., Na^+, Mg^{2+}, Al^{3+}) take the names of the metals from which they are derived (sodium, magnesium, aluminum). When a metal forms more than one ion, these are ordinarily distinguished by indicating the charge of the ion by a Roman numeral in parentheses immediately following the name of the metal. Thus we have

Fe^{2+} iron(II) $\quad$ Fe^{3+} iron(III)

2. Monatomic negative ions are named by adding the suffix *ide* to the stem of the name of the nonmetal from which they are derived. For example,

Br^- bromide $\qquad$ S^{2-} sulfide

3. Polyatomic ions (Table 8.3) are given special names. (A more extensive list of polyatomic ions with a discussion of their nomenclature can be found in Appendix 3.)

You should know the names and formulas of the ions in Table 8.3

Example 8.1 Deduce the formulas of the following ionic compounds, using the Periodic Table and Tables 8.2 and 8.3 to obtain the charges of the ions present:

a. calcium fluoride
b. scandium oxide
c. zinc sulfate
d. ammonium phosphate
e. chromium(III) nitrate

Solution In each case, we obtain the formula from the charges of the cation and anion, realizing that the compound must be electrically neutral.

a. From the positions of the two elements, calcium and fluorine, in the Periodic Table (2A and 7A), we deduce that the ions present are Ca^{2+} and F^-. Two F^- ions are required to balance one Ca^{2+} ion. Formula: CaF_2.
b. Scandium, in Group 3B, forms a +3 ion; oxygen, in 6A, forms a −2 ion. Two Sc^{3+} ions will neutralize three O^{2-} ions. Formula: Sc_2O_3.
c. The ions involved are Zn^{2+} (Table 8.2) and SO_4^{2-} (Table 8.3). One +2 ion balances one −2 ion. Formula: $ZnSO_4$.
d. From Table 8.3, we see that the ions involved are NH_4^+ and PO_4^{3-}. The formula is $(NH_4)_3PO_4$. The parentheses are used to indicate that there are three NH_4^+ ions for every PO_4^{3-} ion.
e. From the name written for the positive ion, we deduce that it has a +3 charge, i.e., Cr^{3+}. In Table 8.3, we locate the nitrate ion, NO_3^-. The formula must be $Cr(NO_3)_3$.

8.2 NATURE OF THE COVALENT BOND

We can readily interpret the ionic bond found in compounds such as NaCl in terms of the strong electrostatic forces to be expected when two oppositely charged ions are brought into contact with one another. The nature of the *covalent* or *electron pair* bond formed when two hydrogen atoms combine to form an H_2 molecule is more difficult to visualize. To emphasize that a pair of electrons is being shared by the two atoms in the molecule, the bond is often represented as

$$H:H \qquad \text{or} \qquad H{-}H$$

with the understanding that the two dots or the straight line drawn between the two hydrogen atoms represent a covalent bond. This simple picture can be misleading if it is taken to imply that the two electrons are located at a fixed position between the two nuclei. A more accurate map of the electron density would resemble that shown in Figure 8.3. At any instant the two electrons may be located at any of various points around the two nuclei. There is, however, a somewhat greater probability of finding the electrons between the two nuclei than at the far ends of the molecule.

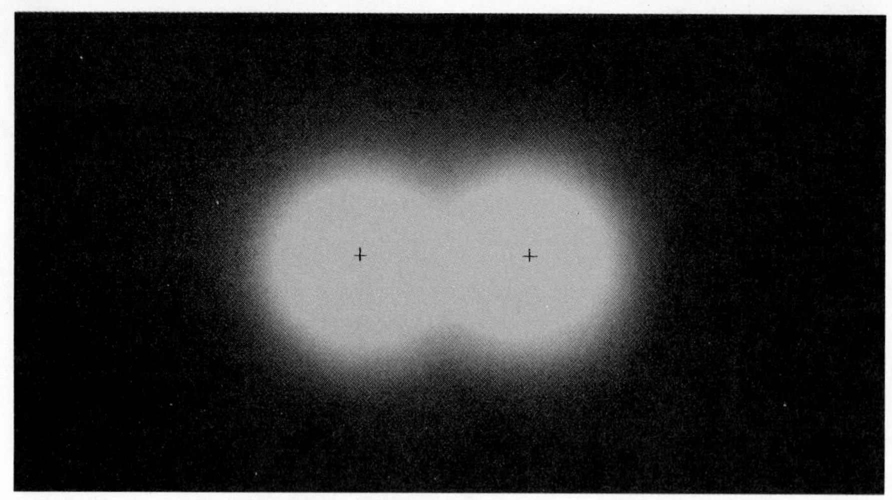

Figure 8.3 Electron density in H_2. The depth of shading is proportional to the probability of finding an electron in a particular region. In chemical bonds there tends to be a concentration of electronic charge between the nuclei.

A question that has long intrigued chemists is: why should the sharing of two electrons between two nuclei result in increased stability? Why, for example, should the H_2 molecule be more stable, by about 436 kJ/mol, than two isolated hydrogen atoms? The first plausible answer to this question was put forth in 1927 by two physicists, W. A. Heitler and T. London. Using the principles of quantum mechanics, they were able to calculate the interaction energy of two hydrogen atoms as a function of the distance between them.

At large distances of separation (far right, Fig. 8.4) there is essentially no interaction between the two hydrogen atoms. As the atoms come closer together (moving to the left in Fig. 8.4), they experience an attraction which leads gradually to an energy minimum. This minimum occurs at an internuclear distance of 0.074 nm; the attractive energy at this point is 436 kJ/mol, the bond energy of the H_2 molecule. At this distance of separation the molecule is in its most stable state. If we attempt to bring the atoms closer together, repulsive forces become increasingly important and the energy curve rises steeply.

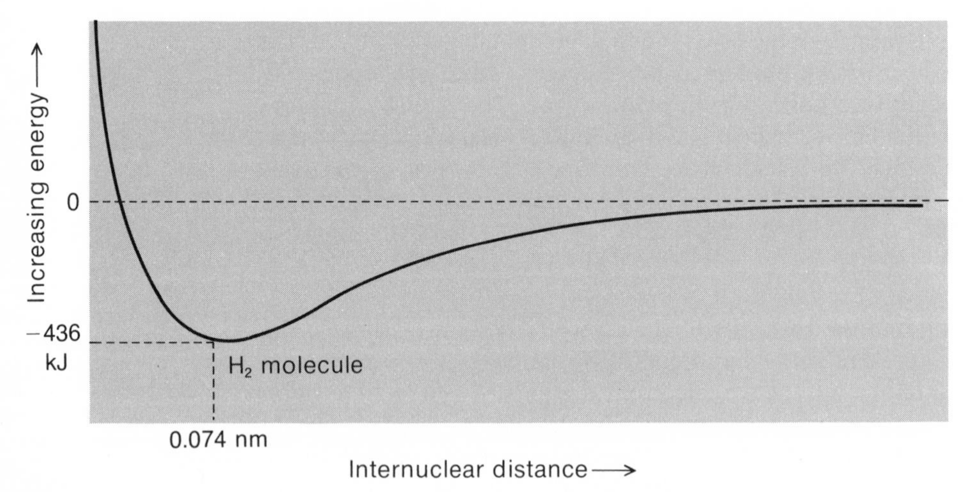

Figure 8.4 Energy of the H_2 molecule as a function of the distance between the two nuclei. The minimum in the curve occurs at the observed internuclear distance. Energy is compared to that of two separated hydrogen atoms.

The existence of the energy minimum shown in Figure 8.4 is responsible for the stability of the hydrogen molecule. The principal factor leading to this minimum is an electrostatic one. In a sense, the high electron density in the region between the two nuclei tends to "shield" these positive charges from one another, thereby contributing to the stability of the molecule. Specifically, at a distance of 0.074 nm, attractive forces between the electron of one atom and the nucleus of the other (electron 1 and nucleus 2, electron 2 and nucleus 1) exceed the repulsive forces between particles of like charge (electron 1 and electron 2, nucleus 1 and nucleus 2). Hence the system in this state—the hydrogen molecule—is more stable than the isolated atoms.

The mathematical equations upon which the Heitler-London model is based cannot be solved exactly for any but the simplest of molecules. However, by certain approximation methods, it is possible to extend the model to rationalize the stability of covalent bonds in a variety of elementary and compound substances. One way to do this, known as the *atomic orbital* or **valence bond** method, was developed in the 1930's by Linus Pauling and J. C. Slater, among others. It predicts that if two atoms are to form a covalent bond, they must each have an unpaired electron to contribute. Looking at the electron configurations of the hydrogen and fluorine atoms,

	1s	2s	2p	
H	(↑)			(unpaired 1s electron)
F	(↑↓)	(↑↓)	(↑↓) (↑↓) (↑)	(unpaired 2p electron)

we would predict, correctly, that two hydrogen atoms, two fluorine atoms, or one hydrogen and one fluorine atom could combine to form the molecules H_2, F_2, and HF. In each molecule, the atoms would be held together by a covalent bond containing one electron from each of the bonded atoms. The two atoms would share both electrons in the bond. Thus, in the HF molecule, the H atom would have the electron configuration

We might say that the bond in HF is formed by the "overlap" of the 1s orbital of H with a 2p orbital of F

	1s	
H atom in HF	(↑↓)	(↓ represents electron contributed by F)

while the fluorine atom would have the configuration

	1s	2s	2p	
F atom in HF	(↑↓)	(↑↓)	(↑↓) (↑↓) (↑↓)	(↓ represents electron from H atom)

From this point of view, *the covalent bond consists of a pair of electrons of opposed spins filling an atomic orbital on both of the bonded atoms* (1s for H, 2p for F).

Throughout most of this chapter, we shall use the atomic orbital approach, in part because of its simplicity, but more important because it can account satisfactorily for the properties of most of the covalently bonded substances with which we deal in general chemistry. It does, however, suffer from certain rather serious deficiencies. In particular, it tends to underemphasize the extent to which the energy levels of electrons in atoms are modified by covalent bond formation.

In the last section of this chapter, we will consider an alternative approach to covalent bonding, known as the **molecular orbital** method, developed originally by F. Hund and R. S. Mulliken around 1930.

8.3 PROPERTIES OF THE COVALENT BOND

Three of the most important characteristics of a covalent bond are its *polarity,* which tells us how the bonding electrons are distributed between the two atoms, its *bond energy,* which is a measure of its strength, and the *length* of the bond, which measures the distance between centers of the two bonded atoms. In this section, we will examine how these properties vary with the nature of the atoms which share the electron pair.

Polar and Nonpolar Covalent Bonds

As we might expect, the two electrons joining the atoms in the H_2 molecule are equally shared by the two nuclei. Stated another way, a bonding electron is as likely to be found in the vicinity of one nucleus as the other. Bonds of this type are described as **nonpolar.** We expect to find nonpolar bonds whenever the two atoms joined are identical, as is the case in H_2 or F_2.

In the HF molecule, the distribution of the bonding electrons is somewhat different from that in H_2 or F_2. Here, the density of the electron cloud is concentrated around the fluorine atom; the bonding electrons, on the average, are shifted toward fluorine and away from hydrogen (Fig. 8.5). Bonds in which the electron distribution is unsymmetrical are referred to as **polar.** The hydrogen fluoride molecule can be described as a **dipole,** with the fluorine atom acting as a negative pole and the hydrogen atom as a positive pole.

Since atoms of two different elements always differ at least slightly in their attraction for electrons, covalent bonds between unlike atoms are always polar. The extent of polarity can be interpreted in terms of the relative *electronegativities* (Fig. 7.4, p. 158) of the two atoms joined by the bond. In the O—F bond, the bonding electrons are only slightly displaced towards fluorine, which is a little more electronegative than oxygen (4.0 vs. 3.5). On the other hand, fluorine has a much higher electronegativity than hydrogen (4.0 vs. 2.1), so the polarity in the H—F bond is much more pronounced than in O—F.

A polar covalent bond may be thought of as being intermediate between a pure (nonpolar) covalent bond, in which the electrons are equally shared, and a pure ionic bond, in which there has been a complete transfer of electrons from one atom to the other. In this sense, we sometimes express bond polarity in terms of *partial ionic character.* The greater the difference in electronegativity between two elements, the more ionic will be the bond between them. The relationship

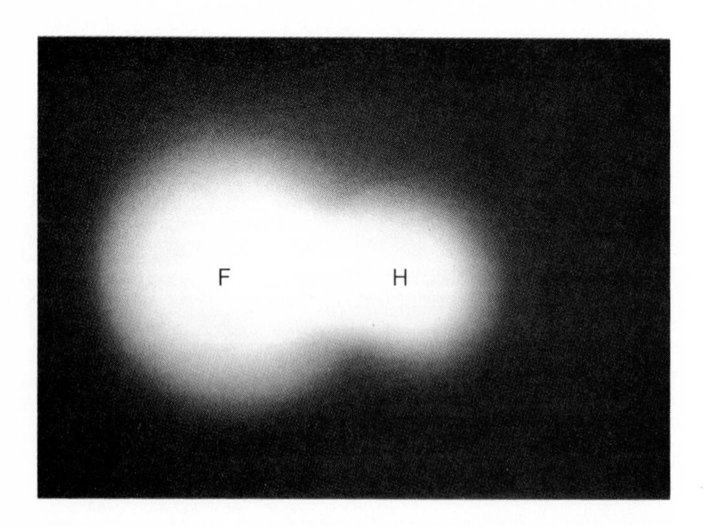

Figure 8.5 Electron density in HF. The electron cloud associated with the bonding electrons is shifted toward the F atom.

between these two variables is shown graphically in Figure 8.6. Notice that a difference of 1.7 electronegativity units corresponds to a bond with approximately 50 per cent ionic character. Such a bond might be described as being halfway between a pure covalent and a pure ionic bond.

It is clearly an oversimplification to refer to a bond between two elements as being "ionic" or "covalent." Consider, for example, the bonding in compounds formed by a 1A or 2A metal with a nonmetal in Group 6A or 7A. The difference in electronegativity ranges from a minimum of 0.6 for the beryllium-tellurium pair to a maximum of 3.3 for cesium and fluorine. The percentage of ionic character in the bonds formed between these pairs of elements shows a corresponding variation, from about 10 per cent for BeTe to 95 per cent for CsF. We might say that the bonding in beryllium telluride is "predominantly covalent," while that in cesium fluoride is "predominantly ionic."

Chemists often call a bond "ionic" if it is more than 50% ionic

Example 8.2 Using only the Periodic Table, arrange the following bonds in order of increasing polarity (ionic character): O—Cl, O—O, O—Na

Solution The first member of the series must be the O—O bond; since the atoms are identical, the bond must be nonpolar. To decide which bond comes next, recall from Chapter 7 that electronegativity increases as we move across in the Table and decreases as we move down. Chlorine lies below but to the right of oxygen; we might expect the two elements to have roughly equal electronegativities. On the other hand, sodium lies far to the left of oxygen in the same period, so should have a much lower electronegativity. The proper sequence is:

$$O{-}O < O{-}Cl < O{-}Na$$

We progress from a nonpolar to a slightly polar to a strongly polar, essentially ionic, bond.

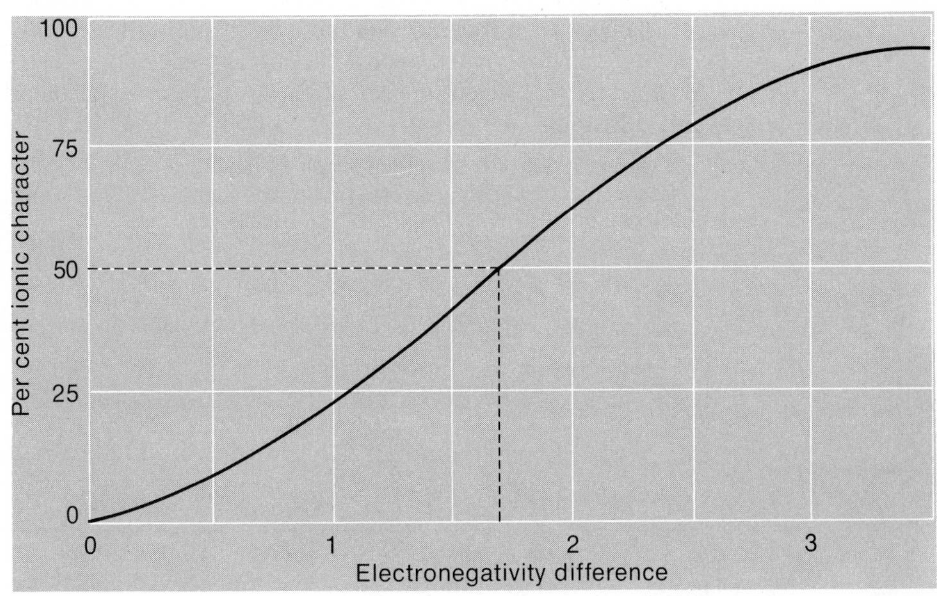

Figure 8.6 Relation between ionic character of a bond and the difference in electronegativity of the bonded atoms.

Bond Distances and Energies

It is possible to estimate the distance between atoms joined by a covalent bond by adding their atomic radii (Appendix 2). For a nonpolar bond, such as that in the F_2 molecule, the value obtained

$$\text{F—F distance} = 2 \text{ radius F} = 0.064 \text{ nm} + 0.064 \text{ nm} = 0.128 \text{ nm}$$

must be exactly equal to the observed internuclear distance, since the atomic radius of a nonmetal is defined as one half the distance between centers of bonded atoms in the elementary substance.

When two unlike atoms, differing in electronegativity, are joined by a covalent bond, the observed internuclear distance is ordinarily smaller than that calculated by taking the sum of the atomic radii. Thus, for HF, we would calculate, using covalent radii,

$$\text{H—F distance} = \text{radius H} + \text{radius F} = 0.037 \text{ nm} + 0.064 \text{ nm} = 0.101 \text{ nm}$$

as compared to an observed distance of 0.092 nm. The simplest way to explain the bond shortening in HF and similar molecules is to assume that the introduction of partial ionic character into a covalent bond strengthens it and, hence, tends to pull the bonded atoms closer together.

A more direct measure of bond strength is its bond energy (Table 4.2, p. 70). Here again, we find that polarity contributes to bond strength. Compare, for example, the observed bond energy for H—F, 565 kJ, to the value we would calculate by taking the average of the two nonpolar bonds H—H and F—F:

Bond polarity tends to strengthen and shorten bonds

$$\frac{436 \text{ kJ} + 153 \text{ kJ}}{2} = 294 \text{ kJ}$$

We see that the H—F bond is some 271 kJ/mol stronger than might be expected if it were nonpolar.

In Table 8.4 we list the "extra bond energy," Y, associated with polarity in the hydrogen halide molecules. Notice that the difference between observed and calculated values decreases as the bond becomes less polar. It has its maximum value, 271 kJ/mol, in the strongly polar HF molecule. In HI, where the two atoms have nearly equal electronegativities, the difference has nearly disappeared; Y is only 3 kJ/mol.

Observations of this sort led Linus Pauling to propose the electronegativity scale discussed in Chapter 7. He suggested the following relationship between Y (kJ/mol) and the difference in electronegativity between two atoms

$$Y = 96(\Delta E.N.)^2$$

TABLE 8.4 OBSERVED AND CALCULATED BOND ENERGIES (kJ/MOL) OF THE HYDROGEN HALIDES

	Bond Energy			
	HF	HCl	HBr	HI
H—X (calc.)*	294	340	314	294
H—X (obs.)	565	431	368	297
Y (obs. − calc.)	271	91	54	3

$$\text{*calc.} = \frac{\text{H—H} + \text{X—X}}{2}$$

In the case of the H—Br bond, where Y = 54, we have

$$(\Delta E.N.)^2 = \frac{54}{96};\ \Delta E.N. = \left(\frac{54}{96}\right)^{1/2} = 0.7$$

To assign absolute values to electronegativities, Pauling arbitrarily took that of fluorine to be 4.0 and related all other values to it.

Multiple Bonds

Ordinarily it is found that the distance between two atoms joined by a covalent bond and the energy required to break it are nearly constant in different molecules. Consider, for example, the first three compounds listed in Table 8.5, ethane, propane, and butane. The measured C—C bond distance and the calculated value of the C—C bond energy are the same in each case.

When we find that one or the other of these properties changes significantly, we suspect a change in bond type. This is the case with the last two compounds listed in Table 8.5 (ethylene, C_2H_4, and acetylene, C_2H_2). The distance between the carbon atoms in ethylene (0.133 nm) is significantly shorter than the C—C bond distance (0.154 nm). At the same time, the bond holding the carbon atoms together in this compound is significantly stronger than the C—C bond. These effects become even more pronounced in acetylene.

This evidence is interpreted to mean that there are **multiple** bonds between the carbon atoms in ethylene and acetylene. Specifically, we say that there is a **double bond** in ethylene, consisting of two pairs of electrons joining the two carbon atoms,

```
   H  H
   |  |
H—C=C—H
```

and a **triple bond** in acetylene (three pairs of electrons)

H—C≡C—H

We always find that double or triple bonds are stronger than single bonds between the same two atoms. Compare, for example, the bond energy of the triple bond in the N_2 molecule to that of the single bond in the N_2H_4 molecule:

```
               H       H
                \ ..  /
:N≡N:            N—N
                /  .. \
               H       H
```

941 kJ/mol for N≡N   159 kJ/mol for N—N

TABLE 8.5 CARBON-CARBON BOND DISTANCE AND ENERGIES

	Bond Distance (nm)	Bond Energy (kJ/mol)
C_2H_6	0.154	347
C_3H_8	0.154	347
C_4H_{10}	0.154	347
C_2H_4	0.133	598
C_2H_2	0.120	820

Hydrazine, N_2H_4, is used as a rocket fuel, in part because its combustion produces the very stable N_2 molecule and hence liberates a large amount of energy.

8.4 LEWIS STRUCTURES; THE OCTET RULE

The concept of the covalent or electron-pair bond was first introduced by the American physical chemist G. N. Lewis in 1916. To rationalize the stability of this bond, he pointed out that *atoms, by sharing electrons, can acquire a stable, noble-gas configuration.* For example, when two hydrogen atoms, each with one electron, combine to form an H_2 molecule,

$$\mathrm{H\cdot + H\cdot \rightarrow (H{:}H)}$$

each hydrogen atom acquires a share in the two electrons and, in that sense, attains the electronic configuration of the noble gas helium (atomic number = 2). Similarly, when two fluorine atoms, each with seven electrons in its outermost principal energy level (n = 2), combine to form the F_2 molecule,

$$\mathrm{{:}\ddot{\underset{..}{F}}\cdot + \cdot\ddot{\underset{..}{F}}{:} \rightarrow ({:}\ddot{\underset{..}{F}}{:}\ddot{\underset{..}{F}}{:})}$$

each atom attains the neon structure with eight electrons in the outermost level. Although Lewis proposed his model for covalent bonding on empirical grounds many years before valence bond theory was developed, Lewis structures are consistent with that theory and indeed furnish the form in which it is used by most chemists.

The structures written above are referred to as *Lewis structures* (or, in the vernacular, "flyspeck formulas"). In writing the Lewis structure for a species, we include only those electrons in outer levels which can participate in covalent bonding, the so-called *valence electrons.* For the A-group elements, **the number of valence electrons of an atom is given by the group number in the Periodic Table:**

Group	1A	2A	3A	4A	5A	6A	7A	8A
No. valence e^-	1	2	3	4	5	6	7	8

In the Lewis structure for a species, a covalent bond may be shown as a pair of electron dots written between the bonded atoms or, more commonly, as a straight line connecting the two atoms. Lewis structures for the simple molecules formed by hydrogen with the nonmetals of the second period are:

Unshared electron pairs are shown as dots

CH_4	NH_3	H_2O	HF
H—C(—H)(—H)—H	H—N̈(—H)—H	H—Ö—H (with unshared pairs on O)	H—F: (with unshared pairs on F)

Note that in each case the central atom (C, N, O, F) is surrounded by 8 valence electrons.

Many of the polyatomic ions listed in Table 8.3 can be assigned simple Lewis structures. For example, the OH^- and NH_4^+ ions can be represented as

$$\mathrm{({:}\ddot{\underset{..}{O}}{-}H)^-} \text{ and } \left[\begin{array}{ccc} & H & \\ & | & \\ H- & N & -H \\ & | & \\ & H & \end{array}\right]^+$$

In both these ions, hydrogen atoms are joined by covalent bonds to nonmetal atoms (O, N). In both ions there are eight valence electrons. In the case of the hydroxide ion, this is one more than the number associated with the neutral atoms ($6 + 1 = 7$), in agreement with the -1 charge of the ion. The $+1$ charge of the NH_4^+ ion is accounted for when we realize that four hydrogen atoms and one nitrogen atom would supply 9 valence electrons ($4 + 5 = 9$), one more than the number present in the ion.

The rule, illustrated above, that atoms in covalently bonded species tend to have noble gas configurations is often referred to as the **octet rule.** Nonmetals, with the exception of hydrogen, achieve a noble gas structure by acquiring an "octet" of electrons. As we shall see later, some stable, covalently bonded species "violate" the octet rule. Nevertheless, we shall find it very useful in suggesting plausible electronic structures for molecules and polyatomic ions.

Writing Lewis Structures

Lewis structures for many molecules and ions are readily written by inspection. However, it may be helpful to follow the general procedure outlined below.

1. *Draw a skeleton structure for the molecule or ion, joining atoms by single bonds.* In some cases, only one arrangement of atoms is possible; in others, experimental evidence must be used to decide between two or more alternative structures.

2. *Count the number of valence electrons.* For a molecule, we simply sum up the valence electrons of the atoms present. For a *polyatomic anion,* electrons are *added* to take into account the negative charge. For a polyatomic *cation,* a number of electrons equal to the positive charge must be *subtracted.*

3. *Deduct two valence electrons for each single bond written in step 1. Distribute the remaining electrons as unshared pairs so as to give each atom eight electrons if possible.* If you find that there are "too few electrons to go around," convert single to multiple bonds. The formation of a double bond compensates for a deficiency of two electrons; a triple bond, a deficiency of four electrons.

The application of these rules and some further guiding principles are illustrated in Example 8.3.

Example 8.3 Draw Lewis structures for:

a. ClO^- b. SO_4^{2-} c. CH_2O

Solution

a. Only one skeleton structure is possible for the hypochlorite ion

$$(Cl—O)^-$$

To obtain the total number of valence electrons we add one (the charge of the ion) to the number contributed by chlorine in Group 7A (7) and oxygen in Group 6A (6).

$$\text{no. of valence electrons} = 7 + 6 + 1 = 14$$

Deducting the two electrons used to make the covalent bond leaves 12. Putting six of these electrons around each atom, we arrive at a reasonable structure for the ClO^- ion which satisfies the octet rule, since both the Cl and O atoms have a share in eight electrons. (The bonding electrons are counted for both atoms.)

$$(:\ddot{\underset{..}{Cl}}—\ddot{\underset{..}{O}}:)^-$$

b. Various skeletons could be written for the sulfate ion. However, in *ions* such as SO_4^{2-}, NO_3^-, and CO_3^{2-} we ordinarily find that *each oxygen atom is bonded to the central, nonmetal atom.* Following this general rule, we write

```
[     O    ]2-
[     |    ]
[  O—S—O  ]
[     |    ]
[     O    ]
```

The number of valence electrons is found by adding the charge of the ion to the total contributed by the sulfur and oxygen atoms:

$$\text{no. of valence electrons} = 6 + 4(6) + 2 = 32$$

Deducting eight electrons for the four covalent bonds in the skeleton structure leaves 24. Putting six electrons around each oxygen atom gives us a plausible Lewis structure for the sulfate ion.

```
[       :Ö:       ]2-
[        |        ]
[  :Ö—S—Ö:  ]
[        |        ]
[       :Ö:       ]
```

c. A reasonable skeleton structure for formaldehyde would be

```
    H
    |
H—C—O
```

The number of valence electrons in this neutral species is simply the total of those contributed by the carbon, hydrogen and oxygen atoms.

$$\text{no. of valence electrons} = 4 + 2(1) + 6 = 12$$

Deducting six electrons for the three covalent bonds in the skeleton structure leaves us only six to work with. We might spend these remaining electrons in either of two ways:

```
    H               H
    |               |
H—C—Ö   or   H—C—Ö:
```

But, whatever we do, we are two electrons shy of the number required to give both carbon and oxygen an octet. To remedy this deficiency, we form a double bond between the carbon and oxygen atoms

```
    H
    |
H—C=Ö:
```

to give a reasonable and, as it happens, the correct structure for the formaldehyde molecule.

There are other structures that could be written for CH_2O which would put eight electrons around each atom. However, this is the only such structure that conforms to the general principle that *carbon,* in all its stable compounds, *forms four covalent bonds.*

Carbon atoms in molecules don't have nonbonded electrons around them

Resonance Forms

In certain cases the Lewis structure does not adequately describe the properties of the ion or molecule that it represents. Consider, for example, the SO_2 molecule, for which we can derive the following structure:

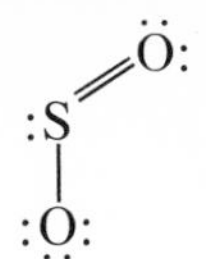

This structure implies that there are two different kinds of sulfur-to-oxygen bonds in SO_2, double and single, yet experiment tells us that there is only one kind of bond in this molecule. In particular, we find the two bond distances are equal, 0.143 nm.

Valence bond theory "explains" this situation by assuming that each of the bonds in SO_2 is a "hybrid," intermediate between a single and double bond. The fact that the observed bond distance is half-way between those expected for a single and double bond lends support to this idea. To express this concept within the framework of Lewis structures, we sometimes write two structures

:S(=Ö:)—Ö: ⟷ :S(—Ö:)=Ö:

with the understanding that the true structure is intermediate between them. In valence bond terminology, these are referred to as *resonance* forms. The concept of resonance is introduced to rationalize the fact that a single Lewis structure does not adequately describe the properties of substances such as sulfur dioxide.

Another species for which it is necessary to invoke the idea of resonance is the nitrate ion. Here three equivalent structures can be written

[:Ö—N(=Ö:)—Ö:]⁻ ⟷ [:Ö—N(—Ö:)=Ö:]⁻ ⟷ [:Ö=N(—Ö:)—Ö:]⁻

to explain the experimental observation that the three nitrogen to oxygen bonds in the NO_3^- ion are identical in all respects.

We will encounter other examples of molecules and ions whose properties can be interpreted in terms of resonance. In this connection it may be well to point out that:

1. Resonance forms do not imply different kinds of molecules. Sulfur dioxide is built up of only one type of molecule, whose structure is assumed to be between those of the two resonance forms.

2. Resonance can be anticipated when it is possible to write two or more Lewis structures which are about equally plausible. In the case of the nitrate ion, the three structures we have written are equivalent. One could, in principle, write many other structures, but none of them would put eight electrons around each atom, so presumably would not make a major contribution to the true structure of the nitrate ion.

3. In writing resonance forms, one can shift only electrons, not atoms. The structure

:Ö—O(=N̈:)—Ö:

could not be a resonance form of the nitrate ion, since the atoms are arranged in a quite different way.

Exceptions to the Octet Rule

All the electrons can't be paired if there is an odd number of them

Although most of the molecules and polyatomic ions that we talk about in general chemistry follow the octet rule, there are some familiar species that do not. Among these are molecules containing an odd number of valence electrons. Nitric oxide, NO, and nitrogen dioxide, NO_2, fall in this category:

$$NO\text{: number of valence electrons} = 5 + 6 = 11$$

$$NO_2\text{: number of valence electrons} = 5 + 6(2) = 17$$

For such *odd electron* species (sometimes called free radicals) it is impossible to write Lewis structures in which each atom obeys the octet rule. In valence-bond terminology, the NO molecule is considered to be a resonance hybrid with the two contributing structures

$$\cdot\ddot{N}{=}\ddot{O}: \leftrightarrow :\ddot{N}{=}\ddot{O}\cdot$$

Several different resonance structures can be written for NO_2, of which the more plausible are of the type

$$\cdot N(=\ddot{O}:)(-\ddot{O}:)$$

Species such as NO and NO_2, in which there are unpaired electrons, are **paramagnetic;** that is, they show a weak attraction toward a magnetic field. Elementary oxygen is also paramagnetic (see Plate 5), which suggests that the conventional Lewis structure

$$:\ddot{O}{=}\ddot{O}:$$

is incorrect, since it requires that all the electrons be paired. The paramagnetism of oxygen could be explained by the structure

$$:\dot{\ddot{O}}{-}\dot{\ddot{O}}:$$

in which there are two unpaired electrons. However, this structure, like the one written previously, is unsatisfactory. In the first place, it does not conform to the octet rule; much more important, it does not agree with experimental evidence. The distance between the two oxygen atoms in O_2 (0.121 nm) is considerably smaller than that ordinarily observed with an O—O single bond (0.148 nm). These properties of oxygen are difficult to explain in terms of valence-bond theory. As we shall see in Section 8.7, the molecular orbital approach leads to a more satisfactory picture of the electron distribution in the O_2 molecule.

There are other species for which Lewis structures written to conform to the octet rule are unsatisfactory. Examples include the fluorides of beryllium and boron, which exist in the vapor as molecules of BeF_2 and BF_3, respectively. Although one could write multiple bonded structures for these molecules in accordance with the octet rule, experimental evidence suggests the structures

$$:\ddot{F}{-}Be{-}\ddot{F}: \quad \text{and} \quad B(-\ddot{F}:)_3$$

in which the central atom is surrounded by four and six valence electrons, respectively, rather than eight.

At the opposite extreme, certain of the halides of the 5A and 6A elements have structures in which the central atom is surrounded by more than eight valence electrons. In PF_5 and PCl_5, the phosphorus atom is joined by single bonds to each of five halogen atoms and consequently must be surrounded by ten bonding electrons. An analogous structure for SF_6 requires that a sulfur atom have 12 valence electrons around it.

8.5 MOLECULAR GEOMETRY

The "geometry" of a simple diatomic molecule can be specified by giving the internuclear distance. With molecules containing more than two atoms, another factor is involved: the angles between the bonds. In such molecules as H_2O and CO_2, it is important to know whether the three atoms are in a straight line, as would be the case if the angle between the bonds were 180°, or arranged in a triangular pattern corresponding to a bond angle of less than 180°.

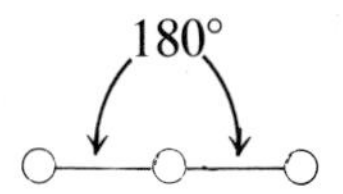

The "bond angle" is the angle between the lines joining the bonded atoms

Electron Pair Repulsion

The major features of the geometry of molecules and polyatomic ions can be predicted by a simple principle first suggested by Sidgwick and Powell in 1940 and developed more recently by R. J. Gillespie of McMaster University in Hamilton, Ontario.

The electron pairs surrounding an atom, as one might expect on the basis of electrostatic repulsion, are oriented to be as far apart as possible.

We will now consider how the concept of maximum distance of separation of electron pairs can be used to predict the geometries of molecules and polyatomic ions in which there are two, three, or four electron pairs around a central atom. Later (Chapter 21) we will consider the geometry of more complex species.

Species with Single Bonds, No Unshared Electron Pairs Around Central Atom

Figure 8.7 shows the geometries to be expected for molecules in which all of the electron pairs (two, three, or four) surrounding a central atom are used to form single bonds. In each case, the electron pairs are directed so as to be as far apart as possible. With two electron pairs (e.g., Be in BeF_2), this leads to a 180° bond angle and hence a **linear** molecule. Three electron pairs around a central atom (B in BF_3) are directed at 120° angles, toward the corners of an **equilateral triangle.** Notice that this molecule is planar. That is, all four atoms are in the same plane.

If the central atom forms four single bonds, the electron pair repulsion principle requires a three-dimensional structure. The four electron pairs are directed toward the corners of a regular **tetrahedron,** a three-dimensional figure with four faces, all of which are equilateral triangles. All the bond angles are 109.5°, the tetrahedral angle. Methane, CH_4, is a classic example of a molecule with this structure.

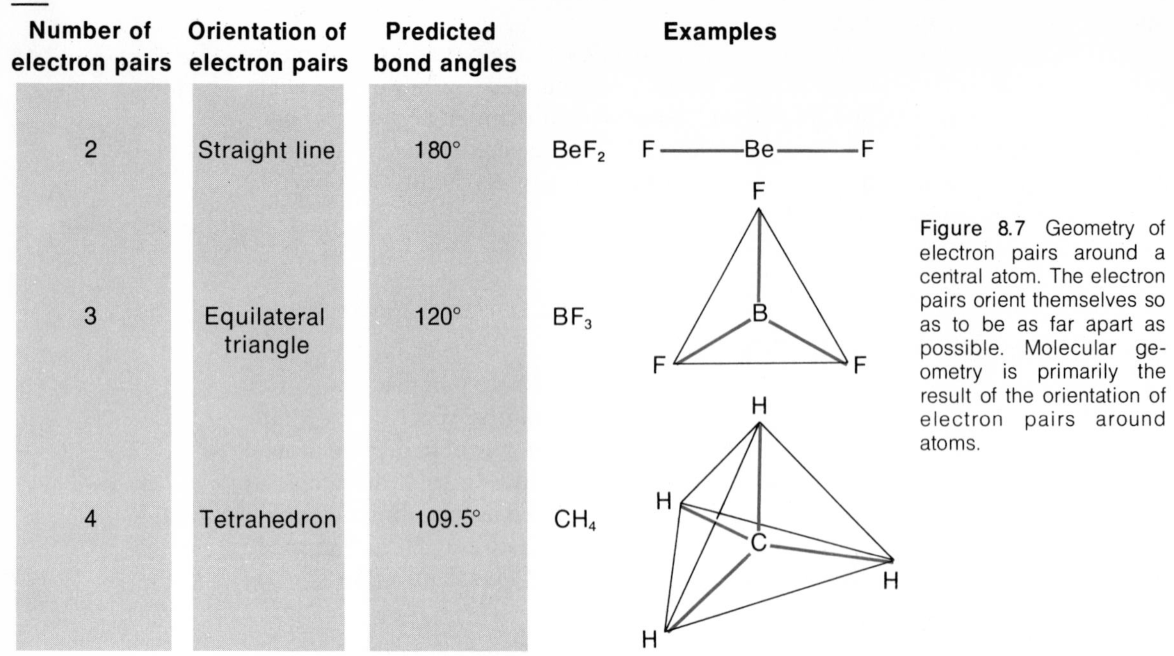

Number of electron pairs	Orientation of electron pairs	Predicted bond angles	Examples
2	Straight line	180°	BeF_2 F——Be——F
3	Equilateral triangle	120°	BF_3
4	Tetrahedron	109.5°	CH_4

Figure 8.7 Geometry of electron pairs around a central atom. The electron pairs orient themselves so as to be as far apart as possible. Molecular geometry is primarily the result of the orientation of electron pairs around atoms.

The tetrahedron is a basic unit in the structure of many organic molecules. Indeed, whenever carbon forms four single bonds, it shows tetrahedral geometry. Many polyatomic ions also show this structure. Referring back to the Lewis structures of NH_4^+ (p. 186) and SO_4^{2-} (p. 188), we see that in both cases the central atom (N, S) is surrounded by four electron pairs, all of which are used to form single bonds. We conclude that in both ions the central atom is at the center of a regular tetrahedron, symmetrically surrounded by four other atoms (H, O).

Species with Unshared Electron Pairs Around Central Atom

The electron pair repulsion principle is readily extended to predict the geometries of molecules or polyatomic ions in which one or more of the electron pairs around the central atom are unshared (sometimes described picturesquely as a "lone pair"). In Figure 8.8, we show the geometries of the NH_3 and H_2O molecules, whose Lewis structures were given on p. 186. Positions of the unshared pairs, shown as charge clouds in Figure 8.8, cannot be determined experimentally. We can visualize the ammonia molecule as a "tetrahedron with one corner missing" or, more simply, as a *pyramid* with the nitrogen atom located above the center of the equilateral triangle formed by the three hydrogen atoms. When two corners of the tetrahedron are removed, as in H_2O, we are left with a *bent* molecule with an oxygen atom at the center.

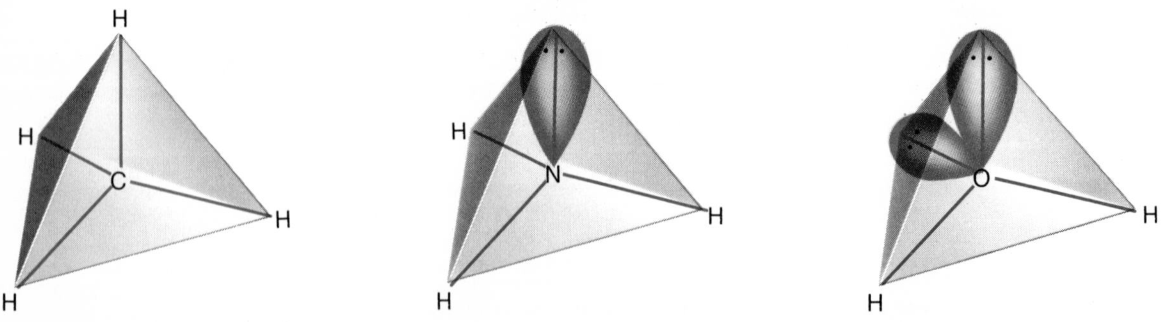

Figure 8.8 Geometries of CH_4, NH_3, and H_2O. When the four pairs of electrons around the central atom are all shared, the molecule is tetrahedral. When one pair is unshared (NH_3), the structure is that of a pyramid, with an equilateral triangle as a base. Two unshared pairs, as in H_2O, lead to a bent molecule.

In general, for any atom obeying the octet rule and forming no multiple bonds, we can assume that the four pairs of electrons around that atom will be directed toward the corners of a regular tetrahedron. The name used to describe the geometry of a simple molecule containing such an atom at its center will depend upon the number of single bonds. If there are four single bonds and no lone pairs, the molecule will be tetrahedral (e.g., CH_4, CH_2Cl_2). If there are three bonds and one lone pair (NH_3), the molecule will be pyramidal. With two bonds and two lone pairs (H_2O, H_2S), the molecule is described as bent. In all cases, the electron pairs around the central atom have a tetrahedral arrangement, but *in describing the geometry of the molecule we consider only the positions of the nuclei.*

This same principle is readily extended to species in which there are three electron pairs about a central atom, one of which is a lone pair (Example 8.4).

Example 8.4 Predict the geometry of a molecule which has the structure

$$:\ddot{\underset{..}{Y}}-\ddot{X}-\ddot{\underset{..}{Y}}:$$

Solution The three pairs of electrons around the central atom should be directed toward the corners of an equilateral triangle. This means that the molecule will be bent, with a bond angle of about 120°, i.e.,

X
Y Y
120°

Experimentally, we find that bond angles in molecules where the central atom has at least one unshared pair of electrons tend to be somewhat smaller than those listed in Figure 8.7. For example, in NH_3 the experimental bond angle is about 107°, a little less than the predicted tetrahedral angle of 109.5°. The effect appears to be slightly greater in H_2O, where the bond angle is 105°. These discrepancies have been attributed to the influence of the unshared pair(s). We might expect the electron cloud formed by the unshared pair in NH_3 to spread out over a greater volume than that of the three pairs connected to hydrogen atoms. This would tend to force the bonding pairs closer to one another and thereby reduce the bond angle. Where there are two pairs of unshared electrons, as in H_2O, this effect is more pronounced.

Species Containing Multiple Bonds

The idea of electron pair repulsion can be extended to predict the geometries of molecules or ions containing double or triple bonds if we assume that *so far as molecular geometry is concerned, a multiple bond behaves as if it were a single electron pair.* Thus, we find that the CO_2 molecule, like BeF_2, is linear

F—Be—F, O═C═O

The unshared electron pairs are not shown here. Can you put them in?

while the SO_3 molecule, like BF_3, is triangular

Applying this principle to the acetylene molecule, where there is a triple bond between the carbon atoms, we predict that each of the bond angles must be 180°, which gives the observed linear structure

$$H{-}C{\equiv}C{-}H$$

The ethylene molecule, with a double bond between the two carbon atoms, has the geometry to be expected if each carbon atom had only three pairs of electrons around it.

```
H       H
 \     /
  C = C
 /     \
H       H
```

The six atoms are located in a plane with bond angles of 120°.

Example 8.5 Predict the geometries of the SO_2 molecule and the NO_3^- ion.

Solution Looking at the Lewis structure of SO_2 on p. 189, we count around the central sulfur atom one double bond, one single bond, and one unshared pair of electrons. Pretending that the double bond is a single electron pair, we predict a bent molecule with a 120° bond angle (the observed angle is 119.5°):

```
     O
    //
  :S
   |
   O
```

The Lewis structure of the NO_3^- ion is given on p. 189. We would expect the nitrogen atom to behave as if it were surrounded by three pairs of electrons. Consequently it should have a structure like that of SO_3 or BF_3. The nitrogen atom is at the center of an equilateral triangle with the three oxygen atoms at the corners of that triangle. The ion is planar.

All of the statements we have made concerning molecular geometry are summarized in a slightly different form in Table 8.6.

Hopefully this discussion has convinced you of the usefulness of the electron pair repulsion principle in predicting molecular geometries. All the examples we have dealt with involve relatively simple species containing only a few atoms, but

TABLE 8.6 GEOMETRIES OF SPECIES IN WHICH A CENTRAL ATOM, X, IS SURROUNDED BY FOUR, THREE, OR TWO ELECTRON PAIRS

Number of Atoms Bonded To X	Number of Lone Pairs Around X	Predicted Bond Angle	Geometry of Species
4	0	109°	tetrahedral
3	1	109°	pyramidal
2	2	109°	bent
3	0	120°	equilateral triangle
2	1	120°	bent
2	0	180°	linear

the same approach can be extended to more complex molecules. You should keep in mind that there is one important "prerequisite" for the use of this method. **You must know or be able to derive the Lewis structure of a species before you can predict its geometry.**

Polarity of Molecules

A polar molecule is one in which there is a separation of positive and negative charge, that is, + and − poles. Any diatomic molecule in which the atoms differ (HF, HCl, . . .) is polar, with a negative pole located at the more electronegative atom. A diatomic molecule such as H_2, in which the two atoms are identical, is nonpolar.

Is O_2 polar?

In molecules containing more than two atoms, we must know the bond angles to decide whether or not the molecule is polar. Consider, for example, the two triatomic molecules shown at the top of Figure 8.9, BeF_2 and H_2O. The linear BeF_2 molecule is nonpolar; the two polar bonds cancel each other. From a slightly different point of view, we might say that the centers of negative and positive charge coincide in this molecule. In contrast, the bent water molecule is polar: the charge centers do not coincide. One can visualize a negative pole located at the oxygen atom with the compensating positive pole located midway between the two hydrogen atoms. Thus the H_2O molecule, like HF and HCl, is a dipole.

Another molecule which is nonpolar despite the presence of polar bonds is CCl_4. The four C—Cl bonds are polar, with the bonding electrons slightly displaced toward the chlorine atoms. However, because of the symmetrical pattern in which the four chlorines are arranged around the carbon atom, the polar bonds

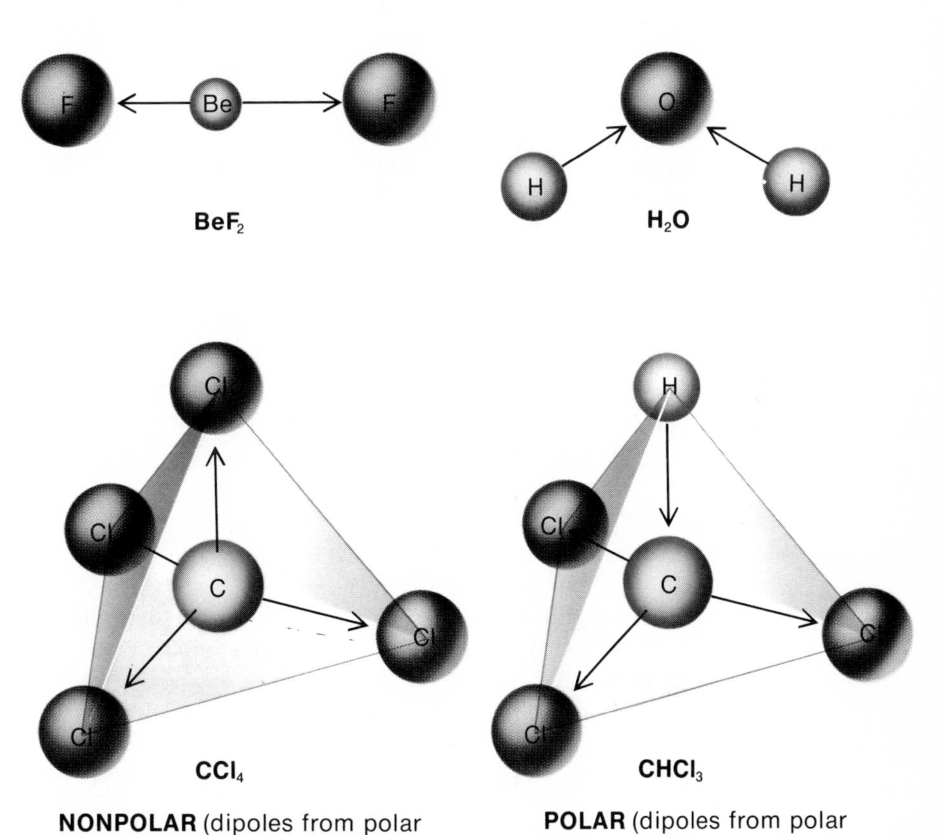

Figure 8.9 The BeF_2 and CCl_4 molecules are nonpolar because the polar bonds cancel each other. This does not happen in the polar H_2O and $CHCl_3$ molecules. (The arrows point to the atom of higher electronegativity.)

cancel each other. If one of the chlorine atoms in CCl_4 is replaced by hydrogen, the symmetry is destroyed and we obtain a polar species, the chloroform molecule, $CHCl_3$.

Example 8.6 Would you expect the SO_2 molecule to be a dipole? the NO_3^- ion?

NH_3 is polar. Why?

Solution In Example 8.5, we decided that the SO_2 molecule must be bent, with a bond angle of 120°. We would expect it to be slightly polar, with a + pole at the less electronegative sulfur atom and a − pole midway between the two oxygens. Referring again to Example 8.5, we decided that the NO_3^- ion has a symmetrical structure with the nitrogen atom at the center of an equilateral triangle and three oxygen atoms at the corners. The three polar bonds cancel each other; the NO_3^- ion is not a dipole.

8.6 HYBRID ATOMIC ORBITALS

Simple valence-bond theory explains covalent bond formation in terms of the sharing of unpaired electrons. From this point of view, it would seem that the number of bonds formed by a given atom would be governed by the number of unpaired electrons in the valence shell. In the case of carbon, whose orbital diagram is

	1s	2s	2p
$_6C$	(⇅)	(⇅)	(↑)(↑)( )

we might predict the formation of two covalent bonds. Yet we know from experience that carbon invariably forms four bonds rather than two. In order to explain this and other discrepancies within the framework of the valence-bond model, it is necessary to invoke a new kind of atomic orbital, the so-called hybrid bond orbital.

*sp*³ *Hybrid Orbitals*

The fact that a carbon atom forms four covalent bonds could be explained by assuming that, prior to reaction, one of the 2s electrons is promoted to the 2p level

	1s	2s	2p
$_6C$	(⇅)	(↑)	(↑)(↑)(↑)

Now that the carbon atom has four unpaired electrons, it can form four covalent bonds by sharing electrons with other atoms such as those of hydrogen.

	1s	2s	2p	
$_6C$	(⇅)	(⇅)	(⇅)(⇅)(⇅)	(CH_4 molecule)

(The colored arrows indicate electrons supplied by hydrogen: the horizontal lines are drawn to enclose the orbitals involved in bond formation.)

There is one fundamental objection to the model we have just described. It implies that two different kinds of bonds are formed. One of these would be an "s" bond while the other three would be "p" bonds. Experimentally it is found that the four bonds formed are identical in all respects. This leads us to believe that the four orbitals used for bond formation by carbon must be equivalent. In valence-bond terminology, we describe this situation by saying that an s and three p orbitals have been *mixed,* or *hybridized,* to give four new bonding orbitals described as **sp^3 hybrids.**

"Promotion of electrons" and "hybridization of orbitals" are two separate phenomena

It should be clearly understood that sp^3 hybrid orbitals have their own unique properties, quite different from those of the orbitals from which they are formed. It can be shown, by arguments based on quantum mechanics, that the four equivalent orbitals are directed toward the corners of a regular tetrahedron. This is, as we have seen, the orientation that keeps four pairs of electrons as far apart as possible.

The fact that the bond angles in ammonia and water are very nearly tetrahedral suggests that the four electron pairs surrounding the central atom in these molecules also occupy sp^3 hybrid orbitals. If the bonds formed by nitrogen in NH_3 or oxygen in H_2O were pure "p" bonds, we would expect them to be oriented at right angles to each other.

Other Types of Hybrid Orbitals; sp and sp^2 Hybrids

There are many other types of hybrid bond orbitals in addition to sp^3. Two of the more important for our purposes are **sp^2** and **sp** hybrids, formed by boron and beryllium, respectively (note the resemblance to sp^3 hybrids with carbon).

	Atom			*Molecule*			
	2s	2p		2s	2p		
Be	(⇅)	( )( )( )	→	(⇅)	(⇅)( )( )	BeF_2	sp hybridization
B	(⇅)	(↑)( )( )	→	(⇅)	(⇅)(⇅)( )	BF_3	sp^2 hybridization
C	(⇅)	(↑)(↑)( )	→	(⇅)	(⇅)(⇅)(⇅)	CH_4	sp^3 hybridization

Just as valence bond theory invokes promotion followed by sp^3 hybridization to explain why carbon forms four equivalent bonds rather than two, so it proposes sp^2 hybridization to explain three bonds rather than one with boron and sp hybridization with beryllium to explain the formation of two bonds rather than none.

Table 8.7 lists the geometries to be expected for the three types of hybrid orbitals that we have considered. You will note the striking similarity to Figure

TABLE 8.7 HYBRID ORBITALS AND THEIR GEOMETRIES*

Number of Bonds	Atomic Orbitals	Hybrid Orbitals	Orientation	Example
2	s, p	sp	linear	BeF_2
3	s, two p	sp^2	equilateral triangle	BF_3
4	s, three p	sp^3	tetrahedron	CH_4

*Two other types of hybrid orbitals (dsp^2, d^2sp^3) are discussed in Chapter 21.

8.7, where we arrived at the same structures by the electron pair repulsion principle. Valence-bond theory correlates molecular geometry with the type of hybridization shown by the central atom. One implies the other; in particular, *if you know the geometry of a molecule (or can deduce it from the electron pair repulsion model), you can predict the type of hybridization that is likely to be involved.*

Hybridization of Atoms Forming Multiple Bonds

We saw in Section 8.5 that the bond angles in C_2H_4 are the same as those in BF_3:

F B—F F 120° H H C=C H H 120°

This suggests that the bonding orbitals used by carbon in C_2H_4 are the same as those used by boron in BF_3: sp^2 hybrids. That is, a carbon atom in C_2H_4 uses three sp^2 hybrid orbitals, two in forming bonds with hydrogen atoms, one with the other carbon atom. The "extra" pair of electrons in the double bond between the two carbon atoms is not hybridized.

Figure 8.10 indicates the shapes of the several orbitals occupied by bonding electrons in the C_2H_4 molecule. Note that:

1. The three hybrid orbitals (Fig. 8.10*A*) have an electron density which is completely symmetrical about a line joining the bonded atoms. If, for example, we were to move out a given distance from any point on the C—C axis, we would encounter the same electron density regardless of the direction in which we moved (up, down, into the plane of the paper, out from the plane of the paper, etc.). Bonds formed from orbitals of this type, which are symmetrical about the bond axis, are often referred to as **sigma (σ) bonds.**

All single bonds are sigma bonds

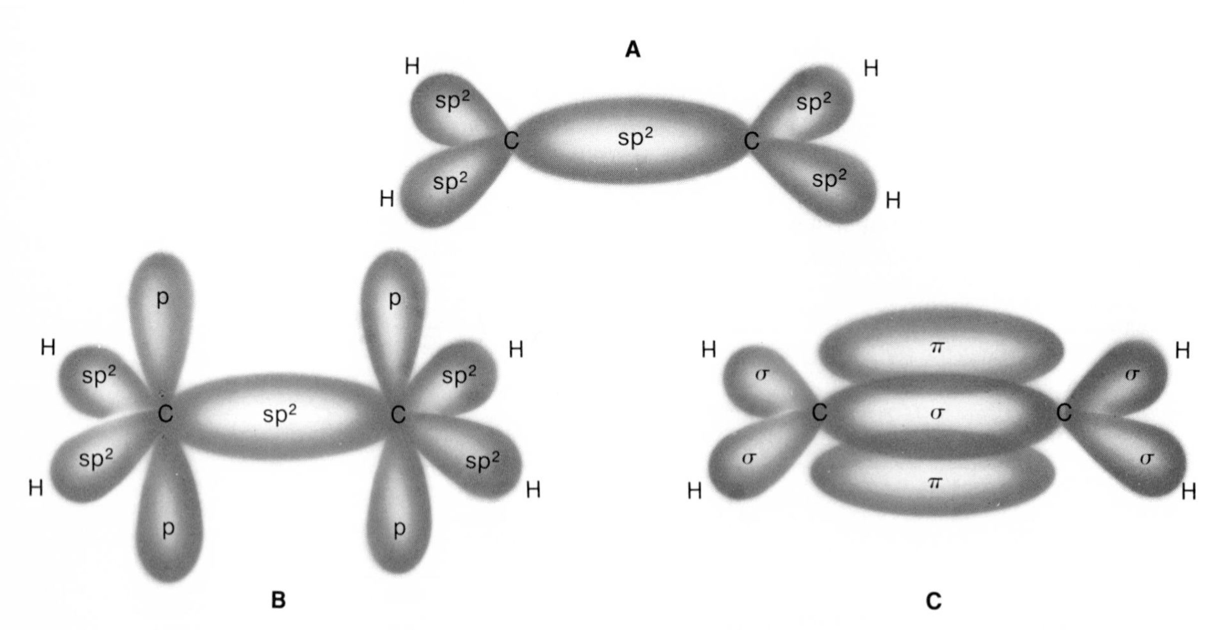

Figure 8.10 Bonding in C_2H_4. In *A* the sigma bond framework is shown. In *B* the p orbitals, one per carbon atom, have been added. The overlap of the p orbitals of the two carbon atoms to form the two lobes of a π bond is shown in *C*, to give the final electronic structure of C_2H_4. For effective overlap the C_2H_4 molecule must be planar. In the molecule there are five σ bonds and one π bond.

2. The orbital occupied by the extra electron pair of the double bond (Fig. 8.10*C*) has an electron density which is *not* symmetrical about the C—C axis. This orbital, formed by combining p orbitals of the two carbon atoms, is composed of two lobes, one above the C—C axis, the other below it. Within these lobes, the electron density is relatively high; in other regions around the C—C axis, it is low. Bonds formed from orbitals such as this, in which the electron density is not symmetrical about a line joining the bonded atoms, are called **pi (π) bonds.**

All multiple bonds contain one sigma bond; the others are pi bonds

The argument which we have just applied to ethylene, C_2H_4, is readily extended to acetylene, C_2H_2. Here we recall that the molecule is linear, analogous to BeF_2:

F—Be—F (180°)   H—C≡C—H (180°)

This suggests that one of the electron pairs between the carbon atoms is hybridized with a C—H bond to give sp hybridization around each carbon atom. The other two electron pairs in the triple bond are not hybridized. Instead, they are located in two different orbitals, oriented at 90° angles to each other. Each of these orbitals forms a π bond, which looks very much like that shown in Figure 8.10*C*.

The situation we have described for C_2H_4 and C_2H_2 applies to any species in which an atom forms multiple bonds. Only one of these bonds can be hybridized with the other electron pairs around the atom to give sp or sp^2 hybridization. The other electron pairs comprising the multiple bonds come from overlapping p orbitals; they are frequently referred to as "π bonds." Thus we might say that there is one π bond in C_2H_4 and two π bonds in C_2H_2.

Example 8.7 Referring to the Lewis structure of SO_2,

:S(=Ö:)—:Ö:

describe the hybridization around the S atom and state the number of π bonds.

Solution In SO_2, we can hybridize *three* electron pairs around sulfur (the lone pair, the single bond, and one of the electron pairs in the double bond). Hence we have sp^2 hybridization. The "extra" electron pair of the double bond is not hybridized and is referred to as a π bond.

8.7 MOLECULAR ORBITALS

The valence-bond model presented in this chapter explains a great many of the structural features of molecules and polyatomic ions. Most important, it accounts, at least qualitatively, for the stability of the covalent bond in terms of an overlap of atomic orbitals. By making use of the concept of hybridization, valence-bond theory succeeds in accounting for the principles of molecular geometry. By introducing the idea of resonance, it is even possible to rationalize the properties of molecules such as SO_2 for which a conventional Lewis structure is inadequate.

A major weakness of the valence-bond approach has been its inability to predict correctly the magnetic properties of many simple molecules. An example previously cited is molecular oxygen. The same problem arises with the B_2

molecule found in boron vapor at high temperatures. This molecule, like O_2, is paramagnetic even though it has an even number, six, of valence electrons.

The deficiencies of valence-bond theory arise from an oversimplification inherent in its approach. It assumes that the electrons in a molecule occupy atomic orbitals of the individual atoms. For example, in the CH_4 molecule, we describe the bonding in terms of the ls orbitals of the H atoms and the four sp^3 hybrid orbitals of the C atom. Clearly this is an approximation, since each electron in CH_4 must really be in an orbital characteristic of the molecule as a whole.

In line with this idea, the molecular orbital theory attempts to treat electron arrangements in molecules in terms of orbitals involving the whole molecule. The molecular orbital approach involves three basic operations:

1. The atomic orbitals associated with isolated atoms are combined to give a new set of molecular orbitals characteristic of the molecule as a whole. In doing this we always find that *the number of molecular orbitals formed is equal to the number of atomic orbitals combined.* For example, when two atoms combine to form a diatomic molecule, two s orbitals, one from each atom, yield two molecular orbitals. Again, six p orbitals, three from each atom, give a total of six different molecular orbitals.

2. Having arrived at a set of molecular orbitals, we attempt to arrange them in order of increasing energy. In principle, the energies of molecular orbitals, like those of atomic orbitals, can be derived by solving the Schrödinger wave equation. In practice, it is impossible to make precise calculations for any but the simplest of molecules. What is actually done is to deduce the relative energies of molecular orbitals from experimental observations on molecules or ions in which those orbitals are used. The spectra and magnetic properties of these species are useful here, but it must be admitted that we sometimes reason after the fact, reshuffling the order of molecular orbitals to explain new experimental evidence.

3. The valence electrons in the molecule are distributed among the available molecular orbitals in much the same way that electrons in atoms are fed into atomic orbitals. In particular, we find that:

a. *Each molecular orbital can hold a maximum of two electrons.*

b. *Electrons go into the lowest molecular orbital available.* A higher orbital starts to fill only when each orbital below it has its quota of two electrons.

c. *Hund's rule is obeyed.* When two orbitals of equal energy are available to 2 electrons, one electron goes into each, giving two half-filled orbitals.

If you read between the lines in the above description of the molecular orbital theory, you can see that it attempts to do for molecules what the electron configuration theory does for atoms. By semi-empirical means, we arrive at a description of the orbitals available to the electrons in a molecule. We then fill these orbitals, using much the same rules as were used to fill sublevels in atoms.

To illustrate molecular orbital theory let us apply it to the diatomic molecules of the elementary substances in the first two periods of the periodic table. We choose these elements as examples because MO theory is simplest to apply here and gives results which are in many ways superior to those of the valence-bond approach.

Hydrogen and Helium; Combination of ls Atomic Orbitals

Molecular orbital (MO) calculations show that the combination of two ls atomic orbitals leads to the formation of two molecular orbitals, one of which has an energy lower than that of the atomic orbitals from which it is formed (Fig. 8.11); placing electrons in this orbital gives a species which is more stable than the two isolated atoms would be. For this reason the lower molecular orbital in Figure 8.11 is referred to as a **bonding orbital.** The other molecular orbital has a higher energy than the corresponding atomic orbitals; electrons entering it find themselves in an unstable, higher energy state. It is referred to as an **antibonding orbital.**

The electron density corresponding to these two molecular orbitals is shown at the right of Figure 8.11. It will be observed that the bonding orbital has a high electron density in the region between the nuclei, which accounts for its stability. In the antibonding orbital, the probability of finding an electron between the nuclei is very small; the electron density is concentrated at the far ends of the "molecule." Since the nuclei are less shielded from each other than they are in the isolated atoms, the antibonding orbital is unstable with respect to the individual atomic orbitals. The electron density in both the bonding and antibonding orbitals is symmetrical with respect to the internuclear axis; both of these are sigma orbitals. In MO notation, these molecular orbitals are designated as σ^b_{1s} and σ^*_{1s}, respectively; the asterisk is used to represent the antibonding orbital.

In the H_2 molecule, where each atom contributes a 1s electron, the σ^b_{1s} orbital is filled, giving a single bond. In contrast, the He_2 molecule, with four 1s electrons to distribute, would fill both the bonding and antibonding orbitals. These would cancel each other and the net number of bonds would be zero

The "2" in the denominator accounts for the fact that a bond contains 2 electrons

$$\text{no. of bonds} = \frac{B - NB}{2} = \frac{2 - 2}{2} = 0$$

where B is the number of electrons in bonding orbitals and NB is the number of electrons in nonbonding orbitals. The stability of He_2 would be no greater than that of two isolated He atoms—small wonder that it does not exist.

Second Period Elements; Combination of 2s and 2p Orbitals

As we have just seen, MO theory predicts correctly the existence of a single bond in the H_2 molecule and the nonexistence of He_2. This is hardly

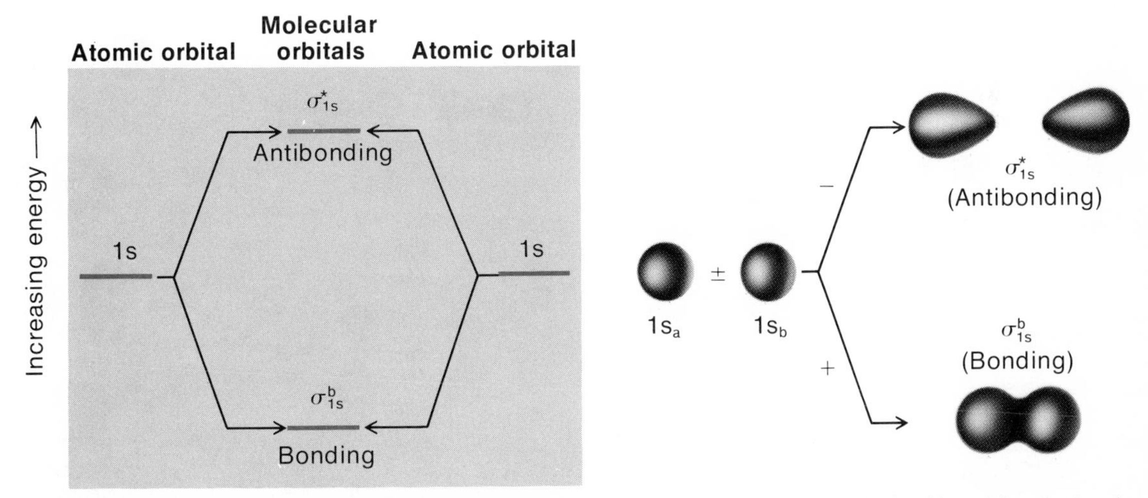

Figure 8.11 Molecular orbitals formed by combining two 1s orbitals. The bonding orbital is obtained by adding the two 1s orbitals on the two atoms, and produces high electron density between the atoms and lower energy for the orbital. The antibonding orbital results if the 1s orbitals are subtracted one from the other, resulting in low electron density between the atoms and a higher energy for the orbital.

astounding; valence-bond theory comes to the same conclusion, as did G. N. Lewis sixty years ago. The situation is somewhat more complicated with the diatomic molecules, real and hypothetical, of the second period elements. Three of these, N_2, O_2, and F_2, are familiar to all of us. The molecules Li_2, B_2, and C_2 are less common but have been observed and studied in the gas phase. In contrast, the molecules Be_2 and Ne_2 are either highly unstable or nonexistent. Let us see what molecular orbital theory predicts as to the structure and stability of these molecules. We start by considering how the atomic orbitals containing the valence electrons of the second period elements (2s and 2p) are used to form molecular orbitals.

Combining two 2s atomic orbitals, one from each atom, gives two molecular orbitals completely analogous to those previously discussed. These orbitals are designated as σ_{2s}^{b} (sigma, bonding, 2s) and σ_{2s}^{*} (sigma, antibonding, 2s).

Let us now consider what happens to the 2p orbitals. You will recall that in an isolated atom there are three such orbitals oriented at right angles to each other. We designate these orbitals arbitrarily as p_x, p_y, and p_z. When two p_x orbitals (one from each atom) combine, they form the two molecular orbitals shown at the upper right of Figure 8.12. These are both sigma orbitals, symmetrical about the x axis. One of them, designated σ_{2p}^{b}, is a bonding orbital, while the other, σ_{2p}^{*}, is antibonding. You will note that these orbitals have "shapes" (electron cloud densities) that resemble those of the sigma molecular orbitals shown in Figure 8.11.

The way in which p_y or p_z atomic orbitals combine with each other is quite different. As you can see from Figure 8.12, two such atomic orbitals, one from each atom, combine to form two pi molecular orbitals. The higher energy orbital, π_{2p}^{*}, where the electron density is concentrated away from the nuclei, is antibonding. The lower energy orbital, π_{2p}^{b}, with a relatively high electron density between the nuclei, is a bonding orbital. There are two such sets of orbitals, one formed by p_y atomic orbitals, the other by p_z atomic orbitals (not shown in the figure).

Summarizing the molecular orbitals available to the valence electrons of the second period elements, we have:

One σ_{2s}^{b} and one σ_{2s}^{*} orbital.
One σ_{2p}^{b} and one σ_{2p}^{*} orbital.
Two π_{2p}^{b} and two π_{2p}^{*} orbitals.

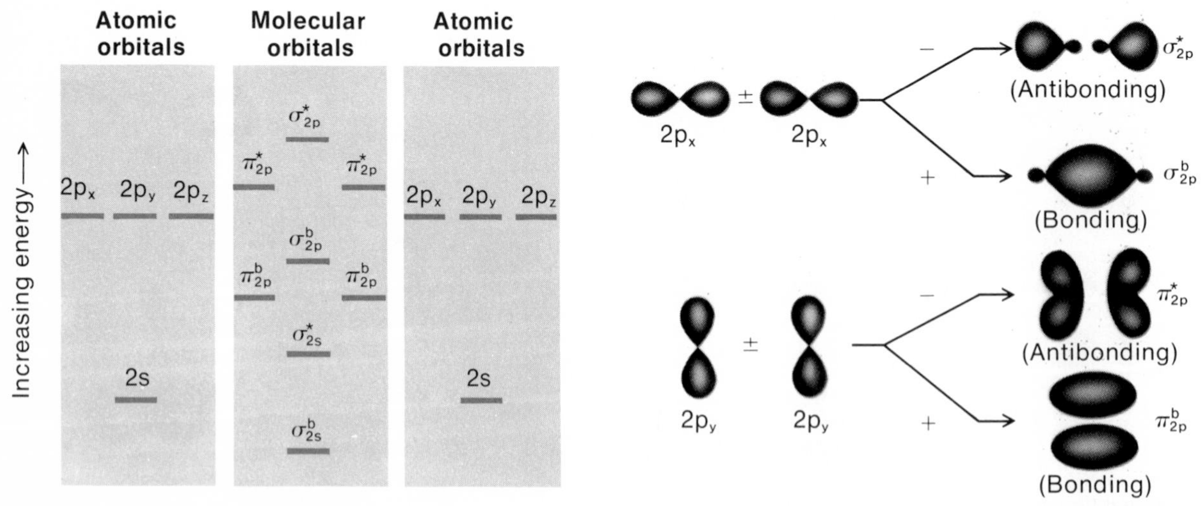

Figure 8.12 Molecular orbitals obtained by combining 2p atomic orbitals on two atoms. There are two π_{2p}^{b} orbitals and two π_{2p}^{*} orbitals because both the $2p_y$ and $2p_z$ orbitals can participate in π orbital formation. The molecular orbitals from the 2s electrons have the structure shown in Figure 8.11.

TABLE 8.8 PREDICTED AND OBSERVED PROPERTIES OF DIATOMIC MOLECULES OF SECOND PERIOD ELEMENTS

	Occupancy of Orbitals							
	σ^b_{2s}	σ^*_{2s}	π^b_{2p}	π^b_{2p}	σ^b_{2p}	π^*_{2p}	π^*_{2p}	σ^*_{2p}
Li_2	(⇅)	( )	( )	( )	( )	( )	( )	( )
Be_2	(⇅)	(⇅)	( )	( )	( )	( )	( )	( )
B_2	(⇅)	(⇅)	(↑)	(↑)	( )	( )	( )	( )
C_2	(⇅)	(⇅)	(⇅)	(⇅)	( )	( )	( )	( )
N_2	(⇅)	(⇅)	(⇅)	(⇅)	(⇅)	( )	( )	( )
O_2	(⇅)	(⇅)	(⇅)	(⇅)	(⇅)	(↑)	(↑)	( )
F_2	(⇅)	(⇅)	(⇅)	(⇅)	(⇅)	(⇅)	(⇅)	( )
Ne_2	(⇅)	(⇅)	(⇅)	(⇅)	(⇅)	(⇅)	(⇅)	(⇅)

	Predicted Properties		Observed Properties	
	No. of Unpaired e^-	No. of Bonds	No. of Unpaired e^-	Bond Energy (kJ/mol)
Li_2	0	1	0	105
Be_2	0	0	0	unstable
B_2	2	1	2	289
C_2	0	2	0	628
N_2	0	3	0	941
O_2	2	2	2	494
F_2	0	1	0	153
Ne_2	0	0	0	nonexistent

Hund's rule requires that there be two unpaired electrons in B_2 and O_2

The relative energies of these orbitals and hence the order in which they are filled in the diatomic molecules of the second period (at least through N_2) are indicated at the left of Figure 8.12.

Based upon this order of filling, we show in Table 8.8 the structures predicted by MO theory for the diatomic molecules Li_2 through Ne_2. Note the excellent agreement between theory and experiment, particularly with regard to the number of unpaired electrons. There is also a general correlation between the predicted number of bonds

$$\text{no. of bonds} = \frac{B - NB}{2}$$

and the bond energy. We would expect a double bond to be stronger than a single bond and a triple bond to be stronger than a double bond.

A major triumph of MO theory is its ability to predict correctly that the O_2 molecule should contain both a double bond and two unpaired electrons; simple valence-bond theory is unable to rationalize this observation. Again, the molecular orbital approach is successful where valence-bond theory fails in explaining the paramagnetism of the B_2 molecule. For the N_2 and F_2 molecules, both theories agree in predicting a triple and single bond, respectively, with no unpaired electrons.

Molecular orbital theory can also be applied quite successfully to predict some of the properties of heteronuclear molecules and ions of the elements of the second period (Example 8.8).

Would the bonding in NO^+ be stronger or weaker than that in NO?

Example 8.8 Using MO theory, predict the electronic structure, number of bonds, and number of unpaired electrons in the NO molecule.

Solution We have 11 valence electrons to account for (five from nitrogen, six from oxygen). Placing these in molecular orbitals in order of increasing energy:

	σ^b_{2s}	σ^*_{2s}	π^b_{2p}	π^b_{2p}	σ^b_{2p}	π^*_{2p}	π^*_{2p}	σ^*_{2p}
NO	(⇅)	(⇅)	(⇅)	(⇅)	(⇅)	(↑)	( )	( )

There are eight electrons in bonding orbitals (two in σ^b_{2s}, four in π^b_{2p}, two in σ^b_{2p}) and three in antibonding orbitals (two in σ^*_{2s}, one in π^*_{2p}). Hence

$$\text{no. bonds} = \frac{8 - 3}{2} = \frac{5}{2}$$

There is one unpaired electron (in the π^*_{2p} orbital).

PROBLEMS

8.1 *Formulas of Ionic Compounds* Give the formulas of

a. aluminum sulfide
b. ammonium sulfate
c. zinc nitrate

8.2 *Bond Polarity* Which of the following bonds should be the least polar, As—Se, As—Te, or Sb—Se?

8.3 *Lewis Structures* Give the Lewis structures of

a. ClO_2^- b. NO_2^- c. PO_4^{3-}

8.4 *Molecular Geometry* Describe the geometries of the species in Problem 8.3.

8.5 *Molecular Polarity* Which of the species in Problem 8.3 are dipoles?

8.6 *Hybrid Bond Orbitals* State the hybrid orbitals used by each central atom in Problem 8.3.

8.7 *Molecular Orbitals* Give the MO structure of NO^+.

8.8 Certain of the following statements are generally valid; others are open to criticism. Which ones would you criticize and why?

a. No transition metal ion contains outer s electrons.
b. Anions are larger than cations.
c. All bonds between unlike atoms are at least slightly polar.
d. Species with unpaired electrons are paramagnetic.

8.9 Predict the formulas of the following ionic compounds:

a. lithium oxide
b. scandium hydroxide
c. manganese(II) phosphate
d. zinc fluoride
e. ammonium nitrate

8.27 Criticize the following statements:

a. Metals in Groups 1A, 2A, and 3A achieve noble gas configurations by losing one, two, and three electrons, respectively.
b. The number of covalent bonds formed by an atom is equal to the number of unpaired electrons in the isolated, gaseous atom.
c. The bond energy of a double bond is twice that of a single bond between the same atoms.
d. The linear molecule X—Y—Z is nonpolar.

8.28 How many moles of ions are there in one mole of

a. ammonium sulfate
b. aluminum nitrate
c. chromium(III) carbonate
d. rubidium selenide
e. magnesium chloride

8.10 Complete and balance the following equations for the formation of ionic compounds:

a. $K(s) + I_2(s) \rightarrow$
b. $Ba(s) + S(s) \rightarrow$
c. $Mg(s) + O_2(g) \rightarrow$
d. $Sc(s) + Cl_2(g) \rightarrow$
e. $Ag(s) + O_2(g) \rightarrow$

8.11 Give the formulas of monatomic ions which have the same electron configuration as

a. He b. Fe^{2+} c. Cl^- d. Sr^{2+}

8.12 If we arbitrarily call a bond "ionic" when the electronegativity difference between the two atoms joined by the bond is greater than 1.7, with what elements does boron form ionic bonds?

8.13 Draw Lewis structures for

a. H_2S
b. SO_3^{2-}
c. CO_3^{2-}
d. ClO_3^-

8.14 Draw Lewis structures for the following species (the skeleton structure is indicated by the way the molecule is written)

a. $H_2C—CHCl$
b. $HO—NO_2$
c. $F_2N—NF_2$

8.15 One of the important components of the upper atmosphere is ozone, O_3. Three possible structures for ozone are

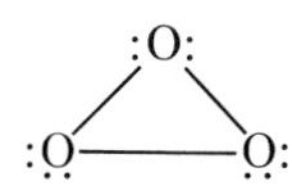

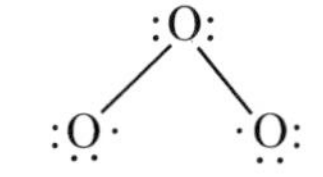

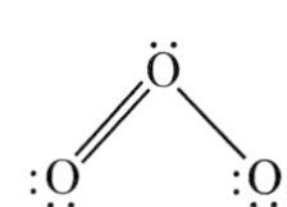

How could you determine experimentally which of these structures is correct?

8.16 Draw all the major resonance forms for

a. SO_3
b. SO_3^{2-}
c. CO_2

8.17 Give the formula of a polyatomic ion which you would expect to have the same Lewis structure as

a. F_2
b. PCl_3
c. SO_2
d. SO_3

8.29 Write balanced equations for

a. the reaction between calcium and nitrogen gas.
b. the decomposition of cesium hydride to the elements.
c. the precipitation of $CaCO_3$ from water solution.
d. the reaction between barium and chlorine.

8.30 Give the electron configurations of

a. Cr^{2+} b. Co^{3+} c. Mg^{2+} d. N^{3-}

8.31 Consider the bonds formed by Mg and Al with N and P. Estimate the % ionic character in each bond.

8.32 Draw Lewis structures for

a. CO_2
b. CN^-
c. CO
d. PH_4^+

8.33 Draw Lewis structures for

a. O—N—O—N—O
b. $H_3C—CHO$
c. Cl_2SO

8.34 The compound $Na_2S_2O_3$ is used as a fixing agent in photography

a. What are the charges of the ions present?
b. Write a Lewis structure for the polyatomic anion.
c. Describe the geometry of the polyatomic anion.

8.35 Which of the following pairs represent resonance forms of the same species?

a. $:\dot{N}=\ddot{O}:$ and $:\ddot{N}=\dot{O}:$

b. $(H)(H)C=C(Cl)(Cl)$ and $(H)(Cl)C=C(Cl)(H)$

c. $H_3C—C(=O)—CH_3$ and $H_3C—CH_2—C(=O)—H$

8.36 Give the formula of a molecule which you would expect to have the same Lewis structure as

a. ClO^-
b. NO_3^-
c. NH_4^+
d. HSO_4^-

8.18 One of the most objectionable compounds in photochemical smog is peroxyacetyl nitrate, PAN. Its skeleton is

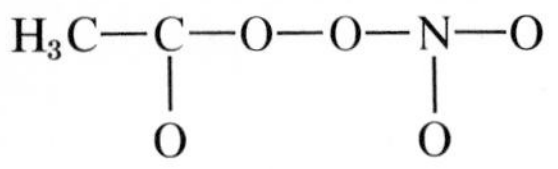

```
H3C—C—O—O—N—O
    |       |
    O       O
```

Write the Lewis structure for this molecule and give all the bond angles.

8.19 For each of the following, draw Lewis structures and describe the geometry.

a. CCl_4
b. BO_3^{3-}
c. SiO_4^{4-}

8.20 Describe the geometries of the species in Problem 8.13.

8.21 Which of the species in Problem 8.19 are dipoles?

8.22 Consider the following Lewis structures. For each, indicate all bond angles and state whether the species is a dipole.

```
          H
          |
a. :Cl—C—H
          |
          H

b. H—O—Cl:

c. (:O=N=O:)+

       H   H
       |   |
d. H—N—N—H
```

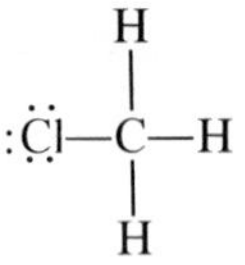

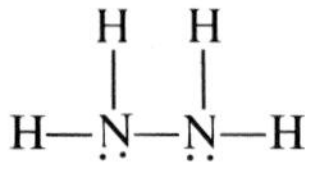

8.23 Indicate the hybrid orbitals used by carbon in

a. CH_4
b. C_2H_4
c. C_2H_2
d. $H_3C—OH$
e. CH_2O

How many π bonds are there in each species?

8.24 Give the formula of a molecule or ion in which an atom of

a. Be forms two bonds, using sp orbitals.
b. S forms two bonds, using sp^3 orbitals.
c. S forms four bonds, using sp^3 orbitals.
d. sulfur forms one π bond.
e. oxygen forms two π bonds.

8.37 A major eye irritant in smog is acrolein, which has the skeleton structure

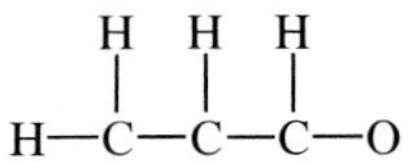

```
   H  H  H
   |  |  |
H—C—C—C—O
```

Draw the Lewis structure for this molecule and indicate all the bond angles.

8.38 Draw Lewis structures and describe the geometry of

a. NF_3
b. NO^-
c. ClO_4^-
d. PO_3^{3-}

8.39 Describe the geometries of the species in Problem 8.32.

8.40 Which of the species in Problem 8.38 are dipoles?

8.41 Draw Lewis structures for species in which a central nitrogen atom is bonded to two other atoms and the bond angle is

a. 109°
b. 120°
c. 180°

Which of these species are dipoles?

8.42 Indicate the hybrid orbitals used by each carbon atom in

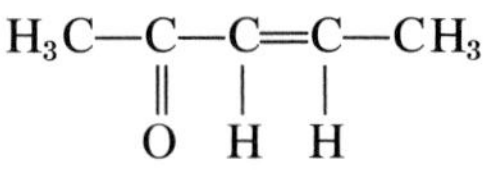

```
H3C—C—C=C—CH3
    ‖  |  |
    O  H  H
```

8.43 Give the formula of a molecule or ion in which an atom of

a. O forms two bonds, using sp^3 orbitals.
b. O forms a π bond.
c. B forms three bonds, using sp^2 orbitals.
d. B forms four bonds, using sp^3 orbitals.
e. nitrogen forms two π bonds.

8.25 Consider the -1 ions formed by adding an electron to the diatomic molecules of the elements of atomic numbers 3 through 9. Using the MO approach, list

a. the number of electrons in each of the 2s and 2p MO's.
b. the number of bonds in each of these ions.

8.26 Using the MO approach, give the number of unpaired electrons in

a. CO
b. C_2^+
c. O_2^{2-}
d. O_2^+

8.44 Suppose that in building up molecular orbitals, the σ^b_{2p} were placed below the π^b_{2p}. Prepare a diagram similar to Table 8.8 based on this assignment. For which species would this change in relative energies affect your prediction of number of bonds? number of unpaired electrons?

8.45 Consider the O_2 molecule, whose MO structure is given in Table 8.8. If successive electrons are removed to give the first, second, third, . . . ionization energies, where would you expect to find the largest jump in ionization energy?

*8.46 A certain compound contains 62.0% C, 10.4% H, and 27.5% O by mass. The vapor at 100°C and 1.00 atm has a density of 1.90 g/dm^3. Suggest two possible Lewis structures for this compound.

*8.47 When an atom forms six bonds, they are directed toward the corners of a regular octahedron (p. 507). What would you predict for the geometry of a molecule in which the central atom is surrounded by six pairs of electrons of which

a. one is a lone pair?
b. two are lone pairs?
c. three are lone pairs?

*8.48 Consider the hypothetical reaction $Na(s) + Cl_2(g) \rightarrow NaCl_2(s)$, where the product contains Na^{2+} and Cl^- ions. Using data in Section 8.1, taking the bond energy in Cl_2 to be 243 kJ/mol, the second ionization energy of Na to be 4520 kJ/mol, and assuming the lattice energy of $NaCl_2$ to be that of $MgCl_2$, -2494 kJ/mol, estimate the heat of formation of $NaCl_2$ and comment upon its stability.

9

PHYSICAL PROPERTIES AS RELATED TO STRUCTURE

The physical properties of pure substances are determined largely by two factors:

—the nature of the structural units (atoms, molecules, or ions) of which the substance is composed.

—the strength of the forces (interatomic, intermolecular, or interionic) between these particles.

In this chapter, we will consider how such physical properties as melting point, boiling point, and electrical conductivity are related to structure. We will distinguish between four types of substances which differ from one another in the nature of their structural units and the forces that hold these units together:

1. Ionic compounds (such as NaCl, MgO, and $CaCO_3$).
2. Molecular substances (such as H_2, CO_2, and H_2O).
3. Macromolecular substances (such as C and SiO_2).
4. Metals (such as Na, Mg, and Fe).

9.1 IONIC COMPOUNDS

In discussing ionic bonding in Chapter 8, we referred briefly to the characteristic properties of ionic compounds. In general we find that:

1. Ionic compounds are solids at room temperature. Their melting points are relatively high, ranging from several hundred degrees Celsius to over 2000°C. This behavior reflects the strong electrostatic forces between oppositely charged ions. Only at high temperatures do the ions acquire sufficient kinetic energy to overcome these attractive forces and attain the freedom of motion characteristic of the liquid state.

2. Ionic compounds are good conductors of electricity either in the molten state or in water solution. In either case, the conducting species are the charged ions (e.g., Na^+, Cl^-) which, in moving through the liquid, carry an electric current. In the solid state, where the ions are restricted to vibration about fixed positions in the crystal lattice, the conductivity is very low.

3. Many (but by no means all) ionic compounds are soluble in the polar solvent water. In contrast, all ionic solids are essentially insoluble in nonpolar organic solvents. We will have more to say about the general topic of solubility in Chapters 12 and 18.

According to Coulomb's Law, the attractive force between oppositely charged particles is directly proportional to their charges and inversely proportional to the square of the distance between them. Applied to ions in a crystal lattice, the relation becomes:

$$f = \text{constant} \times \frac{q_1 \times q_2}{(r_+ + r_-)^2} \quad \begin{matrix} (q_1, q_2 = \text{charges of + and − ions}) \\ (r_+, r_- = \text{radii of + and − ions}) \end{matrix} \tag{9.1}$$

According to Equation 9.1, interionic forces should be greatest for small ions of high charge. The melting points of ionic compounds might be expected to increase with increasing force. If we compare the compounds NaCl and BaO, for which the sums of the ionic radii are nearly the same but the charge products differ by a factor of four, we find that barium oxide has a considerably higher melting point (Table 9.1). The size effect is illustrated by a comparison of MgO and BaO; the smaller radius of the Mg^{2+} ion as compared to the Ba^{2+} ion results in a stronger attraction for the O^{2-} ion and, consequently, a higher melting point.

TABLE 9.1 EFFECT OF IONIC SIZE AND CHARGE ON MELTING POINT

COMPOUND	CHARGES OF IONS	$r_+ + r_-$ (nm)	MELTING POINT (°C)
NaCl	+1, −1	0.095 + 0.181 = 0.276	800
BaO	+2, −2	0.135 + 0.140 = 0.275	1920
MgO	+2, −2	0.065 + 0.140 = 0.205	2800

Decomposition Upon Heating

Binary ionic compounds such as NaCl or MgO ordinarily melt without decomposition to give the corresponding pure liquid. However, certain more complex ionic solids decompose upon heating to give two or more pure substances. This behavior is shown by **hydrates**, in which water molecules are incorporated into the crystal lattice. An example is $BaCl_2 \cdot 2\ H_2O$, which loses water at about 100°C to form anhydrous barium chloride:

$$BaCl_2 \cdot 2\ H_2O(s) \rightarrow BaCl_2(s) + 2\ H_2O(g) \tag{9.2}$$

Certain hydrated salts lose water upon exposure to dry air at room temperature; this process is referred to as **efflorescence**. A familiar example is copper sulfate pentahydrate, $CuSO_4 \cdot 5H_2O$. Blue crystals of this salt lose water to form first the lower hydrates, $CuSO_4 \cdot 3H_2O$ (blue) and $CuSO_4 \cdot H_2O$ (white), and finally, upon heating, anhydrous copper sulfate.

The formulas of hydrates indicate the number of moles of water per mole of compound

Another type of ionic compound which frequently decomposes instead of melting when heated is one containing an oxyanion (e.g., OH^-, CO_3^{2-}). Typically the products of decomposition are a solid metal oxide and a volatile nonmetal oxide. For example, when slaked lime, $Ca(OH)_2$, is heated to about 600°C, it breaks down to quicklime, CaO, and water vapor:

$$Ca(OH)_2(s) \rightarrow CaO(s) + H_2O(g) \tag{9.3}$$

The same type of reaction occurs when limestone, $CaCO_3$, is heated to about 800°C:

$$CaCO_3(s) \rightarrow CaO(s) + CO_2(g) \quad (9.4)$$

9.2 MOLECULAR SUBSTANCES

As a class, molecular substances tend to be volatile, with relatively low melting points and boiling points, usually below 300°C. All substances that are gases at room temperature and virtually all liquids are molecular in nature. Typical examples include elementary hydrogen, H_2 (bp = −253°C), methane, CH_4 (bp = −162°C), and benzene, C_6H_6 (bp = 80°C). If the molecular mass is quite high, the substance may be a solid at room temperature. Examples include naphthalene, $C_{10}H_8$ (mp = 80°C) and elementary iodine, I_2 (mp = 113°C).

Molecular species are ordinarily poor conductors of electricity, either in the liquid state or in solution. Neutral particles such as molecules are incapable of carrying an electric current. A few polar molecules react with water to form ions and hence give a conducting water solution. An example is hydrogen chloride, which undergoes the following reaction when added to water:

$$HCl(g) \rightarrow H^+(aq) + Cl^-(aq) \quad (9.5)$$

Reactions of this type are relatively uncommon; water solutions of most molecular substances (e.g., carbon dioxide, sugar, and methyl alcohol) are very poor conductors.

Interatomic vs. Intermolecular Forces

The generally low melting and boiling points of molecular substances are a direct consequence of the weak forces between molecules. They imply nothing about the forces within a molecule, which are ordinarily quite strong. Consider, for example, elementary hydrogen, which is made up of diatomic molecules. To melt or boil hydrogen, it is necessary only to overcome the attractive forces holding the molecules together–rigidly in the solid, loosely in the liquid state. The low melting point (−259°C) and boiling point (−253°C) of hydrogen reflect the weakness of these forces. The fact that the covalent bond between hydrogen atoms in the H_2 molecule is extremely strong is immaterial; this bond remains intact when hydrogen melts or boils. Very high temperatures are required to break the H—H covalent bond; even at 2400°C, only about 1% of the H_2 molecules are dissociated into atoms.

Trends in Melting and Boiling Points

Of the several structural factors that influence the volatility of molecular substances, the most obvious is molecular mass. Both melting point and boiling point tend to increase with molecular mass. Compare, for example, the four halogens listed in Table 9.2; there is a steady increase in both melting point and boiling point as we go from F_2 (MM = 38) to I_2 (MM = 254). The same trend is evident with the hydrocarbons listed in the table. As molecular mass increases,

TABLE 9.2 EFFECT OF MOLECULAR MASS ON MELTING POINT AND BOILING POINT OF MOLECULAR SUBSTANCES

	HALOGENS			
	F_2	Cl_2	Br_2	I_2
MM	38	71	160	254
mp (°C)	−223	−102	−7	113
bp (°C)	−187	−35	59	184

	HYDROCARBONS (saturated, straight-chain)								
	CH_4	C_2H_6	C_3H_8	C_4H_{10}	C_5H_{12}	C_6H_{14}	C_8H_{18}	$C_{16}H_{34}$	$C_{20}H_{42}$
MM	16	30	44	58	72	86	114	226	282
mp (°C)	−184	−172	−188	−135	−130	−94	−56	20	38
bp (°C)	−162	−88	−42	0	36	69	126	288	345

we go from compounds which are gases at 25°C and 1 atm (CH_4 through C_4H_{10}) to liquids (C_5H_{12} through $C_{16}H_{34}$) to low-melting, paraffin-like solids.

Another factor in addition to molecular mass which affects the volatility of molecular substances is polarity. Compounds built up of polar molecules melt and boil at slightly higher temperatures than nonpolar substances of comparable molecular mass (Table 9.3). The effect of polarity on melting or boiling point is ordinarily small enough to be obscured by differences in molecular mass. For example, in the series $HCl \rightarrow HBr \rightarrow HI$, the boiling point increases steadily with molecular mass despite decreasing polarity. However, in molecular compounds in which hydrogen is bonded to a small, highly electronegative atom (N, O, F), polarity has a much greater effect on volatility. Hydrogen fluoride (*bp* 19°C), despite its low molecular mass, has the highest boiling point of all the hydrogen halides. Water (*bp* 100°C) and ammonia (*bp* −33°C) also have abnormally high boiling points compared to those of the other hydrides of the 6A and 5A elements. In these three cases, the effect of polarity reverses the normal trend to be expected on the basis of molecular mass alone (Fig. 9.1, p. 212).

HF, H_2O and NH_3 are rather special cases

As one might expect, the substances that are most difficult to condense to liquids or solids are those in which the basic structural unit is both light in mass and nonpolar. Such substances include the nonmetallic elements of low molecular mass (H_2, *bp* = −253°C; N_2, *bp* = −196°C; O_2, *bp* = −183°C; F_2, *bp* = −187°C) and the lower members of the noble gas group (He, *bp* = −269°C; Ne, *bp* = −246°C; Ar, *bp* = −186°C).

TABLE 9.3 BOILING POINTS OF POLAR VS. NONPOLAR SUBSTANCES

NONPOLAR			POLAR		
Formula	Molecular Mass	Boiling Point (°C)	Formula	Molecular Mass	Boiling Point (°C)
N_2	28	−196	CO	28	−192
SiH_4	32	−112	PH_3	34	−85
GeH_4	77	−90	AsH_3	78	−55
Br_2	160	59	ICl	162	97

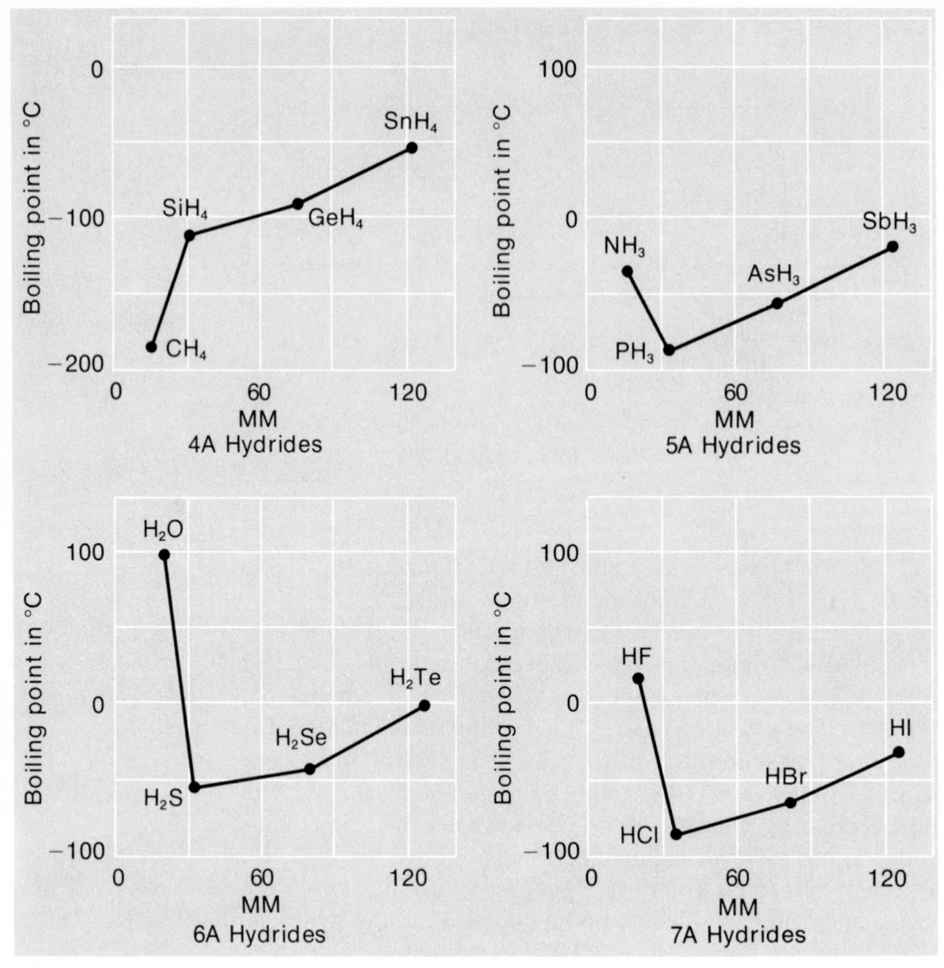

Figure 9.1 Boiling points of some nonmetal hydrides. Note the abnormally high boiling points of NH_3, HF, and particularly, H_2O.

Example 9.1 How would you expect the boiling point of nitric oxide, NO, to compare to

a. the boiling point of N_2O?
b. the average of the boiling points of N_2 and O_2?

Solution

a. Since NO has a molecular mass considerably lower than that of N_2O (30 vs. 44) it should be lower boiling. The observed boiling points are −152°C for NO and −88°C for N_2O.
b. If molecular mass were the only factor, we would expect the boiling point of NO to be halfway between those of N_2(−196°C) and O_2(−183°C). In other words, we would predict a value close to −190°C. However, NO, unlike N_2 or O_2, is a polar molecule. This factor accounts for its higher boiling point, −152°C.

An explanation of these trends in volatility lies in the nature of the forces holding molecular substances together. These forces, for reasons which we shall now consider, increase in magnitude with the size and polarity of the molecule.

DIPOLE FORCES. The effect of polarity on the physical properties of molecular substances is readily explained in terms of the way in which polar molecules line up with respect to one another (Fig. 9.2). The most stable arrangement is one in which the positive pole of one molecule is as close as possible to the negative pole of a neighboring molecule. Under these conditions, there is an electrostatic attraction, referred to as a *dipole force*, between adjacent molecules.

The electrostatic attraction between neighboring molecules within an iodine chloride crystal is similar in origin to that between oppositely charged ions in solid sodium chloride. However, the dipole forces between polar molecules in ICl are an order of magnitude weaker than the ionic bonds in NaCl. In the former case the unequal electronegatives of iodine and chlorine produce only partial + and − charges within the molecule. In NaCl, on the other hand, a complete transfer of electrons leads to ions with full + and − charges.

Dipole forces are also much weaker than polar covalent bonds

When iodine chloride is heated to 27°C the comparatively weak dipole forces are no longer able to hold the molecules in rigid alignment and the solid melts. Dipole forces remain significant in the liquid state, in which the polar molecules are still relatively close to each other. Only in the gas, where the molecules are very far apart, do the electrical forces become negligible. Consequently the boiling points as well as the melting points of polar compounds such as ICl are higher than those of nonpolar substances of comparable molecular mass.

HYDROGEN BONDS. The abnormal properties of hydrogen fluoride, water, and ammonia result from the presence in these substances of an unusually strong type of intermolecular force. This attractive force, exerted between the hydrogen atom of one molecule and the fluorine, oxygen, or nitrogen atom of another, is

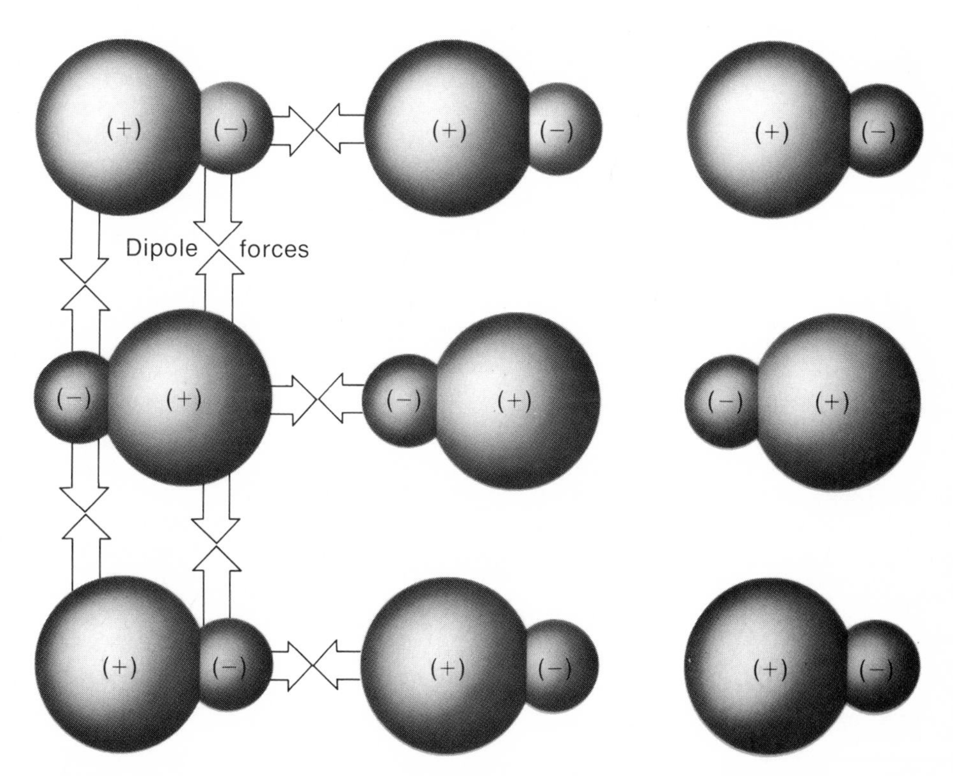

Figure 9.2 Dipole forces in an ICl crystal. The (+) and (−) indicate relatively small positive and negative charges on the iodine and chlorine atoms in the molecules; these charges give rise to dipole forces which contribute to the bonding energy of the lattice.

sufficiently unique to be given a special name, the *hydrogen bond*. There are two reasons why hydrogen bonds are stronger than ordinary dipole forces:

1. The difference in electronegativity between hydrogen (2.1) and fluorine (4.0), oxygen (3.5), or nitrogen (3.0) is great enough to cause the bonding electrons in HF, H_2O, and NH_3 to be markedly displaced from the hydrogen atom. Consequently, the hydrogen atoms in these molecules, insofar as their interaction with neighboring molecules is concerned, behave almost like bare protons. The hydrogen bond is strongest in HF, in which the difference in electronegativity is greatest, and weakest in NH_3, where the difference in electronegativity is relatively small.

2. The small size of hydrogen allows the fluorine, oxygen, or nitrogen atom of one molecule to approach the hydrogen atom in another very closely. It is significant that hydrogen bonding appears to be limited primarily to compounds containing these three elements, all of which have comparatively small atomic radii. The larger chlorine and sulfur atoms, with electronegativities (3.0, 2.8) similar to that of nitrogen, show little or no tendency to form hydrogen bonds in such compounds as HCl and H_2S.

How do we know this?

The energy required to break a mole of hydrogen bonds varies from about 5 to 50 kJ; even though the hydrogen bond is much weaker than a covalent bond, it is the strongest type of intermolecular force. Since many molecules of biological importance contain O—H and N—H bonds, hydrogen bonding is very common in such substances and often has an important influence on their properties. The geometry of proteins and nucleic acids, which are long-chain organic molecules, is fixed by hydrogen bonds between N—H and C=O groups in neighboring chains. The famous "double helix" model for DNA (deoxyribonucleic acid), proposed by Watson and Crick in 1953, was based on such a structure.

Many of the unusual properties of liquid water are a direct consequence of hydrogen bonding. For example, the high specific heat of water relative to other liquids or solids reflects the large amount of energy required to break hydrogen bonds, which become less numerous as the temperature is raised. Even at 100°C, enough hydrogen bonds remain to make the heat of vaporization of water, on a joule per gram basis (2257 J/g) greater than that of any other liquid.

When water freezes to ice, an open, hexagonal pattern of molecules results (Fig. 9.3). Each oxygen atom in the ice crystal is bonded to four hydrogens, two by ordinary covalent bonds at a distance of 0.099 nm and two by hydrogen bonds 0.177 nm in length. The large proportion of "empty space" in the ice structure explains why ice is less dense than liquid water. Indeed, water starts to decrease in density if cooled below 4°C, which indicates that the transition from a closely packed to an open structure occurs gradually over a temperature range rather than taking place abruptly at the freezing point. It is believed that even in water at room temperature, some of the molecules are oriented in an open, ice-like pattern. More and more

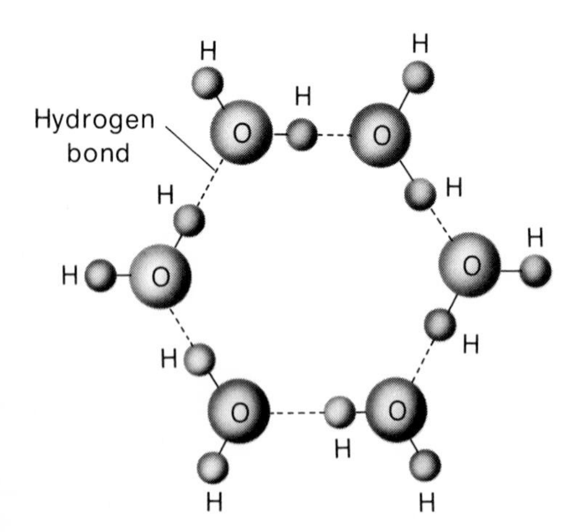

Figure 9.3 Plane projection of the structure of ice. The hexagons produce an open lattice, which accounts for the fact that ice is less dense than water. Hydrogen bonds help stabilize the ice lattice.

molecules assume this pattern as the temperature is lowered. Below 4°C the transition to the open structure predominates over the normal contraction on cooling, and liquid water expands as its temperature is lowered towards 0°C.

DISPERSION (LONDON) FORCES.* The two types of intermolecular forces already discussed can exist only between polar molecules. A different type of attractive force must be postulated to explain the existence of the liquid and solid states of such nonpolar substances as bromine and iodine. Since the melting and boiling points of nonpolar substances tend to increase with molecular mass, we deduce that the magnitude of this force increases with the mass or size of the molecule. The fact that the volatility of polar as well as nonpolar substances decreases with increasing molecular mass suggests that this type of intermolecular force must be common to all molecular substances.

The intermolecular force whose characteristics we have speculated upon is known as a *dispersion force*. Its origin is more difficult to visualize than that of the dipole force or hydrogen bond. Like them, it is basically electrical in nature. However, while hydrogen bonds and dipole forces arise from an attraction between permanent dipoles, dispersion forces are due to what might be called temporary or instantaneous charge separations.

It has been pointed out that, on the average, electrons in a nonpolar molecule such as H_2 are as close to one nucleus as to the other. However, at any given instant, the electron cloud may be concentrated at one end of the molecule (position 1A in Fig. 9.4). A fraction of a second later it may be at the opposite end of the molecule (position 1B). The situation is analogous to that of a person watching a tennis match from a position directly in line with the net. At one instant his eyes are focused on the player to his left; a moment later, they shift to the player on his right. Over a period of time, he looks to one side as often as the other; the "average" position of focus of his eyes is straight ahead.

In H_2, the dispersion forces are very weak

The instantaneous concentration of the electron cloud on one side or the other of the center sets up a temporary dipole in the H_2 molecule. This, in turn, induces a similar dipole in an adjacent molecule. When the electron cloud in the first molecule is at position 1A, the electrons in the second molecule are attracted to 2A. As the first electron cloud shifts to 1B, the electrons of the second molecule are pulled back to 2B. These temporary dipoles, both oriented in the same direction, lead to an attractive force between the molecules; this is the dispersion force.

*These forces are sometimes referred to as van der Waals forces. Properly speaking, van der Waals forces include all the intermolecular forces referred to in this section, as well as one other, the force between a dipole in one molecule and an induced dipole in an adjacent molecule. It is the sum of all these forces which determines the magnitude of the deviation of real gases from ideal behavior (cf. van der Waals equation, Chapter 5).

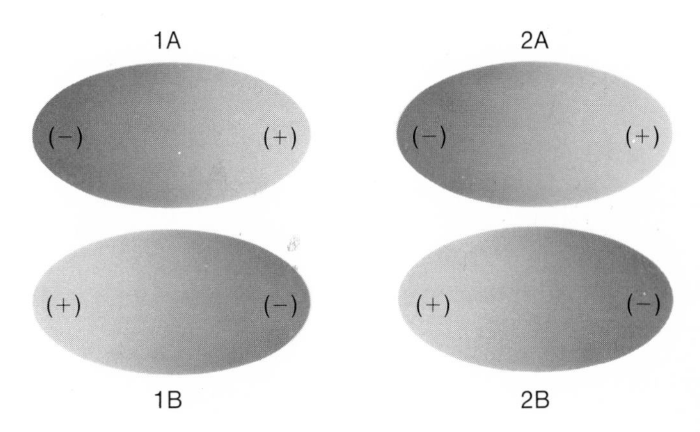

Figure 9.4 Temporary dipoles in H_2 molecules. These small negative and positive charges account for the attractive forces between even nonpolar molecules such as H_2. They are caused by a very slight interaction between the random movements of the electrons in molecules.

The strength of dispersion forces depends upon the ease with which the electronic distribution in a molecule can be distorted or "polarized" by a temporary dipole set up in an adjacent molecule. As one might expect, the ease of polarization depends primarily upon the size of the molecule. Large molecules where the electrons are far removed from the nuclei are more readily polarized than small, compact molecules in which the nuclei are closer to the electrons. In general, molecular size and molecular mass parallel each other; hence, the observation that dispersion forces increase in magnitude with molecular mass.

The fact that dispersion forces are more directly related to the size and shape of a molecule than to its mass enables us to explain certain anomalies (Fig. 9.5). Given two molecules of the same mass but of different shape or size, the electrons in the smaller or more compact molecule will have less freedom to move about, giving the molecule a smaller polarizability and the substance a lower boiling point. The effect can be quite substantial; SF_6, with small, compact, symmetrical molecules, has a relatively low polarizability and a low boiling point, −64°C, whereas decane, $C_{10}H_{22}$, of about the same mass but with a considerably larger volume, boils at 174°C. Although this is an extreme example, it suggests that we should be somewhat cautious in correlating physical properties of molecular substances with their masses. In particular, predictions of trends in melting point or boiling point can be made with confidence only when the molecules involved have reasonably similar structures (e.g., F_2, Cl_2, Br_2, I_2, or CH_4, C_2H_6, C_3H_8, . . .).

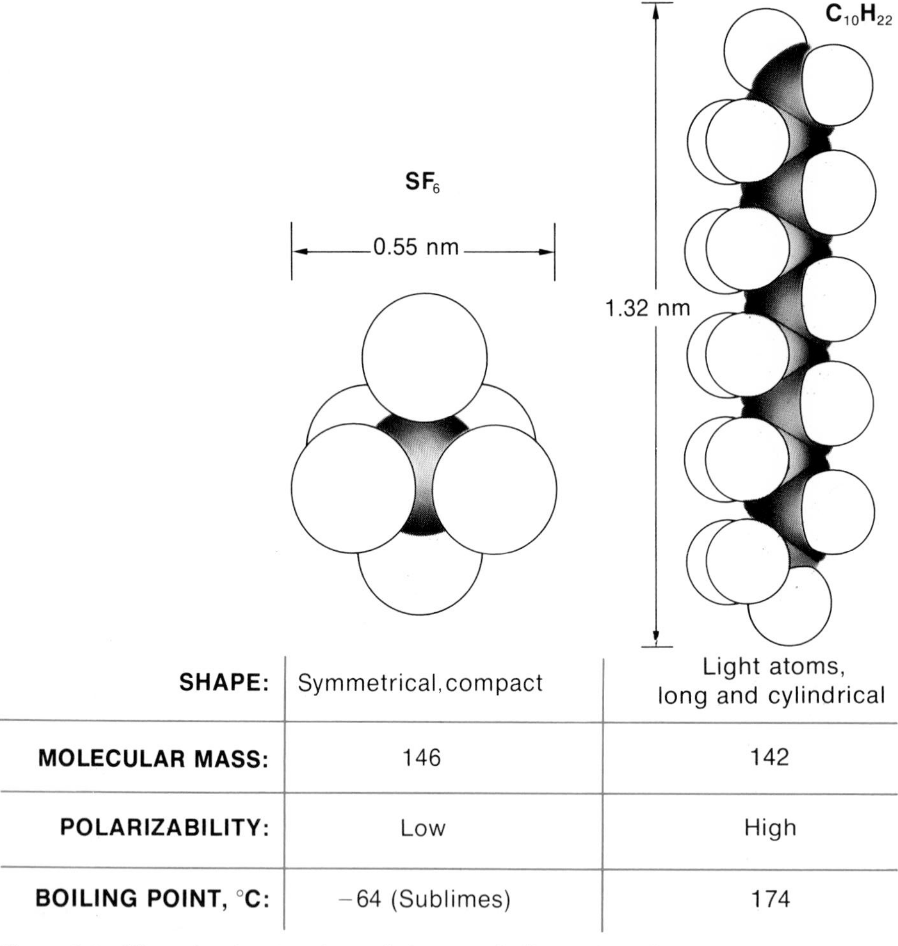

SHAPE:	Symmetrical, compact	Light atoms, long and cylindrical
MOLECULAR MASS:	146	142
POLARIZABILITY:	Low	High
BOILING POINT, °C:	−64 (Sublimes)	174

Figure 9.5 Effect of molecular size and shape on boiling points of substances with similar molecular masses. The more easily electrons can move about in a molecule, the more polarizable the molecule will be, and the greater will be the dispersion forces between the molecules.

Of the three types of intermolecular forces, dispersion forces are most common and, in the majority of cases, most important. They are the only forces of attraction between nonpolar molecules; even in the polar substance HCl, it is estimated that dispersion forces contribute 85 per cent of the total intermolecular force. Only where hydrogen bonding is involved do dispersion forces play a minor role. In water about 80 per cent of the intermolecular attraction can be attributed to hydrogen bonding and only 20 per cent to dispersion forces.

This may seem surprising, but there are a lot of electrons to be polarized in HCl

Example 9.2 What types of intermolecular forces are present in H_2? CCl_4? $CHCl_3$? H_2O?

Solution The H_2 and CCl_4 molecules are nonpolar (Chapter 8). Hence, they are attracted to each other only by dispersion forces. The $CHCl_3$ and H_2O molecules are both polar (Chapter 8). In $CHCl_3$ there are dipole forces as well as dispersion forces. In H_2O, hydrogen bonds as well as dispersion forces are present.

9.3 MACROMOLECULAR SUBSTANCES

Covalent bond formation need not, and often does not, lead to small, discrete molecules. Instead, it can result in structures of the type shown in Figure 9.6, where all the atoms are held together by a network of electron-pair bonds. Substances with this type of structure are referred to as macromolecular; the entire crystal in effect consists of one huge molecule. The compositions of the species shown in Figure 9.6 would perhaps best be described by the formulas X_n and $(XY)_n$, with the understanding that n is a very large number, of the order of Avogadro's number. More commonly, we represent them by the simple formulas X and XY.

Macromolecular substances are invariably high-melting solids with melting points above 1000°C. This general property reflects the strength of the covalent bonds that must be broken to free individual atoms from the crystal lattice. In this respect, macromolecular substances differ markedly from ones containing small, discrete molecules, where only weak intermolecular forces need be overcome to melt the crystal.

Two other typical properties of macromolecular substances are a direct consequence of their structure:

1. They are insoluble in all common solvents. For solution to occur, enough energy would have to be supplied to break most of the covalent bonds in the crystal.

2. They are usually poor electrical conductors, since there are no charged particles available to carry the current.

Many naturally occurring substances are macromolecular in nature; among the most common are the crystalline forms of carbon and silicon dioxide.

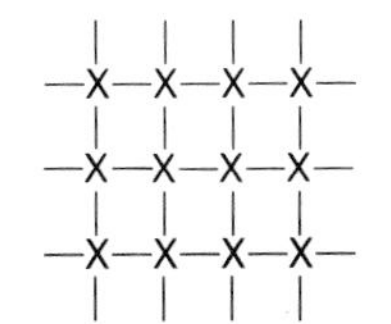

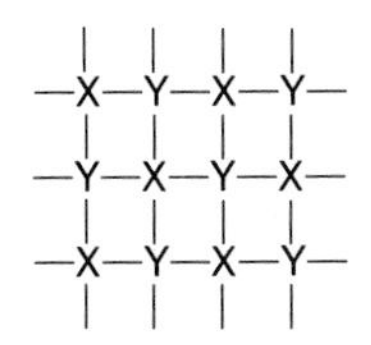

Figure 9.6 Schematic representation of two macromolecular crystals. In a macromolecular substance all atoms in the crystal are covalently linked together, so that the crystal is essentially one molecule.

Allotropic Forms of Carbon

Many elements, of which carbon is one example, can occur in more than one form in the same physical state. The general phenomenon is called *allotropy*; the different modifications are referred to as *allotropes*. The forms may differ in molecular formula, as is the case with the element oxygen, which can exist as O_2 or O_3 molecules in the gas state. Allotropy is most common in the solid state. Here it may result from a difference in the way the molecules are packed in the crystal, as in monoclinic and rhombic sulfur. Alternatively, allotropic forms of a solid element may differ in the way the atoms are bonded to one another. This is the case with the two crystalline forms of the element carbon, graphite and diamond. Both allotropes are macromolecular, which explains why both graphite and diamond have very high melting points, above 3500°C. However, as you can see from Figure 9.7, the bonding patterns in the two crystals are quite different.

Graphite is a moderately good electrical conductor, due to the mobility of electrons in the pi bonds

The graphite crystal is planar with the carbon atoms arranged in a hexagonal pattern. Each carbon atom is bonded to three others, forming one double bond and two single bonds. The forces between the layers in graphite are of the dispersion type and are quite weak. Consequently, the layers can readily slide past one another so that graphite is soft and slippery to the touch.* When you write with a "lead" pencil, which is really made of graphite, thin layers of graphite rub off onto the paper.

In diamond, each carbon atom forms single bonds with four other carbon atoms arranged tetrahedrally around it. The bonds are strong enough (bond energy C—C = 347 kJ/mol) to produce a rugged, three-dimensional lattice. Diamond is one of the very hardest of substances, and is used industrially in cutting tools and quality grindstones.

At room temperature and atmospheric pressure, graphite is the stable form of carbon. Diamond, in principle, should slowly transform to graphite under ordinary

*Studies show that the lubricating properties of graphite disappear under high vacuum. Apparently these properties require the presence of adsorbed H_2O or O_2 molecules, which act much like a film of oil on a metal surface, allowing graphite layers to slide readily past one another.

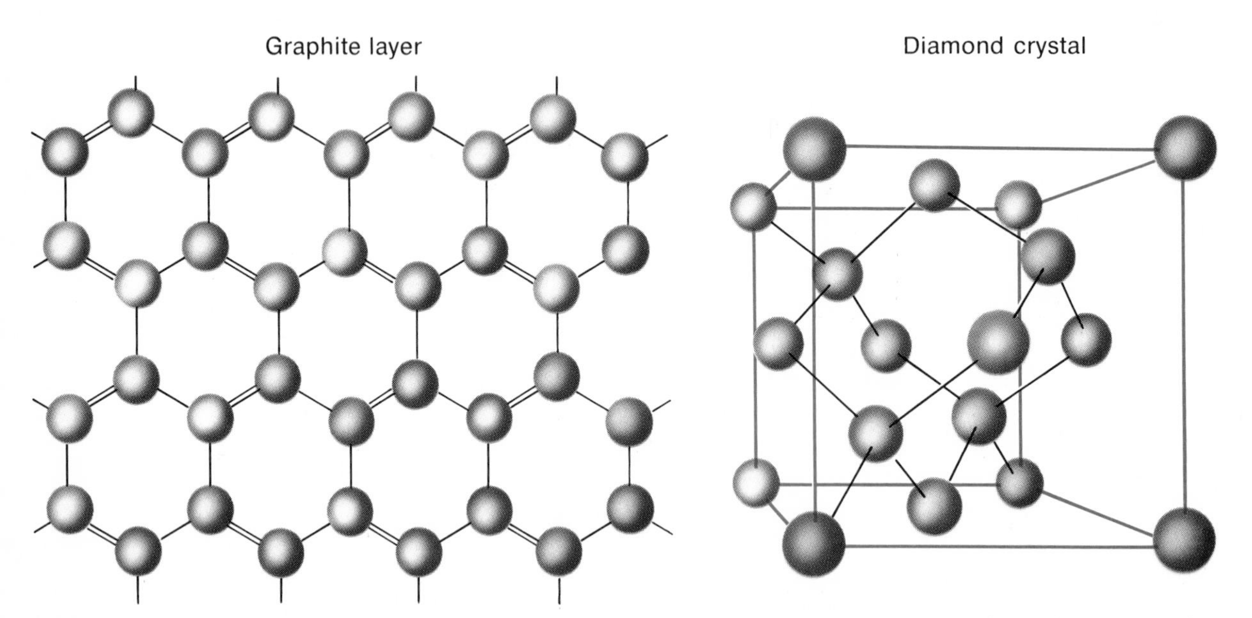

Figure 9.7 Crystal structures of diamond and graphite. In diamond each carbon atom is at the center of a tetrahedron, on each corner of which is another carbon atom. In graphite, the carbon atoms are linked together in planes of hexagons, which form layers which are loosely bound in the solid; graphite would have several resonance structures, of which one is shown here.

conditions. Fortunately for the owners of diamond rings this transition occurs at zero rate unless the diamond is heated to about 1500°C, where the conversion occurs rapidly. For understandable reasons, no one has ever become very excited over the commercial possibilities of this process. The more difficult task of converting graphite to diamond has aroused much greater enthusiasm.

Since diamond has a higher density than graphite (3.51 vs. 2.26 g/cm³), its formation should be favored by high pressures. Theoretically, at 25°C and 8000 atm graphite should turn to diamond. However, under those conditions the reaction has a negligible rate. At higher temperatures it goes faster, but the required pressure goes up too; at 2000°C, a pressure of about 100 000 atm is needed. In 1954 scientists at the General Electric laboratories were able to achieve these high temperatures and pressures and converted graphitic carbon to diamond for the first time. The synthetic diamonds produced by this process were at first quite small, but their size and quality have been improved so that one carat diamonds are now occasionally formed. At present, most of our industrial diamonds are synthetic.

Several more or less amorphous forms of carbon can be prepared; charcoal and carbon black are two of the more familiar. Charcoal can be made by heating wood or other materials with a high carbon content, such as sugars, to a high temperature in the absence of air, driving off water and other volatiles. Carbon black is usually made by burning natural gas under fuel rich conditions and collecting the soot on cool metal plates; this material is used in compounding rubber for automobile tires, where it adds considerable wear resistance. A form of carbon with very high surface area called activated carbon is produced in much the same way as charcoal; the material is very porous and absorbent. Activated carbon is used industrially to clarify sugar solutions and has had some application in public water supply systems, where it is effective in removing odors. In all the amorphous forms of carbon the atoms are arranged in irregular hexagonal patterns similar in many respects to the structure of graphite.

Silicon Dioxide

Silicon forms a strong bond with oxygen (*B.E.* Si—O = 368 kJ/mol) which in turn can bond to another silicon, forming chains of the type

$$\begin{array}{ccccccccc} & | & & | & & | & & | & & | & \\ — & Si & — & O & — & Si & — & O & — & Si & — \\ & | & & | & & | & & | & & | & \end{array}$$

These chains link together to form a macromolecular structure of the type shown in Figure 9.8, p. 220. In the crystal, each silicon atom is at the center of a tetrahedron, bonded to *four* oxygen atoms; each oxygen atom is linked to *two* silicon atoms.

Quartz is the most common form of silicon dioxide, or silica, and is the main component of sea sand. Since the orientation of the oxygen atoms in the lattice is not unique, there are several crystalline modifications of SiO_2, over 20 as a matter of fact, which differ in the manner in which the atoms are actually packed in the crystal. Quartz, like graphite and diamond, has a high melting point, about 1700°C. Unlike most solids, however, it does not melt sharply to a liquid but turns to a viscous mass over a reasonably wide temperature range, first softening at about 1400°C. The viscous fluid probably contains fairly long —Si—O—Si—O— chains, with enough bonds broken so that the material can flow.

Ordinary glass is made by heating a mixture of sand, limestone ($CaCO_3$), and soda ash (Na_2CO_3) to the melting point; carbon dioxide is driven out of the hot mass, which then contains glass, a solution of the oxides of silicon, sodium, and calcium in a mole ratio of about 7:1:1. This material softens at about 600°C and has

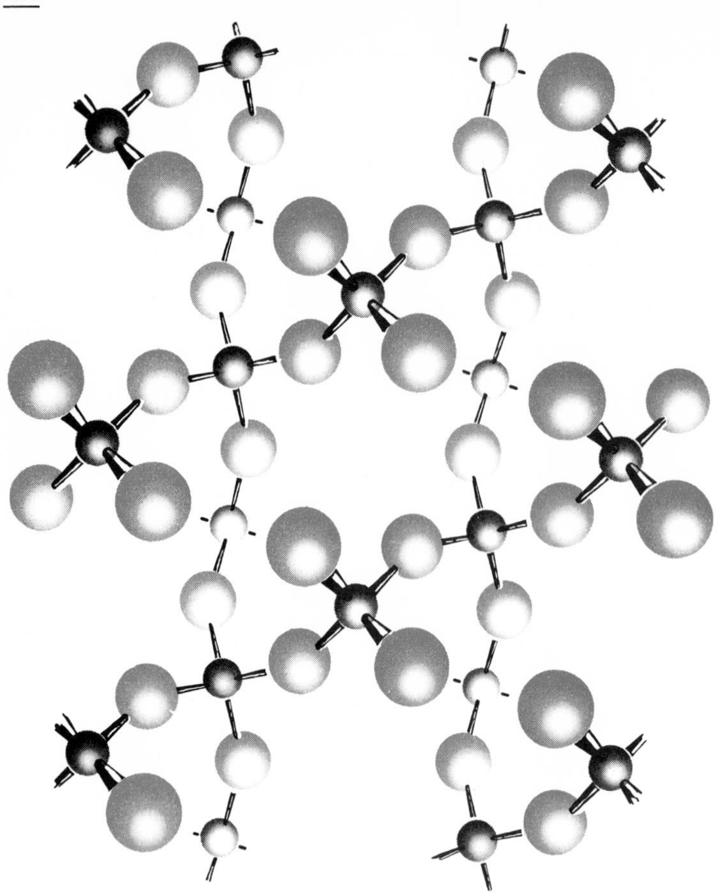

Figure 9.8 Crystal structure of quartz, SiO_2. In SiO_2 every silicon atom (small) is linked tetrahedrally to four oxygen atoms (large).

a wider temperature range of high viscosity than does quartz. Glass does not crystallize on cooling, but retains a disordered structure with many of the properties of a macromolecular substance.

Soft glass, which we have described, is produced in enormous quantities and is used in making bottles and window panes. It is not satisfactory for laboratory beakers and test tubes because it cracks fairly easily when subjected to thermal or mechanical shock. Hard, or borosilicate, glass, called Pyrex or Kimax, is made from a melt of silicon, boron, aluminum, sodium, and potassium oxides and is much more able to withstand both chemical attack and thermal shock. Hard glass softens at about 800°C and is readily worked in the flame of a propane-oxygen torch. Essentially all the glassware used these days in chemical equipment is made from borosilicate glass.

Silicates

The bulk of the rocks, minerals, and soils found in the earth's crust are composed of compounds similar in structure to silica, in that every silicon atom in the crystal is bonded tetrahedrally to four oxygen atoms. These materials are called silicates and are essentially ionic in nature. The negative ion is an oxyanion in which silicon is the central atom. The anion may be a very simple species (e.g., SiO_4^{4-}) or exceedingly complex, depending upon the number of atoms present and the geometric pattern in which these atoms are linked. The negative charge of the silicon oxyanion is balanced by positive charges on the cations of one or more metals (e.g., Na^+, Mg^{2+}, Al^{3+}) which fit into the crystal lattice.

From a structural point of view, the simplest silicates are those containing discrete silicon oxyanions, such as those shown in Figure 9.9. Of these, the simplest is the SiO_4^{4-} anion (compare PO_4^{3-}, SO_4^{2-}, ClO_4^-). This ion is present in the semiprecious stones zircon, $ZrSiO_4$, and the garnets, of which $Ca_3Cr_2(SiO_4)_3$ is typical. In the oxyanion shown at the right of Figure 9.9, six tetrahedra are linked through oxygen atoms at two corners to form a hexagonal ring structure. This is the anion present in emerald, $Be_3Al_2Si_6O_{18}$.

In the $Si_6O_{18}^{6-}$ ion, as in the fibrous minerals (p. 222), 2 of 4 oxygen atoms in a tetrahedron are shared

Most silicate minerals have macromolecular structures in which SiO_4^{4-} tetrahedra are linked together to form networks in one, two, or three dimensions. Two such structures are shown in Figure 9.10, p. 222. The structure at the left, in which tetrahedra are linked together to form an infinite chain, is typical of the *fibrous* minerals such as diopside, empirical formula $CaMgSiO_3$. The silicon-to-oxygen ratio in these minerals is readily derived by noting that two of the four oxygen atoms of each tetrahedron are shared equally with the next tetrahedron in the chain. Thus, for every silicon atom, the number of oxygen atoms is

$$2 + \frac{1}{2}(2) = 3$$

Silicates of this type are particularly dangerous air pollutants, presumably because the long fibers are readily adsorbed on the surface of lung tissue. Over long periods of time, they attack and destroy the tissue, causing the disease known as silicosis. Asbestos, a fibrous silicate with a somewhat more complicated structure than that shown in Figure 9.10, is one of the worst offenders.

The silicate structure shown at the right of Figure 9.10 is typical of *layer* minerals such as talc, $Mg_3(OH)_2Si_4O_{10}$. Here, SiO_4^{4-} tetrahedra are linked together in two dimensions to give an infinite layer. Three of four oxygen atoms are shared with adjacent tetrahedra, so the number of oxygen atoms per silicon atom is

$$1 + \frac{1}{2}(3) = \frac{5}{2}$$

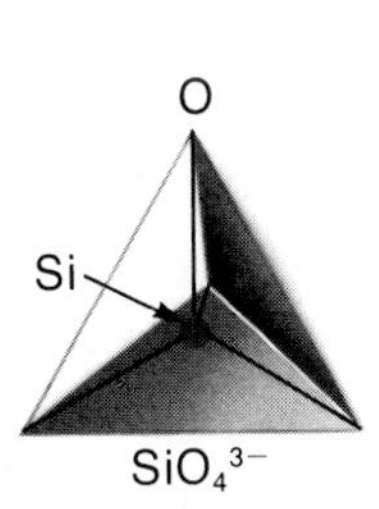

Figure 9.9 Two silicate anions. In solids containing these ions the cations are arranged in voids in the oxygen atom lattices. Garnets have the structure on the left, while emeralds crystallize in the hexagonal lattice on the right.

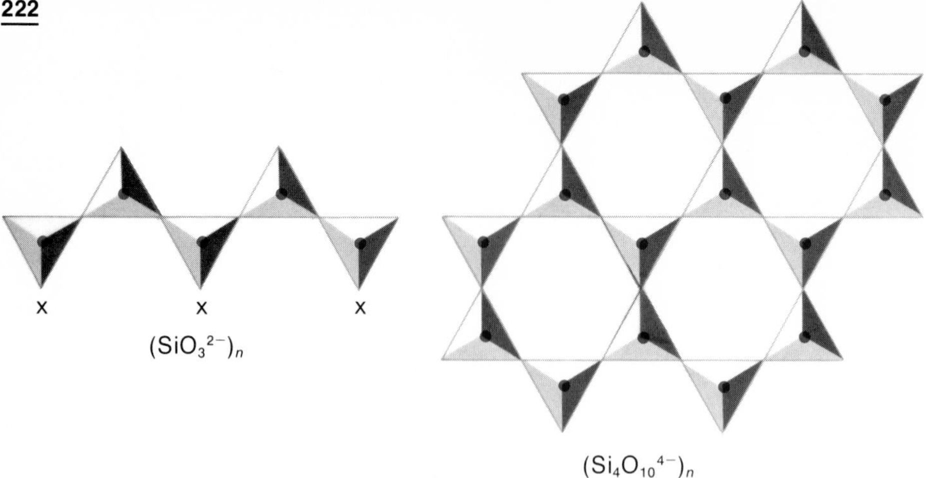

Figure 9.10 Two infinite silicate lattices. Asbestos contains a double anion chain similar to the one shown at the left. The structure of mica is that of the infinite sheet. (This figure and Fig. 9.9 have been redrawn from J. A. Campbell, *Chemical Systems*. San Francisco, W. H. Freeman and Co., 1970, p. 644.)

The sheets are held together only by weak dispersion forces; consequently, they slide past one another rather easily. As a result, talcum powder has a slippery feeling similar to that of graphite. Another common layer mineral, mica, has a somewhat more complicated structure in which cations are located between the layers. The electrostatic attraction between oppositely charged ions makes it more difficult to separate sheets of mica, so it is less slippery to the touch and is not useful as a lubricant.

The simplest three-dimensional silicate network is that shown previously for SiO_2 (Fig. 9.8). Here, all the oxygen atoms in a SiO_4^{4-} tetrahedron are shared and the oxygen-to-silicon ratio is

$$\frac{1}{2}(4) = 2$$

Many common minerals may be considered derivatives of SiO_2 in which some of the silicon atoms have been replaced by aluminum. An example is feldspar, in which an Al atom is substituted for every fourth Si atom, leading to the empirical formula $KAlSi_3O_8$. The zeolite minerals used in water softening (Chapter 13) also have this type of structure. Granite, the chief component of the earth's crust, is a heterogeneous mixture of mica, feldspar, and quartz.

9.4 METALS

In Chapter 7, we described several of the unique properties of metals. For our purposes here, we emphasize those properties which distinguish metals from the other types of substances that we have discussed (ionic, molecular, and macromolecular). Metals as a class are:

1. *Nonvolatile*. With one exception (mercury), all metals are solids at 25°C. They show a wide variation in melting point, ranging from slightly above room temperature (*mp* Cs = 29°C, Ga = 30°C) to several thousand degrees Celsius (*mp* W = 3380°C).

2. *Insoluble in water and organic solvents*. No metals "dissolve" in water in the true sense. As we saw in Chapter 7, a few very active metals react chemically with water to form hydrogen gas. Liquid mercury dissolves many metals, forming

solutions called amalgams. Perhaps the most familiar amalgams are those of silver and gold, formed when ores containing these metals are extracted with mercury.

3. *Excellent conductors of electricity*. Of the four types of substances we have discussed, metals are unique in that they are good conductors in the solid state. The electrical current is carried by electrons moving through the metal under the influence of an external electric field.

The physical and chemical properties of metals suggest a structure in which electrons are relatively mobile. Only if this is true can we explain why metals are good conductors and lose electrons readily to form positive ions. A simple approach to metallic bonding, known as the **electron-sea** model, explains these characteristics of metals in a qualitative way. Here, the metallic lattice is pictured as a regular array of positive ions (i.e., metal atoms minus their valence electrons) anchored in position like bell buoys in a mobile "sea" of electrons. At least to a first approximation, the valence electrons are able to wander through the lattice in much the same way that gas molecules move freely throughout their container. Presumably they would tarry a while in the vicinity of a positively charged metal ion, but they would not be permanently imprisoned between two metal ions. In Figure 9.11 we have attempted to show what a tiny portion of a metal crystal might look like according to the electron-sea model.

This is sometimes referred to as the "rice pudding" model, where the cations are the raisins

This simple picture of metallic bonding offers an obvious explanation of the high electrical conductivity of metals and their ability to lose electrons readily when they react with nonmetals. It can also be used to rationalize many of the other properties of metals referred to in Chapter 7. High thermal conductivity is explained by assuming that heat is transferred through the metal by "collisions" between electrons, which must occur frequently. Again, since electrons are not tied down to a particular bond with a characteristic energy, they should be able to absorb and re-emit light over a wide range of wavelengths. This explains why metal surfaces are such excellent reflectors.

According to the electron-sea model, the strength of the metallic bond is directly related to the charge of the positive ions that occupy lattice positions. Looking at Figure 9.11, we can readily see that the "lattice energy" of magnesium should be greater than that of sodium, since we are dealing with +2 rather than +1 ions, and two electrons per cation rather than one. This is consistent with the fact that magnesium has a considerably higher melting point than sodium (650°C vs. 98°C). Indeed, the general tendency for the melting points of metals to increase as we move across in the Periodic Table (e.g., *mp* K = 64°C, Ca = 850°C, Sc =

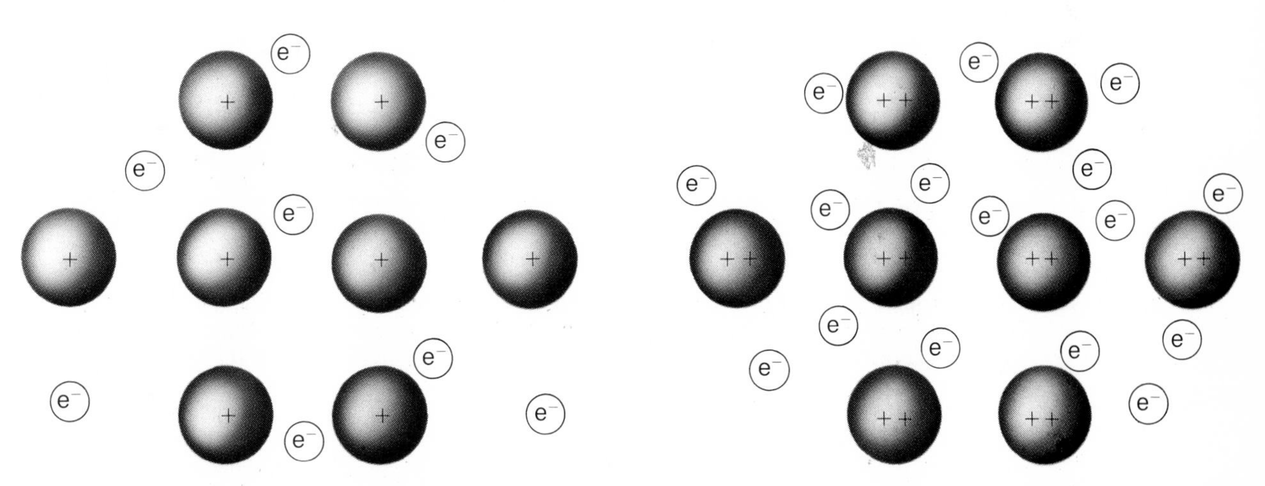

Figure 9.11 Electron-sea model for metallic bonding in sodium (Na^+, e^-) and magnesium (Mg^{2+}, $2e^-$). The hardness of metals increases with the number of electrons available for metallic bonding.

1430°C) can be explained in this way. The strength of the electrostatic forces holding the metal lattice together should increase with the charge of the positive ion (+1, +2, +3).

9.5 SUMMARY

It is possible to tie together many of the ideas expressed in this chapter by means of Table 9.4, which correlates the properties of different types of substances with their structures.

TABLE 9.4 TYPES OF SUBSTANCES; PROPERTIES AS RELATED TO STRUCTURE

Structural Units	Forces Within Units	Forces Between Units	Properties	Examples
1. Ions	——	Ionic bond	High *mp*. Conductors in molten state or water solution. Usually soluble in water, insoluble in organic solvents.	NaCl MgO
2. Molecules				
a. Nonpolar	Covalent bond	Dispersion	Low *mp, bp;* often gas or liquid at 25°C. Nonconductors. Insoluble in water, soluble in organic solvents.	H_2 CCl_4
b. Polar	Covalent bond	Dispersion, dipole, H bond	Similar to nonpolar but generally higher *mp* and *bp,* more likely to be water-soluble.	HCl NH_3
3. Macromolecules	Covalent bond	——	Hard, very high-melting solids. Nonconductors. Insoluble in common solvents.	C SiO_2
4. Cations, mobile electrons	——	Metallic bond	Variable *mp*. Good conductors in solid. Insoluble in common solvents.	Na Fe

Example 9.3 A certain substance is a liquid at room temperature and is insoluble in water. Suggest which of the four categories in Table 9.4 it probably fits into and list additional experiments which could be carried out to confirm your prediction.

Solution The fact that it is a liquid suggests that it is probably molecular in nature. The fact that it is insoluble in water agrees with this classification. To confirm that it is indeed molecular, we might measure the conductivity of the liquid, which should be essentially zero. It should also be soluble in most organic solvents.

PROBLEMS

9.1 *Melting Points of Ionic Compounds* Which should be the higher melting, CaO or KCl?

9.2 *Melting and Boiling Points of Molecular Substances* Arrange in order of increasing boiling point: Br_2, ICl, Cl_2.

9.3 *Intermolecular Forces* List the types of intermolecular forces present in

a. N_2 b. H_2S c. H_2O

9.4 *Types of Substances* A certain solid dissolves in water to form a conducting solution. Upon heating, it decomposes to give off a gas and form another solid. This behavior would be characteristic of:

a. CCl_4 b. graphite c. iron d. NaF e. Li_2CO_3

9.5 Explain in your own words the difference between

a. a molecular and a macromolecular substance.
b. polar covalent bonds and dipole forces.
c. hydrogen bonding and covalent bonding.

9.6 Explain in terms of structure why

a. NaCl has a higher melting point than ICl.
b. SiO_2 has a higher melting point than CO_2.
c. Hg is a better conductor than S.
d. H_2O has a higher boiling point than H_2S.

9.7 Criticize each of the following statements:

a. All high melting substances are ionic.
b. Boiling point increases with molecular mass.
c. Solutions prepared by shaking ionic solids with water are good conductors.

9.8 Classify each of the following as ionic, molecular, or macromolecular:

a. CH_4 b. C c. asbestos
d. $MgCO_3$ e. $CHCl_3$

9.9 For each of the following pairs of molecular substances, choose the higher boiling:

a. HF or HCl
b. O_2 or S_2
c. SiH_4 or PH_3
d. CH_4 or C_2H_6

9.10 For each of the following pairs, choose the higher-boiling member. Explain your reasoning in each case.

a. CH_4 or SiH_4
b. NaCl or CH_4
c. Cr or N_2
d. H_2O or SiO_2

9.17 Distinguish between

a. a discrete silicate anion and a macromolecular silicate anion.
b. metallic bonds and ionic bonds.
c. the forces within structural units in CO_2 and the forces between these units.

9.18 Explain in terms of structure why

a. graphite is much softer than diamond.
b. Cl_2 has a higher boiling point than F_2.
c. NaCl becomes a conductor when it melts.
d. metals are good reflectors of light.

9.19 Which of the following statements are always valid? generally valid? generally invalid?

a. Dispersion forces exist between all molecules.
b. Hydrogen bonds are found in all compounds containing hydrogen.
c. Silicates have macromolecular structures.

9.20 Classify each of the following as molecular, macromolecular, or metallic:

a. sugar
b. brass
c. chromium
d. propane
e. talc

9.21 Which would be the higher-melting substance in each of the following pairs? (The first two pairs are ionic, the others molecular.)

a. NaF or MgO
b. MgO or BaO
c. NH_3 or PH_3
d. PH_3 or SbH_3.

9.22 Follow the directions of Problem 9.10 for the following pairs of substances:

a. garnet or iodine
b. quicklime or benzene
c. mercury or quartz
d. lead or radon

9.11 For which of the following would melting require that ionic bonds be broken? covalent bonds?

a. MgF_2 b. CO_2 c. SiO_2 d. Fe
e. H_2O

9.12 What kind of attractive forces (in addition to dispersion forces) would have to be overcome to

a. melt ice?
b. dissolve NaCl in water?
c. decompose $MgCO_3$ to MgO?
d. dissolve SiO_2 in CCl_4?

9.13 Write a balanced equation for the thermal decomposition of

a. $Mg(OH)_2$ b. LiOH c. Li_2CO_3

9.14 Consider the empirical formulas of the following silicates. Which ones could contain macromolecular anions of the one-dimensional type (infinite chain)? two-dimensional (infinite layer)? three-dimensional?

a. $MgSiO_3$ b. $MgSi_2O_5$
c. Mg_2SiO_4 d. $KAlSiO_4$

9.15 A certain solid is insoluble in water, a nonconductor, and does not melt when heated to 1000°C. Into which of the four categories in Table 9.4 does it probably fall?

9.16 A certain solid is believed to have either a molecular or macromolecular structure. What single experiment would you perform to decide which category it fits into?

9.23 For which of the following would it be necessary only to overcome dispersion forces to boil the substance?

a. HCl b. Cr c. C d. N_2
e. $MgCO_3$

9.24 List the strongest type of bond or intermolecular force in

a. He b. H_2 c. CaF_2
d. asbestos e. Na

9.25 Which of the following ionic compounds would you expect to decompose before melting? What would be the decomposition products?

a. NaF
b. $Al(OH)_3$
c. $CuCl_2 \cdot 5\ H_2O$
d. $Al_2(CO_3)_3$

9.26 Follow the directions of Problem 9.14 for the following silicates:

a. K_4SiO_4
b. $KFeSi_2O_6$
c. $KAlSi_4O_{10}$

9.27 A certain pure liquid is a good conductor of electricity at 0°C. It is insoluble in water and has a very low vapor pressure at 25°C. Identify the substance.

9.28 Describe several simple experiments which would allow you to decide whether a certain substance is metallic or ionic in nature.

*9.29 The silicate mineral asbestos has a structure similar to that of talc except that two chains are linked together through the oxygen atoms marked "x" in Figure 9.10 to form an infinite double chain. Sketch a portion of the asbestos structure and give the empirical formula of the silicate anion.

*9.30 The density of gaseous hydrogen fluoride at 28°C and 1.00 atm is 2.30 g/dm³. What information does this yield about the intermolecular forces in hydrogen fluoride?

10

AN INTRODUCTION TO ORGANIC CHEMISTRY

Having presented in the previous several chapters current theories of atomic structure, chemical bonding, and intermolecular forces, it seems appropriate that we show how these ideas can be applied to a somewhat more descriptive part of chemistry. We select organic chemistry for this purpose, since it both relates well to the theories and is a very important area of chemistry.

Organic chemistry is the study of those substances containing carbon in combination with hydrogen and a few other nonmetals, particularly oxygen, nitrogen, sulfur, and the halogens. In spite of this limitation as to composition, there are far more organic compounds than compounds of all of the other elements combined.

Organic compounds are important in our daily lives in many ways. The human body, for example, consists largely of tissues composed of organic molecules of a wide variety of sizes and structures. These molecules, many of which are synthesized by our bodies, are daily utilized as ingredients in complex reactions which produce additional new molecules, release energy, and, in general, carry on the life process. Similar molecules are found in all plant and animal systems, where they selectively participate in a myriad of reactions controlling growth, maturity, and reproduction.

One particularly important area of organic chemistry is concerned with the identification and development of medicinal products. Although man has for centuries recognized certain plant and animal products as being valuable medicines, developments since 1940 have produced new chemicals that have truly revolutionized the practice of medicine. Sulfa drugs, cortisones, steroids, tranquilizers, and a variety of other medicinal agents, isolated or synthesized by researchers in the last three decades, have made it possible to deal with illness much more effectively than at any time in man's history. If one were to rate the scientific areas of progress in this century which have contributed most to the well-being of mankind, it would be only fair to rank our ability to treat illness with medicines obtained through the knowledge of organic chemistry at the top of the list.

10.1 NATURE OF ORGANIC SUBSTANCES

Organic chemistry owes its existence to the rather unique properties of the carbon atom. Carbon atoms, whose electron configuration is $1s^22s^22p^2$, can, you recall, form four bonds with other atoms by hybridization. These bonds may all be single, or may be a combination of single with double or triple bonds. Carbon is unusual in that it forms strong bonds with a variety of nonmetal atoms (bond energy C—H = 414 kJ/mol, C—O = 351 kJ/mol, C—Cl = 331 kJ/mol, C—N = 293 kJ/mol). More important still is the unique ability of the carbon atom to form strong single and multiple bonds with itself (bond energy C—C = 347 kJ/mol, C═C = 598 kJ/mol, C≡C = 820 kJ/mol). This means that carbon atoms in organic molecules can be linked together in chains of varying length. No other element possesses this capacity to anywhere near the same degree. Silicon and germanium, in the same family as carbon, can also form four bonds, but the relative weakness of Si—Si and Ge—Ge bonds (bond energy Si—Si = 176 kJ/mol) restricts the chemistry of these elements very severely.

Because the electronegativity of carbon is near that of hydrogen and the other nonmetals commonly found in organic compounds, the bonding in these substances is typically covalent and the compounds themselves are molecular. The bonding in organic molecules is almost always consistent with the octet rule, with bond angles given by the electron pair repulsion principle (Chapter 8). In stable organic molecules there are four bonds around each carbon atom; nonbonding electron pairs around carbon are virtually unknown.

Almost the only exception is carbon monoxide, :C≡O:

The physical properties of organic substances are determined by intermolecular attractions resulting from dispersion forces, dipole forces, and hydrogen bonds. Because organic molecules can contain from just a few to many thousands of atoms, with many different compositions, it is not surprising that they occur as gases, as liquids of all ranges of viscosity, and as solids, from the very soft to those nearly as hard as metals.

10.2 HYDROCARBONS

Organic compounds can be classified into groups and subgroups, according to the nature of the covalent bonds and the kinds of atoms present. One very large group includes those substances whose molecules contain only carbon and hydrogen atoms. These substances are called **hydrocarbons,** and, depending on the kinds of carbon bonds present, they can be further classified as alkanes, alkenes, alkynes, or aromatic substances.

Saturated Hydrocarbons: Alkanes

One large and structurally simple subgroup of the hydrocarbons includes those substances whose molecules contain only single carbon-carbon bonds. These substances are called **saturated** hydrocarbons or **alkanes.** In the alkanes the carbon atoms are bonded to each other in chains, which may be long or short, single or branched. Sometimes the alkanes are called paraffins.

The simplest alkanes are methane, CH_4, ethane, C_2H_6, and propane, C_3H_8:

```
    H              H  H              H  H  H
    |              |  |              |  |  |
  H—C—H          H—C—C—H          H—C—C—C—H
    |              |  |              |  |  |
    H              H  H              H  H  H
 methane          ethane            propane
```

Around the carbon atoms in these molecules, and indeed in any saturated hydrocarbon, there are four bonds, involving sp^3 hybrids; they are directed toward the corners of a tetrahedron, as would be expected from the electron pair repulsion rule. This means that in propane, C_3H_8, and in the higher alkanes, the carbon chain is bent or "zigzag" so there is less atom-atom interference than may seem indicated in planar diagrams.

For the smaller alkane molecules we find that all of the possible Lewis structures for a given molecular formula exist. Recognizing that there may be side branches on the main carbon chain, you could easily guess, and correctly, that there are two possible four-carbon alkanes:

```
      H                    H
      |                    |
   H—C—H                H—C—H
      |                    |
   H—C—H                H—C—CH3
      |                    |
   H—C—H                H—C—H
      |                    |
   H—C—H                   H
      |
      H

   butane             2-methylpropane
```

The CH_3 group is called methyl

Example 10.1 Show that the structure of butane is consistent with the octet rule.

Solution Each carbon atom has four valence electrons, each of which is in one of the four bonding sp^3 orbitals around the atom. Each H atom has one valence electron. The carbon atoms share electrons with each other and with hydrogen atoms in such a way as to have eight electrons, with two electrons in each of the four bonds to carbon. Each H atom will share two electrons. These conditions are easily met if we simply put the atoms together in the known structure; in the diagram the electrons contributed by hydrogen are shown as open circles.

```
   H  H  H  H
   ○● ○● ○● ○●
 H○●C●●C●●C●●C●○H
   ○● ○● ○● ○●
   H  H  H  H
```

The two structures drawn for C_4H_{10} illustrate the general fact that in organic chemistry a given molecular formula does not necessarily lead to one unique molecular structure. Although the two four-carbon alkanes, butane and 2-methylpropane, both have the same molecular formula, C_4H_{10}, they are two distinct compounds with their own characteristic physical and chemical properties. Compounds such as these, having the same formula but different molecular structures, are called **isomers.** Isomerism of this and other kinds is very common in organic chemistry and is one of the factors which accounts for the huge number of different organic compounds. For small molecules, it is quite easy to find the number of possible isomers (Example 10.2).

Example 10.2 How many isomers would you expect for molecules with the formula C_5H_{12}?

Solution First sketch the straight-chain structure:

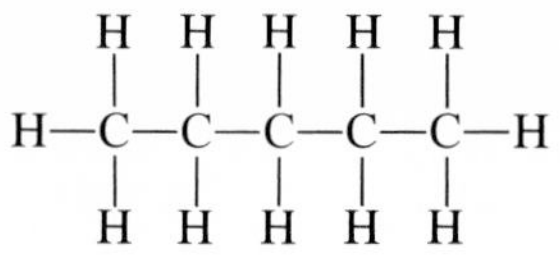

```
    H   H   H   H   H
    |   |   |   |   |
H — C — C — C — C — C — H
    |   |   |   |   |
    H   H   H   H   H
```

I

Counting hydrogens, we find there are 12, as required. Since each carbon atom has four bonds and each hydrogen one, this is a correct Lewis structure.

Having found one correct structure, we need to determine all the *nonequivalent* alternate structures. Two such structures are shown below (only the carbon skeletons are shown).

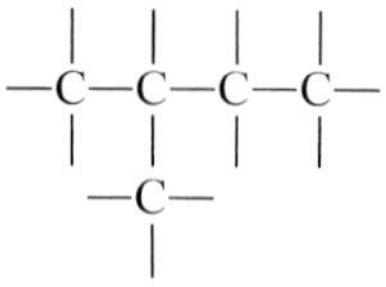

```
                                        |
                                      — C —
  |   |   |   |                         |
— C — C — C — C —                 — C — C — C —
  |   |   |   |                         |
    — C —                             — C —
      |                                 |
```

II 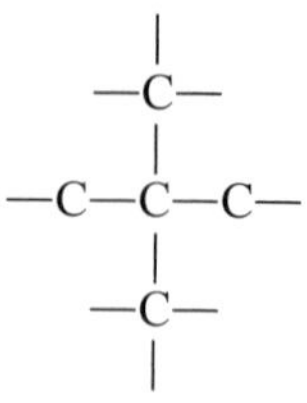 III

In structure II the longest carbon chain consists of only four atoms, so that II and I are clearly different. Similarly, in structure III, there are only three carbon atoms in the longest carbon chain, so it differs from both I and II. Looking at structures I, II, and III you will notice another important difference which distinguishes one from another. In I, no carbon atom is attached to more than 2 other carbon atoms. In II there is one carbon atom that is bonded to 3 other carbons. Finally, in III, the carbon atom in the center is bonded to four other carbon atoms.

At this point, working only with pencil and paper, you might be tempted to draw other structures, such as

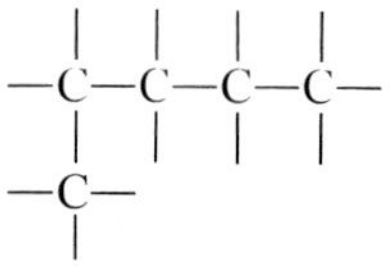

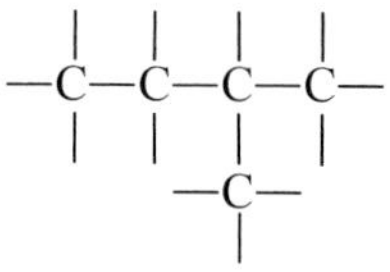

```
  |   |   |   |                 |   |   |   |
— C — C — C — C —             — C — C — C — C —
  |   |   |   |                 |   |   |   |
— C —                                 — C —
  |                                     |
```

However, a few moments' reflection (or access to a molecular model kit) should convince you that these are in fact equivalent to structures written previously. In particular, the first one, like I, has a 5-carbon chain in which no carbon atom is attached to more than two other carbons. The second structure, like II, has a 4-carbon chain with one carbon atom bonded to three other carbons. Structures I, II, and III represent the three possible isomers of C_5H_{12}; there are no others.

As the number of carbon atoms increases, we find that the number of possible structures goes up very rapidly. For example, for the six-carbon alkanes, there are five possibilities, all of which are known; in the structures below we have omitted the H atoms so that you can see the carbon chains more clearly:

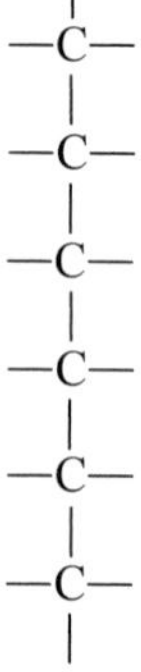

```
  |
— C —
  |         |             |
— C —     — C —         — C —
  |         |   |         |               |              |
— C —     — C — C —     — C —           — C —          — C —
  |         |   |         |   |           |   |        |   |   |
— C —     — C —         — C — C —       — C — C —    — C — C — C —
  |         |             |   |       |   |   |        |   |   |
— C —     — C —         — C —       — C — C —          — C —
  |         |             |           |   |              |
— C —     — C —         — C —           — C —          — C —
  |         |             |               |              |
```

hexane | 2-methylpentane | 3-methylpentane | 2,3-dimethylbutane | 2,2-dimethylbutane

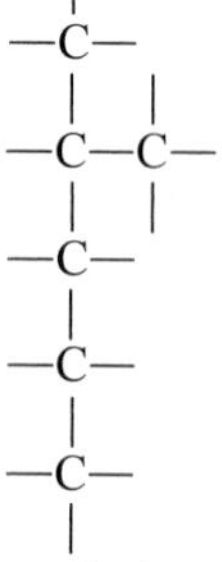

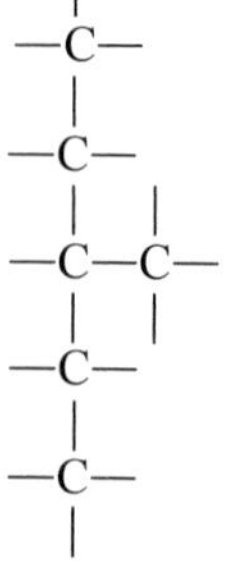

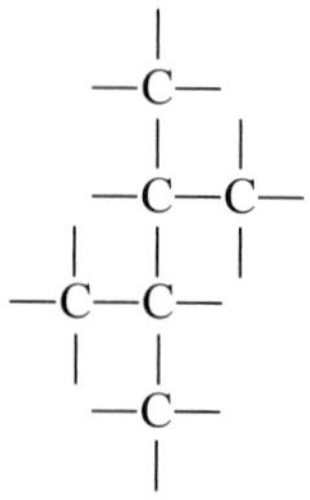

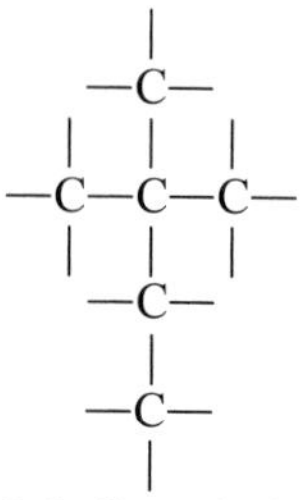

You should note that all of the six-carbon alkanes shown have the same molecular formula, C_6H_{14}.

If one builds space-filling models of the alkanes (Fig. 10.1), it becomes clear that the outer regions of the molecules consist mainly of hydrogen atoms, and that as far as the surface is concerned one molecule looks much like another. There are no well-defined positive and negative centers, so the molecules are relatively nonpolar. They are soluble in each other and in other nonpolar solvents, but are essentially insoluble in water. The alkanes are colorless and relatively odorless. Their melting and boiling points increase with molecular mass, reflecting the fact that intermolecular bonding in the compounds is due to dispersion forces. Methane and ethane are both gases (*bp* −161° C and −88° C, respectively). Hexane is a liquid (*bp* 69° C), whereas eicosane, $C_{20}H_{42}$, a long-chain compound, is a solid (*mp* 38° C). The solids have the waxy properties of the common household paraffin used to seal jelly jars and make candles. The alkanes were originally called paraffins because of the waxy nature of the long-chain members of this class.

The six-carbon alkanes all boil at temperatures in the vicinity of 60° C; however, as the chains become more branched, the boiling points drop noticeably. In light of the fact that the polarizability of compact, globular molecules is less than that of linear ones of the same composition, we would expect smaller dispersion forces between the more branched, compact species. The drift downwards in boiling point as one goes from hexane to 2,2-dimethylbutane thus seems very reasonable, and indeed is illustrative of the magnitude of the effect of molecular shape on boiling point (Table 10.1).

As a group the alkanes are relatively unreactive chemically. Their principal reaction, which accounts for their use as a fuel in heating systems and in the internal combustion engine, is that of burning in air:

$$CH_4(g) + 2\ O_2(g) \rightarrow CO_2(g) + 2\ H_2O(l) \qquad (10.1)$$

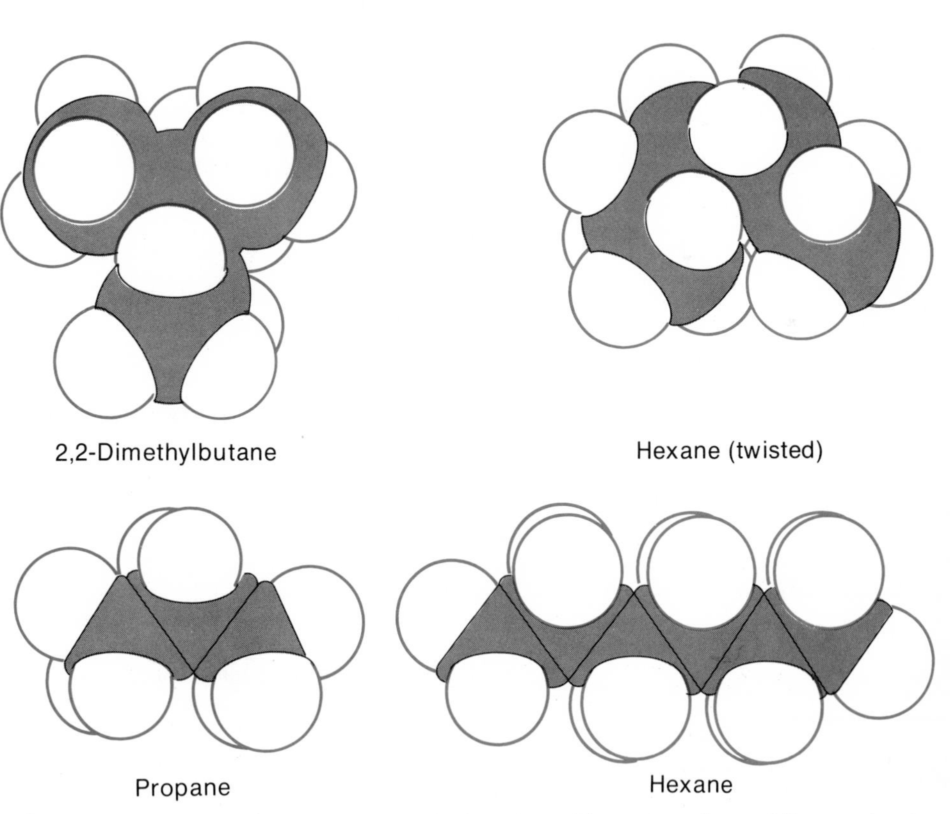

Figure 10.1 Space-filling models of some hydrocarbons. The outer surfaces of these molecules consist mainly of hydrogen atoms.

TABLE 10.1 BOILING POINTS OF SOME SIX-CARBON HYDROCARBONS (°C)*

Compound	Boiling point
Hexane	69
2-Methylpentane	60
3-Methylpentane	64
2,3-Dimethylbutane	58
2,2-Dimethylbutane	50

*See sketches of carbon skeletons on page 230.

$$C_8H_{18}(l) + \frac{25}{2}\,O_2(g) \rightarrow 8\,CO_2(g) + 9\,H_2O(l) \tag{10.2}$$

Closely related to the alkanes is another class of hydrocarbons known as the cycloalkanes. These compounds resemble the alkanes in that they contain only single carbon-carbon bonds; they differ from them in that the carbon atoms are arranged in a ring. The most common cycloalkanes contain rings involving 5 or 6 carbon atoms; such structures have bond angles which are rather close to the tetrahedral angle (109°) and hence involve relatively little strain. However, three- and four-membered rings are known and larger rings are occasionally observed.

The cyclohexane ring is puckered in a "chair" form:

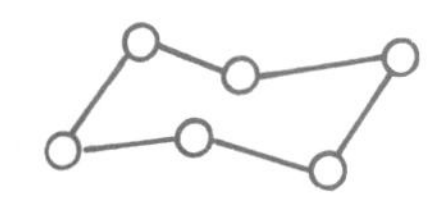

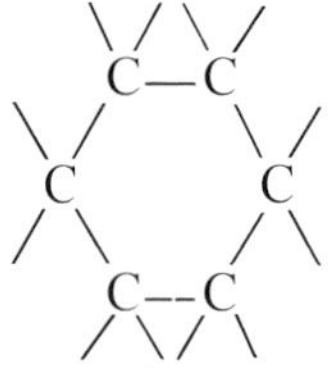

cyclohexane

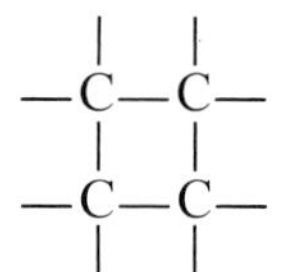

cyclobutane

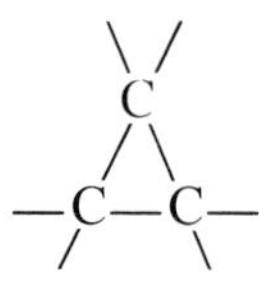

cyclopropane

The molecular formula of a cycloalkane differs from that of a chain alkane with the same number of carbon atoms in one important respect: it contains two fewer hydrogens. Thus we have cyclobutane, C_4H_8, as opposed to butane, C_4H_{10}, and cyclohexane, C_6H_{12}, vs. hexane, C_6H_{14}.

Example 10.3 Show that the formula of a saturated hydrocarbon containing n carbon atoms is C_nH_{2n+2} if there are no rings present, and is C_nH_{2n} if the molecule contains a ring.

Solution Sketch a straight unbranched chain of carbon atoms:

```
  |   |   |   |   |   |   |   |   |   |
—C—C—C—C—C—C—C—C—C—C—
  |   |   |   |   |   |   |   |   |   |
```

Clearly, for this molecule, there are two H atoms at the ends of the bonds above and below each carbon atom, plus one H atom at each end of the chain, making the formula C_nH_{2n+2}. If one makes branched isomers, say by exchanging a CH_3 group from one end with an H atom on the chain, this can be done without changing the formula of the molecule, no matter how many branches there are (or, indeed, how long they are). If, however, the two carbon atoms at the ends of the chain are linked to form a ring, the two end H atoms are lost. This makes the formula of the ring paraffin become C_nH_{2n}.

Even though we have considered the structures of only a few hydrocarbons, they suggest a problem in organic chemistry that cannot be ignored—namely, how does one name organic compounds simply and meaningfully? If one is to write about such substances, characteristic names are needed, for example, for each of the isomers we have described, since it is not always possible or convenient to draw the structures. In the early days of chemistry each newly discovered compound was given a trivial name derived from its source, use, color, or perhaps the name of its discoverer. This worked fine for a while, but as more and more substances were isolated or synthesized it became apparent that some system for naming was needed that reflected structure, rather than the fact that the substance was first found on or in a goat, or in sour milk, or was dark red.

The name "urea" doesn't tell you what the formula of the compound is, but it suggests where it was found originally

By 1930 the system of nomenclature in general use was established by international agreement. If you go on to take a course in organic chemistry, you will be expected to become familiar with the so-called IUPAC names of organic compounds. The names we have given for some of the molecules we have sketched are IUPAC names. Unfortunately for the novice chemist, since the systematic names become quite complicated even for fairly simple species, practicing chemists very frequently use the old trivial names, in spite of their limitations. So, we speak of lactic acid (from milk), instead of 2-hydroxypropanoic acid, and glycerol, instead of 1,2,3-propanetriol, and will probably continue to do so. In this text we will not emphasize nomenclature, and will name substances as a chemist usually would.

If you examine the structures and names of the six-carbon hydrocarbons on page 230, you can see something of the system that is used in generating IUPAC names for hydrocarbons. The name reflects the longest chain of carbon atoms: a six-carbon chain is a hexane; a five-carbon, a pentane; a four-carbon chain is a butane. A CH_3 branch is given the name methyl, and its location is designated by stating the position, in the main chain, of the carbon atom to which it is attached; 2-methylpentane indicates that there is a CH_3 group on the second carbon atom in a five-carbon chain. A C_2H_5 group is given the name ethyl, a C_3H_7 group is called propyl, and a C_4H_9 group is a butyl group.

Unsaturated Hydrocarbons: Alkenes and Alkynes

If one or more of the carbon-carbon bonds in a hydrocarbon is a double or triple bond, that substance is said to be **unsaturated.** If there is one double bond present, the material is called an **alkene** or olefin. If there is one triple bond, the substance is an **alkyne,** or an acetylene, after C_2H_2, the first member of that group of substances.

The alkenes are similar to the alkanes in their number and variety. They are usually made by a process called *cracking*. When a long-chain paraffin is heated in the presence of an appropriate catalyst, carbon-carbon bonds in the chain will tend to rupture, producing an alkene and either hydrogen or another alkane. The bond may break at various places in the chain, so the number of products obtained by cracking a single substance can be quite large. On the following page we have listed some of the alkenes produced by cracking butane, C_4H_{10}. 1-Butene, propylene, and ethylene are the olefinic analogs of the paraffins butane, propane, and ethane.

The bond geometry around a carbon atom which is double-bonded was discussed in Chapter 8. Since three of the electron pairs around carbon form σ bonds by sp^2 hybridization, the bonds all lie in a plane and the bond angles are all about

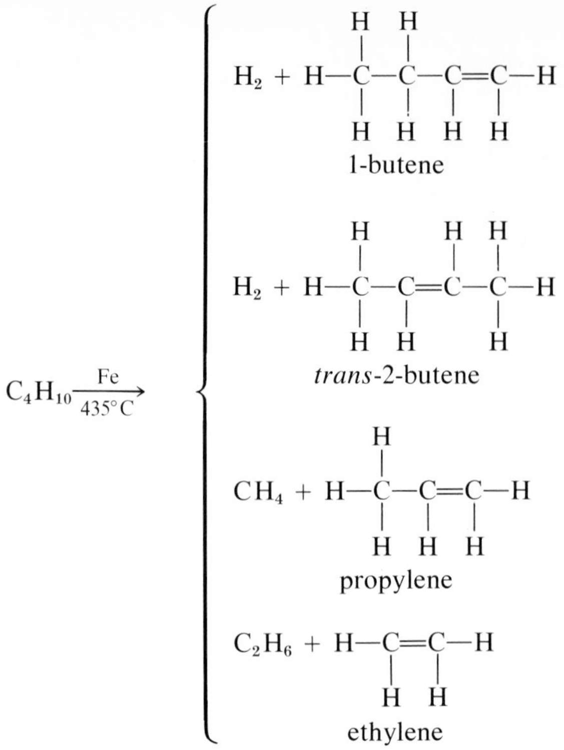

$$C_4H_{10} \xrightarrow[435°C]{Fe} \begin{cases} H_2 + H{-}CH_2{-}CH_2{-}CH{=}CH{-}H & \text{1-butene} \\ H_2 + H{-}CH_2{-}CH{=}CH{-}CH_2{-}H & \textit{trans}\text{-2-butene} \\ CH_4 + H{-}CH_2{-}CH{=}CH{-}H & \text{propylene} \\ C_2H_6 + H{-}CH{=}CH{-}H & \text{ethylene} \end{cases}$$

```
              H  H
              |  |
H2 + H—C—C—C=C—H
       |  |  |  |
       H  H  H  H
         1-butene

       H     H  H
       |     |  |
H2 + H—C—C=C—C—H
       |  |     |
       H  H     H
     trans-2-butene

        H
        |
CH4 + H—C—C=C—H
        |  |  |
        H  H  H
       propylene

C2H6 + H—C=C—H
         |  |
         H  H
       ethylene
```

120°; the double bond consists of one of the σ bonds and the π bond formed by the fourth electron pair.

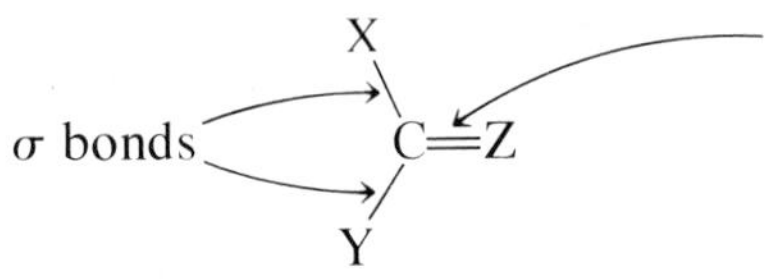

```
             X
              \
σ bonds →      C=Z   ← one σ bond and one π bond
              /
             Y
```

one σ bond and one π bond

X, Y, Z, and C lie in a plane

all bond angles 120°

There is no free rotation about a double bond, so a molecule like 2-butene can exist in two isomeric forms, called **geometric** isomers, in which the two like groups (H atoms or CH_3 groups) attached to double-bonded carbon atoms can exist either close to one another (***cis*** form) or relatively far apart (***trans*** form):

What would you call

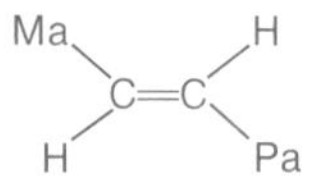

```
Ma      H
  \    /
   C=C        ?
  /    \
H       Pa
```

answer: transparent

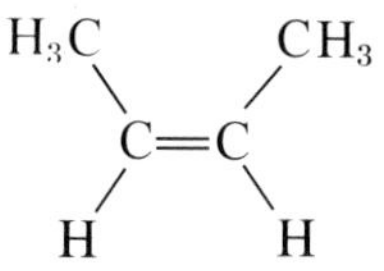

```
H3C      CH3          H3C      H
   \    /                \    /
    C=C                   C=C
   /    \                /    \
  H      H              H      CH3
 cis-2-butene         trans-2-butene
```

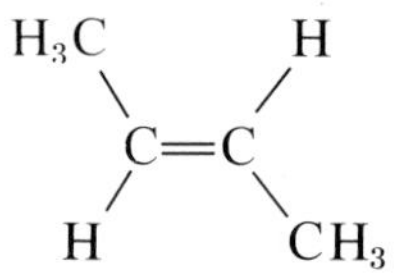

In both isomers all of the carbon atoms and the two single hydrogen atoms lie in a plane.

Alkene hydrocarbons are more reactive chemically than are the alkanes, because of the presence of the double bond. A reaction common to these hydrocarbons is one of addition, in which a molecule such as H_2, HBr, or Br_2 adds directly to the double bond; the resulting saturated compound may contain one or more halogen atoms (see Section 10.5).

$$\underset{\text{propylene}}{\mathrm{H{-}\overset{H}{\underset{H}{C}}{-}\overset{H}{C}{=}\overset{H}{C}{-}H}} + \mathrm{HBr} \rightarrow \underset{\text{2-bromopropane}}{\mathrm{H{-}\overset{H}{\underset{H}{C}}{-}\overset{H}{\underset{Br}{C}}{-}\overset{H}{\underset{H}{C}}{-}H}} \quad (10.3)$$

Example 10.4 Draw all the possible isomers of $C_2H_2Cl_2$, a derivative of ethylene.

Solution When the two chlorine atoms are bonded to the same carbon, only one isomer is possible (structure I below). If one chlorine is bonded to each carbon, we obtain a pair of *cis-trans* isomers (II and III):

$$\underset{\text{I}}{(H)(H)C{=}C(Cl)(Cl)} \qquad \underset{\text{II }(cis)}{(Cl)(H)C{=}C(Cl)(H)} \qquad \underset{\text{III }(trans)}{(H)(Cl)C{=}C(Cl)(H)}$$

Another very important reaction of olefins is that of polymerization, in which the molecules add to each other to form long chains. In Chapter 25 we will discuss polymers produced by this kind of reaction.

Molecules containing a carbon-carbon triple bond are even more reactive than the olefins. Acetylene, C_2H_2, the simplest of these substances, is typically unstable and chemically reactive. Its most familiar use is in the oxyacetylene torch, where it is burned in oxygen to produce a high-temperature flame useful for welding and cutting metals. It undergoes addition reactions with hydrogen or the halogens to produce alkenes or saturated substituted hydrocarbons, many of which are very important industrially. Acetylene itself is a gas. It can be made by reacting calcium carbide, CaC_2, with water, or by the carefully controlled oxidation of methane:

$$CaC_2(s) + 2\ H_2O(l) \rightarrow C_2H_2(g) + Ca(OH)_2(s) \quad (10.4)$$

$$4\ CH_4(g) + 3\ O_2(g) \rightarrow 2\ C_2H_2(g) + 6\ H_2O(g) \quad (10.5)$$

In the methane oxidation the gases are passed very quickly through an electric arc and the products are quenched in water to prevent further oxidation of the acetylene to the much more stable carbon dioxide.

In a molecule containing triple-bonded carbon, the carbon atom will always be attached to two other atoms. You will recall that the geometry of such a group will be linear, as would be expected from the arrangement of the two σ bonds arising from the sp hybrids on carbon. The triple bond is made up from one of those σ bonds and two π bonds from the other two pairs of p electrons on the carbon atoms (Fig. 10.2, at top of next page).

Aromatic Hydrocarbons

There is one group of unsaturated hydrocarbons that is so important as to be classified separately. Substances in this class are called *aromatic*, probably

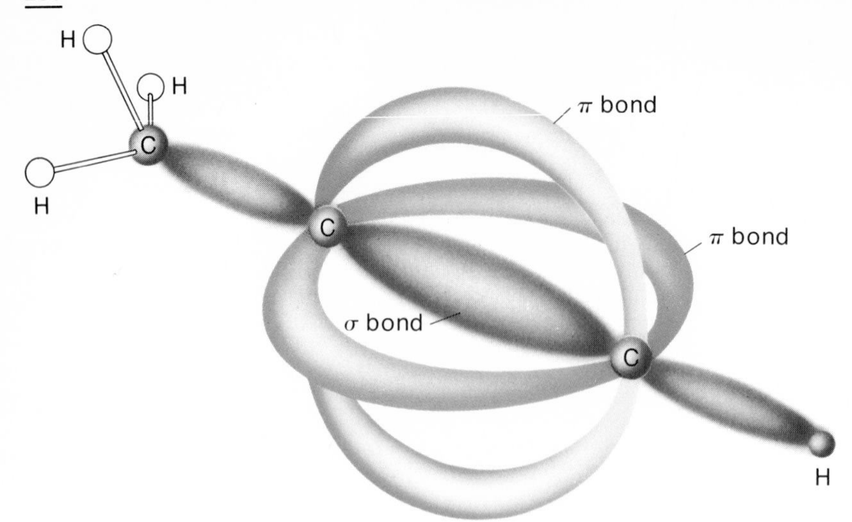

Figure 10.2 Structure of methyl acetylene. The triple carbon-carbon bond consists of a σ bond from the sp hybrids and two π bonds formed by the p_y and p_z electrons on the two carbon atoms.

because many of them have rather strong, often pleasant, odors. All aromatic substances can be considered to be derived from one compound, benzene, C_6H_6:

```
         H
         |
   H     C     H
    \  /   \\ /
     C       C
     ||      |
     C       C
    /  \   // \
   H     C     H
         |
         H
```

Benzene is a most interesting substance, both from a theoretical and practical point of view. Knowing that the bonds around a double-bonded carbon atom tend to form three 120° angles, we would expect that a ring such as that in benzene could form without strain and so might be reasonably stable. Theory would also predict that the molecule be planar, since the atomic geometry around double-bonded carbon atoms is always planar.

Benzene is indeed a planar molecule, and in addition is also a regular hexagon, with all carbon-carbon bonds having equal length. This high degree of symmetry is explained by assuming that benzene has resonance forms similar to those used previously to explain the equal length bonds in molecules like SO_2 and in the NO_3^- ion.

```
         H                          H
         |                          |
   H     C     H              H     C     H
    \  /   \\ /                \ //   \  /
     C       C                   C       C
     ||      |       <-->        |       ||
     C       C                   C       C
    /  \   // \                /  \\   /  \
   H     C     H              H     C     H
         |                          |
         H                          H
```

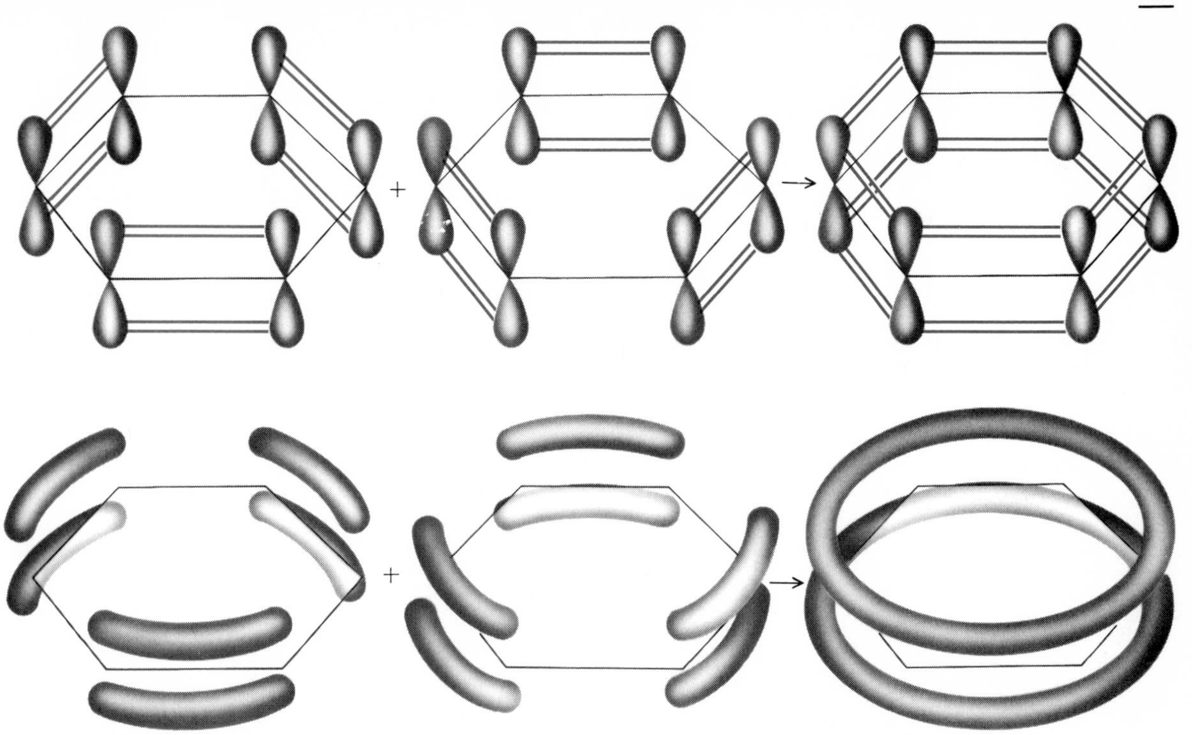

Figure 10.3 π bonding in benzene. Overlap of p orbitals (upper diagrams) forms the π bonds (lower diagrams). Resonance of the π bonds produces doughnut-shaped electron clouds above and below the plane of the molecule.

It is thought that the resonance in benzene stabilizes the molecule

The σ bond framework in benzene arises from sp^2 hybridization on the carbon atoms. The π electrons, one from each carbon atom, appear to be quite mobile in benzene and produce strong π bonding above and below the plane of the molecule, which is thought to stabilize the structure considerably (Fig. 10.3). Molecular orbital theory has been applied to the π electrons in benzene and indicates that the π bonding is as shown at the lower right of Figure 10.3.

Benzene was discovered by Michael Faraday in 1825. It is a liquid which boils at about 80°C, typical of hydrocarbons of similar molecular mass. Its chemical properties are surprising, in that it is much less reactive than, for instance, C_2H_2, although it has the same empirical formula. It does not readily add either hydrogen or the halogens, as would acetylene or typical olefins. Its reaction with a substance like Br_2 must be carried out catalytically and is one of substitution rather than addition, substituting Br atoms for H atoms and producing HBr:

$$C_6H_6(l) + Br_2(l) \xrightarrow[50^\circ C]{Fe} C_6H_5Br(l) + HBr(g) \qquad (10.6)$$

$$C_6H_5Br(l) + Br_2(l) \xrightarrow[50^\circ C]{Fe} C_6H_4\,Br_2(l) + HBr(g) \qquad (10.7)$$

Kekulé, a German chemist, was one of the first to study these reactions, and sought the isomers to be expected from them. On the basis of the number of those isomers, he proposed essentially the structure we now accept.

Example 10.5 Given the known structure of benzene, how many isomers are there for bromobenzene? for dibromobenzene?

Solution For brevity, the benzene molecule may be represented as a hexagon, to which one attaches any atoms which are substituted for hydrogen:

C_6H_6 = [hexagon] C_6H_5Br = [hexagon] Br

In view of the fact that all positions on the benzene ring are equivalent, we would expect only one kind of bromobenzene molecule. Kekulé found only one form and concluded that an open chain, in which of necessity the C atoms would not all be equivalent, was not a correct structure. With the dibromobenzenes, the Br atoms may be on adjacent ring carbons, on two alternate carbon atoms, or on opposite carbons:

Br, Br   Br, Br   Br, Br

There are three isomeric dibromobenzenes, referred to as *ortho-*, *meta-* and *para-*dibromobenzene, in the order written.

There are many aromatic substances, all of which contain one or more benzene rings. In Figure 10.4 we indicate the structures of some of the simpler ones, all of which can be found in coal tar, a gummy black material obtained when soft coal is heated to about 1000°C in the absence of air. (In the drawings the ring car-

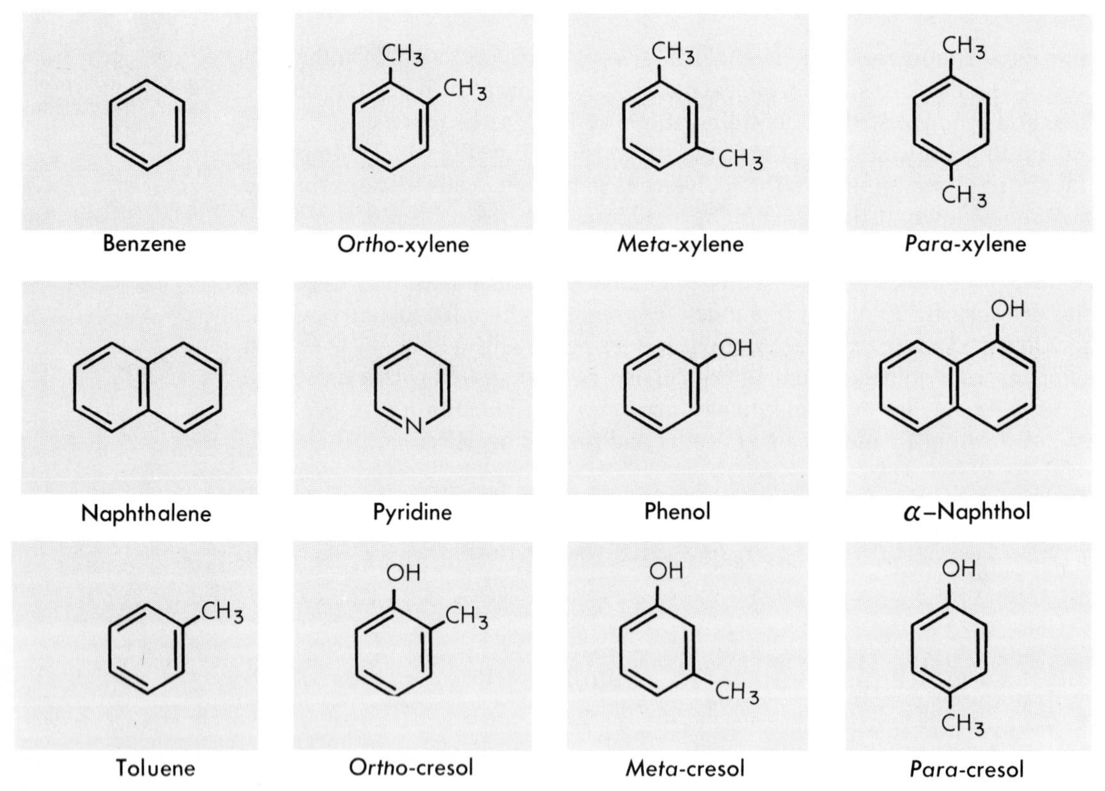

Figure 10.4 Some aromatic substances found in coal tar. These materials are vaporized from coal when it is heated in the absence of air. The prefixes *ortho*, *meta*, and *para* indicate relative positions of disubstituted species on the benzene ring.

bon and hydrogen atoms are omitted.) These compounds are all of commercial importance and are made from petroleum as well as from coal. Benzene, toluene, and the xylenes are used in gasoline and are industrial intermediates. Naphthalene is still used to some extent in mothballs, but its main use is in the manufacture of those polymers known as alkyd resins, which form the film in many automobile enamels. Phenol, sometimes called carbolic acid, was the first antiseptic; it is effective, but unfortunately also corrosive, causing severe skin burns. These days phenol is used in enormous amounts in the manufacture of Bakelite plastics and glues for plywood. The cresols are the main components of the wood preservative known as creosote.

In addition to the substances in Figure 10.4, coal tar also contains many polynuclear hydrocarbons composed of various numbers of benzene rings fused together. Some of these are powerful carcinogens (cancer-producing agents). This property was discovered early in this century after coal tar workers were found to have an abnormally high incidence of skin cancer. One of the more active carcinogens is 1,2-benzopyrene, a yellow crystalline material, a component of cigaret smoke and thought to be a cause of lung cancer, which is prevalent among heavy smokers. Another potent carcinogen is methylcholanthrene which, like 1,2-benzopyrene, contains three coplanar, linear benzene rings, a common feature in the molecules of many of these very dangerous substances.

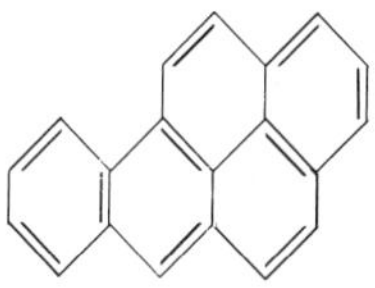

CH_3

1,2-benzopyrene

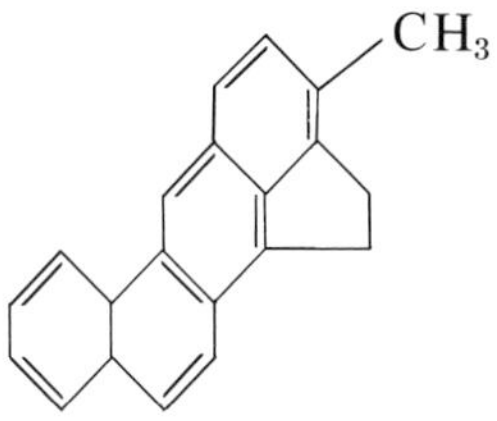

methylcholanthrene

10.3 SOURCES OF HYDROCARBONS

In Chapter 4 we noted that man's main sources of chemical energy are petroleum, natural gas, and coal. We are now in a position to examine how these materials are used to make the fuels and chemicals needed to move our vehicles, heat our homes, and produce the many substances, such as textiles, paints, and plastics, manufactured by the chemical industry.

Petroleum

At present, petroleum and natural gas rank as the main source of not only our fuels but the bulk of the organic chemicals used in industry. Petroleum, or oil, is found in underground deposits in many regions of the earth. Its origin lies in plants and animals which lived on the earth and in the sea many millions of years ago. The residues from these organisms accumulated in certain regions, possibly as a result of geologic conditions, became buried, and were subjected to high pressures and reducing conditions over long periods of time. The resulting material, petroleum, is a complex mixture of hydrocarbons, containing paraffinic chain and ring molecules, aromatic molecules, and small amounts of oxygen- and sulfur-containing substances.

Petroleum has been known for thousands of years, being observed as surface seepages, particularly as oil films on streams and ponds. It had no known use, except as "medicine oil," until about the mid-nineteenth century when the first oil wells were drilled in the United States and Rumania. Crude oil, on being distilled, yielded a fraction known as kerosene, which had immediate commercial importance as a lamp fuel; for many years kerosene was the main product of the petroleum industry.

With the development of the automobile, a lower-boiling fraction obtained in the distillation of petroleum, straight-run gasoline, became of dominant value. Since a 200-ℓ barrel of crude oil yielded only about thirty litres of gasoline, it became a matter of great commercial significance to increase the size of the gasoline fraction. Chemical research on this problem has been very successful and has resulted in (1) more and better gasolines, (2) a knowledge of what substances make up petroleum, (3) a knowledge of what substances make a good gasoline, and (4) much basic knowledge in organic chemistry, including some of the material presented in the previous section of this chapter.

As a result of petroleum research, chemists now are able to manipulate almost at will the end products of the petroleum refining process to meet many different kinds of demands, from home heating gases to jet fuels to road asphalts, in the way which makes best use of the starting material. A byproduct of this research has been the petrochemical industry.

Following a rough distillation of the crude oil into the fractions indicated in Table 10.2, the higher molecular mass fractions are carried through a controlled pyrolysis or *cracking* process, in which they are heated to about 500°C, often under catalytic conditions (see Section 10.2). In this process the molecules suffer a rupture of a carbon-carbon bond, yielding olefins and paraffins of lower molecular mass than the original fraction. The product contains a substantial fraction of substances which boil in the same range as gasoline, thus increasing significantly the yield of gasoline from crude oil. The lighter olefins, particularly ethylene and propylene, have in recent years found a market as raw materials in the plastics industry.

The manipulation of hydrocarbon structures is now a highly sophisticated science

In order to further increase gasoline yield and quality, there are several other procedures used to treat both the original distillate fractions and the products from the cracking step. Since it has been found that branched-chain molecules perform better as motor fuels than do those containing straight chains, the middle distillation fractions are often subjected to a *re-forming* or isomerization process in which the fraction is passed over a solid catalyst such as aluminum chloride at about 200°C. With pentane, for example, one would obtain on isomerization

```
     H   H                          CH3
     |   |                           |
CH3—C———C—CH3                  CH3—C—CH3
     |   |                           |
   H3C   H                          CH3
 2-methylbutane              2,2-dimethylpropane
```

TABLE 10.2 FRACTIONS OBTAINED ON DISTILLATION OF PETROLEUM

FRACTION	BOILING RANGE (°C)	CARBON ATOM CONTENT	DIRECT USE
Gas	below 20	C_1—C_4	gas heating
Petroleum ether	20–60	C_5—C_6	industrial solvent
Light naphtha	60–100	C_6—C_7	industrial solvent
Straight-run gasoline	40–200	C_5—C_{10}	motor vehicle fuel
Kerosene	175–325	C_{11}—C_{18}	jet fuel
Gas oil	275–500	C_{15}—C_{40}	diesel fuel
Lubricating oil	above 400		lubricant
Asphalt	nonvolatile		roofing and road construction

Both of these substances are superior to the unbranched pentane as automobile fuel. Isomerization reactions also occur to some extent during cracking and so make an additional contribution in that step to the fuel quality of the product.

One of the difficulties with the internal combustion engine is that a gasoline-air fuel mixture on ignition has a tendency to detonate, or "knock," rather than burn smoothly. This tendency increases as the compression ratio increases, which is unfortunate since engines with high compression ratios tend to be more efficient. The *octane number* of a gasoline is a measure of its resistance to knock. It is obtained by comparing the knocking characteristics of the gasoline to those of "isooctane" and heptane:

$$\begin{array}{ccccccccc} & & CH_3 & & H & & H & & \\ & & | & & | & & | & & \\ CH_3 & - & C & - & C & - & C & - & CH_3 \\ & & | & & | & & | & & \\ & & CH_3 & & H & & CH_3 & & \end{array}$$

"isooctane"
(2,2,4-trimethylpentane)

$$\begin{array}{ccccccccccccccccc} & & H & & H & & H & & H & & H & & H & & H & & \\ & & | & & | & & | & & | & & | & & | & & | & & \\ H & - & C & - & C & - & C & - & C & - & C & - & C & - & C & - & H \\ & & | & & | & & | & & | & & | & & | & & | & & \\ & & H & & H & & H & & H & & H & & H & & H & & \end{array}$$

heptane

In the system used, isooctane, which resists knocking very well, is assigned an octane number of 100, and heptane, which is very prone to knocking, an octane number of 0. Gasoline with the same knocking properties as a mixture of 90 per cent isooctane and 10 per cent heptane would be rated as "90 octane."

About the best the petroleum industry can economically produce at present using the processes we have described is 90 octane gasoline; this material is a complex mixture of saturated and aromatic hydrocarbons, and is the fuel that is currently sold as "no-lead" gasoline. Premium, or high-test, gasoline contains tetraethyl lead, which improves the octane rating very markedly, and makes 95 octane gasoline possible and practical. Newer cars are restricted to no-lead fuel because the catalytic converters put in the exhaust line to remove air pollutants are easily "poisoned." The switch to no-lead gasoline has required a decrease in the compression ratio in the engines in the newer vehicles. Although these engines are somewhat less efficient, their smaller average size has resulted in a substantial net decrease in fuel consumption per mile.

In the near future, we will probably switch completely to no-lead gasoline

Natural Gas

Above many petroleum deposits there is a gas, called natural gas, which is a mixture of the light paraffins, roughly 75 per cent methane, 15 per cent ethane, 5 per cent propane, and the rest higher hydrocarbons. This material is also found in many places in the earth where there is no petroleum.

In the early days of the petroleum industry the natural gas produced by an oil well was considered to be a nuisance and was usually burned, or flared, at the well; until recently this was still done at some of the oil fields in the Middle East. Gradually it was realized that natural gas, even without treatment, was a very satisfactory heating fuel, and pipelines to cities near gas fields were built to fire home heating systems. Following World War II a network of pipelines was built, so that now nearly every large city in the United States and Canada has natural gas supplied via a large, high-pressure system.

The natural gas used in homes is a mixture of methane and ethane, essentially a perfect fuel for heating or cooking. The propane and butane are removed as liquids by compressing and cooling the gas; this LP (liquefied petroleum) gas has a wide market where pipeline gas is not available. The main market, however, for natural gas is currently the chemical industry, which uses it as a raw material for the production of many important substances, among them ethylene, methyl alcohol, acetylene, and ammonia.

Reserves of natural gas are steadily dropping in the United States, which now imports significant amounts of this fuel (about half of Canada's annual production now goes to the U.S.). At present nonpriority uses of natural gas are being limited or prohibited, so that such applications as grain drying, gas-fired power plants, and industrial process heat are rapidly being converted to use oil or coal. Considerable research is currently being devoted to producing synthetic gas to replace natural gas, by using a more abundant natural source of stored energy, coal.

Coal

Coal is a solid organic material, usually black or dark brown in color, found in layers in many parts of the earth. Like petroleum, coal has its origins in the plant material that lived on the earth about fifty million years ago. Over eons of time this material, under the conditions of pressure and temperature that prevailed, gradually changed its composition, losing hydrogen and oxygen and increasing its relative carbon content. Coal is a complex mixture of many substances; the composition of a given coal sample may vary widely, and will depend on its age, history, and the nature of the living material from which it came.

The aromatic hydrocarbons in coal have a very high C:H ratio

Coal contains mostly carbon, much of which is bound in fused aromatic rings, similar to those in graphite, but with much less extensive structures. In addition to carbon, which makes up about 70 per cent by mass, coal also contains about 6 per cent hydrogen, 10 per cent oxygen, 2 per cent sulfur, 1 per cent nitrogen, and about 10 per cent of other elements, all for the most part in chemical combina-

Figure 10.5 Huge power shovels, like this Gem of Egypt made by the Hanna Coal Company, help make strip mining much more efficient than deep mining. This 7000-tonne monster is as tall as a 12 story building and can pick up more than 200 tonnes of coal in one scoop; that amount will more than fill a railroad car. (Photo courtesy of Consolidated Coal Company.)

tion with carbon. It is not easy to identify the particular substances present in coal in the way that one can do with petroleum; information about such substances is to some extent implied by the composition of coal tar and other material that can be driven from coal by heating it in the absence of air.

Relatively, recoverable coal resources in the world are enormous, of the order of 10^{16} kg. As pointed out in Chapter 4, the energy equivalent of this coal is about 10 times that of the world's resources of petroleum and natural gas. Coal is obtained from both underground and surface mines; it is found in layers, or seams, which vary in thickness from about 1 to 3 m. The technology for coal recovery from both deep and surface, or strip, mines is well developed, with automatic machinery making possible very high worker efficiency (Fig. 10.5). Both types of mining operations have negative effects on the environment, but strip mining in Appalachia in recent years has in some instances been particularly detrimental. Rehabilitation of land used for strip mining does appear to be feasible, at least in areas where the annual rainfall is 25 cm or more, and current mining operations are designed to minimize harmful effects on the environment. There are extensive deposits of coal in the United States and Canada, primarily in the western states and provinces (about 15×10^{14} kg in the U.S. and perhaps as much as 2×10^{14} kg in Canada).

Gas and Gasoline from Coal

In the first third of this century, coal was the main fuel used for heating homes, but it virtually disappeared when oil and natural gas became available. The automatic furnaces possible with oil and gas had real appeal for the houseowner who had experienced for years the chore of shovelling in the coal and taking out the ashes. Now, in view of the world's energy shortage, we would like to move in the opposite direction, substituting coal for natural gas and petroleum. The problem is one of converting coal to either a gaseous or a liquid fuel. The technology for gasification is well developed and it seems likely that in the next decade synthetic gas from coal will gradually replace natural gas for many applications.

Before natural gas became popular, gas for home cooking stoves was made in towns and cities at the local gas works by a very simple process. Air was passed through a column of hot *coke,* the high-carbon residue produced when coal is heated in the absence of air. A strongly exothermic reaction occurred:

$$C(s) + O_2(g) \rightarrow CO_2(g) \qquad \Delta H = -394 \text{ kJ} \qquad (10.8)$$

The temperature of the coke went up to about 1500°C, the air was cut off, and steam was blown through the hot coke, where it reacted as follows:

$$C(s) + H_2O(g) \rightarrow CO(g) + H_2(g) \qquad \Delta H = +131 \text{ kJ} \qquad (10.9)$$

forming what is called *water gas,* which is a relatively good fuel. The above reaction is endothermic, so the temperature of the coke gradually dropped and the procedure was repeated.

A major drawback to the use of water gas is the poisonous nature of CO

Various modifications of this approach have been developed, of which perhaps the most successful is the Lurgi process, developed in Germany (Fig. 10.6). This process is continuous rather than batch, and uses coal directly rather than coke. Pure oxygen and steam are blown under pressure through the hot coal, volatilizing benzene, phenol, and other coal tar chemicals, and producing a good quality fuel gas (40 per cent H_2, 15 per cent CO, 15 per cent CH_4, and 30 per cent CO_2) often called **synthesis gas.**

Although synthesis gas made from coal is useful as a heating fuel, it is even potentially more useful as a raw material for the production of hydrocarbons and

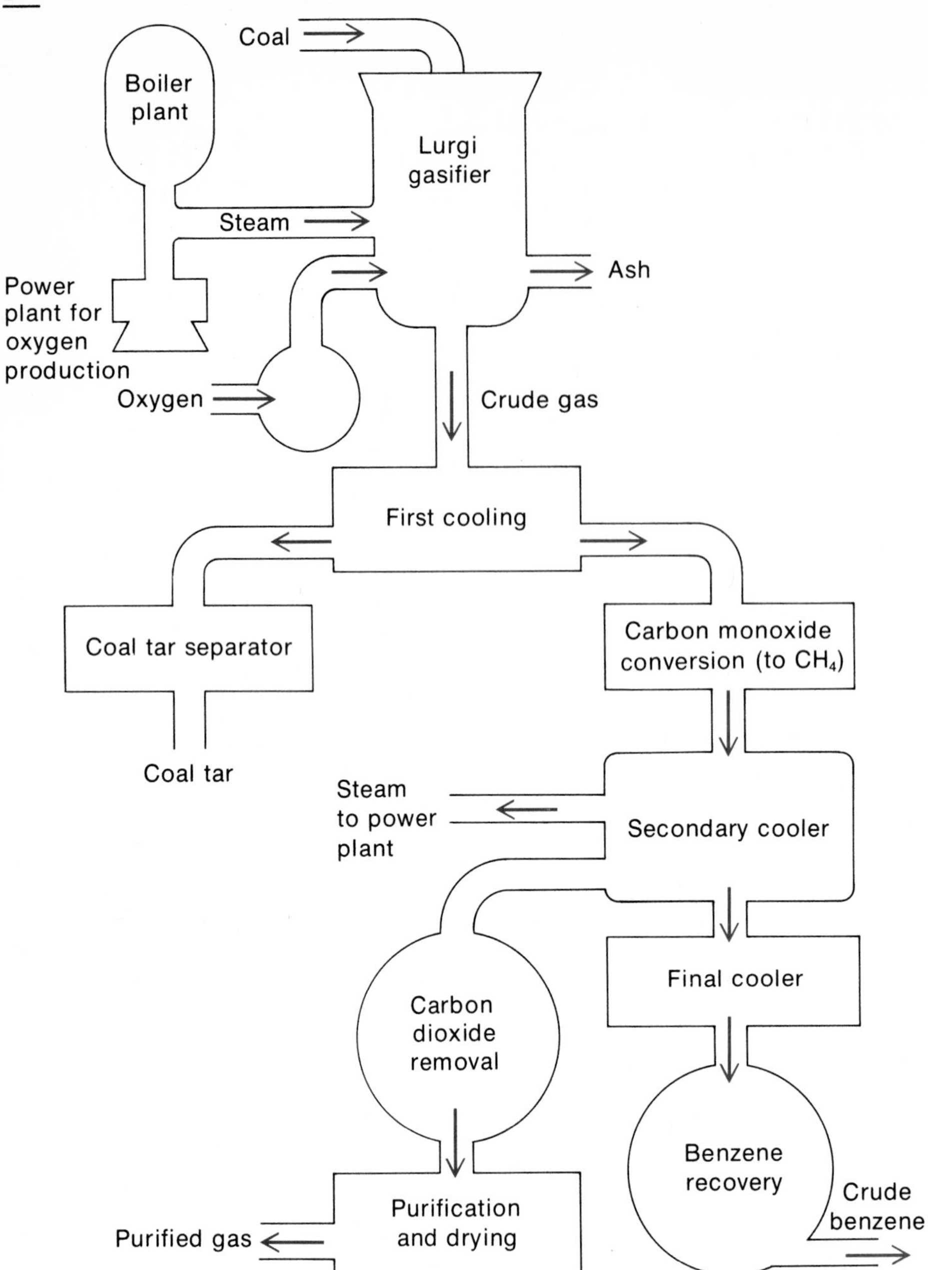

Figure 10.6 Lurgi process for coal gasification. This process operates at elevated pressures, produces H_2, CO, and appreciable amounts of CH_4. Coal tar chemicals are recovered and maximum use is made of available heat.

many other organic chemicals. In the presence of suitable catalysts, carbon monoxide and hydrogen will participate in the following sorts of reactions:

$$CO(g) + 3\ H_2(g) \rightarrow CH_4(g) + H_2O(g) \quad (10.10)$$

$$6\ CO(g) + 13\ H_2(g) \rightarrow C_6H_{14}(g) + 6\ H_2O(g) \quad (10.11)$$

$$8\ CO(g) + 4\ H_2(g) \rightarrow C_4H_8(g) + 4\ CO_2(g) \quad (10.12)$$

At 1 atm pressure and 200°C reactions like these yield a variety of products, mostly straight-chain paraffins and olefins, with chain lengths from about 3 to 20 carbon atoms. About 60 per cent of the product can be classified as gasoline with a relatively low octane number (about 55). Motor fuel in Germany during World War II was made by mixing this material with high-octane gasoline or benzene.

Until recently the cost of coal was such as to make liquid fuels produced by reaction of CO and H_2 noncompetitive with those made from petroleum. Hence, there has been relatively little use of coal gasification processes to make gasoline. An exception is a plant in South Africa, where coal is cheap and petroleum deposits have not been found. This plant takes advantage of modern technology of coal gasification, and in particular operates at elevated pressures, where hydrocarbon yield is increased. It produces synthesis gas by the Lurgi process and then hydrocarbons by CO–H_2 reactions. The raw product is much like petroleum and is treated by many of the processes used in the petroleum industry, yielding high-octane gasoline, diesel fuel, and LP gas in commercial quantities. The efficiency of this plant is such that it can produce one kilogram of hydrocarbons for every five kilograms of coal that are used. It is likely that in the next decade plants of this sort will be built in other countries to help close the gap between fuel needs and production from native petroleum.

Such plants offer a long-term source of (relatively expensive) gasoline

10.4 OXYGEN-CONTAINING COMPOUNDS

The element most commonly encountered in organic compounds, after carbon and hydrogen, is oxygen. In organic compounds it can be present in many different combinations, each of which may give rise to an important class of substances. In this section we will restrict our attention to a few of the simpler classes of organic compounds containing oxygen. Where appropriate, we will show how these substances can be prepared from one well known compound, ethyl alcohol.

Alcohols

An alcohol is a compound in which an OH group has been substituted for a hydrogen atom in a hydrocarbon. The four simple alcohols whose structures are indicated below can be considered to be derived from the hydrocarbons CH_4, C_2H_6, and C_3H_8:

```
    H           H  H            H  H  H           H  H  H
    |           |  |            |  |  |           |  |  |
 H—C—OH      H—C—C—OH       H—C—C—C—OH      H—C—C—C—H
    |           |  |            |  |  |           |  |  |
    H           H  H            H  H  H           H  OH H
```

methyl alcohol   ethyl alcohol   propyl alcohol   isopropyl alcohol

These alcohols, and many others that we could write, have the general formula R—OH, where R is an **alkyl group** (e.g., CH_3—, CH_3CH_2—, $CH_3CH_2CH_2$—).

The most common alcohol is ethyl alcohol or ethanol, C_2H_5OH. This substance is a colorless liquid, about as viscous as water, with a rather faint, sharp odor. It is the potent ingredient of alcoholic beverages, where it is present at various concentrations (about 6 per cent in strong beer, 15 per cent in wine, 40 per cent in distilled spirits). It is important industrially as a solvent and as a chemical intermediate.

As compared with propane, C_3H_8, which has about the same molecular mass, ethyl alcohol has a very high boiling point (78°C vs. −43°C). We attribute this difference to hydrogen bonding associated with the OH group, which increases the attractive forces between neighboring molecules in the liquid. Because of hydrogen bonding the boiling points of all the alcohols are relatively high. This type of bonding also results in a larger viscosity, or resistance to flow, than one usually finds in the hydrocarbons.

Ethyl alcohol can be made by fermentation of sugars or grain. The reaction is carried out in solution in the absence of air and in the presence of certain enzymes, which are natural catalysts, secreted by various yeasts; for glucose, a simple sugar, the reaction would be:

$$C_6H_{12}O_6(aq) \rightarrow 2\ C_2H_5OH(aq) + 2\ CO_2(g) \tag{10.13}$$

Ethanol makes a moderately good motor fuel, and it has been suggested that it might be commercially feasible to produce it by fermentation of organic wastes, such as wood chips.

About 80 per cent of the 1×10^9 kg of industrial ethyl alcohol produced annually in the United States is made synthetically by the catalyzed reaction of water with ethylene, C_2H_4. This is a typical addition reaction, in which the water molecule is split and the fragments add to the double-bonded carbon atoms:

$$H_2C{=}CH_2 + H{-}O{-}H \xrightarrow[\substack{200\ atm \\ 300°C}]{H_3PO_4} H{-}\underset{H}{\overset{H}{C}}{-}\underset{H}{\overset{H}{C}}{-}O{-}H \tag{10.14}$$

The first member of the alcohol series, methyl alcohol, or methanol, CH_3OH, can be made from natural products or synthetically. It is present in the liquid distillate obtained when charcoal is made by heating wood in the absence of air; for this reason methanol is sometimes called wood alcohol. Synthetically, methanol is made from synthesis gas, the mixture of carbon monoxide and hydrogen referred to previously:

$$CO(g) + 2\ H_2(g) \xrightarrow[\substack{250\ atm \\ 350°C}]{ZnO\text{-}Cr_2O_3} CH_3OH(l) \tag{10.15}$$

Methanol is made in larger amounts by synthetic means than any other organic chemical (3×10^9 kg annually in the U.S.). It is used in jet fuels, as a solvent, and to make formaldehyde for resins. Where it is cheap it has had limited use as an additive to gasoline. Methanol is poisonous, causing blindness and death, so although it is an intoxicant, it must not be used in alcoholic beverages.

Ethers

In an alcohol, oxygen is always singly bonded to two different atoms, carbon and hydrogen. In an ether, on the other hand, oxygen is singly bonded to two carbon atoms, as in the structures shown:

```
    H     H          H     H  H           H  H     H  H
    |     |          |     |  |           |  |     |  |
 H—C—O—C—H       H—C—O—C—C—H       H—C—C—O—C—C—H
    |     |          |     |  |           |  |     |  |
    H     H          H     H  H           H  H     H  H
```

dimethyl ether — methyl ethyl ether — diethyl ether

In an ether the two groups bonded to oxygen don't have to be the same, so the general formula of an ether is ordinarily written R—O—R′, with the understanding that R and R′ may be the same or different alkyl groups.

Every ether has at least one alcohol with which it is isomeric; ethyl alcohol, for example, is an isomer of dimethyl ether, both having the molecular formula C_2H_6O. Unlike alcohols, however, ethers exhibit very little hydrogen bonding. They boil at about the same temperature as the paraffins of the same total chain length (*bp* $C_2H_5OC_2H_5 = 35°C$, $C_5H_{12} = 36°C$), and far below the isomeric alcohol (*bp* $C_4H_9OH = 118°C$).

Why not?

As a group ethers are about as unreactive as the saturated hydrocarbons and are not important synthetically. The most common ether is diethyl ether, $C_2H_5OC_2H_5$, which was one of the early anesthetics. It has some use as a solvent, particularly to extract organic substances from water, with which it is only slightly miscible. It is quite volatile and its vapor is notorious for its flammability.

Example 10.6 Draw the structures of all the alcohols and ethers having four carbon atoms, molecular formula $C_4H_{10}O$.

Solution Let us consider the alcohols first. Recall (p. 229) that there are two 4-carbon paraffins (butane and 2-methylpropane) with the molecular formula C_4H_{10}. Each of these leads to two different alcohols, depending upon whether the OH group is attached to a terminal carbon or one in the middle of a chain. The skeleton structures of these four alcohols are

```
                                                  |            |
                                                 —C—          —C—
  |  |  |  |           |  |  |  |             |  |  |       |  |  |
 —C—C—C—C—OH          —C—C—C—C—              —C—C—C—       —C—C—C—OH
  |  |  |  |           |  |  |  |             |  |  |       |  |  |
                             OH                  OH
```

Now for the ethers. The two groups attached to oxygen could both be C_2H_5:

```
  |  |     |  |
 —C—C—O—C—C—
  |  |     |  |
```

Alternatively, we could have a CH_3 and a C_3H_7 group, in which case two structures are possible:

```
                                          |
                                         —C—
  |     |  |  |                     |     |  |
 —C—O—C—C—C—          and          —C—O—C—C—
  |     |  |  |                     |     |  |
```

In summary, there are 7 different compounds with the molecular formula $C_4H_{10}O$; 4 of these are alcohols and 3 are ethers.

The hydroxyl group in alcohols is moderately reactive and can be further oxidized to yield several possible products. In the presence of air and a silver catalyst, ethanol vapor undergoes the following reaction:

$$\mathrm{H{-}\overset{H}{\underset{H}{C}}{-}\overset{H}{\underset{H}{C}}{-}OH + \tfrac{1}{2}\,O_2 \rightarrow H{-}\overset{H}{\underset{H}{C}}{-}\overset{H}{C}{=}O + H_2O} \qquad (10.16)$$

acetaldehyde

The product is an aldehyde and, like all such compounds, it contains a C=O group, called a *carbonyl* group, in which the carbon atom is bonded to at least one hydrogen atom. The general formula of an aldehyde may be written as $\mathrm{R{-}\overset{H}{C}{=}O}$, where R may be an alkyl group or, in the simplest case, an H atom. Thus we have

$$\mathrm{H{-}\overset{H}{C}{=}O} \qquad \mathrm{H_3C{-}\overset{H}{C}{=}O} \qquad \mathrm{H_3C{-}CH_2{-}\overset{H}{C}{=}O}$$

formaldehyde — acetaldehyde — propionaldehyde

Aldehydes have boiling points lower than those of the corresponding alcohols (no hydrogen bonding) but much higher than the saturated hydrocarbons of the same molecular mass (the C=O bond is highly polar).

Aldehydes are relatively reactive chemically; the simplest one, formaldehyde, $H_2C{=}O$, polymerizes explosively, and so is prepared in water solution; the commercial product is called formalin (about 40 per cent $H_2C{=}O$) and has a disagreeable odor, as biology students who have used it as a preservative will attest. The higher molecular mass aldehydes smell much better and are actually used in perfumes. Formaldehyde is by far the most important aldehyde; it is a basic raw material for making Bakelite and melamine resins. Acetaldehyde is primarily used to make acetic acid.

If an alcohol with an OH group on a nonterminal carbon atom is carefully oxidized, the product is a *ketone* rather than an aldehyde:

$$\mathrm{H{-}\overset{H}{\underset{H}{C}}{-}\overset{H}{\underset{OH}{C}}{-}\overset{H}{\underset{H}{C}}{-}H + \tfrac{1}{2}\,O_2 \xrightarrow[1\ min]{450^\circ C} H{-}\overset{H}{\underset{H}{C}}{-}\underset{\|\atop O}{C}{-}\overset{H}{\underset{H}{C}}{-}H + H_2O} \qquad (10.17)$$

isopropyl alcohol (2-propanol) — acetone (a ketone)

Isopropyl alcohol is rubbing alcohol

A ketone, like an aldehyde, contains the carbonyl group, C=O, but in the ketones this group is in the interior of the chain rather than at the end. Examples, in addition to acetone (dimethyl ketone) include

$$\mathrm{H_3C{-}\underset{\|\atop O}{C}{-}CH_2{-}CH_3} \qquad \mathrm{H_3C{-}CH_2{-}\underset{\|\atop O}{C}{-}CH_2{-}CH_3}$$

methyl ethyl ketone — diethyl ketone

The general formula of a ketone is $R{-}\underset{\displaystyle O}{\underset{\|}{C}}{-}R'$, where R and R′ are alkyl groups which may be the same or different.

Acetone, CH_3COCH_3, is the first member of the ketone series and by far the most important. It is a toxic liquid with a rather pleasant odor; it is quite polar and an excellent solvent for a wide variety of organic substances.

Every ketone has at least one aldehyde which is isomeric with it. Note, for example, that acetone and propionaldehyde have the same molecular formula, C_3H_6O. Furthermore, since there is a double bond in the molecule, a ketone or aldehyde has two fewer hydrogen atoms than the alcohol or ether with the same number of carbon atoms (C_3H_8O).

Acids

If ethanol being produced by fermentation is exposed to air it will be oxidized by the bacteria present to acetic acid:

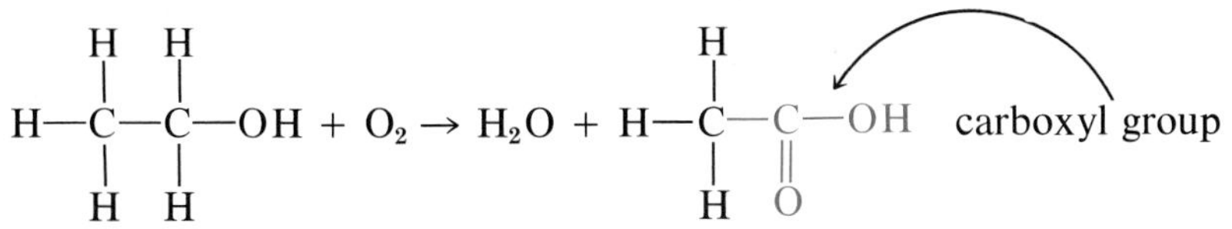

$$H{-}\underset{H}{\overset{H}{C}}{-}\underset{H}{\overset{H}{C}}{-}OH + O_2 \rightarrow H_2O + H{-}\underset{H}{\overset{H}{C}}{-}\underset{\displaystyle O}{\underset{\|}{C}}{-}OH \quad \text{carboxyl group} \qquad (10.18)$$

acetic acid

This reaction, which many amateur wine makers have discovered to their dismay, is used to make vinegar from cider. The pungent, rather disagreeable, odor of vinegar is that of acetic acid.

Acetic acid, like all organic acids, contains the *carboxyl* group, COOH; the general formula of an organic acid may be written as R—COOH. The presence of this very highly polar, hydrogen-bonding group gives organic acids characteristically high boiling points (118°C for acetic acid). Acetic acid is the most important of the organic acids; its main industrial use is in the manufacture of cellulose acetate film. Other organic acids include

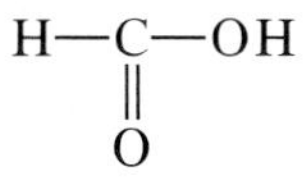

$$H{-}\underset{\displaystyle O}{\underset{\|}{C}}{-}OH \qquad H_3C{-}CH_2{-}\underset{\displaystyle O}{\underset{\|}{C}}{-}OH \qquad H_3C{-}CH_2{-}CH_2{-}\underset{\displaystyle O}{\underset{\|}{C}}{-}OH$$

formic acid 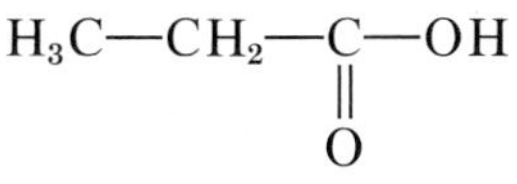 — propionic acid 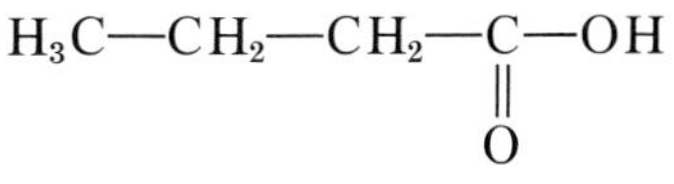 — butyric acid

Butyric acid is found in rancid butter

All of these examples are liquids with sharp penetrating odors (vile might be a better word to describe the odor of butyric acid). In water, all organic acids tend to ionize by the reaction

$$R{-}\underset{\displaystyle O}{\underset{\|}{C}}{-}OH(aq) \rightleftharpoons R{-}\underset{\displaystyle O}{\underset{\|}{C}}{-}O^-(aq) + H^+(aq) \qquad (10.19)$$

The hydrogen ion produced in this reaction makes the solution acidic (Chapter 19).

The product of the reaction between an acid and an alcohol is a substance called an ester. If, for example, acetic acid and ethyl alcohol are mixed, the following reaction occurs:

In naming an ester, the alcohol residue comes first, then the acid

$$CH_3{-}\underset{\underset{O}{\|}}{C}{-}OH + H{-}O{-}CH_2{-}CH_3 \rightarrow CH_3{-}\underset{\underset{O}{\|}}{C}{-}O{-}CH_2{-}CH_3 + H_2O \quad (10.20)$$

ethyl acetate (an ester)

In the reaction, water is eliminated and must be removed if the ethyl acetate is to be formed in good yield. Ethyl acetate is a typical ester; all esters have the general formula R—COOR′, where R is an alkyl group (or H atom in the case of esters of formic acid) and R′ is an alkyl group which may be the same as R or different from it. Other examples include:

$$H{-}\underset{\underset{O}{\|}}{C}{-}O{-}CH_3 \qquad\qquad CH_3{-}\underset{\underset{O}{\|}}{C}{-}O{-}CH_3$$

methyl formate   methyl acetate

Like most esters, ethyl acetate has a pleasant odor and a lower water solubility than either the alcohol (ethyl alcohol) or the acid (acetic acid) from which it is derived. No less than five esters, including ethyl acetate, are present in the vapor that gives pineapple its characteristic odor. Esters are used as industrial solvents and in the manufacture of some common plastics (Plexiglas, Dacron).

TABLE 10.3 CLASSES OF OXYGEN-CONTAINING ORGANIC SUBSTANCES

CLASS	FUNCTIONAL GROUP*	GENERAL FORMULA**	EXAMPLE	NAME
Alcohol	R—OH	$C_nH_{2n+2}O$	$C_2H_5{-}OH$	ethyl alcohol
Ether	R—O—R′	$C_nH_{2n+2}O$, $n \geq 2$	$C_2H_5{-}O{-}C_2H_5$	diethyl ether
Aldehyde	$R{-}\overset{\overset{H}{\mid}}{C}{=}O$	$C_nH_{2n}O$	$CH_3{-}\overset{\overset{H}{\mid}}{C}{=}O$	acetaldehyde
Ketone	$R{-}\underset{\underset{O}{\|}}{C}{-}R'$	$C_nH_{2n}O$, $n \geq 3$	$CH_3{-}\underset{\underset{O}{\|}}{C}{-}CH_3$	acetone
Acid	$R{-}\underset{\underset{O}{\|}}{C}{-}OH$	$C_nH_{2n}O_2$	$CH_3{-}\underset{\underset{O}{\|}}{C}{-}OH$	acetic acid
Ester	$R{-}\underset{\underset{O}{\|}}{C}{-}O{-}R'$	$C_nH_{2n}O_2$, $n \geq 2$	$CH_3{-}\underset{\underset{O}{\|}}{C}{-}O{-}CH_3$	methyl acetate

*R and R′ are alkyl groups (CH_3, C_2H_5, . . .); in formaldehyde, formic acid, and the esters of formic acid, R is H.

**Formulas assume no C=C or C≡C bonds and no rings.

In Table 10.3 we have summarized the structures and formulas of the oxygen-containing substances that we have discussed. Each class has its characteristic functional group and structure. Note that the general formulas are such that the six classes form three pairs that are isomeric with each other; they are: (1) alcohols and ethers; (2) aldehydes and ketones; (3) acids and esters.

Example 10.7 Given substances with the following molecular formulas

a. C_3H_6O
b. C_3H_8O
c. CH_2O_2

decide, using Table 10.3, whether the substance could be an alcohol, ether, aldehyde, ketone, acid, or ester. Draw structures for all the molecules involved.

Solution

a. The molecular formula is that of an aldehyde or ketone (n = 3; 2n = 6). With only three carbon atoms, the chain must be unbranched. There is one aldehyde and one ketone:

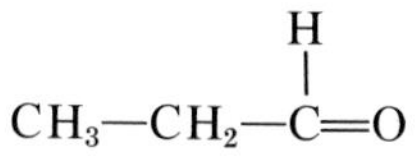

$$CH_3—CH_2—\overset{\overset{\displaystyle H}{|}}{C}=O \qquad\qquad CH_3—\underset{\underset{\displaystyle O}{\|}}{C}—CH_3$$

aldehyde (end C=O) 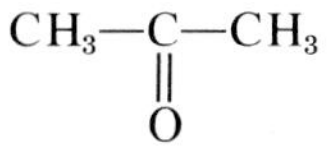 ketone (inside C=O)

b. The formula is that of an alcohol or ether (n = 3; 2n + 2 = 8). There are two alcohols and only one ether:

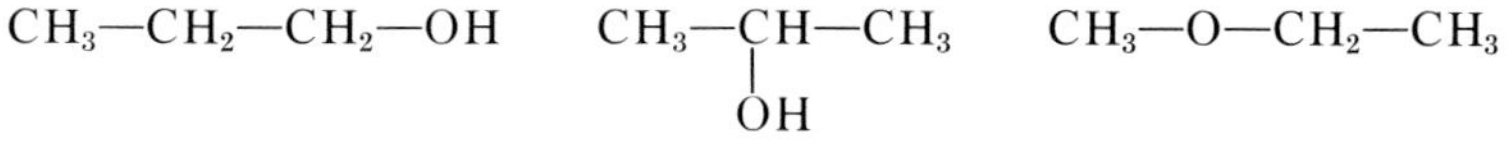

$$CH_3—CH_2—CH_2—OH \qquad CH_3—\underset{\underset{\displaystyle OH}{|}}{C}H—CH_3 \qquad CH_3—O—CH_2—CH_3$$

c. The formula fits that of an acid or ester (2 oxygen atoms). Since there must be at least two carbon atoms for an ester, the substance can only be an acid (formic acid):

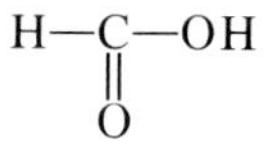

$$H—\underset{\underset{\displaystyle O}{\|}}{C}—OH$$

Soaps and Detergents

Animal and vegetable fats are esters of long-chain organic acids, called fatty acids, and various alcohols, of which glycerol is the most important. Like all esters, fats can be split, or saponified, back to the parent alcohol and acid by being heated with sodium hydroxide solution; the product is called a soap.

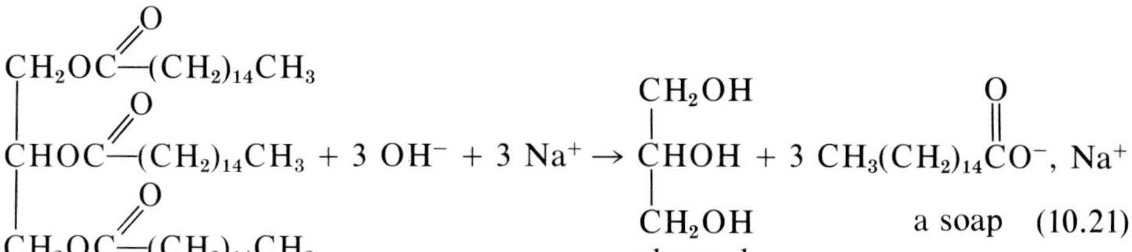

$$\begin{array}{l} CH_2OC(=O)—(CH_2)_{14}CH_3 \\ | \\ CHOC(=O)—(CH_2)_{14}CH_3 \\ | \\ CH_2OC(=O)—(CH_2)_{14}CH_3 \end{array} + 3\ OH^- + 3\ Na^+ \rightarrow \begin{array}{l} CH_2OH \\ | \\ CHOH \\ | \\ CH_2OH \end{array} + 3\ CH_3(CH_2)_{14}\overset{\overset{\displaystyle O}{\|}}{C}O^-,\ Na^+ \qquad (10.21)$$

a typical fat glycerol a soap

A *soap* is the sodium salt of a fatty acid. It obtains its useful cleaning characteristics from the combination of a long-chain hydrocarbon, which has good solvent action on other hydrocarbons, with the ionic COO^- group and its high water solubility. The main difficulty with soap is that in hard water, which typically contains calcium ions, the soap precipitates as an insoluble calcium salt. This creates various problems, particularly in laundering clothes.

Research by soap chemists to improve the solubility of cleaning agents resulted, in about 1950, in the development of substances called **detergents.** Detergents are also made from fats, but following saponification the fatty acids themselves are recovered. These are reduced with hydrogen to alcohols, which are made into inorganic esters by reaction with sulfuric acid:

$$\underset{\text{long-chain alcohol}}{R{-}O{-}H} + H{-}O{-}\overset{\displaystyle O}{\underset{\displaystyle O}{\overset{|}{\underset{|}{S}}}}{-}O{-}H \rightarrow \underset{\text{an alkylsulfuric acid}}{R{-}O{-}\overset{\displaystyle O}{\underset{\displaystyle O}{\overset{|}{\underset{|}{S}}}}{-}O{-}H} + H_2O \quad (10.22)$$

By reacting the remaining acid hydrogen with sodium hydroxide one obtains a detergent. Although there are some difficulties associated with detergents, they are now widely used in washing machines and dishwashers (see Fig. 13.6 and Section 13.3). The annual worldwide production of detergents is now about 8×10^9 kg as compared to 6×10^9 kg of soap!

10.5 SOME HALOGEN-CONTAINING ORGANIC COMPOUNDS

In discussing the chemical reactions of hydrocarbons we mentioned that it is possible to introduce one or more halogen atoms into an organic molecule by an addition reaction between an alkene and a hydrogen halide:

$$C_2H_4(g) + HCl(g) \rightarrow C_2H_5Cl(l) \quad (10.23)$$

or a halogen:

$$C_2H_4(g) + Cl_2(g) \rightarrow C_2H_4Cl_2(l) \quad (10.24)$$

The products, which are called **alkyl halides,** can be considered to be derivatives of hydrocarbons in which one or more hydrogen atoms have been replaced by halogen atoms (compare C_2H_5Cl and $C_2H_4Cl_2$ to C_2H_6).

The alkyl halides are moderately polar, but exhibit no hydrogen bonding, so they boil at temperatures only slightly higher than those of hydrocarbons of about the same molecular mass (*bp*, °C: $C_2H_5Cl = 12$, $C_4H_{10} = 0$). These substances are fairly reactive chemically and are useful, and common, reagents in organic synthesis. They are good solvents for many organic substances, but are not themselves soluble in water.

One can prepare alkyl halides by many methods, including the direct reaction of the elementary halogen with saturated hydrocarbons. This reaction is carried out in the gas phase, and when once started it occurs very rapidly and tends to produce polysubstituted products. With ethane and chlorine, the simplest reaction would be:

$$C_2H_6(g) + Cl_2(g) \xrightarrow{300°C} C_2H_5Cl(s) + HCl(g) \tag{10.25}$$

but many other products are formed, including $ClCH_2CH_2Cl$, CH_3CHCl_2, $ClCH_2CHCl_2$, and so on. The reaction is used industrially where a pure product is not required. The various species produced are separated by fractional distillation, sometimes not very effectively if they have similar boiling points.

Several chlorinated hydrocarbons are quite well known. Chloroform, $CHCl_3$, is a low-boiling liquid, once commonly used as an anesthetic but now known to be a carcinogen. Carbon tetrachloride, CCl_4, was at one time used in fire extinguishers and as a dry-cleaning fluid. However, CCl_4 reacts with air at high temperatures to form $COCl_2$, phosgene, which is a poisonous gas, so modern fire extinguishers use other substances such as CO_2. Tetrachloroethylene, $Cl_2C{=}CCl_2$, has in recent years replaced CCl_4 in dry cleaning, mainly because of its lower toxicity and volatility. Freon, CCl_2F_2, used in refrigerators and aerosol cans because of its low boiling point, is quite inert, but its use is currently being curtailed because of the possibility that it might deplete ozone concentrations in the upper atmosphere.

Rather surprisingly, there are almost no halogen-containing substances found in living organisms. It appears that, in spite of their relatively high concentrations in seawater, halide ions were not accepted by living systems as suitable components during the course of evolution. This may explain in part why many common insecticides and herbicides are chlorinated compounds. Unfortunately, these substances frequently have detrimental effects on the environment, so their use for some purposes has now been prohibited (see Section 13.3).

PROBLEMS

10.1 *Hydrocarbons*

a. Draw the structures of two alkanes which have the molecular formula C_4H_{10}.
b. Draw the structure of the alkene of molecular formula C_3H_6.
c. Give the molecular formula of the acetylenic hydrocarbon with 3 carbon atoms.

10.2 *Petroleum* Explain why distillation of 200 ℓ of petroleum yields only about 30 ℓ of gasoline. How is it possible to increase the yield of gasoline?

10.3 *Oxygen-Containing Compounds*

a. Assign the molecules shown to one of the following classes: alcohol, ether, aldehyde, ketone, acid, ester.

```
    H  H  H            H  H     H             H  H
    |  |  |            |  |     |             |  |
  H—C—C—C—H          H—C—C—C—C—H           H—C—C—C=O
    |  |  |            |  |  ‖  |             |  |  |
    H OH  H            H  H  O  H             H  H  OH
```

b. Which of the classes listed in (a) could have the molecular formula $C_2H_4O_2$?

10.4 *Halogen-Containing Compounds* Suggest two different methods of preparing ethyl bromide, C_2H_5Br, starting with a hydrocarbon.

10.5 Draw Lewis structures for molecules with the following formulas:

a. C_3H_8
b. CH_3OH
c. C_3H_4

10.6 How many alkanes have the molecular formula C_7H_{16}? Sketch their structures.

10.7 Draw the structures of two molecules of formula C_3H_6.

10.8 2-pentene, C_5H_{10}, has a double bond between the second and third carbon atoms in the chain. Sketch its *cis* and *trans* isomers.

10.9 Classify the following as alkanes, alkenes or alkynes (assume there are no rings):

a. C_4H_8
b. C_4H_6
c. C_8H_{18}
d. C_2H_4

10.10 How many different tribromobenzenes could one hope to isolate?

10.11 Give all the bond angles in the toluene molecule, whose structure is given in Figure 10.4.

10.12 Using the following bond energies: C=C = 598, C—H = 414, C—C = 347 kJ/mol, estimate ΔH for

$$C_2H_6(g) + C_2H_4(g) \rightarrow C_4H_{10}(g)$$

10.13 In the water-gas process, Reactions 10.8 and 10.9, how many moles of carbon need to be burned to furnish the heat needed to form a mole of CO and a mole of H_2?

10.14 Noting that each carbon atom forms four bonds, put in the hydrogen atoms in the structure of 1,2-benzopyrene, p. 239, and give the molecular formula.

10.15 Synthesis gas, CO and H_2, can be used to form a variety of substances, depending on the catalyst used. Taking water to be the other product, write a balanced equation for the reaction that uses synthesis gas to make

a. C_3H_8
b. C_2H_5OH

10.16 Draw the complete structure of the molecules implied by the following formulas:

a. $CH_3CHBrCCl_2CH_3$
b. $CH_3COCH_2CH_3$
c. $CH_2OHCOOH$

10.25 Draw structures that satisfy the octet rule for molecules with the following formulas:

a. C_2H_5OH
b. CH_3OCH_3
c. C_4H_8

10.26 How many isomeric cyclic species have the molecular formula C_5H_{10}? Sketch their structures.

10.27 Draw the structures of two molecules of formula C_4H_6.

10.28 Would you expect a molecule with the formula $C_3H_4Cl_2$ to have *cis* and *trans* isomers? Explain.

10.29 Write the molecular formula of a hydrocarbon containing 6 carbon atoms which is

a. an alkane
b. an alkene
c. an alkyne
d. an aromatic

10.30 Suppose the three double bonds in benzene were fixed in position, i.e., there were two distinctly different kinds of bonds. How many dibromobenzenes would there be?

10.31 Cyclobutadiene, C_4H_4, should have a ring structure similar to benzene, but has not been isolated in pure form. Suggest an explanation for its instability.

10.32 Using bond energies given in Table 4.2, Chapter 4, estimate ΔH for the reaction in the gas state between formic acid and methyl alcohol to give methyl formate and water.

10.33 Using heats of formation given in Table 4.1, Chapter 4, calculate ΔH for the following reaction involving synthesis gas:

$$3\ CO(g) + 7\ H_2(g) \rightarrow C_3H_8(g) + 3\ H_2O(l)$$

10.34 Put hydrogen atoms in the proper places in the methylcholanthrene molecule, whose structure is given on p. 239. What is its molecular formula?

10.35 Taking carbon dioxide to be the other product, write a balanced equation for the formation of each of the following from synthesis gas:

a. C_2H_4
b. C_2H_5OH

10.36 Draw the complete structures of the molecules implied by the following formulas:

a. $CH_2OHCHOHCH_2OH$
b. $CH_2{=}CHCH_2CH_2OH$
c. CH_3CHO

10.17 Draw skeleton structures of all the alcohols with the molecular formula $C_5H_{12}O$.

10.18 Draw structures for all the saturated acids and esters that contain four carbon atoms.

10.19 Indicate the class of organic compound having the formula:

a. $CH_3—O—C_2H_5$
b. $CH_3—C≡C—H$
c. C_3H_7CHO
d. $C_2H_5COC_2H_5$

10.20 Identify each of the following as an alcohol, ether, aldehyde, ketone, acid, or ester:

a. $C_6H_{14}O$
b. C_2H_4O
c. $C_3H_6O_2$
d. C_3H_6O

10.21 Calculate the mass percentage of carbon in a compound containing 2 carbon atoms which is

a. an alkane
b. an olefin
c. an acid
d. an alcohol

10.22 Arrange the following substances in order of increasing boiling point:

a. C_2H_6
b. CH_3COOH
c. C_2H_4
d. C_2H_5OH

10.23 Draw the structure of a fat which, upon treatment with NaOH, gives glycerol and an organic acid containing 12 carbon atoms.

10.24 One difficulty with preparing alkyl halides by direct reaction of a halogen with a paraffin is that too many products are obtained. How many different dichlorinated isomers could be made from propane, C_3H_8?

10.37 Draw skeleton structures of all molecules having the formula C_4H_8O (no C=C bonds or rings). Classify each as an aldehyde or ketone.

10.38 Draw the structures of all the oxygen-containing compounds (Table 10.3) that contain 3 carbon atoms and indicate to what class each belongs.

10.39 Identify the class of compound to which the following substances belong:

a. $CH_3CH_2CHOHCH_3$
b. HCOOH
c. $CH_3CH=CHCH_3$
d. CH_3COOCH_3

10.40 Follow the directions of Problem 10.20 for:

a. CH_2O
b. $C_{12}H_{26}O$
c. $C_{12}H_{24}O_2$
d. CH_4O

10.41 Arrange all of the possible 2-carbon compounds containing no elements other than C, H, and O in order of decreasing mass percentage of carbon.

10.42 Arrange the following in order of increasing boiling point.

a. butane
b. butyl alcohol
c. 2-methylpropane
d. diethyl ether

10.43 Draw the structure of a detergent (sodium salt) having 12 carbon atoms.

10.44 How many isomers with the formula $C_4H_7Cl_3$ could one obtain by reacting C_4H_{10} with Cl_2?

*10.45 C_8H_8 is called cyclooctatetraene. Assuming the carbon skeleton of the molecule forms a regular octagon, and that cyclooctatetraene exhibits the same kind of resonance as does benzene, how many dibromo derivatives would you expect the substance to have? (Actually, C_8H_8 is not planar and does not behave as a typical aromatic compound.)

*10.46 In the acetone molecule, CH_3COCH_3, which atoms lie in a plane? What bond angle is associated with The C—C=O group? the C—C—H group? the C—C—C group?

*10.47 Noting the composition of the product gas in the Lurgi process (15% CO, 15% CH_4, 30% CO_2, 40% H_2), write a balanced equation for the overall reaction that occurs, starting with carbon, oxygen, and steam. What is ΔH for this reaction per mole of carbon consumed? Comment on the significance of the sign of ΔH. Where does the heat come from?

*10.48 A sample of an organic compound which contains carbon and hydrogen and possibly oxygen is burned in air. A 10.0-mg sample produces 22.0 mg of CO_2 and 12.0 mg of H_2O. Under less strenuous conditions, the compound reacts with O_2 to form a ketone. Identify the compound.

11

LIQUIDS AND SOLIDS; PHASE CHANGES

In Chapter 5, we discussed the laws governing the physical behavior of gases and the interpretation of these laws in terms of the kinetic theory. In succeeding chapters we have had frequent occasion to refer to substances in the liquid and solid states. We shall now discuss the properties of these two condensed states of matter. In doing so, we shall be particularly interested in:

1. The particle structure of liquids and solids and its influence upon their physical properties.
2. The equilibria between the gaseous, liquid and solid phases of a pure substance.

11.1 NATURE OF THE LIQUID STATE

The structure of liquids is less well established than that of gases or solids. Despite a great deal of research in this area, we still do not have a clear picture of the way in which molecules are arranged in even the most common liquid, water. We do, however, have a reasonably detailed knowledge of the average distances between atoms or molecules in a liquid. Moreover, we can estimate with considerable accuracy the magnitude of the forces between particles in a liquid. It is these two aspects of liquid structure which we shall now consider.

A gas at 1 atm is mostly empty space

At ordinary temperatures and pressures, the molecules in a liquid are much closer together than they are in a gas (Fig. 11.1). In the liquid there is very little free volume between molecules; in the gas, on the other hand, only a small fraction of the total volume is occupied by the molecules themselves. The close contact between molecules in a liquid as compared to a gas explains many of the differences in physical properties between these two states of matter. In particular, we see why the volume of a liquid is much less sensitive to pressure and temperature than that of a gas.

Since the molecules in a liquid are much closer together than those in a gas, attractive forces between molecules are considerably stronger. Energy has to be absorbed to overcome these forces, separate the molecules, and thereby convert the liquid to its vapor. To boil water we apply heat from an outside source such as a Bunsen burner or a hot plate. When a liquid evaporates at room temperature, heat is absorbed from the surroundings. You feel cold when you come out of a shower or a swimming pool because the evaporation of moisture draws heat from the body surface and in doing so lowers its temperature.

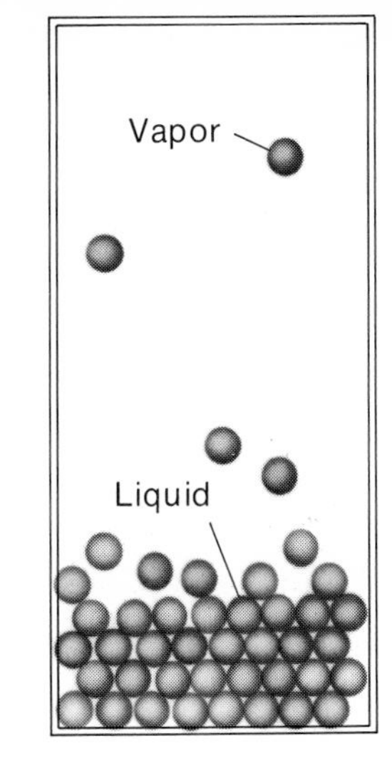

Figure 11.1 In a vapor at one atmosphere the molecules are about ten times as far apart as they are in the liquid. The volume of a liquid is roughly equal to the volume of its molecules.

The amount of heat required to vaporize a mole of liquid against a constant external pressure (e.g., in an open container) is referred to as the **molar heat of vaporization,** ΔH_{vap}. The heat of vaporization is a measure of the strength of the intermolecular forces in a liquid. Compare, for example, the heat of vaporization of water, approximately 40 kJ/mol, to that of methane, 9 kJ/mol. The difference between these two quantities reflects the strong hydrogen bonds in liquid water which are absent in liquid methane.

Since a molecule at the surface of a liquid has fewer neighbors than one in the interior, there is an unbalanced attractive force tending to pull it into the body of the liquid. This explains why liquids in contact with gases tend to achieve as small a surface as possible. Raindrops falling through air, or gas bubbles rising through a liquid, take on the shape of a sphere, thereby achieving the smallest possible surface area for a given volume.

The work required to expand the surface of a liquid by unit area is referred to as its **surface tension,** γ. This quantity is expressed in joules per square metre ($\equiv$N/m). The surface tension, like the heat of vaporization, is a measure of the strength of intermolecular forces. Water, in which these forces are comparatively strong, has a high surface tension, 0.073 J/m^2 at 20°C, considerably greater than that of most organic liquids (0.02 to 0.03 J/m^2). This explains, at least in part, why benzene or ethyl alcohol tend to "wet" or spread out on solid surfaces more readily than water. The wetting ability of water can be increased by adding a soap or detergent which drastically lowers the surface tension. Mercury, with a surface tension even higher than that of water, is a notoriously poor wetting agent. When spilled on a laboratory bench or on the floor, mercury forms tiny spherical drops which are difficult to retrieve.

11.2 LIQUID-VAPOR EQUILIBRIUM

All of us are familiar with the phase change referred to as vaporization, in which a liquid is converted to its vapor. In an open container, this process continues until all the liquid has evaporated; water in a flower pot or a beaker "disappears" by vaporization into the atmosphere. In a closed container the situation is quite different; vaporization occurs only until *equilibrium* is reached between liquid and vapor.

To understand what is meant by liquid-vapor equilibrium, let us consider what happens from a molecular viewpoint when a volatile liquid is placed in a closed container (Fig. 11.2). At any given instant, the molecules in the liquid will have a wide range of velocities. Some of them will be virtually motionless; most will be moving at a velocity close to the average. A few molecules will have very high velocities and hence sufficient kinetic energy to overcome intermolecular forces and escape into the vapor. Occasionally, one of these molecules may collide with the surface and re-enter the liquid. At first, the movement of molecules is primarily in one direction, from liquid to vapor. Gradually, however, as the concentration of molecules in the vapor builds up, the rate of condensation approaches the rate of vaporization. Eventually, we reach a position of **dynamic equilibrium** at which these two rates become equal; the number of molecules condensing in unit time exactly balances the number vaporizing. From this point on, the concentration of molecules in the gas phase has a certain, fixed value which does not change with time.

Vapor Pressure

The process which we have just described cannot be followed visually. We cannot see molecules escape from the liquid nor can we see their concentration build up to its equilibrium value. It is possible, however, to follow the process by making use of the apparatus shown in Figure 11.3. At the beginning of the experiment, the flask is empty, the stopcock is open to the atmosphere, and the mercury levels in the two arms of the manometer are at the same height. A few drops of liquid are squeezed out of the dropper bulb into the flask and the stopcock is closed. As vaporization occurs, molecules leaving the liquid increase the pressure in the space above it. This increase in pressure registers on the manometer; the mercury level in the left arm of the U tube falls, while that in the right arm rises. The pressure steadily increases until equilibrium is reached; at that point it becomes constant and the mercury levels stop moving.

At equilibrium, both the liquid and vapor must be present

The pressure of vapor in equilibrium with a pure liquid at a given temperature is referred to as the **vapor pressure** of the liquid. The apparatus shown in

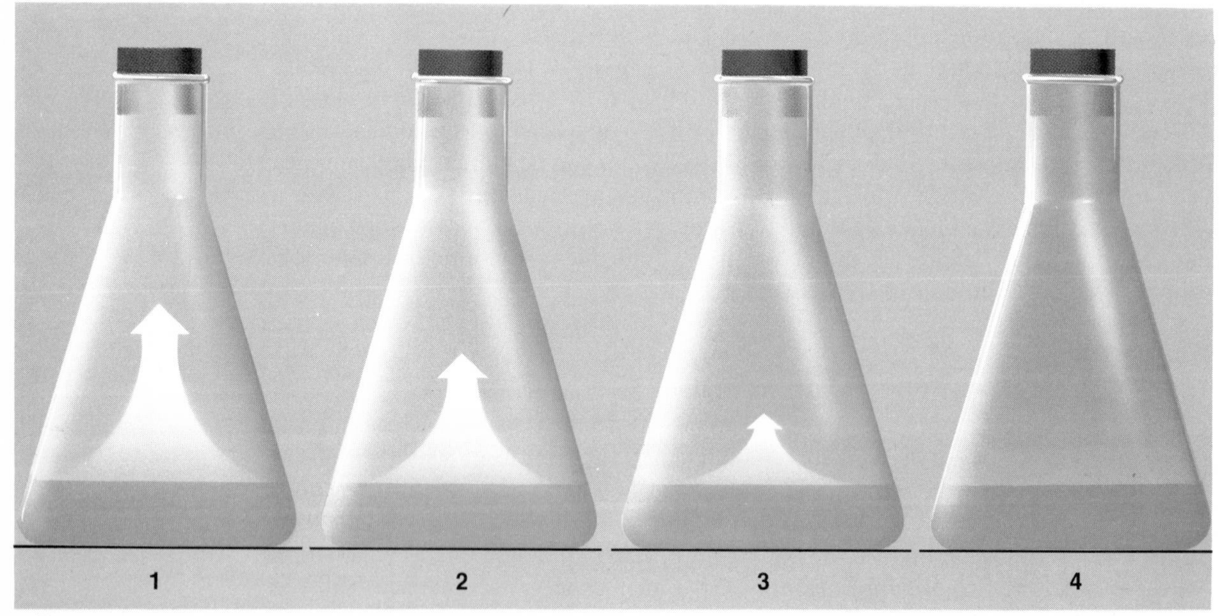

Figure 11.2 Initially, a liquid exposed to a vacuum will evaporate rapidly (1 and 2), but after a short time the rate of vaporization of molecules will become equal to their rate of condensation, and the concentration, and pressure, of the vapor molecules will approach an equilibrium, constant, value (3 and 4).

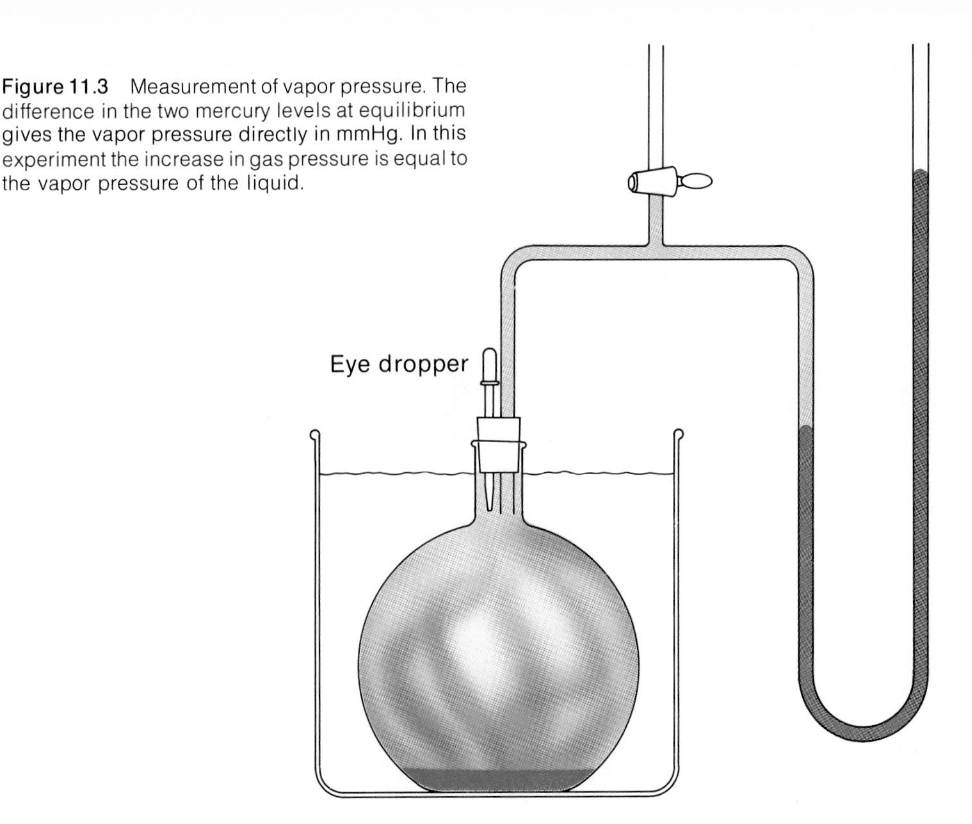

Figure 11.3 Measurement of vapor pressure. The difference in the two mercury levels at equilibrium gives the vapor pressure directly in mmHg. In this experiment the increase in gas pressure is equal to the vapor pressure of the liquid.

Figure 11.3 can be used to measure vapor pressure. If we place liquid benzene in the flask and allow it to come to equilibrium with its vapor at 25°C, we find that the difference in mercury levels is 92 mm; it follows that the vapor pressure of liquid benzene at 25°C is:

$$92\ \text{mmHg} \times \frac{101.3\ \text{kPa}}{760\ \text{mmHg}} = 12.3\ \text{kPa}$$

INDEPENDENCE OF VOLUME. As our discussion implies, a liquid at a given temperature has a fixed vapor pressure characteristic of the substance. It is important to realize that *so long as both liquid and vapor are present, the pressure exerted by the vapor will be independent of the volume of the container.* To see what this statement means, consider Figure 11.4, which shows the vaporization of liquid benzene taking place at 25°C in a container whose volume can be varied by raising the piston. We place a small amount of benzene in the container, raise the piston to the level shown in A, and allow equilibrium to be established. Vaporization occurs until the concentration of molecules in the vapor becomes constant at a pressure of 12.3 kPa. If the equilibrium is disturbed by raising the piston to the level shown in B, the pressure will drop momentarily. However, liquid will quickly evaporate to establish the original concentration of molecules in the vapor and, hence, the original vapor pressure, 12.3 kPa. This process will be repeated each time we raise the piston, as more and more liquid vaporizes. Eventually, all the liquid will have been converted to vapor (Fig. 11.4C). If we continue to raise the piston beyond this point, no more molecules can vaporize to compensate for the increase in volume; the concentration of molecules in the vapor will drop (Fig. 11.4D) and the pressure will decrease in accordance with Boyle's Law.

The vapor pressure of a liquid depends only on its temperature

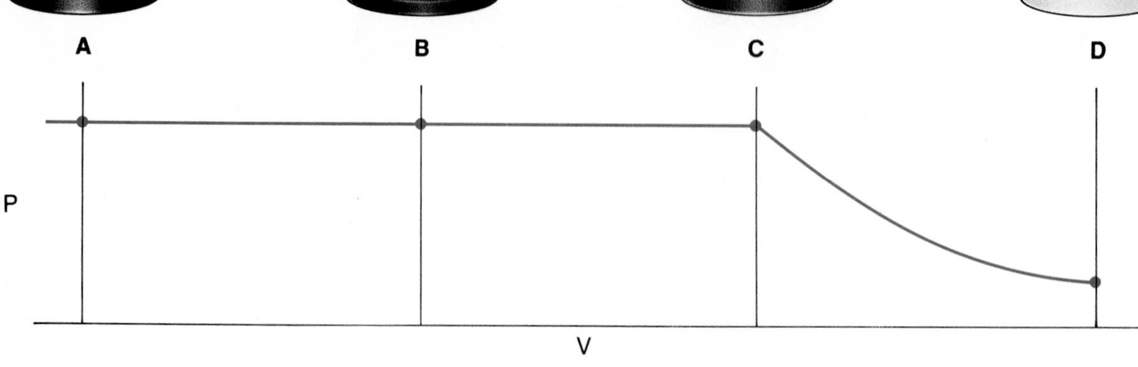

Figure 11.4 The pressure of a vapor in equilibrium with a liquid is independent of the volume of the container as long as there is liquid present (A and B). When all the liquid is vaporized (C), a further increase in volume (D) decreases the pressure in accordance with Boyle's Law. Since System C consists only of gas, its volume can be calculated from the Ideal Gas Law if the temperature and amount of sample are known.

Example 11.1 Suppose that in the experiment just described we start with 0.100 mol of benzene at 25°C (vp = 12.3 kPa).

a. To what value must the volume be increased in order for the liquid to just disappear (point C in Fig. 11.4)?
b. What will be the pressure of benzene vapor when the volume is 12.0 dm^3? 30.0 dm^3?

Solution

a. The liquid will "disappear" when the 0.100 mol of benzene originally present is completely vaporized. Rephrasing the question, we are asked to calculate the volume occupied by 0.100 mol of benzene vapor at 25°C and 12.3 kPa.

From the Ideal Gas Law, we have:

Note the difference between vapor pressure, observed at equilibrium, and gas pressure, observed when *only* the gas is present

$$V = \frac{nRT}{P} = \frac{(0.100 \text{ mol})[8.31 \text{ kPa} \cdot \text{dm}^3/(\text{mol} \cdot \text{K})](298 \text{ K})}{12.3 \text{ kPa}} = 20.1 \text{ dm}^3$$

b. At 12.0 dm^3, which might represent point B in Figure 11.4, both liquid and vapor are present. Hence, P = 12.3 kPa, the equilibrium vapor pressure of benzene at 25°C. At a volume of 30.0 dm^3, which might represent point D in Figure 11.4, the situation is quite different. *Only* vapor is present and its pressure is less than the equilibrium value. Perhaps the simplest way to calculate the pressure at this point is to apply Boyle's Law:

$$P_D = P_C \times \frac{V_C}{V_D} = 12.3 \text{ kPa} \times \frac{20.1 \text{ dm}^3}{30.0 \text{ dm}^3} = 8.24 \text{ kPa}$$

101.3
100
80
60
40
20
0
0 20 40 60 80 100
PRESSURE, kPa
TEMPERATURE, °C
Water

Figure 11.5 Effect of temperature on the vapor pressure of water. As you can see from the diagram at the right, vapor pressure increases ever more rapidly as temperature rises. *Below,* If $\log_{10}P$ is plotted vs $1/T$, where T is the absolute temperature, a straight line is obtained from which the heat of vaporization of the liquid can be calculated.

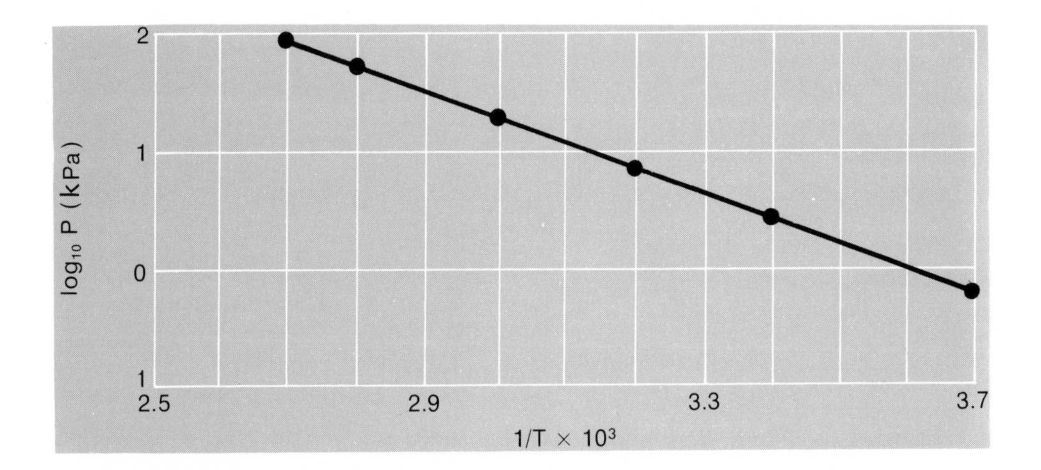

DEPENDENCE ON TEMPERATURE. The vapor pressure of a liquid always increases as temperature rises. Water evaporates more readily on a hot, dry day; stoppers in bottles of volatile liquids such as ether or gasoline may pop out if the temperature increases. The effect of temperature upon the vapor pressure of water is shown in Figure 11.5.

In studying the relationship between two variables such as vapor pressure and temperature, scientists prefer to work with linear functions which fit the general algebraic equation

$$y = Ax + B$$

In this case we can obtain a linear relation by plotting the logarithm of the vapor pressure ($\log_{10} P$) on the y axis vs. the reciprocal of the absolute temperature ($1/T$) on the x axis. Such a plot is shown at the bottom of Figure 11.5; the general equation is

$$\log_{10} P = \frac{A}{T} + B \qquad (11.1)$$

where A is the slope of the straight line and B is the intercept.

The constant A in Equation 11.1 can be related to the heat of vaporization of the liquid. Specifically, it can be shown that:

$$A = -\Delta H_{vap}/(2.30\ R) \qquad (11.2)$$

where ΔH_{vap} is the molar heat of vaporization in joules per mole and R is the gas law constant = 8.31 J/(mol · K). Substituting for A in Equation 11.1:

$$\log_{10} P = \frac{-\Delta H_{vap}}{(2.30)(8.31)T} + B \qquad (11.3)$$

A convenient method of measuring the heat of vaporization of a liquid takes advantage of the linear relationship between $\log_{10} P$ and $1/T$. The vapor pressure is measured at a series of different temperatures. At each temperature we calculate $\log_{10} P$ and $1/T$, plot these points on a graph, and then draw the best straight line through the points. By taking the slope of this line, the constant A is obtained. The heat of vaporization is then calculated using Equation 11.2 (cf. Problem 11.12 at the end of this chapter).

For many purposes it is convenient to have a two-point equation relating the vapor pressure (P_2) at one temperature (T_2) to that (P_1) at another temperature (T_1). Such an equation can be obtained by applying Equation 11.3 at the two temperatures.

at T_2: $$\log_{10} P_2 = \frac{-\Delta H_{vap}}{(2.30)(8.31)T_2} + B$$

at T_1: $$\log_{10} P_1 = \frac{-\Delta H_{vap}}{(2.30)(8.31)T_1} + B$$

Subtracting, we obtain

$$\log_{10} P_2 - \log_{10} P_1 = \frac{-\Delta H_{vap}}{(2.30)(8.31)}\left[\frac{1}{T_2} - \frac{1}{T_1}\right]$$

Rearranging to a somewhat more convenient form

$$\log_{10} \frac{P_2}{P_1} = \frac{\Delta H_{vap}}{(2.30)(8.31)}\left[\frac{T_2 - T_1}{T_2\,T_1}\right] \qquad (11.4)$$

Equation 11.4, in a more general form, is one of the most useful mathematical relationships in all of chemistry. We will meet it again when we deal with chemical equilibrium (Chapter 15) and with rate of reaction (Chapter 16). When applied to the dependence of vapor pressure upon temperature, it is known as the Clausius-Clapeyron equation, honoring the German physicist Rudolph Clausius, who was pre-eminent in the field of thermodynamics in Europe in the last half of the 19th century, and the French engineer B. P. E.

Clapeyron, who first proposed it, in a somewhat different form, in 1834. The Clausius-Clapeyron equation is particularly useful in calculating the vapor pressure of a liquid at one temperature, knowing the vapor pressure at another temperature and the heat of vaporization.

Example 11.2 The vapor pressure of benzene at 50°C is 36.3 kPa. Calculate its vapor pressure at 60°C, given that the heat of vaporization is 30 500 J/mol.

Solution Substituting in Equation 11.4, expressing temperature in K:

$$\log_{10} \frac{P_2}{36.3 \text{ kPa}} = \frac{30\,500(333 - 323)}{(2.30)(8.31)(333)(323)} = +0.148$$

where P_2 is the vapor pressure at 60°C. Since the antilog of +0.148 is 1.41, we have:

$$P_2/36.3 \text{ kPa} = 1.41; P_2 = 1.41 \times 36.3 \text{ kPa} = 51.2 \text{ kPa}$$

Boiling Point

When we heat a liquid in an open container, bubbles form, usually at the base of the container where heat is being applied. The first bubbles that we see are often air, driven out of solution by the increase in temperature. Eventually, however, when a certain temperature is reached, vapor bubbles form throughout the liquid, rise to the surface and break. When this happens, we say that the liquid is boiling.

We find that the temperature at which a liquid boils depends upon the pressure above it. To understand why this is the case, let us refer to Figure 11.6, which shows vapor bubbles rising in a boiling liquid. In order for a vapor bubble to form and grow as it approaches the surface, the pressure within it, P_1, must be at least equal to the pressure above it, P_2. But P_1 is simply the vapor pressure of the liquid. From this argument we conclude that **a liquid boils at a temperature at which its vapor pressure becomes equal to the pressure above its surface.** If this pressure is one standard atmosphere, 101.3 kPa, the liquid boils at a temperature which we call the **normal boiling point.** The vapor pressure of water reaches 101.3 kPa at 100°C; that of benzene becomes equal to one atmosphere at a somewhat lower temperature, 80°C. We refer to 100°C as the normal boiling point of water; 80°C is the normal boiling point of benzene.

As we might expect, the boiling point of a liquid can be reduced by lowering the pressure above it. It is possible to boil water at 25°C by evacuating the space above it with a vacuum pump or even an ordinary water aspirator. When we reach a pressure of 3.2 kPa, the equilibrium vapor pressure at 25°C, the water starts to boil. Chemists often purify a high-boiling compound which may decompose or oxidize at its normal boiling point by boiling at a reduced temperature under vacuum and condensing the vapor.

The boiling point of a liquid depends on how you boil it

If you have been fortunate enough to camp in the high Sierras or the Rockies, or live at a great distance above sea level, you may have noticed that it takes longer to boil potatoes, eggs and other foods at high altitudes. The reduced pressure lowers the temperature at which water boils in an open container, and hence slows down the physical and chemical changes that take place when potatoes or eggs are cooked. In principle, this problem can be solved by using a pressure cooker, in which the pressure developed is high enough to raise the boiling point of water above 100°C. Pressure cookers are

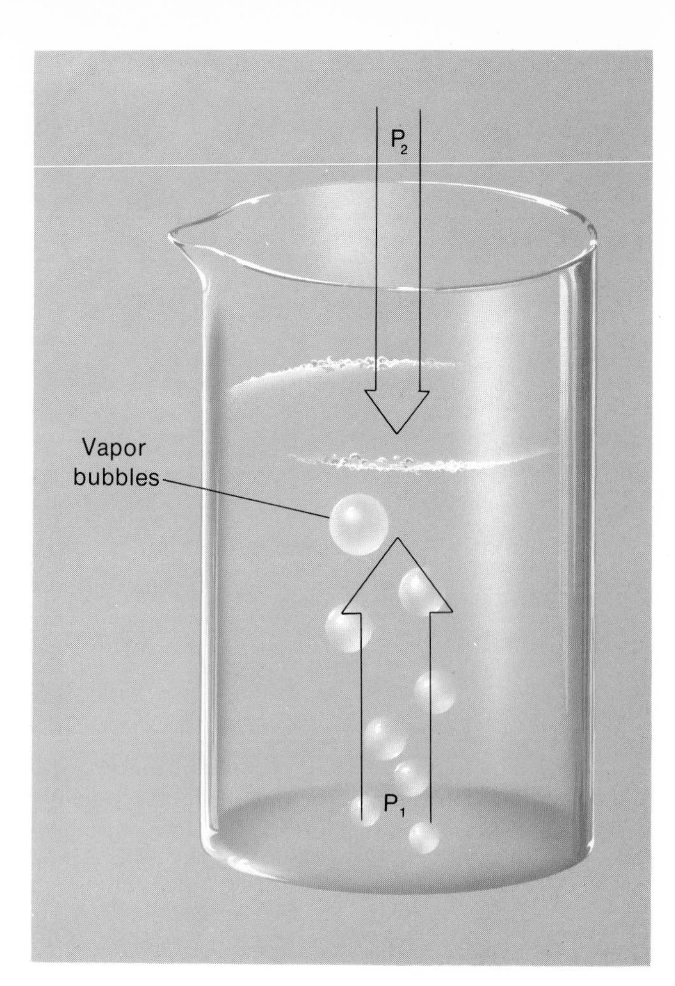

Figure 11.6 A liquid boils when its vapor pressure (P_1) exceeds the pressure above it (P_2).

indeed commonly used in localities such as Cheyenne, Wyoming (elevation 1859 m), and Flagstaff, Arizona (elevation 2101 m), but not by mountain climbers, who have to carry all their equipment on their backs.

We have enough trouble as it is

As we see from Table 11.1, the normal boiling point of a liquid is inversely related to its vapor pressure at room temperature. A volatile liquid such as ether, where the vapor pressure has reached 0.582 atm at 20°C, has to be heated only to 35°C to bring its vapor pressure to one atmosphere. In contrast, mercury (with a very low vapor pressure at room temperature) must reach a much higher temperature before it can boil at 1 atm pressure.

There is a direct correlation between the heat of vaporization of a liquid

TABLE 11.1 PHYSICAL PROPERTIES OF LIQUIDS

	Vapor Pressure at 20°C	Normal Boiling Point	Heat of Vaporization	$\frac{\Delta H_{vap}}{T_b}$
Mercury	0.000 16 kPa	630 K	59 400 J/mol	94
Water	2.3 kPa	373 K	40 600 J/mol	109
Benzene	10 kPa	353 K	30 500 J/mol	86
Bromine	23 kPa	332 K	30 000 J/mol	90
Ether	59 kPa	308 K	25 900 J/mol	84
Ethane	3 600 kPa	184 K	15 500 J/mol	84
Oxygen		90 K	6 820 J/mol	76

and its normal boiling point. For many liquids we find that the heat of vaporization (joules per mole) is approximately 88 times the normal boiling point in K.

$$\frac{\Delta H_{vap}}{T_b} \approx 88 \qquad (11.5)$$

This generalization, known as Trouton's rule, holds to within 5 to 10% for most organic liquids. For water and other liquids in which hydrogen bonding is involved, the heat of vaporization is somewhat higher than that calculated by Trouton's rule.

Critical Temperature and Pressure

Let us imagine an experiment in which a sample of liquid benzene is placed in an evacuated, heavy-walled glass tube (Fig. 11.7), which is then sealed and heated to higher and higher temperatures. The pressure of the vapor rises steadily; to 1 atm at 80°C, to 14 atm at 200°C, to 43 atm at 280°C . . . . Nothing spectacular happens (unless, of course, there happens to be a weak spot in the tube) until we reach 289°C, where the vapor pressure of benzene is 48 atm. Suddenly, as we pass this temperature, the meniscus separating the liquid and vapor phases disappears! The tube now contains only one phase, benzene vapor.

In other experiments, we find that it is impossible to condense benzene vapor at temperatures above 289°C, regardless of how much pressure is applied. Even at pressures as high as 1000 atm, benzene vapor stubbornly refuses to liquefy at 290 or 300°C. This behavior is typical of all substances. There is a temperature, called the **critical temperature,** above which the liquid phase of a pure substance cannot exist. The pressure which must be applied to bring about condensation at that temperature is called the **critical pressure.** Alternatively, one can regard the critical pressure as the vapor pressure of the liquid at its critical temperature.

Table 11.2 lists the critical temperatures of several common substances. The species in the column at the left, all of which have critical temperatures below 25°C, are often referred to as "permanent gases." Applying pressure at room temperature will not condense a permanent gas; it must be cooled as well. The permanent gases are stored and sold under high pressures, often 150 atm or greater; when the valve on the cylinder is opened, the pressure drops as gas escapes, as would be expected from the Ideal Gas Law.

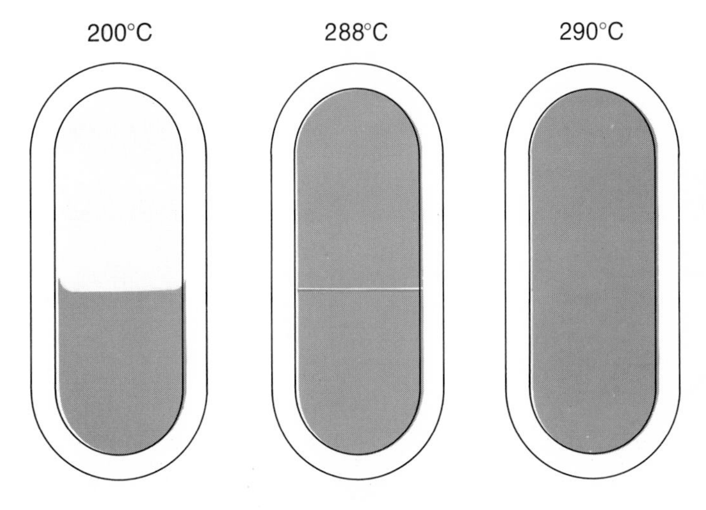

Figure 11.7 When a sample of benzene is heated in a sealed tube, the properties of the liquid and vapor approach one another near the critical temperature, 289°C. As one heats the sample through that temperature the meniscus disappears.

TABLE 11.2 CRITICAL TEMPERATURES

"PERMANENT GASES"		"CONDENSABLE GASES"		"LIQUIDS"	
Helium	−268°C	Carbon dioxide	31°C	Ether	194°C
Hydrogen	−240	Ethane	32	Ethyl alcohol	243
Nitrogen	−147	Propane	97	Benzene	289
Argon	−122	Ammonia	132	Bromine	302
Oxygen	−119	Chlorine	146	Water	374
Methane	−82	Sulfur dioxide	158	Mercury	1460

When you open a CO_2 fire extinguisher, some solid CO_2 is observed. Why?

The gases listed in the center column of Table 11.2 have critical temperatures above 25°C; they are available commercially as liquids in high pressure cylinders. When we open the valve on a cylinder of propane, the gas that escapes is replaced by vaporization of liquid, and the pressure returns to its original value. Only when the liquid has completely vaporized and the tank is almost empty does the gauge pressure drop.

11.3 NATURE OF THE SOLID STATE

Solids tend to crystallize in definite geometric forms which can frequently be seen by the naked eye (Plate 7). In ordinary table salt, we can distinguish small, cubic crystals of sodium chloride. Large, beautifully formed crystals of such minerals as quartz (SiO_2) and fluorite (CaF_2) are found in nature. Even in finely powdered solids, it is ordinarily possible to observe distinct crystal forms under a microscope.

The existence of crystals with distinct geometric forms implies that the particles making up the crystal are arranged in a definite, three-dimensional pattern. The geometry of the pattern may be relatively simple or exceedingly complex. In this section, we will look at some of the simpler ways in which atoms, ions, or molecules can be arranged to form a crystal lattice. First, we will consider a general method of determining crystal structures. This involves studying the extent to which a beam of x-rays of known wavelength is diffracted (bent) when it passes through a crystalline solid.

Crystal Structures from X-Ray Diffraction

Those of you who have had a course in physics may recall that visible light striking a glass plate ruled with a large number of closely spaced parallel lines is broken up by diffraction into a series of beams oriented at definite angles to each other. The basic requirement for diffraction is that the lines be evenly spaced and that the distance between them be about the same as the wavelength of the incident light. In 1912 Max Von Laue, an assistant lecturer in Physics at Munich, suggested that a crystal, which is made up of particles uniformly spaced about 10^{-10} m apart, could act as a diffraction grating for x-rays, which have wavelengths ranging from 10^{-11} to 10^{-9} m. His reasoning was confirmed by two Ph.D. students at Munich, Friedrich and Knipping, who passed a beam of x-rays through a copper sulfate crystal and observed a distinct pattern of spots on a photographic plate (Fig. 11.8).

In 1913 W. H. Bragg, at the University of Leeds, and his son, W. L. Bragg, adapted x-ray diffraction to determine crystal structures. The principle used is embodied in the Bragg equation, which relates the angle of diffraction to the distance between successive layers of atoms or ions in the crystal.

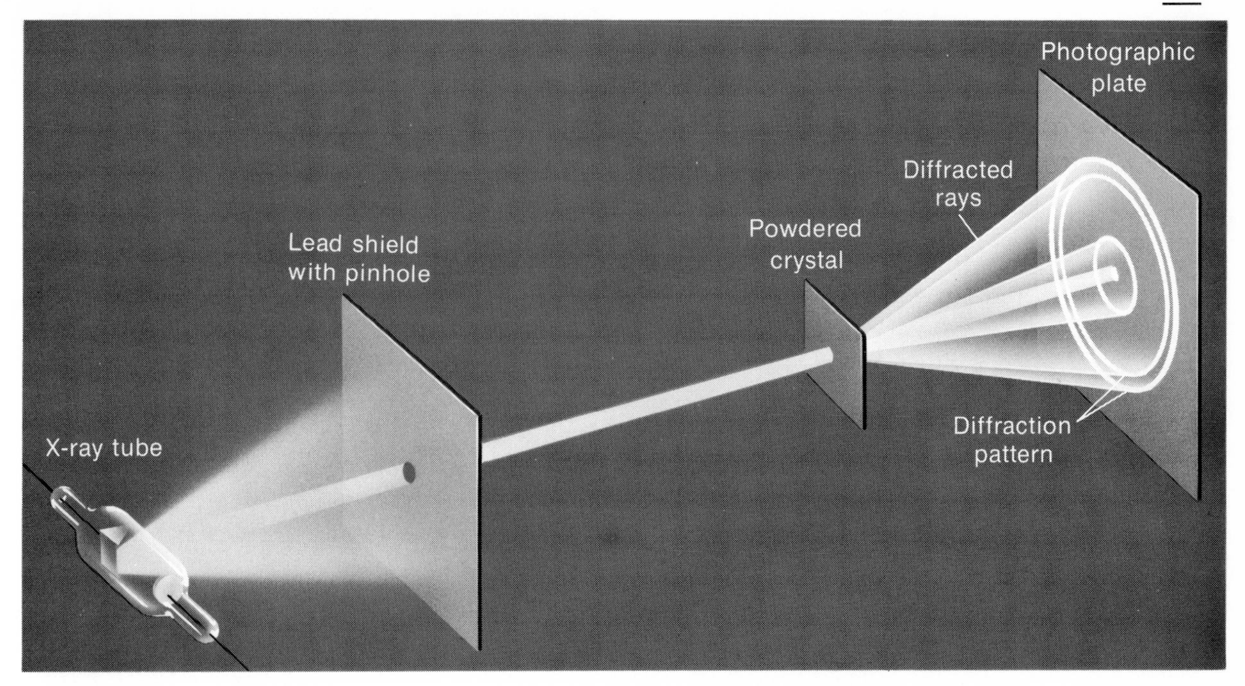

Figure 11.8 X-ray diffraction. If the sample is powdered, as shown in this schematic drawing, monochromatic x-rays are used. From the angles at which the rays are diffracted, one can calculate, using Bragg's Law, the distances between planes of atoms in the crystal.

$$\sin \theta = \frac{n\lambda}{2d} \qquad (11.6)$$

In this equation, θ is the angle between the beam and the layer of atoms, λ is the wavelength of the x-rays, d is the distance between successive layers, and n is an integer (1, 2, 3, . . .) known as the order number of the diffracted beam.

By measuring the angles at which x-rays of known wavelength are diffracted by a crystal, we can calculate from Equation 11.6 the distances between planes of atoms or ions. In this way values of atomic or ionic radii can be obtained. It is more difficult to deduce the geometric pattern in which the particles are arranged. The problem is that an x-ray beam "sees" not one but many different series of layers oriented at various angles to one another.

Despite these difficulties, x-ray crystallographers have been able to unravel the particle structures of a wide variety of crystals, ranging from simple inorganic salts to complex organic molecules. The basic approach is to assume a particular crystal structure and calculate the angles and intensities of the diffracted beams to be expected for that structure. Comparison with experimental data can then suggest refinements which will give better agreement. Prior to the computer age, this was a tedious process. It took Dorothy Hodgkin and her associates at Oxford eight years to work out the structure of vitamin B-12 (molecular formula $C_{63}H_{84}N_{14}O_{14}PCo$). In 1964, nine years after her work was published, she was awarded the Nobel prize in Chemistry. At present, using computers, x-ray crystallographers can determine structures as complex as those of proteins, in which a single molecule may contain a thousand or more atoms (Chapter 25).

Unit Cells

The basic information that comes out of x-ray diffraction studies concerns the dimensions and geometric form of what is known as the **unit cell,** the smallest unit which, repeated over and over again in three dimensions, generates the crystal lattice. Perhaps the simplest unit cell to visualize is the **simple cubic cell,** which consists of eight atoms (or molecules, or ions) located at the corners of a cube. Two other types of cubic cells are shown in Figure 11.9. One of these, in which there is an atom at the center of the cube, is referred to as a **body-centered cubic cell.** The third cell, in which there are atoms at the center of each face of the cube, is called a **face-centered cubic cell.**

In considering the unit cells shown in Figure 11.9, it is important to keep two points in mind:

1. Only in the simple cubic cell do atoms at the corners of the cube touch each other. In both of the other cells, the atoms at the corners are forced slightly apart. In the body-centered cube, atom contact is along a body diagonal. That is, the atom at the center of the cube touches atoms at opposite corners. In the face-centered cube, atoms touch along a face diagonal. The atom at the center of each face is in contact with atoms at opposite corners of the face.

2. Certain of the atoms shown in Figure 11.9 do not belong exclusively to a single unit cell. An atom in the center of a face of a cube is shared by another cube that touches that face. In effect, then, only half of that atom can be assigned to one cell. An atom at the corner of a cube forms a part of eight different cubes that touch at that point. In this sense, only one eighth of a

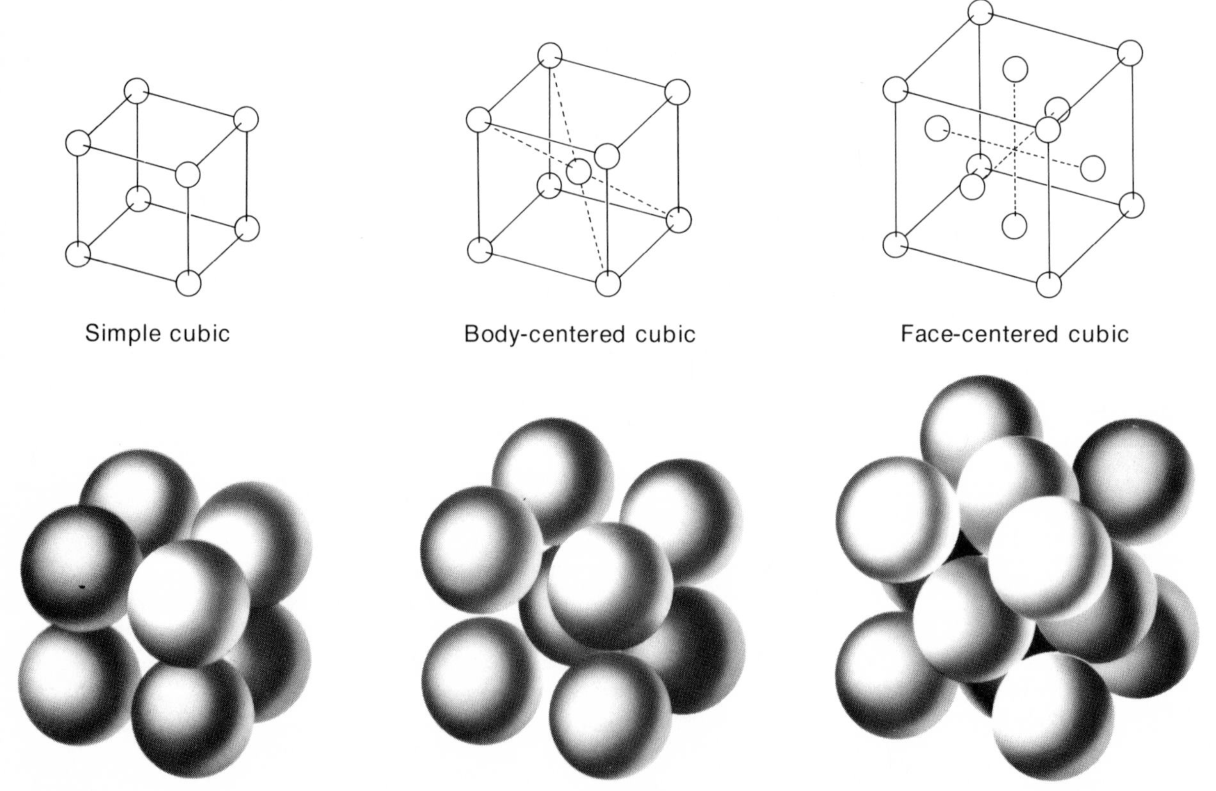

Figure 11.9 Three types of cubic lattices. In the simple cubic lattice, there is an atom at each corner of the cube. In the body-centered lattice, in addition to the eight atoms, there is an atom at the center of the cube. In face-centered lattices, there are, in addition to the eight atoms, atoms at the center of each of the six faces of the cube.

corner atom "belongs" to a particular cell. The number of atoms to be assigned to each of the unit cells in Figure 11.9 is:

Simple cubic: 8 atoms × 1/8 = 1 atom per unit cell
Body-centered cubic: 8 atoms × 1/8 + 1 atom = 2 atoms per unit cell
Face-centered cubic: 8 atoms × 1/8 + 6 atoms × 1/2 = 4 atoms per unit cell

Example 11.3 Copper crystallizes in a structure with a face-centered cubic unit cell 0.363 nm on an edge. Calculate the atomic radius of copper.

Solution It will be helpful to draw a face of the unit cell; points A, B, C, and D locate the centers of the four atoms at the corners of this face.

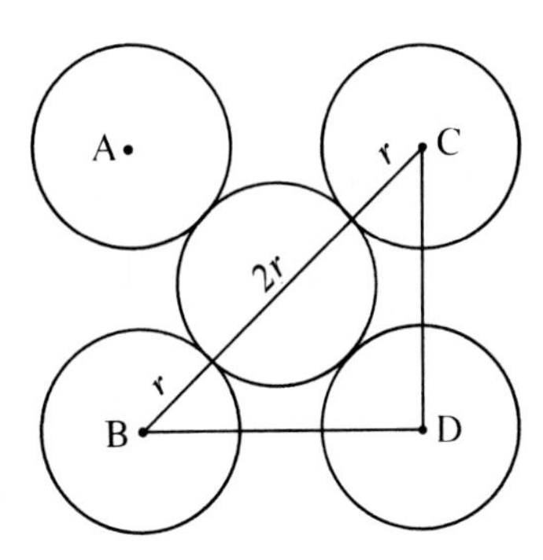

The distances BD and CD are given as 0.363 nm. The distance along the face diagonal, BC, can be calculated from the Pythagorean theorem:

$$(BC)^2 = (BD)^2 + (CD)^2 = 2(0.363\text{ nm})^2$$

$$BC = \sqrt{2}\,(0.363\text{ nm}) = 1.41 \times 0.363\text{ nm} = 0.512\text{ nm}$$

But, from the figure, BC = $4r$, where r is the radius of the metal atom:

$$4r = 0.512\text{ nm};\ r = 0.128\text{ nm}$$

When the atoms in a crystal lattice are all of the same size, as is the case with metals, they tend to pack quite closely together. From this point of view, the simple cubic structure is very unstable, since it contains a relatively large amount of empty space. A body-centered cubic structure is more closely packed, with less "waste space"; about 20 metals have this type of unit cell. A still more efficient way of packing spheres of the same size is the face-centered cubic structure, where the amount of empty space is a minimum. A total of 54 different metals have a structure based on either the face-centered cubic cell or a close relative in which the packing is equally efficient (hexagonal closest packed structure).

Golf balls or oranges pack naturally in an FCC structure

The geometry of ionic crystals, where there are two different kinds of ions, is necessarily more complicated than that of metallic lattices. Nevertheless, it is possible to visualize the packing in certain ionic crystals in terms of the simple unit cells discussed above. An example is the lithium chloride crystal, where the large Cl^- ions (r = 0.181 nm) form a face-centered cubic lattice, with the small Li^+ ions (r = 0.060 nm) fitting into "holes" in the lattice, one Li^+ ion per Cl^- ion (Fig. 11.10). The structure of sodium chloride is similar to that of LiCl, except that the Na^+ ions (r = 0.095 nm) are slightly too large to fit into a

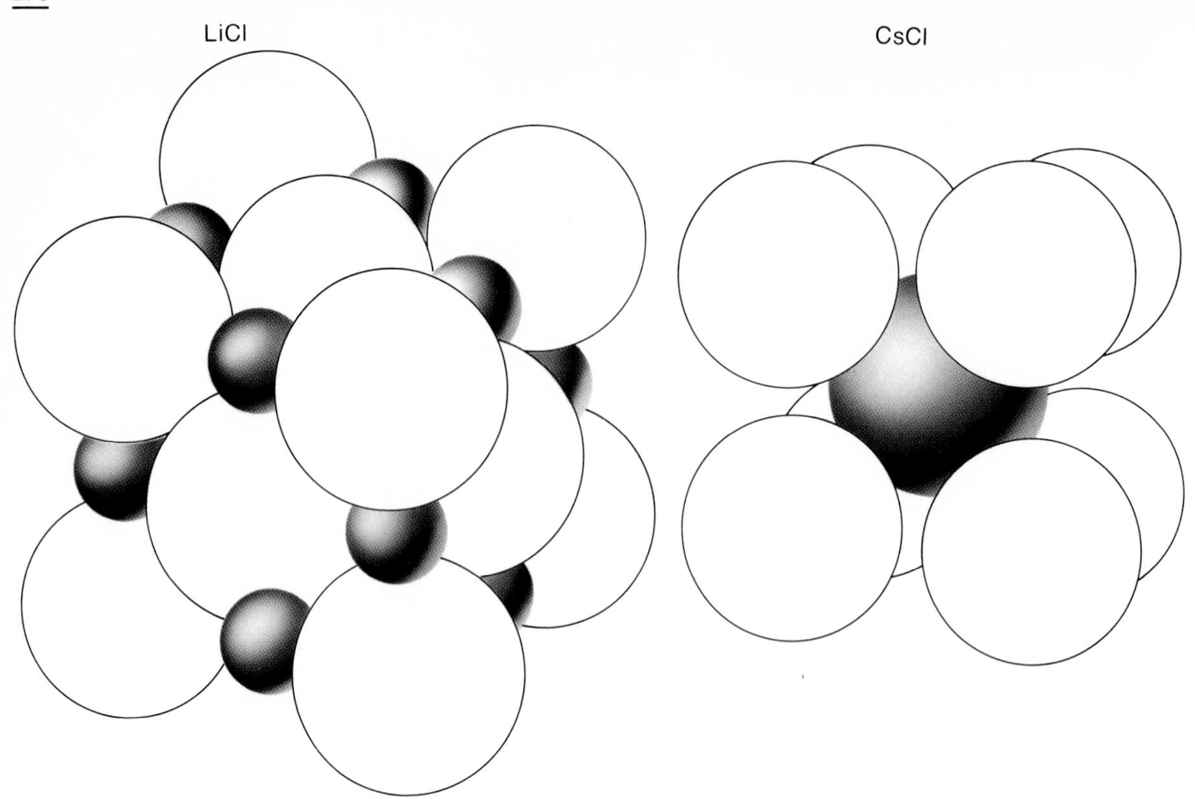

Figure 11.10 In LiCl, where the cation is much smaller than the anion, there is anion-anion contact in a face-centered lattice. In CsCl, the Cs^+ ion at the center of a cube touches eight Cl^- ions at the corners.

close-packed lattice of Cl^- ions touching each other along the faces of a cube. Consequently, the face-centered cubic array of Cl^- ions found in LiCl is slightly expanded in NaCl.

In cesium chloride, the cations are much too large (r of Cs^+ = 0.169 nm) to fit into a face-centered cubic array of Cl^- ions. We find that cesium chloride crystallizes in a quite different structure (Fig. 11.10) in which each Cs^+ ion is located at the center of a cube outlined by Cl^- ions. The Cs^+ ion at the center touches all the Cl^- ions at the corners of the cube; the Cl^- ions do not touch each other.

Defect Crystals

Up to this point in our discussion of the structure of crystalline solids we have considered them to be perfect crystals, with all the atoms or ions lined up in a precise geometric pattern. A perfect crystal, like an ideal gas, is an abstraction; the crystals that we find in nature or prepare in the laboratory always contain imperfections. These defects, even if relatively few in number, can profoundly affect the physical and chemical properties of a solid.

Many different types of crystal defects are possible. One of the most common, at least in ionic crystals, is shown in Figure 11.11. Here, an ion has moved from its regular site to occupy what is known as an *interstitial* position. Ordinarily, as in Figure 11.11, it is the cation that moves; since cations are usually smaller than anions, they can more readily fit into free spaces in the lattice. The sensitivity to visible and UV light that is characteristic of silver bromide and other silver salts used in photography has been explained in terms of this type of defect. When light strikes a silver bromide crystal, enough energy is available to remove an electron from an occasional Br^- ion.

$$Br^- \rightarrow Br + e^-$$

The electron produced is able to migrate through the crystal until it comes in contact with an interstitial Ag^+ ion which it converts to an atom of silver metal.

$$Ag^+ + e^- \rightarrow Ag$$

The silver atoms formed by this mechanism apparently act as nuclei for the formation of grains of silver, which are visible to the naked eye. At first, they are seen as a purplish discoloration at the surface of the silver bromide; on further exposure to light the entire solid turns black.

The incorporation of a small number of foreign atoms or ions into a crystal lattice can have a drastic effect on the properties of the solid. The effect on the electrical conductivity of metalloids such as germanium or silicon is especially striking. Extremely pure samples of germanium and silicon are nonconductors because the valence electrons are tied down in the four covalent bonds that each atom forms with its neighbors. The conductivity increases dramatically when small amounts of arsenic or boron (as little as 0.0001 mol %) are introduced into the crystal. To understand how this comes about, let us consider how the crystal structure of a 4A element might be disturbed by substituting atoms of a 5A or 3A element.

To make a semiconductor, one must start with fantastically pure Ge or Si

An atom of arsenic, which has five valence electrons, can fit into the crystal lattice of germanium or silicon because of the close similarity in atomic radii (0.121 nm vs. 0.122 nm or 0.117 nm). To do so, however, it must give up its valence electron. This electron can move relatively freely through the crystal under the influence of an electrical field, giving rise to an **n-type semiconductor** (current carried by the flow of negative charge). If an atom of boron or other element with three valence electrons is introduced into the lattice, a somewhat different situation arises. An electron deficiency is created at the site occupied by the foreign atom; it is surrounded by seven valence electrons rather than

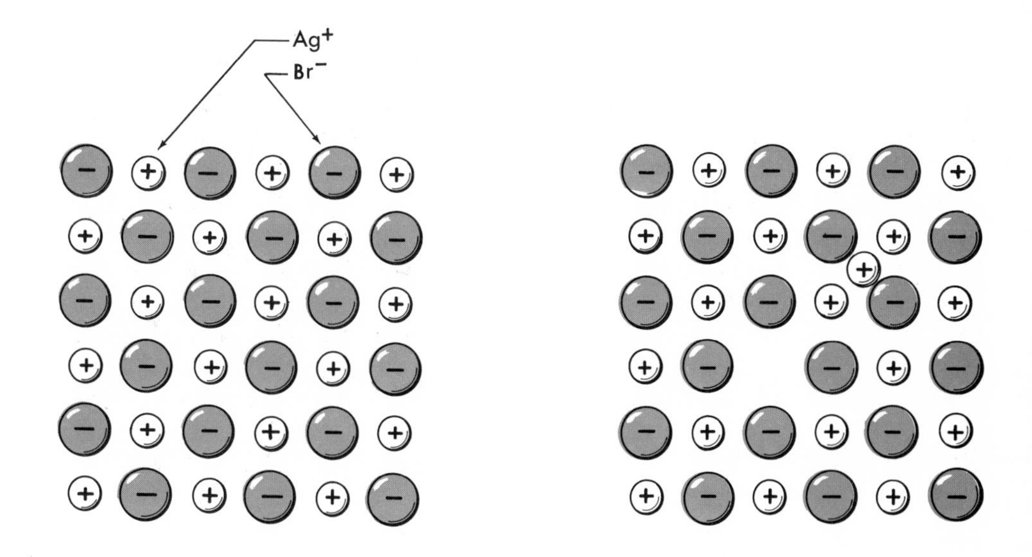

Figure 11.11 Schematic drawing of a common type of defect in ionic crystals (AgBr). The cation occupies an interstitial position, leaving a hole at the point where the cation would ordinarily be.

eight; in this sense, there is a "positive hole" in the lattice. In an electrical field, an electron moves from a neighboring atom to fill this hole. In so doing, it creates an electron deficiency around the atom which it leaves. Conduction in this type of defect crystal, a **p-type semiconductor,** can be thought of as a movement of positive holes through the lattice.

A semiconductor in which there is a junction between an electron-rich and an electron-deficient region acts as a rectifier, capable of converting alternating to direct current. Such an "n-p" junction can be formed by starting with a pure silicon disc and introducing a trace of boron on one side of the disc and a trace of arsenic on the other. These impurities are allowed to diffuse into the silicon at high temperatures until the region of p-type semiconductor meets the region of n-type semiconductor. Electrons flow readily from the electron-rich region containing arsenic to the electron-deficient area created by the presence of boron atoms. Potentials as high as 1000 V are incapable of bringing about electron flow in the opposite direction.

Perhaps the best known of all semiconductor devices is the transistor,

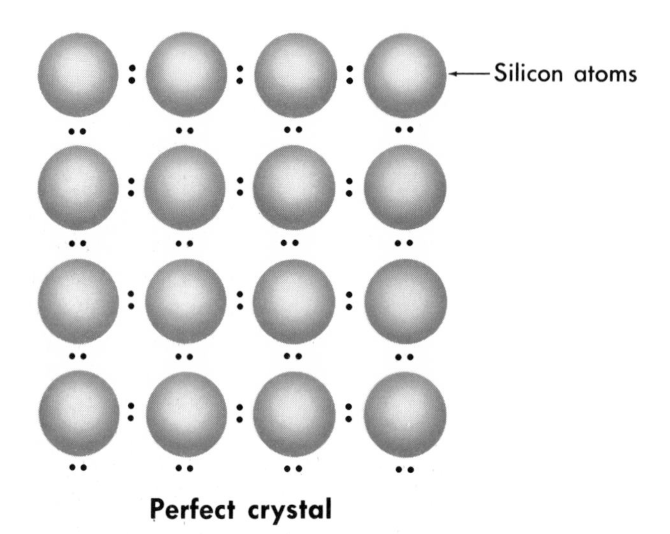

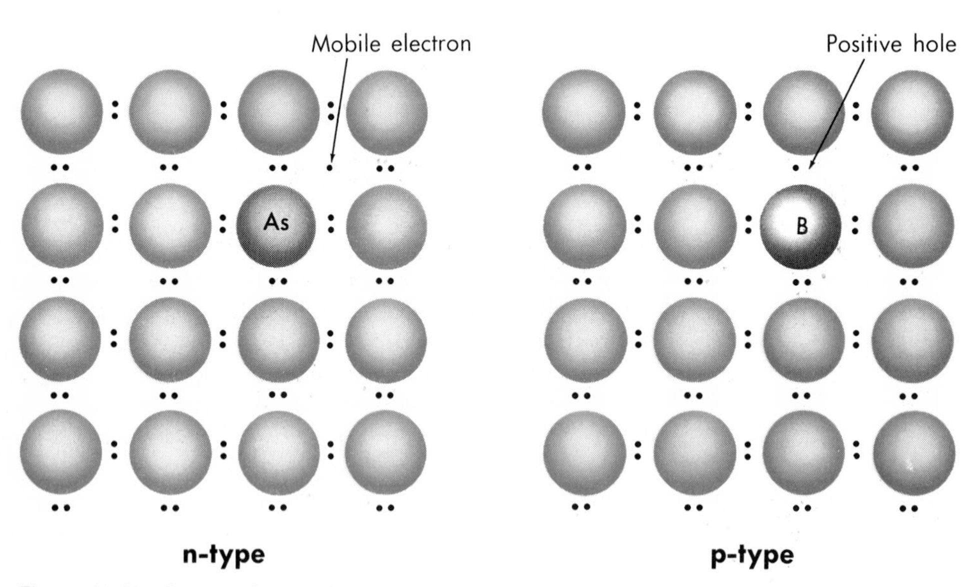

Figure 11.12 Semiconductors derived from silicon. In n-type semiconductors the impurity atoms furnish mobile electrons to the crystal. In p-type semiconductors there is a deficiency of electrons, since the impurity atoms have three rather than four valence electrons.

in which n-p-n or p-n-p junctions are created by forming a sandwich of alternate electron-rich and electron-poor regions. A transistor amplifies an electric current in much the same way as a vacuum tube. The small size of transistors makes them ideal for use in hearing aids, miniature radios, electronic calculators, and computers. Their use in television and radio receivers, in which they have virtually displaced vacuum tubes, reflects their greater durability and lower energy consumption.

A semiconductor whose surface contains an n-p junction can convert radiant energy to electrical energy. This principle is used in the *solar cell,* which has been used as a source of power in space vehicles. In this device, light striking a very pure crystal of germanium or silicon impregnated with an electron-rich impurity such as arsenic ejects some of the loosely held electrons. These electrons are collected at the surface, which is coated with a thin, transparent layer of a p-type semiconductor. The electrons then pass through an electrical circuit where their energy is used to do useful work.

A major drawback to solar cells based on germanium or silicon is an economic one. It appears that their cost will have to be reduced by a factor of at least 50 to make them competitive with conventional sources of electrical energy. This situation has stimulated research on other light-sensitive semiconducting materials which might be less expensive to produce. Two substances now under study are cadmium sulfide, CdS, and gallium arsenide, GaAs.

11.4 PHASE EQUILIBRIA

Earlier in this chapter, we discussed in some detail the equilibrium between a liquid and its vapor. For a pure substance, there are at least two other types of phase equilibria to be considered: solid-liquid and solid-vapor. Many of the important characteristics of all these equilibria can be summarized by what is known as a **phase diagram.** Figure 11.13 shows a portion of the phase diagram for the pure substance water. Pressure is plotted along the vertical axis, temperature along the horizontal axis.

To understand what this diagram implies, let us consider first the significance of the three lines, AB, AC, and AD. *Each of these lines tells us the pressures and temperatures at which two phases are in equilibrium with each other.* Specifically:

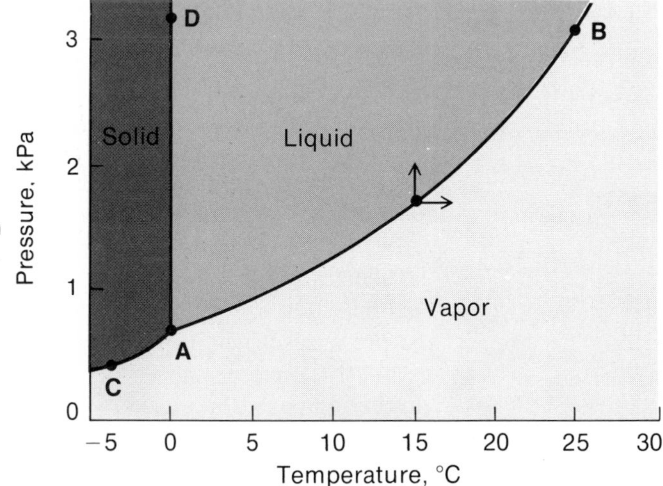

Figure 11.13 Phase diagram for water in the region from −5 to +30°C. The triple point of water occurs at A.

1. Line AB is a portion of the vapor pressure-temperature curve of liquid water. At any temperature and pressure along this line, liquid water is in equilibrium with water vapor. From the curve we see that at point A, these two phases are in equilibrium at 0°C and about 0.6 kPa (more exactly 0.01°C and 0.608 kPa). At B, corresponding to 25°C, the pressure exerted by the vapor in equilibrium with liquid water is about 3.2 kPa. If line AB were extended, we would read an equilibrium pressure of one atmosphere at 100°C, the normal boiling point of water. The line would end at 374°C, the critical temperature of water, where the pressure would be 218 atm.

2. Line AC represents the vapor pressure curve of ice. At any point along this line, such as −3°C and 0.4 kPa (point C) or 0°C and 0.6 kPa (point A), ice and vapor are in equilibrium with each other.

3. Line AD tells us the conditions of temperature and pressure at which liquid water is in equilibrium with ice. We shall have more to say a little later about the nature of this line.

The point A on the phase diagram represents the one point (temperature and pressure) at which all three phases, liquid, solid, and vapor, are in equilibrium with one another. Surprisingly enough, it is called, simply, the *triple point*. For water the triple point temperature is 0.01°C; at this temperature liquid water and ice have the same vapor pressure, 0.608 kPa.

The three areas of the phase diagram, labeled "solid," "liquid," and "vapor" in Figure 11.13, comprise regions of temperature and pressure at which only one phase is present. To show that this is the case, let us consider what happens to an equilibrium mixture of two phases when the pressure or temperature is changed. Specifically, let us start at the point on line AB indicated by a filled circle, where liquid water and vapor are in equilibrium at 15°C and 1.7 kPa. Intuition tells us that if we suddenly increase the pressure on this mixture, holding the temperature at 15°C, condensation will occur. The phase diagram, based on experimental data, confirms our hunch; by increasing the pressure at 15°C we move up into the liquid region (vertical arrow). Suppose, in another experiment, we hold the pressure constant but increase the temperature. It seems reasonable that this change would cause the liquid to vaporize. The phase diagram tells us that our reasoning is correct; an increase in temperature (horizontal arrow) corresponds to a shift into the vapor region.

Superficially at least, the phase diagrams of other pure substances resemble that of water. Nevertheless, differences in the triple point temperature and in the orientation of the solid-liquid line AD cause substances such as iodine, carbon dioxide and benzene to behave quite differently from water when they undergo the phase changes of sublimation or fusion.

Sublimation

The process by which a solid changes directly to its vapor without passing through the liquid state is called sublimation. Examination of Figure 11.13 tells tells us that a *solid will sublime at any temperature below the triple point when the pressure above it is reduced below the equilibrium vapor pressure.* To illustrate what this means, let us consider the conditions under which ice sublimes. On a cold, dry winter day, when the temperature is below 0°C and the pressure of water vapor in the air is less than the equilibrium value (0.608 kPa at 0°C), ice or snow "disappears" by sublimation. The rate of sublimation can be increased by evacuating the space above the ice. This is done commercially in the freeze-drying process for making dehydrated foods. A food such as eggs or lean meat is frozen, put into a vacuum chamber and evacuated to

We lose a lot of snow that way in Minnesota

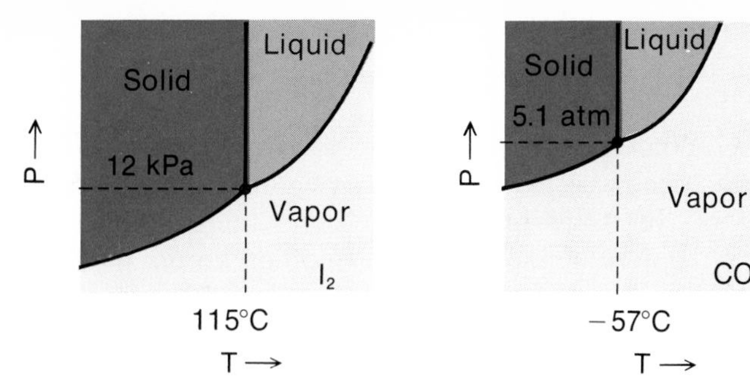

Figure 11.14 Phase diagrams of I_2 and CO_2 near their triple points. Since the triple point of CO_2 occurs at 5.1 atm, it is not possible to melt dry ice in an open container.

pressures of 0.1 kPa or less. Gradually the ice crystals formed on freezing sublime to give a product whose mass is only a fraction of that of the original food.

Iodine sublimes more readily than ice because its triple point pressure, 12 kPa, is much higher. If we heat iodine crystals in a test tube, we can see purple iodine vapor form. The vapor can be made to condense by bringing it into contact with a cold surface (Plate 8) such as a watch glass or tube filled with dry ice. A similar two-step process is used by organic chemists to purify volatile solids such as naphthalene or benzoic acid.

If we wish to sublime iodine by heating it in a test tube, we must be careful to stay below the triple point temperature, 115°C. If the temperature exceeds 115°C, the solid melts. No such problem arises with solid carbon dioxide (dry ice), which has a triple point pressure above one atmosphere (5.1 atm at −57°C) (Fig. 11.14). Liquid CO_2 cannot be prepared by heating dry ice in an open container; regardless of the temperature, the solid passes directly to the vapor.

How *could* you make liquid CO_2?

Fusion

The **melting point** of a solid or, conversely, the **freezing point** of a liquid is taken to be the temperature at which the solid and liquid phases of a pure substance are in equilibrium with each other. Melting points are ordinarily measured in an open container at atmospheric pressure. The melting points that we find recorded in the literature for various substances represent the temperature at which solid and liquid are in equilibrium with each other at an external pressure of 1 atm. For most substances the melting point under these conditions is virtually identical with the triple point. For water, the difference is only 0.01°C.

Although the effect of pressure upon melting point is very small, we are often interested in the direction of this effect. To decide whether the melting point will be increased or decreased by compressing the solid, we need only

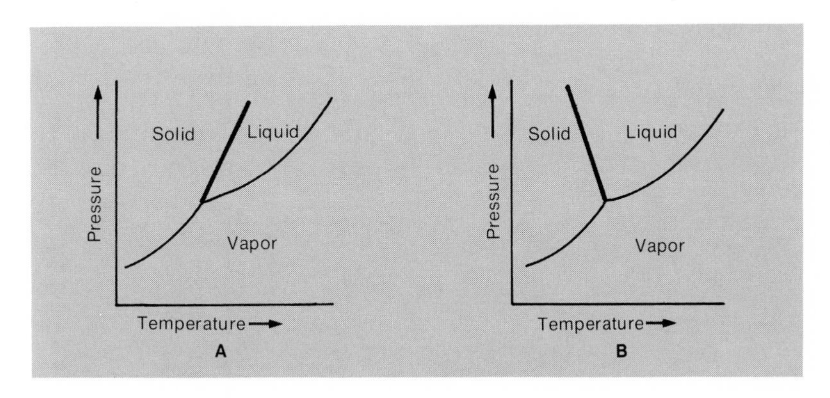

ure 11.15 Effect of pressure on the melting nt of a solid. When the solid is more dense than liquid, the behavior will be as in A, and the lting point will increase with pressure. If the uid is more dense, the melting point will decrease pressure increases, as shown in B. The slopes he heavy lines are highly exaggerated to show effect more clearly.

apply the principle that **an increase in pressure favors the formation of the more dense phase.** We can distinguish between two types of behavior (Fig. 11.15):

1. Where the solid is the more dense phase, as is the case with most substances, we have the situation shown in Figure 11.15*A*. An increase in pressure "favors" the solid; at higher pressures the solid becomes stable at temperatures above the normal melting point. In other words, the melting point is raised by an increase in pressure. The solid-liquid equilibrium line is inclined to the right, shifting away from the y axis as it rises.

2. Where the liquid is the more dense phase, the situation is as indicated in Figure 11.15*B*. The liquid-solid line is inclined to the left, toward the y axis. An increase in pressure favors the formation of liquid; i.e., the melting point is decreased by raising the pressure. One of the few substances that behaves this way is that remarkable substance, water. The fact that an ice cube floats in a glass of water shows that the liquid is the more dense phase. Experimentally, we find that the melting point of ice decreases by about 1°C for every 134 atm of applied pressure. This effect can be demonstrated by suspending two heavy masses from a wire stretched across a block of ice. The pressure exerted by the masses melts a thin layer of ice around the wire. As the wire falls, the pressure above it drops and the ice re-forms. (In time the wire passes completely through the block of ice, which appears to be unchanged.)

ΔH for Phase Changes

Experience tells us that melting a solid and vaporizing a liquid are endothermic processes. An ice cube melting in a glass of water absorbs heat from the water, thereby lowering its temperature. Water in a saucepan on a gas range stops boiling if we turn off the burner, thereby removing the source of heat. The reverse processes, freezing of a liquid or condensation of a vapor, are exothermic. The amounts of heat evolved for a given amount of substance are exactly equal to those absorbed when the phase change occurs in the opposite direction.

This is one reason burns from steam are so severe

The heat flow associated with a phase change taking place at constant pressure is equal to the enthalpy difference between the two phases. In Figure 11.16, we show what happens to the enthalpy of one mole of the substance H_2O as its

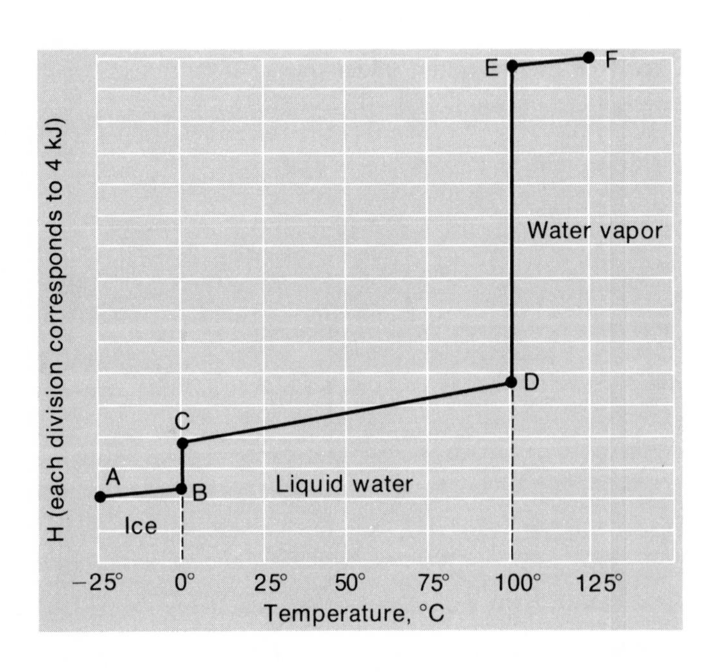

Figure 11.16 Molar enthalpy of water between −25°C and 125°C. Note that it takes a lot more heat to vaporize the liquid (D → E) than to melt the solid (B → C). When a solid, liquid, or gas is heated, its enthalpy always increases.

TABLE 11.3 ENTHALPY CHANGES INVOLVING H_2O

Per Mole		Per Gram	
ΔH_{fus} (at 0°C)	5.99 kJ/mol	ΔH_{fus} (at 0°C)	333 J/g
ΔH_{vap} (at 100°C)	40.63 kJ/mol	ΔH_{vap} (at 100°C)	2257 J/g
Molar heat cap. (s)	37 J/(mol · °C)	Specific heat (s)	2.05 J/(g · °C)
Molar heat cap. (l)	75 J/(mol · °C)	Specific heat (l)	4.18 J/(g · °C)
Molar heat cap. (g)	33 J/(mol · °C)	Specific heat (g)	1.84 J/(g · °C)

Molar heat capacity = *GMM* × specific heat

temperature is raised from −25°C to 125°C. The large vertical jumps in enthalpy at 0°C (BC) and at 100°C (DE) correspond to the **molar heat of fusion** and the **molar heat of vaporization, respectively.** We can also use Figure 11.16 to estimate how the molar enthalpies of the pure phases (ice, liquid water, and water vapor) change with temperature. The slopes of the nearly straight lines AB, CD, and EF give us the *molar heat capacities* [J/(mol · °C)] of ice, liquid water, and water vapor, respectively.

The molar enthalpy changes for H_2O described above are listed in Table 11.3. They can be estimated from Figure 11.16 or calculated from the corresponding quantities per gram, given at the right of the table, by multiplying by 18.0 g/mol. The information given in Table 11.3 is useful in a great many practical calculations, as illustrated by Example 11.4.

Example 11.4 An ice cube weighing 30.0 g melts in water originally at 20.0°C.

a. How much heat is absorbed from the water when the ice cube melts?
b. How many grams of water can be cooled to 0°C by the melting of the ice cube?

Solution

a. This calculation can be made from the molar heat of fusion or, perhaps more simply, from the heat of fusion per gram.

$$Q_{ice} = 30.0\ \text{g} \times 333\frac{\text{J}}{\text{g}} = 9990\ \text{J}$$

b. Since heat is transferred from the water to the ice, Q_{water} is equal in magnitude but opposite in sign to Q_{ice}:

$$Q_{water} = -Q_{ice} = -9990\ \text{J}$$

But, Q_{water} = (mass water in g)(specific heat)(temperature change).

Substituting numbers: −9990 J = (no. g water)[4.18 J/(g · °C)](0.0 − 20.0)°C

$$\text{Solving: no. g water} = \frac{-9990\ \text{J}}{[4.18\ \text{J/(g}\cdot{}^\circ\text{C)}](-20.0^\circ\text{C})} = 120\ \text{g water}$$

Notice that the amount of water cooled from 20°C to 0°C is four times the amount of ice melted (120 g vs. 30 g). This means that if you want to prepare a cold drink starting with water at room temperature, and make sure there is some ice left at equilibrium, you had better add somewhat more than one part of ice per four parts of water.

The amount of heat absorbed when one mole of a solid sublimes is referred to as the **heat of sublimation.** The heat of sublimation is often calculated by making use of Hess's Law which tells us that the enthalpy change depends only upon the final and initial states and is independent of the number of steps. Applied to sublimation, Hess's Law requires that, at a given temperature,

$$\begin{array}{ccc} \Delta H_{subl} = & \Delta H_{fus} + & \Delta H_{vap} \\ (s \rightarrow v) & (s \rightarrow l) & (l \rightarrow v) \end{array} \tag{11.7}$$

For water at 0°C, these quantities are 50.8 kJ/mol, 6.0 kJ/mol, and 44.8 kJ/mol, respectively.*

Our discussion of ΔH of phase changes has focused upon water, but the principles introduced can be applied to any substance. In Table 11.4, we list the enthalpy changes for fusion, vaporization, and sublimation for several different substances. Notice that in each case, $\Delta H_{vap} >> \Delta H_{fus}$; ΔH_{subl}, since it is the sum of the other two enthalpy changes, is the largest of the three.

TABLE 11.4 ΔH (kJ/MOL) FOR PHASE CHANGES

SUBSTANCE		mp(°C)	ΔH_{fus}*	ΔH_{vap}*	ΔH_{subl}*
Mercury	Hg	−39	2.3	56.5	58.8
Water	H_2O	0	6.0	44.8	50.8
Benzene	C_6H_6	5	10.7	34.6	45.3
Naphthalene	$C_{10}H_8$	80	19.3	40.6	59.9
Sodium chloride	NaCl	800	29	180	209

*Values given at the melting point.

11.5 NONEQUILIBRIUM PHASE CHANGES

The phase changes we have considered in this chapter have been assumed to take place under equilibrium conditions. In practice, the phase changes that we observe in the laboratory or in the world around us occur at temperatures and pressures at least slightly removed from equilibrium. Consequently, we often find that systems do not behave in quite the manner predicted by the phase diagram. A liquid being vaporized may not boil smoothly, particularly at low pressures. Instead, it may superheat to temperatures above the calculated boiling point and then "bump" and boil furiously. A very pure liquid on cooling does not start to freeze exactly at its freezing point; it has to be supercooled below that temperature before crystallization occurs.

The tendency of liquids to superheat or supercool reflects the difficulty of building a new phase within the body of the liquid. For a liquid to freeze, some centers, or nuclei, are needed on which crystallization can occur. These centers may be dust particles or small crystals of the substance itself, added to induce crystallization. It is sometimes possible to persuade a supercooled liquid to freeze by stirring vigorously, which may redistribute foreign particles through the body of the liquid. In particularly stubborn cases, scratching the sides of the container is sometimes effective, possibly because it removes tiny particles of glass from the walls and introduces them into the liquid. In a pure liquid, uncontaminated, unstirred and unscratched, crystallization nuclei are absent, and supercooling may be extensive. As an extreme example, very pure water can be cooled to −40°C without freezing.

*The heat of vaporization changes slightly with temperature; compare the value given in Table 11.3 for 100°C. For an explanation, see Problem 11.44.

For a liquid to boil, there must be centers at which bubbles can form. These centers may be microscopic bubbles of dissolved gas, dust particles, or sharp crystal edges or corners. When a liquid is to be distilled, a small quantity of an inert solid such as marble or porcelain is usually added. These solids have sharp edges on which vapor bubbles can form. In the absence of such solids, pronounced superheating can occur. Water that has been thoroughly purged of dust particles and gas bubbles has been heated in an open capillary tube to a temperature of 270°C without boiling.

PROBLEMS

11.1 *Vapor Pressure* A sample of water vapor exerts a pressure of 40.0 kPa at 100°C. If it is cooled to 50°C at constant volume, what will be its pressure, assuming no condensation occurs? Given that the equilibrium vapor pressure at 50°C is 12.3 kPa, decide whether or not condensation will occur.

11.2 *Vapor Pressure vs. Temperature* The vapor pressure of benzene is 0.358 atm at 50.0°C and one atmosphere at 80.0°C. Using Equation 11.4, calculate the heat of vaporization of benzene.

11.3 *Unit Cells* A certain metal with a face-centered cubic unit cell has an atomic radius of 0.140 nm. What is the dimension of the unit cell?

11.4 *Phase Diagrams* Referring to Figure 11.13, state what phase(s) is (are) present at

a. −5°C, 1.3 kPa
b. −5°C, 0.1 kPa
c. 25°C, 3.2 kPa

11.5 *ΔH for Phase Changes* Calculate ΔH for the following process, using data in Table 11.3:

$$2.00 \text{ g } H_2O(g, 100°C) \rightarrow 2.00 \text{ g } H_2O(l, 60°C)$$

11.6 How would you explain to an elementary school pupil why

a. liquids are less compressible than gases?
b. raindrops tend to have a nearly spherical form?
c. a tin can filled with steam at 100°C collapses when it is cooled?
d. his mother uses a pressure cooker when she wants to prepare foods quickly?

11.7 Design an experiment to

a. purify naphthalene by sublimation.
b. determine the atomic radius of a metal.
c. decide whether the melting point of a substance will increase or decrease with pressure.

11.8 Criticize each of the following statements:

a. The Ideal Gas Law can be used to determine how vapor pressure changes with temperature at constant volume.
b. In a cubic unit cell, atoms at the edges of the cube touch each other.
c. A simple way to make a semiconductor is to impregnate silicon with a small amount of germanium.

11.25 How would you explain to a mountain climber

a. how freeze-dried foods are made?
b. why it takes longer at high altitudes to cook food by boiling?
c. why a propane cylinder registers a constant pressure until it is almost empty?
d. why a slightly porous canvas bag keeps water cool?

11.26 Describe an experiment to

a. determine the heat of sublimation of benzene.
b. determine Avogadro's number by a method based on x-ray diffraction.
c. demonstrate superheating.

11.27 Criticize each of the following statements:

a. The Ideal Gas Law can be used to determine how vapor pressure changes with volume at constant temperature.
b. There are nine atoms per unit cell in a body-centered cubic structure.
c. In a crystal of NaCl, 50% of the interstitial ions will be Na^+.

11.9 The density of liquid water at 25°C is 0.998 g/cm^3.

a. What is the volume of one mole of liquid water at this temperature?
b. Using the Ideal Gas Law, calculate the volume of one mole of water vapor at 25°C and the equilibrium vapor pressure, 3.2 kPa.
c. Assuming that water molecules can be treated as spheres with a radius of 0.138 nm, find the volume of a mole of water molecules. ($V = 4\pi r^3/3$)
d. Using the answers to (a), (b), and (c), calculate the percentages of the total volume that are occupied by the molecules in liquid water and in water vapor at 25°C and 3.2 kPa.

11.10 One mole of liquid benzene, C_6H_6, is added to the apparatus shown in Figure 11.4. The apparatus is maintained at 20°C, at which temperature benzene has a vapor pressure of 10.0 kPa.

a. What will be the volume below the piston when the liquid is just vaporized?
b. What will be the pressure of the vapor when the volume is 100 dm^3?
c. What will be the pressure of the vapor when the volume is 300 dm^3?

11.11 The vapor pressure of benzene at various temperatures is as follows:

P (kPa)	10.0	15.7	24.1	35.9
t (°C)	20	30	40	50

A sample of benzene vapor at 50°C and 16.0 kPa pressure is cooled in a stoppered flask. Will it condense when cooled to 40°C? 30°C? 20°C? Estimate the temperature at which condensation will start.

11.12 Applying a graphical method to the data in Problem 11.11, calculate the heat of vaporization of benzene and compare it to the value given in Table 11.1.

11.13 The vapor pressure of bromine at 20°C is 23.3 kPa; its heat of vaporization is 30.0 kJ/mol. Estimate

a. the vapor pressure of bromine at 30°C.
b. the normal boiling point of bromine.

11.14 The normal boiling point of n-hexane is 69°C. Using Trouton's rule, estimate the heat of vaporization. Using this value, estimate the vapor pressure at 25°C.

11.28 The density of iron at 25°C is 7.87 g/cm^3.

a. What is the volume of one mole of solid iron at 25°C?
b. Taking an iron atom to be a sphere of radius 0.126 nm, calculate the volume of a mole of iron atoms.
c. From your answers to (a) and (b), calculate the fraction of "empty space" in an iron crystal at 25°C.

11.29 A mother uses a water vaporizer to raise the humidity in her child's bedroom, which is at 25°C and has a volume of 3.0×10^4 dm^3. Assuming that the air is completely dry to begin with and that no moisture leaves the room,

a. how many grams of water must she put in the vaporizer to ensure that the air becomes saturated with water vapor (vp = 3.2 kPa at 25°C)?
b. what will be the final pressure of water vapor in the room if she puts 800 g of water in the vaporizer?
c. what will be the final pressure of water vapor in the room if she puts 400 g of water in the vaporizer?

11.30 A sample of water vapor at 100°C and 56 kPa pressure is cooled in a container of constant volume.

a. What will be the pressure of the vapor at 85°C (vp water = 58 kPa)?
b. What will be the pressure of the vapor at 70°C (vp water = 31 kPa)?

11.31 Using the data in Appendix 1 for the vapor pressure of water at 20°C, 30°C, 40°C, 50°C, obtain ΔH_{vap} by a graphical method.

11.32 The vapor pressure of a certain liquid is doubled when its temperature rises from 25°C to 35°C. Calculate its heat of vaporization.

11.33 A certain organic compound has a normal boiling point of 80°C; its heat of vaporization is measured to be 444 J/g. Using Trouton's rule, estimate the molecular mass of the compound.

11.15 The normal boiling point of ammonia is −33°C. Its vapor pressure at 25°C is 10 atm. Which of the following statements concerning ammonia must be true?

a. A tank of ammonia at 25°C which has a pressure of 5 atm must contain liquid NH_3.
b. A tank of ammonia at 25°C which has a pressure of 1 atm cannot contain liquid NH_3.
c. The critical temperature of NH_3 must be greater than 0°C.

11.16 When x-rays with a wavelength of 0.0500 nm are diffracted by a certain crystal, the angle of diffraction of the first order beam (n = 1) is found to be 10.0°.

a. Calculate the distance between the layers of atoms responsible for the diffraction.
b. What will be the angle of diffraction for the second order beam (n = 2)?

11.17 Consider the following two-dimensional pattern (the lines indicate bonds, the points atoms).

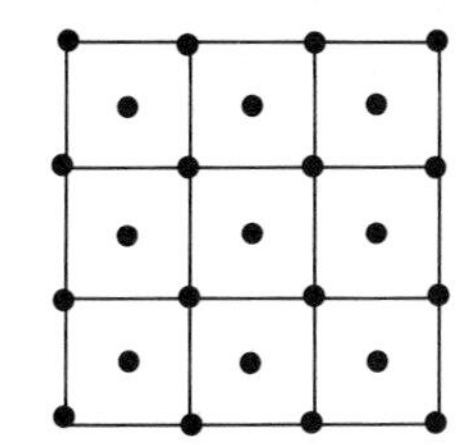

a. What is the unit cell in this structure?
b. How many atoms should be assigned to the unit cell?

11.18 Potassium (at. rad. 0.231 nm) crystallizes in a body-centered cubic structure.

a. How many atoms are there per unit cell?
b. What is the length of one side of the unit cell?
c. What is the volume of the unit cell?
d. What is the molar volume of potassium?

11.19 In the CsCl structure shown in Figure 11.10, what is the distance between centers of the Cl^- ions at the edges of the cube?

11.20 Suppose a small number of atoms of the following elements were introduced into a silicon crystal. Would you expect to form a semiconductor and, if so, would it be of the n- or p-type?

a. Al
b. Sb
c. Se
d. Sn

11.34 The critical point of sulfur dioxide is 157°C, 78 atm. Liquid sulfur dioxide has a vapor pressure of 3.8 atm at 25°C. Which of the following statements must be true?

a. Sulfur dioxide is a gas at 25°C and atmospheric pressure.
b. A tank of sulfur dioxide at 25°C can have a pressure of 5 atm.
c. Sulfur dioxide gas cooled to 150°C and 80 atm pressure will condense.
d. The normal boiling point of sulfur dioxide lies between 25 and 157°C.

11.35 The Braggs used diffraction patterns obtained with sodium chloride to determine the wavelengths of x-rays. The distance between one set of diffraction planes in NaCl is 0.276 nm. Using x-rays given off when copper is bombarded by electrons, the diffraction angles are found to be 16.2°, 33.8°, 56.7° (n = 1, 2, 3). Calculate the wavelength of the x-rays.

11.36 Consider the structure, repeated indefinitely in two dimensions:

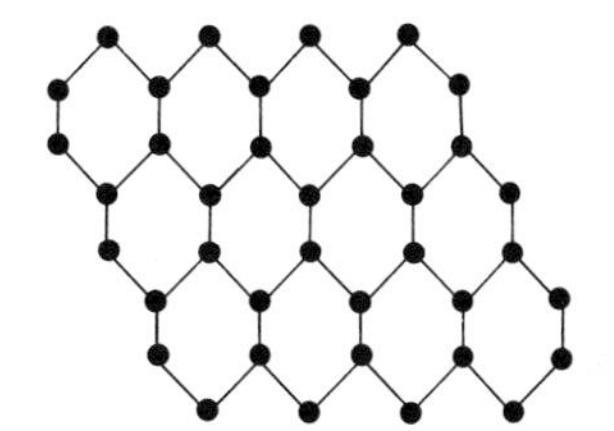

a. How many unit cells are shown?
b. How many atoms should be assigned to a unit cell (each dot is an atom)?

11.37 Gold crystallizes in a face-centered cubic structure. It has an atomic radius of 0.144 nm. What is the dimension of the unit cell?

11.38 In the LiCl structure shown in Figure 11.10, where the Cl^- ions form a face-centered cubic unit cell, what is the length of an edge of the cube? Show that it is possible to fit a Li^+ ion between Cl^- ions along an edge of the cube. Could a Na^+ ion fit?

11.39 From the standpoint of size, which of the atoms in Problem 11.20 would fit most readily into a silicon lattice? a germanium lattice? (See table of atomic radii, Appendix 2.)

11.21 The triple point of benzene is 5°C, 0.028 atm. The density of liquid benzene is 0.894 g/cm³; that of the solid is 1.005 g/cm³. The normal boiling point of benzene is 80°C; its critical point is 289°C, 48 atm. Sketch a phase diagram for benzene in the region 0°C to 300°C.

11.22 Using Figure 11.13, predict the phase(s) that will be present at a pressure of 2.7 kPa if the temperature is

a. 10°C
b. 25°C
c. 22°C
d. −3°C

11.23 Calculate ΔH for the following processes:

a. freezing 6.00 g of $H_2O(l)$.
b. heating 6.00 g of $H_2O(l)$ from 20.0 to 80.0°C.
c. condensing 6.00 g of $H_2O(g)$ at 100°C.

11.24 If an ice cube at 0°C, weighing 5.00 g, is added to 20.0 g of liquid water at 100°C, what will be the final temperature?

11.40 From information given in this chapter, draw a phase diagram for water in the region −10°C to 400°C.

11.41 Using Figure 11.13, predict what, if anything, will happen if liquid water at 10°C and 2.7 kPa is

a. expanded to 0.7 kPa at constant T.
b. compressed to 3.3 kPa at constant T.
c. cooled to −5°C at constant P.
d. heated to 18°C at constant P.

11.42 Using Table 11.4, calculate ΔH for

a. melting 1.00 g of benzene.
b. subliming 12.0 g of Hg(s) at −39°C.
c. vaporizing 2.16 mol of benzene at 5°C.

11.43 How much water can be heated from 20.0 to 100.0°C by condensing 1.00 g of steam, originally at 100.0°C, in the water?

*11.44 The heat of vaporization of a substance represents the difference in enthalpy between the gas and liquid phases, i.e., $\Delta H_{vap} = H_{gas} - H_{liquid}$. For water (Fig. 11.16), ΔH_{vap} at 100°C is given by the height of the vertical line DE, and is equal to 40.6 kJ/mol. ΔH_{vap} at other temperatures can be obtained by extrapolating the straight lines EF and CD and measuring the distance between them at the desired value of T.

a. Following this procedure, show that ΔH_{vap} at 0°C is 44.8 kJ/mol.
b. Calculate ΔH_{vap} at 25°C; 125°C.
c. Looking at the data in Table 11.3, can you explain why ΔH_{vap} decreases with increasing temperature?

*11.45 Calculate the fraction of empty space or "free volume" in a crystal made of simple cubic unit cells; body-centered cubic unit cells; face-centered cubic unit cells. Consider the volume of the atoms assigned to each cell in comparison to the volume of the cell itself.

*11.46 It has been suggested that the pressure exerted on a skate blade is sufficient to melt the ice beneath it and form a thin film of water which makes it easier for the blade to slide over the ice. Assume that the skater weighs 150 lb and that the blade has an area of 0.10 in². Calculate the pressure exerted on the blade (15 lb/in² = 1 atm). From information in the text, calculate the decrease in melting point at this pressure. Comment on the plausibility of this explanation and suggest another mechanism by which the water film might be formed.

12

SOLUTIONS

In previous chapters we have dealt almost entirely with the structures and properties of pure substances. From now on we will be dealing increasingly with solutions, where most chemical reactions occur. In this chapter we will develop the background for a discussion of solution chemistry by studying the structure and physical properties of solutions.

12.1 INTRODUCTION

A solution can be defined as a homogeneous mixture of two or more substances. Seawater, filtered if necessary to remove seaweed, surfboards and sharks, is a solution of various salts in water. "Clean air," a rare commodity these days, is a solution composed of nitrogen, oxygen, argon and small amounts of other gases.

From a structural standpoint, homogeneity implies that the particles of the different species present are of molecular size (~5 nm or less in diameter) and that they are distributed in a more or less random pattern. To illustrate, in a solution formed by adding ethyl alcohol to water, the particles are individual molecules, perhaps 1 nm in diameter. These molecules occur randomly throughout the liquid; there is no tendency for large clusters of alcohol or water molecules to come together in one region of the solution. In a *suspension,* either or both of these criteria are not met. Fog is formed by the clustering of thousands upon thousands of water molecules into droplets large enough to be visible; it is considered to be a suspension of water in air rather than a true solution.

Many salad dressings are liquid-liquid suspensions

Solutions may exist in any of the three states of matter: gas, solid or liquid. Air is the most familiar example of a gaseous solution. Solid solutions are relatively common; many metallic alloys fall into this category. The "nickel" coin is actually a solid solution containing 25 mass per cent of nickel dissolved in copper. Twelve-carat gold is a solid solution containing equal parts by mass of gold and silver (pure gold is 24 carat).

Throughout this chapter we will be dealing primarily with liquid solutions, which are often further subdivided according to the physical states of the pure components. We may have a solution in which both components are liquid (e.g., a martini). One component may be a gas (soda water) or a solid (salt water).

In a gas-liquid or solid-liquid solution, we ordinarily refer to the liquid as the "solvent" and the other component (gas or solid) as the "solute." This choice of words reflects the way in which we visualize the solution process; it is natural to think of carbon dioxide or sodium chloride as "dissolving" in water rather than the reverse. If both components are liquids, the designations solute and solvent are more ambiguous. Frequently the component present in the greater amount is called the solvent. In a solution of 1 g of ethyl alcohol in 100 g of water, we would probably speak of the water as the solvent and the alcohol as the solute.

TABLE 12.1 CONCENTRATIONS OF LABORATORY ACID AND BASE SOLUTIONS

Solute			Moles Solute per dm^3	% Solute (by Mass)	Density (g/cm^3)
Hydrochloric acid	HCl	conc.	12	36	1.18
		dilute	6	20	1.10
Nitric acid	HNO_3	conc.	16	72	1.42
		dilute	6	32	1.19
Sulfuric acid	H_2SO_4	conc.	18	96	1.84
		dilute	3	25	1.18
Ammonia	NH_3*	conc.	15	28	0.90
		dilute	6	11	0.96

*Often labeled "NH_4OH."

If the amounts are more nearly equal, the choice becomes less clear-cut and must be specified.

Several different adjectives are used to indicate in a qualitative way the relative amounts of the components present in a solution. We may describe a solution containing a small amount of solute as being "dilute"; another solution, containing more solute, might be referred to as "concentrated." In a few cases these terms have taken on quantitative significance. Dilute and concentrated solutions of certain acids and bases, labeled as such in the laboratory, have the concentrations specified in Table 12.1.

Relative concentrations of solutions are sometimes expressed in a different way by using the terms "saturated," "unsaturated," and "supersaturated." A **saturated** solution of a solute, X, is one which is in equilibrium with undissolved X. To prepare a saturated solution of sodium chloride in water at 25°C, we could bring an excess of the solid into contact with water at that temperature and stir, vigorously and repeatedly, until no more NaCl goes into solution. The resulting solution, which contains 36.5 g of sodium chloride in 100 g of water, is said to be saturated with sodium chloride. Addition of further solid sodium chloride fails to change its concentration in solution.

An **unsaturated** solution contains a lower concentration of solute than a saturated solution. A solution at 25°C containing less than 36.5 g of sodium chloride in 100 g of water is said to be unsaturated with respect to sodium chloride. Such a solution is not in a state of true equilibrium; if solute is added, it dissolves to approach a saturated solution.

Freshly opened soda pop is a supersaturated solution of CO_2 in water

A **supersaturated** solution represents an unstable situation in which the solution actually contains more than the equilibrium concentration of solute. Supersaturated solutions most commonly arise when a hot, saturated solution of a solid in a liquid is allowed to cool. To illustrate the principle involved, consider sodium acetate, $NaC_2H_3O_2$, which readily forms supersaturated water solutions.

At 20°C a saturated solution contains 46.5 g of sodium acetate in 100 g of water; at higher temperatures the solubility of sodium acetate is considerably greater. If we heat 80 g of this salt with 100 g of water until it is completely dissolved (a temperature of about 50°C is required) and then cool carefully, without shaking or stirring, to 20°C, the excess solute remains in solution. This supersaturated solution can be maintained indefinitely so long as there are no nuclei upon which crystallization can start. If a small seed crystal of sodium acetate is added, crystallization quickly takes place until equilibrium is attained by the formation of a saturated solution.

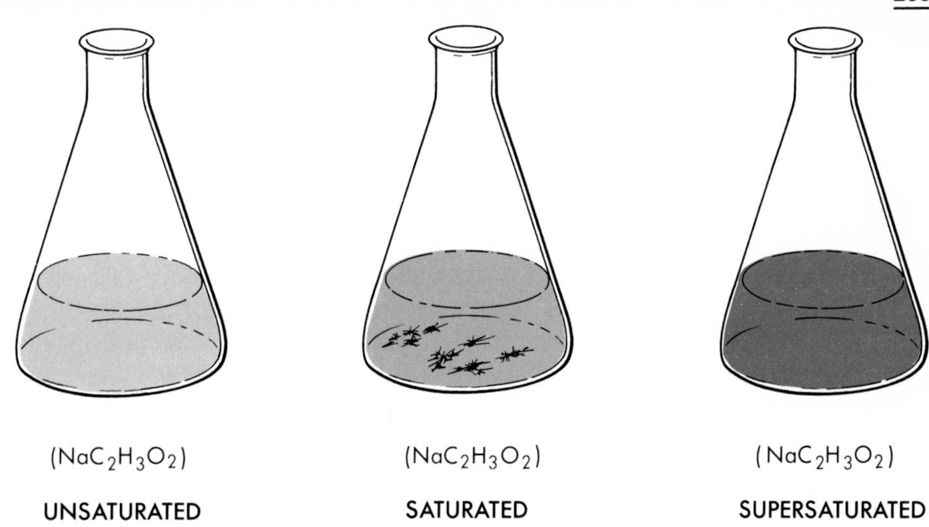

Figure 12.1 Unsaturated, saturated, and supersaturated solutions of sodium acetate. Only the saturated solution is in equilibrium with undissolved solute. If the supersaturated solution is disturbed, say by adding a crystal of sodium acetate, crystallization of excess solute will occur rapidly until equilibrium between solid and solution is established.

12.2 CONCENTRATION UNITS

The properties of solutions depend to a large extent upon the relative amounts of solute and solvent present. For this reason, in any quantitative work involving solutions it is important to specify concentrations. This can be done by stating either the relative amounts of solute and solvent or, alternatively, the amount of one component relative to the total amount (mass or volume) of solution.

One of the simplest ways to specify solution concentrations is to give the **mass percentages** of the different components. We might, for example, describe "concentrated hydrochloric acid" (Table 12.1) by saying that it contains 36% by mass of hydrogen chloride. This tells us that in 100 g of concentrated hydrochloric acid, we can expect to find 36 g of HCl and 64 g of H_2O; in 1.00 g of the acid there will be 0.36 g HCl and 0.64 g H_2O, and so on.

In general, the mass percentage of component A in a solution is given by

$$\% \text{ of A} = \frac{\text{mass A}}{\text{total mass of solution}} \times 100 \tag{12.1}$$

Example 12.1 Using Table 12.1, calculate

a. the number of grams of NH_3 in 240 g of concentrated ammonia.
b. the number of grams of HNO_3 in 10.0 cm^3 of dilute nitric acid.

Solution

a. From the table, we note that the mass percentage of NH_3 is 28%. Rearranging Equation 12.1, we have

$$\text{mass } NH_3 = \frac{\% \text{ of } NH_2}{100} \times \text{total mass of solution}$$
$$= 0.28 \times 240 \text{ g} = 67 \text{ g } NH_3$$

b. Before we can calculate the mass of HNO_3, we must obtain the total mass of 10.0 cm^3 of dilute nitric acid. Noting from Table 12.1 that the density of dilute HNO_3 is 1.19 g/cm^3, we have

$$\text{total mass of solution} = 10.0\ \text{cm}^3 \times 1.19\ \frac{\text{g}}{\text{cm}^3} = 11.9\ \text{g}$$

Now we proceed as in (a) to find the mass of HNO_3:

$$\text{mass } HNO_3 = \frac{\%\ HNO_3}{100} \times \text{total mass of solution}$$
$$= 0.32 \times 11.9\ \text{g} = 3.8\ \text{g } HNO_3$$

In Chapter 3 we stressed the usefulness of the mole concept in discussing the properties of pure substances. In this chapter we shall have occasion to use three different concentration units in which the amount of solute is expressed in moles. The first of these is the **mole fraction;** the mole fraction of component A, which we shall designate as X_A, is given by the equation

$$X_A = \frac{\text{no. of moles of A}}{\text{total no. moles of all components}} \qquad (12.2)$$

and is ordinarily expressed as a decimal (Example 12.2).

Example 12.2 What are the mole fractions of benzene, C_6H_6, and toluene, C_7H_8, in a solution prepared by adding 500 g of benzene to 500 g of toluene?

Solution Since 1 mol C_6H_6 = 6(12.0 g) + 6(1.0 g) = 78.0 g
and 1 mol C_7H_8 = 7(12.0 g) + 8(1.0 g) = 92.0 g

$$\text{no. of moles of } C_6H_6 = 500\ \text{g} \times \frac{1\ \text{mol}}{78.0\ \text{g}} = 6.41\ \text{mol}$$

$$\text{no. of moles of } C_7H_8 = 500\ \text{g} \times \frac{1\ \text{mol}}{92.0\ \text{g}} = 5.43\ \text{mol}$$

Using Equation 12.2:

$$X_{C_6H_6} = \frac{6.41}{6.41 + 5.43} = 0.541$$

$$X_{C_7H_8} = \frac{5.43}{5.43 + 6.41} = 0.459$$

Sometimes we refer to the mole percent of a component in solution, which is 100 times the mole fraction. Thus in the solution referred to in Example 12.2, we would have 54.1 mol % of benzene and 45.9 mol % of toluene.

Molality (m) is defined as the number of moles of solute per kilogram of solvent

$$m = \frac{\text{no. of moles solute}}{\text{no. kilograms solvent}} \qquad (12.3)$$

Example 12.3 Calculate the molality of a solution prepared by dissolving 1.00 g of urea, $CO(NH_2)_2$, in 48.0 g of water.

Solution To obtain the number of moles of solute, we note that

$$1\ \text{mol } CO(NH_2)_2 = 12.0\ \text{g} + 16.0\ \text{g} + 2(14.0\ \text{g}) + 4(1.0\ \text{g}) = 60.0\ \text{g}$$

$$\text{no. moles solute} = 1.00\ \text{g urea} \times \frac{1\ \text{mol urea}}{60.0\ \text{g urea}} = 0.0167\ \text{mol urea}$$

no. kilograms solvent = 48.0/1000 = 0.0480

$$m = \frac{0.0167}{0.0480} = 0.348$$

As Example 12.3 implies, we can prepare a solution of a given molality by dissolving a known mass of solute in a predetermined mass of solvent. The precision with which the concentration is known is limited only by that of the balance used to make the weighings. An advantage of molality as a concentration unit is that it is independent of temperature; a one molal solution prepared at 20°C will retain the same molality at 100°C, provided there is no loss of solute or solvent on heating.

Concentrations of reagents in the general chemistry laboratory are most often specified by stating the number of moles per cubic decimetre (litre) of solution. Using the symbol M to represent this concentration unit* we have:

$$M = \frac{\text{no. moles solute}}{\text{no. cubic decimetres solution}} \quad (12.4)$$

m and M are almost equal in dilute water solution. Why?

Example 12.4

a. How would you prepare 25 dm³ of solution containing 0.10 mol/dm³ of $BaCl_2$, starting with solid $BaCl_2$?
b. What volume of the solution in (a) would you take to get 0.020 mol of $BaCl_2$?

Solution

a. Using Equation 12.4, we can calculate the number of moles of $BaCl_2$ required.

$$\text{no. moles } BaCl_2 = (M\ BaCl_2) \times (\text{no. cubic decimetres solution})$$

$$= 0.10\ \frac{\text{mol}}{\text{dm}^3} \times 25\ \text{dm}^3 = 2.5\ \text{mol}$$

To find out how many grams of $BaCl_2$ we should weigh out, we note that one mole of $BaCl_2$ weighs 137 g + 2(35.5) = 208 g.

$$\text{no. grams } BaCl_2 = 2.5\ \text{mol} \times \frac{208\ \text{g}}{1\ \text{mol}} = 520\ \text{g}$$

It follows that we should weigh out 520 g of $BaCl_2$ and stir with sufficient water to give a final volume of 25 dm³.

b. Solving Equation 12.4 for volume, we have

$$\text{no. cubic decimetres solution} = \frac{\text{no. moles } BaCl_2}{M\ BaCl_2}$$

$$= \frac{0.020\ \text{mol}}{0.10\ \text{mol/dm}^3} = 0.20\ \text{dm}^3\ (\text{i.e., } 200\ \text{cm}^3)$$

One way to make up a solution of the desired concentration in moles per cubic decimetre is to start with the calculated mass of pure solute and add enough water to give the required volume (Example 12.4a). If high accuracy is required, the solu-

*This unit is commonly referred to as the "molarity" of a solution. We will avoid the use of this term because of possible confusion with a new concentration unit which is coming into use in connection with SI: moles per cubic metre (mol/m³). Note that 1 mol/m³ = 0.001 mol/dm³.

tion may be prepared in a volumetric flask (Fig. 1.1, p. 3). Another method, which is often more convenient, is to start with a concentrated solution and dilute it with water. The calculations involved are illustrated in Example 12.5.

Example 12.5 How would you prepare 100 cm^3 of 2.0 *M* HCl, starting with concentrated hydrochloric acid (12 *M*)?

Solution Clearly, the more concentrated solution (12 *M*) should be diluted with water to give a 2.0 *M* solution. The question is: What volume of 12 *M* HCl should we start with to prepare 100 cm^3 of 2.0 *M* HCl? The key to answering this question is to realize that *the number of moles of solute is not changed by dilution.* In the final solution, we want

$$2.0 \frac{\text{mol}}{\text{dm}^3} \times 0.100 \text{ dm}^3 = 0.20 \text{ mol}$$

We must then take a sufficient volume of 12 *M* HCl to give us 0.20 mol. Proceeding as in Example 12.4b:

$$\text{volume } 12\ M \text{ HCl} = \frac{\text{no. moles HCl}}{M \text{ HCl}} = \frac{0.20 \text{ mol}}{12 \text{ mol/dm}^3} = 0.017 \text{ dm}^3 \text{ (17 cm}^3\text{)}$$

We conclude that we should start with 17 cm^3 of 12 *M* HCl and dilute with water to a final volume of 100 cm^3 to give a 2.0 *M* solution.

12.3 PRINCIPLES OF SOLUBILITY

Theory has not yet progressed to the point where we can predict the solubilities of even the simplest solutes in common solvents. The best that we can do is to apply general principles, based on structural considerations, to predict the relative solubilities of different solutes in a common solvent or, vice versa, the relative solubilities of a given solute in a series of solvents. The following brief discussion of solubility principles deals with solutions in the liquid state and is subdivided according to whether the solute is a liquid, a gas or a solid. Solubilities of ionic solutes in water will be discussed in Chapter 18.

Liquid-Liquid

In discussing the solubility of two liquids in each other, it is sometimes stated that "like dissolves like." A more meaningful way of expressing this idea is to say that **liquids** with similar structures and consequently **with intermolecular forces of about the same type and magnitude will be soluble in each other in all proportions.** To illustrate, consider the liquid aliphatic hydrocarbons pentane, C_5H_{12}, and hexane, C_6H_{14}, which are completely miscible with each other. Molecules of these nonpolar substances are held together by dispersion forces of about the same magnitude. The forces between C_5H_{12} molecules in pure liquid pentane are about as strong as those between C_5H_{12} and C_6H_{14} molecules in a solution of pentane in hexane. A pentane molecule readily passes into solution in hexane because it undergoes no significant change in environment in the solution process.

Moderate differences in polarity between solute and solvent seem to have little effect on solubility. Chloroform, $CHCl_3$, which is a polar molecule, and carbon tetrachloride, CCl_4, which is nonpolar, are soluble in each other in all proportions. Moreover, chloroform and carbon tetrachloride show similar solvent properties; both are commonly used to dissolve a variety of organic compounds including hydrocarbons, fats, and greases. This implies that the intermolecular forces in $CHCl_3$ and CCl_4 are nearly equal, despite the considerable difference in polarity. Apparently dipole forces make only a minor contribution in $CHCl_3$; it is about as easy to "break into" the liquid structure of chloroform as it is in carbon tetrachloride.

We commonly observe that nonpolar substances have very small water solubilities. Petroleum, a complex mixture of hydrocarbons, spreads out in a thin film on the surface of a lake or an ocean rather than dissolving in the water. A typical nonpolar hydrocarbon, pentane, has a mole fraction solubility in water of only 0.000 03. This is readily understood in terms of the structure of liquid water. In order to dissolve appreciable quantities of pentane in water, it would be necessary to break the hydrogen bonds holding water molecules together. There is no compensating attractive force between C_5H_{12} and H_2O to supply the energy required to break into the water structure.

Of the relatively few organic liquids which dissolve readily in water, the majority are oxygen-containing compounds of low molecular mass. Two familiar examples are the alcohols containing one and two carbon atoms,

```
      H                        H   H
      |                        |   |
  H—C—O—H                  H—C—C—O—H
      |                        |   |
      H                        H   H
 Methyl alcohol            Ethyl alcohol
```

both of which are soluble in water in all proportions. Methyl and ethyl alcohol each contain an —OH group, as does water. Even more important, both these compounds are known to be hydrogen-bonded in the liquid state. Consequently it is hardly surprising that they dissolve readily in water. One would expect the intermolecular forces between alcohols and water in solution to be roughly comparable to those in the pure liquids.

The presence or absence of H bonds has a pronounced effect on solubility

As the number of carbon atoms in the alcohol molecule increases, we find that the solubility in water decreases. The compound n-butyl alcohol, C_4H_9OH, has a limited solubility in water; its mole fraction in a saturated water solution at 20°C is only about 0.02. Octyl alcohol, $C_8H_{17}OH$, is extremely insoluble (X = 0.0008 in a saturated water solution at 20°C). The same trend is observed with many other types of organic compounds; there is a general tendency for solubility to decrease with an increase in chain length. This effect can be explained most simply in terms of the large number of hydrogen bonds in water that must be broken if a long-chain molecule is to be inserted into the water structure.

Solid-Liquid

In contrast to liquid-liquid pairs, where complete miscibility is common, solids are always found to have limited solubilities in liquids. If we add iodine (*mp* = 115°C) to carbon tetrachloride at 25°C, a saturated solution is formed when the mole fraction of I_2 is only 0.011. On the other hand, bromine, Br_2, which is a liquid at room temperature, is infinitely soluble in carbon tetrachloride; it is impossible to form a saturated solution of these two liquids.

The limited solubility of solid solutes in liquid solvents is explained most simply in terms of a difference in magnitude of intermolecular forces in the two species. The fact that a substance is a solid at 25°C implies that its intermolecular forces are considerably stronger than those in another substance which is a liquid at the same temperature. Thus, we would expect the dispersion forces in $I_2(s)$ to

TABLE 12.2 SOLUBILITIES OF SOLID HYDROCARBONS IN BENZENE AT 25°C*

Solute	Melting Point (°C)	X Solute
Anthracene	218	0.008
Phenanthrene	100	0.21
Naphthalene	80	0.26
Biphenyl	69	0.39

*Mole fraction of solid in saturated solution.

be an order of magnitude greater than those in CCl_4(l). In view of this difference, it is hardly surprising that the solubility of iodine in carbon tetrachloride is relatively low.

Following this line of reasoning, we would predict that the closer a solid is to its melting point, the more closely its intermolecular forces should match those of a liquid, and hence the more soluble it should be. Putting it another way, we would expect low-melting solids to be more soluble at a given temperature than high-melting solids of similar structure. This prediction is confirmed by data such as those in Table 12.2. Of the four hydrocarbon solutes listed, the most soluble in benzene is biphenyl, which is only 44°C below its melting point at the solution temperature, 25°C. The least soluble is the high-melting solid, anthracene, which is nearly 200°C below its melting point at room temperature.

To estimate the relative solubilities of a given solid in different solvents, we take into account the principles discussed earlier in connection with liquid-liquid solubilities. Nonpolar or slightly polar solids are most soluble in solvents of low polarity and least soluble in hydrogen-bonded solvents such as water. An example is the well known pesticide DDT, which has a structure similar to that of carbon tetrachloride or chloroform:

H
|
Cl—⟨benzene ring⟩—C—⟨benzene ring⟩—Cl
|
Cl—C—Cl
|
Cl

DDT

Cl
|
Cl—C—Cl
|
Cl

Carbon tetrachloride

H
|
Cl—C—Cl
|
Cl

Chloroform

Would DDT be soluble in CCl_4?

DDT, like carbon tetrachloride or chloroform, is quite soluble in nonpolar or slightly polar organic solvents. This explains why it tends to concentrate in the fatty tissue of fish, birds and game, often with lethal effects. In contrast, DDT is almost quantitatively insoluble in water; only about 10^{-6} g of DDT dissolves in a cubic decimetre of water. Its low molecular solubility contributes to the persistence of this pesticide in the environment; DDT is not washed out of contaminated soil even by repeated rainfalls.

Gas-Liquid

Following the reasoning outlined for solid-liquid and liquid-liquid solutions, we arrive at the following conclusions regarding the solubilities of gases in liquids:

1. The higher the boiling point of a gas, the more nearly its intermolecular forces will approach in magnitude those of a liquid. Consequently, in comparing the solubilities of different gases in the same solvent, we ordinarily find that the one with the highest boiling point is the most soluble.

2. The best solvent for a given gas will be the one whose intermolecular forces are most similar in nature to those of the gaseous solute.

Both of these principles are illustrated by the solubility data for the noble gases given in Table 12.3. Note the steady increase in solubility with boiling point (He < Ne < Ar < Kr < Xe < Rn) for both solvents. The reduced solubility of all these gases in water as compared to benzene reflects the strong intermolecular hydrogen bonding in water; in the nonpolar solvent benzene, as in the noble gases themselves, the attractive forces between particles are of the dispersion type.

This same trend would hold for other solvents

TABLE 12.3 SOLUBILITY OF THE NOBLE GASES IN BENZENE AND WATER AT 25°C AND 1 ATM*

Gas	Boiling Point (°C)	Solvent: Benzene	Solvent: Water
He	−269	0.76×10^{-4}	0.069×10^{-4}
Ne	−246	1.14×10^{-4}	0.082×10^{-4}
Ar	−186	8.9×10^{-4}	0.25×10^{-4}
Kr	−152	27.3×10^{-4}	0.45×10^{-4}
Xe	−109	110×10^{-4}	0.86×10^{-4}
Rn	−62	310×10^{-4}	1.63×10^{-4}

*Mole fraction of gas in saturated solution.

12.4 EFFECT OF TEMPERATURE AND PRESSURE ON SOLUBILITY

The mutual solubilities of two substances A and B depend not only upon their structures and physical properties, but also upon the external conditions of temperature and pressure. The effects of these two variables upon solubility are readily explained if we regard the solution process as an equilibrium:

$$A + B \rightleftarrows \text{solution}$$

and apply the principles discussed in Chapter 11 concerning the influence of temperature and pressure upon physical equilibria.

Temperature

In any equilibrium, an increase in temperature favors the endothermic process. This means that if heat is absorbed when A dissolves in B

$$A + B \rightleftarrows \text{solution} \qquad \Delta H > 0$$

an increase in temperature will increase the solubility. Conversely, if dissolving A in B evolves heat

$$A + B \rightleftarrows \text{solution} \qquad \Delta H < 0$$

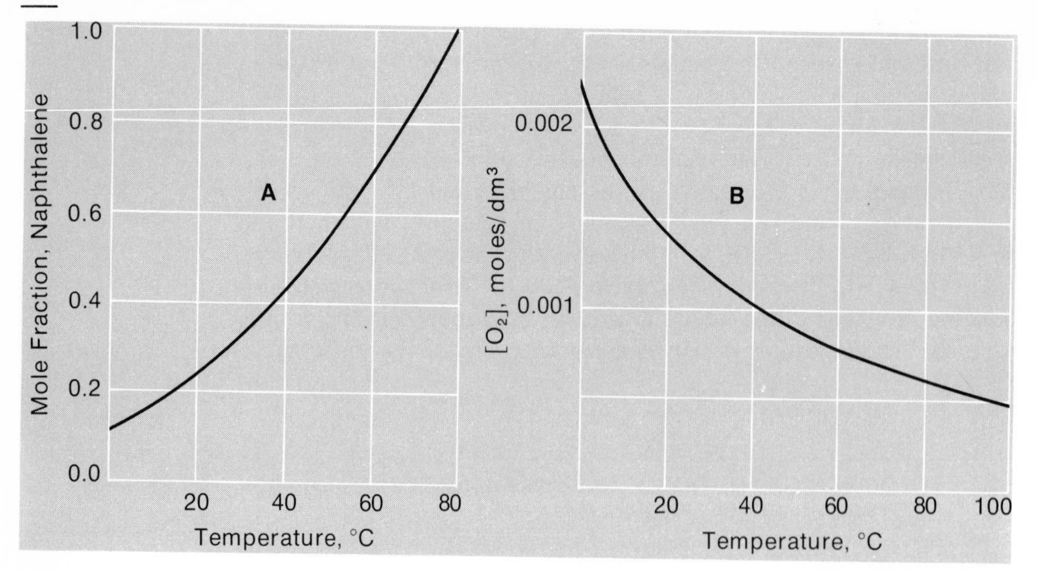

Figure 12.2 *A*, Solubility of naphthalene in benzene. Solubility increases with temperature, becoming complete at the melting point of naphthalene (80°C). *B*, Solubility of O_2 in water at 1 atm. As with most gases, the solubility decreases markedly as the temperature goes up.

an increase in temperature will favor the reverse process; i.e., it will reduce the solubility.

Dissolving a solid in a liquid is almost always an endothermic process, primarily because heat (in an amount about equal to the heat of fusion) must be absorbed to break down the crystal lattice. Consequently, the solubility of a solid in a liquid solvent usually increases with temperature. An example is naphthalene, whose mole fraction solubility in benzene increases from 0.26 at 25°C to 0.54 at 50°C and to 1.0 (i.e., infinite solubility) at its melting point, 80°C (Figure 12.2*A*).

Heat is given off when a gas condenses

Dissolving a gas in a liquid usually (but not always) evolves heat; the enthalpy change approximates that for the condensation of the gas. As a result, we ordinarily find that gas solubility decreases with an increase in temperature. This general rule is followed by all gases dissolving in water; it accounts for the observation that bubbles of air form when a beaker of water is heated. By the time the boiling point has been reached, virtually all of the air dissolved at room temperature is expelled. The reduced solubility of oxygen in water at high temperatures (Figure 12.2*B*) is a major factor in "thermal pollution" of water supplies. The lowered oxygen concentration makes it difficult for fish and other aquatic life to survive in warm water.

Pressure

Pressure has a major effect on solubility only for gas-liquid systems. We find that at moderate pressures the solubility of a gas is directly proportional to its partial pressure in the gas phase over the solution,

$$C_g = kP_g \tag{12.5}$$

where P_g is the partial pressure of the gas over the solution, C_g is its concentration in solution, and k is a constant characteristic of the particular gas-liquid system. A simple kinetic explanation of Equation 12.5 (Henry's Law) is suggested in Figure 12.3.

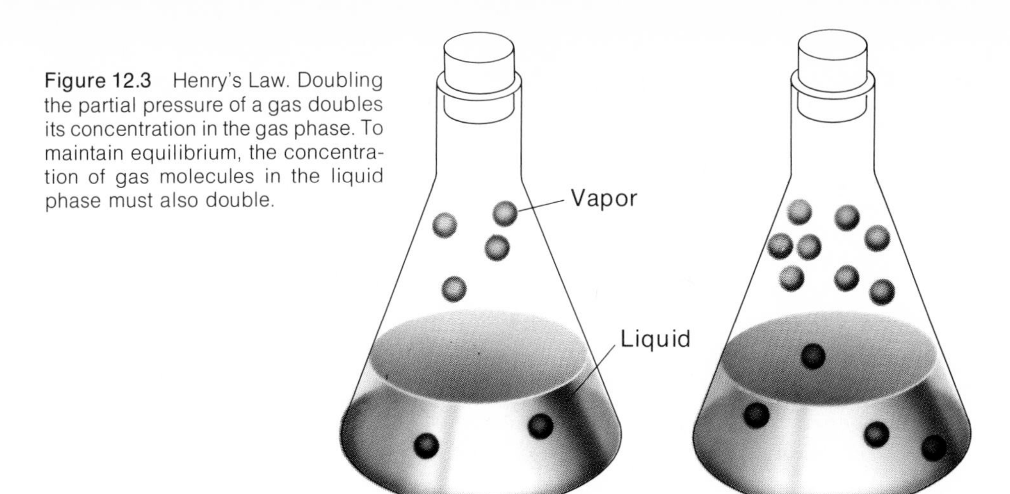

Figure 12.3 Henry's Law. Doubling the partial pressure of a gas doubles its concentration in the gas phase. To maintain equilibrium, the concentration of gas molecules in the liquid phase must also double.

Example 12.6 The solubility of pure oxygen in water at 20°C and 1.00 atm pressure is 1.38×10^{-3} mol/dm³. Calculate the concentration of O_2 at 20°C and a partial pressure of 0.21 atm.

Solution One way to solve this problem is to use the solubility data for pure oxygen to calculate k in Equation 12.5. Knowing k and the partial pressure of O_2, we can then obtain the required concentration of O_2.

$$k = \frac{\text{conc. } O_2}{\text{pressure } O_2} = \frac{1.38 \times 10^{-3}\ \text{mol/dm}^3}{1.00\ \text{atm}} = 1.38 \times 10^{-3}\ \frac{\text{mol/dm}^3}{\text{atm}}$$

At a partial pressure of 0.21 atm

$$\text{conc. } O_2 = 1.38 \times 10^{-3}\ \frac{\text{mol/dm}^3}{\text{atm}} \times 0.21\ \text{atm} = 2.9 \times 10^{-4}\ \text{mol/dm}^3$$

The quantity we have just calculated is the equilibrium concentration of oxygen in water at 20°C saturated with air, where $P_{O_2} = 0.21$ atm.

How do you think this compares to the solubility of O_2 in air?

The influence of partial pressure upon gas solubility is utilized in making carbonated beverages, including beer, sparkling wines, and many soft drinks. So-called "soda-water" is bottled under a carbon dioxide pressure of about 4 atm. When the bottle is opened, the pressure above the liquid drops to 1 atm, and CO_2 bubbles rapidly out of solution. If left uncapped, the liquid is exposed to air in which the partial pressure of CO_2 is very small (0 to 0.001 atm), more carbon dioxide slowly diffuses out of solution, and the beverage becomes "flat." Pressurized containers for whipped cream, shaving cream, etc., work on a similar principle. Opening a valve causes dissolved gas to come out of solution, carrying the liquid with it as a foam.

The excruciatingly painful and sometimes fatal affliction known as the "bends" is another consequence of the effect of pressure on gas solubility. Compressed air breathed by divers working underwater dissolves in the body fluids and tissues. If the individual is suddenly exposed to atmospheric pressure, the excess air comes out of solution as tiny bubbles which impair circulation and affect nerve impulses. One way to minimize this effect is to substitute a mixture of helium and oxygen for the compressed air. Helium, which has a much lower boiling point than nitrogen, is only about one fifth as soluble. Consequently much less gas comes out of solution on decompression.

12.5 COLLIGATIVE PROPERTIES OF NONELECTROLYTE SOLUTIONS

What properties of solutions would *not* be colligative?

Certain properties of solutions are found to depend primarily upon the concentrations of solute particles rather than their nature. These are called **colligative properties;** they include vapor pressure lowering, boiling point elevation, freezing point depression and osmotic pressure. In this chapter we shall consider colligative properties of nonelectrolyte solutions, where the solute particles are molecules. Properties of electrolyte solutions (ionic solutes) are discussed in Chapter 13.

The laws relating colligative properties to solute concentration are presented below as Equations 12.6 to 12.10. They are best regarded as limiting laws, which are approached more and more closely as the solution becomes more dilute. In practice these equations will ordinarily be valid to within, at most, a few per cent at concentrations as high as one molal.

Vapor Pressure Lowering

Experience tells us that the equilibrium vapor pressure of solvent above a solution is lower than that of the pure solvent. Concentrated water solutions of substances such as sugar or urea evaporate more slowly than pure water, reflecting the lowering of the water vapor pressure by the presence of solute. Indeed, if the solute concentration is high enough, water vapor from the atmosphere may condense into the solution, thereby diluting it.

A quantitative study of vapor pressure lowering shows that it is a true colligative property; it is directly proportional to the concentration of solute but independent of the nature of the solute molecule. For example, the vapor pressure of water above a 0.10 *m* solution of either sugar or urea is the same, about 0.0057 kPa less than that of pure water. In a 0.20 *m* solution, the vapor pressure lowering is twice as great, 0.0114 kPa.

The relationship between solvent vapor pressure and concentration can be expressed as

$$P_1 = X_1 P_1^0 \qquad (12.6)$$

where P_1^0 is the vapor pressure of solvent over the solution, X_1 is the mole fraction of solvent in solution and P_1^0 is the vapor pressure of pure solvent at the same

● Solvent molecules
● Solute molecules

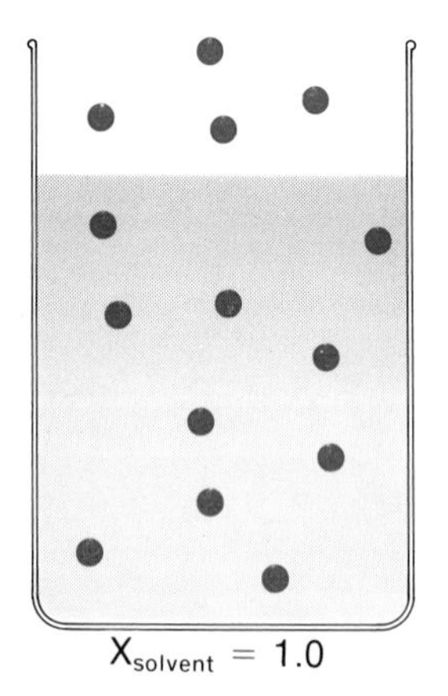

$X_{solvent} = 1.0$

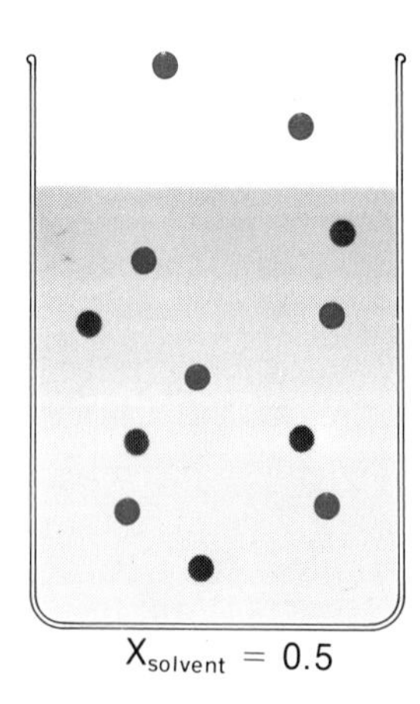

$X_{solvent} = 0.5$

Figure 12.4 Raoult's Law. Adding a solute lowers the concentration of solvent molecules in the liquid phase. To maintain equilibrium, the concentration of solvent molecules in the gas phase, and hence the vapor pressure, must decrease. If its mole fraction in the solution is decreased to 0.5, the vapor pressure of the solvent will decrease by a factor of 2.

temperature. Note that since X_1 in a solution must be less than 1, P_1 must be less than P_1^0 (Fig. 12.4).

We can obtain a direct expression for the vapor pressure lowering by making the substitution $X_2 = 1 - X_1$, where X_2 is the mole fraction of solute.

$X_1 + X_2 = 1$ for a two-component solution

$$P_1 = (1 - X_2)P_1^0$$

Rearranging $$P_1^0 - P_1 = X_2P_1^0$$

The quantity $(P_1^0 - P_1)$, the difference between the solvent vapor pressure in the pure liquid and in solution, is, by definition, the vapor pressure lowering, so

$$VPL = X_2P_1^0 \qquad (12.7)$$

Example 12.7 Calculate the vapor pressure lowering of a solution containing 100 g of sugar, $C_{12}H_{22}O_{11}$, in 500 g of water at 25°C.

Solution In order to use Equation 12.7, we need to know the vapor pressure of pure water (3.168 kPa at 25°C) and X_2, the mole fraction of sugar. Since the molecular masses of sugar and water are 342 and 18.0, respectively, we have

$$X_2 = \frac{\text{no. moles } C_{12}H_{22}O_{11}}{\text{no. moles } C_{12}H_{22}O_{11} + \text{no. moles } H_2O} = \frac{100/342}{100/342 + 500/18.0} = 0.0104$$

From Equation 12.7, we have

$$VPL = 0.0104 \times 3.168 \text{ kPa} = 0.0330 \text{ kPa}$$

(The vapor pressure of water over the solution would be: (3.168 − 0.033) kPa = 3.135 kPa.)

Equation 12.6, known as Raoult's Law, can, in principle at least, be applied to solutions in which both components are volatile. If *both* components follow Raoult's Law, we have

$$P_{tot} = P_1 + P_2 = X_1P_1^0 + X_2P_2^0$$

Solutions which follow this relation are said to be **ideal;** an example of such a solution is that formed by benzene and toluene, two organic liquids of very similar structures. For an equimolar mixture ($X_1 = X_2 = ½$) at 20°C, where the vapor pressures of the pure liquids are 10.0 kPa (benzene) and 3.2 kPa (toluene), we calculate a total pressure of

$$P_{tot} = ½\,(10.0 \text{ kPa}) + ½\,(3.2 \text{ kPa}) = 6.6 \text{ kPa}$$

Experimentally, we find that the total pressure over the solution is 6.7 kPa, slightly greater than the calculated value.

Boiling Point Elevation

A solution of a *nonvolatile** solute invariably boils at a *higher* temperature than the pure solvent. In dilute solutions the boiling point elevation is found to be

*Volatile solutes ordinarily *lower* the boiling point because they contribute to the total vapor pressure over the solution.

TABLE 12.4 MOLAL FREEZING POINT AND BOILING POINT CONSTANTS

Solvent	Freezing Point	k_f	Boiling Point	k_b
Water	0°C	1.86	100°C	0.52
Acetic acid	17	3.90	118	2.93
Benzene	5.50	5.10	80.0	2.53
Cyclohexane	6.5	20.2	81	2.79
Camphor	178	40.0	208	5.95
p-Dichlorobenzene	53	7.1	—	—

directly proportional to solute concentration. The relationship between these two variables can be written in the form

$$\Delta T_b = k_b \times m \tag{12.8}$$

where ΔT_b is the boiling point elevation in degrees Celsius, m is the molality, and k_b is a constant characteristic of the particular solvent. For water, k_b is 0.52°C. In other words, a 1 m water solution of a nonvolatile solute (sugar, urea, ethylene glycol, but *not* methyl alcohol) has a boiling point 0.52°C higher than that of pure water (100.52°C at 1 atm). Organic solvents ordinarily have values of k_b which are considerably larger than that of water (Table 12.4).

The elevation of the boiling point associated with solutions of nonvolatile solutes is readily explained in terms of a lowering of the vapor pressure. Since the solution, at any given temperature, has a vapor pressure *lower* than that of the pure solvent, a *higher* temperature must be reached before the solution boils, that is, before its vapor pressure becomes equal to the external pressure. Figure 12.5 illustrates this reasoning graphically.

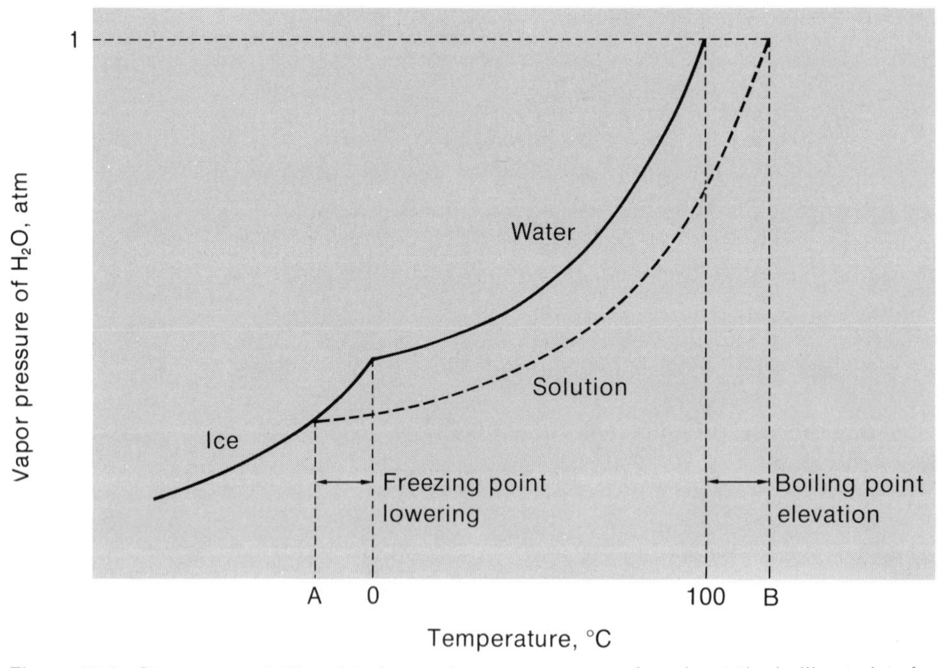

Figure 12.5 Since a nonvolatile solute lowers the vapor pressure of a solvent, the boiling point of a solution will be higher and the freezing point lower than the corresponding points for the pure solvent. Water solutions freeze *below* 0°C at A, and boil *above* 100°C at B.

Solutions containing small amounts of solute freeze or melt at temperatures below that of the pure solvent. For example, a mixture of naphthalene and biphenyl containing only 5 per cent by mass of biphenyl is completely melted at 77°C, some 3°C below the melting point of pure naphthalene (80°C). Organic chemists frequently make use of this behavior to check the purity of a solid which they have prepared in the laboratory. By comparing the melting point of their sample to that of an authentic sample of the pure solid, they get a rough estimate of the amount of impurity present.

The lowering of the freezing point in dilute solution, like the lowering of the vapor pressure and the elevation of the boiling point, is directly proportional to solute concentration. The relationship may be expressed as

$$\Delta T_f = k_f \times m \qquad (12.9)$$

where ΔT_f is the freezing point lowering, m is the molality, and k_f is a constant for a particular solvent. Note the resemblance between the equations for freezing point lowering and boiling point elevation. The constant k_f is ordinarily larger than k_b (cf. Table 12.4). For water, k_f is 1.86°C; for organic solvents, it is commonly a larger number. Camphor, a complex ketone of molecular formula $C_{10}H_{16}O$, has an unusually large freezing point constant, 40.0°C.

The freezing point depression, like the boiling point elevation, is a direct consequence of the lowering of the solvent vapor pressure by the solute (Fig. 12.5). Notice that the freezing point of the solution is taken to be the temperature at which the solvent in solution has the same vapor pressure as the *pure solid solvent*. This implies that it is pure solvent (e.g., ice) which separates when the solution freezes. In practice, this is ordinarily the case.

Icebergs contain very little salt

The use of Equations 12.8 and 12.9 in calculations involving freezing and boiling points of solutions of nonelectrolytes is illustrated by Example 12.8.

Example 12.8 Calculate the freezing point and boiling point, at 1 atm, of a solution of 2.60 g of urea, $CO(NH_2)_2$, in 50.0 g of water.

Solution To use Equations 12.8 and 12.9, we must first calculate the molality. Noting that the molecular mass of urea is 60.0, we have

$$\text{no. moles urea} = 2.60/60.0 = 0.0433$$

$$m = \frac{\text{no. moles urea}}{\text{no. kilograms water}} = \frac{0.0433}{0.0500} = 0.866$$

Hence:

$$\Delta T_b = 0.52°\text{C} \times 0.866 = 0.45°\text{C};\ bp = 100.00°\text{C} + 0.45°\text{C} = 100.45°\text{C}$$

$$\Delta T_f = 1.86°\text{C} \times 0.866 = 1.61°\text{C};\ fp = 0.00°\text{C} - 1.61°\text{C} = -1.61°\text{C}$$

We take advantage of the phenomenon of freezing point lowering when we add antifreeze to a car radiator to prevent the water from freezing. So-called permanent antifreeze uses ethylene glycol, an organic dialcohol which has the structure

```
    H     H
    |     |
 H—C——C—H
    |     |
   OH    OH
```

Ethylene glycol can be used year round as a radiator coolant since it is relatively nonvolatile (*bp* = 197°C) and raises the boiling point of water.

Another organic compound sometimes used as an antifreeze is methyl alcohol, CH_3OH. Methyl alcohol is cheaper than ethylene glycol. Moreover, since its molecular mass is only a little more than half that of ethylene glycol (32 vs. 62), its freezing point lowering per gram is nearly twice as great. The principal disadvantage of methyl alcohol is its volatility (*bp* = 65°C); it lowers the boiling point of water and also tends to distill out of solution.

Osmotic Pressure

If the beakers contained salt solutions of different concentrations, what would happen?

Imagine an experiment in which two beakers, one containing pure water, the other a sugar solution, are placed under a bell jar (Fig. 12.6) As time passes, it is found that the liquid level in the beaker containing the solution rises, while the level of pure water in the other beaker falls. Eventually, by evaporation and condensation, all the water is transferred to the solution; at the end of the experiment, the beaker that contained pure water is empty.

The driving force behind the process just described is the difference in vapor pressure of water in the two beakers. Water moves spontaneously from a region in which its vapor pressure is high (pure water) to a region in which its vapor pressure is low (sugar solution). The air in the bell jar is permeable only to water molecules; the nonvolatile solute is unable to move from one beaker to the other.

The apparatus shown at the right of Figure 12.6 can be used to achieve a result similar to that found in the bell jar experiment. Here, the sugar solution is separated from the pure water by a semipermeable membrane, which may be an animal bladder, a slice of vegetable tissue (a hollow carrot works quite well) or a piece of parchment. This membrane, by a mechanism which is poorly under-

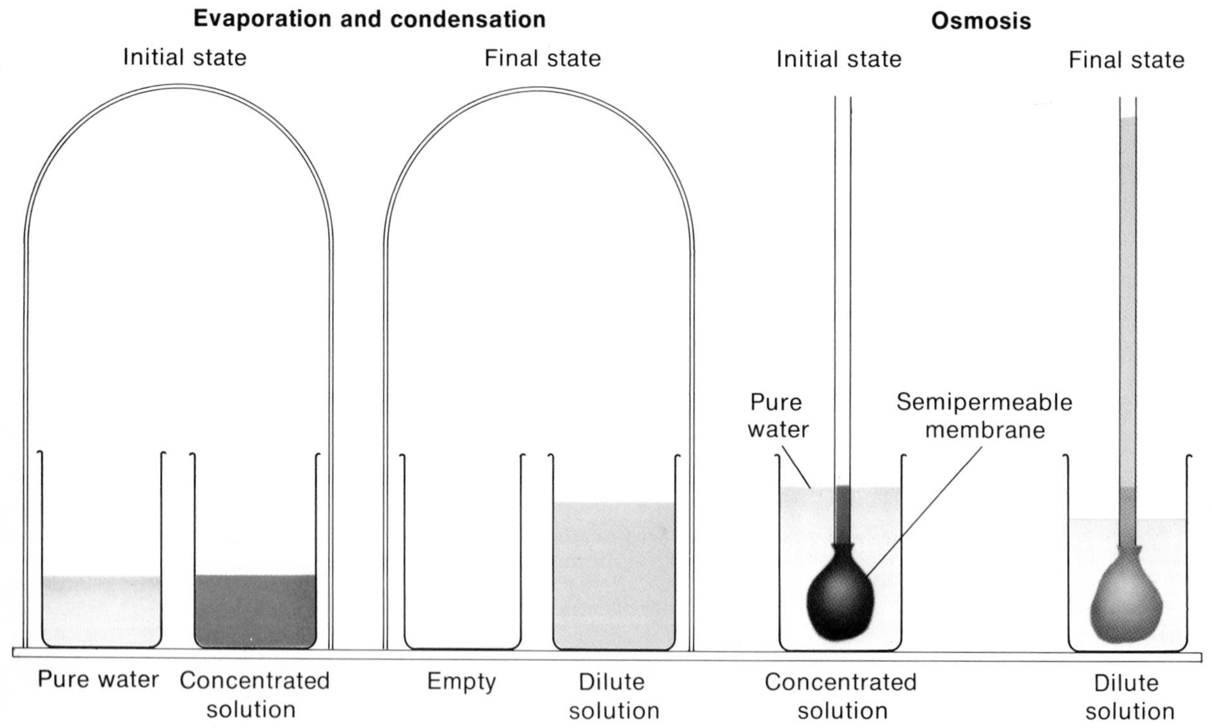

Figure 12.6 Water will tend to move spontaneously into the region where its vapor pressure is lowest. In the bell jar experiment, all the pure water would tend to vaporize and condense in the solution. During osmosis pure water passes through the membrane, diluting the solution.

stood, allows solvent but not solute molecules to pass through it. As before, water moves from a region in which its vapor pressure is high to a region in which its vapor pressure is low. The process, taking place through a membrane permeable only to the solvent, is referred to as **osmosis.**

The passage of water molecules through a membrane into a solution may be prevented by applying pressure to the solution. In Figure 12.6, the process of osmosis stops when the liquid level in the tube reaches a certain height. At this point, the tendency of water to move through the membrane is just balanced by the pressure exerted by the column of liquid.

The external pressure which is just sufficient to prevent osmosis is referred to as the **osmotic pressure** of a solution. Osmotic pressure may be measured precisely in an apparatus such as that shown in Figure 12.7. The inner, porous tube A contains within it a strong, semipermeable membrane consisting of a film of copper ferrocyanide, $Cu_2Fe(CN)_6$. This insoluble film is formed by allowing solutions containing Cu^{2+} and $Fe(CN)_6^{4-}$ ions to diffuse into each other through the walls of the tube. The tube is filled with pure water; the compartment B surrounding the tube is filled with the solution whose osmotic pressure is to be measured. Pressure is applied to the solution at C so as to maintain a constant level D in the capillary attached to the tube containing pure water. This pressure is, by definition, the osmotic pressure.

The osmotic pressure, π, of a dilute solution is a colligative property in that it is directly proportional to the concentration of solute and independent of its nature. The equation relating osmotic pressure to solute concentration is usually written in the form

$$\pi = MRT \qquad (12.10)$$

where π = osmotic pressure in kilopascals, M = concentration in moles per cubic decimetre, R = gas constant [8.31 kPa · dm³/(mol · K)] and T = temperature in K. Notice that since M is equal to the number of moles of solute, n, divided by the volume, V, Equation 12.10 could be written as $\pi = \frac{nRT}{V}$ or $\pi V = nRT$, which has the same form as the Ideal Gas Law.

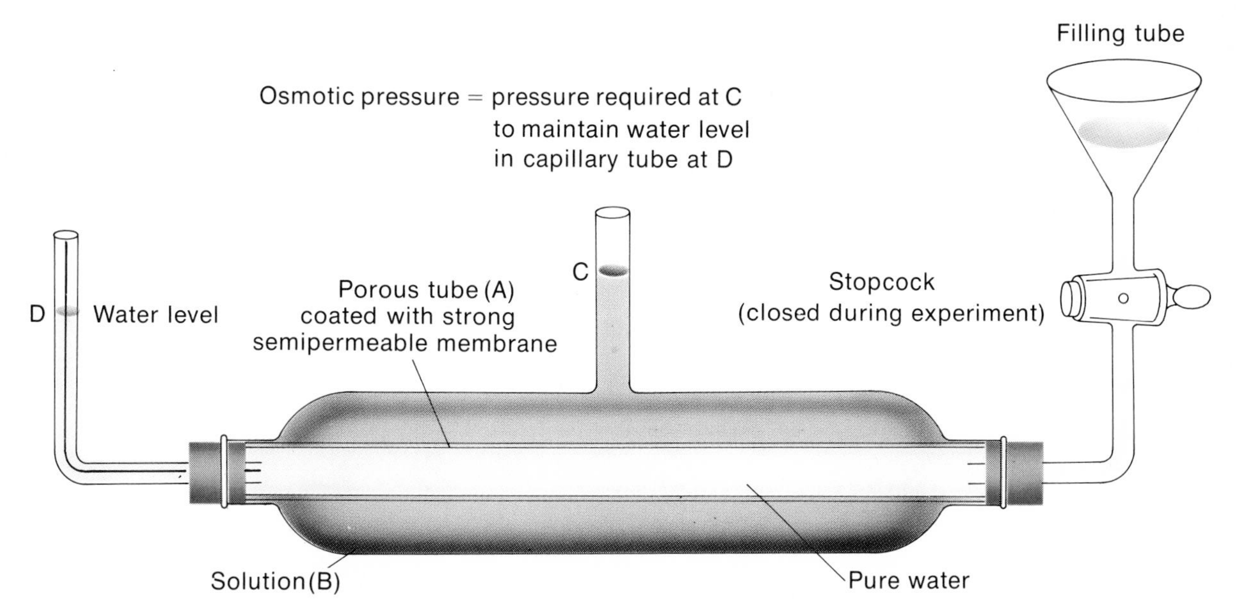

Figure 12.7 Measurement of osmotic pressure. The osmotic pressure of a solution is equal to the excess pressure which must be applied to the solution to keep the pure solvent from diffusing into that solution.

Substitution into Equation 12.10 shows that the osmotic pressure, even in dilute solution, is relatively large. Consider, for example, a 0.10 M solution at 25°C, where we calculate

$$\pi = 0.10 \frac{\text{mol}}{\text{dm}^3} \times 8.31 \frac{\text{kPa} \cdot \text{dm}^3}{\text{mol} \cdot \text{K}} \times 298 \text{ K} = 248 \text{ kPa}$$

A pressure of 248 kPa is equivalent to a column of water 25 m high! This gives some indication of the driving force behind osmosis and the difficulties of measuring osmotic pressure accurately using ordinary membranes.

Osmosis plays a vital role in many biological processes. Nutrient and waste materials pass by osmosis through the cell walls of animal tissues, which show varying degrees of permeability to different solutes. A striking example of a natural osmotic process can be followed by watching under a microscope what happens when red blood cells are placed in pure water. Water passes through the walls to dilute the solution inside the cell, which swells and eventually bursts, releasing its red pigment. If the blood cells are placed in a concentrated sugar solution, water moves in the reverse direction; the cells shrink in size and shrivel up. To avoid these effects, nutrient solutions used in intravenous feeding are carefully adjusted so as to have the same osmotic pressure as blood (about 780 kPa).

The walls of plant as well as animal cells act as semipermeable membranes. Flowers immersed in sugar or salt solution wilt as they are dehydrated by osmosis; if transferred to pure water, they regain their freshness as water moves back into the cells.

Determination of Molecular Masses from Colligative Properties

In Chapter 5 we discussed the determination of molecular masses from gas density data. This method works quite well for many gases and volatile liquids but is useless for substances which decompose on heating, such as sugar or urea. An alternative approach, applicable to a wide variety of nonelectrolytes, involves measuring one of the colligative properties of their solutions. The calculations involved are illustrated in Example 12.9.

Example 12.9 A student determines the molecular mass of a certain nonelectrolyte by weighing out 1.100 g of the solid and dissolving it in 20.0 g of benzene. He measures the freezing point of the solution to be 4.38°C. Knowing that pure benzene has a freezing point of 5.50°C and a k_f of 5.10, calculate the molecular mass of the solute.

Solution Let us first calculate from the data the observed freezing point lowering. Then we can use Equation 12.9 to obtain the molality of the solution. Finally, from the defining equation for molality (12.3), we can calculate the molecular mass of the solute.

$$\Delta T_f = fp \text{ pure benzene} - fp \text{ solution} = 5.50°\text{C} - 4.38°\text{C} = 1.12°\text{C}$$

Applying Equation 12.9, we have

$$\Delta T_f = k_f \times m; \qquad m = \frac{\Delta T_f}{k_f} = \frac{1.12°\text{C}}{5.10°\text{C}} = 0.220$$

But $$m = \frac{\text{no. moles solute}}{\text{no. kilograms solvent}} = \frac{\text{no. grams solute}/GMM \text{ solute}}{\text{no. kilograms solvent}}$$

Rearranging $$GMM \text{ solute} = \frac{\text{no. grams solute}}{(m)(\text{no. kilograms solvent})}$$

Substituting $$GMM = \frac{1.100 \text{ g}}{(0.220)(0.0200)} = 250 \text{ g}$$

It is possible to determine molecular masses by measuring the vapor pressure lowering, boiling point elevation or osmotic pressure. Freezing point lowering is perhaps most commonly used because the effect is comparatively large (compare, for example, the constants for the freezing point lowering, 1.86°C, and the boiling point elevation, 0.52°C, for water solutions). Organic solvents are generally preferred to water because their k_f values are larger. Camphor, with k_f = 40°C, is a popular choice, although its high melting point (178°C) complicates the temperature measurement.

Osmotic pressure measurements are the method of choice for solutes of very high molecular mass, where the concentration in moles per cubic decimetre is ordinarily quite low (Example 12.10).

Example 12.10 The osmotic pressure of a solution prepared by dissolving 1.0 g of hemoglobin in enough water to form 100 cm³ of solution is 0.366 kPa at 20°C. Estimate

a. the concentration of the solution (M).
b. the molecular mass of hemoglobin.

Solution

a. Rearranging Equation 12.10 to solve for M:

$$M = \frac{\pi}{RT}$$

But, π = 0.366 kPa, R = 8.31 kPa · dm³/(mol · K), T = 293 K. Hence

$$M = \frac{0.366}{(8.31)(293)} \frac{\text{mol}}{\text{dm}^3} = 1.50 \times 10^{-4} \frac{\text{mol}}{\text{dm}^3}$$

b. From the defining equation for M:

$$M = \frac{\text{no. moles}}{\text{no. cubic decimetres soln.}} = \frac{\text{no. grams}/GMM}{\text{no. cubic decimetres soln.}}$$

$$1.50 \times 10^{-4} \text{ mol} = \frac{1.0 \text{ g}}{(0.100 \text{ dm}^3)(GMM)}$$

Solving: $GMM = \frac{1.0 \text{ g}}{1.50 \times 10^{-5} \text{ mol}} = 6.7 \times 10^4$ g/mol; $MM \approx 67\,000$

This method can be used to find the *MM* of a synthetic polymer like polyethylene

PROBLEMS

12.1 *Concentration Units* A solution is prepared by dissolving 20.0 g of methyl alcohol, CH_3OH, in 30.0 g of water. Calculate

a. the mass % of CH_3OH.
b. the mole fraction of CH_3OH.
c. the molality of CH_3OH.

12.2 *Volume Concentration* A solution is prepared by dissolving 15.0 g of $CaCl_2$ in water to form 600 cm^3 of solution.

a. What is the concentration of $CaCl_2$ in moles per cubic decimetre?
b. What volume of this solution should be taken to prepare 125 cm^3 of 0.100 M solution?

12.3 *Henry's Law* The solubility of nitrogen in water at 20°C and 1.00 atm is 6.90×10^{-4} mol/dm^3. What is its solubility at 20°C and 0.79 atm, the partial pressure of nitrogen in air?

12.4 *Raoult's Law* Assuming ideal behavior, calculate the partial pressures of hexane and heptane at 20°C over a solution in which the mole fraction of hexane is 0.40. The vapor pressures of pure hexane and heptane at 20°C are 16.3 kPa and 4.8 kPa, respectively.

12.5 *Colligative Properties* A solution is prepared by dissolving 15.0 g of glucose, $C_6H_{12}O_6$, in 200 g of water at 25°C. Calculate

a. the freezing point of the solution (k_f = 1.86°C).
b. the boiling point at 1.00 atm (k_b = 0.52°C).
c. the osmotic pressure (assume $M \approx m$).

12.6 *Molecular Mass* A solution of 10.0 g of a certain solute in 120 g of benzene has a freezing point 4.10°C below that of pure benzene. Calculate the molecular mass of the solute (k_f benzene = 5.10°C).

12.7 How would you explain to a high school chemistry student why

a. the solubility of solids in liquids ordinarily increases with temperature?
b. the boiling point of a sugar solution is greater than that of pure water?
c. gas bubbles form in a canteen of water opened at the top of a mountain?
d. the value of M for a dilute water solution is nearly the same as its molality?

12.8 How would you prepare

a. a supersaturated solution of a solid in water?
b. a supersaturated solution of a gas in water?
c. 0.200 dm^3 of 0.10 M $Ba(NO_3)_2$ solution, starting with the solid?
d. 0.200 dm^3 of 0.10 M $Ba(NO_3)_2$, starting with 1.2 M solution?

12.9 The solution of hydrogen peroxide sold as a bleach and disinfectant contains 3.0% by mass of H_2O_2 and has a density of 1.0 g/cm^3.

a. What is the mass of H_2O_2 in 1.00 cm^3 of this solution?
b. What volume of oxygen, at 25°C and 1.00 atm, is liberated when 1.00 cm^3 of this solution decomposes? The reaction is $H_2O_2(aq) \rightarrow H_2O(aq) + \frac{1}{2}\ O_2(g)$. The label on the bottle says that it forms ten times its own volume of oxygen.

12.10 A water solution contains 15.0 g of sugar, $C_{12}H_{22}O_{11}$, in 120 cm^3. The density of this solution is 1.047 g/cm^3. Calculate

a. M (Equation 12.4).
b. the molality of sugar.
c. the mass per cent of sugar.
d. the mole fraction of sugar.

12.26 How would you explain to a nurse

a. why it is important to control the concentration of solutions injected into the blood?
b. why ethyl alcohol is so much more soluble in water than is benzene?
c. what causes the ailment known as the "bends"?
d. why gas bubbles form in a specimen bottle taken out of a refrigerator?

12.27 You are given a sample of pure sugar, $C_{12}H_{22}O_{11}$, weighing 30.0 g. How would you prepare from it a water solution in which

a. the mass per cent of sugar is 10.5?
b. the mole fraction of sugar is 0.0186?
c. the molality of sugar is 0.100?
d. there is 0.100 mol/dm^3 of sugar?

12.28 Using the data in Table 12.1, calculate

a. the mass of HNO_3 in 100 g of concentrated nitric acid.
b. the mass of HNO_3 in 100 cm^3 of concentrated nitric acid.
c. the volume of water which must be added to 1.00 dm^3 of concentrated HCl to form dilute HCl (neglect the volume change on mixing).

12.29 A solution prepared by dissolving 30.0 g of ethyl alcohol, C_2H_5OH, in 50.0 g of carbon tetrachloride, CCl_4, has a density of 1.28 g/cm^3. Calculate

a. the mass per cent of ethyl alcohol.
b. the mole fraction of ethyl alcohol.
c. the molality of ethyl alcohol.
d. M (Equation 12.4).

12.11 Complete the following table for water solutions.

Solute	g solute	M	V (dm^3)
H_2SO_4	29.0	______	0.650
H_2SO_4	56.6	1.99	______
$NaNO_3$	______	0.154	12.6

12.12 What volume of 0.200 M K_3PO_4 solution can be prepared from 50.0 cm^3 of 0.450 M solution?

12.13 The "proof" of an alcoholic beverage is defined as twice the percentage by volume of alcohol, C_2H_5OH. How many grams of alcohol are there in one litre of 80 proof Scotch whiskey? (d alcohol = 0.80 g/cm^3).

12.14 The solubility of methane in benzene at 25°C and 20.0 kPa is 4.7×10^{-3} mol/dm^3. What is the solubility at 25°C and 300 kPa?

12.15 Which of the following would you expect to be the more soluble in water:

a. $H_3C-\overset{\overset{\displaystyle O}{\|}}{C}-OH(l)$ or $H_3C-\overset{\overset{\displaystyle O}{\|}}{C}-OCH_3(l)$?
b. Ar or He?
c. NaCl or CCl_4?
d. $H_3C-O-CH_3(l)$ or $H_3C-OH(l)$?

12.16 The mole fraction solubilities of $CH_4(g)$, $C_2H_6(g)$, $C_5H_{12}(l)$, and $C_{31}H_{64}(s)$ in benzene at 25°C and 1 atm are 0.0021, 0.0148, 1.00, and 0.0023, respectively. Explain the relative magnitudes of these numbers in terms of solubility principles.

12.17 Calculate the vapor pressure of water over a 10.0 m sugar solution at 25°C (vp water at 25°C = 3.17 kPa).

12.18 The two liquids n-hexane and n-heptane form an ideal solution. Their vapor pressures at 30°C are 25.2 and 7.7 kPa, respectively. What is the total vapor pressure over a solution of these two liquids at 30°C in which the mole fraction of n-hexane is 0.650?

12.19 Calculate the boiling point at 1 atm and the freezing point of the following solutions:

a. 15.0 g of citric acid, $C_6H_8O_7$, in 50.0 g of water
b. 2.82 g of iodoform, CHI_3, in 10.0 g of benzene (Table 12.4)

12.30 Work Problem 12.11, substituting acetone, $H_3C-CO-CH_3$, for the solutes listed.

12.31 Derive a formula for the volume of water, V_w, which must be added to V_c dm^3 of concentrated solution of M_c to give a dilute solution of concentration M_d. Assume no change in volume in dilution.

12.32 A solution prepared by dissolving 16.0 g of I_2 in 100 g of ethyl alcohol, C_2H_5OH, has a density of 0.889 g/cm^3. In this solution which is the larger,

a. mole fraction of I_2 or mass fraction of I_2?
b. m I_2 or M I_2?

12.33 The solubility of carbon dioxide in seawater is 3.0×10^{-2} mol/dm^3 at 25°C and one atmosphere pressure. Calculate the concentration of CO_2 in seawater in equilibrium with air in which the partial pressure of CO_2 is 0.133 kPa.

12.34 Which one of each pair of substances listed in Problem 12.15 would you expect to be the more soluble in hexane?

12.35 The mole fractions of Cl_2, Br_2, and I_2 in a saturated solution in carbon tetrachloride at 20°C and one atmosphere are 0.187, 1.00 and 0.010, respectively. Explain the relative magnitude of these numbers in terms of solubility principles.

12.36 How many grams of urea, $CO(NH_2)_2$, must be added to 100 g of water to give a solution with a vapor pressure 0.100 kPa less than that of pure water at 25°C? at 50°C? (vp water at 25°C = 3.17 kPa, at 50°C = 12.3 kPa.)

12.37 The vapor pressures of benzene and toluene at 100°C are 1.77 atm and 0.73 atm, respectively. The two liquids form an ideal solution. What must be the mole fraction of benzene in a toluene solution which boils at 100°C at 1.00 atm?

12.38 How many grams of the following substances would have to be dissolved in 100 g of water to give a solution freezing at −2.00°C? What would be the boiling points of these solutions at 1 atm?

a. urea, $CO(NH_2)_2$
b. glucose, $C_6H_{12}O_6$

12.20 Isopropyl alcohol, C_3H_8O, is used as an antifreeze in windshield washers. What volume of isopropyl alcohol ($d = 0.79$ g/cm^3) should be added to 1.00 dm^3 of water to give a solution which would not freeze above -20.0°C?

12.21 The osmotic pressure of blood at body temperature, 37.0°C, is about 780 kPa. Taking the solutes in blood to be nonelectrolytes, estimate their total concentration (*M*).

12.22 Of the three solutions: 1.0 *M* sugar in water, 1.0 *M* methyl alcohol in water, 1.0 *M* methyl alcohol in benzene, which will have the same

a. freezing point?
b. boiling point?
c. osmotic pressure?

12.23 A student finds that a solution of 1.62 g of a certain nonelectrolyte in 20.0 g of water freezes at -0.46°C. Calculate the molecular mass of the nonelectrolyte.

12.24 The freezing point of *p*-dichlorobenzene is 53.1°C; its k_f value is 7.10. A solution of 1.20 g of sulfanilamide (one of the sulfa drugs) in 15.0 g of *p*-dichlorobenzene freezes at 49.6°C. What is the molecular mass of sulfanilamide?

12.25 The osmotic pressure of a solution prepared by dissolving 1.00 g of insulin in 100 g of water is 4.32 kPa at 25°C. Calculate the molecular mass of insulin.

12.39 Tetrahydrofuran, C_4H_8O, has been suggested as an antifreeze. How many grams of tetrahydrofuran would have to be added to water to give the same freezing point lowering as one gram of ethylene glycol, $C_2H_6O_2$? If the prices of tetrahydrofuran and ethylene glycol are $3.10/kg and $1.80/kg, respectively, which would be the cheaper antifreeze?

12.40 What is the osmotic pressure of a solution containing 7.18 g of sugar, $C_{12}H_{22}O_{11}$, in 1.00 dm^3 of solution at 20°C?

12.41 If osmosis is responsible for sap rising in a tree, estimate the height in metres to which sap can rise at 25°C if it is 0.20 *M* in sugar and the water outside the sap-bearing tubule contains nonelectrolytes at a concentration of 0.01 *M* (1 kPa = 10.2 cm water).

12.42 Progesterone, a female hormone, is found on analysis to contain 9.5% H, 10.2% O and 80.3% C. A solution of 1.50 g of progesterone in 10.0 g of benzene freezes at 3.07°C. What is the molecular formula of progesterone?

12.43 Nicotine has the empirical formula C_5H_7N. A solution of 0.60 g of nicotine in 12.0 g of water boils at 100.17°C at 1 atm. What is the molecular formula of nicotine?

12.44 A saturated solution of a certain protein in water contains 5.18 g of solute/dm^3. The solution has an osmotic pressure of 0.413 kPa at 20°C. What is the molecular mass of the protein? Estimate the freezing point of the solution.

*12.45 The solubility of air in blood at one atmosphere and body temperature, 37°C, is 6.6×10^{-4} mol/dm^3. If a diver breathes compressed air at 5.00 atm, how many cubic centimetres of air are released from 1.00 cm^3 of blood when he returns to the surface, where the pressure is 1.00 atm?

*12.46 An alternative way to state Henry's Law of gas solubility is to say that the volume of gas which dissolves in a fixed volume of solution is independent of pressure. Demonstrate algebraically that this statement is valid.

*12.47 The water-soluble form of the drug heroin has a molecular mass of 423. A 0.100 g sample of this drug mixed with sugar (*MM* = 342) is added to 1.00 g of water to give a solution freezing at -0.500°C. Estimate the mass per cent of heroin in the sample.

*12.48 A martini, weighing about 150 g, contains 30% by mass of alcohol. About 15% of the alcohol in the martini passes directly into the bloodstream, which, for an adult, has a volume of about 7.0 dm^3. Estimate the concentration of alcohol in the blood, in grams per cubic centimetre, of a person who drinks two martinis before dinner. (A concentration of 0.0030 g/cm^3 is frequently taken as indicative of intoxication in a "normal" adult.)

13

WATER, PURE AND OTHERWISE

Life as we know it depends upon water, which is by far the most abundant substance in plant and animal tissue as well as in most of the world around us. Water accounts for about 70% of the mass of your body; in many growing plants, this percentage is even higher. Over 80% of the earth's surface is covered by water—as a relatively pure liquid in lakes and rivers, as a dilute salt solution in the oceans, or as a nearly pure solid in snowfields, glaciers, and the polar ice caps.

The unusual properties of water, referred to in Chapter 9, have a profound effect on the nature of our environment. Its high specific heat helps to prevent large fluctuations in the surface temperature of the earth. Bodies of water such as oceans and lakes absorb solar heat during the day and release it to the atmosphere at night without undergoing an appreciable temperature change. On the moon, whose surface consists of rocky material with a specific heat only about one fifth that of water, the temperature can vary from 120°C to −150°C.

Another factor holding the earth's temperature relatively constant is the high heat of vaporization of water which, in joules per gram, is greater than that of any other liquid. About one third of the solar energy which reaches the surface of the earth is dissipated by vaporizing water from oceans, lakes, rivers and ice fields. The same mechanism is in large part responsible for maintaining the temperature of our bodies within very narrow limits. Much of the heat generated by metabolism is removed by the evaporation of water through the pores of the skin.

All of these effects can be explained in terms of hydrogen bonding

The peculiar way in which the density of water changes with temperature, reaching its maximum value at 4°C, is of great importance in areas which have cold winter climates. On nights in late autumn when the air temperature drops sharply, the layer of more dense water at the surface of a lake sinks to the bottom, displacing the water beneath it. This circulation process distributes dissolved oxygen and nutrients more or less uniformly throughout the water. Eventually, the lake reaches a stable situation with a bottom layer of water at a temperature of 4°C (39°F), at which aquatic life can survive through the winter. A similar process occurring all year round in the oceans is a major factor in determining the patterns of ocean currents. Seawater at depths below about one kilometre has a nearly constant temperature between 0 and 4°C, regardless of what part of the world it comes from.

The fact that water expands when it freezes has both beneficial and detrimental effects. The damage to plant and animal tissue that accompanies freezing is largely due to this expansion, which causes cell walls to burst. The same process, taking place at the surface of the earth, is responsible for the breakdown of rocky materials to yield fertile soils. Again, if ice were more dense than water it would tend to sink in lakes and rivers, making it much easier for them to freeze solidly in winter, with disastrous consequences to marine life.

The properties of water change, sometimes dramatically, when foreign substances dissolve in it. These substances may enter the water by natural processes, as is the case with the dissolved salts in seawater. Others, including those classed as pollutants, enter as a direct result of man's activities. One of our major objectives in this chapter will be to consider what effect these impurities have on natural bodies of water and how they may be removed. Before doing so, let us take a closer look at some of the special properties of water solutions, which were introduced in a general way in Chapter 12.

13.1 WATER AS A SOLVENT; ELECTROLYTE SOLUTIONS

Although water is sometimes described as the "universal solvent," the phrase is misleading. As pointed out earlier, water does not dissolve detectable amounts of metallic or macromolecular substances. Relatively few molecular substances are appreciably soluble in water. Those that do dissolve (e.g., methanol, CH_3OH; urea, $H_2N—CO—NH_2$; ammonia, NH_3) are those that form hydrogen bonds with water. Only one class of substances, ionic solids, can be described as being generally water-soluble. Indeed, water is one of the very few liquids capable of dissolving, at room temperature, appreciable amounts of a wide variety of ionic compounds (e.g., NaCl, $CaCl_2$, $CuSO_4$, . . .).

In a general way, we can explain the solvent power of water for ionic compounds in terms of its ability to reduce the attractive force between oppositely charged ions. Coulomb's Law, applied to ions in solution, can be written in the form:

$$f = \frac{q_1 \times q_2}{Dr^2} \qquad (13.1)$$

where f is the force of electrostatic attraction between a + and a − ion, q_1 and q_2 are the charges of the ions, r is the distance between them, and D is a characteristic property of the solvent known as its **dielectric constant.** Clearly, the larger the dielectric constant, the smaller will be the attractive force between the ions and the less likely they will be to combine with one another and precipitate out of solution. Water has one of the largest dielectric constants (78.5 at 25°C) of all liquids; this value compares to 2.3 for benzene, 4.8 for chloroform, and 32.6 for methyl alcohol.

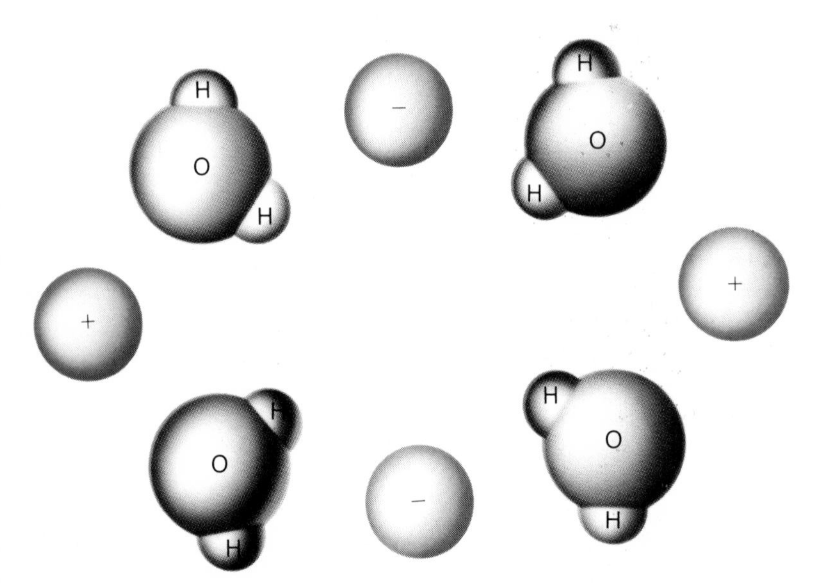

Figure 13.1 Polar water molecules oriented as shown markedly decrease the electrostatic attraction between oppositely charged ions and hence promote solubility. The force between + and − ions in water is only about 1/80 of that which would be present if the ions were similarly arranged in air (D_{air} = 1, D_{water} = 78.5).

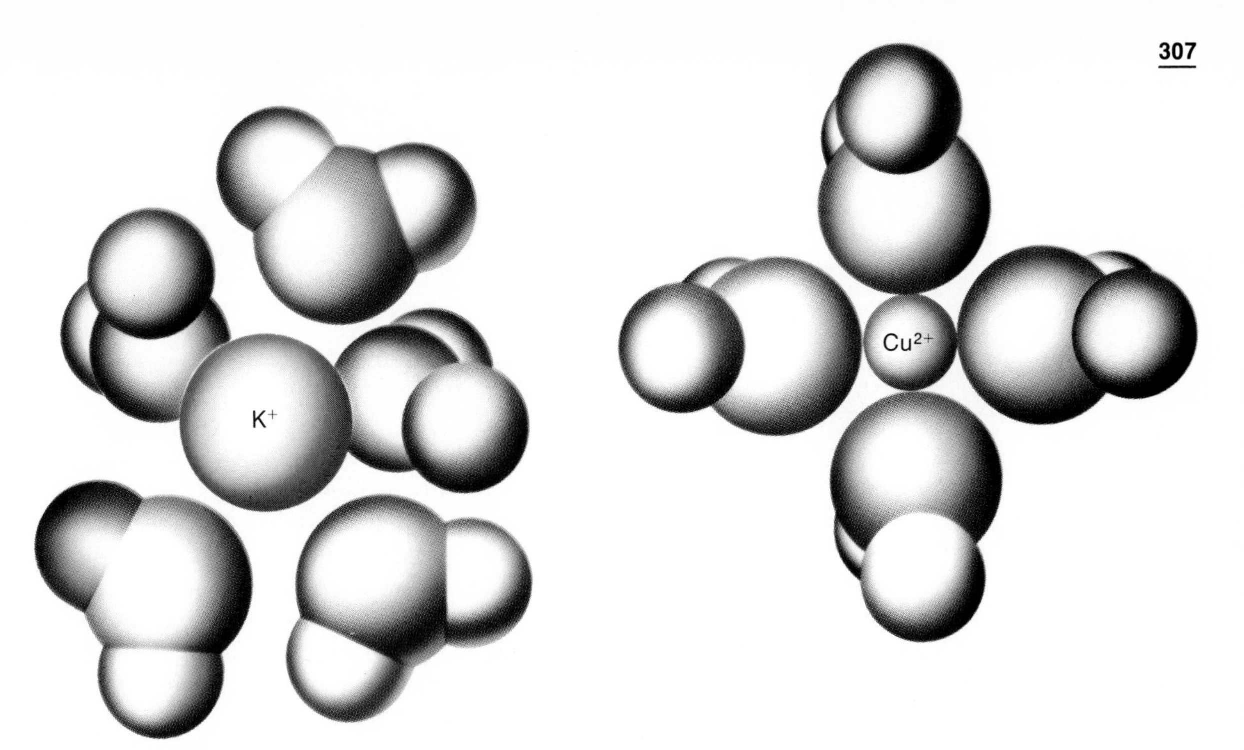

Figure 13.2 Hydrated species formed by K^+ and Cu^{2+} in water solution. The energy evolved in forming these hydrated ions is about equal to the lattice energy of compounds of these ions.

A physical interpretation of the high dielectric constant of water is suggested in Figure 13.1. The small, highly polar water molecule can interject itself between + and − ions in solution. The negative pole of the H_2O molecule (oxygen atom) points to the cation, while the positive pole (hydrogen atoms) points to a neighboring anion. In this way, water molecules effectively "insulate" the ions from one another, thereby making it easier for them to pass from the solid state into solution.

Another factor which promotes the solubility of ionic compounds is the interaction of the dissolved ions with water molecules to form **hydrated ions** (Fig. 13.2). Hydration is a strongly exothermic process; the energy released is large enough to overcome the lattice energy holding the ions to one another in the crystal lattice. The effect is greatest with ions of high "charge density" (charge-to-size ratio) such as Cu^{2+}, where the strongly bonded species $Cu(H_2O)_4^{2+}$ is formed. With larger cations of lower charge, such as K^+, the hydration energy is smaller, but still of the same order of magnitude as the lattice energy. Hydration effects appear to be least significant for singly charged anions such as Cl^- or NO_3^-, which are considerably larger than most cations.

All transition metal cations exist as hydrated species in dilute water solution

Conductivity of Water Solutions

Solutes are often classified into three categories according to the electrical conductivities of their water solutions. Substances which dissolve as molecules and, hence, give nonconducting solutions are classed as *nonelectrolytes*. Examples include methyl alcohol, CH_3OH, and sugar, $C_{12}H_{22}O_{11}$. The properties of nonelectrolyte solutions were discussed in Chapter 12.

Substances which exist in water solution as an equilibrium mixture of ions and molecules are called *weak electrolytes*. An example is hydrogen fluoride; a 0.1 *M* water solution of this compound contains a few H^+ and F^- ions in equilibrium with a large number of HF molecules (cf. Chapter 19).

TABLE 13.1 CONDUCTIVITIES OF 0.10 *M* SALT SOLUTIONS RELATIVE TO 0.10 *M* NaCl (25°C)

1:1 Salts		2:1 Salts		1:2 Salts		3:1 Salts	
LiCl	0.90	$MgCl_2$	1.82	Na_2SO_4	1.69	$LaCl_3$	2.79
NaCl	**1.00**	$CaCl_2$	1.92	K_2SO_4	2.05	$LaBr_3$	2.92
NaBr	1.04	$BaCl_2$	1.97	K_2CO_3	2.02	$PrBr_3$	2.91
NaI	1.02	$CaBr_2$	2.02	K_2CrO_4	2.02	$FeCl_3$	2.51
KCl	1.21	$SrBr_2$	2.04			$Fe(NO_3)_3$	2.77
$NaClO_4$	0.92					$AlBr_3$	2.74
KNO_3	1.13						
$AgNO_3$	1.02						

Strong electrolytes exist almost exclusively as ions in water solution. In an electrical field, these (charged) ions move through the solution, thereby conducting the current. The conductivities of strong electrolytes, at concentrations as low as 0.1 *M*, are at least 100 000 times that of pure water. Compounds which are ionic in the solid state act as strong electrolytes; examples include NaCl (Na^+, Cl^-) and $Ba(OH)_2$ (Ba^{2+}, 2 OH^-). A few molecular species ionize almost completely when added to water. An example is hydrogen chloride, which exists as HCl molecules in the pure gas but is converted almost entirely to ions (H^+, Cl^-) in water.

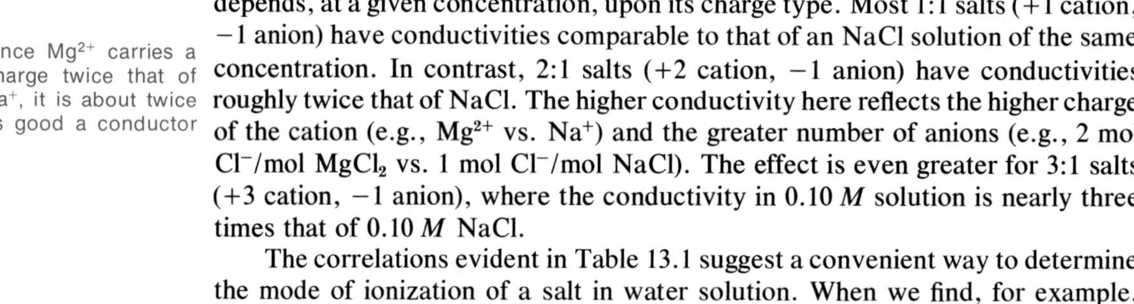

As Table 13.1 indicates, the conductivity of a solution of a strong electrolyte depends, at a given concentration, upon its charge type. Most 1:1 salts (+1 cation, −1 anion) have conductivities comparable to that of an NaCl solution of the same concentration. In contrast, 2:1 salts (+2 cation, −1 anion) have conductivities roughly twice that of NaCl. The higher conductivity here reflects the higher charge of the cation (e.g., Mg^{2+} vs. Na^+) and the greater number of anions (e.g., 2 mol Cl^-/mol $MgCl_2$ vs. 1 mol Cl^-/mol NaCl). The effect is even greater for 3:1 salts (+3 cation, −1 anion), where the conductivity in 0.10 *M* solution is nearly three times that of 0.10 *M* NaCl.

Since Mg^{2+} carries a charge twice that of Na^+, it is about twice as good a conductor

The correlations evident in Table 13.1 suggest a convenient way to determine the mode of ionization of a salt in water solution. When we find, for example, that the compound $KNaSO_4$ has a conductivity in 0.10 *M* solution about halfway

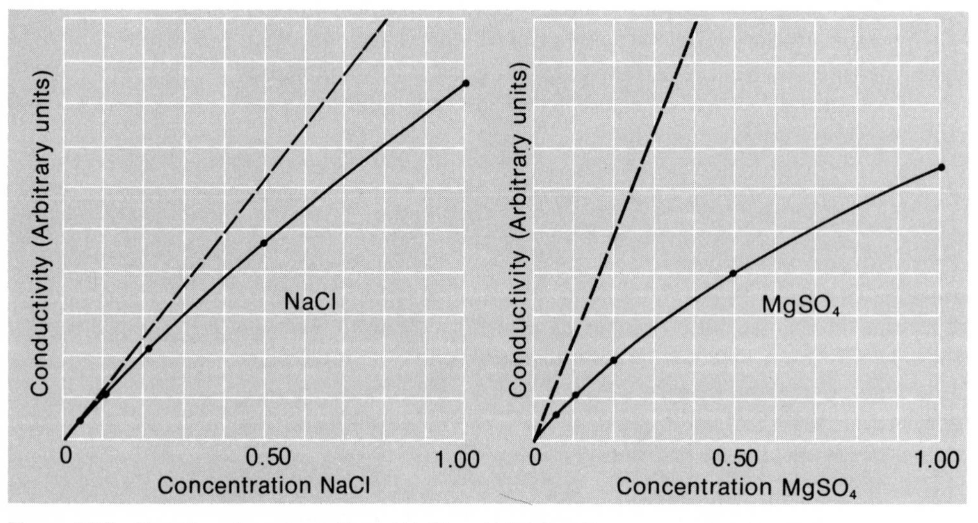

Figure 13.3 Variation of conductivities of NaCl and $MgSO_4$ with concentration. The dotted lines represent the direct proportionality that one would expect if the ions behaved as completely independent particles, as they do at infinite dilution.

between those of Na_2SO_4 and K_2SO_4 but nearly twice that of 0.10 M NaCl, we deduce that this salt must ionize as a 1:2 salt, i.e.,

$$KNaSO_4(s) \rightarrow K^+(aq) + Na^+(aq) + SO_4^{2-}(aq) \qquad (13.2)$$

rather than

$$KNaSO_4(s) \rightarrow K^+(aq) + NaSO_4^-(aq)$$

or in some other manner.

As you might expect, the conductivity of solutions of a given electrolyte varies with concentration. If the ions acted as completely independent particles, we would expect conductivity to be directly proportional to concentration (e.g., doubling the concentration would double the number of ions in a given volume). In practice, we find (Fig. 13.3) that this simple relationship is followed only at very low concentrations. As concentration increases, the conductivity falls below the expected value. The effect becomes more pronounced as the charge of the ions increases (compare the 2:2 salt $MgSO_4$ to the 1:1 salt NaCl).

Colligative Properties of Electrolyte Solutions

We saw in Chapter 12 that in dilute solution the vapor pressure lowering, boiling point elevation, freezing point depression and osmotic pressure are directly proportional to the concentration of solute *particles*. On this basis we would predict that, at a given molality, an electrolyte such as sodium chloride would have a greater effect on these properties than a nonelectrolyte such as sugar. When a mole of sugar dissolves, one mole of solute molecules is obtained; a mole of NaCl, on the other hand, yields two moles of ions. By the same argument, a one molal solution of $CaCl_2$ (three moles of ions per mole of solute) should have a lower vapor pressure, a higher boiling point, a lower freezing point and a greater osmotic pressure than one molal NaCl.

These predictions are confirmed experimentally. At a given molality the vapor pressures of solutions of sugar, sodium chloride and calcium chloride decrease in that order (i.e., sugar > NaCl > $CaCl_2$). Many electrolytes form saturated solutions whose vapor pressures are so low that the solids pick up water (*deliquesce*) when exposed to moist air. This phenomenon occurs with calcium chloride, whose saturated solution has a vapor pressure only 20% that of pure water. If dry $CaCl_2$ is exposed to moist air in which the vapor pressure of water is greater than 20% of the equilibrium value, it deliquesces to form a concentrated aqueous solution. Calcium chloride is sometimes spread on dirt roads in summer to prevent them from becoming dusty and developing a rough, "washboard" surface.

The freezing points of electrolyte solutions, like their vapor pressures, are lower than those of nonelectrolytes at the same concentration. Sodium chloride and calcium chloride are used to melt ice on highways; their aqueous solutions have freezing points as low as −21°C and −55°C, respectively.

Unfortunately, these salts also increase the rate of corrosion of cars that use those highways

The equations for the freezing point lowering, boiling point elevation and osmotic pressure of electrolytes are similar to those given in Chapter 12 for nonelectrolytes except for the introduction of a multiplier i.

$$\Delta T_f = i(1.86°C)m \qquad (13.3)$$

$$\Delta T_b = i(0.52°C)m \qquad (13.4)$$

$$\pi = iMRT \qquad (13.5)$$

TABLE 13.2 FREEZING POINT LOWERINGS OF SOLUTIONS OF NaCl AND $MgSO_4$

	ΔT_f Observed (°C)		i (Calc. from Eq. 13.3)	
Molality	NaCl	$MgSO_4$	NaCl	$MgSO_4$
0.005	0.0182	0.0158	1.96	1.70
0.01	0.0360	0.0301	1.94	1.62
0.02	0.0714	0.0573	1.92	1.54
0.05	0.176	0.132	1.89	1.42
0.10	0.348	0.246	1.87	1.32
0.20	0.685	0.454	1.84	1.22
0.50	1.68	1.00	1.81	1.08

If we regard an electrolyte simply as a source of solute particles which act independently of one another in determining colligative properties, we would predict that i should be equal to *the number of moles of ions per mole of electrolyte;* that is, i would equal 2 for NaCl and $MgSO_4$, 3 for $CaCl_2$ and Na_2SO_4, and so on.

Looking at the data in Table 13.2, we see that the situation is not so simple as the preceding discussion implies. The observed freezing point lowerings of NaCl and $MgSO_4$ are smaller than we would predict from Equation 13.3 with $i = 2$. In other words, for these salts at any finite concentration, the multiplier i is less than 2; it approaches a limiting value of 2 as the solution becomes more and more dilute.

Comparing the numbers in the last two columns of Table 13.2, we see that the observed value of i depends upon:

—*the nature of the electrolyte*. At a given concentration, i is smaller for $MgSO_4$ than for NaCl.

—*the concentration of the electrolyte*. For both NaCl and $MgSO_4$, i decreases as concentration increases.

You will recall that the conductivities of water solutions of these salts showed a similar behavior. Indeed, if we were to plot the observed values of ΔT_f vs. concentration for NaCl and $MgSO_4$, the curves obtained would closely resemble those shown in Figure 13.3.

Despite the complications just noted, we can use measurements of colligative properties to decide upon the nature or extent of ionization of an electrolyte in water (Example 13.1).

Example 13.1 The freezing points of 0.010 m solutions of $NaHCO_3$ and HF are −0.038°C and −0.024°C. Use this information to determine

a. whether $NaHCO_3$ ionizes predominantly as

$$NaHCO_3(s) \rightarrow Na^+(aq) + HCO_3^-(aq); \; i \rightarrow 2$$

or

$$NaHCO_3(s) \rightarrow Na^+(aq) + H^+(aq) + CO_3^{2-}(aq); \; i \rightarrow 3$$

b. the extent to which HF, in 0.010 m solution, ionizes to H^+ and F^-.

Solution

a. Let us calculate the value of i in Equation 13.3.

$$0.038°C = i\,(1.86°C)(0.010)$$

$$i = \frac{0.038}{0.0186} = 2.0$$

Clearly, the first mode of ionization is indicated.

b. Again, we start by calculating i.

$$i = \frac{0.024}{0.0186} = 1.3$$

If HF were not ionized at all, i would be 1. If it were completely ionized to H^+ and F^-, i would be 2. Since the observed value of i is 30% of the way from 1 to 2, we conclude that HF is about 30% ionized in 0.010 m solution.

Structure of Electrolyte Solutions

The deviations between the observed properties of electrolyte solutions and those predicted on the basis of completely independent solute ions have been explained in various ways. Many prominent chemists have worked in this area, starting with Svandte Arrhenius, who in 1903 won a Nobel Prize for his theory of ionization in water solution. He believed that in a solution of NaCl, as in a solution of HF, there was an equilibrium between free ions (Na^+, Cl^-) and molecules (NaCl). According to Arrhenius, the ions were the predominant species but there were enough molecules present to make both the conductivity and the freezing point lowering of NaCl solutions fall below the predicted values based upon complete ionization. This simple picture fell into disfavor when x-ray studies gave no evidence for "molecules" of NaCl in the solid state. Since they did not exist in the solid state, it did not seem reasonable that they would be present in solution.

In 1923 Peter Debye, then at Zurich, later Nobel laureate at Cornell, and a student of his, Erich Hückel, pointed out that, because of electrostatic attraction,

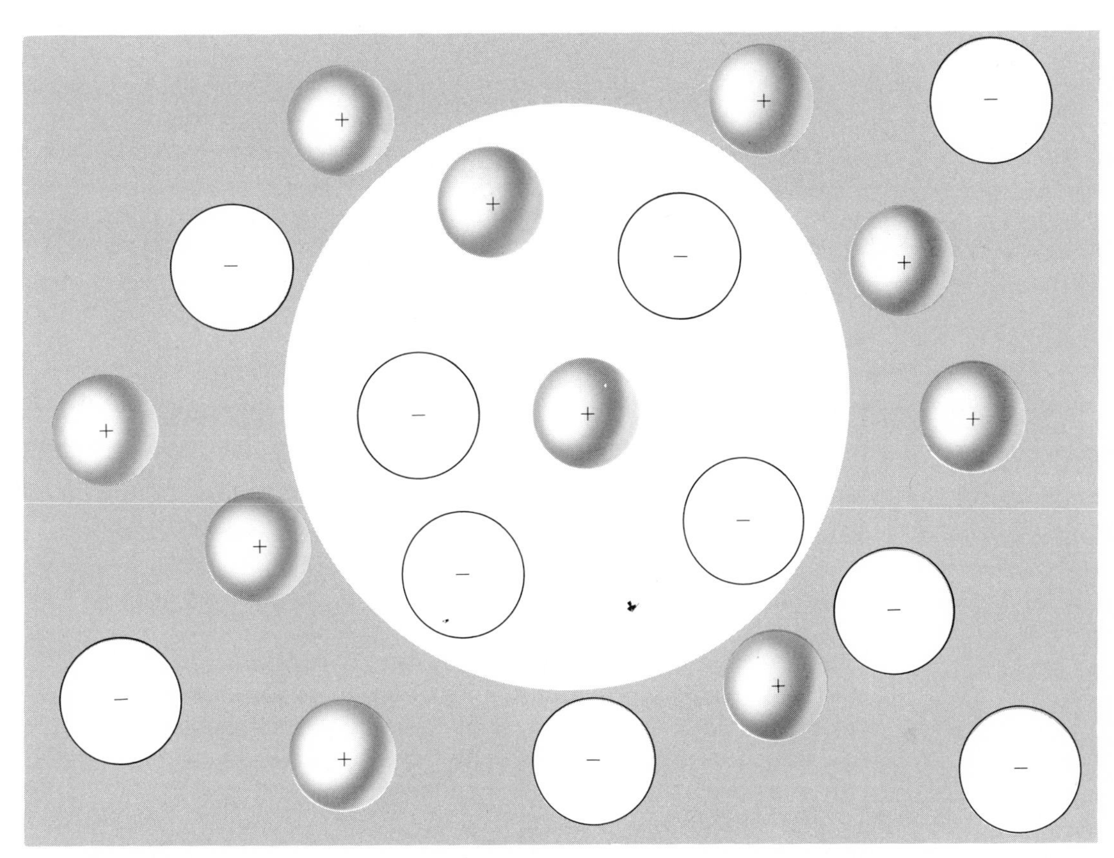

Figure 13.4 Ion atmosphere. An ion on the average is surrounded by more ions of opposite charge than of like charge.

there will be, near a given ion in solution, more ions of opposite than of like charge. In other words, an ion in solution will have associated with it an *ionic atmosphere,* containing an excess of oppositely charged ions (Fig. 13.4). The existence of such an atmosphere prevents ions from acting as completely independent solute particles. One result is to make an ion somewhat less effective than a nonelectrolyte molecule in its influence on colligative properties. Another is to reduce the conductivity of an ion, since it must "drag" the oppositely charged atmosphere with it as it moves through the solution.

Qualitatively at least, the Debye-Hückel approach offers a plausible explanation for many of the properties of electrolyte solutions. Since ion atmosphere effects become more pronounced at high concentrations, we would predict, in agreement with experiment, that deviations from "ideal" behavior should become more pronounced at high concentrations. Moreover, the Debye-Hückel model explains why these deviations should be greater for $MgSO_4$ than for NaCl; the ion atmosphere should be more important around ions of high charge (+2, −2 in $MgSO_4$) than near ions of low charge (+1, −1 in NaCl).

By taking into account the effect of the ion atmosphere, Debye and Hückel were able to derive equations which could be applied to predict the concentration dependence of such properties as conductivity and freezing point lowering. These equations work reasonably well in very dilute solutions, below about 0.01 *m*. Various modifications, notably that of Onsager for the conductivity of electrolyte solutions, have extended their range to as high as 0.1 *m*. In more concentrated solutions, however, all attempts to extend the Debye-Hückel treatment have proved unsatisfactory.

The essential weakness of the ion atmosphere model is that it neglects several other structural factors that operate even in relatively dilute solution. These include:

—*the hydration of ions* (recall Figure 13.2), which effectively ties up some of the water molecules and removes them from their role as solvent. This factor is most important with small cations of high charge such as Mg^{2+} or Cu^{2+}, but probably becomes significant with all ions at high concentrations.

—*changes in the water structure due to the presence of ions.* There is general agreement that ions can disturb the equilibrium between "free" H_2O molecules and "ice-like," hydrogen-bonded clusters of molecules. Precisely how this comes about and exactly what effect it has on the properties of electrolyte solutions is, to say the least, debatable.

—*short-range association between ions of opposite charge, to form "ion pairs."* This idea, first advanced by Niels Bjerrum in 1926, was largely ignored at the time because of the popularity of the Debye-Hückel theory, but is now coming back into favor. Ion pair formation appears to be of minor importance in solutions of 1:1 salts such as NaCl, except perhaps at high concentrations. However, there is general agreement that with magnesium sulfate, where we are dealing with +2 and −2 ions, the principal species, even at concentrations as low as 0.10 *m*, is the $MgSO_4$ ion pair rather than "free" Mg^{2+} and SO_4^{2-} ions.

Ion pairs are common in seawater and affect many of its properties

Unfortunately, despite many years of research, we still have no quantitative theory of electrolyte solutions which incorporates all these factors. We cannot, for example, write a general equation to predict the colligative behavior of simple salts at concentrations above, at best, 0.1 *m*. To be candid about it, our knowledge of the structure of electrolyte solutions has not increased substantially in the past twenty years. We have become more aware of the factors that must be taken into account to develop a comprehensive theory of salt solutions but, so far at least, no one has been able to put them all together.

13.2 NATURAL SOURCES OF WATER

The total amount of water above, below, and on the surface of the earth is estimated to be 1.33×10^{24} kg, about 5% of the total mass of the earth. The way

TABLE 13.3 DISTRIBUTION OF THE EARTH'S WATER

	MASS	% OF TOTAL
Oceans	$13\,000 \times 10^{20}$ g	97.7
Ice and snow	250×10^{20}	1.9
Underground	50×10^{20}	0.4
Lakes	1.2×10^{20}	0.009
Atmosphere	0.17×10^{20}	0.0013
Rivers and streams	0.02×10^{20}	0.0002

in which this water is distributed among various natural sources is indicated in Table 13.3. Notice that most of the earth's water is either contaminated by dissolved salts (the oceans) or relatively inaccessible (ice and snow, underground deposits). Lakes and rivers, which are the primary source of fresh water for most of the world's population, comprise less than 0.01% of its total water supply. However, even this small fraction represents about 4×10^{10} kg per person. In the literal sense, at least, the world is not about to run out of fresh water.

The oceans, which account for most of the earth's water, are a dilute electrolyte solution in which the principal solute species are Na^+ and Cl^- ions. These two ions, along with six others (Mg^{2+}, Ca^{2+}, K^+, SO_4^{2-}, HCO_3^-, Br^-), account for more than 99% of the total dissolved solids in seawater (Table 13.4). At least 50 other inorganic ions and literally thousands of organic molecules are also present, mostly in trace amounts.

The concentration of gold in seawater is about 2×10^{-11} *M*

Analyses of seawater samples from different parts of the world show that the major constituents are present in remarkably constant ratios. For example, the mole ratio of Cl^- to Na^+ is found to be 1.166 ± 0.006 in all of the oceans. The total concentration of dissolved salts is more variable, normally falling in the range of 33 to 37 g/kg seawater.

The total concentration of dissolved salts in river waters (Table 13.5) is, of course, much lower than that in seawater. Interestingly enough, the principal ions found in river waters are not the same as those in seawater. The most common anion in river waters is HCO_3^- rather than Cl^-; the most common cation is usually Ca^{2+} rather than Na^+. Elaborate chemical models of the oceans have been devised to explain these differences. It appears that shellfish play a major role in the depletion of Ca^{2+} and HCO_3^- ions in seawater. Their shells consist of calcium carbonate, $CaCO_3$, formed at the expense of these ions. In contrast, there seems to be no major mechanism for the removal of Na^+ and Cl^- ions from seawater.

Lakes vary enormously in the purity of their water. A high mountain lake fed entirely by rainwater or melting snow may contain as little as 0.01% of dissolved salts. At the other extreme, a few of the world's major lakes are "saltier" than

TABLE 13.4 COMPOSITION OF SEAWATER (mol/dm³)*

CATIONS		ANIONS		NEUTRAL SPECIES	
Na^+	0.4570	**Cl^-**	0.5331	H_3BO_3	0.0004
Mg^{2+}	0.0520	**SO_4^{2-}**	0.0276	N_2	0.0004
Ca^{2+}	0.0101	**HCO_3^-**	0.0023	O_2	0.0002
K^+	0.0097	**Br^-**	0.0008	SiO_2	0.0001
Sr^{2+}	0.0001	CO_3^{2-}	0.0003		

*Values given at 25°C for seawater containing 3.5% by mass of dissolved salts. Species indicated in bold type are considered "major constituents"; all species at concentrations of 1×10^{-4} mol/dm³ or greater are listed.

The Amazon is a "clean" river

TABLE 13.5 CONCENTRATIONS OF IONS IN RIVER WATER (mol/dm^3)

	Hudson at Green Island	Mississippi at Baton Rouge	Colorado at Yuma	Amazon at Obides
Na^+	0.000 21	0.000 48	0.005 41	0.000 07
Mg^{2+}	0.000 20	0.000 31	0.001 24	0.000 02
Ca^{2+}	0.000 80	0.000 85	0.002 35	0.000 14
K^+	0.000 05	0.000 08	0.000 11	0.000 05
Cl^-	0.000 14	0.000 43	0.003 18	0.000 07
SO_4^{2-}	0.000 25	0.000 43	0.003 01	0.000 01
HCO_3^-	0.001 53	0.001 65	0.003 00	0.000 29

seawater. Great Salt Lake in Utah has a total concentration of dissolved salts about six times as great as that of the oceans. Searles Lake in the Mojave desert of California is not, in the ordinary sense, a lake at all. It consists of a vast deposit of solid salts permeated with a saturated brine solution. The brine is a major commercial source of a wide variety of chemicals including salts of the alkali metals, particularly potassium and lithium.

13.3 WATER POLLUTION

During the past decade we have become increasingly aware of the problem of water pollution. Fish kills caused by human and industrial wastes have been reported in rivers as far apart as the Androscoggin in northern New England and the Columbia in the Pacific Northwest. Oil slicks along our beaches and pesticide residues leaching into inland waters have been held responsible for reductions in the population of waterfowl and songbirds. Many of our lakes and rivers are so badly contaminated that they cannot even be used for recreational purposes.

Figure 13.5 (Photograph by Donald Chandler, courtesy of Dr. Arthur D. Harlen, Professor of Zoology, University of Wisconsin.)

Perhaps we should start our discussion of water pollution by explaining what it means. "Polluted water" is, quite simply, water that is unfit for the purpose for which it is intended. It is entirely possible for water to be "polluted" for one purpose yet pure enough for certain other uses. Water containing 0.05 ppm of lead ion (i.e., 0.05 g of $Pb^{2+}/10^6$ g of water) is considered to be unfit for drinking but is perfectly satisfactory for most industrial purposes. Water contaminated with 10% alcohol along with smaller amounts of sugar and other organic compounds would not be used for washing dishes or irrigating asparagus, yet, as a beverage, many people prefer wine to distilled water.

Sources of Pollution

Many of the inorganic cations that we find objectionable in surface and ground waters come from natural sources. Water coming into contact with limestone ($CaCO_3$) or dolomite ($CaCO_3 \cdot MgCO_3$) picks up Ca^{2+} and Mg^{2+} ions which make it too "hard" for many household and industrial uses. Underground deposits of iron compounds are responsible for the presence of Fe^{3+} in certain ground waters. This ion reacts with hot water to form rust (hydrated ferric hydroxide) which deposits as a brown stain on bathtubs and clothing.

$$Fe^{3+}(aq) + 3\ H_2O \rightarrow Fe(OH)_3(s) + 3\ H^+(aq) \qquad (13.6)$$

Another natural pollutant that we see in many of our rivers is silt, which consists of suspended mineral particles resulting from land erosion. One billion tonnes of silt are carried into the oceans each year by our rivers; four times as much settles out along the way, gradually filling lakes and reservoirs. Hoover Dam, built less than fifty years ago, has already lost more than half of its capacity because of deposition of sediment.

Perhaps the most obvious man-made pollutant is human sewage, which began to contaminate water supplies when the flush toilet was introduced around 1800. Many of the epidemics of cholera and typhoid fever which were common during the 19th century were traced to this source. Even today, microorganisms in human waste can be a serious health hazard. Infectious hepatitis is carried in blood and feces by a virus that is highly resistant to ordinary methods of water treatment.

Another development occurred around 1800 that greatly increased the extent of water pollution. This was the Industrial Revolution. Today, it is estimated that industry is the source of approximately two thirds of the total amount of man-made pollutants. Traditionally the worst offenders have been food processors, the pulp and paper industry, metal producers and chemical manufacturers, not necessarily in that order. For years many industries dragged their feet in the area of pollution abatement. Now the situation has changed; many larger corporations are rapidly increasing their efforts in this area, spending several billion dollars a year to control water pollution.

Much, much more water is used in industry than in homes

One of the most serious effects of human and industrial wastes is to increase the **biochemical oxygen demand** ***(BOD)*** of natural water supplies. Aerobic bacteria use oxygen to degrade the complex organic compounds in sewage to simpler and generally unobjectionable species such as CO_2, NO_3^- ions, and SO_4^{2-} ions. This process reduces the amount of dissolved oxygen in the water, sometimes to the point where animal life cannot survive. If the oxygen content drops too low, anaerobic bacteria take over the decomposition process, forming rather noxious pollutants such as CH_4, NH_3, and H_2S.

The *BOD* of polluted water is determined by measuring the amount of oxygen consumed by a sample of known volume. The water sample is first diluted with air-saturated distilled water to ensure an excess of oxygen. The concentration of dissolved oxygen in the diluted sample is determined immediately and again after a period of five days. From the decrease in oxygen concentration, one can calculate the *BOD*, which is usually expressed as

$$BOD = \frac{\text{no. milligrams } O_2 \text{ required}}{\text{no. cubic decimetres sample}} \equiv \text{parts } O_2 \text{ per million parts sample} \quad (13.7)$$

Example 13.2 Calculate the *BOD* of a water supply which contains one gram of urea for every 100 dm^3 of water, assuming that the reaction between urea and oxygen is

$$H_2N{-}\underset{\underset{\displaystyle O}{\|}}{C}{-}NH_2(aq) + 4\ O_2(aq) \rightarrow CO_2(aq) + 2\ NO_3^-(aq) + 2\ H^+(aq) + H_2O$$

Solution From the balanced equation we see that 4 mol O_2 are required to react with 1 mol urea, $H_2N{-}CO{-}NH_2$. But, since 1 mol O_2 weighs 32.0 g and 1 mol urea weighs 60.0 g, we have

$$4(32.0 \text{ g } O_2) \triangleq 1(60.0 \text{ g urea}), \text{ or } 128 \text{ g } O_2 \triangleq 60.0 \text{ g urea}$$

So, to react with 1 g of urea, we need

$$1.00 \text{ g urea} \times \frac{128 \text{ g } O_2}{60.0 \text{ g urea}} = 2.13 \text{ g } O_2 = 2.13 \times 10^3 \text{ mg } O_2$$

From Equation 13.7,

$$BOD = \frac{2.13 \times 10^3 \text{ mg } O_2}{100 \text{ dm}^3} = 21.3 \text{ mg } O_2/\text{dm}^3$$

A *BOD* of 0 to 3 is characteristic of "pure" or nearly pure water. *BOD* values higher than 5 indicate water of doubtful purity. Untreated city sewage typically has a *BOD* of 100 to 400; some industrial wastes have values as high as 10 000. You can appreciate the problems created if these materials are dumped directly into a water supply.

Types of Pollutants

Water pollutants are commonly classified into eight major categories (Table 13.6). Rather than attempting to discuss all these types of pollutants, we will consider only a few examples of current interest.

There are a lot of ways to pollute water

TABLE 13.6 CLASSIFICATION OF WATER POLLUTANTS

TYPE	EXAMPLES
1. Oxygen-demanding wastes	1. Human, animal waste; decaying vegetation
2. Infectious agents	2. Bacteria and viruses
3. Organic molecules	3. Detergents, insecticides, oil
4. Plant nutrients	4. Nitrates, phosphates
5. Inorganic chemicals	5. Hg, Cd^{2+}, Pb^{2+}
6. Heat	6. Water used for cooling in industry
7. Suspended material	7. Silt from land erosion
8. Radioactive substances	8. Fallout products, radioactive waste

DETERGENTS. The structure of a type of detergent widely used in many countries prior to 1965 is shown at the top of Figure 13.6. This detergent had one serious shortcoming from an ecological standpoint—it is not biodegradable; that is, bacteria are unable to chew up the organic anion, breaking it down into simpler species. Consequently the detergent stayed around in water supplies more or less indefinitely, causing some rivers and sewage treatment ponds to become covered with mountains of foam. Fortunately, chemists were able to modify the structure of the anion to give a detergent which is degradable by aerobic (oxygen-consuming) bacteria. The modification was a simple one; the branched side chain was replaced by a straight chain to give the structure shown at the bottom of Figure 13.6. For reasons which we do not understand, aerobic bacteria seem to find straight chains more palatable than branched chains.

The change from branched to straight-chain detergents, which occurred about ten years ago, solved the foam problem. Commercial detergent products can, however, contribute to pollution in another way. In some brands of detergents, as much as one half of the total mass consists of inorganic phosphates (e.g., sodium tripolyphosphate, $Na_5P_3O_{10}$), which form very stable complexes (ion pairs) with cations such as Ca^{2+}, Mg^{2+} and Fe^{3+} that would otherwise interfere with the cleansing action of the detergent. Phosphates are important plant nutrients and as such tend to promote the growth of algae. There is evidence to indicate that detergent phosphates are a factor in the blooms of algae that are choking many of our lakes, notably Lake Erie. There is considerable controversy on this point; it has been shown that in many water systems the critical element in algae growth is carbon or nitrogen rather than phosphorus.

Most of the "low-phosphate" and "nonphosphate" detergents now on the market contain carbonate or silicate salts as a substitute for phosphates. These products are generally less effective cleansing agents; moreover some of them give strongly alkaline solutions that can be hazardous for household use. The best

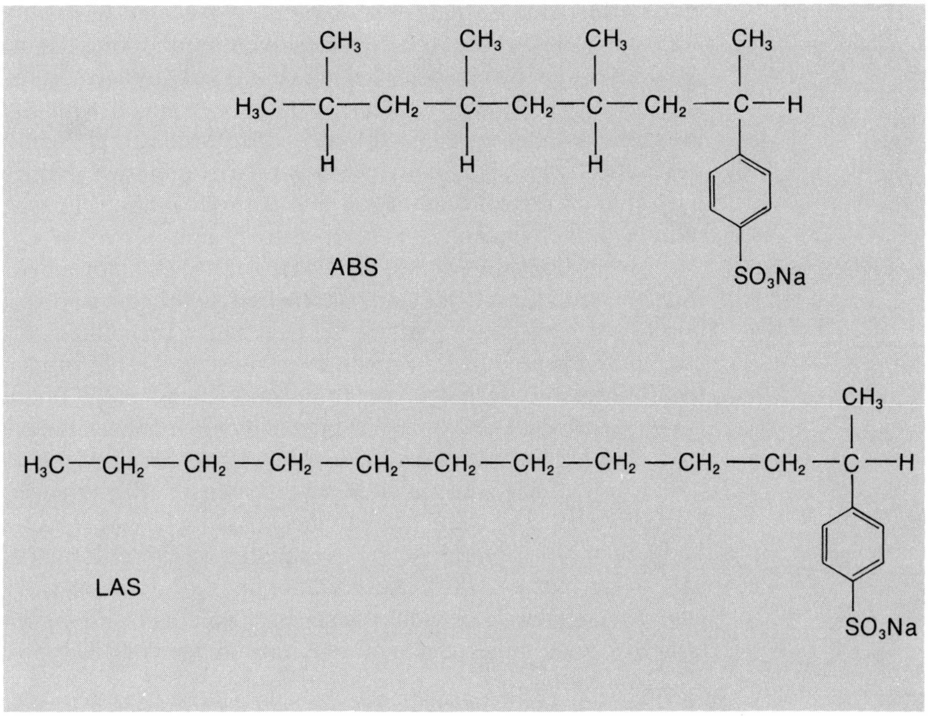

ABS

LAS

Figure 13.6 Bacteria are much more able to degrade a straight chain detergent (linear alkylsulfonate, LAS) than they are a branched chain detergent (alkylbenzenesulfonate, ABS).

long-term solution to the "phosphate" problem probably involves installing advanced sewage treatment systems, which are capable of removing phosphates and other plant nutrients as well.

INSECTICIDES. Prior to World War II the effective ingredients of most insecticides were inorganic compounds of copper, lead and arsenic. These compounds were toxic to nearly all forms of animal life, including man. In 1939 a "safe" organic insecticide was found, *d*ichloro*d*iphenyl*t*richloroethane (DDT):

```
              H
              |
Cl—⟨benzene⟩—C—⟨benzene⟩—Cl
              |
          Cl—C—Cl
              |
              Cl
```

DDT

DDT was found to be effective against a wide variety of disease-bearing and agricultural insects. It was used during World War II to prevent epidemics of typhus and malaria carried by lice and mosquitoes, respectively. After the war, its use in agriculture and forestry increased to the point where 8×10^7 kg were manufactured in 1963. Its production has been dropping ever since.

As early as 1946 a species of housefly was found to be resistant to DDT. This phenomenon has now become quite common; over 100 insect pests are known to have developed immunity in a classic example of Darwinian evolution. To meet this problem a variety of other insecticides were developed, most of which, like DDT, are chlorinated hydrocarbons. Most of these compounds, like DDT, are nontoxic to humans, at least over a short time span.

In many communities, DDT was sprayed routinely for mosquito control

DDT and many of its relatives are persistent insecticides in that they are not readily degraded in the environment. At one time this was considered to be an advantage since these insecticides stayed around long enough to control successive generations of insects. Today, the disadvantages are all too apparent. The stability of chlorinated organic pesticides, coupled with their high solubility in fatty tissues, causes them to concentrate in the food chain. A study of aquatic life in the Long Island marshes showed that minnows eating plankton containing 0.04 ppm of DDT accumulated one part per million; in sea gulls feeding on the minnows, the concentration jumped to 75 ppm.

The toxicity to wildlife of DDT and other chlorinated insecticides is well documented. Insecticides are at least partially responsible for the near extinction of such predatory birds as the bald eagle, the golden eagle and the peregrine falcon. The reproductive capacities of these and other birds are inversely related to DDT uptake. Ospreys, which normally average 2.5 young per year, produce 1.1 offspring at a DDT level of 3 ppm and only 0.5 when the concentration of DDT reaches 5 ppm.

These effects were in large part responsible for government action taken in Canada (1969) and the United States (1972) to severely restrict the use of DDT. More recently, several other insecticides of the chlorinated hydrocarbon type have been banned. With these compounds, the action was taken because they have been shown to produce cancer in certain laboratory animals. There is a concern that, over long periods of time, insecticides of this type could have the same effect on humans.

Several insecticides are now available which decompose rather rapidly in water and, hence, pose less of a long-term threat than DDT. Most of these compounds are organic phosphorus derivatives: parathion is one example.

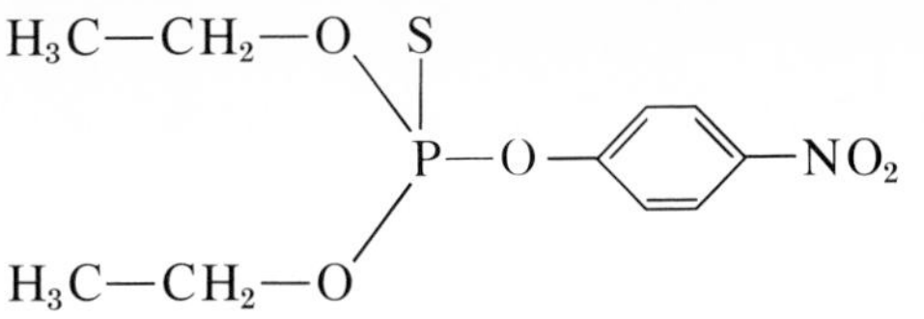

$$H_3C-CH_2-O \quad S$$
$$P-O-C_6H_4-NO_2$$
$$H_3C-CH_2-O$$

parathion

Parathion decomposes readily in water; in twenty days, half of the insecticide is gone. Unfortunately parathion and many related phosphorus insecticides are far more toxic to humans than DDT. In one year, 1958, 100 fatalities in India and over 300 in Japan were attributed to the careless use of parathion.

An insecticide which has been widely used recently to control gypsy moth infestations is carbaryl (Sevin):

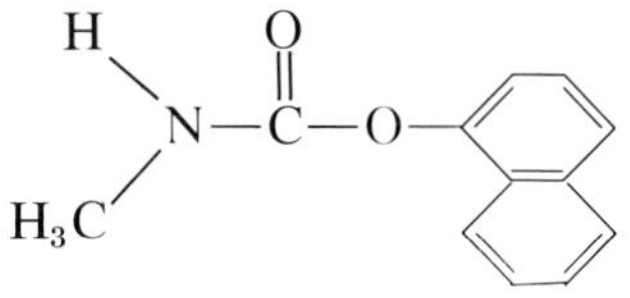

$$H \qquad O$$
$$N-C-O-C_{10}H_7$$
$$H_3C$$

carbaryl

Sevin is not so toxic as DDT and has a less harmful effect on the environment, largely because it decomposes quite rapidly. This factor, however, makes it less effective as an insecticide; repeated sprayings are required.

The problems experienced with DDT and other chemical insecticides have stimulated research on other methods of insect control. One approach which shows considerable promise involves the use of sex attractants. It has been applied in attempts to control the codling moth, the boll weevil, and the gypsy moth. The female insect secretes a tiny amount of a chemical sex attractant during the mating period. In the case of the gypsy moth, the attractant has been identified as

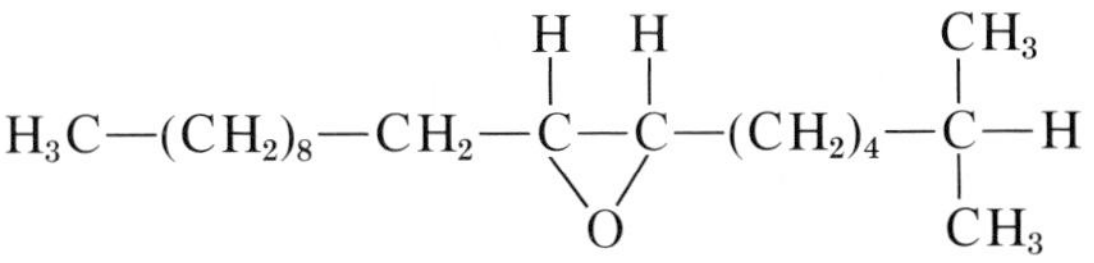

$$H_3C-(CH_2)_8-CH_2-\overset{H}{C}-\overset{H}{C}-(CH_2)_4-\overset{CH_3}{\underset{CH_3}{C}}-H \quad (\text{the two } C\text{'s bridged by } O)$$

Another approach uses "juvenile hormones" which prevent insects from maturing

This complex organic molecule, now produced synthetically in the laboratory, is used to bait traps in infested areas. As little as 2×10^{-5} g is potent enough to attract all the males in the vicinity.

We can hope that this or other nonchemical methods of insect control will ultimately free us from our dependence on insecticides. We cannot forget, however, that in many parts of the world today chemical insecticides are essential to the prevention of disease and famine. The United Nations' World Health Organization has pointed out that the use of DDT freed an estimated billion people from the threat of malaria. It warns that a worldwide ban before effective substitutes are found would be "a disaster to world health."

TRACE ELEMENTS. The presence in our environment of compounds of elements known to be toxic even at very low concentrations has become a major concern in both water and air pollution. As you will note from Table 13.7, several of these elements are introduced into the atmosphere by combustion of fossil fuels. For example, the major source of lead pollution is the burning of gasoline containing the antiknock additive lead tetraethyl, $Pb(C_2H_5)_4$. Other toxic metals,

TABLE 13.7 TOXIC TRACE ELEMENTS IN THE ENVIRONMENT

ELEMENT	MAJOR SOURCES	EFFECTS ON HUMANS
Arsenic	Coal burning, impurity in phosphates, processing of sulfide ores	Large doses cause gastrointestinal disorders; smaller quantities may be carcinogenic
Beryllium	Coal burning, nuclear fuel processing, some rocket fuels	Long exposure even at low levels causes lung damage, often fatal
Cadmium	Electroplating wastes, impurity in all products containing Zn	At low levels (0.2 mg/d) causes hypertension. Degenerative bone disease at high levels
Lead	Leaded gasolines, paints	Nausea, irritability at low levels; large doses cause brain damage
Mercury	Electrochemical industry, certain fungicides	Large doses of organic Hg compounds cause brain damage, often fatal

On the other hand, traces of many elements are beneficial or even essential to living organisms

like cadmium and mercury, enter water supplies as waste products from industrial processes.

The recent development of instrumental methods capable of detecting metals in the part per million or even part per billion range prompted a systematic search for trace elements in water supplies and the food chains that they support. The results have been disturbing. In 1970 Norwald Finreite, a graduate student at the University of Western Ontario, reported mercury concentrations in fish caught in Lake St. Clair to be as high as 7 ppm. This number is to be compared to the "safe" limit set for fish by the Food and Drug Administration of 0.5 ppm, and the concentration of 0.2 ppm later found to be typical of fish in unpolluted waters.

Most of the mercury found in fish is present as an organic derivative, dimethyl mercury, $(CH_3)_2Hg$, a toxic substance which is known to concentrate in the food chain in much the same way as DDT. This compound is synthesized from elementary mercury by certain anaerobic bacteria living at the bottom of lakes and rivers. Dimethyl mercury appears to concentrate in brain tissue and remain there for long periods of time. Symptoms of mercury poisoning arise when the concentration of $(CH_3)_2Hg$ in the brain reaches 5 ppm; 12 ppm is usually fatal.

HEAT. Whenever we attempt to convert heat into useful work, a considerable portion of that heat is discharged into the surroundings. The "surroundings" may be the atmosphere; alternatively, the heat may be discharged to a neighboring body of water. This can create a problem of *thermal pollution* quite as serious as the types of material pollution discussed up to this point. The electrical power industry requires large quantities of cooling water to condense steam that has been used to generate electrical energy. In a conventional power plant, at least 2 kJ of heat are discharged to the water circulated through condensers for every kilojoule of electrical energy produced. In a nuclear power plant, this ratio is even higher, as high as 3 to 1.

The discharge of hot water into a river or lake raises its temperature, typically by 5 to 10°C at the source. Occasionally temperatures as high as 60°C

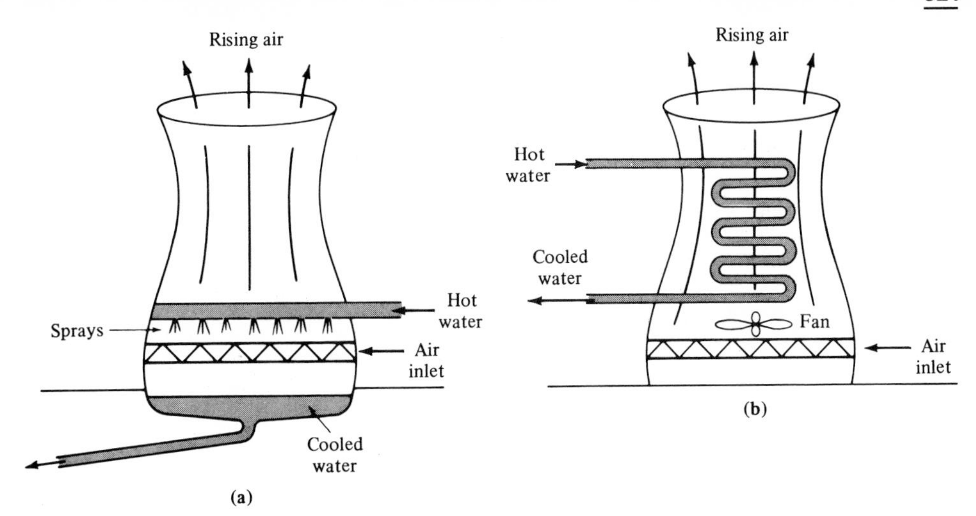

Figure 13.7 Evaporative cooling (a) is more efficient than cooling by heat transfer (b) but tends to create a lot of mist or fog.

have been recorded. Even a relatively small temperature increase can have an adverse effect on aquatic life; salmon and trout cannot live in water much above 25°C. At higher temperatures the rate of metabolism increases, creating a greater demand for oxygen. Less oxygen is available, since its solubility decreases with increasing temperature.

One way to minimize the problem of thermal pollution is to use a cooling tower to reduce the temperature of waste water before it is discharged into a stream. In one type of tower, air is blown through the water to take advantage of the cooling effect of evaporation. The disadvantage of this design is that large quantities of water vapor are introduced into the atmosphere. Not only does this waste water; it may even form clouds or fog banks in the area around the tower. This problem is avoided in towers in which the warm water does not come into direct contact with air. By circulating the water through a maze of tubes over which compressed air is blown, its temperature can be reduced in much the same way as in an ordinary automobile radiator. Unfortunately, this type of tower is about three times as expensive to build and maintain as one of the evaporative type. Even so, the total increase in cost to the consumer's electrical bill is estimated to be less than 5%.

The Second Law of Thermodynamics tells us that no machine can completely convert heat into work; in this sense, "waste" heat is inevitable

Many suggestions have been made for making use of the heated water produced by power plants. In a few cases, it has been used to heat buildings in the vicinity, but the relatively low temperature of the water discourages this practice. Another possibility is to use it to promote the growth of oysters and certain other shellfish that seem to flourish in warm water.

13.4 WATER PURIFICATION

Perhaps you have heard the statement that "running water purifies itself." To some extent this is true for an isolated mountain stream. However, the method is not very practical for purifying the water used by a city such as Chicago or Montreal. Methods of treating enormous quantities of slightly or moderately contaminated water are necessary and are operating in virtually every metropolitan area in the world. In this section we shall consider some of the processes that are used to purify the water we use in our homes and in industry.

Removal of Suspended Matter

The simplest way to clarify muddy water is to allow the suspended material in it to settle. This process, called *sedimentation,* takes place when water is allowed to stand in a reservoir or to flow through a series of settling tanks. Sedimentation alone is often insufficient to remove all the suspended matter; silt particles small enough to be in colloidal suspension cannot be removed in this manner.

The process of *coagulation* is used to remove colloidal particles from water. It involves adding a compound to water which forms a gelatinous precipitate that carries suspended particles down with it. Aluminum salts are among the more popular coagulants; the Al^{3+} ion reacts with water to give a precipitate which can be represented most simply as $Al(OH)_3$.

$$Al^{3+}(aq) + 3\ H_2O \rightarrow Al(OH)_3(s) + 3\ H^+(aq) \qquad (13.8)$$

Millions of kilograms of hydrated aluminum sulfate, $Al_2(SO_4)_3 \cdot 18\ H_2O$, are used annually to clarify municipal and industrial water supplies.

Water which has passed through a coagulation plant is often filtered to remove any remaining suspended material. *Filtration* (Fig. 13.8) can be accomplished by allowing the water to trickle through a layer of sand supported on several layers of successively coarser gravel. A thin layer of precipitated material forms on the sand. Up to a point at least, this added layer of finely divided particles improves the efficiency of the filter. When it becomes too thick, it is removed by backflushing with water.

Disinfection

City water which is to be used for drinking must be treated chemically to remove disease-causing bacteria. Disinfection is also essential for water in municipal swimming pools, although the tolerable bacteria count there can be somewhat

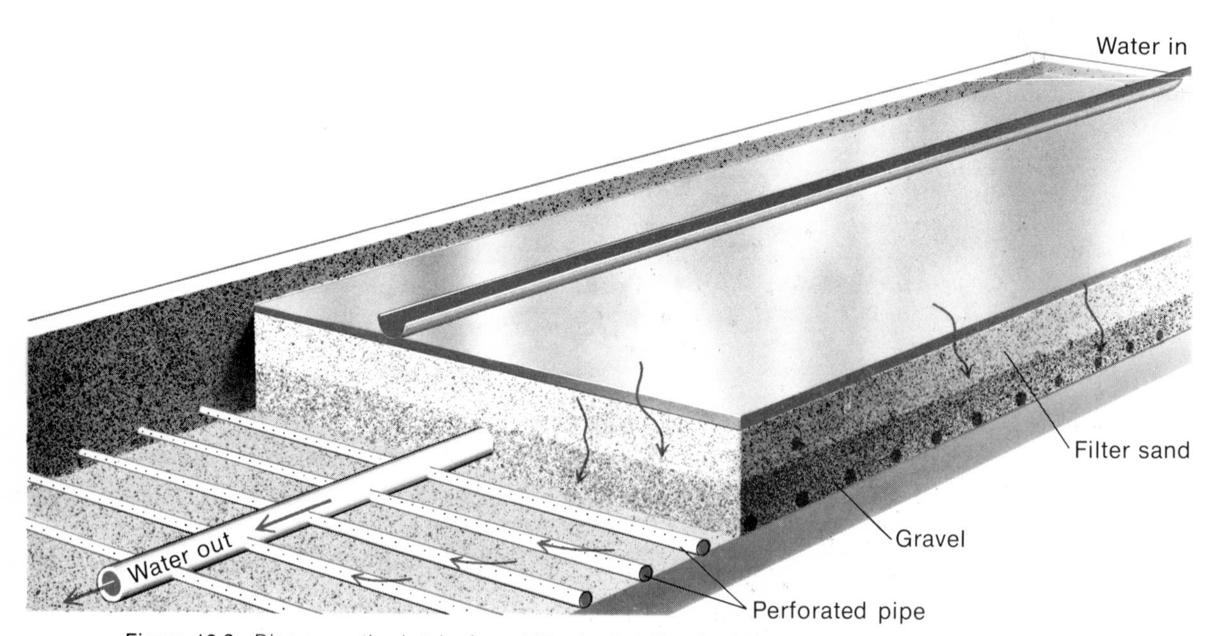

Figure 13.8 Diagrammatic sketch of sand filter bed. A filter bed like this might easily cover a hectare.

higher. The most common disinfectant is chlorine, which kills bacteria, apparently by inhibiting the activity of certain enzymes essential to their metabolism. Large water treatment plants use liquid chlorine from high pressure tanks as a disinfectant. Enough is added to give a residual Cl_2 concentration of 0.2 to 1.0 ppm. This concentration is ordinarily high enough to destroy any bacteria entering the water after it leaves the treatment plant.

Many chemicals other than chlorine can be used to disinfect water supplies. Ozone, O_3, has been used for many years in Canada and Western Europe and is now becoming more popular in the United States. Although somewhat more expensive than chlorine, it is considerably more effective, killing many germs and viruses that are not attacked by chlorine. It leaves no chemical residue and hence does not produce objectionable tastes or odors in drinking water. By the same token, however, ozone does not offer any protection against microorganisms that enter the water after it has been treated.

Removal of Taste and Odor

No matter how safe water may be to drink, we object to it if it has an unpleasant taste or odor. City dwellers accustomed to chlorinated water enjoy drinking from a clear, cold mountain stream. Most of us are turned off by the odor of hydrogen sulfide associated with water from hot mineral springs, despite its supposed medicinal effect.

One way to remove objectionable tastes and odors from water is to pass it through a filter bed containing activated charcoal, a finely divided form of carbon. The truly enormous surface area of this material (as much as 600 m^2/g) enables it to *adsorb* large quantities of various impurities. Most of these are organic compounds held physically on the surface of the carbon. In the case of chlorine, a chemical reaction occurs:

Aeration can be used to remove certain odors, notably that of H_2S

$$2\ Cl_2(aq) + C(s) + 2\ H_2O \rightarrow CO_2(aq) + 4\ H^+(aq) + 4\ Cl^-(aq) \qquad (13.9)$$

Through this reaction, the taste and odor associated with chlorinated drinking water can be eliminated. After the filter bed has been operated for some time (typically about a year), most of the carbon can be reclaimed by heating to about 1000°C to remove adsorbed organic impurities.

Water Softening

"Hard" water, caused by the presence of Ca^{2+} and Mg^{2+} ions, has several undesirable properties. For one thing, it forms a precipitate with soaps, which are sodium salts of long chain organic acids (Chapter 10). A typical soap is sodium stearate, $NaC_{18}H_{35}O_2$. Since the calcium and magnesium salts of stearic acid are insoluble, the following reaction occurs when a sodium stearate soap is used in hard water:

$$M^{2+}(aq) + 2\ C_{18}H_{35}O_2^-(aq) \rightarrow M(C_{18}H_{35}O_2)_2(s);\ M = Ca \text{ or } Mg \qquad (13.10)$$

This reaction continues until nearly all the Ca^{2+} and Mg^{2+} ions are removed; only then does the soap become effective as a cleaning agent.

From an industrial standpoint, a more serious drawback of hard water is the tendency of calcium and magnesium salts to precipitate when water is heated or partially evaporated, as in a steam boiler. The nature of the precipitate depends upon the anion present. A common anion in surface water is the HCO_3^- ion, formed by the reaction of atmospheric CO_2 with water. When water containing Ca^{2+} and HCO_3^- ions is heated, a precipitate of calcium carbonate forms as a result of the reaction sequence:

$$\begin{array}{r} 2\,HCO_3^-(aq) \rightarrow CO_3^{2-}(aq) + CO_2(g) + H_2O \\ \underline{Ca^{2+}(aq) + CO_3^{2-}(aq) \rightarrow CaCO_3(s)} \\ 2\,HCO_3^-(aq) + Ca^{2+}(aq) \rightarrow CaCO_3(s) + CO_2(g) + H_2O \end{array} \quad (13.11)$$

The scale on teakettles and coffeepots is usually calcium carbonate; it can be removed by treating with a weakly acidic solution such as vinegar.

The problem becomes more serious if there are SO_4^{2-} ions in hard water. In this case, heating yields a precipitate of calcium sulfate, one of the few salts which is less soluble at high than at low temperatures.

$$Ca^{2+}(aq) + SO_4^{2-}(aq) \rightarrow CaSO_4(s) \quad (13.12)$$

Calcium sulfate is particularly likely to form in a boiler from which water is being vaporized, thereby increasing the concentration of Ca^{2+} and SO_4^{2-} ions. It deposits as a tightly adherent scale which lowers the heat conductivity of the boiler and may even block the flow of water through the tubes.

We will discuss two of the more common methods used to soften water, the lime-soda and cation exchange processes. Our discussion will deal only with the removal of Ca^{2+} ions, although both processes will remove Mg^{2+} ions as well.

LIME-SODA METHOD. This process, which has been in use for over a hundred years, is still used in many municipal water-softening plants. It involves adding two chemicals, slaked lime, $Ca(OH)_2$, and soda ash, Na_2CO_3, to the hard water. The purpose of adding sodium carbonate is obvious; it acts as a source of CO_3^{2-} ions to precipitate Ca^{2+}

$$\underset{\text{(hard water)}}{Ca^{2+}(aq)} + \underset{\text{(from } Na_2CO_3)}{CO_3^{2-}(aq)} \rightarrow CaCO_3(s) \quad (13.13)$$

The function served by the lime is less apparent. Indeed, it might seem that the addition of $Ca(OH)_2$ to water in an attempt to remove Ca^{2+} ions would be a step in the wrong direction. As a matter of fact, lime is an effective water softener only when there are HCO_3^- ions in the hard water. To understand why this is the case, consider what happens when one mole of $Ca(OH)_2$ is added to water containing two moles of HCO_3^- ions. Three separate reactions are involved:

$$(1) \quad Ca(OH)_2(s) \rightarrow Ca^{2+}(aq) + 2\,OH^-(aq)$$

$$(2) \quad 2\,HCO_3^-(aq) + 2\,OH^-(aq) \rightarrow 2\,CO_3^{2-}(aq) + 2\,H_2O$$

$$(3) \quad \underline{Ca^{2+}(aq) + CO_3^{2-}(aq) \rightarrow CaCO_3(s)}$$

$$Ca(OH)_2(s) + 2\,HCO_3^-(aq) \rightarrow CaCO_3(s) + CO_3^{2-}(aq) + 2\,H_2O \quad (13.14)$$

We note from this reaction sequence that the lime furnishes the OH^- ions required to convert HCO_3^- ions in the water to CO_3^{2-} ions. The net effect of the addition of the lime, as represented by Equation 13.14, is to form an extra mole of free CO_3^{2-} ions. These ions are then capable of removing a mole of Ca^{2+} originally present in the hard water:

$$\underset{\text{(hard water)}}{Ca^{2+}(aq)} + \underset{\text{(from } HCO_3^- \text{ via 13.14)}}{CO_3^{2-}(aq)} \rightarrow CaCO_3(s)$$

Excess $Ca(OH)_2$ will make the water harder rather than softer

Thus, for every mole of $Ca(OH)_2$ added, an extra mole of Ca^{2+} ions is removed from the hard water. It is, of course, extremely important not to add too much $Ca(OH)_2$ to the water; any excess over that required for Reaction 13.14 will tend to increase the Ca^{2+} ion concentration.

The sequence of steps involved in softening hard water by the lime-soda process may be summarized as follows:

1. The water is first analyzed for Ca^{2+} and HCO_3^-.
2. Sufficient lime is added to give one mole of $Ca(OH)_2$ for every two moles of HCO_3^-. Every mole of $Ca(OH)_2$ added removes one mole of Ca^{2+} from the hard water.
3. Any Ca^{2+} remaining in the hard water is removed by adding soda ash, Na_2CO_3, on a 1:1 mole basis.

In practice, the slaked lime and soda ash are added simultaneously. In a water treatment plant designed to remove suspended solids it is convenient to add these chemicals in the coagulation step. The calcium carbonate that precipitates comes down with the suspended material.

Example 13.3 A sample of water is found on analysis to contain 0.0030 mol/dm^3 of Ca^{2+} and 0.0040 mol/dm^3 of HCO_3^-. Calculate the number of moles of $Ca(OH)_2$ and Na_2CO_3 that should be added to 1.00 dm^3 of this water to soften it.

Solution The first step is to add 1 mol $Ca(OH)_2$ for every 2 mol HCO_3^- present:

$$\text{no. moles } Ca(OH)_2 = 0.0040/2 = 0.0020 \text{ mol } Ca(OH)_2$$

This removes an equal amount, 0.0020 mol, of Ca^{2+} from the water, leaving in one cubic decimetre of the water

$$(0.0030 - 0.0020) \text{ mol} = 0.0010 \text{ mol of } Ca^{2+}$$

Consequently, it is necessary to add 0.0010 mol of Na_2CO_3 to the water.

CATION EXCHANGE. The lime-soda method of water softening is impractical for household use; one can hardly expect a homeowner to analyze water, calculate how much of each chemical to add, and then filter off the precipitates. For many industrial purposes, the lime-soda process is also inappropriate, because it usually leaves water supersaturated with $CaCO_3$. This can be particularly serious in a commercial laundry, where the calcium carbonate may end up on the customer's clothing.

In situations such as these, the preferred method of water softening is the cation exchange process, in which Ca^{2+} ions in the hard water are replaced by Na^+ ions. This exchange takes place when hard water is passed through a column containing either a certain type of mineral known as a *zeolite,* which occurs widely in nature, or a man-made "synthetic zeolite."

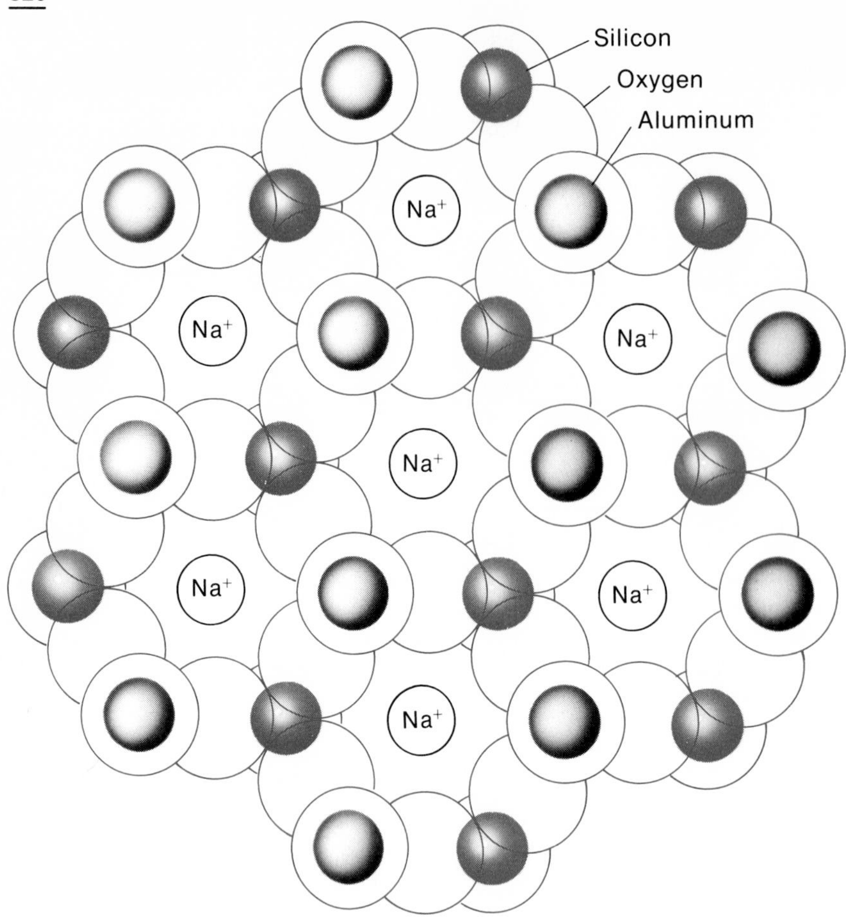

Figure 13.9 Structure of a natural zeolite, $NaAlSiO_4$. The anion, $AlSiO_4^-$, is macromolecular, and the sodium ions are loosely held in holes in the lattice.

To understand the principle of cation exchange, let us examine the structure of a natural zeolite, which has the empirical formula $NaAlSiO_4$ (Fig. 13.9). Notice that atoms of aluminum, silicon and oxygen are bonded together to form a huge "macroanion" similar in structure to a macromolecule such as quartz. The negative charge of the anion is balanced by Na^+ ions trapped in "holes" in the anionic lattice. When pure water flows through the zeolite, nothing spectacular happens; the Na^+ ions cannot leave the crystal since this would create an unbalance of charge. However, if hard water comes in contact with the zeolite, some of the Na^+ ions migrate out of the lattice, being replaced by an equivalent number of Ca^{2+} ions. This process may be represented by the equation

Most home water softening systems use zeolites

$$Ca^{2+}(aq) + 2\ NaZ(s) \rightarrow CaZ_2(s) + 2\ Na^+(aq) \qquad (13.15)$$

where Z stands for a small portion of the macroanion. The effect is to replace a Ca^{2+} (or Mg^{2+}) ion responsible for hardness by two less objectionable Na^+ ions.

After a zeolite column has been used for some time, more and more of the vacancies in the lattice become filled with Ca^{2+} ions. Eventually, an equilibrium is reached beyond which no further exchange of cations will occur. The column can, however, be returned to its original state by flushing with a concentrated solution of sodium chloride. Reaction 13.15 is reversed and the zeolite is ready for reuse.

Modern cation exchangers use synthetic organic resins which have a greater capacity for picking up Ca^{2+} ions than do natural zeolites. One type of resin, known as a "sulfonated polystyrene," has the structure indicated in Figure 13.10.

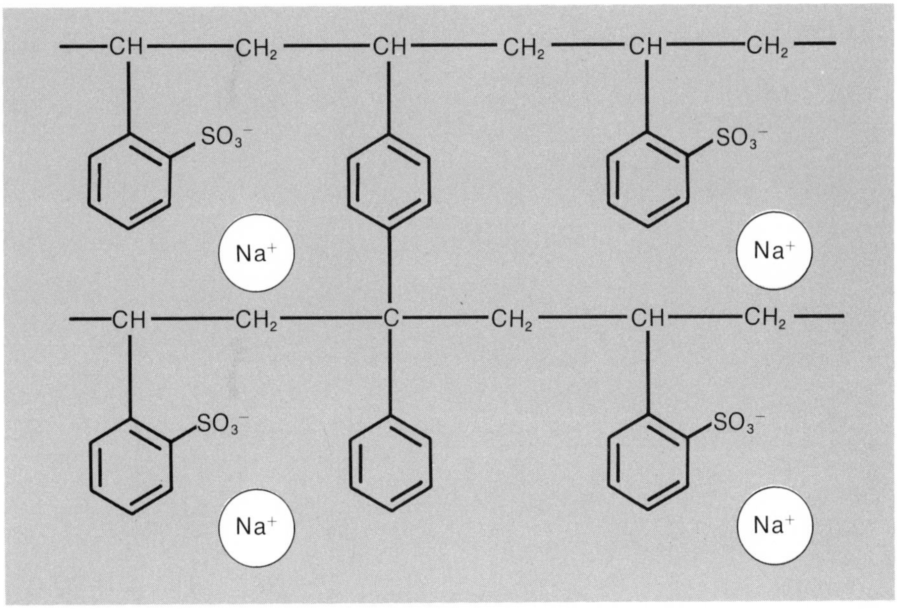

Figure 13.10 Structure of a synthetic cation exchange resin. The Na^+ ions readily exchange with Ca^{2+} or Mg^{2+} ions when hard water is passed through the resin, which is in a column in the form of small beads.

It operates on the same principle as a zeolite; Na^+ ions associated with the resin are exchanged for Ca^{2+} ions in the hard water. A major advantage of this type of cation exchanger is that it can be modified to form part of a unit which will remove *all* the ions from water (p. 328).

Desalination

In certain parts of the world, notably in hot, arid regions bordered by the sea, the most economical way to obtain fresh water is to remove the dissolved salts from seawater. The process of desalination is also used to purify brackish water (i.e., water with a relatively low salt content), which is common in much of the southwestern part of the United States. A variety of methods have been tested for removing all or nearly all of the ions from salt water; we will discuss a few of the more promising of these.

DISTILLATION. The oldest method of desalination, distillation, is the one that is used most widely today. It accounts for about 95% of the total amount of fresh water recovered from the oceans. The price of desalinated water prepared in this way has decreased from \$1 to 6¢/m^3, largely because of improved distillation techniques worked out by chemical engineers.

Solar stills are now being used in some commercial installations

REVERSE OSMOSIS. We saw in Chapter 12 that when a solution is separated by a semipermeable membrane from a sample of pure water, there is a tendency for the water to move through the membrane to the solution side by the process of osmosis. Clearly, ordinary osmosis could hardly be used to desalinate water; the movement is in the wrong direction.

You will recall, however, that osmosis can be prevented by applying, on the solution side, a pressure just equal to the osmotic pressure of the solution. If the applied pressure *exceeds* the osmotic pressure, water moves out of the solution to

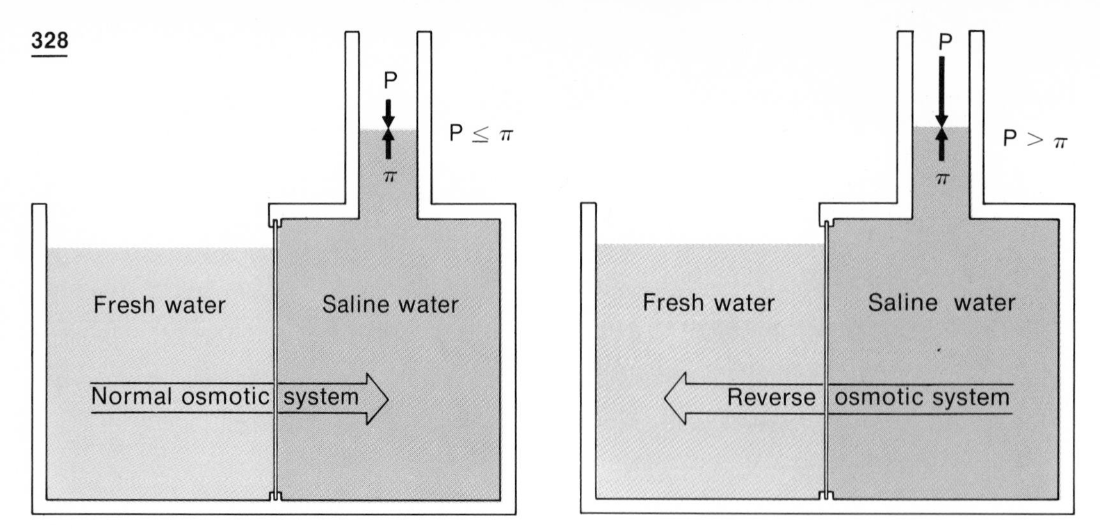

Figure 13.11 Purification of water by reverse osmosis. Sufficient pressure is applied to the salt solution to drive water through the membrane against the force of osmosis.

the pure water side of the membrane. This process, called reverse osmosis, accomplishes the objective of desalination—the extraction of pure water from salt water (Fig. 13.11).

A major problem in reverse osmosis has been to find membranes strong enough to withstand the high pressures required and at the same time be impermeable to the ions in salt water. Recently, synthetic membranes made from nylon or cellulose acetate have been developed to meet these criteria. Desalination plants using such membranes are now producing several million litres per day of fresh water. Since the energy requirement for reverse osmosis is only about 30% of that for distillation, this process seems particularly promising for the future.

ION EXCHANGE. Water of very high purity can be obtained by the ion exchange process shown schematically in Figure 13.12. The column at the left contains a synthetic cation exchange resin similar to that discussed earlier except that the exchangeable cation is H^+ rather than Na^+. The second column is packed with an anion exchange resin in which replaceable OH^- ions are embedded in a cationic network. Referring to the two resins as HR (cation exchanger) and R′OH (anion exchanger), we can write the following equations to describe what happens when water containing a dissolved salt MX is deionized by passing through the two columns:

1st column: $$M^+(aq) + HR(s) \rightarrow MR(s) + H^+(aq)$$

2nd column: $$X^-(aq) + R'OH(s) \rightarrow R'X(s) + OH^-(aq)$$

The H^+ and OH^- ions produced react with each other to form water:

$$H^+(aq) + OH^-(aq) \rightarrow H_2O$$

Adding the three equations just written, we arrive at the overall reaction for the removal of MX:

$$M^+(aq) + X^-(aq) + HR(s) + R'OH(s) \rightarrow MR(s) + R'X(s) + H_2O \quad (13.16)$$

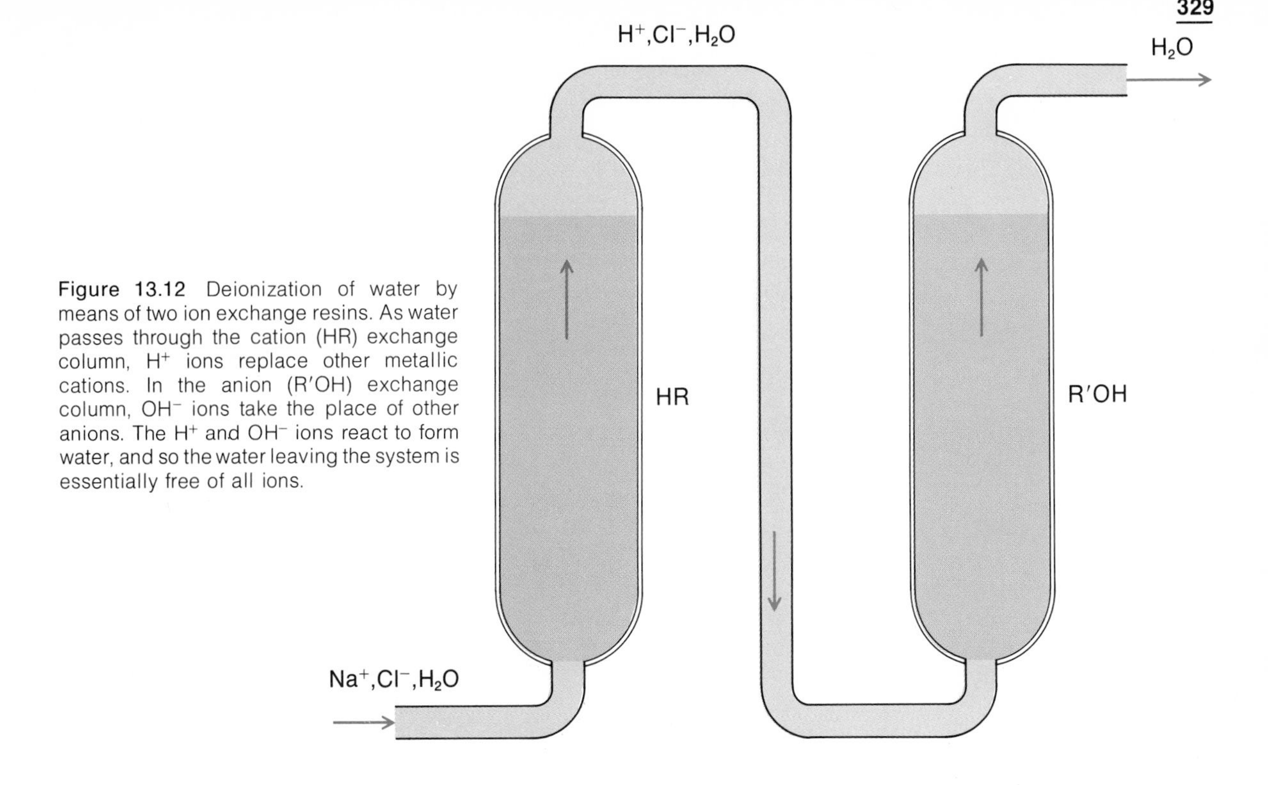

Figure 13.12 Deionization of water by means of two ion exchange resins. As water passes through the cation (HR) exchange column, H^+ ions replace other metallic cations. In the anion (R′OH) exchange column, OH^- ions take the place of other anions. The H^+ and OH^- ions react to form water, and so the water leaving the system is essentially free of all ions.

We see from these equations that there are no ions in solution at the end of the process, which is referred to as *deionization*. In principle at least, any salt present in the water will be removed; M^+ can be any cation (e.g., Na^+) and X^- any anion (e.g., Cl^-). To regenerate the first column, it can be flushed with a strong solution of hydrochloric acid, which acts as a source of H^+ ions. The second column is returned to its original state by flushing with sodium hydroxide.

Example 13.4 Write balanced equations for the removal by ion exchange of dissolved $Ca(NO_3)_2$.

Solution

cation exchange: $Ca^{2+}(aq) + 2\ HR(s) \rightarrow CaR_2(s) + 2\ H^+(aq)$

anion exchange: $2\ NO_3^-(aq) + 2\ R'OH(s) \rightarrow 2\ R'NO_3(s) + 2\ OH^-(aq)$

neutralization: $2\ H^+(aq) + 2\ OH^-(aq) \rightarrow 2\ H_2O$

overall: $Ca^{2+}(aq) + 2\ NO_3^-(aq) + 2\ HR(s) + 2\ R'OH(s) \rightarrow CaR_2(s) + 2\ R'NO_3(s) + 2\ H_2O$

The major drawback to the ion-exchange method is the cost of the chemicals used to regenerate the resins. They are expensive enough to rule out this method for desalination except for brackish waters in which the concentration of salt is at a maximum 300 ppm or less (the concentration in seawater is 35 000 ppm). On the other hand, this method is widely used to purify water in the laboratory. Water which has been passed through deionizing columns has an electrical conductivity less than one tenth of that of ordinary distilled water.

Distilled water has the advantage that it is bacteria-free

PROBLEMS

13.1 *Colligative Properties of Electrolytes* What would you expect to be the limiting value for i in Equation 13.3 for Na_3PO_4? If the freezing point of a 0.10 m solution of Na_3PO_4 is $-0.57°C$, what is the actual value of i at this concentration?

13.2 *BOD* What is the *BOD* of water containing 1.00 g of benzene/10^3 dm^3? The reaction of benzene with oxygen is: $C_6H_6(l) + \frac{15}{2} O_2(aq) \rightarrow 6\ CO_2(aq) + 3\ H_2O$

13.3 *Water Softening* How many moles of $Ca(OH)_2$ and Na_2CO_3 should be added to soften 10 dm^3 of water in which the concentrations of Ca^{2+} and HCO_3^- are 6.0×10^{-4} and 8.0×10^{-4} mol/dm^3, respectively?

13.4 Describe at least one effect, other than those listed in the text, that the high specific heat and heat of vaporization of water have on our environment.

13.5 How would you expect the conductivities of 0.10 M solutions of the following salts to compare to that of 0.10 M NaCl?

a. K_2CrO_4
b. $TlNO_3$
c. $Al(NO_3)_3$

13.6 What would you expect the limiting value of i (Equations 13.3 to 13.5) to be for

a. CH_3OH
b. $AlCl_3$
c. Na_2SO_4
d. $CuSO_4$
e. $KClO_4$

13.7 A certain acid H_2X has a freezing point in 0.010 m solution of $-0.038°C$. Does it ionize predominantly as

$$H_2X(aq) \rightarrow H^+(aq) + HX^-(aq)$$

or

$$H_2X(aq) \rightarrow 2\ H^+(aq) + X^{2-}(aq)?$$

13.8 Explain in your own words why

a. water is such a good solvent for ionic compounds.
b. the conductivity of a salt solution increases with concentration.
c. the value of i in Equations 13.3 to 13.5 is less than 2 for 1:1 salts at finite concentrations.

13.9 Using the data in Table 13.3, estimate the distance in cm by which the average level of the oceans would be raised if 20% of the ice and snow melted. Take the surface area of the oceans to be 3.61×10^8 km^2.

13.10 Using the data in Table 13.4, calculate

a. the mole ratio of Na^+ to K^+ in seawater.
b. the number of grams of Cl^- per cubic decimetre of seawater.
c. the molality of Na^+ in seawater (density = 1.024 g/cm^3).

13.23 List at least two consequences, other than those described in the text, of the fact that ice is less dense than water.

13.24 Arrange the following compounds in order of increasing conductivity in 0.10 M water solution: KNO_3, CH_3OH, HF, $CaCl_2$.

13.25 Assuming that 0.0010 m solutions of the following compounds behave ideally, calculate their freezing points.

a. $H_2N{-}CO{-}NH_2$
b. KNO_3
c. $CaBr_2$
d. $Al_2(SO_4)_3$

13.26 The freezing point of a 0.010 m solution of a certain acid HX is $-0.024°C$. What fraction of the acid is ionized?

13.27 Describe experiments which would show

a. whether K_2CrO_4 is a weak or strong electrolyte.
b. whether K_2CrO_4 ionizes to give K^+ and $KCrO_4^-$ or 2 K^+ and CrO_4^{2-}.
c. whether a salt XY is a 1:1 or a 2:2 electrolyte.

13.28 Using the data in Table 13.3, estimate the amount of heat released in kilojoules if 1% of the water in the atmosphere condensed. Take the heat of vaporization of water to be 44 kJ/mol.

13.29 Using the data in Table 13.4, calculate the total mass of dissolved salts in 1.00 dm^3 of seawater.

13.11 Explain in your own words why

a. seawater has a lower freezing point than fresh water.
b. the concentration of Na^+ in seawater is much higher than that of Ca^{2+}, although the reverse is true in river water.

13.12 The concentration of dissolved oxygen in a sample of waste water decreases from 7.0×10^{-3} g/dm^3 to 4.8×10^{-3} g/dm^3 after five days. Calculate the *BOD* of the sample.

13.13 Twenty kilograms of acetaldehyde, CH_3CHO, is discharged into a water supply which has a total volume of 2.5×10^6 dm^3. By how much does this increase the *BOD* of the water? Assume the following reaction occurs:

$$CH_3CHO(aq) + \tfrac{5}{2}\ O_2(aq) \rightarrow 2\ CO_2(aq) + 2\ H_2O$$

13.14 Write a balanced equation for the reaction of O_2 in the presence of aerobic bacteria with the insecticide Sevin (p. 319); assume the products are CO_2, H_2O, NO_3^-, and H^+.

13.15 Discuss the relative merits of using "high-phosphate" as against "low-phosphate" detergents in a rural home where waste water goes into a septic tank.

13.16 Discuss the advantages and disadvantages of substituting parathion for DDT.

13.17 If the concentration of mercury in a certain brand of canned tuna fish is known to be 0.52 ppm, how many grams of mercury are there in a 50 g can? If all the mercury is present as $Hg(CH_3)_2$, how much of this compound is there in the can?

13.18 How many grams of carbon would be required (Eq. 13.9) to treat one cubic metre of water in which the concentration of chlorine is 0.54 ppm?

13.19 A water supply contains the following ions in the indicated concentrations. Determine the number of moles of $Ca(OH)_2$ and Na_2CO_3 that should be added to soften 1.0 dm^3 of this water.

a. $1.8 \times 10^{-4}\ M\ Ca^{2+}$, $2.6 \times 10^{-4}\ M\ HCO_3^-$
b. $1.8 \times 10^{-4}\ M\ Ca^{2+}$, no HCO_3^-
c. $1.8 \times 10^{-4}\ M\ Ca^{2+}$, $3.6 \times 10^{-4}\ M\ HCO_3^-$

13.20 A principal objective of water softening is the removal of Ca^{2+}; yet, in the lime-soda process, more Ca^{2+} ions are added in the form of $Ca(OH)_2$. Explain why this is done.

13.30 Explain in your own words why

a. the concentration of O_2 in seawater is much more variable than that of N_2.
b. HCO_3^- is usually the most abundant anion in river water but not in seawater.

13.31 The *BOD* of the water in a certain river is 6.2 mg/dm^3. Before reaction, the concentration of O_2 in the river is 8.9×10^{-3} g/dm^3. What is the concentration of O_2 in this water when equilibrium is reached with the *BOD*?

13.32 A certain electroplating company discharges CN^- into a river. An analyst finds that every time a batch of cyanide is released, the *BOD* of the river water increases by 3.0 mg/dm^3. Assuming the reaction

$$2\ CN^-(aq) + \tfrac{5}{2}\ O_2(g) + 2\ H^+(aq) \rightarrow 2\ CO_2(aq) + N_2(g) + H_2O$$

what is the concentration of CN^- in moles per cubic decimetre?

13.33 Write a balanced equation for the aerobic decomposition of the insecticide parathion; assume $P \rightarrow PO_4^{3-}$, $S \rightarrow SO_4^{2-}$, and H^+ ions are formed.

13.34 Suggest at least two ways in which PO_4^{3-} might get into water supplies other than through the use of detergents.

13.35 Suggest several different approaches to reducing the need for chemical insecticides.

13.36 The concentration of Cd^{2+} in a certain water supply is 6.0×10^{-7} mass %. Estimate the molality of Cd^{2+}.

13.37 A column containing 100 g of activated charcoal is used to remove chlorine at a concentration of 0.80 ppm in water. Referring to Equation 13.9, how many cubic decimetres of water can be "dechlorinated" by the column?

13.38 A municipal water supply contains the following ions in the indicated concentrations. Calculate the number of grams of $Ca(OH)_2$ and Na_2CO_3 that should be added to soften 10 dm^3 of this water.

a. $3.0 \times 10^{-4}\ M\ Ca^{2+}$, $4.0 \times 10^{-4}\ M\ HCO_3^-$
b. $3.0 \times 10^{-4}\ M\ Ca^{2+}$, $2.0 \times 10^{-4}\ M\ HCO_3^-$
c. $3.0 \times 10^{-4}\ M\ Ca^{2+}$, $6.0 \times 10^{-4}\ M\ HCO_3^-$

13.39 Suppose (as is seldom the case) that the concentration of HCO_3^- in hard water were more than twice that of Ca^{2+} (e.g., $3.0 \times 10^{-4}\ M$ vs. $1.0 \times 10^{-4}\ M$). Would it then be necessary to add one mole of $Ca(OH)_2$ for every two moles of HCO_3^-? Explain.

13.21 Write balanced equations for

a. the removal of K^+ from water using the zeolite NaZ.
b. the removal of dissolved $FeCl_3$ by passing water through a deionizer containing the resins HR and R′OH.
c. conversion of K_2SO_4 to KNO_3 using an anion exchange resin.

13.22 Considering seawater to be a 3.5% by mass solution of NaCl, calculate

a. the temperature at which salt water starts to freeze (take $i = 1.8$).
b. the freezing point of the solution that remains after 60% of the water has been removed as ice (take $i = 1.7$).

13.40 Write balanced equations for

a. the removal of SO_4^{2-} from water using the anion exchange resin R′Cl.
b. the removal of $Al(NO_3)_3$ from water by use of the resins HR and R′OH.
c. conversion of NaCl to $ZnCl_2$ using a cation exchange resin.

13.41 Taking the concentration of seawater to be 0.53 *M*, calculate the pressure required to extract pure water from seawater by reverse osmosis. (Assume $i = 1.9$, $t = 25°C$.)

*13.42 Each day a certain power plant takes in 2.0×10^6 dm³ of water at 25°C from a river and uses it to condense 4.0×10^4 kg of steam.

a. What is the temperature of the water when it is discharged?
b. If the river into which this water is discharged has a flow of 4.0×10^7 dm³/d, estimate the increase in temperature to be expected downstream.

*13.43 Suppose we wish to remove a *BOD* of 140 from a sample of municipal waste by adding O_2 via air-saturated water. For 1.00 dm³ of waste, how much water should we add? (Solubility pure O_2 in water $= 1.2 \times 10^{-3}$ mol/dm³.) By what factor is the waste water diluted with the air-saturated water?

*13.44 Seawater, like all matter, must be electrically neutral. Use the data in Table 13.4 to show that this is indeed the case.

*13.45 In very dilute solution, the molar conductivity of $MgSO_4$ is about twice that of NaCl, yet the freezing point lowerings are about the same. Explain.

14

SPONTANEITY OF REACTION; ΔG AND ΔS

In discussing thermochemistry in Chapter 4, we concentrated upon reactions taking place at constant temperature and pressure. You will recall that under these conditions the heat flow associated with the reaction is exactly equal to the difference in enthalpy, ΔH, between products and reactants.

In this chapter we return to a discussion of the energy changes accompanying reactions at constant temperature and pressure. This time we will be interested in one crucial question: how can we predict in advance whether a reaction will occur under such conditions, given sufficient time? Putting it another way, how can we decide whether a reaction, at a certain temperature and pressure, will be spontaneous?

We shall find that a knowledge of the sign and magnitude of the enthalpy change, ΔH, will help us to answer this basic question. However, in order to establish a completely general criterion of spontaneity, it will be necessary to introduce two new functions; these are the free energy change, ΔG, and the entropy change, ΔS.

14.1 CRITERIA OF SPONTANEITY. USEFUL WORK

We know from experience that certain processes going on in the world around us are **spontaneous** in the sense that **they occur** "by themselves" **without the exertion of any outside force.** An ice cube melts when we take it out of the refrigerator. Iron exposed to moist air rusts. A mixture of methane and oxygen burns when we set a match to it. Certain other processes are clearly nonspontaneous. No one has ever observed ice cubes forming in a glass of water at 25°C. A hammer left out on the lawn all winter does not lose its coating of rust when spring comes. The products of combustion of natural gas do not recombine to form methane and oxygen. In summary, at room temperature and atmospheric pressure, the following processes are spontaneous, whereas the reverse processes are nonspontaneous:

"Spontaneous" may not be the best word to describe what we mean, but we can't think of a better one

$$H_2O(s) \rightarrow H_2O(l) \qquad (14.1)$$

$$2\ Fe(s) + \tfrac{3}{2}\ O_2(g) + 3\ H_2O(l) \rightarrow 2\ Fe(OH)_3(s) \qquad (14.2)$$

$$CH_4(g) + 2\ O_2(g) \rightarrow CO_2(g) + 2\ H_2O(l) \qquad (14.3)$$

It is not always easy to tell whether a reaction is spontaneous; appearances can be deceiving. A mixture of hydrogen and oxygen can be maintained for long periods of time without any apparent reaction. This might lead us to conclude mistakenly that the reaction

$$H_2(g) + \frac{1}{2} O_2(g) \rightarrow H_2O(l) \qquad (14.4)$$

is nonspontaneous. However, if we bring a lighted match or a platinum foil up to the mixture, Reaction 14.4 occurs, often with explosive violence. This tells us that the formation of liquid water from the elements is a spontaneous process, since it is *capable* of proceeding by itself once it is initiated. More generally, we see that the word "spontaneous" in the sense we are using it here *does not imply that the process occurs rapidly*. Indeed, many spontaneous reactions take place very, very slowly unless the proper catalyst is present (Chapter 16).

It would be very convenient to be able to predict in advance whether a given reaction is potentially spontaneous. Consider, for example, the reaction

$$CO(g) + NO(g) \rightarrow CO_2(g) + \frac{1}{2} N_2(g) \qquad (14.5)$$

If this reaction could be shown to be spontaneous, at let us say 25°C and 1 atm pressure, it might offer an ideal way to remove two of the most serious air pollutants, carbon monoxide and nitric oxide. Certainly it would be worth looking for a suitable catalyst or other means of initiating the reaction. If, on the other hand, it can be shown by calculation that it is impossible to make this reaction occur at any reasonable temperature and pressure, we may as well forget about it and concentrate upon other processes for removing these species from automobile exhaust.

A hundred years ago many chemists felt that they had a general criterion for predicting reaction spontaneity. The prevailing idea, put forth by P. M. Berthelot in Paris and Julius Thomsen in Copenhagen, was that all spontaneous reactions are exothermic. If this were true, all we would have to do to predict reaction spontaneity would be to calculate the enthalpy change, ΔH, and look at its sign. If ΔH turned out to be negative, we could assume that the reaction must be spontaneous; if ΔH were positive, the reaction could not occur by itself.

It turns out that almost all exothermic *chemical reactions* are spontaneous *at 25°C and 1 atm*. For reactions 14.2 to 14.4, ΔH is -791 kJ, -891 kJ, and -286 kJ, respectively; each of these reactions is spontaneous at room temperature and atmospheric pressure. On the other hand, this simple rule fails for many familiar phase changes. An example is the melting of ice (Equation 14.1) which takes place at 25°C and 1 atm even though it is endothermic ($\Delta H = +6.0$ kJ). In another case we find that potassium nitrate dissolves when added to water at room temperature even though the solution process

$$KNO_3(s) \rightarrow K^+(aq) + NO_3^-(aq); \; \Delta H = +35 \text{ kJ} \qquad (14.6)$$

is endothermic.

There is still another basic objection to using the sign of ΔH as a completely general criterion for spontaneity. We often find that endothermic reactions which are nonspontaneous at room temperature become spontaneous when the temperature is raised. Consider, for example, the reaction by which carbon dioxide and quicklime are prepared from limestone

$$CaCO_3(s) \rightarrow CaO(s) + CO_2(g); \; \Delta H = +178.0 \text{ kJ} \qquad (14.7)$$

At 25°C and 1 atm, this reaction is nonspontaneous, as shown by the existence of the white cliffs of Dover over eons of time. However, if the temperature is raised to about 850°C, the limestone decomposes to give off carbon dioxide at 1 atm pressure. If we operate at lower pressures, calcium carbonate decomposes even more readily; at 1×10^{-4} atm, Reaction 14.7 becomes spontaneous at 500°C. We see then that it is possible to make this endothermic reaction spontaneous by increasing the temperature or reducing the pressure, even though ΔH remains virtually constant at +178.0 kJ, nearly independent of both temperature and pressure. Clearly, even though the sign of ΔH is a fairly reliable indicator of reaction spontaneity at room temperature and atmospheric pressure, it cannot be the general criterion that we are looking for.

Nearly a century ago, J. Willard Gibbs, professor of mathematical physics at Yale, showed that the proper criterion for spontaneity of a reaction is its ability to produce useful work.* He proved that **if, at constant temperature and pressure, a reaction can, in principle or in practice, be harnessed to perform useful work, that reaction is spontaneous. If work has to be supplied from the surroundings to make the reaction occur, it cannot be spontaneous.**

Like many scientific truths, this one seems obvious once it is stated

To illustrate the meaning of Gibbs' criterion of spontaneity, we will take as our first example the vaporization of water:

$$H_2O(l, 101°C, 1 \text{ atm}) \rightarrow H_2O(g, 101°C, 1 \text{ atm}) \qquad (14.8)$$

We know from experience that this process is spontaneous; water heated above 100°C in a beaker or saucepan boils to produce steam at atmospheric pressure. Thomas Newcomen demonstrated, when he designed a practical steam engine over two centuries ago, that this process can perform useful work. Notice that when we boil water in an open container in the laboratory or in the kitchen, no useful work is produced; the force of the steam escaping into the atmosphere is not harnessed in any way. The fact that it is *possible* to make this process do work in a machine such as a steam engine proves its spontaneity.

Let us now consider a chemical reaction, the combustion of methane

$$CH_4(g) + 2\ O_2(g) \rightarrow CO_2(g) + 2\ H_2O(l) \qquad (14.3)$$

As pointed out earlier, this reaction occurs spontaneously when we light a Bunsen burner or a gas range. Under these conditions, no useful work is done; all that is observed is the evolution of heat. However, there are at least two ways in which we can persuade this reaction to do work for us. One is to carry out the reaction in an internal combustion engine, thereby generating mechanical energy. Natural gas has been used as a fuel to drive fleets of trucks operated by private companies and government agencies in California. Alternatively the reaction of methane with oxygen can be carried out in a fuel cell, a device that generates electrical energy (Fig. 14.1). Fuel cells using spontaneous reactions such as this were used in the Apollo space program; they are now the subject of a major research program for the production of electrical energy from fuels (Chapter 22).

In contrast to the two processes just discussed, consider the decomposition of water to the elements.

$$H_2O(l) \rightarrow H_2(g) + \tfrac{1}{2}\ O_2(g) \qquad (14.9)$$

This reaction can be made to take place in the electrolysis cell shown in Figure 14.2. However, in order for this to happen, electrical energy must be supplied from an external source, which may be a fuel cell, a lead storage battery, or a

*The adjective "useful" excludes the type of work shown in Figure 4.7, p. 78. The work done in "pushing back" the atmosphere during a volume change accompanying a constant pressure reaction is essentially "useless."

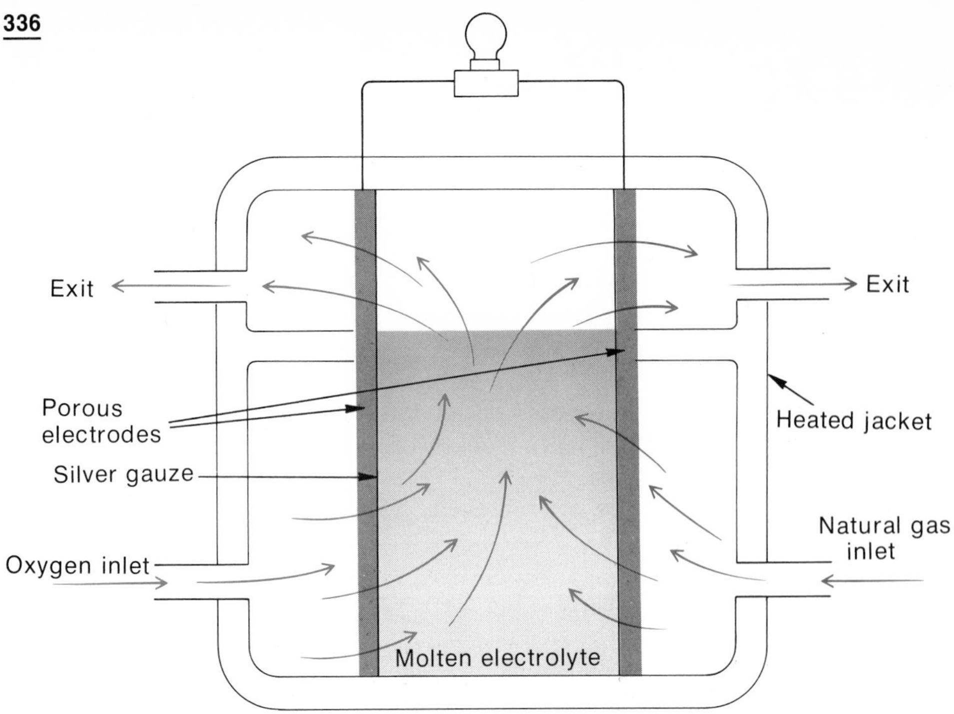

Figure 14.1 Generation of electrical energy in a fuel cell using the reaction between methane and oxygen.

series of dry cells. Since work has to be done to make the reaction go, we conclude that it is nonspontaneous. Notice that "nonspontaneous" does not imply "impossible"; we often find that we can make nonspontaneous reactions occur, but it always costs us work to do so.

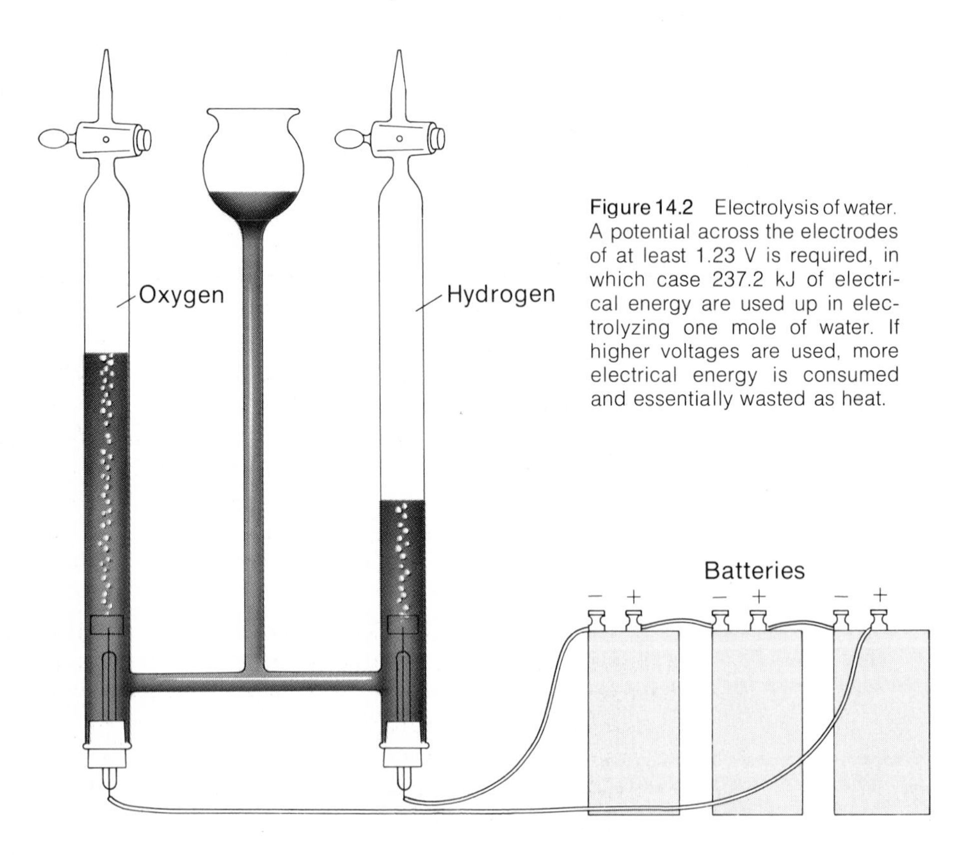

Figure 14.2 Electrolysis of water. A potential across the electrodes of at least 1.23 V is required, in which case 237.2 kJ of electrical energy are used up in electrolyzing one mole of water. If higher voltages are used, more electrical energy is consumed and essentially wasted as heat.

14.2 FREE ENERGY CHANGE, ΔG

The amount of useful work that we can get out of a spontaneous reaction depends to some extent upon the efficiency of the machine used to harness the reaction. When natural gas burns in an internal combustion engine, we seldom get more than 100 to 200 kJ of useful work per mole of methane. In a fuel cell we can do somewhat better, perhaps obtaining as much as 700 kJ of work per mole of methane. We find, however, that there is an upper limit to the amount of work that can be obtained from Reaction 14.3. No matter how ingenious or how economical we are, we can never get more than 818.0 kJ of useful work per mole of methane consumed at 25°C and 1 atm. We may get no work at all, we may get 400 kJ or 800 kJ or even, conceivably, 815 kJ, but no scheme that we can devise will enable us to get more than 818.0 kJ of useful work out of this reaction.

The capacity of a spontaneous reaction, carried out at constant temperature and pressure, to produce useful work can be interpreted in terms of a fundamental property of the substances taking part in the reaction known as their **free energy** and given the symbol ***G***. Specifically, *the amount of useful work that can be obtained from such a reaction is limited by the difference in free energy between products and reactants*. For Reaction 14.3, the free energy of the products (1 mol CO_2 + 2 mol H_2O) is 818.0 kJ *less* than that of the reactants (1 mol CH_4 + 2 mol O_2):

The free energy of a substance reflects its capacity to supply useful work in a reaction

$$CH_4(g) + 2\ O_2(g) \rightarrow CO_2(g) + 2\ H_2O(l);\ \Delta G = -818.0\ \text{kJ}$$

This decrease in free energy for the reaction system places an upper limit, 818.0 kJ, on the amount of useful work that can be produced in the surroundings when the reaction takes place.

The argument that we have gone through can be applied equally well to a nonspontaneous reaction. Consider, for example, Reaction 14.9. Here we find that the free energy of the products (1 mol H_2 + ½ mol O_2) is 237.2 kJ *greater* than that of the reactant (1 mol H_2O):

$$H_2O(l) \rightarrow H_2(g) + \tfrac{1}{2}\ O_2(g);\ \Delta G = +237.2\ \text{kJ}$$

In order to make this reaction go, we must supply at least 237.2 kJ of useful work from the surroundings (e.g., a storage battery) to give the reaction system the required amount of free energy. We can, if we wish, supply 250 or 400 kJ of electrical energy, in which case some of it will be wasted. In no case, however, can we electrolyze a mole of H_2O by furnishing less than 237.2 kJ of useful work.

As our discussion implies, there is a simple relationship between the sign of the free energy change for a reaction and its spontaneity. We can distinguish three cases:

1. **If ΔG is −**, the reaction at constant temperature and pressure is capable of producing useful work and hence is **spontaneous.**

2. **If ΔG is +**, work must be supplied to carry out the reaction at constant temperature and pressure and we say that the reaction is **nonspontaneous.** Alternatively, we can say that *the reverse reaction is spontaneous.*

3. **If ΔG = 0,** the reaction system is at **equilibrium;** there is no driving force tending to make the reaction go in either direction.

Free Energy Change at 1 atm, $\Delta G^{1\ atm}$

The free energy change, ΔG, unlike the enthalpy change, ΔH, can vary considerably with both temperature and pressure. Indeed, this must be the case if the sign of ΔG is to be the criterion of reaction spontaneity. We have seen that there is a shift in the direction in which the reaction

$$CaCO_3(s) \rightarrow CaO(s) + CO_2(g)$$

proceeds as we change the temperature and pressure. At 25°C and 1 atm this reaction is nonspontaneous ($\Delta G > 0$). It can be made spontaneous ($\Delta G < 0$) by raising the temperature or lowering the pressure. In Table 14.1 we have tabulated values of ΔG at various temperatures and pressures. Notice that it changes from +130 kJ at 25°C and 1 atm to −25 kJ at 1000°C and 1 atm and to −174 kJ at 1000°C and 10^{-6} atm (0.1 Pa). In contrast, ΔH for this reaction remains virtually constant at +178 kJ over the entire range of temperatures and pressures.

TABLE 14.1 VARIATION OF ΔG (kJ) WITH TEMPERATURE AND PRESSURE FOR THE REACTION $CaCO_3(s) \rightarrow CaO(s) + CO_2(g)$

t(°C)	P = 1 atm	$P = 10^{-2}$ atm	$P = 10^{-4}$ atm	$P = 10^{-6}$ atm
25	+130	+119	+107	+96
200	+103	+85	+66	+48
400	+71	+50	+28	+7
600	+39	+5	−29	−63
800	+7	−35	−76	−118
1000	−25	−75	−124	−174

This means that we will limit our discussion of spontaneity to reactions occurring at 1 atm

We will discuss later in this chapter the variation of ΔG with temperature. We will not, however, consider further its variation with pressure. In all our calculations we shall work with the **free energy change at one atmosphere pressure,** for which we will write the symbol $\Delta G^{1\,atm}$. You should keep in mind, however, that ΔG and, hence, the direction of spontaneity of reaction might be quite different if we were to allow the pressure to vary.

The "laws of thermochemistry" discussed in Chapter 4 in connection with the enthalpy change for a reaction, ΔH, can be applied equally well to the free energy change, ΔG. Specifically:

1. *ΔG is directly proportional to the amount of substance that reacts or is produced in a reaction.* Knowing that ΔG for the decomposition of one mole of water (18.0 g) at 25°C and 1 atm via Reaction 14.9 is +237.2 kJ, it follows that the free energy change per gram of water is

$$1.00 \text{ g } H_2O \times \frac{237.2 \text{ kJ}}{18.0 \text{ g } H_2O} = +13.2 \text{ kJ}$$

2. *ΔG for a reaction is equal in magnitude but opposite in sign to ΔG for the reverse reaction.* For example, given that

$$H_2O(l) \rightarrow H_2(g) + \tfrac{1}{2}\, O_2(g);\ \Delta G^{1\,atm} \text{ at } 25°C = +237.2 \text{ kJ}$$

it follows that

$$H_2(g) + \tfrac{1}{2}\, O_2(g) \rightarrow H_2O(l);\ \Delta G^{1\,atm} \text{ at } 25°C = -237.2 \text{ kJ}$$

3. *If a reaction can be regarded as the sum of two or more other reactions, ΔG for the overall reaction must be the sum of the free energy changes for the other reactions.* Thus we have:

$$C(s) + \tfrac{1}{2} O_2(g) \rightarrow CO(g);\ \Delta G^{1\,atm} \text{ at } 25°C = -137.3 \text{ kJ}$$

$$CO(g) + \tfrac{1}{2} O_2(g) \rightarrow CO_2(g);\ \Delta G^{1\,atm} \text{ at } 25°C = -257.1 \text{ kJ}$$

$$C(s) + O_2(g) \rightarrow CO_2(g);\ \Delta G^{1\,atm} \text{ at } 25°C = -394.4 \text{ kJ}$$

Free Energies of Formation

We saw in Chapter 4 how changes in enthalpy, ΔH, can be calculated from heats of formation, ΔH_f, of products and reactants. It is possible to carry out analogous calculations for free energy changes, ΔG, in chemical reactions if the free energies of formation, ΔG_f, of products and reactants are available. The free energy of formation of a pure substance is taken to be the free energy change when one mole of the substance is formed from the elements at 1 atm pressure; it is assigned the symbol $\Delta G_f^{1\,atm}$. In Table 14.2 we have listed values of $\Delta G_f^{1\,atm}$ for a variety of compounds at 25°C.

TABLE 14.2 FREE ENERGIES OF FORMATION (kJ/mol) AT 25°C AND 1 ATM

$AgBr(s)$	−95.9	$CO(g)$	−137.3	$H_2O(g)$	−228.6	$NH_4Cl(s)$	−203.9
$AgCl(s)$	−109.7	$CO_2(g)$	−394.4	$H_2O(l)$	−237.2	$NO(g)$	+86.7
$AgI(s)$	−66.3	$C_2H_2(g)$	+209.2	$H_2S(g)$	−33.0	$NO_2(g)$	+51.8
$Ag_2O(s)$	−10.8	$C_2H_4(g)$	+68.1	$HgO(s)$	−58.5	$NiO(s)$	−216.3
$Ag_2S(s)$	−40.3	$C_2H_6(g)$	−32.9	$HgS(s)$	−48.8	$PbBr_2(s)$	−259.8
$Al_2O_3(s)$	−1576.4	$C_3H_8(g)$	−23.5	$KBr(s)$	−379.2	$PbCl_2(s)$	−314.0
$BaCl_2(s)$	−810.9	$CoO(s)$	−213.4	$KCl(s)$	−408.3	$PbO(s)$	−188.5
$BaCO_3(s)$	−1138.9	$Cr_2O_3(s)$	−1046.8	$KClO_3(s)$	−289.9	$PbO_2(s)$	−219.0
$BaO(s)$	−528.4	$CuO(s)$	−127.2	$KF(s)$	−533.1	$Pb_3O_4(s)$	−617.6
$BaSO_4(s)$	−1353.1	$Cu_2O(s)$	−146.4	$MgCl_2(s)$	−592.3	$PCl_3(g)$	−286.3
$CaCl_2(s)$	−750.2	$CuS(s)$	−49.0	$MgCO_3(s)$	−1029	$PCl_5(g)$	−324.6
$CaCO_3(s)$	−1128.8	$CuSO_4(s)$	−661.9	$MgO(s)$	−569.6	$SiO_2(s)$	−805.0
$CaO(s)$	−604.2	$Fe_2O_3(s)$	−741.0	$Mg(OH)_2(s)$	−833.7	$SnCl_4(l)$	−474.0
$Ca(OH)_2(s)$	−896.8	$Fe_3O_4(s)$	−1014.2	$MgSO_4(s)$	−1173.6	$SnO(s)$	−257.3
$CaSO_4(s)$	−1320.3	$HBr(g)$	−53.2	$MnO(s)$	−363.2	$SnO_2(s)$	−519.7
$CCl_4(l)$	−68.6	$HCl(g)$	−95.3	$MnO_2(s)$	−466.1	$SO_2(g)$	−300.4
$CH_4(g)$	−50.8	$HF(g)$	−270.7	$NaCl(s)$	−384.0	$SO_3(g)$	−370.4
$CHCl_3(l)$	−71.5	$HI(g)$	+1.3	$NaF(s)$	−541.0	$ZnO(s)$	−318.2
$CH_3OH(l)$	−166.2	$HNO_3(l)$	−79.9	$NH_3(g)$	−16.6	$ZnS(s)$	−198.3

You will note that $\Delta G_f^{1\,atm}$ is negative for most compounds at 25°C and 1 atm. This implies that the formation of a compound from the elements under these conditions is ordinarily a spontaneous process. Conversely, most compounds are stable with respect to decomposition to the elements at 25°C and 1 atm. There are exceptions; two of the more interesting are NO ($\Delta G_f^{1\,atm}$ = +86.7 kJ/mol) and NO_2 ($\Delta G_f^{1\,atm}$ = +51.8 kJ/mol). In principle, these compounds should decompose to N_2 and O_2 under ordinary conditions. The fact that they stay around long enough to be a major problem in air pollution implies that the rate of decomposition must be extremely slow.

Free energy changes at 1 atm can be calculated from $\Delta G_f^{1\,atm}$ data in a manner completely analogous to that used to find enthalpy changes from ΔH_f data. The relation we use is

$$\Delta G^{1\,atm} = \sum \Delta G_f^{1\,atm} \text{ products} - \sum \Delta G_f^{1\,atm} \text{ reactants} \qquad (14.10)$$

Example 14.1 Find the free energy change $\Delta G^{1\,atm}$ at 25°C for the reaction

$$4\ NH_3(g) + 5\ O_2(g) \rightarrow 4\ NO(g) + 6\ H_2O(l)$$

Solution Applying Equation 14.10 and realizing that the free energy of formation of the elementary substance O_2, like its heat of formation, is zero, we have

$$\begin{aligned}\Delta G^{1\,atm} &= 4\ \Delta G_f^{1\,atm}\ NO(g) + 6\ \Delta G_f^{1\,atm}\ H_2O(l) - 4\ \Delta G_f^{1\,atm}\ NH_3(g)\\ &= 4(86.7\ kJ) + 6(-237.2\ kJ) - 4(-16.6\ kJ)\\ &= -1010.0\ kJ\end{aligned}$$

The reaction is spontaneous at 25°C and 1 atm.

14.3 ENTROPY CHANGE, ΔS

Let us consider once again the reaction

$$CaCO_3(s) \rightarrow CaO(s) + CO_2(g) \tag{14.7}$$

In Figure 14.3 we have plotted the free energy change at 1 atm pressure as a function of absolute temperature, using data taken from Table 14.1. In the same plot we show ΔH, which remains nearly constant at +178.0 kJ over the entire temperature range.

It is clear from the plot that ΔG, unlike ΔH, is strongly dependent upon temperature. Indeed, there is a linear relationship between ΔG and T of the form y = a + bx. The constant, a (the y intercept), is +178.0 kJ. The slope, b, is calculated to be about −0.160 kJ/K. In other words, for this reaction,

$$\Delta G = +178.0\ kJ - 0.160\frac{kJ}{K} \times T$$

We see from the figure that the quantity "178.0 kJ" represents the enthalpy change, ΔH. The quantity "0.160 kJ/K" also represents the change in a fundamental property characteristic of the substances taking part in the reaction. This

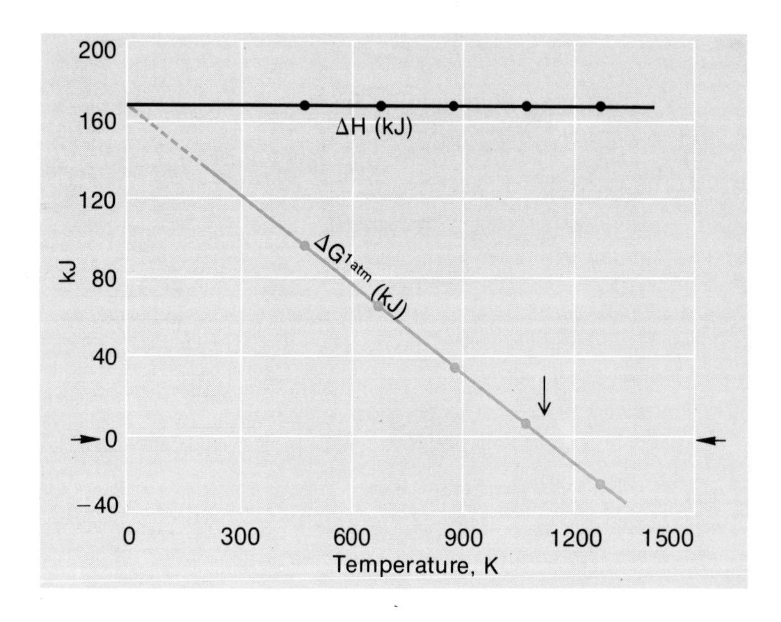

Figure 14.3 ΔH and $\Delta G^{1\,atm}$ as a function of temperature for the reaction: $CaCO_3(s) \rightarrow CaO(s) + CO_2(g)$. Below about 1100 K, $\Delta G^{1\,atm}$ is positive and the reaction is nonspontaneous. Above 1100 K, $CaCO_3$ will decompose spontaneously in a container open to the atmosphere.

property is called the **entropy** and is given the symbol ***S***. Associated with every reaction there is a change in entropy, ΔS, equal to the entropy of the products minus the entropy of the reactants.

$$\Delta S = \sum S_{products} - \sum S_{reactants} \tag{14.11}$$

For reaction 14.7 we can write

$$\Delta S = S \text{ of 1 mol CaO(s)} + S \text{ of 1 mol } CO_2(g) - S \text{ of 1 mol } CaCO_3(s)$$

$$= +0.160 \text{ kJ/K} = +160 \text{ J/K}$$

The analysis that we have just applied to the decomposition of limestone can be generalized to describe the relationship between ΔG and ΔH in any reaction. The general equation is written in the form

$$\Delta G = \Delta H - T\Delta S \tag{14.12}$$

Clearly, spontaneity depends on both ΔH and ΔS as well as T

where T is the absolute temperature in K and ΔS is the difference in entropy between products and reactants.

We shall have a great deal more to say later in this chapter about the usefulness of Equation 14.12. In order to appreciate its significance, we must have a clear physical picture of what is meant by the entropy of a pure substance.

Nature of Entropy

The entropy of a substance, like its enthalpy, free energy, or volume, is one of its characteristic properties. Qualitatively, entropy is a measurement of *disorder* or *randomness*. Substances which are highly disordered have high entropies; low entropy is associated with strongly ordered substances. "Order," in the sense we use it here, describes the extent to which the particles of a substance are confined to a given region of space. In a crystal, where the atoms, molecules, or ions are fixed in position, the entropy is relatively low. When a solid melts, any particle is free to move through the entire liquid volume and the entropy of the substance increases. Upon vaporization the particles acquire still greater freedom to move about and there is a large increase in entropy.

Considering the rather vague way in which we have described entropy, you may be surprised to learn that we can determine absolute values for the molar entropies of pure substances. Actually, it is possible to define entropy very precisely in a thermodynamic sense and measure it by calorimetric means. The details of how this is done are beyond the level of this text, but the results are readily interpreted. In Table 14.3 we have listed the molar entropies of some common solids, liquids, and gases.

TABLE 14.3 MOLAR ENTROPIES (J/K) AT 1 ATM, 25°C

Solids		Liquids		Gases	
C (diamond)	2.4	H_2O	69.9	He	126.0
Si	18.7	Hg	77.4	Ar	154.7
Na	51.0	CH_3OH	126.8	O_2	205.0
Pb	64.9	Br_2	152.3	Cl_2	222.9
NaCl	72.4	C_2H_5OH	160.7	NO_2	241.5
KCl	82.7	$CHCl_3$	202.9	Br_2	245.3
$CaCl_2$	113.8	CCl_4	214.4	N_2O_4	304.3

For a single substance, $S(s) < S(l) < S(g)$

Our simple interpretation of entropy in terms of disorder is well borne out by the data in Table 14.3. We see that solids as a group have lower entropies than liquids, which in turn have lower entropies than gases. Among substances in the same physical state, it is clear that molar entropy goes up with the number of atoms per formula unit; complex molecules have higher entropies than simple ones because there are more atoms to move about. Among solids, those that are very hard like diamond have lower entropies than those that are soft like sodium or lead.

As we might expect, the entropy of a substance increases as the temperature rises. The increased kinetic energy weakens the forces holding the particles together and gives them greater freedom of motion. For simple molecules, the effect is relatively small. The molar entropy of oxygen at 1 atm, $S^{1\ atm}$, increases by less than 4 per cent when the temperature is raised from 25 to 100°C. An increase in pressure decreases the entropy of a substance because it confines the molecules to a smaller volume. For liquids and solids the effect is very small, but for gases it is appreciable. Increasing the pressure on a gas from 1 to 10 atm decreases its molar entropy by about 19 J/K.

ΔS for Phase Changes

When a substance passes from a more ordered to a less ordered state by melting or vaporizing, its entropy increases. We can calculate the entropy change of vaporization at 1 atm pressure by realizing that $\Delta G_{vap}^{1\ atm}$ must be zero at the normal boiling point, since the liquid and vapor are in equilibrium at this temperature and pressure. Setting ΔG equal to zero in Equation 14.12 we obtain

$$\Delta S_{vap}^{1\ atm} = \frac{\Delta H_{vap}}{T_b} \qquad (14.13)$$

where T_b is the normal boiling point in K. Similarly,

$$\Delta S_{fus}^{1\ atm} = \frac{\Delta H_{fus}}{T_f} \qquad (14.14)$$

where T_f is the melting point at 1 atm pressure.

Example 14.2 Use Equations 14.13 and 14.14 to obtain $\Delta S^{1\ atm}$ for the vaporization of liquid benzene and the melting of solid benzene. For benzene, ΔH_{vap} = 30 500 J/mol, ΔH_{fus} = 10 700 J/mol. The boiling point and melting point of benzene at 1 atm are 80°C and 5°C, respectively.

Solution To apply these equations, the temperature must be expressed in K:

$$T_b = 80 + 273 = 353\ K;\ T_f = 5 + 273 = 278\ K$$

Hence:

$$\Delta S_{vap}^{1\ atm} = \frac{30\ 500\ J}{353\ K} = 86.4\ J/K$$

$$\Delta S_{fus}^{1\ atm} = \frac{10\ 700\ J}{278\ K} = 38.5\ J/K$$

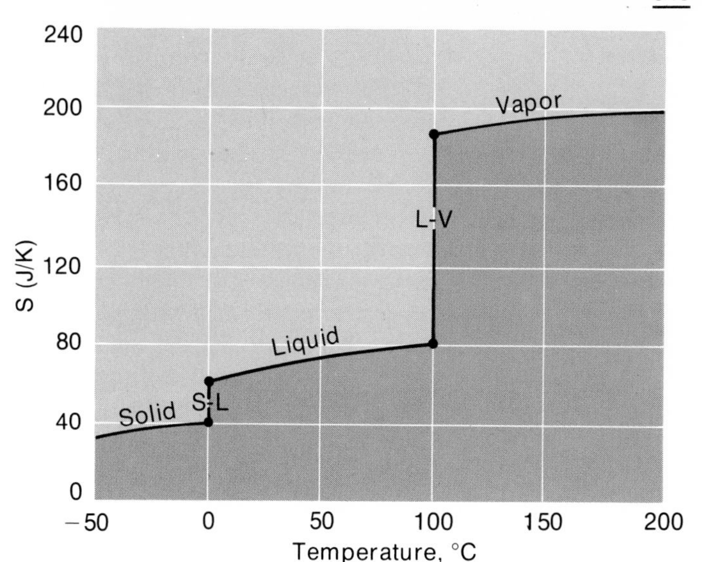

Figure 14.4 Variation with temperature of $S^{1\ atm}$ for one mole of water. Note the large increase in entropy at 0°C (fusion) and the even larger increase at 100°C (vaporization).

You may recall from Chapter 11 that the quantity $\Delta H_{vap}/T_b$ and hence $\Delta S^{1\ atm}$ is about 88 J/K for most liquids. The value just calculated for benzene is remarkably close to that. The entropy change on fusion, as in Example 14.2, is ordinarily much less than that for vaporization. This implies that the "increase in randomness" when a liquid vaporizes is ordinarily much greater than that associated with the melting of a solid (see also the plot of S vs. T for water, Fig. 14.4).

ΔS for Chemical Reactions

A knowledge of the factors that lead to either high or low entropy can help us to predict the sign of ΔS for a chemical reaction. Almost invariably **a reaction which results in an increase in the number of moles of gas is accompanied by an increase in entropy; if the number of moles of gas decreases, ΔS will be a negative quantity.** This rule is illustrated by Reactions 1 to 3 of Table 14.4. In Reaction 4, where there is the same number of moles of gas on both sides of the equation, ΔS is relatively small.

This is true because entropies of gases are so much greater than those of liquids and solids

TABLE 14.4 VALUES OF $\Delta S^{1\ atm}$ (J/K)

Reaction	$\Delta S^{1\ atm}$	Δn_{gas}	Δn_{tot}
1. $CaCO_3(s) \rightarrow CaO(s) + CO_2(g)$	+161	+1	
2. $N_2O_4(g) \rightarrow 2\ NO_2(g)$	+177	+1	
3. $CH_4(g) + 2\ O_2(g) \rightarrow CO_2(g) + 2\ H_2O(l)$	−243	−2	
4. $2\ HCl(g) \rightarrow H_2(g) + Cl_2(g)$	−20	0	
5. $CuSO_4 \cdot 5\ H_2O(s) \rightarrow CuSO_4(s) + 5\ H_2O(l)$	+157		+5
6. $PbBr_2(s) \rightarrow Pb(s) + Br_2(l)$	+56		+1
7. $PbI_2(s) \rightarrow Pb(s) + I_2(s)$	+5		+1
8. $2\ AgI(s) \rightarrow 2\ Ag(s) + I_2(s)$	−26		+1

For reactions in which no gases are involved (Reactions 5 to 8, Table 14.4), we frequently, but not always, find that an increase in the total number of moles leads to an increase in entropy. As Table 14.4 implies, we can be confident that this rule will apply only if Δn_{tot} is large (Reaction 5) or if a liquid is produced at the expense of a solid (Reactions 5 and 6).

The entropy change for a reaction, at least above 25°C, is nearly independent of temperature. This is true despite the fact that the entropy of an *individual substance* invariably increases with temperature. In most cases the increase in entropy of the products just about balances that of the reactants and ΔS does not change appreciably with temperature. Much the same argument can be used, incidentally, to justify neglecting the temperature dependence of ΔH, the enthalpy change of a reaction.*

On the other hand, ΔS for a reaction may vary appreciably with pressure if, as is often the case, there is a change in the number of moles of gas. For this reason among others, we shall work with the **entropy change at one atmosphere pressure, $\Delta S^{1\,atm}$**, in all our calculations.

14.4 THE GIBBS-HELMHOLTZ EQUATION

Now that we have a clearer idea of the nature of entropy, let us examine Equation 14.12 in greater detail. This equation, one of the most useful in all of chemistry, is called the Gibbs-Helmholtz equation in honor of the two men most responsible for its development.

$$\Delta G = \Delta H - T\Delta S \qquad (14.12)$$

Like most fundamental equations, it has a very simple form. In words, it tells us that the driving force for a chemical reaction (ΔG) depends upon two quantities: the energy change due to making and breaking of bonds (ΔH), and the product of the change in randomness (ΔS) times the absolute temperature. The two factors which tend to make ΔG negative and, hence, lead to a spontaneous reaction are:

1. A NEGATIVE VALUE OF ΔH. Exothermic reactions ($\Delta H < 0$) will tend to be spontaneous inasmuch as they contribute to a negative value of ΔG. On the molecular level this means that there will be a tendency to form "strong" bonds at the expense of "weak" ones.

2. A POSITIVE VALUE OF ΔS. If the entropy change is positive ($\Delta S > 0$), the term $-T\Delta S$ will make a negative contribution to ΔG. Consequently there will be a tendency for reactions to be spontaneous if the products are less ordered than the reactants.

If we think about the matter carefully, we realize that there is an inherent tendency, common to inanimate substances (and probably to animate objects like students and professors), to become disorganized. A beaker shatters when we drop it on a stone lab bench, rivers become polluted, and wilderness trails get cluttered with litter. All these processes involve an increase in entropy and are spontaneous, as anyone who tries to reverse them soon discovers.

---

*So far as Equation 14.12 is concerned, there is an added justification for ignoring the temperature dependence of ΔH and ΔS. These two quantities always change in the same direction as temperature changes (e.g., if ΔH becomes more positive, so also does ΔS). This means that the two effects tend to cancel each other and the true value of ΔG at any temperature is about the same as the one we calculate by taking ΔH and ΔS to be constant, independent of temperature.

In many physical processes the entropy increase is the major driving force. An example is the formation of a solution. When nitrogen diffuses into oxygen or benzene dissolves in toluene, the enthalpy change is virtually zero, but ΔS is a positive quantity because the solution is less ordered than the pure substances from which it is formed.

In certain reactions, ΔS is nearly zero and ΔH is the only important component of the driving force for spontaneity. An example is the synthesis of hydrogen fluoride from the elements:

$$\tfrac{1}{2}\ H_2(g) + \tfrac{1}{2}\ F_2(g) \rightarrow HF(g) \qquad (14.15)$$

For this reaction, ΔH is a large negative number, −268.6 kJ, reflecting the fact that the bonds in HF are stronger than those in the H_2 and F_2 molecules. As we might expect for a reaction in which there is no change in the number of moles of gas, $\Delta S^{1\,atm}$ is very small, about 0.0070 kJ/K. The free energy change, $\Delta G^{1\,atm}$, is −270.7 kJ at 25°C, virtually identical with ΔH. Even at very high temperatures, the difference between $\Delta G^{1\,atm}$ and ΔH is relatively small ($\Delta G^{1\,atm} = -282.6$ kJ at 2000 K).

Could HF be decomposed to the elements by heating at 1 atm?

The more common case is one in which both ΔH and $T\Delta S$ make significant contributions to ΔG; both must be considered to decide upon the direction of spontaneity. In particular, their signs and magnitudes will determine what effect, if any, an increase in temperature at constant pressure will have upon reaction spontaneity.

Effect of Temperature on Reaction Spontaneity

When the temperature of a reaction system is increased at constant pressure the direction in which the reaction proceeds spontaneously may or may not change, depending upon the signs of ΔH and ΔS. The four possible situations, deduced from the Gibbs-Helmholtz equation, are summarized in Table 14.5.

Notice that if ΔH and ΔS have opposite signs (I and II), it is impossible by a change in temperature alone to reverse the direction in which the reaction proceeds spontaneously. The two terms ΔH and $-T\Delta S$ reinforce one another, and

TABLE 14.5 EFFECT OF TEMPERATURE ON REACTION SPONTANEITY AT A GIVEN PRESSURE

	ΔH	ΔS	$\Delta G = \Delta H - T\Delta S$	REMARKS
I	−	+	always −	Spontaneous at all T; reverse reaction always nonspontaneous
II	+	−	always +	Nonspontaneous at all T; reverse reaction occurs
III	+	+	+ at low T − at high T	Nonspontaneous at low T; becomes spontaneous as T is raised
IV	−	−	− at low T + at high T	Spontaneous at low T; at high T, reverse reaction becomes spontaneous

ΔG has the same sign at all temperatures. Reactions of this type are relatively uncommon. One important example is the thermal decomposition of carbon monoxide

$$CO(g) \rightarrow C(s) + \tfrac{1}{2} O_2(g) \tag{14.16}$$

for which ΔH is +110.5 kJ and $\Delta S^{1\,atm}$ is −0.0899 kJ/K. Substituting into the Gibbs-Helmholtz equation, we have

$$\begin{aligned} \Delta G^{1\,atm} &= \Delta H - T\Delta S^{1\,atm} \\ &= 110.5 \text{ kJ} + (0.0899 \text{ kJ/K})T \end{aligned}$$

Clearly, $\Delta G^{1\,atm}$ is positive at all temperatures; the reaction cannot take place spontaneously at one atmosphere pressure regardless of temperature. In practical terms this means that it would be futile to hope that the carbon monoxide in automobile exhausts might be eliminated by thermal decomposition to the elements.

It is more common to find that ΔH and ΔS have the same sign (types III and IV). When this happens, the enthalpy and entropy factors oppose each other, ΔG changes sign as the temperature increases, and the direction in which the reaction proceeds spontaneously reverses. At low temperatures ΔH predominates and the exothermic reaction occurs; as the temperature rises the quantity $T\Delta S$ eventually predominates and the reaction which leads to an increase in entropy occurs. In most cases 25°C is a "low" temperature, at least when the pressure is maintained at 1 atm; this explains the empirical rule mentioned earlier that exothermic reactions are usually spontaneous at room temperature and atmospheric pressure.

An example of a general type of reaction for which both ΔH and ΔS have the same sign is the dissociation of a molecule into atoms, e.g.,

$$X_2(g) \rightarrow 2\, X(g) \tag{14.17}$$

At low *T*, species with strong bonds are formed; at high *T*, species with the greatest disorder are favored

ΔH is a positive quantity because heat must be absorbed to break the bond holding the molecule together. ΔS is positive since there is an increase in the number of moles of gas. This means that at sufficiently high temperatures, all molecules, even ones that are very stable such as H_2 and N_2, will break down to free atoms.

Calculation of $\Delta S^{1\,atm}$

If the Gibbs-Helmholtz equation is solved for ΔS we obtain

$$\Delta S = \frac{\Delta H - \Delta G \text{ at } T}{T}$$

Knowing ΔH and ΔG at some temperature T (in K), we can readily obtain ΔS for a reaction. Ordinarily, ΔG is calculated from a table of free energies of formation at 25°C and 1 atm. Under these conditions the above equation becomes

$$\Delta S^{1\,atm} = \frac{\Delta H - \Delta G^{1\,atm} \text{ at 298 K}}{298 \text{ K}} \tag{14.18}$$

Example 14.3 For the vaporization of water at 1 atm pressure

$$H_2O(l, 1\ atm) \rightarrow H_2O(g, 1\ atm)$$

obtain, using data in Tables 4.1 and 14.2,

a. ΔH b. $\Delta G^{1\,atm}$ at 298 K c. $\Delta S^{1\,atm}$

Solution

a. From Table 4.1 (p. 67):

$$\begin{aligned}\Delta H &= \Delta H_f\ H_2O(g) - \Delta H_f\ H_2O(l)\\ &= -241.8\ kJ - (-285.8\ kJ) = +44.0\ kJ\end{aligned}$$

b. From Table 14.2:

$$\begin{aligned}\Delta G^{1\,atm} \text{ at } 298\ K &= \Delta G_f^{1\,atm}\ H_2O(g) - \Delta G_f^{1\,atm}\ H_2O(l)\\ &= -228.6\ kJ - (-237.2\ kJ) = +8.6\ kJ\end{aligned}$$

c. Using Equation 14.18:

$$\Delta S^{1\,atm} = \frac{\Delta H - \Delta G^{1\,atm} \text{ at } 298\ K}{298\ K} = \frac{+44.0\ kJ - 8.6\ kJ}{298\ K} = 0.119\frac{kJ}{K}$$

$$= 119\ J/K$$

Calculation of $\Delta G^{1\,atm}$ at Temperatures Other Than 298 K

Once the values of ΔH and $\Delta S^{1\,atm}$ have been established, $\Delta G^{1\,atm}$ is readily calculated at any temperature from the Gibbs-Helmholtz equation. For the vaporization of water, where we found in Example 14.3 that $\Delta H = +44.0$ kJ and $\Delta S = +0.119$ kJ/K, we have

$$\Delta G^{1\,atm} \text{ at } T = +44.0\ kJ - \frac{0.119\ kJ}{K} \times T$$

In Table 14.6 we list values of $\Delta G^{1\,atm}$ calculated from the above equation at 25° temperature intervals between 0°C (273 K) and 125°C (398 K). For convenience, values of ΔH and $T\Delta S^{1\,atm}$ are also included.

TABLE 14.6 $H_2O(l, 1\ atm) \rightarrow H_2O(g, 1\ atm)$

t(°C)	T(K)	ΔH(kJ)	$T\Delta S$(kJ)	ΔG(kJ)
0	273	+44.0	+32.2	+11.8
25	298	+44.0	+35.2	+8.8
50	323	+44.0	+38.1	+5.9
75	348	+44.0	+41.1	+2.9
100	373	+44.0	+44.0	0.0
125	398	+44.0	+47.0	−3.0

ΔH and ΔS are essentially independent of T, but ΔG varies markedly with T

Above 100°C, the $T\Delta S$ term predominates, ΔG is negative, and water boils spontaneously at 1 atm pressure. At temperatures below 100°C the ΔH term prevails, $\Delta G^{1\,atm}$ is positive, and water in an open beaker refuses to boil. Instead, the reverse reaction, which amounts to the condensation of supercooled steam, occurs. At precisely 100°C, ΔH and $T\Delta S$ becomes equal, ΔG is zero, and the system is at equilibrium, balanced on a knife edge; there is no tendency for reaction to occur in either direction.

Calculation of Temperature at Which $\Delta G^{1\,atm} = 0$

The analysis that we have just carried out suggests a simple way to obtain an important quantity: the temperature at which a reaction is at equilibrium at one atmosphere pressure. All that is required is to determine the temperature at which $\Delta G^{1\,atm}$ becomes zero. To illustrate, let us consider the reaction

$$CaO(s) + SO_3(g) \rightarrow CaSO_4(s) \qquad (14.19)$$

for which $\Delta H = -402.0$ kJ and $\Delta G^{1\,atm} = -345.7$ kJ at 298 K. This reaction offers an approach to removing sulfur trioxide (formed by the combustion of high-sulfur coal) from furnace gases, thereby cutting down on atmospheric pollution. Clearly the reaction is spontaneous at room temperature, but we need to know whether it will remain spontaneous at the high temperatures reached in coal furnaces. Perhaps the most general approach to this problem is to calculate the temperature at which the reaction is at equilibrium at 1 atm pressure.

Example 14.4 Calculate the temperature at which the reaction

$$CaO(s) + SO_3(g) \rightarrow CaSO_4(s)$$

is at equilibrium at one atmosphere pressure.

Solution We need a general expression for $\Delta G^{1\,atm}$ as a function of T. To obtain it, we first evaluale $\Delta S^{1\,atm}$ as in Example 14.3:

$$\Delta S^{1\,atm} = \frac{\Delta H - \Delta G^{1\,atm} \text{ at } 298\text{ K}}{298\text{ K}}$$

$$= \frac{-402.0\text{ kJ} - (-345.7\text{ kJ})}{298\text{ K}} = -0.189\text{ kJ/K}$$

Hence:

$$\Delta G^{1\,atm} = \Delta H - T\Delta S^{1\,atm}$$

$$= -402.0\text{ kJ} + 0.189\,\frac{\text{kJ}}{\text{K}} \times T$$

We now solve for the temperature at which $\Delta G^{1\,atm}$ becomes zero.

$$0 = -402.0 + 0.189\ T$$

$$T = 402.0/0.189 = 2130\text{ K}$$

We conclude from Example 14.4 that Reaction 14.19 is spontaneous at any temperature below 2130 K (1860°C). Since furnace temperatures are ordinarily well below 1000°C, the addition of quicklime, calcium oxide, to coal furnaces would appear to be a practical way of tying up sulfur trioxide and preventing its escape into the air. Indeed, this process is now being used and offers one of the more promising methods of solving this particular problem in the area of air pollution.

A HISTORICAL PERSPECTIVE

Josiah Willard Gibbs (1839–1903)

A century ago chemistry was primarily an empirical science. The outstanding chemists of that era were experimentalists who isolated and characterized new substances. The principles of chemistry were descriptive or correlative in nature, as illustrated by the atomic theory of Dalton and the Periodic Table of Mendeleev. Two theoreticians working in the latter half of the 19th century changed the very nature of chemistry by deriving the mathematical laws that govern the behavior of matter undergoing physical or chemical change. One of these was James Clerk Maxwell, whose contributions to kinetic theory were discussed in Chapter 5. The other was J. Willard Gibbs, professor of mathematical physics at Yale from 1871 until his death in 1903.

In 1876 Gibbs published in the Transactions of the Connecticut Academy of Sciences the first portion of a remarkable paper entitled, "On the Equilibrium of Heterogeneous Substances." When the paper was completed in 1878 (it was 323 pages long), the foundation had been established for the science of chemical thermodynamics. Here for the first time appeared the concepts of maximum work and free energy that we have used in this chapter to discuss the spontaneity of chemical reactions. Included as well were the basic principles of chemical equilibrium to which we shall devote Chapter 15. Sections of the paper went on to apply the laws of thermodynamics to develop principles of phase equilibria (Chapter 11), laws of dilute solutions (Chapter 12), the nature of adsorption at surfaces (Chapter 16), and the mathematical relationships governing energy changes in voltaic cells (Chapter 23).

If Gibbs had never published another paper, this single contribution would have placed him among the greatest theoreticians in the history of science. Generations of experimental scientists have established their reputations by demonstrating in the laboratory the validity of the relationships that Gibbs derived at his desk. Many of these relationships were rediscovered by others; an example is the Gibbs-Helmholtz equation developed in 1882 by Helmholtz, who was completely unaware of Gibbs' work.

In the years that remained to him, Gibbs made substantial contributions in chemistry, astronomy, and mathematics. Among these were two papers published in 1881 and 1884 that established the discipline known today as vector analysis. His last work, published in 1901, was a book entitled "Elementary Principles in Statistical Mechanics." Here Gibbs used the statistical principles that govern the behavior of systems to develop thermodynamic equations that he had derived from an entirely different point of view at the beginning of his career. Here, too, we find the "randomness" interpretation of entropy that has received so much attention in the social as well as the natural sciences.

J. Willard Gibbs is often cited as an example of the "prophet without honor in his own country." His colleagues at New Haven and elsewhere in the United States seem not to have realized the significance of his work until late in his life. During his first ten years as a professor at Yale he received no salary. In 1920 when he was first proposed for the Hall of Fame of Distinguished Americans at New York University, he received nine votes out of a possible 100. Not until 1950 was he elected to that body. Even today the name of J. Willard Gibbs is generally unknown among educated Americans outside of those interested in the natural sciences.

Admittedly, Gibbs himself was largely responsible for the fact that for many years his work did not attract the attention it deserved. He made little effort to publicize it; the *Transactions of the Connecticut Academy of Sciences* was hardly the leading scientific journal of its day. Gibbs was one of those rare individuals who seem to have no inner need for recognition by contemporaries. His satisfaction came from solving a problem in his mind; having done so, he was ready to pass on to other problems without being concerned whether people understood what he had done. His papers are not easy to read; he seldom cites examples to illustrate his abstract reasoning. Frequently the implications of the laws that he derives are left for the reader to grasp on his own. One of his colleagues at Yale confessed many years later that none of the members of the Connecticut Academy of Sciences understood his paper on thermodynamics; as he put it, "We knew Gibbs and took his contributions on faith."

Gibbs achieved recognition in Europe long before his work was generally appreciated in this country. Maxwell, the pre-eminent theoretician of his time, somehow came across a copy of Gibbs' paper on thermodynamics, saw its significance, and referred to it repeatedly in his own publications. Wilhelm Ostwald, who said of Gibbs, "To physical chemistry, he gave form and content for a hundred years," translated the paper into German in 1892. Seven years later, Le Châtelier translated it into French.

Muriel Rukeyser, in her biography of Gibbs, tells an anecdote that reveals a great deal about the man and the scientist. In one of Gibbs' early papers there was a discussion of the phase equilibria between ice, liquid water, and water vapor. In Miss Rukeyser's words, "Again, Willard Gibbs had stopped at the bare idea and left undone that step which might have bridged the gap between himself and his audience. Maxwell added the last personal impression which must have touched and delighted Gibbs more than any other gift." The eminent English theoretician made a plaster model showing graphically the thermodynamic relationships involved and sent it to New Haven. Gibbs took the model to class but never referred to it in his lectures. One day, a student asked where it had come from.

"A friend sent it to me," said Gibbs with his own punishing modesty. "Who is the friend?" the boy asked, knowing very well who it was. But all that Gibbs would say was, "A friend in England."

PROBLEMS

14.1 ΔG Consider the reaction 2 $SO_3(g) \rightarrow 2\ SO_2(g) + O_2(g)$.

a. Using Table 14.2, calculate ΔG at 25°C and 1 atm.
b. Is this reaction spontaneous at 25°C and 1 atm?
c. What is ΔG for the reverse reaction?
d. Calculate ΔG for the decomposition of 1.00 g of SO_3 to SO_2 and O_2 at 25°C and 1 atm. (*MM* SO_3 = 80.1)

14.2 *ΔS for Phase Changes* Calculate ΔS for the vaporization of one mole of water. The heat of vaporization is 2257 J/g at the normal boiling point, 100°C. (*MM* H_2O = 18.0)

14.3 *ΔS for Reactions* Estimate the sign of ΔS for the reaction in Problem 14.1.

14.4 *Gibbs-Helmholtz Equation* Consider the reaction in Problem 14.1, for which you calculated $\Delta G^{1\,atm}$ at 25°C. Calculate

a. ΔH, using Table 4.1, p. 67.
b. $\Delta S^{1\,atm}$
c. $\Delta G^{1\,atm}$ at 500°C
d. T at which $\Delta G^{1\,atm} = 0$

14.5 Which of the following are spontaneous? nonspontaneous?

a. formation of a solution of alcohol in water.
b. melting of ice at −10°C and 1 atm.
c. going to the top of a mountain on a ski lift.
d. separation of oxides of sulfur from stack gases.

14.6 How might the following processes, all of which are nonspontaneous, be accomplished?

a. $H_2O(l) \rightarrow H_2(g) + \frac{1}{2}\ O_2(g)$
b. salt solution $\rightarrow$ NaCl(s) + water
c. $6\ CO_2(g) + 6\ H_2O(l) \rightarrow C_6H_{12}O_6(s) + 6\ O_2(g)$

14.7 Consider the reaction involved in the electrolysis of bauxite ore

$$Al_2O_3(s) \rightarrow 2\ Al(s) + \frac{3}{2}\ O_2(g)$$

for which $\Delta G = +1576$ kJ.

Could this reaction be accomplished by supplying 800 kJ of electrical energy per mole of Al_2O_3? 1600 kJ?

14.8 The oxidation of iron is represented by the equation

$$4\ Fe(s) + 3\ O_2(g) \rightarrow 2\ Fe_2O_3(s);\ \Delta G = -1481\ kJ$$

a. What is the free energy of formation of Fe_2O_3?
b. What is ΔG for the formation of 1.00 g of Fe_2O_3?

14.9 Given

$$C(s) + O_2(g) \rightarrow CO_2(g);\ \Delta G = -394.4\ kJ$$

$$CO(g) + \frac{1}{2}\ O_2(g) \rightarrow CO_2(g);\ \Delta G = -257.1\ kJ$$

Calculate ΔG for $C(s) + \frac{1}{2}\ O_2(g) \rightarrow CO(g)$

14.21 a. Under what conditions is the vaporization of water at 25°C spontaneous? nonspontaneous? neither?
b. The solubility of NaCl in water at 25°C is 6 mol/dm^3. Which of the following processes are spontaneous? nonspontaneous?

$$NaCl(s) \rightarrow NaCl(aq,\ 2\ M)$$
$$NaCl(s) \rightarrow NaCl(aq,\ 8\ M)$$

14.22 Consider the reactions in Problem 14.6. Suggest how useful work might be obtained from the reverse of each of these reactions.

14.23 Consider the decomposition of $KClO_3$

$$KClO_3(s) \rightarrow KCl(s) + \frac{3}{2}\ O_2(g)$$

for which $\Delta G = -118$ kJ.

a. How much useful work can be obtained when one mole of $KClO_3$(s) decomposes?
b. Could this reaction be reversed by doing 50 kJ of work per mole of $KClO_3$? 120 kJ?

14.24 For the decomposition of sodium chloride

$$NaCl(s) \rightarrow Na(s) + \frac{1}{2}\ Cl_2(g);\ \Delta G = +384\ kJ$$

a. What is the free energy of formation of NaCl?
b. How many grams of Na could be produced by supplying 160 kJ of electrical energy?

14.25 For the two reactions:

$$2\ Fe(s) + \frac{3}{2}\ O_2(g) \rightarrow Fe_2O_3(s)$$

$$4\ Fe_2O_3(s) + Fe(s) \rightarrow 3\ Fe_3O_4(s)$$

ΔG is −741 kJ and −79 kJ, respectively. Calculate ΔG_f for Fe_3O_4.

14.10 Consider the phase change

$$H_2O(l) \rightarrow H_2O(g)$$

a. Using Table 14.2, calculate ΔG for this process at 25°C and 1 atm.
b. Explain why the sign of ΔG in (a) is positive.
c. How would you change the temperature (at constant P) to make ΔG negative?
d. How would you change the pressure (at constant T) to make ΔG negative?

14.11 Calculate the free energy change at 25°C and 1 atm for

a. $2\ H_2S(g) + 3\ O_2(g) \rightarrow 2\ SO_2(g) + 2\ H_2O(l)$
b. $Ca(OH)_2(s) + CO_2(g) \rightarrow CaCO_3(s) + H_2O(l)$

Which of these reactions are spontaneous?

14.12 How would you expect the molar entropy of each of the following to compare to that of $Cl_2(g)$?

a. $Cl_2(l)$
b. $F_2(g)$
c. $H_2(g)$
d. $Br_2(g)$
e. $Cl_2O_7(g)$

14.13 The heat of fusion of water is 333 J/g. Calculate ΔS for the melting of one mole of ice.

14.14 Predict the sign of ΔS for

a. crystallization of salt from a supersaturated water solution.
b. $CH_4(g) + 2\ O_2(g) \rightarrow CO_2(g) + 2\ H_2O(l)$
c. a collision between two cars.
d. $Ag^+(aq) + Cl^-(aq) \rightarrow AgCl(s)$
e. the process that occurs when a soft drink bottle is opened.
f. $CO_2(g) \rightarrow CO(g) + \frac{1}{2}\ O_2(g)$

14.15 Each of the following is spontaneous at 25°C and 1 atm. In which cases would you expect the major component of the driving force to be ΔH? ΔS?

a. $H_2(g) + Cl_2(g) \rightarrow 2\ HCl(g)$
b. $2\ Cl(g) \rightarrow Cl_2(g)$
c. water flowing over a dam.
d. benzene + toluene $\rightarrow$ solution

14.16 Consider the following reactions, each at 1 atm.

a. $N_2(g) + O_2(g) \rightarrow 2\ NO(g)$
b. $Mg(s) + Cl_2(g) \rightarrow MgCl_2(s)$
c. $H_2(g) + S(s) \rightarrow H_2S(g)$

For these reactions, ΔH is +90 kJ, −642 kJ, and −20 kJ, respectively. ΔS is +12 J/K, −166 J/K, and +40 J/K, respectively. Which of these reactions are spontaneous at all temperatures? only at high temperatures? only at low temperatures?

14.26 The free energy of formation of ice at 25°C and 1 atm is −236.7 kJ/mol.

a. Using Table 14.2, calculate ΔG for the freezing of one mole of $H_2O(l)$ at 25°C and 1 atm.
b. Explain why the sign of ΔG in (a) is positive.
c. Predict the sign of ΔG for the process in (a) at 10°C; −10°C.
d. If the pressure were increased, would ΔG become more positive or less positive?

14.27 Calculate $\Delta G^{1\ atm}$ at 25°C for

a. $2\ NH_3(g) \rightarrow N_2(g) + 3\ H_2(g)$
b. $4\ NH_3(g) + 7\ O_2(g) \rightarrow 4\ NO_2(g) + 6\ H_2O(l)$
c. $SiO_2(s) + 2\ H_2(g) \rightarrow Si(s) + 2\ H_2O(l)$

Which of these reactions are spontaneous?

14.28 Arrange the following, as best you can, in order of increasing entropy.

a. $LiCl(s)$
b. $Cl_2(g)$
c. $Li(s)$
d. $Ne(g)$
e. $I_2(g)$

14.29 "Dry ice" sublimes at −79°C and 1 atm; its heat of sublimation is 25.9 kJ/mol. Calculate the molar entropy of sublimation; the entropy change per gram of CO_2.

14.30 Predict the sign of ΔS for

a. adsorption of a gas on a solid surface.
b. osmosis.
c. $SnO_2(s) + 2\ H_2(g) \rightarrow Sn(s) + 2\ H_2O(l)$
d. assembling the pages of a book.
e. $C_{14}H_{28}(s) + H_2(g) \rightarrow C_{14}H_{30}(s)$
f. the formation of polyethylene, a polymer of empirical formula $(CH_2)_n$, from ethylene, C_2H_4.

14.31 Give the signs you would expect for ΔG, ΔH, and ΔS for each of the following:

a. electrolysis of water to give $H_2(g)$ and $O_2(g)$.
b. dissolving a small amount of $NaNO_3$ in water (the solution becomes cold).
c. melting ice on a highway by adding salt.

14.32 Comment on the *general* validity of each of the following statements.

a. An exothermic reaction is spontaneous.
b. A reaction for which ΔS is positive is spontaneous.
c. ΔS is positive for a reaction in which there is an increase in the number of moles.
d. If ΔH and ΔS are both positive, ΔG will decrease when the temperature rises.

14.17 For the reaction: $Ag(s) + ½ Cl_2(g) \rightarrow AgCl(s)$, calculate

a. ΔH
b. $\Delta G^{1\,atm}$ at 298 K
c. $\Delta S^{1\,atm}$
d. $\Delta G^{1\,atm}$ at 1000 K

14.18 For the reaction $2\ SO_3(g) \rightarrow 2\ SO_2(g) + O_2(g)$, using data in Tables 4.1 and 14.2, obtain an expression for $\Delta G^{1\,atm}$ as a function of temperature. Prepare a table listing values of $\Delta G^{1\,atm}$ at intervals of 200 K between 200 K and 1200 K.

14.19 Consider the water gas reaction, which has been suggested as one way to get a gaseous fuel from coal:

$$C(s) + H_2O(g) \rightarrow CO(g) + H_2(g)$$

a. Is this reaction spontaneous at 25°C and 1 atm?
b. Is the reaction likely to become spontaneous if the temperature is raised?
c. At what temperature will the system be in equilibrium at 1 atm?
d. What other factors would have to be considered to decide whether this reaction is feasible?

14.20 Two possible ways of producing iron from iron ore are

a. $Fe_2O_3(s) + \frac{3}{2}\ C(s) \rightarrow 2\ Fe(s) + \frac{3}{2}\ CO_2(g)$
b. $Fe_2O_3(s) + 3\ H_2(g) \rightarrow 2\ Fe(s) + 3\ H_2O(g)$

Which of these reactions would proceed spontaneously at the lower temperature?

14.33 For the reaction $MgCO_3(s) \rightarrow MgO(s) + CO_2(g)$, calculate

a. ΔH
b. $\Delta G^{1\,atm}$ at 298 K
c. $\Delta S^{1\,atm}$
d. $\Delta G^{1\,atm}$ at 500 K

14.34 For the reaction $2\ Ag_2O(s) \rightarrow 4\ Ag(s) + O_2(g)$, obtain an expression for $\Delta G^{1\,atm}$ as a function of temperature. Calculate the temperatures at which $\Delta G^{1\,atm} =$ +20 kJ, 0 kJ, and −20 kJ.

14.35 At least on paper, the decomposition of methanol would seem to be a good way to prepare methane:

$$CH_3OH(l) \rightarrow CH_4(g) + ½\ O_2(g)$$

a. Is this reaction spontaneous at 25°C and 1 atm?
b. What is the sign of ΔS for this reaction? What is its magnitude?
c. Above what temperature is this reaction spontaneous?
d. What objection, if any, can you see to producing CH_4 by this reaction? (Consider the other product.)

14.36 It is desired to produce Sn metal from its ore, cassiterite, SnO_2, at as low a temperature as possible. The ore might be heated by itself (to produce O_2), heated with H_2 (to produce water vapor), or heated with carbon (to produce CO_2). Which would you recommend, based on thermodynamic principles?

*14.37 Consider the data in Table 14.1.

a. Show that at any given temperature, the variation of ΔG with pressure is given by the expression

$$\Delta G^P = \Delta G^{1\,atm} + a \log P$$

i.e., that ΔG is a linear function of $\log P$.
b. Evaluate a at 25°C.
c. Calculate ΔS at pressures of 1 atm, 10^{-2} atm, and 10^{-4} atm at 25°C.
d. Show that $\Delta S^P = \Delta S^{1\,atm} - \frac{a}{T} \log P$.

*14.38 Two simple reactions that have been used or suggested for rocket propulsion are

$$H_2(g) + ½\ O_2(g) \rightarrow H_2O(g) \qquad H_2(g) + F_2(g) \rightarrow 2\ HF(g)$$

How much useful work can be obtained per gram of reactant from each of these reactions at 25°C and 1 atm? At 1000°C and 1 atm?

*14.39 The overall reaction that occurs when sugar is metabolized is

$$C_{12}H_{22}O_{11}(s) + 12\ O_2(g) \rightarrow 12\ CO_2(g) + 11\ H_2O(l)$$

For this reaction, ΔH is −5650 kJ and $\Delta G^{1\,atm}$ is −5790 kJ at 25°C.

a. If 30% of the free energy difference is actually converted to useful work, how many kilojoules of work could be obtained when one gram of sugar is metabolized at body temperature, 37.0°C?
b. How many grams of sugar would you have to eat to get the energy to climb a mountain 2.0 km high? ($W = 9.79 \times 10^{-6}\ mh$, where W = work in kilojoules, m is mass in grams, and h is the height in metres)

*14.40 When a copper wire is exposed to air at room temperature, it becomes coated with a black oxide, CuO. If the wire is heated above a certain temperature, the black oxide is converted to a red oxide, Cu_2O. At a still higher temperature, the oxide coating disappears. Explain these observations in terms of the thermodynamics of the reactions

$$2\ CuO(s) \rightarrow Cu_2O(s) + \tfrac{1}{2}\ O_2(g)$$

$$Cu_2O(s) \rightarrow 2\ Cu(s) + \tfrac{1}{2}\ O_2(g)$$

and estimate the temperatures at which the changes occur.

15

CHEMICAL EQUILIBRIUM IN GASEOUS SYSTEMS

In Chapter 14, we showed how the sign of the free energy change, ΔG, can be used to make predictions concerning reaction spontaneity. The calculations we made were subject to one important restriction. Since we worked only with $\Delta G^{1\ atm}$, we could only predict the *direction* in which a reaction will proceed spontaneously when products and reactants are at *one atmosphere pressure*. In this chapter, our goal is more ambitious. We seek to develop principles to predict both the *direction* and *extent* to which a reaction involving gases will proceed when reactants and products are at *any* given pressure or concentration.

The approach we will follow is to study the conditions that apply to a system at chemical equilibrium. To understand what chemical equilibrium implies, consider what happens when a sample of calcium carbonate is introduced into a closed container at 850°C. Some of the calcium carbonate decomposes by way of the reaction

$$CaCO_3(s) \rightarrow CaO(s) + CO_2(g)$$

As the concentration of carbon dioxide builds up in the gas phase, some of the CO_2 molecules react with calcium oxide

$$CaO(s) + CO_2(g) \rightarrow CaCO_3(s)$$

Eventually the rates of these two competing reactions become equal, the concentration of carbon dioxide does not change with time, and we have arrived at equilibrium. Analysis shows that at 850°C, the concentration of CO_2 is 0.0108 mol/dm³. This is true regardless of how much solid calcium carbonate or calcium oxide there is in the container. In other words, we could describe the "position of the equilibrium" in this chemical system at 850°C

$$CaCO_3(s) \rightleftharpoons CaO(s) + CO_2(g) \qquad (15.1)$$

by saying that the equilibrium concentration of gaseous carbon dioxide is 0.0108 mol/dm³.

The chemical equilibrium just described is really very similar to the physical equilibrium between liquid and gaseous water, discussed in Chapter 11:

$$H_2O(l) \rightleftharpoons H_2O(g) \qquad (15.2)$$

You will recall that the position of this equilibrium at any given temperature can be described by citing a single number. We might, for example, say that at 100°C the equilibrium pressure of water vapor is one atmosphere, regardless of how much liquid water is present. Alternatively, we could say that the equilibrium concentration of $H_2O(g)$ at 100°C, as calculated from its pressure using the Ideal Gas Law, is 0.0327 mol/dm^3.

Unfortunately, most chemical systems at equilibrium cannot be described as simply as in Reactions 15.1 or 15.2. In particular, if more than one gaseous species is involved, we cannot specify the equilibrium concentration of one species independent of that of the others. For the system

$$N_2O_4(g) \rightleftharpoons 2\ NO_2(g) \qquad (15.3)$$

the equilibrium concentrations of NO_2 and N_2O_4 do not have fixed values at a particular temperature. Instead, they can take on an infinite number of values, depending upon the way in which the system is prepared. There is, however, a fixed relationship between these two concentrations that can be expressed as a single number at a particular temperature. Let's see if we can discover the nature of this relationship.

15.1 THE N_2O_4-NO_2 EQUILIBRIUM SYSTEM. CONCEPT OF K_c

To study the equilibrium system represented by Reaction 15.3, we might start by admitting pure N_2O_4 into an evacuated container of fixed volume which is surrounded by a bath of boiling water that maintains a constant temperature of 100°C. Let us suppose that we admit one mole of N_2O_4 to a 10.0-dm^3 container, thus establishing its initial concentration at 0.100 mol/dm^3. As time passes, we find that a reddish-brown color develops. This color is characteristic of the NO_2 molecule, an odd-electron species; N_2O_4 is colorless. Eventually the intensity of the color reaches a maximum and then remains constant. This means that NO_2 molecules are being formed at the same rate they are recombining to form N_2O_4. We have arrived at a state of chemical equilibrium and can safely conclude that there will be no further change in the concentrations of NO_2 or N_2O_4.

We now withdraw a sample of the gaseous mixture and analyze it, finding that the equilibrium concentration of N_2O_4 is 0.040 mol/dm^3. In other words, in achieving equilibrium the concentration of N_2O_4 has decreased by 0.060 mol/dm^3:

$$\text{decrease in } N_2O_4 \text{ conc.} = \text{initial conc. } N_2O_4 - \text{equil. conc. } N_2O_4$$
$$= 0.100 \text{ mol/dm}^3 - 0.040 \text{ mol/dm}^3 = 0.060 \text{ mol/dm}^3$$

From the coefficients of Equation 15.3, we see that for every mole of N_2O_4 that decomposes two moles of NO_2 are formed. It follows that the concentration of NO_2 must have increased by 2 × 0.060 mol/dm^3, i.e.,

$$\text{increase in } NO_2 \text{ conc.} = 2 \times \text{decrease in } N_2O_4 \text{ conc.}$$
$$= 2(0.060 \text{ mol/dm}^3) = 0.120 \text{ mol/dm}^3$$

Since we had no NO_2 to start with, its equilibrium concentration must be 0.120 mol/dm³. Thus we see that the "position of the equilibrium" in this experiment could be described by stating that:

$$[N_2O_4] = 0.040 \text{ mol/dm}^3; [NO_2] = 0.120 \text{ mol/dm}^3$$

The square brackets, here and elsewhere throughout this text, are used to represent equilibrium concentrations in moles per cubic decimetre.

The results of this experiment are recorded under the heading of Experiment 1 in Table 15.1. The table also includes the results of two other experiments with the same chemical system at the same temperature, but approaching equilibrium from different points. In Experiment 2, we start with pure NO_2; in Experiment 3, the initial system contains both NO_2 and N_2O_4. Note that, for each experiment, all the numbers can be obtained knowing only the original concentrations of NO_2 and N_2O_4 and the equilibrium concentration of one of these species. In each case, the changes in concentration of the two species are related by the algebraic equation

$$\Delta \text{ conc. } NO_2 = -2(\Delta \text{ conc. } N_2O_4)$$

If the conc. of N_2O_4 *decreases* by x, that of NO_2 *increases* by 2x

which follows directly from Equation 15.3.

Looking at the data in Table 15.1, you might wonder whether these three equilibrium systems have anything in common. Specifically, is there any mathematical relationship between the equilibrium concentrations of NO_2 and N_2O_4 valid for all three experiments? It turns out that there is, although it is not an obvious one. The quotient, $[NO_2]^2/[N_2O_4]$ is the same, about 0.36, in each case:

$$\text{Expt. 1} \quad \frac{[NO_2]^2}{[N_2O_4]} = \frac{(0.120)^2}{0.040} = 0.36$$

$$\text{Expt. 2} \quad \frac{[NO_2]^2}{[N_2O_4]} = \frac{(0.072)^2}{0.014} = 0.37$$

$$\text{Expt. 3} \quad \frac{[NO_2]^2}{[N_2O_4]} = \frac{(0.160)^2}{0.070} = 0.36$$

TABLE 15.1 EQUILIBRIUM MEASUREMENTS IN THE N_2O_4–NO_2 SYSTEM AT 100°C*

		Initial Conc. (mol/dm³)	Change in Conc. (mol/dm³)	Equil. Conc. (mol/dm³)
Expt. 1	NO_2	**0.000**	+0.120	0.120
	N_2O_4	**0.100**	−0.060	**0.040**
Expt. 2	NO_2	**0.100**	−0.028	0.072
	N_2O_4	**0.000**	+0.014	**0.014**
Expt. 3	NO_2	**0.100**	+0.060	0.160
	N_2O_4	**0.100**	−0.030	**0.070**

*The numbers in bold type are those measured experimentally.

Experimentally, this relationship is found to hold for any equilibrium system containing NO_2 and N_2O_4 at 100°C. Regardless of where we start, we eventually arrive at an equilibrium whose position is described by the condition that

$$\frac{[NO_2]^2}{[N_2O_4]} = 0.36 \qquad \text{(at 100°C)}$$

Further experiments with many different systems containing NO_2 and N_2O_4 at various temperatures lead to the general conclusion:

At any given temperature, the quantity

$$\frac{[NO_2]^2}{[N_2O_4]}$$

is a constant, independent of the amounts of N_2O_4 and NO_2 that we start with, the volume of the container, or the total pressure. This constant is referred to as the **equilibrium constant,** K_c, for the reaction

$$N_2O_4(g) \rightleftharpoons 2\ NO_2(g)$$

At 100°C the numerical value of K_c is 0.36; at 150°C it is a considerably larger number, 3.2.

15.2 GENERAL FORM AND PROPERTIES OF K_c

We have seen that for the system

$$N_2O_4(g) \rightleftharpoons 2\ NO_2(g);\ K_c = \frac{[NO_2]^2}{[N_2O_4]}$$

For every gaseous system, an analogous expression can be written. The form of that expression can be deduced from the Law of Chemical Equilibrium:

For any reaction at a given temperature, there is a condition relating the equilibrium concentrations of products and reactants. This condition is expressed by a number, called the **equilibrium constant,** which has the form of a quotient. **The numerator of this quotient is obtained by multiplying together the equilibrium concentrations of products, each raised to a power equal to its coefficient in the balanced equation written to describe the reaction. The denominator is obtained in the same manner, using the equilibrium concentrations of reactants.**

It's easier to apply this law than to state it

To illustrate the application of this law, consider the reactions

$$2\ HI(g) \rightleftharpoons H_2(g) + I_2(g) \qquad (15.4)$$

$$N_2(g) + 3\ H_2(g) \rightleftharpoons 2\ NH_3(g) \qquad (15.5)$$

for which the equilibrium constant expressions take the form:

$$15.4\text{: } K_c = \frac{[H_2] \times [I_2]}{[HI]^2}$$

$$15.5\text{: } K_c = \frac{[NH_3]^2}{[N_2] \times [H_2]^3}$$

The actual values of K_c for reactions like 15.5 and 15.4 are found from experimental data such as that in Table 15.1. What is required is enough information to determine the equilibrium concentrations of all reactants and products. The data may furnish all of these concentrations directly, but more often some of them must be calculated.

Example 15.1 An equilibrium mixture of N_2, H_2, and NH_3 at 300°C is analyzed, and it is found that:

$$[N_2] = 0.25 \text{ mol/dm}^3; \quad [H_2] = 0.15 \text{ mol/dm}^3; \quad [NH_3] = 0.090 \text{ mol/dm}^3$$

Find K_c at 300°C for the reaction: $N_2(g) + 3\ H_2(g) \rightleftharpoons 2\ NH_3(g)$

Solution K_c is found by simply substituting into the equilibrium equation:

$$K_c = \frac{[NH_3]^2}{[N_2] \times [H_2]^3} = \frac{(0.090)^2}{(0.25) \times (0.15)^3} = 9.6$$

Example 15.2 A mole of H_2 and a mole of I_2 are put into a 10-dm^3 container at 520°C and allowed to come to equilibrium. Analysis of the equilibrium system shows that 0.12 mol HI is present. Find K_c at 520°C for the reaction

$$H_2(g) + I_2(g) \rightleftharpoons 2\ HI(g)$$

Solution In this case we can determine the equilibrium concentration of HI immediately, but need to calculate the equilibrium concentrations of H_2 and I_2. Since for every two moles of HI formed, one mole H_2 and one mole I_2 are consumed, to make 0.12 mol HI requires 0.06 mol H_2 and 0.06 mol I_2. Since we started with one mole of both reactants, at equilibrium we have:

$$\text{no. moles } H_2 = 1.00 - 0.06 = 0.94 = \text{no. moles } I_2$$

Now substituting into the equation for K_c, using concentrations:

$$K_c = \frac{[HI]^2}{[H_2] \times [I_2]} = \frac{(0.12 \text{ mol/10 dm}^3)^2}{(0.94 \text{ mol/10 dm}^3) \times (0.94 \text{ mol/10 dm}^3)} = 0.016$$

In writing the expression for K_c for a particular chemical system, there are a couple of important points to keep in mind.

1. *The expression which we write for K_c is dependent upon the equation written to describe the equilibrium system.* To illustrate what this statement implies, consider once again the N_2O_4–NO_2 system, for which we wrote

$$N_2O_4(g) \rightleftharpoons 2\ NO_2(g);\ K_c = \frac{[NO_2]^2}{[N_2O_4]} = 0.36 \text{ (at 100°C)}$$

There are many other equations which could be used to describe this system. In particular, we could have written

$$\tfrac{1}{2}\ N_2O_4(g) \rightleftharpoons NO_2(g)$$

in which case the expression for the equilibrium constant would have been

$$K_c' = \frac{[NO_2]}{[N_2O_4]^{1/2}}$$

Alternatively, we might have written

$$2\ NO_2(g) \rightleftharpoons N_2O_4(g)$$

and arrived at

$$K_c'' = \frac{[N_2O_4]}{[NO_2]^2}$$

There are three different ways of stating the equilibrium condition

Since all of these expressions describe the same chemical system, K_c, K_c', and K_c'' must be related. If you examine the concentration terms in the above equations, it should be apparent that the relationship is:

$$K_c' = (K_c)^{1/2} = (0.36)^{1/2} = 0.60 \text{ (at } 100°\text{C)}$$

$$K_c'' = 1/K_c = 1/0.36 = 2.8 \text{ (at } 100°\text{C)}$$

Each of these three equations is a perfectly valid way of expressing the relationship between the concentrations of NO_2 and N_2O_4 in equilibrium at 100°C. Clearly it would be ambiguous at the very least to say that for this system, at 100°C, "the equilibrium constant is 0.36." The numerical value of K_c is meaningful only when associated with a particular chemical equation.

2. *If a pure liquid or solid is involved in a reaction, its concentration does not appear in the expression for K_c.* For example, for the systems

$$CaCO_3(s) \rightleftharpoons CaO(s) + CO_2(g) \qquad (15.1)$$

$$CO_2(g) + H_2(g) \rightleftharpoons CO(g) + H_2O(l) \qquad (15.6)$$

the equilibrium constants have the form

$$15.1{:}\ K_c = [CO_2]$$

$$15.6{:}\ K_c = \frac{[CO]}{[CO_2] \times [H_2]}$$

where the terms for solids (CaO, $CaCO_3$) and liquids (H_2O) are omitted.

To understand why this simplification is possible, consider Reaction 15.6 taking place at, let us say, 100°C. The equation tells us that liquid water is present in the equilibrium system. As long as this is the case, the pressure of water vapor is fixed and equal to one atmosphere. Alternatively, we could say (p. 356) that the concentration of water vapor in equilibrium with liquid water has a fixed value, about 0.0327 mol/dm^3 at 100°C. This remains true regardless of how much liquid water we have or what the concentrations of the other species may be. In other words, the concentration of H_2O in the gas phase is itself a constant and hence can be, and is, incorporated into K_c.

A property of K_c which we will find very useful in this and succeeding chapters is expressed by the Rule of Multiple Equilibria, which states that:

If a reaction can be expressed as the sum of two or more reactions, K_c for the overall reaction is the product of the equilibrium constants of the individual reactions. That is, if

Reaction 3 = Reaction 1 + Reaction 2

then $$K_c(\text{Reaction 3}) = K_c(\text{Reaction 1}) \times K_c(\text{Reaction 2}) \quad (15.7)$$

To illustrate the application of this rule, consider the reactions at 700°C:

$$SO_2(g) + \tfrac{1}{2}\, O_2(g) \rightleftharpoons SO_3(g);\ K_c = 20 \quad (15.8)$$

$$NO_2(g) \rightleftharpoons NO(g) + \tfrac{1}{2}\, O_2(g);\ K_c = 0.012 \quad (15.9)$$

If we add these two equations, the "½ $O_2(g)$" cancels and we obtain

$$SO_2(g) + NO_2(g) \rightleftharpoons SO_3(g) + NO(g) \quad (15.10)$$

so it follows from the rule that

$$K_c \text{ for } 15.10 = (K_c \text{ for } 15.8)(K_c \text{ for } 15.9) = 20 \times 0.012 = 0.24$$

The validity of the rule can be demonstrated by writing the expressions for K_c for Reactions 15.8 and 15.9 and multiplying:

We will use this rule extensively in later chapters

$$(K_c \text{ for } 15.8)(K_c \text{ for } 15.9) = \frac{[SO_3]}{[SO_2] \times [\cancel{O_2}]^{1/2}} \times \frac{[NO] \times [\cancel{O_2}]^{1/2}}{[NO_2]}$$

$$= \frac{[SO_3] \times [NO]}{[SO_2] \times [NO_2]}$$

Looking at Reaction 15.10, we see that the quotient we have just obtained is precisely equal to K_c for that reaction.

15.3 APPLICATIONS OF K_c

A knowledge of the equilibrium constant for a particular reaction at a given temperature is valuable to the chemist who wishes to carry out the reaction in the laboratory or predict whether it will occur in nature. Often, knowing only the magnitude of K_c, we can predict whether a proposed reaction is likely to be feasible. Consider, for example, two reactions which might serve as methods of "fixing" atmospheric nitrogen (i.e., converting it to a useful compound).

$$N_2(g) + O_2(g) \rightleftharpoons 2\, NO(g);\ K_c = \frac{[NO]^2}{[N_2] \times [O_2]} = 1 \times 10^{-30} \text{ at } 25°C \quad (15.11)$$

$$N_2(g) + 3\, H_2(g) \rightleftharpoons 2\, NH_3(g);\ K_c = \frac{[NH_3]^2}{[N_2] \times [H_2]^3} = 5 \times 10^{8} \text{ at } 25°C \quad (15.5)$$

Looking at the magnitude of K_c for Reaction 15.11, it is clear that this would not be a practical way to fix nitrogen, at least at room temperature. Since K_c is such a tiny number, the equilibrium concentration of NO would be extremely small; virtually all of the N_2 and O_2 would remain unreacted. In contrast, Reaction 15.5 looks a lot more promising; K_c is so large that the equilibrium system will consist mostly of the desired product, ammonia. In other words, the reaction should go virtually to completion. (Unfortunately, Reaction 15.5 takes essentially forever to reach equilibrium at 25°C, but that is another problem which we'll worry about in Chapter 16.)

In general, we can say that if K_c for a reaction is very large, that reaction will tend to proceed far to the right; the system at equilibrium will consist mostly of products. If, on the other hand, K_c is small, very little reaction will occur; the equilibrium system will consist largely of reactants.

The equilibrium constant for a reaction can also be used to make quantitative predictions concerning the direction and extent to which a reaction will occur. We will now examine how this is done.

Prediction of the Direction of Reaction

For the general gas-phase reaction

$$\text{aA(g)} + \text{bB(g)} \rightleftharpoons \text{cC(g)} + \text{dD(g)}$$

where A, B, C, and D represent chemical substances and the small letters a, b, c, and d are their coefficients in the balanced equation, we can write

$$\frac{[\text{C}]^{\text{c}} \times [\text{D}]^{\text{d}}}{[\text{A}]^{\text{a}} \times [\text{B}]^{\text{b}}} = K_c$$

where K_c is the equilibrium constant for the reaction and the square brackets, as always, represent equilibrium concentrations in moles per cubic decimetre. When we carry out a reaction in the laboratory, the original concentration quotient

$$\frac{(\text{orig. conc. C})^{\text{c}} \times (\text{orig. conc. D})^{\text{d}}}{(\text{orig. conc. A})^{\text{a}} \times (\text{orig. conc. B})^{\text{b}}}$$

Note the importance of distinguishing between original and equilibrium concentrations

will seldom be equal numerically to K_c. If it is not, reaction will occur in one direction or the other so as to bring the concentrations of products and reactants to the ratio required at equilibrium. We can distinguish two possibilities:

1. If

$$\frac{(\text{orig. conc. C})^{\text{c}} \times (\text{orig. conc. D})^{\text{d}}}{(\text{orig. conc. A})^{\text{a}} \times (\text{orig. conc. B})^{\text{b}}} < K_c$$

the reaction will proceed from left to right, i.e.,

$$\text{aA(g)} + \text{bB(g)} \rightarrow \text{cC(g)} + \text{dD(g)}$$

In this way, the concentrations of products increase and those of reactants decrease. As this happens, the concentration quotient increases until it becomes equal to K_c, at which point reaction ceases.

2. If

$$\frac{(\text{orig. conc. C})^{\text{c}} \times (\text{orig. conc. D})^{\text{d}}}{(\text{orig. conc. A})^{\text{a}} \times (\text{orig. conc. B})^{\text{b}}} > K_c$$

we conclude that the concentrations of products are "too high" and those of the reactants "too low" to meet the equilibrium condition. Reaction must proceed in the reverse direction, i.e.,

$$\text{cC(g)} + \text{dD(g)} \rightarrow \text{aA(g)} + \text{bB(g)},$$

increasing the concentrations of A and B and simultaneously reducing those of C and D. This reduces the concentration quotient to its equilibrium value given by K_c.

Example 15.3 Nitric oxide, an important pollutant in air, is formed from the elements at high temperatures, such as those obtained when gasoline burns in an automobile engine. At 2000°C, K_c for the reaction

$$N_2(g) + O_2(g) \rightleftharpoons 2\ NO(g)$$

is 0.10. Predict the direction in which the system will move to reach equilibrium at 2000°C if one starts with

a. 1.62 mol N_2 and 1.62 mol O_2 in a 2.0-dm^3 container.
b. 4.0 mol N_2, 1.0 mol O_2, and 0.80 mol NO in a 20-dm^3 container.

Solution In each case, we first calculate the initial concentrations of N_2, O_2, and NO. We then compare the original concentration quotient

$$\frac{(\text{orig. conc. NO})^2}{(\text{orig. conc. } N_2)(\text{orig. conc. } O_2)}$$

to K_c, 0.10, to decide which way reaction will proceed.

In a), we have

orig. conc. N_2 = 1.62 mol/2.0 dm^3 = 0.81 mol/dm^3
orig. conc. O_2 = 1.62 mol/2.0 dm^3 = 0.81 mol/dm^3
orig. conc. NO = 0, since there is no NO present

Hence the original concentration quotient is

$$\frac{(0)^2}{(0.81)(0.81)} = 0 < K_c$$

so the reaction must proceed to the right to produce NO. This would remain true regardless of how much N_2 and O_2 we start with or how small K_c might be. As long as there is no NO present originally, some of it must be formed in order to establish equilibrium.

If we start with pure reactants, we always get at least a little product

We approach b) in exactly the same way.

orig. conc. N_2 = 4.0 mol/20 dm^3 = 0.20 mol/dm^3
orig. conc. O_2 = 1.0 mol/20 dm^3 = 0.050 mol/dm^3
orig. conc. NO = 0.80 mol/20 dm^3 = 0.040 mol/dm^3

$$\text{Hence: } \frac{(\text{orig. conc. NO})^2}{(\text{orig. conc. } N_2)(\text{orig. conc. } O_2)} = \frac{(0.040)^2}{(0.20)(0.050)} = \frac{0.0016}{0.010} = 0.16$$

Since this quotient is larger than K_c, 0.10, the reverse reaction must take place. In this way, conc. NO decreases (i.e., the numerator becomes smaller); conc. N_2 and O_2 increase (the denominator becomes larger). Ultimately, the quotient decreases until it becomes equal to 0.10, at which point equilibrium is established.

Prediction of Extent of Reaction

Knowing the equilibrium constant K_c for a particular reaction and the original concentrations of reactants, it is possible to calculate the extent to which a reaction occurs. The basic approach that we will use involves a three-step process. The first step requires that we express the equilibrium concentrations of all species in terms of a single unknown, x. These concentrations are then substituted into the expression for K_c to give an equation which can be solved for x. Having found x, we can then readily obtain the concentrations of all the species in the equilibrium mixture. This approach is illustrated in Example 15.4.

Example 15.4 Consider the system described in Example 15.3*a*, where we started with concentrations of N_2 and O_2 of 0.81 mol/dm³ and decided that some NO must be formed by the reaction

$$N_2(g) + O_2(g) \rightleftharpoons 2\ NO(g);\ K_c = 0.10 \text{ at } 2000°C$$

Under these conditions what will be the equilibrium concentrations of NO, N_2, and O_2?

Solution We must first express the equilibrium concentrations of N_2, O_2, and NO in terms of a single unknown, x. We will choose x to be the number of moles per cubic decimetre of N_2 that react to achieve equilibrium. The balanced equation tells us that, since 1 mol O_2 reacts with 1 mol N_2, an equal number of moles per cubic decimetre, x, of O_2 must be consumed. Furthermore, since 2 mol NO are formed for every mole of N_2 that reacts, we must form 2x mol/dm³ of NO. Summarizing our reasoning in the form of a table similar to Table 15.1, we have

	ORIG. CONC. (mol/dm³)	CHANGE IN CONC. (mol/dm³)	EQUIL. CONC. (mol/dm³)
N_2	0.81	−x	0.81 − x
O_2	0.81	−x	0.81 − x
NO	0	+2x	2x

We are now ready to substitute equilibrium concentrations into the expression for K_c:

$$K_c = 0.10 = \frac{[NO]^2}{[N_2] \times [O_2]} = \frac{(2x)^2}{(0.81 - x)^2}$$

This is a second-order equation in x; such equations can always be solved by use of the "quadratic formula" (p. 366). In this case, however, we can simplify the arithmetic enormously by noting that the right-hand side of the equation is a perfect square. Taking the square root of both sides, we have

$$(0.10)^{1/2} = 0.32 = \frac{2x}{0.81 - x}$$

which is now readily solved for x:

$$2x = 0.32(0.81 - x) = 0.26 - 0.32x$$

$$2.32x = 0.26;\ x = 0.26/2.32 = 0.11$$

Referring back to the equilibrium table above, we see that

$$[NO] = 2x = 0.22 \text{ mol/dm}^3$$

$$[N_2] = [O_2] = 0.81 - x = 0.81 - 0.11 = 0.70 \text{ mol/dm}^3$$

The calculations that we have just gone through suggest three general comments:

1. There are many different ways in which the unknown may be chosen in a problem of this type. It doesn't matter what quantity we take, provided that we are

consistent in relating equilibrium concentrations to it. In Example 15.4, we might have taken the unknown, y, to be the number of moles per cubic decimetre of NO formed, in which case the concentrations of NO, N_2, and O_2 at equilibrium would have been y, 0.81 − y/2, and 0.81 − y/2, respectively. Had we made this choice, our equations would have looked slightly different, but the final answers would have been the same.

2. The arithmetic involved in equilibrium calculations can be quite simple, as in this case, or somewhat more complex (see Example 15.5, p. 366). The reasoning involved is the same. In any case, it is helpful to set up a table similar to Table 15.1 to summarize the analysis of the problem.

3. At 2000°C, appreciable quantities of nitric oxide are formed from nitrogen and oxygen, which explains why the NO resulting from high-temperature combustion processes is a serious air pollutant.

15.4 EFFECT OF CHANGES IN CONDITIONS UPON THE POSITION OF AN EQUILIBRIUM

Once a system has attained equilibrium, it is possible to change the ratio of products to reactants by changing the external conditions. We will consider three ways in which a chemical equilibrium can be disturbed:

1. Adding (or removing) a reactant or product.
2. Changing the volume of the system.
3. Changing the temperature.

Qualitatively, we can deduce the direction in which an equilibrium will shift when one of these changes is made by applying Le Châtelier's Principle, which states that:

Nature, and societies, seem to oppose changes in the status quo

If a system at equilibrium is altered in any way, the system will shift so as to minimize the effect of the change.

Quantitatively, we can predict the extent of the shift by working with the equilibrium constant, K_c. In the discussion that follows, we will illustrate both of these approaches.

Change in the Number of Moles of Reactants or Products

According to Le Châtelier's Principle, **if we disturb a chemical system at equilibrium by adding a species (reactant or product), the reaction will proceed in such a direction as to consume part of the added species. Conversely, if we remove a species, the system will shift so as to restore part of that species.**

Applying this general rule to the system

$$N_2(g) + O_2(g) \rightleftharpoons 2\ NO(g)$$

we see that if we were to add reactant (N_2 or O_2) or remove some product (NO), reaction would occur in the forward direction to restore equilibrium. If, on the other hand, we were to add product (NO) or remove reactant (N_2 or O_2) the reverse reaction would have to occur to restore the concentration ratio to the value required by K_c.

The extent to which reaction occurs when a system at equilibrium is disturbed by adding or removing a species can be calculated using the equilibrium constant, K_c (Example 15.5).

Example 15.5 Consider the system described in Example 15.4

$$N_2(g) + O_2(g) \rightleftharpoons 2\ NO(g);\ K_c = 0.10$$

where we found $[N_2] = [O_2] = 0.70$ mol/dm^3, and $[NO] = 0.22$ mol/dm^3. Suppose we add enough N_2 to increase its concentration temporarily to 1.00 mol/dm^3. When equilibrium is restored, what will be the equilibrium concentrations of all three species?

Solution Our analysis here is entirely similar to that used in Example 15.4. We will take the "initial concentrations" to be those prevailing immediately after the addition of N_2. As before, we will take our variable, x, to be the number of moles per cubic decimetre of N_2 that reacts to restore equilibrium. The equilibrium table becomes

	ORIG. CONC. (mol/dm^3)	CHANGE IN CONC. (mol/dm^3)	EQUIL. CONC. (mol/dm^3)
N_2	1.00	$-x$	$1.00 - x$
O_2	0.70	$-x$	$0.70 - x$
NO	0.22	$+2x$	$0.22 + 2x$

Substituting into the expression for K_c, we obtain

$$\frac{(0.22 + 2x)^2}{(1.00 - x)(0.70 - x)} = 0.10$$

This time, we cannot solve for x as simply as in Example 15.4, since the denominator of the left-hand side is not a perfect square. Instead, we use the general method of solving a quadratic equation, which involves rearranging to the form

$$ax^2 + bx + c = 0$$

and applying the "quadratic formula"

$$x = \frac{-b \pm \sqrt{b^2 - 4ac}}{2a}$$

To convert the equilibrium equation to the desired form, we proceed as follows:

$$(0.22 + 2x)^2 = 0.10(1.00 - x)(0.70 - x)$$

$$0.0484 + 0.88x + 4x^2 = 0.10(0.70 - 1.70x + x^2) = 0.0700 - 0.17x + 0.10x^2$$

Bringing all the terms to the left-hand side, we obtain

$$3.90x^2 + 1.05x - 0.0216 = 0$$

so $a = 3.90$, $b = 1.05$, $c = -0.0216$, and

$$x = \frac{-1.05 \pm \sqrt{(1.05)^2 + 4(3.90)(0.0216)}}{7.80} = \frac{-1.05 \pm \sqrt{1.44}}{7.80}$$

In solving quadratics of this type, one answer is always physically impossible

$$= \frac{-1.05 \pm 1.20}{7.80} = \frac{0.15}{7.80} \text{ or } \frac{-2.25}{7.80} \text{ (i.e., 0.02 or } -0.29)$$

Of the two answers, only 0.02 is plausible. A value of -0.29 would imply that the reaction proceeds in the opposite direction and that the concentration of NO at equilibrium is negative!

Having found x, we can now calculate the equilibrium concentrations.

$$[N_2] = 1.00 - x = 0.98 \text{ mol/dm}^3$$

$$[O_2] = 0.70 - x = 0.68 \text{ mol/dm}^3$$

$$[NO] = 0.22 + 2x = 0.26 \text{ mol/dm}^3$$

You will note from this example that:

1. Reaction did indeed occur in the forward direction as Le Châtelier's Principle predicted; 0.02 mol/dm^3 of N_2 reacted with an equal quantity of O_2 to produce an additional 0.04 mol/dm^3 of NO.

2. Only a portion of the added nitrogen reacted. Its final concentration, 0.98 mol/dm^3, is intermediate between that before equilibrium was disturbed, 0.70 mol/dm^3, and that immediately afterwards, 1.00 mol/dm^3. This point is further illustrated in the simplified system shown in Figure 15.1.

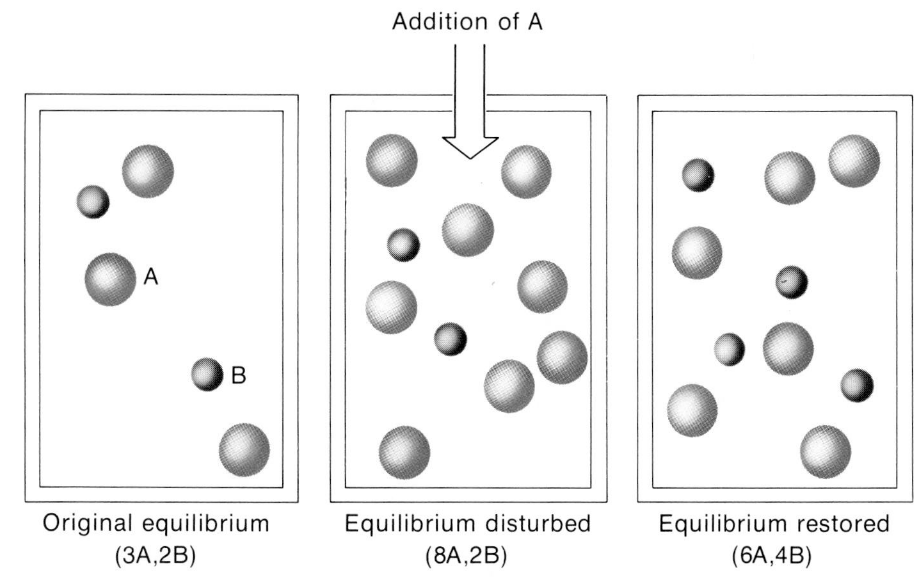

Figure 15.1 Effect of adding reactant to the system $A(g) \rightleftharpoons B(g)$. Of every 5 molecules of A added, 2 are eventually converted to B. Since K_c equals [B]/[A] and equals 2/3 originally (first box), it is necessary that reaction occur until this ratio is re-established (third box, ratio = 4/6 = 2/3).

Changes in Volume

To understand how a change in container volume can change the position of an equilibrium, consider the N_2O_4–NO_2 system:

$$N_2O_4(g) \rightleftharpoons 2\ NO_2(g)$$

When the volume of the system is decreased (Fig. 15.2, p. 368), the number of molecules in a unit volume and hence the total pressure temporarily increase. In order to minimize this effect, some of the NO_2 molecules present combine with each other to form N_2O_4. Since two molecules of NO_2 are required to form one N_2O_4 molecule, this reduces the total number of molecules and hence the pressure. This partially compensates for the pressure increase, as predicted by Le Châtelier's Principle.

We could have arrived at this same conclusion by referring to the form of the equilibrium constant expression for this system:

$$K_c = \frac{[NO_2]^2}{[N_2O_4]}$$

Realizing that the concentration of any species, n/V, is equal to the number

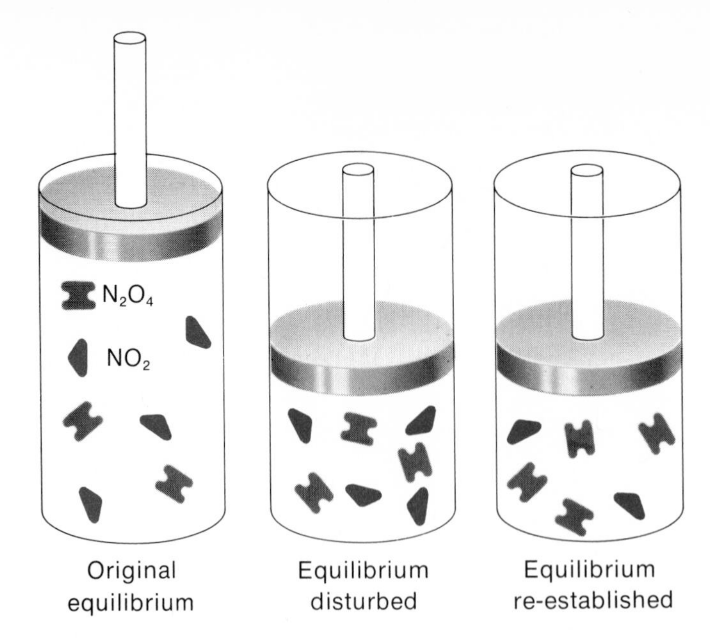

Figure 15.2 Effect of a decrease in volume on the $N_2O_4(g) \rightleftharpoons 2\,NO_2(g)$ system. The immediate effect (middle cylinder) is to crowd the same number of moles of gas into a smaller volume and so increase the total pressure. This is partially compensated for by the conversion of some of the NO_2 to N_2O_4, thereby reducing the total number of moles of gas present.

of moles, n, divided by the total volume, V, we can rewrite this expression as

$$K_c = \frac{(n\ NO_2/V)^2}{(n\ N_2O_4/V)}$$

where $n\ NO_2$ and $n\ N_2O_4$ represent the numbers of moles of these two species at equilibrium in a closed container of volume V. Simplifying, we obtain

$$K_c = \frac{(n\ NO_2)^2}{(n\ N_2O_4) \times V}$$

If V gets smaller, the expression becomes larger than K_c, so the reverse reaction must occur

Now, if we make V smaller, as in Figure 15.2, there is only one way in which the right-hand side of this equation can remain constant. Some NO_2 must be converted to N_2O_4, thereby decreasing ($n\ NO_2$) and increasing ($n\ N_2O_4$).

It is possible, using the value of K_c (0.36 at 100°C), to calculate the extent to which NO_2 is converted to N_2O_4 when the volume of the system is decreased. The results of such calculations are summarized in Table 15.2 for a system where we start with a large volume. As the volume is decreased from 10.0 to 1.0 dm³, more and more of the NO_2 is converted to N_2O_4. Notice that the total number of moles of gas decreases steadily as a result of this conversion.

The analysis we have gone through for the N_2O_4–NO_2 system can be applied to any chemical equilibrium involving gases. **When the position of an equilibrium is disturbed by decreasing the volume, reaction takes place in a direction so as to**

TABLE 15.2 EFFECT OF CHANGE IN VOLUME ON THE EQUILIBRIUM SYSTEM: $N_2O_4(g) \rightleftharpoons 2\,NO_2(g)$; $K_c = 0.36$ at 100°C

V(dm³)	$n\ NO_2$	$n\ N_2O_4$	n_{tot}
10.0	1.20	0.40	1.60
5.0	0.96	0.52	1.48
2.0	0.68	0.66	1.34
1.0	0.52	0.74	1.26

decrease the number of moles of gas (e.g., $2\ NO_2 \rightarrow N_2O_4$). When the volume is increased, the reaction which increases the number of moles of gas (e.g., $N_2O_4 \rightarrow 2\ NO_2$) takes place.

The application of this principle to various reactions involving gases is shown in Table 15.3. Notice particularly that it is the change in the number of moles of *gas* which determines which way the equilibrium shifts when the volume is increased (system 4), and that when there is no change in the number of moles of gas (system 5), a change in volume has no effect upon the position of the equilibrium.

TABLE 15.3 EFFECT OF A CHANGE IN VOLUME UPON THE POSITION OF GASEOUS EQUILIBRIA

SYSTEM	V INCREASES	V DECREASES
1. $N_2O_4(g) \rightleftharpoons 2\ NO_2(g)$	→	←
2. $SO_2(g) + \frac{1}{2}\ O_2(g) \rightleftharpoons SO_3(g)$	←	→
3. $N_2(g) + 3\ H_2(g) \rightleftharpoons 2\ NH_3(g)$	←	→
4. $C(s) + H_2O(g) \rightleftharpoons CO(g) + H_2(g)$	→	←
5. $N_2(g) + O_2(g) \rightleftharpoons 2\ NO(g)$	0	0

Changes in volume of gaseous systems at equilibrium ordinarily result in changes in pressure. For example, if we cut the volume of the N_2O_4–NO_2 system in half, we momentarily increase the pressure by a factor of two. Instead of saying that the shift in the position of the equilibrium comes about because of a change in volume, we might ascribe the shift to the pressure change that accompanies the volume decrease. Specifically, we could say that an *increase in pressure shifts the position of the equilibrium in such a way as to decrease the number of moles of gas* ($2\ NO_2 \rightarrow N_2O_4$).

In discussing the effect of pressure upon the position of an equilibrium, we must be careful to specify how the pressure change comes about. There are many different ways in which we could change the total pressure in the N_2O_4–NO_2 system without changing the volume. We might, for example, add helium, which would increase the total number of moles of gas and hence the total pressure. Since the volume remains unchanged, we see from the expression

$$K_c = \frac{(n\ NO_2)^2}{(n\ N_2O_4) \times V}$$

that the ratio $(n\ NO_2)^2/n\ N_2O_4$ remains constant. In other words, the position of the equilibrium is unaffected by the addition of helium. In general we find that addition of an unreactive gas at constant volume has essentially no effect upon the position of an equilibrium.

Changes in Temperature

So far our attention has been directed toward changes in equilibrium systems at constant temperature, where the equilibrium constant K_c is a fixed number, independent of volume, pressure, or the numbers of moles of reactants or products. If the temperature changes, the magnitude of K_c changes as well. Le Châtelier's Principle predicts that an increase in temperature favors an endothermic process, i.e., a process which absorbs heat and hence tends to minimize the temperature increase. This means that **if the forward reaction is endothermic,**

an increase in temperature will result in a larger value of K_c. An example is the system

$$N_2O_4(g) \rightleftharpoons 2\ NO_2(g)$$

where the forward reaction is endothermic

$$N_2O_4(g) \rightarrow 2\ NO_2(g);\ \Delta H = +58.2\ \text{kJ}$$

For this reaction, K_c increases from about 5×10^{-4} at 0°C to 0.36 at 100°C. Consequently, the concentration of NO_2 relative to that of N_2O_4 increases as the temperature rises (Plate 9).

In contrast, for the system

$$N_2(g) + 3\ H_2(g) \rightleftharpoons 2\ NH_3(g)$$

where the forward reaction is exothermic

$$N_2(g) + 3\ H_2(g) \rightarrow 2\ NH_3(g);\ \Delta H = -92.4\ \text{kJ}$$

K_c decreases as the temperature rises, falling from 5×10^8 at room temperature to about 0.5 at 400°C. In general, we find that **if the forward reaction is exothermic, an increase in temperature leads to a decrease in K_c.**

We are often interested not only in the direction in which the equilibrium constant changes as the temperature increases, but also the extent to which it changes. It is possible to calculate the magnitude of the temperature effect by making use of an equation analogous to the Clausius-Clapeyron equation* introduced in Chapter 11:

$$\log_{10} \frac{K_2}{K_1} = \frac{\Delta H}{(2.30)(8.31)} \left[\frac{T_2 - T_1}{T_2 T_1} \right] \tag{15.12}$$

where K_2 and K_1 are the equilibrium constants at temperatures T_2 and T_1, respectively (K), and ΔH is the enthalpy change in *joules* for the forward reaction.

---

*Strictly speaking, this equation would apply exactly only if ΔH were independent of temperature and, for *gaseous* equilibria, if the expression for K involved partial pressures rather than concentrations in moles per cubic decimetre. For these reasons among others, we cannot expect values of K_c calculated from this equation to be accurate to more than one significant figure.

Example 15.6 In the Haber Process by which elementary nitrogen is converted to ammonia

$$N_2(g) + 3\ H_2(g) \rightleftharpoons 2\ NH_3(g)$$

ΔH is −92.4 kJ. Taking the equilibrium constant to be 5×10^8 at room temperature (25°C), calculate its value at 400°C.

Usually, as here, temperature has a large effect on K_c

Solution We shall take T_2 to be the higher temperature.

$$T_2 = (273 + 400)\ \text{K} = 673\ \text{K};\ T_1 = (273 + 25)\ \text{K} = 298\ \text{K}$$

Substituting these values and that of ΔH in Equation 15.12, we obtain

$$\log_{10} \frac{K_2}{K_1} = \frac{-92\ 400}{(2.30)(8.31)} \left[\frac{673 - 298}{673 \times 298} \right] = -9.0$$

To find the ratio K_2/K_1, we note that the number whose logarithm is −9.0 is 1×10^{-9}.

$$\frac{K_2}{K_1} = 1 \times 10^{-9}; \; K_2 = K_1 \, (1 \times 10^{-9})$$
$$= 5 \times 10^{8} \times 1 \times 10^{-9} = 0.5$$

Noting that the equilibrium constant for this reaction decreases by a factor of 10^9 in going from room temperature to 400°C, we infer that to obtain a high yield of ammonia by the Haber Process, it would be desirable to use as low a temperature as possible. However, as we will see in Chapter 17, a moderately high temperature (about 400°C) is required to make the reaction go at a reasonable rate.

15.5 RELATION BETWEEN THE FREE ENERGY CHANGE AND THE EQUILIBRIUM CONSTANT

In Chapter 14, we pointed out that the free energy change, ΔG, is the fundamental criterion of spontaneity. Reactions occur spontaneously at a given temperature and pressure if ΔG is a negative quantity. In this chapter we have discussed the direction and the extent to which chemical reactions take place spontaneously in terms of the equilibrium constant, K. As you might expect, these two quantities are intimately related. It can be shown by a thermodynamic argument which will not be presented here that

$$\Delta G^0 = -RT \ln K$$
$$= -(2.30)(8.31)\; T \log_{10} K \qquad (\Delta G^0 \text{ in joules}) \qquad (15.13)$$

where ΔG^0 is the so-called "standard free energy change," i.e., the free energy change when all the species involved in the reaction are at unit concentrations, K is the equilibrium constant for the reaction, R is the gas constant, and T is the absolute temperature in K.

Looking at Equation 15.13, we can distinguish three possible situations, depending upon the sign of ΔG^0 for the reaction:

1. If ΔG^0 is a negative quantity, $\log_{10} K$ must be positive, K is greater than one, and the reaction proceeds spontaneously in the forward direction when all species are at unit concentrations.

2. If ΔG^0 is positive, $\log_{10} K$ must be negative, K is less than one, and the reverse reaction is spontaneous when all species are at unit concentrations.

3. If, perchance, $\Delta G^0 = 0$, $\log_{10} K = 0$, $K = 1$, and the reaction is at equilibrium when all species are at unit concentrations.

Equation 15.13 also allows us to give physical meaning to the magnitude of the free energy change. If ΔG^0 is a large negative number, K will be much greater than one and the forward reaction, under ordinary conditions, will go virtually to completion. Conversely, if ΔG^0 is a large positive number, K will be a small fraction and the reverse reaction will go nearly to completion.

Example 15.7 Calculate K for reactions which, at 25°C, have ΔG^0 values of
a. −80.0 kJ
b. +80.0 kJ

Solution Solving Equation 15.13 for log K, we have

$$\log_{10} K = \frac{-\Delta G^0}{(2.30)(8.31)\,T}$$

Remember that ΔG^0 is in joules in Eq. 15.13

At 25°C, $T = 298$, and $\log_{10} K = \frac{-\Delta G^0}{(2.30)(8.31)(298)} = \frac{-\Delta G^0}{5\,700}$

a.
$$\log_{10} K = -\frac{(-80\,000)}{5\,700} = 14.0$$

Taking antilogs, $K = 1 \times 10^{14}$

b.
$$\log_{10} K = \frac{-80\,000}{5\,700} = -14.0$$

Taking antilogs, $K = 1 \times 10^{-14}$

From Example 15.7, we see that values of ΔG^0 of -80 kJ and $+80$ kJ lead to equilibrium constants that are very large in the first case (1×10^{14}) and very small in the second (1×10^{-14}). Indeed, only if ΔG^0 is rather close to zero (perhaps in the range -10 to $+10$ kJ) will the magnitude of K be such that we will have appreciable concentrations of both products and reactants in the equilibrium system. Most of the reactions for which we have made equilibrium calculations in this chapter fall into this latter category.

You may have noticed that throughout this section we have used the general symbol "K" rather than "K_c" to represent the equilibrium constant. Furthermore, we have not specified exactly what we mean by "unit concentrations." There is a rational explanation for this apparent fuzziness. Equation 15.13 is a quite general one which applies to any type of equilibrium, including those in aqueous solution to be discussed in Chapters 18 through 23. The exact meaning of ΔG^0 and K will depend to some extent upon the type of equilibrium being discussed.

For systems of the type discussed in this chapter and the preceding one, Chapter 14, where the species involved are either *gases or pure liquids or solids,* "unit concentration" is ordinarily taken to mean *one atmosphere pressure*. Hence, ΔG^0 becomes identical with the quantity $\Delta G^{1\,atm}$ discussed in Chapter 14. Under these conditions, the "K" involved in Equation 15.13 is not the equilibrium constant K_c which we have discussed in this chapter. Instead, it is a quantity known as K_p, which is the equilibrium constant obtained when the concentrations of all species are expressed as their *partial pressures (in atmospheres).*

$$\Delta G^{1\,atm} \text{ (in joules)} = -(2.30)(8.31)\ T \log_{10} K_p \qquad (15.14)$$

To illustrate the difference between K_c and K_p, consider the reaction

$$N_2(g) + 3\ H_2(g) \rightleftharpoons 2\ NH_3(g) \qquad (15.5)$$

where: $K_c = \frac{[NH_3]^2}{[N_2] \times [H_2]^3}$; $\qquad K_p = \frac{(p\ NH_3)^2}{(p\ N_2) \times (p\ H_2)^3}$

In the expression for K_c, the terms $[NH_3]$, $[N_2]$, and $[H_2]$ represent, as we have seen, equilibrium concentrations in moles per cubic decimetre. In contrast, in the expression for K_p, the terms (p NH_3), (p N_2), and (p H_2) represent equilibrium partial pressures in atmospheres.

Using the Ideal Gas Law, it is possible to relate the concentration of a gaseous species in moles per cubic decimetre to its partial pressure in atmospheres. Going one step further, it is possible to relate the two equilibrium constants, K_c and K_p, for a particular reaction. The general relationship, derived in Problem 15.41 at the end of this chapter, is

$$K_p = K_c\ (0.0821\ T)^{\Delta n_g} \qquad (15.15)$$

where the exponent in this equation, Δn_g, represents the change in the number of moles of gas as the reaction proceeds in the forward direction.

Using Equations 15.14 and 15.15, it is possible to calculate the equilibrium constant discussed in this chapter, K_c, from the quantity $\Delta G^{1\,atm}$ discussed in Chapter 14. Consider, for example, Reaction 15.5. From Table 14.2, p. 339, we see that at 25°C, i.e., 298 K:

$$N_2(g) + 3\ H_2(g) \rightarrow 2\ NH_3(g)$$

$$\Delta G^{1\,atm} = 2\Delta G_f^{1\,atm}\ NH_3(g) = -33.2 \text{ kJ} = -33\,200 \text{ J}$$

Applying Equation 15.14, we have, on solving for $\log_{10} K_p$,

$$\log_{10} K_p = \frac{-\Delta G^{1\,atm}}{(2.30)(1.99)T} = \frac{33\ 200}{(2.30)(8.31)(298)} = 5.9 = 0.9 + 5.0$$

$$K_p = 8 \times 10^5 \text{ (at 25°C)}$$

To use Equation 15.15, we note that for this reaction, $\Delta n_g = -2$ (i.e., 2 mol of gaseous NH_3 are formed from a total of 4 mol of gaseous reactants). So

$$K_p = 8 \times 10^5 = K_c(0.0821\ T)^{-2} = \frac{K_c}{(0.0821 \times 298)^2} = \frac{K_c}{(24.5)^2}$$

Solving: $K_c = 8 \times 10^5 (24.5)^2$

$$= 8 \times 10^5\ (600) = 4800 \times 10^5 \approx 5 \times 10^8 \text{ (at 25°C)}$$

PROBLEMS

15.1 *Expression for K_c* Write the expressions for K_c for

a. $2\ CO_2(g) \rightleftharpoons 2\ CO(g) + O_2(g)$
b. $SnO_2(s) + 2\ CO(g) \rightleftharpoons Sn(s) + 2\ CO_2(g)$

15.2 *Rate of Multiple Equilibria* Show algebraically that K_c for the reaction $SnO_2 \rightleftharpoons Sn(s) + O_2(g)$ is the product of the two expressions written for K_c in Problem 15.1 and explain how this illustrates the rule of multiple equilibria.

15.3 *Determination of K_c* For the system $2\ HI(g) \rightleftharpoons H_2(g) + I_2(g)$, it is found that if one starts with pure HI at a concentration of 0.50 mol/dm³, its concentration at equilibrium is 0.10 mol/dm³. Set up a table analogous to Table 15.1 and use it to calculate K_c for the reaction.

15.4 *Direction of Reaction* Use the value of K_c calculated in Problem 15.3 to decide in which direction reaction will occur when the original concentrations are

a. HI, 0.25 *M;* H_2, 0.00 *M;* I_2, 0.00 *M*
b. HI, 0.10 *M;* H_2, 0.50 *M;* I_2, 0.20 *M*

15.5 *Extent of Reaction* What is the equilibrium concentration of H_2 in Problem 15.4 a?

15.6 *Effect of Changes in Conditions Upon Equilibrium* Consider the system

$$2\ SO_2(g) + O_2(g) \rightleftharpoons 2\ SO_3(g);\ \Delta H = -198.2 \text{ kJ}$$

In which direction will the equilibrium system move if

a. SO_3 is added.
b. SO_3 is removed.
c. the volume is increased.
d. the temperature is increased.
e. the pressure is increased by adding N_2.

15.7 *Change of K_c with T* For the reaction in Problem 15.6, $K_c = 100$ at 1000 K. Calculate K_c at 1200 K.

15.8 *Relation Between ΔG^0 and K* Calculate ΔG^0 for a reaction for which $K = 100$ at 1000 K.

15.9 Write expressions for K_c for each of the following systems:

a. $2\ NO_2(g) \rightleftharpoons 2\ NO(g) + O_2(g)$
b. $CO(g) + \frac{1}{2}\ O_2(g) \rightleftharpoons CO_2(g)$
c. $CH_4(g) + 2\ O_2(g) \rightleftharpoons CO_2(g) + 2\ H_2O(g)$
d. $CH_4(g) + 2\ O_2(g) \rightleftharpoons CO_2(g) + 2\ H_2O(l)$

15.25 Write expressions for K_c for each of the following systems:

a. $2\ NH_3(g) \rightleftharpoons N_2(g) + 3\ H_2(g)$
b. $Al_2O_3(s) + 3\ H_2(g) \rightleftharpoons 2\ Al(l) + 3\ H_2O(g)$
c. $N_2H_4(l) \rightleftharpoons N_2(g) + 2\ H_2(g)$
d. the decomposition of 1 mol NOCl(g) to the elements.

15.10 At 520°C, K_c is 67 for the reaction

$$H_2(g) + I_2(g) \rightleftharpoons 2\ HI(g)$$

What is K_c at the same temperature for

a. $2\ HI(g) \rightleftharpoons H_2(g) + I_2(g)$?
b. $\frac{1}{2}\ H_2(g) + \frac{1}{2}\ I_2(g) \rightleftharpoons HI(g)$?

15.11 Given that, for the reactions

$$SnO_2(s) + 2\ H_2(g) \rightleftharpoons Sn(s) + 2\ H_2O(g)$$

$$CO(g) + H_2O(g) \rightleftharpoons CO_2(g) + H_2(g)$$

K_c is 21 and 0.034, respectively, calculate K_c for the reaction

$$SnO_2(s) + 2\ CO(g) \rightleftharpoons Sn(s) + 2\ CO_2(g)$$

15.12 Consider the equilibrium

$$N_2(g) + 3\ H_2(g) \rightleftharpoons 2\ NH_3(g)$$

At 400°C, the equilibrium concentrations of N_2, H_2, and NH_3 are 0.15, 0.80, and 0.20 mol/dm^3, respectively. Calculate K_c.

15.13 For the reaction in Problem 15.12 at a different temperature, it is found that if one starts with N_2 and H_2 each at a concentration of 0.18 mol/dm^3, the equilibrium concentration of NH_3 is 0.040 mol/dm^3. Calculate K_c.

15.14 When limestone is heated, it decomposes according to the reaction

$$CaCO_3(s) \rightleftharpoons CaO(s) + CO_2(g)$$

It is found that if one starts with 1.25 mol $CaCO_3$ in a 5.0-dm^3 container at 1000°C, 40% of the limestone decomposes.

a. What is $[CO_2]$?
b. What is K_c for the reaction?
c. How many grams of CaO are formed?

15.15 For the reaction

$$2\ NO_2(g) \rightleftharpoons 2\ NO(g) + O_2(g)$$

K_c at a certain temperature is 0.50. Predict the direction in which reaction will occur to establish equilibrium if one starts with

a. conc. NO_2 = conc. NO = conc. O_2 = 0.10 M.
b. 1.23 mol NO_2 and 0.168 mol O_2 in a 14.5-dm^3 container.
c. 2.0 mol of NO_2, 0.40 mol NO, and 0.10 mol O_2 in a 10-dm^3 container.

15.16 For the reaction

$$2\ HI(g) \rightleftharpoons H_2(g) + I_2(g)$$

K_c = 0.016 at 520°C. Calculate the concentrations of all species at equilibrium in a 5.0-dm^3 container starting with

15.26 At 25°C, K_c is 2.2×10^{-3} for the reaction

$$ICl(g) \rightleftharpoons \frac{1}{2}\ I_2(g) + \frac{1}{2}\ Cl_2(g)$$

Calculate K_c for

a. $2\ ICl(g) \rightleftharpoons I_2(g) + Cl_2(g)$
b. $I_2(g) + Cl_2(g) \rightleftharpoons 2\ ICl(g)$

15.27 Given that K_c for the reactions

$$H_2(g) + S(s) \rightleftharpoons H_2S(g)$$

$$S(s) + O_2(g) \rightleftharpoons SO_2(g)$$

is 1.0×10^{-3} and 5.0×10^6, respectively, calculate K_c for the reaction

$$H_2(g) + SO_2(g) \rightleftharpoons H_2S(g) + O_2(g)$$

15.28 Consider the equilibrium

$$2\ SO_2(g) + O_2(g) \rightleftharpoons 2\ SO_3(g)$$

At equilibrium at 700°C, 0.40 mol of SO_2, 0.030 mol of O_2, and 1.00 mol of SO_3 are present in a 2.0-dm^3 container. Calculate K_c.

15.29 For the reaction in Problem 15.28 at a different temperature, it is found that when one starts with pure SO_3 at a concentration of 1.20 mol/dm^3, its equilibrium concentration is 0.42 mol/dm^3. Calculate K_c.

15.30 Consider the equilibrium

$$C(s) + CO_2(g) \rightleftharpoons 2\ CO(g)$$

When this system is at equilibrium at 700°C in a 2.0-dm^3 container, there are 0.10 mol CO, 0.20 mol CO_2, and 0.40 mol C present. When cooled to 600°C, an additional 0.04 mol of C(s) forms. Calculate K_c at 700°C and again at 600°C.

15.31 For the reaction

$$4\ HCl(g) + O_2(g) \rightleftharpoons 2\ H_2O(g) + 2\ Cl_2(g)$$

K_c at a certain temperature is 1.6. Predict the direction in which the system will move to reach equilibrium if one starts with

a. conc. HCl = conc. O_2 = conc. H_2O = 0.20 M.
b. 1.20 mol HCl, 0.60 mol O_2, 1.40 mol H_2O, and 0.80 mol Cl_2 in a 4.0-dm^3 container.

15.32 For the reaction

$$SO_2(g) + NO_2(g) \rightleftharpoons SO_3(g) + NO(g)$$

K_c is 9.0 at 700°C. Calculate the equilibrium concentrations of all species if one starts with

a. 0.50 mol of HI.
b. 0.25 mol H_2, 0.25 mol I_2.
c. 0.50 mol H_2, 0.50 mol I_2.

15.17 In the catalytic converter of an automobile exhaust system, some SO_2 is converted to SO_3 by the reaction

$$SO_2(g) + \frac{1}{2} O_2(g) \rightleftharpoons SO_3(g)$$

If one starts with a concentration of SO_2 of 2.0×10^{-4} mol/dm³ and if the O_2 concentration remains constant at 1.0×10^{-2} mol/dm³, what will be the equilibrium concentration of SO_3? ($K_c = 20$)

15.18 When chlorine gas is heated, it decomposes according to the reaction

$$Cl_2(g) \rightleftharpoons 2\ Cl(g)$$

K_c for this reaction is 1.2×10^{-6} at 1000°C and 3.6×10^{-2} at 2000°C. Starting with a concentration of Cl_2 of 0.10 mol/dm³, what will be the equilibrium concentration of atomic chlorine at

a. 2000°C?
b. 1000°C? (Note that K_c at 1000°C is so small that $[Cl_2]$ will be very nearly equal to its original concentration.)

15.19 Consider the data in Table 15.2. Show by calculation that if one starts with 2.00 mol NO_2 in a 10-dm³ container, the number of moles of NO_2 at equilibrium is indeed 1.20.

15.20 Consider the equilibrium

$$N_2(g) + 3\ H_2(g) \rightleftharpoons 2\ NH_3(g);\ \Delta H = -92.4\ kJ$$

A mixture of these three substances reaches equilibrium at 200°C. Predict the direction in which the system will move to re-establish equilibrium if

a. 1 mol of H_2 is removed.
b. the total pressure is increased by adding H_2.
c. the total pressure is increased by adding He.
d. the volume of the container is reduced.
e. the temperature is raised to 300°C.

15.21 For the reaction

$$CO(g) + H_2O(g) \rightleftharpoons CO_2(g) + H_2(g)$$

equilibrium is established at a certain temperature when the concentrations of CO, H_2O, CO_2, and H_2 are 1.0×10^{-2}, 2.0×10^{-2}, 1.2×10^{-2}, and 1.2×10^{-2} mol/dm³, respectively.

a. Calculate K_c.
b. If enough CO is added to raise its concentration temporarily to 2.0×10^{-2} *M*, what will be the final concentrations of all species?

a. conc. SO_2 = conc. NO_2 = 3.0×10^{-3} *M*.
b. conc. SO_2 = conc. NO_2 = conc. SO_3 = conc. NO = 3.0×10^{-3} *M*.

15.33 The concentrations of CO and CO_2 in a sample taken from an automobile exhaust are 4.0×10^{-5} and 4.0×10^{-4} mol/dm³, respectively. If this exhaust were passed through an afterburner at 1600°C in which the concentration of O_2 is maintained at a constant value of 4.0×10^{-4} mol/dm³, what would be the equilibrium concentration of CO? At 1600°C, K_c for the reaction

$$CO(g) + \frac{1}{2} O_2(g) \rightarrow CO_2(g)$$

is about 1×10^4.

15.34 For the reaction

$$N_2(g) + O_2(g) \rightarrow 2\ NO(g)$$

$K_c = 1 \times 10^{-30}$ at 25°C and 0.10 at 2000°C. Starting with 0.040 mol/dm³ of N_2 and 0.010 mol/dm³ of O_2, calculate the equilibrium concentration of NO

a. at 2000°C.
b. at 25°C. (Note that K_c is very small at this temperature. The equilibrium concentrations of N_2 and O_2 will be virtually equal to their original concentrations.)

15.35 Proceeding as in Problem 15.19, check by calculation the values for the numbers of moles of NO_2 and N_2O_4 given in Table 15.2 when the volume is 1.0 dm³.

15.36 Consider the equilibrium

$$2\ HBr(g) \rightleftharpoons H_2(g) + Br_2(g);\ \Delta H = +74.4\ kJ$$

Predict the direction in which the equilibrium will shift if the pressure is increased by

a. compressing the system.
b. adding H_2.
c. adding HBr.
d. adding Ar.
e. raising the temperature.

15.37 Water gas, a commercial fuel, is made by the reaction of hot coke with steam:

$$C(s) + H_2O(g) \rightleftharpoons CO(g) + H_2(g)$$

When equilibrium is established at 800°C, the concentrations of CO, H_2, and H_2O are 4.0×10^{-2}, 4.0×10^{-2}, and 1.0×10^{-2} mol/dm³, respectively.

a. Calculate K_c.
b. What will be the final concentrations of CO and H_2 if enough steam is added to raise its concentration temporarily to 4.0×10^{-2} *M*?

15.22 Consider the reaction

$$SnO_2(s) + 2\ H_2(g) \rightleftharpoons Sn(s) + 2\ H_2O(g)$$

for which $\Delta H = +97.1$ kJ. At 27°C, K_c for this reaction is 1×10^{-11}. Estimate the value of K_c at 227°C.

15.23 For a certain reaction at 300 K, $K = 6.0 \times 10^2$. What is ΔG^o?

15.24 Consider the general reaction

$$A + B \rightleftharpoons C + D$$

a. How does the standard free energy change for the reverse reaction, ΔG^o_R, compare to that for the forward reaction, ΔG^o_F?
b. Using the relation in (a) and Equation 15.13, show that the equilibrium constant for the reverse reaction, K_R, must be the reciprocal of that for the forward reaction, K_F.

15.38 Consider the Haber process

$$N_2(g) + 3\ H_2(g) \rightleftharpoons 2\ NH_3(g)$$

Taking $K_c = 5 \times 10^8$ at 25°C and the heat of formation of NH_3 to be -46.2 kJ/mol, estimate the temperature at which $K_c = 1$.

15.39 Calculate K at 500 K for reactions for which ΔG^o is $+10.0$ kJ; -10.0 kJ.

15.40 Consider the reactions:

$$A \rightarrow B; \quad \Delta G^o_1$$

$$B + C \rightarrow D; \quad \Delta G^o_2$$

a. Relate the standard free energy change, ΔG^o_3, for the reaction

$$A + C \rightarrow D$$

to ΔG^o_1 and ΔG^o_2.
b. Using the relation in (a) and Equation 15.13, demonstrate the validity of the Rule of Multiple Equilibria, i.e., $K_3 = K_2 \times K_1$.

*15.41

a. Using the Ideal Gas Law, show that the partial pressure of a gas in atmospheres is equal to (0.0821 T) times its concentration in mol/dm³.
b. For the general gas phase reaction

$$aA(g) + bB(g) \rightleftharpoons cC(g) + dD(g)$$

$$\text{show that } K_p = \frac{(pC)^c(pD)^d}{(pA)^a(pB)^b} = K_c(0.0821\ T)^{\Delta n_g}$$

$$\text{where } \Delta n_g = (c + d) - (a + b)$$

*15.42 For the reaction $2\ SO_2(g) + O_2(g) \rightleftharpoons 2\ SO_3(g)$

a. Calculate $\Delta G^{1\ atm}$ at 300 K; at 1000 K (recall the discussion in Chapter 14).
b. Calculate K_p at 300 K and 1000 K, using Equation 15.14.
c. Calculate K_c at 300 K and 1000 K, using Equation 15.15.

*15.43 Sufficient N_2O_4 is put into a 10.0-dm³ flask to establish an initial pressure of 100 kPa at 25°C. Part of the N_2O_4 dissociates to NO_2; at equilibrium at 25°C, the total pressure is 117 kPa. Calculate K_c for $N_2O_4(g) \rightleftharpoons 2\ NO_2(g)$ at 25°C.

*15.44 It has been suggested that the generation of nitric oxide in the upper atmosphere by the combustion of fuel in a supersonic transport might destroy the layer of ozone which protects us from ultraviolet light. The reaction is

$$NO(g) + O_3(g) \rightarrow NO_2(g) + O_2(g)$$

a. Given that the free energy of formation of ozone is $+163.6$ kJ/mol at 298 K and 1 atm and using other data from Table 14.2, Chapter 14, calculate $\Delta G^{1\ atm}$, K_p, and K_c for this reaction at 25°C.
b. Assuming that the concentrations of NO, O_3, and O_2 in the upper atmosphere before reaction occurs are 2×10^{-9}, 1×10^{-9}, and 2×10^{-3} mol/dm³, respectively, calculate the equilibrium concentration of O_3. (Assume there is no NO_2 present originally.)
c. According to your calculations, what percentage of the O_3 would be destroyed? (Note that you are assuming that the rate of reaction is great enough to quickly establish equilibrium, which may not be the case.)

16

RATES OF REACTION

A chemist, in studying a reaction, frequently must be concerned with several factors. To maximize the yield of a particular product he can, as we have seen, apply thermodynamic principles and use free energy changes and equilibrium constants to establish the equilibrium conditions. If the reaction proceeds reasonably rapidly, this may give enough information to set the reaction conditions. If, however, the reaction is relatively slow, taking perhaps a week instead of a few minutes or an hour to reach equilibrium, he will need to study the rate, or kinetics, of the reaction before carrying it out in the laboratory.

Although some reactions, particularly those that involve small gas molecules or ions in solution, proceed rapidly, there are many spontaneous reactions that occur very slowly indeed. Consider, for example, the reaction

$$CO(g) + NO(g) \rightarrow CO_2(g) + \tfrac{1}{2} N_2(g)$$

We can calculate that ΔG for this reaction, at 25°C and 1 atm, is −343.8 kJ, indicating that the reaction is spontaneous under these conditions. Indeed, K_c at 25°C is so large ($\sim 10^{60}$) that these two toxic gases, even at the low concentrations found in automobile exhausts, should combine almost completely to form carbon dioxide and elementary nitrogen. Unfortunately, however, this reaction takes place so slowly under ordinary conditions that it does not offer a practical method of removing carbon monoxide and nitric oxide from air on our freeways.

Many of the most familiar substances in our environment are unstable from a thermodynamic viewpoint. The fossil fuels—coal, petroleum, and natural gas—should, according to thermodynamic calculations, be converted to carbon dioxide and water upon exposure to air. The same is true of the organic compounds that make up the living cells of our bodies. Life persists because some reactions that are in principle spontaneous occur at an infinitesimal rate under the conditions of temperature and pressure that prevail on the earth's surface.

It's sobering to realize that people are thermodynamically unstable

We conclude that there is no direct correlation between the rate of a reaction and the thermodynamic driving force as expressed by the free energy change or the equilibrium constant. In order to predict how rapidly a reaction will occur, we must become familiar with a different set of principles which fall in the area of *chemical kinetics*. We shall develop and discuss certain of these principles in this chapter, where our primary concern will be with reactions involving gases. In later chapters we shall have more to say about reaction rates in water solution.

16.1 MEANING OF REACTION RATE

In order to discuss reaction rate in a meaningful way, we must know precisely what is meant by the term. The rate of reaction is a positive quantity that tells us how the concentration of a reactant or product changes with time. To see what this statement means, consider the reaction between carbon monoxide and nitrogen dioxide

$$CO(g) + NO_2(g) \rightarrow CO_2(g) + NO(g) \qquad (16.1)$$

The rate of this reaction can be taken to be the change in concentration per unit time of one of the products, let us say, carbon dioxide

$$\text{rate} = \frac{\text{change in conc. } CO_2}{\text{time interval}} = \frac{\Delta \text{ conc. } CO_2}{\Delta t} \qquad (16.1a)$$

Alternatively, the rate could be expressed in terms of the disappearance of a reactant

The minus sign is needed to make the rate positive

$$\text{rate} = \frac{-\Delta \text{ conc. CO}}{\Delta t} \qquad (16.1b)$$

Notice that, because of the 1:1 stoichiometry of Equation 16.1, these two expressions for the rate of reaction are equivalent. For every mole of CO_2 formed, one mole of CO disappears. If, in one second, the concentration of CO_2 were to increase by 0.020 mol/dm^3 (Δ conc. CO_2 = +0.020 mol/dm^3), that of CO would have to decrease by the same amount (Δ conc. CO = −0.020 mol/dm^3). The rate, calculated from either 16.1a or 16.1b, would be +0.020 mol/(dm$^3 \cdot$ s).

Notice that the units of reaction rate will be those of concentration divided by time. We shall consistently express concentrations in moles per cubic decimetre (mol/dm^3). Time, on the other hand, may be given in seconds (s), minutes (min), hours (h), days (d), or years (a). Thus, units of reaction rate may be:

$$\text{mol/(dm}^3 \cdot \text{s), mol/(dm}^3 \cdot \text{min), mol/(dm}^3 \cdot \text{a), . . .}$$

Let us consider how we might experimentally determine the rate of the reaction between CO and NO_2, as expressed by Equation 16.1b. Clearly we need to know how the concentration of CO changes with time. To find out, we might start by introducing 0.10 mol CO and 0.10 mol NO_2 into a 1.0-dm^3 container at 400°C. Every ten seconds we withdraw a small sample of gas from the reaction mixture, cool it quickly to stop the reaction, and analyze for CO. The results of such an experiment are listed in Table 16.1 and shown graphically in Figure 16.1.

You will notice, from both the table and the figure, that the concentration of carbon monoxide drops rapidly in the initial stages of the reaction. As the reaction proceeds, the concentration of CO decreases more and more slowly. Clearly the rate of reaction is itself a function of time. Over the first 10-s period, the average rate is

$$\frac{-\Delta \text{ conc. CO}}{\Delta t} = \frac{-(0.067 - 0.100)}{10 - 0} = \frac{0.033}{10} = 0.0033 \frac{\text{mol}}{\text{dm}^3 \cdot \text{s}}$$

TABLE 16.1 RATE OF REACTION $CO(g) + NO_2(g) \rightarrow CO_2(g) + NO(g)$

(Data taken at 400°C with initial concentrations of CO = NO_2 = 0.10 mol/dm^3)

Conc. CO	0.100	0.067	0.050	0.040	0.033	. . .	0.017	. . .	0.002
Time (s)	0	10	20	30	40	. . .	100	. . .	1000

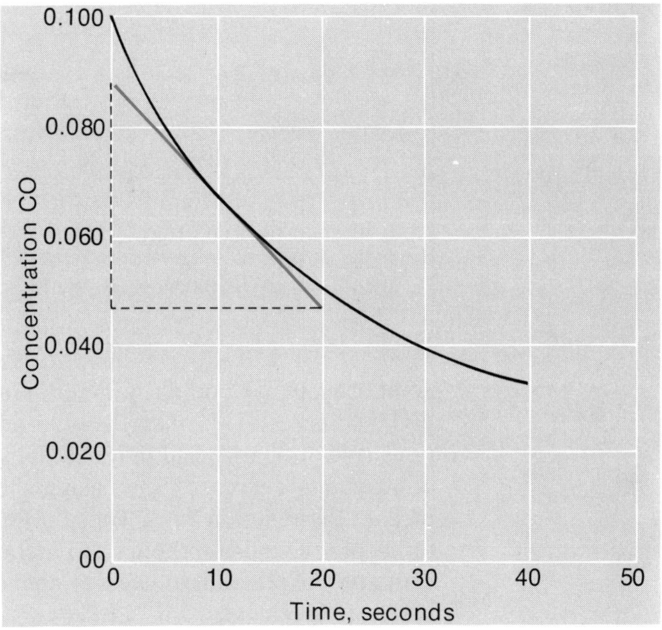

Figure 16.1 Variation of concentration of CO with time in the reaction $CO(g) + NO_2(g) \rightarrow CO_2(g) + NO(g)$. The rate at any point is equal to minus the slope of the tangent to the curve. At $t = 10$, slope $= \Delta$ conc. $CO/\Delta t = -0.044/20 = -0.0022$ mol/(dm³ · s). The rate is 0.0022 mol/(dm³ · s).

In the period between 10 and 20 s, the average rate is considerably smaller

$$\frac{-\Delta \text{ conc. CO}}{\Delta t} = \frac{-(0.050 - 0.067)}{10} = \frac{0.017}{10} = 0.0017 \frac{\text{mol}}{\text{dm}^3 \cdot \text{s}}$$

If we require the rate of reaction at a particular instant, let us say at $t = 10$ s, we might guess that it would be halfway between these two rates, i.e.,

$$\frac{0.0033 + 0.0017}{2} = 0.0025 \frac{\text{mol}}{\text{dm}^3 \cdot \text{s}}$$

A better way to estimate the rate at this point would be to draw a tangent to the curve of Figure 16.1 at $t = 10$ s and find its slope. If we do this carefully, we find an instantaneous rate of 0.0022 mol/(dm³ · s).

A somewhat more sophisticated way to obtain the rate of reaction at a given time uses the principles of calculus. The instantaneous rate of reaction, in derivative notation, is given by

$$\text{rate} = \frac{-\text{d (conc. CO)}}{\text{d}t}$$

This relationship suggests that to find the rate at $t = 10$ s we could fit the data in Table 16.1 to an algebraic equation relating conc. CO to time, obtain a general expression for the derivative, and evaluate it at $t = 10$ s. We shall not pursue this approach further because experience indicates that manipulations of this type are not familiar to students in general chemistry, even those who have had an elementary course in calculus.

16.2 DEPENDENCE OF REACTION RATE UPON CONCENTRATION

In discussing the reaction of carbon monoxide with nitrogen dioxide, we pointed out that the rate of reaction decreases with time. From a slightly different

TABLE 16.2 INITIAL RATES OF REACTION, mol/(dm$^3 \cdot$ s)
$CO(g) + NO_2(g) \rightarrow CO_2(g) + NO(g)$ at 400°C

Series 1			Series 2			Series 3		
Conc. CO	Conc. NO_2	Rate	Conc. CO	Conc. NO_2	Rate	Conc. CO	Conc. NO_2	Rate
0.10	0.10	0.005	0.10	0.20	0.010	0.10	0.30	0.015
0.20	0.10	0.010	0.20	0.20	0.020	0.20	0.30	0.030
0.30	0.10	0.015	0.30	0.20	0.030	0.30	0.30	0.045
0.40	0.10	0.020	0.40	0.20	0.040	0.40	0.30	0.060

point of view, we could say that the rate decreases as the concentration of CO or NO_2 decreases. This observation is generally valid for a variety of chemical reactions. We ordinarily find that reactions proceed more slowly as the concentrations of reactants decrease. Increasing the concentration of reactants ordinarily increases the reaction rate. These observations can be interpreted quite simply in terms of the collision theory of reaction rates (Section 16.5). The higher the concentration of reactant molecules, the more frequently they will collide and be converted to products.

One way to study the effect of concentration upon reaction rate is to obtain the initial rate (i.e., the rate at $t = 0$) as a function of the concentration of a particular reactant, holding the concentrations of all other reactants constant. We might, for example, conduct a series of experiments in which we measure the initial rate of the CO-NO_2 reaction at different concentrations of CO, holding the concentration of NO_2 constant. Data for three such series are presented in Table 16.2.

What would happen to the rate if both conc. CO and conc. NO_2 were doubled?

Looking at the vertical columns in Table 16.2 we observe that the rate is directly proportional to the concentration of CO. If, for example, the concentration of CO is doubled (e.g., from 0.10 to 0.20 mol/dm^3), the rate also doubles (from 0.005 to 0.010 mol/(dm$^3 \cdot$ s) in series 1, from 0.010 to 0.020 mol/(dm$^3 \cdot$ s) in series 2, and so forth). In a similar way we can deduce the effect of NO_2 concentration upon rate by examining the horizontal rows of data in Table 16.2. Notice, for example, that when the concentration of CO is held constant at 0.10 mol/dm^3 (first horizontal row) the rate increases in direct proportion to the concentration of NO_2. We conclude that the rate of this reaction is directly proportional to the concentrations of both CO and NO_2 or that

$$\text{rate} = k\,(\text{conc. CO})(\text{conc. NO}_2) \qquad (16.2)$$

The constant of proportionality k in Equation 16.2 is referred to as the **rate constant** for the reaction. For a particular reaction, k is a function only of temperature; it is independent of the concentrations of reactants. It can be calculated from the observed rate at known reactant concentrations. Substituting in Equation 16.2 at a point where the concentrations of CO and NO_2 are both 0.10 mol/dm^3, we have

$$0.005\,\frac{\text{mol}}{\text{dm}^3 \cdot \text{s}} = k\left(0.10\,\frac{\text{mol}}{\text{dm}^3}\right)\left(0.10\,\frac{\text{mol}}{\text{dm}^3}\right)$$

Solving:
$$k = \frac{0.005\ \text{mol/(dm}^3 \cdot \text{s)}}{(0.10\ \text{mol/dm}^3)^2} = 0.5\,\frac{\text{dm}^3}{\text{mol} \cdot \text{s}}$$

Given the value of k, we can calculate the rate of reaction at 400°C for any given concentrations of CO and NO_2:

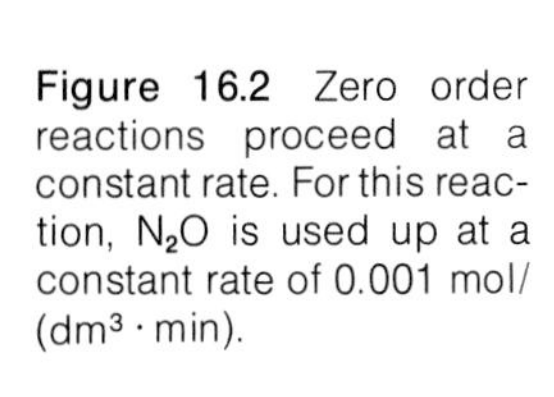

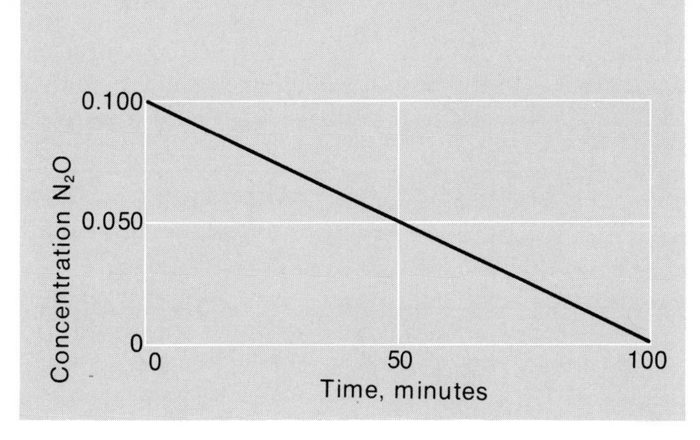

Figure 16.2 Zero order reactions proceed at a constant rate. For this reaction, N_2O is used up at a constant rate of 0.001 mol/($dm^3 \cdot$ min).

dm^3, we could express the concentration of N_2O in terms of time by the following equation:

$$\text{conc. } N_2O = 0.100 - 0.001t \qquad (16.4a)$$

where t is the elapsed time in minutes. After 40 min, 0.060 mol/dm^3 would be left, and after 100 min the N_2O would be all gone. A graph of conc. N_2O against time would yield a straight line (Fig. 16.2).

For the general zero order reaction, we obtain very similar relations:

$$A \rightarrow \text{products; rate} = -\frac{\Delta \text{ conc. A}}{\Delta t} = k(\text{conc. A})^0 = k$$

or:

$$\Delta \text{ conc. A} = -k\Delta t \qquad (16.4b)$$

But, Δ conc. A = (conc. of A at time t) − (conc. A at time zero)

$$= \text{conc. A} - (\text{conc. A})_0$$

and, $\Delta t = t - 0 = t$

So, from 16.4b, we have: conc. A − (conc. A)$_0$ = $-kt$, or

$$\text{conc. A} = (\text{conc. A})_0 - kt \qquad (16.4)$$

In a zero order reaction, all the reactant is consumed in a finite time

First Order Reactions

Most reactions, as we have noted, are not zero order, but have rates that depend on reactant concentrations. First order reactions are very common; a classic example is the decomposition of dinitrogen pentoxide:

$$2\ N_2O_5(g) \rightarrow 4\ NO_2(g) + O_2(g) \qquad \text{rate} = k(\text{conc. } N_2O_5)$$

At 67°C the rate constant for this reaction is 0.35 min^{-1}.

Given, say, an initial concentration of N_2O_5 of 1.0 mol/dm^3, we would like to find an equation like 16.4a, expressing conc. N_2O_5 in terms of the time. Since the rate is proportional to conc. N_2O_5, the rate decreases as conc. N_2O_5 decreases, making the relationship between concentration and time a more complicated expression than with zero order reactions (see Table 16.3). The problem really requires the calculus for a rigorous solution (see p. 384), but the final equation one obtains is quite simple, and if we work with $\log_{10}$ conc. N_2O_5 instead of conc. N_2O_5, is similar in form to Equation 16.4a:

$$\log_{10} \text{conc. } N_2O_5 = \log_{10} 1.0 - 0.35t/2.30 \qquad (16.5a)$$

Given Equation 16.5a we could calculate the concentration of N_2O_5 at any time t (Fig. 16.3). (In this case it would take infinite time to use up all the N_2O_5, since the rate approaches zero as conc. N_2O_5 goes to zero.)

TABLE 16.3 CONCENTRATION-TIME DATA FOR THE REACTION $2\ N_2O_5(g) \rightarrow 4\ NO_2(g) + O_2(g)$ at 67°C

t (min)	0	1	2	3	4	5
Conc. N_2O_5	1.000	0.705	0.497	0.349	0.246	0.173
Log conc. N_2O_5	0.000	−0.152	−0.301	−0.457	−0.609	−0.762

For the general first order reaction,

$$B \rightarrow \text{products; rate} = -\frac{\Delta \text{ conc. B}}{\Delta t} = k \text{ conc. B}$$

$$\log_{10} \text{conc. B} = \log_{10}(\text{conc. B})_0 - kt/2.30 \qquad (16.5b)$$

where $(\text{conc. B})_0$ is the initial concentration and t is the elapsed time.

The first order equation is usually written in a slightly different and somewhat more convenient form. Rearranging Equation 16.5b, we obtain

$$\log_{10}(\text{conc. B})_0 - \log_{10}(\text{conc. B}) = \frac{kt}{2.30}$$

or:

$$\log_{10} \frac{(\text{conc. B})_0}{(\text{conc. B})} = \frac{kt}{2.30} \qquad (16.5)$$

Derivation of Equation 16.5 involves using the calculus to integrate the rate equation. Letting X equal conc. B, we can say

$$\text{rate} = -\frac{dX}{dt} = kX$$

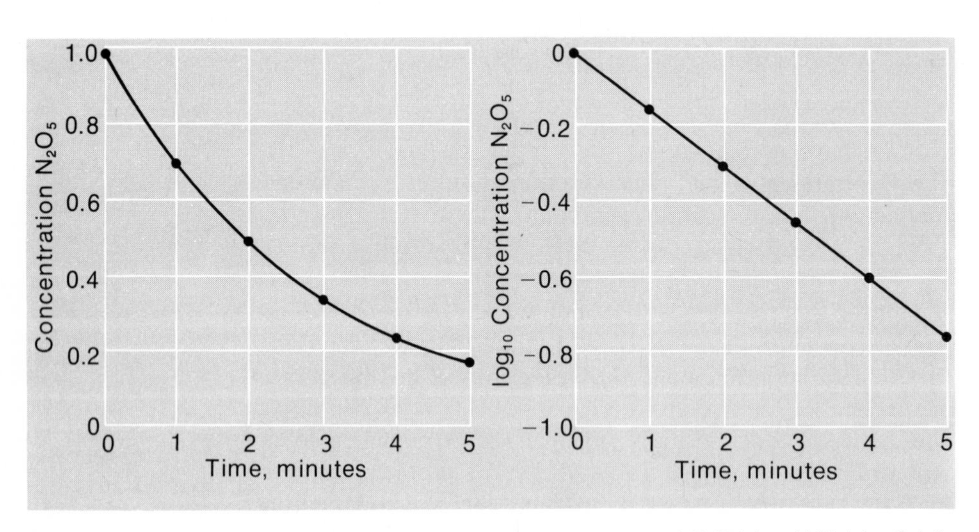

Figure 16.3 Variation of concentration of N_2O_5 with time in the reaction $2\ N_2O_5(g) \rightarrow 4\ NO_2(g) + O_2(g)$. For this or any other first order reaction a plot of log (conc.) vs. time is linear.

Rearranging to separate variables, we get the equation $\frac{dX}{X} = -k\,dt$

Integrating from $t = 0$ to t and from X_0 to X:

$$\int_{X_0}^{X} \frac{dX}{X} = -k\int_0^t dt \qquad \ln X - \ln X_0 = -kt$$

where ln X represents the natural logarithm of X (base e). Natural logarithms are easily converted to ordinary (base 10) logarithms: $\ln X = 2.30 \log_{10}X$

$$\log_{10}X - \log_{10}X_0 = -\frac{kt}{2.30} \qquad \log_{10}\frac{X_0}{X} = \frac{kt}{2.30} \qquad (16.5)$$

For a first order reaction, Equation 16.5 can be used to calculate the concentration of a reactant remaining after a given time (Example 16.2a), the time required for reactant concentration to drop to a certain level (Example 16.2b), or the time required for a given fraction of a sample to react (Example 16.2c).

Example 16.2 The decomposition of hydrogen peroxide to water and oxygen

$$2\ H_2O_2(l) \rightarrow 2\ H_2O(l) + O_2(g)$$

is a first order reaction with a rate constant of $0.0410\ \text{min}^{-1}$.

a. If we start with a 0.500 *M* H_2O_2 solution, what will be its concentration after 10.0 min?
b. How long will it take for the concentration of H_2O_2 to drop from 0.500 *M* to 0.100 *M*?
c. How long will it take for one-half of a sample of H_2O_2 to decompose?

Solution

a. Substituting numbers into Equation 16.5, letting X = conc. H_2O_2:

$$\log_{10}\frac{0.500}{X} = \frac{0.0410 \times 10.0}{2.30} = 0.178$$

The antilog of 0.178 is 1.51, so $\frac{0.500}{X} = 1.51$; solving, $X = 0.33\ \text{mol/dm}^3$.

b. Solving Equation 16.5 for t, we obtain:

$$t = \frac{2.30}{k}\log_{10}\frac{X_0}{X} = \frac{2.30}{0.0410}\log_{10}\frac{0.500}{0.100} = 56.1 \log_{10} 5.00$$

But, $\log_{10} 5.00 = 0.699$, so $t = 56.1(0.699) = 39.2$ min.

c. When half of the sample has decomposed, $X = X_0/2$; $X_0 = 2X$; $X_0/X = 2$
Using the equation from part b:

$$t = \frac{2.30}{k}\log_{10} 2 = \frac{2.30(0.301)}{k} = \frac{0.693}{k}$$

With $k = 0.0410\ \text{min}^{-1}$, we have $t = \frac{0.693}{0.0410}$ min = 16.9 min.

The analysis of Example 16.2c shows that *the time required for a given fraction of a reactant to decompose via a first order reaction is independent of concentration.* Specifically, the time required for one half the sample to decompose, often referred to as the **half-life** of the reaction, is

$$t_{1/2} = \frac{0.693}{k} \qquad (16.6)$$

Thus for the reaction discussed in Example 16.2, where $k = 0.0410\ \text{min}^{-1}$, $t_{1/2} \approx 17$ min, and the initial concentration of H_2O_2 is 0.500 mol/dm³:

after 17 min, conc. H_2O_2 = 0.250 mol/dm³ (1/2 left)
after 34 min, conc. H_2O_2 = 0.125 mol/dm³ (1/4 left)
after 51 min, conc. H_2O_2 = 0.0625 mol/dm³ (1/8 left)
after 68 min, conc. H_2O_2 = 0.0312 mol/dm³ (1/16 left)

Notice from Equation 16.6 that the half-life, $t_{1/2}$, is inversely proportional to the rate constant, k. A "fast" reaction, for which k is large, will have a short half-life. Conversely, a "slow" reaction (small value of k) will be characterized by a comparatively long half-life.

Second Order Reactions

For the general second order reaction:

$$\text{C} \rightarrow \text{products; rate} = k(\text{conc. C})^2$$

it is possible to derive the following relation between the concentration of reactant and time:

$$\frac{1}{\text{conc. C}} = \frac{1}{(\text{conc. C})_0} + kt \qquad (16.7)$$

Given Equations 16.4, 16.5, and 16.7, you can probably see that the quantities which depend in a linear way on time for zero, first, and second order reactions are, respectively, the concentration (X), its logarithm (log X), and its reciprocal (1/X). This means that, given data for concentration in terms of time, one can find the order of a reaction by simply plotting those quantities against time and noting which graph produces a straight line.

Example 16.3 The following data were obtained for the gas phase decomposition of hydrogen iodide.

t (h)	0	2	4	6
conc. HI	1.00	0.50	0.33	0.25

In this reaction zero, first, or second order in HI?

Solution It will be useful to prepare a table in which we list X, log X, and 1/X as a function of time, letting X = conc. HI.

t	X	$\log_{10}$X	1/X
0	1.00	0.00	1.0
2	0.50	−0.30	2.0
4	0.33	−0.48	3.0
6	0.25	−0.60	4.0

If it is not obvious from the table above that the only linear plot will be that of 1/X vs. t, that point should be clear from Figure 16.4. We conclude that we are dealing with a second order reaction.

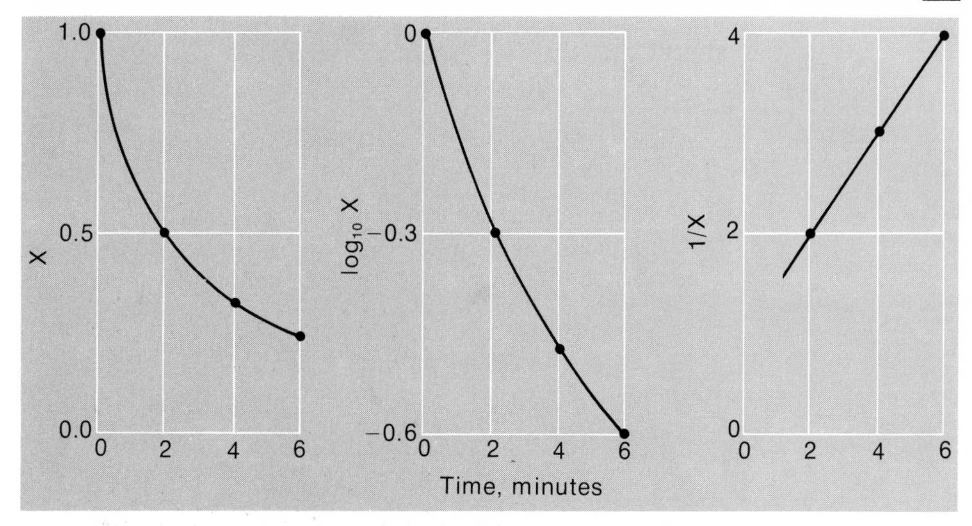

Figure 16.4 Decomposition of HI. If a plot of X vs. *t* is linear, the reaction is zero order. If log X vs. *t* is linear, the reaction is first order. If 1/X vs. *t* is linear, as here, the reaction is second order.

16.3 DEPENDENCE OF REACTION RATE UPON TEMPERATURE

The rates of most chemical reactions increase as the temperature rises. A person in a hurry to prepare dinner employs this principle in using a pressure cooker to cook potatoes, apples, or a pot roast (hopefully, not all at the same time). By storing the leftovers in a refrigerator, we slow down the chemical reactions responsible for food spoilage. As a general and very approximate rule, it is often stated that an increase in temperature of 10°C doubles the reaction rate. If this rule holds, foods should cook twice as fast in a pressure cooker at 110°C as in an open saucepan and deteriorate four times as rapidly at room temperature (25°C) as they do in a refrigerator at 5°C.

From all this, it doesn't necessarily follow that chemists are good cooks

Qualitatively, we can explain the effect of temperature upon reaction rate by recalling from Chapter 5 that raising the temperature greatly increases the fraction of molecules having high kinetic energies. These are the molecules that are most likely to react when they collide (Section 16.5). To derive a quantitative relationship between reaction rate and temperature, let us consider the reaction between CO and NO_2 discussed earlier. If we determine the rate constant of this reaction over a series of temperatures between 600 and 800 K, the data in Table 16.4 are obtained.

TABLE 16.4 TEMPERATURE DEPENDENCE OF THE RATE CONSTANT FOR THE REACTION $CO(g) + NO_2(g) \rightarrow CO_2(g) + NO(g)$

TEMPERATURE (K)	k [$dm^3/(mol \cdot s)$]
600	0.028
650	0.22
700	1.3
750	6.0
800	23

As usual, we search for a linear relationship between some function of the rate constant, k, and temperature. It turns out that a plot of $\log_{10}k$ vs. $1/T$ is a straight line (Fig. 16.5). This tells us that

$$\log_{10}k = A - \frac{B}{T} \tag{16.8}$$

where A and B are constants for a particular reaction that can be determined from the slope and intercept of the plot. The general validity of Equation 16.8 was first demonstrated by the Swedish physical chemist Svandte Arrhenius in 1887.

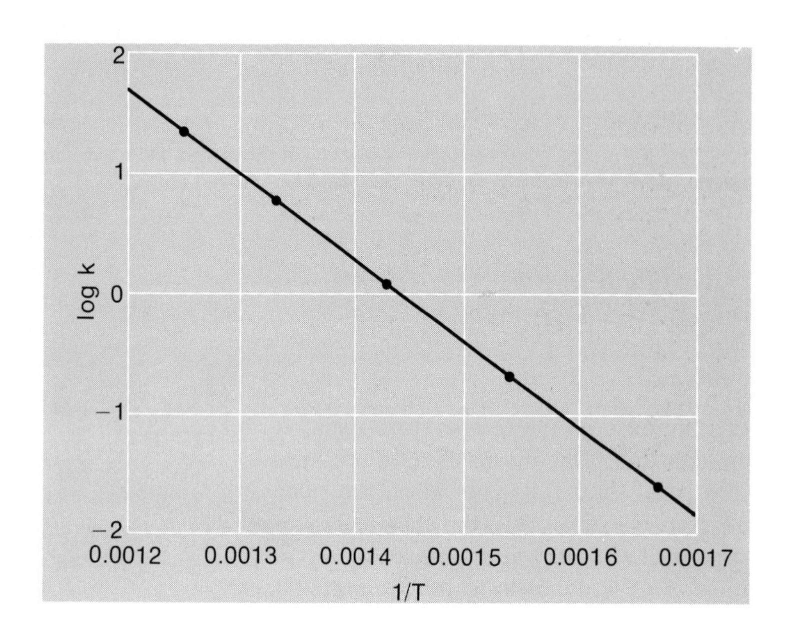

Figure 16.5 When log k is plotted against $1/T$ for the CO-NO_2 reaction, a straight line is obtained. Note that every time log k increases by 1, k increases by a factor of ten. The rate of this reaction changes by about a factor of 1000 in the temperature range shown (600–800 K).

The constant B in Equation 16.8 is directly related to an important parameter known as the **activation energy,** E_a, for the reaction. It turns out that $B = E_a/2.30\,R$, where R is the gas law constant, 8.31 J/K. Making this substitution in Equation 16.8, we obtain

$$\log_{10}k = A - \frac{E_a}{(2.30)(8.31)\,T} \tag{16.9}$$

Equation 16.9 can be used directly to obtain the activation energy for a reaction from a plot of $\log_{10}k$ vs. $1/T$ (slope $= -E_a/(2.30)(8.31)$. Alternatively, it can be manipulated (cf. Equations 11.3 and 11.4, Chapter 11) to give the following relationship between the rate constants, k_2 and k_1, at two different temperatures T_2 and T_1:

$$\log_{10}\frac{k_2}{k_1} = \frac{E_a}{(2.30)(8.31)}\left(\frac{T_2 - T_1}{T_2\,T_1}\right) \tag{16.10}$$

where E_a, the activation energy, is expressed in joules. The usefulness of this relation is illustrated in Example 16.4.

Example 16.4

a. The activation energy for the reaction 2 $NO_2(g) \rightarrow 2\ NO(g) + O_2(g)$ is 1.14×10^5 J. At 600 K, $k = 0.75\ dm^3/(mol \cdot s)$. Calculate k at 700 K.

b. What must be the value of E_a if the rate constant for a reaction is to double when the temperature increases from 27 to 37°C?

Solution

a. Applying Equation 16.10 with $T_2 = 700$ K, $T_1 = 600$ K.

$$\log_{10}\frac{k_2}{k_1} = \frac{1.14 \times 10^5}{(2.30)(8.31)}\left(\frac{700 - 600}{700 \times 600}\right) = 1.42$$

Taking antilogs, $\frac{k_2}{k_1}$ = antilog 1.42 = 26

$$k_2 = 26k_1 = 26 \times 0.75\ dm^3/(mol \cdot s) = 20\ dm^3/(mol \cdot s)$$

b. If $k_2/k_1 = 2.00$, then $\log_{10} k_2/k_1 = \log_{10} 2 = 0.301$
Substituting in Equation 16.10:

$$0.301 = \frac{E_a(310 - 300)}{(2.30)(8.31)(310)(300)}$$

Solving, $E_a = 53\ 500$ J

Note that if E_a were appreciably greater than 53.5 kJ, k would more than double for a 10° rise in temperature; if E_a were smaller than 53.5 kJ, k would increase by less than a factor of two. Clearly the empirical rule that a temperature increase of 10°C doubles the reaction rate is at best a crude approximation.

The activation energy associated with a reaction can be given a simple physical interpretation. In order for a reaction to occur between stable molecules, a certain amount of energy must be absorbed to weaken the bonds holding the reactant molecules together. The quantity E_a represents the energy required to bring the reactants to the point where they can rearrange to form products. As we might expect, very fast reactions are characterized by small activation energies. Looking at Equation 16.9, we see that if E_a is small, the constant A, which is a positive number, will predominate and k will be relatively large.

Expanding upon this physical picture, we can say that stable molecules, before being converted to products, must pass through an unstable, high energy, intermediate species. This transient, highly reactive species is referred to as an *activated complex*. Its exact nature is difficult to determine; in the reaction between CO and NO_2 it might be a "pseudomolecule" made up of CO and NO_2 molecules in close contact. We might visualize the path of reaction somewhat as follows:

Often, electrons in reactant molecules must be promoted to higher orbitals

O≡C + O—N(=O) → O≡C···O···N(=O) → O=C=O + N=O

reactants — activated complex — products

The dotted lines represent "partial bonds" in the activated complex; the N—O bond in the NO_2 molecule has been partially broken and a new bond between carbon and oxygen atoms has started to form.

According to this picture, we interpret the activation energy to be the difference in energy between the activated complex and the reactant molecules. This

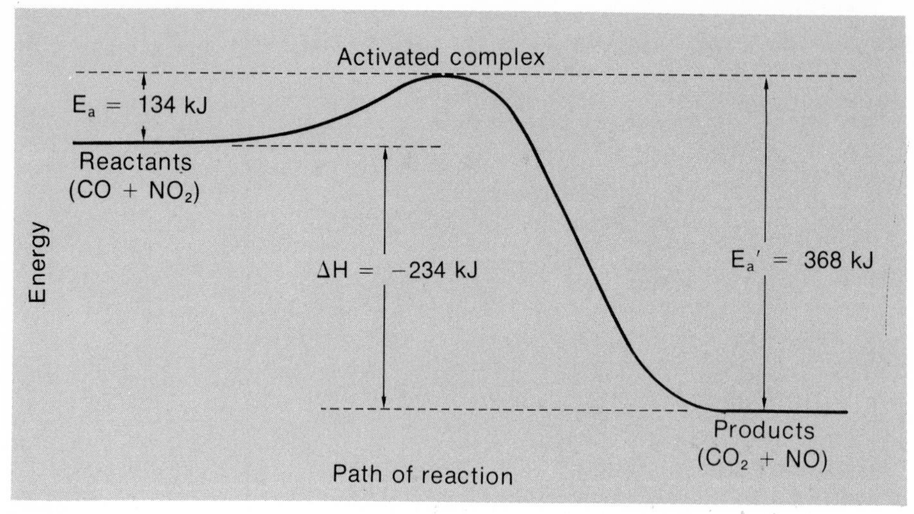

Figure 16.6 Concept of activation energy. During the reaction step, about 134 kJ, the activation energy E_a, must be furnished to the reactants for each mole of CO that reacts. This energy activates each CO-NO_2 complex to the point where reaction can proceed.

E_a is an energy barrier which can prevent spontaneous reactions from occurring rapidly

concept is shown schematically in Figure 16.6 for the CO—NO_2 reaction. Notice that the activated complex has an energy 134 kJ greater than that of the reactants, CO and NO_2, and 368 kJ greater than that of the products. In other words, the activation energies for the forward reaction, E_a, and the reverse reaction, E_a' are, respectively,

$$E_a = 134 \text{ kJ}: E_a' = 368 \text{ kJ}$$

The difference between these two quantities represents the energy (or enthalpy) difference between products and reactants

$$CO(g) + NO_2(g) \rightarrow CO_2(g) + NO(g): \Delta H = -234 \text{ kJ}$$

In general, for any reaction

$$\Delta H = E_a - E_a' \tag{16.11}$$

For an exothermic reaction ($\Delta H < 0$), the activation energy for the reverse reaction, E_a', must be greater than that for the forward reaction, E_a. If, on the other hand, the products have a higher enthalpy than the reactants ($\Delta H > 0$), $E_a > E_a'$.

16.4 CATALYSIS

It has long been known that certain substances called *catalysts* can increase the rate of a reaction without being consumed by it. A familiar example of a reaction which is extremely susceptible to catalysis is the decomposition of hydrogen peroxide:

$$H_2O_2(aq) \rightarrow H_2O(l) + \tfrac{1}{2} O_2(g) \tag{16.12}$$

This occurs rather slowly under ordinary conditions (recall Example 16.2) but takes place almost instantaneously if a pinch of manganese dioxide, MnO_2, is added. All the MnO_2 can be recovered when the reaction is over, indicating that it is a true catalyst. Another reaction of this type is the decomposition of nitrous oxide

$$N_2O(g) \rightarrow N_2(g) + \tfrac{1}{2}\, O_2(g) \qquad (16.13)$$

which can be accelerated by bringing the N_2O into contact with a metal such as gold.

The effectiveness of a catalyst in increasing the rate of a reaction is explained in terms of a lowering of the activation energy (Fig. 16.7). Compare, for example, the values of E_a for Reaction 16.13: 250 kJ for the uncatalyzed reaction vs. 120 kJ on a gold surface. The reduction in activation energy is achieved by providing an alternative pathway of lower energy for the reaction. The decomposition of N_2O on gold involves chemical adsorption of the gas on the metal surface, with the formation of a weak bond between the oxygen of the N_2O molecule and a gold atom. This, in turn, weakens the bond joining nitrogen to oxygen, making it easier for the molecule to break apart.

$$N{\equiv}N{-}O + Au \rightarrow N{\equiv}N\cdots O\cdots Au \rightarrow N{\equiv}N + O\cdots Au \;\; (O\cdots Au \rightarrow \tfrac{1}{2}\, O_2 + Au)$$

As Figure 16.7 indicates, the presence of a catalyst does not affect the relative energies of reactants and products. The overall enthalpy and free energy changes as well as the equilibrium constant are the same for the catalyzed and uncatalyzed reactions. We conclude that a catalyst can be effective only if we are dealing with a spontaneous reaction. It would be futile to search for a catalyst to bring about a nonspontaneous reaction; ΔG and K would still be unfavorable and the reaction could not occur.

Catalysts can speed up reactions, but can't change the position of an equilibrium

Many reactions which take place slowly, if at all, under ordinary laboratory conditions occur readily in the body in the presence of catalysts called enzymes, which are proteins with high molecular weights. An example is the combustion of sugar

$$C_{12}H_{22}O_{11}(s) + 12\, O_2(g) \rightarrow 12\, CO_2(g) + 11\, H_2O(l)$$

which is difficult to bring about directly, as, for example, by heating a sample of sugar in a test tube. In the body, sugar is metabolized at 37°C in a complex series of biochemical reactions which give carbon dioxide and water as end products. Each step in the sequence is catalyzed by a particular enzyme specifically adapted for that purpose.

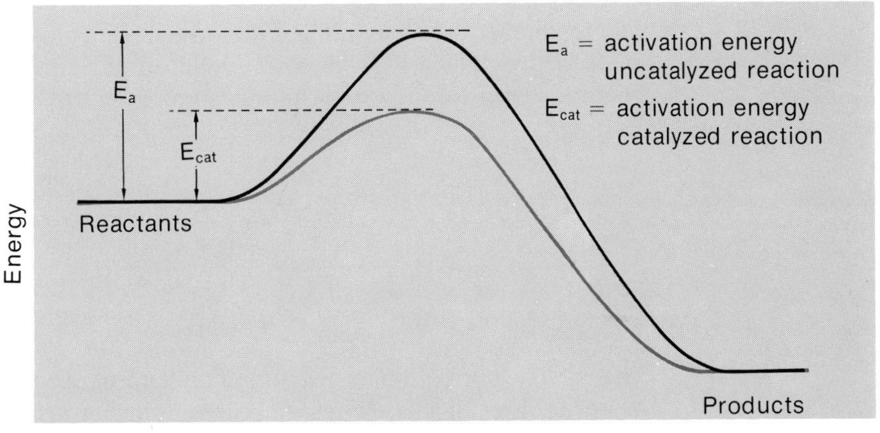

Figure 16.7 By changing the path by which a reaction occurs, a catalyst can lower the activation energy that is required, and so speed up the reaction.

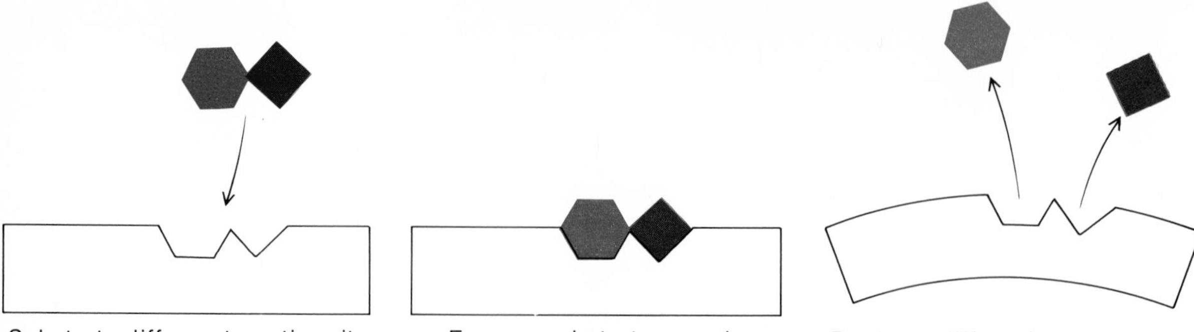

Figure 16.8 In enzyme catalysis, the substrate appears to fit, in a "lock and key" arrangement, on the enzyme. After adsorption, the enzyme configuration often changes, which assists in cleaving the crucial substrate bond, and thereby increases the rate of reaction.

Enzymes are very specific catalysts for one or at most a few reactions

The effectiveness of an enzyme in catalyzing biochemical reactions can be interpreted crudely in terms of the "lock and key" analogy shown in Figure 16.8. The reactant, known in biochemical jargon as the "substrate," fits into a specific site on the enzyme surface, where it is held in position by intermolecular forces. The substrate-enzyme complex can then either decompose or react with another species such as a water molecule. The validity of this model gains some support from the fact that the kinetics of enzyme-catalyzed reactions resemble those found in surface catalysis where a similar model is well established.

Enzyme activity is diminished in the presence of certain substances known as inhibitors. One way in which an inhibitor can operate is to occupy sites on an enzyme molecule which are supposed to be reserved for the substrate. Frequently inhibitors have geometries closely resembling those of the substrates they replace. For example, the metabolism of citric acid is inhibited by its close relative, fluorocitric acid, which can presumably fit into the same slot in an enzyme.

$$\begin{array}{rcl} & \mathrm{H} & \\ & | & \\ \mathrm{H}- & \mathrm{C} & -\mathrm{COOH} \\ & | & \\ \mathrm{HO}- & \mathrm{C} & -\mathrm{COOH} \\ & | & \\ \mathrm{H}- & \mathrm{C} & -\mathrm{COOH} \\ & | & \\ & \mathrm{H} & \end{array} \qquad\qquad \begin{array}{rcl} & \mathrm{F} & \\ & | & \\ \mathrm{H}- & \mathrm{C} & -\mathrm{COOH} \\ & | & \\ \mathrm{HO}- & \mathrm{C} & -\mathrm{COOH} \\ & | & \\ \mathrm{H}- & \mathrm{C} & -\mathrm{COOH} \\ & | & \\ & \mathrm{H} & \end{array}$$

citric acid fluorocitric acid

In another case we find that the insecticide parathion inhibits the hydrolysis of the acetylcholine ion, which is produced at the end of a nerve cell when an impulse passes through it on its way to or from the brain.

$$\begin{array}{rcl} & \mathrm{S} & \\ & | & \\ \mathrm{C_2H_5{-}O}- & \mathrm{P} & -\mathrm{O}-\langle\bigcirc\rangle-\mathrm{NO_2} \\ & | & \\ & \mathrm{O{-}C_2H_5} & \end{array} \qquad \left[\begin{array}{rcl} & \mathrm{CH_3} & \qquad\qquad\qquad\quad \mathrm{O} \\ & | & \qquad\qquad\qquad\quad \| \\ \mathrm{H_3C}- & \mathrm{N} & -\mathrm{CH_2}-\mathrm{CH_2}-\mathrm{O}-\mathrm{C}-\mathrm{CH_3} \\ & | & \\ & \mathrm{CH_3} & \end{array}\right]^{+}$$

parathion acetylcholine ion

The presence of parathion leads to a buildup of acetylcholine ions at nerve cell junctions. This blocks the transmission of further nerve impulses, leading to unconsciousness and, eventually, death. Presumably the parathion operates by occupying the active site on the enzyme molecule responsible for breaking down the acetylcholine ion.

Until quite recently we knew almost nothing about the molecular structure of enzymes; the "lock and key" model was little more than a convenient way to rationalize the kinetics of enzyme-controlled reactions. Within the past decade it has been possible to establish by x-ray crystallography the structures and configurations of some of the simpler enzymes. Even more remarkably, the structures of some enzyme-substrate complexes have been determined. Research in this area led in 1969 to the first laboratory synthesis of an enzyme, ribonuclease, for which Drs. Stein and Moore of Rockefeller University and Anfinsen of NIH won the 1972 Nobel prize in chemistry.

16.5 COLLISION THEORY OF REACTION RATES

As we have seen, kinetic studies can furnish us with information about the effects of concentration, temperature, and catalysts on reaction rates. One of the goals of chemical kinetics is to explain these effects on a theoretical basis, developing as best we can a *mechanism,* or path, for the reaction that is consistent with its observed order, activation energy, and rate constant at a given temperature. If we understand a reaction mechanism we may be able to increase the rate or, if necessary, slow it down so that it will not interfere with another, more desirable, reaction.

One of the earliest, and most successful, approaches to an explanation of reaction rates is the collision theory, which is based on the very plausible idea that reactions occur as a result of collisions between reacting species. The theory proposes that the rate of the reaction

$$A + B \rightarrow \text{products}$$

will be proportional to the number of A-B collisions that occur in unit volume of the reaction mixture per second. For a mixture of A and B at ordinary pressures, this number, which is called the collision number, Z, is enormous, of the order of 10^{30} collisions/($dm^3 \cdot s$), and is much, much larger than the reaction rate, at least for nonexploding mixtures.

If all collisions were effective, reaction would occur instantaneously

There are at least two reasons why, for a bimolecular process such as

$$CO(g) + NO_2(g) \rightarrow CO_2(g) + NO(g),$$

the rate of reaction is ordinarily much smaller than the collision number, Z. In the first place, in order for reaction to occur, the colliding molecules must have a total kinetic energy at least equal to the activation energy, E_a (134 kJ/mol). Collisions between low energy molecules will be ineffective since they simply cannot provide sufficient energy to push the reaction over the energy barrier. Moreover, the two reactant molecules must be favorably oriented with respect to one another if they are to react upon collision. If, when CO and NO_2 molecules collide, the carbon and nitrogen atoms come in contact with one another (Fig. 16.9, p. 394), it is unlikely that the necessary transfer of an oxygen atom will take place.

Taking account of these two factors, we can write our rate expression in the following way

$$\text{rate} = f \times p \times Z \qquad (16.14)$$

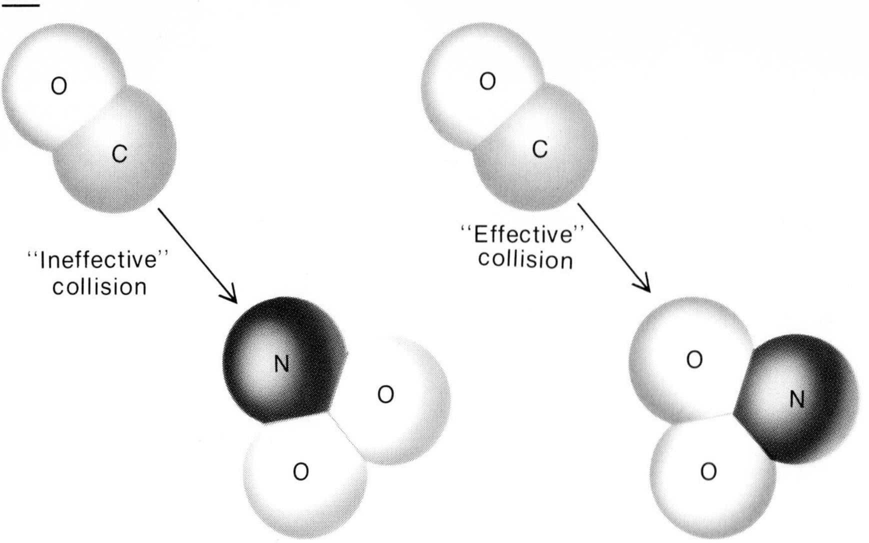

Figure 16.9 For a collision to result in reaction, not only must it involve enough energy, but also the molecules must be properly oriented.

where p is the probability that the colliding molecules will be favorably oriented for reaction and f is the fraction of molecules that possess sufficient energy to react upon collision. Since both p and f are less than unity, we can expect the rate of reaction to be significantly less than the collision number Z.

It is difficult to compare experimentally measured reaction rates with those predicted by Equation 16.14 because the factor p can seldom be estimated accurately. We can, however, use this equation to rationalize the temperature dependence of reaction rate. You may recall from our discussion of the Maxwell distribution of molecular energies in Chapter 5 that the fraction of molecules having energies equal to or greater than a certain energy, for example, E_a, is

$$f = e^{-E_a/RT} = 10^{-E_a/2.30RT} \tag{16.15}$$

where e is the base of natural logarithms, R is the gas constant, and T the absolute temperature. Making this substitution in Equation 16.14, we obtain

$$\text{rate} = p \times Z \times 10^{-E_a/2.30RT}$$

Taking the logarithm of both sides of this equation and substituting for R its value 8.31 J/K, we obtain

$$\log_{10} \text{rate} = \log_{10}(p \times Z) - \frac{E_a}{(2.30)(8.31)T} \tag{16.16}$$

Since the rate constant k is equal to the rate at unit concentrations, we see that this equation becomes identical with Equation 16.9 if we take $A = \log_{10}(p \times Z)$. We conclude that the collision theory satisfactorily explains the dependence of reaction rate upon temperature. Note that the effect of temperature on reaction rate is primarily due to the influence on the fraction of molecules with enough kinetic energy to react, and is not the result of its effect on Z. (Z, indeed, increases slowly with temperature, being proportional to molecular speed through Equation 5.13, p. 111, but this in general influences the rate only slightly.)

Having dealt with the effect of temperature on reaction rate, let us return to Equation 16.14 and see what the collision theory has to say about the effect of concentration on rate. Neither f nor p would be expected to vary with concentration, the former being only a function of temperature and the latter dependent only on the geometrical arrangement of the colliding molecules. The collision number Z does, however, depend on concentration, and for a reaction between A and B would be expected to increase in direct proportion to their concentrations. Doubling the concentration of A or B would double Z; doubling both concentrations would increase Z by a factor of four. In general we can express Z in the following way:

$$Z = C \times (\text{conc. A}) \times (\text{conc. B})$$

where C is a constant that depends on the speeds and sizes of the molecules involved in the reaction. Using this equation for Z in Equation 16.14, we obtain

$$\text{rate} = f \times p \times C \times (\text{conc. A}) \times (\text{conc. B}) = k(\text{conc. A})(\text{conc. B})$$

where the rate constant k equals $f \times p \times C$.

Collision theory, then, predicts that a reaction between A and B will be first order in A, first order in B, and second order overall. Some reactions of this sort, indeed, are second order, but many others are not, so this prediction must be considered a serious deficiency of the theory. The general features of the theory are, however, very reasonable; it deals with the role of temperature very well, but fails in its simple form to explain reaction orders properly.

By simple collision theory, all reactions would be second order

16.6 REACTION MECHANISMS

A reaction mechanism, as we mentioned, is a description of the path, or sequence of steps, by which the reaction occurs. If, as postulated by the collision theory, the reaction occurs in one step by virtue of a collision between reacting molecules, the mechanism would simply be the equation for the reaction. This would appear to be the case in the formation of hydrogen iodide:

$$H_2(g) + I_2(g) \rightarrow 2\ HI(g) \qquad \text{rate} = k(\text{conc. } H_2)(\text{conc. } I_2)$$

Experimentally, the reaction is second order overall, which is consistent with a mechanism involving a simple collision between H_2 and I_2 molecules (see, however, Example 16.5, p. 397).

In order to explain a reaction that is not second order, it is usually necessary to formulate a mechanism that involves two or more steps, the net result of which is the overall reaction. In recent years it has become apparent that multistep reaction mechanisms are the rule, rather than the exception.

One of the early applications of a multistep mechanism was to explain first order reactions, which the collision theory could not handle. For the general reaction we have

$$A \rightarrow \text{product, P} \qquad \text{rate} = k(\text{conc. A})$$

The observed order can be obtained by the following reaction mechanism:

1. As a result of collisions between A molecules, some activated, unstable molecules, which we will indicate as A*, are formed. These molecules reach a state of chemical equilibrium with ordinary A molecules:

$$\text{Step I:} \quad 2\ A \rightleftarrows A^* + A \qquad K_c = \frac{(\text{conc. } A^*)(\text{conc. A})}{(\text{conc. A})^2} = \frac{(\text{conc. } A^*)}{(\text{conc. A})}$$

2. A small fraction of the A* molecules decomposes to form P, at a rate that is proportional to the concentration of A*:

Step II: $A^* \rightarrow P$ rate $= k'$(conc. A*)

where k' is the rate constant for Step II. Since P is formed only in this step, the rate of the reaction is the rate of Step II. To express the rate in terms of A, rather than A*, whose concentration is not known, we resort to the equilibrium condition:

$$(\text{conc. A}^*) = K_c(\text{conc. A})$$

Substituting this value in the rate equation we get

$$\text{rate} = k'(\text{conc. A}^*) = k'K_c(\text{conc. A}) = k(\text{conc. A})$$

where $k'K_c$ equals k, the experimentally observed rate constant for the reaction. Clearly the reaction is first order by the mechanism made up of Step I plus Step II.

Reaction mechanisms are often rather complex and can be confusing at first sight. They can be interpreted in various ways, but we shall limit our use of them to predicting the reaction order which they imply. In making the prediction, we observe several rules with respect to the individual steps in the mechanism. Specifically:

1. The rate equation for any *step* is written in such a way that the order for each reactant is equal to the coefficient of that reactant in that step. For example, for the following steps in a mechanism, the rate expressions would be as shown:

$CH_3 + HCl \rightarrow CH_4 + Cl$	rate $= k(\text{conc. } CH_3)(\text{conc. HCl})$
$2\ NO + O_2 \rightarrow 2\ NO_2$	rate $= k(\text{conc. NO})^2(\text{conc. } O_2)$
$NO_2 \rightarrow NO + O$	rate $= k(\text{conc. } NO_2)$

2. In many mechanisms one of the steps has a much slower rate than any of the others. It turns out that the rate of the overall reaction, which is what is measured experimentally, will be determined by the rate of that slow step. To see why this is the case, consider the following possible mechanism for the reaction $H_2(g) + Br_2(g) \rightarrow 2\ HBr(g)$:

Step A:	$Br_2 \rightleftarrows 2\ Br$ (fast)
Step B:	$Br + H_2 \rightarrow HBr + H$ (slow)
Step C:	$H + Br_2 \rightarrow HBr + Br$ (fast)

The equilibrium in Step A is set up rapidly and furnishes active Br atoms to the mixture. Product is formed in each of the other two steps, with Step C occurring much more rapidly than B. This means that every time an H atom is formed in Step B it reacts essentially instantaneously with Br_2 in Step C to form HBr and another Br atom. Consequently, Step B, the slow step, is **rate determining;** that is, the rate of the overall reaction, the formation of HBr, is fixed by the rate at which B occurs. (The actual overall rate is not *equal* to the rate of Step B, but is twice that rate, since for every HBr molecule formed in Step B another is produced simultaneously in C.) By an argument similar to the one we have used here we can show that for every reaction with one slow step, the overall rate will be determined by the rate of that step, and so will have an order consistent with that step. The situation is analogous to that in a relay race in which there is one slow runner (B) and a fast runner (C). The time required to reach the finish line will be determined primarily by B, particularly if he is much slower than C.

3. Frequently, reactive intermediates are present as reactants in the rate determining step. Since their concentrations are not known experimentally, they must be eliminated from the predicted rate equation if it is to be compared with the observed one. In the reaction just discussed, the Br atoms are reactive intermediates, present at low, unknown concentration. Their concentration, which would appear in the rate equation for Step B, must be eliminated in developing the final predicted rate equation. This could be done through use of the equilibrium condition implied in Step A. The final rate equation, which must include only measurable concentrations, should agree with the observed rate law if the mechanism is a reasonable one.

The following example shows how one applies these ideas in testing a proposed reaction mechanism.

Example 16.5 Although its observed rate equation is consistent with simple collision theory, an alternative mechanism has been proposed for the reaction

$$H_2(g) + I_2(g) \rightarrow 2\ HI(g)$$

Step A: $I_2 \rightleftarrows 2\ I$ (fast)
Step B: $H_2 + 2\ I \rightarrow 2\ HI$ (slow)

Show that this mechanism leads to the rate law for the reaction, which is observed to be rate $= k(\text{conc. } H_2)(\text{conc. } I_2)$

Solution Since Step B occurs slowly, and is the only one making the product, the rate of that step and the rate of the overall reaction must be equal. Writing the rate of Step B according to the rules we have cited, we obtain:

$$\text{For Step B: rate} = k_B(\text{conc. } H_2) \times (\text{conc. I})^2 = \text{rate of overall reaction}$$

The presence of a rate-determining step in a mechanism greatly simplifies the algebra

This rate equation contains conc. I, where I is a reactive iodine atom, whose concentration is too small to be measured. By using the equilibrium condition required by Step A, we can express the concentration of iodine atoms in terms of that of the reactant, I_2, iodine molecules, which can be measured experimentally:

$$\text{For Step A: } K_c = \frac{(\text{conc. I})^2}{(\text{conc. } I_2)} \text{ or } (\text{conc. I})^2 = K_c(\text{conc. } I_2)$$

We now eliminate $(\text{conc. I})^2$ from the rate equation for Step B, obtaining

$$\text{rate} = k_B(\text{conc. } H_2) \times K_c(\text{conc. } I_2) = k_BK_c(\text{conc. } H_2)(\text{conc. } I_2)$$

as the predicted rate law. Taking k_BK_c to equal k, the observed rate constant, we see that this mechanism indeed predicts a rate law consistent with the observed one.

The result of Example 16.5 is illustrative of one of the limitations of mechanism analysis, namely that *the fact that a mechanism agrees with the observed rate expression does not prove that it is correct.* For the HI formation reaction, both simple collision theory and the mechanism in the example show agreement with the observed kinetics. At this time it appears that the mechanism proposed in the example is the correct one.

Chain Reactions

Many gas phase reactions are initiated by the formation, at very low concentrations, of an extremely reactive species which sets off a series of reactions leading to the formation of products. Such processes are referred to as **chain reactions;**

typically they occur very rapidly after a short induction period to allow for the formation of the reactive species. An important example of a chain reaction is the formation of hydrogen chloride from the elements. A mixture of hydrogen and chlorine stored at room temperature in the dark shows no evidence of reaction over long periods of time. However, if the mixture is exposed to ultraviolet light or heated to 200°C, a vigorous reaction occurs. The first step in this reaction, referred to as **chain initiation,** involves the reversible dissociation of a chlorine molecule into atoms:

$$Cl_2 \rightleftharpoons 2\ Cl \qquad (16.17a)$$

The chlorine atoms formed are extremely reactive toward hydrogen molecules:

$$Cl + H_2 \rightarrow HCl + H \qquad (16.17b)$$

This reaction forms another highly reactive species, a hydrogen atom, which attacks a chlorine molecule:

$$H + Cl_2 \rightarrow HCl + Cl \qquad (16.17c)$$

For every Cl atom formed in the initiation step, 10^6 HCl molecules may be produced

In this way, the chlorine atoms are regenerated very quickly and can react with more H_2 molecules; the **chain propagation,** represented by Equations 16.17b and 16.17c, occurs *over and over again* until the H_2 and Cl_2 are almost completely converted to HCl.

The hydrogen and chlorine atoms, which act as **chain carriers,** can be consumed by reaction with each other:

$$H + Cl \rightarrow HCl;\ Cl + Cl \rightarrow Cl_2;\ H + H \rightarrow H_2 \qquad (16.17d)$$

These processes represent **chain termination,** since they break the chain mechanism.

As you can well imagine, kinetic studies of chain reactions are not easy to carry out in the laboratory. Frequently a chemist trying to establish the order of a chain reaction finds that it is over before he has had time to make any measurements. In extreme cases the only result of his efforts may be shattered glassware and broken or badly bent apparatus. The H_2-Cl_2 reaction is particularly notorious in this respect.

Surface Reactions

When a reaction occurs at a solid surface rather than in the gas phase, the activation energy and the rate expression ordinarily change. This implies a difference in mechanism in the two types of reactions. Mechanisms of surface reactions can be quite complicated; we shall consider only the simplest possible process (Fig. 16.10) in which a molecule adsorbed on the surface of a solid catalyst decomposes to give two or more product molecules. An example of such a reaction is the decomposition of N_2O on gold.

$$N_2O(g) \xrightarrow{Au} N_2(g) + \tfrac{1}{2}\ O_2(g)$$

As we might expect, the rate of the surface reaction is directly proportional to the number of adsorbed molecules or, from a slightly different point of view, to the fraction θ of the surface covered by these molecules. Therefore,

$$\text{rate} = k\theta$$

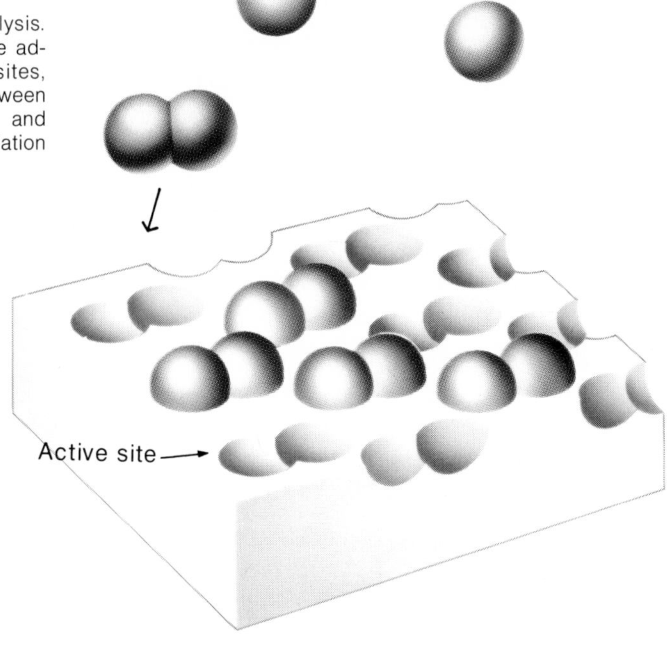

Figure 16.10 Surface catalysis. Reactant gas molecules are adsorbed on the active sites, weakening key bonds between atoms in those molecules and speeding up the dissociation reaction.

In order to deduce the order of the reaction with respect to gaseous reactant (e.g. N_2O), we must relate θ to concentration in the gas phase. There are two limiting situations of interest:

1. At low concentrations of N_2O it is found that the fraction of the surface covered by adsorbed molecules is directly proportional to their concentration in the gas phase. So, we can say that

$$\theta = \text{b(conc. } N_2O) \qquad \text{and} \qquad \text{rate} = k\text{b(conc. } N_2O)$$

You can see that the reaction under such conditions is first order.

2. At high concentrations the surface becomes saturated with reactant molecules. Further increases in concentration have no effect on the amount of gas adsorbed, since essentially all the active sites are already occupied. Or, in other words, at high concentrations of N_2O,

$$\theta = 1 \qquad \text{and} \qquad \text{rate} = k\theta = k$$

Here the rate of decomposition is constant and the reaction is, therefore, zero order. This is the sort of situation present in many solid-catalyzed reactions and explains the fact that they often appear to be zero order.

Most zero order reactions can be explained this way

It is possible to explain these experimental observations of order by a more theoretical argument than we have presented. In 1918 Irving Langmuir at the General Electric Laboratories derived the following equation for θ as a function of concentration:

$$\theta = \frac{\text{b(conc. R)}}{1 + \text{b(conc. R)}} \tag{16.18}$$

This expression is called the Langmuir adsorption isotherm, and fits the observed behavior of many gases adsorbed on solids. The constant b depends on the nature of the gas R, the solid surface used, and the temperature. At low concentrations the second term in the denominator of Equation 16.18 is very small, so the denominator is essentially equal to 1, and θ is proportional to concentration. At high concentrations the denominator becomes about equal to b(conc. R) and θ approaches unity.

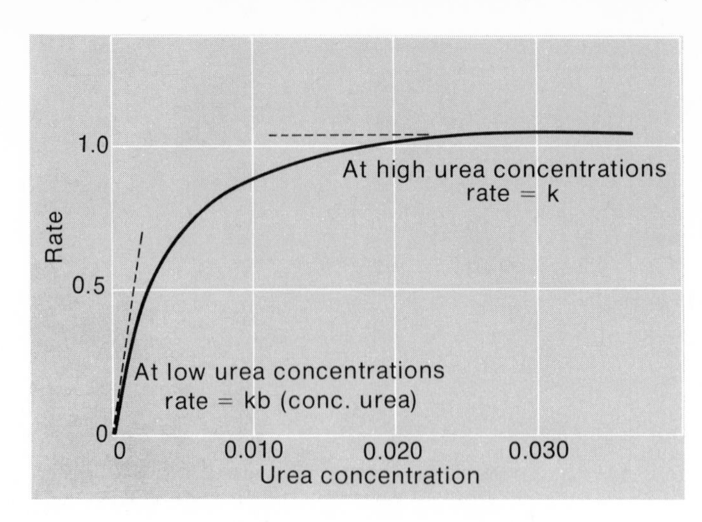

Figure 16.11 Rate of enzyme-catalyzed reaction as a function of substrate concentration. At low concentrations the reaction is first order; it approaches zero order at high concentrations, since the enzyme surface becomes saturated.

On the basis of this discussion we would expect that the rates of reactions catalyzed by adsorption would be first order at low reactant concentrations, gradually shifting to zero order at high concentrations. This is indeed the behavior exhibited by the decomposition of N_2O on gold and many other reactions occurring on solid surfaces. Many enzyme-catalyzed reactions show similar kinetic properties. In Figure 16.11 we show the rate of hydrolysis of urea as catalyzed by the enzyme urease.

$$H_2N{-}\underset{\underset{O}{\|}}{C}{-}NH_2(aq) + H_2O \xrightarrow{\text{urease}} 2\ NH_3(aq) + CO_2(g)$$

At low urea concentrations the rate is proportional to concentration as expected, while at high concentrations the enzyme becomes saturated and the rate becomes constant (zero order). The functional form of the curve in Figure 16.11 is the same as that in Equation 16.18, supporting the idea that the mechanisms for solid-catalyzed gas reactions and enzymatic reactions are very similar.

PROBLEMS

16.1 *Meaning of Reaction Rate* In a study of the reaction $CH_3CHO(g) \rightarrow CH_4(g) + CO(g)$, it is found that the concentration of CH_3CHO goes from 0.0266 mol/dm^3 to 0.0248 mol/dm^3 in 5.4 min. What is the average rate of the reaction over that period?

16.2 *Rate Expressions* The rate constant for the reaction $CO(g) + NO_2(g) \rightarrow CO_2(g) + NO(g)$ at 400°C is 0.50 dm^3/(mol · s), and the reaction is first order with respect to both CO and NO_2.

a. What is the overall order of the reaction?
b. What is the rate of the reaction when conc. CO = 0.025 mol/dm^3 and conc. NO_2 = 0.040 mol/dm^3?

16.3 *First Order Reaction* At 67°C the first order rate constant for the reaction: $2\ N_2O_5(g) \rightarrow 4\ NO_2(g) + O_2(g)$ is 0.35 min^{-1}. How long would it take for conc. N_2O_5 to go from 0.160 to 0.020 mol/dm^3 at 67°C?

16.4 *Dependence of Rate Upon Temperature* The rate constant for the reaction: $CO(g) + NO_2(g) \rightarrow CO_2(g) + NO(g)$ is 0.220 at 650 K and 23.0 at 800 K in dm^3/(mol · s). What is the activation energy in kilojoules?

16.5 *Reaction Mechanism* The observed rate law for a reaction between A_2 and B_2 is

$$\text{rate} = \text{constant} \times \text{conc. } A_2 \times (\text{conc. } B_2)^{1/2}$$

Show that the following mechanism is consistent with this rate law:

$$B_2 \rightleftharpoons 2B$$
$$A_2 + B \rightarrow \text{products; (slow)}$$

16.6 Express the rate of the reaction

$$C_2H_6(g) \rightarrow C_2H_4(g) + H_2(g)$$

a. in terms of Δconc. C_2H_6.
b. in terms of Δconc. C_2H_4.

16.7 One of the reactions involved in smog formation is known to be that between ozone and nitric oxide: $O_3(g) + NO(g) \rightarrow O_2(g) + NO_2(g)$. This reaction has been shown to be first order in both ozone and nitric oxide with a rate constant of 1.2×10^7 dm^3/(mol · s). Calculate the concentration of NO_2 formed per second in polluted air where the concentrations of ozone and nitric oxide are both 5×10^{-8} mol/dm^3. From the magnitude of your answer, would you expect the conversion of NO to NO_2 to occur rapidly or slowly?

16.8 For the reaction: A(g) → B(g), the following rate data were obtained with rates in mol/(dm^3 · s), conc. in mol/dm^3:

conc. A	0.0800	0.0600	0.0400	0.0200
rate	0.0040	0.0030	0.0020	0.0010

a. What is the order of the reaction?
b. What is the rate constant?
c. What would be the rate of reaction with conc. A = 0.240 mol/dm^3?

16.9 For a reaction A(g) → products, the rate is 0.025 mol/(dm^3 · min) when conc. A = 0.40 mol/dm^3. What will be the rate when conc. A = 1.00 mol/dm^3 if the reaction is

a. zero order in A?
b. first order in A?
c. second order in A?

16.10 For the decomposition of N_2O_5 at 45°C the following data were obtained:

t (s)	0	200	400	600	800
conc. (*M*)	1.64	1.45	1.28	1.13	1.00

a. By plotting log conc. vs. *t*, show that the reaction is first order.
b. From the graph, find the rate constant, *k*.
c. Using the rate law, calculate the rate at *t* = 200 s and compare to the average rate in the interval 0 to 400 s.

16.11 Find the order (0, 1, or 2) for the reaction X(g) → products, given:

time (min)	0	15	30	45
conc. X (*M*)	0.800	0.600	0.478	0.398

16.21 Express the rate of the reaction

$$2\ HI(g) \rightarrow H_2(g) + I_2(g)$$

a. in terms of Δconc. H_2.
b. in terms of Δconc. HI, if you want the same rate as in (a).

16.22 One of the major eye irritants in smog is formaldehyde, CH_2O, which may be formed by the reaction of ozone with ethylene: $O_3(g) + C_2H_4(g) \rightarrow 2\ CH_2O(g) + O(g)$, which is known to be second order with a rate constant of about 2×10^3 dm^3/(mol · s). The concentrations of O_3 and C_2H_4 in heavily polluted air are estimated to be about 5×10^{-8} and 1×10^{-8} mol/dm^3, respectively. What is the rate of production of formaldehyde in mol/(dm^3 · s)? How long will it take to build up a formaldehyde concentration of 1×10^{-8} mol/dm^3, the threshold above which eye irritation becomes noticeable? (Assume constant concentrations of O_3 and C_2H_4.)

16.23 For the reaction A(g) + B(g) → products, the following rate data were obtained [initial rates in mol/(dm^3 · s)]:

conc. A	conc. B	initial rate
0.500 *M*	0.400 *M*	6.00×10^{-3}
0.250 *M*	0.400 *M*	1.50×10^{-3}
0.250 *M*	0.800 *M*	3.00×10^{-3}

a. What is the order with respect to A? B?
b. What is the rate constant?

16.24 The rate of a certain reaction: B(g) → products is 0.0050 mol/(dm^3 · s) when conc. B = 0.200 mol/dm^3. What is the rate constant *k* if the reaction is

a. zero order in B?
b. first order in B?
c. second order in B?

16.25 For a certain first order reaction, the following data were obtained:

t (min)	0	20	40	60
conc. (*M*)	0.0600	0.0475	0.0378	0.0300

a. From a plot of conc. vs. time, estimate the concentration at 80 min.
b. From a plot of log conc. vs. time, estimate the concentration at 80 min.
c. Which result would you expect to be more accurate? Explain.

16.26 Find the order (0, 1, or 2) and the rate constant for the reaction D(g) → products from the following rate data:

time (s)	0	20	40	60
conc. D (*M*)	0.600	0.400	0.267	0.178

16.12 The first order rate constant for the decomposition of ethyl chloride, C_2H_5Cl, at 700 K is 2.50×10^{-3} min^{-1}. Starting with a concentration of 0.200 mol/dm^3, calculate

a. conc. C_2H_5Cl after one hour; one day.
b. the time required for conc. C_2H_5Cl to fall to one half its original value.

16.13 In the first order decomposition of cyclobutane at 750 K, it is found that 25% of a sample has decomposed in 80 s. How long would it take for 50% of the sample to decompose?

16.14 The activation energy of a certain enzyme-catalyzed reaction is 50.0 kJ. By what per cent is the rate of this reaction increased if you have a fever of 40.0°C, assuming a normal temperature of 37.0°C?

16.15 In the reaction A(g) + B(g) → products, how will f, p, and Z in Equation 16.14 be affected if

a. conc. A is doubled and conc. B is tripled?
b. the temperature rises from 127°C to 147°C, taking E_a = 60 kJ?

(Note that Z depends to some extent upon temperature, since it is proportional to average molecular speed.)

16.16 The activation energy for the reaction $CO(g) + NO_2(g) \rightarrow CO_2(g) + NO(g)$ is about 134 kJ. Show how this value can be obtained from Figure 16.5.

16.17 The activation energy for the reaction $H_2(g) + Cl_2(g) \rightarrow 2\ HCl(g)$ is about 155 kJ. Using data in Table 4.1, calculate the activation energy for the reverse reaction and prepare a diagram similar to Figure 16.6 for this reaction.

16.18 How do you explain the fact that the reaction $CO(g) + NO_2(g) \rightarrow CO_2(g) + NO(g)$

a. occurs slowly at room temperature, even though $\Delta G = -222$ kJ?
b. occurs rapidly at high temperatures?
c. occurs more slowly at low pressures than at high?

16.19 The reaction $CO(g) + Cl_2(g) \rightleftharpoons COCl_2(g)$ has the rate law: rate = constant × (conc. CO)(conc. $Cl_2)^{3/2}$. Show that this rate law is consistent with the mechanism

$$Cl_2 \rightleftharpoons 2\ Cl \text{ (fast)}$$
$$Cl + CO \rightleftharpoons COCl \text{ (fast)}$$
$$Cl_2 + COCl \rightarrow COCl_2 + Cl \text{ (slow)}$$

16.27 The half-life of nitromethane, CH_3NO_2, at 500 K is 650 s. For this first order reaction

a. find the rate constant.
b. determine the time required for conc. CH_3NO_2 to drop from 0.050 M to 0.0125 M.
c. the concentration of CH_3NO_2 one hour after the time elapsed in (b).

16.28 For the first order decomposition of acetone at 650 K it was found that the concentration was 0.0300 M after 200 min and 0.0200 M after 400 min.

a. What is the rate constant?
b. What was the initial concentration?

16.29 Raw milk will sour in about four hours at 28°C but will last up to 48 h in a refrigerator at 5°C. Assuming the rate to be inversely related to souring time, what is the activation energy for the reaction involved in the souring of milk?

16.30 Use Equation 16.15 to find the fraction of molecules in a gas at 300 K with energy equal to 40 kJ/mol or more. Find the fraction at 360 K. Calculate the ratio of the rates of reaction at 360 K and 300 K if E_a is 40 kJ, assuming that only f in Equation 16.14 varies significantly with temperature.

16.31 The rate constant for the decomposition of CH_3CHO is 0.0105 $dm^3/(mol \cdot s)$ at 700 K. If the activation energy is 188 kJ, find the rate constant at 800 K.

16.32 For the reaction $2\ O_3(g) \rightarrow 3\ O_2(g)$, the activation energy is about 117 kJ. Taking the heat of formation of $O_3(g)$ to be +142 kJ/mol, draw a diagram similar to that in Figure 16.6 for the decomposition of O_3.

16.33 Explain briefly but clearly why

a. all collisions between reactant molecules do not lead to reaction.
b. a reaction A(g) + B(g) → products is not necessarily second order overall.
c. the slow step in a mechanism determines the overall rate of a reaction.

16.34 For the reaction $H_2(g) + Br_2(g) \rightarrow 2\ HBr(g)$, the observed rate law under certain conditions is:

$$\text{rate} = \text{constant} \times \text{conc. } H_2 \times (\text{conc. } Br_2)^{1/2}$$

Show that this law is consistent with the mechanism suggested on p. 396.

16.20 At low temperatures, the rate law for the reaction $CO(g) + NO_2(g) \rightarrow CO_2(g) + NO(g)$ is: rate = constant × (conc. $NO_2)^2$. Which of the following mechanisms is consistent with this rate law?

a. $CO + NO_2 \rightarrow CO_2 + NO$
b. $2\ NO_2 \rightleftharpoons N_2O_4$ (fast)
   $N_2O_4 + 2\ CO \rightarrow 2\ CO_2 + 2\ NO$ (slow)
c. $2\ NO_2 \rightarrow NO_3 + NO$ (slow)
   $NO_3 + CO \rightarrow NO_2 + CO_2$ (fast)
d. $2\ NO_2 \rightarrow 2\ NO + O_2$ (slow)
   $2\ CO + O_2 \rightarrow 2\ CO_2$ (fast)

16.35 For the decomposition of ozone, the mechanism is believed to be

$$O_3 \rightleftharpoons O_2 + O \text{ (fast)}$$
$$O + O_3 \rightarrow 2\ O_2 \text{ (slow)}$$

Find the rate law associated with this mechanism, expressing it in terms of the concentrations of O_3 and O_2.

***16.36** For a certain enzyme-catalyzed reaction the rate is measured as a function of substrate concentration with the following results [rate in mol/(dm³ · min)]:

conc. S (*M*)	0.050	0.100	0.200	0.500	1.00	1.50	2.00
rate	0.026	0.046	0.075	0.120	0.150	0.164	0.171

From a plot similar to that in Figure 16.11, determine k and b in the equation

$$\text{rate} = \frac{k\text{b (conc. S)}}{1 + \text{b(conc. S)}}$$

***16.37** The ozone present in polluted air is formed by the two step mechanism:

$$NO_2(g) \xrightarrow{k_1} NO(g) + O(g) \qquad \text{(first order)}$$

$$O(g) + O_2(g) \xrightarrow{k_2} O_3(g) \qquad \text{(second order)}$$

It is known that $k_1 = 6.0 \times 10^{-3}\ s^{-1}$ and $k_2 = 1.0 \times 10^6\ dm^3/(mol \cdot s)$. The concentrations of NO_2 and O_2 in polluted air are about 3.0×10^{-9} and 1.0×10^{-2} mol/dm³, respectively. One can assume that the concentration of atomic oxygen reaches a "steady state," a low constant concentration at which point it is being consumed in the second reaction at the same rate as it is being produced in the first.

a. Calculate the steady state concentration of O(g) in polluted air.
b. Calculate the rate of formation of O_3 in polluted air.
c. If the rate of formation of O_3 remains constant, how long would it take for its concentration to build up to one part per million in air at 25°C and 1 atm? Under such conditions air contains about 0.041 mol/dm³.

***16.38** Suggest how one might distinguish experimentally between the two possible mechanisms suggested in the text for the H_2-I_2 reaction. Note that both mechanisms lead to the same rate expression. The second mechanism is discussed in an article by J. H. Sullivan in the *Journal of Chemical Physics,* **46**:73, 1967.

***16.39** In a first order reaction, let us suppose that a quantity, X, of reactant is added at regular intervals of time, Δt. At first the amount of reactant in the system builds up; eventually, however, it levels off at a "saturation value" given by the expression:

$$\text{saturation value} = \frac{X}{1 - 10^{-a}}, \text{ where } a = 0.30\frac{\Delta t}{t_{1/2}}$$

This analysis applies to the intake of mercury into the body, where one takes in a certain amount each day. The half-life for elimination of mercury appears to be about 70 d.

Suppose that a person eats 50 g of fish containing 0.50 g Hg/10⁶ g fish each day. Using the above equation, calculate the number of grams of Hg in his body at "saturation," and compare to the value at which symptoms of mercury poisoning appear, about 0.014 g. How many grams of fish would he need to eat each day to reach the toxic limit?

17

THE ATMOSPHERE

Life on this planet depends upon the presence of the relatively thin layer of air that surrounds it. Even though the atmosphere accounts for only about 0.0001% of the total mass of the earth, it is the reservoir from which we draw oxygen essential to metabolism, carbon dioxide for photosynthesis, and nitrogen, whose compounds are essential to plant growth. Our climate is governed by the movement of water vapor from the earth's surface into the atmosphere and back again.

Even trace constituents of the atmosphere can have beneficial or detrimental effects on the delicate balance of life. Small amounts of ozone at an elevation of about 30 km absorb most of the harmful ultraviolet radiation of the sun. On the other side of the ledger, as little as 0.2 ppm of ozone near the earth's surface promotes the photochemical reactions responsible for smog formation.

In this chapter we will consider some of the physical and chemical properties of the components of the atmosphere. With that background, we will discuss the general problem of air pollution, how it comes about, and how it can be controlled. Throughout the chapter, we will have frequent occasion to apply the principles of chemical thermodynamics (Chapter 14), gaseous equilibrium (Chapter 15), and chemical kinetics (Chapter 16).

17.1 THE COMPOSITION OF THE ATMOSPHERE

Air has 5 principal components

In Table 17.1 are given the mole fractions of the various components of the atmosphere. Two species are omitted: water vapor, whose mole fraction may vary from 0.02 in the tropics to 0.0005 in polar regions, and suspended particles (e.g., dust, smoke), which vary greatly both in concentration and in chemical composition.

TABLE 17.1 COMPOSITION OF CLEAN, DRY AIR AT SEA LEVEL

COMPONENT	MOLE FRACTION	COMPONENT	MOLE FRACTION	COMPONENT	MOLE FRACTION
N_2	0.780 8	Ne	1.82×10^{-5}	SO_2	$<1 \times 10^{-6}$
O_2	0.209 5	He	5.24×10^{-6}	O_3	$<1 \times 10^{-7}$
Ar	0.009 34	CH_4	2×10^{-6}	NO_2	$<2 \times 10^{-8}$
CO_2	0.003 14	Kr	1.14×10^{-6}	I_2	$<1 \times 10^{-8}$
		H_2	5×10^{-7}	NH_3	$<1 \times 10^{-8}$
		N_2O	5×10^{-7}	CO	$<1 \times 10^{-8}$
		Xe	8.7×10^{-8}	NO	$<1 \times 10^{-8}$

TABLE 17.2 CONCENTRATION UNITS FOR GASES AS RELATED TO MOLE FRACTION (X)

Volume % = mole % = 100 X	Parts per million (ppm) = $10^6\ X$
Mass % = volume % × $\dfrac{MM}{\text{aver. } MM^*}$	Parts per billion (ppb) = $10^9\ X$
Partial pressure = X (total pressure)	
mol/dm^3 = $\dfrac{\text{partial pressure (kPa)}}{(8.31)\ T}$	

*The average molecular mass of dry air is 29.0.

The composition of air, indeed that of any gaseous mixture, can be expressed in a number of ways. Several of the more common concentration units used for solutions of gases are related to mole fraction in Table 17.2.

Note that, for any gaseous solution:

—Volume % is identical with mole %.

—Mass % can be taken to be the mass of the component per hundred grams of gaseous solution; alternatively, it can be calculated from volume % as indicated in Table 17.2.

—The relations given in Table 17.2 for the partial pressure of a gaseous component and for its concentration in moles per cubic decimetre are based on the use of the Ideal Gas Law (Chapter 5).

$\frac{n}{V} = \frac{P}{RT}$

—Parts per million or parts per billion, commonly used to express the concentrations of minor components, are based on *mole fractions* for *gaseous* solutions. For example, when we say that a species in air is present at a concentration of "one part per million," we mean that 1 mol of that species would be found in 10^6 mol of air (or one molecule per million molecules of air). In contrast, ppm or ppb in *liquid* solution are based on *mass fractions*. When we say that a water supply contains one part per million of lead, we mean that there is one gram of lead in 10^6 g of water.

Example 17.1 Using the information in Tables 17.1 and 17.2, calculate

a. the volume %, mass %, partial pressure at 100.0 kPa total pressure, and concentration in moles per cubic decimetre at 25°C of N_2 in clean, dry air.

b. the concentration of krypton in parts per million and parts per billion.

Solution

a. Volume % = 100(0.7808) = 78.08

$$P_{N_2} = 0.7808 \times 100.0 \text{ kPa} = 78.08 \text{ kPa}$$

$$\text{Mass \%} = 78.08 \times \frac{28.01}{29.0} = 75.4$$

$$M = \frac{78.08}{(8.31)(298)} = 0.0315$$

(Note that O_2, MM = 32, and Ar, AM = 40, raise the average molecular mass of air above that of N_2.)

b. $10^6(1.14 \times 10^{-6}) = 1.14$
$10^9(1.14 \times 10^{-6}) = 1140$

The two major components of air, nitrogen and oxygen, are obtained by the fractional distillation (Chapter 1) of liquid air. One method of liquefying air is shown in Figure 17.1. Notice that cooling is accomplished by two quite different methods. In part, compressed air is cooled by passing it through a heat exchanger around which a cold fluid (e.g., water, ammonia, or liquid air, depending on the temperature desired) is circulated. Further cooling is achieved by allowing the compressed air to expand suddenly, typically from about 200 atm to 20 atm. When the gas molecules move farther apart, heat must be absorbed to overcome the attractive forces between these molecules. The absorption of heat lowers the kinetic energy and hence the temperature of the air. The cold air is recirculated several times until it finally reaches a low enough temperature to start condensing.

When liquid air is allowed to warm up slowly, nitrogen (bp = 77 K) evaporates first, leaving a residue which is mostly liquid oxygen (bp = 90 K). Of the noble gases, helium (bp = 4 K) and neon (bp = 27 K) pass off in the nitrogen fraction. Most of the argon (bp = 87 K) and all the heavier noble gases are combined with the liquid oxygen, from which they can be removed and purified by repeated fractionation. Neon is extracted from the nitrogen fraction in a similar way. Helium is obtained from natural gas, where its concentration is much higher than in air.

17.2 NITROGEN

We will begin our discussion of the properties of the principal components of the atmosphere by considering the most abundant species, elementary nitrogen.

Chemical Properties

At room temperature and atmospheric pressure, elementary nitrogen fails to

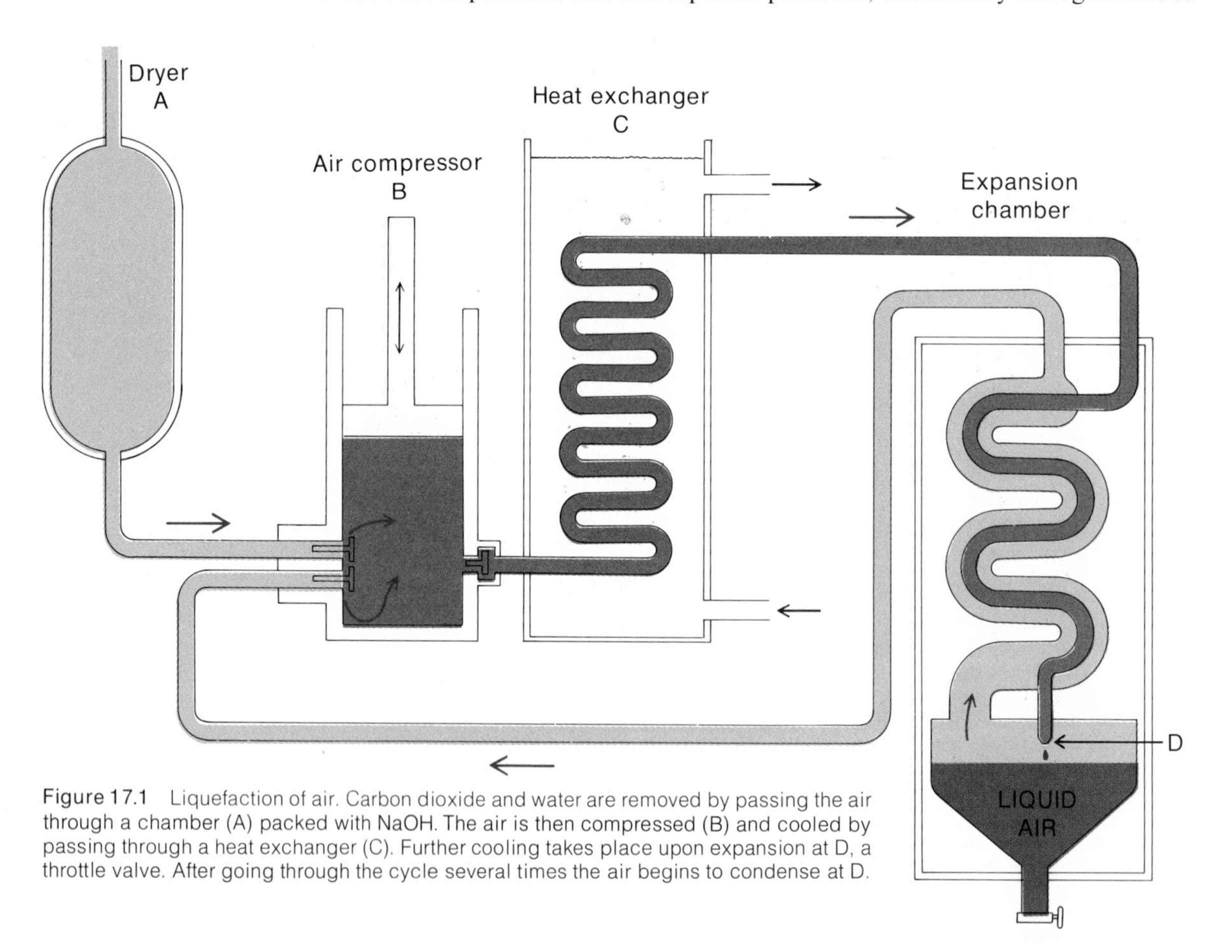

Figure 17.1 Liquefaction of air. Carbon dioxide and water are removed by passing the air through a chamber (A) packed with NaOH. The air is then compressed (B) and cooled by passing through a heat exchanger (C). Further cooling takes place upon expansion at D, a throttle valve. After going through the cycle several times the air begins to condense at D.

react with any other element. Its inertness can be attributed to the strength of the triple bond holding the N_2 molecule together

$$:N \equiv N:(g) \rightarrow 2\ :\dot{N}\cdot(g); \quad \Delta H = +941 \text{ kJ}$$

The high stability of the bond makes it difficult to form binary compounds of nitrogen from the standpoint of reaction rate (high activation energy) and/or reaction spontaneity (unfavorable ΔH_f, ΔG_f).

At high temperatures, nitrogen reacts with a few very reactive metals such as lithium and magnesium.

$$6\ Li(s) + N_2(g) \rightarrow 2\ Li_3N(s) \qquad (17.1)$$

$$3\ Mg(s) + N_2(g) \rightarrow Mg_3N_2(s) \qquad (17.2)$$

These compounds contain the nitride ion, $(:\ddot{N}:)^{3-}$. Nitrogen also reacts with boron and aluminum at high temperatures to form compounds with empirical formulas BN and AlN, but here the bonding is predominantly covalent rather than ionic. Boron nitride and aluminum nitride have macromolecular structures similar to those of graphite and diamond.

Nitrogen reacts directly with only two nonmetals, oxygen and hydrogen.

$$N_2(g) + O_2(g) \rightarrow 2\ NO(g) \qquad (17.3)$$

$$N_2(g) + 3\ H_2(g) \rightarrow 2\ NH_3(g) \qquad (17.4)$$

The reaction with oxygen is spontaneous only at temperatures above 2000°C; the product, nitric oxide, should decompose to the elements upon cooling ($\Delta G_f^{1\ atm}$ NO = +86.7 kJ/mol at 25°C). Ammonia, NH_3, is considerably more stable than nitric oxide ($\Delta G_f^{1\ atm}$ NH_3 = −16.6 kJ/mol at 25°C). However, as we shall see, it can be formed at a reasonable rate only at relatively high temperatures in the presence of special catalysts.

The uses of elementary nitrogen reflect its chemical inertness at ordinary temperatures. Liquid nitrogen is used as a coolant in preference to liquid air because the gas given off on evaporation does not support combustion. Certain foods, including instant coffee, peanuts, and potato chips, are packed under nitrogen to prevent oxidation and loss of flavor. Nitrogen gas under pressure is used to force flammable liquid fuels and their oxidizers into rocket motors.

Nitrogen Fixation

Combined nitrogen in the form of protein is essential to both plant and animal life. The key step in the nitrogen cycle in nature is the fixation of elementary nitrogen, i.e., its conversion to compounds such as ammonia or nitric acid. Small amounts of elementary nitrogen are converted to nitric oxide by lightning discharges; subsequent reaction with oxygen and water forms nitric acid, HNO_3.

A more important natural process of nitrogen fixation involves the action of certain bacteria found in the roots of plants such as peas, beans, clover, and alfalfa. These bacteria contain enzymes that catalyze the conversion of elementary nitrogen to ammonia. Although the mechanism of this process is not thoroughly understood, it appears that two different enzymes are involved. Both contain iron atoms; one of the enzymes contains molybdenum as well. It is generally believed that molecular nitrogen forms a weak complex with the metal atoms of the enzymes; this is then converted chemically to a more stable ammonia complex. Recently, this process has been simulated in the laboratory, using certain compounds of another transition metal, titanium.

Faced with an expanding population and inadequate food supplies, man has attempted for centuries to speed up the nitrogen cycle in nature by adding nitrogen compounds directly to the soil. Prior to this century the only way to do this was to add "organic nitrogen" (i.e., manure). In 1908 Fritz Haber in Germany argued that it would be feasible to fix atmospheric nitrogen by reacting it with hydrogen to form ammonia.

$$N_2(g) + 3\ H_2(g) \rightleftharpoons 2\ NH_3(g);\ \Delta H = -92.4\ \text{kJ} \qquad (17.4)$$

His research was supported by German industrialists who were interested in using the ammonia to produce nitric acid, the starting material for such explosives as nitroglycerine and trinitrotoluene, TNT. In 1913 the first large-scale ammonia plant went into production. Throughout World War I the Haber process produced sufficient quantities of ammonia to make Germany independent of foreign supplies of nitrates, which were cut off by the British blockade.

The Haber process is now the main source of fixed nitrogen in the world. Its feasibility depends upon choosing experimental conditions under which nitrogen will react rapidly with hydrogen to give a reasonably high yield of ammonia. At room temperature and atmospheric pressure, the position of the equilibrium strongly favors the formation of ammonia ($K_c = 5 \times 10^8$) but the rate of reaction is virtually zero. Equilibrium can be reached more rapidly by increasing the temperature. However, since Reaction 17.4 is exothermic ($\Delta H < 0$), high temperatures reduce the equilibrium constant and hence the yield of ammonia (Table 17.3). High pressures, on the other hand, have a favorable influence on both the rate of the reaction and the position of the equilibrium. An increase in pressure makes it possible for the system to attain equilibrium more rapidly because it increases the concentrations of the gases involved. It also increases the relative amount of ammonia present at equilibrium, since the formation of NH_3 results in a decrease in the number of moles of gas (4 mol → 2 mol).

Much of Haber's research was devoted to finding a catalyst that would enable Reaction 17.4 to be carried out at a reasonable rate without going to excessively high temperatures. In a modern ammonia synthesis plant the catalyst used is a specially prepared mixture of iron, potassium oxide, and aluminum oxide. The synthesis is carried out at a temperature of 400 to 450°C and pressures of 200 to 600 atm. Ammonia ($bp = -33$°C) is condensed out as a liquid from the gaseous mixture; unreacted hydrogen and nitrogen are recycled to raise the yield of ammonia.

The annual production of ammonia by the Haber process in the United States and Canada is about 16×10^9 kg. Most of it is used as a fertilizer, either directly, in

Note the large effect of pressure on the yield of NH_3

TABLE 17.3 EFFECT OF TEMPERATURE AND PRESSURE ON THE YIELD OF AMMONIA IN THE HABER PROCESS ($[H_2] = 3\ [N_2]$)

		MOLE % NH_3 IN EQUILIBRIUM MIXTURE				
°C	K_c	10 atm	50 atm	100 atm	300 atm	1000 atm
200	650	51	74	82	90	98
300	9.5	15	39	52	71	93
400	0.5	4	15	25	47	80
500	0.08	1	6	11	26	57
600	0.014	0.5	2	5	14	13

the form of ammonium salts such as $(NH_4)_2SO_4$ or $(NH_4)_3PO_4$, or after conversion to urea, $H_2N—CO—NH_2$, by the reaction

$$2\ NH_3(g) + CO_2(g) \rightarrow H_2N—CO—NH_2(s) + H_2O(l) \qquad (17.5)$$

A relatively small portion of the ammonia is used to make nitric acid by the Ostwald process, which involves the following series of reactions:

$$4\ NH_3(g) + 5\ O_2(g) \rightarrow 4\ NO(g) + 6\ H_2O(g) \quad (1000°C,\ \text{Pt catalyst}) \quad (17.6)$$

$$2\ NO(g) + O_2(g) \rightarrow 2\ NO_2(g) \quad (\text{NO allowed to cool in air}) \quad (17.7)$$

$$3\ NO_2(g) + H_2O(l) \rightarrow 2\ HNO_3(l) + NO(g) \quad (NO_2\ \text{passed into water}) \quad (17.8)$$

The hydrogen used as a reactant in the Haber process accounts for virtually the entire cost of the ammonia formed. You may be surprised to learn that most of the hydrogen produced in the United States and Canada comes, not from the electrolysis of water (which is too expensive), but from the reaction of steam at about 1000°C with hydrocarbons, the most important of which is methane, the principal constituent of natural gas:

Some H_2 is obtained as a by-product in the cracking of petroleum to give gasoline

$$CH_4(g) + H_2O(g) \rightarrow CO(g) + 3\ H_2(g) \qquad (17.9)$$

This reaction is nonspontaneous at room temperature ($\Delta G^{1\ atm} = +142.1$ kJ at 25°C) but becomes spontaneous at high temperatures ($\Delta H = +206.1$ kJ, $\Delta S^{1\ atm} = +215$ J/K).

Until quite recently, the price of the hydrogen produced by Reaction 17.9 had been stable for several decades, at about 2¢ a kilogram. The sharp rise in the price of natural gas changed this situation, leading directly to an increase in the cost of the ammonia produced by the Haber process. This in turn was largely responsible for a big jump in fertilizer prices. The effect has been particularly severe for many of the underdeveloped countries, which must have an abundant supply of cheap fertilizer if they are to avert mass starvation. Two approaches are possible: one is to find a cheaper way to make hydrogen, the other is to find a way to fix nitrogen which does not involve hydrogen as a reactant.

17.3 OXYGEN

Elementary oxygen is much more reactive than nitrogen, reflecting both its higher electronegativity (3.0 vs. 2.5 for nitrogen) and the fact that the bond in the O_2 molecule is considerably weaker than that in N_2. Among the elements, only the noble gases, the halogens, and a few inactive metals (Pd, Au, Pt) fail to react directly with oxygen.

More than half of the 18×10^9 kg of oxygen produced annually in the United States and Canada is used in the manufacture of iron and steel. In the *basic oxygen process* for making steel, pure oxygen is blown over the surface of molten iron to lower the percentage of carbon by burning it to CO_2. At the same time, impurities such as silicon and phosphorus are converted to oxides and removed. The whole process takes only about 1 to 2 h as compared to 8 to 10 h for the older, open hearth process which it is rapidly displacing.

TABLE 17.4 OXIDES OF THE METALS OF THE FIRST TRANSITION SERIES

CATION	Sc	Ti	V	Cr	Mn	Fe	Co	Ni	Cu	Zn
		TiO_2	V_2O_5	CrO_3	Mn_2O_7					
			VO_2		MnO_2					
M^{3+}	Sc_2O_3	Ti_2O_3	V_2O_3	Cr_2O_3	Mn_2O_3	Fe_2O_3	Co_2O_3			
2 M^{3+}, M^{2+}					Mn_3O_4	Fe_3O_4	Co_3O_4			
M^{2+}		TiO	VO	CrO	MnO	FeO	CoO	NiO	CuO	ZnO
M^{+}									Cu_2O	

Reaction with Metals

IONIC OXIDES. When oxygen reacts with a metal, the product is most often an ionic solid containing the oxide ion, O^{2-}. The reactions with lithium and calcium are typical:

$$4\ Li(s) + O_2(g) \rightarrow 2\ Li_2O(s) \tag{17.10}$$

$$2\ Ca(s) + O_2(g) \rightarrow 2\ CaO(s) \tag{17.11}$$

Many transition metals form more than one oxide (Table 17.4). Most frequently, the metal cation carries a charge of +3 (Cr_2O_3, Fe_2O_3) or +2 (NiO, ZnO). Manganese, iron, and cobalt form oxides of formula M_3O_4, in which four O^{2-} ions with a total charge of −8 are balanced by one +2 and two +3 ions.

When a metal forms more than one ionic oxide, it is ordinarily the higher oxide (i.e., the one containing the cation of highest charge) that is formed at low temperatures. For example, powdered copper, exposed to air at room temperature, is very slowly converted to black copper(II) oxide

$$2\ Cu(s) + O_2(g) \rightarrow 2\ CuO(s) \tag{17.12}$$

This oxide can be converted to copper(I) oxide, Cu_2O, by heating to about 900°C

$$2\ CuO(s) \rightarrow Cu_2O(s) + \tfrac{1}{2}\ O_2(g) \tag{17.13}$$

This behavior is readily understood in terms of the thermodynamic principles introduced in Chapter 14. For Reaction 17.13, $\Delta G^{1\,atm}$ at 298 K = +108.0 kJ, so CuO does not convert to Cu_2O under these conditions. However, since $\Delta S^{1\,atm}$ for this reaction is also positive (+120 J/K), $\Delta G^{1\,atm}$ becomes smaller as the temperature is increased. At about 900°C (cf. Problem 17.9), $\Delta G^{1\,atm}$ goes to zero; above this temperature, Cu_2O is the stable oxide.

Many ionic oxides, notably those of the 1A and 2A metals, react with water, to form the corresponding metal hydroxide in water solution:

Many metal oxides (e.g., Al_2O_3, Fe_2O_3) are too insoluble to react

$$Li_2O(s) + H_2O \rightarrow 2\ Li^+(aq) + 2\ OH^-(aq) \tag{17.14}$$

$$CaO(s) + H_2O \rightarrow Ca^{2+}(aq) + 2\ OH^-(aq) \tag{17.15}$$

In each case, the reaction is that between an O^{2-} ion and an H_2O molecule to give two OH^- ions. In electron-dot notation:

$$(:\ddot{\underset{\cdot\cdot}{O}}:)^{2-} + H{-}\ddot{\underset{\cdot\cdot}{O}}{-}H \rightarrow 2\ (:\ddot{\underset{\cdot\cdot}{O}}{-}H)^-$$

Any water solution containing a high concentration of hydroxide ion has basic properties (Chapter 19). For this reason, compounds which react with water to form OH^- ions are often referred to as *basic anhydrides*.

COVALENT OXIDES. A few transition metals form oxides in which the bonding is predominantly covalent rather than ionic. Titanium dioxide, TiO_2, and manganese dioxide, MnO_2, have macromolecular structures analogous to SiO_2. The highest oxide of manganese, Mn_2O_7, is a reddish liquid that freezes below −20°C, indicating a molecular structure. None of these compounds react with water to form hydroxide ions.

PEROXIDES AND SUPEROXIDES. Although the most common anion formed by oxygen is O^{2-}, the element can, under certain conditions, form diatomic anions with the Lewis structures indicated below:

$$(:\ddot{\underset{..}{O}}-\ddot{\underset{..}{O}}:)^{2-} \qquad\qquad (:\ddot{\underset{..}{O}}-\ddot{\underset{.}{O}}:)^{-}$$

peroxide — superoxide

The peroxide ion is formed directly when sodium or barium reacts with oxygen at room temperature:

$$2\ Na(s) + O_2(g) \rightarrow Na_2O_2(s)\ \text{(sodium peroxide)} \qquad (17.16)$$

$$Ba(s) + O_2(g) \rightarrow BaO_2(s)\ \text{(barium peroxide)} \qquad (17.17)$$

(Note that in Na_2O_2, two Na^+ ions are required to balance one O_2^{2-} ion; in BaO_2, one Ba^{2+} balances one O_2^{2-}). Under the same conditions, potassium gives the superoxide (one K^+ ion per O_2^- ion):

$$K(s) + O_2(g) \rightarrow KO_2(s)\ \text{(potassium superoxide)} \qquad (17.18)$$

Metal peroxides react vigorously with water to form a solution containing hydrogen peroxide, H_2O_2 (CAUTION: Concentrated aqueous solutions of H_2O_2 are dangerous to work with), and the corresponding metal hydroxide:

Metal peroxides and superoxides are also dangerous to work with

$$Na_2O_2(s) + 2\ H_2O \rightarrow 2\ Na^+(aq) + 2\ OH^-(aq) + H_2O_2(aq) \qquad (17.19)$$

The reaction of superoxides with water is similar except that an additional product, molecular oxygen, is formed:

$$2\ KO_2(s) + 2\ H_2O \rightarrow 2\ K^+(aq) + 2\ OH^-(aq) + H_2O_2(aq) + O_2(g) \qquad (17.20)$$

Example 17.2 Give the formulas of the oxide, peroxide, and superoxide of cesium, and write balanced equations for their reactions with water.

Solution Since cesium is in Group 1A, its cation must have a +1 charge. The formulas are as follows:

oxide: Cs_2O (two Cs^+, one O^{2-})

peroxide: Cs_2O_2 (two Cs^+, one O_2^{2-})

superoxide: CsO_2 (one Cs^+, one O_2^-)

The reactions with water are entirely analogous to 17.14, 17.19, and 17.20:

$$Cs_2O(s) + H_2O \rightarrow 2\ Cs^+(aq) + 2\ OH^-(aq)$$

$$Cs_2O_2(s) + 2\ H_2O \rightarrow 2\ Cs^+(aq) + 2\ OH^-(aq) + H_2O_2(aq)$$

$$2\ CsO_2(s) + 2\ H_2O \rightarrow 2\ Cs^+(aq) + 2\ OH^-(aq) + H_2O_2(aq) + O_2(g)$$

Reaction with Nonmetals

In nonmetal oxides, the bonding is predominantly covalent rather than ionic. Many of the most familiar nonmetal oxides, including CO, CO_2, SO_2, and SO_3, are molecular in nature. A few, notably those of silicon (SiO_2) and boron (B_2O_3) are macromolecular.

Most nonmetals (e.g. C, S) can form more than one oxide. In order to obtain a pure product, it is necessary to choose conditions which favor the formation of one oxide at the expense of the other. Normally, the higher oxide forms when the nonmetal burns at ordinary temperatures in an excess of oxygen. For example, phosphorus and carbon form P_4O_{10} and CO_2 when ignited in air.

$$P_4(s) + 5\ O_2(g) \rightarrow P_4O_{10}(s) \quad \text{low } T\text{, excess } O_2 \qquad (17.21)$$

$$C(s) + O_2(g) \rightarrow CO_2(g) \quad \text{low } T\text{, excess } O_2 \qquad (17.22)$$

Formation of the lower oxide (P_4O_6, CO) is favored by high temperatures or a limited supply of oxygen; carbon monoxide is formed under precisely these conditions when a fuel-rich mixture burns in an automobile engine.

The reaction of sulfur with oxygen is a particularly important one, since it is the first step in the preparation of sulfuric acid. When sulfur is heated in air, it catches fire, forming a choking gas made up principally of sulfur dioxide.

$$S(l) + O_2(g) \rightarrow SO_2(g) \qquad (17.23)$$

Up to 2% of sulfur trioxide is formed simultaneously. To prepare this compound in good yield, sulfur dioxide is reacted further with oxygen.

$$SO_2(g) + \tfrac{1}{2}\ O_2(g) \rightleftharpoons SO_3(g);\ \Delta H = -99.1\ \text{kJ} \qquad (17.24)$$

The rate of this reaction and the equilibrium yield of sulfur trioxide depend upon temperature and pressure. The principles involved here are the same as those discussed previously in connection with the Haber process. At low temperatures, the equilibrium constant for the formation of sulfur trioxide is large (Table 17.5), but equilibrium is reached very slowly. As the temperature is raised, the rate increases, but since the reaction is exothermic, the yield of sulfur trioxide drops off. High pressures would tend to increase both the yield of sulfur trioxide and the rate of reaction.

TABLE 17.5 EFFECT OF TEMPERATURE ON THE EQUILIBRIUM $SO_2(g) + \tfrac{1}{2}\ O_2(g) \rightleftharpoons SO_3(g)$

t(°C)	400	500	600	700	800
K_c	2300	400	70	20	7
Mol % SO_3*	96	88	66	40	20

*Under the condition that $[O_2] = \tfrac{1}{2}[SO_2]$, and $P_{tot} = 1$ atm.

In the so-called *contact process* for the manufacture of sulfuric acid, the equilibrium represented by Equation 17.24 is reached rapidly by passing sulfur dioxide and oxygen, at atmospheric pressure, over a solid catalyst at a temperature of about 650°C. The equilibrium mixture is then recycled at a lower temperature, 400 to 500°C, to increase the yield of sulfur trioxide. The two catalysts which have proved most effective are vanadium pentoxide, V_2O_5, and finely divided platinum. The sulfur trioxide produced can be converted to sulfuric acid by reaction with water.

$$SO_3(g) + H_2O \rightarrow H_2SO_4(aq) \qquad (17.25)$$

This reaction is ordinarily carried out in sulfuric acid solution; if sulfur trioxide is added directly to water, sulfuric acid is formed as a fog of tiny particles which are difficult to recover. In 1975, about 100×10^9 kg of H_2SO_4 were manufactured in the world, more than for any other chemical.

Reaction 17.25 is typical of molecular nonmetal oxides, many of which react with water to form acids. Compare the reactions

$$CO_2(g) + H_2O \rightarrow H_2CO_3(aq) \qquad (17.26)$$

$$P_4O_{10}(s) + 6\ H_2O \rightarrow 4\ H_3PO_4(aq) \qquad (17.27)$$

These nonmetal oxides and many others, like SO_3, are referred to as *acid anhydrides*.

17.4 THE NOBLE GASES

All the noble gases (Periodic Group 8A) except radon are found in the atmosphere. Argon is by far the most abundant, accounting for more than 99.7% of the total noble gas content of air. Small amounts of helium and radon are found in association with radioactive minerals (Chapter 24). Natural gas, from which helium is extracted commercially, sometimes contains as much as six per cent by volume of that gas.

Helium is also used to provide an inert atmosphere for welding Al and Mg

Helium, because of its low density (about 1/7 that of air) and chemical inertness, is used in dirigibles and in synthetic atmospheres to make breathing easier for patients suffering from certain types of lung disorders. Smaller quantities are used as a carrier gas in gas chromatography and as a liquid coolant to achieve very low temperatures ($bp = -269°C$). Argon and, more recently, krypton are used to provide an inert atmosphere in light bulbs and thereby retard decomposition of the tungsten filament. In illuminated "neon signs" a high voltage is passed through a glass tube containing neon at a very low pressure. The red glow emitted under these conditions corresponds to an intense line in the neon spectrum at 640 nm.

The first person to isolate one of the noble gases was the English scientist Sir Henry Cavendish. He subjected a sample of atmospheric nitrogen to repeated electrical discharges in the presence of oxygen, thereby forming oxides of nitrogen which were dissolved out in water. Cavendish reported that a bubble of gas, about 1% of the original volume, remained undissolved. Nothing came of his observation for over a hundred years. In 1892 Lord Rayleigh, Professor of Physics at the Cavendish Laboratory in Cambridge, found that nitrogen prepared by removing oxygen from air had a density about 0.5% greater than that of nitrogen prepared synthetically by the decomposition of ammonium nitrite, NH_4NO_2. A Scotsman, Sir William Ramsey, separated the heavier impurity from atmospheric nitrogen;

TABLE 17.6 PROPERTIES OF BINARY COMPOUNDS OF THE NOBLE GASES

	ΔH_f	*$\Delta G_f^{1\,atm}$ at 298 K
XeF_2	−109 kJ	−75 kJ
XeF_4	−218	−136
XeF_6	−293	−172
XeO_3	+418	——
KrF_2	+59	——

*Calculated by extrapolation from higher temperatures.

from a study of its emission spectrum, he identified it as a new element which he called argon, from the Greek word for "lazy." Over the next five years, Rayleigh and Ramsey, working together, isolated the remaining members of the noble gas family. Helium had previously been detected in the solar spectrum in 1868, but was unknown on earth.

Up until about ten years ago the noble gases were more frequently referred to as "inert gases," since they were believed to be completely unreactive toward other elements and compounds. In 1962 Neil Bartlett, a 29 year old chemist doing research on platinum-fluorine compounds at the University of British Columbia, isolated a reddish solid which he showed to be $O_2^+(PtF_6)^-$. Realizing that the ionization energy of xenon (1130 kJ/mol) is virtually identical with that of molecular oxygen (1110 kJ/mol), he prepared by an analogous method a substance which he showed to be $Xe^+(PtF_6)^-$. The success of Bartlett's experiments opened up a new era in noble gas chemistry. Within a few months a group at the Argonne National Laboratories near Chicago prepared the first stable binary compound of a noble gas, XeF_4, by direct reaction of the elements under pressure at 400°C. Xenon tetrafluoride is a white molecular solid with a melting point of 140°C.

Since 1962 several different compounds of xenon with fluorine and oxygen have been reported. The binary compounds of xenon whose existence is well established are listed in Table 17.6. The three fluorides of xenon are thermodynamically stable, as indicated by their negative free energies of formation. On the other hand, the oxide, XeO_3, is unstable and decomposes violently on the slightest provocation. The lighter noble gases are much less reactive than xenon: only one binary compound of krypton, KrF_2, has been isolated. There have been attempts in many laboratories to prepare fluorides or oxides of argon, but all results so far have been negative.

Compounds containing Xe-N bonds have recently been prepared

17.5 CARBON DIOXIDE

The percentage of carbon dioxide in the air is too small for it to be extracted profitably. Instead, it is obtained commercially by the combustion of coke or other carbon-containing fuels,

$$C(s) + O_2(g) \rightarrow CO_2(g) \qquad (17.28)$$

by heating limestone or other carbonate minerals,

$$CaCO_3(s) \rightarrow CaO(s) + CO_2(g) \qquad (17.29)$$

or as a by-product of fermentation processes, e.g.,

$$C_6H_{12}O_6(aq) \rightarrow 2\ C_2H_5OH(aq) + 2\ CO_2(g) \qquad (17.30)$$

We are familiar with the solid form of carbon dioxide (Dry Ice) and its use as a refrigerant. A common type of fire extinguisher contains liquid CO_2 at its equilibrium vapor pressure, about 59 atm at 20°C. When the cylinder valve is opened, the liquid boils. The amount of heat absorbed in this process is great enough to lower the temperature of the escaping gas to the point (−78°C) where it forms a finely divided "snow" of solid carbon dioxide. Since pure CO_2 is a nonconductor of electricity, extinguishers of this type are particularly useful in dealing with electrical fires. They cannot be used to put out fires of burning magnesium or aluminum, as Londoners found out to their dismay during the fire-bombings in World War II, since these metals react exothermically with the carbon dioxide.

$$2\ \mathrm{Mg}(s) + \mathrm{CO_2}(g) \rightarrow 2\ \mathrm{MgO}(s) + \mathrm{C}(s);\ \Delta H = -810.1\ \mathrm{kJ} \qquad (17.31)$$

The average concentration of carbon dioxide in the atmosphere has increased by about 10% in the past century. Increased consumption of fossil fuels is primarily responsible for this trend. Extensive land clearing, which reduces the amount of CO_2 consumed in photosynthesis, is also a factor. It is estimated that by the year 2050, unless steps are taken to restore the CO_2 balance, its concentration may be twice what it is today.

From time to time, fears are expressed that increasing concentrations of CO_2 in the atmosphere may eventually produce significant changes in the world's climate. Carbon dioxide molecules absorb infrared radiation given off by the earth, thereby acting as an insulating blanket to prevent heat from escaping into outer space. From 1850 to 1950, worldwide temperatures rose by an average of about 0.5°C. To what extent this effect is correlated with an increase in CO_2 concentration is hard to say. Many other factors could be involved. Interestingly enough, temperatures appear to have dropped slightly since 1950, despite a rising carbon dioxide level. This cooling trend may be due to increased amounts of suspended particles, formed by fuel combustion and industrial processes. These particles reflect solar radiation back into space, thus offsetting the "greenhouse effect" of CO_2.

Direct heating of air from combustion of fuels may be a factor

17.6 WATER VAPOR

The concentration of water vapor in the air is often expressed in terms of the *relative humidity*, which is defined as

$$R.H. = \frac{P_{H_2O}}{P^0_{H_2O}} \times 100\% \qquad (17.32)$$

where P_{H_2O} is the partial pressure of water vapor in the air and $P^0_{H_2O}$ is the equilibrium vapor pressure of water at the same temperature. On a day when the temperature is 25°C ($P^0_{H_2O}$ = 3.17 kPa) and the partial pressure of water vapor is 2.60 kPa, the relative humidity is

$$\frac{2.60}{3.17} \times 100\% = 82.0\%$$

Since the equilibrium vapor pressure of water increases with rising temperature, Equation 17.32 implies that the relative humidity will drop when cold air is warmed up (Example 17.3).

In Minnesota, humidifiers are commonly used in houses in winter

Example 17.3 On a cold winter day, when the outside temperature is 0°C, the partial pressure of water in the air is 0.40 kPa. Calculate the relative humidity of the outside air and the relative humidity of the same air when it is brought into a house and warmed to 20°C.

Solution The equilibrium vapor pressure of water is 0.61 kPa at 0°C and 2.34 kPa at 20°C (Appendix 1). Consequently

$$R.H.\ \text{outside} = \frac{0.40}{0.61} \times 100\% = 66\%$$

$$R.H.\ \text{inside} = \frac{0.40}{2.34} \times 100\% = 17\%$$

Conversely, if a warm air mass is suddenly cooled, the relative humidity increases, perhaps to 100% or more. When this happens, liquid water condenses; this is precisely the way in which clouds are formed in the atmosphere. Clouds consist of many billions of tiny droplets of liquid water, on the average perhaps 0.01 mm in diameter. Droplets of this size are too small to fall to the earth's surface as rain. The growth of small water droplets at temperatures above 0°C is ordinarily a very slow process. The rate of growth is increased in the presence of dust particles, which act as nuclei upon which small droplets can condense. This explains why volcanic eruptions are often followed by rainstorms.

More frequently ice crystals formed in the colder upper regions of clouds act as nuclei for precipitation. In principle, ice crystals should form at 0°C; in practice, because of supercooling, they seldom develop unless the temperature drops to at least −15°C. To stimulate the formation of ice crystals, clouds are sometimes seeded with Dry Ice (Fig. 17.2). The sublimation of solid carbon dioxide absorbs enough heat from the cloud to reduce the temperature below that required for ice crystal formation. Another substance that is frequently used is finely divided silver iodide, which has a crystal structure similar to that of ice. The presence of silver iodide tends to prevent supercooling and, hence, allows ice crystals to form at temperatures close to 0°C.

Figure 17.2 Cloud seeding with Dry Ice. The particles of Dry Ice sublime rapidly in the cloud, cooling the fine water droplets to the point where they freeze and act as nuclei for condensation of other droplets in the cloud.

Silver iodide has also been used with some success in seeding hurricanes. The objective here is to add so much silver iodide that an enormous number of tiny ice crystals form. These crystals are too small to bring about precipitation, but the heat evolved in their formation tends to dissipate the storm clouds at the eye of the hurricane. One difficulty is that hurricane seeding may simply shift the storm off course without weakening it appreciably. For this reason, seeding experiments are limited to storms that are far removed from populated areas.

17.7 THE UPPER ATMOSPHERE

The species that we find in air near the surface of the earth are largely stable atoms and molecules such as N_2, O_2, Ar, CO_2, and H_2O. Above about 30 km, the situation is quite different. Exploration of the upper atmosphere by rockets and satellites equipped with small scale mass spectrometers and other sensitive instruments reveal additional species such as those listed in Table 17.7. Some of these are neutral particles (oxygen, hydrogen, and nitrogen atoms; OH radicals) formed by the dissociation of molecules. Others are positive ions formed by the ionization of diatomic molecules (NO^+, O_2^+, N_2^+) or atoms (O^+, H^+). At high altitudes certain of these species make up an appreciable fraction of the total atmosphere. For example, at a height of 120 km, there are about as many free oxygen atoms as O_2 molecules.

Not too many of either, since the pressure is so low (5×10^{-8} atm)

All of the reactions listed in Table 17.7 are endothermic; they require either the breaking of a chemical bond or the removal of an electron. In the upper atmosphere, sunlight serves as the source of energy for such reactions. The wavelengths of light given in Table 17.7 are calculated from the Einstein equation (Chapter 6) in the form:

$$\lambda_{max}(\text{nm}) = \frac{1.196 \times 10^5}{\Delta E\ (\text{kJ/mol})} \tag{17.33}$$

You will note that all of the wavelengths listed are in the ultraviolet region of the spectrum, below 400 nm. Very little of this high-energy radiation reaches the surface of the earth, so reactions such as those listed in Table 17.7 are very unlikely in the air around us. Above 30 km, significant amounts of radiation in the near-UV (200 to 400 nm) are available and species such as O, H, and OH begin to appear. At still higher altitudes, where the air is thinner, sunlight contains enough radiation in the far UV (<200 nm) to bring about highly endothermic processes, forming species such as N, O_2^+, and N_2^+.

TABLE 17.7 FORMATION OF HIGH-ENERGY SPECIES IN THE UPPER ATMOSPHERE

Molecular Fragments			Cations		
Reaction	ΔE (kJ/mol)	λ_{max}* (nm)	Reaction	ΔE (kJ/mol)	λ_{max}* (nm)
$NO_2 \rightarrow NO + O$	+305	392	$NO \rightarrow NO^+ + e^-$	+920	130
$O_2 \rightarrow O + O$	+494	242	$O_2 \rightarrow O_2^+ + e^-$	+1109	108
$H_2O \rightarrow H + OH$	+502	238	$O \rightarrow O^+ + e^-$	+1314	91.0
$NO \rightarrow N + O$	+632	189	$H \rightarrow H^+ + e^-$	+1314	91.0
$N_2 \rightarrow N + N$	+941	127	$N_2 \rightarrow N_2^+ + e^-$	+1510	79.2

*Longest wavelength of light which can supply the energy indicated.

These species can persist since there are so few collisions that conversion to stable molecules is slow

The existence in the upper atmosphere of molecular fragments and ions is responsible for many phenomena that affect our lives. To cite just one example, the presence of free electrons and positive ions, mainly NO^+ and O_2^+, produces a conducting layer in the region above 50 km that makes it possible to transmit radio waves over long distances.

Ozone

The component of the upper atmosphere which has received the most attention in recent years is ozone, an allotropic form of elementary oxygen which has the molecular formula O_3. Some of the properties of this substance are listed in Table 17.8. From a practical standpoint, the most important of these is the capacity of ozone to absorb ultraviolet light; 95 to 99% of sunlight in the wavelength range 200 to 300 nm is absorbed by ozone in the upper atmosphere. If this radiation were to reach the surface of the earth, it could have several adverse effects. By a conservative estimate, a decrease in O_3 concentration of only 5% would produce an additional 8000 cases of skin cancer annually in the United States alone.

TABLE 17.8 PROPERTIES OF O_2 AND O_3

	O_2	O_3
Melting point (°C)	−218	−192
Boiling point (°C)	−183	−112
Lewis Structure	see discussion in Chapter 8	:Ö=Ö—Ö: (bent)
UV absorption	very weak	strong in region 200 to 300 nm

Most of the ozone in the earth's atmosphere is located in a relatively narrow layer extending from an altitude of 20 km to 40 km. Even here, its concentration is quite small, reaching a maximum of about 10 ppm at 30 km. The ozone in this layer is formed by collisions between O_2 molecules and highly reactive oxygen atoms:

$$O_2(g) + O(g) \rightarrow O_3(g) \qquad (17.34)$$

Ozone molecules formed by Reaction 17.34 decompose by several mechanisms, of which the most important is

$$O_3(g) + O(g) \rightarrow 2\ O_2(g) \qquad (17.35)$$

This reaction takes place at a rather slow rate by direct collision between an O_3 molecule and an O atom. Alternatively, it can occur more rapidly by a two-step process in which a trace constituent of the upper atmosphere acts as a catalyst. Two such catalysts which have received a great deal of attention recently are listed in Table 17.9.

The fact that nitric oxide acts as a catalyst for ozone decomposition suggests that any appreciable increase in NO concentration in the upper atmosphere could deplete the ozone layer. This possibility was one factor which influenced the U.S. Senate in 1971 to cut off funds for the development of the supersonic transport (SST). It was suggested that these planes, burning jet fuel with air at high temperatures, might produce significant amounts of nitric oxide by the reaction mentioned earlier:

$$N_2(g) + O_2(g) \rightarrow 2\ NO(g) \qquad (17.3)$$

TABLE 17.9 MECHANISM FOR THE CATALYTIC DECOMPOSITION OF OZONE

	CATALYST	
	NO Molecule	Cl Atom
Mechanism	$NO + O_3 \rightarrow NO_2 + O_2$ $NO_2 + O \rightarrow NO + O_2$	$Cl + O_3 \rightarrow ClO + O_2$ $ClO + O \rightarrow Cl + O_2$
Overall reaction	$O_3 + O \rightarrow 2\ O_2$	$O_3 + O \rightarrow 2\ O_2$

Since the SST has a cruising altitude of 20 km, at the lower edge of the ozone layer, the NO formed would come into direct contact with O_3.

More recently, attention has been focused on the Cl-catalyzed decomposition of ozone. At the time this mechanism was discovered, in 1973, it was believed to be unimportant because there was no known source of free chlorine atoms in the upper atmosphere. Less than a year later, it was suggested that two halogenated methane derivatives might serve as such a source. The compounds $CFCl_3$ and CF_2Cl_2, which are widely used as refrigerants and aerosol propellants, decompose to form chlorine atoms when exposed to UV radiation at 200 nm.

$$CFCl_3(g) \rightarrow CFCl_2(g) + Cl(g) \quad (17.36)$$

$$CF_2Cl_2(g) \rightarrow CF_2Cl(g) + Cl(g) \quad (17.37)$$

Light at this wavelength is readily available at the upper edge of the ozone layer; significant amounts of $CFCl_3$ and CF_2Cl_2 (but not, at least to this point, Cl atoms) have been detected at this altitude.

17.8 AIR POLLUTION

Since the discovery of fire, man has polluted the atmosphere with noxious gases and soot. When coal began to be used as a fuel in the 14th century, the problem became one of public concern. Increased fuel consumption by industry, concentration of population in urban areas, and the advent of motor vehicles have, over the years, made the problem worse. Today the principal cause of pollution in our atmosphere is the gasoline engine.

Table 17.10 gives some idea of the relative amounts and toxic effects of the five major air pollutants. Note that while there are relatively few suspended particles in the air, their high toxicity makes them one of the most dangerous pollutants.

TABLE 17.10 MAJOR AIR POLLUTANTS

	RELATIVE AMOUNT*	TOXICITY FACTOR†	TOXIC EFFECT**
1. Carbon monoxide (CO)	1.00	1	1
2. Hydrocarbons	0.24	2	0.5
3. Nitrogen oxides (NO, NO_2)	0.15	80	12
4. Sulfur oxides (SO_2, SO_3)	0.23	30	7
5. Suspended particles	0.17	100	17

*Based on data for United States, 1970.
†Estimated from total effect on environment.
**Relative amount × toxicity factor.

There are, of course, a great many other substances which act as contaminants in air. Radioactive species produced by nuclear fallout qualify, as do toxic gases released accidentally into the atmosphere. Substances, such as freon, which threaten the existence of the ozone layer (p. 419) might be included. In general, any substance whose addition to the atmosphere produces a measurable effect on man or his environment can be classified as a pollutant. In this chapter, we will limit our attention to the pollutants listed in Table 17.10.

Sources and Effects of Pollutants

In Table 17.11 are listed the principal sources in our society of the five major air pollutants. We should emphasize that natural processes are important sources of many of these substances. For example, the amount of methane produced by the anaerobic decay of organic matter in swamps and marshes far exceeds the total amount of hydrocarbons listed in Table 17.11. However, since pollutants from natural sources are ordinarily spread out over wide areas, they usually pose no threat to health. In contrast, man-made pollutants tend to be concentrated in areas of high population density and hence are more harmful.

TABLE 17.11 SOURCES OF AIR POLLUTANTS IN THE UNITED STATES (1970)*

	SUSPENDED PARTICLES	SO_2, SO_3	NO, NO_2	HYDRO-CARBONS	CO
Transportation (primarily automobiles)	0.6	0.9	10.6	17.7	100.0
Fuel combustion for heat and electricity	6.2	24.1	9.1	0.5	0.7
Industry	11.9	5.5	0.2	5.0	10.4
Solid waste disposal	1.3	0.1	0.4	1.8	6.5
All others	3.1	0.3	0.4	6.5	15.3
	23.1	30.9	20.7	31.5	132.9

*Millions of tonnes per year.

SUSPENDED PARTICLES. The finely divided solids and tiny drops of liquids found suspended in the atmosphere vary greatly in size. Particles in cigarette smoke have diameters as small as 10^{-6} cm; dust particles produced by a cement kiln may be as large as 10^{-2} cm. Industrial processes involving cutting, grinding, and spraying are major sources of this type of pollutant. The combustion of coal produces large amounts of unburned carbon (soot) and finely divided inorganic material (fly ash) which are emitted to the atmosphere. Fuel oil is cleaner-burning; natural gas is even better in this respect. A properly adjusted burner using natural gas produces virtually no smoke or soot.

Reduced visibility (Fig. 17.3) caused by absorption and scattering of light is one of the more obvious effects of high concentrations of suspended particles. Soiled clothing and soot-covered buildings are other symptoms that are all too familiar to the city-dweller. Many of the suspended particles resulting from industrial processes are known to be harmful to health. Black lung disease, common among coal miners, and silicosis, an occupational hazard of granite cutters, are caused by the deposition of tiny particles of carbonaceous material or silicon dioxide in the lungs. Over a period of years, these materials attack lung tissue and increase susceptibility to such diseases as tuberculosis and lung cancer. Similar effects are produced by asbestos, which is widely used in brake linings and as a fire-proofing and insulating material. Among a group of asbestos workers in New York City it was found that one death in five was due to lung cancer.

SULFUR OXIDES. Most of the coal burned in heating and power plants in the United States and Canada contains from 1 to 3% sulfur, mostly in the form of minerals such as pyrites, FeS_2. Upon combustion, the sulfur is converted to sulfur dioxide:

Steel mills use tremendous quantities of coal and in some cities are a main source of air pollution

$$4\ FeS_2(s) + 11\ O_2(g) \rightarrow 2\ Fe_2O_3(s) + 8\ SO_2(g) \qquad (17.38)$$

Most hydrocarbon fuels contain very little sulfur. An exception is the residual oil left after the distillation of petroleum. This tarry material, which is used in many industrial and municipal heating plants, contains from 1 to 2% sulfur. The sulfur, present largely as organic compounds, is converted to sulfur dioxide upon combustion. Metallurgical processes involving the roasting of sulfide ores (Chapter 7) are another major source of SO_2.

Sulfur dioxide at concentrations as low as 0.3 ppm can cause acute injury to plants. Most healthy adults can tolerate considerably higher SO_2 levels without apparent adverse effects. However, individuals who suffer from chronic respiratory diseases such as bronchitis or asthma are much more sensitive. Increasing the concentration of SO_2 from 0.1 to 0.2 ppm can cause them to start coughing and experience severe difficulties in breathing.

From equilibrium considerations we might expect sulfur dioxide in the air to be converted to sulfur trioxide.

$$SO_2(g) + \tfrac{1}{2}\ O_2(g) \rightarrow SO_3(g);\ K_c = 6 \times 10^{12} \text{ at } 25°C$$

Figure 17.3 New York City smog. Most of the large cities in the world have at least occasional accumulations of smog such as that shown here. (Courtesy of Planned Parenthood—World Population.)

However, as pointed out previously, this reaction takes place very slowly in the absence of a catalyst. Ordinarily, less than 1% of the SO_2 in polluted air is converted to SO_3. Suspended solids in the atmosphere, such as those found in coal smoke, can raise this fraction to 5% or more by catalyzing this reaction. The further reaction of sulfur trioxide with water in the air or one's lungs (Reaction 17.25) produces sulfuric acid.

It appears that many if not all of the adverse effects of sulfur oxides in the atmosphere are caused by the sulfuric acid produced when SO_3 reacts with water. This is certainly the case with building materials such as limestone or marble ($CaCO_3$), which are readily attacked by H_2SO_4:

$$CaCO_3(s) + H_2SO_4(aq) \rightarrow CaSO_4(s) + CO_2(g) + H_2O \qquad (17.39)$$

The calcium sulfate formed is soluble enough to be gradually leached away by rain water. This process is responsible for the deterioration of such structures as the Acropolis in Athens, which has suffered more damage in this century than in the preceding 20 centuries. Another effect which can be attributed to sulfuric acid is the deterioration of the paper in books and documents. Curiously enough, manuscripts printed before 1750 are almost immune to attack by sulfur oxides in the atmosphere. At about that time modern methods of papermaking were introduced; they leave traces of metal oxides which catalyze the conversion of SO_2 to SO_3 and hence to sulfuric acid.

NITROGEN OXIDES. The high-temperature combustion of fuels is the principal source of this type of air pollutant. Detectable amounts of nitric oxide are formed from the elements via Reaction 17.3 at temperatures of 1200°C or

Figure 17.4 Marble statues sometimes become unrecognizable as a result of attack by oxides of sulfur. (From Wagner, R. H., *Environment and Man*, W. W. Norton & Co., New York, 1971.)

greater (the temperature reached in an internal combustion engine can exceed 2000°C). In air, NO is slowly converted to NO_2 by oxygen:

$$2\ NO(g) + O_2(g) \rightarrow 2\ NO_2(g);\ k = 7 \times 10^{-3}\ dm^6/(mol^2 \cdot s) \text{ at } 25°C \quad (17.40)$$

and more rapidly by ozone:

$$NO(g) + O_3(g) \rightarrow NO_2(g) + O_2(g);\ k = 1.2 \times 10^7\ dm^3/(mol \cdot s) \text{ at } 25°C \quad (17.41)$$

Perhaps the most harmful effect of oxides of nitrogen in the atmosphere is the key role that they play in the formation of *photochemical smog*, which gained its reputation originally in Los Angeles but has now become common in cities as widely separated as Montreal, Denver, and Washington, D.C. Smog formation is initiated by the photochemical decomposition of nitrogen dioxide (Table 17.7)

$$NO_2(g) \rightarrow NO(g) + O(g) \quad (17.42)$$

which occurs on exposure to light at the edge of the visible region (392 nm). The oxygen atoms produced can react with NO by the reverse of Reaction 17.42, but most of them react with the more abundant species O_2

$$O(g) + O_2(g) \rightarrow O_3(g) \quad (17.43)$$

to produce ozone, a major component of photochemical smog. Both of these very reactive species, oxygen atoms and ozone molecules, attack hydrocarbons, primarily of the olefin type. A variety of products are eventually formed, including alcohols, ketones, and aldehydes. The eye irritation which is a familiar feature of photochemical smog is due to the presence of compounds such as formaldehyde and acrolein:

```
    H                        H  H
    |                        |  |
  H—C=O               H₂C=C—C=O
formaldehyde             acrolein
```

Unsaturated hydrocarbons can also react directly with either NO or NO_2 to produce such compounds as

```
H₃C—O—N—O        and      H₃C—C—O—O—N—O
      ||                      ||       ||
      O                       O        O
 methyl nitrate           peroxyacetyl nitrate (PAN)
```

both of which are powerful eye irritants.

HYDROCARBONS. These organic compounds enter the atmosphere directly as vented gases from petroleum refineries or by evaporation from the fuel tanks of automobiles. A more important source is automobile exhaust, which contains significant amounts of unburned hydrocarbons unless special precautions are taken to ensure complete combustion. Most of these hydrocarbons are of the saturated type (e.g., CH_4, C_2H_6, . . . C_8H_{18}) and appear to be relatively harmless. A few of them, however, are olefins produced by the "cracking" of saturated hydrocarbons such as octane:

$$\underset{\text{octane}}{C_8H_{18}(l)} \rightarrow 2\ \underset{\text{ethylene}}{C_2H_4(g)} + \underset{\text{butene}}{C_4H_8(g)} + H_2(g) \quad (17.44)$$

As pointed out, unsaturated hydrocarbons such as ethylene and butene can react with species such as O, NO, or O_3 to produce some of the most noxious components of photochemical smog.

CARBON MONOXIDE. The incomplete combustion of fossil fuels produces carbon monoxide as well as unburned hydrocarbons. The principal culprit is again the automobile. If sufficient oxygen is present, all of the carbon atoms in the fuel are converted to CO_2:

$$C_8H_{18}(l) + 12.5\ O_2(g) \rightarrow 8\ CO_2(g) + 9\ H_2O(l) \qquad (17.45)$$

However, if the fuel mixture is too "rich", i.e., contains too much fuel and too little air, considerable amounts of CO may be formed:

$$C_8H_{18}(l) + 11.5\ O_2(g) \rightarrow 6\ CO_2(g) + 2\ CO(g) + 9\ H_2O(l) \qquad (17.46)$$

In principle, the carbon monoxide should be converted to CO_2 in the atmosphere

$$CO(g) + \tfrac{1}{2}\ O_2(g) \rightarrow CO_2(g):\ K_c = 3 \times 10^{45} \text{ at } 25°C \qquad (17.47)$$

but this reaction, like the conversion of SO_2 to SO_3, is ordinarily quite slow.

Carbon monoxide taken into the lungs reduces the ability of the blood to transport oxygen through the body. It does this by forming a complex with the hemoglobin of the blood which is more stable than that formed by oxygen (Chapter 21).

$$CO(g) + Hem \cdot O_2(aq) \rightleftharpoons O_2(g) + Hem \cdot CO(aq):\ K_c = 210 \qquad (17.48)$$

$$(Hem = \text{hemoglobin})$$

A person suffering from acute or chronic carbon monoxide poisoning has to breathe more air to deliver a given amount of oxygen to body cells. If as much as 10 to 20% of hemoglobin is tied up by carbon monoxide, the resulting strain on the heart and lungs can be fatal. Even if the concentration of $Hem \cdot CO$ is only 2% of that of $Hem \cdot O_2$, mental activity can be impaired to the point where simple problems in arithmetic become difficult to carry out. One can calculate (Problem 17.20) that this ratio will be reached at equilibrium when the concentration of CO in the air is about 20 ppm.

Reducing Air Pollution

HYDROCARBONS, CO, OXIDES OF NITROGEN. Referring back to Table 17.11, we see that the automobile is the major source of these three air pollutants. In 1970, the average vehicle in use in the United States and Canada produced about 130 g of carbon monoxide, 18 g of unburned hydrocarbons, and 6 g of oxides of nitrogen per kilometre traveled. In the past several years, strenuous efforts have been made to reduce these emissions. The problem is a complex one for several reasons. In particular, attempts to reduce the level of one of these pollutants may lead to:

—*higher levels of another pollutant*. Emissions of carbon monoxide and unburned hydrocarbons can be reduced by raising the burning temperature so as to achieve complete combustion to carbon dioxide and water:

$$CO(g) + \tfrac{1}{2}\ O_2(g) \rightarrow CO_2(g)$$

$$C_8H_{18}(l) + \frac{25}{2}\ O_2(g) \rightarrow 8\ CO_2(g) + 9\ H_2O(g)$$

However, as the temperature rises, so does the level of oxides of nitrogen, produced by the endothermic reaction

$$N_2(g) + O_2(g) \rightarrow 2\ NO(g);\ \Delta H = +180.8\ kJ$$

—*reduced fuel economy and/or performance* (see discussion below). The "energy crisis" of the 1970's has made this factor particularly important. To what extent this has been used as an excuse by car manufacturers and government agencies to justify the relaxation of clean air standards is debatable.

Prior to 1975, most automobile manufacturers attempted to reduce hydrocarbon and CO emissions by relatively minor changes in engine design with the objective of ensuring complete combustion of the fuel. This was accomplished by increasing the ratio of air to fuel, modifying the design of the combustion chamber to obtain better mixing, and retarding the spark that ignites the fuel-air mixture. These changes were not enthusiastically received by many drivers because they tend to reduce gas mileage and make starting more difficult, particularly in cold weather.

Starting with 1975 model cars, a different approach has been followed to reduce emissions of hydrocarbons and CO. This involves inserting into the exhaust system a "catalytic converter" (Fig. 17.5). This device contains from 1 to 3 g of a mixture of two heavy metals, platinum and palladium, imbedded in an aluminum oxide base. In the presence of the catalyst, CO and hydrocarbons in the exhaust are converted to carbon dioxide and water. The catalytic system, as compared to ones previously used, gives better gas economy and smoother performance. On the other hand, it appears to have at least two possible disadvantages. In the first place, Pt-Pd mixtures are very efficient catalysts for the conversion of SO_2 in the exhaust to sulfuric acid (Reactions 17.24, 17.25), which is a serious air pollutant. Moreover, the catalyst is readily "poisoned," particularly by the $Pb(C_2H_5)_4$–$C_2H_4Br_2$ mixture added to gasoline to raise its octane number. Cars equipped with catalytic converters must use unleaded gasoline, which is somewhat more expensive and may itself pose a health hazard because of the increased fraction of aromatic hydrocarbons present, some of which are carcinogenic (Chapter 10).

Some recent research suggests Br rather than Pb is the culprit

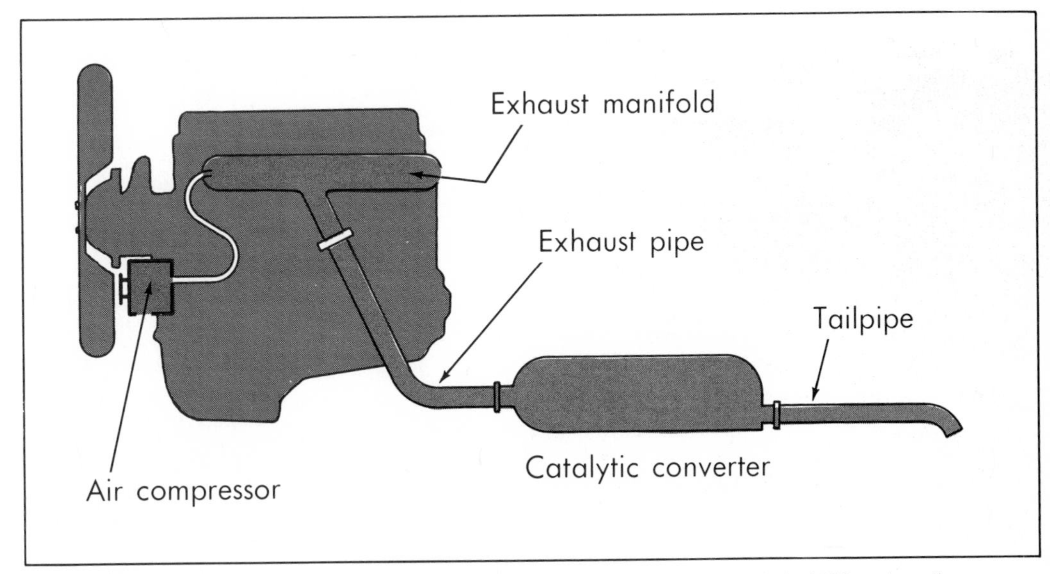

Figure 17.5 Catalytic converters, used on essentially all U.S. automobiles made in 1975, reduce the concentrations of CO and hydrocarbons in the car exhaust by facilitating their oxidation to CO_2 and water vapor.

Several different approaches have been studied for reducing exhaust emissions of oxides of nitrogen. One currently in use involves recirculating from 10 to 20% of the exhaust gases. This lowers the combustion temperature and hence cuts down on the formation of NO from nitrogen and oxygen. Such a system has one inherent disadvantage. Unless the amount of exhaust gases recirculated is controlled very precisely, there are adverse effects on both performance and fuel economy. Another method is to introduce a second catalyst into the exhaust system to promote the decomposition of NO, perhaps by the reaction

$$2\ CO(g) + 2\ NO(g) \rightarrow 2\ CO_2(g) + N_2(g) \qquad (17.49)$$

This reaction is spontaneous ($\Delta G^{1\ atm}$ at 25°C = −688 kJ) but comparatively slow at the "parts per million" level at which CO and NO are present in automobile exhaust.

SULFUR OXIDES (SO_2, SO_3). The problem here is one of cleaning up emissions from power and heating plants, which account for nearly 80% of the sulfur oxides produced by man. An obvious way to prevent SO_2 emissions is to remove sulfur compounds from fuel before it is burned. Pyrites, FeS_2, can be separated from coal by "washing" with a concentrated solution of calcium chloride (d = 1.35 g/cm^3). The coal (d = 1.2 g/cm^3) floats in this solution, while pyrites (d = 4.9 g/cm^3) sinks to the bottom. With finely divided coal, the process of flotation (Chapter 7) can be used. Neither of these methods is effective with organic sulfur compounds, which account for nearly half of the sulfur content of coal and virtually all of that in fuel oil.

A different approach to this problem is to add a chemical to react with the sulfur oxides after they are formed. The most common method uses limestone, $CaCO_3$, which may be added to stack gases either as a dry powder or as a slurry in water. The reaction is

$$CaCO_3(s) + SO_2(g) + \tfrac{1}{2}\ O_2(g) \rightarrow CaSO_4(s) + CO_2(g) \qquad (17.50)$$

Unfortunately, the calcium sulfate formed tends to adhere to the surface of the limestone particles, thereby stopping the reaction. Many other chemicals have been tested but none appears to be entirely satisfactory.

SUSPENDED PARTICLES. Several different methods are used to remove smoke particles and dust from smokestack gases. Large particles (diameter > 0.005

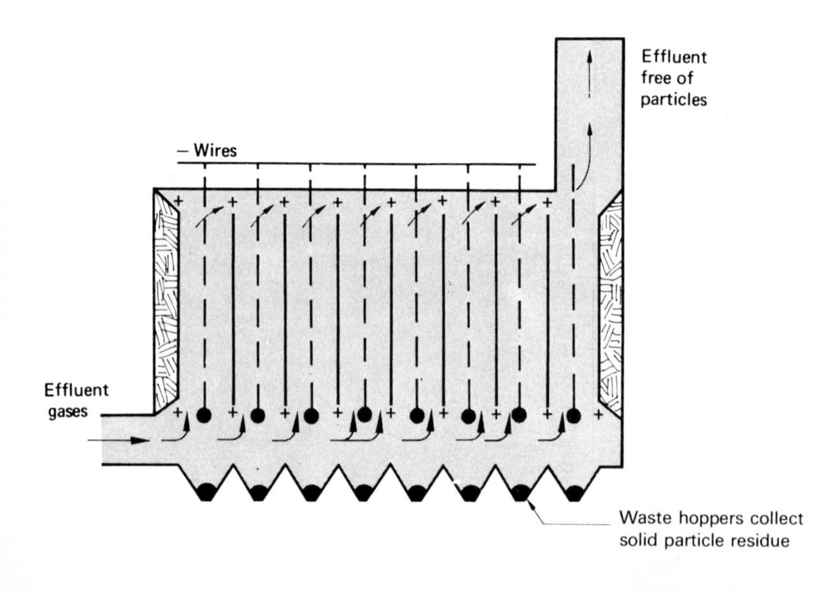

Figure 17.6 Schematic drawing of electrostatic precipitator. Particles pick up charge as they pass between plates and are precipitated on plates of opposite charge. The potential across the plates is about 50 000 V. (From G. Gordon and W. Zoller, *Chemistry in Modern Perspective*, Addison-Wesley Publishing Co., Reading, Mass., 1975.)

cm) can be removed quite easily by simply allowing them to settle by gravity in a large chamber. This approach is impractical for small particles, which settle at a much slower rate. Particles with diameters as small as 0.0001 cm can be removed by passing the stack gas through an electrostatic precipitator of the type shown in Figure 17.6. This device consists of a series of plates which are charged to high voltages, alternately + and −. Particles approaching a given plate tend to acquire its charge. They are then attracted to the surface of the next plate, from which they fall into the hopper below.

PROBLEMS

17.1 *Concentration Units* The mole fraction of Ar in dry air is 9.34×10^{-3}. Using Table 17.2, calculate its concentration in

a. volume %.
b. mass %.
c. moles per cubic decimetre at 25°C and a total pressure of one atmosphere.
d. parts per million.

17.2 *Chemistry of Oxygen* Write balanced equations to represent

a. the formation of aluminum oxide from the elements.
b. the formation of barium peroxide from the elements.
c. the reaction of barium peroxide with water.
d. the reaction of carbon dioxide with water.

17.3 *Relative Humidity* The equilibrium vapor pressure of water at 20°C is 2.34 kPa. If the relative humidity at this temperature is 52%, what is the partial pressure of water in the air?

17.4 Using Tables 17.1 and 17.2, calculate the concentration of CO_2 in air in

a. volume %.
b. mass %.
c. moles per cubic decimetre at 25°C and P_{tot} = 100 kPa.
d. parts per million.

17.5 Explain in your own words why

a. air cools when its pressure drops.
b. liquid N_2 is safer to use than liquid air.
c. pure O_2 gives a higher flame temperature than air.
d. argon is cheaper than neon or krypton.
e. CO_2 is more effective than water in putting out electrical fires.

17.6 Describe by means of chemical equations what happens when

a. lithium at high temperatures comes in contact with nitrogen gas.
b. lithium at room temperature reacts with oxygen.
c. barium oxide is added to water.
d. ammonia burns in air at 1000°C.
e. $SO_3(g)$ dissolves in water to form an acidic solution.

17.7 Discuss the effect of changes in temperature and applied pressure on the position of the following equilibria and the rate at which they are attained:

a. $N_2(g) + O_2(g) \rightleftarrows 2\ NO(g)$
b. $N_2(g) + 3\ H_2(g) \rightleftarrows 2\ NH_3(g)$

17.22 The concentration of CO in polluted air is 11.2 ppm. Calculate its concentration in

a. mole fraction.
b. partial pressure (P_{tot} = one atmosphere).
c. moles per cubic decimetre (25°C).
d. grams per cubic metre.

17.23 Would it be feasible, at least in principle, to

a. omit the heat exchanger (Fig. 17.1) in the liquefaction of air?
b. fix nitrogen by passing it over heated lithium?
c. use high pressures in the preparation of SO_3 from SO_2?
d. seed clouds by spraying with a liquid with a high heat of vaporization?

17.24 Describe by means of chemical equations how you would prepare

a. $H_2O_2(aq)$ from $Na_2O_2(s)$.
b. $HNO_3(l)$ from $NH_3(g)$.
c. urea from ammonia.
d. $Cu_2O(s)$ from $CuO(s)$.
e. $CO(g)$ and $H_2(g)$ from $CH_4(g)$.
f. $H_3PO_4(aq)$ from $P_4O_{10}(s)$.

17.25 Discuss how one might increase the rate and/or the yield of product in the following reactions:

a. $SO_2(g) + \frac{1}{2}\ O_2(g) \rightarrow SO_3(g)$
b. $Xe(g) + 2\ F_2(g) \rightarrow XeF_4(s)$

17.8 Write Lewis structures for:

a. O_2^{2-}
b. O_2^-
c. O^{2-}
d. CO

17.9 Using data on heats of formation in Table 4.1, Chapter 4 and free energies of formation, Table 14.2, Chapter 14, show that the temperature at which the reaction $2\ CuO(s) \rightarrow Cu_2O(s) + \frac{1}{2}\ O_2(g)$ becomes spontaneous is approximately 900°C.

17.10 Consider the data in Table 17.3.

a. From the table, determine the number of moles of NH_3, H_2, and N_2 in an equilibrium mixture containing one mole of gas at 200°C and 50 atm. Note that $[H_2] = 3[N_2]$.
b. Using the Ideal Gas Law, calculate the volume occupied by one mole of gas under these conditions.
c. Combining your answers to (a) and (b), calculate the equilibrium concentrations in mol/dm³ of NH_3, H_2, and N_2.
d. From your answer to (c), calculate K_c at 200°C and compare to the value given in the table.

17.11 Describe the effects which might result from

a. an increase in the concentration of CO_2 in the atmosphere.
b. an increase in the concentration of NO in the upper atmosphere.
c. a rapid drop in temperature on a day when the relative humidity is 80%.

17.12 What are the basic raw materials used in the commercial preparation of

a. NH_3?
b. H_2SO_4?
c. urea?

17.13 Calculate the longest wavelength of a photon that can bring about the decomposition of CCl_2F_2 (Reaction 17.37), using the bond energy table in Chapter 4.

17.14 The rate constant for the decomposition of ozone by the mechanism: $O_3(g) + O(g) \rightarrow 2\ O_2(g)$, is 7.8×10^5 dm³/(mol · s). Calculate the rate of decomposition of ozone when the concentrations of O_3 and O are 1.0×10^{-8} and 4.2×10^{-13} mol/dm³, respectively.

17.15 The rate constant for the NO-catalyzed decomposition of O_3 (Table 17.9) is 2.4×10^9 dm³/(mol · s), as compared to 7.8×10^5 dm³/(mol · s) for the direct decomposition (Problem 17.14). The second step in the catalyzed decomposition is rate-determining. If the concentration of NO_2 in the upper atmosphere is 1/2000 that of O_3, what is the ratio of the rates of the two reactions?

17.26 Write Lewis structures for:

a. N^{3-}
b. OH^-
c. O_3
d. O

17.27 Using data from Tables 4.1 and 14.2, calculate the temperature at which the reaction $CH_4(g) + H_2O(g) \rightarrow CO(g) + 3\ H_2(g)$ becomes spontaneous at 1 atm pressure.

17.28 Consider the data in Table 17.5.

a. Given that the mole % of SO_3 at 800°C is 20 and that $[O_2] = \frac{1}{2}\ [SO_2]$, calculate the mole % of all three gases in the equilibrium mixture at this temperature.
b. Using the Ideal Gas Law, find the number of moles of gas in a 1.00-dm³ container at 1.00 atm and 800°C.
c. Combining your results in (a) and (b), find the equilibrium concentrations of all three gases in a 1.00-dm³ container at 1.00 atm and 800°C.
d. From your answers to (c), find K_c at 800°C and compare to the value given in the table.

17.29 Describe the possible effects of

a. increasing the concentration of suspended particles in the air.
b. decreasing the concentration of O_3 in the upper atmosphere.
c. increasing the temperature on a day when the relative humidity is 80%.

17.30 What is (are) the basic source(s) of

a. the neon used in advertising signs?
b. the nitric acid on the lab shelf?
c. the NO_2 which promotes the formation of photochemical smog?

17.31 Calculate the maximum energy in kcal/mole of a bond that can be broken by absorbing red light at 680 nm.

17.32 If ozone is being produced by the mechanism $O_2(g) + O(g) \rightarrow O_3(g)$ at the same rate it is decomposing by the reaction in Problem 17.14, and if the concentrations of O_2 and O are 1.3×10^{-4} and 4.2×10^{-13} mol/dm³, respectively, what is the rate constant for the above reaction?

17.33 It is estimated that the amount of O_3 in the upper atmosphere decomposing by the NO-catalyzed reaction is four times that decomposing by the direct reaction. Using the rate constants given in Problem 17.15, estimate the ratio of the concentrations of NO_2 and O_3.

17.16 Taking the concentrations of NO, O_3, and O_2 in polluted air to be 4×10^{-8}, 2×10^{-8}, and 1.0×10^{-2} mol/dm^3, respectively, calculate and compare the rates of Reactions 17.40 and 17.41 under these conditions.

17.17 Taking the total volume of air in the Los Angeles basin to be 6000 km^3 and assuming the maximum permissible concentration of CO to be 4×10^{-7} mol/dm^3, calculate

a. the total mass in grams of CO permissible in the LA basin.
b. the number of vehicle kilometres required to produce this CO, assuming the average vehicle emits 130 g CO/km.

17.18 Suggest why

a. the concentration of suspended particles is *not* expressed in parts per million.
b. unleaded gasoline is more expensive than gasoline containing $Pb(C_2H_5)_4$.
c. an estimate of the amount of photochemical smog can be obtained by measuring the O_3 concentration.
d. smog in Los Angeles ordinarily builds up in the early morning hours.

17.19 Consider Reaction 17.45.

a. How many grams of O_2 are required to burn one gram of C_8H_{18}?
b. Taking the mass % of O_2 in air to be 21, calculate the number of grams of air required per gram of octane. (This is approximately the "stoichiometric ratio," i.e., the air/fuel ratio required for complete combustion.

17.20 Consider Reaction 17.48. Suppose the concentration of Hem·CO is 2.0% of that of Hem·O_2. Calculate

a. the concentration ratio of CO/O_2 required.
b. the mole fraction of CO in air necessary, taking that of O_2 to be 0.20.
c. the concentration of CO required in parts per million.

17.21 Explain why

a. increasing the air/fuel ratio decreases emissions of CO from cars but ordinarily increases NO emissions.
b. recycling exhaust gases reduces NO emissions.
c. catalytic converters are probably not an ultimate answer to cleaning up automobile emissions.
d. adding $CaCO_3$ to stack gases is not an entirely satisfactory way to reduce SO_2 emissions.

17.34 Taking k for the reaction $2\,CO(g) + 2\,NO(g) \rightarrow 2\,CO_2(g) + N_2(g)$ to be 3×10^{-27} $dm^6/(mol^2 \cdot s)$, calculate the rate of this reaction when the concentrations of CO and NO are 4×10^{-6} and 4×10^{-9} mol/dm^3, respectively. (The reaction is first order in CO and second order in NO.)

17.35 Repeat the calculation of Problem 17.17 for hydrocarbons, taking the maximum permissible concentration to be 1×10^{-8} mol/dm^3, the average molecular mass to be 100, and the average emission of hydrocarbons to be 18 g/km.

17.36 Explain why

a. oxides of sulfur in the air cause damage to marble statues.
b. emission of NO from automobiles can be reduced by lowering the air/fuel ratio.
c. unsaturated hydrocarbons are more dangerous air pollutants than saturated ones.
d. aromatic hydrocarbons are more dangerous air pollutants than saturated ones.

17.37 Repeat the calculations of Problem 17.19 for Reaction 17.46, which represents incomplete combustion.

17.38 Using the equilibrium constant given with Reaction 17.48 and taking the concentration of O_2 in air to be 8.2×10^{-3} mol/dm^3, calculate

a. the concentration of CO in mol/dm^3 required to convert 10% of the hemoglobin in the blood to the CO complex.
b. the corresponding concentration of CO in parts per million. Take the total pressure to be 1.00 atm and $t = 27°C$.

17.39 Suggest what effect each of the following changes might have on the level of pollutants in the air:

a. Using electrically powered cars.
b. Extracting metals from oxide rather than sulfide ores (e.g., ZnO vs. ZnS).
c. Converting power plants from coal to nuclear fuel.
d. Replacing gasoline with alcohol, C_2H_5OH, which gives a lower combustion temperature.

*17.40 A certain type of coal contains 3.0% by mass of sulfur.

a. What mass of sulfur dioxide will be produced by burning 100 g of this coal?
b. What volume of SO_2 at 25°C and one atmosphere will be produced?
c. If 1.00 m^3 of air is used to burn the coal, what will be the concentration of SO_2 in the stack gas in parts per million?

*17.41 Most of the ozone in the upper atmosphere exists in a layer at an altitude between 20 and 40 km.

a. Calculate the total volume in litres of this layer, taking the radius of the earth to be 6.4×10^3 km and using the following relation for the difference in volume between two concentric spheres of radii r_1 and r_2: $\Delta V \approx 4\pi r_1^2(r_2 - r_1)$
b. Taking the average pressure of air in this layer to be 0.013 atm and the temperature to be −20°C, calculate the total number of moles of gas in the layer.
c. Assuming the average concentration of ozone in the layer to be 7 ppm, what is the total mass in grams of the ozone in the upper atmosphere?

18

PRECIPITATION REACTIONS

Most of the reactions considered to this point have been ones that take place in the gas phase or at the solid-gas interface. As important as such reactions are, reactions taking place in aqueous solution are perhaps of even greater significance. In the oceans, on rain-soaked fields and mountains, and in living organisms occur a multitude of chemical reactions that influence our world and our lives in important ways.

In this and succeeding chapters, we will study a variety of reactions taking place in water solution. We begin by looking at what is perhaps the simplest of such reactions, that of precipitation. When certain solutions of electrolytes are mixed, such as nickel chloride, $NiCl_2$, and sodium hydroxide, NaOH, the positive ion of one solution combines with the negative ion of the other solution to form an insoluble solid. We will develop principles to predict under what conditions such reactions will occur, what their products will be, and how they can be represented by chemical equations.

18.1 NET IONIC EQUATIONS

If 0.1 *M* solutions of $NiCl_2$ and NaOH are mixed, a green gelatinous precipitate forms. In principle, the green precipitate might be either sodium chloride, NaCl, resulting from the interaction of Na^+ ions of the NaOH solution with Cl^- ions of the $NiCl_2$ solution, or nickel hydroxide, $Ni(OH)_2$, formed when Ni^{2+} ions of the $NiCl_2$ solution come in contact with OH^- ions of the NaOH solution. Experience enables us to make a choice between these two possibilities: we know that sodium chloride, ordinary table salt, is neither green nor insoluble. By elimination, we deduce that nickel hydroxide, $Ni(OH)_2$, must be the product of the reaction.

Having deduced the nature of the reaction, it is now possible to write an equation for it. The product is solid nickel hydroxide, $Ni(OH)_2$; the reactants are Ni^{2+} and OH^- in aqueous solution. Consequently the balanced equation is

$$Ni^{2+}(aq) + 2\ OH^-(aq) \rightarrow Ni(OH)_2(s) \qquad (18.1)$$

The solid that forms in such a reaction is called a precipitate

This equation, representing a reaction between ions in water solution, is an example of a **net ionic equation**. We will be writing such equations extensively throughout the rest of this text. In writing net ionic equations, we follow the same principles as in writing any type of equation. In particular:

1. *The identity of the product must be established before the equation can be written*. You will note that in writing Equation 18.1, we essentially worked back-

wards. After deciding that the product must be $Ni(OH)_2$, it followed that the reactants must be a Ni^{2+} ion and two OH^- ions.

2. *Species which do not take part in the reaction are not included in the equation*. In Equation 18.1, neither the Na^+ ions from the NaOH solution nor the Cl^- ions from the $NiCl_2$ solution are included. They are "spectator ions" in the sense that they take no part in the reaction, being present in solution before and after precipitation.

3. *The physical states of reactants and products are indicated in the equation*. The symbol (aq) is written after Ni^{2+} and OH^- to remind you that these ions are originally present in water solution. To indicate that $Ni(OH)_2$ is an insoluble solid, we write (s) after it.

Sometimes, two precipitation reactions occur when two solutions are mixed. Suppose, for example, a solution of barium hydroxide, $Ba(OH)_2$, is added to a solution of nickel sulfate, $NiSO_4$. Both of the possible products, $Ni(OH)_2$ and $BaSO_4$, are insoluble in water. Hence, we would predict that both of these compounds should precipitate. Experiment confirms this deduction; if we look at the precipitate under a microscope, it is possible to distinguish white crystals of $BaSO_4$ from green particles of $Ni(OH)_2$. In representing this double precipitation reaction, we write two net ionic equations, since two entirely different reactions are taking place:

Precipitation reactions usually occur instantaneously; $E_a \approx 0$

$$Ni^{2+}(aq) + 2\ OH^-(aq) \rightarrow Ni(OH)_2(s) \qquad (18.1)$$

and

$$Ba^{2+}(aq) + SO_4^{2-}(aq) \rightarrow BaSO_4(s) \qquad (18.2)$$

Net ionic equations such as those written above can be given a quantitative meaning by following the principles outlined in Chapter 3. Example 18.1 illustrates how this is done.

Example 18.1 When 300 cm³ of 0.10 *M* Na_2SO_4 is added to 200 cm³ of 0.20 *M* $BaCl_2$ solution, a white precipitate forms.

a. Determine the number of moles of each ion originally present.
b. Write a net ionic equation for the precipitation reaction.
c. Determine the number of moles of each ion remaining in solution after precipitation.

Solution

a. Let us start by calculating the number of moles of Na_2SO_4 and $BaCl_2$ originally present:

$$\text{no. moles } Na_2SO_4 = 0.300\ dm^3 \times 0.10\frac{mol}{dm^3} = 0.030\ mol$$

$$\text{no. moles } BaCl_2 = 0.200\ dm^3 \times 0.20\frac{mol}{dm^3} = 0.040\ mol$$

But, noting from the formulas of the salts that 1 mol Na_2SO_4 produces 2 mol Na^+ and 1 mol SO_4^{2-}, while 1 mol $BaCl_2$ yields 1 mol Ba^{2+} and 2 mol Cl^-, we have

$$2 \times 0.030 = 0.060\ mol\ Na^+$$

$$1 \times 0.030 = 0.030\ mol\ SO_4^{2-}$$

$$1 \times 0.040 = 0.040\ mol\ Ba^{2+}$$

$$2 \times 0.040 = 0.080\ mol\ Cl^-$$

b. In principle, there are two possible precipitates, NaCl and $BaSO_4$. Knowing that NaCl is soluble and $BaSO_4$ insoluble, we deduce that the equation must be

$$Ba^{2+}(aq) + SO_4^{2-}(aq) \rightarrow BaSO_4(s)$$

c. Looking at the answers in (a) and the equation just written, we see that SO_4^{2-} is the limiting reagent. The 0.030 mol SO_4^{2-} will be essentially all consumed. This will require that 0.030 mol Ba^{2+} react, leaving 0.040 − 0.030 = 0.010 mol of that ion in solution. Since the Na^+ and Cl^- ions do not take part in the reaction, their amounts will remain unchanged. Summarizing this reasoning in the form of a table:

	Original	Change	Final
No. moles Na^+	0.060		0.060
No. moles SO_4^{2-}	0.030	−0.030	0.000
No. moles Ba^{2+}	0.040	−0.030	0.010
No. moles Cl^-	0.080		0.080

Na^+ and Cl^- ions are "spectators"

Notice that all the Na^+ and Cl^- ions remain in solution. Part of the Ba^{2+} and virtually all the SO_4^{2-} react. Actually, as we shall see in Section 18.3, equilibrium considerations require that a trace of SO_4^{2-} be left in solution, but the number of moles of SO_4^{2-} is much less than 0.001.

18.2 SOLUBILITIES OF IONIC COMPOUNDS

Solubility data obtained from an experimental study of a limited number of precipitation reactions can be used to predict results of a great many other reactions. For example, having established that nickel hydroxide is insoluble in water, we can predict that mixing solutions of $Ni(NO_3)_2$ and NaOH, $NiSO_4$ and KOH, or $NiBr_2$ and $Ca(OH)_2$ will result in the formation of a precipitate of $Ni(OH)_2$, according to Equation 18.1. Again, if it is established through experiment that sodium chloride and potassium nitrate are both soluble in water, it follows that no precipitation reaction will occur when solutions of $NaNO_3$ and KCl are mixed. It might appear that all we need to predict the results of possible precipitation reactions is a table of solubilities in which every ionic solid is neatly classified as soluble or insoluble. Unfortunately there are difficulties associated with any attempt to set up such a simple classification scheme.

Ionic solids do not fall neatly into the categories "soluble" and "insoluble" with a sharp dividing line between them. Instead, they cover an enormous range of solubility. One of the most soluble salts known is lithium chlorate, $LiClO_3$, which dissolves to the extent of 35 mol/dm³ at room temperature. We can safely predict that this compound will not be the product of a precipitation reaction. At the other extreme is mercuric sulfide, HgS; one can calculate that a saturated solution should contain only about 10^{-26} mol/dm³ of Hg^{2+} and S^{2-} ions. Quite clearly, we can expect to get a precipitate of mercuric sulfide whenever water solutions containing Hg^{2+} and S^{2-} ions are mixed, even if the solutions are extremely dilute. Lead chloride, $PbCl_2$, is an example of a compound of intermediate solubility. When equal volumes of 0.1 *M* $Pb(NO_3)_2$ and 0.1 *M* NaCl are mixed, the solubility of $PbCl_2$ is exceeded and it precipitates. If, on the other hand, the solutions mixed are somewhat more dilute, say 0.04 *M*, no precipitate is formed. Compounds such as lead chloride are difficult to classify as soluble or insoluble; it is perhaps begging the question to classify them as slightly soluble.

Bearing these qualifications in mind, we can classify the more common ionic solids on the basis of their solubility behavior according to the rules outlined in Table 18.1. The use of solubility rules to predict the results of precipitation reactions is illustrated by Example 18.2.

You will find this table very useful in predicting the results of precipitation reactions

TABLE 18.1 SOLUBILITY RULES

NO_3^-	All nitrates are soluble.
Cl^-	All chlorides are soluble except AgCl, Hg_2Cl_2, and $PbCl_2$.*
SO_4^{2-}	All sulfates are soluble except $CaSO_4$,* $SrSO_4$, $BaSO_4$, Hg_2SO_4, $HgSO_4$, $PbSO_4$, and Ag_2SO_4.*
CO_3^{2-}	All carbonates are insoluble except those of the 1A elements and NH_4^+.
OH^-	All hydroxides are insoluble except those of the 1A elements, $Sr(OH)_2$ and $Ba(OH)_2$. ($Ca(OH)_2$ is slightly soluble.)
S^{2-}	All sulfides except those of the 1A and 2A elements and NH_4^+ are insoluble.

*Insoluble compounds are those which precipitate upon mixing equal volumes of solutions 0.1 *M* in the corresponding ions. Compounds which fail to precipitate at concentrations slightly below 0.1 *M* are starred.

Example 18.2 Write balanced equations for the reactions, if any, that occur when equal volumes of 0.1 *M* solutions of the following compounds are mixed:

a. $AgNO_3$ and Na_2CO_3 b. Na_2SO_4 and KOH c. $CuSO_4$ and $Ba(OH)_2$

Solution In each case we first write down the formulas of the two possible precipitates and then decide, on the basis of the solubility rules, whether one or both of these compounds will precipitate.

a. Possible precipitates: Ag_2CO_3, $NaNO_3$. Table 18.1 indicates that Ag_2CO_3 is insoluble (Ag is not a 1A element!), whereas $NaNO_3$ is soluble (all nitrates are soluble). Consequently, we have

$$2\ Ag^+(aq) + CO_3^{2-}(aq) \rightarrow Ag_2CO_3(s)$$

b. Possible precipitates: NaOH, K_2SO_4. From the solubility rules, it is clear that both of these compounds are soluble. No precipitation reaction occurs.

c. Possible precipitates: $Cu(OH)_2$, $BaSO_4$. Of these compounds, $BaSO_4$ is specifically listed as insoluble; it can be deduced that $Cu(OH)_2$ is also insoluble. Two precipitation reactions occur simultaneously:

$$Ba^{2+}(aq) + SO_4^{2-}(aq) \rightarrow BaSO_4(s)$$

and

$$Cu^{2+}(aq) + 2\ OH^-(aq) \rightarrow Cu(OH)_2(s)$$

Unfortunately there is no simple theory which enables us to explain the solubility rules listed in Table 18.1. The difficulties involved in predicting relative solubilities of ionic compounds are inherent in the nature of the solution process, in which solvent water molecules are intimately involved. Normally when a solid dissolves in a liquid, we expect both the enthalpy change, ΔH, and the entropy change, ΔS, to be positive, as pointed out in Chapters 12 and 14. When an ionic solute dissolves in water, this may not be the case. The formation of hydrated ions (Chapter 13) evolves heat, which tends to make ΔH negative, and orients water molecules in fixed positions around ions, which tends to decrease the entropy. These complications, among others, make it difficult to predict from first principles the relative solubilities of different ionic compounds in water. The fact that we have to depend upon empirical rules to predict solubilities is another illustration of our basic lack of knowledge concerning the structure of electrolyte solutions and, indeed, of water itself.

In most cases we find that the solubilities of ionic compounds are inversely related to the charge densities (charge to size ratio) of their ions. Compounds containing ions of low charge density (e.g., large ions of +1 or −1 charge) tend to be more soluble than those containing highly charged small ions. We find, for example, that virtually all the compounds of K^+ (charge = +1, r = 0.133 nm) are soluble in water, while many Ca^{2+} (charge = +2, r = 0.099 nm) salts such as CaF_2, $CaCO_3$, and $CaSO_4$ are relatively insoluble. Again virtually all the salts containing NO_3^- are soluble. In contrast, many compounds containing anions of higher charge (e.g., CO_3^{2-}) or smaller size (e.g., OH^-) are insoluble in water.

18.3 SOLUBILITY EQUILIBRIA

Qualitative Aspects: Common Ion Effect

When silver acetate is dissolved in pure water to form a saturated solution, an equilibrium is established between the solid and its ions in solution.

$$AgC_2H_3O_2(s) \rightleftharpoons Ag^+(aq) + C_2H_3O_2^-(aq) \qquad (18.3)$$

The concentrations of Ag^+ and $C_2H_3O_2^-$ ions under these conditions are equal to each other and have a fixed value, independent of the amount of silver acetate or water used in preparing the saturated solution. At 20°C the equilibrium concentrations of Ag^+ and $C_2H_3O_2^-$ in a solution prepared by dissolving silver acetate in pure water are about 0.045 mol/dm³.

It is possible to change the relative concentrations of Ag^+ and $C_2H_3O_2^-$ ions in equilibrium with $AgC_2H_3O_2(s)$ in various ways. In particular, we can do this by adding a solution of a salt containing one of these ions. We might, for example, add a concentrated solution of $AgNO_3$. The added Ag^+ ions disturb the equilibrium represented by Equation 18.3; the reverse reaction occurs to precipitate more silver acetate. This behavior is referred to as the *common ion effect*. When equilibrium is reestablished, the concentration of $C_2H_3O_2^-$ ion is lower than it was originally, perhaps 0.020 M (orig. conc. = 0.045 M), while that of Ag^+ ion is higher, perhaps 0.10 M (orig. conc. = 0.045 M). The experiment which we have described is shown in Plate 10.

The common ion effect explains why the solubility of a solid such as $AgC_2H_3O_2$ is lower in a solution of a salt containing a common ion (Ag^+ or $C_2H_3O_2^-$) than in pure water. We find, for example, that if we add silver acetate to a 0.10 M solution of $AgNO_3$, only about 0.02 mol/dm³ of the solid dissolves, as compared to 0.045 mol/dm³ in pure water. In effect, the presence of Ag^+ ions in the solution reduces the extent to which Reaction 18.3 takes place. The same effect would be observed if $AgC_2H_3O_2(s)$ were added to a solution of sodium acetate or any other salt containing $C_2H_3O_2^-$ ions.

Similarly, $CaSO_4$ is less soluble in Na_2SO_4 solution than in water

Quantitative Treatment: Solubility Product Constant

As the above discussion implies, there is an inverse relationship between the equilibrium concentrations of ions in contact with the electrolyte from which they are derived. Any process which shifts the equilibrium so as to increase the concentration of one of these ions decreases the concentration of the other. The quantitative relationship between these concentrations can be introduced by considering a slightly soluble 1:1 electrolyte, MX, in equilibrium with its ions in aqueous solution:

$$MX(s) \rightleftharpoons M^+(aq) + X^-(aq) \quad (18.4)$$

For this reaction the equilibrium constant expression takes the form

$$K_{sp} = [M^+] \times [X^-] \quad (18.5)$$

Some solid MX must be present, however, if equilibrium is to exist

where the square brackets, as always, are used to denote equilibrium concentrations in moles per cubic decimetre, of the ions M^+ and X^- in the solution. The concentration of solid MX does not appear in the equilibrium expression. (Recall the discussion of gas-solid equilibrium in Section 15.2, Chapter 15.) For example, for silver acetate we have

$$[Ag^+] \times [C_2H_3O_2^-] = 2 \times 10^{-3} \text{ (at 25°C)} = K_{sp} \text{ of } AgC_2H_3O_2$$

The quantity K_{sp} appearing in Equation 18.5 is a particular type of equilibrium constant referred to as the *solubility product constant*. For any given electrolyte K_{sp} has a fixed value independent of the concentrations of the individual ions in equilibrium with the solid.

The concept of the solubility product constant is readily extended to electrolytes of any valence type. For an electrolyte MX_2,

$$MX_2(s) \rightleftharpoons M^{2+}(aq) + 2\ X^-(aq) \quad (18.6)$$

we can write

$$K_{sp} = [M^{2+}] \times [X^-]^2$$

Thus, we have, for example,

$$K_{sp} \text{ of } PbCl_2 = [Pb^{2+}] \times [Cl^-]^2 = 1.7 \times 10^{-5}$$

Again, for a compound such as bismuth sulfide, Bi_2S_3,

$$Bi_2S_3(s) \rightleftharpoons 2\ Bi^{3+}(aq) + 3\ S^{2-}(aq) \quad (18.7)$$

$$K_{sp} = [Bi^{3+}]^2 \times [S^{2-}]^3 = 1 \times 10^{-72}$$

In the general case, the solubility product principle may be stated as follows. **In any water solution in equilibrium with a slightly soluble ionic compound, the product of the concentrations of the ions, each raised to a power equal to its coefficient in the net ionic equation for the solution process, has a constant value.**

In principle at least, the numerical value of K_{sp} can be calculated from the measured solubility, or vice versa (Example 18.3).

Example 18.3

a. The measured solubility of $AgC_2H_3O_2$ at 20°C is 0.045 mol/dm^3. Calculate K_{sp} for this salt.
b. K_{sp} of the mineral fluorite, CaF_2, is known to be 2×10^{-10}. Estimate the solubility, in moles per cubic decimetre, of this mineral.

Solution

a. From Equation 18.3, it is evident that for every mole of $AgC_2H_3O_2$ that dissolves, a mole of Ag^+ and a mole of $C_2H_3O_2^-$ enter the solution. Hence, when 0.045 mol/dm³ of solid dissolves, the equilibrium concentrations of Ag^+ and $C_2H_3O_2^-$ must both be 0.045 *M*.

$$K_{sp} = [Ag^+] \times [C_2H_3O_2^-] = (0.045) \times (0.045) = 2.0 \times 10^{-3}$$

b. From the equation:

$$CaF_2(s) \rightarrow Ca^{2+}(aq) + 2\ F^-(aq)$$

we see that for every mole of CaF_2 that dissolves, one mole of Ca^{2+} and two moles of F^- enter the solution. Consequently, if we let

$$S = \text{solubility } CaF_2 \text{ (moles per cubic decimetre)}$$

then:

$$[Ca^{2+}] = S, \text{ and } [F^-] = 2\ S$$

Substituting in the expression:

$$K_{sp} = [Ca^{2+}] \times [F^-]^2 = 2 \times 10^{-10}$$

$$(S)(2\ S)^2 = 2 \times 10^{-10};\ 4\ S^3 = 2 \times 10^{-10};\ S^3 = 0.5 \times 10^{-10} = 50 \times 10^{-12}$$

Solving,

$$S = (50)^{1/3} \times 10^{-4} = 3.7 \times 10^{-4} \text{ mol/dm}^3$$

You must distinguish carefully between *solubility* and *solubility product constant*

In practice, the observed solubility of an ionic compound in water is usually somewhat greater than the value calculated from K_{sp} by the procedure illustrated in Example 18.3*b*. This occurs because some of the solid dissolves to form species in solution other than the simple ions which appear in the expression for K_{sp}. In the particular case of calcium fluoride, some of the calcium in solution is in the form of the "ion pair," CaF^+, rather than the simple Ca^{2+} ion. This effect is particularly important when both ions have relatively high charges (e.g., +2 and −2). With calcium sulfate, for example, about half of the calcium in solution is in the form of the ion pair, $CaSO_4$. This means that the solubility of this solid is about twice as great as would be calculated from K_{sp}, where it is assumed that the only calcium-containing species in solution is the Ca^{2+} ion.

It is also commonly found that the solubility of an ionic solid is greater in a solution of a "foreign" electrolyte (i.e., one having no ions in common with the solid) than it is in pure water. For example, the solubility of AgCl in 0.10 *M* KNO_3 solution is about 30% greater than it is in water. This effect can be explained, at least in part, by the "ion atmosphere" model of electrolyte solutions discussed in Chapter 13. The solution equilibrium

$$AgCl(s) \rightleftharpoons Ag^+(aq) + Cl^-(aq)$$

is shifted to the right if the Ag^+ and Cl^- ions in solution are able to form such an atmosphere (i.e., if they can surround themselves with ions of opposite charge). Adding foreign ions such as K^+ and NO_3^- promotes the formation of this atmosphere and hence makes the solution process more spontaneous.

In this text, we will neglect effects such as those just discussed and assume that the solubility product expression is followed quantitatively. This is an approximation ordinarily followed by chemists simply because it is not feasible to include these effects except under special conditions.

TABLE 18.2 SOLUBILITY PRODUCT CONSTANTS AT 25°C

Acetates	$AgC_2H_3O_2$	2×10^{-3}	Iodides	AgI	1×10^{-16}
				PbI_2	1×10^{-8}
Bromides	AgBr	5×10^{-13}			
	$PbBr_2$	5×10^{-6}	Sulfates	$BaSO_4$	1.0×10^{-10}
				$CaSO_4$	3×10^{-5}
Carbonates	$BaCO_3$	2×10^{-9}		$PbSO_4$	1×10^{-8}
	$CaCO_3$	5×10^{-9}			
	$MgCO_3$	2×10^{-8}	Sulfides	Ag_2S	1×10^{-49}
				CdS	1×10^{-27}
Chlorides	AgCl	1.6×10^{-10}		CoS	1×10^{-21}
	Hg_2Cl_2	1×10^{-18}		CuS	1×10^{-36}
	$PbCl_2$	1.7×10^{-5}		FeS	1×10^{-18}
				HgS	1×10^{-52}
Chromates	Ag_2CrO_4	2×10^{-12}		MnS	1×10^{-13}
	$BaCrO_4$	2×10^{-10}		NiS	1×10^{-21}
	$PbCrO_4$	2×10^{-14}		PbS	1×10^{-28}
				ZnS	1×10^{-23}
Fluorides	BaF_2	2×10^{-6}			
	CaF_2	2×10^{-10}			
	PbF_2	4×10^{-8}			
Hydroxides	$Al(OH)_3$	5×10^{-33}			
	$Cr(OH)_3$	1×10^{-30}			
	$Fe(OH)_2$	1×10^{-15}			
	$Fe(OH)_3$	5×10^{-38}			
	$Mg(OH)_2$	1×10^{-11}			
	$Zn(OH)_2$	5×10^{-17}			

We can use solubility product constants such as those listed in Table 18.2 to:

1. Calculate the concentration of one ion in equilibrium with another in solution (Example 18.4).
2. Determine whether or not a precipitate will form when two solutions are mixed (Example 18.5).

Example 18.4 Mg^{2+} ions in seawater are extracted by adding calcium hydroxide to precipitate $Mg(OH)_2$ ($K_{sp} = 1 \times 10^{-11}$). If enough hydroxide ions are added to form a precipitate and make the final concentration of $OH^- = 1 \times 10^{-3}$ *M:*

a. What is the equilibrium concentration of Mg^{2+} at this point?
b. What percentage of the Mg^{2+} originally present remains in the seawater (original conc. $Mg^{2+} = 5 \times 10^{-2}$ *M*)?

Solution

a. From the equation:

$$Mg(OH)_2(s) \rightleftarrows Mg^{2+}(aq) + 2\ OH^-(aq)$$

$$K_{sp} = [Mg^{2+}] \times [OH^-]^2 = 1 \times 10^{-11}$$

Since $[OH^-] = 1 \times 10^{-3}$ *M*, we have:

$$[Mg^{2+}] \times (1 \times 10^{-3})^2 = 1 \times 10^{-11}$$

$$[Mg^{2+}] = \frac{1 \times 10^{-11}}{1 \times 10^{-6}} = 1 \times 10^{-5}$$

b. Originally, the concentration of Mg^{2+} was 5×10^{-2} *M;* it is now 1×10^{-5} *M*

$$\%Mg^{2+} \text{ remaining} = \frac{1 \times 10^{-5}}{5 \times 10^{-2}} \times 100 = 0.02\%$$

In other words 99.98% of the Mg^{2+} has been precipitated out of the seawater as $Mg(OH)_2$.

Example 18.5 Hard water containing Ca^{2+} ions frequently precipitates calcium sulfate, $CaSO_4$. When the concentration of Ca^{2+} is 0.01 *M*, will a precipitate of $CaSO_4$ form if

a. enough SO_4^{2-} is added to make its concentration 0.001 *M*?
b. 400 cm³ of 0.02 *M* Na_2SO_4 is added to 800 cm³ of the hard water?

Solution

a. To decide whether or not a precipitate forms, we must compare the actual concentration product to that required for equilibrium ($K_{sp} = 3 \times 10^{-5}$)

$$\text{conc. } Ca^{2+} \times \text{conc. } SO_4^{2-} = (0.01) \times (0.001) = 1 \times 10^{-5}$$

Since the concentration product is smaller than K_{sp}, we conclude that there are not enough ions in the solution to establish equilibrium. In other words, no $CaSO_4$ forms.

b. This problem is analogous to (a), except for one important factor. We must calculate the concentrations of Ca^{2+} and SO_4^{2-} *after* the two solutions are mixed, since the volume available to the ions has increased. To obtain the concentration of Ca^{2+}, we note that the 800 cm³ of solution 0.01 *M* in Ca^{2+} has, in effect, been diluted to 1200 cm³ by the addition of the Na_2SO_4 solution. Consequently the concentration of Ca^{2+} has decreased by a factor of 800/1200:

$$\text{conc. } Ca^{2+} = 0.01 \times \frac{800}{1200} = 0.007\ M$$

Similarly: $$\text{conc. } SO_4^{2-} = 0.02 \times \frac{400}{1200} = 0.007\ M$$

Consequently: $$(\text{conc. } Ca^{2+}) \times (\text{conc. } SO_4^{2-}) = (7 \times 10^{-3})(7 \times 10^{-3}) = 49 \times 10^{-6} = 5 \times 10^{-5}$$

Clearly, the concentration product, 5×10^{-5}, is greater than K_{sp}, 3×10^{-5}. Solid $CaSO_4$ will precipitate until the concentrations of Ca^{2+} and SO_4^{2-} drop to the point at which their product becomes equal to 3×10^{-5}.

The general principle, illustrated by this example, is that a **precipitate will form when two solutions are mixed if, and only if, the product of the ion concentrations, each raised to the power given by its coefficient in the solubility equation, exceeds K_{sp}.** Thus, for $CaSO_4$, we will obtain a precipitate if

$$(\text{conc. } Ca^{2+}) \times (\text{conc. } SO_4^{2-}) > K_{sp} = 3 \times 10^{-5}$$

while for $PbCl_2$ a precipitate will form if

$$(\text{conc. } Pb^{2+}) \times (\text{conc. } Cl^{-})^2 > K_{sp} = 1.7 \times 10^{-5}$$

If the product is less than K_{sp}, no solid will come down and equilibrium will not be established.

18.4 PRECIPITATION REACTIONS IN ANALYTICAL CHEMISTRY

Quantitative Analysis

Precipitation reactions are a useful tool to the analyst who wants to know the percentage of a particular element or compound in a mixture. In the branch of analytical chemistry known as **gravimetric analysis**, a weighed sample is first brought into solution by adding a suitable solvent. A reagent is then added which precipitates only the species being analyzed for. To complete the analysis, the precipitate is separated and weighed. The calculations involved in this type of analysis are illustrated in Example 18.6.

Example 18.6 A sample of insecticide weighing 1.200 g is analyzed for arsenic by dissolving it in acid and adding hydrogen sulfide in excess. When this is done, 0.260 g of arsenic sulfide, As_2S_3, is obtained. Calculate the percentage of arsenic in the insecticide.

Solution From the formula As_2S_3, we see that

$$2 \text{ mol As} \simeq 1 \text{ mol } As_2S_3$$

$$149.8 \text{ g As} \simeq 246.0 \text{ g } As_2S_3$$

The mass of arsenic in 0.260 g of As_2S_3 must then be

$$0.260 \text{ g } As_2S_3 \times \frac{149.8 \text{ g As}}{246.0 \text{ g } As_2S_3} = 0.158 \text{ g As}$$

Knowing the mass of As and the total mass of the sample, we can readily calculate the per cent of As in the sample.

$$\%\text{As} = \frac{\text{mass As}}{\text{mass sample}} \times 100 = \frac{0.158 \text{ g}}{1.200 \text{ g}} \times 100 = 13.2\% \text{ As}$$

In gravimetric and volumetric analysis, precipitation reactions must go essentially to completion

Another way to determine the percentage of a substance in a mixture is to measure the volume of a reagent of known concentration required to react exactly with it. Analyses based on volume measurements of this type fall in the general area of **volumetric analysis**. They may involve precipitation (Example 18.7) or other types of reactions.

Example 18.7 The percentage of Cl^- ion in a powdered bleach is found by measuring the volume of 0.184 *M* $AgNO_3$ required to react with it. It is found that 15.0 cm^3 of $AgNO_3$ is required to react with a sample weighing 0.800 g. Calculate the per cent of Cl^- in the bleach.

Solution Knowing the volume and concentration of the $AgNO_3$ solution, we can readily calculate the number of moles of $AgNO_3$ added

$$\text{no. moles } AgNO_3 = 0.0150 \text{ dm}^3 \times 0.184 \frac{\text{mol}}{\text{dm}^3} = 0.002\ 76 \text{ mol}$$

But this must be the number of moles of Ag^+ (1 mol $Ag^+ \simeq$ 1 mol $AgNO_3$) and also the number of moles of Cl^- (1 mol Ag^+ reacts with 1 mol Cl^-)

$$\text{no. moles } Cl^- = 0.002\ 76 \text{ mol}$$

Having calculated the number of moles of Cl^-, we next obtain the number of grams of Cl^- and finally its percentage

$$\text{no. grams } Cl^- = 0.002\ 76 \text{ mol } Cl^- \times \frac{35.45 \text{ g } Cl^-}{1 \text{ mol } Cl^-} = 0.0978 \text{ g } Cl^-$$

$$\%Cl^- = \frac{\text{grams } Cl^-}{\text{grams sample}} \times 100 = \frac{0.0978 \text{ g}}{0.800 \text{ g}} \times 100 = 12.2\% \text{ } Cl^-$$

In this method for the determination of chloride, as in all volumetric analyses, it is essential to know the exact point at which reaction is complete. In principle, we could do this by noting the point at which the precipitate of silver chloride stops forming. In practice, it is more convenient to add a substance known as an *indicator*, which changes color when the equivalence point is reached, that is, when chemically equivalent quantities of precipitating agent and sample are present. A few drops of potassium chromate solution added to a chloride sample serves as a suitable indicator for titration with silver ions. At the equivalence point, when essentially all the chloride ion has been removed as silver chloride, a dark red precipitate of silver chromate, Ag_2CrO_4, forms. Since this compound requires a somewhat higher concentration of Ag^+ to precipitate than does silver chloride, Ag_2CrO_4 does not form as long as an appreciable amount of chloride ion remains in solution.

The experimental setup for the quantitative determination of Cl^- ion by this method, referred to as a Mohr titration, is shown in Plate 11.

Qualitative Analysis

The qualitative detection of ions in a mixture is commonly accomplished by a scheme of analysis in which precipitation reactions play a major role. The general procedure in such a scheme is illustrated in Example 18.8.

Example 18.8 Develop a scheme based on the information given in Table 18.1 to separate and identify the ions Ag^+, Cu^{2+}, and Ca^{2+} in a water solution.

Solution We look first for a reagent to precipitate one ion, leaving the others in solution. Clearly, Cl^- qualifies, since it precipitates Ag^+ but not Cu^{2+} or Ca^{2+}. The first step in the scheme might then be the addition of Cl^- in the form of a solution of hydrochloric acid. Formation of a precipitate would indicate Ag^+; if no precipitate forms, Ag^+ must be absent.

We next choose a reagent to distinguish between Cu^{2+} and Ca^{2+}. We see from Table 18.1 that S^{2-} is a suitable choice. Addition of a solution of sodium sulfide would precipitate Cu^{2+} if it is present, leaving Ca^{2+} in solution.

At this point, we need only test for Ca^{2+}. This could be done by adding a solution of Na_2CO_3, which will precipitate Ca^{2+} as $CaCO_3$. Failure to obtain a precipitate at this point would show Ca^{2+} to be absent. In summary

$Ag^+(aq)$, $Cu^{2+}(aq)$, $Ca^{2+}(aq)$

↓ $Cl^-(aq)$

$AgCl(s)$ | $Cu^{2+}(aq)$, $Ca^{2+}(aq)$

↓ $S^{2-}(aq)$

$CuS(s)$ | $Ca^{2+}(aq)$

↓ $CO_3^{2-}(aq)$

$CaCO_3(s)$

In a qual scheme, we remove ions one at a time, usually starting with the ion which forms the fewest soluble compounds

The scheme developed in Example 18.8 illustrates two important points concerning qualitative analysis:

1. The order in which reagents are added is crucial. It would have been useless to have added CO_3^{2-} to the original solution since it gives a precipitate with all three cations.

2. All reactions must be essentially quantitative. Any Ag^+ left after the first step would interfere with the tests for Cu^{2+} and Ca^{2+}, since both Ag_2S and Ag_2CO_3 are insoluble.

Several different analytical schemes have been devised to separate and identify the 25 or more ions commonly included in a laboratory course in qualitative analysis. Needless to say, these schemes are a great deal more complex than that outlined in Example 18.8. Many other types of reactions in addition to precipitation are used. We shall consider later how acid-base reactions (Chapter 20), complex-ion formation reactions (Chapter 21), and oxidation-reduction reactions (Chapter 23) are utilized in qualitative analysis. At this stage, it may be useful to give a brief outline of one of the simpler schemes of cation analysis.

The cations are first separated into five major groups on the basis of the solubilities of their compounds.

GROUP 1 (Ag^+, Pb^{2+}, Hg_2^{2+}). Separated as insoluble chlorides by addition of dilute HCl. All other cations remain in solution, since their chlorides are soluble.

GROUP 2 (Cu^{2+}, Bi^{3+}, Cd^{2+}, Hg^{2+}, As^{3+}, Sb^{3+}, Sn^{4+}). All these ions form extremely insoluble sulfides (e.g., K_{sp} CuS $= 1 \times 10^{-36}$). The sulfides are precipitated by adding H_2S in acidic solution, in which the concentration of S^{2-} is very small.

GROUP 3 (Al^{3+}, Cr^{3+}, Fe^{3+}, Zn^{2+}, Ni^{2+}, Co^{2+}, Mn^{2+}). These sulfides are somewhat more soluble than those of Group 2 (e.g., K_{sp} MnS $= 1 \times 10^{-13}$). The ions are precipitated by adding H_2S in basic solution, where there is a relatively high concentration of S^{2-}. Under these conditions, Al^{3+} and Cr^{3+} precipitate as the hydroxides, the other ions as sulfides.

GROUP 4 (Mg^{2+}, Ca^{2+}, Sr^{2+}, Ba^{2+}). May be precipitated as carbonates.

GROUP 5 (Na^+, K^+, NH_4^+). Special reagents are ordinarily used to identify these ions. The violet color imparted to a flame by K^+ is a useful test for that ion.

Once the cations have been separated into groups, the problem becomes one of further subdividing the groups. We shall not attempt at this point to describe the chemistry involved in these separations. A single example may, however, be instructive. Lead can be separated from the other Group 1 ions by boiling the chloride precipitate with water, filtering the hot mixture, and testing the hot filtrate with CrO_4^{2-}. The formation of a yellow precipitate of lead chromate, $PbCrO_4$, which is much less soluble than $PbCl_2$, indicates the presence of lead.

$$PbCl_2(s) \rightarrow Pb^{2+}(aq) + 2\ Cl^-(aq)$$

$$Pb^{2+}(aq) + CrO_4^{2-}(aq) \rightarrow PbCrO_4(s)$$

$$PbCl_2(s) + CrO_4^{2-}(aq) \rightarrow 2\ Cl^-(aq) + PbCrO_4(s) \qquad (18.9)$$

18.5 PRECIPITATION REACTIONS IN INORGANIC PREPARATIONS

Precipitation reactions are a convenient source of insoluble ionic compounds. Barium sulfate can be prepared by adding a slight excess of sulfuric acid to a solution of a barium salt

$$Ba^{2+}(aq) + SO_4^{2-}(aq) \rightarrow BaSO_4(s) \qquad (18.10)$$

The precipitate is filtered, washed to remove foreign ions, and dried at 100°C to give a product of high purity. A similar technique is used to prepare the insoluble silver halides (AgCl, AgBr, AgI), used in photography, from silver nitrate. As noted earlier, an important industrial process, the extraction of magnesium from seawater, depends upon a precipitation reaction. Calcium hydroxide is added to precipitate magnesium hydroxide, thereby separating Mg^{2+} from the more abundant Na^+ ions.

$$Mg^{2+}(aq) + 2\ OH^-(aq) \rightarrow Mg(OH)_2(s) \qquad (18.11)$$

In addition to their use in preparing insoluble inorganic compounds, precipitation reactions offer a means of converting one soluble ionic compound to another. Suppose, for example, we wish to convert Na_2CO_3 to NaOH. One way to do this is to add a solution of $Ca(OH)_2$. A precipitate of $CaCO_3$ forms, leaving Na^+ and OH^- ions in solution. Evaporation of this solution gives solid sodium hydroxide.

The spectator ions Na^+ and OH^- are included here because they are required for the final step (Eq. 18.12)

Precipitation:

$$2\ Na^+(aq) + CO_3^{2-}(aq) + Ca^{2+}(aq) + 2\ OH^-(aq) \rightarrow CaCO_3(s) + 2\ Na^+(aq) + 2\ OH^-(aq)$$

Evaporation:

$$2\ Na^+(aq) + 2\ OH^-(aq) \rightarrow 2\ NaOH(s) \qquad (18.12)$$

About one fifth of the 10^{10} kg of sodium hydroxide prepared annually in the world is made by this two-step sequence.

Reviewing the reaction sequence that we have just gone through, we see that the basic principle is a simple one. To convert a compound MX (e.g., Na_2CO_3) to another compound, MY (e.g., NaOH), we add a reagent ($Ca(OH)_2$) containing the desired anion, Y, and a cation which forms a precipitate with the anion we want to get rid of, X. By filtering off this precipitate, we effectively substitute Y for X. Applying this principle in another case, we see that metal chlorides, MCl, can be converted to the corresponding nitrates, MNO_3, by adding $AgNO_3$. After filtering off the insoluble AgCl, we are left with M^+ and NO_3^- ions in solution; evaporation gives $MNO_3(s)$.

By a simple extension of this principle, it is possible to "exchange cations" in ionic compounds. For example, to convert $BaCl_2$ to $MgCl_2$, we first add a solution of $MgSO_4$:

$$Ba^{2+}(aq) + 2\ Cl^-(aq) + Mg^{2+}(aq) + SO_4^{2-}(aq) \rightarrow BaSO_4(s) + Mg^{2+}(aq) + 2\ Cl^-(aq)$$

The precipitated barium sulfate is then filtered off and the solution evaporated to bring the Mg^{2+} and Cl^- ions out of solution:

$$Mg^{2+}(aq) + 2\ Cl^-(aq) \rightarrow MgCl_2(s) \qquad (18.13)$$

PROBLEMS

18.1 *Net Ionic Equations* Using Table 18.1, write a net ionic equation for the reaction that takes place when 0.200 dm^3 of 0.10 M $Ba(OH)_2$ solution is mixed with 0.300 dm^3 of 0.20 M $FeCl_2$ solution.

18.2 *Mole Relations* For the reaction described in Problem 18.1, give the number of moles of each ion present before and after reaction.

18.3 *K_{sp} and Solubility*

a. K_{sp} for $CaCO_3$ is 5×10^{-9}. Calculate the solubility of $CaCO_3$ in water in moles per cubic decimetre.
b. The solubility of $Zn(OH)_2$ in water is 2.3×10^{-6} mol/dm^3. Calculate K_{sp}.

18.4 *Applications of K_{sp}* K_{sp} of CdS is 1×10^{-27}.

a. What is $[S^{2-}]$ in equilibrium with 0.01 M Cd^{2+}?
b. Will a precipitate form when conc. $Cd^{2+} = 0.010\ M$, and conc. $S^{2-} = 1 \times 10^{-21}\ M$?

18.5 *Precipitation Reactions in Analysis* Addition of excess H_2SO_4 to a sample weighing 1.000 g gives 0.466 g of $BaSO_4$. What is the percentage of Ba in the sample? (*AM* Ba = 137; *FM* $BaSO_4$ = 233)

18.6 *Conversion of Salts* What reagent would you use to convert $MgCl_2$ to $Mg(NO_3)_2$? K_2SO_4 to KCl?

18.7 Write net ionic equations for the precipitation reactions that occur when solutions of the following are mixed:

a. NaCl and $Pb(NO_3)_2$
b. $Al(NO_3)_3$ and KOH
c. $Ni(NO_3)_2$ and KCl
d. $Ba(OH)_2$ and $FeSO_4$

18.8 Explain why $CaSO_4$ is

a. more soluble in water than in 1 M H_2SO_4.
b. somewhat more soluble in KNO_3 solution than in pure water.
c. somewhat more soluble in water than predicted from its K_{sp} value.
d. much less soluble in water than is NaCl.

18.9 400 cm³ of 0.120 M $MgCl_2$ solution is added to 200 cm³ of 0.180 M KOH solution.

a. Calculate the number of moles of each ion present before reaction.
b. Write a balanced net ionic equation for the reaction that occurs.
c. Calculate the number of moles of each ion present after reaction.

18.10 Criticize the following statement: In any water solution, the product of the concentrations of Ag^+ and Cl^- is given by the K_{sp} value, 1.6×10^{-10}.

18.11 Complete the following table (K_{sp} AgBr = 5.0×10^{-13}) and plot the points on a graph.

$[Ag^+]$	$[Br^-]$
1.0×10^{-12}	______
2.0×10^{-12}	______
______	0.20
______	0.10

18.12 What is the solubility of AgBr (the number of moles per cubic decimetre that dissolves) in 0.30 M NaBr?

18.13 Using K_{sp} values given in Table 18.2, calculate the solubility, in moles per cubic decimetre, of

a. $BaCO_3$
b. PbF_2

18.14 The solubility of PbI_2 in pure water is about 1.3×10^{-3} mol/dm³. Estimate its solubility product constant.

18.26 Write net ionic equations for the precipitation reactions that occur when solutions of the following are mixed.

a. Na_2CO_3 and $BaCl_2$
b. BaS and $ZnSO_4$
c. KNO_3 and $CuCl_2$
d. $SbCl_3$ and Na_2S

18.27 You are given a beaker containing a saturated solution of $PbCl_2$ in pure water. Suggest two reagents which you might add to this solution to give a precipitate of $PbCl_2$. Can you suggest two ways in which you might increase the solubility of $PbCl_2$?

18.28 0.200 dm³ of 0.150 M $Al_2(SO_4)_3$ solution is added to 0.300 dm³ of 0.100 M NaOH. Calculate

a. the number of moles of $Al(OH)_3$ formed.
b. the number of moles of each ion present before and after reaction.
c. the final concentration, in moles per cubic decimetre, of each ion.

18.29 Criticize the following statement: Since K_{sp} of Ag_2CrO_4 is smaller than that of AgCl (2×10^{-12} vs. 1.6×10^{-10}), Ag_2CrO_4 must be less soluble in water than AgCl.

18.30 Complete the following table:

$[Ag^+]$	$[CrO_4^{2-}]$
2×10^{-6}	______
1×10^{-4}	______
______	4×10^{-6}
______	2×10^{-4}

(K_{sp} Ag_2CrO_4 = 2×10^{-12})

18.31 What is the solubility of $PbCl_2$ in 2.0 M HCl?

18.32 Calculate the solubility of

a. $Mg(OH)_2$, in moles per cubic decimetre.
b. CoS, in grams per cubic decimetre.

18.33 The solubility of $CaCO_3$ in pure water is about 0.007 g/dm³. Estimate the value of K_{sp} for $CaCO_3$.

18.15 Decide whether a precipitate will form when

a. enough Ag^+ is added to tap water, in which conc. $Cl^- = 2 \times 10^{-4}$ M, to make conc. $Ag^+ = 1 \times 10^{-5}$ M.
b. 200 cm³ of 0.010 M $Pb(NO_3)_2$ solution is mixed with 300 cm³ of 0.080 M HCl.

18.16 The concentration of Pb^{2+} in a certain water supply is 2×10^{-5} M. If it is desired to remove 80% of the Pb^{2+} by adding CrO_4^{2-}, what must be $[CrO_4^{2-}]$?

18.17 Ground water which comes in contact with the mineral fluorite becomes saturated with CaF_2.

a. What is the concentration of F^- in the water?
b. Will a precipitate form if enough Ba^{2+} is added to make its conc. 1×10^{-3} M?

18.18 Hydrogen sulfide is slowly added to a solution 0.10 M in both Co^{2+} and Fe^{2+}.

a. Which ion precipitates first?
b. What is the concentration of the first ion when the second precipitate starts to form?

18.19 If Cl^- is slowly added to a solution 0.20 M in Pb^{2+}

a. has a precipitate formed when conc. $Cl^- = 5 \times 10^{-3}$ M?
b. at what conc. of Cl^- does a precipitate start to form?
c. what percentage of the Pb^{2+} remains when conc. $Cl^- = 2.0 \times 10^{-2}$ M?

18.20 An analyst determines the percentage of lead in an insecticide by precipitating it as PbS. If he obtains 0.150 g of precipitate from a sample weighing 1.098 g, what is the percentage of lead?

18.21 A student determines the per cent of Cl^- in a mixture by weighing out 0.400 g, dissolving in 20 cm³ of water, and titrating with 15.6 cm³ of 0.120 M $AgNO_3$ solution. What is the per cent of Cl^- in the mixture?

18.22 Develop a scheme, based on Table 18.1, to separate and identify the ions in a mixture containing

a. Ag^+, Cu^{2+}, Ba^{2+}
b. Cl^-, OH^-, SO_4^{2-}

18.23 To analyze an unknown which may contain SO_4^{2-}, Cl^-, and S^{2-}, a student adds Pb^{2+} to one portion and Ba^{2+} to another. In both cases, he gets a precipitate. On this basis, what can he say about the identity of the unknown?

18.34 Will a precipitate form when

a. enough Pb^{2+} is added to a solution 0.020 M in I^- to make conc. $Pb^{2+} = 1.0 \times 10^{-3}$ M?
b. 500 cm³ of 0.010 M $BaCl_2$ solution is added to 300 cm³ of 0.010 M NaF?

18.35 The concentration of Mg^{2+} ion in seawater is 5×10^{-2} M. What is the $[OH^-]$ necessary to remove 90% of the Mg^{2+}?

18.36 Fluoridation of water supplies may add about one part per million of F^- ion, i.e., 1 g $F^-/10^6$ g of water.

a. What is the concentration of F^- in this water in moles per cubic decimetre?
b. Will a precipitate form when hard water in which conc. $Ca^{2+} = 2 \times 10^{-4}$ M is fluoridated?

18.37 A solution is 0.10 M in Cl^- and 0.10 M in CrO_4^{2-}. If Ag^+ ions are slowly added, what precipitate forms first? What is the concentration of the first ion when the second ion starts to precipitate? On the basis of your answers to these questions, suggest why CrO_4^{2-} ion might serve as a suitable indicator for the titration of Cl^- with Ag^+.

18.38 Enough Pb^{2+} is added to 2.0 dm³ of solution originally 1×10^{-4} M in SO_4^{2-} to form a precipitate and give a final Pb^{2+} conc. of 1.0×10^{-3} M.

a. What is $[SO_4^{2-}]$ at that point?
b. What percentage of the SO_4^{2-} is left?
c. How many moles of Pb^{2+} had to be added to do this?

18.39 An analyst wishes to test body tissue for arsenic by precipitation as As_2S_3. If the tissue contains about 0.020% As and if he must get at least 50 mg of As_2S_3, how large a sample must he use?

18.40 A student finds that 21.4 cm³ of 0.100 M NaOH is required to precipitate the Mg^{2+} in a 20.8 cm³ sample of seawater. If the precipitate formed is pure $Mg(OH)_2$, what is the conc. of Mg^{2+} in seawater?

18.41 Develop a scheme, based on Table 18.1, to separate and identify

a. Cu^{2+}, Ca^{2+}, Pb^{2+}
b. S^{2-}, Cl^-, OH^-

18.42 To analyze an unknown which may contain Ag^+, Ni^{2+}, and Na^+, a student adds first HCl and then HNO_3. In neither case does she get a precipitate. What can she say about the identity of the unknown?

18.24 A student is given a bottle containing a solid which may be either of two compounds. What test should she make to decide whether the compound is

a. NaCl or $CaCO_3$?
b. $Ba(NO_3)_2$ or KNO_3?
c. Na_2CO_3 or Na_2SO_4?

18.25 Describe in some detail how the following conversions might be accomplished:

a. $CuCl_2$ to CuS
b. $CuCl_2$ to $Cu(NO_3)_2$
c. $CuCl_2$ to $CuCO_3$

18.43 A student is given a bottle which may contain either of two compounds. How should he proceed to find out which is present,

a. NaCl or CH_4?
b. NaCl or $CaCl_2$?
c. Na_2CrO_4 or Na_2SO_4?

18.44 You have a sample of RbCl weighing 1.000 g which you wish to convert to pure $RbNO_3$.

a. What reagent should you add?
b. How many moles of the reagent should you add?
c. How would you obtain $RbNO_3(s)$?

*18.45 The white pigment in certain paints contains Pb^{2+} ions, which react with H_2S to form black, insoluble PbS. If the concentration of H_2S in air is 15 ppm, how many cubic decimetres of air at 25°C and one atmosphere must come in contact with one gram of Pb^{2+} to convert half of it to PbS? (Consult Table 17.1, Chapter 17, for the appropriate conversion factors.)

*18.46 The inside of a teakettle is coated with 10.0 g of $CaCO_3$. If the teakettle is washed with 1.00 dm^3 of pure water, what fraction of the precipitate is removed? How many successive 1.00-dm^3 portions of water would have to be used to remove half of the $CaCO_3$?

*18.47 The concentrations of various cations in seawater are:

ion	Na^+	Mg^{2+}	Ca^{2+}	Al^{3+}	Fe^{3+}
concentration (*M*)	0.46	0.050	0.01	4×10^{-7}	2×10^{-7}

a. At what conc. of OH^- does $Mg(OH)_2$ start to precipitate?
b. At this concentration, will any of the other ions precipitate?
c. If enough OH^- is added to precipitate 50% of the Mg^{2+}, what percentage of each of the other ions will be precipitated?
d. Under the conditions in (c), what mass of precipitate will be obtained from a litre of seawater?

*18.48 When $CaSO_4$ dissolves in water, the following equilibria are established:

$$CaSO_4(s) \rightleftharpoons CaSO_4(aq) \qquad K_1 = 6 \times 10^{-3}$$
$$CaSO_4(aq) \rightleftharpoons Ca^{2+}(aq) + SO_4^{2-}(aq) \qquad K_2 = 5 \times 10^{-3}$$

a. What are the equilibrium concentrations of $CaSO_4(aq)$, Ca^{2+}, and SO_4^{2-}?
b. What is the total number of moles of $CaSO_4$ that dissolve in 1.00 dm^3 of water?
c. How does your answer in (b) compare to that calculated directly from K_{sp} of $CaSO_4 = 3 \times 10^{-5}$?

19

ACIDS AND BASES

In the world of the chemist the most important solvent, as you have probably already recognized, is water. Water is so common a solvent for all sorts of reactions that we often forget that in many reactions its role is as important as that of the solutes. In one group of reactions in particular, water has an especially significant part to play; these reactions involve those species known as acids and bases.

19.1 THE DISSOCIATION OF WATER; NATURE OF ACIDS AND BASES

One might realistically say that acids and bases exist because of one of the key properties of water, namely, its dissociation to ions according to the equation:

$$H_2O \rightleftarrows H^+(aq) + OH^-(aq) \tag{19.1}$$

In any aqueous solution the above equilibrium between water and its ions, H^+ and OH^-, is, as we shall see, an exceedingly important one. The reaction to equilibrium is very rapid in both directions; it does not go very far to the right, so that at any moment only a relatively few H_2O molecules are dissociated. Following the general rules of the law of equilibrium, we would formulate the equilibrium expression for Reaction 19.1 as

$$K_c = \frac{[H^+] \times [OH^-]}{[H_2O]}$$

In aqueous solution the concentration of water is typically very large, about 55 M, and remains essentially constant for dilute solutions. Rather than carry its concentration in calculations, we invariably include it in K_c; under such conditions the equilibrium expression takes the form

$$K_w = [H^+] \times [OH^-] = 1.0 \times 10^{-14} \tag{19.2}$$

The dissociation constant of water is given the symbol K_w, and is, as you can see, very small, being about 1.0×10^{-14} at 25°C.

By Equation 19.2 any water solution at 25°C will contain H^+ and OH^- ions at such concentrations that their product will be 1.0×10^{-14}. If one is dealing with pure water, it is easy to calculate the concentrations of these two ions, since by Equation 19.1 they will form in equal amounts:

in pure H_2O $\quad [H^+] = [OH^-] \quad [H^+] \times [OH^-] = [H^+]^2 = 1.0 \times 10^{-14}$

$$[H^+] = 1.0 \times 10^{-7}\ M = [OH^-]$$

TABLE 19.1 RELATIONS BETWEEN $[H^+]$, $[OH^-]$ AND *pH* IN AQUEOUS SOLUTIONS

Solution	No. 1	No. 2	No. 3	No. 4	No. 5	No. 6	No. 7	No. 8	No. 9
$[H^+]$	10^0	10^{-2}	10^{-4}	10^{-6}	10^{-7}	10^{-8}	10^{-10}	10^{-12}	10^{-14}
$[OH^-]$	10^{-14}	10^{-12}	10^{-10}	10^{-8}	10^{-7}	10^{-6}	10^{-4}	10^{-2}	10^0
pH	0	2	4	6	7	8	10	12	14
	Acidic				Neutral	Basic			

A solution in which $[H^+]$ equals $[OH^-]$ is said to be *neutral.* Such solutions turn out to be relatively rare, but in principle at least, pure water is neutral.

Knowing $[H^+]$, one can always describe a solution as being acidic, basic or neutral

Ordinarily the concentrations of hydrogen and hydroxide ions in a solution are not equal. A water solution in which the hydrogen ion concentration is larger than the hydroxide ion concentration is said to be *acidic.* On the other hand, a water solution in which the hydroxide ion concentration is larger than that of hydrogen ion is *basic.* As the concentration of one ion goes up, that of the other must go down proportionally so that Equation 19.2 will be satisfied. This means that any solution in which $[H^+]$ is greater than $1.0 \times 10^{-7}\,M$ is acidic; a solution in which $[OH^-]$ is greater than $1.0 \times 10^{-7}\,M$ is basic. In Table 19.1 we have indicated some possible combinations of concentrations of these two ions. Solutions 1 through 4 would be classified as acidic; solutions 6 through 9 are progressively more basic.

19.2 *pH*

We have seen that it is possible to describe quantitatively the acidity or basicity of a water solution by specifying the concentration of H^+ ion. In 1909 Sörensen proposed an alternative method of accomplishing this purpose, making use of a term known as *pH,* defined as follows:

$$pH \equiv -\log_{10}[H^+] = \log_{10} 1/[H^+] \qquad (19.3)$$

A great many important solutions have a *pH* between 1 and 14, corresponding to a variation in hydrogen ion concentration of 10^{-1} to $10^{-14}\,M$. For such solutions it is sometimes convenient to express acidity in terms of *pH* rather than H^+ ion concentration, thereby avoiding the use of either small fractions or negative exponents (Table 19.2).

TABLE 19.2 *pH* OF SOME COMMON LIQUIDS

Lemon juice	2.2–2.4	Saliva, human	6.5–7.5
Vinegar	3.0	Blood, human	7.3–7.5
Tomato juice	3.5	Urine, human	4.8–8.4
Beer	4–5	Seawater	8.3
Cheese	4.8–6.4	Wine	2.8–3.8
Cow's milk	6.3–6.6	Drinking water	6.5–8.0

It follows from the definition that *the lower the pH, the more acidic a solution is;* conversely, the higher the *pH*, the more basic is the solution. Just as it is possible to differentiate between neutral, basic, and acidic solutions on the basis of the concentrations of H^+ or OH^- ions, so one can make this distinction in terms of *pH:*

neutral solution: $[H^+] = 10^{-7}\,M,\ pH = 7.0$
acidic solution: $[H^+] > 10^{-7}\,M,\ pH < 7.0$
basic solution: $[H^+] < 10^{-7}\,M,\ pH > 7.0$

Example 19.1 illustrates the calculation of the *pH* of water solutions and the interrelationships between *pH*, $[H^+]$, and $[OH^-]$.

Example 19.1 Calculate
a. the *pH* of a solution 0.020 *M* in H^+.
b. the *pH* of a solution 2.5×10^{-3} *M* in OH^-.
c. the $[H^+]$ in seawater.

Solution The conversion of $[H^+]$ to *pH* and vice versa involves a simple use of logarithms.

a. $[H^+] = 0.020\,M = 2.0 \times 10^{-2}\,M$

$$pH = -\log_{10}[H^+] = -\log_{10}(2.0 \times 10^{-2}) = -(0.3 - 2.0) = 1.7$$

b. Before calculating the *pH*, we need to know $[H^+]$. We can obtain this by making use of the relation: $[H^+] \times [OH^-] = 1.0 \times 10^{-14}$.

Could the *pH* of a solution be negative?

$$[H^+] = \frac{1.0 \times 10^{-14}}{[OH^-]} = \frac{1.0 \times 10^{-14}}{2.5 \times 10^{-3}} = 0.40 \times 10^{-11} = 4.0 \times 10^{-12}$$

$$pH = -\log_{10}(4.0 \times 10^{-12}) = -(0.6 - 12.0) = 11.4$$

c. From Table 19.2 we note that the *pH* of seawater is 8.3.

$$-\log_{10}[H^+] = 8.3$$

$$\log_{10}[H^+] = -8.3 = 0.7 - 9.0$$

Taking antilogs, $[H^+] = 5 \times 10^{-9}\,M$

(If you are not very familiar with the use of logarithms you should consult Appendix 4.)

The *pH*, $[H^+]$, or $[OH^-]$ of a water solution may be determined experimentally in a number of different ways, one of which involves the use of acid-base indicators, which undergo a color change over a rather narrow *pH* range (Plate 12). Litmus, which is red in acidic and blue in basic solution, was one of the first indicators and is still widely used. By the judicious use of one or more of the indicators listed in Table 19.3, p. 450, it is possible to bracket quite accurately the *pH* or $[H^+]$ of a solution. For example, if one finds that a solution gives a red (basic) color with phenolphthalein but a yellow (acidic) color with alizarine yellow, its *pH* must be approximately 10.

A universal indicator made by combining several acid-base indicators may be used to determine, within about one unit, the *pH* of any water solution. This mixture of indicators shows an entire spectrum of colors ranging from deep red in strongly acidic solution to deep blue in strongly basic solution (Plate 13). A similar

TABLE 19.3 TYPICAL ACID-BASE INDICATORS

NAME	pH INTERVAL	ACID COLOR (LOW pH)	BASE COLOR (HIGH pH)
Methyl violet	0.0–1.6	yellow	violet
Methyl yellow	2.9–4.0	red	yellow
Methyl orange	3.1–4.4	red	yellow
Methyl red	4.8–6.2	red	yellow
Bromthymol blue	6.0–8.0	yellow	blue
Thymol blue	8.0–9.6	yellow	blue
Phenolphthalein	8.2–10.0	colorless	pink
Alizarine yellow	10.1–12.0	yellow	red

principle is used to prepare pH paper, used to test the acidity of biological fluids and for soil testing. Strips of paper impregnated with a mixture of indicators can be designed to give gradations of color over a wide or narrow pH range.

In a more qualitative way, the acidity or basicity of a solution can be established by some very simple tests. Dilute acidic solutions are characteristically sour; lemon juice, vinegar, and rhubarb get at least part of their taste from the acids they contain. It wouldn't be advisable to test-taste solutions which are highly acidic, since there are various undesirable side effects, to say the least. Such solutions, however, will react with zinc or magnesium metal, evolving hydrogen gas. They will also react readily with carbonates, in which case bubbles of carbon dioxide are given off.

Basic solutions typically have a bitter taste and feel slippery. A solution of lye, sodium hydroxide, feels very slippery, but this is because it is dissolving a surface layer of skin. Probably a safer test for bases is to add a small amount of a solution of a magnesium salt such as $MgCl_2$. The formation of a white precipitate of magnesium hydroxide indicates the solution is basic.

19.3 STRONG ACIDS AND BASES

In order for a solute to make a solution acidic it must somehow release H^+ ion to the solution. The simplest kind of substance which can do this is one like HCl, which on being dissolved in water undergoes the following reaction:

$$HCl(aq) \rightarrow H^+(aq) + Cl^-(aq) \qquad (19.4)$$

For this reaction, we can assume K_c is infinite

Strong acids, of which HCl is a classic example, ionize essentially completely in water in a reaction like 19.4, producing hydrogen ion and an anion. In a solution of a strong acid there are assumed to be no acid molecules, but only hydrogen ions and the anions resulting from the dissociation. In a 0.5 M HCl solution, made by dissolving a half mole of HCl(g) per litre of solution, $[H^+]$ is 0.5 M, $[Cl^-]$ is 0.5 M, and [HCl] is virtually *zero*. In solution, then, HCl behaves as a strong electrolyte. It conducts the electric current very well and has the colligative properties of an ionized substance.

There are only a few strong acids; the six most important ones are listed in Table 19.4. Since all these acids are very soluble in water, it is possible to use them to prepare solutions in which $[H^+]$ is very high. Concentrated HCl is about 12 M and concentrated HNO_3 is about 16 M. Sulfuric acid is furnished commercially as essentially pure H_2SO_4; that liquid contains almost no free water, is a very strong dehydrating agent and a good oxidizing agent, as well as having acid

TABLE 19.4 THE STRONG ACIDS AND BASES

ACIDS		BASES
Hydrochloric acid	HCl	Hydroxides of the 1A metals:
Hydrobromic acid	HBr	$LiOH$, $NaOH$, KOH, $RbOH$, $CsOH$
Hydriodic acid	HI	Hydroxides of the 2A metals:
Nitric acid	HNO_3	$Mg(OH)_2$, $Ca(OH)_2$, $Sr(OH)_2$, $Ba(OH)_2$
Perchloric acid	$HClO_4$	
Sulfuric acid	H_2SO_4	

properties. HCl, HNO_3, and H_2SO_4 are all very important industrial chemicals; each has its own characteristic properties, but they share in common the capacity to furnish high concentrations of hydrogen ion in water solutions. The "work-horse" acids of the chemistry laboratory are 6 *M* HCl, 6 *M* HNO_3, and 3 *M* H_2SO_4, sometimes referred to as "dilute" acids.

Analogous to the strong acids are the **strong bases,** of which $NaOH$ is the most common example. Sodium hydroxide dissolves very readily in water to give a solution containing Na^+ and OH^- ions.

"Strong" means that an acid or base is completely dissociated; nothing else is implied

$$NaOH(s) \rightarrow Na^+(aq) + OH^-(aq) \qquad (19.5)$$

In solution $NaOH$ is effectively 100 per cent ionized, as are all the strong bases. In 0.8 *M* $NaOH$ solution, made by dissolving 0.8 mol $NaOH/dm^3$ of solution, $[Na^+]$ is 0.8 *M*, $[OH^-]$ is 0.8 *M*, and $[NaOH]$, the concentration of undissociated $NaOH$, is just about zero.

There are not many strong bases; most of them are listed in Table 19.4. They are essentially limited to the hydroxides of the 1A and 2A metals. Of these bases, only $NaOH$ and KOH are commonly used in the chemistry laboratory. All compounds of lithium, rubidium, and cesium are expensive; the 2A hydroxides have limited solubilities. Calcium hydroxide is often used in industry when a strong base in needed and high solubility is not critical.

Using stock solutions of strong acids and bases it is possible to prepare, by dilution, acidic and basic solutions of nearly any concentration.

Example 19.2 How would you prepare, from 1 *M* HCl and 1 *M* $NaOH$ stock solutions, 1.0 dm^3 of a solution in which $[H^+]$ is

a. 4×10^{-3} *M*?
b. 5×10^{-13} *M*?

Solution

a. Since HCl is completely ionized, the solution to be made will have to be 4×10^{-3} *M* HCl. 1.0 dm^3 of such a solution will contain 4×10^{-3} mol HCl. We need to find the volume of 1 *M* HCl which will contain that number of moles; we use the relation:

$$\text{no. moles HCl} = M \times V;\ V = \frac{4 \times 10^{-3}\ \text{mol}}{1\ \text{mol/dm}^3} = 4 \times 10^{-3}\ \text{dm}^3 = 4\ \text{cm}^3$$

We measure out 4 cm^3 1 *M* HCl and dilute to a volume of 1.0 dm^3 mixing the final solution well.

b. We cannot prepare this solution by dilution of 1 *M* HCl with water, since the best we can do by that approach is to get essentially pure water, which has a $[H^+]$ of 10^{-7} *M*. Since the solution is basic, we make it from the stock 1 *M* $NaOH$.

We first need to find $[OH^-]$, using the expression for K_w:

$$[H^+] \times [OH^-] = K_w = 1.0 \times 10^{-14}$$

$$[OH^-] = \frac{1.0 \times 10^{-14}}{5 \times 10^{-13}} = 0.2 \times 10^{-1}\ M = 2 \times 10^{-2}\ M$$

To make 1.0 dm^3 of this solution, we need 2×10^{-2} mol NaOH. By the procedure used in part a, we find that we would require 20 cm^3 1 *M* NaOH. We measure out that volume of the basic stock solution and dilute to 1.0 dm^3 with water, remembering to mix well.

19.4 WEAK ACIDS

Most acids and bases are not strong, i.e., completely ionized in water, but fall in the category called weak. A *weak acid* or a *weak base* is characterized by the fact that in solution its reaction to form H^+ or OH^- ion proceeds only to a small extent.

A typical example of a weak acid is hydrofluoric acid, made by dissolving hydrogen fluoride, HF, in water. Rather than dissociating completely, like HCl, HF in solution ionizes only partially:

$$HF(aq) \rightleftarrows H^+(aq) + F^-(aq) \qquad (19.6)$$

In the final solution the concentrations of H^+ and F^- ions are low as compared with the concentration of undissociated HF molecules. For example, in 0.1 *M* HF solution, only about 8 per cent of the HF exists in ionized form. This property of weak acids makes their behavior considerably different from that of strong acids.

Species Which Behave as Weak Acids

There are many substances which in water solution behave as weak acids. It is convenient to classify them in three categories:

1. MOLECULAR SPECIES CONTAINING AN IONIZABLE PROTON. These include a small multitude of organic acids, of which acetic acid, the acid component of vinegar, $HC_2H_3O_2$, is the most common example. All the organic acids contain at least one acidic hydrogen atom and dissociate in a manner similar to acetic acid:

$$HC_2H_3O_2(aq) \rightleftarrows H^+(aq) + C_2H_3O_2^-(aq)$$

Distilled water is weakly acidic because of dissolved CO_2

Inorganic weak acids are also fairly common and include many molecular species, such as carbonic acid, H_2CO_3, sulfurous acid, H_2SO_3, hydrogen sulfide, H_2S, and hydrocyanic acid, HCN, as well as HF. Many of these acids are solutions of a gas in water; in some cases the gas is an oxide. Carbonic acid and sulfurous acid are simply aqueous solutions of CO_2 and SO_2, respectively. In other weak acids the solute gas does not contain oxygen, as with H_2S, HCN, and HF. Some weak acids contain two or more ionizable hydrogen atoms and dissociate in two or more steps, the first of which always occurs to a greater extent than the second or third. Carbonic acid is an acid of this sort:

$$H_2CO_3(aq) \rightleftarrows H^+(aq) + HCO_3^-(aq) \qquad \text{first ionization}$$

$$HCO_3^-(aq) \rightleftarrows H^+(aq) + CO_3^{2-}(aq) \qquad \text{second ionization}$$

2. ANIONS CONTAINING AN IONIZABLE PROTON. A few salts such as $NaHSO_4$ which contain an anion (HSO_4^-) with an ionizable hydrogen show acidic properties in water solution. These anions ionize by a reaction like that of the HCO_3^- ion.

3. CATIONS. A salt such as ammonium chloride, NH_4Cl, is weakly acidic because of the reaction

$$NH_4^+(aq) \rightleftarrows H^+(aq) + NH_3(aq)$$

Most metal cations (indeed, all except those of the 1A and 2A metals) are weak acids. In water, cations such as Al^{3+} and Zn^{2+} are hydrated; the following reversible reactions produce enough protons to make solutions of $AlCl_3$ or $Zn(NO_3)_2$ acidic to litmus:

$$Al(H_2O)_6^{3+}(aq) \rightleftharpoons H^+(aq) + Al(H_2O)_5(OH)^{2+}(aq)$$

$$Zn(H_2O)_4^{2+}(aq) \rightleftharpoons H^+(aq) + Zn(H_2O)_3(OH)^+(aq)$$

Experimentally, one can distinguish between strong and weak acids by measuring:

a. The conductivities of their water solutions. Strong acids behave as strong electrolytes; indeed, a 0.10 *M* solution of hydrochloric acid, HCl, has a considerably higher conductivity than 0.10 *M* NaCl. In contrast, solutions of molecular acids such as hydrofluoric acid or acetic acid are relatively poor electrical conductors because there are few ions present.

b. The colligative properties of their water solutions. The freezing point of a 0.10 *M* solution of HCl is almost exactly the same as that of 0.10 *M* NaCl. In contrast, the freezing point of a 0.10 *M* solution of acetic acid is comparable to that of 0.10 *M* solutions of nonelectrolytes such as sugar or urea.

c. The *pH* of their water solutions. A 0.10 *M* solution of perchloric acid has a *pH* of 1, indicating complete dissociation. The *pH* of a 0.10 *M* solution of $HC_2H_3O_2$ is about 2.9, indicating that relatively few of the acetic acid molecules are ionized.

The equilibria between species in solutions of weak acids can be handled quantitatively in a manner analogous to that developed previously in connection with gases (Chapter 15) and slightly soluble salts (Chapter 18). In particular, it is possible to derive an expression for the equilibrium constant, or the ionization constant, for the dissociation of a weak acid in water. We shall now consider how this equilibrium constant, symbol K_a, is expressed, how it is determined experimentally, and how it is used in practical calculations.

Expression for K_a

Consider the dissociation of the weak acid HX:

$$HX(aq) \rightleftarrows H^+(aq) + X^-(aq) \qquad (19.7)$$

Formulating the expression for the equilibrium constant, K_a, for this reaction according to the rules we have used previously, we obtain

$$K_a = \frac{[H^+] \times [X^-]}{[HX]} \qquad (19.8)$$

The equilibrium constant, K_a, is called the **ionization constant** or **acid dissociation constant** of the weak acid HX.

To illustrate the form taken by K_a for various weak acids, consider the three species HF, NH_4^+, and H_2CO_3:

Note that the concentration of undissociated acid appears in the expression for K_a

$$HF(aq) \rightleftarrows H^+(aq) + F^-(aq) \qquad K_a = \frac{[H^+] \times [F^-]}{[HF]}$$

$$NH_4^+(aq) \rightleftarrows H^+(aq) + NH_3(aq) \qquad K_a = \frac{[H^+] \times [NH_3]}{[NH_4^+]}$$

$$H_2CO_3(aq) \rightleftarrows H^+(aq) + HCO_3^-(aq) \qquad K_a = \frac{[H^+] \times [HCO_3^-]}{[H_2CO_3]}$$

Experimental Determination of K_a

It is possible to measure the ionization constants of weak acids by several different methods. One of the procedures is illustrated in the following example.

Example 19.3 Some of the common organic acids were originally detected by their odors and were given names which indicated their source. A case in point is caproic acid, found in the skin secretions of goats (L. *caper,* goat). Caproic acid, $CH_3(CH_2)_4COOH$, is similar to acetic acid but has a longer carbon chain; only the last hydrogen in the formula is acidic, so we can call the acid HCap.

The concentration of hydrogen ion in a solution prepared by adding 0.030 mol HCap to 1.0 dm³ of water was measured and found to be 6.5×10^{-4} *M*. Find K_a for caproic acid.

Solution The dissociation of caproic acid would occur as follows:

$$HCap(aq) \rightleftarrows H^+(aq) + Cap^-(aq) \qquad K_a = \frac{[H^+] \times [Cap^-]}{[HCap]}$$

To find the numerical value of K_a, we must know the equilibrium concentrations of H^+, Cap^-, and HCap. That of H^+ has been determined to be 6.5×10^{-4} *M*. From the equation for the dissociation of HCap, it is evident that one mole of caproate ion, Cap^-, is formed for every mole of H^+ ion.

$$[H^+] = [Cap^-] = 6.5 \times 10^{-4}\ M$$

To obtain the equilibrium concentration of HCap, we note that for every mole of H^+ formed, a mole of HCap must dissociate. Basing our calculation on 1.0 dm³ of solution, which would contain 0.030 mol caproic acid originally, we can see that [HCap] at equilibrium can be found by subtracting $[H^+]$ from the original concentration of HCap:

$$\begin{aligned}[HCap] &= 3.0 \times 10^{-2}\ M - 6.5 \times 10^{-4}\ M \\ &= 3.0 \times 10^{-2}\ M - 0.065 \times 10^{-2}\ M \\ &= 2.9 \times 10^{-2}\ M\end{aligned}$$

Substituting:

$$K_a = \frac{[H^+] \times [Cap^-]}{[HCap]} = \frac{(6.5 \times 10^{-4}) \times (6.5 \times 10^{-4})}{2.9 \times 10^{-2}} = 1.5 \times 10^{-5}$$

In Table 19.5 (p. 460) we have listed the ionization constants of some of the more common weak acids. The values of these constants vary widely, reflecting the fact that some weak acids are stronger than others. The larger the dissociation constant, the greater the tendency to ionize to produce hydrogen ion, and the larger the fraction of weak acid that is dissociated. We find that acetic acid, $HC_2H_3O_2$($K_a = 1.8 \times 10^{-5}$) is weaker than hydrofluoric acid ($K_a = 7.0 \times 10^{-4}$) but stronger than boric acid ($K_a = 5.8 \times 10^{-11}$). This means that in 0.5 *M* solutions of these three acids, the per cent dissociation of HF would be largest and that of H_3BO_3 would be smallest.

Use of K_a in Calculations

Examples 19.4 and 19.5 illustrate how ionization constants can be used in calculations involving solutions of weak acids.

Example 19.4 Calculate the $[H^+]$ in a 1.0 *M* solution of $HC_2H_3O_2$.

Solution From Table 19.5 we have

$$HC_2H_3O_2(aq) \rightleftarrows H^+(aq) + C_2H_3O_2^-(aq)$$

$$K_a = \frac{[H^+] \times [C_2H_3O_2^-]}{[HC_2H_3O_2]} = 1.8 \times 10^{-5}$$

Let x represent the unknown equilibrium concentration of H^+. In 1.0 dm^3 of solution, x will therefore also equal the number of moles of H^+ formed by the reaction. It is evident from the equation for the dissociation of acetic acid that a mole of acetate ions is produced for every mole of H^+ formed. Neglecting the small amount of H^+ produced from the water, the concentration of $C_2H_3O_2^-$ must also be x. The formation of x moles of H^+ consumes an equal number of moles of acetic acid; the equilibrium concentration of $HC_2H_3O_2$ must then be 1.0 − x. Substituting,

$$\frac{(x)(x)}{1.0 - x} = 1.8 \times 10^{-5}$$

This equation could be rearranged to the form $x^2 + bx + c = 0$ and solved for x, using the quadratic formula. Such a procedure is tedious and, in this case, unnecessary. Since $HC_2H_3O_2$ is a weak acid, only slightly dissociated in water, the equilibrium concentration of $HC_2H_3O_2$, 1.0 − x, must be very nearly equal to its *original* concentration before dissociation, namely, 1.0 *M*. Making this approximation, we obtain

$$\frac{x^2}{1.0} = 1.8 \times 10^{-5};\ x^2 = 1.8 \times 10^{-5} = 18 \times 10^{-6}$$

Solving for x by extracting the square root of both sides, we have

$$x = [H^+] = 4.2 \times 10^{-3}$$

The fact that the concentration of H^+, 0.0042, is so *much less* than the original concentration of $HC_2H_3O_2$ (only about 0.42% of it) justifies the approximation made earlier, i.e., 1.0 − x = 1.0. In general, the expression for K_a is rarely valid to better than ±5 per cent. Consequently, in the expression

1.0 − 0.0042 = 1.0, to two significant figures

$$K_a = \frac{x^2}{a - x}$$

in which x = $[H^+]$ and a = concentration of weak acid prior to dissociation, one is justified in setting (a − x) equal to a, provided this approximation does not introduce an error of more than about 5 per cent. In practice it is ordinarily simplest to make the approximation, calculate x, and compare to a. If the value of x thus obtained is less than 5 per cent of a, the approximation is valid. If x is greater than about 5 per cent of a, one can go back to the original equation and solve by means of the quadratic formula or, alternatively, by the method of successive approximations (Example 19.5).

Example 19.5 Lactic acid, $CH_3CHOHCOOH$, gets its name from sour milk, from which it was first isolated in 1780 (L. *lactis*, milk). K_a for HLac is 8.4×10^{-4}. Find the $[H^+]$ in a sample of sour milk containing 0.100 *M* HLac as its acid component.

Solution Lactic acid dissociates according to the equation:

$$HLac(aq) \rightleftarrows H^+(aq) + Lac^-(aq) \qquad K_a = \frac{[H^+] \times [Lac^-]}{[HLac]} = 8.4 \times 10^{-4}$$

Letting $[H^+] = x$, $[Lac^-]$ will also equal x, $[HLac] = 0.100 - x$, and we obtain

$$\frac{(x)(x)}{0.100 - x} = 8.4 \times 10^{-4}$$

If we make the same approximation as before, i.e., $0.10 - x \approx 0.10$,

$$x^2 = 8.4 \times 10^{-5} = 84 \times 10^{-6} \qquad x = 9.2 \times 10^{-3} \approx [H^+]$$

We note that in this case the calculated concentration of H^+ is more than 5 per cent of the original concentration of undissociated acid;

$$\frac{9.2 \times 10^{-3}}{0.10} \times 100 = 9.2\%$$

This means that about 9.2% of the HLac originally present is dissociated, so that 0.100 *M* was really not a very good value to use for [HLac]. A much better approximation would be obtained by recognizing that we should, and now can, use $0.100 - x$, where x is the value just calculated, as [HLac]:

If $$[H^+] \approx 9.2 \times 10^{-3}$$

then $$[HLac] = 0.100 - 9.2 \times 10^{-3} = 0.100 - 0.0092 = 0.091\ M$$

Substituting this value for [HLac] into the expression for K_a, we have

$$\frac{x^2}{0.091} = 8.4 \times 10^{-4}$$

$$x^2 = 7.6 \times 10^{-5} = 76 \times 10^{-6} \qquad x = 8.7 \times 10^{-3}\ M = [H^+]$$

You could, of course, use the quadratic formula, but that is tedious at best (try it!)

This value of $[H^+]$ is closer to the true $[H^+]$ in the solution, since 0.091 *M* is a better approximation to the equilibrium concentration of HLac than was 0.100 *M*. If we are still not satisfied, we can attempt a further improvement, using the value of $[H^+]$ we just calculated to obtain a still better value for [HLac]. If we do this, we find that [HLac] keeps the value 0.091 *M;* that is, if

$$[H^+] = 8.7 \times 10^{-3}\ M$$

then $$[HLac] = 0.100 - 0.0087 = 0.091\ M$$

This means that if we were to solve again for $[H^+]$ we would get the same answer. In other words, "we have gone about as far as we can go."

This method of *successive approximations* is a very useful one for working problems of this type involving equilibria of all sorts. Usually it is found that the first approximation, (i.e., $a - x = a$), is excellent. Occasionally it is desirable to make a further refinement by the method illustrated here.

The extent of dissociation of a weak acid depends not only upon the acid dissociation constant but also on the concentration of the acid in the solution. Taking acetic acid as a typical example of a weak acid, we find that a 0.1 *M* solution of such an acid will be about 1 per cent dissociated, so that in the solution $[H^+]$

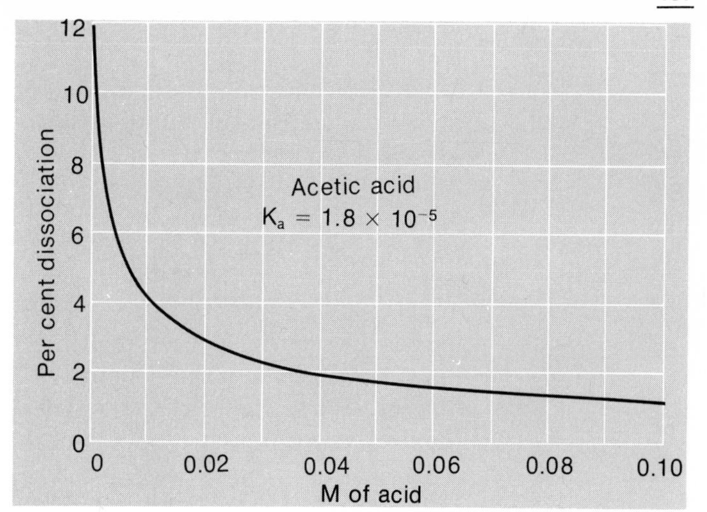

Figure 19.1 The per cent dissociation of a weak acid increases as the acid is made more dilute. Acetic acid is about 4% dissociated in its 0.01 *M* solution, but only about 1.3% dissociated in a 0.1 *M* solution.

is about 0.001 *M* and the *pH* is about 3. As the concentration of the acid decreases, its per cent of dissociation goes up. The actual degree of dissociation can be determined by the methods in Examples 19.4 and 19.5, and in Figure 19.1 we have plotted the results of such calculations. Note that for most concentrations of acetic acid ordinarily encountered, the per cent dissociation is quite low, so that the approximation in Example 19.4 is valid.

19.5 WEAK BASES

There is a rather large family of substances called *weak bases,* paralleling the weak acids, which react in solution to create appreciable concentrations of hydroxide ion. The behavior of weak bases is analogous to that of the weak acids. A weak base on being added to water reacts in such a way that the solution becomes basic, but the concentration of hydroxide ion is much lower than it would be if the reaction to form OH^- ion went to completion.

The most commonly encountered weak base is ammonia, NH_3, made by dissolving NH_3 gas in water. The solution is basic because ammonia in aqueous solution reacts in the following way:

$$NH_3(aq) + H_2O \rightleftarrows NH_4^+(aq) + OH^-(aq) \qquad (19.9)$$

The reaction does not tend to proceed very far to the right, so that in ammonia solutions $[OH^-] \ll [NH_3]$; it is for this reason that we classify NH_3 as a weak base.

Ammonia has several properties that increase its usefulness as a base in the chemistry laboratory. It is very soluble in water, so its solutions can be quite concentrated; the stock solutions are 6 *M* (dilute) and 17 *M* (concentrated). Since NH_3 is volatile, it can be driven from the solution by boiling, if necessary. Ammonia also forms stable complexes with many metal cations (Chapter 21), which makes it an important reagent in qualitative analysis.

Solutions of NH_3 are often labeled NH_4OH, although there is no evidence for the existence of such a molecule

Anions that Behave as Weak Bases

Although ammonia is perhaps the most important of the weak bases, many other species react with water in a similar way. Returning to Equation 19.9, we might say that the reason ammonia solutions are basic is that NH_3 has the capacity to remove H^+ ions from solvent water molecules, forming NH_4^+ ions and the OH^- ions that give the solution its basic character. Any species which, like NH_3, can remove H^+ ions from water will, of necessity, also produce a basic solution. A great many anions fall into this category.

As an example of a basic anion consider the acetate ion, $C_2H_3O_2^-$, derived from acetic acid, $HC_2H_3O_2$. The fact that acetic acid is ionized to only a small extent in water solution

$$HC_2H_3O_2(aq) \rightleftharpoons H^+(aq) + C_2H_3O_2^-(aq)$$

implies that there is a strong tendency for the reverse reaction to occur – i.e., for the $C_2H_3O_2^-$ ion to pick up a proton. This means that if we add a salt containing this ion, such as $NaC_2H_3O_2$, to water, the acetate ion will tend to remove a proton from the water molecule by the following reaction:

$$C_2H_3O_2^-(aq) + H_2O \rightleftharpoons HC_2H_3O_2(aq) + OH^-(aq) \qquad (19.10)$$

We would not expect Reaction 19.10 to go very far to the right, since water is very reluctant to give up hydrogen ions. However, the reaction does go to a small extent; it produces enough OH^- ions to make the solution weakly basic ($pH \approx 9$).

A reaction analogous to 19.10 occurs with *all anions derived from weak acids.* We find that solutions of NaF and Na_2CO_3 are basic because of the reactions:

$$F^-(aq) + H_2O \rightleftharpoons HF(aq) + OH^-(aq) \qquad (19.11)$$

$$CO_3^{2-}(aq) + H_2O \rightleftharpoons HCO_3^-(aq) + OH^-(aq) \qquad (19.12)$$

Notice the similarity between these equations and Equations 19.9 and 19.10. In each case a weak base (NH_3, $C_2H_3O_2^-$, F^-, CO_3^{2-}) abstracts a proton from a water molecule to form a weak acid (NH_4^+, $HC_2H_3O_2$, HF, HCO_3^-); the OH^- ions formed at the same time make the solution basic ($pH > 7$).

Because of its basic character, the acetate ion is sometimes referred to as the **conjugate base** of acetic acid. The terminology is reversible; we say that acetic acid is the **conjugate acid** of the acetate ion. All weak acid, weak base pairs can be related in this way; F^- ion is the conjugate base of HF, while the ammonium ion, NH_4^+, is the conjugate acid of ammonia, NH_3.

What is the conjugate base of HCO_3^-? the conjugate acid?

Expression for K_b

For reactions such as 19.9 and 19.10 we can formulate an equilibrium constant expression in the usual way:

$$NH_3(aq) + H_2O \rightleftharpoons NH_4^+(aq) + OH^-(aq); \; K_b = \frac{[NH_4^+] \times [OH^-]}{[NH_3]}$$

$$C_2H_3O_2^-(aq) + H_2O \rightleftharpoons HC_2H_3O_2(aq) + OH^-(aq); \; K_b = \frac{[HC_2H_3O_2] \times [OH^-]}{[C_2H_3O_2^-]}$$

Notice that here, as usual, the concentration of water (which is approximately constant at 55 *M*) is incorporated into the equilibrium constant. The constant, K_b, called a **base dissociation constant,** is ordinarily a small number. For NH_3, it is 1.8×10^{-5}, for $C_2H_3O_2^-$, $K_b = 5.6 \times 10^{-10}$. Comparing these two numbers, we conclude that the acetate ion is a considerably weaker base than the ammonia molecule.

Calculations involving K_b are carried out in a manner quite similar to those for K_a (compare Examples 19.6 and 19.4).

Example 19.6 Calculate $[OH^-]$ in a 0.20 M solution of NH_3 ($K_b = 1.8 \times 10^{-5}$).

Solution From the equation $NH_3(aq) + H_2O \rightleftharpoons NH_4^+(aq) + OH^-(aq)$, we see that NH_4^+ and OH^- ions are produced in equal amounts. Moreover, one mole of NH_3 is consumed for every mole of OH^- produced. Hence, if we let $[OH^-] = x$, it follows that:

$$[NH_4^+] = [OH^-] = x; [NH_3] = 0.20 - x$$

Substituting into the expression for K_b:

$$K_b = \frac{[NH_4^+] \times [OH^-]}{[NH_3]}; \quad 1.8 \times 10^{-5} = \frac{x^2}{0.20 - x}$$

Assuming x is relatively small, i.e., $0.20 - x \approx 0.20$

$$x^2 = 0.20(1.8 \times 10^{-5}) = 3.6 \times 10^{-6}$$

$$x = [OH^-] = 1.9 \times 10^{-3}\,M$$

We see that x is indeed small; the per cent ionization of 0.20 M NH_3 is only about 1%

$$\frac{1.9 \times 10^{-3}}{0.20} \times 100 = 1.0\%$$

so the approximation is justifiable.

The pH of 0.1 M $NaC_2H_3O_2$ would be calculated the same way

Relation Between K_b and K_a

It is possible to measure dissociation constants of weak bases in the laboratory by procedures very much like those used for weak acids. In practice, this is seldom necessary. Instead, we take advantage of a simple mathematical relationship between K_b for a weak base and K_a for its conjugate weak acid. This relationship can be derived by adding together the equations for the dissociation of the weak acid HA and the weak base A^-:

(1)	$HA(aq) \rightleftharpoons H^+(aq) + A^-(aq)$	$K_I = K_a$ of HA
(2)	$A^-(aq) + H_2O \rightleftharpoons HA(aq) + OH^-(aq)$	$K_{II} = K_b$ of A^-
(3)	$H_2O \rightleftharpoons H^+(aq) + OH^-(aq)$	$K_{III} = K_w$

Since Equation (1) + Equation (2) = Equation (3) we have, according to the Law of Multiple Equilibria (Chapter 15):

$$K_I \times K_{II} = K_{III}$$

or:

$$(K_a \text{ of HA}) \times (K_b \text{ of } A^-) = K_w = 1.0 \times 10^{-14} \qquad (19.13)$$

Equation 19.13 allows us to calculate K_b for any weak base if we know K_a for its conjugate weak acid (Example 19.7).

Example 19.7 Lactic acid, HLac, is known to have a dissociation constant K_a equal to 8.4×10^{-4}. Find K_b for the lactate ion, Lac^-.

Solution The lactate ion is the conjugate base of lactic acid, since it is produced by the ionization of the parent acid. Therefore,

$$K_b = K_w/K_a = \frac{1.0 \times 10^{-14}}{8.4 \times 10^{-4}} = 0.12 \times 10^{-10} = 1.2 \times 10^{-11}$$

In Table 19.5 we list the values of K_a and K_b for many of the common weak acids and their conjugate bases. The acids are arranged in order of decreasing values of K_a. Since K_b is inversely proportional to K_a, it follows that the bases are in order, from top to bottom, of increasing values of K_b. By looking at the table, you can see immediately that nitrous acid, HNO_2, is a stronger acid than is acetic acid, $HC_2H_3O_2$, since it lies above acetic acid. Conversely, the acetate ion, $C_2H_3O_2^-$, is a stronger base than the nitrite ion, NO_2^-, since it lies below NO_2^- in the table.

As pointed out earlier some acids, of which H_2CO_3 is typical, dissociate in more than one step. Notice from the table that K_a for the first step ($H_2CO_3 \rightarrow H^+ + HCO_3^-$) is considerably larger, by a factor of 10^4, than K_a for the second step ($HCO_3^- \rightarrow H^+ + CO_3^{2-}$). Consequently, virtually all of the H^+ ions in a solution of carbonic acid, or other acids of this type, come from the first dissociation.

TABLE 19.5 DISSOCIATION CONSTANTS OF WEAK ACIDS AND BASES

(Increasing Acid Strength ↑ — Increasing Basic Strength ↓)

	Acid	K_a	Base	K_b
Sulfurous acid	H_2SO_3	1.7×10^{-2}	HSO_3^-	5.9×10^{-13}
Hydrogen sulfate ion	HSO_4^-	1.2×10^{-2}	SO_4^{2-}	8.3×10^{-13}
Phosphoric acid	H_3PO_4	7.5×10^{-3}	$H_2PO_4^-$	1.3×10^{-12}
Hydrofluoric acid	HF	7.0×10^{-4}	F^-	1.4×10^{-11}
Nitrous acid	HNO_2	4.5×10^{-4}	NO_2^-	2.2×10^{-11}
Formic acid	$HCHO_2$	1.8×10^{-4}	CHO_2^-	5.6×10^{-11}
Benzoic acid	$HC_7H_5O_2$	6.6×10^{-5}	$C_7H_5O_2^-$	1.5×10^{-10}
Acetic acid	$HC_2H_3O_2$	1.8×10^{-5}	$C_2H_3O_2^-$	5.6×10^{-10}
Propionic acid	$HC_3H_5O_2$	1.4×10^{-5}	$C_3H_5O_2^-$	7.1×10^{-10}
Carbonic acid	H_2CO_3	4.2×10^{-7}	HCO_3^-	2.4×10^{-8}
Hydrogen sulfide	H_2S	1×10^{-7}	HS^-	1×10^{-7}
Dihydrogen phosphate ion	$H_2PO_4^-$	6.2×10^{-8}	HPO_4^{2-}	1.6×10^{-7}
Hydrogen sulfite ion	HSO_3^-	5.6×10^{-8}	SO_3^{2-}	1.8×10^{-7}
Hypochlorous acid	$HClO$	3.2×10^{-8}	ClO^-	3.1×10^{-7}
Boric acid	H_3BO_3	5.8×10^{-10}	$H_2BO_3^-$	1.7×10^{-5}
Ammonium ion	NH_4^+	5.6×10^{-10}	NH_3	1.8×10^{-5}
Hydrocyanic acid	HCN	4.0×10^{-10}	CN^-	2.5×10^{-5}
Hydrogen carbonate ion	HCO_3^-	4.8×10^{-11}	CO_3^{2-}	2.1×10^{-4}
Hydrogen phosphate ion	HPO_4^{2-}	1.7×10^{-12}	PO_4^{3-}	5.9×10^{-3}
Hydrogen sulfide ion	HS^-	1×10^{-15}	S^{2-}	1×10^{1}

For the acids: $HX(aq) \rightleftarrows H^+(aq) + X^-(aq)$ $\qquad K_a = \frac{[H^+] \times [X^-]}{[HX]}$

For the bases: $X^-(aq) + H_2O \rightleftarrows HX(aq) + OH^-(aq)$ $\qquad K_b = \frac{[HX] \times [OH^-]}{[X^-]}$

19.6 ACID-BASE PROPERTIES OF SALT SOLUTIONS

A salt MX can be considered to be the product of the reaction of the base MOH with the acid HX

Having completed the previous two sections, you could probably correctly decide that both H_2SO_4 and HNO_2 solutions would be acidic and that solutions of NaOH and NH_3 would be basic. You might have a bit more trouble making similar decisions regarding solutions of salts such as Na_2SO_4 or NH_4NO_3, since we have not specifically discussed their acid-base properties. Using the principles we have developed, however, we can predict the acid-base character of almost any salt solution.

We can assume that salts in water solution are completely ionized, so that the acid-base properties of the solutions will be those of the ions present. Some ions in solution are acidic, some basic, and some are neutral. The neutral ions are perhaps easiest to understand, so let's work with them first.

Neutral Ions

For an ion in solution to affect the *pH*, it must react in some way so as to generate H^+ or OH^- ions. If it does not react, it will be neutral. Relatively few ions fall in this category; the more common ones are listed in Table 19.6. Notice that:

—*the neutral anions are those derived from strong acids* (compare Table 19.4).

—*the neutral cations are those derived from strong bases* (compare Table 19.4). There is a very simple explanation for this relationship. Consider, for example, the chloride ion, produced by ionization of hydrogen chloride in solution:

$$HCl(aq) \rightarrow H^+(aq) + Cl^-(aq) \qquad K_a \rightarrow \infty$$

Since HCl is a strong acid, its ionization in solution is complete; K_a for the acid can be taken to be infinite. This means that in solution, the Cl^- ion has *no tendency to combine with H^+ ions* from any source, including water, and so does not itself influence the *pH* of the solution in any way; it is neutral. The same kind of argument applies to sodium ion, produced by the ionization of sodium hydroxide, a strong base. Since the base ionizes completely to Na^+ and OH^- ions, *K* for the dissociation is essentially infinite, so there is no tendency for Na^+ ion to combine with OH^- ion, whatever its source. Sodium ion is, therefore, neutral. Similar arguments apply to all ions (other than H^+ and OH^-) produced by the ionization of a strong acid or strong base.

Basic Anions

As we noted in Section 19.5, anions formed by the ionization of weak acids are themselves weak bases. There is a small army of such anions; those listed in Table 19.6 are only typical examples. In contrast, there are no common basic cations.

Acidic Ions

As mentioned earlier, all cations except those derived from the 1A and 2A metals act as weak acids in water solution. Included in this category are NH_4^+, Al^{3+}, and essentially all the transition metal ions, which are far too numerous to list separately in Table 19.6.

TABLE 19.6 ACID-BASE PROPERTIES OF SOME COMMON IONS IN WATER SOLUTION

	Neutral		Basic		Acidic
Anion	Cl^- Br^- I^-	NO_3^- ClO_4^- SO_4^{2-}	$C_2H_3O_2^-$ F^- CO_3^{2-} S^{2-} PO_4^{3-}	CN^- NO_2^- HCO_3^- HS^- HPO_4^{2-}	HSO_4^-
Cation	Li^+ Na^+ K^+	Mg^{2+} Ca^{2+} Ba^{2+}	none		Al^{3+} NH_4^+ transition metal ions

Anions containing ionizable protons can show both acidic and basic character. The hydrogen carbonate ion, HCO_3^-, is typical. It can ionize via the reaction

$$HCO_3^-(aq) \rightleftarrows H^+(aq) + CO_3^{2-}(aq) \qquad K_a = 4.8 \times 10^{-11}$$

tending to make its solution acidic. Since it is the conjugate base of H_2CO_3, carbonic acid, it also has basic properties:

$$HCO_3^-(aq) + H_2O \rightleftarrows H_2CO_3(aq) + OH^-(aq) \qquad K_b = 2.4 \times 10^{-8}$$

0.1 *M* $NaHCO_3$ has a *pH* of 8.3

Since K_b for HCO_3^- is larger than its value of K_a, a solution containing HCO_3^- will tend to be slightly basic. This is true of most other ions of this type; solutions containing HS^-, HPO_4^{2-}, and $H_2BO_3^-$ are all weakly basic. One striking exception is the HSO_4^- ion, which is a relatively strong acid

$$HSO_4^-(aq) \rightleftarrows H^+(aq) + SO_4^{2-}(aq) \qquad K_a = 1.2 \times 10^{-2}$$

The principles we have just discussed can be used to predict not only the acid-base properties of individual ions but also those of the salts from which they are derived. Whether a particular salt solution will be neutral, basic, or acidic depends upon the net effect of the cation and anion present (Example 19.8).

Example 19.8 Classify each of the following 0.1 *M* solutions as acidic, neutral, or basic: $NaNO_3$, KCN, Na_2CO_3, NH_4Br, $AlCl_3$.

Solution

$NaNO_3$	neutral cation, neutral anion	neutral solution
KCN	neutral cation, basic anion	basic solution
Na_2CO_3	neutral cation, basic anion	basic solution
NH_4Br	acidic cation, neutral anion	acidic solution
$AlCl_3$	acidic cation, neutral anion	acidic solution

19.7 GENERAL THEORIES OF ACIDS AND BASES

Chemical theorists have always been interested in the fundamental properties of acids and bases. Acids and bases have been known since the beginnings of chemistry, and the fact that their behavior is so closely linked to water, the solvent, tended to confuse matters. So far in this chapter we have considered an acid to be a substance which in water solution produces an excess of H^+ ion. A base was similarly defined to be a substance which, directly or indirectly, forms excess OH^- ion in solution. This approach to acids and bases is indeed a very practical one and was first proposed by Arrhenius in 1884.

The Arrhenius concept defines acids and bases in terms of the species they produce on addition to water. This approach tends to minimize to some extent the role that the solvent plays in acid-base systems, and in so doing sacrifices some rather useful insights into the relationship between acids and bases. An alternate, more general picture, suggested independently by Brönsted in Denmark and Lowry in England in 1923, shows the relationships more clearly.

Brönsted-Lowry Concept

According to the Brönsted-Lowry idea, an acid-base reaction is one in which there is a *proton transfer* from one species to another. The species which gives up, or *donates,* the proton is referred to as an acid; the molecule or ion which *accepts* the proton is a *base*.

To illustrate the application of the Brönsted-Lowry concept, let us re-examine as an example of an acid-base reaction the ionization of HF in water solution:

$$HF(aq) \rightleftarrows H^+(aq) + F^-(aq) \qquad K_a = 7.0 \times 10^{-4} \qquad (19.14)$$

We immediately run into a snag, since although HF appears to be the proton donor, there isn't any obvious acceptor. According to Brönsted and Lowry, in such a reaction a proton acceptor must be present. In this case the acceptor is the water molecule itself. Recognizing that, we can write the reaction for the ionization in the following way:

$$\underset{\text{acid}}{HF(aq)} + \underset{\text{base}}{H_2O} \rightleftarrows H_3O^+(aq) + F^-(aq) \qquad (19.14a)$$

K_a for Eq. 19.14a = K_a for Eq. 19.14

This equation satisfies the Brönsted-Lowry picture; it tells us that in the reaction the acid, HF, the proton donor, reacts with the base, water, the proton acceptor, to form a hydrated proton and a fluoride ion.

One advantage in writing the reaction in this way is that it shows more clearly than Equation 19.14 why the ionization occurs at all. Certainly the breaking of the bond in HF takes energy; in fact, the bond-breaking reaction is so endothermic that if it were the only change that occurred, K_a would be virtually zero. We would, however, expect that the H^+ ion produced in solution would react strongly with water to form a hydrated species, and Equation 19.14a emphasizes that hydration. (We have written the formula of the hydrated proton as H_3O^+, although it is very likely that there are several hydrates, such as $H_5O_2^+$ or $H_9O_4^+$, present in the acidic solution.)

Another feature of the Brönsted-Lowry approach becomes clear when we write Reaction 19.14a in the reverse direction:

$$\underset{\text{acid}}{H_3O^+(aq)} + \underset{\text{base}}{F^-(aq)} \rightleftarrows \underset{\text{acid}}{HF(aq)} + \underset{\text{base}}{H_2O} \qquad (19.14a')$$

In this reaction the fluoride ion is clearly a base, since it accepts a proton from the acid, H_3O^+. Using the Arrhenius approach we showed that F^- ion was a base by a rather different argument; here its role as a base is clear.

It is possible to look at all acid-base reactions in terms of the Brönsted-Lowry idea. In every such reaction an acid reacts with a base to form another acid and another base which are conjugate to the original ones. We might write several other examples, taken from earlier portions of this chapter. In each case the acid can give a proton *to* water, and the base has the capacity to extract a proton *from* water.

$$\underset{\text{acid}}{HC_2H_3O_2(aq)} + \underset{\text{base}}{H_2O} \rightleftarrows \underset{\text{acid}}{H_3O^+(aq)} + \underset{\text{base}}{C_2H_3O_2^-(aq)}$$

$$\underset{\text{acid}}{NH_4^+(aq)} + \underset{\text{base}}{H_2O} \rightleftarrows \underset{\text{acid}}{H_3O^+(aq)} + \underset{\text{base}}{NH_3(aq)}$$

$$\underset{\text{base}}{F^-(aq)} + \underset{\text{acid}}{H_2O} \rightleftarrows \underset{\text{acid}}{HF(aq)} + \underset{\text{base}}{OH^-(aq)}$$

$$\underset{\text{base}}{NH_3(aq)} + \underset{\text{acid}}{H_2O} \rightleftarrows \underset{\text{acid}}{NH_4^+(aq)} + \underset{\text{base}}{OH^-(aq)}$$

On examining the above reactions, one notes that water, according to the Brönsted-Lowry approach, can act as both an acid and a base. There are some other species which can act as both acid and base; that is, they are capable of either donating or accepting a proton. The HCO_3^- ion is an example; in reaction with sodium hydroxide solution it is an acid, whereas with hydrofluoric acid it behaves as a base:

$$\underset{\text{acid}}{HCO_3^-(aq)} + \underset{\text{base}}{OH^-(aq)} \rightleftarrows \underset{\text{acid}}{H_2O} + \underset{\text{base}}{CO_3^{2-}(aq)}$$

$$\underset{\text{base}}{HCO_3^-(aq)} + \underset{\text{acid}}{HF(aq)} \rightleftarrows \underset{\text{acid}}{H_2CO_3(aq)} + \underset{\text{base}}{F^-(aq)}$$

An important advantage of the Brönsted-Lowry concept of acids and bases is that it can be applied to solutions in which the solvent is not water. In liquid ammonia, for example, sodium hydride ionizes and participates in the following acid-base reaction:

$$\underset{\text{acid}}{NH_3} + \underset{\text{base}}{H^-} \rightarrow \underset{\text{acid}}{H_2} + \underset{\text{base}}{NH_2^-}$$

Here the ammonia molecule acts as an acid, donating a proton to the base, H^-.

The Lewis Concept

We have seen that the Brönsted-Lowry picture represents a considerable extension of the Arrhenius concept of acids and bases. However, the Brönsted-Lowry picture is restricted in one important respect: it can be applied only to reactions involving a proton transfer. In particular, in order to act as a Brönsted-Lowry acid, a species must contain an ionizable hydrogen atom.

The Lewis acid-base concept, first proposed by the American physical chemist G. N. Lewis in 1923, removes this restriction. The Lewis concept considers an *acid* to be a species that can *accept* an *electron pair;* a *base* is a substance that can *donate* an *electron pair.* According to the Lewis concept, any reaction which leads to the formation of a coordinate covalent bond (both electrons furnished by same species) is an acid-base reaction.

From a structural point of view, the Lewis concept of a base does not differ in any essential way from the Brönsted concept. In order for a species to accept a proton and thereby act as a Brönsted base it must possess an unshared pair of electrons. Consider, for example, the NH_3 molecule, the H_2O molecule, and the F^- ion, all of which can act as Brönsted bases:

$$\text{H—}\ddot{\text{N}}\text{—H (with H below N)}, \qquad \text{H—}\ddot{\underset{\cdot\cdot}{\text{O}}}\text{—H}, \qquad (:\ddot{\underset{\cdot\cdot}{\text{F}}}:)^-$$

Each of these species contains an unshared pair of electrons that is utilized in accepting a proton to form the NH_4^+ ion, the H_3O^+ ion, or the HF molecule.

$$\left[\begin{array}{c}\text{H}\\ \text{H:}\underset{\cdot\cdot}{\text{N}}\text{:H}\\ \text{H}\end{array}\right]^+ \qquad \left[\begin{array}{c}\text{H:}\ddot{\underset{\cdot\cdot}{\text{O}}}\text{:H}\\ \text{H}\end{array}\right]^+ \qquad \text{H:}\ddot{\underset{\cdot\cdot}{\text{F}}}\text{:}$$

Clearly, NH_3, H_2O, and F^- can also be Lewis bases since they possess an unshared electron pair which can be donated to an acid. We see then that the Lewis concept does not significantly change the number of species which can behave as bases.

On the other hand, the Lewis concept greatly increases the number of species which can be considered to be acids. The substance which accepts an electron pair and therefore acts as a Lewis acid can be a proton:

In each case, the Lewis base furnishes the pair of electrons which forms the bond with H^+

$$\underset{\text{acid}}{H^+(aq)} + \underset{\text{base}}{H_2O} \rightarrow H_3O^+(aq)$$

$$\underset{\text{acid}}{H^+(aq)} + \underset{\text{base}}{NH_3(aq)} \rightarrow NH_4^+(aq)$$

It can equally well be a cation, such as Zn^{2+}, which is capable of forming a coordinate covalent bond with a Lewis base:

$$\underset{\text{acid}}{Zn^{2+}(aq)} + \underset{\text{base}}{4\ H_2O} \rightarrow Zn(H_2O)_4^{2+}(aq)$$

$$\underset{\text{acid}}{Zn^{2+}(aq)} + \underset{\text{base}}{4\ NH_3} \rightarrow Zn(NH_3)_4^{2+}(aq)$$

Another important class of Lewis acids comprises molecules containing an incomplete octet of electrons. A classic example is boron trifluoride, BF_3, which reacts readily with ammonia, accepting a pair of electrons:

$$\underset{\text{acid}}{\begin{array}{c}\text{F}\\ |\\ \text{F—B}\\ |\\ \text{F}\end{array}} + \underset{\text{base}}{\begin{array}{c}\text{H}\\ |\\ \text{:N—H}\\ |\\ \text{H}\end{array}} \rightarrow \begin{array}{c}\text{F}\quad\text{H}\\ |\quad\ |\\ \text{F—B—N—H}\\ |\quad\ |\\ \text{F}\quad\text{H}\end{array}$$

Although the Lewis model for acids and bases has certain advantages, in most cases when a chemist speaks of an acid or base he uses the terms in the Arrhenius or Brönsted-Lowry sense. The term "Lewis acid" is more restricted and generally taken to refer to a species like BF_3, which is a powerful electron pair acceptor. Lewis acids of this sort are often encountered in organic chemistry, where they serve as catalysts for many reactions.

A HISTORICAL PERSPECTIVE

Gilbert Newton Lewis (1875–1946)

The Lewis concept of acids and bases, like the Lewis structures discussed in Chapter 8, was the product of the American physical chemist, G. N. Lewis. Born in Massachusetts, Lewis grew up in Nebraska, then came back East to obtain his B.S. (1896) and Ph.D. (1899) at Harvard. Although he stayed on for a few years as an instructor, Lewis seems never to have been happy at Harvard. A precocious student and an intellectual rebel, he was repelled by the highly traditional atmosphere that prevailed in the chemistry department there in his time. Many years later, he refused an honorary degree from his alma mater.

After leaving Harvard, Lewis made his reputation at M.I.T., where he was promoted to full professor in only four years. In 1912, he moved across the country to the University of California at Berkeley as Dean of the College of Chemistry and department chairman. He remained there for the rest of his life. Under his guidance, the chemistry department at Berkeley became perhaps the most prestigious in the country. Among the faculty and graduate students that he attracted were five future Nobel Prize winners: Harold Urey in 1934, William Giauque in 1949, Glenn Seaborg in 1951, Willard Libby in 1960, and Melvin Calvin in 1961.

In administering the chemistry department at Berkeley, Lewis demanded excellence in both research and teaching. Virtually the entire staff was involved in the general chemistry program; at one time eight full professors carried freshman sections. Several department members became leaders of chemical education in America. Among them is Joel Hildebrand, who came to California in 1913 and was still active in teaching and research sixty years later.

Like so many physical chemists, G. N. Lewis maintained throughout his life a fascination with chemical thermodynamics. His Ph.D. thesis was in this area, as were all his early publications. Many of the standard free energies of formation listed in Table 14.2, Chapter 14, and the standard electrode potentials given in Table 23.1, Chapter 23, are based on data obtained by Lewis and his students. In 1923 he published with Merle Randall a text entitled *Thermodynamics and the Free Energy of Chemical Substances*. More than fifty years later, a revised edition of that text is still widely used in graduate courses in chemistry.

Lewis' interest in chemical bonding and structure dates from 1902. In attempting to explain "valence" to a class at Harvard, he devised an atomic model to rationalize the octet rule. His model was deficient in many respects; for one thing, Lewis visualized cubic atoms with electrons located at the corners. Perhaps this explains why his ideas of atomic structure were not published until 1916. In that year, Lewis conceived of the electron-pair bond. This concept and its implications were elaborated upon in a book that he published in 1923, *Valence and the Structure of Atoms and Molecules* (recently reprinted by Dover Publications). Here, in Lewis' characteristically lucid style, we find many of the basic principles of covalent bonding that are accepted today. Here too is the Lewis definition of acids and bases as electron-pair acceptors and donors. Curiously enough, this general approach to acid-base reactions seems to have been virtually ignored for many years until revived in a paper published by Lewis in 1938.

The years from 1923 to 1938 were relatively unproductive for G. N. Lewis so far as his own research was concerned. The applications of the electron-pair bond came largely in the areas of organic and quantum chemistry; in neither of these fields did Lewis feel at home. In the early '30's, he published a series of relatively minor papers dealing with the properties of deuterium. Then, in 1939, he began to publish in the field of photochemistry. Of approximately 20 papers in this area, several were of fundamental importance, comparable in quality to the best work of his early years. Retired officially in 1945, Lewis died a year later while carrying out an experiment on fluorescence.

PROBLEMS

19.1 *$[H^+]$, $[OH^-]$, pH* The concentration of H^+ in a certain solution is 2.0×10^{-4} *M*. Find $[OH^-]$ and the *pH*.

19.2 *Determination of K_a* The *pH* of a 0.20 *M* solution of HA is 3.0. Find K_a of HA.

19.3 *Calculations Involving K_a* What is $[H^+]$ in a 0.10 *M* solution of a weak acid with an ionization constant of 4.0×10^{-5}?

19.4 *K_b and K_a* Calculate K_b for the conjugate base of the acid described in Problem 19.3.

19.5 *Strong and Weak Acids and Bases* Classify each of the following as a strong acid, strong base, weak acid, or weak base:

a. NaOH b. HF c. NH_4^+ d. NH_3 e. F^- f. HI

19.6 *Acid-Base Properties of Salts* Classify each of the following solutions as acidic, basic or neutral and explain by writing net ionic equations:

a. $NaC_2H_3O_2$ b. $ZnCl_2$ c. KNO_3 d. NH_4Br

19.7 *Brönsted-Lowry Acids and Bases* In the reaction $HCO_3^-(aq) + H_3O^+(aq) \rightleftharpoons H_2CO_3(aq) + H_2O$, which species behave as Brönsted acids? Brönsted bases?

19.8 Write a phrase defining

a. a strong acid.
b. a weak base.
c. an acidic solution.
d. a Brönsted base.

19.28 What is meant by each of the following statements?

a. Nitric acid, HNO_3, is a strong acid.
b. K_a for $HC_2H_3O_2$ is equal to 1.8×10^{-5}.
c. K_b for CN^- is equal to 2.5×10^{-5}.
d. A 0.3 *M* NaF solution is basic.

19.9 Find the *pH* of solutions with the following $[H^+]$:

a. 1×10^{-2} *M*
b. 0.000 10 *M*
c. 4.0×10^{-5} *M*
d. 6.2×10^{-10} *M*

Which solutions are acidic?

19.29 Find the *pH* of solutions with the following $[H^+]$:

a. 0.1 *M*
b. 10 *M*
c. 7.0×10^{-3} *M*
d. 8.2×10^{-9} *M*

Which of the solutions are basic?

19.10 Calculate $[H^+]$ and $[OH^-]$ in solutions with the following *pH*:

a. 4.0 b. 8.52 c. 0.00 d. 12.60

19.30 Find $[H^+]$ and $[OH^-]$ in solutions having the following *pH*:

a. 9.0 b. 3.20 c. −1.05 d. 7.46

Which solutions are acidic?

19.11 Find $[H^+]$, $[OH^-]$, and the *pH* of the following solutions:

a. 0.30 *M* HBr
b. 0.50 *M* KOH
c. a solution made by dissolving 100 g NaOH in water to make 500 cm^3 of solution.
d. a solution made by diluting 10 cm^3 6.0 *M* HCl to 300 cm^3 with water.

19.31 Find $[H^+]$, $[OH^-]$, and the *pH* of the following solutions:

a. 0.80 *M* NaOH
b. 0.60 *M* HCl
c. a solution made by diluting 8.0 cm^3 6.0 *M* KOH to a volume of 480 cm^3 with water.
d. a solution made by dissolving 75 g HCl in water to make 2.0 dm^3 of solution.

19.12 Write balanced net ionic equations for the reactions which make each of the following substances behave as acids in water solution:

a. HI b. H_2CO_3 c. $CuCl_2$ d. $SO_2(g)$

19.13 Indicate by means of net ionic equations why water solutions of each of the following have a *pH* greater than 7:

a. NH_3
b. $KC_2H_3O_2$
c. NaOH
d. $NaHCO_3$

19.14 Butyric acid, $CH_3CH_2CH_2COOH$, is one of the decomposition products in rancid butter and has a well-deserved malodorous reputation. In a well-vented hood a student measured the *pH* of 0.400 dm^3 of a solution containing 0.20 mol butyric acid and found it to be 2.50. Find K_a for butyric acid from this information.

19.15 Find the $[H^+]$, the *pH*, and the per cent dissociation of the acid in

a. 0.25 *M* HBr.
b. 0.25 *M* HOCl.

19.16 Chloroacetic acid, $CH_2ClCOOH$, has a K_a of 1.4×10^{-3}. Using the method of successive approximations, find $[H^+]$ in 0.100 *M* $CH_2ClCOOH$.

19.17 Write the net ionic equation for the reaction that makes KCN in water solution behave as a base.

a. What is K_b for the reaction?
b. Find the *pH* of 0.20 *M* KCN.

19.18 Find the value of K_b for the conjugate base of each of the following acids:

a. gallic acid, present in tea; $K_a = 3.9 \times 10^{-5}$
b. mandelic acid, obtained from bitter almonds; $K_a = 1.4 \times 10^{-4}$

19.19 State whether solutions of the following substances in water would be acidic, basic, or neutral:

a. NH_4Cl b. H_3PO_4 c. Na_3PO_4
d. KNO_3 e. $KHCO_3$ f. NaCN

19.20 Write net ionic equations to explain the acidity or basicity of the various solutions listed in Problem 19.19.

19.21 Arrange the following 0.1 *M* solutions in order of increasing *pH:*

KOH Na_2CO_3 NaOCl NH_3
HOCl H_2SO_4 HCl Na_2SO_4

19.32 Write balanced net ionic equations showing why each of the following substances produces an acidic solution in water:

a. H_3PO_4 b. $CO_2(g)$ c. HNO_2 d. $AlCl_3$

19.33 Write net ionic equations to explain why the following species act as weak bases in water solution.

a. CH_3NH_2 (similar to NH_3)
b. NO_2^-
c. HPO_4^{2-}
d. CHO_2^-

19.34 Formic acid, HCOOH, is the irritant in nettles and ants, and is the simplest organic acid. Sketch the molecular structure, referring back to Chapter 10 if you need to. Which H atom is acidic? If the *pH* of a solution of 5.0 g HCOOH in 1.00 dm^3 of water is 2.30, estimate the dissociation constant K_a for formic acid.

19.35 Find the $[H^+]$, *pH*, and the per cent dissociation of the acid in

a. 0.60 *M* HCl.
b. 0.60 *M* NH_4NO_3.

19.36 A chloropropionic acid, $CH_3CH_2ClCOOH$, has a K_a of 1.5×10^{-3}. Using the method of successive approximations, find $[H^+]$ in 0.60 *M* $CH_3CH_2ClCOOH$.

19.37 Write the net ionic equation for the reaction that makes solutions of sodium benzoate, $NaC_7H_5O_2$, basic. (Take HBz as benzoic acid, and Bz^- as the benzoate ion.)

a. What is K_b for the reaction?
b. Find the *pH* of 0.50 *M* NaBz.

19.38 Find the value of K_a for the conjugate acid of the following organic bases:

a. dimethylamine, used to make the insecticide, Sevin; $K_b = 5.2 \times 10^{-4}$
b. aniline, an important dye intermediate; $K_b = 3.8 \times 10^{-10}$

19.39 State whether aqueous solutions of the following substances would be acidic, basic, or neutral:

a. $Al(NO_3)_3$ b. HOCl c. NaOCl
d. NH_4NO_3 e. Na_2CO_3 f. $NaHSO_4$

19.40 Write net ionic equations for the reactions in water of the various species in Problem 19.39.

19.41 Arrange the following 0.1 *M* solutions in order of increasing *pH:*

H_2CO_3 KBr HI NH_3
KCN NaOH NH_4Br

19.22 For each of the following reactions, indicate the Brönsted acids and bases:

a. $H_3O^+(aq) + HSO_3^-(aq) \rightleftarrows H_2SO_3(aq) + H_2O$
b. $HF(aq) + OH^-(aq) \rightleftarrows F^-(aq) + H_2O$
c. $NH_4^+(aq) + H_2O \rightleftarrows NH_3(aq) + H_3O^+(aq)$

19.23 Find K for each of the reactions in Problem 19.22, using Table 19.5 and, if necessary, the Multiple Equilibria Rule, predict whether the equilibrium state would tend to favor reactants or products.

19.24 Which of the following species can act as Brönsted acids? Brönsted bases? Lewis acids? Lewis bases?

a. CO_3^{2-} b. H_3O^+ c. Fe^{3+} d. $H_2PO_4^-$

19.25 Two solutions are needed in the stockroom, each with a volume of 10 dm^3 and a pH equal to 11.00. How many moles of each solute would it take if one solution is to be made with NaOH and the other with $NH_3(K_b = 1.8 \times 10^{-5})$?

19.26 Give two examples of

a. an acid that is stronger than HF.
b. a salt which dissolves in water to yield a neutral solution.
c. a salt which dissolves in water to yield an acidic solution.
d. an oxide that would dissolve in water to form a basic solution.
e. a base that is stronger than NH_3.

19.27 For each of the solutions listed, indicate the main solute species present (0.10 M or greater) and find the pH of the solution.

a. 0.10 M NaBr
b. 0.60 M NaF
c. 0.50 M NH_4NO_3
d. 0.30 M LiOH

19.42 For each of the following reactions, indicate the Brönsted acids and bases:

a. $CN^-(aq) + H_2O \rightleftarrows HCN(aq) + OH^-(aq)$
b. $HCO_3^-(aq) + H_3O^+(aq) \rightleftarrows H_2CO_3(aq) + H_2O$
c. $HC_2H_3O_2(aq) + HS^-(aq) \rightleftarrows C_2H_3O_2^-(aq) + H_2S(aq)$

19.43 Find K for each of the reactions in Problem 19.42, using Table 19.5, and if necessary, the Multiple Equilibria Rule. Which of these reactions will proceed to the right if all solute reactants are 0.1 M?

19.44 Which of the following species can act as Brönsted acids? Brönsted bases? Lewis acids? Lewis bases?

a. H_2O b. NO_3^- c. OCl^- d. HSO_4^-

19.45 You are asked to make two solutions, both with a pH of 2.00 and with a volume of 6.00 dm^3. How many moles of each solute will it take if one of the solutions is to be made with HCl and the other with chloroacetic acid, HClAc, whose K_a is 1.4×10^{-3}?

19.46 Give two examples of

a. a base that is weaker than ammonia.
b. an acid that is weaker than H_2S.
c. a salt which dissolves in water to yield a basic solution.
d. an acid that is made by dissolving an oxide in water.
e. a strong acid.

19.47 For each of the solutions below, indicate the main species present (0.10 M or greater) and find the pH of the solution.

a. 0.20 M HI
b. 0.10 M NaOCl
c. 0.60 M KNO_3
d. 0.050 M $Ba(OH)_2$

*19.48 Using information given in Table 19.5 and the Law of Multiple Equilibria, determine the equilibrium constant for the reaction

$$H_2S(aq) \rightleftharpoons 2\,H^+(aq) + S^{2-}(aq)$$

What is the concentration of S^{2-} in a solution 0.10 M in H_2S which has a pH of 3.0?

*19.49 Silver hydroxide, AgOH, is insoluble in water ($K_{sp} = 1 \times 10^{-8}$). Describe a simple qualitative experiment which would enable you to determine whether AgOH is a strong or weak base.

*19.50 It is found that 0.20 M solutions of the three sodium salts NaX, NaY, and NaZ have pH's of 7.0, 8.0, and 9.0, respectively. Arrange the acids HX, HY, and HZ in order of increasing strength. Where you can, find K_a for the acids.

*19.51 Among the important biochemical substances are the amino acids, of which glycine is an example:

$$H_2N{-}CH_2{-}\underset{\underset{\displaystyle O}{\|}}{C}{-}OH$$

In each amino acid there is a weak acid group —COOH and a weak basic group $—NH_2$. Into what species would glycine tend to convert in very acidic solution? What species would it form in very basic solution? Given that K_a for —COOH and K_b for $—NH_2$ are about equal, glycine in water can be shown to exist as yet another species, called a zwitterion. Suggest the structure of this species.

20

ACID-BASE REACTIONS

In the previous chapter we were concerned for the most part with the acidic or basic properties of solutions containing one solute, which was either a strong or weak acid or base. If we mix two such solutions, one acidic and one basic, an acid-base reaction occurs. To find the properties of the final solution we must be able to write the net ionic equation for the reaction that occurs, determine its associated equilibrium constant, and apply the principles of equilibrium, much as we did in dealing with precipitation reactions.

20.1 TYPES OF ACID-BASE REACTIONS

Reactions of Strong Acids and Bases

Consider the reaction that takes place when a solution of a strong acid like HCl, HNO_3, or H_2SO_4 is mixed with that of a strong base like NaOH or KOH. If, say, nitric acid is mixed with sodium hydroxide, the H^+ ions in the completely ionized acid solution will react with the OH^- ions in the ionized base:

Why does $K = 1/K_w$?

$$H^+(aq) + OH^-(aq) \rightleftarrows H_2O \qquad K = 1/K_w = 1.0 \times 10^{14} \qquad (20.1)$$

This is the reaction that occurs when a solution of any strong acid is mixed with a solution of any strong base. It is often called a *neutralization* reaction, since if just enough base is added to react with all the acid, the solution becomes neutral. If, for example, exactly 200 cm^3 0.500 M HNO_3 are mixed with exactly 200 cm^3 0.500 M NaOH, the H^+ ion from the acid will be just neutralized by the OH^- ion from the base; the final solution will contain only the nitrate ions from the acid and sodium ions from the base and will have a pH of 7.

Since K for Reaction 20.1 is enormous, the reaction between H^+ and OH^- ions is, for all practical purposes, *quantitatively complete*. Of course, it is not necessary that equal amounts of acid and base be mixed for a reaction to occur. In many cases an excess of one or the other will be used, and the final solution will contain a salt and the reagent present in excess. Mixing 200 cm^3 0.500 M HNO_3 with 300 cm^3 0.500 M NaOH would result in a solution equivalent to one made from $NaNO_3$ and NaOH, since the NaOH is in excess. Under such circumstances the solution would be basic and, in fact, it is easy to show that in that solution $[OH^-]$ will be 0.10 M.

Example 20.1 Find $[OH^-]$ in a solution made by mixing 200 cm³ 0.500 *M* HNO_3 with 300 cm³ 0.500 *M* NaOH.

Solution When the solutions are mixed the reaction which occurs is

$$H^+(aq) + OH^-(aq) \rightleftarrows H_2O$$

In the acid:

$$\text{no. moles } H^+ = M_{HNO_3} V_{HNO_3} = 0.500 \frac{\text{mol } H^+}{\text{dm}^3} \times 0.200 \text{ dm}^3 = 0.100 \text{ mol}$$

In the base:

$$\text{no. moles } OH^- = M_{NaOH} V_{NaOH} = 0.500 \text{ mol } OH^-/\text{dm}^3 \times 0.300 \text{ dm}^3 = 0.150 \text{ mol}$$

Assuming complete reaction, 0.100 mol H^+ in the acid will be neutralized by 0.100 mol OH^- from the base, forming 0.100 mol H_2O and leaving an excess of 0.050 mol OH^- in the 500 cm³ of mixed solution.

$$[OH^-] = \frac{\text{no. moles } OH^-}{\text{volume solution}} = \frac{0.050}{0.500} = 0.10\ M$$

Reactions of Weak Acids with Strong Bases

When a solution of a weak acid is mixed with that of a strong base, an acid-base reaction again occurs. Indeed, H^+ ions from the dissociated part of the weak acid react with OH^- ions from the base. This depletes the hydrogen ion concentration and more weak acid dissociates to restore equilibrium. The net change which occurs is the disappearance of HX, the weak acid, by reaction with OH^- ion to form water and the X^- ion:

$$HX(aq) + OH^-(aq) \rightleftarrows H_2O + X^-(aq) \qquad K = \frac{1}{K_b \text{ of } X^-} \qquad (20.2)$$

Note the similarity of this reaction to Reaction 20.1. We take HX as a reactant, rather than H^+, since in solution the weak acid is primarily in the form of HX molecules.

While Reaction 20.1 is the reverse of the reaction by which water ionizes, Reaction 20.2 is the reverse of the reaction of a basic anion with water:

$$X^-(aq) + H_2O \rightleftharpoons HX(aq) + OH^-(aq)$$

So, just as the equilibrium constant for Equation 20.1 is $1/K_w$, that for Equation 20.2 is $1/K_b$, the reciprocal of the base dissociation constant of the X^- ion. Since K_b is usually very small, its reciprocal is very large, and Reaction 20.2, like 20.1, tends to go quantitatively to the right.

If equivalent amounts of HX and OH^- are mixed, the final solution will be that of a salt, as before, but in this case it will be that of a salt of a *weak* acid and will not be neutral. If, for example, we mix 20 cm³ 1.0 *M* $HC_2H_3O_2$, acetic acid, with 20 cm³ 1.0 *M* NaOH, the final solution will contain the salt, sodium acetate. Since acetate ion behaves as a weak base, the solution will be basic. In general, at the equivalence point of the reaction between any weak acid and a strong base (i.e., when equivalent amounts of acid and base have been mixed), the solution will be basic.

Example 20.2 A solution of $HC_2H_3O_2$ is mixed with a solution of NaOH.

a. Write the net ionic equation for the reaction which occurs.
b. Calculate K for this reaction.
c. If a solution containing 0.1 mol $HC_2H_3O_2$ is mixed with a solution containing 0.1 mol NaOH, describe the nature of the solution obtained.

Solution

a. Referring to Equation 20.2, HX becomes $HC_2H_3O_2$, so the reaction is

$$HC_2H_3O_2(aq) + OH^-(aq) \rightleftarrows H_2O + C_2H_3O_2^-(aq)$$

b. K for the reaction is equal to $1/K_b$ for the $C_2H_3O_2^-$ ion. In Table 19.5, p. 460, we find that K_b for $C_2H_3O_2^-$ is 5.6×10^{-10}. Therefore,

$$K = \frac{1}{5.6 \times 10^{-10}} = 1.8 \times 10^9$$

c. Since K_b is very large, the reaction produces just about 0.1 mol $C_2H_3O_2^-$ ion and uses up just about all of the $HC_2H_3O_2$ and OH^- ion. The solution contains 0.1 mol Na^+ ion from the original NaOH solution. So we have as the main species in the final solution, 0.1 mol Na^+ ion and 0.1 mol $C_2H_3O_2^-$ ion. This solution is completely equivalent to one of equal volume made by dissolving 0.1 mol $NaC_2H_3O_2$, sodium acetate, in water. The solution would be slightly basic.

With weak acids containing more than one acidic hydrogen, reaction with OH^- ions will occur stepwise. An important example is the reaction of carbonic acid, H_2CO_3, with a solution of a strong base such as NaOH:

$$H_2CO_3(aq) + OH^-(aq) \rightleftarrows H_2O + HCO_3^-(aq) \qquad (20.3)$$

$$HCO_3^-(aq) + OH^-(aq) \rightleftarrows H_2O + CO_3^{2-}(aq) \qquad (20.4)$$

H_2SO_3 and H_3PO_4 would behave similarly

The ultimate product depends on the relative quantities of acid and base which are used. *The first reaction goes nearly to completion before the second one starts.* If one mole of OH^- ions is added per mole of H_2CO_3, the product is the HCO_3^- ion. Addition of another mole of OH^- removes the second proton to give the carbonate ion, CO_3^{2-}. Reaction 20.4 would occur directly if a solution of a strong base were added to a solution made by dissolving sodium hydrogen carbonate, $NaHCO_3$, in water.

Reactions of Strong Acids and Weak Bases

If a solution of a weak base, such as NH_3, is mixed with a solution of a strong acid, such as HCl, the net overall reaction would be:

$$H^+(aq) + NH_3(aq) \rightleftarrows NH_4^+(aq) \qquad K = \frac{1}{K_a \text{ of } NH_4^+} \qquad (20.5)$$

This reaction is also similar to Reaction 20.1. In this case we need to write NH_3 as the basic species, rather than OH^-, since in an ammonia solution the main species present is NH_3, not OH^-. Reaction 20.5 is the reverse of the ionization reaction of the weak acid NH_4^+. Therefore, the equilibrium constant for Reaction 20.5 is $1/K_a$, where K_a is the acid dissociation constant for NH_4^+. Since K_a for

NH_4^+ is very small, K for Reaction 20.5 is very large, and, like the other two types of acid-base reactions we have discussed, tends to go essentially to completion. All reactions between solutions of strong acids and weak bases are similar to the H^+-NH_3 reaction; all produce a weak acid as a product, and all have as an equilibrium constant the reciprocal of the ionization constant of that acid.

Example 20.3 A solution of sodium acetate, $NaC_2H_3O_2$, is mixed with a solution of nitric acid, HNO_3.

a. Write the net ionic equation for the reaction which occurs.
b. Determine the equilibrium constant for this reaction.
c. Describe the nature of the solution obtained by mixing 1.0 dm³ of 0.1 *M* $NaC_2H_3O_2$ with 1.0 dm³ of 0.1 *M* HNO_3.

Solution

a. In sodium acetate solution, the solute is completely ionized, and the species present are Na^+ ion, which is neutral, and $C_2H_3O_2^-$ ion, which is a weak base. Nitric acid is a strong acid, so its solution contains NO_3^- ion, which is neutral, and H^+ ion. The acid-base reaction which takes place is therefore

$$C_2H_3O_2^-(aq) + H^+(aq) \rightleftarrows HC_2H_3O_2(aq)$$

b. The reaction is the reverse of the one by which acetic acid, $HC_2H_3O_2$, ionizes, so K equals $1/K_a$ for acetic acid. From Table 19.5, K_a equals 1.8×10^{-5}. Therefore,

$$K = \frac{1}{1.8 \times 10^{-5}} = 5.6 \times 10^4$$

c. The original solutions contained 0.1 mol Na^+ and 0.1 mol $C_2H_3O_2^-$, and 0.1 mol H^+ and 0.1 mol NO_3^-. In the reaction, essentially all of the $C_2H_3O_2^-$ and H^+ combine to form 0.1 mol $HC_2H_3O_2$. The final solution is equivalent to one of equal volume made by dissolving 0.1 mol $NaNO_3$ and 0.1 mol $HC_2H_3O_2$ in water. It is weakly acidic.

The *pH* of the solution would be that of 0.05 *M* $HC_2H_3O_2$

If an anion is capable of acquiring two protons, the reaction with a strong acid occurs in two steps; an example of such a species is the carbonate ion:

$$H^+(aq) + CO_3^{2-}(aq) \rightleftarrows HCO_3^-(aq) \qquad K = 1/K_a \text{ of } HCO_3^- \quad (20.6)$$

$$HCO_3^-(aq) + H^+(aq) \rightleftarrows H_2CO_3(aq) \rightleftarrows CO_2(g) + H_2O \quad K = 1/K_a \text{ of } H_2CO_3 \quad (20.7)$$

Addition of excess acid to a solution containing CO_3^{2-} ion may lead to the evolution of CO_2 gas if the initial carbonate ion concentration is high enough. Carbon dioxide may also be formed if sodium hydrogen carbonate is added to an acidic solution. This is, of course, what happens in the stomach when bicarbonate of soda, $NaHCO_3$, is taken to relieve acid indigestion.

In working with the various kinds of acid-base reactions, you would be well advised to take advantage of their similarities. In each case an acid is reacting with a base, but where the acid or the base is weak, we write the net ionic equation in terms of that weak acid or base rather than in terms of H^+ or OH^-, respectively. Since the net ionic equations for the various reactions are the reverse of equations we have previously investigated, the equilibrium constants for those reactions are the reciprocals of the constants for those previous reactions. In general the values of K for acid-base reactions are large, so the reactions may be considered to go nearly to completion. In Table 20.1 we have summarized the pertinent relationships for the three sorts of acid-base reactions we have considered.

TABLE 20.1 CHARACTERISTICS OF ACID-BASE REACTIONS

TYPE	EXAMPLE	K	SPECIES AT EQUIVALENCE POINT
Strong acid + strong base HCl-NaOH	$H^+(aq) + OH^-(aq) \rightleftarrows H_2O$	$\frac{1}{K_w} = 1.0 \times 10^{14}$	Na^+, Cl^- neutral solution
Weak acid + strong base $HC_2H_3O_2$-NaOH	$HC_2H_3O_2(aq) + OH^-(aq) \rightleftarrows H_2O + C_2H_3O_2^-(aq)$	$\frac{1}{K_b} = 1.8 \times 10^9$	Na^+, $C_2H_3O_2^-$ basic solution
Weak base + strong acid NH_3-HCl	$NH_3(aq) + H^+(aq) \rightleftarrows NH_4^+(aq)$	$\frac{1}{K_a} = 1.8 \times 10^9$	NH_4^+, Cl^- acidic solution

Reactions of Strong Acids with Solids

Unlike the previous acid-base reactions, all of which occur between solutions, there is an important class of reactions in which the source of the base is a solid. Many of the compounds of the metals contain anions which are basic; a majority of these compounds are insoluble in water. These include the hydroxides, the carbonates, phosphates, and sulfides of most of the metals except those in Group IA. These substances dissolve in water to only a small extent to achieve equilibrium with their ions in solution as discussed in Chapter 18. Barium carbonate is a typical example:

$$BaCO_3(s) \rightleftarrows Ba^{2+}(aq) + CO_3^{2-}(aq) \qquad K = K_{sp} = 2 \times 10^{-9}$$

As you can see from the value of the solubility product constant, not much $BaCO_3$ would dissolve in a liter of water. Faced with the problem of preparing a solution in which $[Ba^{2+}]$ was, say, 0.2 M, from solid $BaCO_3$, you might well decide it was impossible. However, recalling that for equilibrium of the solid with its ions it is only necessary that

$$[Ba^{2+}] \times [CO_3^{2-}] = 2 \times 10^{-9}$$

You can see that if you could make $[CO_3^{2-}]$ small enough, $[Ba^{2+}]$ could, at equilibrium be 0.2 M or even larger. So, barium carbonate would, indeed, dissolve in a solution as long as in that solution, $[CO_3^{2-}]$ was kept very low, of the order of 1×10^{-8} M.

It turns out that by adding a strong acid we can indeed lower the $[CO_3^{2-}]$ to a very small value. Recall that carbonate ion is a weak base, and will react with strong acids as in Equations 20.6 and 20.7, forming HCO_3^- or H_2CO_3. These reactions act to deplete $[CO_3^{2-}]$ and thus enhance the solubility of $BaCO_3$. The quantitative effect of acids on the solubility of $BaCO_3$ is perhaps most easily seen by writing the net ionic equation for the reaction of the solid with a strong acid:

$$BaCO_3(s) + 2\,H^+(aq) \rightleftarrows Ba^{2+}(aq) + H_2CO_3(aq) \qquad (20.8)$$

The equilibrium constant for Reaction 20.8 is a very large number, about 1×10^8 (see Example 20.4). This means that this reaction, like all of those we have studied in this section, goes essentially to completion. In other words, by using excess acid, it is possible to bring large amounts of $BaCO_3$ into solution.

Example 20.4 Apply the Law of Multiple Equilibria to calculate the equilibrium constant for Reaction 20.8.

Solution We can regard Reaction 20.8 as the sum of two reactions:

$$(1)\ BaCO_3(s) \rightleftharpoons Ba^{2+}(aq) + CO_3^{2-}(aq); \qquad K_1$$

$$(2)\ 2\ H^+(aq) + CO_3^{2-}(aq) \rightleftharpoons H_2CO_3(aq); \qquad K_2$$

$$BaCO_3(s) + 2\ H^+(aq) \rightleftharpoons Ba^{2+}(aq) + H_2CO_3(aq); \ K$$

Applying the Law of Multiple Equilibria:

$$K = K_1 \times K_2$$

Examination of Reaction (1) shows that it is the solubility product expression for $BaCO_3$. Hence:

$$K_1 = K_{sp}\ BaCO_3 = 2 \times 10^{-9}$$

This example illustrates the usefulness of the Law of Multiple Equilibria

To evaluate K_2, we note that Equation (2) is the sum of two equations previously considered, 20.6 and 20.7:

$$(2a)\ H^+(aq) + CO_3^{2-}(aq) \rightleftharpoons HCO_3^-(aq); \quad K = 1/K_a \text{ of } HCO_3^-$$

$$(2b)\ H^+(aq) + HCO_3^-(aq) \rightleftharpoons H_2CO_3(aq); \quad K = 1/K_a \text{ of } H_2CO_3$$

$$(2)\ \ 2\ H^+(aq) + CO_3^{2-}(aq) \rightleftharpoons H_2CO_3(aq); \ K_2$$

$$\text{So: } K_2 = \frac{1}{K_a\ HCO_3^-} \times \frac{1}{K_a\ H_2CO_3} = \frac{1}{4.8 \times 10^{-11}} \times \frac{1}{4.2 \times 10^{-7}} = \frac{1}{2.0 \times 10^{-17}} = 5 \times 10^{16}$$

Finally, multiplying K_1 by K_2, we obtain the equilibrium constant for the reaction we are interested in, Reaction 20.8:

$$K = K_1 \times K_2 = (2 \times 10^{-9})(5 \times 10^{16}) = 1 \times 10^8$$

By analyses similar to the one we have just carried out, we find that carbonates in general, like $BaCO_3$, are soluble in excess strong acid:

$$MgCO_3(s) + 2\ H^+(aq) \rightleftharpoons Mg^{2+}(aq) + H_2CO_3(aq); \ K = 1 \times 10^9$$

as are most hydroxides:

$$Mg(OH)_2(s) + 2\ H^+(aq) \rightleftharpoons Mg^{2+}(aq) + 2\ H_2O; \ K = 1 \times 10^{17}$$

and many sulfides:

$$MnS(s) + 2\ H^+(aq) \rightleftharpoons Mn^{2+}(aq) + H_2S(aq); \ K = 1 \times 10^9$$

The only exceptions are some very insoluble sulfides which will be discussed in Section 20.5. Summarizing, we can say that: **Insoluble inorganic compounds containing basic anions tend to be soluble in solutions of strong acids.** The use of *pH* control in affecting the solubilities of substances which are insoluble in water is a beautiful example of how the chemist, by using the laws of equilibrium, can drive a reaction such as 20.8 to the right or left as the situation demands.

There are, as you may know, some insoluble salts which contain anions which are essentially neutral. Among the most common of these are some containing Cl^- or SO_4^{2-} ions, such as AgCl, $BaSO_4$, and $PbSO_4$. If you attempt to dissolve these substances in acid, you find that for the most part nothing happens; they remain insoluble. The reason for this becomes apparent if you carry out the preceding development substituting, for example, $PbSO_4$ for $BaCO_3$. An equation like 20.8 results. However, since K_a for H_2SO_4 is very, very large (sulfuric acid is a strong acid), K for the reaction of $PbSO_4$ with H^+ is just about zero, so lead sulfate fails to dissolve in acid. A similar situation applies for AgCl; HCl, like H_2SO_4, is a strong acid. In general, water-insoluble salts of strong acids cannot be dissolved by adding acid.

20.2 ACID-BASE TITRATIONS

Since the equilibrium constants for many acid-base reactions are very large, these reactions are often applied in quantitative analysis. In particular, they are used in titrations to determine concentrations of acids and bases in solution and to aid in the analysis of mixtures. The general procedure is similar to that used with titrations involving precipitation reactions. We measure with a buret the volume of a standardized acidic or basic solution which is required to just react with a measured volume of a solution of the base or acid whose concentration we wish to determine. The problem of knowing when stoichiometrically equivalent amounts of acid and base have been mixed is handled in general by the use of suitable indicators or a *pH* meter. Procedures for carrying out acid-base titrations have been developed to the point where they are among the most precise methods of chemical analysis. They are used in nearly every laboratory, be it in a university, power plant, or brewery. Examples 20.5 and 20.6 indicate the calculations which are involved in some typical titrations.

Example 20.5 It is found that 22.3 cm^3 of 0.240 M NaOH is required to react with a 50.0-cm^3 sample of vinegar, a solution of acetic acid in water. Calculate the concentration of acetic acid in the vinegar.

Solution From the equation for the reaction of OH^- ions with acetic acid

$$HC_2H_3O_2(aq) + OH^-(aq) \rightarrow C_2H_3O_2^-(aq) + H_2O$$

This is the easiest way to find the concentration of an acidic solution

it is evident that the reactants are consumed in a 1:1 mole ratio. That is,

$$\text{no. moles } OH^- = \text{no. moles } HC_2H_3O_2$$

But

$$\text{no. moles of } OH^- = (M\ \text{NaOH})(\text{volume NaOH})$$

$$\text{no. moles } HC_2H_3O_2 = (M\ HC_2H_3O_2)(\text{volume } HC_2H_3O_2 \text{ solution})$$

Hence

$$(M\ \text{NaOH})(\text{volume NaOH}) = (M\ HC_2H_3O_2)(\text{volume } HC_2H_3O_2 \text{ solution})$$

$$0.240\ \frac{\text{mol}}{\text{dm}^3}(0.0223\ \text{dm}^3) = (M\ HC_2H_3O_2)(0.0500\ \text{dm}^3)$$

Solving,

$$M\ HC_2H_3O_2 = 0.240\ \frac{\text{mol}}{\text{dm}^3} \times \frac{0.0223}{0.0500} = 0.107\ M$$

Example 20.6 A chemist synthesizes a substance which he believes is barbituric acid, $MM = 128.1$, a precursor for many sleeping tablets. Barbituric acid has one acidic hydrogen, $K_a = 9.8 \times 10^{-5}$. To help with the identification, he titrates a 0.5000-g crystalline sample with 0.1000 M NaOH and finds that at the equivalence point he has added 39.10 cm³ of the base. Is the sample likely to be barbituric acid?

Solution From the data, we can calculate the MM of the compound. One mole of the acid will react with one mole of OH^-. Therefore,

$$\text{no. moles acid} = \text{no. moles OH}^- \text{ added}$$

$$\frac{\text{no. grams acid}}{GMM} = M_{NaOH} \times V_{NaOH} = \frac{0.100 \text{ mol}}{1 \text{ dm}^3} \times 0.039\ 10 \text{ dm}^3$$

$$= 3.910 \times 10^{-3} \text{ mol}$$

$$GMM = \frac{\text{no. grams acid}}{\text{no. moles OH}^-} = \frac{0.5000 \text{ g}}{3.910 \times 10^{-3} \text{ mol}} = 127.9 \text{ g/mol}$$

Since the molecular mass of the sample is very nearly that reported for barbituric acid, the material may well be barbituric acid. Further evidence, such as melting point and infrared spectrum, would have to be obtained to confirm the identification.

Normality: Gram Equivalent Masses of Acids and Bases

The concentrations of solutions used in titrations are sometimes expressed in terms of normality rather than moles per cubic decimetre. For any solute A its normality in solution is defined as:

$$\text{Normality of A} = \frac{\text{no. } GEM \text{ of A}}{\text{no. cubic decimetres solution}}; \quad N_A = \frac{\text{no. } GEM \text{ of A}}{V_A} \qquad (20.9)$$

One gram equivalent mass, *GEM,* or one equivalent of A, is that mass of A which, in a reaction between A and B, will react with one *GEM* of the species B. This approach makes for very simple calculations of titration results, once the normality of A or B has been established. For the reaction mixture at the equivalence point,

$$\text{no. } GEM \text{ A added} = \text{no. } GEM \text{ B added}$$

Using Equation 20.9:

$$N_A \times V_A = N_B \times V_B \qquad (20.10)$$

Knowing the normality N_A and the volume of the solution of A required to react exactly with a measured volume of the solution of B, the normality of B is very easily found. Equation 20.10 can be applied to any titration reaction, including acid-base reactions.

The only difficulty in using normalities of solutions is to establish a simple rational definition for the amount of substance in one gram equivalent mass. For acids and bases the *GEM* is defined as follows:

1 *GEM* acid = mass of acid which reacts with one mole of OH^-
1 *GEM* base = mass of base which reacts with one mole of H^+

By this definition,

1 *GEM* HCl = 36.5 g HCl = 1 mol HCl

1 *GEM* NaOH = 40 g NaOH = 1 mol NaOH
1 *GEM* H_2SO_4 = 49 g H_2SO_4 = ½ mol H_2SO_4
1 *GEM* $Ca(OH)_2$ = 37 g $Ca(OH)_2$ = ½ mol $Ca(OH)_2$

Hence, for a given solution, N and M are simply related; for example,

1.0 N HCl = 1.0 M HCl	0.5 N NaOH = 0.5 M NaOH
1.0 N H_2SO_4 = 0.5 M H_2SO_4	0.1 N $Ca(OH)_2$ = 0.05 M $Ca(OH)_2$

In some cases the relation between these two concentration units is not as apparent. It is sometimes the case that a solution with a certain value of M may have several possible normalities, depending on the reaction in which the reagent participates. If, for instance, we react phosphoric acid, H_3PO_4, with a base, we can stop the titration when we have removed one, or two, or three protons:

I	$H_3PO_4(aq) + OH^-(aq) \rightleftarrows H_2PO_4^-(aq) + H_2O$	$N_{H_3PO_4} = M_{H_3PO_4}$
II	$H_3PO_4(aq) + 2\ OH^-(aq) \rightleftarrows HPO_4^{2-}(aq) + 2\ H_2O$	$N_{H_3PO_4} = M_{H_3PO_4} \times 2$
III	$H_3PO_4(aq) + 3\ OH^-(aq) \rightleftarrows PO_4^{3-}(aq) + 3\ H_2O$	$N_{H_3PO_4} = M_{H_3PO_4} \times 3$

In Reaction I, 1.0 dm^3 of 1 M H_3PO_4 would react with one mole OH^- ion and, hence, would contain one equivalent, and would be 1 N; in Reaction II, 1.0 dm^3 of the *same solution* reacts with *two* moles of OH^- ion and so would be 2 N H_3PO_4; in Reaction III, the same solution would be 3 N H_3PO_4.

The fact that the normality of a given reagent *depends on the reaction in which it participates* is a serious deficiency of the whole idea of normality. In recent years some chemists have suggested that concentrations never be expressed in normalities because of the ambiguities we have just noted. It is certainly true that concentrations are less ambiguous when expressed in moles per cubic decimetre, and indeed M is the preferred concentration unit. However, in cases where one is working with a particular reaction under routine conditions it may be worthwhile, because of the simplicity and general validity of Equation 20.10, to use normalities.

Acid-Base Indicators

The end point of an acid-base titration is established by observing a change in color in an acid-base indicator, which is added in small amount before carrying out the titration. The color of the indicator, which is ordinarily an organic dye, depends upon the H^+ ion concentration in the solution being titrated. In this sense, it "indicates" the rather drastic change in pH that ordinarily occurs near the end point of an acid-base titration.

To illustrate how an indicator works, let us consider a typical example, bromthymol blue. Like most acid-base indicators, bromthymol blue is a weak acid; it has a dissociation constant of about 1×10^{-7}. Using HIn to represent the molecular acid and In^- the conjugate base formed when bromthymol blue dissociates in water, we have:

$$\underset{\text{yellow}}{HIn(aq)} \rightleftharpoons H^+(aq) + \underset{\text{blue}}{In^-(aq)} \qquad K_a = \frac{[H^+] \times [In^-]}{[HIn]} = 1 \times 10^{-7}$$

The essential characteristic of bromthymol blue, like that of all other acid-base indicators, is that *the two species HIn and In^- have different colors.* In this case, the HIn molecule is yellow and the In^- ion is blue.

Let us consider what color bromthymol blue will have in solutions of different

pH. Depending on the $[H^+]$, the ratio $[In^-]/[HIn]$ will take on whatever value is required to satisfy the above equation. Solving that equation for the ratio, we get

$$\frac{[In^-]}{[HIn]} = \frac{1 \times 10^{-7}}{[H^+]} \tag{20.11}$$

Using Equation 20.11, we can distinguish three possibilities:

1. If, in a particular solution, $[H^+]$ is 10^{-6} or greater ($pH \leq 6$), most of the indicator will be in the form of the HIn molecule and the solution will appear yellow. Thus, at pH 6:

The pH is that of the solution and is not influenced by the presence of the indicator

$$\frac{[In^-]}{[HIn]} = \frac{1 \times 10^{-7}}{1 \times 10^{-6}} = 0.1$$

and there will be only one In^- ion (blue) for every ten HIn molecules (yellow).

2. If $[H^+] = 10^{-7}$ ($pH = 7$), equal amounts of In^- and HIn will be present

$$\frac{[In^-]}{[HIn]} = \frac{1 \times 10^{-7}}{1 \times 10^{-7}} = 1$$

and the solution will appear green (equal amounts of blue and yellow).

3. If $[H^+]$ is 10^{-8} or less ($pH \geq 8$), the In^- ion will predominate and the solution will appear blue.

$$\frac{[In^-]}{[HIn]} = \frac{1 \times 10^{-7}}{1 \times 10^{-8}} = 10$$

In practice, bromthymol blue changes color gradually from yellow to blue as the pH goes from 6 to 8. Like most indicators, it has a range of about 2 pH units over which it changes color. We say that bromthymol blue has an **end point** at pH 7 since at that pH the change in color is most easily detected.

The end point observed with any particular indicator will depend upon the magnitude of its acid dissociation constant, K_a. The $[H^+]$ at the end point will equal the K_a of the indicator. Methyl red ($K_a = 1 \times 10^{-5}$) has an end point at pH 5; it changes color from red to yellow as the pH goes from 4 to 6. Phenolphthalein ($K_a = 1 \times 10^{-9}$) changes from colorless to pink in the pH range 8–10 (recall Plate 12).

In carrying out an acid-base titration, we try to select an indicator which changes color at or very near the pH of the equivalence point of the reaction; i.e., the point at which equal quantities of acid and base have been added. A successful titration, one with good quantitative results, is possible only if the **end point** as established by the indicator actually occurs at the **equivalence point** in the titra-

TABLE 20.2 COLORS OF ACID-BASE INDICATORS

INDICATOR	K_a	$pH =$ 3	4	5	6	7	8	9	10	11
Bromthymol blue	1×10^{-7}	y	y	y	y	gr	b	b	b	b
Methyl red	1×10^{-5}	r	r	o	y	y	y	y	y	y
Phenolphthalein	1×10^{-9}	c	c	c	c	c	c	p	p	p

y = yellow r = red c = colorless p = pink o = orange gr = green

$$\frac{[In^-]}{[HIn]} = \frac{K_a}{[H^+]}$$

If $[H^+] \ll K_a$, color of In^- ion predominates.
If $[H^+] \gg K_a$, color of HIn predominates.

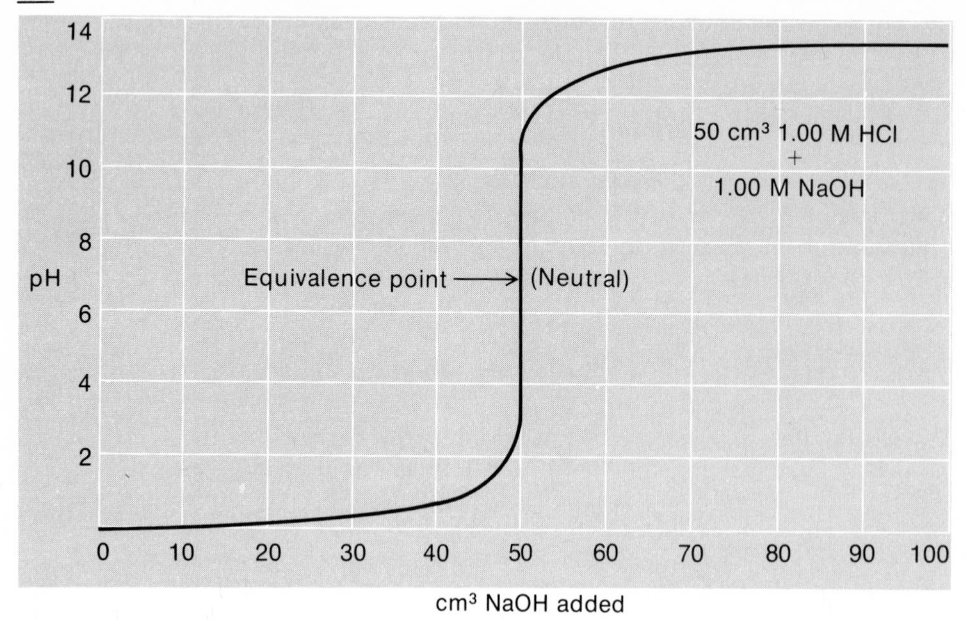

Figure 20.1 Titration of a strong acid with a strong base. At the equivalence point the *pH* changes very rapidly on addition of base, more than 6 *pH* units for one drop of base.

tion. Since the *pH* at the equivalence point in an acid-base reaction depends on the relative strengths of the acid and base involved, one cannot always use the same indicator for different kinds of acid-base titrations.

1. TITRATION OF A STRONG ACID WITH A STRONG BASE. If a solution of a strong base like NaOH is slowly added to a solution of a strong acid like HCl, the *pH* of the solution will be initially low and will gradually rise as the acid is neutralized (Fig. 20.1). In the vicinity of the equivalence point the *pH* changes very rapidly with added base. At the equivalence point itself the change in *pH* is extremely rapid, going up by perhaps 6 *pH* units on addition of one drop of base. The actual equivalence point is at *pH* 7, since at that point the solution contains only a neutral salt, such as NaCl. Bromthymol blue would serve well as an indicator of the end point, but since the *pH* of the solution near the end point is so sensitive to added base, one can equally well use phenolphthalein (end point at *pH* 9) or methyl red (end point at *pH* 5) in the titration. The *pH* of the solution at a few points during the titration is calculated in the following example.

Example 20.7 50.00 cm³ of 1.000 *M* HCl is titrated with 1.000 *M* NaOH. Find the *pH* of the solution after the following volumes of 1.000 *M* NaOH have been added:

a. 49.99 cm³
b. 50.00 cm³
c. 50.01 cm³

Solution During the titration the following reaction occurs:

$$H^+(aq) + OH^-(aq) \rightleftarrows H_2O \qquad K = 1.0 \times 10^{14}$$

a. At this point, we have:

$$(50.00 - 49.99)\ cm^3 = 0.01\ cm^3 = 1 \times 10^{-5}\ dm^3$$

of unneutralized 1.000 *M* HCl in a total volume of almost exactly 100 cm^3 (0.100 dm^3) of solution. Consequently,

$$[H^+] = \frac{1 \times 10^{-5}\ dm^3 \times 1.000\ mol/dm^3}{0.100\ dm^3} = 1 \times 10^{-4}\ M;\ pH = 4.0$$

b. The 50.00 cm^3 of 1 *M* NaOH exactly neutralizes the 50.00 cm^3 of 1 *M* HCl, to give a solution of NaCl. Since both Na^+ and Cl^- are neutral species, the *pH* is that of pure water, 7. This is the equivalence point in the titration. Note that only 0.01 cm^3 of base is required at the equivalence point to move the *pH* from 4 to 7. This volume is much less than that of 1 drop of reagent.

This approach can be used to find the *pH* at any point during the titration

c. We are now past the equivalence point and OH^- is in excess. Specifically, we have

$$(50.01 - 50.00)\ cm^3 = 0.01\ cm^3 = 1 \times 10^{-5}\ dm^3$$

of unneutralized 1.000 *M* NaOH in a total volume just slightly greater than 100 cm^3 (0.100 dm^3) of solution.

$$[OH^-] = \frac{1 \times 10^{-5}\ dm^3 \times 1.000\ mol/dm^3}{0.100\ dm^3} = 1 \times 10^{-4}\ M$$

$$[H^+] = \frac{1 \times 10^{-14}}{1 \times 10^{-4}} = 1 \times 10^{-10}\ M;\ pH = 10.0$$

Clearly, a small excess of base drives the *pH* from 7 to 10. Since the *pH* around the equivalence point is so sensitive to added base, any indicator with an end point between 4 and 10 would be satisfactory. Phenolphthalein is often used.

2. TITRATION OF A WEAK ACID WITH A STRONG BASE. If a solution of a weak acid is titrated with a strong base (Fig. 20.2), the change of *pH* as base is added differs in some important ways from that observed with a strong acid. Consider, for example, what happens when a strong base such as NaOH is added to a solution of acetic acid:

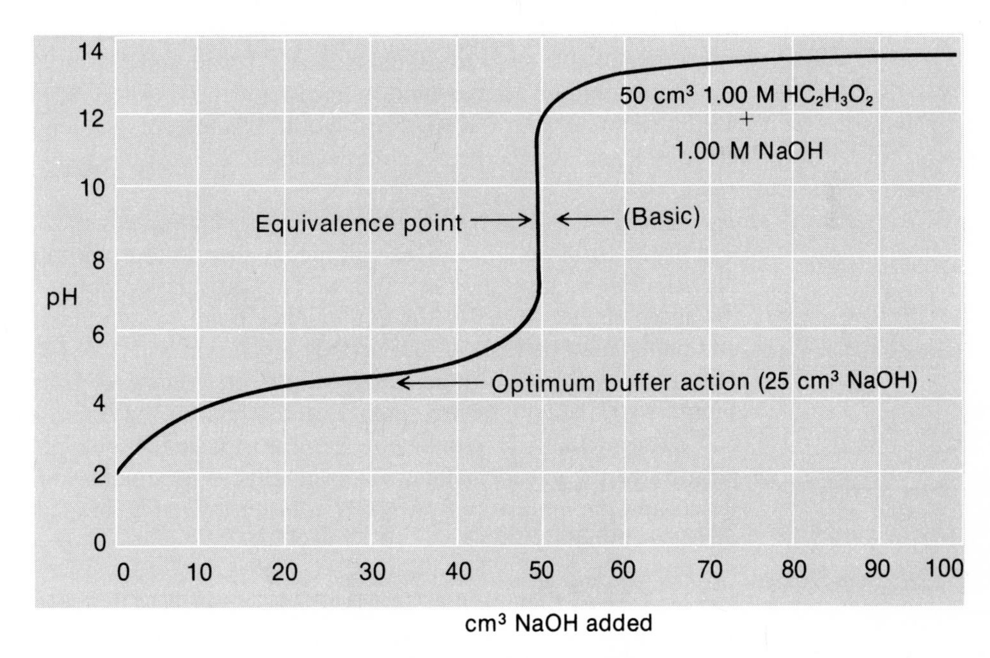

Figure 20.2 Titration of a weak acid with a strong base. The steep portion of the curve is shorter than when a strong acid is titrated; the acid-base indicator should have an end point at about *pH* 9.

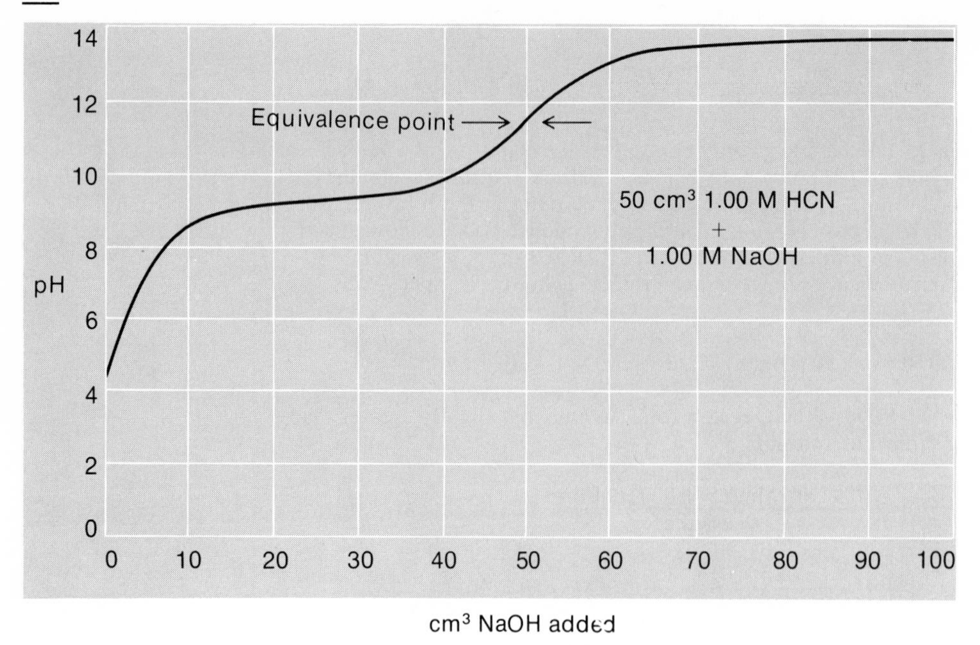

Figure 20.3 Titration of a very weak acid with a strong base. The steep portion of the curve is too short to allow accurate determination of the equivalence point with an acid-base indicator.

$$HC_2H_3O_2(aq) + OH^-(aq) \rightleftarrows H_2O + C_2H_3O_2^-(aq) \qquad (20.12)$$

Since the acid is weak, the pH at the beginning of the titration will be typically higher by a few units than with a strong acid of the same molarity. The pH rises more rapidly in the early stages of the titration than it did before and goes through a somewhat smaller change in the immediate vicinity of the equivalence point. At the equivalence point the solution contains only sodium acetate, which we have seen previously is basic. These factors make the choice of indicator more crucial. In the case of the titration of acetic acid, the equivalence point occurs at about pH 9, which is the pH of dilute $NaC_2H_3O_2$ solution. Phenolphthalein (end point at pH 9) would be an excellent indicator for the titration. If one attempted to use methyl red (end point at pH 5), it would change color much too early, when the titration was only about 65 per cent complete, hardly a good time to stop adding base.

With a weak acid having a very small ionization constant, the beginning portion of the titration curve rises still higher than it does with acetic acid. This has the effect of making the detection of the equivalence point more difficult, since the sharp change of pH at the equivalence point, which makes possible a precise end point, may be lost. If one titrates a solution of HCN ($K_a = 4 \times 10^{-10}$) with NaOH he observes a dependence of pH on volume of base added like that in Figure 20.3; the equivalence point is actually at pH 11.5, but an indicator with an end point at that pH would not show a rapid color change at the equivalence point since the pH is changing so gradually. It is actually impossible to precisely determine the molarity of an HCN solution by using an acid-base indicator in a direct titration.

3. TITRATION OF A WEAK BASE WITH A STRONG ACID. When a strong acid such as HCl is added to a solution of ammonia, the acid-base reaction is

$$NH_3(aq) + H^+(aq) \rightleftarrows NH_4^+(aq) \qquad (20.5)$$

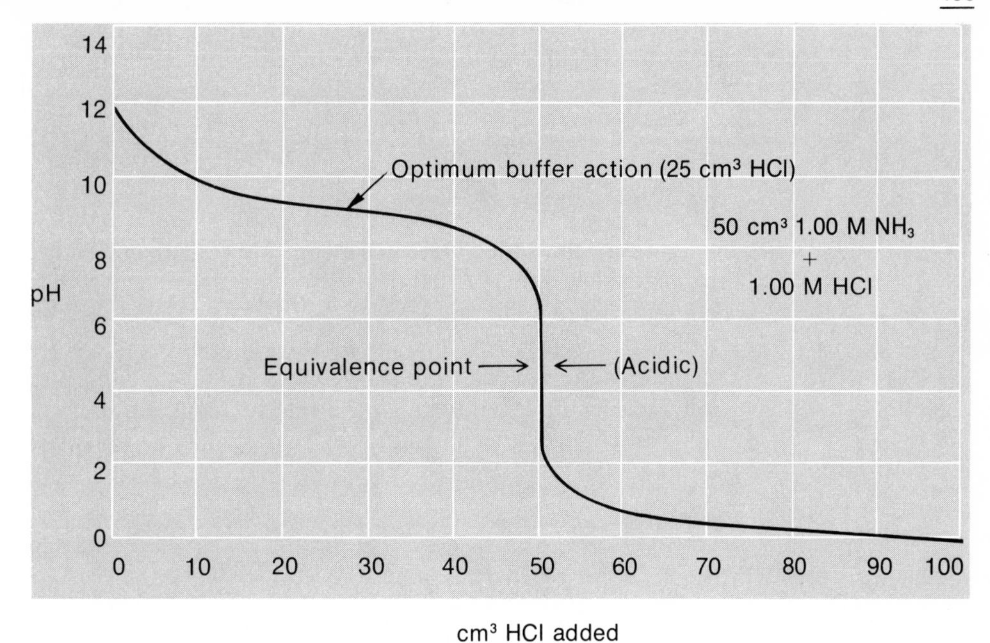

Figure 20.4 Titration of a weak base with a strong acid. The acid-base indicator should change color at about *pH* 5.

The NH_3 solution is initially basic, but with a lower *pH* than one would have with a strong base of equal concentration (Fig. 20.4). The *pH* drops as acid is added, with the steepest drop occurring as we go through the equivalence point. The solution at the equivalence point contains only ammonium chloride, and is acidic due to the presence of NH_4^+ ion. The *pH* at the equivalence point is about 5, so methyl red (end point at *pH* 5) would be a good end point indicator. If the base being titrated is too weak, the *pH* changes too slowly near the equivalence point. It would be very difficult to titrate the acetate ion ($K_b = 5.6 \times 10^{-10}$) in a solution of $NaC_2H_3O_2$, with hydrochloric acid, since the end point would not be sharp.

Why not use phenolphthalein in this titration?

20.3 BUFFERS

It is a well-known fact that the *pH* of blood and many other body fluids is relatively insensitive to the addition of acid or base. If one adds 0.01 mol of HCl or NaOH to 1 dm^3 of blood, its *pH* changes by less than 0.1 unit from its normal value of 7.4. In contrast, the addition of these same quantities of acid or base to pure water changes its *pH* by about 5 units (from 7 to 2 with 0.01 mol HCl and from 7 to 12 with 0.01 mol NaOH).

A solution whose *pH* changes relatively little on addition of acid or base is said to be *buffered*. For buffer action a solution must ordinarily contain two species, one capable of reacting with H^+ ions and the other with OH^- ions; in addition, the species in the buffer must not react with each other. Usually a buffer consists of a mixture of an acid and its conjugate base.

One of the simplest buffers can be prepared by adding acetic acid ($HC_2H_3O_2$) to a solution of sodium acetate (Na^+ ions, $C_2H_3O_2^-$ ions). If a strong base is added to the resulting mixture, the OH^- ions will react with acetic acid molecules:

$$OH^-(aq) + HC_2H_3O_2(aq) \rightleftharpoons H_2O + C_2H_3O_2^-(aq)$$

As we have seen, this reaction is nearly quantitative, so the added OH^- ions are almost completely removed. Addition of a strong acid to the buffer results in reaction of the H^+ ions of the acid with acetate ions of the sodium acetate:

$$H^+(aq) + C_2H_3O_2^-(aq) \rightleftarrows HC_2H_3O_2(aq)$$

In either case the added OH^- or H^+ ions are consumed and are unable to produce the drastic change in *pH* that occurs when a strong acid or base is added to water or an unbuffered solution.

The *pH* of a buffer depends on the value of K_a of the weak acid it contains and the relative concentrations of that acid and its conjugate base. Addition of strong acid or base will change the relative concentrations and so will cause a small *pH* change, the magnitude of which can be calculated as illustrated in Example 20.8.

Example 20.8 1.0 dm^3 of buffered solution contains 0.10 mol $HC_2H_3O_2$ and 0.10 mol $C_2H_3O_2^-$.

a. Calculate the *pH* of the buffer.
b. Calculate the *pH* after addition of 0.01 mol H^+ ion.
c. Calculate the *pH* after addition of 0.01 mol OH^- ion.

Solution

a. In the buffer the *pH* is established by the equilibrium between acetic acid and acetate ion:

$$K_a = \frac{[H^+] \times [C_2H_3O_2^-]}{[HC_2H_3O_2]} = 1.8 \times 10^{-5} \qquad [H^+] = 1.8 \times 10^{-5} \times \frac{[HC_2H_3O_2]}{[C_2H_3O_2^-]}$$

In the buffer $[C_2H_3O_2^-]$ and $[HC_2H_3O_2]$ are both equal to 0.1 *M*, so

$$[H^+] = K_a = 1.8 \times 10^{-5}\ M \qquad pH = 4.74$$

b. When 0.01 mol H^+ is added, this will react with 0.01 mol $C_2H_3O_2^-$ to form 0.01 mol $HC_2H_3O_2$. This means that in the final solution there are 0.09 mol $C_2H_3O_2^-$ ion and 0.11 mol $HC_2H_3O_2$, since we started with 0.10 mol of each. Therefore

Note that the H^+ or OH^- added is consumed; $[H^+]$ is calculated from the expression for K_a

$$[H^+] = 1.8 \times 10^{-5} \times \frac{[HC_2H_3O_2]}{[C_2H_3O_2^-]} = 1.8 \times 10^{-5} \times \frac{0.11}{0.09} = 2.2 \times 10^{-5}$$

$$pH = 4.66$$

The addition of 0.01 mol H^+ to the buffer lowers the *pH* by 0.08 unit.

c. In this case, addition of 0.01 mol OH^- results in its reaction with 0.01 mol $HC_2H_3O_2$, producing 0.01 mol $C_2H_3O_2^-$. In the final solution there will be 0.09 mol $HC_2H_3O_2$ and 0.11 mol $C_2H_3O_2^-$.

$$[H^+] = 1.8 \times 10^{-5} \times \frac{[HC_2H_3O_2]}{[C_2H_3O_2^-]} = 1.8 \times 10^{-5} \times \frac{0.09}{0.11} = 1.5 \times 10^{-5}$$

$$pH = 4.82$$

Again we find that the *pH* changes by less than 0.1 unit.

The buffer we have considered in our example is approximately the same as is obtained during the titration of $HC_2H_3O_2$ with NaOH in Figure 20.2, p. 481, at the stage when 25 cm^3 of NaOH have been added. At that point the acid is half neutralized, and $[HC_2H_3O_2]$ equals $[C_2H_3O_2^-]$. You will note that at that point in the titration the *pH* is changing relatively slowly with added base; a buffer takes advantage of the fact that a mixture containing roughly equal amounts of weak acid and its conjugate base has a relatively stable *pH*, at least as compared to the unbuffered system, which in Figure 20.2 corresponds to the point where 50 cm^3 NaOH have been added. There are two properties of buffers that are nicely shown by Figure 20.2.

1. The capacity of a buffer to absorb H^+ or OH^- ions is limited. If enough strong acid or base is added, all the conjugate base or weak acid in the buffer will be used up and the strong acid or base will then take over the system. Relating our example to Figure 20.2, the addition of 0.01 mol H^+ or OH^- is equivalent to removing or adding about 2.5 cm^3 of the NaOH solution. If we should add ten times that amount of NaOH (25 cm^3) we would exhaust the buffer and would go to a *pH* of about 9. Similarly, addition of HCl in amount sufficient to react with all the acetate ion would lower the *pH* to about 2.9, the value calculated for 0.1 *M* acetic acid.

2. The *pH* range of a buffer is limited, usually to about one unit above or one unit below the optimum value, where the weak acid and its conjugate base have equal concentrations. Referring again to Figure 20.2, you can see that in order to change the acetate-acetic acid buffer from its optimum *pH* of about 4.74 to about 5.7 we would have to add roughly 22 cm^3 more of the NaOH solution. Such a solution would still be rated as a buffer but would contain very little $HC_2H_3O_2$, since most of it has been neutralized. This means that the capacity of that buffer to absorb OH^- ion is very limited. Similarly, if one wished to lower the *pH* of the buffer by one unit, he would make the buffer by adding roughly 2.5 cm^3 of NaOH to 25 cm^3 of $HC_2H_3O_2$ of the same molarity. This buffer would absorb OH^- ion very well, but would have very low capacity for H^+ ion. For these reasons one tries to find buffer systems where K_a of the acid is about equal to the concentration of H^+ desired.

What would be the optimum *pH* of an NH_3-NH_4^+ buffer ($K_a = 5.6 \times 10^{-10}$)?

There are many common buffers, some of which are used by the body in maintaining the *pH* of its many fluids. In the latter class are the H_2CO_3–HCO_3^- buffer and a phosphate buffer, $H_2PO_4^-$–HPO_4^{2-}. In the blood the *pH* is maintained for the most part by a rapid-acting H_2CO_3–HCO_3^- buffer. This buffer has a *pH* of about 6.4 if the concentrations of H_2CO_3 and HCO_3^- are equal. The actual *pH* of blood is kept close to 7.4 by the fact that the ratio $[H_2CO_3]/[HCO_3^-]$ is kept at about 0.05; this gives a calculated *pH* of about 7.7, but this is lowered a bit by other buffers, particularly the phosphate buffer, which has a normal *pH* of 7.2. The capacity of the H_2CO_3–HCO_3^- buffer in the blood is greatest for acid, as is desirable since acid, particularly lactic acid, is produced by exercise and must be removed quickly if the *pH* in the body cells is not to drop so low as to kill the cells. Carbon dioxide itself is also produced in metabolism and must be continuously absorbed by the blood and vented through the lungs. There are several crucial equilibria related to the mechanics of transporting oxygen to the tissues and CO_2 to the lungs, and these depend in a remarkable way on the proper blood *pH*. Without a simple buffer available, *pH* control in the blood would be impossible.

Buffers have many applications in analytical and physical chemistry. Electrophoresis, used to separate and purify proteins, depends on very accurate *pH* control and is always carried out in a buffer. Many reactions involving organic and inorganic systems are studied in buffers, since their rates depend markedly on *pH*. One of the most simple and accurate methods for determining the ionization constant of a weak acid involves preparation of a buffer containing equal amounts of the acid and its conjugate base.

Example 20.9 A student is instructed to determine the ionization constant of an acid by the following procedure. A sample of the weak acid HA is dissolved in water to give 50 cm^3 of solution. The solution is then split into two equal 25-cm^3 portions. One of these portions is neutralized with sodium hydroxide and then mixed with the unneutralized portion. The *pH* of the final mixture is measured and found to be 4.0. Show how K_a for the acid can be determined by this experiment.

Solution By definition, $K_a = \frac{[H^+] \times [A^-]}{[HA]}$

By the design of the experiment, half the HA molecules originally present are converted to A^- by reaction with NaOH. The final mixture is a buffer, in which $[HA] = [A^-]$.

When the solutions of HA and A^- are mixed, there is essentially no reaction between these species

In the expression for K_a, $\frac{[A^-]}{[HA]}$ equals 1, so $K_a = [H^+]$.

From the measured *pH* of 4.0, it follows that:

$$[H^+] = 1.0 \times 10^{-4} = K_a$$

Note that one advantage of this method is that you do not have to know the concentration of the acid in the original solution.

It may be helpful to review at this point the several ways in which we have used the expression for K_a, i.e.,

$$K_a = \frac{[H^+] \times [A^-]}{[HA]}$$

to solve practical problems in this chapter. We have applied this expression to:

—determine the color of an indicator; that is, to calculate the ratio $[A^-]/[HA]$, knowing K_a and the concentration of H^+ in the solution.

—determine the concentration of H^+ in a buffer, knowing K_a and the ratio $[A^-]/[HA]$.

—determine K_a, knowing $[H^+]$ and the value of the ratio $[A^-]/[HA]$.

20.4 APPLICATION OF ACID-BASE REACTIONS IN INORGANIC SYNTHESES

Acid-base reactions are used very frequently in preparing inorganic compounds. They lend themselves to the preparation of a variety of salts; in addition, acid-base reactions can be applied in the laboratory to prepare small amounts of certain volatile acids and bases.

Preparation of Salts

One of the most common industrial methods for the preparation of metallic salts takes advantage of the reaction between strong acids and solids containing basic anions. Let us assume, for example, that we wish to make some pure crystalline $CuCl_2$ and that we have available a solution containing some Cu^{2+} ion in the presence of a different anion (e.g., SO_4^{2-}, NO_3^-).

Our procedure is to precipitate the hydroxide of copper by adding a solution of NaOH:

$$Cu^{2+}(aq) + 2\ OH^-(aq) \rightleftarrows Cu(OH)_2(s)$$

The solid product can be separated from the rest of the solution and washed with distilled water to remove any soluble contaminants, leaving essentially pure $Cu(OH)_2$. To the solid we add hydrochloric acid, HCl, slowly, with stirring, to bring the solid back into solution.

$$Cu(OH)_2(s) + 2\ H^+(aq) \rightleftarrows Cu^{2+}(aq) + 2\ H_2O$$

As soon as all the solid has dissolved, we stop adding acid and should have, at that point, a solution containing only Cu^{2+} ions and Cl^- ions from the acid. We boil the solution down and recover pure crystalline copper(II) chloride as a dihydrate:

$$Cu^{2+}(aq) + 2\ Cl^-(aq) + 2\ H_2O \rightarrow CuCl_2 \cdot 2\ H_2O(s)$$

By essentially the same approach one can prepare salts of all the metallic cations which form insoluble hydroxides. The anions of choice are introduced by simply selecting the proper acid—nitric acid for nitrates, sulfuric for sulfates, and so on.

Preparation of Volatile Acids or Bases

A convenient way to prepare small quantities of ammonia is to react an ammonium salt or a solution containing NH_4^+ ions with a strong base like sodium hydroxide:

$$NH_4^+(aq) + OH^-(aq) \rightarrow NH_3(g) + H_2O \qquad (20.13)$$

The solution must be heated to decrease the solubility of NH_3 and hence drive it out of solution.

Several volatile weak acids are easily prepared by treating one of their salts or solutions thereof with a strong, nonvolatile acid such as sulfuric acid. Carbon dioxide, sulfur dioxide, and hydrogen sulfide are frequently made in the laboratory by this kind of reaction:

$$CO_3^{2-}(aq) + 2\ H^+(aq) \rightleftarrows H_2CO_3(aq) \rightleftarrows CO_2(g) + H_2O \qquad (20.14)$$

$$SO_3^{2-}(aq) + 2\ H^+(aq) \rightleftarrows H_2SO_3(aq) \rightleftarrows SO_2(g) + H_2O(aq) \qquad (20.15)$$

$$FeS(s) + 2\ H^+(aq) \rightleftarrows H_2S(g) + Fe^{2+}(aq) \qquad (20.16)$$

TABLE 20.3 VOLATILITY OF ACIDS FROM WATER SOLUTION

ACID	BOILING POINT (°C)	
Hydrogen sulfide	100	H_2S escapes on warming
Carbonic acid	100	CO_2 escapes on warming
Sulfurous acid	100	SO_2 escapes on warming
Hydrochloric acid	110	constant boiling mixture (20% HCl)
Hydrofluoric acid	120	constant boiling mixture (35% HF)
Nitric acid	121	constant boiling mixture (68% HNO_3)
Hydrobromic acid	126	constant boiling mixture (47% HBr)
Hydriodic acid	127	constant boiling mixture (57% HI)
Phosphoric acid	213	$2\ H_3PO_4 \rightarrow H_4P_2O_7 + H_2O$
Sulfuric acid	330	$H_2SO_4 \rightarrow SO_3 + H_2O$

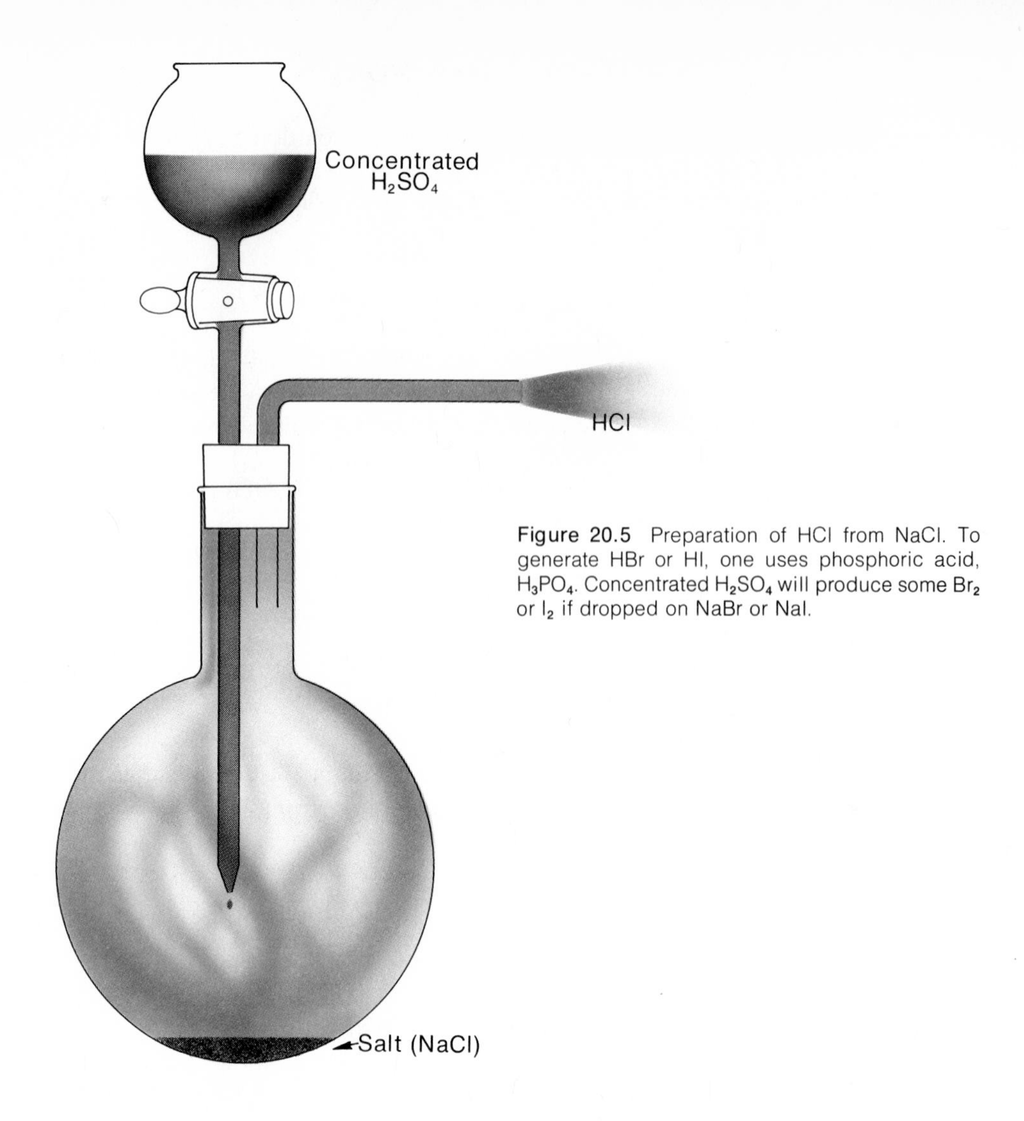

Figure 20.5 Preparation of HCl from NaCl. To generate HBr or HI, one uses phosphoric acid, H_3PO_4. Concentrated H_2SO_4 will produce some Br_2 or I_2 if dropped on NaBr or NaI.

The usefulness of these reactions depends on the acid formed being volatile, not on its being weak. Indeed, one can prepare volatile strong acids by reactions entirely analogous to these. For example, hydrogen chloride can be formed by allowing sulfuric acid to drop on sodium chloride (Fig. 20.5):

$$NaCl(s) + H_2SO_4(l) \rightarrow NaHSO_4(s) + HCl(g)$$

20.5 APPLICATIONS OF ACID-BASE REACTIONS IN QUALITATIVE ANALYSIS

Acid-base reactions are widely used in qualitative analysis for either of two different purposes. An acid-base reaction may be used to test for a specific ion by converting that ion into a volatile, easily detected product. Alternatively, such a reaction is often applied to separate one ion from another, where advantage is taken of the ability of acidic solutions to dissolve certain water-insoluble salts while leaving others unaffected.

Tests for Specific Ions

Some of the reactions in the previous section work very well for the identification of certain ions commonly found in solution. Reaction 20.13 is used for detection of NH_4^+ ion. The NH_3 formed is detected either by its odor or by its effect on moist litmus; of all the gases commonly found in the analytical laboratory, only NH_3 is basic to litmus.

Reactions 20.14, 20.15, and 20.16 are used in anion analysis to test for carbonates, sulfites, and sulfides. Carbon dioxide has no odor but is easily detected by passing it into a solution of barium hydroxide, producing a white precipitate of barium carbonate:

$$Ba^{2+}(aq) + 2\ OH^-(aq) + CO_2(g) \rightarrow BaCO_3(s) + H_2O$$

Is this an acid-base reaction?

Sulfur dioxide has the characteristic odor of burning sulfur and also will decolorize a purple solution of $KMnO_4$. Hydrogen sulfide is hard to miss, since its odor is bad and characteristic; it will also darken a piece of filter paper moistened with lead nitrate solution:

$$Pb^{2+}(aq) + H_2S(g) \rightarrow PbS(s) + 2\ H^+(aq)$$

Separation of Ions

The use of an acid-base reaction to separate one ion from another is illustrated by a procedure commonly used in anion analysis to distinguish between CO_3^{2-} and SO_4^{2-}. These ions are first precipitated as the barium salts. The mixed precipitate of $BaCO_3$ and $BaSO_4$ is then treated with hydrochloric acid. The carbonate is brought into solution by Reaction 20.8, while the sulfate is unaffected. (Why is the sulfate immune to acid?)

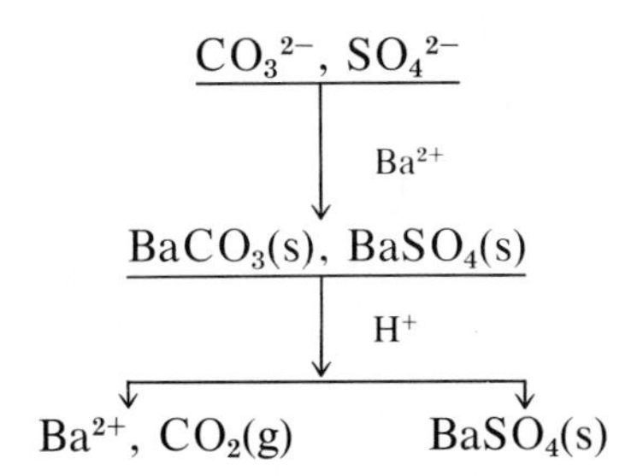

A more subtle application of this principle is involved in the reaction of hydrochloric acid with a mixed precipitate of CuS and ZnS. The zinc sulfide dissolves in acid while copper sulfide does not. The separation is more commonly carried out in reverse; that is, by adding hydrogen sulfide in acidic solution to a mixture of Zn^{2+} and Cu^{2+} ions. CuS is precipitated while Zn^{2+} remains in solution.

Zn^{2+}, Cu^{2+}

↓ H_2S, H^+ (0.3 *M*)

Zn^{2+} | $CuS(s)$

As noted in Chapter 18, ions that form water-insoluble sulfides can be separated into two groups in this manner. Those ions which form extremely insoluble sulfides (Group 2 = Cu^{2+}, Bi^{3+}, As^{3+}, Sb^{3+}, Sn^{2+}, Sn^{4+}, Cd^{2+}, Hg^{2+}) are precipitated by hydrogen sulfide at a pH of 0.5, while those sulfides which are somewhat more soluble (Co^{2+}, Fe^{2+}, Fe^{3+}, Ni^{2+}, Zn^{2+}, Mn^{2+} in Group 3) are precipitated only after the solution is made basic.

The general principles behind this method of separation were discussed in the section on reactions of strong acids with basic solids. The reaction by which a solid sulfide MS dissolves in a solution of a strong acid is

$$MS(s) + 2\,H^+(aq) \rightleftarrows M^{2+}(aq) + H_2S(aq) \qquad (20.17)$$

The solubility of the solid (mol/dm³) will equal $[M^{2+}]$. The equilibrium constant K for the reaction is as follows:

This equation was obtained by the procedure used in Example 20.4

$$K = \frac{K_{sp}}{1 \times 10^{-22}} = \frac{[H_2S] \times [M^{2+}]}{[H^+]^2} \qquad (20.18)$$

where K_{sp} is the solubility product constant of the sulfide and 1×10^{-22} is the product of the two ionization constants of H_2S (i.e., K_a H_2S $\times$ K_a HS^-; recall Example 20.4).

Equation 20.18 must be satisfied in a system containing a solid sulfide MS in equilibrium with M^{2+}, H_2S, and H^+. In this system we can control $[H^+]$ with strong acid and maintain $[H_2S]$ at about 0.1 M by saturating the solution with hydrogen sulfide gas. Rearranging Equation 20.18 to the form

$$[M^{2+}] = \frac{K_{sp}}{1 \times 10^{-22}} \times \frac{[H^+]^2}{[H_2S]} \qquad (20.18a)$$

we can readily pick out two factors which will affect the solubility, $[M^{2+}]$:

1. SOLUBILITY PRODUCT CONSTANT OF THE SULFIDE. $[M^{2+}]$ varies directly with the value of K_{sp} so, as we might expect, the more soluble the sulfide is in water, the more soluble it will be in acid. Zinc sulfide, ZnS ($K_{sp} = 1 \times 10^{-23}$) is more soluble in water than is copper sulfide ($K_{sp} = 1 \times 10^{-36}$), so we would expect it to be more soluble in acid.

2. ACID CONCENTRATION. Since $[M^{2+}]$ is proportional to $[H^+]^2$, the solubility of the metal sulfide will be very sensitive to $[H^+]$ and indeed will increase sharply with increasing acidity of the solution.

It is possible to apply these principles quantitatively, as illustrated in Example 20.10.

Example 20.10

a. Find K (Equation 20.18) for the reactions of CuS and ZnS with acid.
b. Find the solubility of CuS and ZnS, i.e., the equilibrium concentrations of Cu^{2+} and Zn^{2+}, in a solution 0.3 M in H^+ and saturated with H_2S (conc. H_2S = 0.1 M), using Equation 20.18a.

Solution

a. For CuS, $K_{sp} = 1 \times 10^{-36}$, so $K = \frac{1 \times 10^{-36}}{1 \times 10^{-22}} = 1 \times 10^{-14}$

For ZnS, $K_{sp} = 1 \times 10^{-23}$, so $K = \frac{1 \times 10^{-23}}{1 \times 10^{-22}} = 1 \times 10^{-1} = 0.1$

Clearly, K for the reaction of ZnS with acid is much larger than that for CuS.

b. From Equation 20.18a, we have:

$$[M^{2+}] = K \times \frac{[H^+]^2}{[H_2S]}, \text{ where } K = \frac{K_{sp}}{10^{-22}}$$

Using the values of K calculated in part (a):

CuS: $$[Cu^{2+}] = (1 \times 10^{-14})\frac{(0.3)^2}{0.1} = 1 \times 10^{-14}$$

ZnS: $$[Zn^{2+}] = (0.1)\frac{(0.3)^2}{0.1} = 0.1$$

We conclude that the solubility of CuS under these conditions is negligibly small. On the other hand, the solubility of ZnS is rather large, about 0.1 mol/dm^3. It should then be possible to separate Cu^{2+} from Zn^{2+} by adding H_2S in acidic solution ($[H^+] = 0.3\ M$); CuS but not ZnS would precipitate. Notice, however, that if $[H^+]$ were somewhat lower, say $0.03\ M$, the solubility of ZnS would be much lower

$$[Zn^{2+}] = \frac{0.1\ (0.03)^2}{0.1} = 0.001$$

and the separation might not be achieved. This is why, in analyses of this kind, *pH* control is very important.

20.6 AN INDUSTRIAL APPLICATION OF ACID-BASE REACTIONS: THE SOLVAY PROCESS

A commercially important process in which acid-base reactions play an important part is the so-called Solvay process for the manufacture of sodium hydrogen carbonate, $NaHCO_3$, and sodium carbonate, Na_2CO_3. The principal economic advantage of the Solvay process is that it uses as raw materials sodium chloride and limestone, both of which are cheap and available from naturally occurring deposits.

Preparation of $NaHCO_3$

The first product formed in the Solvay process is sodium hydrogen carbonate, $NaHCO_3$. To prepare this compound, advantage is taken of its comparatively low solubility at temperatures near the freezing point of water. Carbon dioxide is bubbled through a concentrated brine solution saturated with ammonia and maintained at a temperature of approximately 0°C. Under these conditions, finely divided crystals of sodium hydrogen carbonate precipitate. The equation for the overall reaction may be written:

$$CO_2(g) + H_2O + NH_3(g) + Na^+(aq) + Cl^-(aq) \rightarrow NaHCO_3(s) + NH_4^+(aq) + Cl^-(aq) \qquad (20.19)$$

The sodium hydrogen carbonate is filtered from the solution of ammonium chloride. It is possible to prepare $NaHCO_3$ in this manner because its solubility at 0°C (0.82 mol/dm^3) is considerably less than that of any of the other possible products: NH_4Cl (5.5 mol/dm^3), NaCl (6.1 mol/dm^3), or NH_4HCO_3 (1.5 mol/dm^3).

A reaction such as 20.19, involving as it does a large number of species, can best be understood if we break it down into a series of relatively simple steps. In this case, three such steps may be considered:

1. Bubbling carbon dioxide through water establishes the equilibrium

$$CO_2(g) + H_2O \rightleftarrows H_2CO_3(aq) \rightleftarrows H^+(aq) + HCO_3^-(aq) \qquad (20.19a)$$

The concentration of HCO_3^- ion produced by this reaction (0.000 18 *M*) is far below that required to precipitate $NaHCO_3$ (about 0.10 *M*).

2. To increase the concentration of HCO_3^- ions in solution, it is necessary to add a reagent that will shift the equilibrium in 20.19a to the right. This may be accomplished by adding ammonia, which removes H^+ ions by converting them to NH_4^+ ions:

$$NH_3(g) + H^+(aq) \rightarrow NH_4^+(aq) \qquad (20.19b)$$

Adding (a) + (b): $CO_2(g) + NH_3(g) + H_2O \rightarrow NH_4^+(aq) + HCO_3^-(aq)$

3. Under these conditions the HCO_3^- ions are present at a sufficiently high concentration to be precipitated by the sodium ions of the sodium chloride solution:

$$Na^+(aq) + Cl^-(aq) + HCO_3^-(aq) \rightarrow NaHCO_3(s) + Cl^-(aq) \qquad (20.19c)$$

Adding all three equations, and cancelling $HCO_3^-(aq)$, we get Equation 20.19:

$$CO_2(g) + H_2O + NH_3(g) + Na^+(aq) + Cl^-(aq) \rightarrow NaHCO_3(s) + NH_4^+(aq) + Cl^-(aq) \qquad (20.19)$$

Preparation of Na_2CO_3 from $NaHCO_3$

The greater part of the sodium hydrogen carbonate prepared by the Solvay process is converted to the more widely used salt, sodium carbonate. This conversion is accomplished by heating $NaHCO_3$ to about 300°C; at this temperature carbon dioxide and water vapor are given off and a white residue of sodium carbonate remains:

$$2\ NaHCO_3(s) \rightarrow Na_2CO_3(s) + CO_2(g) + H_2O(g) \qquad (20.20)$$

The carbon dioxide formed is recycled to prepare more $NaHCO_3$ by Reaction 20.19.

Preparation of CO_2 and Recovery of NH_3

For every mole of $NaHCO_3$ or Na_2CO produced by the Solvay process, one mole of CO_2 is consumed. The carbon dioxide is produced by heating limestone:

$$CaCO_3(s) \rightarrow CaO(s) + CO_2(g) \qquad (20.21)$$

The economics of the Solvay process, like those of all industrial operations, depend upon the effective use of all the products. In particular, it is important that the calcium oxide produced by Reaction 20.21 be utilized. Furthermore, the solution of ammonium chloride remaining after the precipitation of sodium hydrogen carbonate is far too valuable to be discarded. Indeed, if the ammonia consumed in Reaction 20.19 were not recovered, the Solvay process would be economically impossible, since the ammonia costs more than the sodium hydrogen carbonate is worth. Fortunately these two problems—the utilization of the CaO produced in Reaction 20.21 and the recovery of the NH_3 used in 20.19—can be solved simultaneously. It was pointed out earlier that ammonia can be formed from ammonium salts by heating with a base:

$$2\ NH_4^+(aq) + 2\ OH^-(aq) \rightarrow 2\ NH_3(g) + 2\ H_2O$$

Calcium oxide, upon addition to water, gives a moderately basic solution:

$$CaO(s) + H_2O \rightarrow Ca^{2+}(aq) + 2\ OH^-(aq)$$

Consequently, if the solution of NH_4Cl remaining after the precipitation of $NaHCO_3$ is heated with CaO, the reaction

$$CaO(s) + 2\ NH_4^+(aq) + 2\ Cl^-(aq) \rightarrow Ca^{2+}(aq) + 2\ Cl^-(aq) + 2\ NH_3(g) + H_2O \quad (20.22)$$

occurs, liberating ammonia and leaving as a final by-product a solution of calcium chloride.

Unfortunately, there is no market for all the $CaCl_2$ formed

Summary of the Solvay Process

The raw materials used in the Solvay process are sodium chloride, water, and limestone. The first product formed is sodium hydrogen carbonate, $NaHCO_3$,

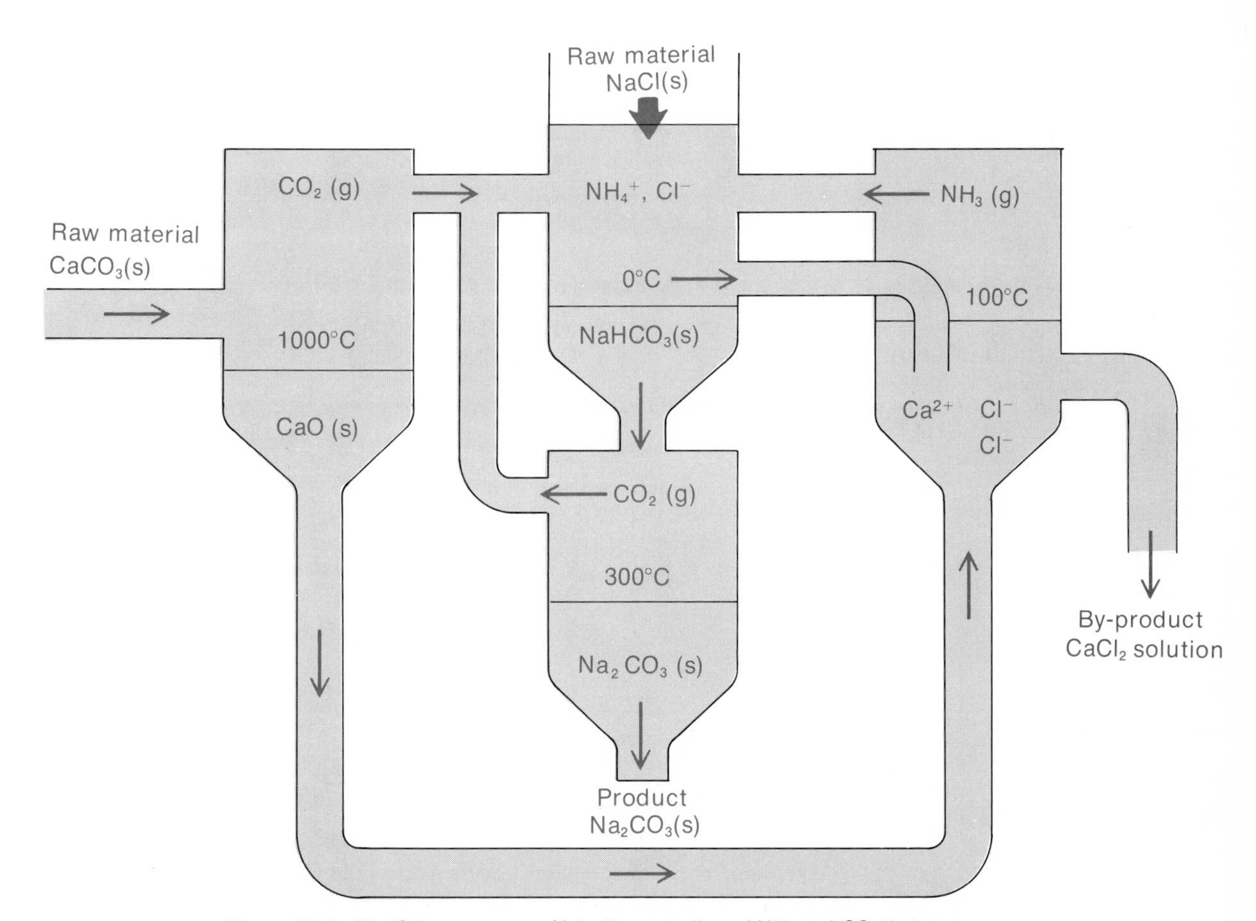

Figure 20.6 The Solvay process. Note the recycling of NH_3 and CO_2 that occurs.

which precipitates at 0°C from a sodium chloride solution saturated with CO_2 (produced by the thermal decomposition of $CaCO_3$) and NH_3 (recovered by heating the solution remaining with CaO). The $NaHCO_3$ may be sold for use in medicine, as a baking powder, and so on, or it may be converted to Na_2CO_3 by heating. More than 7×10^9 kg of sodium carbonate are produced annually in the United States, making it rank tenth among the industrial chemicals. A fairly strong, nonvolatile base, it is used in the manufacture of glass, paper, soap, and many other chemicals.

Sodium carbonate is sometimes called washing soda, because in the days when soap was used for laundry purposes the carbonate was added to the wash water to improve its solvent properties. Washing soda is not a serious pollutant, since carbonates occur naturally in surface waters. The main difficulty with carbonates as cleaning agents is that they form insoluble $CaCO_3$ with hard water, and this gums up the wash pretty badly. Recently, washing soda has had something of a comeback in non-phosphate detergents because of concern for our environment.

In the past decade some localized deposits of nearly pure trona, $Na_2CO_3 \cdot NaHCO_3 \cdot 2\,H_2O$, discovered in 1938 near Green River, Wyoming, have been very extensively mined and currently furnish more than half the sodium carbonate needs of the United States. It is quite possible that this source will soon replace the Solvay process for two reasons. First, mining a product tends to be cheaper than producing it by a chemical process. Second, the calcium chloride formed as a by-product in the Solvay process has never found sufficient use as a chemical. There is some demand for it for ice removal on highways and as a drying agent in the laboratory, but in the past much of it has been disposed of by spraying the solution on fields and letting it evaporate in the air. This has created serious pollution problems in nearby lakes and rivers, which in some cases have necessitated the closing of Solvay process plants.

PROBLEMS

20.1 *Equations for Acid-Base Reactions* Write balanced net ionic equations for the reaction between:

a. solutions of KOH and $HClO_4$.
b. solutions of KOH and $HC_2H_3O_2$.
c. solutions of HCl and NH_3.
d. $ZnCO_3(s)$ and a solution of HCl.

20.2 *Equilibrium Constants for Acid-Base Reactions* Calculate K for the reactions in Problem 20.1a, b, and c. (K_b $C_2H_3O_2^-$ = K_a NH_4^+ = 5.6×10^{-10})

20.3 *Acid-Base Titrations* In a titration it is found that 50 cm³ of 0.30 *M* NaOH will just neutralize 15 cm³ of an HCl solution. What is the concentration *(M)* of the HCl?

20.4 *Acid-Base Indicators* Methyl red has a K_a of 1×10^{-5}; the acid form, HIn, is red and its conjugate base, In^-, is yellow.

Complete the following table

pH	3	5	7
$[In^-]$ $[HIn]$	—	—	—
Color	—	—	—

20.5 *Buffers* A buffer is prepared by mixing 0.20 mol NH_4Cl with 0.40 mol NH_3 (K_a NH_4^+ = 5.6×10^{-10}). What is the *pH* of the buffer?

20.6 *Qualitative Analysis* Write net ionic equations to describe how you would test for the following ions in aqueous solution:

a. CO_3^{2-} b. S^{2-} c. NH_4^+

20.7 The net ionic equations for acid-base reactions differ, depending on whether the acids and bases are strong or weak. Explain why this is so.

20.28 For what kind of acid-base reaction is the net ionic equation written as

$$H^+(aq) + OH^-(aq) \rightarrow H_2O?$$

Cite an acid-base reaction which would not have this as its net ionic equation.

20.8 Write net ionic equations for the reactions, if any, which occur when solutions of the following substances are mixed. Do not include ions which do not participate.

a. HCl and KOH
b. NH_3 and HI
c. HF and KCN
d. KOH and NH_3
e. HNO_3 and $NaHCO_3$
f. NaH_2PO_4 and NaOH

20.29 Write net ionic equations for the reactions, if any, which occur when the following solutions are mixed. Do not include ions unless they react.

a. 0.5 *M* NaOH and 0.5 *M* HF
b. 0.2 *M* HCl and 0.1 *M* $NaC_2H_3O_2$
c. 0.1 *M* NH_4Cl and 0.4 *M* KOH
d. 0.3 *M* NH_3 and 0.1 *M* $NaC_2H_3O_2$
e. 0.2 *M* HNO_3 and 0.05 *M* $Ba(OH)_2$
f. 0.1 *M* $NaNO_3$ and 0.1 *M* HCl

20.9 Calculate *K* for each of the following reactions:

a. $HOCl(aq) + OH^-(aq) \rightarrow OCl^-(aq) + H_2O$
b. $F^-(aq) + H^+(aq) \rightarrow HF(aq)$
c. $HOCl(aq) + F^-(aq) \rightarrow HF(aq) + OCl^-(aq)$

Which of these reactions would occur to the greatest extent? the smallest extent?

20.30 Calculate *K* for the following reactions:

a. $HCO_3^-(aq) + OH^-(aq) \rightarrow CO_3^{2-}(aq) + H_2O$
b. $HCO_3^-(aq) + H^+(aq) \rightarrow H_2CO_3(aq)$
c. $CO_3^{2-}(aq) + 2\ H^+(aq) \rightarrow H_2CO_3(aq)$

20.10 Calculate *K* for:

a. $MgCO_3(s) + 2\ H^+(aq) \rightarrow Mg^{2+}(aq) + H_2CO_3(aq)$
b. $Mg(OH)_2(s) + 2\ H^+(aq) \rightarrow Mg^{2+}(aq) + 2\ H_2O$

20.31 Calculate *K* for

a. $MnS(s) + 2\ H^+(aq) \rightarrow Mn^{2+}(aq) + H_2S(aq)$
b. $Al(OH)_3(s) + 3\ H^+(aq) \rightarrow Al^{3+}(aq) + 3\ H_2O$

20.11 Although both $PbCO_3$ and $PbSO_4$ are insoluble in water, $PbCO_3$ is much more soluble in nitric acid than is $PbSO_4$. Explain.

20.32 Both AgCl and Ag_2CrO_4 are insoluble in water. In nitric acid, however, Ag_2CrO_4 is quite soluble while AgCl is essentially insoluble. Explain this difference in behavior.

20.12 A solution is made by mixing 200 cm^3 0.20 *M* NaOH with 100 cm^3 0.10 *M* HNO_3. Find the number of moles of each ion and its concentration in the solution after reaction is complete. What is the *pH* of the solution?

20.33 A solution is prepared by dissolving 0.10 mol NaOH in 200 cm^3 of 1.0 *M* $HC_2H_3O_2$. Assuming complete reaction, find $[HC_2H_3O_2]$ and $[C_2H_3O_2^-]$ in the solution. Then, using K_a for acetic acid, find the (small) concentration of H^+ in the solution.

20.13 In standardizing an NaOH solution, a student found that 38.46 cm^3 of the base neutralized exactly 32.33 cm^3 0.1064 *M* HCl. Find the concentration *(M)* of the NaOH.

20.34 In titrating an H_2SO_4 solution, it was found that 23.66 cm^3 of 0.2137 *M* NaOH would neutralize 22.04 cm^3 of the acid. Find the concentration *(M)* of the H_2SO_4.

20.14 NaOH solutions are often standardized by titration against weighed samples of potassium hydrogen phthalate, KHPh, an organic crystalline compound with one acidic hydrogen. This material has a high molar mass, 204.22 g/mol, and can be obtained in very pure form. A 2.6328-g sample of KHPh required 38.54 cm^3 NaOH to reach a phenolphthalein end point. What was the concentration of NaOH?

20.35 One can find the gram equivalent mass (*GEM*) of an acid by titration against a standardized NaOH solution. A 1.48-g crystalline sample of citric acid, found in oranges and lemons, was found to require 43.65 cm^3 of 0.527 *M* NaOH for complete neutralization. Find the *GEM* of citric acid.

20.15 In each of the following reactions the underlined reagent is 0.10 *M*. What is its normality?

a. $\underline{HC_2H_3O_2(aq)} + OH^-(aq) \rightleftarrows H_2O + C_2H_3O_2^-(aq)$
b. $\underline{H_2CO_3(aq)} + OH^-(aq) \rightleftarrows HCO_3^-(aq) + H_2O$
c. $\underline{H_2CO_3(aq)} + 2\ OH^-(aq) \rightleftarrows CO_3^{2-}(aq) + 2\ H_2O$
d. $\underline{NH_3(aq)} + H^+(aq) \rightleftarrows NH_4^+(aq)$

20.36 In each of the following reactions the underlined reagent is 0.30 *N*. What is its *M* value?

a. $\underline{H_2SO_3(aq)} + OH^-(aq) \rightleftarrows HSO_3^-(aq) + H_2O$
b. $\underline{H_2SO_3(aq)} + 2\ OH^-(aq) \rightleftarrows SO_3^{2-}(aq) + 2\ H_2O$
c. $\underline{H_3PO_4(aq)} + 3\ OH^-(aq) \rightleftarrows PO_4^{3-}(aq) + 3\ H_2O$
d. $\underline{HClO(aq)} + OH^-(aq) \rightleftarrows H_2O + OCl^-(aq)$

20.16 Phenol red, one of the common acid-base indicators, has a K_a equal to 1×10^{-8}. Its acid form is yellow, while its conjugate base is red. What color would the indicator have in a solution of pH 6? 7? 8? 9? 12? Would phenol red be a good indicator to use to find the equivalence point in an NaOH–HBr titration?

20.17 Given three acid-base indicators—methyl orange (end point at pH 4), bromthymol blue (end point at pH 7), and phenolphthalein (end point at pH 9)—which would you select for the following titrations? Why?

a. HBr with NaOH
b. HBr with NH_3
c. $HC_2H_3O_2$ with KOH
d. $NaC_2H_3O_2$ with HCl

20.18 50.00 cm³ 0.1000 M HCl is titrated with 0.1000 M NaOH. Calculate the pH of the solution after the following volumes of NaOH have been added:

a. 00.00 cm³
b. 25.00 cm³
c. 49.90 cm³
d. 50.00 cm³
e. 50.01 cm³
f. 75.00 cm³

20.19 Which of the following pairs of solutions could be used to prepare a buffer?

a. NaOH and HCl
b. HNO_2 and $NaNO_2$
c. HNO_2 and NaOH
d. Na_3PO_4 and HCl

20.20 A buffer solution is made by adding 0.20 mol formic acid and 0.40 mol sodium formate to 1.00 dm³ of water.

a. Find the pH of the buffer. (K_a formic acid $= 1.8 \times 10^{-4}$)
b. What is the pH after 0.02 mol HCl is added?
c. What is the pH after 0.02 mol NaOH is added?
d. What is the pH after 0.20 mol NaOH is added?

20.21 Blood is buffered mainly by the H_2CO_3–HCO_3^- system, in which the ratio of $[HCO_3^-]$ to $[H_2CO_3]$ is about 20:1.

a. What would the pH of blood be if this were the only buffer? (K_a $H_2CO_3 = 4.2 \times 10^{-7}$)
b. Is this buffer more effective against addition of acid than of base? Why?
c. What would the pH of the buffer become if the relative amount of H_2CO_3 were increased so that the concentration ratio was 2:1?

20.22 A student dissolves 0.50 g of a weak acid in 10 cm³ of water. He divides the solution into two equal parts, titrates one of them with NaOH to the equivalence point, and adds this solution to the other portion. The measured pH is 5.3. Estimate K_a.

20.37 Bromcresol green, a well-known acid-base indicator, has a K_a value of 1×10^{-5}. Its acid form is yellow, and its basic form is blue. What color would bromcresol green be in a solution of pH 1? 3? 4? 5? 7? 10? For which titration would bromcresol green be most satisfactory: NaOH and HCl, NaOH and NH_4Cl, or NH_3 and HNO_3?

20.38 Given the acid-base indicators in Problem 20.17, which would you choose for the following titrations? Why?

a. H_2SO_4 with KOH
b. KCN with HCl
c. NH_3 with HNO_3
d. HF with NaOH

20.39 50.00 cm³ of 0.2000 M NaOH is titrated with 0.2000 M HCl. Find the pH of the solution after the following volumes of HCl have been added:

a. 0.00 cm³
b. 25.00 cm³
c. 49.99 cm³
d. 50.00 cm³
e. 50.10 cm³
f. 100.00 cm³

20.40 Which of the following pairs of solutions could be used to make a buffer?

a. H_2CO_3 and NaOH
b. NaH_2PO_4 and Na_2HPO_4
c. $HC_2H_3O_2$ and NaOH
d. NH_3 and HCl

20.41 A buffer is made by mixing 500 cm³ 0.80 M NH_3 with 500 cm³ 0.80 M NH_4Cl.

a. Find the pH of the buffer. (K_a $NH_4^+ = 5.6 \times 10^{-10}$)
b. Find the pH after adding 0.02 mol HCl.
c. Find the pH after adding 0.02 mol NaOH.
d. Find the pH after adding 0.50 mol HCl.

20.42 One of the functions of the H_2CO_3–HCO_3^- buffer in the blood is to quickly remove lactic acid, HLac, from tissues after it is produced by active exercise. (K_a HLac $= 8.4 \times 10^{-4}$)

a. Find K for $HLac(aq) + HCO_3^-(aq) \rightarrow H_2CO_3(aq) + Lac^-(aq)$.
b. In normal blood, $[H_2CO_3] = 0.0014\ M$ and $[HCO_3^-] = 0.027\ M$. Find the pH.
c. Find the pH of blood after the addition of 0.0050 mol/dm³ of HLac.

20.43 Suppose that in Problem 20.22, one third rather than one half of the original solution were titrated and the measured pH was 6.4. What is K_a for the acid?

20.23 Describe in words how you might prepare

a. NH_4I from NH_3.
b. $CuSO_4$ from $Cu(NO_3)_2$.
c. $Ni(OH)_2$ from $NiCO_3$.
d. CO_2 from $CaCO_3$.

20.24 Write net ionic equations to show how, using acid-base reactions, you could separate the following ions:

a. SO_4^{2-}, SO_3^{2-}
b. CO_3^{2-}, NO_3^-
c. NH_4^+, Mg^{2+}
d. Cl^-, S^{2-}

20.25 How could you distinguish between the following solids, all of which are white?

a. $NaCl$, Na_2CO_3
b. NH_4Cl, Na_2SO_4
c. $NaHSO_4$, $NaHCO_3$
d. KCN, KBr

20.26 CdS ($K_{sp} = 1 \times 10^{-27}$) is the most soluble of the Group II sulfides.

a. Find K for the reaction of CdS with acid.
b. Find the solubility of CdS in a solution that is 0.1 M in H_2S and in which H^+ is 0.3 M.
c. What fraction of the Cd^{2+} in a solution in which its concentration is 0.1 M will precipitate under the conditions in (b)?

20.27 In the Solvay process,

a. why is ammonia used?
b. why is the reaction carried out at 0°C?
c. what is the source of the pollution problem?

20.44 Describe in some detail how you could make

a. NH_3 from NH_4Cl.
b. $CdSO_4$ from $CdCl_2$.
c. $CuCl_2$ from CuO.
d. $NaHCO_3$ from NaOH.

20.45 How could you use acid-base or precipitation reactions to separate and identify the following ions in a mixture?

$$SO_4^{2-},\ CO_3^{2-},\ Cl^-,\ NH_4^+$$

Write net ionic equations for each reaction involved.

20.46 A sample consists of one or both members of the following pairs. How could you determine its composition?

a. $NaHCO_3$, $CuSO_4$
b. Na_2CO_3, $Ba(NO_3)_2$
c. NH_4Cl, $CaCO_3$

20.47 Using Table 18.2, predict the order in which Co^{2+}, Fe^{2+}, Mn^{2+}, Ni^{2+}, and Zn^{2+} would precipitate from a solution saturated with H_2S if the cations were all 0.1 M and the pH of the solution was very slowly increased from 0 to 9. At pH 7, with $[H_2S] = 0.1\ M$, what would be the equilibrium concentration of Fe^{2+}?

20.48 The overall reaction in the Solvay process for the preparation of Na_2CO_3 is:

$$2\ Na^+(aq) + CaCO_3(s) \rightarrow Na_2CO_3(s) + Ca^{2+}(aq)$$

Write balanced equations for each step of the process and add them to give this equation as the net result.

*20.49 50.00 cm³ of 0.2000 M $HC_2H_3O_2$ ($K_a = 1.8 \times 10^{-5}$) is titrated with 0.2000 M NaOH. Find the pH of the solution after the following volumes of NaOH have been added:

a. 0.00 cm³ b. 25.00 cm³ c. 49.90 cm³ d. 50.00 cm³ e. 50.01 cm³ f. 100.00 cm³

*20.50 In a titration of 50 cm³ 1.00 M $HC_2H_3O_2$ with 1.00 M NaOH, a student used bromcresol green as an indicator ($K_a = 1 \times 10^{-5}$). What volume of NaOH would it take to reach the end point with this indicator? What would be a better indicator for this titration?

*20.51 Determine the pH at which CoS ($K_{sp} = 1 \times 10^{-21}$) will begin to precipitate from a solution 0.1 M in both Co^{2+} and H_2S.

*20.52 Ammonium chloride solutions are slightly acidic, so they are better solvents than water for insoluble substances such as $Mg(OH)_2$. Find the solubility of $Mg(OH)_2$ in 0.2 M NH_4Cl and compare with the solubility in water. Hint: Find K for the reaction $Mg(OH)_2(s) + 2\ NH_4^+(aq) \rightarrow Mg^{2+}(aq) + 2\ NH_3(aq) + 2\ H_2O$.

21

COMPLEX IONS; COORDINATION COMPOUNDS

If white, anhydrous copper sulfate, $CuSO_4$, is exposed to ammonia gas, a deep blue crystalline product is formed. Chemical analysis shows that this product contains four moles of ammonia per mole of copper sulfate; its formula might be written as $Cu(NH_3)_4SO_4$. The positive ion in this compound consists of a Cu^{2+} ion bonded to four ammonia molecules, i.e., $Cu(NH_3)_4{}^{2+}$. The reaction between the Cu^{2+} ion and the NH_3 molecules is represented in electron dot notation as

$$Cu^{2+} + 4\,:\!\begin{array}{l} \ \ \diagup H \\ N\!-\!H \\ \ \ \diagdown H \end{array} \rightarrow \left[\begin{array}{ccccc} & & H\ H\ H & & \\ & & \diagdown | \diagup & & \\ H & & N & & H \\ & \diagdown & | & \diagup & \\ H\!-\!N & - & Cu & - & N\!-\!H \\ & \diagup & | & \diagdown & \\ H & & N & & H \\ & & \diagup | \diagdown & & \\ & & H\ H\ H & & \end{array}\right]^{2+}$$

$$\underset{\text{colorless}}{Cu^{2+}} + 4\ NH_3 \rightarrow \underset{\text{deep blue}}{Cu(NH_3)_4{}^{2+}} \qquad (21.1)$$

The nitrogen atom of each NH_3 molecule contributes a pair of unshared electrons to form a covalent bond with the Cu^{2+} ion. This bond and others like it, where both electrons are contributed by the same atom, is referred to as a **coordinate** covalent bond. Clearly, there are four such bonds in the $Cu(NH_3)_4{}^{2+}$ ion.

In one sense, Reaction 21.1 resembles that between a proton and an ammonia molecule to form an ammonium ion:

$$H^+ + :\!\begin{array}{l} \ \ \diagup H \\ N\!-\!H \\ \ \ \diagdown H \end{array} \rightarrow \left[\begin{array}{c} H \\ | \\ H\!-\!N\!-\!H \\ | \\ H \end{array}\right]^{+}$$

$$H^+ + NH_3 \rightarrow NH_4{}^+$$

In both cases, the ammonia molecule is acting as a Lewis base, donating the pair of electrons required to form the coordinate covalent bond. The Cu^{2+} ion, like the H^+ ion, acts as a Lewis acid in accepting a pair of electrons.

The $Cu(NH_3)_4^{2+}$ ion is commonly referred to as a complex ion. In the broadest sense, a complex ion may be considered to be a charged species consisting of more than one atom. Strictly speaking, such oxyanions as nitrate (NO_3^-) and sulfate (SO_4^{2-}) can be classified as complex ions. However, we will use the term in a more restricted sense to refer to a **charged species in which a metal atom is joined by coordinate covalent bonds to neutral molecule(s) and/or negative ions.** Species such as $Cu(H_2O)_4^{2+}$ and $Zn(H_2O)_3(OH)^+$, encountered in previous chapters, are properly considered to be complex ions.

The metal atom in a complex ion (Cu, Zn, . . .) is referred to as the **central atom.** The molecules (NH_3, H_2O, . . .) or anions (OH^-, Cl^-, . . .) attached to the central atom are known as coordinating groups or **ligands.** The number of bonds (e.g., 2, 4, 6) formed by the central atom is called its **coordination number.**

Compounds containing a complex ion such as $Cu(NH_3)_4^{2+}$ or $Zn(H_2O)_3(OH)^+$ are referred to as **coordination compounds.** Typical examples include

$[Cu(NH_3)_4]SO_4$ $\qquad$ $[Zn(H_2O)_3(OH)]Cl$

$[Cu(NH_3)_4]Cl_2$ $\qquad$ $[Zn(H_2O)_3(OH)]NO_3$

In writing the formulas of coordination compounds, we will identify the complex ion by writing brackets around it, as in the examples above. It is understood that the species within the brackets (4 NH_3 molecules in one case, 3 H_2O molecules, and an OH^- ion in the other) are bonded directly to the central metal atom. Species written outside the brackets (e.g., SO_4^{2-}, Cl^-, NO_3^-) are present as free ions analogous to those in simple salts such as $CuSO_4$ and $CuCl_2$. The nomenclature of coordination compounds is discussed in Appendix 3.

21.1 STRUCTURES OF COORDINATION COMPOUNDS. CHARGES OF COMPLEX IONS.

In Table 21.1, we have indicated the structures of five different coordination compounds formed by Pt^{2+}. The identities of the complex species in these compounds can be established from experimental observations on the properties of their water solutions. Two types of experiments are particularly valuable:

1. *Conductivity measurements.* We find that solutions of compounds 1, 2, 4, and 5 conduct an electric current. The conductivities of 0.1 *M* solutions of $[Pt(NH_3)_4]Cl_2$ and $K_2[PtCl_4]$ are about the same as those of simple 2:1 or 1:2 salts such as $CaCl_2$ or K_2SO_4, implying that they have similar structures. The conductivities of 0.1 *M* solutions of compounds 2 and 4 are about the same as those of salts such as KCl or KNO_3, implying that they consist of +1 and −1 ions. Compound 3 is quite insoluble in water and appears to be a nonelectrolyte analogous to such molecular species as methanol or sugar.

TABLE 21.1 COMPLEXES FORMED BY Pt^{2+} WITH NH_3 AND Cl^-

SALT	COMPLEX	CHARGE OF COMPLEX	ANALOGOUS SIMPLE SALT
1. $[Pt(NH_3)_4]Cl_2$	$Pt(NH_3)_4^{2+}$	+2	$CaCl_2$
2. $[Pt(NH_3)_3Cl]Cl$	$Pt(NH_3)_3Cl^+$	+1	KCl
3. $[Pt(NH_3)_2Cl_2]$	$Pt(NH_3)_2Cl_2$	0	—
4. $K[Pt(NH_3)Cl_3]$	$Pt(NH_3)Cl_3^-$	−1	KNO_3
5. $K_2[PtCl_4]$	$PtCl_4^{2-}$	−2	K_2SO_4

Cl$^-$ ions bonded directly to Pt^{2+} are not precipitated by Ag^+

2. *Behavior toward $AgNO_3$.* When silver nitrate is added to a cold solution of compounds 3, 4, or 5, no precipitate forms. This suggests that the Cl^- ions in these compounds are tightly bound within the complex. Compound 1 gives two moles of AgCl per mole of Pt, implying two free Cl^- ions. When silver nitrate is added to compound 2, only one mole of AgCl is formed per mole of Pt. This suggests that only one of the Cl^- ions is free; the other is bonded directly to the central metal atom.

Once the formula of a complex has been established, we can readily obtain its charge in a formal way by taking *the algebraic sum of the charge of the metal ion and those of the ligands.* Thus, for the complexes of Pt^{2+} listed in Table 21.1, realizing that NH_3 is a neutral molecule and that the Cl^- ion has a charge of -1, we obtain:

1. $+2 + 4(0) = +2$
2. $+2 + 3(0) + 1(-1) = +1$
3. $+2 + 2(0) + 2(-1) = 0$
4. $+2 + 1(0) + 3(-1) = -1$
5. $+2 + 4(-1) = -2$

Example 21.1 The Pt^{4+} ion forms a coordination compound with the formula $Pt(NH_3)_4Cl_4$. Treatment of one mole of this compound with $AgNO_3$ gives two moles of AgCl. Deduce the structure of the compound, give the charge of the complex ion, and suggest a simple salt which would have a similar conductivity in water solution.

Solution The information given implies that there are only two free Cl^- ions in the compound. The other two Cl^- ions and all the NH_3 molecules must be tied up in the complex, bonded directly to Pt^{4+}. Thus the structure must be

$$[Pt(NH_3)_4Cl_2]Cl_2$$

The charge of the complex ion is readily derived, knowing that we started with a Pt^{4+} ion, that an NH_3 molecule has zero charge, and that a Cl^- ion has a -1 charge.

$$+4 + 4(0) + 2(-1) = +2$$

The compound is analogous to a simple 2:1 salt such as $CaCl_2$ and hence would be expected to show similar conductivity behavior.

Species such as $[Pt(NH_3)_2Cl_2]$ or $[Zn(H_2O)_2(OH)_2]$ in which the charge of the metal ion (+2) is exactly balanced by the total charge of the ligands (−2) are called neutral complexes. Compounds of this type are typically insoluble in water. We find, for example, that when enough OH^- ions are added to a solution of a Zn^{2+} salt to give an appreciable concentration of the neutral complex $Zn(H_2O)_2(OH)_2$, a precipitate of "hydrated zinc hydroxide" forms. Many of the transition metal hydroxides precipitate out of aqueous solution, not as simple unhydrated species (e.g., $Zn(OH)_2$) but as hydrated compounds with compositions close to that of a neutral complex.

The industrial method of refining nickel (the Mond process) involves forming a neutral complex between a nickel atom and four carbon monoxide molecules. When carbon monoxide at 60°C is passed over nickel containing iron and cobalt impurities, the following reaction occurs:

$$Ni(s) + 4\ CO(g) \rightarrow Ni(CO)_4(g) \qquad (21.2)$$

Under these conditions, none of the impurities react. The so-called nickel carbonyl can be separated from excess CO by cooling; at 43°C it condenses to a colorless liquid. To recover the nickel, the neutral complex is heated to 200°C to reverse Reaction 21.2.

21.2 COMPOSITION OF COMPLEX IONS

Central Metal Atom

In general, we find a correlation between the charge density (charge to size ratio) of a metal ion and its ability to form complex ions. Cations of high charge and small size, such as the +2 and +3 ions of the transition metals, form a large number of stable complex ions. The poorest complex formers are large metal ions of +1 charge, specifically such alkali metal cations as Na^+ and K^+.

In water solution, cations of the transition metals exist as complex ions

Among the transition metals which form more than one cation, it is usually the ion of higher charge which is the better complex former. We find, for example, that the complexes formed by Cr^{3+}, Fe^{3+}, and Co^{3+} are both more numerous and more stable than those of the +2 ions of these metals, Cr^{2+}, Fe^{2+}, and Co^{2+}.

The correlation between charge density and complexing ability implies that electrostatic forces between a metal ion and the unpaired electrons of a ligand are a major factor in complex ion formation. We shall have more to say on this topic when we discuss bonding in complex ions in Section 21.4. It is well to point out, however, that factors other than charge density must be considered in assessing the complexing ability of a metal ion. The Ag^+ ion (r = 0.126 nm) forms many more stable complexes than either Na^+ (r = 0.095 nm) or Ca^{2+} (r = 0.099 nm), both of which have higher charge densities.

Ligands. Chelating Agents

In principle, any molecule or anion with an unshared pair of electrons can donate them to a metal ion to form a coordinate covalent bond. Thus we would expect such species as the ammonia molecule, the water molecule, or the fluoride ion to be good coordinating agents:

```
     H            H
    /            /
 :N—H        :O:          [:F:]⁻
    \            \
     H            H
```

In contrast, species such as the methane molecule and the ammonium ion, which have no unshared electron pairs, do not act as ligands.

```
    H                 [    H    ]+
    |                 [    |    ]
 H—C—H                [ H—N—H   ]
    |                 [    |    ]
    H                 [    H    ]
```

In practice, the electron pair donor in complex ion formation is ordinarily an atom of one of the more electronegative elements (C, N, O, S, F, Cl, Br, I). Hundreds of different ligands containing one or more of these atoms are known. Among the ligands most frequently encountered in general chemistry are the NH_3 and H_2O molecules and the OH^-, Cl^-, and CN^- ions (Table 21.2).

The relative abilities of different ligands to coordinate with metal ions depend upon a great many factors. One of the most important of these is the basicity of

TABLE 21.2 A FEW COMPLEX IONS INVOLVING SIMPLE LIGANDS

H_2O	NH_3	OH^-	Cl^-	CN^-
$Cu(H_2O)_4^{2+}$	$Ag(NH_3)_2^+$	$Zn(OH)_4^{2-}$	$AgCl_2^-$	$Ag(CN)_2^-$
$Cr(H_2O)_6^{3+}$	$Cu(NH_3)_4^{2+}$	$Al(OH)_6^{3-}$	$CuCl_4^{2-}$	$Ni(CN)_4^{2-}$
$Co(H_2O)_6^{3+}$	$Zn(NH_3)_4^{2+}$	$Cr(OH)_6^{3-}$	$AlCl_4^-$	$Zn(CN)_4^{2-}$
$Ni(H_2O)_6^{2+}$	$Co(NH_3)_6^{3+}$		$PtCl_4^{2-}$	$Fe(CN)_6^{3-}$
$Zn(H_2O)_6^{2+}$	$Ni(NH_3)_6^{2+}$		$PtCl_6^{2-}$	$Fe(CN)_6^{4-}$

the ligand. It is perhaps not too surprising to find that molecules or ions which have a strong attraction for a proton are among the better coordinating agents. The species NH_3, OH^-, and CN^-, all of which are strong bases in the Brönsted-Lowry or Lewis sense, form stable complexes with a wide variety of transition metal ions. The ClO_4^- ion, which shows no tendency to acquire a proton in water solution, is a notoriously poor coordinating agent; the NO_3^- and HSO_4^- ions, both derived from strong acids, form relatively few stable complexes. It should be pointed out, however, that the Cl^- ion, which does not act as a base in water, forms stable complexes with many transition metal ions (cf. Table 21.2).

One of the first ligands to be studied extensively was the ethylenediamine molecule:

$$\begin{array}{ccccccccc} & & & & H & & H & & \\ & & & & | & & | & & \\ H & - & \ddot{N} & - & C & - & C & - & \ddot{N} & - & H \\ & & | & & | & & | & & | & & \\ & & H & & H & & H & & H & & \end{array}$$

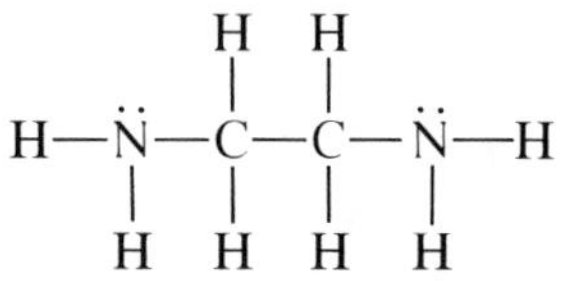

This molecule, containing two nitrogen atoms, each with an unshared pair of electrons, forms two coordinate covalent bonds with metal atoms. These bonds are extremely stable, as shown by the fact that the addition of ethylenediamine to a solution containing the $Cu(NH_3)_4^{2+}$ ion results in the displacement of the four NH_3 molecules by two $H_2N{-}CH_2{-}CH_2{-}NH_2$ molecules.

Cu²⁺ still has a coordination number of 4 in $Cu(en)_2^{2+}$

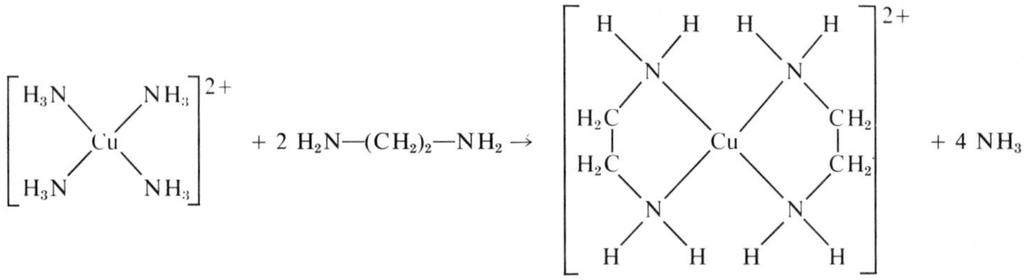

$$\left[\begin{array}{ccc} H_3N & & NH_3 \\ & Cu & \\ H_3N & & NH_3 \end{array}\right]^{2+} + 2\ H_2N{-}(CH_2)_2{-}NH_2 \rightarrow \left[\begin{array}{ccccc} H\ \ H & & & & H\ \ H \\ N & & & & N \\ H_2C & & Cu & & CH_2 \\ | & & & & | \\ H_2C & & & & CH_2 \\ N & & & & N \\ H\ \ H & & & & H\ \ H \end{array}\right]^{2+} + 4\ NH_3$$

In writing a chemical equation to represent this reaction, we frequently abbreviate the ethylenediamine molecule as "en":

$$Cu(NH_3)_4^{2+}(aq) + 2\ en(aq) \rightarrow Cu(en)_2^{2+}(aq) + 4\ NH_3(aq) \qquad (21.3)$$

The fact that ethylenediamine is a powerful coordinating agent is due in part to the high stability of the five-membered rings that it forms with metal ions. Methylenediamine, $H_2N{-}CH_2{-}NH_2$, is not nearly as good a complexing ligand.

Another factor which tends to make reactions such as 21.3 spontaneous is the increase in the number of solute molecules (2 en $\rightarrow$ 4 NH_3). As we saw in Chapter 14, a reaction in which the number of molecules increases usually results in an increase in entropy. A positive value of ΔS tends to make ΔG negative ($\Delta G = \Delta H - T\Delta S$), corresponding to a spontaneous reaction.

A great many molecules and anions in addition to ethylenediamine form more than one bond with a metal ion. Ligands which act in this way are known as chelating agents: the complexes formed are called **chelates** from the Greek word meaning a crab's claw. Two anions which can act as chelating agents are the oxalate ion, $C_2O_4^{2-}$ and the carbonate ion, CO_3^{2-}. (The numbers (1) and (2) in the diagram below indicate the atoms involved in chelate formation.)

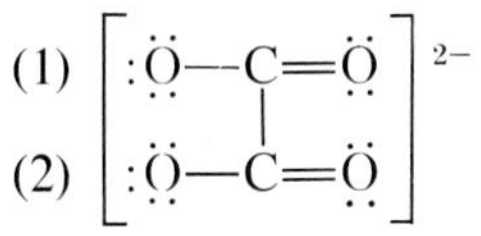

(1) :Ö—C=Ö (2) :Ö—C=Ö, overall charge 2−

$C_2O_4^{2-}$

(1) :Ö, (2) :Ö, both bonded to C=Ö, overall charge 2−

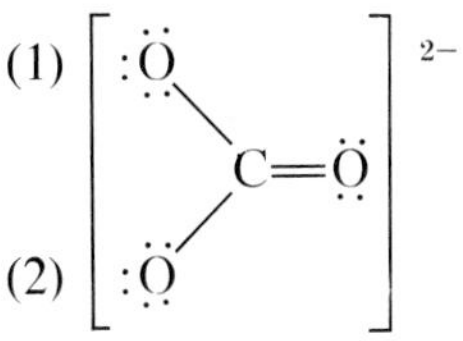

CO_3^{2-}

Chelating agents are abundant in nature and play an important role in many processes essential to plant and animal life. Certain species of soybeans are known to synthesize and secrete organic chelating agents that extract iron from insoluble compounds in the soil, thereby making it available for metabolic processes in the plant. Mosses and lichens growing on boulders use a similar process to obtain the metal ions they need for growth.

Many important natural products are chelates in which a central metal atom is bonded into a large organic molecule. In chlorophyll, the green coloring matter of plants, the central atom is magnesium; in the essential vitamin B_{12}, it is cobalt. The structure of heme, the pigment responsible for the red color of blood, is shown in Figure 21.1. We see that there is an Fe^{2+} ion at the center surrounded by four nitrogen atoms at the corners of a square. A fifth coordination position around the iron is occupied by a high molecular mass organic molecule (globin) which, in combination with heme, gives the proteinaceous material that we refer to as hemoglobin. The composition of the globin molecule is extremely important. A relatively minor variation in a tiny subsection of the molecule is responsible for the disease called sickle cell anemia, in which misshapen red blood cells clog capillaries, causing blood clots and depriving tissues of oxygen.

In hemoglobin, the heme molecule lies in a crevice of the large globin group

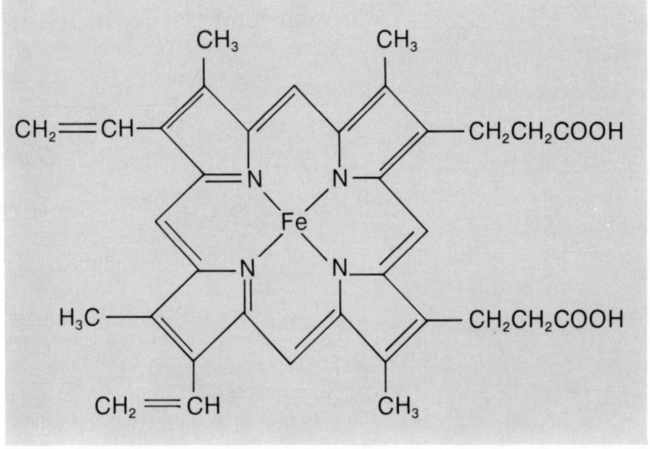

CH_3, CH_3, $CH_2{=}CH$, CH_2CH_2COOH, N, N, Fe, N, N, H_3C, CH_2CH_2COOH, $CH_2{=}CH$, CH_3

Figure 21.1 Structure of heme. Fe^{2+} ion is at the center of an octahedron, surrounded by four nitrogen atoms, a globin molecule, and a water molecule.

The sixth coordination position in hemoglobin appears to be occupied by a water molecule which can be replaced reversibly by oxygen to give a derivative known as oxyhemoglobin.

$$\text{Hemoglobin} + O_2 \rightleftharpoons \text{Oxyhemoglobin} + H_2O$$

The oxyhemoglobin complex needs to be stable, but not too stable

The position of this equilibrium is sensitive to the pressure of oxygen. In the lungs, where the blood is saturated with air (partial pressure $O_2 = 0.2$ atm), the hemoglobin is almost completely converted to the oxygenated form. In the tissues serviced by arterial blood, the partial pressure of oxygen drops and the oxyhemoglobin breaks down to release elementary oxygen essential for the combustion of food. By this reversible process, hemoglobin acts as an oxygen carrier, absorbing oxygen in the lungs and liberating it to the tissues.

Unfortunately, hemoglobin forms a complex with carbon monoxide that is considerably more stable than oxyhemoglobin. The equilibrium constant for the reaction

$$\text{Hemoglobin}\cdot O_2(aq) + CO(g) \rightleftharpoons \text{Hemoglobin}\cdot CO(aq) + O_2(g) \quad (21.4)$$

is about 200 at body temperature. Consequently, the carbon monoxide complex is formed preferentially in the lungs even at CO concentrations as low as one part per thousand. When this happens, the flow of oxygen to the tissues is cut off, resulting eventually in muscular paralysis and death.

Recently, chemists have synthesized Fe^{2+} complexes which have structures similar to that of heme. These complexes, like heme, absorb O_2 reversibly in a 1:1 mole ratio. Such complexes might possibly serve as substitutes for hemoglobin in synthetic blood and hence might be useful in treating such diseases as sickle cell anemia. So far, however, the "synthetic hemes" which have been prepared are unstable at body temperature (37°C).

Coordination Number

As may be seen from Table 21.3, the most common coordination number shown by metals in complex ions is 6. A coordination number of 4 is less common, while a coordination number of 2 is restricted largely to the +1 ions of the 1B elements. Odd coordination numbers (1, 3, 5) are very rare.

Certain metal ions show only one coordination number regardless of the ligands involved in the complex ion. For example, Pt^{2+} invariably forms four bonds, giving complex ions such as $Pt(NH_3)_4^{2+}$ and $PtCl_4^{2-}$. Similarly, Cr^{3+} and Co^{3+} always show a coordination number of 6. Other metal ions show variable coordination numbers, depending upon the nature of the ligand. The Ni^{2+} ion, for

TABLE 21.3 COORDINATION NUMBER AND GEOMETRY OF COMPLEX IONS*

COORDINATION NO.	GEOMETRY	EXAMPLES
2	linear	Cu^+, $\mathbf{Ag^+}$, Au^+
4	square planar	Cu^{2+}, Ni^{2+}, $\mathbf{Pt^{2+}}$, Pd^{2+}
4	tetrahedral	Al^{3+}, Ni^{2+}, Co^{2+}, Zn^{2+}, Cd^{2+}
6	octahedral	Al^{3+}, $\mathbf{Cr^{3+}}$, $\mathbf{Fe^{2+}}$, $\mathbf{Fe^{3+}}$, Co^{2+} $\mathbf{Co^{3+}}$, Ni^{2+}, Cu^{2+}, Zn^{2+}, Cd^{2+}, $\mathbf{Pt^{4+}}$

*Ions which show only one coordination number are in heavy type.

example, has two different coordination numbers, 4 and 6, shown respectively in the complex ions $Ni(CN)_4^{2-}$ and $Ni(H_2O)_6^{2+}$. Aluminum (Al^{3+}) and zinc (Zn^{2+}) behave similarly.

21.3 GEOMETRY OF COMPLEX IONS

Coordination Number = 2

Complex ions in which two ligands are coordinated to a central metal atom are invariably linear. The structures of the $Ag(NH_3)_2^+$, $Ag(CN)_2^-$ and $Au(CN)_2^-$ ions may be represented as follows:

$$(H_3N{-}Ag{-}NH_3)^+,\ (N{\equiv}C{-}Ag{-}C{\equiv}N)^-,\ (N{\equiv}C{-}Au{-}C{\equiv}N)^-$$

Coordination Number = 4

For a complex in which the metal ion forms four bonds, two different geometries are possible. The four coordinating groups may be located at the corners of a square to give what is known, redundantly, as a *square planar complex*. In other cases, the four bonds are directed toward the corners of a regular tetrahedron *(tetrahedral complex)*. The complexes of Pt^{2+} are of the square planar type, whereas the four-coordinated complexes of Zn^{2+} are tetrahedral (Fig. 21.2).

Certain square complexes can exist in two different forms with quite different properties. Consider, for example, the complex $[Pt(NH_3)_2Cl_2]$. Two forms of this compound, differing in absorption spectrum, water solubility, and chemical reactivity, have been prepared. One of these, made by reacting ammonia with the $PtCl_4^{2-}$ ion, has a structure in which the two ammonia molecules are located at adjacent corners of a square. In the other form, prepared by reacting the $Pt(NH_3)_4^{2+}$ ion with hydrochloric acid, the two ammonia molecules are located at opposite corners of the square:

Tetrahedral complexes do not show geometrical isomerism

H₃N Cl / Pt / H₃N Cl	H₃N Cl / Pt / Cl NH₃
Cis	Trans

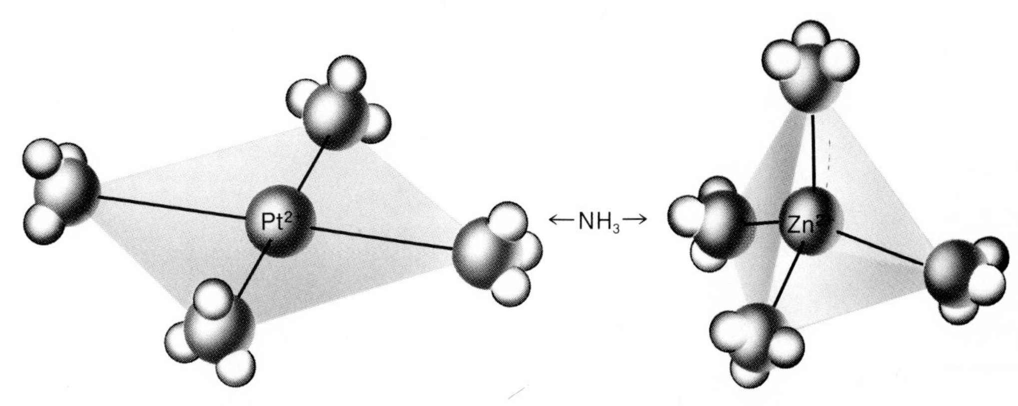

Figure 21.2 Structure of $Pt(NH_3)_4^{2+}$ (square planar) and $Zn(NH_3)_4^{2+}$ (tetrahedral). Geometrical isomers can exist for square planar complexes but not for tetrahedral complexes.

The two forms of $Pt(NH_3)_2Cl_2$ are called **geometrical isomers.** From a structural standpoint, they differ only in the spatial arrangement of the groups coordinated about the central atom. The form in which like groups are as close together as possible is called the ***cis*** isomer; the form in which like groups are far apart is referred to as the ***trans*** isomer. Geometrical isomerism can occur with any square planar complex of general formula Ma_2b_2 or Ma_2bc, in which M refers to the central atom and a, b, c represent ligands.

Assigning a *cis* or *trans* configuration to a particular isomer of a complex ion is by no means a simple experimental problem. Physical methods are perhaps most reliable for this purpose. X-ray diffraction studies can be used to determine the position of one ligand relative to another, but this method is tedious at best. A simpler approach is to use paper chromatography. At least for octahedral complexes (see below) it is found that the *trans* isomer moves more rapidly than the *cis*. Presumably this means that two ligands such as Cl^- ions can bond more strongly to the paper when they are in a *cis* position.

Historically chemical methods were widely used to assign *cis* and *trans* configurations. When the *cis* isomer of $[Pt(NH_3)_2Cl_2]$ is reacted with oxalate ions in solution, two Cl^- ions are displaced by a $C_2O_4{}^{2-}$ ion to form a chelate.

$$\begin{matrix} H_3N & & Cl \\ & Pt & \\ H_3N & & Cl \end{matrix} + \left[\begin{matrix} O-C=O \\ | \\ O-C=O \end{matrix}\right]^{2-} \rightarrow \begin{matrix} H_3N & & O-C=O \\ & Pt & | \\ H_3N & & O-C=O \end{matrix} + 2\ Cl^- \qquad (21.5)$$

Chelation cannot occur with the *trans* isomer since the oxalate ion cannot stretch out to attach itself to platinum at two points *trans* to each other. Unfortunately this approach leads to misleading results with many complex ions where changes of configuration can occur in water solution.

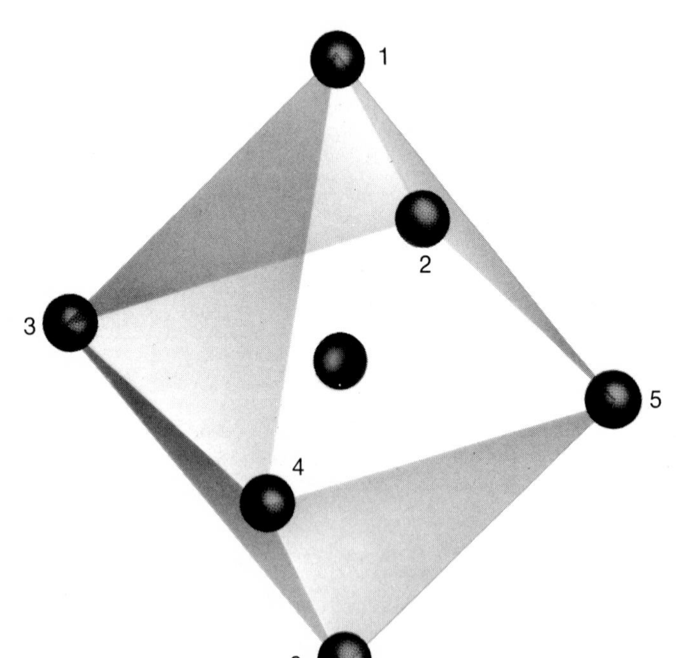

Figure 21.3 Regular octahedron. In this highly symmetric structure all of the six ligands surrounding the central atom occupy completely equivalent sites.

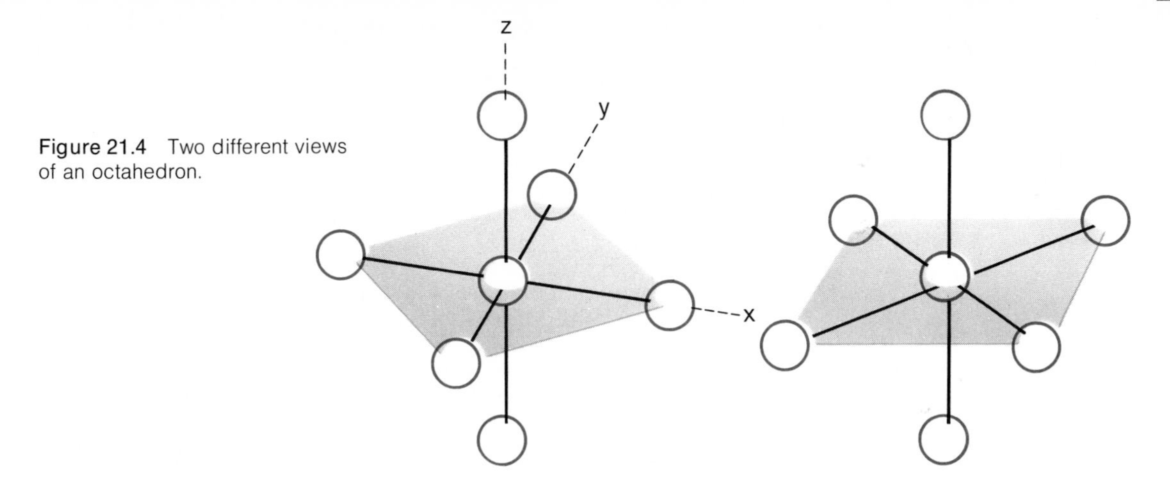

Figure 21.4 Two different views of an octahedron.

Coordination Number = 6

The six groups surrounding the metal ion in such complexes as $Fe(CN)_6^{3-}$ and $Co(NH_3)_6^{3+}$ are located at the corners of a regular octahedron, a figure with six corners and eight faces, all of which are equilateral triangles (Fig. 21.3). The metal ion is located at the center of the octahedron.

The spatial distribution of ligands in octahedral complexes is often shown by skeleton structures such as those in Figure 21.4. The drawing at the left emphasizes that the six ligands are located along three axes at right angles to each other (x, y, and z axes), *equidistant from the metal atom at the center*. The skeleton structure at the right of Figure 21.4 is easier to draw and serves to emphasize another characteristic of an octahedral complex; it can be visualized as a derivative of a square planar complex in which the two additional ligands are located along a line perpendicular to the square at its center.

The skeleton structure is often indicated simply as

Geometrical isomerism can occur in octahedral as well as square complexes. Notice from Figure 21.3 that for any given position of a ligand in an octahedral complex, there are four equivalent positions equidistant from the first, and one at a greater distance. If, for example, we choose position 1 as a point of reference, groups located at 2, 3, 4, and 5 will be equidistant from it, while a group at position 6 will be farther away. We may refer to positions 1 and 2, 1 and 3, 1 and 4, or 1 and 5 as being *cis* to each other while positions 1 and 6 are *trans*. Consequently, an ion such as $Co(NH_3)_4Cl_2^+$ can exist in two isomeric forms, one in which the two chloride ions are in a *cis* relationship to each other and another in which they are in a *trans* configuration (Fig. 21.5). These two isomers, like those of $[Pt(NH_3)_2Cl_2]$,

Cl
H$_3$N Cl
Co
H$_3$N NH$_3$
NH$_3$

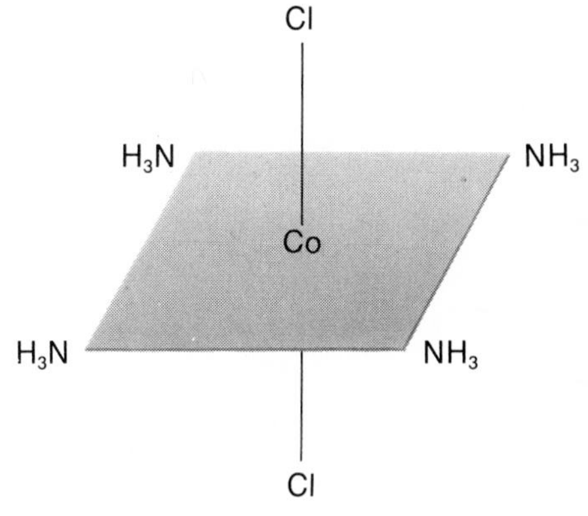

Figure 21.5 *Cis* and *trans* isomers of the $[Co(NH_3)_4Cl_2]^+$ complex ion. Note that the two Cl atoms are closer in the *cis* isomer (left) than in the *trans* isomer (right).

have different physical and chemical properties. The most striking difference is their color; *trans*-$[Co(NH_3)_4Cl_2]Cl$ is a bright green solid, while *cis*-$[Co(NH_3)_4Cl_2]Cl$ is violet.

Example 21.2 How many isomers are possible for the neutral complex $[Co(NH_3)_3Cl_3]$?

Solution It is best to approach this problem systematically. We might start by putting two NH_3 molecules in *trans* positions, perhaps at the "top" and "bottom" of the octahedron (Fig. 21.6a). We then ask ourselves: In how many spatially different positions can we place the third NH_3 molecule? A moment's reflection should convince you that there is really only one choice. All four of the remaining positions are equivalent in that they are *cis* to the two groups that we have already located. Choosing one of these positions arbitrarily, we get our first isomer (Fig. 21.6b).

It's easy to overestimate the number of isomers

To see if there are other isomers we start again, this time locating two NH_3 molecules *cis* to each other (Fig. 21.6c). If we were to place the third NH_3 molecule at one of the other corners of the square, we would simply reproduce the first isomer. (Remember that the symmetry of a regular octahedron requires that the distance across the diagonal of the square be the same as that from "top" to "bottom.") We are left with two equivalent positions. Placing the third NH_3 molecule arbitrarily at the "top" we arrive at a second isomer (Fig. 21.6d), distinctly different from the first because all three NH_3 molecules are *cis* to one another.

We have now exhausted in a logical manner all the possibilities for geometrical isomerism, finding two isomers. There are no others.

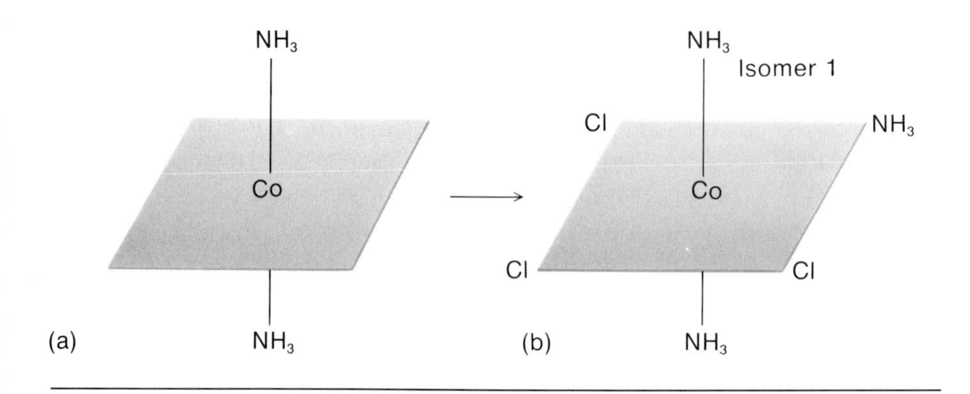

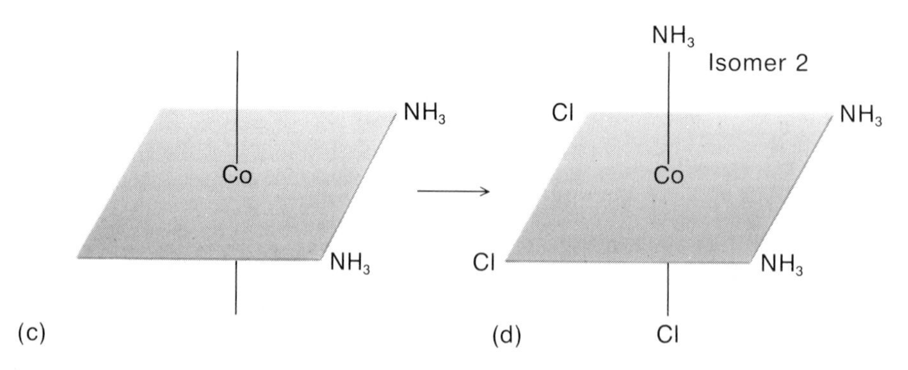

Figure 21.6 Isomers of $Co(NH_3)_3Cl_3$ (Example 21.2). In drawing isomeric structures one must be careful to check to see if two apparently different structures are not actually equivalent.

The complexes formed by Co^{3+} with NH_3 molecules and Cl^- ions played an important part in the development of the structural theory of coordination compounds. By the last decade of the 19th century, four such complexes were known. Their formulas were written as

1. $CoCl_3 \cdot 6\ NH_3$ (orange-yellow)
2. $CoCl_3 \cdot 5\ NH_3$ (violet)
3. $CoCl_3 \cdot 4\ NH_3$ (green)
4. $CoCl_3 \cdot 3\ NH_3$ (green)

Treatment of any of these compounds with hydrochloric acid fails to remove the ammonia, suggesting that the NH_3 molecules are strongly bonded to cobalt. When silver nitrate is added to a water solution of (1), all of the chlorine is precipitated immediately as AgCl. With compound (2), on the other hand, only ⅔ of the chlorine is precipitated and, with (3), only ⅓. Compound (4) when treated with a cold solution of silver nitrate fails to give any precipitate at all.

In 1893, Alfred Werner, a 26-year-old instructor at Zurich, reflecting on the properties of these and analogous compounds of Cr^{3+} and Pt^{4+}, proposed the basic structural theory of coordination compounds presented in this chapter. Specifically, he suggested that all of these complexes have an octahedral structure with six ligands attached directly to cobalt. Enclosing the complexes within square brackets, we would write their formulas as

Seems simple, now that we understand the principles

1. $[Co(NH_3)_6]Cl_3$
2. $[Co(NH_3)_5Cl]Cl_2$
3. $[Co(NH_3)_4Cl_2]Cl$
4. $[Co(NH_3)_3Cl_3]$

Only those Cl^- ions outside the coordination sphere are precipitated by silver nitrate.

Werner's ideas, reputed to have come to him in a dream, were revolutionary and hotly disputed at the time. To check their validity, Werner studied the conductivities of water solutions of compounds (1) through (4). He found, as predicted, that compounds (1) through (3) had conductivities comparable to simple 3:1, 2:1, and 1:1 salts respectively (e.g., $Al(NO_3)_3$, $Mg(NO_3)_2$, $NaNO_3$). Compound (4) behaved essentially as a nonelectrolyte.

Werner realized that the octahedral structure he was proposing required that compounds (3) and (4) show geometrical isomerism. Isolating their isomers proved a formidable task indeed. Not until 1907 was Werner able to prepare the *cis* isomer of $[Co(NH_3)_4Cl_2]Cl$, which had a violet color and certain other properties quite different from those of the green *trans* isomer, compound (3). Even today, only one form of $[Co(NH_3)_3Cl_3]$ is known. However, the two isomers of the analogous compound, $[Co(NH_3)_3(NO_2)_3]$, have been isolated.

In 1913, in recognition of his outstanding contributions to inorganic chemistry, Alfred Werner received the Nobel prize. Six years later he died at the age of 52.

21.4 ELECTRONIC STRUCTURE OF COMPLEX IONS

The *atomic orbital* or *valence-bond* approach presented in Chapter 8 can be extended to describe the electronic structures and rationalize the geometries of complex ions. Alternatively, one can use the molecular orbital approach discussed briefly in Chapter 8. Still another model, based on what is known as *crystal-field theory,* has been developed to explain certain of the physical and chemical properties of coordination compounds which are difficult to understand in terms of the valence-bond model. Here we shall describe only the valence-bond and crystal-field treatments.

Valence-Bond (Atomic Orbital) Model

You may recall that the valence-bond model was applied in Section 8.6, Chapter 8, to derive the electronic structures of simple molecules such as BeF_2, BF_3, and CH_4. It can also be applied in a quite straightforward way to the bonding

TABLE 21.4 VALENCE BOND MODEL APPLIED TO COMPLEX IONS

Coordination Number	Geometry	Hybrid Orbitals Occupied by Ligand Electrons	Example
2	linear	sp	$Cu(NH_3)_2^+$
4	tetrahedral	sp^3	$Zn(NH_3)_4^{2+}$
4	square, planar	dsp^2	$Ni(CN)_4^{2-}$
6	octahedral	d^2sp^3	$Cr(NH_3)_6^{3+}$

in complex ions. The basic assumption of the model is that **electron pairs donated by the ligands enter hybrid orbitals associated with the central metal atom.** The particular orbitals which these electron pairs enter depend upon the coordination number and the geometry of the complex. The possible sets of hybrid bonding orbitals are listed in Table 21.4.

Of the types of hybridization listed in Table 21.4, sp and sp^3 hybrid orbitals were discussed extensively in Chapter 8, pp. 196–197. In square planar complexes the four pairs of bonding electrons enter a d orbital, an s orbital, and two p orbitals (hence, "dsp^2"). For complex ions formed by metals in the first transition series, such as $Ni(CN)_4^{2-}$, these would be one 3d, one 4s, and two 4p orbitals. The four individual orbitals are "hybridized" to give four equivalent orbitals having the same energy. In octahedral complexes, the six pairs of bonding electrons occupy two d orbitals, an s orbital, and three p orbitals (d^2sp^3). With Cr^{3+} or other ions derived from metals in the first transition series, these are two 3d, one 4s, and three 4p orbitals.

Orbital diagrams for the four complex ions listed in Table 21.4 are shown in Figure 21.7. Here, the hybrid orbitals are enclosed by horizontal lines. The bonding electrons within these orbitals, all of which are contributed by the ligands, are shown as colored arrows. The electrons shown in black are those contributed by the metal ion itself; none of these are involved in bonding. To save space, only those electrons beyond the argon core are shown; the inner 18 electrons of each metal ion are represented by the notation [Ar].

The orbital diagrams shown in Figure 21.7 can be derived in a rather simple way. To illustrate the reasoning involved, consider the $Cr(NH_3)_6^{3+}$ ion. With a coordination number of 6, we deduce that the 6 pairs of electrons contributed by the NH_3 molecules must enter two 3d, one 4s, and three 4p orbitals. These are the electrons shown as colored arrows in the figure. To indicate that the two electrons in an orbital must have opposite spins, we draw one arrow "up" and the

Complex Ion		3d	4s	4p
$Cu(NH_3)_2^+$	[Ar]	(↑↓)(↑↓)(↑↓)(↑↓)(↑↓)	(↑↓)	(↑↓)( )( )
$Zn(NH_3)_4^{2+}$	[Ar]	(↑↓)(↑↓)(↑↓)(↑↓)(↑↓)	(↑↓)	(↑↓)(↑↓)(↑↓)
$Ni(CN)_4^{2-}$	[Ar]	(↑↓)(↑↓)(↑↓)(↑↓)(↑↓)	(↑↓)	(↑↓)(↑↓)( )
$Cr(NH_3)_6^{3+}$	[Ar]	(↑)(↑)(↑)(↑↓)(↑↓)	(↑↓)	(↑↓)(↑↓)(↑↓)

Figure 21.7 Orbital diagrams of complex ions given in Table 21.4. Electron pairs furnished by the ligands are shown in color.

other "down." After locating the bonding electrons, we pause to consider how many electrons will be contributed to the diagram by the Cr^{3+} ion. Since chromium has atomic number 24, there must be 24 electrons in the neutral Cr atom. The Cr^{3+} ion would have three fewer electrons, or 21. Of these, $21 - 18 = 3$ are beyond the argon core. Summarizing:

total no. of e^- in Cr = 24
total no. of e^- in Cr^{3+} = $24 - 3 = 21$
no. of e^- beyond [Ar] in Cr^{3+} = $21 - 18 = 3$

With three empty 3d orbitals available, we would expect the three electrons to half-fill each orbital and to have the same spins (recall Hund's rule, p. 142). Thus, we arrive at the orbital diagram for $Cr(NH_3)_6^{3+}$ shown in Figure 21.7.

We should emphasize that *all* of the octahedral complexes of Cr^{3+} have the same structure, that just derived for $Cr(NH_3)_6^{3+}$. This would be the case, for example, for the $Cr(OH)_6^{3-}$ ion and for the chelate, $Cr(en)_3^{3+}$.

The orbital diagrams of the other complex ions given in Table 21.4 can be derived in an entirely analogous fashion. Example 21.3 offers a further illustration of the principles involved.

Example 21.3 Write an orbital diagram for $Fe(CN)_6^{4-}$.

Solution Since the coordination number is 6, we expect the bonding electrons from the CN^- ligands to enter d^2sp^3 hybrid orbitals. So, we put them there:

3d					4s	4p		
( )	( )	( )	(⇅)	(⇅)	(⇅)	(⇅)	(⇅)	(⇅)

We must now decide how many electrons are contributed by the central metal ion. To do this, we must first deduce its charge. Since the complex ion has a charge of −4 and each CN^- ion has a charge of −1, we must be dealing with a +2 ion of iron, i.e.,

$$Fe^{2+} + 6\ CN^- \rightarrow Fe(CN)_6^{4-}$$

In the +2 ion of Fe (at. no. = 26), there must be 24 e^-; 6 of these are located beyond the argon core:

It's best to put the ligand electrons in first and leave them there

total no. e^- in Fe^{2+} = $26 - 2 = 24$
no. of e^- beyond [Ar] in Fe^{2+} = $24 - 18 = 6$

We have just enough room in the empty 3d orbitals to accommodate these 6 electrons, one pair to an orbital. Hence the completed orbital diagram is

	3d					4s	4p		
$Fe(CN)_6^{4-}$ [Ar]	(⇅)	(⇅)	(⇅)	(⇅)	(⇅)	(⇅)	(⇅)	(⇅)	(⇅)

The valence-bond approach has proven extremely useful in correlating the geometries of complex ions with their electronic structures. It has also been quite successful in explaining certain of the properties of complex ions, notably their magnetic behavior. The structures shown in Figure 21.7 imply that the complex ions $Cu(NH_3)_2^+$, $Zn(NH_3)_4^{2+}$, and $Ni(CN)_4^{2-}$ should all be *diamagnetic* since they have no unpaired electrons. In contrast, the $Cr(NH_3)_6^{3+}$ ion, with 3 unpaired electrons, should be *paramagnetic*. These predictions are confirmed experimentally.

The first three complexes are weakly repelled by a magnetic field; $Cr(NH_3)_6^{3+}$, on the other hand, is attracted into the field with a force that corresponds to 3 unpaired electrons.

With certain complex ions, however, the observed magnetic behavior does not agree with the structure predicted by valence bond theory. An example is the $Fe(H_2O)_6^{2+}$ ion. From the arguments we have gone through, its orbital diagram should be identical with that of the $Fe(CN)_6^{4-}$ ion shown in Example 21.3. Consequently, it should be diamagnetic (no unpaired electrons), like the $Fe(CN)_6^{4-}$ ion. Experimentally, it is found to be paramagnetic, with 4 unpaired electrons. Although it is possible to rationalize this observation within the framework of valence-bond theory,* there seems to be no convincing explanation as to why the CN^- ion should form one type of complex with Fe^{2+} and the H_2O molecule a quite different type.

In another area, valence-bond theory has been notably ineffective in explaining what is perhaps the most striking characteristic of coordination compounds: their brilliant colors. A quite different approach, known as the crystal-field model, has been much more successful in this area.

Crystal-Field Model

In the crystal field model, the ligand electrons are not shared with the metal ion

The crystal-field model considers that the attractive forces holding a complex ion together are primarily electrostatic rather than covalent. In this approach, unlike the valence-bond model, it is assumed that the ligands do not feed electrons into the orbitals of the central metal atom. Instead, their only effect is to change the relative energies of these orbitals through electrostatic interactions. To illustrate how these interactions arise, let us consider a specific example, the formation of the $Fe(CN)_6^{4-}$ ion:

$$Fe^{2+} + 6\ CN^- \rightarrow Fe(CN)_6^{4-}$$

In the bare Fe^{2+} ion, all the 3d orbitals have the same energy. Consequently, the six 3d electrons of Fe^{2+} are spread out among them in accordance with Hund's rule.

	3d
Fe^{2+}	(↑↓) (↑) (↑) (↑) (↑)

However, when six CN^- ions approach the Fe^{2+} ion to form an octahedral complex, geometric considerations suggest that these orbitals should be split into two groups of different energies. To understand why this is the case, let us examine the orientation of the electron density clouds associated with the five d orbitals (Fig. 21.8).

Imagine now six CN^- ions approaching these orbitals, two along the z axis (north and south), two along the x axis (east and west), and two along the y axis.

---

*One "explanation" is that the d orbitals occupied by ligand electrons in $Fe(H_2O)_6^{2+}$ are those in the 4th principal energy level rather than the 3rd. The electrons in the Fe^{2+} ion (6 beyond Ar) could then spread over all the 3d orbitals, giving the structure shown below

		3d	4s	4p	4d
$Fe(H_2O)_6^{2+}$	[Ar]	(↑↓) (↑) (↑) (↑) (↑)	(↑↓)	(↑↓) (↑↓) (↑↓)	(↑↓)(↑↓)( )( )( )

which does indeed have four unpaired electrons.

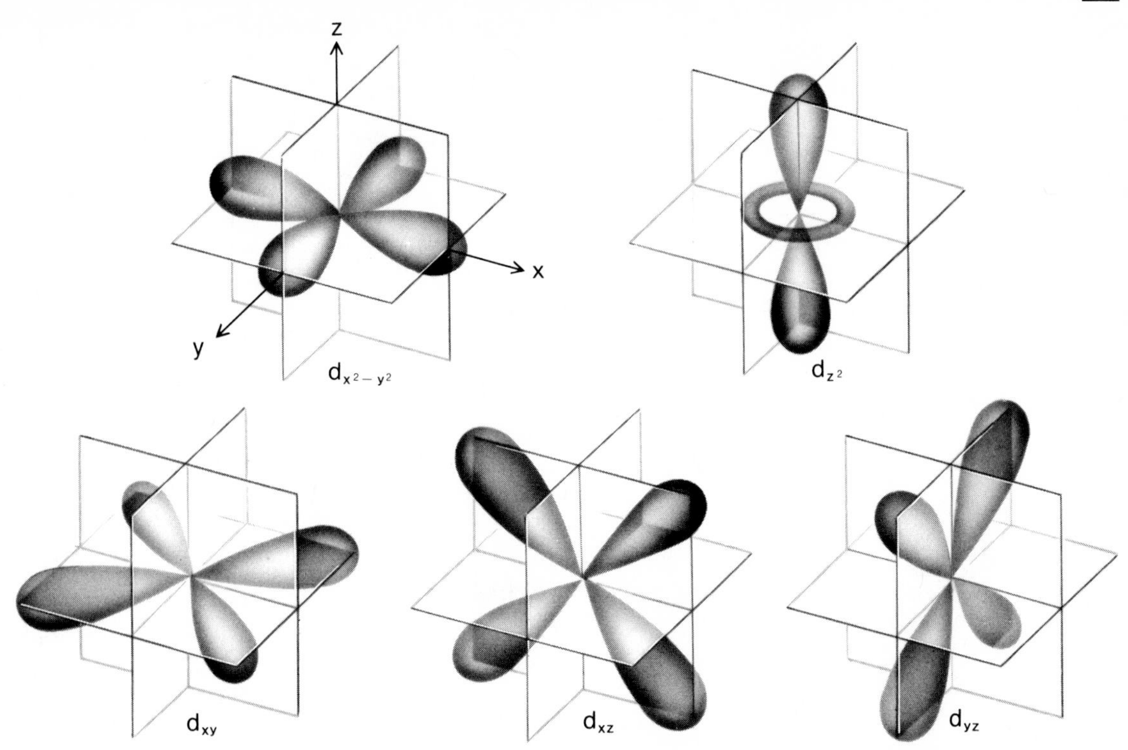

Figure 21.8 Spatial orientation of d orbitals. Note that the $d_{x^2-y^2}$ and d_{z^2} orbitals are oriented toward ligands approaching the corners of an octahedron.

When they move in close to either of the two orbitals labeled "d_{z^2}" or "$d_{x^2-y^2}$", a strong electrostatic repulsion is set up because the electron density in these orbitals is concentrated along the z, x, and y axes. In contrast, the electrons in the other three orbitals (d_{xy}, d_{xz}, d_{yz}) are less strongly repelled since their densities are concentrated between the axes rather than along them. In other words, in the "octahedral field" created by the approach of CN^- ions, the five 3d orbitals of Fe^{2+}, or indeed of any transition metal ion, are split into two groups. One group, comprising the d_{xy}, d_{yz}, and d_{xz} orbitals, is lower in energy than the other group of two orbitals, d_{z^2} and $d_{x^2-y^2}$. Schematically, we have the situation shown in Figure 21.9.

As a result of the splitting of d orbitals, it seems reasonable to postulate a rearrangement of the electronic structures of Fe^{2+} when it forms a complex with

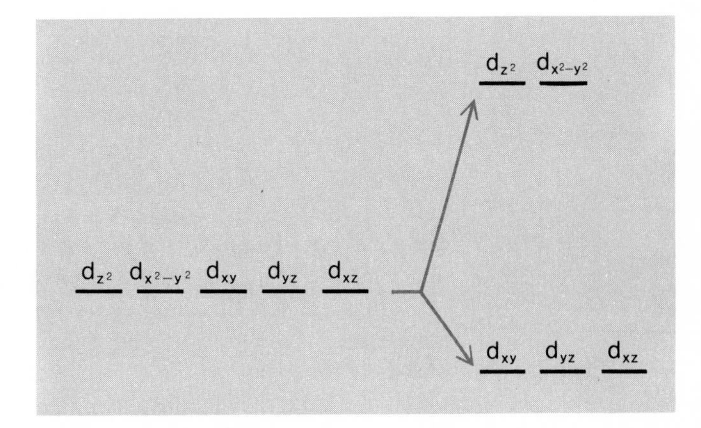

Figure 21.9 Splitting of d orbitals in an octahedral field. Two of the orbitals ($d_{x^2-y^2}$, d_{z^2}) are raised in energy, while three (d_{xy}, d_{yz}, d_{xz}) are lowered.

CN^-. The six 3d electrons of the Fe^{2+} will tend to pair in the three lower energy levels, in "violation" of Hund's rule.

Fe^{2+} (in $Fe(CN)_6^{4-}$)

( )( )

(↑↓)(↑↓)(↑↓)

The extent to which the energies of the d orbitals are split when a complex ion is formed depends upon how strongly the ligands interact with the electrons in these orbitals. The neutral H_2O molecule repels electrons less than the negatively charged CN^- ion and hence gives a smaller energy separation. As a result, the electron distribution in the Fe^{2+} ion is retained in $Fe(H_2O)_6^{2+}$. In this case, the energy to be saved by pairing electrons is not great enough to overcome the tendency for electrons to remain unpaired.

Fe^{2+} (in $Fe(H_2O)_6^{2+}$)

(↑)(↑)

(↑↓)(↑)(↑)

The energy difference between the two sets of levels is large in $Fe(CN)_6^{4-}$ and small in $Fe(H_2O)_6^{2+}$

Looking at the structures written for Fe^{2+} in the $Fe(CN)_6^{4-}$ and $Fe(H_2O)_6^{2+}$ complexes, we see that the crystal field model explains why the first complex is diamagnetic and the second paramagnetic with 4 unpaired electrons. **"Low-spin"** (smaller number of unpaired electrons) and **"high-spin"** (larger number of unpaired electrons) complexes analogous to these are observed with many transition metal ions (Example 21.4).

Example 21.4 Using the crystal-field model, derive the structure of the Co^{2+} ion in low-spin and high-spin octahedral complexes.

Solution We first determine the number of d electrons available. Since the atomic number of cobalt is 27, there are $27 - 2 = 25$ electrons in Co^{2+}. Of these, 18 are accounted for by the argon core, leaving 7 in the 3d orbitals. In a low-spin complex, formed with strongly interacting ligands, 6 of these electrons are crowded into the lower three energy levels, leaving only 1 unpaired electron. In a high-spin complex, where the splitting is small, Hund's rule is followed and there are 3 unpaired electrons.

low-spin	high-spin
(↑)( )	(↑)(↑)
(↑↓)(↑↓)(↑↓)	(↑↓)(↑↓)(↑)

Both types of complexes are known. The $Co(CN)_6^{4-}$ ion is of the low-spin type; $Co(H_2O)_6^{2+}$ is a high-spin complex.

Crystal-field theory offers a simple explanation of the fact that complexes formed by metal ions with unfilled d orbitals are colored. The difference in energy between the two sets of d levels turns out in many cases to be equal in magnitude to the energy of a photon of light in the visible region. This means that d electrons in complexes can jump from the lower level to the upper one by absorbing energy from light. When this happens with white light, some of the component wavelengths are removed and the light reflected from the sample is colored. Complexes of ions such as Zn^{2+}, in which all the d orbitals are filled, are colorless.

The extent of d-level splitting determines the wavelength of the light absorbed

TABLE 21.5 COLORS OF COMPLEX IONS OF Co^{3+}

COMPLEX	COLOR OBSERVED	COLOR ABSORBED	APPROXIMATE WAVELENGTH (NM) ABSORBED
$Co(NH_3)_6^{3+}$	yellow	violet	430
$Co(NH_3)_5NCS^{2+}$	orange	blue	470
$Co(NH_3)_5H_2O^{3+}$	red	blue-green	500
$Co(NH_3)_5Cl^{2+}$	purple	yellow-green	530
trans $Co(NH_3)_4Cl_2^+$	green	red	680

and hence the color that we see when we look at the complex. This effect is illustrated in Table 21.5 and Plate 14. We see that when we substitute for NH_3 such ligands as SCN^-, H_2O, or Cl^-, which produce smaller d orbital splittings, the light absorbed shifts to longer wavelengths (lower energy). On the basis of these and other observations, it is possible to arrange various ligands in order of decreasing tendency to split the d orbitals of a metal ion. An abbreviated version of such a *spectrochemical* series is

$$CN^- > NO_2^- > en > NH_3 > SCN^- > H_2O > F^- > Cl^-$$

Lest it be supposed that the crystal field theory can explain all the properties of complex ions, we should point out at least one of its deficiencies. If one thinks of the bonding in complex ions as being primarily electrostatic, it is hard to explain why certain molecules which are only slightly polar, such as CO, can be effective coordinating agents. In order to explain this, it is necessary to modify the crystal field theory to take into account covalent as well as ionic bonding. A more sophisticated version of the electrostatic approach, which we shall not attempt to describe here, known as **ligand field theory,** has been developed with this in mind.

21.5 RATE OF COMPLEX ION FORMATION

When an inorganic chemist wishes to prepare a particular complex ion, he most frequently uses a substitution reaction in which one ligand is replaced by another. This process is often carried out in aqueous solution. For example, to prepare the $Cu(NH_3)_4^{2+}$ ion, we might add aqueous ammonia to a solution of copper sulfate or copper nitrate containing the $Cu(H_2O)_4^{2+}$ cation. A step-wise reaction occurs,

In this reaction, the ligands exchange very rapidly

$$Cu(H_2O)_4^{2+}(aq) + NH_3(aq) \rightarrow Cu(H_2O)_3(NH_3)^{2+}(aq) + H_2O$$
$$Cu(H_2O)_3(NH_3)^{2+}(aq) + NH_3(aq) \rightarrow Cu(H_2O)_2(NH_3)_2^{2+}(aq) + H_2O$$
$$Cu(H_2O)_2(NH_3)_2^{2+}(aq) + NH_3(aq) \rightarrow Cu(H_2O)(NH_3)_3^{2+}(aq) + H_2O$$
$$Cu(H_2O)(NH_3)_3^{2+}(aq) + NH_3(aq) \rightarrow Cu(NH_3)_4^{2+}(aq) + H_2O$$

$$Cu(H_2O)_4^{2+}(aq) + 4\ NH_3(aq) \rightarrow Cu(NH_3)_4^{2+}(aq) + 4\ H_2O \qquad (21.6)$$

with the final product being the $Cu(NH_3)_4^{2+}$ ion. As ammonia is added, the light blue color of $Cu(H_2O)_4^{2+}$ changes almost instantaneously to the deeper blue, almost purple color of $Cu(NH_3)_4^{2+}$.

Complex ions differ greatly in the rate at which they undergo substitution reactions. Consider, for example, what happens when nitric acid is added to two different solutions, one containing the $Cu(NH_3)_4^{2+}$ ion, the other the analogous Co^{3+} complex, $Co(NH_3)_6^{3+}$. In the first case, substitution takes place instantaneously to generate $Cu(H_2O)_4^{2+}$.

$$\underset{\text{purple}}{Cu(NH_3)_4^{2+}(aq)} + 4\ H^+(aq) + 4\ H_2O \rightarrow \underset{\text{blue}}{Cu(H_2O)_4^{2+}(aq)} + 4\ NH_4^+(aq) \tag{21.7}$$

In contrast, days or even weeks may go by before we detect any color change or other evidence for reaction in the solution containing the Co^{3+} complex. This is true even though the reaction

$$Co(NH_3)_6^{3+}(aq) + 6\ H^+(aq) + 6\ H_2O \rightarrow Co(H_2O)_6^{3+}(aq) + 6\ NH_4^+(aq) \tag{21.8}$$

is thermodynamically spontaneous; the equilibrium constant for Reaction 21.8 is of the order of 10^{20}.

Complexes such as those of Cu^{2+} which exchange ligands virtually instantaneously are described as **labile.** Complexes such as $Co(NH_3)_6^{3+}$ which undergo substitution reactions at a rate which is measurable by ordinary techniques are referred to as nonlabile or **inert.** There is no sharp dividing line between these two categories: inert complexes can have half-lives ranging from a few minutes to several years.

Among the relatively few cations which consistently form inert complexes are Cr^{3+}, Co^{3+}, Pt^{2+}, and Pt^{4+}. It is no coincidence that complexes of these ions were among the first to be studied in the laboratory and that much of our knowledge about the structures and properties of complex ions is based upon their behavior in aqueous solution. The great advantage of working with inert complexes is that they retain their identity in solution at least long enough for their chemical and physical properties to be studied (Plate 15). Once formed, they are relatively easy to separate from solution as pure compounds.

21.6 COMPLEX ION EQUILIBRIA

In the preceding section we compared the (kinetic) labilities of different complex ions. It is equally important to be able to compare their (thermodynamic) stabilities. One way to do this is to measure the equilibrium constant for the reaction that occurs when a compound containing a particular complex ion is added to water. In doing this, it is customary to treat the reaction as if it were a simple *dissociation,* ignoring intermediate steps. Thus, for the equilibrium involved when a compound containing $Ag(NH_3)_2^+$ ions is added to water, we write

$$Ag(NH_3)_2^+(aq) \rightleftharpoons Ag^+(aq) + 2\ NH_3(aq);\ K_d = \frac{[Ag^+] \times [NH_3]^2}{[Ag(NH_3)_2^+]} \tag{21.9}$$

where K_d represents the dissociation constant of the complex ion.

The fact that the equilibrium constant for Reaction 21.9 is a very small number, 4×10^{-8}, means that $Ag(NH_3)_2^+$ dissociates to only a small extent when added to water. Looking at it another way, even in quite dilute solutions of ammonia, the concentration of Ag^+ will be low relative to that of $Ag(NH_3)_2^+$ (Example 21.5).

Example 21.5 Calculate the ratio of the concentration of Ag^+ to that of $Ag(NH_3)_2^+$ in a solution prepared by adding a small amount of $AgNO_3$ to a large volume of 0.02 *M* NH_3.

Solution Substituting in the expression for K_d (Equation 21.9):

$$K_d = \frac{[Ag^+] \times (0.02)^2}{[Ag(NH_3)_2^+]} = 4 \times 10^{-8}$$

Solving:

$$\frac{[Ag^+]}{[Ag(NH_3)_2^+]} = \frac{4 \times 10^{-8}}{(2 \times 10^{-2})^2} = \frac{4 \times 10^{-8}}{4 \times 10^{-4}} = 1 \times 10^{-4}$$

In other words, even in this very dilute solution only 1/10 000 of the silver is present as free Ag^+; 99.99% is in the form of the ammonia complex.

The stabilities of different complexes of the same cation are inversely related to their dissociation constants; the smaller the value of K_d, the more stable the complex. When we find that the equilibrium constant for the reaction

$$Ag(S_2O_3)_2^{3-}(aq) \rightleftharpoons Ag^+(aq) + 2\ S_2O_3^{2-}(aq);\ K_d = \frac{[Ag^+] \times [S_2O_3^{2-}]^2}{[Ag(S_2O_3)_2^{3-}]} \quad (21.10)$$

Note the analogy with weak acids; we could say that HCN ($K_a = 4 \times 10^{-10}$) is more "stable" than HF ($K_a = 7 \times 10^{-4}$)

is only 1×10^{-13}, we deduce that the $Ag(S_2O_3)_2^{3-}$ complex is even more stable than $Ag(NH_3)_2^+$ ($K_d = 4 \times 10^{-8}$). This complex is formed in photography when the developed film is "fixed" by washing with sodium thiosulfate, $Na_2S_2O_3$. The thiosulfate complex is sufficiently stable to bring into solution unreacted silver bromide remaining on the film.

TABLE 21.6 DISSOCIATION CONSTANTS OF COMPLEX IONS

MC_2		MC_4		MC_6	
$AgCl_2^-$	1×10^{-6}	$CdCl_4^{2-}$	4×10^{-3}	$Cr(OH)_6^{3-}$	1×10^{-38}
$Ag(NH_3)_2^+$	4×10^{-8}	$Cu(NH_3)_4^{2+}$	2×10^{-13}	$Co(NH_3)_6^{2+}$	1×10^{-5}
$Ag(S_2O_3)_2^{3-}$	1×10^{-13}	$Cu(CN)_4^{2-}$	1×10^{-25}	$Co(NH_3)_6^{3+}$	1×10^{-35}
$Ag(CN)_2^-$	1×10^{-21}	$Ni(CN)_4^{2-}$	1×10^{-31}	$Fe(CN)_6^{4-}$	1×10^{-24}
$CuCl_2^-$	1×10^{-5}	$Zn(NH_3)_4^{2+}$	1×10^{-9}	$Fe(CN)_6^{3-}$	1×10^{-31}
$Cu(NH_3)_2^+$	2×10^{-11}	$Zn(OH)_4^{2-}$	3×10^{-16}		
		$Zn(CN)_4^{2-}$	1×10^{-17}		

21.7 COMPLEX IONS IN ANALYTICAL CHEMISTRY

Qualitative Analysis

Reactions involving the formation of complex ions are widely used in qualitative analysis for either of two purposes. The ability of a metal ion to form a colored complex or a precipitate with a particular complexing agent may be used as a specific test for that ion. Alternatively, two ions may be separated from one another by adding a complexing agent that forms a complex with only one of them. Frequently this is accomplished by adding a complexing agent to a mixture of insoluble solids, selectively bringing one of them into solution.

TESTS FOR SPECIFIC IONS. An extremely sensitive test for the Cu^{2+} ion in water solution involves its ability to form a deep blue complex with ammonia. The color of the $Cu(NH_3)_4^{2+}$ ion is much more intense than that of the light blue $Cu(H_2O)_4^{2+}$ ion; it can be detected at concentrations of Cu^{2+} as low as 10^{-4} *M*. Certain other ions interfere with this test; nickel, for example, also forms a deep blue complex ion with ammonia.

The Fe^{3+} ion in water solution is readily detected by adding a solution of potassium thiocyanate, KSCN, to give the blood-red color characteristic of the $FeSCN^{2+}$ complex ion (more exactly, $Fe(H_2O)_5SCN^{2+}$).

$$Fe(H_2O)_6^{3+}(aq) + SCN^-(aq) \rightarrow Fe(H_2O)_5SCN^{2+}(aq) \qquad (21.11)$$

Chelating agents, because of their ability to form extremely stable complexes, are widely used to test for specific cations. Dimethyl glyoxime

```
H3C—C——C—CH3
    ‖   ‖
HO—N    N—OH
   ¨    ¨
```

is one such compound. It uses the unshared electron pairs on the two nitrogen atoms to form chelates with many metal ions. The formation of a red, insoluble complex of dimethyl glyoxime with Ni^{2+} is a very sensitive test for that ion.

SEPARATION OF IONS. Cations are often separated from one another by taking advantage of differences in their tendencies to form complex ions with a particular ligand. To illustrate the method, consider the two ions Fe^{3+} and Al^{3+}, which are ordinarily precipitated in the same group in cation analysis. If sodium hydroxide is added to a solution containing these ions, they precipitate as $Fe(OH)_3$ and $Al(OH)_3$. However, as more sodium hydroxide is added, the aluminum hydroxide dissolves to form a complex ion

$$Al(OH)_3(s) + 3\ OH^-(aq) \rightarrow Al(OH)_6^{3-}(aq) \qquad (21.12)$$

Iron(III) hydroxide fails to dissolve and is thus separated from Al^{3+}.

Many transition metal ions, like Al^{3+}, form sufficiently stable complex ions with OH^- to go into solution in concentrated sodium hydroxide. Among the hydroxides which dissolve in a strong base are $Zn(OH)_2$, $Cr(OH)_3$ and, to a lesser extent, $Cu(OH)_2$. Compounds such as these, which are capable of reacting with OH^- ions as well as H^+ ions are said to be **amphoteric.**

Another complexing agent which is frequently used to separate metal ions is the ammonia molecule, NH_3. In the analysis of the Group 1 cations, advantage is taken of the stability of the $Ag(NH_3)_2^+$ complex to separate silver from mercury. Treatment of a precipitate containing AgCl with dilute ammonia leads to the reaction

$$AgCl(s) + 2\ NH_3(aq) \rightarrow Ag(NH_3)_2^+(aq) + Cl^-(aq)$$

Any labile NH_3 complex can be decomposed by adding H^+

bringing the silver into solution in the form of the complex ion. To confirm the presence of Ag^+, one can add nitric acid to the solution. The hydrogen ions from the acid destroy the complex by converting NH_3 molecules to NH_4^+ ions.

$$Ag(NH_3)_2^+(aq) + Cl^-(aq) + 2\ H^+(aq) \rightarrow AgCl(s) + 2\ NH_4^+(aq)$$

SOLUBILITY OF SOLIDS IN COMPLEXING AGENTS. The extent to which a solid dissolves in a complexing agent depends upon several factors. To illustrate the principles involved, let us consider a general reaction in which a silver halide, AgX (X = Cl, Br, or I) reacts with a complexing agent, C, which might be NH_3,

$S_2O_3^{2-}$, or any other ligand that forms a complex with Ag^+. To evaluate the equilibrium constant, K, for the reaction

$$AgX(s) + 2\ C(aq) \rightleftharpoons AgC_2^+(aq) + X^-(aq);\ K = ?$$

we note that this equation can be regarded as the sum of two other equations:

The approach here is similar to that used in dealing with the solubility of solids in acids (Chapter 20)

$$(1)\ AgX(s) \rightleftharpoons Ag^+(aq) + X^-(aq) \qquad K_1 = K_{sp} \text{ of } AgX$$

$$(2)\ Ag^+(aq) + 2\ C(aq) \rightleftharpoons AgC_2^+(aq) \qquad K_2 = 1/K_d \text{ of } AgC_2^+$$

$$AgX(s) + 2\ C(aq) \rightleftharpoons AgC_2^+(aq) + X^-(aq);\ K = K_1 \times K_2 = \frac{K_{sp}\ AgX}{K_d\ AgC_2^+} \quad (21.13)$$

From Equation 21.13, we see that the magnitude of K and hence the solubility of AgX in the complexing agent will be increased by:

1. *High solubility of AgX in water* (i.e., large value of K_{sp}). Thus we would predict that AgCl ($K_{sp} = 1.6 \times 10^{-10}$) would be more soluble in any complexing agent than AgI ($K_{sp} = 1 \times 10^{-16}$).

2. *High stability of the complex AgC_2^+* (i.e., small value of K_d, the equilibrium constant for the dissociation of the complex). All silver salts will be more soluble in $S_2O_3^{2-}$ (K_d $Ag(S_2O_3)_2^{3-} = 1 \times 10^{-13}$) than in NH_3 (K_d $Ag(NH_3)_2^+ = 4 \times 10^{-8}$).

From the equation written for the reaction of a silver halide with a complexing agent, we notice a third factor which will tend to enhance solubility:

3. *High concentration of the complexing agent.* The greater the concentration of C, the farther the equilibrium will shift to the right and the greater will be the amount of solid that goes into solution. Experimentally, we find that silver bromide is considerably more soluble in concentrated ammonia (15 *M*) than in dilute ammonia (6 *M*).

Example 21.6

a. Write an equation for the reaction that occurs when silver bromide is treated with aqueous ammonia.
b. Calculate K for the reaction (K_{sp} AgBr $= 5 \times 10^{-13}$; K_d $Ag(NH_3)_2^+ = 4 \times 10^{-8}$).
c. Determine the solubility of AgBr in 6 *M* NH_3; in 15 *M* NH_3.

Solution

a. $AgBr(s) + 2\ NH_3(aq) \rightarrow Ag(NH_3)_2^+(aq) + Br^-(aq)$

b. $K = \dfrac{K_{sp} \text{ of } AgBr}{K_d \text{ of } Ag(NH_3)_2^+} = \dfrac{5 \times 10^{-13}}{4 \times 10^{-8}} = 1 \times 10^{-5}$

c. From the equation for the reaction, we see that for every mole of AgBr that dissolves, one mole of $Ag(NH_3)_2^+$ and one mole of Br^- is formed. If we let S represent the solubility of silver bromide in ammonia, then

$$S = [Ag(NH_3)_2^+] = [Br^-]$$

The expression for the equilibrium constant for the solution process is

$$K = \frac{[Ag(NH_3)_2^+] \times [Br^-]}{[NH_3]^2} = 1 \times 10^{-5}$$

Substituting: $1 \times 10^{-5} = \frac{S \times S}{[NH_3]^2}$, or $S^2 = [NH_3]^2 \times 1 \times 10^{-5}$

In 6 M NH_3: $S^2 = 36 \times 10^{-5} = 3.6 \times 10^{-4}$; $S \approx 2 \times 10^{-2}$ M

In 15 M NH_3: $S^2 = 225 \times 10^{-5} = 22.5 \times 10^{-4}$; $S \approx 5 \times 10^{-2}$ M

Calculations entirely similar to those just carried out indicate that AgCl is "soluble" in 6 M NH_3 ($S \approx 0.4$ M) while AgI is insoluble even in 15 M NH_3 ($S <$ 0.001 M). Consequently, it is possible to separate a mixture of the three silver halides by first treating with dilute NH_3, in which only AgCl is appreciably soluble, and then with concentrated NH_3, which dissolves AgBr but not AgI.

Quantitative Analysis

The reaction of a metal ion with a complexing agent bears at least a superficial resemblance to the reaction of an H^+ ion with an OH^- ion. Comparing the two equations

$$Cu^{2+}(aq) + 4\ NH_3(aq) \rightarrow Cu(NH_3)_4^{2+}(aq)$$

$$H^+(aq) + OH^-(aq) \rightarrow H_2O$$

we note that in both cases a positive ion is converted to an extremely stable, covalently bonded species. It might seem, then, that we could determine the concentration of a metal ion by titrating with a complexing agent in much the same way that an acid is titrated with a base.

In practice, it is seldom possible to use ordinary complexing agents to analyze quantitatively for metal ions in solution. The difficulty is that, as previously noted, the formation of a metal complex is a stepwise process. If we add ammonia to a solution of a copper salt, the hydrated Cu^{2+} ion is not converted directly to the $Cu(NH_3)_4^{2+}$ ion. Instead, intermediate species containing one, two, or three ammonia molecules are formed. As the concentration of ammonia increases, the various equilibria gradually shift to lower the Cu^{2+} ion concentration. There is no sharp change in "free" Cu^{2+} ion concentration analogous to the abrupt change in H^+ ion concentration that we observe at the equivalence point of an acid-base titration. The end point, instead of being sharp and precise, is drawn out and diffuse.

Analytical chemists have developed a series of reagents that react with metal ions to give extremely stable 1:1 complexes and, hence, are suitable for metal-ion titrations. These substances are chelating agents; the best known is the sodium salt of ethylenediaminetetraacetic acid, commonly called **EDTA.** The anion of this salt has the following structure:

```
     O                                   O
     ||                                  ||
 ⁻O—C—CH₂                           CH₂—C—O⁻
  (1)     \  ..                 ..  /        (5)
            N—CH₂—CH₂—N
          / (3)           (4) \
 ⁻O—C—CH₂                           CH₂—C—O⁻
  (2) ||                                 ||  (6)
      O                                  O
```

A single EDTA anion can attach itself to a metal ion through as many as six different atoms (numbered 1 to 6 in the foregoing structural formula), filling all its coordination requirements. Difficulties inherent in stepwise complex formation are thereby avoided.

To illustrate how an EDTA titration may be carried out, let us consider the use of this chelating agent in the determination of Fe^{3+}. The reaction that occurs may be represented as:

$$Fe^{3+}(aq) + EDTA^{4-}(aq) \rightarrow Fe(EDTA)^{-}(aq) \quad (21.14)$$

The SCN^- ion, which forms a blood-red complex with Fe^{3+}, can be used as an indicator. As EDTA is added, the thiocyanate complex of iron(III) is converted to the more stable EDTA complex. At the equivalence point, the SCN^- ions attached to iron are quantitatively displaced and the color changes from deep red to yellow. Knowing the concentration of the EDTA solution and the volume which must be added to reach the end point, we can readily calculate the amount of Fe^{3+} present.

The uses of EDTA are not confined to analytical chemistry. It is an effective antidote for heavy metal poisoning. Children who have ingested toxic amounts of lead compounds, which were at one time a major component of the pigment in ordinary house paint, are treated by an intramuscular injection of a solution of EDTA. This brings Pb^{2+} into solution as a complex ion and, hence, leads to its elimination from the body. EDTA has also been used to remove radioactive isotopes of metallic elements, notably plutonium, from body tissues. One of the most dangerous of these isotopes, strontium-90, can be eliminated with a derivative of EDTA known as "BAETA," which is unusual in that it forms a more stable chelate with strontium than with calcium.

PROBLEMS

21.1 *Structures of Coordination Compounds* Cr^{3+} forms the coordination compound $Cr(NH_3)_4Br_3$, in which the complex ion has an octahedral geometry. Identify the complex ion, give its charge, and cite a simple salt which would show similar conductivity behavior in water solution.

21.2 *Geometry of Complexes* Using Table 21.3, sketch structures of $Cu(H_2O)_2^+$, $Zn(H_2O)_2Br_2$, $Cu(H_2O)_2Br_2$, and $Fe(H_2O)_5Br^{2+}$. Include all isomers.

21.3 *Atomic Orbital Model* Write the orbital diagram for $Fe(H_2O)_5Br^{2+}$.

21.4 *Crystal Field* Indicate the high-spin and low-spin forms for Fe^{3+}.

21.5 *Complex-Ion Equilibria* What is the ratio: $[Zn^{2+}]/[Zn(NH_3)_4^{2+}]$ in dilute NH_3 (6*M*)? (K_d of $Zn(NH_3)_4^{2+} = 1 \times 10^{-9}$).

21.6 Consider the complex ion $Fe(H_2O)_4(OH)_2^+$.

a. Identify the ligands and give their charges.
b. What is the charge of the central ion?
c. What is the formula of the nitrate salt of this ion?

21.7 Match each compound at the left with the one at the right which would approach it most closely in molar conductivity.

a. $[Co(NH_3)_6]Cl_3$	CH_3OH
b. $[Co(NH_3)_5Cl]Cl$	KNO_3
c. $K_2[PtCl_4]$	$Mg(NO_3)_2$
d. $[Ag(NH_3)_2]NO_3$	$Al(NO_3)_3$
e. $[Cr(H_2O)_3Cl_3]$	$Al_2(SO_4)_3$
f. $Na[Co(NH_3)_2Cl_4]$	$MgSO_4$

21.25 Consider the coordination compound:

$$[Fe(H_2O)_5Cl]Br.$$

a. Identify the ligands and their charges.
b. What is the charge of the central ion?
c. What would happen if this compound were treated with $AgNO_3$ at 0°C?

21.26 Match the following compounds as in Problem 21.7.

a. $[Zn(NH_3)_4]SO_4$	CH_3OH
b. $[Co(NH_3)_5Cl]Cl_2$	KNO_3
c. $[Co(NH_3)_6]_2(CO_3)_3$	$Mg(NO_3)_2$
d. $Li[Pt(NH_3)Cl_3]$	$Al(NO_3)_3$
e. $[Pt(NH_3)_2Cl_2]$	$Al_2(SO_4)_3$
f. $[Co(NH_3)_5CO_3]Br$	$MgSO_4$

21.8 Each of the following compounds has a molar conductivity comparable to NaCl. Identify the complex ion present in each and state the number of moles of AgCl formed at 0°C per mole of compound.

a. $Cr(NH_3)_4Cl_3$
b. $Pd(NH_3)KCl_3$
c. $Pt(en)NH_3Cl_4$
d. $NaPt(OH)_3H_2O$

21.9 Which of the following would you expect to be effective chelating agents?

a. H_2O
b. hydrogen peroxide, HO—OH
c. $H_2N—CH_2—CH_2—CH_2—NH_2$
d. $(CH_3)_2N—NH_2$

21.10 What is the percentage by mass of iron in heme (Figure 21.1)?

21.11 Sketch the geometry of

a. $Cr(H_2O)_3Cl_3$
b. $Zn(H_2O)_3(OH)^+$ (tetrahedral)
c. *cis*-$Ni(H_2O)_2Cl_2$
d. *trans*-$Fe(H_2O)_4Cl_2^+$
e. $Co(en)_3^{3+}$
f. *cis*-$Cr(en)_2Cl_2^+$

21.12 Draw structures for all isomers of the following:

a. $Cr(H_2O)_3Cl_3$
b. $Co(NH_3)_2(en)Cl_2^+$
c. $Fe(H_2O)_4(OH)_2$
d. $Pt(en)Cl_2$ (square planar)
e. $Zn(en)Cl_2$ (tetrahedral)

21.13 Draw as many structures as possible for complex ions which form compounds of formula $Co(NH_3)_4Cl_2Br$.

21.14 Using the valence-bond model, draw orbital diagrams to indicate the electronic structure around the central ion in

a. $Co(NH_3)_6^{3+}$
b. $Co(en)_3^{3+}$
c. $Co(en)_3^{2+}$
d. $Fe(CN)_6^{4-}$
e. $Fe(CN)_6^{3-}$
f. $Mn(H_2O)_6^{2+}$

21.15 Using the valence-bond model, draw orbital diagrams for square planar, tetrahedral, and octahedral complexes of Ni^{2+}.

21.16 Using crystal-field theory, draw electronic structures for the high-spin and low-spin forms of the octahedral complexes listed in Problem 21.14.

21.27 Give the formulas, using only H_2O or Cl^- as ligands, of

a. an octahedral complex of Ni^{2+} which is a nonelectrolyte.
b. two square planar complexes of Ni^{2+} with molar conductivities similar to NaCl.
c. the complex ion in $NiCl_2 \cdot 5\ H_2O$.
d. the complex ion in $NiCl_2 \cdot 3\ H_2O$.

21.28 Consider the following complexes, all of which have either square planar or octahedral geometry. In which is CO_3^{2-} acting as a chelating agent?

a. $Co(NH_3)_5CO_3^+$
b. $Co(NH_3)_4CO_3^+$
c. $Pt(en)CO_3$
d. $Pt(en)NH_3CO_3$

21.29 Every time you breathe, you take in about 400 cm³ of air (21% O_2 by volume) at approximately 20°C and one atmosphere pressure. If all the oxygen in this air entered the bloodstream, how many molecules of hemoglobin would be involved?

21.30 Sketch the geometry of

a. $Cr(NH_3)_5Br^{2+}$
b. $Cu(H_2O)_3Cl^+$ (square planar)
c. *cis*-$Co(NH_3)_4BrCl^+$
d. *trans*-$Ni(H_2O)_2Cl_2$
e. *trans*-$Cr(en)_2Cl_2^+$
f. $Pt(en)_2^{2+}$ (square planar)

21.31 How many isomers are there for complexes with the following formulas (M = metal atom; A, B, C = ligands):

a. MA_4BC
b. MA_3B_2C
c. $MA_2B_2C_2$
d. MA_2BC (tetrahedral)
e. MA_2BC (square planar)

21.32 How many different octahedral complexes of Co^{3+} can you write using only ethylenediamine and/or Cl^- as ligands?

21.33 Follow the directions of Problem 21.14 for the following complexes:

a. $Ag(NH_3)_2^+$
b. $Cu(NH_3)_4^{2+}$
c. $Co(NH_3)_4Cl_2^+$
d. $Co(NH_3)_6^{2+}$
e. $Mn(H_2O)_6^{3+}$

21.34 Using the valence-bond model, list all the cations (+1, +2, or +3) in the first transition series which could form complexes with a noble-gas structure (36 e⁻). Consider coordination numbers of 2, 4, and 6.

21.35 Consider transition metal ions which have from one to ten d electrons. Which of these can have "high-spin" and "low-spin" forms in octahedral complexes?

21.17 Based upon Table 21.5, what do you predict to be the color of the $Co(en)_3^{3+}$ ion?

21.18 Consider the reaction $Co(NH_3)_5CN^{2+}(aq) + H_2O \rightarrow Co(NH_3)_5H_2O^{3+}(aq) + CN^-(aq)$. Suggest at least three different ways in which you might measure the rate of this reaction.

21.19 Using the data in Table 21.6, calculate the ratio: $[Zn^{2+}]/[Zn(OH)_4^{2-}]$ at *pH* 1, 7, and 10.

21.20 Using Tables 21.6 and 18.2, calculate K for the reaction $AgCl(s) + Cl^-(aq) \rightarrow AgCl_2^-(aq)$. How would you expect the solubility of AgCl in 1 M HCl to compare to that in 1 M NH_3?

21.21 Consider the reaction $AgBr(s) + 2\ S_2O_3^{2-}(aq) \rightarrow Ag(S_2O_3)_2^{3-}(aq) + Br^-(aq)$, where $K = 5$. What is the solubility of AgBr in 0.1 M $Na_2S_2O_3$?

21.22 Write balanced equations to explain why

a. silver iodide dissolves in NaCN.
b. zinc hydroxide dissolves in NH_3.
c. AgBr dissolves in NH_3 but reprecipitates when the solution is acidified (two equations).

21.23 Suggest a suitable reagent to bring each of the following compounds into solution. (Consider acid-base reactions as well as complex ion formation.)

a. $Zn(OH)_2$
b. AgBr
c. $Cr(OH)_3$
d. $Fe(OH)_3$

21.24 How many cm^3 of 0.20 M EDTA should be injected into the soil to bring the iron in one gram of Fe_2O_3 into solution?

21.36 Drying agents which are used to remove water vapor from air are often coated with $CoCl_2$, which is blue when anhydrous and red or pink when hydrated. Explain this color change in terms of crystal-field splitting.

21.37 Devise a lecture demonstration to illustrate the lability of the $Cu(NH_3)_4^{2+}$ ion as compared to the inertness of the $Pt(NH_3)_4^{2+}$ ion.

21.38 At what concentration of $S_2O_3^{2-}$ would 99% of the Ag^+ in a solution be converted to $Ag(S_2O_3)_2^{3-}$?

21.39 Calculate K for the reaction $CuS(s) + 4\ NH_3(aq) \rightarrow Cu(NH_3)_4^{2+}(aq) + S^{2-}(aq)$. Comment on the effectiveness of aqueous ammonia as a solvent for CuS.

21.40 Explain how, by measuring the solubility of AgCl in 1 M KSCN, it would be possible to calculate the dissociation constant for the $Ag(SCN)_2^-$ ion.

21.41 Write balanced equations to explain why

a. a color change occurs when HCl is added to a solution of $Cu(NH_3)_4^{2+}$.
b. Zn^{2+} can be separated from Fe^{3+} by treatment with excess OH^- (two equations).
c. addition of NaSCN to a solution of $FeCl_3$ gives a blood-red color.

21.42 How would you accomplish the following?

a. $Zn(OH)_2(s) \rightarrow Zn(NH_3)_4^{2+}(aq)$
b. $Zn(NH_3)_4^{2+}(aq) \rightarrow Zn(H_2O)_4^{2+}(aq)$
c. $Zn(H_2O)_4^{2+}(aq) \rightarrow Zn(OH)_2(s)$
d. $Zn(OH)_2(s) \rightarrow Zn(OH)_4^{2-}(aq)$

21.43 If a child eats 100 g of paint containing 10% lead, how many grams of EDTA (Na salt) should he receive to bring the lead into solution?

*21.44 Consider the equilibrium $Zn(NH_3)_4^{2+}(aq) + 4\ OH^-(aq) \rightarrow Zn(OH)_4^{2-}(aq) + 4\ NH_3(aq)$.

a. Calculate K for this reaction.
b. What is the ratio $[Zn(NH_3)_4^{2+}]/[Zn(OH)_4^{2-}]$ in a solution 1 M in NH_3 ($K_b = 1.8 \times 10^{-5}$)?

*21.45 A solution containing Ni^{2+} and Al^{3+} ions is treated with aqueous ammonia. A colored precipitate forms at first; as more ammonia is added, part of the precipitate dissolves to form a deep blue solution. The precipitate that remains is white; upon treatment with excess OH^-, it forms a clear solution. If acid is slowly added to this solution, a white precipitate forms, which dissolves as more acid is added. Write balanced net ionic equations for each reaction that took place.

*21.46 In the $Ti(H_2O)_6^{3+}$ ion, the splitting between the d levels is 230 kJ/mol. What is the color of this ion?

22
OXIDATION AND REDUCTION: ELECTROCHEMICAL CELLS

Many chemical reactions involve the transfer of electrons from one species to another. A simple example is the formation of Na^+ and F^- ions from the corresponding atoms. The electron transfer is perhaps most simply shown by using electron-dot notation (Chapter 8):

$$Na\cdot + \cdot\ddot{\underset{..}{F}}: \rightarrow Na^+ + \left(:\ddot{\underset{..}{F}}:\right)^- \quad (22.1)$$

It is clear that in this process a sodium atom has transferred an electron to a fluorine atom.

$$Na\cdot \rightarrow Na^+ + e^- \quad (22.1a)$$

$$\cdot\ddot{\underset{..}{F}}: + e^- \rightarrow \left(:\ddot{\underset{..}{F}}:\right)^- \quad (22.1b)$$

Any reaction between atoms which leads to the formation of ions may be analyzed similarly. For example, the reaction between magnesium and oxygen atoms

$$\cdot Mg\cdot + :\dot{\underset{\cdot}{O}}: \rightarrow Mg^{2+} + \left(:\ddot{\underset{..}{O}}:\right)^{2-} \quad (22.2)$$

may be broken down into two half-reactions:

$$\cdot Mg\cdot \rightarrow Mg^{2+} + 2\ e^- \quad (22.2a)$$

and

$$:\dot{\underset{\cdot}{O}}: + 2\ e^- \rightarrow \left(:\ddot{\underset{..}{O}}:\right)^{2-} \quad (22.2b)$$

Processes such as 22.1a and 22.2a, which involve the **loss of electrons**, are referred to as **oxidation** half-reactions; atoms of sodium and magnesium lose electrons to become positively charged ions. The processes represented by 22.1b and 22.2b, which involve the **gain of electrons**, are referred to as **reduction** half-reactions; in gaining electrons atoms of fluorine and oxygen are said to be reduced. The overall Reactions 22.1 and 22.2, in which oxidation and reduction occur simultaneously, are called **oxidation-reduction** reactions or simply ''redox'' reactions.

Since no electrons are created or destroyed, in an oxidation-reduction reaction there can be no net gain or loss of electrons. For example, when lithium reacts with oxygen, two lithium atoms are oxidized to Li^+ for every oxygen atom reduced to O^{2-}.

All electrons produced by oxidation of one species are used up in the reduction of another

oxidation: $$2\ Li\cdot \rightarrow 2\ Li^+ + 2\ e^- \quad (22.3a)$$

reduction: $$:\dot{\underset{\cdot}{O}}: + 2\ e^- \rightarrow \left(:\ddot{\underset{\cdot\cdot}{O}}:\right)^{2-} \quad (22.3b)$$

overall reaction: $$2\ Li\cdot + :\dot{\underset{\cdot}{O}}: \rightarrow 2\ Li^+ + \left(:\ddot{\underset{\cdot\cdot}{O}}:\right)^{2-} \quad (22.3)$$

We shall shortly find that it is possible to broaden the meaning of the terms oxidation and reduction so that they can be applied to a wide variety of reactions, including many in which no ionic species are involved. To accomplish this, it is necessary to introduce a new concept—oxidation number.

22.1 OXIDATION NUMBER

The chemical equation written for the reaction between hydrogen and fluorine

$$\tfrac{1}{2}\ H_2(g) + \tfrac{1}{2}\ F_2(g) \rightarrow HF(g) \quad (22.4)$$

resembles that for the reaction of sodium with fluorine

$$Na(s) + \tfrac{1}{2}\ F_2(g) \rightarrow NaF(s)$$

Indeed, the two reactions themselves have much in common. In both there is an exchange of electrons between atoms; the major difference lies in the extent to which electrons are transferred. In Reaction 22.4 the product is a molecule, HF, rather than a pair of ions (Na^+, F^-). The valence electron of hydrogen is shared with fluorine rather than being transferred to it. This distinction, however, is one of degree rather than kind. The electrons in the H—F covalent bond are displaced strongly toward fluorine. So far as "electron bookkeeping" is concerned, it would seem reasonable to assign these electrons to the fluorine atom

$$H\,\Big|\,:\ddot{\underset{\cdot\cdot}{F}}:$$

By assigning electrons in this way we have, in a sense, given a -1 charge to fluorine which now has one more valence electron (8) than an isolated fluorine atom (7). The hydrogen atom, stripped of its valence electron by this assignment, has in effect acquired a $+1$ charge.

The accounting system that we have just illustrated is widely used in inorganic chemistry. The concept of **oxidation number** is introduced to refer to the charge an atom would have if, as here, the bonding electrons were assigned arbitrarily to the more electronegative element. In the HF molecule, hydrogen is said to have an oxidation number of $+1$, fluorine an oxidation number of -1. In water the bonding electrons are assigned to the more electronegative oxygen atom:

$$H\,\Big|\,:\ddot{\underset{\cdot\cdot}{O}}:\,\Big|\,H$$

This gives oxygen an oxidation number of -2 (8 valence e^- vs. 6 e^- in the neutral atom) and hydrogen an oxidation number of $+1$ (0 e^- vs. 1 e^- in the neutral atom).

In a nonpolar covalent bond, the bonding electrons are split evenly between the two atoms:

$$:\ddot{\underset{..}{F}}\cdot \Big| \cdot\ddot{\underset{..}{F}}: \qquad \text{oxidation no. F} = 0$$

It should be emphasized that the oxidation number of an atom in a covalently bonded substance is an artificial concept. Unlike the charge of an ion, the oxidation number of an element cannot be determined experimentally. The hydrogen atom in the HF or H_2O molecule does not carry a full positive charge; its oxidation number of +1 in these molecules may be regarded as a "pseudocharge."

Rules for Assigning Oxidation Numbers

In principle, oxidation numbers could be determined in any species by assigning bonding electrons in the manner just described. However, such a method would be tedious at best and would require that we know the Lewis structure of the species. In practice, oxidation numbers are ordinarily obtained in a much simpler way, applying certain arbitrary rules. Essentially, there are four such rules:

1. **The oxidation number of an element in an elementary substance is 0.** For example, the oxidation number of chlorine in Cl_2 or of phosphorus in P_4 is zero.

2. **The oxidation number of an element in a monatomic ion is equal to the charge of that ion.** In the ionic compound NaCl, sodium has an oxidation number of +1, chlorine an oxidation number of −1. The oxidation numbers of aluminum and oxygen in Al_2O_3 (Al^{3+}, O^{2-} ions) are +3 and −2, respectively.

3. **Certain elements have the same oxidation number in all or almost all their compounds.** The 1A metals always exist as +1 ions in their compounds and, hence, are assigned an oxidation number of +1. By the same token, the 2A elements always have oxidation numbers of +2 in their compounds. Fluorine, the most electronegative of all elements, has an oxidation number of −1 in all of its compounds.

Oxygen is ordinarily assigned an oxidation number of −2 in its compounds. (An exception arises in compounds containing the peroxide ion, O_2^{2-}, where the oxidation number of oxygen is −1.)

Hydrogen in its compounds ordinarily has an oxidation number of +1. (The only exception is in metal hydrides such as NaH and CaH_2, where hydrogen is present as the H^- ion and hence is assigned an oxidation number of −1.)

4. **The sum of the oxidation numbers of all the atoms in a neutral species is 0; in an ion, it is equal to the charge of that ion.** The application of this very useful principle is illustrated in Example 22.1.

Example 22.1 What is the oxidation number of selenium in Na_2Se? of manganese in MnO_4^-?

Solution For Na_2Se, knowing that the oxidation number of sodium must be +1, we have:

$$2(+1) + \text{oxid no. Se} = 0; \text{ oxid no. Se} = -2$$

In the MnO_4^- ion, taking the oxidation number of oxygen to be −2 and realizing that the sum must be −1:

$$\text{oxid no. Mn} + 4(-2) = -1; \text{ oxid no. Mn} = +7$$

The common oxidation numbers of the elements are tabulated in Figure 22.1. It may be helpful to point out some general principles that are perhaps hidden in this maze of numbers.

1. The metallic elements show only positive oxidation numbers in the compounds they form with nonmetals. This reflects the fact that metals tend to lose rather than to gain electrons when they react with nonmetals. Negative oxidation numbers are found with only a few elements, all of which are strongly electronegative nonmetals. For these elements the lowest oxidation number is equal to the charge of their monatomic anion (N^{3-}, O^{2-}, F^-).

2. As pointed out previously, the 1A and 2A metals show only one oxidation number (1A = +1, 2A = +2). In contrast, transition metals commonly show a variety of oxidation numbers. As we move down a given group in the transition series, higher oxidation numbers become more stable. For example, nickel rarely shows an oxidation number greater than +2, while the lower members of this

Can you think of a compound in which N has an oxidation number of −3?

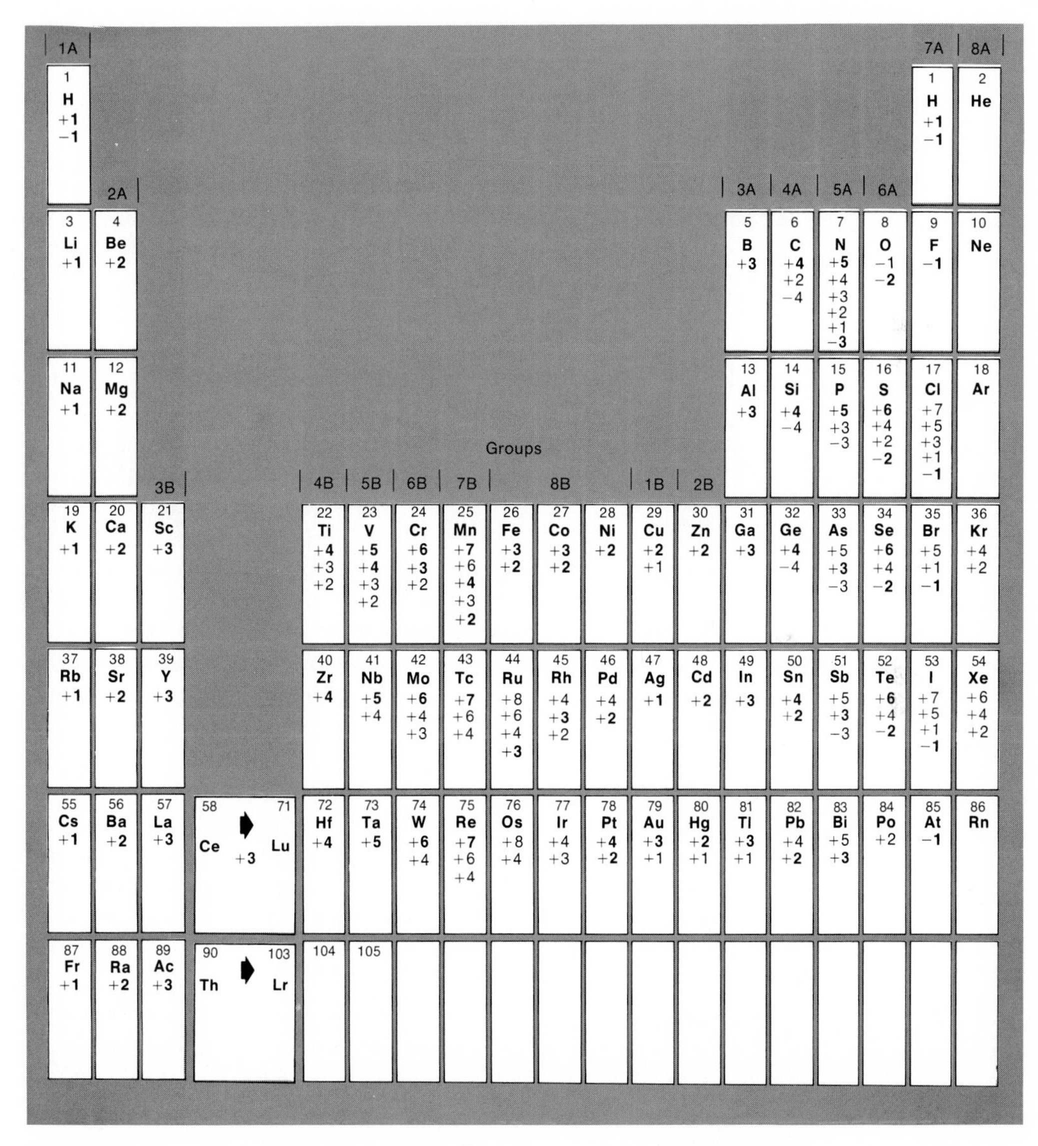

Figure 22.1 Oxidation states of the elements. The most common or stable states are shown in heavy type. Some elements have as many as six oxidation states.

group, Pd and Pt, form a great many stable compounds in which their oxidation number is +4.

3. With some exceptions, the maximum oxidation number of an element is given by its group number in the Periodic Table. In a few families the elements achieve this oxidation number by forming monatomic ions (1A, 2A, 3B). More frequently, elements show their maximum oxidation numbers only in compounds in which they are covalently bonded to oxygen or another highly electronegative element. For example, sulfur has a +6 oxidation number in the species SO_3, H_2SO_4, SO_4^{2-}, and SF_6. We find "+7 chlorine" in the perchlorate ion, ClO_4^-, in which the bonding electrons have all been arbitrarily assigned to oxygen.

$$\left[\begin{array}{ccc} & :\ddot{\underset{..}{O}}: & \\ :\ddot{\underset{..}{O}}: & \boxed{Cl} & :\ddot{\underset{..}{O}}: \\ & :\ddot{\underset{..}{O}}: & \end{array}\right]^-$$

Oxidation and Reduction: A Working Definition

The concept of oxidation number leads directly to a working definition of the terms oxidation and reduction. **Oxidation** is defined as an **increase in oxidation number, reduction** as a **decrease in oxidation number.** Reactions in which one element increases in oxidation number at the expense of another are referred to as oxidation-reduction reactions. Two simple examples follow:

$$2\ Al(s) + 3\ Cl_2(g) \rightarrow 2\ AlCl_3(s) \quad \begin{array}{r} \text{Al oxidized (oxidation no. } 0 \rightarrow +3) \\ \text{Cl reduced (oxidation no. } 0 \rightarrow -1) \end{array} \quad (22.5)$$

$$4\ As(s) + 5\ O_2(g) \rightarrow 2\ As_2O_5(s) \quad \begin{array}{r} \text{As oxidized (oxidation no. } 0 \rightarrow +5) \\ \text{O reduced (oxidation no. } 0 \rightarrow -2) \end{array} \quad (22.6)$$

These definitions are compatible, of course, with our earlier interpretation of oxidation and reduction in terms of loss and gain of electrons. An element which loses electrons inevitably increases in oxidation number; the gain of electrons always results in a decrease in oxidation number. By defining oxidation and reduction in terms of changes in oxidation number, we greatly simplify the electron bookkeeping in redox reactions. For example, analysis of the reaction

$$HCl(g) + HNO_3(l) \rightarrow NO_2(g) + \tfrac{1}{2}\ Cl_2(g) + H_2O(l) \qquad (22.7)$$

in terms of oxidation numbers reveals immediately that chlorine is oxidized (oxidation no. = −1 in HCl, 0 in Cl_2) while nitrogen is reduced (oxidation no. = +5 in HNO_3, +4 in NO_2). It is much more difficult to decide precisely which atoms are "losing" or "gaining" electrons.

In a redox reaction, the oxidizing *agent* is *reduced;* the reducing *agent* is *oxidized*

In discussing oxidation-reduction reactions the phrases **oxidizing agent** and **reducing agent** are frequently used to refer to the species responsible for oxidation and reduction. We speak of Cl_2 in Reaction 22.5 as an oxidizing agent, since it brings about the oxidation of aluminum from 0 to +3. By the same token, Al in this reaction is a reducing agent, being responsible for the reduction of chlorine from an oxidation number of 0 to −1. Similarly:

Reaction	Oxidizing agent	Reducing agent
22.6	$O_2(g)$	As(s)
22.7	$HNO_3(l)$	HCl(g)

22.2 BALANCING OXIDATION-REDUCTION EQUATIONS

Many of the equations that we write to represent oxidation-reduction reactions are simple enough to be balanced by inspection. Such is clearly the case, for example, with Equations 22.4 through 22.6. Frequently, however, we are confronted with a rather more complicated redox reaction where the coefficients in the balanced equation are by no means obvious. We will now consider a general approach to balancing such equations, particularly suited to redox reactions taking place in aqueous solution.

Half-Equation Method

A straightforward way of balancing an oxidation-reduction equation in aqueous solution involves, as a first step, breaking the equation down into an oxidation half-equation and a reduction half-equation. The two half-equations are balanced separately and then combined in such a way as to arrive at an overall equation in which there is no net change in the number of electrons. To illustrate the steps involved, let us start with a relatively simple reaction, that which occurs when a direct electric current is passed through a water solution of $FeCl_3$. The products are the elementary substances iron and chlorine. The unbalanced equation for the reaction is

You may be able to balance this equation by inspection, but it illustrates the approach used with more complex equations

$$Fe^{3+}(aq) + Cl^-(aq) \rightarrow Fe(s) + Cl_2(g)$$

To balance this equation, we proceed as follows:

1. *Split the equation into two half-equations,* one oxidation and one reduction:

reduction: $$Fe^{3+}(aq) \rightarrow Fe(s) \qquad (1a)$$

oxidation: $$Cl^-(aq) \rightarrow Cl_2(g) \qquad (1b)$$

2. *Balance these half-equations, first with respect to mass and then with respect to charge:*

Equation 1a is balanced insofar as mass is concerned, since there is one atom of Fe on both sides. The charges, however, are unbalanced: the Fe atom on the right has 0 charge while the Fe^{3+} ion on the left has a charge of +3. To correct this, we add three electrons to the left of 1a, arriving at

$$Fe^{3+}(aq) + 3\ e^- \rightarrow Fe(s) \qquad (2a)$$

Equation 1b must first be balanced with respect to mass by providing two Cl^- ions to give one molecule of Cl_2:

$$2\ Cl^-(aq) \rightarrow Cl_2(g)$$

To balance charges, two electrons must be added to the right, giving a charge of −2 on both sides:

$$2\ Cl^-(aq) \rightarrow Cl_2(g) + 2\ e^- \qquad (2b)$$

3. Having arrived at two balanced half-equations, *combine them so as to make the number of electrons gained in reduction equal to the number lost in oxidation.*

In Equation 2a, three electrons are gained; in 2b, two electrons are given off. To arrive at a final equation in which no electrons appear, multiply 2a by 2, 2b by 3, and add:

2 × 2a: $$2\ Fe^{3+}(aq) + \not{6}\ e^- \rightarrow 2\ Fe(s) \qquad (3a)$$

3 × 2b: $$6\ Cl^-(aq) \rightarrow 3\ Cl_2(g) + \not{6}\ e^- \qquad (3b)$$

$$2\ Fe^{3+}(aq) + 6\ Cl^-(aq) \rightarrow 2\ Fe(s) + 3\ Cl_2(g) \qquad (3)$$

The equation just balanced corresponds to the simplest type of oxidation-reduction reaction in which only two elements are involved, the one that is reduced and the one that is oxidized. The equation is more difficult to balance and the method just described more pertinent when atoms of elements other than those being oxidized or reduced are involved in the overall reaction. The method we have outlined is readily extended to such cases.

For oxidation-reduction reactions in water solution, the two most common "extra" elements (elements whose atoms undergo no change in oxidation number) are hydrogen and oxygen. Compounds containing these elements take part in a great many oxidation-reduction reactions, often without any change in the oxidation number of H (+1) or oxygen (−2).

To illustrate the balancing of an oxidation-reduction equation in which "extra" elements appear, consider the reaction that occurs between chloride and permanganate ions in acidic solution. Experimental evidence indicates that this reaction can best be represented by the equation

$$MnO_4^-(aq) + H^+(aq) + Cl^-(aq) \rightarrow Mn^{2+}(aq) + Cl_2(g) + H_2O$$

Note that the two elements that undergo a change in oxidation number are manganese (+7 → +2) and chlorine (−1 → 0). Neither hydrogen nor oxygen change oxidation number, yet atoms of these elements participate in the reaction. The oxygen atoms tied up originally in the MnO_4^- ion end up as H_2O molecules; the H^+ ions meet the same fate.

To balance this equation, we proceed as follows:

1. Separate into two half-equations:

oxidation: $$Cl^-(aq) \rightarrow Cl_2(g) \qquad (1a)$$

reduction: $$MnO_4^-(aq) + H^+(aq) \rightarrow Mn^{2+}(aq) + H_2O \qquad (1b)$$

2. Half-equation 1a is readily balanced as before, giving us

$$2\ Cl^-(aq) \rightarrow Cl_2(g) + 2\ e^- \qquad (2a)$$

To balance 1b, we first make sure that there are the same number of Mn atoms on both sides, one. Next, the oxygen is balanced by writing a coefficient of 4 in front of the H_2O on the right to account for the four oxygens in the MnO_4^- ion:

$$MnO_4^-(aq) + H^+(aq) \rightarrow Mn^{2+}(aq) + 4\ H_2O$$

To complete the mass balance, the number of hydrogen atoms on the two sides must be equalized. The four H_2O molecules on the right contain eight hydrogen atoms; there must then be eight H^+ ions on the left:

$$MnO_4^-(aq) + 8\ H^+(aq) \rightarrow Mn^{2+}(aq) + 4\ H_2O$$

Finally, the charges must be balanced; at the moment there is a charge of +2 on the right and +7 on the left (−1 + 8). To balance, five electrons are added to the left:

$$MnO_4^-(aq) + 8\ H^+(aq) + 5\ e^- \rightarrow Mn^{2+}(aq) + 4\ H_2O \qquad (2b)$$

3. Half-equations 2a and 2b are now combined as usual so as to eliminate electrons from the final equation. To do this, multiply 2a by 5 and 2b by 2, producing 10 e^- on both sides:

$$5 \times 2a: \quad 10\ Cl^-(aq) \rightarrow 5\ Cl_2(g) + \cancel{10}\ e^- \qquad (3a)$$

$$2 \times 2b: \quad 2\ MnO_4^-(aq) + 16\ H^+(aq) + \cancel{10}\ e^- \rightarrow 2\ Mn^{2+}(aq) + 8\ H_2O \qquad (3b)$$

$$2\ MnO_4^-(aq) + 16\ H^+(aq) + 10\ Cl^-(aq) \rightarrow 2\ Mn^{2+}(aq) + 8\ H_2O + 5\ Cl_2(g) \quad (3)$$

We frequently have occasion to write balanced equations for oxidation-reduction reactions taking place in basic solution. For such reactions, it would be inappropriate to write equations in which H^+ ions appear, since this ion is present in only very small concentrations in basic solution. Instead, hydrogen in such equations should be in the form of OH^- ions or H_2O molecules. A simple way to accomplish this is to eliminate any H^+ ions appearing in the half-equations, "neutralizing" them by adding an equal number of OH^- ions to both sides. To illustrate, consider the oxidation, in basic solution, of iodide by permanganate ions:

$$I^-(aq) + MnO_4^-(aq) \rightarrow I_2(aq) + MnO_2(s) \text{ (basic solution)}$$

One can proceed exactly as in the foregoing example, to obtain the half-equations

$$\text{oxidation:} \quad 2\ I^-(aq) \rightarrow I_2(aq) + 2\ e^- \qquad (2a)$$

$$\text{reduction:} \quad MnO_4^-(aq) + 4\ H^+(aq) + 3\ e^- \rightarrow MnO_2(s) + 2\ H_2O \qquad (2b)$$

The H^+ ions appearing in the reduction half-equation must now be removed to obtain an equation valid in basic solution. To do this, four OH^- ions are added to both sides, and water is formed on the left by combining H^+ with OH^- ions.

Alternatively, you could add OH^- ions to the final equation

$$MnO_4^-(aq) + 4\ H^+(aq) + 3\ e^- \rightarrow MnO_2(s) + 2\ H_2O$$

$$+\ 4\ OH^-(aq) \rightarrow \qquad +\ 4\ OH^-(aq)$$

$$MnO_4^-(aq) + 4\ H_2O + 3\ e^- \rightarrow MnO_2(s) + 2\ H_2O + 4\ OH^-(aq)$$

Eliminating two water molecules from each side, we arrive at

$$MnO_4^-(aq) + 2\ H_2O + 3\ e^- \rightarrow MnO_2(s) + 4\ OH^-(aq) \qquad (2b')$$

for the reduction half-reaction in basic solution. To obtain the overall equation, we proceed as before, combining 2a and 2b′ in such a way as to make the electron gain equal the electron loss:

$$3 \times 2a: \quad 6\ I^-(aq) \rightarrow 3\ I_2(aq) + \not{6}\ e^- \qquad (3a)$$

$$2 \times 2b': 2\ MnO_4^-(aq) + 4\ H_2O + \not{6}\ e^- \rightarrow 2\ MnO_2(s) + 8\ OH^-(aq) \qquad (3b)$$

$$6\ I^-(aq) + 2\ MnO_4^-(aq) + 4\ H_2O \rightarrow 3\ I_2(aq) + 2\ MnO_2(s) + 8\ OH^-(aq) \qquad (3)$$

Oxidation Number Method of Balancing Equations

The **half-equation** method just described is by no means the only method of balancing oxidation-reduction equations. We have stressed this particular method because it is valuable in studying other aspects of oxidation-reduction reactions. Of the various other methods which can be used to balance redox equations, we shall mention only one—the **oxidation number** method.

To illustrate the application of this method, consider the equation referred to earlier:

$$MnO_4^-(aq) + Cl^-(aq) + H^+(aq) \rightarrow Mn^{2+}(aq) + Cl_2(g) + H_2O$$

To balance this equation by the oxidation number method, we proceed as follows:

1. Find the oxidation number of each element on both sides of the equation, thereby determining which elements have undergone oxidation and reduction:

	Oxid No. Reactants	Oxid No. Products	
Mn	+7	+2	reduced
O	−2	−2	
Cl	−1	0	oxidized
H	+1	+1	

2. By adjusting the coefficients of the species being oxidized and reduced, make the total increase in oxidation number equal to the total decrease.

In this case, each Mn atom undergoes a decrease in oxidation number of five units; each Cl atom increases in oxidation number by one unit. To make the increase in oxidation number equal to the decrease, there must be five Cl atoms oxidized for every Mn reduced. Thus

$$MnO_4^-(aq) + 5\ Cl^-(aq) \rightarrow Mn^{2+}(aq) + 5/2\ Cl_2(g)$$

or, multiplying through by two to eliminate fractional coefficients,

$$2\ MnO_4^-(aq) + 10\ Cl^-(aq) \rightarrow 2\ Mn^{2+}(aq) + 5\ Cl_2(g) \qquad (a)$$

Note that for Equation a, the total increase in oxidation number of Cl is $10 \times 1 = 10$; the total decrease in oxidation number of Mn $= 2 \times 5 = 10$.

3. Having determined the coefficients of the species being oxidized and reduced, balance the number of atoms of the remaining elements in the usual manner.

Here, starting with Equation a, oxygen is balanced first. The presence of two MnO_4^- ions on the left, containing a total of eight oxygen atoms, requires that there be eight H_2O molecules, each with one oxygen atom, on the right:

$$2\ MnO_4^-(aq) + 10\ Cl^-(aq) \rightarrow 2\ Mn^{2+}(aq) + 5\ Cl_2(g) + 8\ H_2O \qquad (b)$$

Finally, to balance the hydrogen, 16 H^+ ions must be added to the left:

$$2\ MnO_4^-(aq) + 10\ Cl^-(aq) + 16\ H^+(aq) \rightarrow 2\ Mn^{2+}(aq) + 5\ Cl_2(g) + 8\ H_2O \qquad (c)$$

The final balanced equation is, as it should be, identical to that previously derived by the half-equation method.

By means of a device known as an electrolytic cell, it is possible to use electrical energy to bring about a nonspontaneous oxidation-reduction reaction. To understand how such a cell operates, let us consider the generalized cell diagram shown in Figure 22.2.

The storage battery at the left provides a source of direct electric current; it could be replaced by a simple dry cell or a DC generator. From the terminals of the battery, indicated by + and − signs, two wires lead to the electrolytic cell. This consists of two electrodes, A and C, dipping into a liquid containing ions M^+ and X^-.

By a mechanism which we shall consider in Section 22.4, the battery acts as an electron pump, pushing electrons into C and removing them from A. In order to maintain electrical neutrality, some process must take place within the cell so as to consume electrons at C and liberate them at A. This process is an oxidation-reduction reaction. At electrode C, known as the **cathode,** an ion or molecule undergoes **reduction** by accepting electrons. At the **anode,** A, electrons are produced by the **oxidation** of an ion or molecule. The overall cell reaction is the sum of the two half-reactions occurring at the electrodes. While electrolysis is proceeding, there is a steady flow of ions to the two electrodes. Positive ions (*cations*) move toward the *cathode;* negative ions (*anions*) move toward the *anode*.

In an *electrolytic* cell, the cathode is −, the anode is +

Commercial Cells

In principle, any oxidation-reduction reaction, no matter how nonspontaneous, may be brought about in an electrolytic cell if the applied voltage is great enough. Many of our most important metals and industrial chemicals are prepared by electrolytic processes of this type. We shall consider three of these:

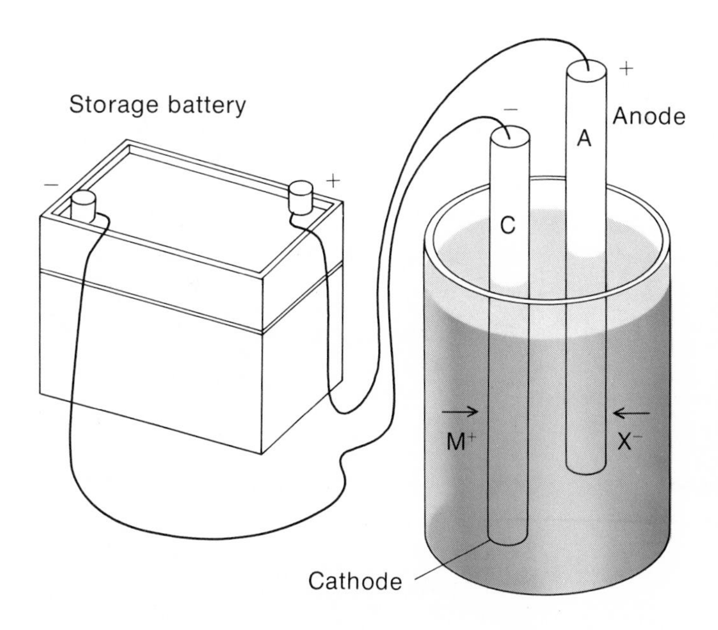

Figure 22.2 Schematic diagram of an electrolytic cell. Reduction occurs at the cathode (C). Oxidation occurs at the anode (A). During electrolysis the cations move toward the cathode, the anions toward the anode.

the production of sodium metal by the electrolysis of molten sodium chloride, the preparation of aluminum from bauxite ore (Al_2O_3), and the production of chlorine, hydrogen, and sodium hydroxide by the electrolysis of a water solution of sodium chloride.

Na FROM NaCl. The so-called Downs cell, used commercially to electrolyze molten sodium chloride, is shown in Figure 22.3. The half-reactions occurring in the cell are particularly simple. At the cylindrical iron *cathode,* sodium ions are reduced to sodium metal:

$$Na^+ + e^- \rightarrow Na(l) \qquad (22.11a)$$

For every sodium ion reduced at the cathode, a chloride ion is oxidized at the graphite *anode:*

$$Cl^- \rightarrow \tfrac{1}{2}\, Cl_2(g) + e^- \qquad (22.11b)$$

The total cell reaction, obtained by summing 22.11a and 22.11b, is

$$NaCl(l) \rightarrow Na(l) + \tfrac{1}{2}\, Cl_2(g) \qquad (22.11)$$

Reaction 22.11 is nonspontaneous; at the operating temperature, 600°C, $\Delta G^{1\,atm}$ is +323 kJ. At least this much energy, in the form of electrical work, must be supplied to decompose one mole of sodium chloride. If the products of the cell reaction, elementary sodium and chlorine, are allowed to come into contact with each other, they will combine spontaneously to give the starting material, sodium chloride. To prevent this, the electrodes in the Downs cell are separated by a circular iron screen, which allows the migration of ions but prevents direct contact between the products of electrolysis.

About 2×10^7 kg of sodium are made annually in the United States by the electrolysis of molten sodium chloride. Its principal use is in the synthesis of organic compounds, particularly tetraethyl lead, the antiknock ingredient of motor fuels. Molten sodium is used as a coolant to remove heat from nuclear reactors; the excellent heat conductivity of sodium and relatively wide liquid range (*mp* = 98°C, *bp* = 880°C) makes it ideal for this purpose. The chlorine formed simultaneously in the Downs process is a valuable by-product. There is, however, a cheaper way of making chlorine electrolytically (see p. 536).

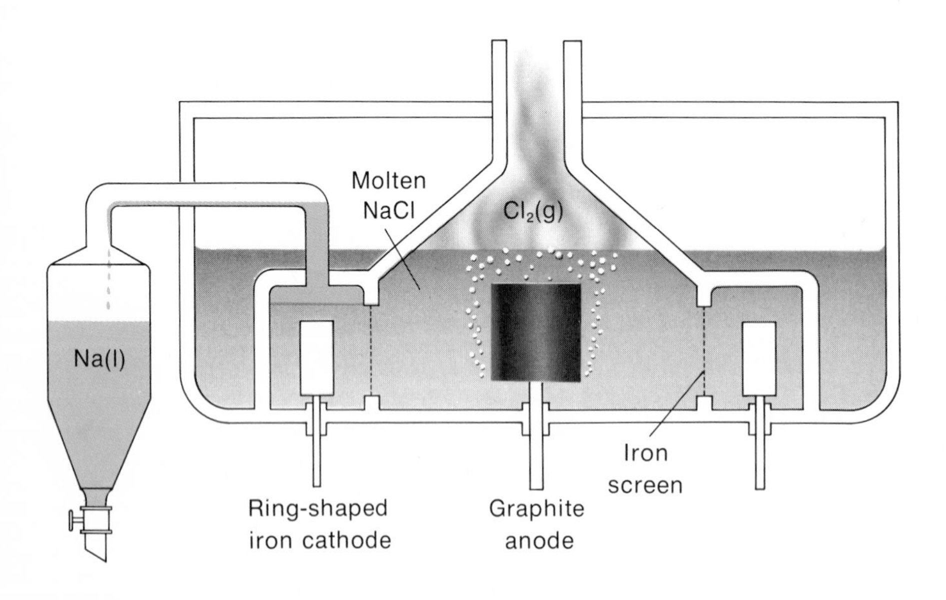

Figure 22.3 Electrolysis of molten sodium chloride. The iron screen is used to prevent sodium and chlorine from coming into contact with each other.

Al FROM Al_2O_3. Aluminum is the third most abundant element in the earth's crust. Its importance as a structural material is indicated by the fact that about 10^{10} kg of aluminum are produced annually (half in the United States and Canada), an amount greater than that of any other metal except copper and iron. Yet, from 1828, when aluminum was first isolated by Wöhler, until 1886, when the Hall electrolytic process for its manufacture was developed, the metal remained little more than a scientific curiosity. Over that entire period, the price of aluminum never fell below $20 a kilogram; today it sells for about $1 a kilogram.

The electrolytic process by which aluminum is produced today from bauxite ore (aluminum oxide) was worked out by Charles Hall, a graduate student at Oberlin College in Ohio. After trying a great many materials, he found that the mineral cryolite, Na_3AlF_6, could be used in the molten state as a solvent for Al_2O_3. The use of cryolite makes it possible to reduce the temperature of electrolysis from 2000°C, the melting point of pure Al_2O_3, to about 1000°C. Curiously enough, within a few weeks of the time that Hall produced his first aluminum, a young Frenchman, Heroult, independently worked out an almost identical process for its manufacture.

Some graduate students are more productive than others

The cell used to produce aluminum from aluminum oxide is shown schematically in Figure 22.4. The bauxite ore, before being admitted to the cell, is purified by a two-step process. It is first dissolved away from impurities (principally oxides of iron) by heating under pressure with a concentrated solution of sodium hydroxide

$$Al_2O_3(s) + 6\ OH^-(aq) + 3\ H_2O \rightarrow 2\ Al(OH)_6^{3-}(aq) \qquad (22.12)$$

and then, after filtering, reprecipitated by adding a weak acid, carbon dioxide, to reverse Reaction 22.12.

$$2\ Al(OH)_6^{3-}(aq) + 6\ CO_2(g) \rightarrow Al_2O_3(s) + 6\ HCO_3^-(aq) + 3\ H_2O \qquad (22.13)$$

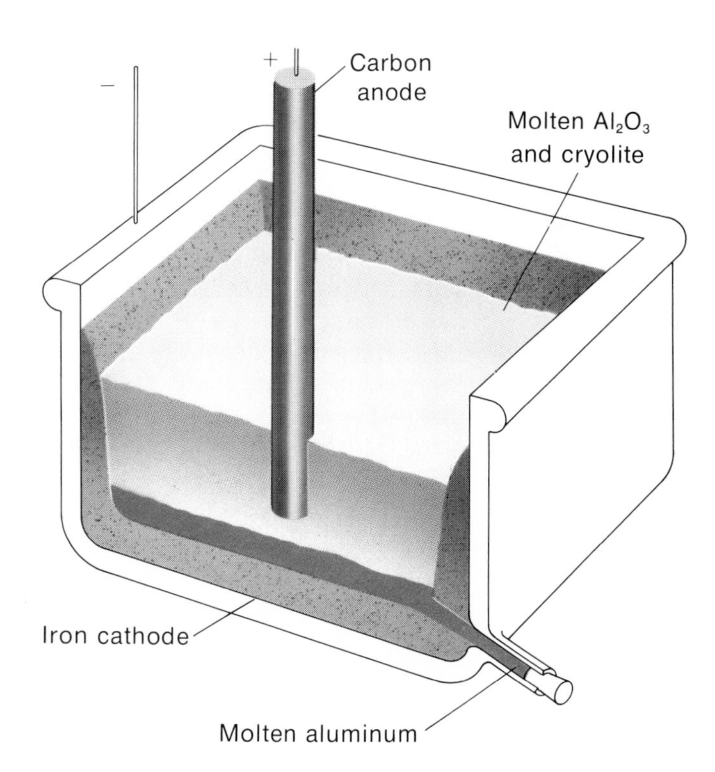

Figure 22.4 Electrolytic preparation of aluminum. Aluminum, being more dense than cryolite, collects at the bottom of the cell and so is protected from oxidation by the air.

The purified ore is then placed in an electrically heated cell, mixed with cryolite,* and melted. The iron wall of the cell serves as the cathode at which Al^{3+} ions are reduced to form molten aluminum. The anodes, retractable carbon rods, are attacked by the oxygen produced in the cell to form a mixture of carbon dioxide and carbon monoxide. The two half-reactions occurring at the electrodes may be represented most simply as

cathode: $$2\ Al^{3+} + 6\ e^- \rightarrow 2\ Al(l) \quad (22.14a)$$

anode: $$3\ O^{2-} \rightarrow \tfrac{3}{2}\ O_2(g) + 6\ e^- \quad (22.14b)$$

$$2\ Al^{3+} + 3\ O^{2-} \rightarrow 2\ Al(s) + \tfrac{3}{2}\ O_2(g) \quad (22.14)$$

The production of 1 kg of aluminum consumes about 2 kg of aluminum oxide, 0.6 kg of anodic carbon, 0.1 kg of cryolite, and about 7.2×10^4 kJ of electrical energy.

Cl_2, NaOH, AND H_2 FROM NaCl(aq). It is ordinarily less expensive and more convenient to carry out an electrolysis in water solution than in the molten state. However, in the presence of water molecules we may obtain quite different products than we get with a melt. As an example, let us consider the electrolysis of a concentrated aqueous solution of sodium chloride, using platinum or graphite electrodes. The anode reaction is the same as that in the Downs cell, i.e.,

$$2\ Cl^-(aq) \rightarrow Cl_2(g) + 2\ e^- \quad (22.15a)$$

The cathode reaction, however, is quite different. Bubbles of hydrogen are given off at this electrode and the solution surrounding it becomes strongly basic. We deduce that water molecules rather than sodium ions are reduced.

$$2\ H_2O + 2\ e^- \rightarrow H_2(g) + 2\ OH^-(aq) \quad (22.15b)$$

The overall cell reaction for the electrolysis of an aqueous solution of sodium chloride is obtained by summing the two half-reactions

$$2\ H_2O + 2\ Cl^-(aq) \rightarrow Cl_2(g) + H_2(g) + 2\ OH^-(aq) \quad (22.15)$$

Notice that one effect of the cell reaction is to replace Cl^- ions by an equal number of OH^- ions. Evaporation of the solution remaining after electrolysis yields solid sodium hydroxide, mixed with some unreacted sodium chloride. Much of the sodium hydroxide and virtually all the chlorine made industrially are prepared by this process; hydrogen is an important by-product.

Reactions analogous to 22.15 occur with many other salts in water solution. For example, electrolysis of a KCl solution gives the same products at the electrodes. In general, whenever the cation of the salt is very difficult to reduce to the metal, as is the case with Na^+ or K^+, water molecules are reduced preferentially at the cathode by Reaction 22.15b.

Most transition metal ions (e.g., Cu^{2+}, Ni^{2+}) are reduced to the metal

When aqueous solutions of certain salts are electrolyzed, it is found that the product at the anode is oxygen gas, produced by the oxidation of water. This commonly happens when the anion of the salt is very difficult to oxidize, as in the

*A mixture of the fluorides of aluminum, sodium, and calcium is now used in place of cryolite. This mixture gives a solution with aluminum oxide which has a lower melting point and density than that obtained with cryolite. The lower density facilitates the separation of the molten aluminum, which sinks to the bottom of the cell.

case of the F^- ion. Consider what happens when a water solution of sodium fluoride is electrolyzed. The anode reaction is

$$H_2O \rightarrow \frac{1}{2}\ O_2(g) + 2\ H^+(aq) + 2\ e^- \quad (22.16a)$$

since H_2O molecules are more readily oxidized than F^- ions. The cathode reaction is identical with 22.15b

$$2\ H_2O + 2\ e^- \rightarrow H_2(g) + 2\ OH^-(aq) \quad (22.16b)$$

since H_2O molecules are easier to reduce than Na^+ ions. The overall reaction, obtained by adding 22.16a to 22.16b and simplifying, is

$$H_2O \rightarrow H_2(g) + \frac{1}{2}\ O_2(g) \quad (22.16)$$

We see that the net result is simply the electrolysis of water to give hydrogen and oxygen; the Na^+ and F^- ions take no part in the reaction.

The electrolysis reaction represented by Equation 22.16 has been proposed as the basis of a ''hydrogen economy'' of the future. As pointed out in Chapter 4, we are rapidly running out of the fossil fuels upon which our present economy is based. All of the potential energy sources of the future—nuclear, solar, geothermal, . . .—involve the generation of electric power at sites far removed from centers of population density. Transmitting electrical energy over long distances by conventional means (e.g., high-voltage transmission lines) is both expensive and wasteful. In principle at least it might be cheaper and more efficient to use the electrical energy where it is generated to electrolyze water. The hydrogen formed could then be pumped through natural gas pipelines to areas where the demand for fuel is great. By burning the hydrogen with air, thereby reversing Reaction 22.16, we could recover as heat virtually all of the energy absorbed in the electrolysis. Alternatively, the hydrogen might be used to regenerate electrical energy, using a fuel cell (Section 22.4).

If solar cells become practical, the sunlight absorbed during the day could be used to produce hydrogen by electrolysis

As a synthetic fuel, hydrogen has one important advantage over the natural fuels now in use. Its combustion product, water, is nonpolluting, in contrast to such substances as CO, unburned hydrocarbons, and SO_2 produced by the combustion of fossil fuels. Indeed, hydrogen is such a clean fuel that it might be possible to use it in a home-heating furnace without a flue, thereby eliminating the need for a chimney. On the other hand, hydrogen has a heating value per unit volume only about 1/3 of that of natural gas. Perhaps more serious, mixtures of hydrogen with air are flammable over wider limits than is the case with natural gas.

Hydrogen has even been suggested as a fuel for motor vehicles. A major drawback here is that the only method now available for storing hydrogen is as a gas under high pressure. An 80-litre tank full of hydrogen at 100 atm pressure would supply only enough energy to drive about 40 km, so frequent refueling would be necessary.

Electroplating

One of the most important applications of electrolytic cells is in the process of electroplating, in which a thin layer of metal (seldom as thick as 0.002 cm) is deposited on an electrically conducting surface. Electroplating is used for many different purposes. Sometimes the reason is to increase the value or improve the appearance of an object, as is the case in gold and silver plating. Chromium plating is designed to provide an attractive shiny surface with improved wearing proper-

TABLE 22.1 ELECTROPLATING PROCESSES

METAL	ANODE	ELECTROLYTE	APPLICATION
Cu	Cu	20% $CuSO_4$, 3% H_2SO_4	electrotype
Ag	Ag	4% AgCN, 4% KCN, 4% K_2CO_3	tableware, jewelry
Au	Au, C, Ni-Cr	3% AuCN, 19% KCN, Na_3PO_4 buffer	jewelry
Cr	Pb	25% CrO_3, 0.25% H_2SO_4	automobile parts
Ni	Ni	30% $NiSO_4$, 2% $NiCl_2$, 1% H_3BO_3	base plate for Cr
Zn	Zn	6% $Zn(CN)_2$, 5% NaCN, 4% NaOH, 1% Na_2CO_3, 0.5% $Al_2(SO_4)_3$	galvanized steel
Sn	Sn	8% H_2SO_4, 3% Sn, 10% cresol-sulfuric acid	"tin" cans

ties. Metals such as zinc or tin are plated over steel to protect it against corrosion. In Table 22.1, we list some of the components of electrolytic cells used to plate various metals.

Copper is also purified electrolytically by this process

One of the simpler electroplating processes is that used with copper (Fig. 22.5). Here, as in most electroplating cells, the metal to be plated is used as the anode (Cu) and the electrolyte contains an ion (Cu^{2+}) derived from that metal. As copper is plated out at the cathode, it goes into solution at the anode

anode: $$Cu(s) \rightarrow Cu^{2+}(aq) + 2\,e^-$$

cathode: $$Cu^{2+}(aq) + 2\,e^- \rightarrow Cu(s)$$

thereby maintaining a constant concentration of Cu^{2+} in the electrolyte solution around the electrodes. Sulfuric acid is added to prevent the solution from becoming basic (Reaction 22.15b), which would contaminate the plated copper with such

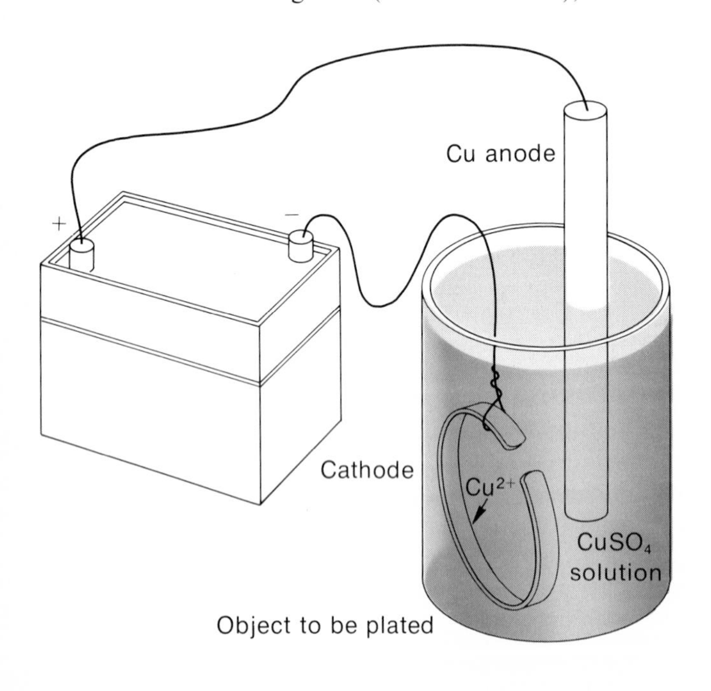

Figure 22.5 Electroplating of copper. The anode is made of pure copper to keep the concentration of Cu^{2+} constant. In this sort of cell there is no net reaction, simply the transfer of copper from one place to another.

compounds as CuO or $Cu(OH)_2$. Traces of organic materials such as glue or gelatin are also added to the solution because it has been found empirically that their presence leads to a more adherent plate.

You will notice from Table 22.1 that metal cyanides are used in many electroplating processes. The CN^- ion acts as a complexing agent to lower the concentration of free metal ion. This tends to prevent the cation from plating too rapidly, which would give a rough or brittle coating. In silver plating the electrode reactions are

Electroplating is still somewhat of an art

anode: $Ag(s) + 2\ CN^-(aq) \rightarrow Ag(CN)_2^-(aq) + e^-$

cathode: $Ag(CN)_2^-(aq) + e^- \rightarrow Ag(s) + 2\ CN^-(aq)$

Solutions containing cyanide ions are extremely toxic: many cases of water pollution have arisen from careless discharge of spent electrolyte by electroplating plants. The problem is magnified by the use of metal cyanides to extract silver and gold from low grade ores. Fish and livestock kills have been traced to streams contaminated by cyanide solutions used in mining operations.

Quantitative Aspects of Electrolysis

From an economic standpoint, one of the most important aspects of an electrochemical process is the relationship between the quantity of electricity passed through the cell and the amounts of substances produced by oxidation and reduction at the electrodes. The nature of this relationship is readily derived from the half-equation for the electrode process.

Half-equation	Quantity of Charge		Amount of Product
$Na^+ + e^- \rightarrow Na$	1 mol e^-	$\triangleq$	1 mol Na = 23.0 g Na
$Mg^{2+} + 2\ e^- \rightarrow Mg$	2 mol e^-	$\triangleq$	1 mol Mg = 24.3 g Mg
$Al^{3+} + 3\ e^- \rightarrow Al$	3 mol e^-	$\triangleq$	1 mol Al = 27.0 g Al

From these relations we see that the passage of one mole of electrons through an electrolytic cell would produce

1 mol (23.0 g) of Na or ½ mol (12.2 g) of Mg or ⅓ mol (9.0 g) of Al

As pointed out in Chapter 2, the charge of an electron, in SI units, is 1.602×10^{-19} C. It follows that the charge carried by a mole of electrons is, in coulombs,

$$6.022 \times 10^{23}\ e^- \times 1.602 \times 10^{-19} \frac{C}{e^-} = 9.65 \times 10^4\ C$$

or

$$1\ mol\ e^- = 96\ 500\ C \tag{22.17}$$

The constant in this equation, 96 500 C/mol e^-, is often referred to as the *Faraday constant* after Michael Faraday, the English scientist who first studied the quantitative aspects of electrochemistry over a century ago (p. 547).

In calculating amounts of substances produced in electrolytic cells, we frequently use a unit of current flow, the ampere (A). When a current of one ampere flows through an electrical circuit, one coulomb passes a given point in the circuit in one second. The number of coulombs passing through a cell can be calculated by multiplying the rate of flow in amperes by the elapsed time in seconds

$$\text{no. of coulombs} = (\text{no. of amperes}) \times (\text{no. of seconds}) \tag{22.18}$$

The relationships which we have just presented can be used in many practical calculations dealing with electrolytic cells, as in Example 22.2.

Example 22.2 Chromium metal can be plated from an acidic solution containing CrO_3 (cf. Table 22.1).

a. Write a balanced half-equation for the plating process.
b. How many grams of chromium will be plated by 20 500 C?
c. How long will it take to plate one gram of chromium using a current of 10.0 A?

Solution

a. The unbalanced half-equation is

$$CrO_3(aq) \rightarrow Cr(s)$$

Proceeding as indicated in Section 22.2, we first add three H_2O molecules to the right to balance oxygen. This requires that we add 6 H^+ to the left side of the equation to balance hydrogen. Finally, to balance the charge, we add 6 e^- to the left.

$$CrO_3(aq) + 6\ H^+(aq) + 6\ e^- \rightarrow Cr(s) + 3\ H_2O$$

b. From the coefficients of the balanced half-equation, we see that 6 mol electrons are equivalent to 1 mol chromium (52.0 g Cr):

$$6 \text{ mol } e^- \simeq 52.0 \text{ g Cr}$$

To find the number of grams of chromium plated, we need only convert the 20 500 C given to moles of electrons, using Equation 22.17, and then to grams of chromium.

$$\text{no. grams Cr} = 20\ 500 \text{ C} \times \frac{1 \text{ mol } e^-}{96\ 500 \text{ C}} \times \frac{52.0 \text{ g Cr}}{6 \text{ mol } e^-}$$
$$= 1.84 \text{ g Cr}$$

c. The indicated path here is to first convert grams of chromium to moles of electrons and then to coulombs. From the number of coulombs and the number of amperes (10.0), we can readily calculate the time in seconds, using Equation 22.18.

$$\text{no. of coulombs} = 1.00 \text{ g Cr} \times \frac{6 \text{ mol } e^-}{52.0 \text{ g Cr}} \times \frac{96\ 500 \text{ C}}{1 \text{ mol } e^-}$$
$$= 11\ 100 \text{ C}$$

$$\text{no. of seconds} = \frac{\text{no. of coulombs}}{\text{no. of amperes}} = \frac{11\ 100}{10.0}$$
$$= 1110 \text{ s, or about 18.5 min}$$

We should point out that in (b) and (c) we have, in effect, assumed a 100% yield of chromium in the electrolytic process. In practice we cannot expect to obtain this; some of the electrons will be wasted in such side reactions as

$$2\ H^+(aq) + 2\ e^- \rightarrow H_2(g)$$

As a result, if we were to carry out in the laboratory the process indicated in (b), we would obtain considerably less than 1.84 g of Cr. In (c) we would find that the time required to plate a gram of chromium would exceed 18.5 min.

The *gram equivalent mass* of a substance involved in an oxidation-reduction reaction is taken to be the mass in grams that is equivalent to one mole of electrons. If the half-equation for the oxidation or reduction is known, the gram equivalent mass can be calculated immediately. For example, for the process referred to in Example 22.2, we see from the half-equation

$$CrO_3(aq) + 6\ H^+(aq) + 6\ e^- \rightarrow Cr(s) + 3\ H_2O$$

$$6\ \text{mol}\ e^- \simeq 52.0\ \text{g Cr}$$

that the gram equivalent mass of chromium must be 52.0 g/6 = 8.67 g. In the laboratory, gram equivalent mass can be determined by measuring the mass of a substance produced by a known quantity of electricity (Example 22.3).

Example 22.3 In the electrolysis of a water solution of indium sulfate, it is found that 5.71 g of In is plated out when a current of 4.00 A is applied for one hour. Assuming a 100% yield of indium, what is its gram equivalent mass?

Solution Let us first find the number of coulombs consumed, convert this to moles of electrons, and then calculate how much indium would be produced by one mole of electrons.

$$\text{no. of coulombs} = (\text{no. of amperes}) \times (\text{no. of seconds}) = 4.00 \times 3600 = 14\,400$$

$$\text{no. of moles electrons} = \frac{14\,400}{96\,500} = 0.149$$

$$\text{i.e., } 0.149\ \text{mol}\ e^- \simeq 5.71\ \text{g In}$$

$$GEM\ \text{In} = 1\ \text{mol}\ e^- \times \frac{5.71\ \text{g In}}{0.149\ \text{mol}\ e^-} = 38.3\ \text{g In}$$

You will notice that the gram equivalent mass of indium is almost exactly 1/3 of its gram atomic mass (*AM* In = 114.8). This implies that the indium cation in solution carries a +3 charge, i.e., that the half-equation for its reduction is

$$In^{3+}(aq) + 3\ e^- \rightarrow In(s)$$

Experiments of this sort offer one of the simplest means of determining the charge of a simple monatomic ion in aqueous solution or, by implication, in a solid ionic compound.

Conversely, if the charge of the ion is known, the *AM* of the metal can be calculated very accurately

22.4 VOLTAIC CELLS

In an electrolytic cell electrical energy is supplied to bring about a nonspontaneous oxidation-reduction reaction. A voltaic (galvanic) cell is designed to achieve the opposite effect; a spontaneous redox reaction serves as a source of electrical energy. All of us are familiar with certain types of voltaic cells, such as the "dry cell" that is used in flashlights and the lead storage battery that supplies electrical energy to an automobile. To understand how a voltaic cell operates, we shall consider first some very simple cells which are easily constructed in the general chemistry laboratory.

Simple Voltaic Cells: The Zn-Cu^{2+} Cell

When a piece of zinc is added to a beaker containing a water solution of copper sulfate, a spontaneous oxidation-reduction reaction takes place.

$$Zn(s) + Cu^{2+}(aq) \rightarrow Zn^{2+}(aq) + Cu(s) \qquad (22.19)$$

Experimentally we observe that a spongy, reddish brown deposit of copper forms on the surface of the zinc; the blue color of the aquated Cu^{2+} ion fades as it is replaced by the colorless Zn^{2+} ion. The temperature of the solution rises as a result of the heat evolved. From the equation we see that the reaction amounts to electron transfer from a zinc atom to a Cu^{2+} ion.

To design a cell that uses Reaction 22.19 as a source of electrical energy, we must make the electron transfer occur indirectly; that is, the electrons given up by zinc atoms must pass through a circuit and do electrical work before they reduce Cu^{2+} ions to copper atoms. Two different ways of accomplishing this are shown in Figure 22.6. Let us first concentrate on the salt-bridge cell shown at the left of the figure and trace the flow of electrical current through it.

1. At the zinc *anode*, electrons are produced by the *oxidation* half-reaction

$$Zn(s) \rightarrow Zn^{2+}(aq) + 2\ e^- \qquad (22.19a)$$

This electrode, which "pumps" electrons into the external circuit, is ordinarily marked as the negative pole of the cell.

2. Electrons generated by Reaction 22.19a move through the external circuit (left to right in Figure 22.6*A*). This part of the circuit may be a simple resistance wire, a light bulb, an electric motor, an electrolytic cell, or some other device that consumes electrical energy.

3. Electrons pass from the external circuit to the copper *cathode* where they are used in the *reduction* of Cu^{2+} ions in the surrounding solution

$$Cu^{2+}(aq) + 2\ e^- \rightarrow Cu(s) \qquad (22.19b)$$

The copper electrode, which "pulls" electrons from the external circuit, is considered to be the positive pole of the cell.

4. To complete the circuit, ions must move through the aqueous solutions in the cell. As Reactions 22.19a and 22.19b proceed, a surplus of positive ions (Zn^{2+})

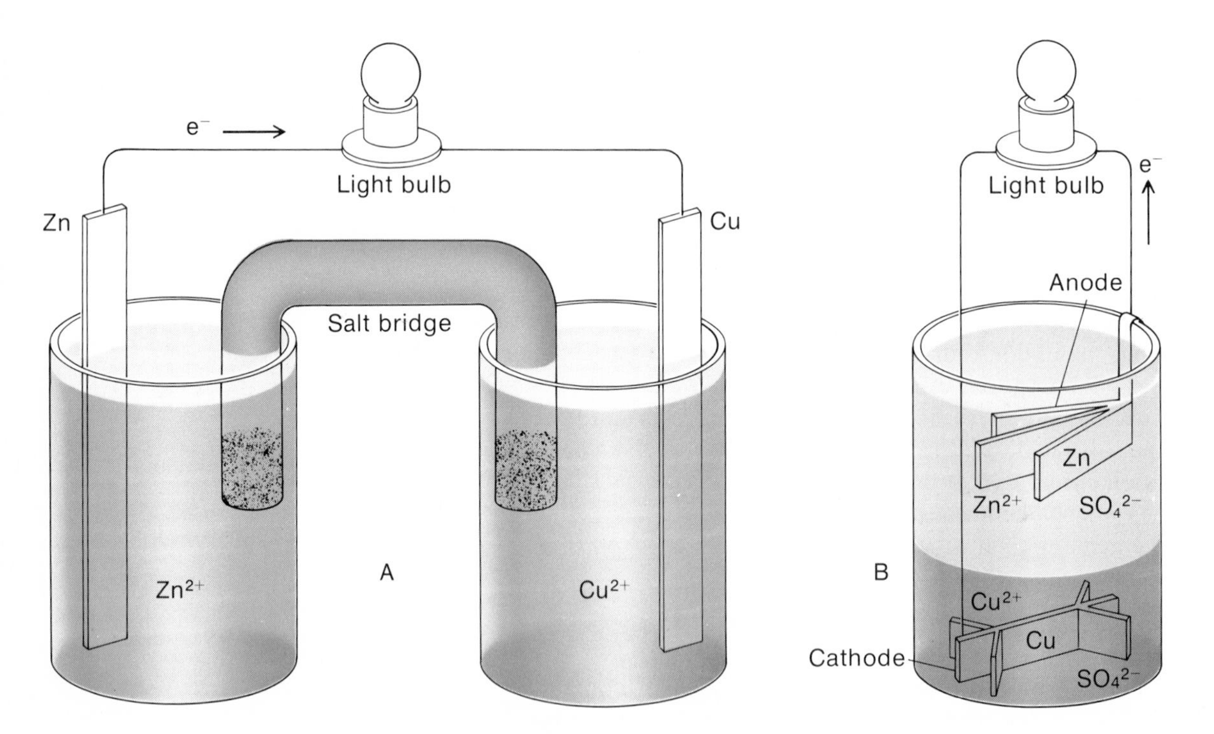

Figure 22.6 Two different cells used to obtain electrical energy from the reaction: $Zn(s) + Cu^{2+}(aq) \rightarrow Zn^{2+}(aq) + Cu(s)$. Note that in neither cell are the reactants, Zn and Cu^{2+}, allowed to come in direct contact with each other.

tends to build up around the zinc electrode. The region around the copper electrode tends to become deficient in positive ions as Cu^{2+} ions are discharged. To maintain electrical neutrality, cations must move toward the copper cathode or, alternatively, anions must move toward the zinc anode. In practice, both migrations occur.

In the cell shown in Figure 22.6*A*, movement of ions occurs through a *salt bridge*. In its simplest form a salt bridge may consist of an inverted U-tube, plugged with glass wool at each end. The tube is filled with a solution of a salt which takes no part in the electrode reactions; potassium nitrate, KNO_3, is frequently used. As current is drawn from the cell, K^+ ions move from the salt bridge into the copper half-cell to compensate for the Cu^{2+} ions discharged at the cathode. At the same time, NO_3^- ions move into the zinc half-cell to neutralize the Zn^{2+} ions formed at the anode.

The salt bridge allows current to flow, but prevents Cu^{2+} ions from coming in contact with Zn

Another way of setting up the Zn-Cu^{2+} cell is illustrated in the so-called "gravity" or Daniell cell shown in Figure 22.6*B*. To form this cell, enough copper sulfate solution is added to the jar to cover the copper electrode. A more dilute, less dense solution of zinc sulfate is then carefully poured over the copper sulfate. So long as the cell is not subjected to vibrations, the boundary between the layers may be maintained over long periods of time. Cells of this design were once used extensively to operate telegraph relays, doorbells, and other stationary electrical apparatus. Since their internal resistance is much lower than that of salt-bridge cells, much larger currents can be drawn from them.

Cells similar in design to that shown in Figure 22.6*A* can be set up to take advantage of the spontaneity of many different oxidation-reduction reactions. We can, for example, devise cells in which the following reactions serve as a source of electrical energy:

$$Ni(s) + Cu^{2+}(aq) \rightarrow Ni^{2+}(aq) + Cu(s) \qquad (22.20)$$

$$Zn(s) + 2\ H^+(aq) \rightarrow Zn^{2+}(aq) + H_2(g) \qquad (22.21)$$

In each case the apparatus consists of two half cells, each containing an electrode dipping into a solution of an appropriate electrolyte (e.g., Ni in $NiSO_4$). In the H_2-H^+ half cell (Fig. 22.7), hydrogen gas is bubbled over a specially prepared platinum electrode. At the anode of each cell, atoms of the element having the greatest tendency to lose electrons (Ni, Zn) are oxidized. The electrons released travel through the external circuit to the cathode (Cu, Pt) where they reduce the cations (Cu^{2+}, H^+) around that electrode. The circuit is completed by the movement of ions through the salt bridge; cations move to the cathode, anions to the anode.

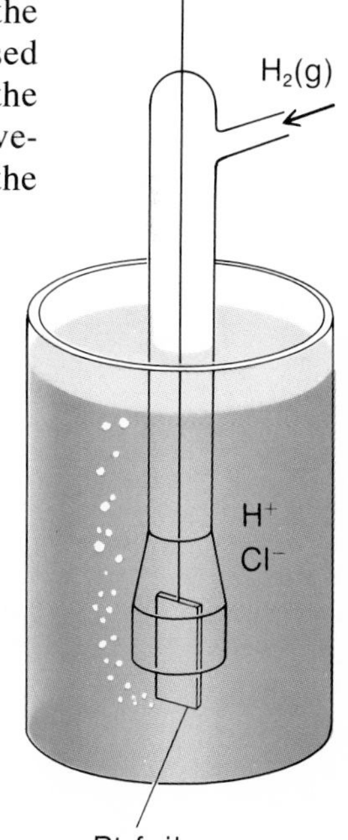

Figure 22.7 An H_2-H^+ half cell. When in combination with a Zn-Zn^{2+} half cell, electrons on the surface of the platinum foil reduce H^+ ions to hydrogen gas: $2\,H^+(aq) + 2e^- \rightarrow H_2(g)$. The H_2-H^+ cell is often used as a reference against which other electrode systems are compared.

Commercial Cells

As we shall see in Chapter 23, salt-bridge cells can provide us with some very practical information concerning the spontaneity and extent of oxidation-reduction reactions. However, their high internal resistance makes them unsuitable for commercial use. If we attempt to draw an appreciable amount of current from a salt-bridge cell, its voltage drops off sharply. There are on the market today many different types of voltaic cells, all of which are capable of supplying a comparatively large current, at least for a short time.

DRY (LECLANCHÉ) CELL. The construction of the ordinary dry cell used in flashlights and small appliances is shown in Figure 22.8. The zinc wall of the cell serves as the anode; the graphite rod passing through the center of the cell is the cathode. The space between the electrodes is filled with a moist paste containing manganese dioxide, carbon black, and ammonium chloride. When the cell is being used to generate energy, the half-reaction at the anode is

$$Zn(s) \rightarrow Zn^{2+}(aq) + 2\ e^- \tag{22.22a}$$

At the cathode, manganese dioxide is reduced to the species $MnO(OH)$, in which manganese is in the $+3$ oxidation state:

$$MnO_2(s) + NH_4^+(aq) + e^- \rightarrow MnO(OH)(s) + NH_3(aq) \tag{22.22b}$$

For the overall reaction:

$$Zn(s) + 2\ MnO_2(s) + 2\ NH_4^+(aq) \rightarrow Zn^{2+}(aq) + 2\ MnO(OH)(s) + 2\ NH_3(aq) \tag{22.22}$$

If too large a current is drawn from the cell, the ammonia formed by Reaction 22.22b forms a gaseous, insulating layer around the carbon electrode. In normal operation this condition is prevented by the migration of Zn^{2+} ions to the cathode, where they react with ammonia molecules to form complex ions such as $Zn(NH_3)_4^{2+}$, $Zn(NH_3)_2(H_2O)_2^{2+}$, and so on.

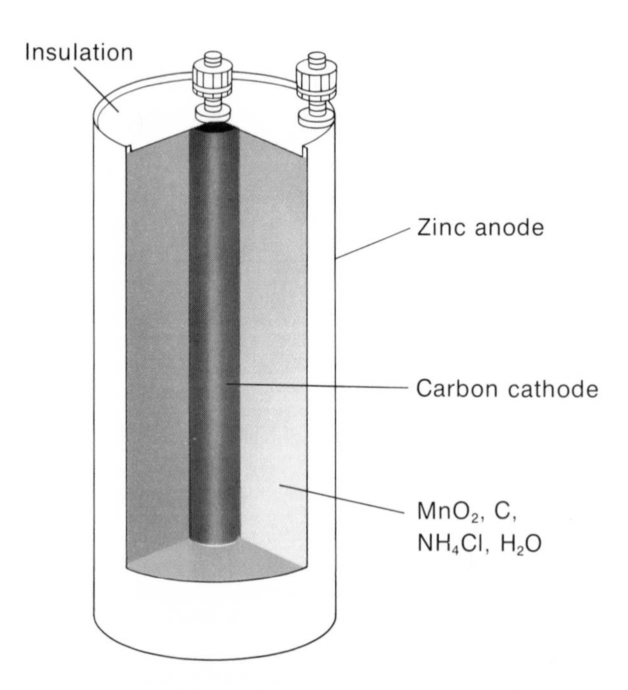

Figure 22.8 Section of a Zn-MnO_2 dry cell. A dry cell like this has a potential of 1.5 V and will deliver a current of 0.5 A for about 6 h.

LEAD STORAGE BATTERY. The 12-V storage battery used in automobiles consists of six voltaic cells of the type shown in Figure 22.9 connected in series. A group of lead plates, the grids of which are filled with spongy gray lead, forms the anode of the cell. The multiple cathode consists of a group of plates of similar design filled with lead dioxide. These two series of plates, alternating with each other throughout the cell, are immersed in a water solution of sulfuric acid, which acts as the electrolyte.

When a lead storage battery is supplying current, the lead in the anode grids is oxidized to Pb^{2+} ions, which immediately precipitate on the plates as lead sulfate, $PbSO_4$. At the cathode the lead dioxide is reduced to Pb^{2+} ions, which also precipitate as $PbSO_4$.

$$Pb(s) + SO_4^{2-}(aq) \rightarrow PbSO_4(s) + 2\ e^- \qquad (22.23a)$$

$$PbO_2(s) + 4\ H^+(aq) + SO_4^{2-}(aq) + 2\ e^- \rightarrow PbSO_4(s) + 2\ H_2O \qquad (22.23b)$$

$$Pb(s) + PbO_2(s) + 4\ H^+(aq) + 2\ SO_4^{2-}(aq) \rightarrow 2\ PbSO_4(s) + 2\ H_2O \qquad (22.23)$$

Deposits of lead sulfate formed by Reactions 22.23a and 22.23b slowly build up on the plates, partially covering and replacing the lead and lead dioxide. As the cell discharges, the concentration of sulfuric acid decreases; for every mole of lead reacting, two moles of H_2SO_4 (4 H^+, 2 SO_4^{2-}) are replaced by two moles of water. The state of charge of a storage battery can be checked by measuring the density of the electrolyte. A low density indicates a low sulfuric acid concentration and, hence, a partially discharged cell.

A lead storage battery, unlike an ordinary dry cell, can be restored to its original condition by passing a direct current through it in the reverse direction. While a storage battery is being charged, it acts as an electrolytic cell; the half-reactions represented by equations 22.23a and 22.23b are reversed:

If the $PbSO_4$ falls off the plates, you've got a real problem

$$2\ PbSO_4(s) + 2\ H_2O \rightarrow Pb(s) + PbO_2(s) + 4\ H^+(aq) + 2\ SO_4^{2-}(aq) \qquad (22.24)$$

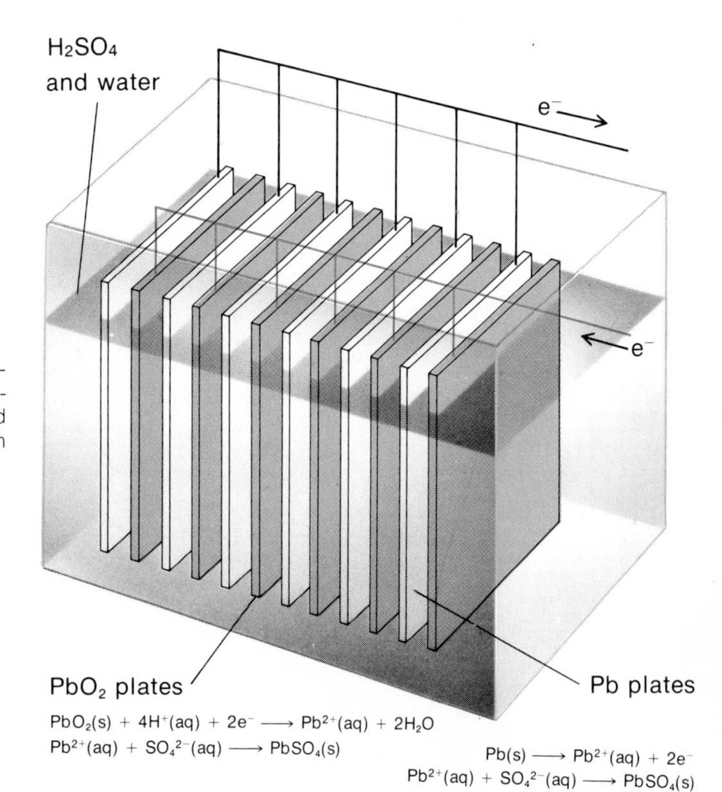

Figure 22.9 Lead storage battery. Two advantages of the lead storage battery are ability to deliver large amounts of energy for a short time and ability to be recharged. A disadvantage is its high mass/energy ratio.

The electrical energy required to bring about Reaction 22.24 may be furnished by a direct-current generator, as in older automobiles, or by an alternator equipped with a rectifier (in modern cars).

FUEL CELLS. The commercial voltaic cells that we have discussed are useful for special purposes where the relatively high cost of such chemicals as Zn, MnO_2, Pb, and PbO_2 is not a crucial factor. From a theoretical point of view, they are interesting in that they convert chemical energy very efficiently into electrical energy. You may recall from Chapter 14 that the free energy change is a measure of the useful work that can be obtained from a reaction. In a well-designed voltaic cell, the electrical work approaches $-\Delta G$ for the cell reaction.

Most of our electrical energy today is produced by generators driven by steam turbines operating on the heat produced by combustion of coal, oil, or natural gas. Here, the conversion of chemical into electrical energy is indirect in that the chemical energy is first converted to heat which is then used to make steam. The indirect process is both theoretically and practically less efficient than the direct conversion that occurs in a voltaic cell. The best power plants convert only about 30 to 40 per cent of the heat of combustion of a fuel into electrical energy. The remainder is dissipated to the air or to bodies of water, where it contributes to thermal pollution.

Scientists and engineers have long speculated on the possibility of converting the chemical energy of fuels directly to electrical energy in a type of voltaic cell known as a fuel cell. In principle there is no reason why this cannot be done. The combustion of a fuel, like the reaction of Zn with Cu^{2+}, is a spontaneous oxidation-reduction reaction and, hence, should serve as a source of electrical energy. To illustrate the intriguing possibilities of a fuel cell, consider the reaction

$$H_2(g) + \frac{1}{2}\, O_2(g) \rightarrow H_2O(l);\ \Delta H = -286\ \text{kJ},\ \Delta G^{1\ atm}\ \text{at}\ 25°C = -237\ \text{kJ}$$

If this reaction were used as a source of heat in a power plant, we would be unlikely to obtain more than about 100 kJ of electrical work per mole of hydrogen burned. In a fuel cell we could, in principle, extract as much as 237 kJ/mol of hydrogen; a more realistic estimate might be 200 kJ. Looking at the data in Table 22.2, we see two advantages of the fuel cell. Not only does it produce about twice as much electrical energy for a given amount of fuel, but at the same time we reduce the amount of waste heat, which contributes to thermal pollution, by a factor of more than two.

One type of H_2-O_2 fuel cell is shown in Figure 22.10. The two gases enter the cell through porous carbon electrodes surrounded by an aqueous solution of potassium hydroxide. The half-reactions at the electrodes are

anode: $$H_2(g) + 2\ OH^-(aq) \rightarrow 2\ H_2O + 2\ e^- \quad (22.25a)$$

cathode: $$\frac{1}{2}\, O_2(g) + H_2O + 2\ e^- \rightarrow 2\ OH^-(aq) \quad (22.25b)$$

$$H_2(g) + \frac{1}{2}\, O_2(g) \rightarrow 2\ H_2O \quad (22.25)$$

TABLE 22.2 ENERGY CONVERSION FOR THE REACTION $H_2(g) + \frac{1}{2}\, O_2(g) \rightarrow H_2O(l)$

	ELECTRICAL ENERGY PER MOLE H_2	WASTE HEAT PER MOLE H_2
Steam turbine	100 kJ	186 kJ
Fuel cell	200 kJ	86 kJ

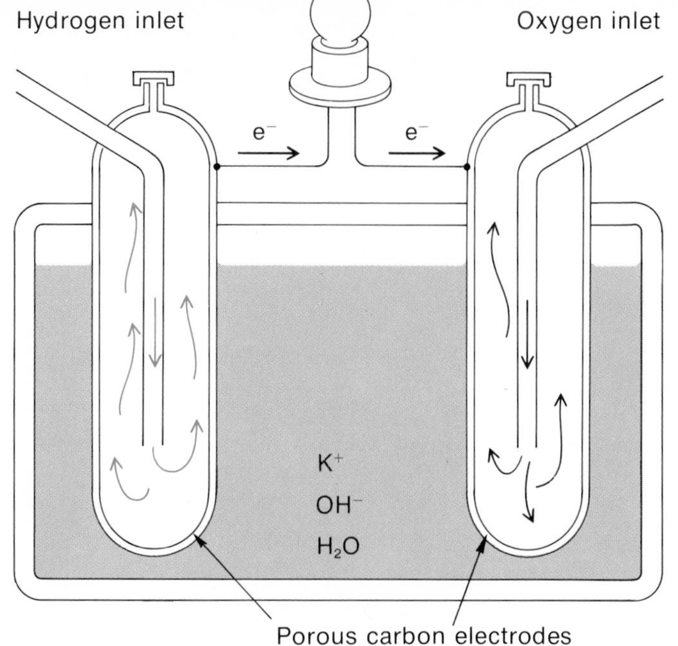

Figure 22.10 Schematic drawing of a hydrogen-oxygen fuel cell. Although several fuel cells of this sort have been developed, none as yet has proved to be practical for ordinary uses.

A fuel cell of this type, suitably modified to operate at zero gravity, was used as a source of electrical energy in the Apollo spacecraft.

Another type of fuel cell uses CH_4 and O_2

In spite of extensive industrial research on these devices, at present we have no commercially practical high-energy fuel cells. The prototypes that have been made have been subject to all sorts of technical problems, ranging from leaks to excessive corrosion to catalytic failures. No large-capacity cells have been produced, and in the small cells, slow reaction rates have been a limiting factor. At the present time, the development of a practical fuel cell in the moderate or high energy range is one of the most important needs of our society. Fuel cells offer one of the best methods of making more efficient use of fossil fuels and, at the same time, greatly decreasing thermal pollution from power generation.

A HISTORICAL PERSPECTIVE

Michael Faraday (1791–1867)

In the first half of the 19th century, the physical sciences were dominated by a series of remarkable Englishmen. In 1808, John Dalton (1766–1844) proposed his atomic theory of matter. At about the same time, Sir Humphry Davy (1778–1829) isolated six new chemical elements (Na, K, Mg, Ca, Sr, and Ba) by electrolysis of their molten carbonates or oxides. Later on, two English physicists, James Joule (1818–1889) and Lord Kelvin (1824–1907) laid the foundations for the laws of thermodynamics.

Probably the greatest experimental scientist of the 19th century and certainly the most prolific was Michael Faraday, who was born and lived almost all his life in what is now greater London. The son of a blacksmith, he had no formal education beyond the rudiments of reading, writing, and arithmetic. Apprenticed to a bookbinder at the age of 13, Faraday educated himself by reading virtually every book that came into the shop. One that particularly impressed him was a textbook, *Conversations in Chemistry,* written by Mrs. Jane Marcet. Within a few years he was carrying out simple experiments in his home laboratory and attending lectures given by Davy at the Royal Institution. Anxious to escape a life of drudgery as a tradesman, Faraday wrote out a copy of these lectures and submitted it to Davy with a request for employment. Shortly afterwards, a vacancy arose and he was hired as a laboratory assistant.

Davy quickly recognized Faraday's talents and as time passed allowed him to work more and more independently. In his years with Davy, Faraday published papers covering almost every field of chemistry. They included studies on the condensation of gases (he was the first to liquefy ammonia), the reaction of silver compounds with ammonia, and the isolation of several organic compounds, the most important of which was benzene. In 1816, Faraday began a series of lectures at the Royal Institution which were brilliantly successful. In 1825 he succeeded Davy as director of the laboratory. As Faraday's reputation grew, it was said that, "Humphry Davy's greatest discovery was Michael Faraday." Perhaps it was witticisms of this sort that led to an estrangement between master and protégé. Late in his life, Davy opposed Faraday's nomination as a Fellow of the Royal Society and is reputed to have cast the only vote against him.

In personality and temperament, Davy and Faraday were poles apart. To Sir Humphry Davy, science was a fascinating hobby which happened to bring him fame and fortune. He pursued it at his leisure, enjoying an active social life and writing mediocre poetry in his spare time. To Michael Faraday, science was an obsession; one of his biographers describes him as a "work maniac." An observer (Faraday had no students) said of him ". . . if he had to cross the laboratory for anything, he did not walk, he ran; the quickness of his perception was equalled by the calm rapidity of his movements." In 1839, he suffered a nervous breakdown, the result of overwork. For much of the rest of his life, Faraday was in poor health. He gradually gave up more and more of his social engagements but continued to do research at the same pace as before.

Faraday developed the laws of electrolysis between 1831 and 1834. In the summer of 1833, he showed that a given amount of electrical current passed through a solution of H_2SO_4 produces the same amount of H_2 regardless of the concentration of the acid, the size of the electrodes, or the voltage applied. He concluded that the amount of a substance produced by electrolysis must be directly proportional to the current passed and independent of other factors. In mid-December of that same year, he began a series of determinations of the electrochemical gram equivalent masses of metals, including tin, lead, and zinc. Despite taking a whole day off for Christmas, he managed to complete these experiments, write up the results of three years' work, and get his paper published in the Philosophic Transactions of the Royal Society in early January, 1834. In this paper, Faraday introduced the basic vocabulary of electrochemistry, using for the first time the terms "anode," "cathode," "ion," "electrolyte," and "electrolysis."

Although we have emphasized Faraday's work in chemistry, his greatest contributions were to physics in the fields of electricity and magnetism. In 1821, he discovered that a wire through which a current is flowing will rotate about a magnetic pole, and so produced the first electric motor. Ten years later, he showed that an electric potential could be created by rotating a copper disc between the poles of a permanent magnet, and thereby discovered the principle governing all modern electrical generators. These two contributions made possible the controlled use of electricity by mankind and were crucial steps in the development of modern society. No less a scientist than Albert Einstein ranked Faraday, along with Newton, Galileo, and Maxwell, as one of the greatest physicists of the ages.

PROBLEMS

22.1 *Oxidation Number* Give the oxidation numbers of each atom in

a. Na_3PO_4 b. C_4H_{10} c. NO_3^-

22.2 *Balancing Redox Equations* Balance the following equation:

$$Fe(s) + H^+(aq) + SO_4^{2-}(aq) \rightarrow Fe^{3+}(aq) + SO_2(g) + H_2O$$

22.3 *Laws of Electrolysis* Consider the electrolysis of a water solution of NaCl (Equation 22.15). How many grams of Cl_2 are produced in two hours using a current of 10.0 A?

22.4 *Electrolytic and Voltaic Cells* Sketch diagrams of

a. a cell in which molten $CaCl_2$ is electrolyzed.
b. a voltaic cell in which the reaction is $Pb(s) + 2\ Ag^+(aq) \rightarrow Pb^{2+}(aq) + 2\ Ag(s)$. In both cells, label anode and cathode and indicate the direction of flow of electrons and ions.

22.5 Give the oxidation number of each atom in

a. OsO_4 b. K_2CO_3 c. C_2H_6O
d. PO_3^{3-} e. CrF_6^{3-} f. HSO_3^-

22.6 Consider the oxidation numbers of nitrogen listed in Figure 22.1. Give the formula of a molecule in which nitrogen shows each of these oxidation numbers.

22.7 Using Figure 22.1, predict the formulas of three different compounds containing the elements Rb, Re, and O (consider only the −2 oxid no. for O).

22.8 Based on Figure 22.1, which of the following species can act as either oxidizing or reducing agents?

a. ClO_3^- b. ClO_4^- c. Cl^- d. Cl_2

22.9 For each of these reactions

a. $2\ Fe(s) + 3\ Cl_2(g) \rightarrow 2\ Fe^{3+}(aq) + 6\ Cl^-(aq)$
b. $Cl_2(g) + 2\ OH^-(aq) \rightarrow Cl^-(aq) + ClO^-(aq) + H_2O$

give the oxidation number of each atom and identify the oxidizing agent and reducing agent.

22.10 Consider the following half-equations, all for reactions in acidic solution:

$Co(s) \rightarrow Co^{3+}(aq)$
$SO_3^{2-}(aq) \rightarrow H_2S(g)$
$Sn^{4+}(aq) \rightarrow Sn^{2+}(aq)$
$NO(g) \rightarrow NO_3^-(aq)$

a. Classify each of these as oxidation or reduction half-reactions.
b. Balance each half-equation.
c. Write as many balanced redox equations as possible by combining these half-equations.

22.25 Give the oxidation number of each atom in

a. As_2O_3 b. CaC_2O_4 c. PF_5
d. NO_2^- e. $Cr_2O_7^{2-}$ f. O_2^-

22.26 Give the formula of an anion in which chlorine shows each of the oxidation numbers given in Figure 22.1.

22.27 Using Figure 22.1 and Table 21.3, p. 504, give the formulas and charges of all the complexes containing Pt, H_2O molecules, and Cl^- ions.

22.28 Which of the following anions could not act as an oxidizing agent (i.e., could not be reduced)?

a. Br^- b. O_2^{2-} c. H^- d. NO_3^-

22.29 For the reaction

$$MnO_4^-(aq) + 8\ H^+(aq) + 5\ Cl^-(aq) \rightarrow Mn^{2+}(aq) + \tfrac{5}{2}\ Cl_2(g) + 4\ H_2O$$

identify

a. the element oxidized.
b. the element reduced.
c. the oxidizing agent.
d. the reducing agent.

22.30 Consider the following half-equations, all for reactions in acidic solution:

$Cd^{2+}(aq) \rightarrow Cd(s)$
$Pb(s) + SO_4^{2-}(aq) \rightarrow PbSO_4(s)$
$NO_3^-(aq) \rightarrow NO_2(g)$
$Br^-(aq) \rightarrow Br_2(l)$

a. Classify each of these as oxidation or reduction processes.
b. Balance each half-equation.
c. Write as many balanced redox equations as possible by combining these half-equations.

22.11 Follow the directions of Problem 22.10 for the following half-equations, all for reactions in basic solution:

$$Cl_2(g) \rightarrow Cl^-(aq)$$
$$NO(g) \rightarrow NO_3^-(aq)$$
$$Ni(s) \rightarrow Ni(OH)_2(s)$$

22.12 Balance the following equations:

a. $Al(s) + Cu^{2+}(aq) \rightarrow Al^{3+}(aq) + Cu(s)$
b. $Cl^-(aq) + NO_3^-(aq) + H^+(aq) \rightarrow Cl_2(g) + NO_2(g) + H_2O$
c. $H_2O_2(aq) + MnO_4^-(aq) + H^+(aq) \rightarrow O_2(g) + Mn^{2+}(aq) + H_2O$
d. $As(s) + NO_3^-(aq) + H_2O \rightarrow AsO_4^{3-}(aq) + NO(g) + H^+(aq)$

22.13 Write balanced equations for each of the following:

a. $P_4(s) \rightarrow PH_3(g) + HPO_3^{2-}(aq)$; acid
b. $P_4(s) \rightarrow PH_3(g) + HPO_3^{2-}(aq)$; base
c. $Re(s) + NO_3^-(aq) \rightarrow ReO_4^-(aq) + NO(g)$; acid
d. $MnO_2(s) + BiO_3^-(aq) \rightarrow MnO_4^{2-}(aq) + Bi^{3+}(aq)$; base

22.14 When moist air comes in contact with iron, corrosion takes place to form $Fe(OH)_3(s)$. Write balanced half-equations for the oxidation and reduction reactions and combine them to give the balanced overall equation.

22.15 To remove traces of CO from air, it is passed over solid I_2O_5, forming CO_2 and I_2.

a. Write a balanced equation for the reaction.
b. How many grams of I_2O_5 are required to remove one gram of CO?
c. What volume (dm^3) of air at 25°C and 1.00 atm containing 2.0×10^{-3} mol % CO can be purified by one gram of I_2O_5?

22.16 Consider the reaction

$$I_2(s) + 2\ Fe^{2+}(aq) \rightarrow 2\ I^-(aq) + 2\ Fe^{3+}(aq)$$

for which $\Delta G = +46$ kJ.

a. Could this reaction take place in a voltaic cell? an electrolytic cell?
b. Sketch a cell in which this reaction could take place, label anode and cathode, and indicate the flow of current through the circuit.

22.17 In the electrolysis of a water solution of KI, $I_2(s)$ is formed at one electrode, $H_2(g)$ and OH^- ions at the other. Write a

a. half-equation for the anode reaction.
b. half-equation for the cathode reaction.
c. balanced equation for the overall reaction.

22.31 Follow the directions of Problem 22.30 for the following half-equations, all for reactions in basic solution:

$$Br_2(l) \rightarrow Br^-(aq)$$
$$Br_2(l) \rightarrow BrO_3^-(aq)$$
$$O_2(g) \rightarrow OH^-(aq)$$

22.32 Balance the following equations:

a. $I_2(s) + S_2O_3^{2-}(aq) \rightarrow I^-(aq) + S_4O_6^{2-}$
b. $Ag_2S(s) + NO_3^-(aq) + H^+(aq) \rightarrow Ag^+(aq) + S(s) + NO(g) + H_2O$
c. $C_2H_6O(aq) + Cr_2O_7^{2-}(aq) + H^+(aq) \rightarrow C_2H_4O(aq) + Cr^{3+}(aq) + H_2O$
d. $Zn(s) + NO_3^-(aq) + H^+(aq) \rightarrow Zn^{2+}(aq) + NH_4^+(aq) + H_2O$

22.33 Write balanced equations for each of the following:

a. $Cl_2(g) \rightarrow ClO_4^-(aq) + Cl^-(aq)$; acid
b. $Cl_2(g) \rightarrow ClO_4^-(aq) + Cl^-(aq)$; base
c. $AuCl_4^-(aq) + AsH_3(g) \rightarrow H_3AsO_3(aq) + Au(s) + Cl^-(aq)$; acid
d. $Am^{3+}(aq) + S_2O_8^{2-}(aq) \rightarrow AmO_2^+(aq) + SO_4^{2-}(aq)$; base

22.34 In gold plating, metallic gold is formed from the $Au(CN)_2^-$ ion. At the same time, $O_2(g)$ is formed. The plating solution is basic. Write equations for the two half-reactions and the overall reaction.

22.35 In the Marsh test for arsenic, Zn is added to the sample in acidic solution. AsO_4^{3-} ions are reduced to $AsH_3(g)$; Zn^{2+} ions are formed.

a. Write a balanced equation for the reaction.
b. How many moles of H^+ are required to react with one gram of AsO_4^{3-}?
c. What volume of AsH_3, measured at 25°C and 1.00 atm, is produced from one milligram of AsO_4^{3-}?

22.36 Magnesium metal can be produced by the electrolysis of molten $MgCl_2$.

a. Write equations for the anode and cathode reactions.
b. Sketch a cell in which this reaction could take place, labeling anode and cathode.
c. Predict the sign of ΔG for the cell reaction.

22.37 When a water solution of $CuSO_4$ is electrolyzed, Cu(s) is formed at one electrode, $O_2(g)$ and H^+ ions at the other. Write a

a. half-equation for the anode reaction.
b. half-equation for the cathode reaction.
c. balanced equation for the overall reaction.

22.18 Explain why

a. it is cheaper to produce $Cl_2(g)$ by electrolysis of aqueous rather than molten NaCl.
b. cryolite is added to the electrolyte in the cell used to make Al.
c. large quantities of NaCN are used in electroplating.
d. the anode in a copper plating cell is made of copper.

22.19 In the electrolysis of an aqueous solution of NaCl, 2.61×10^{22} electrons pass through a cell.

a. How many moles of electrons are used?
b. How many coulombs does this represent?
c. How many grams of H_2, Cl_2, and OH^- are produced, assuming 100% yield?

22.20 A current of 2.00 A is drawn from a dry cell for a period of ten minutes. Assuming that the only reaction taking place is 22.22, p. 544,

a. how many grams of zinc are consumed?
b. what volume of ammonia gas, at 25°C and 1.00 atm, would be produced if the solution were heated?

22.21 A spoon with an area of 4.00 cm² is plated with silver from a $Ag(CN)_2^-$ solution, using a current of 0.500 A for two hours.

a. How many grams of Ag are plated if the current efficiency is 80%?
b. What is the thickness of the silver plate formed (*d* Ag = 10.5 g/cm³)?

22.22 A solution of a certain cerium salt deposits 1.00 g of Ce when 2.75×10^3 C are passed through an electrolytic cell.

a. Calculate the *GEM* of cerium.
b. What is the charge of the cerium cation?

22.23 Design salt-bridge cells in which the following reactions occur. In each case, label the anode and cathode and indicate the direction of flow of current through the circuit.

a. $Ni(s) + 2\ Ag^+(aq) \rightarrow Ni^{2+}(aq) + 2\ Ag(s)$
b. $Pb(s) + PbO_2(s) + 4\ H^+(aq) + 2\ SO_4^{2-}(aq) \rightarrow 2\ PbSO_4(s) + 2\ H_2O$
c. $H_2(g) + 2\ Fe^{3+}(aq) \rightarrow 2\ H^+(aq) + 2\ Fe^{2+}(aq)$

22.38 What changes would be observed if

a. acid were added to the cell used to electrolyze aqueous NaCl?
b. acid were added to a cell used to electroplate silver?
c. a strong acid were added to the $Al(OH)_6^{3-}$ complex ion formed in the purification of bauxite?
d. a bar of iron were used as the anode in the electrolysis of Al_2O_3?

22.39 A certain cell is producing aluminum at the rate of 1.00 kg/d. Assuming a yield of 100%,

a. how many electrons must pass through the cell in one day?
b. what is the current passing through the cell?
c. how much oxygen is being produced simultaneously?

22.40 A cell used to electrolyze NaCl solution is filled with 1.00 dm³ of 2.00 *M* NaCl. A current of two amperes is passed through this solution for eight hours. Assuming that the only reaction that occurs is 22.15, p. 536,

a. what volume of $Cl_2(g)$ is produced at 25°C and 1.00 atm?
b. what fraction of the Cl^- ions has been converted to OH^- ions?

22.41 It is desired to plate a coin, which has a diameter of 2.50 cm and a thickness of 0.15 cm, with a layer of gold 0.0020 cm thick.

a. How many grams of gold are required? (*d* = 19.3 g/cm³)
b. How long will it take to plate the coin from AuCN, using a current of 0.200 A and assuming 100% yield?

22.42 A solution of a platinum salt deposits 2.44 g of Pt when 4800 C are passed through the cell. Assuming that the only reaction that occurs at the cathode is the reduction of a Pt-containing cation to the metal,

a. calculate the *GEM* of platinum.
b. determine the oxidation number of the platinum in the solution species.

22.43 Follow the directions of Problem 22.23 in sketching voltaic cells in which the following reactions occur.

a. $Zn(s) + Cd^{2+}(aq) \rightarrow Zn^{2+}(aq) + Cd(s)$
b. $2\ AuCl_4^-(aq) + 3\ Cu(s) \rightarrow 2\ Au(s) + 8\ Cl^-(aq) + 3\ Cu^{2+}(aq)$
c. $Cl_2(g) + 2\ I^-(aq) \rightarrow 2\ Cl^-(aq) + I_2(s)$

22.24 A lead storage battery is used to electrolyze a water solution of NaCl. Sketch the two cells, labeling anode and cathode, and indicate the direction of flow of current through all parts of the circuit.

22.44 A dry cell is used to electroplate copper, using Cu electrodes. Sketch the two cells, labeling anode and cathode, and indicate the direction of flow of current through all parts of the circuit.

*22.45 Using appropriate data and equations from Chapters 4, 5, and 14, calculate

a. the minimum amount of electrical energy required to form one mole of $H_2(g)$ by electrolysis of H_2O at 25°C and 1.00 atm.
b. the amount of heat that could be obtained by burning one mole of $H_2(g)$ at 25°C and 1.00 atm.
c. the heating value of $H_2(g)$ in kJ/dm^3 at 25°C and 1.00 atm.
d. the heating value of $CH_4(g)$ in kJ/dm^3 at 25°C and 1.00 atm.

*22.46 Using appropriate conversion factors and the definition $1\ V \cdot C = 1\ J$, calculate the value of the Faraday constant in $kcal/(V \cdot mol\ e^-)$.

*22.47 About twenty kilowatt hours of electrical energy is required to produce one kilogram of aluminum from bauxite ore.

a. Using the conversion factor $1\ kWh = 3.60 \times 10^6\ J$, calculate the amount of energy in joules required to produce one gram of Al.
b. In recycling, the major expenditure of energy is for heating aluminum to its melting point (660°C) and melting it. Taking the specific heat of Al to be $0.90\ J/(g \cdot °C)$ and its heat of fusion to be 10.7 kJ/mol, estimate the energy requirement for recycling one gram of aluminum.
c. Comparing your answers to (a) and (b), comment on the energy saving to be expected by recycling aluminum cans.

23

OXIDATION-REDUCTION REACTIONS: SPONTANEITY AND EXTENT

Voltaic cells such as those described in Section 22.4 are of interest to chemists for reasons that go beyond their practical importance as a source of electrical energy. The property of a voltaic cell which is of particular concern to us is its voltage, or "potential," which is a measure of the driving force behind the reaction taking place in the cell. By measuring cell voltages, it is possible to decide whether a given oxidation-reduction reaction will proceed in the laboratory or in the world around us. Moreover, knowing the cell voltage, we can go one step further and predict to what extent the reaction will occur.

23.1 STANDARD VOLTAGES

In a properly designed cell, we find that the voltage measured at a given temperature, let us say 25°C, depends upon two factors: the nature of the cell reaction and the concentrations of the various species (ions or molecules) taking part in the reaction. We will take the **standard voltage, E^0,** for a given cell reaction to be that obtained when *all ions or molecules in solution are at a concentration of one mole per cubic decimetre and all gases are at a partial pressure of one atmosphere*. To illustrate, consider the Zn-H^+ cell referred to in Chapter 22. We find that when the pressure of hydrogen gas is 1 atm and the concentrations of both Zn^{2+} and H^+ are 1 *M*, the cell voltage is +0.76 V. This quantity, +0.76 V, is referred to as the standard cell voltage and given the symbol E^0.

$$Zn(s) + 2\ H^+(aq, 1\ M) \rightarrow Zn^{2+}(aq, 1\ M) + H_2(g, 1\ atm);\ E^0 = +0.76\ V$$

You will recall that any redox reaction can be split into two half-reactions, an oxidation and a reduction. For the above reaction we have

oxidation: $Zn(s) \rightarrow Zn^{2+}(aq, 1\ M) + 2\ e^-$

reduction: $2\ H^+(aq, 1\ M) + 2\ e^- \rightarrow H_2(g, 1\ atm)$

It has been found that, with half-reactions such as these, one can associate standard voltages which denote in a quantitative way the tendency for the reaction to occur at standard concentrations. The standard voltage corresponding to the reduction half-reaction is given the symbol E^0_{red}; that associated with the oxidation half-reaction is written as E^0_{ox}. The standard cell voltage, E^0, is always the sum of these two quantities. For the Zn-H^+ cell, we have

$Zn(s) \rightarrow Zn^{2+}(1\ M) + 2\ e^-$	$E^0_{ox}(Zn \rightarrow Zn^{2+})$
$2\ H^+(1\ M) + 2\ e^- \rightarrow H_2(1\ atm)$	$E^0_{red}(H^+ \rightarrow H_2)$
$Zn(s) + 2\ H^+(1\ M) \rightarrow Zn^{2+}(1\ M) + H_2(1\ atm)$	$E^0 = E^0_{ox} + E^0_{red} = +0.76\ V$

Given the standard voltage of a cell such as this one, it would be useful to be able to calculate the standard voltages for the two half-reactions, since these could be used with other cells in which these half-reactions occur. The problem is that we have two unknowns, E^0_{ox} and E^0_{red}, and only one known, E^0. No matter what we do, we cannot resolve this dilemma. Consequently, we resort to a procedure used with enthalpies or free energies of formation: we refer all values to that of a reference system taken arbitrarily as zero. Specifically, we choose the standard hydrogen reduction as a reference and assign it a voltage of zero, i.e.,

We need two half-cells to measure a voltage

$$2\ H^+(aq, 1\ M) + 2\ e^- \rightarrow H_2(g, 1\ atm);\ E^0_{red}\ (H^+ \rightarrow H_2) = 0.00\ V$$

On the basis of this convention and knowing that E^0 for the Zn-H^+ cell is +0.76 V, it immediately follows that the standard voltage for the oxidation of zinc must be +0.76 V, i.e.,

$$Zn(s) \rightarrow Zn^{2+}(aq, 1\ M) + 2\ e^-;\ E^0_{ox}\ (Zn \rightarrow Zn^{2+}) = +0.76\ V$$

As soon as one voltage has been established, others can readily be determined from measurements on properly designed cells. Suppose, for example, we wish to determine the standard voltage for the reduction of Cu^{2+} to Cu. One way to do this is to set up the cell shown in Figure 22.6, p. 542, using 1 M solutions of Zn^{2+} and Cu^{2+}, and measure the voltage, which we find to be 1.10 V.

$$Zn(s) + Cu^{2+}(aq, 1\ M) \rightarrow Zn^{2+}(aq, 1\ M) + Cu(s);\ E^0 = +1.10\ V$$

In this cell zinc is being oxidized and Cu^{2+} ions reduced. Hence

$$E^0_{ox}\ (Zn \rightarrow Zn^{2+}) + E^0_{red}\ (Cu^{2+} \rightarrow Cu) = +1.10\ V$$

Since the standard voltage for the oxidation of zinc must be the same here as in the Zn-H^+ cell, +0.76 V,

$$+0.76\ V + E^0_{red}\ (Cu^{2+} \rightarrow Cu) = 1.10\ V$$

$$E^0_{red}\ (Cu^{2+} \rightarrow Cu) = 1.10\ V - 0.76\ V = +0.34\ V$$

TABLE 23.1 STANDARD POTENTIALS IN WATER SOLUTION AT 25°C

		STANDARD POTENTIAL (VOLTS)
$Li^+(aq) + e^-$	$\rightarrow Li(s)$	−3.05
$K^+(aq) + e^-$	$\rightarrow K(s)$	−2.93
$Ba^{2+}(aq) + 2\ e^-$	$\rightarrow Ba(s)$	−2.90
$Ca^{2+}(aq) + 2\ e^-$	$\rightarrow Ca(s)$	−2.87
$Na^+(aq) + e^-$	$\rightarrow Na(s)$	−2.71
$Mg^{2+}(aq) + 2\ e^-$	$\rightarrow Mg(s)$	−2.37
$Al^{3+}(aq) + 3\ e^-$	$\rightarrow Al(s)$	−1.66
$Mn^{2+}(aq) + 2\ e^-$	$\rightarrow Mn(s)$	−1.18
$Zn^{2+}(aq) + 2\ e^-$	$\rightarrow Zn(s)$	−0.76
$Cr^{3+}(aq) + 3\ e^-$	$\rightarrow Cr(s)$	−0.74
$Fe^{2+}(aq) + 2\ e^-$	$\rightarrow Fe(s)$	−0.44
$Cr^{3+}(aq) + e^-$	$\rightarrow Cr^{2+}(aq)$	−0.41
$Cd^{2+}(aq) + 2\ e^-$	$\rightarrow Cd(s)$	−0.40
$PbSO_4(s) + 2\ e^-$	$\rightarrow Pb(s) + SO_4^{2-}(aq)$	−0.36
$Tl^+(aq) + e^-$	$\rightarrow Tl(s)$	−0.34
$Co^{2+}(aq) + 2\ e^-$	$\rightarrow Co(s)$	−0.28
$Ni^{2+}(aq) + 2\ e^-$	$\rightarrow Ni(s)$	−0.25
$AgI(s) + e^-$	$\rightarrow Ag(s) + I^-(aq)$	−0.15
$Sn^{2+}(aq) + 2\ e^-$	$\rightarrow Sn(s)$	−0.14
$Pb^{2+}(aq) + 2\ e^-$	$\rightarrow Pb(s)$	−0.13
$2\ H^+(aq) + 2\ e^-$	$\rightarrow H_2(g)$	0.00
$AgBr(s) + e^-$	$\rightarrow Ag(s) + Br^-(aq)$	0.10
$S(s) + 2\ H^+(aq) + 2\ e^-$	$\rightarrow H_2S(aq)$	0.14
$Sn^{4+}(aq) + 2\ e^-$	$\rightarrow Sn^{2+}(aq)$	0.15
$Cu^{2+}(aq) + e^-$	$\rightarrow Cu^+(aq)$	0.15
$SO_4^{2-}(aq) + 4\ H^+(aq) + 2\ e^-$	$\rightarrow SO_2(g) + 2\ H_2O$	0.20
$Cu^{2+}(aq) + 2\ e^-$	$\rightarrow Cu(s)$	0.34
$Cu^+(aq) + e^-$	$\rightarrow Cu(s)$	0.52
$I_2(s) + 2\ e^-$	$\rightarrow 2\ I^-(aq)$	0.53
$Fe^{3+}(aq) + e^-$	$\rightarrow Fe^{2+}(aq)$	0.77
$Hg_2^{2+}(aq) + 2\ e^-$	$\rightarrow 2\ Hg(l)$	0.79
$Ag^+(aq) + e^-$	$\rightarrow Ag(s)$	0.80
$2\ Hg^{2+}(aq) + 2\ e^-$	$\rightarrow Hg_2^{2+}(aq)$	0.92
$NO_3^-(aq) + 4\ H^+(aq) + 3\ e^-$	$\rightarrow NO(g) + 2\ H_2O$	0.96
$AuCl_4^-(aq) + 3\ e^-$	$\rightarrow Au(s) + 4\ Cl^-(aq)$	1.00
$Br_2(l) + 2\ e^-$	$\rightarrow 2\ Br^-(aq)$	1.07
$O_2(g) + 4\ H^+(aq) + 4\ e^-$	$\rightarrow 2\ H_2O$	1.23
$MnO_2(s) + 4\ H^+(aq) + 2\ e^-$	$\rightarrow Mn^{2+}(aq) + 2\ H_2O$	1.23
$Cr_2O_7^{2-}(aq) + 14\ H^+(aq) + 6\ e^-$	$\rightarrow 2\ Cr^{3+}(aq) + 7\ H_2O$	1.33
$Cl_2(g) + 2\ e^-$	$\rightarrow 2\ Cl^-(aq)$	1.36
$ClO_3^-(aq) + 6\ H^+(aq) + 5\ e^-$	$\rightarrow \frac{1}{2}\ Cl_2(g) + 3\ H_2O$	1.47
$Au^{3+}(aq) + 3\ e^-$	$\rightarrow Au(s)$	1.50
$MnO_4^-(aq) + 8\ H^+(aq) + 5\ e^-$	$\rightarrow Mn^{2+}(aq) + 4\ H_2O$	1.52
$PbO_2(s) + SO_4^{2-}(aq) + 4\ H^+(aq) + 2\ e^-$	$\rightarrow PbSO_4(s) + 2\ H_2O$	1.68
$H_2O_2(aq) + 2\ H^+(aq) + 2\ e^-$	$\rightarrow 2\ H_2O$	1.77
$Co^{3+}(aq) + e^-$	$\rightarrow Co^{2+}(aq)$	1.82
$F_2(g) + 2\ e^-$	$\rightarrow 2\ F^-(aq)$	2.87
BASIC SOLUTION		
$Zn(OH)_4^{2-}(aq) + 2\ e^-$	$\rightarrow Zn(s) + 4\ OH^-(aq)$	−1.22
$Fe(OH)_2(s) + 2\ e^-$	$\rightarrow Fe(s) + 2\ OH^-(aq)$	−0.88
$2\ H_2O + 2\ e^-$	$\rightarrow H_2(g) + 2\ OH^-(aq)$	−0.83
$Fe(OH)_3(s) + e^-$	$\rightarrow Fe(OH)_2(s) + OH^-(aq)$	−0.56
$S(s) + 2\ e^-$	$\rightarrow S^{2-}(aq)$	−0.48
$Cu(OH)_2(s) + 2\ e^-$	$\rightarrow Cu(s) + 2\ OH^-(aq)$	−0.36
$CrO_4^{2-}(aq) + 4\ H_2O + 3\ e^-$	$\rightarrow Cr(OH)_3(s) + 5\ OH^-(aq)$	−0.12
$NO_3^-(aq) + H_2O + 2\ e^-$	$\rightarrow NO_2^-(aq) + 2\ OH^-(aq)$	0.01
$Ag_2O(s) + H_2O + 2\ e^-$	$\rightarrow 2\ Ag(s) + 2\ OH^-(aq)$	0.34
$ClO_4^-(aq) + H_2O + 2\ e^-$	$\rightarrow ClO_3^-(aq) + 2\ OH^-(aq)$	0.36
$O_2(g) + 2\ H_2O + 4\ e^-$	$\rightarrow 4\ OH^-(aq)$	0.40
$ClO_3^-(aq) + 3\ H_2O + 6\ e^-$	$\rightarrow Cl^-(aq) + 6\ OH^-(aq)$	0.62
$ClO^-(aq) + H_2O + 2\ e^-$	$\rightarrow Cl^-(aq) + 2\ OH^-(aq)$	0.89

Standard half-cell voltages are ordinarily obtained from a list of *standard potentials* such as that given in Table 23.1. **The potentials listed give us directly the standard voltages for reduction half-reactions.** For example, since the standard potentials for $Zn^{2+} \rightarrow Zn$ and $Cu^{2+} \rightarrow Cu$ are −0.76 V and +0.34 V, respectively, we see immediately that

$$Zn^{2+}(aq) + 2\,e^- \rightarrow Zn(s) \qquad E^0_{red} = -0.76\ V$$
$$Cu^{2+}(aq) + 2\,e^- \rightarrow Cu(s) \qquad E^0_{red} = +0.34\ V$$

Standard voltages for oxidation half-reactions are obtained by changing the sign of the standard potential listed in Table 23.1. Thus we have

$$Zn(s) \rightarrow Zn^{2+}(aq) + 2\,e^- \qquad E^0_{ox} = +0.76\ V$$
$$Cu(s) \rightarrow Cu^{2+}(aq) + 2\,e^- \qquad E^0_{ox} = -0.34\ V$$

In general, standard voltages for forward and reverse reactions (oxidation and reduction) are equal in magnitude but opposite in sign.

Calculation of E^0 from E^0_{red} and E^0_{ox}

As pointed out earlier (p. 554), the standard voltage of any cell is the algebraic sum of the standard voltages of the two half-reactions, reduction and oxidation, taking place in the cell. That is,

$$E^0 = E^0_{red} + E^0_{ox} \qquad (23.1)$$

This simple relationship makes it possible, using Table 23.1, to calculate the standard voltages of more than 3000 different voltaic cells. Example 23.1 illustrates how this is done.

Example 23.1 Using Table 23.1, calculate the standard voltage of a voltaic cell in which the reaction is $2\,Ag^+(aq) + Cu(s) \rightarrow 2\,Ag(s) + Cu^{2+}(aq)$

Solution Splitting the cell reaction into two half-reactions, finding the appropriate standard voltages from the table, and adding:

reduction:	$2\,Ag^+(aq) + 2\,e^- \rightarrow 2\,Ag(s)$;	$E^0_{red} = +0.80\ V$
oxidation:	$Cu(s) \rightarrow Cu^{2+}(aq) + 2\,e^-$;	$E^0_{ox} = -0.34\ V$
		$E^0 = +0.46\ V$

Notice that:

—the calculated E^0 is positive, as it must be for any voltaic cell where a spontaneous redox reaction is taking place.

—E^0_{red} for Ag^+ is taken directly from Table 23.1, where we find

$$Ag^+(aq) + e^- \rightarrow Ag(s);\ E^0_{red} = +0.80\ V$$

We do *not* multiply the voltage by two just because two Ag^+ ions happen to appear in the balanced overall equation. E^0, E^0_{red}, or E^0_{ox} is a measure of the driving force behind a reaction; it is independent of the number of electrons transferred.

It would be nice if you could double the voltage of a cell by multiplying the cell equation by 2, but it doesn't work

Similar calculations can be made in connection with electrolytic as well as voltaic cells. By adding the proper half-reaction voltages, we can calculate the minimum applied voltage necessary, at standard concentrations, to bring about a nonspontaneous oxidation-reduction reaction in an electrolytic cell.

Example 23.2 Calculate the minimum voltage which must be applied in the electrolysis of an aqueous solution of sodium chloride (Reaction 22.15, Chapter 22), assuming standard concentrations.

Solution Writing the electrode reactions with the half-reaction voltages

$$2\ Cl^-(aq) \rightarrow Cl_2(g) + 2\ e^- \qquad E^0_{ox} = -1.36\ V$$
$$2\ H_2O + 2\ e^- \rightarrow H_2(g) + 2\ OH^-(aq) \qquad E^0_{red} = -0.83\ V$$
$$E^0 = -2.19\ V$$

Since the calculated cell voltage is negative, the reaction as written is nonspontaneous and energy must be put into the cell to make the reaction take place. For this reaction, where E^0 is -2.19 V, we must apply a voltage of at least 2.19 V across the electrodes of the cell. This could be done with a 6- or 12-V storage battery or indeed with any direct current source capable of developing a voltage of 2.19 V or more.

Ease of Reduction and Oxidation

Standard voltages for half-reactions measure the relative tendencies of different species to be reduced or oxidized. The more positive the voltage, the more spontaneous is the corresponding half-reaction. A large negative voltage signifies a half-reaction that is difficult to bring about. As an example, consider three species chosen from the left column of Table 23.1:

$$Mg^{2+}(aq) + 2\ e^- \rightarrow Mg(s) \qquad E^0_{red} = -2.37\ V$$
$$2\ H_2O + 2\ e^- \rightarrow H_2(g) + 2\ OH^-(aq) \qquad E^0_{red} = -0.83\ V$$
$$Cu^{2+}(aq) + 2\ e^- \rightarrow Cu(s) \qquad E^0_{red} = +0.34\ V$$

The signs and magnitudes of the standard voltages for these three species tell us that ease of reduction increases in the order $Mg^{2+} < H_2O < Cu^{2+}$. This order is confirmed experimentally. When we pass a direct current through a solution of $CuSO_4$, as in an electroplating cell, Cu^{2+} ions rather than H_2O molecules are reduced. In contrast with copper, magnesium metal cannot be prepared by electrolysis in aqueous solution: if we pass a direct current through a water solution of $MgCl_2$, H_2O molecules rather than Mg^{2+} ions are reduced. As another example, consider the three oxidizable species:

$$2\ I^-(aq) \rightarrow I_2(s) + 2\ e^- \qquad E^0_{ox} = -0.53\ V$$
$$2\ H_2O \rightarrow O_2(g) + 4\ H^+(aq) + 4\ e^- \qquad E^0_{ox} = -1.23\ V$$
$$2\ F^-(aq) \rightarrow F_2(g) + 2\ e^- \qquad E^0_{ox} = -2.87\ V$$

As these voltages imply, iodine but not fluorine can be prepared by oxidation of the corresponding anion in water solution. If we attempt to prepare fluorine gas from aqueous sodium fluoride, either by electrolytic or chemical oxidation, we succeed only in oxidizing water molecules; O_2 rather than F_2 is the product.

From a slightly different point of view, values of E^0_{red} and E^0_{ox} tell us the relative strengths of various oxidizing and reducing agents. A strong oxidizing agent is a species which is readily reduced and hence has a large positive standard voltage for reduction (E^0_{red}). Such species are located in the lower part of the left column of Table 23.1 ($F_2 > Co^{3+} > H_2O_2$). The halogens listed above fluorine decrease in oxidizing power in the order $Cl_2 > Br_2 > I_2$.

The species listed in the right column of Table 23.1 are all capable of acting as reducing agents, at least in principle. Strong reducing agents are readily oxidized; they have large positive values of E^0_{ox}. Species of this type can be found at the upper part of the right column. They include the alkali metals (Li > K > Na),

Group 2A metals (Ba > Ca > Mg), aluminum, and some of the transition metals (Zn > Fe > Cd). Hydrogen gas (E^0_{ox} = 0) is a moderately good reducing agent. Species below H_2 in the right column, all of which have negative values of E^0_{ox}, are relatively weak reducing agents.

Example 23.3 Consider the following species: MnO_4^-, I^-, NO_3^-, Fe^{2+}, Ca. Using Table 23.1, classify each species as an oxidizing or reducing agent. Arrange the oxidizing agents in order of increasing strength; do the same with the reducing agents.

Solution We start by realizing that oxidizing agents (species that can be reduced) are located in the left-hand column. Scanning that column, we find the following oxidizing agents:

$$Fe^{2+}: E^0_{red}\ (Fe^{2+} \rightarrow Fe) = -0.44\ V$$
$$NO_3^-: E^0_{red}\ (NO_3^- \rightarrow NO) = +0.96\ V$$
$$MnO_4^-: E^0_{red}\ (MnO_4^- \rightarrow Mn^{2+}) = +1.52\ V$$

Reducing agents (species that can be oxidized) are found in the right-hand column. Here we locate

$$Ca: E^0_{ox}\ (Ca \rightarrow Ca^{2+}) = +2.87\ V$$
$$I^-: E^0_{ox}\ (I^- \rightarrow I_2) = -0.53\ V$$
$$Fe^{2+}: E^0_{ox}\ (Fe^{2+} \rightarrow Fe^{3+}) = -0.77\ V$$

(Note that Fe^{2+} can act as either an oxidizing agent, in which case it is reduced to Fe, or a reducing agent, where it is oxidized to Fe^{3+}.)

Finally, noting that the more positive the voltage the stronger the oxidizing or reducing agent, we see that

oxidizing agents: $Fe^{2+} < NO_3^- < MnO_4^-$
reducing agents: $Fe^{2+} < I^- < Ca$

23.2 SPONTANEITY AND EXTENT OF REDOX REACTIONS

We have pointed out that a voltaic cell is one in which a spontaneous oxidation-reduction reaction occurs. The converse is also true: any reaction that can occur in a voltaic cell to produce a positive voltage must be spontaneous. To decide whether a given reaction is capable of taking place under a particular set of conditions, all we have to do is to calculate the voltage associated with it. If the calculated voltage is positive, the reaction must be spontaneous. If we calculate a negative voltage, the reaction cannot go by itself; the reverse reaction is spontaneous. A practical application of this simple principle is shown in Example 23.4.

Example 23.4 Using the standard potentials listed in Table 23.1, decide whether

a. Fe(s) will be oxidized to Fe^{2+} by treatment with 1 *M* hydrochloric acid (HCl).
b. Cu(s) will be oxidized to Cu^{2+} by treatment with 1 *M* hydrochloric acid.
c. Cu(s) will be oxidized to Cu^{2+} by treatment with 1 *M* nitric acid (HNO_3).

Solution

a. In order for iron to be oxidized, some species must be reduced. In hydrochloric acid, the only reducible species is the H^+ ion. Looking up the appropriate potentials:

$$\begin{array}{lr} Fe(s) \rightarrow Fe^{2+}(aq) + 2\ e^- & E^0_{ox} = +0.44\ V \\ 2\ H^+(aq) + 2\ e^- \rightarrow H_2(g) & E^0_{red} = 0.00\ V \\ \hline Fe(s) + 2\ H^+(aq) \rightarrow Fe^{2+}(aq) + H_2(g) & E^0 = +0.44\ V \end{array}$$

Since the calculated voltage is positive, we deduce that the reaction should occur.

If species can react in a voltaic cell (E^0 positive), they will react if put into direct contact

In the laboratory, we find that it does. Iron filings dropped into hydrochloric acid slowly dissolve with the evolution of $H_2(g)$.

b. Proceeding in the same way:

$$\begin{array}{lr} Cu(s) \rightarrow Cu^{2+}(aq) + 2\,e^- & E^0_{ox} = -0.34\ V \\ 2\,H^+(aq) + 2\,e^- \rightarrow H_2(g) & E^0_{red} = 0.00\ V \\ \hline Cu(s) + 2\,H^+(aq) \rightarrow Cu^{2+}(aq) + H_2(g) & E^0 = -0.34\ V \end{array}$$

We find, as predicted, that no reaction occurs when copper is added to 1 M hydrochloric acid.

c. In HNO_3, there is another potential oxidizing agent, the NO_3^- ion. Combining the proper half-equations:

$$\begin{array}{lr} 3[Cu(s) \rightarrow Cu^{2+}(aq) + 2\,e^-] & E^0_{ox} = -0.34\ V \\ 2[NO_3^-(aq) + 4\,H^+(aq) + 3\,e^- \rightarrow NO(g) + 2\,H_2O] & E^0_{red} = +0.96\ V \\ \hline 3\,Cu(s) + 2\,NO_3^- + 8\,H^+(aq) \rightarrow 3\,Cu^{2+}(aq) + 2\,NO(g) + 4\,H_2O & E^0 = +0.62\ V \end{array}$$

As predicted, nitric acid does indeed oxidize copper metal to Cu^{2+}; the reduction product is one of the lower oxidation states of nitrogen (e.g., NO or NO_2) rather than $H_2(g)$.

From this example, we see that any metal with a positive value of E^0_{ox} should react with 1 M acid to produce hydrogen gas. Experimentally, this prediction is confirmed. Metals above hydrogen in Table 23.1 are capable of displacing hydrogen from acids, although in many cases the reaction takes place very slowly. Metals with negative standard oxidation voltages such as copper ($E^0_{ox} = -0.34$ V) and silver ($E^0_{ox} = -0.80$ V) do not reduce H^+ ions to $H_2(g)$. Consequently, they fail to react with dilute solutions of an acid such as HCl. They can, however, react with an acid such as HNO_3, where the anion is a strong oxidizing agent (E^0_{red} $NO_3^- = +0.96$ V).

Strictly speaking, the conclusions that we arrive at in Example 23.4 are limited to reactions carried out at standard concentrations (1 M for species in solution, 1 atm for gases) since they are based on values of the standard voltage, E^0. While it is important to know whether a given oxidation-reduction reaction will take place at standard concentrations, this information alone is not sufficient if we wish to carry out the reaction in the laboratory. For one thing, it is highly unlikely that we will want to carry out the reaction under such conditions that all reactants and products are at standard concentrations. Moreover, we would like to know not only whether a particular reaction will take place, but also the extent to which it will occur.

To answer these questions, we need to know the magnitude of the equilibrium constant for the reaction. Fortunately it is possible to calculate K for an oxidation-reduction reaction from the corresponding E^0 value. To obtain a relationship between these two quantities, it is convenient to work through the standard free energy change for the reaction, ΔG^0.

Relation Between E^0 and ΔG^0

You will recall from Chapter 14 that a spontaneous reaction is characterized by a *negative* free energy change. As we have seen in this chapter, a spontaneous oxidation-reduction reaction is always associated with a *positive* voltage. These two statements, taken together, imply a direct relationship between the free energy decrease ($-\Delta G$) and the voltage (E). This relationship can be obtained readily from thermodynamic arguments and is:

$$\Delta G = -nFE$$

where ΔG is the free energy change in kilojoules, n is the number of moles of electrons transferred in the reaction, F is the value of the Faraday constant, 96.5 kJ/V, and E is the voltage.

We are particularly interested in applying this relationship when reactants and products are at their standard concentrations (conc. of 1 M for species in solution, partial pressure of gases 1 atm). Under these conditions we can write

$$\Delta G^0 = -nFE^0 = -96.5\ nE^0 \qquad (23.2)$$

where ΔG^0 is the standard free energy change in kilojoules and E^0 is the standard voltage.

Example 23.5 Using Equation 23.2, calculate ΔG^0 for the reaction

$$Zn(s) + 2\ H^+(aq) \rightarrow Zn^{2+}(aq) + H_2(g)$$

Solution We have previously shown that E^0 for this reaction is +0.76 V. The number of moles of electrons transferred is clearly two (2 mol e^- required to reduce 2 mol H^+; 2 mol electrons released by the oxidation of one mole of Zn). Consequently,

n is easily found from the half-equations

$$\Delta G^0 = -96.5(2)(+0.76)\ \text{kJ} = -147\ \text{kJ}$$

Relation Between E^0 and K

As we pointed out in Chapter 15, the standard free energy change, ΔG^0, is related to the equilibrium constant through the equation

$$\Delta G^0 = -2.30\ RT \log_{10} K \qquad (23.3)$$

where R is the gas constant and T the absolute temperature. Combining Equations 23.2 and 23.3 we obtain

$$nFE^0 = 2.30\ RT \log_{10} K$$

or

$$\log_{10} K = \frac{nFE^0}{2.30\ RT}$$

Substituting the values for the constants F and R in the proper units (96.5 kJ/V and 8.31×10^{-3} kJ/K) and taking T to be 298 K (i.e., 25°C, the temperature at which cell voltages are most commonly measured and tabulated) we obtain

$$\log_{10} K = \frac{nE^0}{0.0591} \quad \text{(at 25°C)} \qquad (23.4)$$

Using Equation 23.4 we can calculate, from the standard cell voltage, $\log_{10} K$ and hence K for any redox reaction taking place in water solution. In the expression for K, concentrations of species in solution are expressed in moles per cubic decimetre. For gases taking part in the reaction, partial pressures in atmospheres are used. Terms for pure solids, pure liquids, or solvent water molecules do not appear in the equilibrium expression.

TABLE 23.2 RELATION BETWEEN E^0 AND K ($n = 2$)

E^0	Log K	K	E^0	Log K	K	E^0	Log K	K
+1.00	+34	10^{34}	+0.10	+3.4	2500	−0.20	−6.8	2×10^{-7}
+0.80	+27	10^{27}	+0.05	+1.7	50	−0.40	−14	10^{-14}
+0.60	+20	10^{20}	0.00	0.0	1	−0.60	−20	10^{-20}
+0.40	+14	10^{14}	−0.05	−1.7	0.02	−0.80	−27	10^{-27}
+0.20	+6.8	6×10^{6}	−0.10	−3.4	0.0004	−1.00	−34	10^{-34}

In Table 23.2 we list values of K calculated from Equation 23.4, corresponding to various values of E^0, for the particular case where $n = 2$. It turns out that if E^0 is greater than about 0.2 V, K will be so large, of the order of 10^7 or greater, that the reaction will go virtually to completion under ordinary conditions. This behavior is characteristic of such reactions as

$$H_2O_2(aq) + 2\ H^+(aq) + 2\ Fe^{2+}(aq) \rightarrow 2\ H_2O + 2\ Fe^{3+}(aq);\ E^0 = +1.00\ V;\ K = 10^{34}$$

Addition of excess hydrogen peroxide to a solution containing Fe^{2+} ions quantitatively oxidizes them to Fe^{3+}. In contrast, we expect that reactions for which the calculated E^0 value is less than about −0.2 V will show little tendency to take place. For example, the reaction

For this reason, most redox reactions go to completion or not at all

$$2\ H^+(aq) + 2\ Fe^{2+}(aq) \rightarrow H_2(g) + 2\ Fe^{3+}(aq);\ E^0 = -0.77\ V;\ K = 10^{-26}$$

does not occur to any detectable extent; iron(II) salts such as $FeSO_4$ or $FeCl_2$ are stable in acidic solution, at least in the absence of dissolved oxygen.

We are particularly interested in working with equilibrium constants for redox reactions where E^0 is a relatively small number, perhaps falling in the range +0.2 to −0.2 V. For such reactions we can expect to find appreciable concentrations of both reactants and products in the equilibrium mixture.

The equilibrium constant for an oxidation-reduction reaction, like any equilibrium constant, can be used to determine

—the extent to which the reaction occurs, starting with pure reactants at known concentrations (Example 23.6b).

—the direction in which the reaction will proceed spontaneously at any given concentrations of reactants and products (Example 23.6c).

Example 23.6 Consider the reaction: $Ag^+(aq) + Fe^{2+}(aq) \rightleftarrows Ag(s) + Fe^{3+}(aq)$

a. Calculate K for this reaction.
b. If enough $AgNO_3$ is added to a solution originally 0.10 M in Fe^{2+} to make the final $[Ag^+] = 1.0\ M$, what will be the equilibrium concentration of Fe^{3+}?
c. Determine the direction in which the reaction will proceed if the concentrations of Fe^{3+}, Fe^{2+}, and Ag^+ are 1.0 M, 0.10 M, and 0.10 M, respectively.

Solution

a. We first obtain E^0 from standard voltages

$$E^0 = E^0_{red}\ (Ag^+ \rightarrow Ag) + E^0_{ox}\ (Fe^{2+} \rightarrow Fe^{3+}) = +0.80\ V - 0.77\ V = 0.03\ V$$

To calculate K, we note that for this reaction, $n = 1$

$$\log_{10} K = \frac{(1)(0.03)}{0.059} = 0.5;\ K = 3$$

b. The expression for K is: $K = \frac{[Fe^{3+}]}{[Ag^{+}] \times [Fe^{2+}]} = 3$

In the reaction, one mole of Fe^{2+} is consumed for every mole of Fe^{3+} formed. Hence, if we let $[Fe^{3+}] = x$, then $[Fe^{2+}] = 0.10 - x$. From the statement of the problem, $[Ag^{+}] = 1.0$. Making these substitutions

$$\frac{x}{0.10 - x} = 3; \quad x = 0.07$$

These calculations are entirely analogous to those carried out in Chapter 15

we see that about 70% of the Fe^{2+} ions are converted to Fe^{3+}. Certainly Ag^{+} would not be the best reagent to use if we wished to quantitatively oxidize Fe^{2+} ions.

c. To decide upon the direction in which the system will move to establish equilibrium, we follow the reasoning described on p. 362, Chapter 15, comparing the actual concentration quotient to that which must exist at equilibrium:

$$\frac{(\text{conc. } Fe^{3+})}{(\text{conc. } Ag^{+}) \times (\text{conc. } Fe^{2+})} = \frac{1.0}{(0.10)(0.10)} = 100$$

Since 100 is much larger than the equilibrium constant ($K = 3$), the *reverse* reaction must occur to establish equilibrium. Some of the Fe^{3+} ions react to form Fe^{2+} and Ag^{+} ions until the concentration ratio drops to the equilibrium value, 3.

Notice that at standard concentrations (1.0 M), the forward reaction is spontaneous ($E^0 = +0.03$ V). By reducing the concentrations of the two reactants to 0.10 M, the direction of the reaction is reversed.

23.3 EFFECT OF CONCENTRATION ON VOLTAGE

As pointed out earlier in this chapter, the voltage of a cell depends not only upon the nature of reactants and products but also upon their concentrations. To illustrate the effect of concentration upon voltage, consider the Zn-Cu^{2+} cell shown in Figure 22.6, p. 542. If we start with 1 M concentrations of Zn^{2+} and Cu^{2+}, the measured voltage should be the standard value, 1.10 V.

$$Zn(s) + Cu^{2+}(aq, 1\ M) \rightarrow Zn^{2+}(aq, 1\ M) + Cu(s); \ E^0 = 1.10\ V \quad (23.5)$$

Consider now what happens when we operate this cell to produce electrical energy. Reaction 23.5 occurs, thereby increasing the concentration of Zn^{2+} and decreasing that of Cu^{2+}. Equilibrium considerations suggest that these changes will reduce the driving force for the reaction, making it less spontaneous.

Since the cell voltage is a measure of the spontaneity of the cell reaction, we might expect that voltage to fall off as product (Zn^{2+}) builds up and reactant (Cu^{2+}) disappears. Experimentally, we find that this prediction is confirmed. As time passes, the voltage decreases steadily. Eventually, the ratio (conc. Zn^{2+})/(conc. Cu^{2+}) becomes equal to the equilibrium constant for the cell reaction ($K = 10^{37}$), and equilibrium is reached. When this happens, the driving force for the reaction disappears, the voltage becomes zero, and we say that the cell is "dead."

There are many other ways in which we might increase the ratio: (conc. Zn^{2+})/(conc. Cu^{2+}) in the Zn-Cu^{2+} cell. We might, for example, add a soluble zinc salt to the zinc half-cell. Alternatively, we could reduce the concentration of Cu^{2+} in the copper half-cell by diluting with water or adding a reagent such as S^{2-} ions, which form a precipitate with Cu^{2+}. Regardless of how we do it, we always find that increasing the ratio (conc. Zn^{2+})/(conc. Cu^{2+}) decreases the cell voltage. Con-

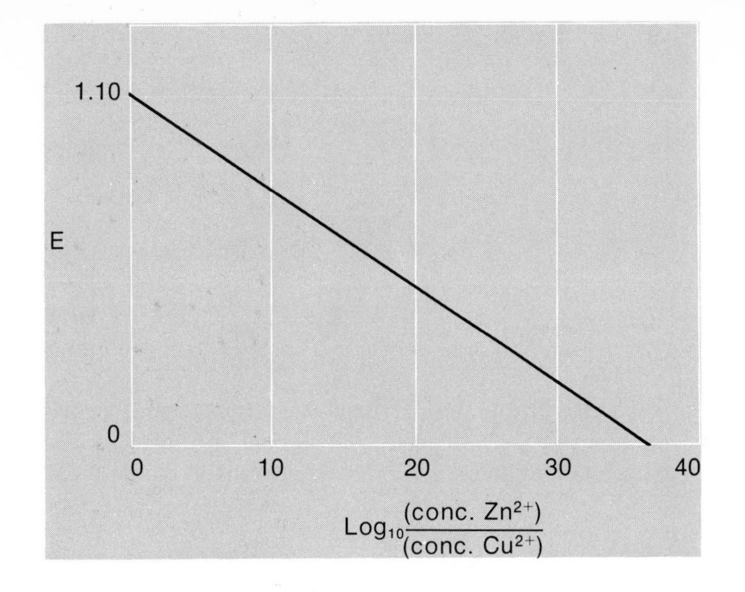

Figure 23.1 Dependence of voltage of cell: $Zn(s) + Cu^{2+}(aq) \rightarrow Zn^{2+}(aq) + Cu(s)$ on concentrations of the two cations. As the cell produces electrical energy, conc. Zn^{2+} increases and conc. Cu^{2+} decreases. The voltage gradually drops from 1.10 V (E^0), becoming zero when equilibrium is reached.

versely if this ratio is decreased, perhaps by adding water or S^{2-} ions to the zinc half-cell so as to lower the concentration of Zn^{2+}, the cell voltage increases.

The behavior just described is typical of all voltaic cells. In general,

—if the concentration of products increases relative to that of reactants, the cell reaction becomes less spontaneous and the voltage decreases.

—if the concentration of products decreases relative to that of reactants, the cell reaction becomes more spontaneous and the voltage increases.

Increasing product concentrations tends to drive the reaction to the left and so decrease the voltage

The Nernst Equation

From Figure 23.1, we can deduce the mathematical form of the relationship between the voltage, *E*, of the Zn-Cu^{2+} cell and the ratio (conc. Zn^{2+})/(conc. Cu^{2+}). Clearly *E* is a linear function of the logarithm of that ratio. That is:

$$E = A + B \log_{10} \frac{(\text{conc. } Zn^{2+})}{(\text{conc. } Cu^{2+})}$$

where A, the y intercept, is 1.10 V and B, the slope, is readily calculated to be about −0.03. Thus for this cell we have

$$E = 1.10 - 0.03 \log_{10} \frac{(\text{conc. } Zn^{2+})}{(\text{conc. } Cu^{2+})}$$

The equation just written is a special form of a more general relation known as the **Nernst equation.** For the general oxidation-reduction reaction at 25°C

$$aA + bB \rightarrow cC + dD$$

in which A, B, C, and D are species whose concentrations can be varied and a, b, c, and d are the corresponding coefficients of the balanced equation, the Nernst equation has the form

$$E = E^0 - \frac{0.0591}{n} \log_{10} \frac{(\text{conc. C})^c(\text{conc. D})^d}{(\text{conc. A})^a(\text{conc. B})^b} \qquad (23.6)$$

(E = cell voltage, E^0 = standard voltage, n = no. moles e^- transferred in reaction.)

Thus, for the cell reactions

$$Zn(s) + 2\ Ag^+(aq) \rightarrow Zn^{2+}(aq) + 2\ Ag(s);\ E^0 = 1.56\ V$$

$$E = 1.56 - \frac{0.0591}{2} \log_{10} \frac{(\text{conc. } Zn^{2+})}{(\text{conc. } Ag^+)^2}$$

and for $\quad Zn(s) + 2\ H^+(aq) \rightarrow Zn^{2+}(aq) + H_2(g) \qquad E^0 = 0.76\ V$

$$E = 0.76 - \frac{0.0591}{2} \log_{10} \frac{(\text{conc. } Zn^{2+})(\text{pressure } H_2)}{(\text{conc. } H^+)^2} \tag{23.7}$$

You will note that, in these equations, terms for the solids taking part in the cell reaction (e.g., Zn, Ag) do not appear in the expression for E. Indeed, the cell voltage is independent of the size or shape of the solid electrodes used.

Example 23.7 Calculate the voltage of a cell in which the following reaction occurs:

$$Zn(s) + 2\ H^+(aq, 0.001\ M) \rightarrow Zn^{2+}(aq, 1\ M) + H_2(g, 1\ atm)$$

Solution Substituting in Equation 23.7:

$$E = 0.76 - \frac{0.0591}{2} \log_{10} \frac{(1)(1)}{(10^{-3})^2}$$

$$= 0.76 - \frac{0.0591}{2} \log_{10} 10^6 = 0.76 - \frac{(0.0591)(6)}{2} = +0.58\ V$$

The Nernst equation can also be applied to determine the effect of changes in concentration on the voltage of an individual half-cell. For example, for the half-reaction

$$MnO_4^-(aq) + 8\ H^+(aq) + 5\ e^- \rightarrow Mn^{2+}(aq) + 4\ H_2O;\ E^0_{red} = 1.52\ V$$

we can write:

$$E_{red} = 1.52 - \frac{0.0591}{5} \log_{10} \frac{(\text{conc. } Mn^{2+})}{(\text{conc. } MnO_4^-)(\text{conc. } H^+)^8}$$

We shall have more to say about this application in Section 23.4.

Use of the Nernst Equation to Determine Concentration of Ions

From the standpoint of chemistry, the most important application of the Nernst equation is in the experimental determination of the concentrations of species in solution. To illustrate what is involved, let us refer back to Equation 23.7. We used this equation, in Example 23.7, to calculate a cell voltage given the concentrations of all the species involved in the reaction. We can turn this procedure around; if we measure the cell voltage and know the concentrations of all but one species, we can use the equation to obtain the "unknown" concentration (Example 23.8).

Example 23.8 If the measured voltage of the Zn-H^+ cell is +0.46 V when the cell reaction is

$$Zn(s) + 2\,H^+(aq) \rightarrow Zn^{2+}(aq,\ 1\ M) + H_2(g,\ 1\ atm),$$

what must be the concentration of H^+?

Solution Applying Equation 23.7:

$$0.46 = 0.76 - \frac{0.0591}{2}\log_{10}\frac{1}{(\text{conc. } H^+)^2}$$

But, since $\log 1/x^2 = -\log x^2 = -2 \log x$, we can write

$$0.46 = 0.76 - \frac{0.0591}{2}(-2 \log_{10} \text{conc. } H^+) = 0.76 + 0.0591 \log_{10}(\text{conc. } H^+)$$

$$\text{Solving: } \log_{10}(\text{conc. } H^+) = \frac{0.46 - 0.76}{0.0591} = -5.1 = 0.9 - 6.0;\ \text{conc. } H^+ = 8 \times 10^{-6}\ M$$

We see from Example 23.8 that the Zn-H^+ cell offers a simple and quite precise method of measuring the concentration of H^+ or the *pH* of a solution. Indeed, it is with cells of this type that *pH* is ordinarily measured in the laboratory. In Figure 23.2 we show a schematic diagram of an instrument known as a *pH* meter which is specially designed for this purpose. The three essential components of the *pH* meter are:

1. A vacuum-tube voltmeter or potentiometer, capable of measuring voltages to at least ±0.01 V.
2. A reference half-cell of known voltage.
3. A half-cell whose voltage depends on the concentration of H^+. This consists of a metal electrode dipping into a solution of known *pH* separated by a *thin, fragile* glass membrane from the solution whose *pH* is to be determined. The voltage across this *glass electrode,* and hence the cell voltage itself, is a linear function of the *pH* of the solution outside the membrane.

A *pH* meter must always be calibrated against a buffer of accurately known *pH*

During the past decade *specific ion electrodes* somewhat similar in design to the glass electrode have been developed to analyze for a variety of cations and anions. One of the first to be used extensively was a fluoride ion electrode which is sensitive to F^- at concentrations as low as 0.1 ppm and, hence, is ideal for monitoring fluoridated water supplies. An electrode which is specific for Cl^-

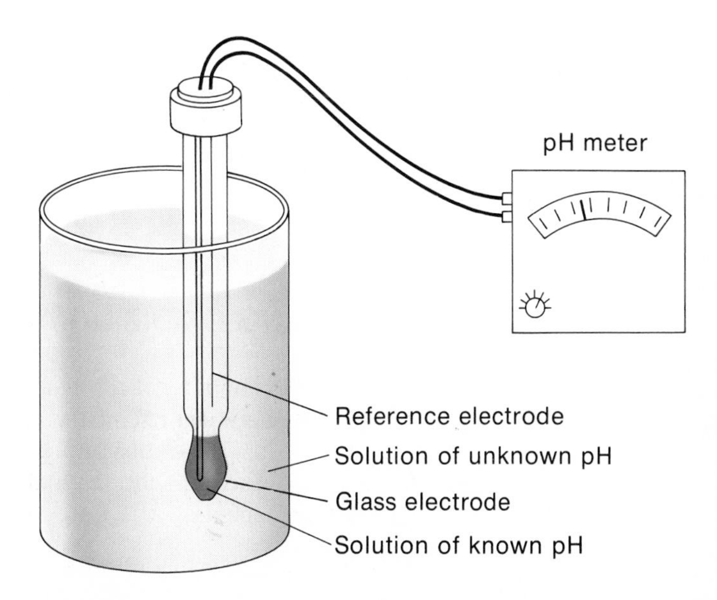

Figure 23.2 Measurement of *pH*. A *pH* meter is a voltmeter with an extremely high resistance. When operating, the meter draws a negligible amount of current from the cell, and so measures the maximum voltage the cell can produce. With a *pH* meter, under optimum conditions, one can measure the *pH* of a solution to ±0.001 unit.

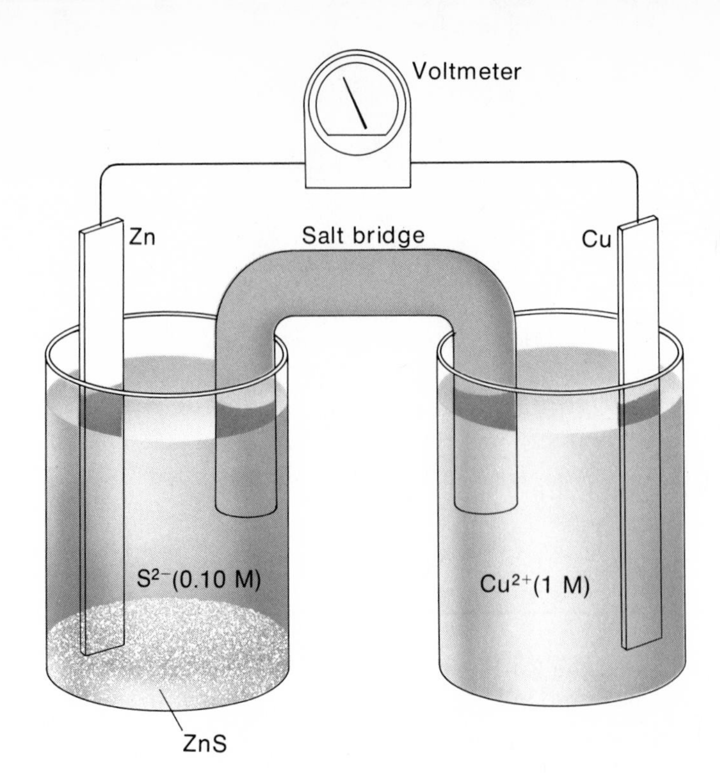

Figure 23.3 Determination of K_{sp} of ZnS, using a Zn-Cu^{2+} cell. Enough Na_2S is added to the zinc half cell to precipitate most of the Zn^{2+} ions and bring the S^{2-} concentration to a known value, such as 0.10 *M*. By measuring the cell voltage under these conditions and using the Nernst equation, one can calculate the concentration of Zn^{2+} in equilibrium with ZnS(s). From that value and the known concentration of S^{2-} ion, the solubility product can be determined.

ions is now being used to diagnose for cystic fibrosis. Attached directly to the skin, it detects the abnormally high concentrations of sodium chloride in sweat that are a characteristic symptom of this disorder. Diagnoses that used to require an hour or more can now be carried out in a few minutes; as a result, large numbers of children can be rapidly and routinely screened.

The general approach illustrated by Example 23.8 can be applied to find ionization constants of weak acids, dissociation constants of complex ions, or solubility product constants of slightly soluble salts. For example, to determine K_{sp} of zinc sulfide, we could start with a standard Zn-Cu^{2+} cell and add enough S^{2-} ions to the zinc half-cell to precipitate virtually all of the Zn^{2+} ions (Fig. 23.3). Applying the Nernst equation to the cell reaction:

Most K_{sp} values were determined this way

$$Zn(s) + Cu^{2+}(aq, 1\ M) \rightarrow Zn^{2+}(aq) + Cu(s)$$

$$E = 1.10 - \frac{0.0591}{2} \log_{10} \frac{(\text{conc. } Zn^{2+})}{1}$$

From the measured cell voltage, the (very low) concentration of Zn^{2+} could be calculated. Multiplying by the equilibrium concentration of S^{2-} in the zinc half-cell would then yield K_{sp} for ZnS.

$$[Zn^{2+}] \times [S^{2-}] = K_{sp}$$

23.4 STRONG OXIDIZING AGENTS

Oxidation-reduction reactions are of considerable importance in industrial, inorganic, and analytical chemistry. They are used to synthesize compounds, to bring otherwise insoluble substances into solution, and to analyze for a variety of ions. Many of these applications make use of one of a relatively small group of strong oxidizing agents.

A "strong" oxidizing agent, as we have noted, is a species which tends to pick up electrons readily; in more quantitative terms, it is an ion or molecule that has a large positive standard reduction voltage. Looking at Table 23.1, we see that most such species fall into one of two categories:

1. Molecules of nonmetals (Br_2, O_2, Cl_2, F_2).
2. Oxyanions (NO_3^-, $Cr_2O_7^{2-}$, ClO_3^-, MnO_4^-) in acid solution.

In this section we shall examine some of the oxidation-reduction reactions in which three of these species, the Cl_2 molecule, the NO_3^- ion, and the $Cr_2O_7^{2-}$ ion, participate. Some of the reactions that we will be talking about are ones that are frequently carried out in the general chemistry laboratory; others are used in industrial processes. Hopefully, they will serve to illustrate the principles concerning spontaneity and extent of redox reactions that we have emphasized in this chapter.

Chlorine

As is indicated by the magnitude of its standard reduction voltage (E^0_{red} = +1.36 V), elementary chlorine is a powerful oxidizing agent. It is particularly effective in oxidizing organic compounds, which accounts for its use in water purification (Chapter 13). Perhaps the most familiar reactions in which chlorine acts as an oxidizing agent are those involving bromide and iodide ions.

$$Cl_2(g) + 2\,Br^-(aq) \rightarrow 2\,Cl^-(aq) + Br_2(l);\ E^0 = +1.36\text{ V} - 1.07\text{ V} = +0.29\text{ V} \quad (23.8)$$

$$Cl_2(g) + 2\,I^-(aq) \rightarrow 2\,Cl^-(aq) + I_2(s);\ E^0 = +1.36\text{ V} - 0.53\text{ V} = +0.83\text{ V} \quad (23.9)$$

These reactions are frequently used to test for the presence of Br^- or I^- ions, as shown in Plate 16. Addition of chlorine to a solution containing either of these ions gives the free halogens Br_2 or I_2. If the water solution is then shaken with a small amount of a nonpolar organic solvent such as carbon disulfide or carbon tetrachloride, the free halogens enter the organic layer, to which they impart their characteristic colors, reddish brown (bromine) or violet (iodine).

Why add the organic solvent? why a "small amount"?

Bromine is prepared commercially from seawater, in which it occurs as Br^- ions, by oxidation with chlorine (Reaction 23.8). About 2×10^8 kg of bromine are produced annually (most of it in the United States) by this method. The concentration of iodide ions in seawater is so low (conc. $I^- = 4 \times 10^{-7}\,M$ vs. conc. $Br^- = 8 \times 10^{-4}\,M$) that it is not feasible to produce iodine in this way. In oil well brines, where the concentration of I^- is considerably higher, Reaction 23.9 can be carried out; most of the iodine we use is prepared by oxidation of the iodides in these brines with chlorine.

Chlorine, unlike oxygen, forms many compounds in which it has a positive oxidation number. These compounds are ordinarily formed by a **disproportionation** reaction in which elementary chlorine is simultaneously oxidized and reduced. An example of one such reaction is that which occurs when chlorine is added to water.

$$Cl_2(g) + H_2O \rightleftharpoons HOCl(aq) + H^+(aq) + Cl^-(aq);\ K = 3 \times 10^{-5} \quad (23.10)$$

The resulting solution, called *chlorine water,* contains equimolar amounts of the weak acid HOCl (hypochlorous acid) and the strong acid HCl. Half the chlorine (oxid state = 0) is reduced to Cl^- ions (oxid state = −1) while the remainder is oxidized to HOCl (oxid state Cl = +1).

The position of the equilibrium in Reaction 23.10 is strongly affected by the concentration of H^+ ions. In basic solution, in which the concentration of H^+ ions is low, chlorine is much more soluble than in pure water. The overall reaction that takes place when chlorine is bubbled through a solution of sodium hydroxide maintained at room temperature is

$$Cl_2(g) + 2\ OH^-(aq) \rightarrow ClO^-(aq) + Cl^-(aq) + H_2O \qquad (23.11)$$

Why would it be dangerous to add acid to a household bleach?

The solution formed by the reaction of chlorine with sodium hydroxide via Reaction 23.11 is sold under various trade names as a household bleach and disinfectant. It is prepared commercially by electrolyzing a stirred water solution of sodium chloride. Recall that the electrolysis of an NaCl solution gives Cl_2 molecules and OH^- ions (Chapter 22); stirring ensures that these species react with each other. The active ingredient of the resulting solution is the hypochlorite ion, a relatively potent oxidizing agent:

$$ClO^-(aq) + H_2O + 2\ e^- \rightarrow Cl^-(aq) + 2\ OH^-(aq) \qquad E^0_{red} = +0.89\ V$$

The reaction of chlorine with a hot, concentrated solution of sodium or potassium hydroxide is quite different from that observed at room temperature. Any ClO^- ions formed decompose on heating to ClO_3^- and Cl^- ions; the net reaction is

$$3\ Cl_2(g) + 6\ OH^-(aq) \rightarrow ClO_3^-(aq) + 5\ Cl^-(aq) + 3\ H_2O \qquad (23.12)$$

Potassium chlorate is a powerful oxidizing agent (E^0_{red} ClO_3^- = +1.47 V) and can react violently with easily oxidized materials, including many organic substances. It is sometimes used as a source of oxygen in the general chemistry laboratory.

$$2\ KClO_3(s) \rightarrow 2\ KCl(s) + 3\ O_2(g) \qquad (23.13)$$

This reaction, in the absence of a catalyst, takes place very slowly below a temperature of 400°C. In the temperature range 350–400°C, the principal products are potassium perchlorate, $KClO_4$, and potassium chloride.

$$4\ KClO_3(s) \rightarrow 3\ KClO_4(s) + KCl(s) \qquad (23.14)$$

Oxyanions, NO_3^- and $Cr_2O_7^{2-}$

Although oxyanions differ greatly in their oxidizing strength, they have certain characteristics in common. In particular,

1. *Oxyanions are stronger oxidizing agents in acidic than in neutral or basic solution.* We find, for example, that concentrated nitric acid is a powerful oxidizing agent, capable of oxidizing both copper and silver, neither of which reacts with H^+ ions alone:

$$Cu(s) + 2\ NO_3^-(aq) + 4\ H^+(aq) \rightarrow Cu^{2+}(aq) + 2\ NO_2(g) + 2\ H_2O \quad (23.15)$$

$$Ag(s) + NO_3^-(aq) + 2\ H^+(aq) \rightarrow Ag^+(aq) + NO_2(g) + H_2O \qquad (23.16)$$

In contrast, salts such as KNO_3, containing the NO_3^- ion, are ineffective oxidizing agents in neutral solution. Again, a solution prepared by adding sulfuric acid to potassium dichromate is frequently used as an oxidizing agent to clean laboratory glassware; it removes greases and oils which are impervious to a solution containing only $K_2Cr_2O_7$.

We can readily explain this relationship between oxidizing strength and acidity by examining the half-equations for the reduction of NO_3^- and $Cr_2O_7^{2-}$ ions.

$$NO_3^-(aq) + 2\ H^+(aq) + e^- \rightarrow NO_2(g) + H_2O$$

$$Cr_2O_7^{2-}(aq) + 14\ H^+(aq) + 6\ e^- \rightarrow 2\ Cr^{3+}(aq) + 7\ H_2O$$

We see that, in both cases, the H^+ ion is involved as a reactant. Increasing its concentration should then make the oxidation more spontaneous; i.e., it should make the reduction voltage a larger positive number (Example 23.9).

Example 23.9 E^0_{red} for the half-reaction

$$NO_3^-(aq) + 2\ H^+(aq) + e^- \rightarrow NO_2(g) + H_2O$$

is +0.78 V. Using the Nernst equation, calculate E_{red} in 10 *M* H^+ and in neutral solution, assuming all other species to be at unit concentrations.

Solution Applying the Nernst equation to this half-reaction with conc. NO_3^- = 1 *M* and pNO_2 = 1 atm, we have

$$E_{red} = +0.78 - \frac{0.0591}{1} \log_{10} \frac{1}{(\text{conc. } H^+)^2}$$

simplifying: $$E_{red} = +0.78 + 2(0.0591) \log_{10} (\text{conc. } H^+)$$

Notice that, as predicted, the reduction voltage increases as the solution becomes more acidic. In 10 *M* acid, conc. H^+ = 10, $\log_{10}$ (conc. H^+) = 1:

$$E_{red} = +0.78 + 2(0.0591)(1) = +0.90\ V$$

In neutral solution, conc. $H^+ = 1 \times 10^{-7}$, $\log_{10}$ (conc. H^+) = −7

$$E_{red} = +0.78 + 2(0.0591)(-7) = -0.05\ V$$

Clearly, the NO_3^- ion in neutral solution will be a weak oxidizing agent.

In neutral solution, the NO_3^- ion is very stable

2. *Oxyanions can be reduced to a variety of species, depending upon the experimental conditions.* Table 23.3 indicates some of the species to which the NO_3^- and $Cr_2O_7^{2-}$ ions can be reduced.

Examining Table 23.3, we can count some 12 different species to which the NO_3^- ion might be reduced and four possible reduction products for $Cr_2O_7^{2-}$. (The actual number of species is somewhat greater; we have left out some of the more

TABLE 23.3 OXIDATION STATES OF N AND Cr

	NITROGEN			CHROMIUM	
	Acidic Solution	Basic Solution		Acidic Solution	Basic Solution
+5	NO_3^-	NO_3^-	+6	$Cr_2O_7^{2-}$	CrO_4^{2-}
+4	$NO_2(g)$	$NO_2(g)$	+3	Cr^{3+}	$Cr(OH)_3(s)$*
+3	HNO_2	NO_2^-	+2	Cr^{2+}	$Cr(OH)_2(s)$
+2	$NO(g)$	$NO(g)$			
+1	$N_2O(g)$	$N_2O(g)$			
0	$N_2(g)$	$N_2(g)$			
−1	NH_3OH^+	NH_2OH			
−2	$N_2H_5^+$	N_2H_4			
−3	NH_4^+	$NH_3(g)$			

*In strongly basic solution, complex ions such as $Cr(OH)_6^{3-}$ will form.

exotic ones.) It might seem futile to hope that we could predict in advance which species would be formed in a particular redox reaction. There are, however, some guiding principles that we can use to predict the position of:

a. *Equilibria between acidic and basic species within a given oxidation state.* If we know the equilibrium constant for the appropriate acid-base reaction, this prediction can be made with confidence. Consider, for example, the two species of nitrogen in the +3 oxidation state, HNO_2 and NO_2^-. Their concentrations can be related through the dissociation constant for the weak acid, HNO_2:

$$HNO_2(aq) \rightleftharpoons H^+(aq) + NO_2^-(aq);\ K_a = 4.5 \times 10^{-4}$$

By rearranging the expression for K_a we obtain:

$$\frac{[HNO_2]}{[NO_2^-]} = \frac{[H^+]}{K_a} = \frac{[H^+]}{4.5 \times 10^{-4}}$$

We see that at H^+ ion concentrations greater than 4.5×10^{-4} *M*, the ratio $[HNO_2]/[NO_2^-]$ will be greater than one; under these conditions the weak acid HNO_2 will be the predominant species. At H^+ ion concentrations less than 4.5×10^{-4} *M*, the principal species present will be the weak base, NO_2^-.

A particularly interesting case of acid-base equilibrium is offered by the +6 state of chromium

Note that 23.17 is an acid-base reaction, not redox

$$\underset{\text{yellow}}{2\ CrO_4^{2-}(aq)} + 2\ H^+(aq) \rightleftharpoons \underset{\text{red}}{Cr_2O_7^{2-}(aq)} + H_2O \qquad (23.17)$$

Here, we can determine the position of the equilibrium visually by noting the color of the solution. At a *pH* greater than 7, the CrO_4^{2-} ion predominates; as the solution is made acidic, the yellow color changes to red, indicating the formation of $Cr_2O_7^{2-}$. These changes are shown in Plate 17.

b. *Equilibria between different oxidation states.* The reduction of +6 chromium almost always leads to the +3 state. Only by using an excess of a powerful reducing agent such as zinc can we reach the relatively unstable +2 state of chromium.

$$Cr^{3+}(aq) + e^- \rightarrow Cr^{2+}(aq);\ E^0_{red} = -0.41\ V$$

In the case of the NO_3^- ion, it is a great deal more difficult to predict which species will be produced on reduction because of the multiplicity of lower oxidation states, ranging from +4 to −3. Indeed, whenever we write an equation showing the reduction of NO_3^- to a single species such as $NO_2(g)$ or $NO(g)$, we are oversimplifying reality; nearly always we get a mixture of reduction products in varying proportions.

When we carry out oxidation-reduction reactions with concentrated nitric acid, nitrogen dioxide, NO_2, can always be detected among the products. For example, treatment of copper metal with concentrated nitric acid yields copious brown fumes of NO_2 in accordance with Reaction 23.15. The formation of NO_2 is readily explained on a kinetic basis. Even though we know very little about the detailed mechanism of the reduction process, it seems certain that the first step is a one-electron transfer to go from the +5 state (NO_3^-) to the +4 state (NO_2).

In principle, reduction of NO_3^- should proceed past NO_2 to lower oxidation states. The standard voltages for the reduction of NO_2 to HNO_2 (+1.08 V), NO (+1.04 V), and N_2 (+1.36 V) are all more positive than that for the reduction of NO_3^- to NO_2 (+0.78 V). However, if NO_2 is produced rapidly, as is the case when concentrated acid is used, it escapes from solution rather than undergoing further reduction. With dilute acid, where reaction occurs more slowly, the situation is quite different; the NO_2 formed in the first step stays around long enough to be further reduced. Possible overall half equations for the reduction of dilute HNO_3 include:

$$NO_3^-(aq) + 4\ H^+(aq) + 3\ e^- \rightarrow NO(g) + 2\ H_2O \qquad (23.18)$$

$$NO_3^-(aq) + 3\ H^+(aq) + 2\ e^- \rightarrow HNO_2(aq) + H_2O \qquad (23.19)$$

The latter reaction can occur in the stomach (pH 1 to 3) where certain bacteria act as the reducing agent. The HNO_2 produced may react with hemoglobin of the blood, converting it from an iron(II) to an iron(III) complex which cannot act as an oxygen carrier. Infant mortality from this source (blue babies) had led to restrictions on the concentration of nitrate in drinking water.

23.5 OXYGEN; THE CORROSION OF IRON

Of all oxidizing agents, elementary oxygen is the most abundant and, in many ways, the most important. Its presence in air insures that all water supplies including reagent solutions used in the laboratory will ordinarily be saturated with atmospheric oxygen. We often forget this and are puzzled by such phenomena as

—the formation of white or yellow precipitates when hydrogen sulfide is used in qualitative analysis

$$\tfrac{1}{2}\ O_2(g) + H_2S(g) \rightarrow S(s) + H_2O \qquad (23.20)$$

—the yellow color that solutions of NaI or KI acquire upon standing

$$\tfrac{1}{2}\ O_2(g) + 2\ I^-(aq) + 2\ H^+(aq) \rightarrow I_2(aq) + H_2O \qquad (23.21)$$

—the cloudiness that develops in a solution of tin(II) chloride

$$\tfrac{1}{2}\ O_2(g) + Sn^{2+}(aq) + H_2O \rightarrow SnO_2(s) + 2\ H^+(aq) \qquad (23.22)$$

From an economic standpoint, the most important redox reaction involving dissolved oxygen is the corrosion of iron and steel. It is estimated that the annual cost to the United States of corrosion of ferrous metals exceeds five billion dollars. We see the results of corrosion all around us in junk piles and auto graveyards. Perhaps as much as 20 per cent of all the iron produced each year goes to replace products whose usefulness has been destroyed by rust.

In order to understand the mechanism by which iron corrodes, let us consider what happens when a sheet of iron is exposed to a neutral water solution containing an electrolyte such as sodium chloride. The iron tends to oxidize according to the half-reaction

Iron won't rust in dry air

$$Fe(s) \rightarrow Fe^{2+}(aq) + 2\ e^- \qquad (23.23a)$$

For this reaction to take place, some other species must be reduced simultaneously. A reasonable possibility would be the reduction of H^+ ions to elementary hydrogen (E^0_{red} H^+ = 0.00 V). This does occur in strongly acidic solution, but cannot take place at an appreciable rate in neutral solution, in which the concentration of H^+ ions is only 10^{-7} M. Instead, oxygen molecules dissolved in the solution are reduced:

$$\tfrac{1}{2}\ O_2(g) + H_2O + 2\ e^- \rightarrow 2\ OH^-(aq) \qquad (23.23b)$$

Adding these two half-equations and noting that iron(II) hydroxide is insoluble in water, we obtain for the primary corrosion reaction,

$$Fe(s) + \frac{1}{2} O_2(g) + H_2O \rightarrow Fe(OH)_2(s) \quad (23.23)$$

There is a great deal of evidence to suggest that Reaction 23.23 represents the first step in the corrosion of iron or steel. However, iron(II) hydroxide is ordinarily further oxidized in the reaction

$$2\ Fe(OH)_2(s) + \frac{1}{2} O_2(g) + H_2O \rightarrow 2\ Fe(OH)_3(s) \quad (23.24)$$

The final product—the loose, flaky deposit that we call rust—has the reddish brown color of iron(III) hydroxide, $Fe(OH)_3$.

A significant clue to the mechanism of corrosion is the experimental observation that the oxidation half-reaction (23.23a) and the reduction half-reaction (23.23b) do not occur at the same location. If we look at a nail extracted from an old building, we frequently find that the rust is concentrated near the head of the nail, which has been in contact with moist air. The most serious pitting, often amounting to disintegration, is found along the shank of the nail, which is embedded in the wood. These observations lead us to believe that oxidation (23.23a) is occurring along a surface that may be some distance away from the point at which oxygen is being reduced (23.23b).

By connecting a voltmeter to two different points on the surface, you can measure a voltage

The fact that oxidation and reduction half-reactions take place at different locations suggests that corrosion occurs by an electrochemical mechanism. The surface of a piece of corroding iron may be visualized as consisting of a series of tiny voltaic cells. At *anodic areas,* iron is oxidized to Fe^{2+} ions; at *cathodic areas,* elementary oxygen is reduced to OH^- ions. Electrons are transferred through the iron, which acts like the external conductor of an ordinary voltaic cell. The electrical circuit is completed by the flow of ions through the water solution or film covering the iron. The fact that rust ordinarily accumulates at cathodic areas suggests that it is primarily Fe^{2+} ions which move through the solution, from anode (23.23a) to cathode (23.23b).

Many of the characteristics of corrosion are most readily explained in terms of an electrochemical mechanism. A perfectly dry metal surface is not attacked by oxygen; iron exposed to dry air does not corrode. This seems plausible if corrosion occurs through a voltaic cell, which requires a water solution through which ions can move to complete the circuit. The fact that corrosion occurs more readily in seawater than in fresh water has a similar explanation. The dissolved salts in seawater supply the ions necessary for the conduction of current.

The existence of discrete cathodic and anodic areas on a piece of corroding

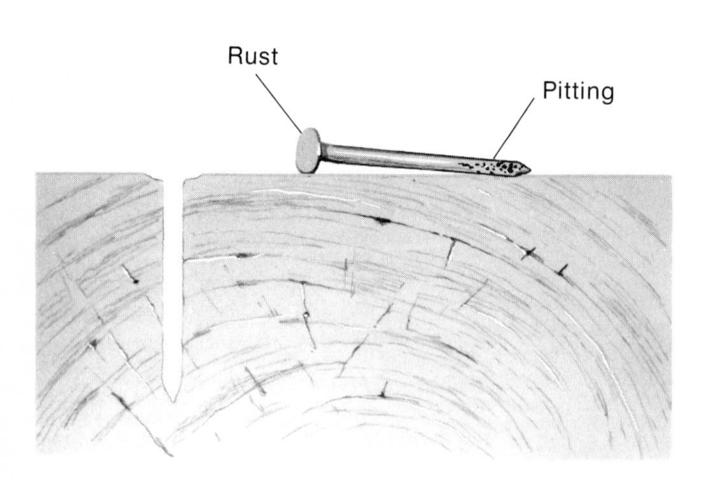

Figure 23.4 Corrosion of an iron nail driven into wood. Rust collects near the head of the nail, but pitting occurs along its length.

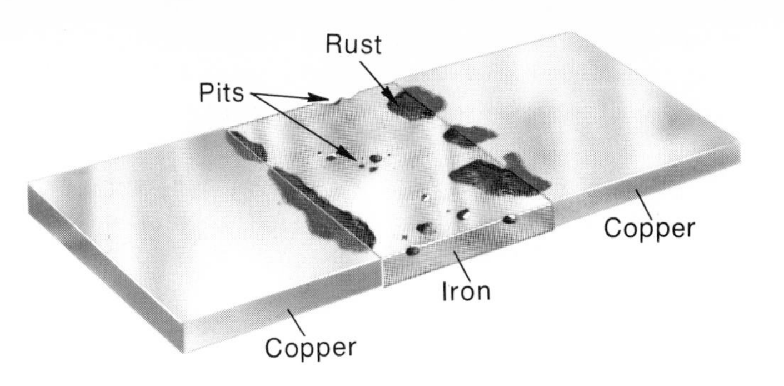

Figure 23.5 Corrosion of iron in contact with copper. Corrosion commonly occurs at points where dissimilar metals are connected.

iron requires that adjacent surface areas differ chemically from each other. There are several ways in which one small area on a piece of iron or steel can become anodic or cathodic with respect to an adjacent area. Two of the most important are:

1. *The presence of impurities at scattered locations along the metal surface.* A tiny crystal of a less active metal such as copper or tin embedded in the surface of the iron acts as a cathode at which oxygen molecules are reduced. The iron atoms in the vicinity of these impurities are anodic and undergo oxidation to Fe^{2+} ions. This effect can be demonstrated on a large scale by immersing in water an iron plate which has been partially copper plated (Fig. 23.5). At the interface between the two metals, a voltaic cell is set up in which the iron is anodic and the copper cathodic. A thick deposit of rust forms at the interface. The formation of rust inside an automobile bumper where the chromium plate stops is another example of this phenomenon.

2. *Differences in oxygen concentration along the metal surface.* To illustrate the effect, consider what happens when a drop of water adheres to the surface of a piece of iron exposed to the air (Fig. 23.6). The metal around the edges of the drop is in contact with water containing a high concentration of dissolved oxygen. The water touching the metal beneath the center of the drop is depleted in oxygen, since it is cut off from contact with air. As a result, a small oxygen-concentration cell is set up. The area around the edge of the drop, where the oxygen concentration is high, becomes cathodic; oxygen molecules are reduced there via Reaction 23.23b. Directly beneath the drop is an anodic area where the iron is oxidized. A particle of dirt on the surface of an iron object can act in much the same way as a drop of water to cut off the supply of oxygen to the area beneath it and thereby establish anodic and cathodic areas. This explains why garden tools left covered with soil are particularly susceptible to rusting.

Iron or steel objects can be protected from corrosion in several different ways. These include:

1. *Covering the surface with a protective coating.* This may be a layer of paint which cuts off access to moisture and oxygen. Under more severe conditions, it may be desirable to cover the surface of the iron or steel with a layer of another metal. Metallic plates, applied electrically (Cr, Ni, Cu, Ag, Zn, Sn) or by immersion at high temperatures (Zn, Sn), are ordinarily more resistant to heat and chemi-

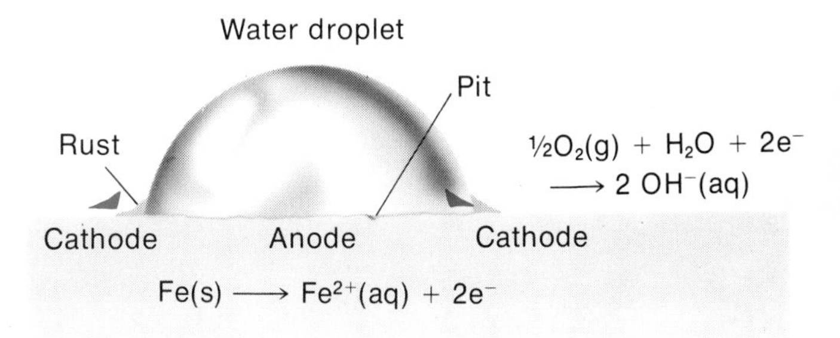

Figure 23.6 Corrosion of iron under a drop of water. The Fe^{2+} ions migrate toward the edge of the drop, where they precipitate as $Fe(OH)_2$, which is then further oxidized by air to $Fe(OH)_3$.

cal attack than the organic coating left when paint dries. If the plating metal is more active than iron (e.g., Zn) it, rather than iron, will be oxidized if the surface is broken. However, if the plating metal is less active than iron, there is a danger that cracks in its surface may enhance the corrosion of the iron or steel. This problem can arise with "tin cans," which are made by applying a layer of tin over a steel base. If the food in the can contains citric acid, some of the tin plate may dissolve, exposing the steel beneath.* When the can is opened, exposing the interior to the air, rust forms spontaneously on the iron surrounding the breaks in the tin surface. A thin coating of lacquer is ordinarily applied over the tin to prevent corrosive effects of this type.

2. *Bringing the object into electrical contact with a more active metal* such as magnesium or zinc. Under these conditions, the iron becomes cathodic and, hence, is protected against rusting; the more active metal serves as a sacrificial anode in a large-scale corrosion cell. This method of combating corrosion, known as *cathodic protection,* is particularly useful for steel objects such as cables or pipelines that are buried under soil or water (Fig. 23.7).

The same result can be achieved by using an external electrical source to give the iron a negative potential

23.6 REDOX REACTIONS IN ANALYTICAL CHEMISTRY

Qualitative Analysis

At certain stages of the standard schemes of cation and anion analysis, oxidizing or reducing agents are used for one of three different purposes:

1. *To bring a sample into solution.* Solids which are insoluble in both water and dilute nonoxidizing acids can frequently be brought into solution with concentrated nitric acid. It reacts with inactive metals such as copper or silver (Equations 23.15 and 23.16) and with many metal sulfides such as those of copper and bismuth

$$CuS(s) + 2\ NO_3^-(aq) + 4\ H^+(aq) \rightarrow Cu^{2+}(aq) + S(s) + 2\ NO_2(g) + 2\ H_2O \quad (23.25)$$

$$Bi_2S_3(s) + 6\ NO_3^-(aq) + 12\ H^+(aq) \rightarrow 2\ Bi^{3+}(aq) + 3\ S(s) + 6\ NO_2(g) + 6\ H_2O \quad (23.26)$$

to form a solution containing the corresponding metal ion.

*Tin forms an extremely stable complex with citrate ion and, hence, is attacked more readily by citric acid than by many stronger inorganic acids.

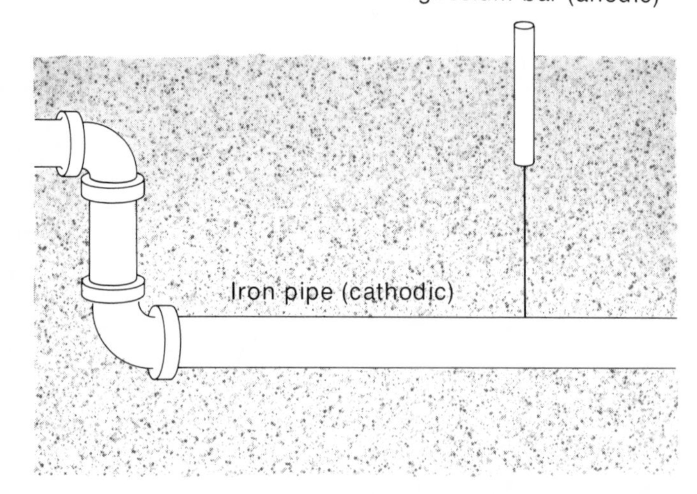

Figure 23.7 Cathodic protection of buried iron pipe. A magnesium or zinc bar is oxidized instead of the iron.

2. *To identify a particular ion.* The mercuric ion, Hg^{2+}, may be tested for by putting its solution on a copper penny or a piece of copper wire; if Hg^{2+} is present, a shiny deposit of metallic mercury forms.

$$Hg^{2+}(aq) + Cu(s) \rightarrow Hg(s) + Cu^{2+}(aq) \qquad (23.27)$$

Another test is to add tin(II) chloride, which reduces Hg^{2+} to the insoluble chloride of mercury(I), Hg_2Cl_2

$$2\ Hg^{2+}(aq) + Sn^{2+}(aq) + 2\ Cl^-(aq) \rightarrow Hg_2Cl_2(s) + Sn^{4+}(aq) \qquad (23.28)$$

3. *To separate ions from one another.* The Cr^{3+} ion can be separated from Al^{3+} and Zn^{2+} by first treating with hydrogen peroxide in basic solution to oxidize chromium from the +3 to the +6 state

$$2\ Cr^{3+}(aq) + 3\ H_2O_2(aq) + 10\ OH^-(aq) \rightarrow 2\ CrO_4^{2-}(aq) + 8\ H_2O \qquad (23.29)$$

and then adding Ba^{2+} to precipitate yellow, insoluble barium chromate

$$Ba^{2+}(aq) + CrO_4^{2-}(aq) \rightarrow BaCrO_4(s) \qquad (23.30)$$

Quantitative Analysis

A species that is readily oxidized can be determined quantitatively by titration with an oxidizing agent in much the same way that a base is titrated with an acid. An oxidizing agent that is frequently used in redox titrations is potassium permanganate, $KMnO_4$.

To illustrate the use of MnO_4^- ion, let us consider a specific redox titration, the determination of Fe^{2+} ions with MnO_4^-.

$$MnO_4^-(aq) + 8\ H^+(aq) + 5\ Fe^{2+}(aq) \rightarrow Mn^{2+}(aq) + 4\ H_2O + 5\ Fe^{3+}(aq) \qquad (23.31)$$

$$E^0 = E^0_{red}\ MnO_4^- + E^0_{ox}\ Fe^{2+} = (+1.52 - 0.77)V = 0.75\ V$$

The large positive E^0 value for this reaction means that the equilibrium constant is large enough ($K = 10^{64}$) to make the reaction go essentially to completion. In the titration, we start with a known volume of an acidified solution containing Fe^{2+} ions and add from a buret a solution of potassium permanganate of known concentration. At the instant the MnO_4^- ions are added, the solution takes on the pink or purple color characteristic of that ion. As the MnO_4^- ions are used up by Reaction 23.31, the color fades almost immediately. However, when all the Fe^{2+} ions have been titrated, i.e., at the equivalence point, the addition of one or two drops of excess MnO_4^- produces a permanent pink color. The volume of titrant necessary to reach this end point is recorded and the concentration of Fe^{2+} ions calculated as indicated in Example 23.10.

Example 23.10 A 20.0-cm^3 sample containing Fe^{2+} ions requires 18.0 cm^3 of 0.100 *M* $KMnO_4$ solution for complete reaction. Calculate the concentration of Fe^{2+} ions in the solution.

Solution Let us first calculate the number of moles of MnO_4^- added. Then, using Equation 23.31, we can calculate the number of moles of Fe^{2+} ion in the sample.

Finally, knowing the volume of the sample, we can calculate the concentration of Fe^{2+}.

$$\text{no. moles } MnO_4^- = 0.100\ \frac{\text{mol}}{\text{dm}^3} \times 0.0180\ \text{dm}^3 = 0.001\ 80\ \text{mol}$$

According to Equation 23.31

$$1\ \text{mol } MnO_4^- \simeq 5\ \text{mol } Fe^{2+}$$

Hence,

$$\text{no. moles } Fe^{2+} = 0.001\ 80\ \text{mol } MnO_4^- \times \frac{5\ \text{mol } Fe^{2+}}{1\ \text{mol } MnO_4^-} = 0.009\ 00\ \text{mol } Fe^{2+}$$

$$\text{conc. } Fe^{2+} = \frac{0.009\ 00\ \text{mol}}{0.0200\ \text{dm}^3} = 0.450\ M$$

PROBLEMS

23.1 *Calculation of Standard Voltage* What is E^0 for a cell in which the reaction is $Sn^{2+}(aq) + Cl_2(g) \rightarrow Sn^{4+}(aq) + 2\ Cl^-(aq)$?

23.2 *Use of E^0* Which of the following species can be oxidized by O_2 in acid solution (standard concentrations)?

a. Cl^- b. Fe^{2+} c. Co^{2+}

23.3 *Relation between E^0, ΔG^0, and K* Calculate ΔG^0 and K at 25°C for the reaction in Problem 23.1.

23.4 *Nernst Equation* Consider the reaction in Problem 23.1. Calculate

a. E when conc. Sn^{4+} = conc. Cl^- = conc. Sn^{2+} = 0.10 M, pCl_2 = 1 atm.
b. conc. Sn^{2+} if E = 1.09 V when conc. Sn^{4+} = conc. Cl^- = 1.0 M, pCl_2 = 1 atm.

23.5 *Redox Titrations* For the reaction $MnO_4^-(aq) + 8\ H^+(aq) + 5\ I^-(aq) \rightarrow Mn^{2+}(aq) + 4\ H_2O + \frac{5}{2}\ I_2(s)$, calculate the concentration of I^- in a solution if 25.0 cm^3 of 0.200 M $KMnO_4$ is required to react with 10.0 cm^3 of solution.

23.6 Calculate E^0 for each of the following voltaic cells:

a. $Pb(s) + 2\ Ag^+(aq) \rightarrow Pb^{2+}(aq) + 2\ Ag(s)$
b. $O_2(g) + 4\ Fe^{2+}(aq) + 4\ H^+(aq) \rightarrow 2\ H_2O + 4\ Fe^{3+}(aq)$
c. Cd-Cd^{2+} half-cell and a Zn-Zn^{2+} half-cell.

23.7 What is the minimum voltage that must be applied to achieve the following reactions in an electrolytic cell?

a. $Ni^{2+}(aq) + 2\ Cl^-(aq) \rightarrow Ni(s) + Cl_2(g)$
b. electrolysis of a water solution of KI to give I_2, OH^-, and H_2.

23.8 Suppose E^0_{red} ($H^+ \rightarrow H_2$) were set at 0.50 V instead of 0.00 V. What would be

a. E^0_{ox} of H_2?
b. E^0_{red} ($Cu^{2+} \rightarrow Cu$)?
c. E^0 for the cell $Zn(s) + Cu^{2+}(aq) \rightarrow Zn^{2+}(aq) + Cu(s)$?

23.24 Calculate E^0 for each of the following voltaic cells:

a. $MnO_2(s) + 4\ H^+(aq) + 2\ I^-(aq) \rightarrow Mn^{2+}(aq) + 2\ H_2O + I_2(s)$
b. $H_2(g) + 2\ OH^-(aq) + S(s) \rightarrow 2\ H_2O + S^{2-}(aq)$
c. a Ag-Ag^+ and a Au-$AuCl_4^-$ half-cell.

23.25 Which of the following electrolyses can be accomplished with a single dry cell (1.5 V)?

a. $Cd^{2+}(aq) + 2\ Cl^-(aq) \rightarrow Cd(s) + Cl_2(g)$
b. electrolysis of water to give H_2 and O_2.

23.26 Suppose E^0_{red} of Ag^+ instead of that of H^+ were set equal to zero. What would be

a. E^0_{red} H^+?
b. E^0_{ox} Ca?
c. E^0 for the cell $Zn(s) + Cu^{2+}(aq) \rightarrow Zn^{2+}(aq) + Cu(s)$?

23.9 Predict the products obtained upon electrolysis of 1 *M* solutions of the following (consider the possibility of forming H_2 and/or O_2 from water):

a. $CuBr_2$ b. AlI_3 c. $AgNO_3$

23.10 Using Table 23.1, classify each of the following as reducing or oxidizing agents and indicate the strongest in each category:

Ag^+, Fe^{2+}, Cl^-, Co^{3+}, O_2

23.11 Decide what reaction, if any, will occur when the following are mixed (standard concentrations).

a. SO_4^{2-}, H^+, H_2S
b. Fe^{2+}, H^+, Cl^-
c. Cu^{2+}, Ag^+
d. H_2, Ca^{2+}, Cr^{3+}

23.12 Explain, in terms of standard voltages, why

a. copper is oxidized by nitric acid but not by hydrochloric acid.
b. Sn^{2+} and Fe^{3+} ions are not found in the same solution.
c. fluorine cannot be prepared by electrolysis of a water solution.

23.13 Calculate ΔG^0 and K for each of the following:

a. $PbO_2(s) + Pb(s) + 4\ H^+(aq) + SO_4^{2-}(aq) \rightarrow 2\ PbSO_4(s) + 2\ H_2O$
b. $O_2(g) + 4\ Cl^-(aq) + 4\ H^+(aq) \rightarrow 2\ H_2O + 2\ Cl_2(g)$

23.14 Consider the reaction $Sn(s) + Pb^{2+}(aq) \rightarrow Sn^{2+}(aq) + Pb(s)$.

a. What is K for this reaction?
b. In which direction will reaction take place if conc. $Pb^{2+} = 0.10\ M$, conc. $Sn^{2+} = 1.0\ M$?
c. If excess tin is added to a solution 1 *M* in Pb^{2+}, what will be $[Sn^{2+}]$?

23.15 Copper(I) salts tend to disproportionate in water solution to copper(II) salts and copper metal, i.e.,

$$2\ Cu^+(aq) \rightarrow Cu^{2+}(aq) + Cu(s)$$

If one started with a 1.0 *M* solution of Cu^+, what would be its concentration at equilibrium?

23.16 A voltaic cell consists of a copper electrode immersed in 1 *M* $CuSO_4$ connected via a salt bridge to a solution of 1 *M* $AgNO_3$ in which is immersed a silver electrode. By what amount will the voltage change if

a. the concentration of Cu^{2+} is decreased to 0.0010 *M*?
b. enough Cl^- is added to precipitate AgCl ($K_{sp} = 1.6 \times 10^{-10}$) and make conc. $Cl^- = 1.0\ M$?
c. the area of the copper electrode is doubled?

23.27 Of the cations listed in Table 23.1, list three which would be reduced to the free metal upon electrolysis in water solution and three which would not.

23.28 Arrange the following species in order of increasing strength as oxidizing agents:

H_2O_2, Zn^{2+}, MnO_4^-, $AuCl_4^-$

23.29 Predict what reaction, if any, will occur when oxygen gas is bubbled through acidic water solutions of the following (standard concentrations):

a. AgF
b. $FeSO_4$
c. $ZnBr_2$

23.30 Explain the following observations in terms of standard voltages:

a. Tin added to a solution of a Sn^{2+} salt protects it from oxidation.
b. The Cu^+ ion is unstable in water solution.
c. The red color of a solution of $K_2Cr_2O_7$ fades when $FeSO_4$ is added.

23.31 For the reaction:

$$3\ A(s) + 2\ B^{3+}(aq) \rightarrow 3\ A^{2+}(aq) + 2\ B(s)$$

the values of $[B^{3+}]$ and $[A^{2+}]$ are 0.020 *M* and 0.0050 *M*, respectively. Calculate K, E^0, and ΔG^0 for this reaction.

23.32 Consider the reaction $MnO_2(s) + 4\ H^+(aq) + 2\ Cl^-(aq) \rightarrow Mn^{2+}(aq) + 2\ H_2O + Cl_2(g)$.

a. Calculate K.
b. What will be the equilibrium conc. of HCl if $[Mn^{2+}] = 1.0\ M$, $pCl_2 = 1$ atm?

23.33 Consider the reaction

$$Fe^{2+}(aq) + 2\ Cr^{2+}(aq) \rightarrow Fe(s) + 2\ Cr^{3+}(aq)$$

If 1 *M* Fe^{2+} is mixed with 2 *M* Cr^{2+}, what will be the equilibrium concentration of Cr^{3+}?

23.34 In a cell in which iron is corroding, what would you expect to be the voltage difference between two points which differ only in oxygen concentration if the partial pressure of oxygen at one point is 1.0 atm and, at the other point, 0.010 atm? The reaction is:

$$Fe(s) + \tfrac{1}{2}\ O_2(g) + H_2O \rightarrow Fe(OH)_2(s)$$

23.17 Calculate the voltages of cells in which the following reactions occur:

a. $Zn(s) + Cu^{2+}(0.10\ M) \rightarrow Zn^{2+}(0.0010\ M) + Cu(s)$
b. $Cu(s) + Cu^{2+}(1\ M) \rightarrow Cu^{2+}(2 \times 10^{-4}\ M) + Cu(s)$

23.18 A half-cell consisting of a silver electrode dipping into a solution of silver nitrate is connected to one in which a copper electrode is in contact with a 0.10 *M* solution of $Cu(NO_3)_2$. Excess HBr is added to the silver half-cell to give a precipitate of AgBr and make $Br^- = 0.10\,M$. Under these conditions the voltage is 0.22 V and the Ag electrode is the anode. Calculate K_{sp} for AgBr.

23.19 Explain how a *pH* meter could be used to determine the ionization constant of acetic acid.

23.20 Describe in words how you could prepare the following species from Cl_2:

a. HClO
b. ClO^-
c. ClO_3^-
d. ClO_4^-

23.21 Write balanced equations to represent

a. the reaction of Zn with dilute HNO_3 to form Zn^{2+} and NH_4^+.
b. the half-reaction for the reduction of NO_3^- to HNO_2.
c. the reaction of gold with aqua regia, a mixture of HCl and HNO_3. Products include $AuCl_4^-$ and $NO_2(g)$.

23.22 Explain why

a. NO_2 is commonly formed in the reduction of HNO_3.
b. a color change occurs when acid is added to a solution of K_2CrO_4.
c. corrosion usually occurs faster in salt water than in fresh water.
d. rust deposits usually form at cathodic rather than anodic areas.

23.23 Potassium permanganate reacts with $C_2O_4^{2-}$ ions to form $CO_2(g)$.

a. Write a balanced equation for the reaction.
b. If 21.6 cm³ of 0.150 *M* $KMnO_4$ is required to titrate 24.9 cm³ of a solution of sodium oxalate, what is the concentration of $C_2O_4^{2-}$?

23.35 Calculate the voltages of cells in which the following reactions occur:

a. $3\ Ag(s) + NO_3^-(10\ M) + 4\ H^+(10\ M) \rightarrow 3\ Ag^+(0.10\ M) + NO(1\ atm) + 2\ H_2O$
b. $Ag^+(0.001\ M) + Ag(s) \rightarrow Ag^+(1\ M) + Ag(s)$

23.36 The voltage for the half-reaction

$$Au^{3+}(aq) + 3\ e^- \rightarrow Au(s)$$

is reduced from 1.50 V to 1.00 V when enough Cl^- is added to make conc. $AuCl_4^-$ = conc. $Cl^- = 1\ M$. Calculate *K* for the reaction

$$AuCl_4^-(aq) \rightarrow Au^{3+}(aq) + 4\ Cl^-(aq)$$

23.37 Explain how a *pH* meter could be used to determine the dissociation constant of water.

23.38 Describe how you could prepare the following species from $Cr_2O_7^{2-}$:

a. CrO_4^{2-}
b. Cr^{3+}
c. $Cr(OH)_3$
d. Cr

23.39 Write balanced equations for

a. the reaction of NiS(s) with conc. HNO_3 to form Ni^{2+}, S, and $NO_2(g)$.
b. the reaction between O_2 and I^- ions in acidic solution.
c. the half-reaction that occurs beneath a water drop on a piece of iron.
d. the reaction of Hg^{2+} ions with Sn^{2+} in the presence of Cl^-.

23.40 Explain why

a. corrosion of a pipeline can be prevented by contact with a bar of zinc.
b. a solution of $SnCl_2$ often turns cloudy on standing.
c. copper dissolves in conc. HNO_3 but not in conc. HCl.
d. in a window screen corrosion is most severe where the wires meet.

23.41 It is possible to determine iodide ions in solution by titration with $KMnO_4$.

a. Write a balanced equation for the reaction.
b. If 18.9 cm³ of 0.100 *M* $KMnO_4$ is required to titrate 17.0 cm³ of a solution of sodium iodide, what is the concentration of I^-?

*23.42 10.0 cm³ of blood from a patient were treated with $C_2O_4^{2-}$ to precipitate calcium oxalate, CaC_2O_4. The precipitate was dissolved in strong acid and titrated with 1.20 cm³ of 0.100 *M* $KMnO_4$. What is the concentration of Ca^{2+}, in grams per cubic decimetre, in the blood sample?

*23.43 Given that

$$Ag^+(aq) + e^- \rightarrow Ag(s);\ E^0_{red} = +0.80\ V$$
$$AgI(s) + e^- \rightarrow Ag(s) + I^-(aq);\ E^0_{red} = -0.15\ V$$

calculate the solubility product constant of AgI.

*23.44 The standard reduction voltage of O_2 in acidic solution is +1.23 V. That is, for the half-reaction $O_2(g) + 4\,H^+(aq) + 4\,e^- \rightarrow 2\,H_2O$, $E^0_{red} = +1.23$ V. Calculate the standard reduction voltage in basic solution and compare to the value listed in Table 23.1.

*23.45 Given that K for the reaction

$$Cl_2(g) + H_2O \rightarrow Cl^-(aq) + H^+(aq) + HOCl(aq)$$

is 3×10^{-5}, apply the Rule of Multiple Equilibria to evaluate K for

$$Cl_2(g) + 2\,OH^-(aq) \rightarrow OCl^-(aq) + Cl^-(aq) + H_2O$$

24

NUCLEAR REACTIONS

The "ordinary chemical reactions" that we have discussed up to this point are ones which involve changes in the outer electronic structure of atoms or molecules. In contrast, there is a large class of processes, called nuclear reactions, which are the result of changes taking place within atomic nuclei. Nuclear reactions differ from ordinary chemical reactions in several important respects. In particular:

1. In ordinary reactions, the different isotopes of an element show virtually identical chemical properties; in nuclear reactions they behave quite differently. Consider, for example, the two isotopes of carbon, $^{12}_{6}C$ and $^{14}_{6}C$. The chemical properties of these isotopes are very similar. Their nuclear properties differ considerably; the $^{12}_{6}C$ nucleus is extremely stable while the $^{14}_{6}C$ nucleus decomposes spontaneously.

2. The nuclear reactivity of an element is essentially independent of its state of chemical combination. In the nuclear chemistry of radium, it makes little difference whether we deal with the element itself or one of its compounds. The radium atom in elementary radium and the Ra^{2+} ion in $RaCl_2$ behave similarly from a nuclear standpoint. In ordinary chemical reactions, on the other hand, the radium atom and the Ra^{2+} ion behave quite differently, since they have different outer electronic structures.

In discussing nuclear reactions or writing equations to represent them, we shall not ordinarily be concerned with what happens to the electrons outside the nucleus. Even though the species taking part in these reactions are atoms, molecules, or ions, the reactions themselves occur within the nucleus.

3. Nuclear reactions frequently involve the conversion of one element to another. Whenever a nuclear process results in a change in the number of protons in the nucleus, a new element of different atomic number is formed. In contrast, elements taking part in ordinary chemical reactions retain their identity.

4. Nuclear reactions are accompanied by energy changes which exceed, by several orders of magnitude, those associated with ordinary chemical reactions. The energy evolved when one gram of radium undergoes radioactive decay (Section 24.1) is about 500 000 times as great as that given off when an equal amount of radium reacts with chlorine to form radium chloride. Still larger amounts of energy are given off in nuclear fission (Section 24.4) and nuclear fusion (Section 24.5).

24.1 RADIOACTIVITY

Certain unstable nuclei decompose spontaneously, giving off energy in the process. A few such nuclei occur in nature; their decomposition is referred to as natural radioactivity. Many more unstable nuclei have been prepared in the laboratory; the process by which they decompose is referred to as induced radioactivity. We will be concerned with the nature of the reactions that occur when radioactive nuclei decompose ("decay") and with the effects that the products of these reactions have on the surroundings.

Natural Radioactivity

This phenomenon was discovered, almost accidentally, by a French scientist, Henri Becquerel, in 1896. He found that a certain uranium salt, potassium uranyl sulfate, $K_2UO_2(SO_4)_2$, gave off a powerful type of high-energy radiation which was capable of blackening a photographic plate. Curiously enough, this radiation seems never to have been detected previously, even though the element uranium had been isolated more than a hundred years before.

Becquerel was able to show that the rate at which radiation was emitted from a uranium salt was directly proportional to the amount of uranium present. There was one apparent exception to this rule: a certain uranium ore known as pitchblende gave off radiation at a rate nearly four times as great as one would calculate on the basis of its uranium content. In July of 1898 Marie and Pierre Curie, colleagues of Becquerel at the Sorbonne, were able to isolate from a tonne of pitchblende ore a fraction of a gram of a new element which was more intensely radioactive than uranium. They named this element polonium, after Marie Curie's native country. Six months later the Curies isolated still another, intensely radioactive, previously unknown element, radium. The Nobel Prize for physics in 1903 was awarded jointly to Henri Becquerel and Marie and Pierre Curie; eight years later, Madame Curie received an unprecedented second Nobel Prize, this time in chemistry.

According to one of the authors, Madame Curie's maiden name was Marya Sklodowska

The radiation emitted by naturally radioactive elements can be separated by an electrical or magnetic field into three distinct parts (Fig. 24.1).

1. **Alpha radiation**, which consists of a stream of positively charged particles (alpha particles) that carry a charge of +2 and have a mass of 4 on the atomic mass

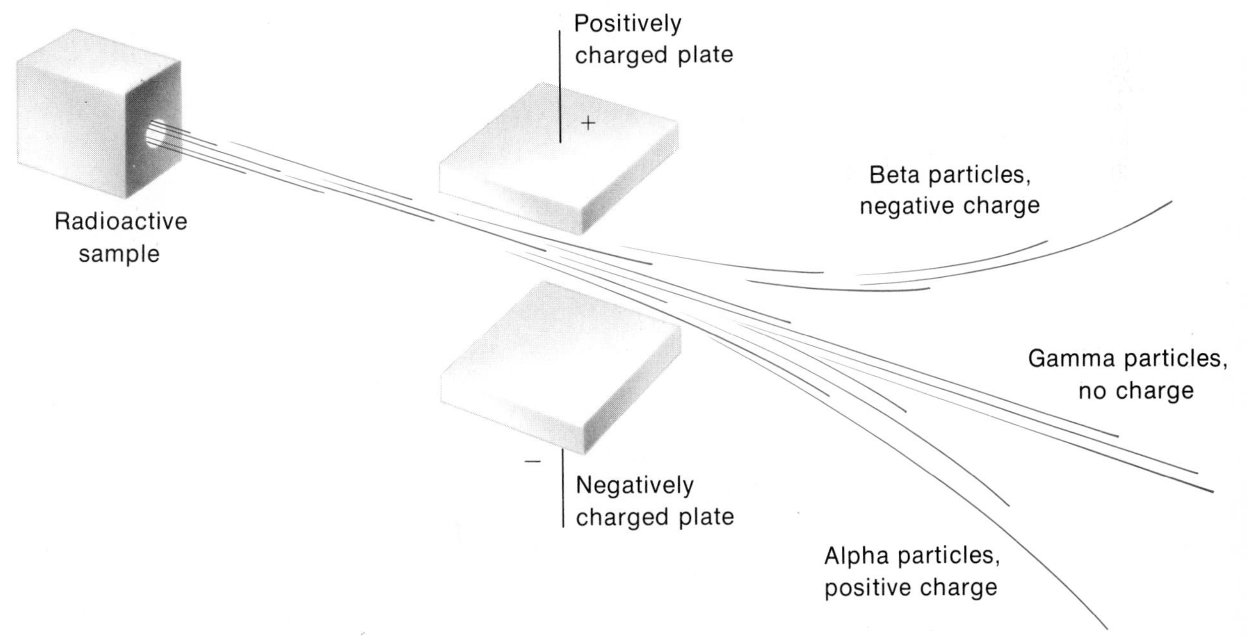

Figure 24.1 The direction in which beta particles are deflected shows that they are negatively charged. Alpha particles move so as to indicate that they carry a positive charge. Gamma rays are undeflected and so must be uncharged.

scale. These particles are identical with the nuclei of ordinary helium atoms (at. no. = 2, mass no. = 4).

When an alpha particle is ejected from the nucleus, there is a decrease of two units in atomic number and a decrease of four in mass number. For example, the loss of an alpha particle by the nucleus of an ordinary uranium atom (at. no. 92, mass no. 238) gives an isotope of thorium with an atomic number of 90 and a mass number of 234. This nuclear reaction may be represented by the equation

$$^{238}_{92}U \rightarrow ^{4}_{2}He + ^{234}_{90}Th \tag{24.1}$$

Note that here, as in all nuclear equations, there is a balance of both atomic number (90 + 2 = 92) and mass number (4 + 234 = 238) on the two sides.

2. **Beta radiation**, which is made up of a stream of negatively charged particles (beta particles) that have all the properties of electrons. The ejection of a beta particle (mass $\cong$ 0, charge = −1) results from the transformation of a neutron (mass = 1, charge = 0) at the surface of the nucleus into a proton (mass = 1, charge = +1). Consequently, beta-emission leaves the mass number unchanged but increases the atomic number by one unit. An example of beta-emission is the spontaneous radioactive decay of thorium-234 (90 protons, 144 neutrons) to protactinium-234 (91 protons, 143 neutrons):

$$^{234}_{90}Th \rightarrow ^{0}_{-1}e + ^{234}_{91}Pa \tag{24.2}$$

In effect, an electron has an atomic number of −1

The symbol $^{0}_{-1}e$ is written to represent a beta particle (electron).

3. **Gamma radiation**, electromagnetic radiation of very short wavelength (λ = 0.0005 to 0.1 nm), i.e., high-energy photons. The emission of gamma radiation accompanies virtually all nuclear reactions as the result of an energy change within the nucleus, whereby an unstable, excited nucleus resulting from alpha- or beta-emission gives off a photon and drops to a lower, more stable energy state. Since gamma-emission changes neither the atomic number nor the mass number, we shall frequently neglect it in writing nuclear equations.

Reactions 24.1 and 24.2 represent only the first two steps in a natural radioactive series which leads ultimately to a stable isotope of lead, $^{206}_{82}Pb$. There are a total of 14 steps in this process (see Problem 24.7); all of the intermediate products are unstable and decay relatively rapidly. The two steps that follow Reactions 24.1 and 24.2 are given in Example 24.1.

Example 24.1 The isotope formed in Reaction 24.2 is unstable, decomposing by emission of a beta particle (Reaction 1). The product of this reaction is also unstable; it decomposes by alpha emission (Reaction 2). Write balanced nuclear equations for Reactions 1 and 2.

Solution The unbalanced equation for Reaction 1 is

$$^{234}_{91}Pa \rightarrow ^{0}_{-1}e + ?$$

where the question mark indicates the isotope formed by beta emission. We see that in order to balance mass and charge, this isotope must have a mass number of 234 and an atomic number of 92 (i.e., 92 − 1 = 91). From the Periodic Table, we find this to be an isotope of uranium (at. no. = 92). Hence the balanced equation is

$$^{234}_{91}Pa \rightarrow ^{0}_{-1}e + ^{234}_{92}U \tag{1}$$

When $^{234}_{92}U$ loses an alpha particle, $^{4}_{2}He$, the mass number decreases by 4 and the atomic number by 2. The isotope produced must have a mass number of 234 −

4 = 230 and an atomic number of 92 − 2 = 90 (thorium, symbol Th). The balanced nuclear equation is

$$^{234}_{92}U \rightarrow {}^{4}_{2}He + {}^{230}_{90}Th \qquad (2)$$

Induced Radioactivity. Bombardment Reactions

During the past fifty years, about 1500 radioactive isotopes have been prepared synthetically. The number of such isotopes per element ranges from 1 (hydrogen and helium) to a maximum of 36 (indium). They are all prepared by bombardment reactions in which a stable (i.e., nonradioactive) nucleus is converted to one that undergoes radioactive decay. A typical example is the nuclear reaction that occurs when the stable isotope of aluminum, $^{27}_{13}Al$, absorbs a neutron to form $^{28}_{13}Al$. The latter is unstable, decomposing by electron emission to give a stable isotope of silicon, $^{28}_{14}Si$. The two steps involved in the process are

$$\text{neutron bombardment: } {}^{27}_{13}Al + {}^{1}_{0}n \rightarrow {}^{28}_{13}Al \qquad (24.3)$$

$$\text{radioactive decay: } {}^{28}_{13}Al \rightarrow {}^{28}_{14}Si + {}^{0}_{-1}e \qquad (24.4)$$

The first radioactive isotopes to be made in the laboratory were prepared in 1934 by Irene (daughter of Marie and Pierre) Curie and her husband, Frederic Joliot. They achieved this by bombarding certain stable isotopes with high-energy alpha particles. One reaction was

$$^{27}_{13}Al + {}^{4}_{2}He \rightarrow {}^{30}_{15}P + {}^{1}_{0}n \qquad (24.5)$$

The product, phosphorus-30, is radioactive. It decays by emitting a particle called a **positron,** which has the same mass as an electron, but a charge of +1 rather than −1.

$$^{30}_{15}P \rightarrow {}^{30}_{14}Si + {}^{0}_{1}e \qquad (24.6)$$

Although positron emission is never observed in natural radioactivity, it is a common mode of decay among radioactive isotopes produced in the laboratory. You will note from Equation 24.6 that the result of positron emission is the conversion of a proton in the nucleus to a neutron (15 p, 15 n in P-30; 14 p, 16 n in Si-30). Consequently, it commonly occurs with "light" isotopes, i.e., nuclei which have too few neutrons to be stable. An example is carbon-11, which decays by emitting a positron:

$$^{11}_{6}C \rightarrow {}^{11}_{5}B + {}^{0}_{1}e \qquad (24.7)$$

In contrast, the "heavy" isotope of the same element, carbon-14, decays by electron emission:

$$^{14}_{6}C \rightarrow {}^{14}_{7}N + {}^{0}_{-1}e \qquad (24.8)$$

The bombarding particles used to induce nuclear reactions ordinarily fall into one of two categories:

1. NEUTRONS. Since a neutron experiences no coulombic repulsion when it approaches a nucleus, it need have only a very small kinetic energy to initiate a nuclear reaction. The only important source of neutrons today is the nuclear reactor, described in Section 24.4.

2. POSITIVE IONS (PROTONS, DEUTERONS, α-PARTICLES, . . .). When cations are accelerated to very high velocities, they can acquire sufficient energy to overcome the coulombic repulsion of the target nucleus and bring about a nuclear reaction. Nuclear physicists and engineers have built several different types of instruments to form focused beams of high energy positive ions. One of the earliest of these, the cyclotron, designed originally by E. O. Lawrence at the University of California, is shown schematically in Figure 24.2.

One of the most interesting applications of bombardment reactions is in the preparation of the so-called transuranium elements. In the past forty years, 14 elements with atomic numbers greater than uranium (93 through 106) have been synthesized. Much of this work has been done by a group at the University of California at Berkeley under the direction of Glenn Seaborg and, more recently, Albert Ghiorso. In the past decade, a Russian group led by G. N. Flerov has been active in the field.

Some of the reactions used to prepare transuranium elements are listed in Table 24.1. Neutron bombardment is effective for the lower members of the series (elements 93 through 95), but the yield of product decreases exponentially with increasing atomic number. To form the heavier transuranium elements, it is necessary to bombard appropriate targets with high-energy positive ions. By using relatively heavy bombarding particles such as carbon-12, one can achieve a considerable increase in atomic number.

Enormous energies are required to cause nuclear reactions by $^{12}_{6}C$ bombardment

With few exceptions, the isotopes of the transuranium elements have very short half-lives. Moreover, most of them, especially those of very high atomic number, have been formed in extremely minute quantities, amounting in some cases to only a few atoms. One of the greatest achievements of the scientists working in this field has been their ability to work out methods of studying the chemical and physical properties of submicrogram amounts of these elements. Both chemical and physical evidence indicate that the transuranium elements up to atomic number 103 are filling out a second rare-earth series by completing the 5f subshell. Element 104 would then be the first member of a new transition series, falling below hafnium in the Periodic Table.

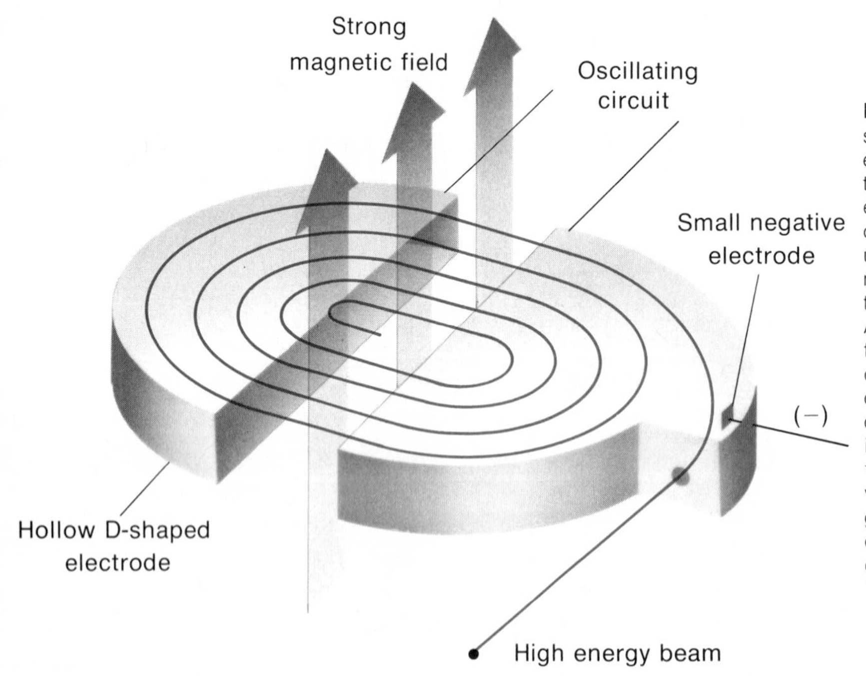

Figure 24.2 The cyclotron consists of two oppositely charged, evacuated "dees" placed between the poles of a powerful electromagnet. Positive ions, originating at the center, enter the upper dee, which is originally at a negative potential. They pass through this dee in a curved path. At the instant they reenter the central corridor, the polarity of the dees is reversed, and the particles enter the lower dee at an increased velocity. The procedure is repeated over and over; the particles move at higher and higher velocities in paths of greater and greater radius. Eventually they are deflected from the periphery of one of the dees to strike the target.

TABLE 24.1 SYNTHESIS OF TRANSURANIUM ELEMENTS

Element	Reaction
NEUTRON BOMBARDMENT	
Neptunium, Plutonium	$^{238}_{92}U + ^{1}_{0}n \rightarrow ^{239}_{92}U \rightarrow ^{239}_{93}Np + ^{0}_{-1}e$
	$^{239}_{93}Np \rightarrow ^{239}_{94}Pu + ^{0}_{-1}e$
Americium	$^{239}_{94}Pu + 2\ ^{1}_{0}n \rightarrow ^{241}_{94}Pu \rightarrow ^{241}_{95}Am + ^{0}_{-1}e$
POSITIVE ION BOMBARDMENT	
Curium	$^{239}_{94}Pu + ^{4}_{2}He \rightarrow ^{242}_{96}Cm + ^{1}_{0}n$
Californium	$^{242}_{96}Cm + ^{4}_{2}He \rightarrow ^{245}_{98}Cf + ^{1}_{0}n$
or:	$^{238}_{92}U + ^{12}_{6}C \rightarrow ^{246}_{98}Cf + 4\ ^{1}_{0}n$
Element 104*	$^{249}_{98}Cf + ^{12}_{6}C \rightarrow ^{257}_{104}? + 4\ ^{1}_{0}n$
Element 105*	$^{249}_{98}Cf + ^{15}_{7}N \rightarrow ^{260}_{105}? + 4\ ^{1}_{0}n$
Element 106*	$^{249}_{98}Cf + ^{18}_{8}O \rightarrow ^{263}_{106}? + 4\ ^{1}_{0}n$

*The equations given represent reactions used by the group at Berkeley. The names of elements 104 through 106 have not been established. For 104 and 105, the Berkeley group has suggested rutherfordium and hahnium, honoring Ernest Rutherford and Otto Hahn, discoverer of nuclear fission. The Russian group prefers the names kurchatovium and bohrium, after the Russian physicist I. V. Kurchatov and Niels Bohr.

Example 24.2 Dr. Seaborg has suggested that by using very heavy nuclei as bombarding particles, it may be possible to synthesize elements of atomic numbers much higher than any now known. One reaction that he has speculated upon is that of $^{48}_{20}Ca$ with $^{244}_{94}Pu$. Assuming that the product nucleus is an isotope of element 114 containing 174 neutrons, write a balanced nuclear equation for the reaction.

Solution The unbalanced equation, from the information given, is

$$^{244}_{94}Pu + ^{48}_{20}Ca \rightarrow ^{288}_{114}X + \_\_\_\_$$

Notice that atomic number is already balanced (114 on both sides). The other product must then be a neutron, $^{1}_{0}n$. Since there is a deficiency of 4 in mass number on the right-hand side (288 vs. 292), four neutrons must be formed. The balanced equation is

$$^{244}_{94}Pu + ^{48}_{20}Ca \rightarrow ^{288}_{114}X + 4\ ^{1}_{0}n$$

Interaction of Radiation with Matter

The alpha, beta, and gamma rays given off during radioactive decay lose their energy when they pass through matter by transferring it to atoms, molecules, or ions with which they collide. These collisions may be elastic, in the sense that only kinetic energy is transferred, thereby raising the temperature of the exposed material. The increased kinetic energy of the target particles is eventually translated into heat, which is given off to the surroundings.

Frequently the interaction of radiation with matter results in inelastic collisions in which the potential energy of the target species is raised. A common type is one in which an electron is excited to a higher energy level. When the electron drops back to its ground state, energy is given off as electromagnetic radiation which may be visible light (λ = 400 to 800 nm), ultraviolet light (λ = 10 to 400 nm), or x-rays (λ = 0.005 to 10 nm). Radium salts give off an intense luminescence, which is easily visible in the dark and can even be detected in daylight with a sample containing more than 0.1 g of radium. The dials of luminous watches are painted with a mixture containing a tiny amount of a radium salt and a substance such as zinc sulfide or anthracene which fluoresces upon exposure to radiation. An instrument commonly used to detect and measure radiation, the *scintillation counter*, uses this same principle. Light produced by radiation striking an organic solid or solution activates a photoelectric cell within the counter.

In another type of inelastic collision, an electron is removed from an atom or molecule to form a positively charged ion. The ionizing ability of alpha, beta, and gamma radiation is utilized in several instruments used to study radioactivity. In the *cloud chamber* (Fig. 24.3), the ions produced serve as nuclei upon which tiny water droplets condense. The path of an alpha or beta particle through the chamber can be observed by following the condensation trail produced in the vapor. The Geiger-Müller counter (Fig. 24.4) amplifies the electric current produced by a flow of electrons and positive ions to oppositely charged electrodes. The current pulse resulting from each ionization is amplified and detected by means of an electrically activated counting device.

γ-rays are the most penetrating but α-radiation is most damaging

The harmful effect of radiation upon human beings is caused by its ability to ionize and ultimately destroy the organic molecules of which body cells are composed. The extent of damage depends upon two factors: the energy and the type of radiation. The former is now expressed in *grays;* a gray corresponds to the absorption of one joule per kilogram of tissue. The total biological effect of radiation, expressed in *rems,* is found by multiplying the number of grays by an appropriate factor for the particular type of radiation

$$\text{no. of rems} = n\ (\text{no. of grays})$$

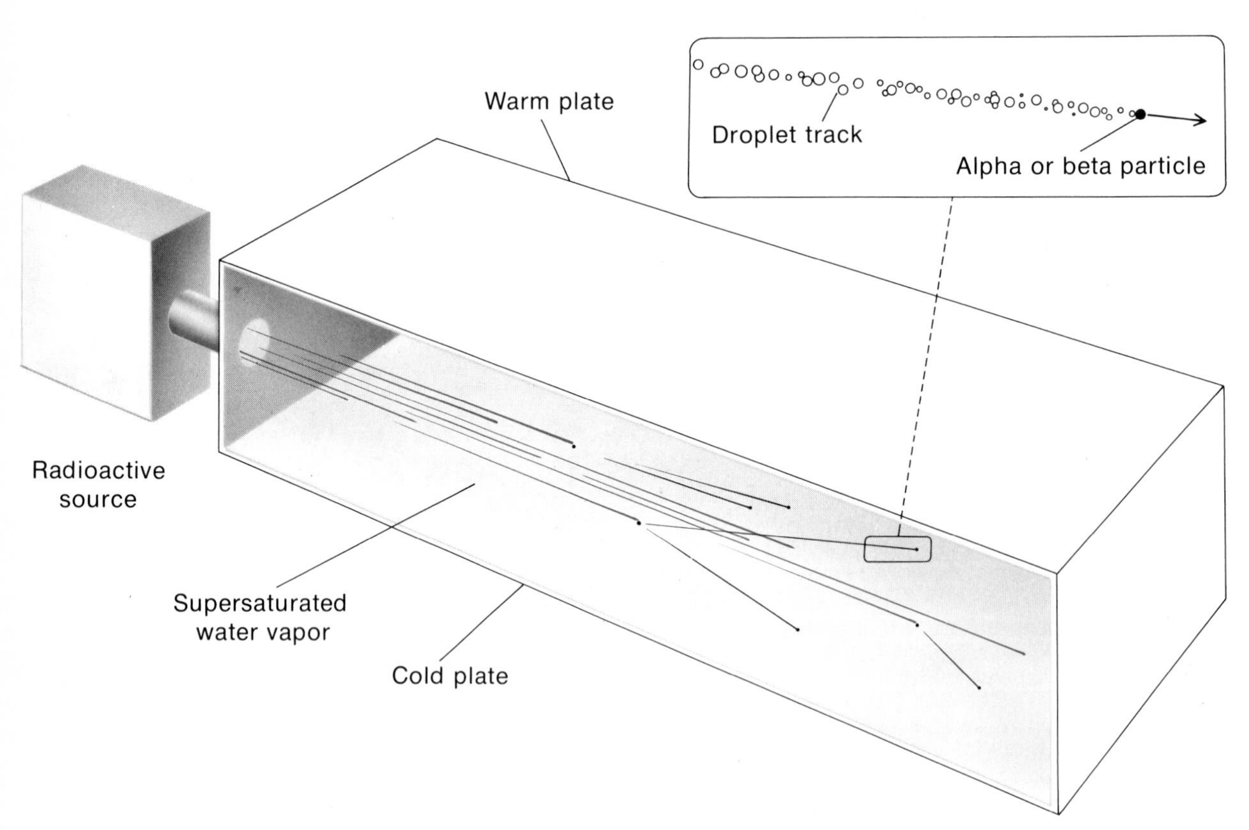

Figure 24.3 Cloud chamber. Supersaturated vapor condenses as tiny droplets on ions produced by emitted particle as it moves through the vapor. The track that is observed consists of those droplets.

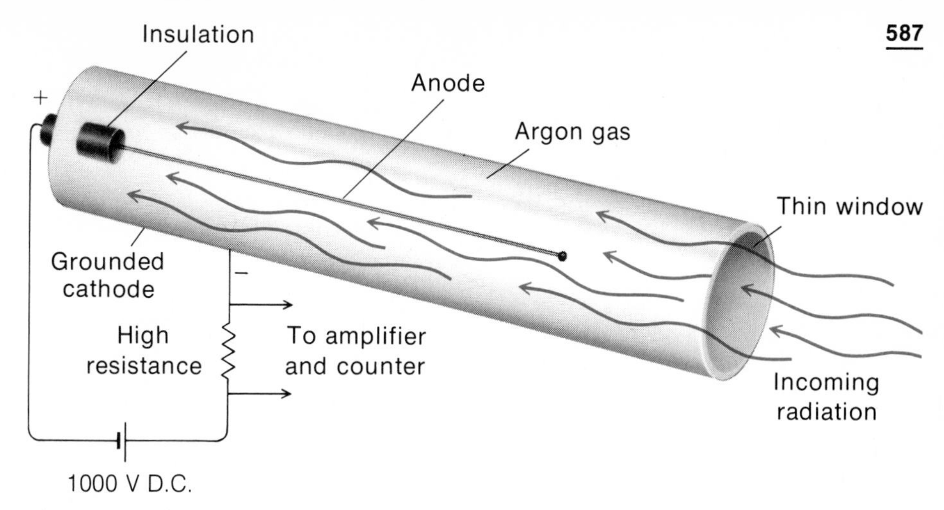

Figure 24.4 Geiger-Müller counter. Ions produced by radiation cause an electrical discharge, which activates a counter.

where n is 100 for β, γ, and X radiation and 1000 for α radiation or high-energy neutrons. Table 24.2 lists some of the effects to be expected when a human being is exposed to a single dose of radiation at various levels. By way of comparison, a single chest x-ray amounts to an exposure of 0.2 rem, a gastrointestinal tract examination about 22 rems.

Small doses of radiation repeated over long periods of time can have extremely serious consequences. Many of the early workers in the field of radioactivity developed cancer as a result of chronic overexposure to radiation. Cases are known in which cancers developed as long as forty years after initial exposure. Recent studies have shown an abnormally large number of cases of leukemia among the survivors of the nuclear bombs detonated at Hiroshima and Nagasaki.

It has long been known that radiation can produce mutations in plants and animals by bringing about changes in chromosomes. There is every reason to suppose that similar genetic effects can arise in human beings as well. Statistical surveys of the children of American radiologists indicate an increased frequency of congenital defects. This is confirmed by studies which have been made of children born to the survivors of Nagasaki and Hiroshima. Perhaps the most disturbing aspect of this problem is that there appears to be no lower limit or "tolerance level" below which the genetic effects of radiation become negligible. Even a small increase in the background level of radiation (now estimated to range from 0.08 to 0.2 rem per year) can be expected to produce a proportional increase in undesirable mutations. For this reason, among others, the Federal Research Council has set an upper limit of 0.17 rem per year above background as the maximum dosage to which major population groups can be exposed. This compares to 0.001 rem per year resulting from the operation of nuclear power plants.

Note that 0.001 rem per year is far below background

TABLE 24.2 EFFECT OF EXPOSURE TO A SINGLE DOSE OF RADIATION

Dose (rems)	Probable Effect
0 to 25	No observable effect
25 to 50	Small decrease in white blood cell count
50 to 100	Lesions, marked decrease in white blood cells
100 to 200	Nausea, vomiting, loss of hair
200 to 500	Hemorrhaging, ulcers, possible death
500+	Fatal

Uses of Radioactive Isotopes

A large number of radioactive isotopes now available have been applied in many areas of basic and applied research. A few of these are discussed below.

MEDICINE. The high-energy radiation given off by radium was used for many years in the treatment of cancer to destroy or arrest the growth of malignant tissue. More recently a radioactive isotope of cobalt, cobalt-60, which is cheaper than radium and gives off even more powerful radiation, has been used for this purpose. Certain types of cancer can be treated internally with the aid of radioactive isotopes. If a patient suffering from cancer of the thyroid (malignant goiter) drinks a solution of sodium iodide containing radioactive iodine (^{131}I or ^{128}I), the iodine moves preferentially to the thyroid gland, where the radiation destroys the malignant cells without affecting the rest of the body.

Trace amounts of radioactive samples injected into the blood stream can be used to detect circulatory disorders. For example, by injecting a sodium chloride solution containing a small amount of radioactive sodium into the leg of a patient and measuring the build-up of radiation in the foot, a physician can quickly find out whether the circulation in that area is abnormal. This may help him to decide whether amputation is necessary and, if so, where it should be done.

INDUSTRY. The frictional wear of piston rings can be monitored by making a test ring slightly radioactive by neutron bombardment and measuring the activity of the iron dust in the lubricating oil that circulates around the piston. The rate of corrosion of steel and the locations at which it is most severe can also be measured by a similar technique.

The thickness of very thin sheets of metal, paper, or plastic can be checked by interposing them between a radioactive source and a detector such as a Geiger counter. If the fraction of the radiation which passes through a sheet is known, it is possible to estimate its thickness quite accurately. Thin spots where the sheet might break down in use can also be detected.

CHEMISTRY. One of the earliest applications of radioactive isotopes in chemistry was to demonstrate the validity of the concept of dynamic equilibrium. You will recall that in our discussion of chemical equilibria (Chapter 15) we defined the equilibrium state as one in which forward and reverse reactions are proceeding at the same rate. For example, in a saturated solution of lead chloride

$$PbCl_2(s) \rightleftharpoons Pb^{2+}(aq) + 2\ Cl^-(aq)$$

the fact that the concentration of Pb^{2+} does not change with time is explained by assuming that the two opposing processes, solution and precipitation, are occurring at the same rate. The availability of a radioactive isotope of lead, Pb-212, suggested a way to check this model. A small amount of this isotope, in the form of $Pb(NO_3)_2$, was injected into a solution saturated with $PbCl_2$. It was found that within a very short time some of the radioactive lead appeared in the lead chloride, indicating that exchange between the solid and the solution was indeed occurring.

Chemists studying the rates and mechanisms of reaction frequently use radioactive isotopes to trace the path of a particular element as it passes through various steps from reactant to final product. Organic chemists in particular have learned a great deal about the mechanism of some rather complex reactions by using carbon-14 as a tracer. One of these is the extremely important natural process of photosynthesis. The overall reaction may be represented as

$$6\ CO_2(g) + 6\ H_2O(l) \rightarrow C_6H_{12}O_6(s) + 6\ O_2(g) \qquad (24.9)$$

This reaction proceeds through a series of steps in which successively more complex organic molecules are synthesized, leading ultimately to carbohydrates of which glucose, $C_6H_{12}O_6$, is typical. To study the path of this reaction, plants are exposed to carbon dioxide containing carbon-14. At various time intervals, the plants are killed and analyzed to determine which organic compounds contain carbon-14 and hence must be early products of the photosynthesis process. Research along these lines by Melvin Calvin at the University of California at Berkeley led to a Nobel Prize in chemistry in 1961.

Unfortunately several of the more common elements do not have radioactive isotopes sufficiently long-lived to be used as tracers. An important example is oxygen; $^{15}_{8}O$, the longest lived radioactive isotope, has a half-life of only 2 min. It is possible to follow reaction mechanisms involving oxygen atoms by enriching with oxygen-18, a stable isotope which is quite rare in nature (abundance = 0.20%). This isotope is readily detected by using a mass spectrometer (Chapter 2) which distinguishes it from the common isotope, oxygen-16.

Inorganic chemists have used oxygen-18 as a tracer to determine the lability or inertness of complex ions. If a sample containing the cation $Fe(H_2O)_6^{3+}$ in which the water molecules are enriched in oxygen-18 is added to ordinary water it is found that exchange with the solvent takes place instantaneously. Within a fraction of a second, we reach an equilibrium distribution of $^{18}_{8}O$ between complex and water. In contrast, this same process takes place much more slowly with the inert complex $Cr(H_2O)_6^{3+}$. The half-life for the reaction

$$Cr(H_2O)_5(H_2O^*)^{3+}(aq) + \underset{(* = ^{18}_{8}O)}{H_2O} \rightarrow Cr(H_2O)_6^{3+}(aq) + H_2O^* \qquad (24.10)$$

is found to be 40 h.

24.2 RATE OF RADIOACTIVE DECAY

The rate at which a radioactive sample decays can be measured by counting the number of particles given off in unit time. Modern instruments for measuring radioactivity such as scintillation counters and Geiger counters do this automatically. In interpreting data obtained with such instruments, it is ordinarily necessary to correct for the "background" radiation given off by naturally radioactive species in the environment.

One generalization which emerges from rate studies with radioactive isotopes is that the rate of decay is essentially independent of temperature. From an experimental standpoint, this eliminates the need for precise temperature control in such studies. From a theoretical point of view, it implies that the activation energy for radioactive decay is zero.

First Order Rate Law

Radioactive decay is one of the most important examples of the first order rate process discussed in Chapter 16. You will recall that for such reactions the rate equation is

$$\log_{10} \frac{X_0}{X} = \frac{kt}{2.30} \qquad (24.11)$$

Here, X_0 is the amount of radioactive substance at zero time (i.e., when the counting process starts) and X is the amount remaining after time *t*. The first order rate constant, *k*, is characteristic of the isotope undergoing radioactive decay.

The half-life is the time required for half the sample to decay

Decay rates of radioactive isotopes are most often expressed in terms of their half-lives, $t_{1/2}$, rather than the first order rate constant, k. As noted in Chapter 16, these two quantities are related by the equation

$$k = \frac{0.693}{t_{1/2}} \tag{24.12}$$

The application of these equations to radioactive decay processes is illustrated in Example 24.3.

Example 24.3 The common radioactive isotope of radium, $^{226}_{88}Ra$, has a half-life of 1620 a.* Calculate

a. the first order rate constant for the decay of radium-226.
b. the fraction of a sample of this isotope which will remain after 100 a.

Solution

a. $k = \frac{0.693}{1620\ a} = 4.28 \times 10^{-4}\ a^{-1}$

b. $\log \frac{X_0}{X} = \frac{(4.28 \times 10^{-4}\ a^{-1})}{2.30}\ 100\ a = 0.0186$

taking antilogs: $X_0/X = 1.044$

The fraction *remaining* is

$$\frac{X}{X_0} = \frac{1}{1.044} = 0.956 \text{ or } 95.6\%$$

This problem illustrates a safety hazard inherent in any attempt to "dispose" of a sample of a relatively long-lived isotope. A sample of a radium salt cannot simply be discharged into the environment in the vain hope that it will decompose rapidly; the level of radiation would be virtually unchanged a century from now. This factor has to be taken into account in any proposed system of radioactive waste disposal.

Qualitatively, half-lives can be interpreted in terms of the level of radiation associated with the corresponding isotopes. Since uranium-238 has a very long half-life (4.5×10^9 a), it gives off radiation very slowly at a rate that remains nearly constant over long periods of time. At the opposite extreme is the isotope polonium-214, which decays by α-emission with a half-life of 1.6×10^{-4} s. Within a second, virtually all the radiation associated with this isotope has been dissipated. Needless to say, isotopes such as these produce a tremendously high level of radiation during their brief existence.

Age of Rocks

Certain radioactive isotopes act as "natural clocks" which can help us to determine the time at which various rock deposits solidified or, in other words, to estimate their age. To understand how this can be done, consider a uranium-bearing rock, formed billions of years ago by solidification from a molten mass. Once the rock became solid, the products of radioactive decomposition of uranium were no longer able to diffuse away and, hence, were incorporated into the rock. Over a long period of time these products, all of which have relatively short half-lives, were

*The SI symbol "a" represents year.

ultimately converted to lead-206. The overall equation for the decay process can be written

$$^{238}_{92}U \rightarrow ^{206}_{82}Pb + 8\,^{4}_{2}He + 6\,_{-1}^{0}e; \qquad t_{1/2} = 4.5 \times 10^{9}\ a \qquad (24.13)$$

Knowing the half-life for this process, it should then be possible, by measuring the ratio of lead-206 to uranium-238 in the rock today, to calculate the time that has elapsed since the rock solidified. If we should find, for example, that equal numbers of atoms of these two isotopes were present, we would infer that the rock must be about 4.5×10^{9} (4.5 billion) years old.

This method of estimating the age of mineral deposits assumes, among other things, that none of the lead-206 has become separated from the parent uranium-238. It is possible to check the validity of this assumption by referring to other "radioactive clocks" which operate in nature. One of these is the β-decay of rubidium-87, which has a half-life of 5.7×10^{10} a.

$$^{87}_{37}Rb \rightarrow ^{87}_{38}Sr + _{-1}^{0}e; \qquad t_{1/2} = 5.7 \times 10^{10}\ a \qquad (24.14)$$

Ages of rocks determined by these and other radioactive methods range from 3 to 4.5×10^{9} a; the latter number is often taken as an approximate value for the age of the earth. Interestingly enough, analyses of rock samples taken from the moon's surface indicate ages in this same range. This would seem to eliminate the once prevalent idea that the moon was torn from the earth's surface by a cataclysmic event a long time after the earth solidified.

Age of Organic Material

During the 1950's Professor W. F. Libby and others worked out a method based upon the decay rate of a naturally occurring isotope, carbon-14, for determining the age of organic matter. This method can be applied to objects from a few hundred up to fifty thousand years old. It has been used, for example, to check the authenticity of canvases of Renaissance painters and to determine the age of relics left by prehistoric cavemen.

Samples older than this have too little C-14 to measure accurately

Carbon-14 is produced in the atmosphere by the interaction of neutrons from cosmic radiation with ordinary nitrogen atoms:

$$^{14}_{7}N + ^{1}_{0}n \rightarrow ^{14}_{6}C + ^{1}_{1}H$$

The carbon-14 formed by this nuclear reaction is eventually incorporated into the carbon dioxide of the air. A steady-state concentration, amounting to about one atom of carbon-14 for every 10^{12} atoms of carbon-12, is established in atmospheric CO_2. A living plant, taking in carbon dioxide, has this same $^{14}C/^{12}C$ ratio, as do plant-eating animals or human beings.

When a plant or animal dies, the intake of radioactive carbon stops. Consequently the radioactive decay of carbon-14

$$^{14}_{6}C \rightarrow ^{14}_{7}N + _{-1}^{0}e \qquad (\text{half-life} = 5720\ a)$$

takes over and the ratio of $^{14}C/^{12}C$ drops. By measuring this ratio and comparing it to that in living plants, one can estimate the time at which the plant or animal died (Example 24.4).

This method depends on a constant $^{14}C/^{12}C$ ratio in the air over the years

Example 24.4 A tiny piece of paper taken from the Dead Sea scrolls, believed to date back to the 1st century, A.D., was found to have a $^{14}C/^{12}C$ ratio 0.795 times that in a plant living today. Estimate the age of the scrolls.

Solution Knowing the half-life of carbon-14 ($t_{1/2}$ = 5720 a), we can calculate the first order rate constant from Equation 24.12. Then, using Equation 24.11, we can obtain the elapsed time

$$k = \frac{0.693}{5720 \text{ a}} = 1.21 \times 10^{-4} \text{ a}^{-1}$$

$$\log \frac{X_0}{X} = \frac{1.21 \times 10^{-4} \text{ a}^{-1} \times t}{2.30}$$

But, $X = 0.795\ X_0$, so: $\log \frac{X_0}{X} = \log \frac{1.000}{0.795} = \log 1.26 = 0.100$

Hence, $0.100 = \frac{1.21 \times 10^{-4} \text{ a}^{-1} \times t}{2.30}$, $t = 1900$ a

24.3 MASS-ENERGY RELATIONS

We pointed out at the beginning of this chapter that the energy change associated with nuclear reactions is greater by several orders of magnitude than that which accompanies ordinary chemical reactions. The energy change can be calculated from Einstein's equation,

$$\Delta E = \Delta mc^2 \qquad (24.15)$$

where Δm is the change in mass,* ΔE is the change in energy, and c is the velocity of light. If one substitutes for c the value 3.00×10^8 m/s, Equation 24.15 gives directly the relation between the energy change in *joules* and the mass change in *kilograms:*

$$\Delta E \text{ (in joules)} = 9.00 \times 10^{16} \times \Delta m \text{ (in kilograms)}$$

In calculations dealing with nuclear reactions, we will ordinarily be interested in obtaining the energy change in *kilojoules* corresponding to a mass change in grams. For these units, since 1 J = 10^{-3} kJ and 1 kg = 10^3 g, we have

$$\Delta E \text{ (in kilojoules)} = 9.00 \times 10^{10} \times \Delta m \text{ (in grams)} \qquad (24.16)$$

Using Equation 24.16 along with the appropriate nuclear masses (Table 24.3), it is possible to calculate the energy change accompanying a nuclear reaction.

Energy changes in ordinary chemical reactions are of the order of 40 kJ/g or less. For example, in the combustion of petroleum, about 46 kJ of heat are

*Specifically, Δm = mass of products − mass of reactants; ΔE = energy of products − energy of reactants. In spontaneous nuclear reactions, the products weigh less than the reactants (Δm negative); in this case, the energy of the product is less than that of the reactants (ΔE negative), and energy is evolved to the surroundings.

TABLE 24.3 NUCLEAR MASSES ON THE C-12 SCALE*

	AT. NO.	MASS NO.	MASS		AT. NO.	MASS NO.	MASS
n	0	1	1.008 67	Br	35	79	78.899 2
H	1	1	1.007 28		35	81	80.897 1
	1	2	2.013 55		35	87	86.902 8
	1	3	3.015 50	Rb	37	89	88.890 9
He	2	3	3.014 93	Sr	38	90	89.886 4
	2	4	4.001 50	Mo	42	99	98.884 9
Li	3	6	6.013 48	Ru	44	106	105.882 9
	3	7	7.014 36	Ag	47	109	108.878 9
Be	4	9	9.009 99	Cd	48	109	108.878 6
	4	10	10.011 34		48	115	114.879 3
B	5	10	10.010 19	Sn	50	120	119.874 7
	5	11	11.006 56	Ce	58	144	143.881 6
C	6	11	11.008 14		58	146	145.886 5
	6	12	11.996 71	Pr	59	144	143.880 7
	6	13	13.000 06	Sm	62	152	151.885 3
	6	14	13.999 95	Eu	63	157	156.891 4
O	8	16	15.990 52	Er	68	168	167.894 1
	8	17	16.994 74	Hf	72	179	178.904 8
	8	18	17.994 77	W	74	186	185.910 7
F	9	18	17.996 01	Os	76	192	191.918 7
	9	19	18.993 46	Au	79	196	195.923 1
Na	11	23	22.983 73	Hg	80	196	195.921 9
Mg	12	24	23.978 45	Pb	82	206	205.929 5
	12	25	24.979 25		82	207	206.930 9
	12	26	25.976 00		82	208	207.931 6
Al	13	26	25.979 77	Po	84	210	209.936 8
	13	27	26.974 39		84	218	217.962 8
	13	28	27.974 77	Rn	86	222	221.970 3
Si	14	28	27.969 24	Ra	88	226	225.977 1
S	16	32	31.963 29	Th	90	230	229.983 7
Cl	17	35	34.959 52	Pa	91	234	233.993 4
	17	37	36.956 57	U	92	233	232.989 0
Ar	18	40	39.952 50		92	235	234.993 4
K	19	39	38.953 28		92	238	238.000 3
	19	40	39.953 58		92	239	239.003 8
Ca	20	40	39.951 62	Np	93	239	239.001 9
Ti	22	48	47.935 88	Pu	94	239	239.000 6
Cr	24	52	51.927 34		94	241	241.005 1
Fe	26	56	55.920 66	Am	95	241	241.004 5
Co	27	59	58.918 37	Cm	96	242	242.006 1
Ni	28	59	58.918 97	Bk	97	245	245.012 9
Zn	30	64	63.912 68	Cf	98	248	248.018 6
	30	72	71.911 28	Es	99	251	251.025 5
Ge	32	76	75.903 80	Fm	100	252	252.027 8
As	33	79	78.902 88		100	254	254.033 1

*Note that these are *nuclear masses*. The masses of the corresponding atoms can be calculated by adding the masses of each extranuclear electron (0.000 549). For example, for an *atom* of $^{4}_{2}He$ we have

$$4.001\ 50 + 2(0.000\ 549) = 4.002\ 60$$

Similarly, for an atom of $^{12}_{6}C$:

$$11.996\ 71 + 6(0.000\ 549) = 12.000\ 00$$

Example 24.5 For the radioactive decay of radium $^{226}_{88}Ra \rightarrow {}^{222}_{86}Rn + {}^{4}_{2}He$, calculate ΔE in kilojoules when

a. one mole of radium decays.
b. one gram of radium decays.

Solution

a. We first calculate Δm for the reaction and then obtain ΔE from Equation 24.16. For the decay of one mole of Ra:

$$\Delta m = \text{mass of 1 mol } ^{4}_{2}\text{He} + \text{mass of 1 mol } ^{222}_{86}\text{Rn} - \text{mass of 1 mol } ^{226}_{88}\text{Ra}$$
$$= 4.0015 \text{ g} + 221.9703 \text{ g} - 225.9771 \text{ g}$$
$$= -0.0053 \text{ g}$$

(Note that since Δm is extremely small, it is necessary to know the masses of products and reactants very accurately to obtain the mass difference to 2 significant figures.)

$$\Delta E \text{ (in kJ)} = 9.00 \times 10^{10} \times (-0.0053) = -4.8 \times 10^{8} \text{ kJ}$$

b. Since one mole of radium weighs 226 g, we have

$$\Delta E = 1 \text{ g Ra} \times \frac{(-4.8 \times 10^{8} \text{ kJ})}{226 \text{ g Ra}} = -2.1 \times 10^{6} \text{ kJ}$$

evolved per gram of fuel burned. Looking at Example 24.5b, we see that ΔE for the radioactive decay of radium is about 50 000 times as great as that for typical non-nuclear processes.

Nuclear Stability; Binding Energy

Using Table 24.3, it is possible to compare the masses of nuclei to those of the individual neutrons and protons of which the nuclei consist. When we do this, we always find that the nuclear mass is less than that of the constituent particles. Consider, for example, the $^{2}_{1}H$ nucleus:

$$\begin{aligned} \text{mass of 1 mol protons} &= 1.007\ 28 \text{ g} \\ \text{mass of 1 mol neutrons} &= \underline{1.008\ 67 \text{ g}} \\ &\ 2.015\ 95 \text{ g} \end{aligned} \qquad \text{mass of 1 mol } ^{2}_{1}\text{H} = 2.013\ 55 \text{ g}$$

In this case, there is a decrease in mass of (2.015 95 g − 2.013 55 g) = 0.002 40 g when one mole of deuterons is formed from protons and neutrons.

The mass decrement is a positive quantity

Similar calculations for other nuclei lead to the **mass decrements** listed in Table 24.4. These quantities refer in each case to the decrease in mass (in grams) associated with the formation of one mole of nuclei from the appropriate number of protons and neutrons. That is, for the reaction to form the nucleus of an element Z,

$$\text{x protons} + \text{y neutrons} \rightarrow {}^{x+y}_{x}Z$$

$$\text{mass decrement} = x(1.007\ 28 \text{ g}) + y(1.008\ 67 \text{ g}) - \text{mass of 1 mol Z}$$

TABLE 24.4 BINDING ENERGIES OF VARIOUS NUCLEI

NUCLEUS	MASS DECREMENT (g/mol)	BINDING ENERGY (kJ/mol)	BINDING ENERGY (kJ/g)
$^{2}_{1}H$	0.002 40	2.16×10^{8}	1.07×10^{8}
$^{3}_{2}He$	0.008 30	7.47	2.48
$^{7}_{3}Li$	0.042 16	37.9	5.41
$^{10}_{5}B$	0.069 56	62.6	6.26
$^{12}_{6}C$	0.098 99	89.1	7.43
$^{27}_{13}Al$	0.241 63	217	8.06
$^{59}_{27}Co$	0.555 63	500	8.49
$^{64}_{30}Zn$	0.600 50	540	8.46
$^{99}_{42}Mo$	0.915 0	824	8.33
$^{157}_{63}Eu$	1.382 2	1244	7.93
$^{196}_{80}Hg$	1.666 2	1500	7.65
$^{238}_{92}U$	1.935 3	1742	7.32

Since the formation of a nucleus from protons and neutrons always involves a decrease in mass, any nucleus must be stable toward decomposition into these particles. Consider, for example, the deuterium nucleus. In order to break up one mole of these nuclei into protons and neutrons, 0.002 40 g of mass would have to be added. This would require the absorption of a tremendous amount of energy. The amount of energy which would have to be absorbed to decompose nuclei into protons and neutrons is referred to as the **binding energy** of the nucleus. The binding energy, in kilojoules per mole, is readily calculated from the mass decrement by using the conversion factor 9.00×10^{10} kJ $\triangleq$ 1 g of mass. For the $^{2}_{1}H$ nucleus:

$$\text{binding energy (kJ/mol)} = \text{mass decrement (g/mol)} \times 9.00 \times 10^{10} \text{ kJ/g}$$

$$= 0.002\ 40 \frac{\text{g}}{\text{mol}} \times 9.00 \times 10^{10} \frac{\text{kJ}}{\text{g}}$$

$$= 2.16 \times 10^{8} \text{ kJ/mol}$$

We interpret this result to mean that the decomposition of a mole of deuterons into protons and neutrons would require the absorption of 2.16×10^{8} kJ. Conversely, 2.16×10^{8} kJ would be evolved if a mole of protons and a mole of neutrons were to combine to form $^{2}_{1}H$.

Notice from Table 24.4 that binding energy in kilojoules per mole increases steadily with nuclear mass. We might expect this to be the case; the more particles there are in the nucleus, the greater should be the total energy required to break it apart. To get a better idea of the relative stabilities of different nuclei, we calculate the binding energy per unit mass, i.e., per *gram* of nuclei. For the $^{2}_{1}H$ nucleus, which has a molar mass of 2.014 g,

Why is binding energy per *gram* a better measure of relative stability?

$$\text{binding energy (kJ/g)} = \text{binding energy (kJ/mol)} \times \frac{1 \text{ mol}}{2.014 \text{ g}}$$

$$= 2.16 \times 10^8 \frac{\text{kJ}}{\text{mol}} \times \frac{1 \text{ mol}}{2.014 \text{ g}}$$

$$= 1.07 \times 10^8 \text{ kJ/g}$$

Looking at the last column of Table 24.4, we see that the binding energy in kilojoules per gram is relatively small for very light isotopes such as ${}^2_1H$ and ${}^3_2He$. It rises to a maximum of about 8.5×10^8 for isotopes of intermediate mass number such as cobalt-59 and then falls off slowly to about 7×10^8 for the very heavy isotopes such as uranium (Fig. 24.5).

If we take the binding energy per gram to be a measure of the relative stability of a nucleus, it is clear from Figure 24.5 that the most stable nuclei are those of intermediate mass such as ${}^{59}_{27}Co$, located near the broad maximum of the curve. Nuclei of very heavy elements such as uranium, located at the far right, should be energetically unstable with respect to *fission* into smaller nuclei. Even greater amounts of energy should be obtainable from the *fusion* of very light nuclei such as ${}^2_1H$ at the far left of the curve.

24.4 NUCLEAR FISSION

Shortly before World War II several groups of scientists were studying the products obtained by bombarding uranium with neutrons, hoping to discover new elements with atomic numbers greater than 92. In 1938 Hahn and Strassman in Germany isolated from the products a compound of a Group 2A element, which they originally believed to be radium (at. no. 88). Subsequently they were able to show conclusively that this element was barium (at. no. 56), indicating that a uranium atom had been split into fragments. Hahn's first reaction to this discovery was one of disbelief; he later stated, in January of 1939, "As chemists, we should replace the symbol Ra . . . by Ba . . . [but], as nuclear chemists, closely associated with physics, we cannot decide to take this step in contradiction to all previous experience in nuclear physics."

If Hahn was reluctant to admit the possibility of an entirely new type of nuclear reaction, a former colleague of his, Lisa Meitner, was not. In a letter published

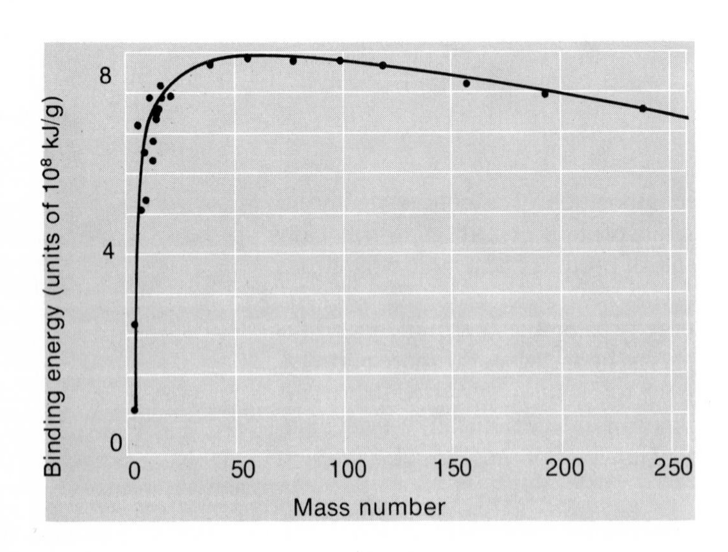

Figure 24.5 Plot of binding energy per gram vs. mass number. Very light and very heavy nuclei are relatively unstable.

with O. R. Frisch in January of 1939, she stated: "At first sight, this result seems very hard to understand. . . . On the basis, however, of present ideas about the behavior of heavy nuclei, an entirely different picture of these new disintegration processes suggests itself. . . . It seems possible that the uranium nucleus . . . may, after neutron capture, divide itself into nuclei of roughly equal size." This revolutionary suggestion was quickly substantiated by experiments carried out in laboratories all over the world.

With the outbreak of World War II, interest in nuclear fission was focused on the enormous amount of energy that could be released in this process. At Los Alamos, in the mountains of New Mexico, a group of scientists under the direction of J. Robert Oppenheimer worked feverishly to produce the fission or "atomic" bomb. Many of the members of this group were exiles from Nazi Germany; they were spurred on by the fear that Hitler would obtain the bomb first. Their work led to the explosion of the first atomic bomb in the New Mexico desert at 5:30 A.M. 1945 7 16. Less than a month later (1945 8 6) the world learned of this new weapon when another bomb was exploded over Hiroshima, killing 70 000 people and completely devastating an area of 10 km^2. Three days later Nagasaki and its inhabitants met a similar fate. On August 14th, the Japanese surrendered and World War II was over.

This research was one of the best kept secrets of all time

The Fission Process (U-235)

Several isotopes of the heavy elements are capable of undergoing fission if bombarded by neutrons of sufficiently high energy. In practice, attention has centered upon two particular isotopes, $^{235}_{92}U$ and $^{239}_{94}Pu$, both of which can be split into fragments by low-energy neutrons. The bomb exploded over Hiroshima used uranium-235; the Nagasaki bomb contained plutonium-239.

Since most of the available data for nuclear fission has to do with uranium-235, our discussion will concentrate upon that isotope. It makes up only about 0.7% of naturally occurring uranium. The more abundant isotope, uranium-238, does not undergo the fission reaction. During World War II, several different processes were studied for the separation of these two isotopes. The most successful technique was that of gaseous effusion (Chapter 5), using the volatile compound UF_6 (bp = 56°C).

FISSION PRODUCTS. When a uranium-235 atom undergoes fission, it splits into two unequal fragments and a number of neutrons and beta particles. The fission process is complicated by the fact that different uranium-235 atoms split up in many different ways. For example, while one atom of $^{235}_{92}U$ is splitting to give isotopes of rubidium (at. no. 37) and cesium (at. no. 55), another may break up to give isotopes of bromine (at. no. 35) and lanthanum (at. no. 57), while still another atom yields isotopes of zinc (at. no. 30) and samarium (at. no. 62).

$$^{1}_{0}n + ^{235}_{92}U \rightarrow ^{90}_{37}Rb + ^{144}_{55}Cs + 2\,^{1}_{0}n \qquad (24.17)$$

$$^{1}_{0}n + ^{235}_{92}U \rightarrow ^{87}_{35}Br + ^{146}_{57}La + 3\,^{1}_{0}n \qquad (24.18)$$

$$^{1}_{0}n + ^{235}_{92}U \rightarrow ^{72}_{30}Zn + ^{160}_{62}Sm + 4\,^{1}_{0}n \qquad (24.19)$$

Fission of a macroscopic sample of uranium-235, containing billions of billions of atoms, gives a large number of products; more than 200 isotopes of 35 different elements have been identified among the fission products of uranium-235.

Since the stable neutron-to-proton ratio near the middle of the periodic table, where the fission products are located, is considerably smaller (~1.2) than that of uranium-235 (1.55), the immediate products of the fission process contain too many neutrons for stability. Isotopes such as $^{90}_{37}Rb$ and $^{144}_{55}Cs$ are radioactive, decaying by electron emission. In the case of rubidium-90, three steps are required to reach a stable nucleus.

$$^{90}_{37}Rb \rightarrow {}^{90}_{38}Sr + {}^{0}_{-1}e; \; t_{1/2} = 2.8 \text{ min}$$

$$^{90}_{38}Sr \rightarrow {}^{90}_{39}Y + {}^{0}_{-1}e; \; t_{1/2} = 29 \text{ a}$$

$$^{90}_{39}Y \rightarrow {}^{90}_{40}Zr + {}^{0}_{-1}e; \; t_{1/2} = 64 \text{ h}$$

The radiation hazard associated with nuclear fallout arises from the formation of radioactive isotopes such as these. One of the most dangerous of these isotopes is strontium-90, which, in the form of strontium carbonate, is readily incorporated into the bones of animals and human beings.

You will notice from Equations 24.17 to 24.19 that two to four neutrons are produced in the fission process for every one consumed. This creates the possibility of a chain reaction. Once a few atoms of uranium-235 split, the neutrons produced can bring about the fission of many more uranium-235 atoms, which in turn yield more neutrons, capable of splitting more uranium atoms, and so on. This is, of course, precisely what happens in the atomic bomb; the energy evolved in successive fissions escalates to give, within a few seconds, a tremendous explosion.

In order for nuclear fission to result in a chain reaction, the uranium-235 or plutonium-239 sample must be large enough so that most of the neutrons are captured internally. If the sample is too small, most of the neutrons produced will escape from the surface, thereby breaking the chain. The *critical mass* of uranium-235 required to maintain a nuclear chain reaction in a bomb appears to be about 40 kg. In the Hiroshima bomb, the critical mass was achieved by using a conventional explosive to fire one piece of uranium-235 into another.

FISSION ENERGY. The evolution of energy accompanying nuclear fission is directly related to the decrease in mass that takes place. One can calculate (Example 24.6, p. 601) that about 80 000 000 kJ of energy is given off for every gram of U-235 that reacts. This is about 40 times as great as the energy change for simple nuclear reactions such as the radioactive decay of radium (Example 24.5). Compared to ordinary chemical reactions, it is truly enormous. The heat of combustion of coal is only about 30 kJ/g; the energy given off when TNT explodes is still smaller, about 2.8 kJ/g. Putting it another way, the fission of one gram of uranium-235 produces as much energy as the combustion of 2700 kg of coal or the explosion of 30 tonnes of TNT.

Nuclear Reactors

Within the past twenty years nuclear reactors, which convert the heat produced by the fission of U-235 into electrical energy, have come into widespread use throughout the world. The type of reactor which is most common in the United States and Europe is the "pressurized light water" reactor shown in Figure 24.6. Water at a pressure of 150 atm is passed through the reactor to absorb the heat given off by the fission process. The water, coming out of the reactor core at a temperature of 320° C, circulates through a closed loop containing a heat exchanger. A second stream of water at a low pressure passes through the heat exchanger and is converted to steam at 270° C. This steam is used to drive a turbogenerator which produces electrical energy.

Reactors of this type have one inherent disadvantage; the water used as a heat exchanger absorbs a large fraction of the neutrons produced by fission. In order to

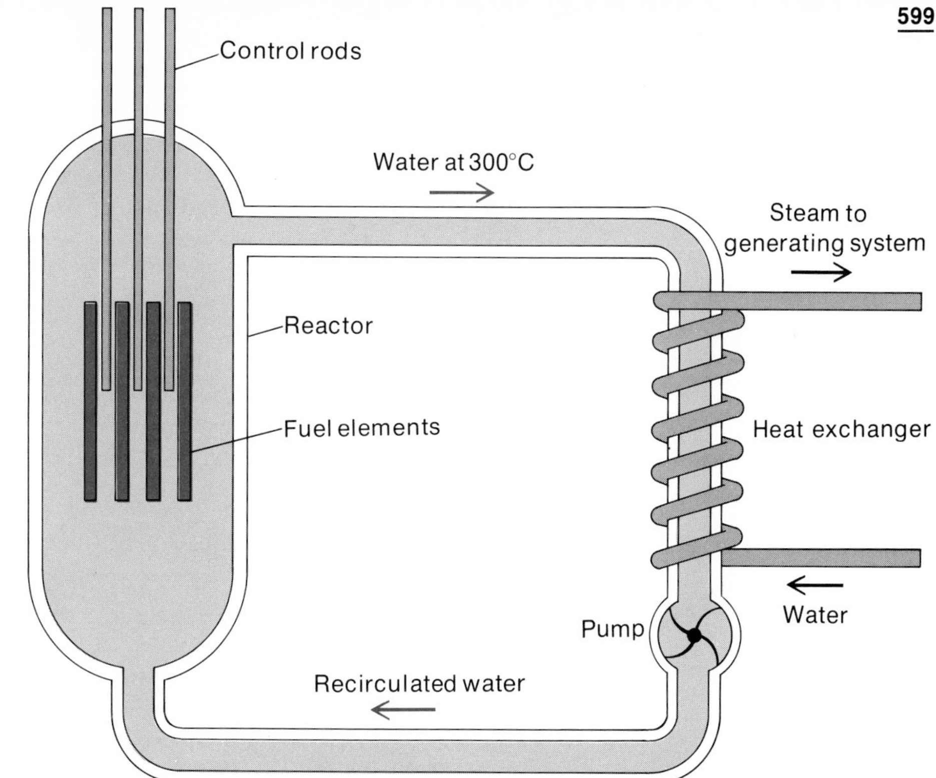

Figure 24.6 Nuclear reactor of the pressurized water type. The control rods are made of a material such as cadmium, which absorbs neutrons very effectively. The rate of fission is carefully monitored and controlled.

sustain the chain reaction, the fuel elements must contain a relatively high percentage of U-235, about five times as great as that in natural uranium. One way to avoid the costly process of isotope separation is to substitute "heavy water" (D_2O), which has very little tendency to absorb neutrons, for ordinary water. This practice is followed in the CANDU reactor, developed in Canada and now being exported to many third-world countries, which uses natural uranium (0.72% U-235) as a fuel.

As reserves of petroleum and natural gas dwindle, fission reactors have received increased attention as a source of electrical energy. In addition to their relatively low fuel cost, these reactors possess another advantage, a greatly reduced level of air pollution. A typical power plant burning coal or oil discharges into the air each year several hundred thousand tonnes of sulfur dioxide, oxides of nitrogen, and ash particles, while a nuclear plant emits practically nothing to the atmosphere.

The most serious problem associated with nuclear reactors is the safety hazard posed by the radioactive products of the fission process. The amounts of such lethal isotopes as Sr-90 and Cs-137 that accumulate in a few weeks' time are equivalent to those produced by the explosion of an atomic bomb of the size dropped on Hiroshima. If the reactor were broken open by a fire, explosion, or earthquake, the fallout could raise the radiation over the surrounding area to dangerous levels. Despite all the precautions that are taken to prevent such an accident, there can be no absolute guarantee that it will never happen. Fears have been expressed that it might come about from leaks in the cooling system, resulting in over-heating and melting of the fuel elements. There is considerable controversy as to whether the elaborate emergency cooling units installed on all reactors are sufficiently fail-safe to prevent this from happening.

So far at least, no such accidents have occurred

As more and more nuclear reactors come into use, it becomes imperative that we develop methods for safe, long-term storage of the radioactive wastes that they produce. So far these wastes have been stored, as concentrated aqueous solutions, in huge underground storage tanks. These tanks are heavily shielded and cooled to absorb the heat given off by radioactive decay, which in some cases would be great enough to keep the solutions at the boiling point. At present, some 500 million litres of liquid waste is in storage. The plan is to evaporate this material to obtain solids, which can be stored more safely, perhaps in abandoned salt mines thousands of feet below the surface of the earth. If this method proves to be practical, it could greatly alleviate the problem of storage of radioactive wastes.

Breeder Reactors

From a long-range standpoint, nuclear reactors using uranium-235 will not be able to satisfy our future power needs. Before the end of this century, our supplies of this relatively rare isotope will be depleted and we will have to turn to other sources of energy. One possibility is to convert the more abundant isotope, uranium-238, into plutonium-239

$$^{238}_{92}U + ^{1}_{0}n \rightarrow ^{239}_{94}Pu + 2\ ^{0}_{-1}e \quad (24.20)$$

which can then undergo fission by reactions such as

$$^{239}_{94}Pu + ^{1}_{0}n \rightarrow ^{90}_{38}Sr + ^{147}_{56}Ba + 3\ ^{1}_{0}n \quad (24.21)$$

Notice that this two-step reaction sequence produces more neutrons than it consumes. Consequently, it leads to a chain reaction similar to the fission of uranium-235, producing comparable amounts of energy. By using this process, we could make use of nearly all the uranium found in nature rather than a tiny fraction of that element. So-called **breeder reactors** based on Reactions 24.20 and 24.21 could satisfy our energy needs for a century or more.

The scientific feasibility of the U-238, Pu-239 cycle was demonstrated many years ago. The bomb exploded over Nagasaki contained plutonium-239 made from uranium-238. The first nuclear reactor to produce electrical energy in this country was of the breeder type. Nevertheless, twenty-five years later, we do not yet have a breeder reactor in commercial operation. A host of technological problems and safety hazards have prevented such reactors from becoming competitive with other methods of generating electrical energy. Among other factors, there is considerable concern about the properties of plutonium-239, which is highly radioactive. The radiation given off when this isotope decays is so powerful that microgram amounts have produced cancer in experimental animals.

Opinions differ greatly as to whether breeder reactors should be built

Demonstration breeder reactors, capable of producing relatively small amounts of electricity, have been in operation for some time in Britain, France, and the USSR. One such reactor, which will cost upwards of half a billion dollars, is now under construction in the United States. This reactor is expected to be in operation by 1980. Another five to ten years will be required to build even more costly breeder reactors capable of producing electrical energy on a commercial scale.

24.5 NUCLEAR FUSION

From Figure 24.5 we see that very light isotopes such as those of hydrogen are potentially unstable with respect to fusion into heavier isotopes of elements such as helium. Indeed, from the form of the nuclear stability curve, we would predict that the energy available from nuclear fusion should be considerably greater than that given off in the fission of an equal mass of a heavy element. This is indeed the case; we can calculate that the energy evolved per gram of reactants in a simple fusion process such as

$$^{2}_{1}H + ^{2}_{1}H \rightarrow ^{4}_{2}He \qquad (24.22)$$

is about three to ten times as great as in the fission of uranium-235 (Example 24.6).

Example 24.6 Calculate the amount of energy evolved, in kilojoules per gram of reactants, in

a. a fusion reaction: $^{2}_{1}H + ^{2}_{1}H \rightarrow ^{4}_{2}He$
b. a fission reaction: $^{235}_{92}U \rightarrow ^{90}_{38}Sr + ^{144}_{58}Ce + ^{1}_{0}n + 4\ ^{0}_{-1}e$

Solution

a. We first calculate the change in mass per mole of product, using Table 24.3.

$$\Delta m = 4.001\ 50\ g - 2(2.013\ 55\ g) = -0.025\ 60\ g$$

Converting to kilojoules:

$$\Delta E = -2.56 \times 10^{-2}\ g \times 9.00 \times 10^{10}\ kJ/g = -2.30 \times 10^{9}\ kJ$$

Since 4.03 g of deuterium are involved, ΔE per gram of reactant is

$$\Delta E = \frac{-2.30 \times 10^{9}\ kJ}{4.03} = -5.75 \times 10^{8}\ kJ$$

b. Proceeding as in (a), we find that, per mole of uranium reacting

$$\Delta m = 89.8864\ g + 143.8816\ g + 1.0087\ g + 4(0.000\ 55\ g) - 234.9934\ g$$
$$= -0.2145\ g$$

Hence, for one mole of U-235:

$$\Delta E = -0.2145\ g \times 9.00 \times 10^{10}\ kJ/g = -1.93 \times 10^{10}\ kJ$$

For one gram of U-235

$$\Delta E = \frac{-1.93 \times 10^{10}\ kJ}{235} = -8.21 \times 10^{7}\ kJ$$

Comparing the answers to (a) and (b), we conclude that the fusion reaction produces about 7 times as much energy per gram of starting material (57.5×10^{7} kJ vs. 8.21×10^{7} kJ) as does the fission reaction. This factor varies from about 3 to 10, depending upon the particular reactions chosen to represent the fusion and fission processes.

As an energy source, nuclear fusion possesses several additional advantages over nuclear fission. For one thing, fusion is a "clean" process in the sense that the products are stable isotopes such as $^{4}_{2}He$, rather than the hazardous radioactive isotopes formed by fission. Equally important, light isotopes suitable for fusion are far more abundant than the heavy isotopes required for fission. We can calculate, for example (Problem 24.39), that the fusion of only 2×10^{-13} per cent of the deuterium ($^{2}_{1}H$) in seawater would meet the total annual energy requirements of the world.

Unfortunately fusion processes, unlike neutron-induced fission, have very high activation energies. In order to overcome the electrostatic repulsion between two deuterium nuclei and cause them to react, they have to be accelerated to velocities that are about 10^{6} m/s, about 10 000 times greater than ordinary molecular velocities at room temperature. The corresponding temperature for fusion, as calculated from kinetic theory (Problem 24.36), is of the order of 10^9°C. In the hydrogen bomb, temperatures of this magnitude were achieved by using a fission reaction to trigger nuclear fusion. If fusion reactions are to be used to generate electricity, it will be necessary to develop equipment in which very high temperatures can be maintained long enough to allow fusion to occur and give off energy. In any conventional container, the reactant nuclei would quickly lose their high kinetic energies by collisions with the walls.

One fusion reaction currently under study is a two-step process involving deuterium and lithium as the basic starting materials:

$$^{2}_{1}H + ^{3}_{1}H \rightarrow ^{4}_{2}He + ^{1}_{0}n$$
$$^{6}_{3}Li + ^{1}_{0}n \rightarrow ^{4}_{2}He + ^{3}_{1}H$$
$$^{2}_{1}H + ^{6}_{3}Li \rightarrow 2\,^{4}_{2}He \qquad (24.23)$$

This particular process is attractive because it has a somewhat lower activation energy than other fusion reactions. Within the past few years, promising results have been obtained with this reaction using "magnetic bottles" in which the reactant nuclei are confined in very strong magnetic fields. So far, prototype models

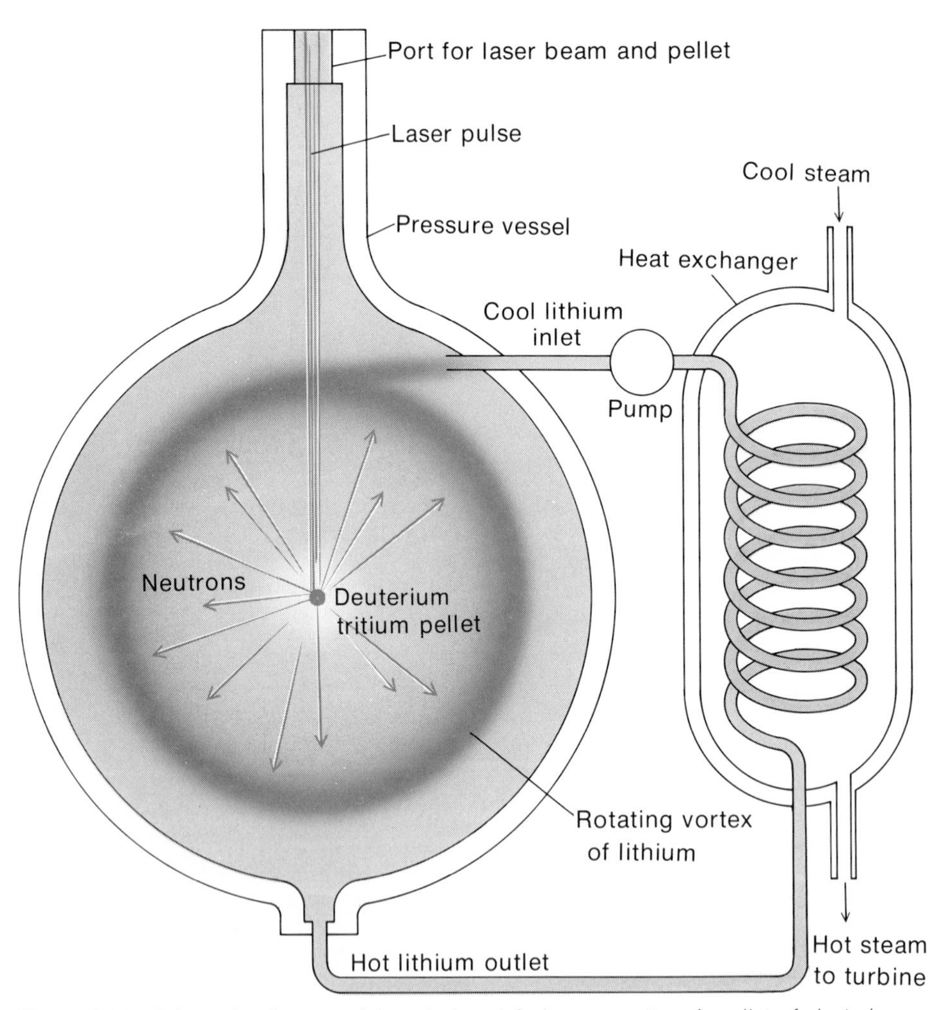

Figure 24.7 Schematic diagram of laser-induced fusion apparatus. A pellet of deuterium-tritium undergoes fusion on absorbing energy from a laser pulse. Neutrons from the fusion reaction react with molten lithium, liberating energy and regenerating tritium (Reaction 24.23).

using this principle have been able to sustain Reaction 24.23 for 0.02 s at best. In order to achieve a net evolution of energy, it will be necessary to extend this time to at least one second.

Another approach to nuclear fusion is shown in Figure 24.7. Tiny glass pellets (about 0.1 mm in diameter) filled with frozen deuterium and tritium are illuminated by a powerful laser beam. In principle at least, the neutrons produced should be able to react with lithium to complete Reaction 24.23 and give off energy. Unfortunately, the economic feasibility of this approach is extremely dubious. The glass pellets are expensive to produce and the laser beam has a very short life expectancy. Whatever approach is used, it seems safe to say that we will not obtain electrical energy on a commercial scale from fusion reactors before the end of this century.

It has not yet been demonstrated that fusion is a practical source of energy

Perhaps the ultimate irony of our time is the fact that we have made so little use of the energy produced in a nuclear fusion process that has been going on since the universe was formed. The energy given off by the sun and other stars results from fusion reactions in which ordinary hydrogen is converted to helium. One mechanism which has been suggested for this process is

$$^{1}_{1}H + ^{1}_{1}H \rightarrow ^{2}_{1}H + ^{0}_{1}e$$

$$^{2}_{1}H + ^{1}_{1}H \rightarrow ^{3}_{2}He;$$

$$^{3}_{2}He + ^{1}_{1}H \rightarrow ^{4}_{2}He + ^{0}_{1}e;$$

$$4\ ^{1}_{1}H \rightarrow ^{4}_{2}He + 2\ ^{0}_{1}e; \quad \Delta E = -6.0 \times 10^{8}\ kJ/g\ reactant \qquad (24.24)$$

Each day, processes such as this, occurring within the sun at a temperature of perhaps 10^{9}°C, produce enormous quantities of energy, of which about 6×10^{18} kJ reaches the surface of the earth. This is roughly equivalent to the total amount of energy that man has consumed since the beginning of time.

PROBLEMS

24.1 *Nuclear Equations* Complete the following nuclear equation:

$$^{96}_{44}Ru + ^{4}_{2}He \rightarrow \_\_\_\_ + ^{1}_{0}n$$

24.2 *Rate of Radioactive Decay* The half-life for the decay of Th-234 is 24.5 d. Calculate the rate constant and the fraction left after 60 d.

24.3 *Mass-Energy Relations* Using Table 24.3, calculate ΔE in kilojoules for the decay of one gram of polonium-210: $^{210}_{84}Po \rightarrow ^{4}_{2}He + ^{206}_{82}Pb$.

24.4 Explain why

a. a U-238 atom and the corresponding +2 ion behave the same way in nuclear reactions.
b. it is frequently found that a beam of radiation from a radioactive sample is split into three beams by an electrical field.
c. $^{16}_{7}N$ decays by electron emission while $^{13}_{7}N$ emits a positron.
d. in the reactor shown in Figure 24.6, the water passing through the core is not converted directly to steam.

24.20 Explain how

a. a U-238 atom and a U-235 atom can have essentially identical chemical properties.
b. a Geiger-Müller counter works.
c. elements of very high atomic number are prepared in the laboratory.
d. radioactive isotopes can be used to demonstrate the dynamic nature of chemical equilibria.

24.5 Predict what would happen if

a. a human being were exposed to radioactive fallout amounting to 1000 rems.
b. ammonia containing N-16 atoms were added to a solution containing $Cu(NH_3)_4^{2+}$.
c. the C-14 method were used to estimate the age of a dinosaur that lived millions of years ago.

24.6 Write balanced nuclear equations for

a. loss of an alpha particle by $^{230}_{90}Th$.
b. loss of an electron by $^{214}_{82}Pb$.
c. fusion of two C-12 nuclei to give another nucleus and a neutron.
d. fission of U-235 to give $^{141}_{56}Ba$, another nucleus, and an excess of two neutrons.

24.7 A certain natural radioactive series starts with uranium-238 and ends with lead-206. Each step in the series involves the loss of either an alpha- or a beta-particle. In the entire series, how many α-particles are given off? how many β-particles?

24.8 The symbolism $^{19}_{9}F(\alpha, n)$ is used to indicate the nuclear reaction in which an alpha particle collides with an F-19 nucleus to form another isotope and emit a neutron. Write nuclear equations for

a. the above reaction.
b. $^{106}_{46}Pd(n, p)$; p = proton.
c. $(\alpha, n)\ ^{12}_{6}C$.

24.9 Cobalt-60, which is widely used in treating cancer patients, has a half-life of 5.26 a.

a. Calculate the rate constant for the decay of this isotope.
b. If a hospital purchases 20 mg of this isotope, how much will remain after ten years?

24.10 The level of radioactivity associated with a certain isotope is measured with a Geiger counter. Over a one-week period the level drops from 420 counts/min to 261 counts/min. Calculate the rate constant for the decay process and the half-life of the isotope.

24.11 Estimate the age of rocks in which

a. the mole ratio of uranium-238 to lead-206 is 1.00.
b. the mole ratio of uranium-238 to lead-206 is 1.05.
c. the mass ratio (in grams) of uranium-238 to lead-206 is 1.05.

Take the half-life of uranium-238 to be 4.5×10^9 a.

24.21 Describe an experiment to determine

a. the half-life of radium (~1600 a).
b. the age of the canvas used for a painting, supposedly in 1500 A.D.
c. the nature of the particle given off when Ca-48 decays.
d. the mass change when radium decomposes by loss of an alpha particle.

24.22 Write balanced nuclear equations for

a. α emission resulting in formation of $^{233}_{91}Pa$.
b. loss of a positron by $^{85}_{39}Y$.
c. fusion of two C-12 nuclei to give Na-23 and another particle.
d. fission of Pu-239 to give $^{130}_{50}Sn$, another nucleus, and an excess of three neutrons.

24.23 A certain radioactive series starts with $^{237}_{93}Np$ and ends with $^{209}_{83}Bi$. How many alpha particles are emitted? how many beta particles?

24.24 Write balanced nuclear equations for the following processes:

a. $^{96}_{42}Mo(d, n)$; d = deuteron
b. $(\alpha, 2n)\ ^{211}_{85}At$
c. $^{10}_{5}B(n, \gamma)$
d. $^{45}_{21}Sc(n, p)$

24.25 Strontium-90, one of the most hazardous isotopes produced in nuclear fission, has a half-life of 29 a.

a. Calculate the rate constant for the decay of strontium-90.
b. What fraction of a strontium-90 sample will remain after 100 a?

24.26 A health officer monitoring the decay of a radioactive sample obtains the following readings on a Geiger-Müller counter:

8:00 A.M. May 6	240	8:00 A.M. May 9	203
11:00 A.M. May 7	225	9:00 A.M. May 12	172

Using a graphical method, estimate the time at which the number of counts will be reduced to 50.

24.27 One way of dating rocks is to determine the relative amounts of potassium-40 and argon-40; the decay of ^{40}K has a half-life of 1.27×10^9 a. Analysis of a certain lunar sample gives the following results in mole ratios:

$^{40}Ar/^{40}K = 4.13$	$(t_{1/2}\ ^{40}K = 1.27 \times 10^9$ a)
$^{206}Pb/^{238}U = 0.66$	$(t_{1/2}\ ^{238}U = 4.5 \times 10^9$ a)
$^{87}Sr/^{87}Rb = 0.041$	$(t_{1/2}\ ^{87}Rb = 5.7 \times 10^{10}$ a)

Using these data, obtain the best possible value for the age of the sample. Can you suggest why the K-Ar method gives a low result?

24.12 The radioactive isotope $^{3}_{1}H$ is produced in nature in much the same way as carbon-14. It has a half-life of 12.3 a. Estimate the age of a sample of Scotch whiskey which has a tritium content 0.65 times that of the water in the area in which the whiskey was produced.

24.13 A paper company is accused of polluting a river with organic wastes from paper manufacture. It maintains that the wastes came from a tanker that spilled oil in the river. Suggest how radioactive dating techniques might be used to settle the controversy.

24.14 Consider the reaction $^{222}_{86}Rn \rightarrow ^{4}_{2}He + ^{218}_{84}Po$.

a. Calculate Δm in grams when one mole of Rn-222 reacts.
b. Calculate ΔE in kilojoules when one mole of Rn-222 reacts; one gram of Rn-222.

24.15 Using Table 24.3, calculate the mass in grams of one mole of O-16 *atoms*.

24.16 Show by calculation whether the following reaction is spontaneous:

$$^{1}_{1}H + ^{18}_{8}O \rightarrow ^{19}_{9}F$$

24.17 Consider the fission reaction $^{235}_{92}U + ^{1}_{0}n \rightarrow ^{144}_{58}Ce + ^{87}_{35}Br + ^{0}_{-1}e + 5\ ^{1}_{0}n$. How many kilojoules of energy are given off per gram of U-235? How many tonnes of TNT would have to be detonated to produce the same amount of energy ($\Delta E = -2.8$ kJ/g)?

24.18 For Mo-99, calculate the mass decrement (g/mol); the binding energy (kJ/mol) and kJ/g.

24.19 Compare the energies given off per gram of reactant in the two fusion processes:

$$^{2}_{1}H + ^{3}_{1}H \rightarrow ^{4}_{2}He + ^{1}_{0}n$$

$$^{2}_{1}H + ^{6}_{3}Li \rightarrow 2\ ^{4}_{2}He$$

24.28 A sample of a funeral shroud taken from an ancient Egyptian tomb is found to have a carbon-14 content 0.560 times that in a living plant. Estimate the age of the linen from which the shroud was made.

24.29 One objection to the C-14 method is that it assumes that the C-14/C-12 ratio in the atmosphere many years ago was the same as it is now. Suggest two different ways in which the validity of this assumption might be checked.

24.30 Consider the alpha decay of $^{245}_{97}Bk$. How much energy, in kilojoules, is given off when one gram of berkelium decays?

24.31 Using Table 24.3, calculate the mass in grams of one mole of $^{24}_{12}Mg^{2+}$ ions.

24.32 Will Al-26 decay spontaneously by positron emission? Si-28? Show by calculation.

24.33 Consider the fission reaction $^{239}_{94}Pu + ^{1}_{0}n \rightarrow ^{144}_{58}Ce + ^{90}_{38}Sr + 2\ ^{0}_{-1}e + 6\ ^{1}_{0}n$. How many grams of Pu-239 would have to react to produce one kilojoule of energy? how many atoms?

24.34 Repeat the calculations of Problem 24.18 for B-10.

24.35 Taking Equation 24.24 to represent the reaction that produces the sun's energy, how many grams of hydrogen would have to be fused to provide the 6.0×10^{18} kJ of energy that reaches the earth each day?

*24.36 It is possible to obtain an estimate of the activation energy for fusion by calculating the energy required to bring two deuterons close enough together to form an alpha particle. This energy can be estimated by using Coulomb's Law: $E = q_1q_2/r$, where q_1 and q_2 are the charges of the deuterons (1.52×10^{-14} esu), r is the radius of the helium nucleus ($\sim 1 \times 10^{-14}$ m), and E is the energy in joules.

a. Estimate E in joules per α particle.
b. Using the equation $E = mv^2/2$, estimate the velocity that a deuteron must have if a collision between two of them is to supply the activation energy for fusion (note that m must be in kilograms).
c. Using the equation $v = (3RT/KMM)^{1/2}$, estimate the temperature that would have to be reached to achieve fusion [$R = 8.31$ J/(mol · K)].

*24.37 Plutonium-239 decays by the reaction $^{239}_{94}Pu \rightarrow ^{235}_{92}U + ^{4}_{2}He$, with a rate constant of 5.5×10^{-11}/min. In a one-gram sample of Pu-239

a. how many grams decompose in 10 min?
b. how much energy in joules is given off in 10 min?
c. what radiation dosage in rems is received by a 60-kg man exposed to a gram of plutonium for 10 min (no. of rems = 1000 × no. of grays)?

*24.38 Natural uranium consists of three isotopes, U-234 (0.0052%), U-235 (0.719%), and U-238 (99.276%). All three isotopes are radioactive, decaying by alpha emission with half-lives of 2.32×10^5 a, 7.07×10^8 a, and 4.51×10^9 a, respectively. What fraction of the alpha particles given off by uranium comes from each isotope?

*24.39 Consider the reaction

$$2\,{}^{2}_{1}H \rightarrow {}^{4}_{2}He$$

a. Calculate ΔE in kilojoules per gram of deuterium fused.
b. How much energy is potentially available from the fusion of all the deuterium atoms in seawater? The percentage of deuterium in water is about 0.015%. The total mass of water in the oceans is 1.3×10^{27} g.
c. What fraction of the deuterium in the oceans would have to be consumed to supply the annual energy requirement of the world (2.2×10^{17} kJ)?

25

POLYMERS: NATURAL AND SYNTHETIC

In previous chapters we have discussed the chemical and physical properties of many kinds of substances. For the most part, these materials were made up of either small molecules or simple ions. In this chapter we will be concerned with an important class of compounds containing large molecules, namely those substances we call polymers.

A polymer is a compound which consists of a large number of small molecular units, called monomers, combined together chemically. A typical polymer molecule contains a chain of monomers several thousand units long, which may or may not have a number of shorter branches. The monomer units may or may not be identical; a given polymer may contain one, two, or many different monomer components.

In the polymer, which is usually a solid, the chains may be roughly parallel, perhaps held together by hydrogen bonds, and forming fibers with high tensile strength and flexibility. On the contrary, the chains may be tangled, which would tend to make a strong film or bulk solid. The properties of the polymer will depend upon the nature of the monomers it contains; its softening or melting point, resistance to solvents and chemical attack, and flexibility and elasticity may vary over wide limits.

Natural polymers produced by plants and animals are essential to all forms of life. The cell membranes of plants and the woody structures of trees are composed in large part of cellulose, a polymeric carbohydrate (Section 25.3). Your body and mine contain thousands of polymers, some simple and some very complex, which make up our tissues, our blood, and our skin. Many of these fall in the class of organic polymers known as proteins (Section 25.4). Natural polymers also have many practical applications in our daily lives. For centuries, wearing apparel has been made from such products as cotton (largely cellulose), wood, silk, and leather (proteinaceous materials).

Since about 1930, a wide variety of synthetic polymers (Sections 25.1 and 25.2) have been made available by the chemical industry. Examples include such well-known textile materials as Dacron and nylon. Most synthetic polymers, like their natural counterparts, are high molecular mass organic compounds. They contain, in addition to carbon and hydrogen, such elements as oxygen, nitrogen, sulfur, and the halogens. Their molecular structures are considerably simpler than those of most natural polymers; ordinarily they contain at most two different types of monomer units.

More industrial chemists work with polymers than with any other type of substance

The reactions used to synthesize polymers in the laboratory fall into two categories:

—*addition* reactions, in which monomers add to one another to give a polymer whose molecular mass is the sum of those of the monomer units from which it is formed.

—*condensation* reactions, in which successive monomer units combine by splitting out a small molecule such as H_2O.

25.1 SYNTHETIC POLYMERS: ADDITION TYPE

Addition polymers are made from monomer units containing multiple bonds. Typically, the monomer is an alkene or a derivative of an alkene with the general formula

$$H_2C{=}CHR$$

where R may be (in the simplest case) a hydrogen atom or, more commonly, a halogen atom or organic functional group. To illustrate the process involved, let us consider the simplest alkene, ethylene, $H_2C{=}CH_2$. You may recall from Chapter 10 that the double bond in ethylene is fairly reactive and will add such species as H_2, Cl_2, or HCl, to form compounds ($H_3C{-}CH_3$, $ClH_2C{-}CH_2Cl$, $H_3C{-}CH_2Cl$), in which there are single carbon-carbon bonds. Ethylene can also undergo an addition reaction with itself to form a polymer called polyethylene:

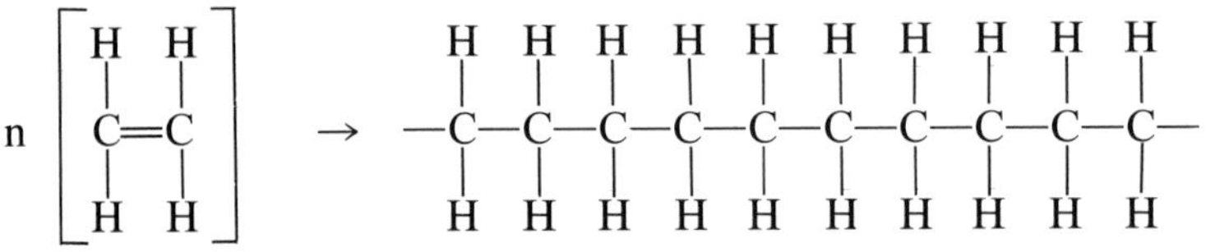

$$n\left[\begin{array}{cc} H & H \\ | & | \\ C & {=}C \\ | & | \\ H & H \end{array}\right] \rightarrow \begin{array}{cccccccccc} H & H & H & H & H & H & H & H & H & H \\ | & | & | & | & | & | & | & | & | & | \\ -C- & C- & C- & C- & C- & C- & C- & C- & C- & C- \\ | & | & | & | & | & | & | & | & | & | \\ H & H & H & H & H & H & H & H & H & H \end{array} \tag{25.1}$$

Ethylene — Section of a polyethylene molecule

In this polymer, the chain consists of carbon atoms, all singly bonded; the bonding around each carbon atom is tetrahedral, so the chain is zigzag. Depending on how the polymer molecule is made, it may contain essentially only straight chains, roughly 2000 units long, or it may have a number of short branches. In the unbranched form, which is called linear polyethylene, neighboring chains are oriented in a regular pattern, at least on a molecular scale. This gives the polymer a significant amount of crystallinity, and hence a relatively high density, melting point, and rigidity. Branched polyethylene is less crystalline and is a rather soft, waxy material.

Polymer chains made from derivatives of ethylene, i.e., monomers having the general formula $H_2C{=}CHR$, may take on any of three different structures:

$-CH_2-CHR-CH_2-CHR-CH_2-CHR-CH_2-CHR-$ head-to-tail

$-CHR-CH_2-CH_2-CHR-CHR-CH_2-CH_2-CHR-$ head-to-head, tail-to-tail

$-CH_2-CHR-CHR-CH_2-CHR-CH_2-CH_2-CHR-$ random

Example 25.1 Sketch a portion of the polymer derived from vinyl chloride, $H_2C{=}CHCl$, assuming it is a

a. head-to-tail polymer.
b. head-to-head, tail-to-tail polymer.

Solution

a. Since in the head-to-tail structure the monomer units are all oriented in the same direction on entering the chain, there will be a Cl atom on every other carbon atom:

$$n\left[\begin{array}{cc} H & H \\ | & | \\ C & {=}C \\ | & | \\ H & Cl \end{array}\right] \rightarrow \begin{array}{cccccccccccc} & H & H & H & H & H & H & H & H & H & H & \\ & | & | & | & | & | & | & | & | & | & | & \\ - & C- & C- & C- & C- & C- & C- & C- & C- & C- & C & - \\ & | & | & | & | & | & | & | & | & | & | & \\ & H & Cl & H & Cl & H & Cl & H & Cl & H & Cl & \end{array}$$

b. In this case, successive monomer units are oriented in opposite ways, so Cl atoms occur in pairs on adjacent carbon atoms:

$$\begin{array}{cccccccccccc} & H & H & H & H & H & H & H & H & H & H & \\ & | & | & | & | & | & | & | & | & | & | & \\ - & C- & C- & C- & C- & C- & C- & C- & C- & C- & C & - \\ & | & | & | & | & | & | & | & | & | & | & \\ & H & Cl & Cl & H & H & Cl & Cl & H & H & Cl & \end{array}$$

In practice it is found that polyvinyl chloride is a head-to-tail polymer of the type shown in Example 25.1a. This is typical of the structure of most addition polymers.

Rather surprisingly, the structure of polyvinyl chloride is not completely established by knowing that the monomer units are arranged head-to-tail. There are, in fact, several nonidentical ways that the chlorine atoms can be attached to the chain, as shown in Figure 25.1 for the general monomer $H_2C{=}CHR$.

The properties of a polymer like this depend on its configuration

In the *isotactic* configuration, the R group (e.g., Cl atom) is always in the same direction with respect to the chain, say always to the right if the chain were observed from above. In the *syndiotactic* form, the R groups alternate positions down the chain, one to the left, one to the right, one to the left, and so on. In the third structure, which has no symmetry of this sort, the R groups are randomly arranged, and the structure is called *atactic*.

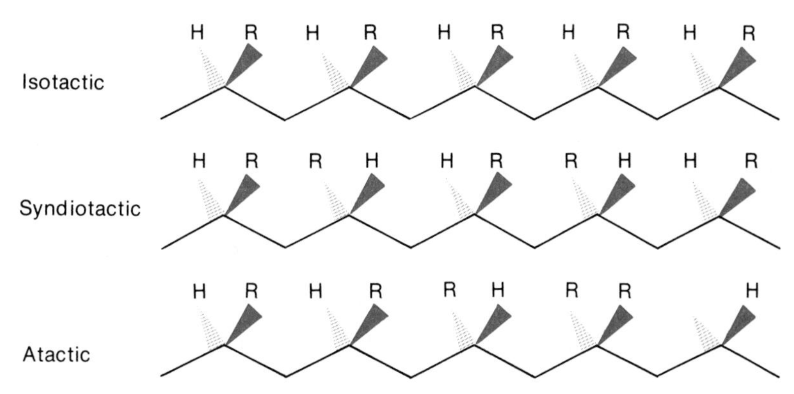

Figure 25.1 Three different structures for an addition polymer of the head-to-tail type. The solid line represents a carbon chain, with each lower vertex being a CH_2 group; the plane of the chain is that of the page. Depending on polymerization conditions, the polymer obtained can have any of the three structures shown.

Example 25.2 Ordinarily vinyl chloride forms a syndiotactic molecule on polymerization. Sketch a portion of the polymer molecule.

Solution For vinyl chloride, R equals Cl, and in the syndiotactic form the Cl groups alternate their positions down the chain. The structure would be

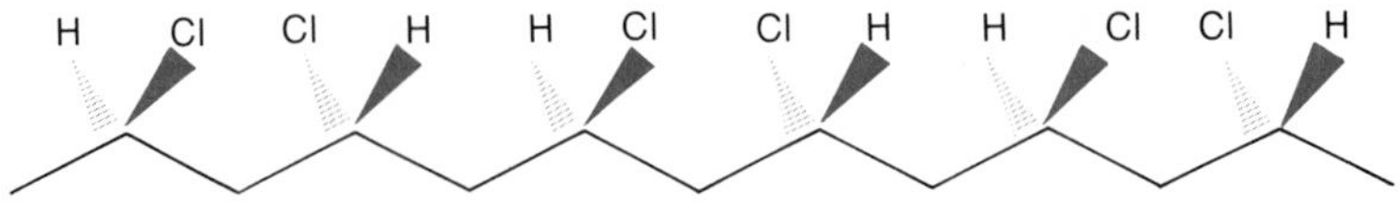

H Cl Cl H H Cl Cl H H Cl Cl H

(There is a carbon atom at each vertex of the chain; the two hydrogen atoms bonded to every other carbon atom are not shown.)

In the early days of the polymer industry, the structures of the molecules that were formed on polymerization were not known. Various catalysts were used to initiate polymerization; some would produce useful polymers from a given monomer, others made poor ones. In about 1955, Guilio Natta, an Italian chemist, discovered some catalysts that enabled him to make both pure isotactic and pure syndiotactic polypropylene, both of which have more desirable properties than the atactic (random) modification, which was the form produced by polymerization processes in use at that time. These catalysts contain a reducing agent and a transition metal salt; a common one is $Al(C_2H_5)_3 \cdot TiCl_4$. Natta catalysts proved to be extremely useful for controlling polymer configurations; Karl Ziegler, for example, used them in Germany to make the first linear polyethylene. In 1963 these two men shared the Nobel Prize in Chemistry for their discoveries.

There are now at least 50 synthetic polymers available commercially

In Table 25.1 we have listed some of the more common addition polymers, along with their uses. Although these polymers vary in their properties, they are similar in many respects. We will briefly examine one of the most common addition polymers, polyvinyl chloride, which is made and processed by methods typical of those used with many polymers.

Polyvinyl Chloride: Preparation and Properties

Vinyl chloride, like many monomers, is a gas at room temperature (*bp* = $-14°C$). Polymerization is usually accomplished in a suspension of the monomer in water at about 50°C. The reaction is initiated by adding a reactive species called a *free radical,* a substance containing an unpaired electron. Such a species is formed by the decomposition of benzoyl peroxide:

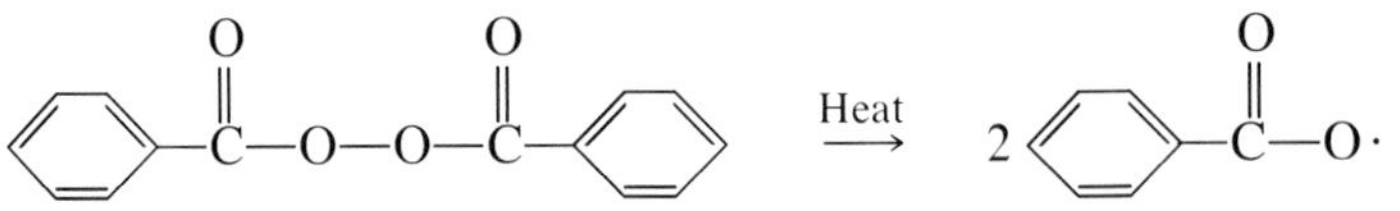

$$\mathrm{C_6H_5{-}\overset{\overset{\displaystyle O}{\|}}{C}{-}O{-}O{-}\overset{\overset{\displaystyle O}{\|}}{C}{-}C_6H_5} \xrightarrow{\text{Heat}} 2\,\mathrm{C_6H_5{-}\overset{\overset{\displaystyle O}{\|}}{C}{-}O\cdot} \qquad (25.2)$$

benzoyl peroxide — a free radical, X·

The free radical, X·, reacts readily with vinyl chloride, forming another free radical:

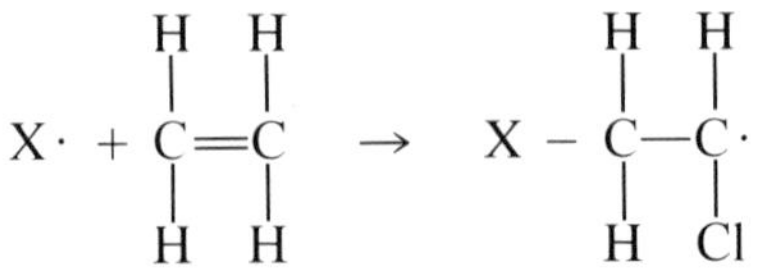

$$X\cdot + \begin{array}{cc} H & H \\ | & | \\ C & = C \\ | & | \\ H & H \end{array} \rightarrow X - \begin{array}{cc} H & H \\ | & | \\ C & - C\cdot \\ | & | \\ H & Cl \end{array} \qquad (25.3)$$

TABLE 25.1 SOME COMMON ADDITION POLYMERS

MONOMER	NAME	POLYMER	USES	AMT PRODUCED IN 1975 (millions of kilograms)
$H_2C{=}CH_2$	ethylene	polyethylene	bags, coatings, toys	3000
$H_2C{=}C(H)(CH_3)$	propylene	polypropylene	beakers, milk cartons	1000
$H_2C{=}C(H)(Cl)$	vinyl chloride	polyvinyl chloride, PVC	floor tile, raincoats, pipe, phonograph records	1600
$H_2C{=}C(H)(CN)$	acrylonitrile	polyacrylonitrile, PAN	rugs; Orlon, Acrilan, and Dynel are copolymers with vinyl chloride or other polar monomers	300
$H_2C{=}C(H)(C_6H_5)$	styrene	polystyrene	cast articles requiring a transparent plastic	1200
$H_2C{=}C(CH_3)(O{-}C({=}O){-}CH_3)$	methyl methacrylate	polymethyl methacrylate, Plexiglas, Lucite	high quality transparent cast objects	20
$F_2C{=}CF_2$	tetrafluoroethylene	Teflon	gaskets, insulation, bearings, pan coatings	10

The product is able to add another vinyl chloride molecule, and then another and another . . . to produce eventually a polymer molecule containing about 25 000 monomer units. As previously pointed out, nearly all of these are arranged head-to-tail in the syndiotactic configuration, with very few branches. The chains are finally terminated when two long chain radicals combine.

The polymerization process, as is the case with most addition polymers, is highly exothermic (Example 25.3). One of the advantages of carrying out the reaction in water suspension is the relative ease of temperature control.

Example 25.3 From information given in the preceding discussion, estimate, for polyvinyl chloride

a. the molecular mass of the polymer.
b. ΔH per mole of monomer reacted (use bond energies).

Solution

a. For vinyl chloride, $H_2O{=}CHCl$, the molecular mass is readily calculated to be 62.5. With 25 000 monomer units, the polymer would have a molecular mass of about $62.5 \times 25\ 000 = 1.6 \times 10^6$.

b. Every time a molecule of monomer is added to the chain, one carbon-carbon double bond is broken and two carbon-carbon single bonds are formed. ΔH per mole of monomer can be estimated from bond energies taken from Tables 4.2 and 8.5. Recalling that bond-breaking is endothermic and bond-making is exothermic:

$$\Delta H = B.E.\ C{=}C - 2\ B.E.\ C{-}C = 598 - 2(347) = -96 \text{ kJ/mol of monomer}$$

Typically, the mole ratio of water to monomer in the polymerization mixture is about 10:1, so 96 kJ is liberated into about 180 g of water. This is considerably more than would be required to bring the water to its boiling point, so the batch must be cooled continuously as polymerization proceeds.

The product formed by suspension polymerization consists of small granules (~0.1 mm in diameter); these are separated by centrifuging and then dried. The pure polymer is quite hard, tough, and very resistant to water. It can, however, be dissolved in hot organic solvents, forming viscous solutions. It can be melted and extruded or molded into water pipe or other rigid articles. For many uses, polyvinyl chloride is softened and made more flexible by addition of substances called plasticizers; these materials are typically nonvolatile esters and fairly good solvents for the polymer. Plasticizers act by decreasing the dispersion forces holding the chains together. Some softening can also be accomplished by polymerizing with another monomer, such as vinyl acetate, to make what is called a copolymer. Depending on the amount of softening introduced, polyvinyl chloride is useful for making such diverse materials as garden hose, wire insulation, phonograph records, floor tile, luggage, shower curtains, raincoats, and liners for swimming pools.

One of the drawbacks of polyvinyl chloride is that it is not stable at high temperatures; even at 100°C it decomposes slowly, evolving HCl. Most polymers are not very useful at elevated temperatures, the prime exception being Teflon, made from tetrafluoroethylene (Table 25.1). Whereas all the other polymers will burn to some degree and can be degraded chemically, Teflon is extremely inert, resisting chemical attack by all known reagents except molten alkali metals. It is a useful plastic from about −70°C to 250°C, a much larger range than can be claimed for other polymers, which tend to become very brittle when cold and usually melt below 200°C. Teflon is a moderately hard, white, waxy solid, tough enough to be machined to close tolerances. Its inertness and temperature stability make it *the* substance of choice for making gaskets and seals which are subject to high temperatures or caustic reagents. Its high electrical resistance and flexibility make it ideal for use in coaxial cables. Its most mundane application is as a nonstick coating for pots and pans, where its desirable properties are put to the test daily by thousands of amateur cooks.

Another drawback is that the monomer is carcinogenic

Rubber

The first known and still the most important of the addition polymers is that material we call rubber. Rubber occurs naturally and is found in the latex, or sap, of the *Hevea* or rubber tree, which is grown mainly in Ceylon and Indonesia. The latex is collected and treated with acetic acid to precipitate the raw rubber, which coagulates, is pressed dry, and rolled into sheets for shipment.

Raw rubber is a gummy, light-colored substance, which becomes sticky when warmed and brittle when cold. In 1834 Charles Goodyear discovered that if natural rubber is heated with sulfur a reaction occurs, which produces a new material with the properties we normally associate with rubber, namely elasticity, flexibility, and resistance to abrasion. Ordinarily commercial rubber is prepared by this process, called vulcanization; rubber also usually contains certain additives, particularly carbon blacks, which are both less expensive than rubber and greatly improve its resistance to wear and tear.

It is now known that natural rubber is a polymeric substance, containing in its molecules many units of a monomer called isoprene, 2-methyl-1,3-butadiene, attached to one another in the following manner:

$$2n\left[\begin{array}{c} CH_3 \qquad\quad H \\ C{-}C \\ CH_2 \qquad\quad CH_2 \end{array}\right] \rightarrow \left[\begin{array}{c} CH_3 \qquad\quad H \qquad\quad CH_3 \qquad\quad H \\ C{=}C \qquad\qquad\qquad C{=}C \\ {-}CH_2 \qquad\quad CH_2{-}CH_2 \qquad\quad CH_2{-} \end{array}\right]_n$$

Isoprene

Natural rubber (poly-*cis*-isoprene)

the CH_2 groups are *cis* with respect to the double bond

In the rubber molecule, the monomer units are all arranged head-to-tail and all CH_2 groups are in the *cis* position with respect to the double bonds; the average molecular mass is about 400 000.

During the vulcanization process the long polymer chains are cross-linked by reaction with fairly short chains of sulfur atoms, from one to about six atoms long; H_2S is eliminated and relatively little unsaturation is lost. As a result of cross-linking, solubility in organic solvents goes down and rigidity tends to go up. Ordinary rubber is produced by vulcanization with about 3% sulfur; hard rubber results if the amount of sulfur is increased to 30%. Too much cross-linking gives a rigid, three-dimensional network, with its associated hardness and brittleness. In ordinary rubber the chains are relatively easy to align and stretch; the cross links

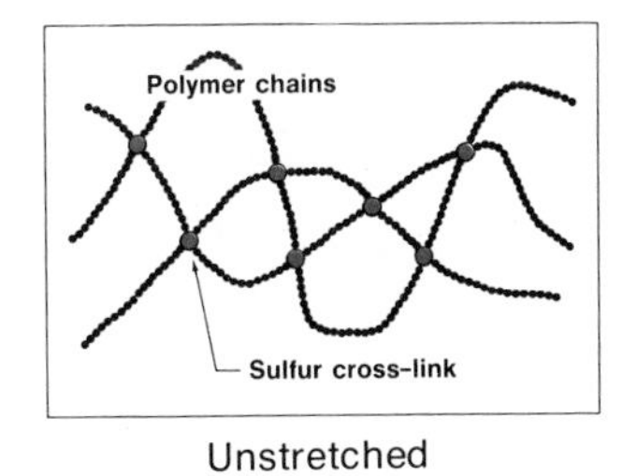

Unstretched

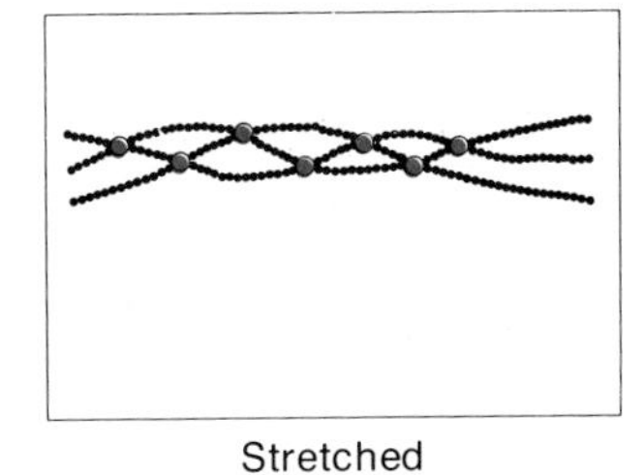

Stretched

Figure 25.2 Schematic drawing showing cross-linking in rubber molecules. The sulfur cross-links are a few atoms long and prevent the flexible chains from moving with respect to each other. When tension is released, the chains assume their previous, more random, arrangement. (From Jones, M. M., Netterville, J. T., et al., *Chemistry, Man, and Society,* 2nd ed., Philadelphia, W. B. Saunders Co., 1976, p. 343.)

serve to let the polymer "remember" where the chains were and so return to their original configuration when tension is released. Without such cross links, the chains slip past one another upon stretching, resulting in very little elasticity.

Since about 1930, several synthetic rubbers have been produced commercially. Wallace Carothers, working at the laboratories of the Du Pont Company in Wilmington, Delaware, discovered one of the most important of these, called neoprene, which is made from the monomer chloroprene:

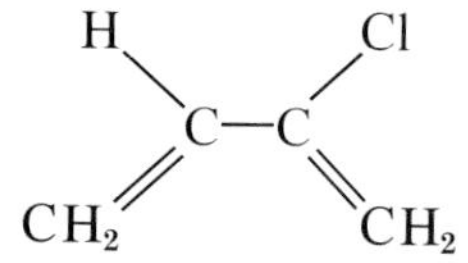

$$\begin{array}{ccc} H & & Cl \\ & \diagdown C{-}C \diagup & \\ CH_2 \diagup\!\!\diagup & & \diagdown\!\!\diagdown CH_2 \end{array}$$

Chloroprene

Neoprene for many purposes is superior to natural rubber, but is considerably more expensive. It is used in large amounts where rubbers that are resistant to organic solvents, water, and oxidation are required. Although it makes an excellent tire rubber, you have more likely encountered it in gasoline fuel lines or shoe soles, where its desirable properties warrant its use.

During World War II supplies of natural rubber from southeast Asia were cut off, making it imperative that the United States produce synthetic rubber in large volume on short notice. A copolymer containing one part of styrene and three parts of 1,3-butadiene, $CH_2{=}CHCH{=}CH_2$, was found to perform satisfactorily, and, in 1945, 600 000 tonnes of this synthetic rubber (Buna S, GRS, and SBR are some of its common names) were produced. It is still more widely used than any other synthetic rubber mainly as tire tread material.

In 1955, research chemists at the Goodyear and Firestone companies discovered that if isoprene is polymerized with Ziegler-Natta catalysts, the product obtained is essentially identical to natural rubber. This material is now produced commercially in relatively small amounts. The monomer 1,3-butadiene, when polymerized with the same catalysts, forms a cheaper rubber with similar properties. Although production of this type of synthetic rubber did not start until 1964, it is now second only to styrene-butadiene rubber in amount manufactured.

25.2 SYNTHETIC POLYMERS: CONDENSATION TYPE

Having recognized the rather simple principle that compounds containing one or more carbon-carbon double bonds will tend to form addition polymers, one might wonder whether other classes of compounds discussed in Chapter 10 might also react to produce polymers. A clue to a source of such compounds is afforded by the reaction of an alcohol with an acid to form an ester:

$$\underset{\text{an alcohol}}{R{-}OH} + \underset{\text{an acid}}{HO{-}\overset{\overset{O}{\|}}{C}{-}R'} \rightarrow \underset{\text{an ester}}{R{-}O{-}\overset{\overset{O}{\|}}{C}{-}R'} + H_2O \qquad (25.4)$$

In the process the two organic molecules are condensed into one, and water is eliminated. If the alcohol used contains two OH groups rather than one, and the acid has two COOH groups, the ester formed will still have a reactive group at each end of the molecule:

$$\underset{\text{a dihydroxy alcohol}}{HO-R-OH} + \underset{\text{a dicarboxylic acid}}{HO-\overset{O}{\overset{\|}{C}}-R'-\overset{O}{\overset{\|}{C}}-OH} \rightarrow \underset{\text{ester with active end groups}}{HO-R-O-\overset{O}{\overset{\|}{C}}-R'-\overset{O}{\overset{\|}{C}}-OH} + H_2O \quad (25.5)$$

Consequently, further condensation can occur with the final result being a long-chain polymeric ester:

$$-O-R-O-\overset{O}{\overset{\|}{C}}-R'-\overset{O}{\overset{\|}{C}}-O-R-O-\overset{O}{\overset{\|}{C}}-R'-\overset{O}{\overset{\|}{C}}-$$

section of a polyester molecule

The **polyester** shown above is a typical condensation polymer. Looking at the polymerization reaction, you will note two important differences between addition and condensation polymers:

1. When a condensation polymer is formed, a small molecule, usually water, is eliminated for every monomer unit added to the polymer chain. In contrast, in addition polymerization, monomer units simply add to one another.

2. To form a condensation polymer, one must start with monomers containing two (or more) functional groups (e.g., —OH, COOH). Usually, but not always, two different monomers are involved. In addition polymers, the key property is the presence of one or more double bonds in the monomer, and typically only one monomer is used.

There are several polyesters of commercial importance. Perhaps the most familiar is the fiber known as Dacron; this is made from the simplest dihydroxy alcohol, ethylene glycol, $HO-CH_2-CH_2-OH$, where R is CH_2-CH_2, and the dicarboxylic acid in which R′ is a benzene ring, ⌬. This polymer forms very strong fibers, and can also be made into film, where it is called Mylar.

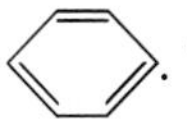

Example 25.4 Kodel is a polyester made from the alcohol and acid shown below. Draw a portion of the Kodel molecule.

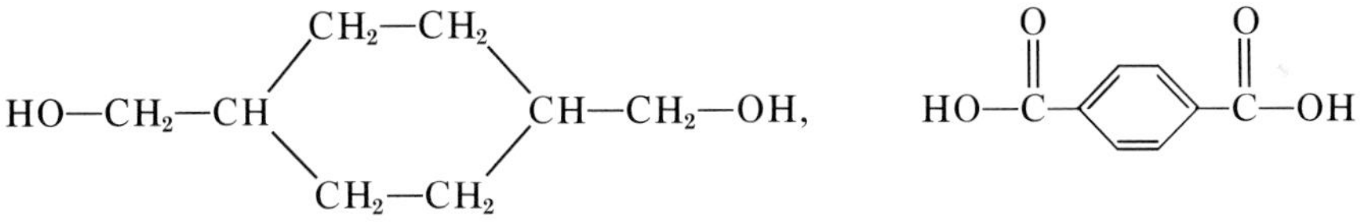

$$HO-CH_2-CH\langle{}^{CH_2-CH_2}_{CH_2-CH_2}\rangle CH-CH_2-OH, \qquad HO-\overset{O}{\overset{\|}{C}}-⌬-\overset{O}{\overset{\|}{C}}-OH$$

Solution Noting that R is

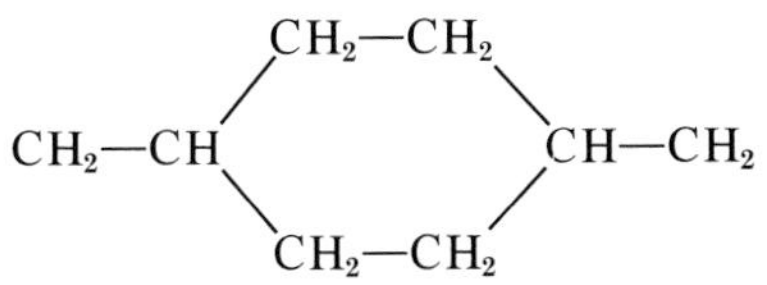

$$CH_2-CH\langle{}^{CH_2-CH_2}_{CH_2-CH_2}\rangle CH-CH_2$$

and R′ is 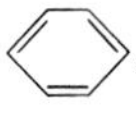 ⌬, we can proceed immediately to draw the repeating unit in the Kodel polymer:

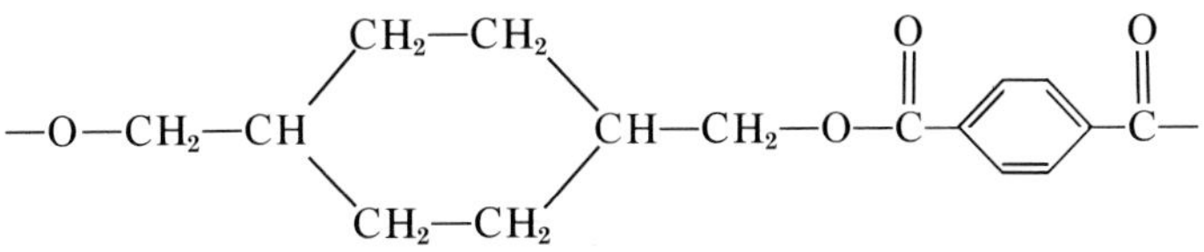

$$-O-CH_2-CH\langle{}^{CH_2-CH_2}_{CH_2-CH_2}\rangle CH-CH_2-O-\overset{O}{\overset{\|}{C}}-⌬-\overset{O}{\overset{\|}{C}}-$$

Polyesters such as Dacron and Kodel, like many addition polymers, are *thermoplastic* (soften on heating) and moderately soluble in hot organic solvents, forming very viscous solutions. The polyesters are somewhat less resistant to chemical attack than most addition polymers in that the esterification reaction can be reversed fairly easily, particularly by hot aqueous NaOH.

Another important class of condensation polymers is that of the **polyamides.** These are made by a reaction that is quite similar to the one used in producing polyesters. Instead of reacting an acid with an alcohol, one reacts it with an analogous substance called an *amine,* which contains an NH_2 rather than an OH group. The acid-amine reaction produces, on elimination of water, a compound known as an *amide:*

This reaction is somewhat oversimplified

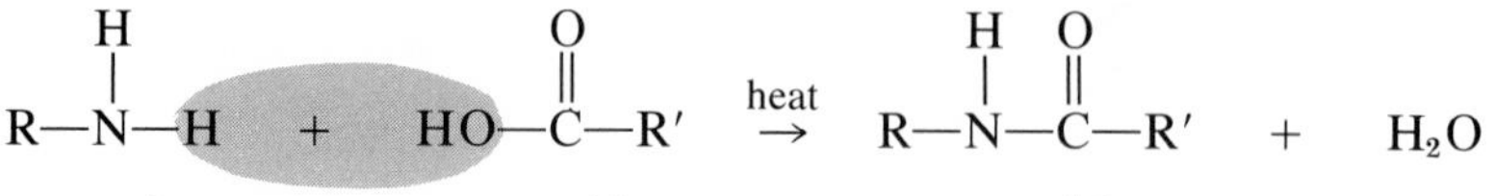

$$\underset{\text{an amine}}{\mathrm{R{-}\overset{\displaystyle H}{\overset{|}{N}}{-}H}} + \underset{\text{an acid}}{\mathrm{HO{-}\overset{\displaystyle O}{\overset{\|}{C}}{-}R'}} \xrightarrow{\text{heat}} \underset{\text{an amide}}{\mathrm{R{-}\overset{\displaystyle H}{\overset{|}{N}}{-}\overset{\displaystyle O}{\overset{\|}{C}}{-}R'}} + H_2O \qquad (25.6)$$

Now, reasoning by analogy with the approach used to make polyesters, you should be able to show how we might produce a polyamide.

Example 25.5

a. Using R and R′, sketch the general structure of an amine and an acid which could be used to make a polyamide.
b. Sketch a portion of the polyamide molecule.

Solution

a. Both the amine and the acid must have two reacting groups. The diamine and the dicarboxylic acid would have the general structures

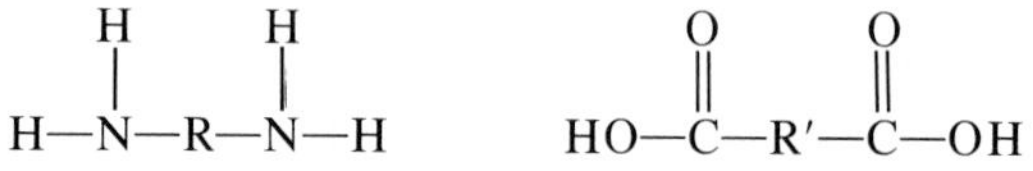

$$\underset{\text{a diamine}}{\mathrm{H{-}\overset{\displaystyle H}{\overset{|}{N}}{-}R{-}\overset{\displaystyle H}{\overset{|}{N}}{-}H}} \qquad \underset{\text{a dicarboxylic acid}}{\mathrm{HO{-}\overset{\displaystyle O}{\overset{\|}{C}}{-}R'{-}\overset{\displaystyle O}{\overset{\|}{C}}{-}OH}}$$

b. By reacting the amine with the acid, we form the amide:

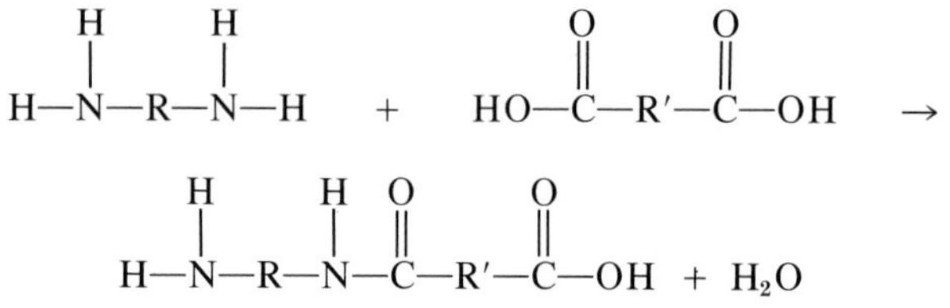

$$\mathrm{H{-}\overset{\displaystyle H}{\overset{|}{N}}{-}R{-}\overset{\displaystyle H}{\overset{|}{N}}{-}H} + \mathrm{HO{-}\overset{\displaystyle O}{\overset{\|}{C}}{-}R'{-}\overset{\displaystyle O}{\overset{\|}{C}}{-}OH} \rightarrow$$

$$\underset{\text{amide with reactive end groups}}{\mathrm{H{-}\overset{\displaystyle H}{\overset{|}{N}}{-}R{-}\overset{\displaystyle H}{\overset{|}{N}}{-}\overset{\displaystyle O}{\overset{\|}{C}}{-}R'{-}\overset{\displaystyle O}{\overset{\|}{C}}{-}OH}} + H_2O$$

This molecule could continue to add amine and acid, alternately, yielding ultimately

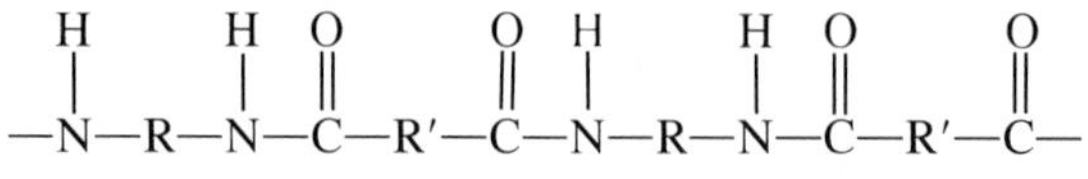

$$\underset{\text{section of a polyamide}}{\mathrm{{-}\overset{\displaystyle H}{\overset{|}{N}}{-}R{-}\overset{\displaystyle H}{\overset{|}{N}}{-}\overset{\displaystyle O}{\overset{\|}{C}}{-}R'{-}\overset{\displaystyle O}{\overset{\|}{C}}{-}\overset{\displaystyle H}{\overset{|}{N}}{-}R{-}\overset{\displaystyle H}{\overset{|}{N}}{-}\overset{\displaystyle O}{\overset{\|}{C}}{-}R'{-}\overset{\displaystyle O}{\overset{\|}{C}}{-}}}$$

Wallace Carothers, the discoverer of neoprene, found that the polyamide formed when R is $(CH_2)_6$ and R′ is $(CH_2)_4$ had very interesting properties. That material is now known as Nylon 66, the two 6's indicating the number of carbon atoms in the monomeric acid and amine.

Nylon is an ideal polymer in many ways. Perhaps its main virtue lies in its ability to form strong but moderately elastic fibers, which are essentially perfect for the manufacture of women's stockings. The reason that nylon makes good fibers appears to be, in part at least, that hydrogen bonding can occur between polymer chains, with the oxygen atom from a carbonyl group interacting with an H atom from an N—H group on an adjacent chain (Fig. 25.3). A typical fiber will contain a large number of parallel molecular chains. In bulk form nylon is also an excellent polymer; it is tough, somewhat waxy, and suitable for use in gears and other high quality parts requiring resistance to wear.

The successful development of nylon encouraged the chemical industry to spend a lot more money on research

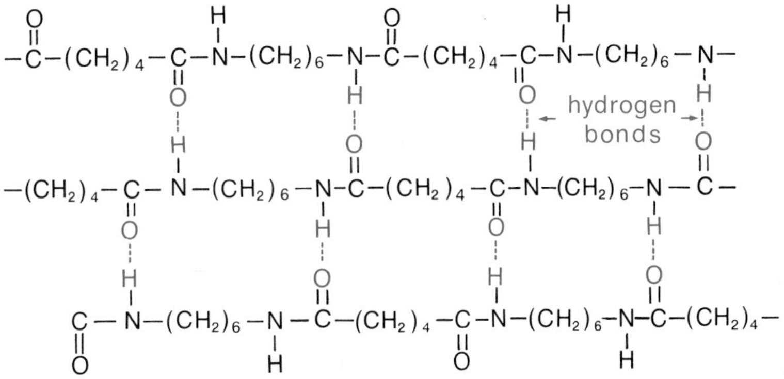

```
  O                  H                  O                     H
  ||                 |                  ||                    |
 -C-(CH2)4-C-N-(CH2)6-N-C-(CH2)4-C-N-(CH2)6-N-
           ||         |           ||         |
           O          H           O          H
           :          :           :  hydrogen:
           H          O           H   bonds  O
           |          ||          |          ||
 -(CH2)4-C- N-(CH2)6-N-C-(CH2)4-C-N-(CH2)6-N-C-
         ||          |          ||         |
         O           H          O          H
         :           :          :          :
         H           O          H          O
         |           ||         |          ||
        C-N-(CH2)6-N-C-(CH2)4-C-N-(CH2)6-N-C-(CH2)4-
        ||         |          ||         |
        O          H          O          H
```

Figure 25.3 Hydrogen bonding in nylon. A nylon fiber will contain many molecular chains which are hydrogen bonded as shown. For maximum bonding the chains go in opposite directions.

Silicones

So far we have been discussing polymers which are purely organic, containing chains either made entirely from carbon atoms, which is the case with all addition polymers, or from carbon and oxygen or nitrogen, as with the polyesters and polyamides. Most polymers, natural and synthetic, are indeed organic, but there is one important class, the silicones, in which silicon is present in the chain.

In Chapter 10 we noted that silicon was one of the few elements that can form long chains, and that in such chains silicon was always combined with oxygen, the chain being of the form —Si—O—Si—O—Si—O—. Silicates all contain such chains, which provide the strength we observe in many rocks and minerals.

Compounds containing silicon and elements other than oxygen are not very numerous and typically react vigorously with water. One of the more important of such compounds is dimethyldichlorosilane, $(CH_3)_2SiCl_2$; addition of this compound to water produces the following reaction:

$$(CH_3)_2SiCl_2(aq) + 2\ H_2O \rightarrow (CH_3)_2Si(OH)_2(aq) + 2\ H^+(aq) + 2\ Cl^-(aq)$$

The silicon-containing product, with two hydroxyl groups on a single silicon atom, is not very stable, and readily combines with itself, eliminating water, and ultimately forming a condensation polymer called a **silicone**:

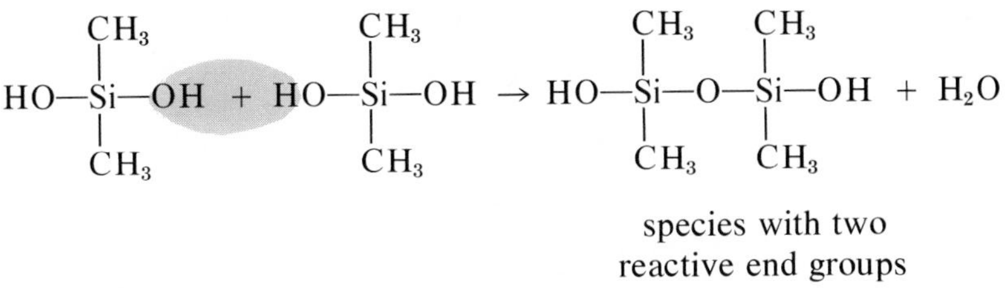

$$HO{-}\overset{CH_3}{\underset{CH_3}{Si}}{-}OH + HO{-}\overset{CH_3}{\underset{CH_3}{Si}}{-}OH \rightarrow HO{-}\overset{CH_3}{\underset{CH_3}{Si}}{-}O{-}\overset{CH_3}{\underset{CH_3}{Si}}{-}OH + H_2O \qquad (25.7)$$

species with two reactive end groups

$$n\left[\begin{array}{c} CH_3 \\ | \\ HO{-}Si{-}OH \\ | \\ CH_3 \end{array}\right] \rightarrow \begin{array}{c} CH_3 \quad\;\; CH_3 \quad\;\; CH_3 \quad\;\; CH_3 \\ | \qquad\quad | \qquad\quad | \qquad\quad | \\ {-}O{-}Si{-}O{-}Si{-}O{-}Si{-}O{-}Si{-} \\ | \qquad\quad | \qquad\quad | \qquad\quad | \\ CH_3 \quad\;\; CH_3 \quad\;\; CH_3 \quad\;\; CH_3 \end{array} + n\,H_2O \qquad (25.8)$$

section of a silicone polymer

A silicone made with methyl groups, as in the above chain, will be a nonvolatile viscous oil, much more stable and resistant to oxidation at high temperatures than petroleum-based oils. Silicone oils are used in oil diffusion pumps and for making high-temperature baths for laboratory use.

Silly putty is intermediate between silicone rubber and silicone oil

Cross-linking the silicone chains produces a silicone rubber, which is not wet by water, does not take a permanent set on compression, and is elastic at both high and low temperatures. These rubbers are used for making gaskets, sealants, and electrical insulation. The common silicone rubber bathtub caulking available in hardware stores contains acetate groups bonded to the silicone chain. In the presence of moisture, acetic acid is liberated and hydroxyl groups replace acetate on the chains. The chains slowly cross-link, or vulcanize, as water is eliminated and oxygen links between chains are formed.

25.3 NATURAL POLYMERS: CARBOHYDRATES

Among the polymers found in living organisms are several that fall into that class of organic compounds called **carbohydrates.** The carbohydrates are compounds of carbon, hydrogen, and oxygen. Their name is derived from the fact that carbohydrates usually have molecular formulas that can be written as $C_n(H_2O)_m$. Some common examples of polymeric carbohydrates are those substances we call cellulose and starch. Hydrolysis of these materials yields a compound called glucose, $C_6H_{12}O_6$, which, like many carbohydrates, is a sugar. Glucose is found free as well as in combined forms, and is present in honey and many fruits and is the sugar found in our blood. Because of its wide occurrence, glucose, either free or combined, is probably the most abundant of all the organic substances.

The chemical bonding in glucose can be represented in two dimensions as

$$\begin{array}{c} H \quad\; H \quad\; H \quad\; H \quad\; H \quad\; H \\ | \qquad | \qquad | \qquad | \qquad | \qquad | \\ O{=}C_1{-}C_2{-}C_3{-}C_4{-}C_5{-}C_6{-}H \\ \quad | \qquad | \qquad | \qquad | \qquad | \\ \quad OH \;\; OH \;\; OH \;\; OH \;\; OH \end{array}$$

The glucose molecule is typical of the simple sugars. It contains a carbon chain with one carbonyl, C=O, group and hydroxyl groups on each carbon atom except the one in the carbonyl group. Since the carbonyl group is on a terminal carbon atom, glucose is an aldehyde (Chapter 10). It is called an *aldohexose,* since it is an aldehyde (aldo), has six carbon atoms (hex), and is a sugar (ose). Ribulose, another simple sugar, contains five carbon atoms and is a ketone, and would be called a *ketopentose*. All sugars are either aldehydes or ketones and typically have hydroxyl groups on all carbon atoms except the one in the carbonyl group. Most simple sugars contain between three and six carbon atoms, with six and five carbons by far the most common.

Animals obtain a substantial part of their energy by oxidation of glucose, the overall reaction being:

$$C_6H_{12}O_6(aq) + 6\ O_2(g) \rightarrow 6\ CO_2(g) + 6\ H_2O \qquad \Delta H = -2880\ \text{kJ} \qquad (25.9)$$

The actual biological oxidation occurs in the cells in a very complex way, involving well over twenty steps, all of which are catalyzed by enzymes. The energy released in some of the steps is stored in various high-energy compounds, which serve to furnish the energy required in later steps. It turns out that a person can convert about 30% of the energy available from Reaction 25.9 into the mechanical work required for such activities as running or lifting weights. Considering that the energy conversion occurs essentially at a constant temperature (about 37° C), this is truly a fantastic efficiency; we at present have no machines that could get any work at all from this reaction under the conditions under which it occurs in an organism.

Looking at the two-dimensional projection of the glucose molecule shown on p. 618, you might suppose that it would be the only aldohexose. It isn't; there are actually 16 different isomeric aldohexoses, all with the same molecular formula and the same bonding as in glucose. The reason that this is so is that glucose exhibits a type of isomerism we have not encountered previously. Certain of the carbon atoms in glucose are *asymmetric*, which means that they are attached to four different groups. As indicated in Figure 25.4, the presence of such an atom in an organic molecule leads to the existence of two different isomers which are mirror images of each other, and which cannot be converted one to the other without breaking chemical bonds.

In glucose, there are four asymmetric carbon atoms (Nos. 2, 3, 4, and 5; carbon atom 2, for example, is bonded to the H, OH, CHO, and $CHOHCHOHCHOHCH_2OH$ groups). Each time we add an asymmetric carbon atom to a molecule, the number of isomers doubles, so there are 2 × 2 × 2 × 2, or 16, isomeric aldohexoses, sometimes called stereoisomers to distinguish them from the structural isomers discussed in Chapters 10 and 21.

As you can well imagine, it was by no means an easy task to determine which one of these isomers corresponded to naturally-occurring glucose. This problem was solved by the great German chemist Emil Fischer between about 1885 and 1900. Fischer made good use of the fact that solutions of sugars and other substances containing asymmetric carbon atoms rotate the plane of a transmitted beam of polarized light, a property which is called optical activity. By synthesizing longer-chain sugars from shorter ones and by degrading longer-chain sugars to shorter ones of known configuration, Fischer and a host of industrious graduate students were able to establish the structures of all of the naturally-occurring aldohexoses, including glucose. For his work, Fischer received the Nobel Prize in 1902.

The carbon chain in glucose is flexible and can orient itself in various ways due to the free rotation that can occur around each carbon-carbon single bond. A

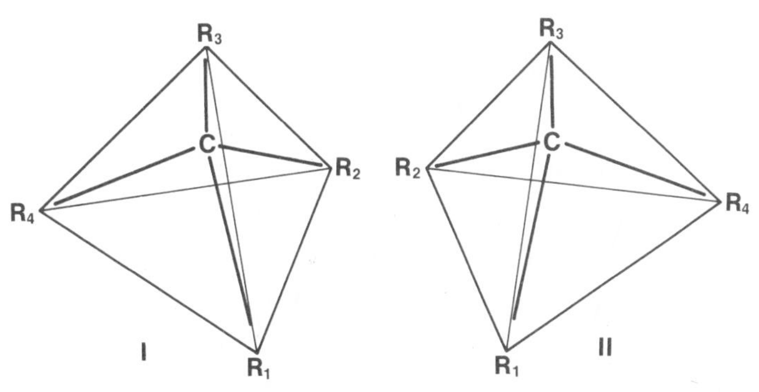

Figure 25.4 Optical isomers. The sketches show the two mirror-image structures to be associated with an asymmetric carbon atom. The two structures would be those of two different substances, which are called optical isomers. There will be 2^n isomers of this sort associated with a molecule in which there are n asymmetric carbon atoms.

particularly important conformation, shown below, leads readily to the formation of a six-membered ring:

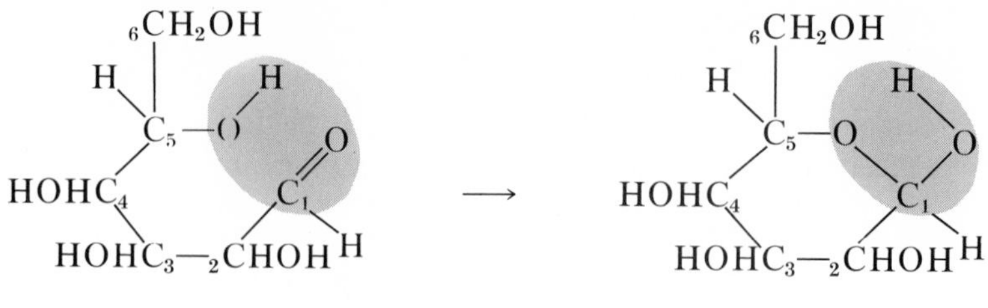

linear form → ring form (25.10)

In the following sketches we show the three-dimensional character of the ring form of glucose. The ring is not planar, but takes the so-called ''chair'' form. If you make a model of this molecule, you will find that the bonds to the H and OH groups from the ring carbons are essentially either perpendicular to the plane of the ring (axial) or lie in the plane (equatorial). In β-glucose, all of the OH groups are equatorial. In α-glucose the OH group on carbon atom 1 is axial; all the others are equatorial. Both forms exist in water solution, with β about twice as likely to be present as α at equilibrium at 25°C. The linear form exists in solution, but has a very low concentration.

When the ring forms, the OH group on C_1 can go into either the α or β position

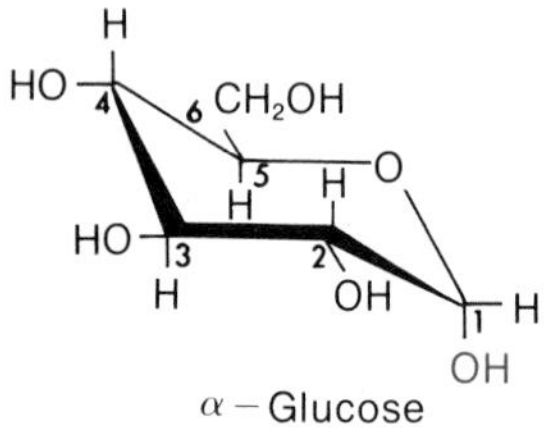

α – Glucose

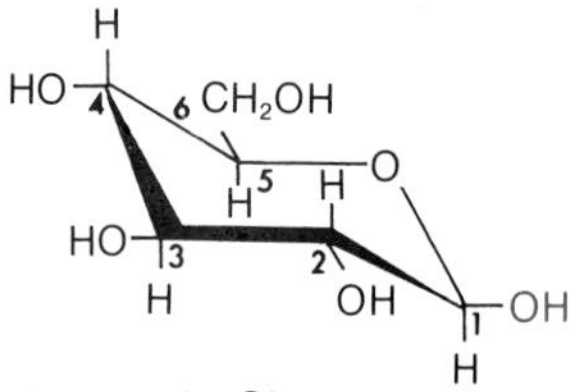

β – Glucose

Combined Forms of Glucose

Although glucose exists in simple form in our blood, it ordinarily is found in living things in various combined forms. The simplest combination occurs in the maltose molecule, which contains two glucose groups connected by an oxygen bond:

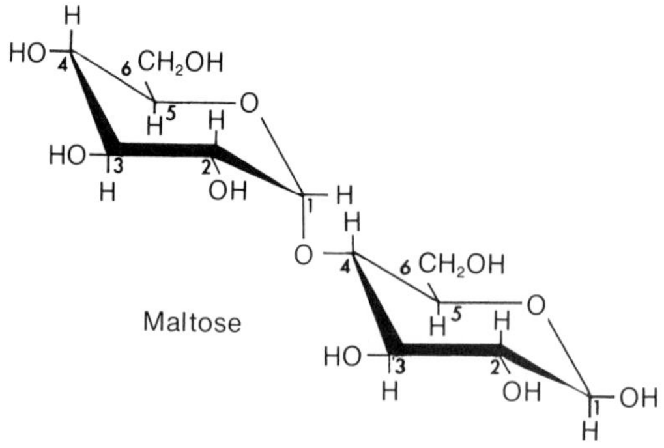

Maltose

If an aqueous maltose solution is heated under acidic conditions or is treated with the enzyme maltase, the maltose molecule reacts with water (hydrolyzes), and cleaves to yield two glucose molecules:

$$\underset{\text{Maltose}}{C_{12}H_{22}O_{11}(aq)} + H_2O \xrightarrow[\text{maltase}]{\text{Acid or}} \underset{\text{Glucose}}{2\ C_6H_{12}O_6(aq)} \qquad (25.11)$$

Because two simple sugar molecules are produced on hydrolysis, maltose is called a **disaccharide.** There are several important disaccharides, but the most common one is sucrose, also known as cane or beet sugar, or just sugar:

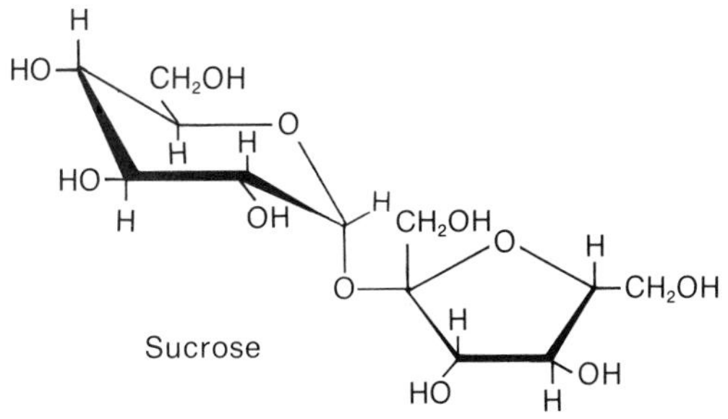

Sucrose

Sucrose on hydrolysis in acid yields glucose and fructose, a ketohexose found in honey and fruit juices.

A disaccharide can be considered to be the product of a condensation reaction between two simple sugars, in which the sugars become linked by an oxygen bridge and water is eliminated. If this sort of condensation reaction were to occur many times, linking many simple sugars together, one would obtain a polymer called a **polysaccharide.** There are many naturally-occurring polysaccharides containing chains of sugar units linked by oxygen bridges. Two of the simplest, and most important, are those substances we call starch and cellulose.

Starch

Starch is a polysaccharide in which, as we noted earlier, the monomer unit is glucose. Starch is found in many plants, where it is stored in the roots and seeds. It is present in large amounts in corn and potatoes, and these are its main commercial sources. Starch granules, as found in peeled potatoes or hulled corn, swell and soften on being heated in water, forming a colloidal suspension. If this material is treated with acid or the proper enzymes, the polysaccharide chains hydrolyze, ultimately yielding glucose. In starch the glucose rings are attached as in maltose, through an oxygen bridge which is in the α form on C_1 on one ring to C_4 on the next. Starch molecules consist of long chains, containing roughly 1000 glucose units. Some of the chains are branched, with side chains occurring roughly every 25 glucose units on the main chain.

In our bodies glucose is stored as a polysaccharide known as glycogen, which is very similar in structure to starch except that it is more highly branched. Glycogen and starch serve animals and plants respectively as the main source of glucose.

Example 25.6 What is the simplest formula of starch?

Solution Starch may be regarded as a condensation polymer of glucose, whose formula is $C_6H_{12}O_6$. Every time a glucose unit is added to the polysaccharide chain a molecule of water is eliminated, so the unit added is not $C_6H_{12}O_6$, but rather $C_6H_{10}O_5$. This is the simplest formula of starch.

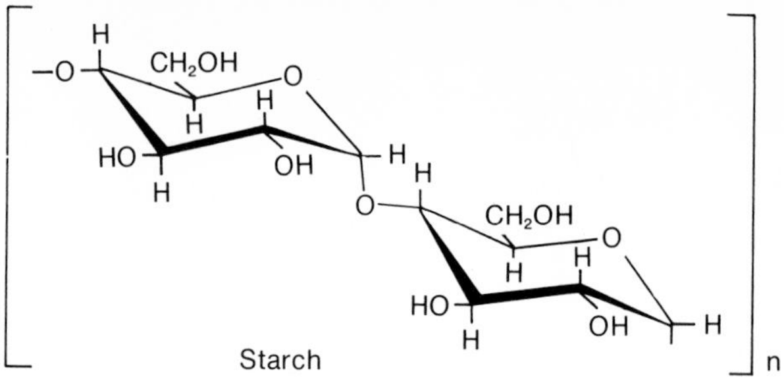

Starch

Figure 25.5 Starch structure. In starch the linkage between glucose rings is through an oxygen bond that is in the α position on the ring to the left of the bond. This structure, along with branching of the chain, prevents formation of long, strong fibers in starch. Potatoes and corn would tend to have molecules with this structure.

Cellulose

We can't digest cellulose, but many animals and some insects can

Cellulose is the substance which constitutes the cell membrane of most plants. About 50% of the woody structure of trees is cellulose, as is about 98% of cotton fiber. Cellulose, like starch, is a polysaccharide, and like starch can be degraded by acid completely to glucose. Starch and cellulose differ in the manner in which the glucose units are linked in the polymer chains. In starch the oxygen bridge between rings is in the α position; with cellulose, the bridge is in the β position. The chains in cellulose are quite long and essentially unbranched, with a molecular mass of about a million (roughly 10 000 glucose units). The molecular structure of cellulose, unlike that of starch, allows for strong hydrogen bonding between the polymer chains. This gives cellulose fibers good resistance to water and high strength (cotton has a tensile strength greater than that of steel).

Because it has such strong fibers, cellulose has many uses. Cotton is used to make thread for fabrics, and these days is usually woven as blends with synthetic fibers (unfortunately, in the authors' opinion). Cotton also is used in some high quality papers and to make some synthetic plastic materials. Most industrial cellulose is obtained from wood. In the process, wood chips are treated with hot sodium hydroxide solution or other chemicals, which dissolve some of the wood components and partially degrade the polymer chains. The insoluble fraction, called wood pulp, consists of impure cellulose, and can be used directly to make paper or can be treated further to make various plastics.

Since cellulose contains hydroxyl groups, it will react with acids to produce esters. The most important cellulose esters are the nitrate and the acetate. These substances do not hydrogen bond, and so are soluble in hydrocarbon solvents.

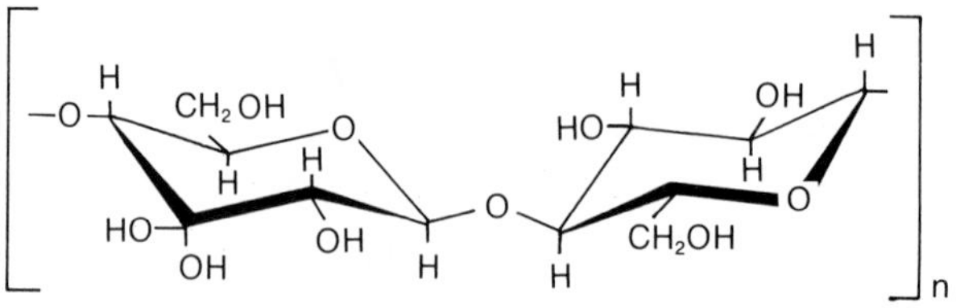

Cellulose

Figure 25.6 Cellulose structure. The bonding between glucose rings in cellulose is through oxygen bridges in the β position for each ring to the left of the bridge. This structure allows for ordered hydrogen bonding between chains and formation of long strong fibers. Cotton and wood fiber would have structures like that shown.

The nitrate can be used to make films, but its main application is in explosives; in the form called guncotton it is used in making smokeless powder, the main propellant in shotgun and rifle cartridges. The first commercial plastic, called celluloid, was first produced in about 1870, and was made from cellulose nitrate, camphor, and ethyl alcohol, which served as a solvent. It had many disadvantages, such as high flammability, nonresistance to many solvents, and noticeable odor, but for about fifty years it was the only significant plastic. Cellulose acetate is used to make photographic film and synthetic fibers.

An interesting alternative method for making synthetic fibers from cellulose involves dissolving cellulose with an appropriate reagent, forcing the solution through a small hole into a bath, and regenerating the cellulose chains by reaction with the bath liquid. The solvent used is carbon disulfide, which reacts with cellulose to form a rather unstable product which is soluble in dilute sodium hydroxide; filaments of essentially pure cellulose form when the hydroxide solution is passed through a spinneret into a solution of sodium hydrogen sulfate. The polymer produced is called rayon, and was one of the early synthetic fibers; its popularity has decreased with the advent of other synthetic polymers which have better properties. The polymer can also be formed into the film called cellophane, for many years essentially the only available transparent film.

25.4 NATURAL POLYMERS: PROTEINS

Among the most important, and abundant, of the substances which make up a living organism are those compounds called proteins. In our own bodies proteins have many functions. Some are fibrous and are the main components of our muscles, hair, and skin. Others are found in body fluids and serve as carriers of inorganic or organic compounds. A large number of proteins are biological catalysts called enzymes; these facilitate and control many biochemical reactions.

Protein is an important component of most foods. It is abundant in lean meats and vegetables like beans and peas, and is present in smaller amounts in nearly everything we eat. In our digestive system proteins are broken down into their components, which are then reassembled in our bodies into other proteins which our system requires. In this section we will look at the structure of proteins.

Nature of Proteins

Proteins, like the polysaccharides starch and cellulose, are polymeric substances. They range in molecular mass from about 6000 for insulin to over a million for some of the more complex enzymes. All proteins, when subjected to treatment with strong acid, will depolymerize into their monomeric components, all of which are **α-amino acids.** α-Amino acids are organic acids in which an NH_2 group is attached to the carbon atom adjacent to a carboxyl group, as indicated in the sketch below:

$$\begin{array}{c} \quad NH_2 \quad\; O \\ \quad | \quad\;\; // \\ R-C-C \\ \quad | \quad\;\; \backslash \\ \quad H \quad\;\; OH \end{array}$$

an α-amino acid

The symbol R represents an organic radical, which may range from an H atom to a large aliphatic or aromatic group. There are about 20 α-amino acids which

TABLE 25.2 THE COMMON α-AMINO ACIDS

(R groups are indicated by shaded areas)

HYDROPHOBIC R GROUPS (decrease solubility in water)

Valine—Val

Leucine—Leu

Isoleucine—Ile

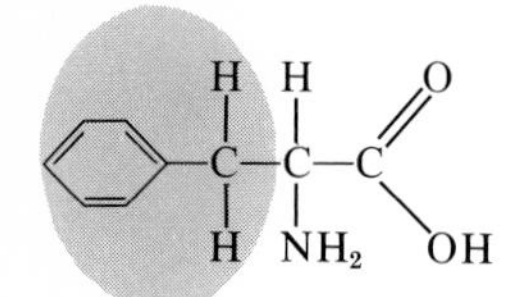

Phenylalanine—Phe

Tryptophan—Trp

Methionine—Met

HYDROPHILIC R GROUPS (enhance water solubility)

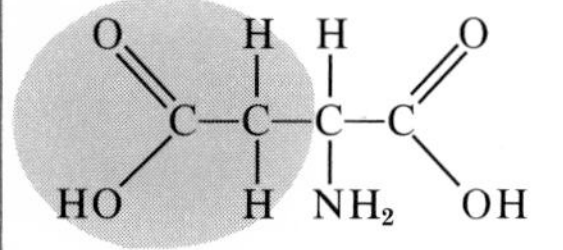

Aspartic Acid—Asp

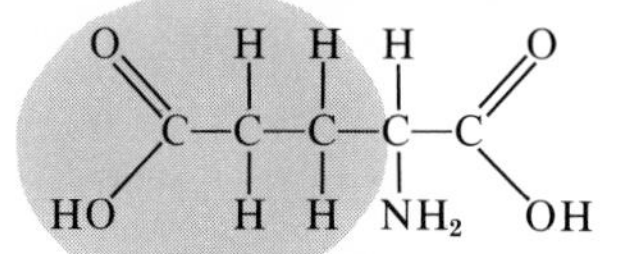

Glutamic Acid—Glu

Lysine—Lys

Arginine—Arg

Histidine—His

SLIGHTLY HYDROPHILIC R GROUPS (little influence on water solubility)

Glycine—Gly

Alanine—Ala

Serine—Ser

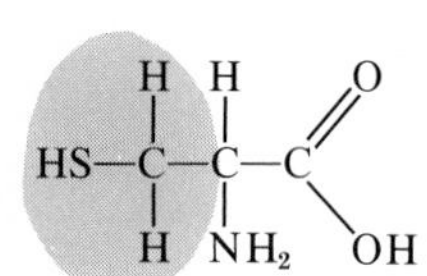

Cysteine—Cys

Threonine—Thr

Tyrosine—Tyr

Asparagine—Asn

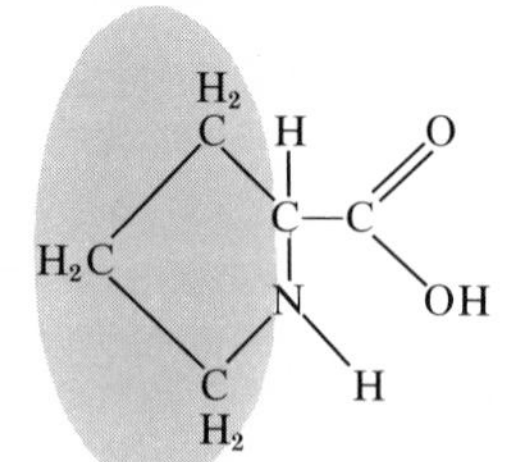

Proline—Pro

Glutamine—Gln

are obtained by breaking down protein molecules, and in Table 25.2 we have listed their structures.

As you can see from the table, the α-amino acids have a wide variety of structures. By virtue of the acidic —COOH and basic $—NH_2$ groups they all contain, their solutions have properties which are markedly influenced by *pH*. At high *pH* (basic solution), the —COOH group loses its proton to become $—COO^-$, giving a negative charge. On the other hand, at low *pH* (acid solution), the $—NH_2$ group picks up a proton to become $—NH_3^+$, and the acid becomes positively charged. At some intermediate *pH*, called the isoelectric point, the molecule is uncharged. The *pH* at which this occurs varies considerably from one amino acid to another. Since the charge on a species affects its water solubility and tendency to adsorb on a solid, the various amino acids can be separated by chromatographic methods at carefully controlled *pH* values.

```
   NH2  O                 NH3+  O
   |   //                 |    //
R—C—C                 R—C—C
   |   \                  |    \
   H    O-                H     OH
```

basic form (anion) | acidic form (cation)

Each α-amino acid has a reactive —COOH group *and* a reactive $—NH_2$ group

Two amino acids may combine by reaction of the acid group on one molecule with the basic group on another:

```
                                                                  peptide
                                                                  linkage
    H  O              H     O                         H  O    H     O
    |  ||             |     ||                        |  ||   |     ||
H—N—C—C—OH + H—N—C——C—OH → H2O + H—N—C—C—N—C——C—OH   (25.12)
 |  |              |  |                          |  |     |  |
 H  H              H  CH3                        H  H     H  CH3
```

Glycine | Alanine | Glycylalanine, a dipeptide (Gly-Ala)

As you can see, this reaction is completely analogous to that between an acid and an amine to form an amide (Section 25.2). The product, called a *dipeptide*, still contains active $—NH_2$ and —COOH groups, either of which can react with another amino acid to form a **tripeptide.** This kind of condensation reaction clearly can continue, ultimately producing a polymer called a *polypeptide*. Those substances we call proteins are all polypeptides. Proteins consist of long chains of amino acid residues linked together by peptide bonds (shown in color):

```
    H  O     H  O     H  O     H  O
    |  ||    |  ||    |  ||    |  ||
—N—C—C—N—C—C—N—C—C—N—C—C—
 |  |     |  |     |  |     |  |
 H  R1    H  R2    H  R3    H  R4
```

Section of a protein molecule

Proteins differ from the other polymers we have discussed up to this point in that they may contain up to 20 different monomer units. This means that there is a multitude of possible proteins (Example 25.7).

Example 25.7 How many different tripeptides could be made from the amino acids in Table 25.2?

Solution In the tripeptide, each of the three amino acids can be any one of the 20 listed in the table. This means that there are 20 possibilities for the first position, 20 for the second, and 20 for the third. Since these choices are independent, the number of tripeptides will be

$$20 \times 20 \times 20 = 8000$$

(Note that the order in the chain is significant; glycyl-alanyl-histidine is a different molecule from histidyl-alanyl-glycine. Ordinarily, we would abbreviate these names to gly-ala-his and his-ala-gly respectively, with the residue having the free $—NH_2$ group listed first.)

Protein Structure Determination

The problem of establishing the detailed structure of proteins has been of interest to biochemists for many years. It is only in the past twenty-five years, however, that experimental techniques adequate to the task have been available, with some of the more impressive results being obtained only quite recently.

The procedure for determining the structure of a protein is complex, but reasonably straightforward. One first must prepare a sample of the pure protein; this is accomplished by a combination of techniques including chromatography, selective precipitation, and electrophoresis (a method which separates species on the basis of the charges they carry). Each technique fractionates the sample on the basis of the different properties of its components, hopefully yielding a pure protein as one of the products.

AMINO ACID CONTENT. Having obtained a pure sample of protein, one then establishes the relative numbers of moles of each of the amino acids it contains. This is done by subjecting the sample to complete hydrolysis (in hot 6 *M* HCl), breaking all the peptide bonds and freeing the individual amino acids. The hydrolyzed sample may be analyzed by column chromatography or by electrophoresis, ultimately producing a chromatogram of the type shown in Figure 25.7.

From such a chromatogram one can establish the relative amounts of the different amino acids present (Column A in Table 25.3). Given the molecular mass of the protein, perhaps from osmotic pressure measurements, one can then determine the number of moles of each amino acid present in a mole of protein (Column B). The reasoning involved is indicated in Example 25.8.

From the analyses of these and other proteins we conclude that there is no clear or simple pattern for the relative amounts of the amino acid residues in these substances. Some proteins, notably the fibrous ones, are simpler in structure than

Example 25.8 The molecular mass of insulin was found by various methods to be about 6000. Find the number of serine residues in the insulin molecule, using data from column A in Table 25.3.

Solution The analysis shows that there is about 0.497 mol serine produced by hydrolysis of 1000 g of insulin. Since the gram molecular mass of insulin is about 6000 g, there will be about 6×0.497, or three moles serine per mole of insulin and, therefore, three serine residues per insulin molecule.

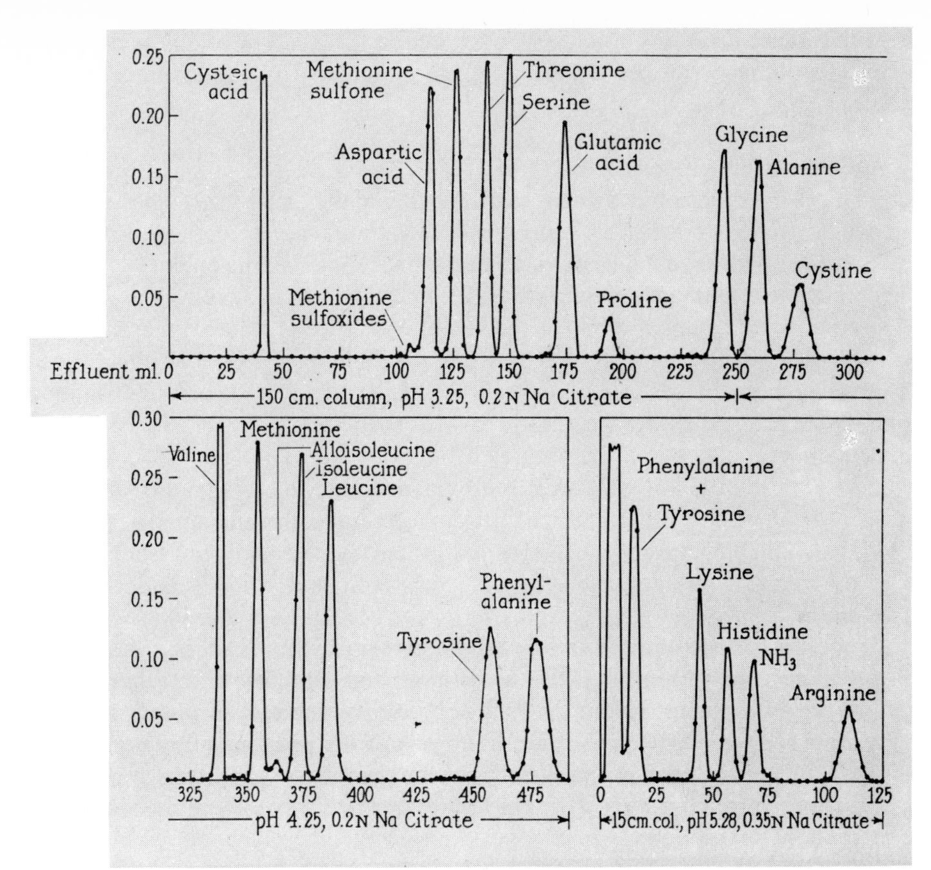

Figure 25.7 Chromatogram showing amino acid analyses of some proteins. The elution of the different acids would be carried out under different *pH* conditions, to take advantage of the variation of isoelectric points of the acids with their structures. (From Moore et al.: Ann. Chem. *30*:1186, 1958.)

TABLE 25.3 AMINO ACID COMPOSITION OF SOME PROTEIN MOLECULES

Amino Acid	Insulin (bovine) A*	Insulin (bovine) B*	Cytochrome c (horse) A	Cytochrome c (horse) B	β-Lacto-globulin A	β-Lacto-globulin B	Human Serum Albumin A	Human Serum Albumin B
Glycine	0.572	4	0.903	12	0.200	7	0.213	15
Alanine	0.505	3	0.455	6	0.793	30	0.046	3
Valine	0.661	5	0.230	3	0.484	18	0.657	45
Leucine	1.008	6	0.461	6	1.178	44	0.837	58
Isoleucine	0.211	1	0.459	6	0.448	17	0.130	9
Proline	0.252	1	0.302	4	0.457	17	0.443	31
Phenylalanine	0.492	3	0.306	4	0.234	9	0.472	33
Tyrosine	0.690	4	0.298	4	0.204	8	0.259	18
Tryptophan	0.000	0	0.075	1	0.094	3	0.010	1
Serine	0.497	3	0.000	0	0.386	14	0.318	22
Threonine	0.175	1	0.756	10	0.433	16	0.386	27
Cysteine	1.030	6	0.150	2	0.141	6	0.262	20
Methionine	0.000	0	0.153	2	0.215	8	0.087	6
Arginine	0.176	1	0.146	2	0.165	6	0.356	25
Histidine	0.336	2	0.232	3	0.103	4	0.226	16
Lysine	0.176	1	1.428	19	0.795	29	0.865	58
Aspartic acid	0.510	3	0.604	8	0.860	32	0.673	46
Glutamic acid	1.262	7	0.901	12	1.298	48	1.155	80

*A = mol amino acid/1000 g protein; B = no. amino acid residues per molecule.

So far no regularities have been observed in the composition of proteins others, but there appears to be no way to predict the amount of any given amino acid in a protein, given the amounts of any other acid in the same or another protein.

AMINO ACID SEQUENCE IN PROTEINS. Once the composition of the protein has been determined, the next step is to find the order in which the various amino acid residues occur on the protein chain, by no means an easy task. Many proteins contain more than one chain; when such is the case, the chains may or may not be linked by chemical bonds. If they are not, they often can be separated from one another by the same methods used to obtain a pure protein. When the chains are linked chemically, it is always through disulfide (—S—S—) bonds between two cysteine molecules; these bonds can be cleaved easily and selectively and the chains then resolved as before and dealt with separately.

Sanger found the sequence in insulin by this method

Sequence analysis has been carried out by various means. In the earliest method the protein chain was broken down randomly into fragments of varying length and the fragments separated chromatographically and analyzed chemically. The chain structure was established by putting the analyzed fragments together on paper, overlapping where such was indicated, and ultimately finding the sequence consistent with all fragments, something like solving a jigsaw puzzle. In the modern method one treats the whole protein with a reagent which reacts with only the NH_2 group on the end of the chain; addition of another reagent removes the amino acid residue on that end, leaving the rest of the chain intact. The residue removed is separated from the rest of the protein and identified, and the procedure repeated until the sequence in the whole chain has been determined. This approach has been highly automated and has facilitated sequence determination to the point where amino acid sequences in peptides containing up to 20 residues are readily established.

The first protein for which the amino acid sequence was determined was insulin, whose structure is

```
                                   ┌──────────S—S──────────┐
                                   │                       │                                         HOOC—Asn
                                   │                       │                                               │
NH2—Gly—Ile—Val—Glu—Gln—Cys—Cys—Ala—Ser—Val—Cys—Ser—Leu—Tyr—Gln—Leu—Glu—Asn—Tyr—Cys
                              │                                                                            │
                              S                                                                            S
                              │                                                                           /
                              S                                                                          S
                              │                                                                          │
NH2—Phe—Val—Asn—Gln—His—Leu—Cys—Gly—Ser—His—Leu—Val—Glu—Ala—Leu—Tyr—Leu—Val—Cys—Gly─┐
                                                                                                          │
                                           HOOC—Ala—Lys—Pro—Thr—Tyr—Phe—Phe—Gly—Arg—Glu─┘
```

Insulin contains two chains, one with 21 residues and the other with 30, with three disulfide bridges; the molecular mass of insulin is calculated to be 5733. For his research on the amino acid sequence in insulin, Frederick Sanger was awarded the Nobel Prize in 1953.

The amino acid sequences of a large number of protein chains have now been determined. Among the larger protein molecules which have been studied are hemoglobin, which has four chains with about 150 residues each, and chymotrypsinogen, an enzyme precursor, with 245 residues (Table 25.4).

Conformation of Protein Molecules

A protein molecule, unlike a molecule of polyethylene, contains many groups which tend to interact with other groups, either on the same or on adjacent

TABLE 25.4 CHARACTERISTICS OF SOME PROTEINS

	Molecular Mass	No. of Residues	No. of Chains
Bovine insulin	5 733	51	2
Human hemoglobin	64 500	574	4
Human serum albumin	68 500	550	1
Bovine chymotrypsinogen	22 600	245	3
Horse γ globulin	149 900	~1 250	4
Bovine glutamate dehydrogenase	1 000 000	~8 300	40
Horse cytochrome c	13 370	104	1

molecules. These interactions may involve hydrogen bonding between O atoms on C=O groups and H atoms in nearby N—H groups, forces between charged R groups on the chain, or solvent-R group interactions. All three kinds of interactions can be quite strong and may affect the properties of a protein. In particular, these interactions determine in large measure the conformation of the protein, the manner in which the units in the chain are arranged in three dimensions.

By far the most important experimental method used to establish protein conformation is x-ray diffraction. The x-ray diffraction pattern of a crystal depends on the positions of all atoms in the crystal and can be used to determine those positions. In recent years it has become possible to apply x-ray diffraction techniques to proteins containing upwards of 200 amino acid residues. From a small crystal of such a protein one can obtain about 50 000 spots in an x-ray diffraction pattern; the positions and intensities of these spots are fed into a computer, which can handle the millions upon millions of calculations necessary to relate those spots to the positions of the atoms that caused them. The protein conformations we shall discuss were all determined by x-ray diffraction methods.

There are two ways in which a protein chain can be twisted so as to maximize hydrogen bonding between carbonyl oxygen atoms and amido hydrogen atoms;

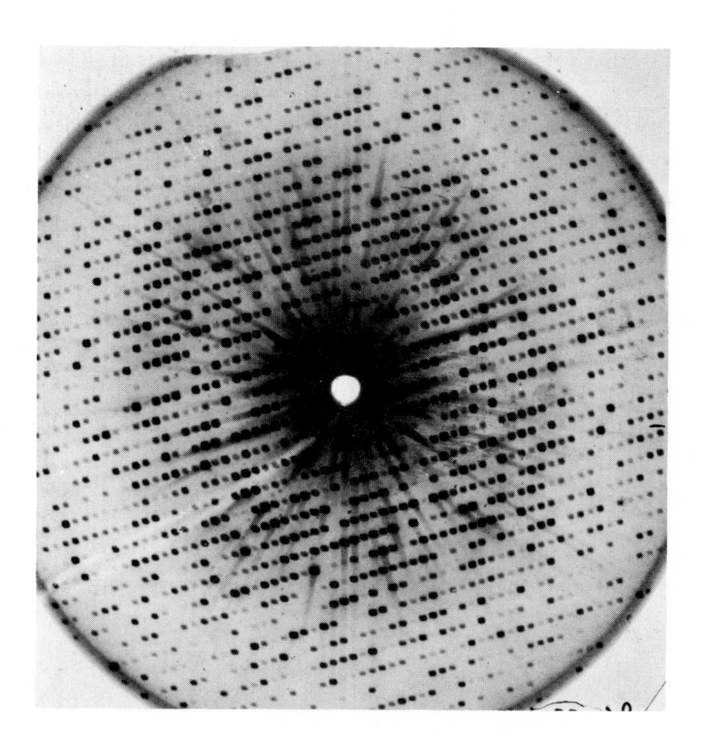

Figure 25.8 X-ray diffraction pattern from whale myoglobin. Even though protein molecules are large, they can often be crystallized. The diffraction pattern may contain several thousand points, whose positions reflect, indirectly, the amino acid sequence and conformation of the protein chain. (From F. C. Kendrew, Scientific American, December, 1961.)

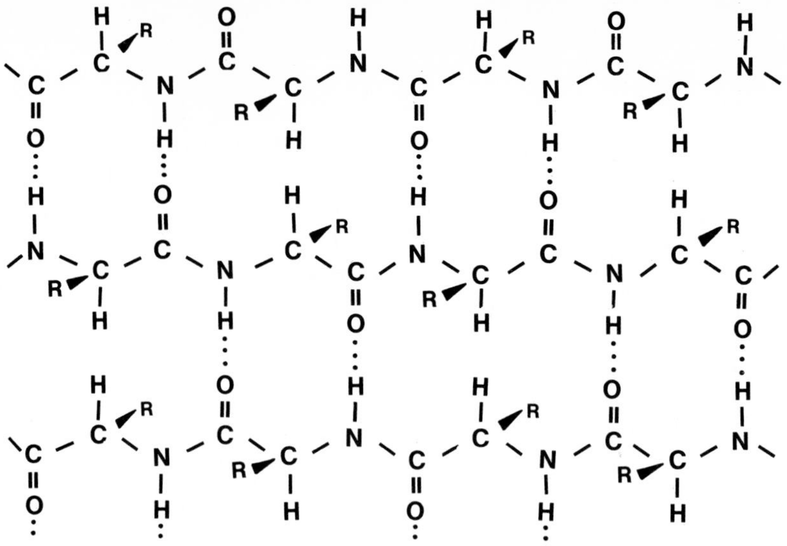

Figure 25.9 β-Keratin structure of silk fiber. When the R groups on the amino acid residues are small, the protein chains can hydrogen bond in the roughly planar structure shown. Since the bonding around the N and noncarbonyl carbon atoms is tetrahedral, the sheet formed has a pleated structure. In the drawing, the hydrogen bonds are shown by dots.

both of these are observed in natural substances. If the R groups from the amino acid residues are small, as are the H atoms in glycine or the CH_3 groups in alanine, then it is possible for the chains to form a sheet as in Figure 25.9. The sheet shown lies in the plane of the paper and is made up from many parallel protein chains held together by hydrogen bonds. As you can see by looking at the middle chain, full hydrogen bonding to adjacent chains is possible at every amino acid residue, making for a very stable fiber. The structure shown is called the pleated sheet, or β-keratin, and is found in silk fiber.

In Figure 25.9 it is clear that there tend to be some R groups in close proximity. In silk the amino acid residues are mainly those of glycine and alanine, where R = H and CH_3, respectively, and it is possible for these small groups to rotate somewhat, all in the same direction, so as to bring alternate R groups in the chain above and below the plane of the sheet, minimizing the steric forces between those groups.

In most proteins the R groups are too bulky to allow for formation of sheets such as are found in the silk structure. Pauling and Corey showed that an alternate structure that also maximized H bonding was one in which the protein chain took the form of a helical coil called an **α-helix.** We have indicated the form of the coil in Figure 25.10, showing first the schematic arrangement of atoms that make up the chain and then a more complete structure to show the positions at which H bonding can occur. Note that the R groups are all on the outside of the helix, where they have the most room, and that the actual structure of the helix is nearly independent of the nature of these groups. The dimensions of the helix correspond quite well to those observed in such natural fibrous proteins as wool, hair, skin, feathers, fingernails, and horn; the structure of these substances is now accepted as being for the most part that of the α-helix.

Among the more interesting protein structures are those exhibited by the so-called globular proteins, which exist as roughly spherical entities in which the chain weaves its way around the sphere. In Figure 25.11 (p. 632) is shown a typical globular protein structure, that of cytochrome c isolated from horse heart. In the drawing each ball represents an α-carbon atom and its associated R group, with the atoms in the peptide linkage denoted by a bond; the actual number of *atoms* in the molecule is about 10 *times* the number of balls shown.

The conformation of cytochrome c is typical of that of the other globular proteins which have been determined. The hydrophilic groups are located in the outer regions, hydrophobic groups are mostly in the center, and the path of the chain is such as to minimize the molecular volume. For these reasons, these proteins have been said to resemble oily drops. In most proteins the observed atomic arrangement is surprisingly stable. If the chain is made to uncoil, by changing the *pH* or raising the temperature slightly, it will in many cases revert to its former configuration when normal conditions are restored. If one actually synthesizes a protein, as has been done for a few proteins the size of cytochrome c, the chains spontaneously take on the natural conformation when the *pH* and temperature have their normal values. Even when more than one chain is present in the protein, the mixture is often able to find the proper conformation. Interchain bonds may stabilize the structures of complex proteins but do not appear to be necessary to bring the chains into the natural arrangement. The observed structure is mainly due to the tendency for hydrophilic and hydrophobic groups to be located, for

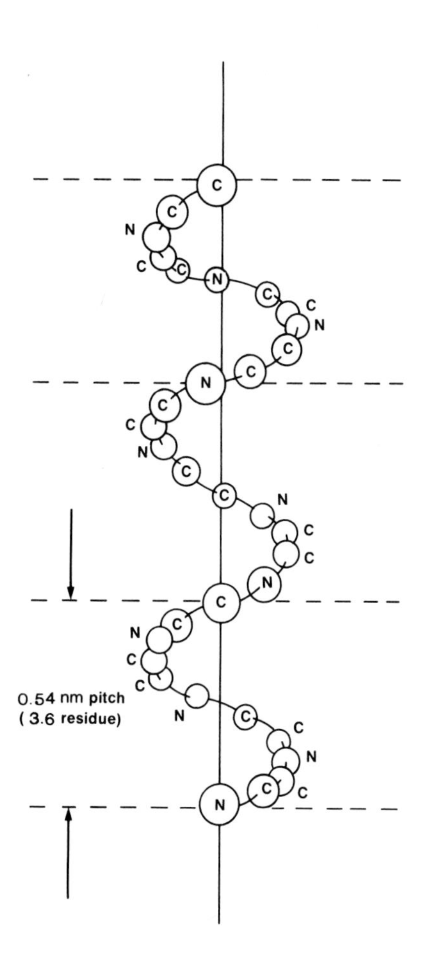

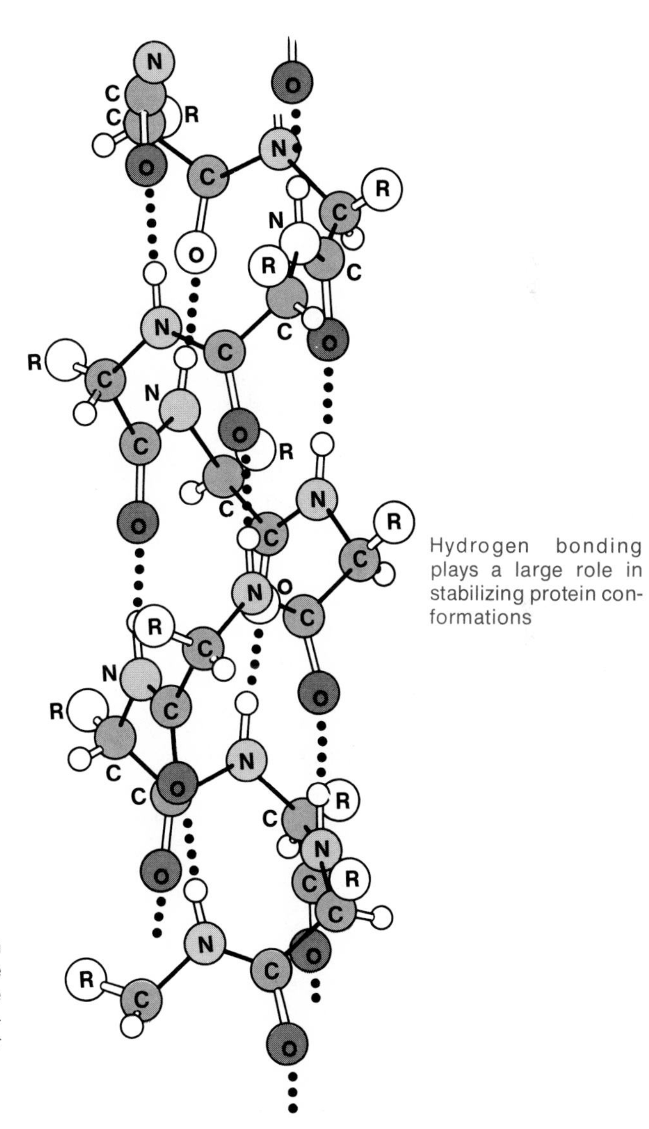

Figure 25.10 α-Helix structure of proteins. The main atom chain in the helix is shown schematically on the left. The sketch on the right more nearly represents the actual positions of the atoms and shows where intrachain hydrogen bonding occurs. Wool and many other fibrous proteins have the α-helix structure.

Hydrogen bonding plays a large role in stabilizing protein conformations reasons of thermodynamic stability, in the outer and inner regions, respectively. This arrangement allows for hydrogen bond formation between hydrophilic groups, minimizes repulsive forces between groups of like charge, and minimizes the free energy of interaction of water and the hydrophobic groups.

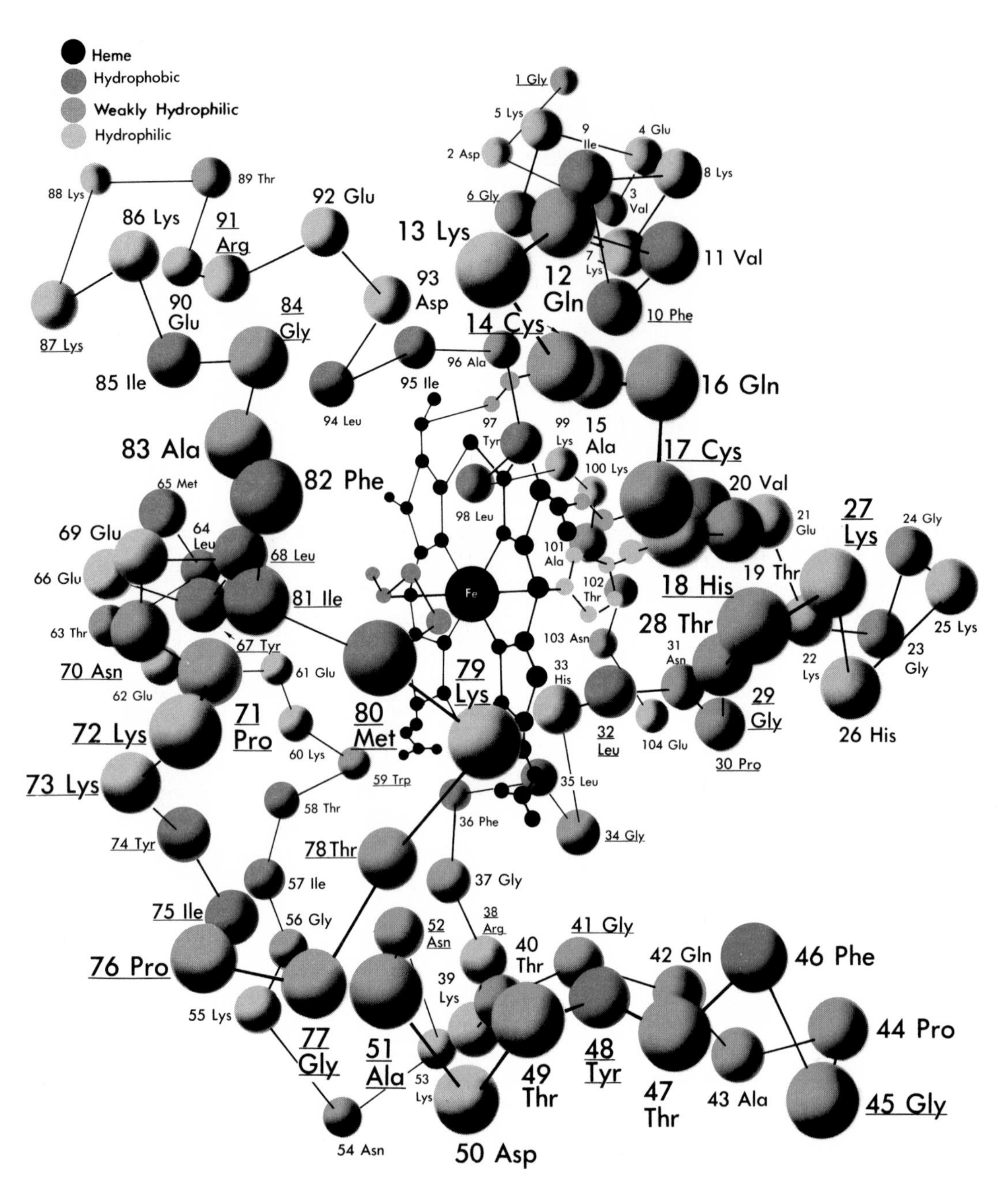

Figure 25.11 Conformation of the cytochrome c molecule. The molecule is roughly spherical, with a single chain beginning at the top of the sketch and ending a little below and right of center. In the drawing each ball represents an α-carbon atom and its associated R group. Sheltered within the cytochrome c molecule is the group called heme (mentioned in Chapter 21), which contains an Fe atom and is vital for the electron-carrying properties of this protein. The conformation of the molecule is typical of that of enzymes, with hydrophobic groups mainly on the inside, hydrophilic groups on the outside, and a compact arrangement of the chain.

In Chapter 16 we discussed briefly the properties of enzymes, the catalysts involved in most of the chemical reactions that occur in living systems. Since there are many such reactions, there are many enzymes. Well over a thousand have so far been identified, and more than a hundred have been isolated as pure crystals. As far as is now known, all enzymes are globular proteins, similar in general structure and conformation to cytochrome c.

All enzymes seem to work by very similar mechanisms. Groups on the surface of the enzyme have the proper arrangement for interaction with a particular reactant molecule, forming a complex in which a particular bond in the reactant is weakened and finally broken. Another enzyme may interact with two molecules in such a way that a bond tends to form between the two. The enzyme surface is sufficiently complex that it interacts with only one, or a few, molecules, making for catalysis of a specific reaction; the "lock and key" analogy is an obvious one, but it is really not too obvious which is the lock and which is the key. The fact that enzyme action is limited to rather narrow ranges of pH and temperature is easily understood in light of the fact that the conformation, and hence the surface, of globular proteins changes dramatically when these substances are heated or exposed to highly acidic or basic media.

In a few cases, it has been possible to determine the conformations of enzyme-reactant complexes

It appears that, once they were established by natural selection, enzymes maintained their general structures through many stages of evolution. Cytochrome c, an enzyme used in one step in oxygen metabolism, is found in all living systems that use oxygen, from man, to birds, to insects, and even to primitive yeasts. In all of these species, cytochrome c is much the same, differing in only 12 out of 104 residues between man and pigeon, and in only 40 places between man and yeast.

In this chapter we have been able to give you only a brief look at some of the many polymers that are observed in living organisms. In particular we have had to defer any discussion of the nucleic acids DNA and RNA to later courses in organic and biochemistry. It seems likely that during your lifetimes some of the main advances in chemistry will be in those areas dealing with the biochemical properties of the large molecules, where chemists are hoping to find solutions to such problems as hereditary disease and cancer.

PROBLEMS

25.1 *Addition Polymers* Styrene, which has the structure $H_2C{=}\overset{H}{C}{-}C_6H_5$ (benzene ring), forms a head-to-tail addition polymer.

a. Sketch a portion of the polystyrene molecule.
b. Calculate ΔH for polymerization per mole of styrene.

25.2 *Condensation Polymers* One of the simplest polyesters would be made from ethylene glycol, $HO{=}CH_2{-}CH_2{-}OH$, and oxalic acid, $HO{-}\underset{\|}{\overset{}{C}}{-}\underset{\|}{C}{-}OH$ (each C double-bonded to O). Sketch a portion of the polymer chain obtained from these monomers.

25.3 *Carbohydrates* Cellulose consists of about 10 000 $C_6H_{10}O_5$ units linked together.

a. What are the percentages by mass of C, H, and O in cellulose?
b. What is the molecular mass of cellulose?

25.4 *Proteins* Sketch the tetrapeptide obtained from four molecules of the α-amino acid glycine (Table 25.2).

25.5 What is meant by each of the following terms?

a. addition polymer
b. isotactic structure
c. peptide bond
d. vulcanization

25.6 Given a sample of polystyrene and a sample of phenol, $HO-C_6H_5$, a solid organic compound, describe several tests you might perform to decide which was which.

25.7 How would you explain to a young science student how to decide whether a given compound might make a useful monomer for addition polymerization? condensation polymerization?

25.8 Acrylonitrile, $H_2C{=}CHCN$, is a commercially important monomer. Sketch a section of the polyacrylonitrile molecule, a head-to-tail polymer.

25.9 Sketch a section of a head-to-tail polyacrylonitrile molecule with a syndiotactic structure.

25.10 Draw the structures of the monomers that would be used to make the following polymers:

a. $$-\overset{Br}{\underset{H}{C}}-\overset{H}{\underset{H}{C}}-\overset{Br}{\underset{H}{C}}-\overset{H}{\underset{H}{C}}-\overset{Br}{\underset{H}{C}}-\overset{H}{\underset{H}{C}}-$$

b. $$-\underset{H}{N}-\overset{H}{\underset{H}{C}}-\underset{\|O}{C}-\underset{H}{N}-\overset{H}{\underset{H}{C}}-\underset{\|O}{C}-$$

c. $$-\overset{O\|}{C}-(CH_2)_4-\overset{O\|}{C}-\overset{H}{N}-(CH_2)_6-\overset{H}{N}-$$

25.11 Describe the polymers in Problem 25.10 as being addition or condensation polymers or proteins. (More than one term may apply.)

25.12 Lexan is a very rugged polyester, in which the monomers can be taken to be carbonic acid,

$$HO-\underset{\|O}{C}-OH, \quad \text{and } HO-C_6H_4-\overset{CH_3}{\underset{CH_3}{C}}-C_6H_4-OH$$

Sketch a section of the Lexan chain.

25.24 Define each of the following terms:

a. condensation polymer
b. asymmetric carbon atom
c. α-amino acid
d. aldopentose

25.25 Given two benzene solutions, one of styrene, the other of polystyrene, both containing the same mass % of solute, which would you expect to have the higher

a. osmotic pressure?
b. freezing point?
c. viscosity?

25.26 Which of the following monomers could form an addition polymer? a condensation polymer?

a. C_2H_6
b. C_2H_4
c. $HO-CH_2-CH_2-OH$
d. $HO-CH_2-CH_3$

25.27 Sketch a portion of a head-to-head, tail-to-tail polymer made from acrylonitrile.

25.28 Show the isotactic structure of the polymer chain in Problem 25.9.

25.29 What monomers would be used to make the following polymers:

a. $$-\overset{O\|}{C}-C_6H_4-\overset{O\|}{C}-O-(CH_2)_2-O-$$

b. $$-\overset{H}{\underset{Cl}{C}}-\overset{H}{\underset{Cl}{C}}-\overset{H}{\underset{Cl}{C}}-\overset{H}{\underset{Cl}{C}}-\overset{H}{\underset{Cl}{C}}-\overset{H}{\underset{Cl}{C}}-$$

c. $$-\overset{O\|}{C}-\overset{H}{N}-(CH_2)_5-\overset{O\|}{C}-\overset{H}{N}-(CH_2)_5-$$

25.30 Describe the polymers in Problem 25.29 as being addition or condensation polymers or proteins.

25.31 Nylon 6 is made from a single monomer which in linear form would be

$$H_2N-(CH_2)_5-\underset{\|O}{C}-OH$$

By using as an analogy the reaction by which a protein forms, draw a portion of the chain in Nylon 6.

25.13 Neoprene, made from the chloroprene monomer, $H_2C{=}CH{-}CCl{=}CH_2$, is a rubber with mainly a *trans* structure. Draw a section of the neoprene molecule.

25.14 Taking the carbon atom in Figure 25.4 to be carbon atom 2 in glucose, put the four different groups attached to that atom on the corners of the tetrahedron. How many different ways can that be done?

25.15 How many asymmetric carbon atoms are there in the following molecules?

a.
$$\begin{array}{ccccccccc} & & H & & H & & H & & H \\ & & | & & | & & | & & | \\ O & = & C & - & C & - & C & - & C & - & H \\ & & & & | & & | & & | \\ & & & & OH & & OH & & OH \end{array}$$

b. $CH_2OH{-}CHOH{-}CH_2OH$

c. H O H; OH OH H CH_2OH; H OH

25.16 What is cellulose? starch? glycogen? How are they similar? How do they differ in structure?

25.17 Estimate ΔH for the formation of one mole of the ring form of glucose ($B.E.$ C—O = 351 kJ/mol, $B.E.$ C=O = 715 kJ/mol).

25.18 Sketch the form in which leucine would exist in acid solution; in basic solution.

25.19 A tripeptide contains a glycine, valine, and lysine residue. Sketch a possible form of the tripeptide, including all atoms. Name it in the shorthand form used for showing amino acid sequences.

25.20 Given the data in column A of Table 25.3 and the fact that the molecular mass of human serum albumin is about 68 500, find the number of valine residues in that albumin molecule.

25.21 A short polypeptide on complete hydrolysis yielded the following numbers of moles of the listed amino acids per hundred grams of polypeptide:

Ala 0.086	Gly 0.258	Val 0.172
Ile 0.086	Pro 0.344	Ser 0.602

What is the lowest molecular mass that would be reasonable for this polypeptide? How many residues would there be in the chain? (Hint: What is the smallest possible number of any of the residues listed?)

25.32 Gutta percha, a rubber-type polymer used for golf ball covers, has the same monomer as in natural rubber, isoprene. However, in gutta percha the CH_2 groups are all *trans* with respect to the double bonds. Sketch a portion of the gutta percha chain.

25.33 Draw the structure of the simplest carbohydrate you can think of which has an asymmetric carbon atom.

25.34 Note the asymmetric carbon atoms, if any, in the following molecules:

a.
$$\begin{array}{ccccccc} & & H & & H & & \\ & & | & & | & & \\ CH_3 & - & C & - & C & - & C & - & OH \\ & & | & & | & & \| \\ & & OH & & NH_2 & & O \end{array}$$

b.
$$\begin{array}{ccccc} CH_3 & - & C & - & CH_2OH \\ & & \| & & \\ & & O & & \end{array}$$

c. $CH_3{-}CHOH{-}CHOH{-}CH_3$

25.35 Cellulose is useful for making paper, whereas starch is not. Suggest an explanation for this difference.

25.36 Using bond energies, estimate ΔH for the hydrolysis of maltose to glucose.

25.37 Sketch the form in which lysine would exist in acid solution; in basic solution. Note that lysine can have a larger charge than leucine under some conditions.

25.38 How many tripeptides could one make from glycine, valine, and lysine, using any number of each amino acid?

25.39 Given the data in Table 25.3, use the simplest procedure you can devise to find the molecular mass of cytochrome c.

25.40 A 1.00-mg sample of a pure enzyme yielded on hydrolysis 0.0165 mg of leucine and 0.0248 mg isoleucine. What is the minimum possible molecular mass of the enzyme? (GMM leucine = GMM isoleucine = 131 g)

25.22 How would you explain the fact that the β-keratin structure, present in silk, is much less common in proteins than the α-helix structure?

25.23 Why is enzymatic catalysis so much more sensitive to temperature and pH than is catalysis that does not involve enzymes?

25.41 In β-keratin, how many hydrogen bonds are formed per amino acid residue?

25.42 Would you expect the rate constant k for an enzyme-catalyzed reaction to have a dependence on temperature that follows the Arrhenius equation (Equation 16.10)? Explain.

*25.43 Using bond energies in Table 4.2 estimate ΔH for protein formation, per mole of amino acid added to the chain. Does the value seem reasonable?

*25.44 One of the earliest, and still one of the most important, polymers is the material known as Bakelite. Bakelite is a condensation polymer of phenol, C_6H_5—OH, and formaldehyde, $H_2C{=}O$. Formaldehyde will react with phenol to produce the following compounds when the ratio is 1:1

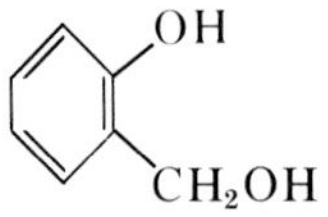

OH, CH_2OH (ortho) and HO, CH_2OH (para)

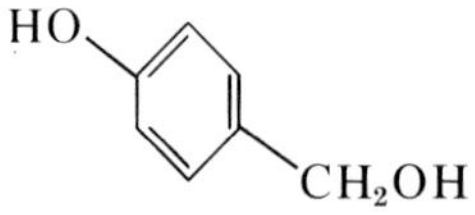

These species, which can be taken to be the monomers, on being heated condense with each other and themselves, eliminating water (formed from the OH groups on CH_2OH and ring hydrogen atoms) and linking benzene rings with CH_2 groups. If the phenol-formaldehyde ratio is 1:1, a linear polymer forms. If the ratio is 1:2, two $H_2C{=}O$ react with each ring, and the chains cross-link at every benzene ring, forming the infusible, insoluble, hard, brittle solid we know as Bakelite. Sketch the linear chain polymer and the cross-linked, essentially macromolecular solid.

APPENDIX

1

CONSTANTS, PROPERTIES OF WATER, AND UNIT CONVERSIONS

CONSTANTS

Acceleration of gravity	= 9.806 m/s
Avogadro's number	= 6.022×10^{23}
Electronic charge	= 1.602×10^{-19} C
Faraday constant	= 9.6487×10^{4} C/mol e^-
Gas constant	= 8.314 kPa · dm³/(mol · K)
	= 8.314 J/(mol · K)
	= 8.314 m² · kg/(s² · mol · K)
Planck's constant	= 6.625×10^{-34} J/s
Velocity of light	= 2.998×10^{8} m/s
π	= 3.142
e	= 2.718
ln x	= 2.303 log x
2.303 R	= 19.15 J/(mol · K)
2.303 RT (at 25°C)	= 5709 J/mol

EQUILIBRIUM CONSTANTS

Acid-base dissociation	Table 19.5, p. 460
Solubility product constants	Table 18.2, p. 438
Complex-ion dissociation	Table 21.6, p. 517

THERMODYNAMIC QUANTITIES

Bond energies	Table 4.2, p. 70
Free energies of formation	Table 14.2, p. 339
Heats of formation	Table 4.1, p. 67
Standard potentials	Table 23.1, p. 555

PROPERTIES OF WATER

Density (g/cm^3)	0.999 87	1.000 00	0.997 07	0.958 38
t (°C)	0	4	25	100

ΔH_{fus} = 5.99 kJ/mol at 0°C

ΔH_{vap}(kJ/mol)	44.8	44.0	40.6
t (°C)	0	25	100

Vapor pressure (kilopascals)

t (°C)	vp	t (°C)	vp	t (°C)	vp	t (°C)	vp
0	0.61	21	2.49	35	5.63	92	75.59
5	0.87	22	2.64	40	7.37	94	81.45
10	1.23	23	2.81	45	9.59	96	87.67
12	1.40	24	2.98	50	12.33	98	94.30
14	1.60	25	3.17	55	15.73	100	101.32
16	1.82	26	3.36	60	19.92	102	108.78
17	1.94	27	3.57	65	25.00	104	116.67
18	2.06	28	3.78	70	31.16	106	125.04
19	2.20	29	4.01	80	47.34	108	133.92
20	2.34	30	4.24	90	70.10	110	143.27

CONVERSION FACTORS

Energy	1 cal = 4.184 J
	1 eV/atom = 96.49 kJ/mol
	1 J/particle = 6.022×10^{20} kJ/mol
	1 g mass = 9.00×10^{10} kJ
	(see also Table 1.2, p. 11)
Length, volume, mass	Table 1.1, p. 9
Pressure	Table 1.2, p. 11

SI UNITS

Base Units

The International System of Units or *Système International* (SI), emphasized throughout this text, was adopted by the 11th General Conference of Weights and Measures in 1960. It is constructed from seven base units, each of which represents a particular physical quantity (Table I).

TABLE I SI BASE UNITS

Physical Quantity	Name of Unit	Symbol
1. Length	metre	m
2. Mass	kilogram	kg
3. Time	second	s
4. Temperature	kelvin	K
5. Amount of substance	mole	mol
6. Electric current	ampere	A
7. Luminous intensity	candela	cd

Of the seven units listed in Table I, the first five are particularly useful in general chemistry. They are defined as follows.

1. The *metre* was redefined in 1960 to be equal to 1 650 763.73 wavelengths of a certain line in the orange-red region of the emission spectrum of krypton-86.
2. The *kilogram* represents the mass of a platinum-iridium block kept at the International Bureau of Weights and Measures at Sevres, France.
3. The *second* was redefined in 1967 as the duration of 9 192 631 770 periods of a certain line in the microwave spectrum of cesium-133.
4. The *kelvin* is 1/273.16 of the temperature interval between the absolute zero and the triple point of water.
5. The *mole* is the amount of substance which contains as many entities as there are atoms in exactly 0.012 kg of carbon-12.

Prefixes Used with SI Units

Decimal fractions and multiples of SI units are designated by using the prefixes listed in Table II. Those which are most commonly used in general chemistry are underlined.

TABLE II SI PREFIXES

Factor	Prefix	Symbol	Factor	Prefix	Symbol
10^{12}	tera	T	10^{-1}	<u>deci</u>	d
10^{9}	giga	G	10^{-2}	<u>centi</u>	c
10^{6}	mega	M	10^{-3}	<u>milli</u>	m
10^{3}	<u>kilo</u>	k	10^{-6}	micro	μ
10^{2}	hecto	h	10^{-9}	<u>nano</u>	n
10^{1}	deca	da	10^{-12}	pico	p
			10^{-15}	femto	f
			10^{-18}	atto	a

Derived Units

In the International System of Units, all physical quantities are represented by appropriate combinations of the base units listed in Table I. To choose a particularly simple example, the SI unit for volume, the cubic metre, represents the volume of a cube one metre on an edge. Again, in SI, the density of a substance can be expressed by dividing its mass in kilograms by its volume in cubic metres. A list of the derived units most frequently used in general chemistry is given in Table III.

TABLE III SI DERIVED UNITS

Physical Quantity	Name of Unit	Symbol	Definition
Area	square metre	m^2	
Volume	cubic metre	m^3	
Density	kilogram per cubic metre	kg/m^3	
Force	newton	N	$kg \cdot m/s^2$
Pressure	pascal	Pa	N/m^2
Energy	joule	J	$kg \cdot m^2/s^2$
Electric charge	coulomb	C	$A \cdot s$
Electric potential difference	volt	V	$J/(A \cdot s)$

Perhaps the least familiar of these units to the beginning chemistry student are the ones used to represent force, pressure, and energy.

The *newton* is defined as the force required to impart an acceleration of one metre per second squared to a mass of one kilogram (recall that Newton's second law can be stated as: force = mass × acceleration).

The *pascal* is defined as the pressure exerted by a force of one newton acting on an area of one square metre (recall that pressure = force/area).

The *joule* is defined as the work done when a force of one newton ($kg \cdot m/s^2$) acts through a distance of one metre (recall that work = force × distance).

Conversions Between SI and Other Units

In Table IV are listed appropriate conversion factors for translating units from other systems to the International System.

TABLE IV CONVERSION FACTORS

QUANTITY	SI UNIT	OTHER UNIT	CONVERSION FACTOR
Energy	joule	calorie	1 cal = 4.184 J
		erg	1 erg = 10^{-7} J
Force	newton	dyne	1 dyn = 10^{-5} N
Length	metre	ångström	1 Å = 10^{-10} m = 10^{-8} cm = 10^{-1} nm
Mass	kilogram	pound	1 lb = 0.453 592 37 kg
Pressure	pascal	bar	1 bar = 10^5 Pa
		atmosphere	1 atm = 1.01 325 × 10^5 Pa
		mmHg	1 mmHg = 133.322 Pa
		lb/in^2	1 lb/in^2 = 6894.8 Pa
Temperature*	kelvin	Celsius	1°C = 1 K
		Fahrenheit	1°F = $\frac{5}{9}$ K
Volume	cubic metre	litre	1 ℓ = 1 dm^3 = 10^{-3} m^3
		gallon (U.S.)	1 gal (U.S.) = 3.7854 × 10^{-3} m^3
		gallon (U.K.)	1 gal (U.K.) = 4.5641 × 10^{-3} m^3
		cubic inch	1 in^3 = 1.6387 × 10^{-6} m^3

*Temperatures in degrees Celsius (t_c) or degrees Fahrenheit (t_f) can be converted to temperatures in kelvins (T) by using the equations:

$$t_c = T - 273.15; \quad t_f = \frac{9}{5}T - 459.67$$

Reference Tables for Non-SI Units

Throughout this text, we have given thermodynamic quantities in kJ/mol. For the convenience of those readers, particularly in the United States, who are more familiar with kilocalories, we list in Tables V to VII certain thermodynamic quantities (enthalpies of formation, free energies of formation, bond energies), in kcal/mol. Table VIII lists vapor pressures of water in mmHg (1 mmHg = 0.1333 kPa) as a function of temperature.

TABLE V HEATS OF FORMATION (KCAL/MOL) AT 25°C AND 1 ATM

Compound	ΔH	Compound	ΔH	Compound	ΔH	Compound	ΔH
AgBr(s)	−23.8	C_2H_2(g)	+54.2	H_2O_2(l)	−44.8	NH_3(g)	−11.0
AgCl(s)	−30.4	C_2H_4(g)	+12.5	H_2S(g)	−4.8	NH_4Cl(s)	−75.4
AgI(s)	−14.9	C_2H_6(g)	−20.2	H_2SO_4(l)	−193.9	NH_4NO_3(s)	−87.3
Ag_2O(s)	−7.3	C_3H_8(g)	−24.8	HgO(s)	−21.7	NO(g)	+21.6
Ag_2S(s)	−7.6	n-C_4H_{10}(g)	−29.8	HgS(s)	−13.9	NO_2(g)	+8.1
Al_2O_3(s)	−399.1	n-C_5H_{12}(l)	−41.4	KBr(s)	−93.7	NiO(s)	−58.4
$BaCl_2$(s)	−205.6	C_2H_5OH(l)	−66.4	KCl(s)	−104.2	$PbBr_2$(s)	−66.3
$BaCO_3$(s)	−291.3	CoO(s)	−57.2	$KClO_3$(s)	−93.5	$PbCl_2$(s)	−85.9
BaO(s)	−133.4	Cr_2O_3(s)	−269.7	KF(s)	−134.5	PbO(s)	−52.1
$Ba(OH)_2$(s)	−226.2	CuO(s)	−37.1	KOH(s)	−101.8	PbO_2(s)	−66.1
$BaSO_4$(s)	−350.2	Cu_2O(s)	−39.8	$MgCl_2$(s)	−153.4	Pb_3O_4(s)	−175.6
$CaCl_2$(s)	−190.0	CuS(s)	−11.6	$MgCO_3$(s)	−266	PCl_3(g)	−73.2
$CaCO_3$(s)	−288.5	$CuSO_4$(s)	−184.0	MgO(s)	−143.8	PCl_5(g)	−95.4
CaO(s)	−151.9	Fe_2O_3(s)	−196.5	$Mg(OH)_2$(s)	−221.0	SiO_2(s)	−205.4
$Ca(OH)_2$(s)	−235.8	Fe_3O_4(s)	−267.0	$MgSO_4$(s)	−305.5	$SnCl_2$(s)	−83.6
$CaSO_4$(s)	−342.4	HBr(g)	−8.7	MnO(s)	−92.0	$SnCl_4$(l)	−130.3
CCl_4(l)	−33.3	HCl(g)	−22.1	MnO_2(s)	−124.5	SnO(s)	−68.4
CH_4(g)	−17.9	HF(g)	−64.2	NaBr(s)	−86.0	SnO_2(s)	−138.8
$CHCl_3$(l)	−31.5	HI(g)	+6.2	NaCl(s)	−98.2	SO_2(g)	−71.0
CH_3OH(l)	−57.0	HNO_3(l)	−41.4	NaF(s)	−136.0	SO_3(g)	−94.5
CO(g)	−26.4	H_2O(g)	−57.8	NaI(s)	−68.8	ZnO(s)	−83.2
CO_2(g)	−94.1	H_2O(l)	−68.3	NaOH(s)	−102.0	ZnS(s)	−48.5

TABLE VI FREE ENERGIES OF FORMATION (KCAL/MOL) AT 25°C, 1 ATM

$AgBr(s)$	−22.9	$CO(g)$	−32.8	$H_2O(g)$	−54.6	$NH_4Cl(s)$	−48.7
$AgCl(s)$	−26.2	$CO_2(g)$	−94.3	$H_2O(l)$	−56.7	$NO(g)$	+20.7
$AgI(s)$	−15.9	$C_2H_2(g)$	+50.0	$H_2S(g)$	−7.9	$NO_2(g)$	+12.4
$Ag_2O(s)$	−2.6	$C_2H_4(g)$	+16.3	$HgO(s)$	−14.0	$NiO(s)$	−51.7
$Ag_2S(s)$	−9.6	$C_2H_6(g)$	−7.9	$HgS(s)$	−11.7	$PbBr_2(s)$	−62.1
$Al_2O_3(s)$	−376.8	$C_3H_8(g)$	−5.6	$KbR(s)$	−90.6	$PbCl_2(s)$	−75.0
$BaCl_2(s)$	−193.8	$CoO(s)$	−51.0	$KCl(s)$	−97.6	$PbO(s)$	−45.1
$BaCO_3(s)$	−272.2	$Cr_2O_3(s)$	−250.2	$KClO_3(s)$	−69.3	$PbO_2(s)$	−52.3
$BaO(s)$	−126.3	$CuO(s)$	−30.4	$KF(s)$	−127.4	$Pb_3O_4(s)$	−147.6
$BaSO_4(s)$	−323.4	$Cu_2O(s)$	−35.0	$MgCl_2(s)$	−141.6	$PCl_3(g)$	−68.4
$CaCl_2(s)$	−179.3	$CuS(s)$	−11.7	$MgCO_3(s)$	−246	$PCl_5(g)$	−77.6
$CaCO_3(s)$	−269.8	$CuSO_4(s)$	−158.2	$MgO(s)$	−136.1	$SiO_2(s)$	−192.4
$CaO(s)$	−144.4	$Fe_2O_3(s)$	−177.1	$Mg(OH)_2(s)$	−199.3	$SnCl_4(l)$	−113.3
$Ca(OH)_2(s)$	−214.3	$Fe_3O_4(s)$	−242.4	$MgSO_4(s)$	−280.5	$SnO(s)$	−61.5
$CaSO_4(s)$	−315.6	$HBr(g)$	−12.7	$MnO(s)$	−86.8	$SnO_2(s)$	−124.2
$CCl_4(l)$	−16.4	$HCl(g)$	−22.8	$MnO_2(s)$	−111.4	$SO_2(g)$	−71.8
$CH_4(g)$	−12.1	$HF(g)$	−64.7	$NaCl(s)$	−91.8	$SO_3(g)$	−88.5
$CHCl_3(l)$	−17.1	$HI(g)$	0.3	$NaF(s)$	−129.3	$ZnO(s)$	−76.1
$CH_3OH(l)$	−39.7	$HNO_3(l)$	−19.1	$NH_3(g)$	−4.0	$ZnS(s)$	−47.4

TABLE VII BOND ENERGIES (KCAL/MOL)

SINGLE BONDS

	H	C	N	O	S	F	Cl	Br	I
H	104	99	93	111	81	135	103	88	71
C		83	70	84	62	116	79	66	57
N			38	53	—	65	48	58	—
O				33	—	44	49	48	48
S					54	68	61	51	—
F						37	61	61	—
Cl							58	52	50
Br								46	43
I									36

MULTIPLE BONDS

Bond	Energy	Bond	Energy
C=C	143	C≡C	196
N=N	100	N≡N	225
O=O	96	C≡N	213
C=N	147		
C=O	171		
C=S	114		

TABLE VIII VAPOR PRESSURE OF WATER IN MILLIMETRES OF MERCURY

t (°C)	*vp*	*t* (°C)	*vp*	*t* (°C)	*vp*	*t* (°C)	*vp*
0	4.58	21	18.65	35	42.2	92	567.0
5	6.54	22	19.83	40	55.3	94	610.9
10	9.21	23	21.07	45	71.9	96	657.6
12	10.52	24	22.38	50	92.5	98	707.3
14	11.99	25	23.76	55	118.0	100	760.0
16	13.63	26	25.21	60	149.4	102	815.9
17	14.53	27	26.74	65	187.5	104	875.1
18	15.48	28	28.35	70	233.7	106	937.9
19	16.48	29	30.04	80	355.1	108	1004.4
20	17.54	30	31.82	90	525.8	110	1074.6

APPENDIX

2

ATOMIC AND IONIC RADII*

Element	Atomic Number	Atomic Radius in nm	Ionic Radius in nm
H	1	0.037	(−1) 0.208
He	2	0.05	
Li	3	0.152	(+1) 0.060
Be	4	0.111	(+2) 0.031
B	5	0.088	
C	6	0.077	
N	7	0.070	
O	8	0.066	(−2) 0.140
F	9	0.064	(−1) 0.136
Ne	10	0.070	
Na	11	0.186	(+1) 0.095
Mg	12	0.160	(+2) 0.065
Al	13	0.143	(+3) 0.050
Si	14	0.117	
P	15	0.110	
S	16	0.104	(−2) 0.184
Cl	17	0.099	(−1) 0.181
Ar	18	0.094	
K	19	0.231	(+1) 0.133
Ca	20	0.197	(+2) 0.099
Sc	21	0.160	(+3) 0.081
Ti	22	0.146	
V	23	0.131	
Cr	24	0.125	(+3) 0.064
Mn	25	0.129	(+2) 0.080
Fe	26	0.126	(+2) 0.075
Co	27	0.125	(+2) 0.072
Ni	28	0.124	(+2) 0.069
Cu	29	0.128	(+1) 0.096
Zn	30	0.133	(+2) 0.074
Ga	31	0.122	(+3) 0.062

*Radii are taken from Pauling, Linus: "The Nature of the Chemical Bond," 3rd ed., Ithaca, New York, Cornell University Press, 1960. Those of the noble gas atoms are from *Journal of Physical Chemistry, 69*:596, 1965.

Element	Atomic Number	Atomic Radius in nm	Ionic Radius in nm
Ge	32	0.122	
As	33	0.121	
Se	34	0.117	(−2) 0.198
Br	35	0.114	(−1) 0.195
Kr	36	0.109	
Rb	37	0.244	(+1) 0.148
Sr	38	0.215	(+2) 0.113
Y	39	0.180	(+3) 0.093
Zr	40	0.157	
Nb	41	0.143	
Mo	42	0.136	
Tc	43		
Ru	44	0.133	
Rh	45	0.134	
Pd	46	0.138	
Ag	47	0.144	(+1) 0.126
Cd	48	0.149	(+2) 0.097
In	49	0.162	(+3) 0.081
Sn	50	0.140	
Sb	51	0.141	
Te	52	0.137	(−2) 0.221
I	53	0.133	(−1) 0.216
Xe	54	0.130	
Cs	55	0.262	(+1) 0.169
Ba	56	0.217	(+2) 0.135
La	57	0.187	(+3) 0.115
Ce	58	0.182	(+3) 0.101
Pr	59	0.182	(+3) 0.100
Nd	60	0.182	(+3) 0.099
Pm	61		
Sm	62		
Eu	63	0.204	(+2) 0.097
Gd	64	0.179	(+3) 0.096
Tb	65	0.177	(+3) 0.095
Dy	66	0.177	(+3) 0.094
Ho	67	0.176	(+3) 0.093
Er	68	0.175	(+3) 0.092
Tm	69	0.174	(+3) 0.091
Yb	70	0.193	(+3) 0.089
Lu	71	0.174	(+3) 0.089
Hf	72	0.157	
Ta	73	0.143	
W	74	0.137	
Re	75	0.137	
Os	76	0.134	
Ir	77	0.135	
Pt	78	0.138	
Au	79	0.144	(+1) 0.137
Hg	80	0.155	(+2) 0.110
Tl	81	0.171	(+3) 0.095
Pb	82	0.175	
Bi	83	0.146	
Po	84	0.165	
At	85		
Rn	86	0.14	
Fr	87		
Ra	88	0.220	
Ac	89	0.20	
Th	90	0.180	
Pa	91		
U	92	0.14	

APPENDIX

3

NOMENCLATURE OF INORGANIC COMPOUNDS

In Chapter 8 we discussed briefly the system of nomenclature that is used with ionic compounds. Here we will review that system in somewhat greater detail and then examine how three other important classes of inorganic compounds (binary compounds of the nonmetals, oxyacids, and coordination compounds) are named.

IONIC COMPOUNDS

The names of ionic compounds are derived from those of the ions of which they are composed. We shall first consider the nomenclature of individual ions and then the names of the compounds they form.

Positive Ions

Monatomic positive ions take the names of the metal from which they are derived:

Na^+	sodium	Ca^{2+}	calcium	Al^{3+}	aluminum

When a metal forms more than one ion, it is necessary to distinguish between these ions. The accepted practice today is to indicate the charge of the ion by a Roman numeral in parentheses immediately following the name of the metal:

Fe^{2+}	iron(II)	Cu^+	copper(I)	Sn^{2+}	tin(II)
Fe^{3+}	iron(III)	Cu^{2+}	copper(II)	Sn^{4+}	tin(IV)

An earlier method, still widely used, adds to the stem of the Latin name of the metal the suffixes *-ous* or *-ic*, representing the lower and higher charges respectively:

Fe^{2+}	ferrous	Cu^+	cuprous	Sn^{2+}	stannous
Fe^{3+}	ferric	Cu^{2+}	cupric	Sn^{4+}	stannic

The only polyatomic cations to be considered here are:

NH_4^+	ammonium	Hg_2^{2+}	mercury(I) or mercurous

Negative Ions

Monatomic negative ions are named by adding the suffix *-ide* to the stem of the name of the nonmetal from which they are derived:

N^{3-}	nitride	O^{2-}	oxide	F^-	fluoride	H^-	hydride
		S^{2-}	sulfide	Cl^-	chloride		
		Se^{2-}	selenide	Br^-	bromide		
		Te^{2-}	telluride	I^-	iodide		

The nomenclature of polyatomic anions is more complex. The names of some of the more common oxyanions are:

NO_3^-	nitrate	ClO_2^-	chlorite	MnO_4^-	permanganate
NO_2^-	nitrite	ClO^-	hypochlorite	CrO_4^{2-}	chromate
SO_4^{2-}	sulfate	CO_3^{2-}	carbonate	$Cr_2O_7^{2-}$	dichromate
SO_3^{2-}	sulfite	PO_4^{3-}	phosphate	OH^-	hydroxide
ClO_4^-	perchlorate	BO_3^{3-}	borate		
ClO_3^-	chlorate	SiO_4^{4-}	silicate		

Notice that when a nonmetal such as nitrogen or sulfur forms two different oxyanions, the suffix *-ate* is used with the anion containing the larger number of oxygen atoms (NO_3^-, SO_4^{2-}), while the suffix *-ite* is used with the anion containing the smaller number of oxygen atoms (NO_2^-, SO_3^{2-}). With elements such as chlorine, which forms more than two oxyanions, the prefixes *per-* (largest number of oxygen atoms) and *hypo-* (smallest number of oxygen atoms) are used as well.

Oxyanions that contain hydrogen as well as nonmetal and oxygen atoms are properly named as illustrated in the following examples:

HCO_3^-	hydrogen carbonate	HPO_4^{2-}	hydrogen phosphate
HSO_4^-	hydrogen sulfate	$H_2PO_4^-$	dihydrogen phosphate

Compounds

The name of the positive ion is given first, followed by the name of the negative ion. Examples are

$CaCl_2$	calcium chloride
$FeBr_2$	iron(II) bromide
$(NH_4)_2SO_4$	ammonium sulfate
$Fe(ClO_4)_3$	iron(III) perchlorate
$NaHCO_3$	sodium hydrogen carbonate

In practice, compounds containing metal atoms, regardless of the type of bonding involved, are ordinarily named as if they were ionic. For example, the compounds $AlCl_3$ and $SnCl_4$, in both of which the bonding is primarily covalent, are named as follows:

$AlCl_3$	aluminum chloride	$SnCl_4$	tin(IV) chloride

BINARY COMPOUNDS OF THE NONMETALS

When a pair of nonmetals form only one compound, that compound may be named quite simply. The name of the element whose symbol appears first in the formula (the element of lower electronegativity) is written first. The second portion of the name is formed by adding the suffix *-ide* to the stem of the name of the second nonmetal. Examples include

HCl	hydrogen chloride
H_2S	hydrogen sulfide
NF_3	nitrogen fluoride

If more than one binary compound is formed by a pair of nonmetals, as is most often the case, the Greek prefixes, *di* = two, *tri* = three, *tetra* = four, *penta* = five, *hexa* = six, and so on, are used to designate the number of atoms in each element. Thus, for the oxides of nitrogen we have

N_2O_5	dinitrogen pentoxide*
N_2O_4	dinitrogen tetroxide*
NO_2	nitrogen dioxide
N_2O_3	dinitrogen trioxide
NO	nitrogen oxide
N_2O	dinitrogen oxide

A great many of the best-known binary compounds of the nonmetals have acquired common names which are widely and, in some cases, exclusively used. These include

H_2O	water	PH_3	phosphine
H_2O_2	hydrogen peroxide	AsH_3	arsine
NH_3	ammonia	NO	nitric oxide
N_2H_4	hydrazine	N_2O	nitrous oxide

OXYACIDS

The names of some of the more common oxygen acids are:

HNO_3	nitric acid	$HClO_4$	perchloric acid	H_2CO_3	carbonic acid
HNO_2	nitrous acid	$HClO_3$	chloric acid	H_3PO_4	phosphoric acid
H_2SO_4	sulfuric acid	$HClO_2$	chlorous acid	H_3BO_3	boric acid
H_2SO_3	sulfurous acid	$HClO$	hypochlorous acid	H_4SiO_4	silicic acid

It is of interest to compare the names of these oxyacids to those of the corresponding oxyanions listed previously. Note that oxyanions whose names end in *-ate* are derived from acids whose names end in *-ic*. Compare, for example, CO_3^{2-} (carbon*ate*) and H_2CO_3 (carbon*ic* acid); NO_3^- (nitr*ate*) and HNO_3 (nitr*ic* acid); ClO_4^- (perchlor*ate*) and $HClO_4$ (perchlor*ic* acid). Oxyanions whose names end in *-ite* are derived from acids whose names end in *-ous*. Thus we have NO_2^- (nitr*ite*) and HNO_2 (nitr*ous* acid); ClO^- (hypochlor*ite*) and $HClO$ (hypochlor*ous* acid).

---

*Note that in this case the *a* is dropped from the prefixes *penta* and *tetra* in the interests of euphony.

COORDINATION COMPOUNDS

The nomenclature of compounds containing complex ions in which a metal atom is held by coordinate covalent bonds to two or more ligands (Chapter 21) is perhaps more involved than that of any other type of inorganic compound. Several rules are required, the more pertinent of which are as follows:

a. As in simple ionic compounds, the cation is named first, followed by the anion.

b. If there is more than one ligand of a particular type attached to the central atom, Greek prefixes are used to indicate the number of these ligands. Where the name of the ligand itself is complex (e.g., ethylenediamine), the number of such ligands is indicated by the prefixes *bis-* or *tris-* instead of *di-* or *tri-* and the name of the ligand is enclosed in parentheses.

c. In naming a complex ion, the names of anionic ligands are written first, followed by those of neutral ligands, and finally that of the central metal atom. This is exactly the reverse of the order in which the groups are listed in the formula of the complex ion: the symbol of the central atom is written first, followed by the formulas of neutral ligands and then those of negatively charged ligands. In writing the formula of a coordination compound, the formula of the complex ion is often set off by brackets.

d. The names of anionic ligands are modified by substituting the suffix *-o* for the usual ending. Thus we have

Cl^-	chloro	CO_3^{2-}	carbonato
OH^-	hydroxo	CN^-	cyano

The names of neutral ligands are ordinarily not changed. Two important exceptions are:

H_2O	aquo	NH_3	ammine

e. The charge of the metal ion is indicated by a Roman numeral following the name of the metal. If the complex is an anion, the suffix *-ate* is added, often to the Latin stem of the name of the metal.

Examples are

$[Co(NH_3)_6]Cl_3$	hexamminecobalt(III) chloride
$[Co(en)_3](NO_3)_3$	tris(ethylenediamine)cobalt(III) nitrate
$[Cr(NH_3)_4Cl_2]Cl$	dichlorotetramminechromium(III) chloride
$[Pt(H_2O)_3Cl]Br$	chlorotriaquoplatinum(II) bromide
$K_3[Fe(CN)_6]$	potassium hexacyanoferrate(III)
$K_4[Fe(CN)_6]$	potassium hexacyanoferrate(II)

The last two compounds are often referred to as potassium ferricyanide and potassium ferrocyanide, respectively.

APPENDIX

4

EXPONENTS AND LOGARITHMS

(This material is taken from "Elementary Mathematical Preparation for General Chemistry," W. L. Masterton and E. J. Slowinski, W. B. Saunders Co., 1974.)

EXPONENTIAL NOTATION

In chemistry we frequently deal with very large or very small numbers. In one gram of the element carbon there are

$$50\ 150\ 000\ 000\ 000\ 000\ 000\ 000$$

atoms of carbon. At the opposite extreme, the mass of a single carbon atom is

$$0.000\ 000\ 000\ 000\ 000\ 000\ 000\ 019\ 94\ \text{g}$$

Numbers such as these are not only difficult to write; they are very awkward to work with. Imagine how tedious it would be to find the mass of 2150 carbon atoms by carrying out the operation

$$\begin{array}{r} 0.000\ 000\ 000\ 000\ 000\ 000\ 000\ 019\ 94\ \text{g} \\ \times\ 2150 \\ \hline \end{array}$$

To simplify operations of this type, we use what is known as **exponential notation.** To express a number in exponential notation, we write it in the form

$$C \times 10^n$$

where C is a number between 1 and 10 (e.g., 1, 2.62, 5.8) and n is a positive or negative integer (e.g., 1, −1, −3). To find n, we count the number of places that the decimal point must be moved to give the coefficient, C. If the decimal point must be moved to the *left,* n is a *positive* integer; if it must be moved to the *right,* n is a *negative* integer. Thus we have

$26.23 = 2.623 \times 10^1$ (decimal point moved 1 place to left)

$$5609 = 5.609 \times 10^3 \quad \text{(decimal point moved 3 places to left)}$$
$$0.0918 = 9.18 \times 10^{-2} \quad \text{(decimal point moved 2 places to right)}$$

Numbers written in exponential notation can be given a very simple interpretation. Recognizing that $10^1 = 10$, $10^3 = 1000$, and $10^{-2} = 1/100 = 0.01$, we could express the three exponentials written above as

$$2.623 \times 10^1 = 2.623 \times 10$$
$$5.609 \times 10^3 = 5.609 \times 1000$$
$$9.18 \times 10^{-2} = 9.18 \times 0.01$$

MULTIPLICATION AND DIVISION. A major advantage of exponential notation is that it simplifies the processes of multiplication and division. To *multiply*, we *add exponents:*

$$10^1 \times 10^2 = 10^{1+2} = 10^3;\ 10^6 \times 10^{-4} = 10^{6+(-4)} = 10^2$$

To *divide*, we *subtract* exponents:

$$10^3/10^2 = 10^{3-2} = 10^1;\ 10^{-3}/10^6 = 10^{-3-6} = 10^{-9}$$

To multiply one exponential number by another, we first multiply the coefficients in the usual manner and then add exponents. To divide one exponential number by another, we find the quotient of the coefficients and then subtract exponents. Examples:

$$(5.00 \times 10^4) \times (1.60 \times 10^2) = (5.00 \times 1.60) \times (10^4 \times 10^2) = 8.00 \times 10^6$$

$$(6.01 \times 10^{-3})/(5.23 \times 10^6) = \frac{6.01}{5.23} \times \frac{10^{-3}}{10^6} = 1.15 \times 10^{-9}$$

It often happens that multiplication or division yields an answer which is not in standard exponential notation. Thus we might have

$$(5.0 \times 10^4) \times (6.0 \times 10^3) = (5.0 \times 6.0) \times 10^4 \times 10^3 = 30 \times 10^7$$

The product is not in standard exponential notation since the coefficient, 30, does not lie between 1 and 10. To remedy this situation, we rewrite the coefficient as 3.0×10^1 and then add exponents

$$30 \times 10^7 = (3.0 \times 10^1) \times 10^7 = 3.0 \times 10^8$$

In another case

$$0.526 \times 10^3 = (5.26 \times 10^{-1}) \times 10^3 = 5.26 \times 10^2$$

RAISING TO POWERS AND EXTRACTING ROOTS. To raise an exponential number to a power or extract a root, we make use of the rules

$$(10^n)^a = 10^{na};\ (10^n)^{1/a} = 10^{n/a}$$

That is:

$$(10^{-2})^3 = 10^{-6};\ (10^{-2})^{1/2} = 10^{-2/2} = 10^{-1}$$

Applying these rules to numbers expressed in exponential notation, we have

$$(2.0 \times 10^{-2})^3 = (2.0)^3 \times (10^{-2})^3 = 8.0 \times 10^{-6}$$

$$(4.0 \times 10^{-2})^{1/2} = (4.0)^{1/2} \times (10^{-2})^{1/2} = 2.0 \times 10^{-1}$$

Here, as in multiplication and division, we operate on the coefficient and exponential terms separately.

Extracting the square root of an exponential number poses a special problem when the exponent is not an even number. Consider, for example,

$$(4.0 \times 10^5)^{1/2}$$

Following the procedure described above, we would obtain

$$(4.0)^{1/2} \times (10^5)^{1/2} = 2.0 \times 10^{5/2}$$

The answer is not in standard exponential notation; indeed $10^{5/2}$ is an awkward expression to work with because it does not readily translate into an ordinary number.

In cases of this type, we rewrite the exponential number to make the exponent an even number. To do this, we may divide the exponential by 10 and multiply the coefficient by 10.

$$(4.0 \times 10^5)^{1/2} = (40 \times 10^4)^{1/2}$$

Now, proceeding in the usual manner, we obtain

$$(40 \times 10^4)^{1/2} = (40)^{1/2} \times (10^4)^{1/2} = 6.3 \times 10^2$$

A similar principle is used in extracting cube roots.

$$(2.0 \times 10^{-5})^{1/3} = (20 \times 10^{-6})^{1/3} = 20^{1/3} \times 10^{-2} = 2.7 \times 10^{-2}$$

(Roots of numbers such as 40 and 20 may be found from tables, from a slide rule, or by using a table of logarithms as described later.)

ADDITION AND SUBTRACTION. Occasionally we may find it necessary to add or subtract two exponential numbers. These processes are extremely simple if both exponents are the same. For example:

$$2.02 \times 10^7 + 3.16 \times 10^7 = (2.02 + 3.16) \times 10^7 = 5.18 \times 10^7$$

$$4.23 \times 10^{-5} - 1.61 \times 10^{-5} = (4.23 - 1.61) \times 10^{-5} = 2.62 \times 10^{-5}$$

If the exponents differ, one of the numbers must be rewritten to make them the same. To add 5.04×10^8 to 3.0×10^7, we might express the latter as a number times 10^8

$$5.04 \times 10^8 + 3.0 \times 10^7 = 5.04 \times 10^8 + 0.30 \times 10^8 = 5.34 \times 10^8$$

LOGARITHMS

We have seen that the processes of multiplication, division, raising to a power, and extracting a root are simplified by expressing the quantities involved in exponential notation. However, even when this is done, a considerable amount of arithmetic may still be required. Suppose, for example, we wish to multiply 6.02×10^{23} by 1.99×10^{-24}. Combining the exponential terms, we obtain

$$(6.02 \times 10^{23}) \times (1.99 \times 10^{-24}) = (6.02 \times 1.99) \times 10^{-1}$$

but we still have to carry out the tedious operation of multiplying 6.02 by 1.99.

We could achieve a further simplification by expressing numbers such as 6.02 and 1.99 as powers of 10; the operation of multiplication could then be replaced by the more rapid process of addition. It is possible to accomplish this by making use of a table of common logarithms, such as that given on pp. A.22–A.23. A common logarithm is simply a power of 10; specifically, *it is the power to which 10 must be raised to give a particular number*. Thus we have

$$\log 100 = \log 10^2 = 2$$

$$\log 0.001 = \log 10^{-3} = -3$$

Again, when we find that the logarithms of 6.02 and 1.99 are, to four significant figures, 0.7796 and 0.2989, respectively, we deduce that

$$6.02 = 10^{0.7796};\ 1.99 = 10^{0.2989}$$

FINDING THE LOGARITHM OF A NUMBER. The four-place table of logarithms on pp. A.22–A.23 allows us to determine directly the logarithm of any three-digit number between 1 and 10. To find the logarithm of 6.02, we follow down the column at the far left of the table until we come to 6.0 and then move across to the column headed "2," reading off the logarithm of 6.02 as 0.7796. Similarly, we find the logarithm of 6.03 to be 0.7803, that of 6.04 to be 0.7810, and so on.

We can readily use the table to estimate accurately the logarithms of four-digit numbers. Suppose, for example, we wish to find the logarithm of 6.023. Since this number is 3/10 of the way between 6.02 and 6.03, its logarithm should be about 3/10 of the way between that of 6.02 (0.7796), and that of 6.03 (0.7803). Expressing this reasoning in the form of an equation

$$\begin{aligned}\log 6.023 &= \log 6.02 + 0.3\ (\log 6.03 - \log 6.02)\\ &= 0.7796 + 0.3\ (0.0007)\\ &= 0.7796 + 0.0002 = 0.7798\end{aligned}$$

The logarithm of a number less than 1 or greater than 10 can be found by writing it in exponential notation and applying the general rule

$$\log (C \times 10^n) = \log C + n$$

For example, to find the logarithms of 60.2 and 0.0602, we first write these numbers as 6.02×10^1 and 6.02×10^{-2}. Then:

$$\log (6.02 \times 10^1) = \log 6.02 + 1 = 0.7796 + 1 = 1.7796$$

$$\log (6.02 \times 10^{-2}) = \log 6.02 - 2 = 0.7796 - 2 = -1.2204$$

FINDING THE NUMBER CORRESPONDING TO A GIVEN LOGARITHM. The operation of finding an antilogarithm (number corresponding to a given logarithm) is simply the inverse of finding the logarithm of a number. To start with a simple example, let us find the number whose logarithm is 0.4997. We locate

0.4997 in the body of the table, noting that it falls in the horizontal column labeled "3.1" under the vertical column labeled "6." We deduce that

$$0.4997 = \log 3.16$$

By the same procedure, we find the antilogarithm of 0.5011 to be 3.17.

Frequently the logarithm we are working with does not appear directly in the table. Suppose, for example, we wish to obtain the antilogarithm of 0.5000. Even though we cannot locate 0.5000 directly, we can bracket it between 0.4997, the logarithm of 3.16, and 0.5011, the logarithm of 3.17. Since 0.5000 is 3/14 or 0.2 of the way between 0.4997 and 0.5011, its antilogarithm must be about 0.2 of the way between 3.16 and 3.17. That is

$$0.5000 = \log 3.162$$

The process just described can be applied where the logarithm falls between 0 and 1, corresponding to a number between 1 and 10. If we need to find the number associated with a logarithm greater than 1 (e.g., 6.4997) or less than 0 (e.g., −0.6021), we first *rewrite the logarithm in the form of a decimal fraction (mantissa) plus or minus a whole number (characteristic).* For example, to find the number whose logarithm is 6.4997, we rewrite this quantity as 0.4997 + 6. Now, we look up the mantissa, 0.4997, in the table, finding its antilogarithm to be 3.16. The antilogarithm of the characteristic, 6, is 10^6. Consequently, the number we are looking for must be 3.16×10^6. Summarizing

$$6.4997 = 0.4997 + 6 = \log (3.16 \times 10^6)$$

With a negative logarithm, we proceed in the same manner. To find the number whose logarithm is −0.6021, we first rewrite this quantity as a decimal fraction, between 0 and 1, minus a whole number. A moment's reflection should convince you that the result is 0.3979 − 1. That is

$$-0.6021 = 0.3979 - 1$$

Having cleared this hurdle, we find from the table that the antilogarithm of 0.3979 is 2.50; that of −1 must be 10^{-1}. The number we are looking for is 2.50×10^{-1}. In another case, to find the number whose logarithm is −3.4128, we proceed as follows

$$-3.4128 = 0.5872 - 4 = \log (3.865 \times 10^{-4})$$

OPERATIONS INVOLVING LOGARITHMS. Since logarithms are exponents, the rules governing the use of exponents apply here as well. Specifically:

Multiplication:

$\log (xy) = \log x + \log y;$ $\log (6.02 \times 1.99) = \log 6.02 + \log 1.99$

Division:

$\log (x/y) = \log x - \log y;$ $\log (9.17/2.62) = \log 9.17 - \log 2.62$

Raising to a power:

$\log (x^n) = n \log x$ $\log (2.00)^3 = 3 \log 2.00$

Extracting a root: $\log (x^{1/n}) = \frac{1}{n} \log x$ $\log (3.00^{1/2}) = \frac{1}{2} \log 3.00$

These relationships enable us to simplify the calculations involved in these four operations. Thus, to multiply 6.02×1.99, we could look up the logarithms of the two numbers, add them together, and look up the antilogarithm of the sum.

$$\log 6.02 + \log 1.99 = 0.7796 + 0.2989 = 1.0785 = \log 1.20 \times 10^1$$

Hence: $$6.02 \times 1.99 = 12.0$$

In another case, to divide 9.17 by 2.62, we subtract logarithms and look up the antilog of the difference:

$$\log 9.17 - \log 2.62 = 0.9624 - 0.4183 = 0.5441 = \log 3.50$$

Hence: $$9.17/2.62 = 3.50$$

NATURAL LOGARITHMS. For calculation purposes, common logarithms are most convenient since our number system is based upon multiples of 10. However, certain of the equations we use in general chemistry involve a different type of logarithm, taken to the base e, where, to four significant figures,

$$e = 2.718 \ldots$$

Logarithms to the base e are referred to as **natural** logarithms and are often written as "ln"; i.e.,

$$\log_e x \equiv \ln x; \qquad \log_{10} x \equiv \log x$$

Tables of natural logarithms, while available, are seldom used. To find the natural logarithm of a number, we first look up the common logarithm and then make use of the relation

$$\ln x = 2.303 \log x$$

To find the natural logarithm of 4.160, we first find the base 10 logarithm from our table to be 0.6191 and then multiply by 2.303:

$$\ln 4.160 = 2.303 \times 0.6191 = 1.426$$

Occasionally we are required to evaluate expressions such as $e^{0.2500}$, where the base of natural logarithms, e, is raised to a power. To do this, we note that if we let $x = e^{0.2500}$:

$$\ln x = 0.2500 = 2.303 \log x$$

$$\log x = 0.2500/2.303 = 0.1086; \qquad x = 1.284 = e^{0.2500}$$

ELECTRONIC CALCULATORS

Many of the operations that we have described in the sections *Exponential Notation* and *Logarithms* can be carried out more or less automatically by electronic calculators selling for less than $50. You will find such a calculator extremely useful for homework problems and for quizzes and exams (if allowed by your instructor). However, you should be aware of some of its limitations. In particular, a calculator can not be used as a crutch to avoid learning the mathematical operations necessary for success in general chemistry. This point was brought home to one of the authors when he found that, on an examination question which led to the equation

$$\log x = 111$$

a large number of students complained that they couldn't evaluate x because their calculator couldn't deal with antilogs above 100!

PROBLEMS

1 Express the following numbers in standard exponential notation:

a. 2712
b. 0.124
c. 31.6×10^{-3}
d. 0.0045×10^{2}
e. 0.000 000 000 000 000 000 000 019 94

2 Carry out the following operations, expressing the answers in standard exponential notation.

a. $(2.61 \times 10^{4}) \times (3.14 \times 10^{-2})$
b. $(5.18 \times 10^{4}) \times (4.91 \times 10^{-9})$
c. $\dfrac{6.29 \times 10^{-2}}{3.35 \times 10^{6}}$
d. $(4.17 \times 10^{-3})/(8.76 \times 10^{-4})$
e. $6.24 \times 10^{-3} + 1.4 \times 10^{-4}$
f. $5.29 \times 10^{3} - 1.61 \times 10^{2}$

3 Find the common logarithms of the following numbers:

a. 6.92
b. 6.92×10^{3}
c. 4.179×10^{2}
d. 0.007 12
e. 3.022×10^{-5}

4 Find the numbers whose common logarithms are

a. 0.8426
b. 0.4703
c. 0.0053
d. 3.2172
e. 2.9000
f. −6.4178
g. −2.1929
h. −22.0100

5 Evaluate each of the following, giving your answers in standard exponential notation.

a. $(2.16 \times 10^{4})^{1/2}$
b. $(6.92 \times 10^{-3})^{1/2}$
c. $(3.40 \times 10^{2})^{0.34}$
d. ln 12.60
e. $\ln (9.10 \times 10^{-2})$
f. $\log_{10} e$
g. $e^{0.6204}$
h. $e^{-1.2520}$

6 Making use of the definition: $pH = -\log_{10}$ (conc. H^+), calculate the pH of solutions in which conc. H^+ is:

a. 1.00×10^{-7}
b. 6.24×10^{-6}
c. 5.1×10^{-9}
d. 12

7 Determine the conc. H^+ in solutions in which the pH is:

a. 5.00
b. 2.361
c. 9.78
d. 7.7

8 In a solution saturated with H_2S, the following relation holds:

$$(\text{conc. } H^+)^2 \times (\text{conc. } S^{2-}) = 1 \times 10^{-23}$$

What is conc. S^{2-} in such a solution which has a pH of 4.5?

9 The standard free energy change of a reaction, ΔG^o (in joules) is related to the equilibrium constant, K, by the expression: $\Delta G^o = -RT \ln K$, where R = 8.31 J/K and T is the temperature in K. Calculate ΔG^o at 298 K for a reaction for which $K = 1.60 \times 10^{-4}$.

ANSWERS

1. 2.712×10^{3}, 1.24×10^{-1}, 3.16×10^{-2}, 4.5×10^{-1}, 1.994×10^{-23}

2. 8.20×10^{2}, 2.54×10^{-4}, 1.88×10^{-8}, 4.76, 6.38×10^{-3}, 5.13×10^{3}

3. 0.8401, 3.8401, 2.6211, −2.1475, −4.5197

4. 6.960, 2.953, 1.012, 1.649×10^{3}, 7.943×10^{2}, 3.821×10^{-7}, 6.413×10^{-3}, 9.772×10^{-23}

5. 1.47×10^{2}, 8.32×10^{-2}, 7.26, 2.534, −2.397, 4.343×10^{-1}, 1.860, 2.860×10^{-1}

6. 7.000, 5.205, 8.29, −1.08

7. 1.0×10^{-5}, 4.36×10^{-3}, 1.7×10^{-10}, 2×10^{-8}

8. 1×10^{-14}

9. 21.6 kJ

TABLE OF LOGARITHMS

	0	1	2	3	4	5	6	7	8	9
1.0	.0000	.0043	.0086	.0128	.0170	.0212	.0253	.0294	.0334	.0374
1.1	.0414	.0453	.0492	.0531	.0569	.0607	.0645	.0682	.0719	.0755
1.2	.0792	.0828	.0864	.0899	.0934	.0969	.1004	.1038	.1072	.1106
1.3	.1139	.1173	.1206	.1239	.1271	.1303	.1335	.1367	.1399	.1430
1.4	.1461	.1492	.1523	.1553	.1584	.1614	.1644	.1673	.1703	.1732
1.5	.1761	.1790	.1818	.1847	.1875	.1903	.1931	.1959	.1987	.2014
1.6	.2041	.2068	.2095	.2122	.2148	.2175	.2201	.2227	.2253	.2279
1.7	.2304	.2330	.2355	.2380	.2405	.2430	.2455	.2480	.2504	.2529
1.8	.2553	.2577	.2601	.2625	.2648	.2672	.2695	.2718	.2742	.2765
1.9	.2788	.2810	.2833	.2856	.2878	.2900	.2923	.2945	.2967	.2989
2.0	.3010	.3032	.3054	.3075	.3096	.3118	.3139	.3160	.3181	.3201
2.1	.3222	.3243	.3263	.3284	.3304	.3324	.3345	.3365	.3385	.3404
2.2	.3424	.3444	.3464	.3483	.3502	.3522	.3541	.3560	.3579	.3598
2.3	.3617	.3636	.3655	.3674	.3692	.3711	.3729	.3747	.3766	.3784
2.4	.3802	.3820	.3838	.3856	.3874	.3892	.3909	.3927	.3945	.3962
2.5	.3979	.3997	.4014	.4031	.4048	.4065	.4082	.4099	.4116	.4133
2.6	.4150	.4166	.4183	.4200	.4216	.4232	.4249	.4265	.4281	.4298
2.7	.4314	.4330	.4346	.4362	.4378	.4393	.4409	.4425	.4440	.4456
2.8	.4472	.4487	.4502	.4518	.4533	.4548	.4564	.4579	.4594	.4609
2.9	.4624	.4639	.4654	.4669	.4683	.4698	.4713	.4728	.4742	.4757
3.0	.4771	.4786	.4800	.4814	.4829	.4843	.4857	.4871	.4886	.4900
3.1	.4914	.4928	.4942	.4955	.4969	.4983	.4997	.5011	.5024	.5038
3.2	.5051	.5065	.5079	.5092	.5105	.5119	.5132	.5145	.5159	.5172
3.3	.5185	.5198	.5211	.5224	.5237	.5250	.5263	.5276	.5289	.5302
3.4	.5315	.5328	.5340	.5353	.5366	.5378	.5391	.5403	.5416	.5428
3.5	.5441	.5453	.5465	.5478	.5490	.5502	.5514	.5527	.5539	.5551
3.6	.5563	.5575	.5587	.5599	.5611	.5623	.5635	.5647	.5658	.5670
3.7	.5682	.5694	.5705	.5717	.5729	.5740	.5752	.5763	.5775	.5786
3.8	.5798	.5809	.5821	.5832	.5843	.5855	.5866	.5877	.5888	.5899
3.9	.5911	.5922	.5933	.5944	.5955	.5966	.5977	.5988	.5999	.6010
4.0	.6021	.6031	.6042	.6053	.6064	.6075	.6085	.6096	.6107	.6117
4.1	.6128	.6138	.6149	.6160	.6170	.6180	.6191	.6201	.6212	.6222
4.2	.6232	.6243	.6253	.6263	.6274	.6284	.6294	.6304	.6314	.6325
4.3	.6335	.6345	.6355	.6365	.6375	.6385	.6395	.6405	.6415	.6425
4.4	.6435	.6444	.6454	.6464	.6474	.6484	.6493	.6503	.6513	.6522
4.5	.6532	.6542	.6551	.6561	.6571	.6580	.6590	.6599	.6609	.6618
4.6	.6628	.6637	.6646	.6656	.6665	.6675	.6684	.6693	.6702	.6712
4.7	.6721	.6730	.6739	.6749	.6758	.6767	.6776	.6785	.6794	.6803
4.8	.6812	.6821	.6830	.6839	.6848	.6857	.6866	.6875	.6884	.6893
4.9	.6902	.6911	.6920	.6928	.6937	.6946	.6955	.6964	.6972	.6981
5.0	.6990	.6998	.7007	.7016	.7024	.7033	.7042	.7050	.7059	.7067
5.1	.7076	.7084	.7093	.7101	.7110	.7118	.7126	.7135	.7143	.7152
5.2	.7160	.7168	.7177	.7185	.7193	.7202	.7210	.7218	.7226	.7235
5.3	.7243	.7251	.7259	.7267	.7275	.7284	.7292	.7300	.7308	.7316
5.4	.7324	.7332	.7340	.7348	.7356	.7364	.7372	.7380	.7388	.7396
5.5	.7404	.7412	.7419	.7427	.7435	.7443	.7451	.7459	.7466	.7474
5.6	.7482	.7490	.7497	.7505	.7513	.7520	.7528	.7536	.7543	.7551
5.7	.7559	.7566	.7574	.7582	.7589	.7597	.7604	.7612	.7619	.7627
5.8	.7634	.7642	.7649	.7657	.7664	.7672	.7679	.7686	.7694	.7701
5.9	.7709	.7716	.7723	.7731	.7738	.7745	.7752	.7760	.7767	.7774

TABLE OF LOGARITHMS (*Continued*)

	0	1	2	3	4	5	6	7	8	9
6.0	.7782	.7789	.7796	.7803	.7810	.7818	.7825	.7832	.7839	.7846
6.1	.7853	.7860	.7868	.7875	.7882	.7889	.7896	.7903	.7910	.7917
6.2	.7924	.7931	.7938	.7945	.7952	.7959	.7966	.7973	.7980	.7987
6.3	.7993	.8000	.8007	.8014	.8021	.8028	.8035	.8041	.8048	.8055
6.4	.8062	.8069	.8075	.8082	.8089	.8096	.8102	.8109	.8116	.8122
6.5	.8129	.8136	.8142	.8149	.8156	.8162	.8169	.8176	.8182	.8189
6.6	.8195	.8202	.8209	.8215	.8222	.8228	.8235	.8241	.8248	.8254
6.7	.8261	.8267	.8274	.8280	.8287	.8293	.8299	.8306	.8312	.8319
6.8	.8325	.8331	.8338	.8344	.8351	.8357	.8363	.8370	.8376	.8382
6.9	.8388	.8395	.8401	.8407	.8414	.8420	.8426	.8432	.8439	.8445
7.0	.8451	.8457	.8463	.8470	.8476	.8482	.8488	.8494	.8500	.8506
7.1	.8513	.8519	.8525	.8531	.8537	.8543	.8549	.8555	.8561	.8567
7.2	.8573	.8579	.8585	.8591	.8597	.8603	.8609	.8615	.8621	.8627
7.3	.8633	.8639	.8645	.8651	.8657	.8663	.8669	.8675	.8681	.8686
7.4	.8692	.8698	.8704	.8710	.8716	.8722	.8727	.8733	.8739	.8745
7.5	.8751	.8756	.8762	.8768	.8774	.8779	.8785	.8791	.8797	.8802
7.6	.8808	.8814	.8820	.8825	.8831	.8837	.8842	.8848	.8854	.8859
7.7	.8865	.8871	.8876	.8882	.8887	.8893	.8899	.8904	.8910	.8915
7.8	.8921	.8927	.8932	.8938	.8943	.8949	.8954	.8960	.8965	.8971
7.9	.8976	.8982	.8987	.8993	.8998	.9004	.9009	.9015	.9020	.9026
8.0	.9031	.9036	.9042	.9047	.9053	.9058	.9063	.9069	.9074	.9079
8.1	.9085	.9090	.9096	.9101	.9106	.9112	.9117	.9122	.9128	.9133
8.2	.9138	.9143	.9149	.9154	.9159	.9165	.9170	.9175	.9180	.9186
8.3	.9191	.9196	.9201	.9206	.9212	.9217	.9222	.9227	.9232	.9238
8.4	.9243	.9248	.9253	.9258	.9263	.9269	.9274	.9279	.9284	.9289
8.5	.9294	.9299	.9304	.9309	.9315	.9320	.9325	.9330	.9335	.9340
8.6	.9345	.9350	.9355	.9360	.9365	.9370	.9375	.9380	.9385	.9390
8.7	.9395	.9400	.9405	.9410	.9415	.9420	.9425	.9430	.9435	.9440
8.8	.9445	.9450	.9455	.9460	.9465	.9469	.9474	.9479	.9484	.9489
8.9	.9494	.9499	.9504	.9509	.9513	.9518	.9523	.9528	.9533	.9538
9.0	.9542	.9547	.9552	.9557	.9562	.9566	.9571	.9576	.9581	.9586
9.1	.9590	.9595	.9600	.9605	.9609	.9614	.9619	.9624	.9628	.9633
9.2	.9638	.9643	.9647	.9652	.9657	.9661	.9666	.9671	.9675	.9680
9.3	.9685	.9689	.9694	.9699	.9703	.9708	.9713	.9717	.9722	.9727
9.4	.9731	.9736	.9741	.9745	.9750	.9754	.9759	.9763	.9768	.9773
9.5	.9777	.9782	.9786	.9791	.9795	.9800	.9805	.9809	.9814	.9818
9.6	.9823	.9827	.9832	.9836	.9841	.9845	.9850	.9854	.9859	.9863
9.7	.9868	.9872	.9877	.9881	.9886	.9890	.9894	.9899	.9903	.9908
9.8	.9912	.9917	.9921	.9926	.9930	.9934	.9939	.9943	.9948	.9952
9.9	.9956	.9961	.9965	.9969	.9974	.9978	.9983	.9987	.9991	.9996

APPENDIX 5

ANSWERS TO PROBLEMS

CHAPTER 1

1.1 13 cm^3 **1.2** 470 km; 4.70×10^{14} nm **1.3** −64°C, −83°F

1.18 a. See p. 19. b. See p. 17. c. See p. 8.

1.19 a. 3.80 g + 9.19 g = 12.99 g
b. Simple distillation OK.
c. Would require that °F = a°C.
d. Could be a mixture.

1.20 a. *Fp* would drop steadily
b. Both substances would crystallize out.
c. Absorbs at 450 nm; gives red color.

1.21 a. Use pipet or buret.
b. Heat to high *T* to obtain C, H; burn in air to two products.
c. Fractional distillation or VPC.

1.22 a. 8.38×10^{-24} cm^3 b. 11.1 g/cm^3 c. Empty space between atoms.

1.23 9.931 cm^3 **1.24** 54 g

1.25 a. 36.5 g/100 g water b. 0.0127 atm c. 5.19×10^4 J

1.26 a. 1081 K b. 11.2 cm^3 c. 0.667 kg/m^3

1.27 Must contain bromine; cool to 0°C to separate. Cool further to check for methane, oxygen.

1.28 4.5×10^{-4} g Hg **1.29** 1.2×10^{17} t **1.30** ~650 nm

1.31 43 g SA, 32 g TA **1.32** 1.19 km^2 **1.33** 15.9

1.34 C, D, E **1.35** 0.14

CHAPTER 2

2.1 a. 3.973, 7.945 b. 7.945/3.973 = 2.000

2.2 a. 9, 9, 10 b. $^{121}_{50}$Sn **2.3** a. 4.68 b. 468 **2.4** 10.82

2.5 a. 1.86×10^{-22} g b. 1.50×10^{23} **2.6** a. 26.04 g b. 3.01×10^{23}

2.23 a. Determine percentages of elements in different compounds of two elements.
b. Use mass spectrometer as with He.
c. Use mass spectrometer as with methane.

2.24 a. H_2O lighter than Cl.
b. No molecules in NaCl.
c. Only if derived from same atom.
d. Only if charges are equal.

2.25 a, b **2.26** $\dfrac{3\,AM \text{ M}/64.0}{2\,AM \text{ M}/48.0} = \dfrac{3}{2} \times \dfrac{48}{64} = \dfrac{9}{8}$

2.27 a. 94 b. 145 c. 94 d. 94, 91

2.28 a. 13 *GAM* H, 7.2 *GAM* C b. 1.8 **2.29** 12.01

2.30 5.4% Si-29, 2.3% Si-30 **2.31** 4.01×10^{15}

2.32 a. 3.06×10^{22} b. 4.67×10^{-23} g **2.33** d < a < e < c < b

2.34 3.67 g/cm^3; *AM* 7 times greater, *AR* about the same **2.35** Pb

2.36 3×10^6 **2.37** 7.5×10^{15}

2.38 $\frac{61.30}{38.70/2} \times 16.0 = 50.68$; $AM \approx 25/1.0 = 25$; $AM = 50.68/2 = 25.34$

2.39 14.0 **2.40** 6.03×10^{23} **2.41** 5.93×10^{23} **2.42** 1.1

2.43 0.009% **2.44** 18.0 g, 36.0 g, 54.0 g, 72.0 g

CHAPTER 3

3.1 CH_2 **3.2** C_2H_4 **3.3** a. 0.349 b. 123 g

3.4 60.0, 4.48, 35.6 **3.5** $4\ PH_3(g) + 8\ O_2(g) \rightarrow P_4O_{10}(s) + 6\ H_2O(l)$

3.6 a. 4.64 b. 2.67 c. 22.6 **3.7** a. O_2 b. 7.10 g c. 73.2%

3.26 a. 52.2, 44.9, 2.83 b. 1.91×10^3 **3.27** 80.0

3.28 a. KNO_2 b. C_5H_7N c. $CrCl_3N_3H_9$ **3.29** 6

3.30 CHCl **3.31** b, d, e **3.32** UF_6 **3.33** SnO_2, SnO

3.34 a. 6.71×10^{-3} b. 30.4 g c. 1.21×10^{22}, 3.68×10^{23}

3.35 a. 3.3×10^{24} g

3.36 a. $B_2H_6(l) + 3\ O_2(g) \rightarrow B_2O_3(s) + 3\ H_2O(l)$
b. $2\ Ag^+(aq) + CO_3^{2-}(aq) \rightarrow Ag_2CO_3(s)$
c. $4\ C_3H_5N_3O_9(l) \rightarrow 12\ CO_2(g) + 6\ N_2(g) + 10\ H_2O(l) + O_2(g)$

3.37 a. 26.9, 1.68, 135 b. 0.750, 0.750, 60.1 c. 3.49, 55.8, 279

3.38 a. 40.55 b. 0.364 c. 2.12 d. 1.49 **3.39** 160 dm^3

3.40 a. $CaO(s) + CO_2(g) \rightarrow CaCO_3(s)$ b. 2.30×10^4 g

3.41 a. 230 b. 23 c. 2.6 **3.42** a. PBr_3 b. 71.3 g c. 36.5

3.43 33 g, 21 g **3.44** 35.9

3.45 a. $2\ C(s) + O_2(g) \rightarrow 2\ CO(g)$; $Fe_2O_3(s) + 3\ CO(g) \rightarrow 2\ Fe(s) + 3\ CO_2(g)$
b. 0.430

3.46 2.1 cm^3

3.47 a. CoO, Co_3O_4
b. $CoCO_3(s) \rightarrow CoO(s) + CO_2(g)$; $6\ CoO(s) + O_2(g) \rightarrow 2\ Co_3O_4(s)$

CHAPTER 4

4.1 a. −14.9 kJ b. 0.0404 g **4.2** −2599.0 kJ **4.3** 0

4.4 a. 500 J b. 0.76 J/(g · °C) **4.5** a. 51 700 J b. −1550 kJ

4.6 −1.98 kJ **4.7** a. −2220 kJ b. −3 c. −2212 kJ

4.24 a. 0.451 kJ b. 2.39 g **4.25** a. 19.0 kJ b. 14.5 kJ

4.26 a. 37.9 kJ b. 3.79×10^4 kJ **4.27** −519.7 kJ

4.28 a. 9.51 kJ b. −871.6 kJ **4.29** −3800 kJ

4.30 a. +71 kJ b. +159 kJ c. +243 kJ; +122 kJ **4.31** 23.8°C

4.32 1900 kJ **4.33** 1430 J/°C **4.34** b, c

4.35 a. 146 J b. 29 J c. 110 J

4.36 a. −89.5 kJ b. +3 c. −97.0 kJ

4.37 a. 1.5×10^8 kJ; 3.7×10^8 kJ b. 3.1×10^{16} kJ c. 2.8×10^{16} kJ

4.38 a. 11 b. Slightly below VW beetle. **4.39** See discussion pp. 82–83.

4.40 a. −0.091 cm^3 b. -9.2×10^{-3} J c. 333 J d. 0.0028%

4.41 2.7, 18, 28, 23 kJ/g; 65, 430, 670, 550 kcal **4.42** 555°C

CHAPTER 5

5.1 a. 5.00 kPa b. 264 cm^3 c. 15 d. 2.81 g/dm^3

5.2 0.67 atm

5.3 a. Raise T to 1192 K; 596 K b. $CO_2 < Ar < O_2 < N_2 < H_2O$

5.23 Variation in atmospheric pressure. **5.24** 0.250 cm^3
5.25 163.4 kPa **5.26** 1.73×10^6 m^3 **5.27** 70°C
5.28 38.7 kPa **5.29** 22.4 dm^3, 24.5 dm^3
5.30 0.0642 g/dm^3, 11.2 g/dm^3 **5.31** CH, 78.1, C_6H_6 **5.32** 6.40 g
5.33 2:1 **5.34** 0.036 g **5.35** 40.1
5.36 a. 8.08×10^{-3}, 1.45×10^{-2} b. 20.0 kPa, 35.9 kPa c. 55.9 kPa
5.37 30.47 dm^3; 0.49%
5.38 a. $u_{O_2} = 0.935\, u_{N_2}$ b. same c. $p_{O_2} = 0.27\, p_{N_2}$ **5.39** 8100 K
5.40 142 s **5.41** a–c are straight lines through origin; d is hyperbola.
5.42 1.67×10^5 dm^3; 3.3×10^5 dm^3
5.43 Very few molecules present; little energy transferred to thermometer.
5.44 $E = 3R\Delta T/2 = 12.5$ J
5.45 No H_2; O_2 on borderline; no gases of $MM > 32$ originally present.
5.46 44.01; 12.01

CHAPTER 6

6.1 a. Discrete energies b. 4.949×10^{-19} J
6.2 1.634×10^{-18} J; 984 kJ/mol **6.3** $1s^22s^22p^3$
6.4
1s 2s 2p
(⇅) (⇅) (↑)(↑)(↑); three unpaired electrons
6.5 1, 0, 0, 1/2; 1, 0, 0, −1/2; 2, 0, 0, 1/2; 2, 0, 0, −1/2; 2, 1, +1, 1/2; 2, 1, 0, 1/2; 2, 1, −1, 1/2
6.22 a. p. 129 b. p. 125 c. p. 124 d. p. 126 **6.23** 437 nm
6.24 Eighth line
6.25 $\Delta E = \dfrac{B(n^2 - 4)}{4n^2}$; $\lambda = \dfrac{4hcn^2}{B(n^2 - 4)} = 364.6 \dfrac{n^2}{n^2 - 4}$ (in nm)
6.26 -1.961×10^{-17} J; 11 800 kJ/mol **6.27** 5.5×10^{-70} J; 3×10^{31}; yes
6.28 3×10^{-13} J; yes in all three cases.
6.29 2; one orbital with room for two electrons.
6.30 Fill sublevels in order of increasing energy; gives number of electrons in each sublevel.
6.31 $1s^22s^22p^63s^23p^6$; $1s^22s^22p^6$; $1s^22s^22p^63s^23p^1$; $1s^22s^22p^63s^23p^64s^23d^6$
6.32 Ground: c; excited: b, d, e, f; wrong: a.
6.33

	1s	2s	2p	3s	3p	4s	3d
Ti	(⇅)	(⇅)	(⇅)(⇅)(⇅)	(⇅)	(⇅)(⇅)(⇅)	(⇅)	(↑)(↑)( )( )( )
Si	(⇅)	(⇅)	(⇅)(⇅)(⇅)	(⇅)	(↑)(↑)( )		
Mn	(⇅)	(⇅)	(⇅)(⇅)(⇅)	(⇅)	(⇅)(⇅)(⇅)	(⇅)	(↑)(↑)(↑)(↑)(↑)

Unpaired electrons: Ti, 2; Si, 2; Mn, 5.
6.34 Ar^{2+}; $1s^22s^22p^63s^23p^4$
6.35 1, 0, 0, 1/2; 1, 0, 0, −1/2; 2, 0, 0, 1/2; 2, 0, 0, −1/2; 2, 1, 1, 1/2; 2, 1, 1, −1/2; 2, 1, 0, 1/2; 2, 1, 0, −1/2; 2, 1, −1, 1/2
6.36 b. $\mathbf{m}_\ell$ must be 0 c. ℓ must be 0 or 1, $\mathbf{m_s}$ 1/2 or −1/2
d. $\mathbf{m_s}$ must be 1/2 or −1/2 e. ℓ must be 0 or 1 f. $\mathbf{m}_\ell$ must be 0
6.37 I, II, III, V; I, II, IV, VI; I, III, IV, V; II, III, IV, VI; I, III, V, VI; II, IV, V, VI
6.38 No **6.39** 3.67×10^{-20} J vs. 3.37×10^{-19} J required
6.40 $1s^21p^42s^2$ **6.41** a. 3, 9 b. 12 c. $1s^32s^32p^2$; $1s^32s^32p^93s^2$

CHAPTER 7

7.1 $6s^26p^4$ **7.2** 6°C, 30°C
7.3 P less metallic, higher ionization energy, higher *EN*, smaller radius.
7.4 $In_2(SO_4)_3$

7.20 a. $Cs(s) + H_2O(l) \rightarrow CsOH(s) + 1/2\ H_2(g)$
$Cs(s) + 1/2\ Cl_2(g) \rightarrow CsCl(s)$
b. $Ba(s) + 2\ H_2O(l) \rightarrow Ba(OH)_2(s) + H_2(g)$
$Ba(s) + Cl_2(g) \rightarrow BaCl_2(s)$
c. $La(s) + 3\ H_2O(l) \rightarrow La(OH)_3(s) + 3/2\ H_2(g)$
$La(s) + 3/2\ Cl_2(g) \rightarrow LaCl_3(s)$

7.21 a. N, P, As, Sb, Bi b. Li, Na, K, Rb, Cs, Fr c. Ga, Ge, As, Se

7.22 $7s^27p^2$ **7.23** a. Larger b. Greater c. Less d. Lower

7.24 a. As > Se > S b. S > Se > As c. S > Se > As d. As > Se > S

7.25 Half-filled p level gives extra stability to N.

7.26 a. Metals at left of period; electrons with same n value ineffective in shielding.
b. Has one electron in outermost level, needs one electron to fill that level.
c. More intervening elements.

7.27 1.8 **7.28** Mg distinctly below curve.

7.29 a. $Mg_3(BiO_4)_2$ b. FrBr c. Au_2SO_4 d. K_2WS_4

7.30 Sb_2O_5, Sb_2O_3

7.31 a. Easier to reduce Al^{3+}. b. Cu more reactive than Au.
c. HgO decomposes to elements more readily than ZnO.

7.32 a. $NiS(s) + 3/2\ O_2(g) \rightarrow NiO(s) + SO_2(g)$
$NiO(s) + CO(g) \rightarrow Ni(s) + CO_2(g)$
b. $BaCO_3(s) \rightarrow BaO(s) + CO_2(g)$
c. $3\ BaO(s) + 2\ Al(s) \rightarrow 3\ Ba(s) + Al_2O_3(s)$

7.33 Growth rate much higher for Cu. **7.34** 5×10^9 kg

7.35 Compounds of Fr: FrCl, FrBr, FrI, Fr_2S, FrOH
Compounds of At: NaAt, $MgAt_2$, HAt
Fr will have lowest ionization energy, highest metallic character, lowest *EN*, largest atomic radius of all elements.
Atomic properties of At will resemble those of Te; will melt at about 200°C, boil at about 300°C, be a black solid.

7.36 Si; $SiCl_4$ **7.37** 33 a

CHAPTER 8

8.1 a. Al_2S_3 b. $(NH_4)_2SO_4$ c. $Zn(NO_3)_2$ **8.2** As-Te

8.3 a. $(:\ddot{O}-\ddot{Cl}-\ddot{O}:)^-$ b. $(:\ddot{O}-\ddot{N}=\ddot{O}:)^-$ c. $(:\ddot{O}-P(:\ddot{O}:)_2-\ddot{O}:)^{3-}$ (P bonded to four O atoms, one above and one below)

8.4 a. Bent b. Bent c. Tetrahedral **8.5** a, b

8.6 a. sp^3 b. sp^2 c. sp^3 **8.7** σ^b_{2s} (⇅) σ^*_{2s} (⇅) π^b_{2p} (⇅) π^b_{2p} (⇅) σ^b_{2p} (⇅)

8.27 a. Metals below Al in 3A would not acquire noble-gas configuration by losing $3e^-$.
b. Not generally true (hybridization).
c. Less than twice (Table 8.5).
d. Polar; X and Z different.

8.28 a. 3 b. 4 c. 5 d. 3 e. 3

8.29 a. $3\ Ca(s) + N_2(g) \rightarrow Ca_3N_2(s)$
b. $2\ CsH(s) \rightarrow 2\ Cs(s) + H_2(g)$
c. $Ca^{2+}(aq) + CO_3^{2-}(aq) \rightarrow CaCO_3(s)$
d. $Ba(s) + Cl_2(g) \rightarrow BaCl_2(s)$

8.30 a. $[Ar]3d^4$ b. $[Ar]3d^6$ c. [Ne] d. [Ne]

8.31 Mg—N = 55%; Mg—P = 20%; Al—N = 40%; Al—P = 10%

8.32 a. $:\ddot{O}{=}C{=}\ddot{O}:$ b. $(:C{\equiv}N:)^-$ c. $:C{\equiv}O:$ d. 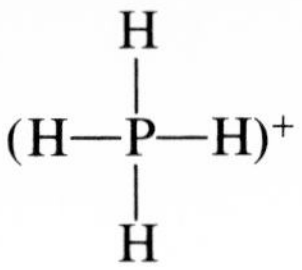

8.33 a. $:\ddot{O}{=}\ddot{N}{-}\ddot{O}{-}\ddot{N}{=}\ddot{O}:$ b. 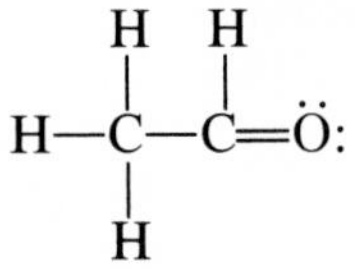 c. 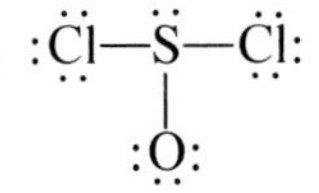

8.34 a. Na^+, $S_2O_3^{2-}$ b. 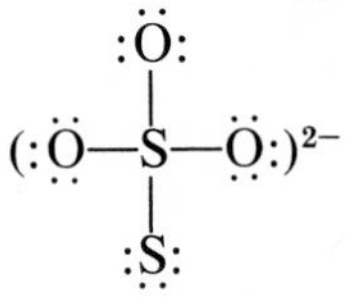 c. Tetrahedral

8.35 a **8.36** a. Cl_2 b. SO_3 c. CH_4 d. $HClO_4$

8.37 Two double bonds, all 120° angles.

8.38 a. 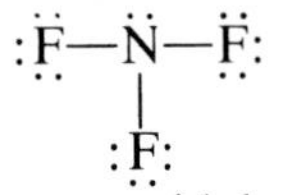 b. $(:\ddot{N}{=}\ddot{O}:)^-$ c. 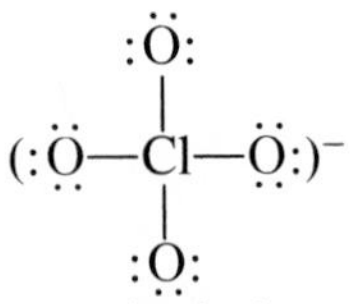 d. 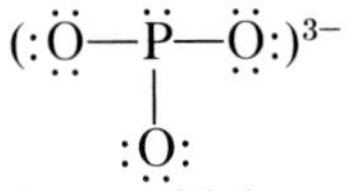

pyramidal linear tetrahedral pyramidal

8.39 a. Linear b. Linear c. Linear d. Tetrahedral **8.40** a, b, d

8.41 a. $(:\ddot{O}{-}\ddot{N}{-}\ddot{O}:)^{3-}$ b. $(:\ddot{O}{-}\ddot{N}{=}\ddot{O}:)^-$ c. $(:\ddot{O}{=}N{=}\ddot{O}:)^+$; a and b are dipoles.

8.42 Left to right: sp^3, sp^2, sp^2, sp^2, sp^3

8.43 a. H_2O b. CH_2O c. BF_3 d. BF_4^- e. N_2

8.44 Unpaired electrons: B_2, C_2 **8.45** Between O_2^{2+} and O_2^{3+}

8.46 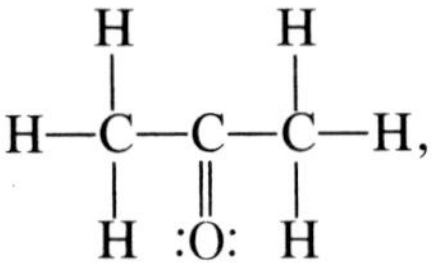, 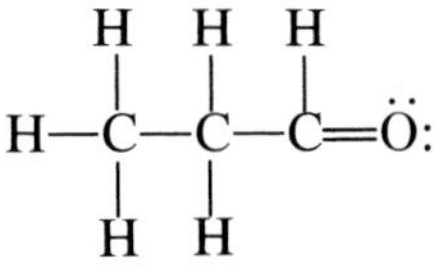

8.47 a. Square pyramid b. Square or pyramid c. Triangle or pyramid

8.48 +2144 kJ

CHAPTER 9

9.1 CaO **9.2** $Cl_2 < Br_2 < ICl$

9.3 a. Dispersion b. Dispersion, dipole c. Dispersion, H bonds

9.4 e

9.17 a. See Figures 9.9, 9.10. b. Cation + electron vs. cation + anion c. Covalent bonds vs. dispersion forces

9.18 a. Weak forces between layers b. Higher *MM* c. Ions free to move d. See discussion, Section 9.4.

9.19 a. Always valid. b. Only if N, O, or F present. c. Valid unless anion is discrete.

9.20 a. Molecular b. Metallic c. Metallic d. Molecular e. Macromolecular anion

9.21 a. MgO b. MgO c. NH_3 d. SbH_3

9.22 a. Garnet b. Quicklime c. Quartz d. Lead **9.23** d

9.24 a. Dispersion b. Covalent bonds c. Ionic bonds d. Covalent and ionic bonds e. Metallic bonds

9.25 b. Al_2O_3, H_2O c. $CuCl_2$, H_2O d. Al_2O_3, CO_2

9.26 b. 1 c. 2 or 3 **9.27** Hg

9.28 Test conductivity of solid, water solubility. **9.29** $Si_4O_{11}^{6-}$

9.30 $MM = 57$; polymerized through H bonding.

10.1 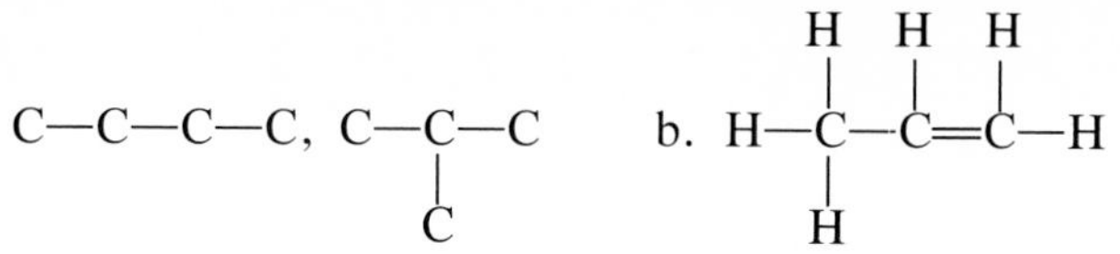 c. C_3H_4

10.2 See discussion, p. 240. **10.3** a. Alcohol, ketone, acid b. Acid

10.4 $C_2H_4(g) + HBr(g) \rightarrow C_2H_5Br(l)$; $C_2H_6(g) + Br_2(g) \rightarrow C_2H_5Br(l) + HBr(g)$

10.25 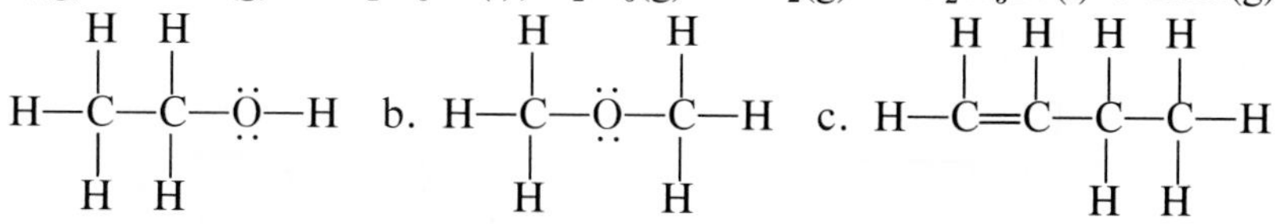

10.26 4 **10.27** $H_3C{-}C{\equiv}C{-}CH_3$, $H{-}C{\equiv}C{-}CH_2{-}CH_3$

10.28 Yes, if there is one Cl bonded to each C.

10.29 a. C_6H_{14} b. C_6H_{12} c. C_6H_{10} d. C_6H_6 **10.30** 4

10.31 Would require sp^2 hybridization with 90° bond angles. **10.32** 0

10.33 −629.7 kJ **10.34** $C_{21}H_{16}$

10.35 a. $4\ CO(g) + 2\ H_2(g) \rightarrow 2\ CO_2(g) + C_2H_4(g)$
b. $3\ CO(g) + 3\ H_2(g) \rightarrow CO_2(g) + C_2H_6O(l)$

10.36 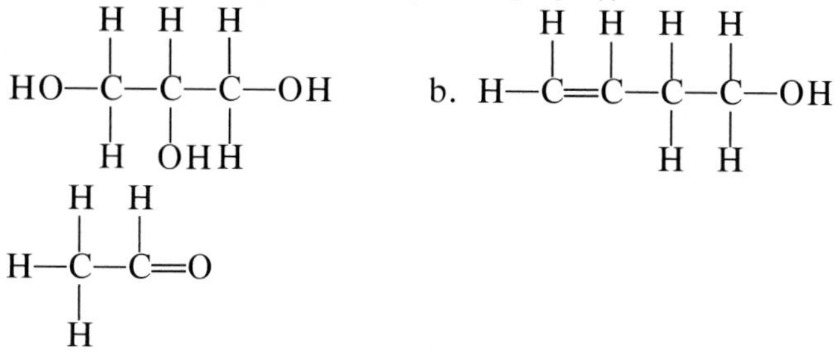

10.37 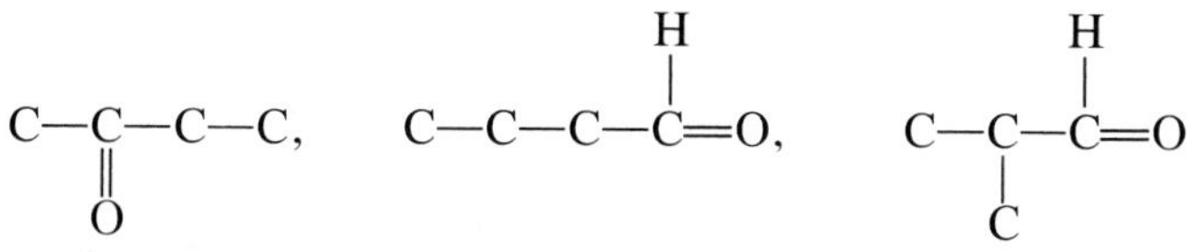
ketone aldehyde aldehyde

10.38 Alcohols: $H_3C{-}CH_2{-}CH_2{-}OH$, $H_3C{-}CHOH{-}CH_3$
Ether: $H_3C{-}O{-}CH_2{-}CH_3$
Aldehyde: 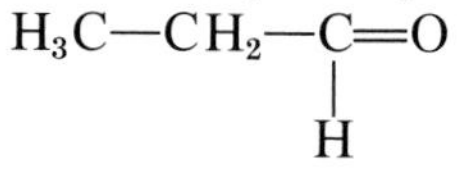

Ketone: 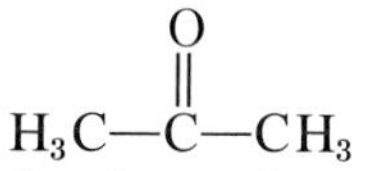
Acid: 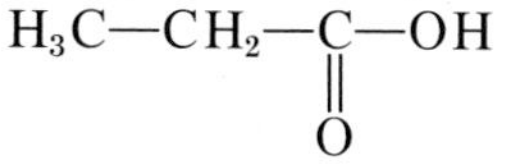

Esters: 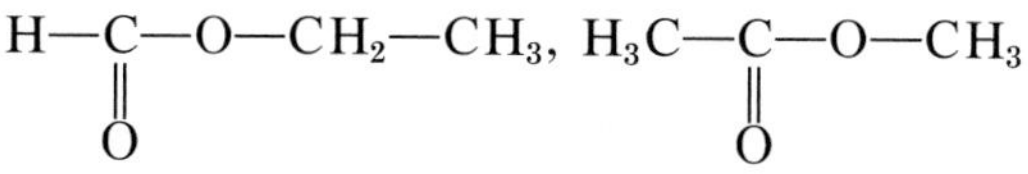

10.39 a. Alcohol b. Acid c. Alkene d. Ester

10.40 a. Aldehyde b. Alcohol or ether c. Acid or ester d. Alcohol

10.41 $C_2H_2 > C_2H_4 > C_2H_6 > C_2H_4O > C_2H_6O > C_2H_4O_2$

10.42 $c < a < d < b$

10.43 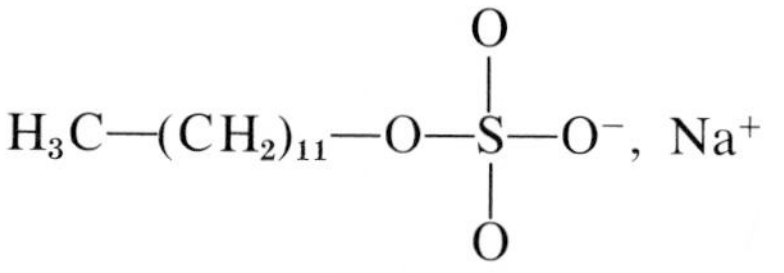

10.44 14 **10.45** 4

10.46 Structure: two C atoms single-bonded to a central C, which is double-bonded to O (C=O); in plane; 120°; 109°; 120°

10.47 +39.1 kJ; combustion of C to CO_2 **10.48** $H_3C-C(H)(OH)-CH_3$

CHAPTER 11

11.1 34.6 kPa; condensation occurs **11.2** 32 400 J/mol
11.3 0.397 nm **11.4** a. Solid b. Vapor c. Liquid-vapor
11.5 −4848 J
11.25 a. See discussion, Section 11.4.
b. Water boils at lower T.
c. Liquid vaporizes to replace escaping gas.
d. Evaporation absorbs heat.
11.26 a. Measure ΔH_{fus}, ΔH_{vap}, and add.
b. Measure type and dimension of unit cell, calculate volume per atom, compare to known molar volume.
c. Heat very pure liquid in clean container.
11.27 a. *Vp* independent of volume. b. Not all atoms belong entirely to cell.
c. >50%; Na^+ smaller than Cl^-.
11.28 a. 7.09 cm^3 b. 5.04 cm^3 c. 0.289
11.29 a. 700 b. 3.2 kPa c. 1.8 kPa
11.30 a. 54 kPa b. 31 kPa **11.31** 43.6 kJ
11.32 52.8 kJ **11.33** 70 **11.34** a, c **11.35** 0.154 nm
11.36 a. 12 b. 2 **11.37** 0.409 nm
11.38 0.513 nm; Na^+ could not fit **11.39** Se
11.40 Triple point at 0°C, 0.0060 atm; normal *bp* at 100°C, 1 atm; critical point at 374°C, 218 atm; s-1 line bends backwards.
11.41 a. Vaporizes c. Freezes
11.42 a. +0.137 kJ b. +3.51 kJ c. +74.7 kJ **11.43** 6.75 g
11.44 b. 44 kJ, 40 kJ c. Molar heat capacity lower for vapor
11.45 0.476, 0.320, 0.260 **11.46** 0.75°C; transfer of heat through blade

CHAPTER 12

12.1 a. 40.0 b. 0.273 c. 20.8 **12.2** a. 0.225 b. 0.0556 dm^3
12.3 5.5×10^{-4} mol/dm^3 **12.4** 6.5, 2.9 kPa
12.5 a. −0.776°C b. 100.22°C c. 1030 kPa **12.6** 104
12.26 a. Must have same π to avoid flow of water.
b. Similar intermolecular forces.
c. Air coming out of solution when P drops.
d. Air coming out of solution when T rises.
12.27 a. Add 256 g water. b. Add 83.3 g water. c. Add 877 g water.
d. Add enough water to give final volume of 877 cm^3.
12.28 a. 72 g b. 102 g c. 1.0 dm^3
12.29 a. 37.5% b. 0.667 c. 13.0 d. 10.4
12.30 0.768, 0.490, 113 g **12.31** $V_w = \dfrac{V_c(M_c - M_d)}{M_d}$
12.32 a. Mass fraction b. Molality

12.33 3.9×10^{-5} mol/dm^3
12.34 a. $H_3C—COOCH_3$ b. Ar c. CCl_4 d. $H_3C—O—CH_3$
12.35 Gases and solids less soluble than liquids. **12.36** 10.9 g; 2.73 g
12.37 0.26 **12.38** a. 6.45 g b. 19.4 g; 100.56°C
12.39 Ethylene glycol **12.40** 51.1 kPa **12.41** 48 m
12.42 $C_{21}H_{30}O_2$ **12.43** $C_{10}H_{14}N_2$ **12.44** 30 500; -3.16×10^{-4}°C
12.45 0.066 cm^3 **12.46** $n = kP = PV/RT;\ V = kRT =$ constant
12.47 41% **12.48** 1.9×10^{-3}

CHAPTER 13

13.1 4; 3.1 **13.2** 3.07 **13.3** 4.0×10^{-3}, 2.0×10^{-3} mol
13.23 Icebergs float, increase in P melts ice.
13.24 $CH_3OH < HF < KNO_3 < CaCl_2$
13.25 a. −0.0019°C b. −0.0037°C c. −0.0056°C d. −0.0093°C
13.26 30%
13.27 a. Measure conductivity. b. Compare conductivity to NaCl, Na_2SO_4.
c. Compare conductivity to NaCl, $MgSO_4$.
13.28 4.2×10^{17} kJ **13.29** 34.34 g
13.30 a. O_2 used by marine life. b. Converted to CO_3^{2-}, ppted as $CaCO_3$.
13.31 2.7×10^{-3} g/dm^3 **13.32** 7.5×10^{-5}
13.33 $C_{10}H_{14}PSNO_5(aq) + 15\ O_2(g) \rightarrow PO_4^{3-}(aq) + SO_4^{2-}(aq) + NO_3^-(aq) + 6\ H^+(aq) + 10\ CO_2(g) + 4\ H_2O$
13.34 Fertilizers, insecticides **13.35** See discussion, Section 13.3.
13.36 5.4×10^{-8} **13.37** 1.5×10^6 dm^3
13.38 a. 0.15, 0.11 b. 0.074, 0.21 c. 0.22
13.39 No; add 1 mol $Ca(OH)_2$ per mole Ca^{2+}.
13.40 a. $2\ R'Cl(s) + SO_4^{2-}(aq) \rightarrow R'_2SO_4(s) + 2\ Cl^-(aq)$
b. $Al^{3+}(aq) + 3\ NO_3^-(aq) + 3\ HR(s) + 3\ R'OH(s) \rightarrow AlR_3(s) + 3\ R'NO_3(s) + 3\ H_2O$
c. $2\ Na^+(aq) + 2\ Cl^-(aq) + ZnR_2(s) \rightarrow 2\ NaR(s) + Zn^{2+}(aq) + 2\ Cl^-(aq)$
13.41 2500 kPa **13.42** a. 37°C b. 0.60°C
13.43 Dilute to 1/18 original volume.
13.44 No. moles + charge = 0.591; no. moles − charge = 0.592.
13.45 Same no. moles ions, but twice the charge.

CHAPTER 14

14.1 a. +140.0 kJ b. No c. −140.0 kJ d. +0.874 kJ
14.2 109 J/K **14.3** +
14.4 a. +198.2 kJ b. +0.195 kJ/K c. +47.5 kJ d. 1020 K
14.21 a. Below equil. *vp*; above equil. *vp*; at equil. *vp*. b. First is spontaneous.
14.22 a. Fuel cell b. Make use of osmotic pressure to raise weight.
c. Metabolism in body
14.23 a. 118 kJ b. No; yes **14.24** a. −384 kJ/mol b. 9.58
14.25 −1014 kJ/mol
14.26 a. +0.5 kJ b. nonspontaneous at 25°C, 1 atm c. +, −
d. more +
14.27 a. +33.2 kJ b. −1149.6 kJ (spont) c. +330.6 kJ
14.28 $c < a < d < b < e$ **14.29** 134 J/K; 3.05 J/K
14.30 a. − b. + c. − d. − e. − f. −
14.31 a. All + b. ΔG −, ΔH +, ΔS + c. ΔG −, ΔH +, ΔS +
14.32 a. Not always true b. Not always true
c. Not always true (only for n_g) d. True

14.33 a. +118 kJ b. +65 kJ c. +0.178 kJ/K d. +29 kJ
14.34 310 K, 460 K, 610 K
14.35 a. No b. +; 0.162 kJ/K c. 1010 K d. Methane burns
14.36 Second reaction (third is cheaper).
14.37 b. 5.7 kJ c. +0.161 kJ/K, +0.198 kJ/K, +0.238 kJ/K
d. Calc. and obs. values agree closely at 10^{-2} and 10^{-4} atm.
14.38 12.7 kJ, 13.5 kJ; 10.3 kJ, 13.9 kJ
14.39 a. 5.1 b. 275 **14.40** 1200 K; 2450 K

CHAPTER 15

15.1 a. $\frac{[CO]^2 \times [O_2]}{[CO_2]^2}$ b. $\frac{[CO_2]^2}{[CO]^2}$
15.2 $K_c = [O_2] = \frac{[CO]^2 \times [O_2]}{[CO_2]^2} \times \frac{[CO_2]^2}{[CO]^2}$
15.3 4.0 **15.4** a. → b. ← **15.5** 0.10
15.6 a. ← b. → c. ← d. ← e. 0 **15.7** 2
15.8 −38 200 J
15.25 a. $\frac{[N_2] \times [H_2]^3}{[NH_3]^2}$ b. $\frac{[H_2O]^3}{[H_2]^3}$ c. $[H_2]^2 \times [N_2]$
d. $\frac{[N_2]^{1/2} \times [O_2]^{1/2} \times [Cl_2]^{1/2}}{[NOCl]}$
15.26 a. 4.8×10^{-6} b. 2.1×10^5 **15.27** 2.0×10^{-10} **15.28** 420
15.29 0.74 **15.30** 0.025, 8×10^{-4} **15.31** a. → b. ←
15.32 a. $[SO_3] = [NO] = 2.2 \times 10^{-3}$; $[SO_2] = [NO_2] = 8 \times 10^{-4}$
b. $[SO_3] = [NO] = 4.5 \times 10^{-3}$; $[SO_2] = [NO_2] = 1.5 \times 10^{-3}$
15.33 2×10^{-6} **15.34** a. 5.2×10^{-3} b. 2×10^{-17}
15.35 $[NO_2] = 0.52$, $[N_2O_4] = 0.74$
15.36 a. 0 b. ← c. → d. 0 e. →
15.37 a. 0.16 b. 0.059 **15.38** 640 K
15.39 9.0×10^{-2}, 11
15.40 a. $\Delta G_3^0 = \Delta G_1^0 + \Delta G_2^0$ b. $\log K_3 = \log K_1 + \log K_2$; $K_3 = K_1 \times K_2$
15.41 a. $P = \frac{nRT}{V} = (\text{conc.})RT$
b. $pC^c = [C]^c \times (0.0821\ T)^c$, etc. $K_p = K_c \times (0.0821\ T)^{\Delta n_g}$
15.42 a. −140.0 kJ, −4.2 kJ b. 3×10^{24}, 1.7 c. 8×10^{25}, 140
15.43 5.5×10^{-3}
15.44 a. −198.5 kJ; 8×10^{34} b. 3×10^{-38} c. Essentially all

CHAPTER 16

16.1 3.3×10^{-4} mol/(dm³ · min)
16.2 a. Second order b. 5.0×10^{-4} mol/(dm³ · s) **16.3** 5.9 min
16.4 134 kJ **16.5** rate $= kK_c^{1/2}$ (conc. A_2) (conc. B_2)$^{1/2}$
16.21 a. $\frac{\Delta \text{ conc. } H_2}{\Delta t}$ b. $-\frac{1}{2} \frac{\Delta \text{ conc. HI}}{\Delta t}$
16.22 1×10^{-12}; 10^4 s
16.23 a. Second order in A, first order in B b. 0.0600 dm³/(mol · s)
16.24 a. 0.0050 mol/(dm³ · s) b. 0.025 s^{-1} c. 0.12 dm³/(mol · s)
16.25 a. ~0.024 *M* b. 0.0238 *M* c. b **16.26** 1; 0.020 s^{-1}
16.27 a. 1.07×10^{-3} s^{-1} b. 1300 s c. 2.6×10^{-4} *M*
16.28 a. 2.02×10^{-3} min^{-1} b. 0.0450 *M* **16.29** 75 kJ

16.30 1.0×10^{-7}, 1.5×10^{-6}, 15
16.31 0.60 $dm^3/(mol \cdot s)$ **16.32** 401 kJ
16.33 a. Molecules may lack sufficient energy or may not be oriented properly.
b. May occur in several steps.
c. See discussion, p. 396.
16.34 Rate = $kK^{1/2}$(conc. H_2)(conc. Br_2)$^{1/2}$
16.35 Rate = $k\dfrac{(\text{conc. } O_3)^2}{(\text{conc. } O_2)}$ **16.36** $k = 0.20$, b = 3.0
16.37 a. 1.8×10^{-15} *M* b. 1.8×10^{-11} mol/($dm^3 \cdot s$) c. 2300 s
16.38 Test for I atoms. **16.39** 0.0025 g; 280 g

CHAPTER 17

17.1 a. 0.934 b. 1.29 c. 3.82×10^{-4} d. 9340
17.2 a. $4\ Al(s) + 3\ O_2(g) \rightarrow 2\ Al_2O_3(s)$ b. $Ba(s) + O_2(g) \rightarrow BaO_2(s)$
c. $BaO_2(s) + 2\ H_2O \rightarrow Ba^{2+}(aq) + 2\ OH^-(aq) + H_2O_2(aq)$
d. $CO_2(g) + H_2O \rightarrow H_2CO_3(aq)$
17.3 1.2 kPa
17.22 a. 1.12×10^{-5} b. 1.12×10^{-5} atm c. 4.58×10^{-7}
d. 2.02×10^{-2} g/m^3
17.23 a. No b. Yes c. Yes d. Yes
17.24 a. $Na_2O_2(s) + 2\ H_2O \rightarrow 2\ Na^+(aq) + 2\ OH^-(aq) + H_2O_2(aq)$
b. Equations 17.6–17.8 c. Equation 17.5
d. $2\ CuO(s) \rightarrow Cu_2O(s) + 1/2\ O_2(g)$ e. Equation 17.9
f. $P_4O_{10}(s) + 6\ H_2O \rightarrow 4\ H_3PO_4(aq)$
17.25 Increase rate by increasing *T* or *P* or by adding catalyst; increase yield by increasing *P* or decreasing *T*.
17.26 a. $(:\ddot{\underset{..}{N}}:)^{3-}$ b. $(:\ddot{\underset{..}{O}}—H)^-$ c. 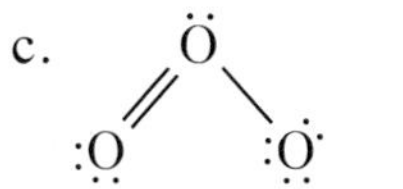 d. $\cdot\ddot{\underset{..}{O}}\cdot$
17.27 959 K
17.28 a. $SO_3 = 20$, $O_2 = 27$, $SO_2 = 53$ b. 0.0114
c. $[SO_3] = 0.0023$, $[O_2] = 0.0031$, $[SO_2] = 0.0060$ d. 6.9
17.29 a. Lowers visibility and perhaps *T* b. Increased exposure to UV
c. Lower relative humidity
17.30 a. Air b. Air, natural gas, water c. Air **17.31** 176 kJ
17.32 60 $dm^3/(mol \cdot s)$ **17.33** 1.3×10^{-3}
17.34 2×10^{-49} mol/($dm^3 \cdot s$) **17.35** a. 6×10^9 g b. 3×10^8
17.36 a. Convert $CaCO_3$ to (soluble) $CaSO_4$.
b. Lower combustion *T* forms less NO.
c. More reactive. d. More reactive, carcinogenic.
17.37 a. 3.23 g b. 15 g **17.38** a. 4.3×10^{-6} b. 105
17.39 a. Should lower pollution, at least along roadways.
b., c. Lower oxides of S. d. More CO, hydrocarbons; less NO_x.
17.40 a. 6.0 g b. 2.3 dm^3 c. 2300
17.41 a. 1.0×10^{22} dm^3 b. 6.3×10^{18} c. 2×10^{15} g

CHAPTER 18

18.1 $Fe^{2+}(aq) + 2\ OH^-(aq) \rightarrow Fe(OH)_2(s)$
18.2

	Ba^{2+}	OH^-	Fe^{2+}	Cl^-
Before	0.020	0.040	0.060	0.12
After	0.020	——	0.040	0.12

18.3 a. 7×10^{-5} b. 4.9×10^{-17} **18.4** a. 1×10^{-25} b. Yes
18.5 27.4 **18.6** $AgNO_3$; $BaCl_2$
18.26 a. $Ba^{2+}(aq) + CO_3^{2-}(aq) \rightarrow BaCO_3(s)$
b. $Ba^{2+}(aq) + SO_4^{2-}(aq) \rightarrow BaSO_4(s)$; $Zn^{2+}(aq) + S^{2-}(aq) \rightarrow ZnS(s)$
d. $2\ Sb^{3+}(aq) + 3\ S^{2-}(aq) \rightarrow Sb_2S_3(s)$
18.27 HCl, $Pb(NO_3)_2$; raise *T*, add KNO_3
18.28 a. 0.0100
b.

	Al^{3+}	SO_4^{2-}	Na^+	OH^-
Before	0.0600	0.0900	0.0300	0.0300
After	0.0500	0.0900	0.0300	——

c. Twice the final no. of moles
18.29 Not same valence type; calculated solubility Ag_2CrO_4 is greater.
18.30 0.5, 2×10^{-4}, 7×10^{-4}, 1×10^{-4} **18.31** 4.2×10^{-6}
18.32 a. 1.4×10^{-4} b. 3×10^{-9} **18.33** 5×10^{-9}
18.34 a. Yes b. No **18.35** 4×10^{-5}
18.36 a. 5×10^{-5} b. No **18.37** AgCl; 4×10^{-5}
18.38 a. 1×10^{-5} b. 10% c. 2.2×10^{-3} **18.39** 150 g
18.40 0.0514 *M*
18.41 a. Cl^-, then S^{2-}, then CO_3^{2-} b. Mg^{2+}, then Ni^{2+}, then Ag^+
18.42 No Ag^+
18.43 a. Note if solid is present. b. Add CO_3^{2-}. c. Add Ag^+, note color.
18.44 a. $AgNO_3$ b. 8.271×10^{-3} c. Filter, evaporate.
18.45 3900 **18.46** 7×10^{-4}, ~700
18.47 a. 1.4×10^{-5} b. Al^{3+}, Fe^{3+} c. ~100% d. 1.5 g
18.48 a. 6×10^{-3}, 5×10^{-3}, 5×10^{-3} b. 1.1×10^{-2} c. ~twice as great

CHAPTER 19

19.1 5.0×10^{-11}, 3.70 **19.2** 5×10^{-6} **19.3** 2.0×10^{-3}
19.4 2.5×10^{-10}
19.5 a. SB b. WA c. WA d. WB e. WB f. SA
19.6 a. Basic: $C_2H_3O_2^-(aq) + H_2O \rightarrow HC_2H_3O_2(aq) + OH^-(aq)$
b. Acidic: $Zn(H_2O)_4^{2+}(aq) \rightarrow Zn(H_2O)_3(OH)^+(aq) + H^+(aq)$
c. Neutral d. Acidic: $NH_4^+(aq) \rightarrow NH_3(aq) + H^+(aq)$
19.7 BA: H_3O^+, H_2CO_3; BB: HCO_3^-, H_2O
19.28 a. Completely dissociated in water to H^+, NO_3^-.
b. $\dfrac{[H^+] \times [Ac^-]}{[HAc]} = 1.8 \times 10^{-5}$ c. $\dfrac{[OH^-] \times [HCN]}{[CN^-]} = 2.5 \times 10^{-5}$
d. $pH > 7$
19.29 a. 1.0 b. −1.0 c. 2.15 d. 8.09 (basic)
19.30 a. 1×10^{-9}, 1×10^{-5} b. 6.3×10^{-4}, 1.6×10^{-11} (acidic)
c. 11, 9.1×10^{-16} (acidic) d. 3.5×10^{-8}, 2.9×10^{-7}
19.31 a. 1.2×10^{-14}, 0.80, 13.92 b. 0.60, 1.7×10^{-14}, 0.22
c. 1.0×10^{-13}, 0.10, 13.00 d. 1.0, 1.0×10^{-14}, 0.00
19.32 a. $H_3PO_4(aq) \rightarrow H^+(aq) + H_2PO_4^-(aq)$
b. $CO_2(g) + H_2O \rightarrow H_2CO_3(aq) \rightarrow H^+(aq) + HCO_3^-(aq)$
c. $HNO_2(aq) \rightarrow H^+(aq) + NO_2^-(aq)$
d. $Al(H_2O)_6^{3+}(aq) \rightarrow Al(H_2O)_5(OH)^{2+}(aq) + H^+(aq)$
19.33 a. $CH_3NH_2(aq) + H_2O \rightarrow CH_3NH_3^+(aq) + OH^-(aq)$
b. $NO_2^-(aq) + H_2O \rightarrow HNO_2(aq) + OH^-(aq)$
c. $HPO_4^{2-}(aq) + H_2O \rightarrow H_2PO_4^-(aq) + OH^-(aq)$
d. $CHO_2^-(aq) + H_2O \rightarrow HCHO_2(aq) + OH^-(aq)$
19.34 2.3×10^{-4} **19.35** a. 0.60, 0.22, 100 b. 1.8×10^{-5}, 4.74, 0.0030
19.36 2.9×10^{-2} **19.37** a. 1.5×10^{-10} b. 8.92

19.38 a. 1.9×10^{-11} b. 2.6×10^{-5}

19.39 a. A b. A c. B d. A e. B f. A

19.40 a. $Al(H_2O)_6^{3+}(aq) \rightarrow Al(H_2O)_5(OH)^{2+}(aq) + H^+(aq)$
b. $HOCl(aq) \rightarrow H^+(aq) + OCl^-(aq)$
c. $OCl^-(aq) + H_2O \rightarrow HOCl(aq) + OH^-(aq)$
d. $NH_4^+(aq) \rightarrow NH_3(aq) + H^+(aq)$
e. $CO_3^{2-}(aq) + H_2O \rightarrow HCO_3^-(aq) + OH^-(aq)$
f. $HSO_4^-(aq) \rightarrow H^+(aq) + SO_4^{2-}(aq)$

19.41 $HI < H_2CO_3 < NH_4Br < KBr < NH_3 < KCN < NaOH$

19.42 a. BA: H_2O, HCN; BB: CN^-, OH^-
b. BA: H_3O^+, H_2CO_3; BB: HCO_3^-, H_2O
c. BA: $HC_2H_3O_2$, H_2S; BB: HS^-, $C_2H_3O_2^-$

19.43 a. 2.5×10^{-5} b. 2.4×10^6 c. 180 (b and c will go to right)

19.44 a. BA, BB, LB b. BB, LB c. BB, LB d. BA, BB, LB

19.45 0.06, 0.50

19.46 a. F^-, $C_2H_3O_2^-$ b. H_3BO_3, NH_4^+ c. NaF, $NaC_2H_3O_2$
d. H_2SO_3, H_2SO_4 e. HCl, HBr

19.47 a. H^+, I^-, 0.70 b. Na^+, OCl^-, 10.25 c. K^+, NO_3^-, 7.00
d. OH^-, 13.00

19.48 1×10^{-17} **19.49** Determine *pH* of $AgNO_3$ solution.

19.50 HX > HY > HZ; K_a of HY = 2×10^{-3}, K_a of HZ = 2×10^{-5}

19.51

```
     H                                              H
     |                                              |
 (H—N—CH2—C—OH)+,  (H2N—CH2—C—O)−,  H—N—CH2—C—O
     |      ||                  ||             |      ||
     H      O                   O              H      O
```

CHAPTER 20

20.1 a. $H^+(aq) + OH^-(aq) \rightarrow H_2O$
b. $HC_2H_3O_2(aq) + OH^-(aq) \rightarrow C_2H_3O_2^-(aq) + H_2O$
c. $H^+(aq) + NH_3(aq) \rightarrow NH_4^+(aq)$
d. $2\ H^+(aq) + ZnCO_3(s) \rightarrow Zn^{2+}(aq) + CO_2(g) + H_2O$

20.2 1.0×10^{14}, 1.8×10^9, 1.8×10^9 **20.3** 1.0

20.4 10^{-2}, 1, 10^2; red, orange, yellow **20.5** 9.55

20.6 a. $CO_3^{2-}(aq) + 2\ H^+(aq) \rightarrow CO_2(aq) + H_2O$
b. $S^{2-}(aq) + 2\ H^+(aq) \rightarrow H_2S(g)$
c. $NH_4^+(aq) + OH^-(aq) \rightarrow NH_3(g) + H_2O$

20.28 Strong acid, strong base; HF-NaOH or NH_3-HCl

20.29 a. $HF(aq) + OH^-(aq) \rightarrow H_2O + F^-(aq)$
b. $H^+(aq) + C_2H_3O_2^-(aq) \rightarrow HC_2H_3O_2(aq)$
c. $NH_4^+(aq) + OH^-(aq) \rightarrow NH_3(g) + H_2O$
e. $H^+(aq) + OH^-(aq) \rightarrow H_2O$

20.30 a. 4.8×10^3 b. 2.4×10^6 c. 5.0×10^{16}

20.31 a. 1×10^9 b. 5×10^9

20.32 $Ag_2CrO_4(s) + H^+(aq) \rightarrow 2\ Ag^+(aq) + HCrO_4^-(aq)$ **20.33** 1.8×10^{-5}

20.34 0.1147 **20.35** 64.3 g

20.36 a. 0.30 b. 0.15 c. 0.10 d. 0.30

20.37 Y, Y, Y, G, B, B; NaOH + HCl **20.38** a. Any b. MO c. MO d. PP

20.39 a. 13.30 b. 12.82 c. 9.30 d. 7.00 e. 3.70 f. 1.18

20.40 a, b, c, d **20.41** a. 9.25 b. 9.21 c. 9.30 d. 1.00

20.42 a. 2.0×10^3 b. 7.66 c. 6.92 **20.43** 2×10^{-7}

20.44 a. Heat with NaOH.
b. Ppt. with NaOH, filter, neutralize with H_2SO_4, evaporate.

c. Dissolve in excess HCl, evaporate.
d. Bubble excess CO_2 through NaOH solution; evaporate.

20.45 Add acid: $2\ H^+(aq) + CO_3^{2-}(aq) \rightarrow CO_2(aq) + H_2O$
Add Ba^{2+}: $Ba^{2+}(aq) + SO_4^{2-}(aq) \rightarrow BaSO_4(s)$
Add Ag^+: $Ag^+(aq) + Cl^-(aq) \rightarrow AgCl(s)$
Heat with OH^-: $NH_4^+(aq) + OH^-(aq) \rightarrow NH_3(aq) + H_2O$

20.46 a. Add acid to test for HCO_3^-; add H_2S to test for Cu^{2+}.
b. Add acid to test for CO_3^{2-}; add SO_4^{2-} to test for Ba^{2+}.
c. Shake with water to test for $CaCO_3$; heat with OH^- to test for NH_4^+.

20.47 1×10^{-9}

20.48 2(Equation 20.19) + Equation 20.20 + Equation 20.21 + Equation 20.22

20.49 a. 2.72 b. 4.74 c. 7.44 d. 8.89 e. 9.3 f. 12.82

20.50 32; PP **20.51** 1.5 **20.52** 0.07 vs. 1.4×10^{-4}

CHAPTER 21

21.1 $[Cr(NH_3)_4Br_2]^+$; NaBr

21.2 Linear, tetrahedral, square planar (two isomers), octahedral

21.3
3d: (⇅)(⇅)(↑)(⇅)(⇅) 4s: (⇅) 4p: (⇅)(⇅)(⇅)

21.4
(↑)(↑) ( )( )
(↑)(↑)(↑) (⇅)(⇅)(↑)
high low

21.5 8×10^{-13}

21.25 a. 5 H_2O, 1 Cl^- b. +2 c. 1 mol AgBr ppts.

21.26 a. $MgSO_4$ b. $Mg(NO_3)_2$ c. $Al_2(SO_4)_3$ d. KNO_3
e. CH_3OH f. KNO_3

21.27 a. $Ni(H_2O)_4Cl_2$ b. $[Ni(H_2O)_3Cl]Cl$, $K[Ni(H_2O)Cl_3]$
c. $Ni(H_2O)_5Cl^+$ d. $Ni(H_2O)_3Cl^+$

21.28 b, c **21.29** 2.1×10^{21}

21.30 a. Octahedral
b. Cl^- at one corner, H_2O molecules at others
c. Co at center; Br at top, NH_3 at bottom; H_3N, Cl, H_3N, NH_3 in the plane
d. Ni at center; H_2O and Cl above, Cl and OH_2 below
e. Co at center; Cl at top and bottom; en, en in the plane
f. en Pt en

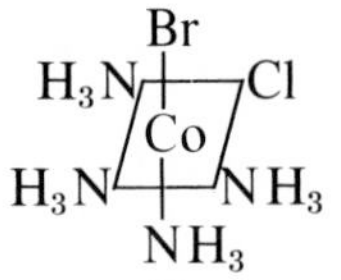

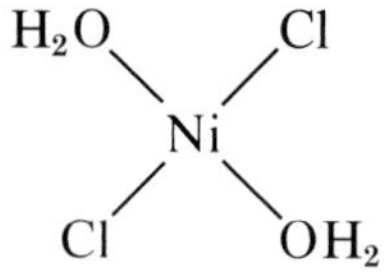

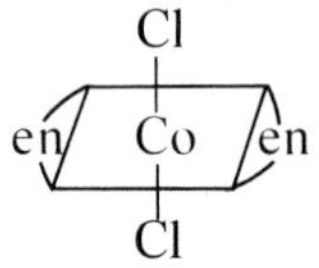

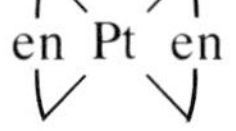

21.31 a. 2 b. 3 c. 5 d. 1 e. 2 **21.32** 5

21.33
a. [Kr] 4d: (⇅)(⇅)(⇅)(⇅)(⇅) 5s: (⇅) 5p: (⇅)( )( )
b. [Ar] 3d: (⇅)(⇅)(⇅)(⇅)(⇅) 4s: (⇅) 4p: (⇅)(⇅)(↑)
c. [Ar] (⇅)(⇅)(⇅)(⇅)(⇅) (⇅) (⇅)(⇅)(⇅)
d. [Ar] (⇅)(⇅)(⇅)(⇅)(⇅) (⇅) (⇅)(⇅)(⇅) 4d: (↑)
e. [Ar] (⇅)(↑)(↑)(⇅)(⇅) (⇅) (⇅)(⇅)(⇅)

21.34 MnA_6^+, FeA_6^{2+}, CoA_6^{3+}, CuA_4^+, ZnA_4^{2+} **21.35** 4, 5, 6, 7

21.36 Cl^- in complex gives blue color, red complex contains H_2O.

21.37 Add acid, note rate of color change.

21.38 3×10^{-6} **21.39** 5×10^{-24}

21.40 $K_d = \dfrac{K_{sp}\ AgCl \times [SCN^-]^2}{S^2}$

21.41 a. $Cu(NH_3)_4^{2+}(aq) + 4\ H^+(aq) + 4\ H_2O \rightarrow Cu(H_2O)_4^{2+}(aq) + 4\ NH_4^+(aq)$
b. $Zn^{2+}(aq) + 4\ OH^-(aq) \rightarrow Zn(OH)_4^{2-}(aq)$
$Fe^{3+}(aq) + 3\ OH^-(aq) \rightarrow Fe(OH)_3(s)$
c. $Fe^{3+}(aq) + SCN^-(aq) \rightarrow FeSCN^{2+}(aq)$

21.42 a. Excess conc. NH_3 b. Excess acid c. NaOH dropwise
d. Excess NaOH

21.43 18 **21.44** a. 3×10^6 b. $\sim 10^3$

21.45 $Ni^{2+}(aq) + 2\ NH_3(aq) + 2\ H_2O \rightarrow Ni(OH)_2(s) + 2\ NH_4^+(aq)$
$Al^{3+}(aq) + 3\ NH_3(aq) + 3\ H_2O \rightarrow Al(OH)_3(s) + 3\ NH_4^+(aq)$
$Ni(OH)_2(s) + 6\ NH_3(aq) \rightarrow Ni(NH_3)_6^{2+}(aq) + 2\ OH^-(aq)$
$Al(OH)_3(s) + 3\ OH^-(aq) \rightarrow Al(OH)_6^{3-}(aq)$
$Al(OH)_6^{3-}(aq) + 3\ H^+(aq) \rightarrow Al(OH)_3(s) + 3\ H_2O$
$Al(OH)_3(s) + 3\ H^+(aq) \rightarrow Al^{3+}(aq) + 3\ H_2O$

21.46 Absorbs at 520 nm; red

CHAPTER 22

22.1 a. +1, +5, −2 b. −5/2, +1 c. +5, −2

22.2 $2\ Fe(s) + 3\ SO_4^{2-}(aq) + 12\ H^+(aq) \rightarrow 2\ Fe^{3+}(aq) + 3\ SO_2(g) + 6\ H_2O$

22.3 26.4 g

22.4 a. Inert electrodes; Ca^{2+} discharged at cathode, Cl^- at anode b. Pb anode, Ag cathode

22.25 a. +3, −2 b. +2, +3, −2 c. +5, −1 d. +3, −2
e. +6, −2 f. −1/2

22.26 ClO_4^-, ClO_3^-, ClO_2^-, ClO^-, Cl^-

22.27 $Pt(H_2O)_6^{4+}$, $Pt(H_2O)_5Cl^{3+}$, $Pt(H_2O)_4Cl_2^{2+}$, $Pt(H_2O)_3Cl_3^+$, $Pt(H_2O)_2Cl_4$,
$Pt(H_2O)Cl_5^-$, $PtCl_6^{2-}$, $Pt(H_2O)_4^{2+}$, $Pt(H_2O)_3Cl^+$, $Pt(H_2O)_2Cl_2$,
$Pt(H_2O)Cl_3^-$, $PtCl_4^{2-}$

22.28 a, c **22.29** a. Cl b. Mn c. MnO_4^- d. Cl^-

22.30 a. R, O, R, O
b. $Cd^{2+}(aq) + 2\ e^- \rightarrow Cd(s)$
$Pb(s) + SO_4^{2-}(aq) \rightarrow PbSO_4(s) + 2\ e^-$
$NO_3^-(aq) + 2\ H^+(aq) + e^- \rightarrow NO_2(g) + H_2O$
$2\ Br^-(aq) \rightarrow Br_2(l) + 2\ e^-$
c. $Cd^{2+}(aq) + Pb(s) + SO_4^{2-}(aq) \rightarrow Cd(s) + PbSO_4(s)$
$Cd^{2+}(aq) + 2\ Br^-(aq) \rightarrow Cd(s) + Br_2(l)$
$2\ NO_3^- + 4\ H^+(aq) + Pb(s) + SO_4^{2-}(aq) \rightarrow 2\ NO_2(g) + 2\ H_2O + PbSO_4(s)$
$2\ NO_3^-(aq) + 4\ H^+(aq) + 2\ Br^-(aq) \rightarrow 2\ NO_2(g) + 2\ H_2O + Br_2(l)$

22.31 a. R, O, R
b. $Br_2(l) + 2\ e^- \rightarrow 2\ Br^-(aq)$
$Br_2(l) + 12\ OH^-(aq) \rightarrow 2\ BrO_3^-(aq) + 6\ H_2O + 10\ e^-$
$O_2(g) + 2\ H_2O + 4\ e^- \rightarrow 4\ OH^-(aq)$
c. $3\ Br_2(l) + 6\ OH^-(aq) \rightarrow 5\ Br^-(aq) + BrO_3^-(aq) + 3\ H_2O$
$2\ Br_2(l) + 4\ OH^-(aq) + 5\ O_2(g) \rightarrow 4\ BrO_3^-(aq) + 2\ H_2O$

22.32 a. $I_2(s) + 2\ S_2O_3^{2-}(aq) \rightarrow 2\ I^-(aq) + S_4O_6^{2-}(aq)$
b. $3\ Ag_2S(s) + 2\ NO_3^-(aq) + 8\ H^+(aq) \rightarrow 6\ Ag^+(aq) + 3\ S(s) + 2\ NO(g) + 4\ H_2O$
c. $3\ C_2H_6O(aq) + Cr_2O_7^{2-}(aq) + 8\ H^+(aq) \rightarrow 3\ C_2H_4O(aq) + 2\ Cr^{3+}(aq) + 7\ H_2O$
d. $4\ Zn(s) + NO_3^-(aq) + 10\ H^+(aq) \rightarrow 4\ Zn^{2+}(aq) + NH_4^+(aq) + 3\ H_2O$

22.33 a. $4\ Cl_2(g) + 4\ H_2O \rightarrow 7\ Cl^-(aq) + ClO_4^-(aq) + 8\ H^+(aq)$
b. $4\ Cl_2(g) + 8\ OH^-(aq) \rightarrow 7\ Cl^-(aq) + ClO_4^-(aq) + 4\ H_2O$
c. $2\ AuCl_4^-(aq) + AsH_3(g) + 3\ H_2O \rightarrow 2\ Au(s) + 8\ Cl^-(aq) + H_3AsO_3(aq) + 6\ H^+(aq)$
d. $Am^{3+}(aq) + S_2O_8^{2-}(aq) + 4\ OH^-(aq) \rightarrow AmO_2^+(aq) + 2\ SO_4^{2-}(aq) + 2\ H_2O$

22.34 4 $Au(CN)_2^-(aq)$ + 4 $OH^-(aq) \rightarrow$ 4 Au(s) + 8 $CN^-(aq)$ + $O_2(g)$ + 2 H_2O
22.35 a. 4 Zn(s) + $AsO_4^{3-}(aq)$ + 11 $H^+(aq) \rightarrow$ 4 $Zn^{2+}(aq)$ + $AsH_3(g)$ + 4 H_2O
b. 0.0791 c. 0.176 cm^3
22.36 a. 2 $Cl^- \rightarrow Cl_2(g)$ + 2 e^-; Mg^{2+} + 2 $e^- \rightarrow$ Mg(s)
b. Inert cathode at which Mg^{2+} is reduced; inert anode at which Cl^- is oxidized. c. +
22.37 a. 2 $H_2O \rightarrow O_2(g)$ + 4 $H^+(aq)$ + 4 e^- b. $Cu^{2+}(aq)$ + 2 $e^- \rightarrow$ Cu(s)
c. 2 H_2O + 2 $Cu^{2+}(aq) \rightarrow O_2(g)$ + 4 $H^+(aq)$ + 2 Cu(s)
22.38 a. Cathode reaction would be 2 $H^+(aq)$ + 2 $e^- \rightarrow H_2(g)$
b. $Ag(CN)_2^-$ converted to Ag^+, HCN
c. Would get $Al^{3+}(aq)$ d. Oxidized to Fe_2O_3
22.39 a. 6.69×10^{25} b. 124 c. 889 g
22.40 a. 7.29 dm^3 b. 29.8% **22.41** a. 0.42 g b. $\sim 10^3$ s
22.42 a. 49.1 g b. +4
22.43 a. Zn anode, Cd cathode; electrons move from Zn to Cd, anions to Zn, cations to Cd.
b. Cu anode, Au cathode; electrons move from Cu to Au, anions move to Cu, cations to Au.
c. Cl_2 reduced at Pt cathode, I^- oxidized at Pt anode.
22.44 dry cell: Zn anode, C cathode
plating cell: Cu anode (Cu^{2+} formed); Cu cathode (Cu plates)
22.45 a. 237.2 kJ b. 285.8 kJ c. 11.7 d. 36.5 **22.46** 23.1
22.47 a. 7.20×10^4 J b. 970 J c. ~99%

CHAPTER 23

23.1 1.21 V **23.2** b **23.3** −234 kJ, 8×10^{40}
23.4 a. 1.27 V b. 8×10^{-5} *M* **23.5** 2.50 *M*
23.24 a. 0.70 V b. 0.35 V c. 0.20 V **23.25** Neither (at std. conc.)
23.26 a. −0.80 V b. 3.67 V c. 1.10 V
23.27 Cu^{2+}, Ag^+, Au^{3+}; Li^+, K^+, Ba^{2+}
23.28 $Zn^{2+} < AuCl_4^- < MnO_4^- < H_2O_2$
23.29 b. $Fe^{2+} \rightarrow Fe^{3+}$ c. $Br^- \rightarrow Br_2$
23.30 a. $Sn^{4+}(aq)$ + Sn(s) $\rightarrow$ 2 $Sn^{2+}(aq)$
b. 2 $Cu^+(aq) \rightarrow Cu^{2+}(aq)$ + Cu(s)
c. $Cr_2O_7^{2-}(aq)$ + 6 $Fe^{2+}(aq)$ + 14 $H^+(aq) \rightarrow$ 2 $Cr^{3+}(aq)$ + 6 $Fe^{3+}(aq)$ + 7 H_2O
23.31 3.1×10^{-4}, −0.035 V, +20 kJ **23.32** a. 4×10^{-5} b. 5.4 *M*
23.33 0.4 *M* **23.34** 0.030 V **23.35** a. 0.32 V b. −0.18 V
23.36 10^{-25} **23.37** Measure $[H^+]$ in 1.0 *M* NaOH
23.38 a. Add base. b. Reduce in acidic soln. c. Make soln. in (b) basic.
d. Electrolysis.
23.39 a. NiS(s) + 2 $NO_3^-(aq)$ + 4 $H^+(aq) \rightarrow Ni^{2+}(aq)$ + S(s) + 2 $NO_2(g)$ + 2 H_2O
b. $O_2(g)$ + 4 $H^+(aq)$ + 4 $I^-(aq) \rightarrow$ 2 H_2O + 2 $I_2(s)$
c. Fe(s) $\rightarrow Fe^{2+}(aq)$ + 2 e^-
d. 2 $Hg^{2+}(aq)$ + 2 $Cl^-(aq)$ + $Sn^{2+}(aq) \rightarrow Hg_2Cl_2(s)$ + $Sn^{4+}(aq)$
23.40 a. Zn oxidizes, rather than Fe.
b. Sn^{2+} oxidized to Sn^{4+}, which ppts as SnO_2.
c. NO_3^- acts as oxidizing agent. d. O_2 concentration cell set up.
23.41 a. 2 $MnO_4^-(aq)$ + 16 $H^+(aq)$ + 10 $I^-(aq) \rightarrow$ 2 $Mn^{2+}(aq)$ + 5 $I_2(s)$ + 8 H_2O
b. 0.556 *M*
23.42 1.20 **23.43** 10^{-16} **23.44** +0.40 V **23.45** 9×10^{15}

CHAPTER 24

24.1 $^{99}_{46}Pd$ **24.2** 0.0283/d; 0.18 **24.3** -2.5×10^6 kJ

24.20 a. Same number protons in nucleus. b. See p. 586. c. See p. 585. d. See p. 588.

24.21 a. Measure level of radioactivity, calculate k, then $t_{1/2}$. b. C-14 method c. Mass spectrometer d. Measure ΔE, calc. Δm.

24.22 a. $^{237}_{93}Np \rightarrow {}^{4}_{2}He + {}^{233}_{91}Pa$ b. $^{85}_{39}Y \rightarrow {}^{0}_{1}e + {}^{85}_{38}Sr$ c. $2\ {}^{12}_{6}C \rightarrow {}^{23}_{11}Na + {}^{1}_{1}H$ d. $^{239}_{94}Pu + {}^{1}_{0}n \rightarrow {}^{130}_{50}Sn + 4\ {}^{1}_{0}n + {}^{106}_{44}Ru$

24.23 7, 4

24.24 a. $^{96}_{42}Mo + {}^{2}_{1}H \rightarrow {}^{1}_{0}n + {}^{97}_{43}Tc$ b. $^{209}_{83}Bi + {}^{4}_{2}He \rightarrow 2\ {}^{1}_{0}n + {}^{211}_{85}At$ c. $^{10}_{5}B + {}^{1}_{0}n \rightarrow {}^{0}_{0}\gamma + {}^{11}_{5}B$ d. $^{45}_{21}Sc + {}^{1}_{0}n \rightarrow {}^{1}_{1}H + {}^{45}_{20}Ca$

24.25 a. 0.024/a b. 0.091 **24.26** June 3

24.27 3.3×10^9 a; Ar escapes **24.28** 4790 a

24.29 Compare to historical records; use other radioactive dating methods.

24.30 2.5×10^6 **24.31** 23.983 94 g **24.32** Yes; no

24.33 1.41×10^{-8}; 3.55×10^{13} **24.34** 0.069 56; 6.26×10^9; 6.26×10^8

24.35 1.0×10^{10} g **24.36** a. 2×10^{-14} b. $\sim 2 \times 10^6$ m/s c. $\sim 10^9$°C

24.37 a. 5.5×10^{-10} g b. -1.2 J c. 20

24.38 49%, 2%, 49%

24.39 a. -5.72×10^8 b. 1.1×10^{32} c. 2.0×10^{-15}

CHAPTER 25

25.1 a. $-CH_2-CHR-CH_2-CHR-$ (each C bearing H above; below: H, R, H, R), where R = phenyl (benzene ring) b. -96 kJ

25.2 $-O-CH_2-CH_2-O-C(=O)-C(=O)-O-$

25.3 a. 44.5, 6.23, 49.4 b. 1.6×10^6

25.4 $H_2N-CH_2-C(=O)-NH-CH_2-C(=O)-NH-CH_2-C(=O)-NH-CH_2-C(=O)-OH$

25.24 a. See p. 615. b. Four different groups bonded. c. $H_2N-\overset{|}{\underset{|}{C}}-COOH$ d. 5-carbon sugar that is an aldehyde.

25.25 a. S b. PS c. PS **25.26** b. A c. C

25.27 $-C-C-C-C-$ (above: H H H H; below: H, CN, CN, H)

25.28 Zigzag chain with substituents: H, CN; H, CN; H, CN; H, CN

25.29 a. $HO-C(=O)-C_6H_4-C(=O)-OH$ and $HO-(CH_2)_2-OH$

b. $Cl-CH=CH-Cl$ (H above each C)

c. $H_2N{-}(CH_2)_5{-}C({=}O){-}OH$

25.30 a. C b. A c. C **25.31** See 25.29c.

25.32
H_3C $CH_2{-}CH_2$ H
C=C C=C
$-H_2C$ H H_3C CH_2-

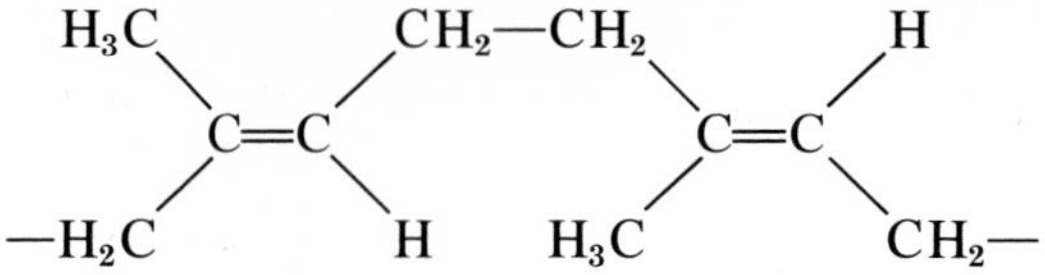

25.33 $O{=}C(H){-}C(H)(OH){-}C(H)(OH){-}H$

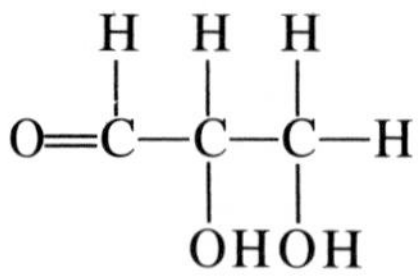

25.34 a. 2 b. 0 c. 2 **25.35** See p. 622. **25.36** 0

25.37 $(H_2N{-}(CH_2)_4{-}C(H)(NH_3){-}COOH)^+$ or $(H_3N{-}(CH_2)_4{-}C(H)(NH_3){-}COOH)^{2+}$;

$H_2N{-}(CH_2)_4{-}C(H)(NH_2){-}COO^-$

25.38 27 **25.39** 13 300 **25.40** 1.59×10^4 **25.41** 1

25.42 Not necessarily; structure changes with *T*. **25.43** -17 kJ

25.44 Linear:
CH_2 CH_2
OH OH OH

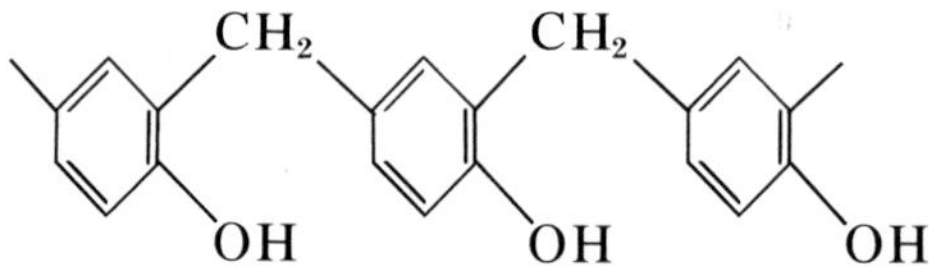

Bakelite:
OH OH OH
CH_2 CH_2
CH_2 CH_2 CH_2
CH_2 CH_2
OH OH OH

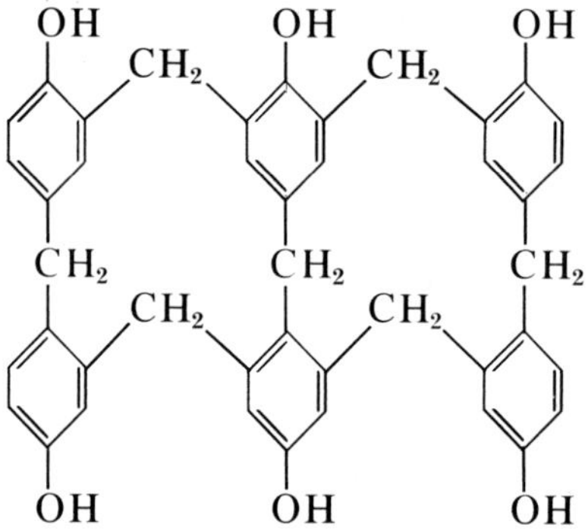

GLOSSARY

Terms in this glossary are part of the basic chemical vocabulary; they are defined in the context used in this text. If you do not find the term you seek here, look for it in the Index, which will refer you to pages where the term is used.

A

absolute zero—the lowest possible temperature at which matter might exist; 0 K, −273.15°C.

acid—a substance which on being dissolved in water produces a solution in which $[H^+]$ is greater than 10^{-7} *M*. Examples: HCl, HNO_3, H_2CO_3, CH_3COOH.

acid dissociation constant—K_a; the equilibrium constant for the following reaction of an acid HA: $HA(aq) \rightleftarrows H^+(aq) + A^-(aq)$.

$$K_a = \frac{[H^+][A^-]}{[HA]}$$

activated complex—a species, formed by collision of energetic particles, which can react to form products or other intermediates.

addition polymer—a polymer produced by reaction of a monomer, usually a derivative of ethylene, adding to itself; no other product is formed.

alcohol—a substance containing an OH group attached to a hydrocarbon chain. Examples: C_2H_5OH, ethyl alcohol; C_4H_9OH, butyl alcohol.

aldehyde—a substance containing a carbonyl, C=O, group at the end of a hydrocarbon chain. Example: $C_2H_5—C(H)=O$, propionaldehyde.

alkali metal—a metal in Group IA. Examples: Li, Na, K.

alkaline (basic)—having a $[OH^-]$ which is greater than 10^{-7} *M*.

alkane—a hydrocarbon containing only single carbon-carbon bonds. Examples: C_2H_6, C_6H_{14}.

alkene—a hydrocarbon which contains one carbon-carbon double bond. Examples: $CH_3CH{=}CH_2$; $H_2C{=}CH_2$.

alkyl group—a hydrocarbon group such as $CH_3—$, $C_2H_5—$, or $C_3H_7—$.

alkyne—a hydrocarbon containing one carbon-carbon triple bond. Example: $HC{\equiv}CH$.

allotrope—one of two or more forms of an elementary substance. Examples: O_2 and O_3 are allotropic forms of oxygen; graphite and diamond are allotropes of carbon.

alpha (α) particle—a helium nucleus, He^{2+} ion.

amalgam—a solution of a metal in mercury.

amphoteric—capable of reacting with both H^+ and OH^- ions, usually an insoluble hydroxide. Examples: $Al(OH)_3$, $Zn(OH)_2$.

anhydride—a substance derived from another by removal of water. Examples: SO_3 is the anhydride of H_2SO_4; CaO is the anhydride of $Ca(OH)_2$.

anion—a species carrying a negative charge. Examples: Cl^-, CO_3^{2-}, $H_2PO_4^-$.

anode—an electrode at which oxidation occurs. Example: If, at a copper electrode, the reaction that occurs is $Cu(s) \rightarrow Cu^{2+}(aq) + 2\,e^-$, then the copper metal is behaving as an anode.

aqua regia—a mixture of concentrated hydrochloric and nitric acids.

aromatic substance—an organic compound containing a benzene ring. Examples: Benzene, C_6H_6, ; toluene, C_7H_8, $—CH_3$; naphthalene, $C_{10}H_8$,

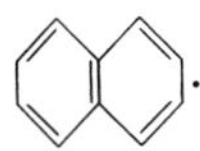

atmosphere, standard—unit of pressure; equal to the pressure exerted by a column of mercury 760 mm high, about 101.3 kPa.

atom—smallest particle of an element; matter is composed of atoms in various chemical combinations. Example: The N atom is the smallest particle of the element nitrogen. Two nitrogen atoms combine to form an N_2 molecule, the smallest particle which has the properties of nitrogen as it is ordinarily found.

atomic mass—a number which reflects the relative mass of an average atom of an element; based on the atomic mass of a ^{12}C isotope taken to be exactly 12. Example: Since the atomic mass of sulfur is about 32 and that of helium is about 4, a sulfur atom weighs about 8 times as much as a helium atom.

atomic number—a number equal to the number of electrons around the nucleus of an atom of an element; also the number of protons in the nucleus of that atom. Example: The atomic number of carbon is 6; there are six electrons outside the nucleus of a C atom and six protons in the nucleus of that atom.

atomic radius—the radius of an atom, taken to be one half the distance between two nuclei in the ordinary form of the elementary substance. Example: The radius of the Cl atom is 0.099 nm, since the internuclear distance in the Cl_2 molecule is 0.198 nm.

aufbau principle—the rule stating that electrons enter energy levels in an atom in order of increasing energy, filling one sublevel before moving into the next.

Avogadro's Law—a principle stating that equal volumes of gases at the same temperature and pressure contain equal numbers of molecules.

Avogadro's number—6.022×10^{23}; the number of units in a mole.

B

baking soda—sodium hydrogen carbonate, $NaHCO_3$.

balanced equation—an equation for a chemical reaction in which the reactants and products contain equal numbers of each kind of atom participating in the reaction. Example: The equation $CH_4(g) + 2\ O_2(g) \rightarrow CO_2(g) + 2\ H_2O(l)$ is balanced since both reactants and products contain one C, four H, and four O atoms.

base—a substance which on dissolving in water produces a solution in which $[OH^-]$ is greater than $10^{-7}\ M$. Examples: NaOH, Na_2CO_3, NH_3.

base dissociation constant—K_b; the equilibrium constant for the following reaction of the base X^-: $X^-(aq) + H_2O \rightleftarrows HX(aq) + OH^-(aq)$.

$$K_b = \frac{[HX][OH^-]}{[X^-]}$$

body-centered cubic—a crystalline structure in which the unit cell is a cube with one atom at each of its corners and the same kind of an atom at its center.

boiling point—that temperature of a liquid at which its vapor pressure equals the applied pressure; a liquid will tend to form bubbles and vaporize at its boiling point; usually reported at one atmosphere pressure (101.3 kPa).

bond—a linkage between two atoms.

bond energy—enthalpy change ΔH associated with a reaction in which a bond is broken. Example: For the reaction $HCl(g) \rightarrow H(g) + Cl(g)$, $\Delta H = 431$ kJ; the bond energy, *B.E.*, of the H—Cl bond is 431 kJ/mol of bond.

bonding orbital—an orbital associated with two atoms in which the energy of its two electrons is less than the energies of those electrons in the separated atoms. The presence of a populated bonding orbital between two atoms stabilizes the bond between them.

Boyle's Law—a relation stating that when a gas sample is compressed at a constant temperature, the product of the pressure and the volume remains constant.

brine—a solution of a salt, usually NaCl, in water.

Brönsted-Lowry acid—a species which donates a proton to another species. Example: In the reaction $HX(aq) + H_2O \rightleftarrows H_3O^+(aq) + X^-(aq)$, HX behaves as a Brönsted-Lowry acid in that it donates a proton to H_2O.

Brönsted-Lowry base—a species that accepts a proton from another species. Example: In the reaction just given, H_2O behaves as a Brönsted-Lowry base, since it accepts a proton from HX.

buffer—a solution which resists change of *pH* more effectively than would a solution of a strong acid or base having the same *pH;* usually contains a weak acid and its conjugate base. Example: A solution containing 0.5 *M* H_2CO_3 and 0.5 *M* $NaHCO_3$ is a buffer with a *pH* of about 6.4; the *pH* is relatively resistant to change caused by addition of small amounts of either H^+ or OH^- ions.

C

calorie—a unit of energy equal to the amount of heat required to raise the temperature of one gram of water by one degree Celsius (strictly speaking from 14.5 to 15.5°C); equal in energy to 4.184 J.

capillary action—the movement of a liquid caused by its interaction with the surface of a very small tube, or capillary. Example: Water rises by capillary action in a small tube immersed in a beaker of water.

carbonate ion—CO_3^{2-}.

catalyst—a substance which affects the rate of a reaction without being used up itself. Example: A piece of platinum foil can act as a catalyst for the combustion of methane in air.

cathode—an electrode at which reduction occurs. Example: If, at a silver electrode, the reaction that occurs is $Ag^+(aq) + e^- \rightarrow Ag(s)$, then the silver metal is serving as a cathode.

cation—an ion having a positive charge. Examples: Fe^{2+}, K^+, and NH_4^+ are all cations.

caustic soda—NaOH, sodium hydroxide.

Celsius degree—a unit of temperature, based on there being 100° between the freezing and boiling points of water; ultimately defined by means of a gas volume thermometer, the absolute temperature scale, and the 100° interval noted above.

centi—prefix on a metric unit indicating a multiple of 10^{-2}. Example: one centimetre = 10^{-2} m.

chain reaction—a type of chemical reaction occurring in steps in which the product of a late step serves as a reactant in an earlier step, thereby allowing a reaction when, once begun, to continue.

Charles' Law—a relation stating that the volume of a gas sample at constant pressure is directly proportional to its absolute temperature.

chelating agent—a complexing ligand that can form more than one bond with a central ion. Example: Ethylenediamine, $H_2N—CH_2—CH_2—NH_2$, en, is a chelating agent which can form two bonds with a metal ion; its complex ion with Cu^{2+}, coordination number 4, has the formula $Cu(en)_2^{2+}$.

chemical equation—an expression which qualitatively and quantitatively describes the reactants and products of a chemical reaction as to their nature and amount. Example: $N_2(g) + 3\ H_2(g) \rightarrow 2\ NH_3(g)$, a chemical equation, tells us that one mole of nitrogen gas reacts with three moles of hydrogen gas to form two moles of ammonia gas.

***cis* isomer**—a geometric isomer in which two identical bonded atoms or groups are relatively close to one another. Example: In square planar $Pt(NH_3)_2Cl_2$, the *cis* isomer has the structure

$$\begin{array}{c} Cl \\ | \\ Cl—Pt—NH_3 \\ | \\ NH_3 \end{array}$$

; the *trans* isomer has the structure

$$\begin{array}{c} NH_3 \\ | \\ Cl—Pt—Cl \\ | \\ NH_3 \end{array}$$

.

coagulation—a change in state in which a solute species forms semisolid clumps and comes out of solution. Example: Milk coagulates if left warm too long.

colligative property—a physical property of a solution which depends on the concentration, but not the kind, of solute particles. Example: The vapor pressure depression of a solution depends on the mole fraction of the solute but not on the nature of the solute, and so is a colligative property.

common ion effect—if to a solution containing ions a solute is added which furnishes one of the ions originally present, the common ion effect will change some of the properties of the solution. Example: The solubility of NaCl in water is decreased by addition of 6 *M* HCl.

complex ion—an ion containing a central metallic cation to which two or more groups are attached by coordinate covalent bonds. Example: In the $Ag(NH_3)_2^+$ complex ion the electrons in the coordinate covalent bonds between Ag^+ and NH_3 are furnished by the NH_3 molecules.

compound—a chemical substance containing more than one kind of atom.

concentration—refers to relative amounts of solute and solvent in a solution; may be stated in many ways, such as per cent solute by mass, or mole fraction, but very often is given in terms of moles of solute per cubic decimetre of solution. Example: In 6 *M* NaOH, there are 6 mol NaOH/dm³ solution.

condensation—conversion of a gas to a liquid or solid.

conductivity—a term referring to the relative ease with which a sample will transmit electricity or heat (should specify which). Example: Since a much larger electrical current will flow through an aluminum rod at a given voltage than through a glass rod of the same shape, the electrical conductivity of aluminum is much greater than that of glass.

configuration—a structure, geometrical arrangement. Example: Carbon tetrachloride molecules have a tetrahedral configuration.

conjugate—refers to related acids and bases, often in connection with Brönsted-Lowry description. Example: In the reaction $HNO_2(aq) + H_2O(aq) \rightleftarrows H_3O^+(aq) + NO_2^-(aq)$, HNO_2 and H_3O^+ behave as acids, while H_2O and NO_2^- act as bases. The NO_2^- ion is the conjugate base of the acid HNO_2, and H_2O is the conjugate base of the H_3O^+ ion. Or, HNO_2 is the conjugate acid of the NO_2^- ion and H_3O^+ ion in the conjugate acid of H_2O.

conversion factor—a ratio, numerically equal to one, by which a quantity is multiplied to obtain an equivalent quantity; often expressed as the equation from which the ratio can be obtained. Example: From the conversion factor, 1 in = 2.54 cm, one can set up the ratio 2.54 cm/1 in, which can be used to convert a distance in inches into its equivalent in centimetres.

coordinate covalent bond—a covalent bond in which the electrons are furnished by only one of the bonded atoms; most commonly encountered in complex ions. Example: In the $Zn(OH)_4^{2-}$ ion, the electrons in the bonds between Zn^{2+} and the OH^- ions are all furnished by the hydroxide ions; these bonds are, therefore, coordinate covalent bonds.

Coulomb's Law—a relation expressing force between charged particles; $F = \frac{q_1 q_2}{Dr^2}$, where F is the force between two particles having charges q_1 and q_2 and r is the distance between them. If q_1 and q_2 have the same sign, force is repulsive; otherwise it is attractive. In a medium having dielectric constant D, the force is decreased by the factor D; D is 1 in a vacuum.

covalent bond—a chemical link between two atoms, produced by shared electrons in the region between the atoms. Example: In the H_2O molecule there is a covalent bond between the O atom and each H atom; each bond contains two electrons, one furnished by the H atom and one by the O atom; both atoms share the electrons in the bond.

critical temperature—the highest temperature at which a substance can exhibit liquid-vapor equilibrium. Equilibrium pressure at that point is called the critical pressure. Above that temperature, liquid cannot be condensed from the vapor at any pressure. Example: Since water has a critical temperature of 374°C, above 374°C one cannot have liquid water in equilibrium with its vapor.

crystal—a sample of matter in which the component atoms or ions are arranged in a regular geometric pattern.

D

Dalton's Law—a relation stating that the total pressure of a gas mixture is equal to the sum of the partial pressures of its components.

density—a property of a sample equal to its mass per unit volume. Example: Since the density of mercury is 13.5 g/cm³, 1.00 cm³ of Hg weighs 13.5 g.

desiccant—a drying agent.

deuterium—a heavy isotope of hydrogen, ^2_1H.

diamagnetic—a descriptive term indicating that a substance does not contain unpaired electrons and so is not attracted into a magnetic field. Example: Since all of the electrons in NH_3 molecules are paired, NH_3 is diamagnetic.

diamond—one of the crystalline forms of carbon.

diffusion—a process by which one substance, by virtue of the kinetic properties of its particles, will gradually mix with another. Example: $H_2S(g)$ prepared in a test tube will slowly diffuse into the surrounding air.

dilute—refers to a solution containing a relatively small amount of solute; opposite of concentrated.

dipole force—refers to attractive force between molecules possessing separate positive and negative poles. Example: Since the HCl molecule has positive and negative ends, there will be dipole forces between neighboring HCl molecules.

dispersion force—an attractive force between molecules which arises because of presence of temporary dipoles. Usually increases with *MM*.

dissociation—separation into two or more species; usually applied to weak acids or bases or complex ions. Example: The dissociation of acetic acid in water to form H^+ ions and acetate ions only occurs to a small extent.

distillation—a procedure in which a liquid is vaporized under conditions where the evolved vapor is later condensed and collected.

double bond—a chemical bond involving two electron pairs.

ductility—ability of a solid to retain strength on being forced through an orifice; characteristic of metals.

E

E—see *energy*.

effusion—the movement of a gas through a capillary or porous solid into another gaseous region or vacuum.

electrode—a general name for anode or cathode.

electrolysis—the passage of a direct electric current through a solution containing ions, producing chemical changes at the electrodes.

electrolyte—a substance which exists as ions in water solution. Example: NaCl (Na^+ and Cl^- in water solution).

electron—the negatively charged component of atoms; exists in a roughly spherical cloud around atomic nucleus; carries 1 unit of negative charge and has a very low mass.

electron cloud—a region of negative charge around an atomic nucleus; associated with an atomic orbital.

electron configuration—a statement of the populations of the electronic energy sublevels in an atom. Example: Since the electron configuration of the Li atom is $1s^22s$, there are two electrons in the 1s sublevel and one electron in the 2s sublevel.

electron spin—a property of an electron loosely related to its spin around an axis. Only two spin states are allowed, usually described by quantum number $\mathbf{m_s}$, which can assume the values $+1/2$ and $-1/2$.

electronegativity—a property of an atom which increases with its tendency to attract the electrons in a bond. Example: Since the Cl atom is more electronegative than the H atom, in the HCl molecule the bonding electrons will be closer to Cl.

electropositive elements—those elements which tend to lose electrons easily. Example: Sodium.

electrostatic forces—the forces between particles caused by their electric charges.

element—a general name given to each of the 106 different atoms. Example: Sulfuric acid, H_2SO_4, contains three elements, hydrogen, sulfur, and oxygen; this is equivalent to saying that in H_2SO_4 there are three different kinds of atoms, H, S, and O.

elementary substance—ordinarily observed form of a chemical substance containing only one kind of atom. Examples: Elementary oxygen consists of O_2 molecules; elementary sodium is metallic Na.

empirical formula—an expression which furnishes relative numbers of atoms of the elements in a chemical substance; expressed as the lowest possible set of integers. Often called the simplest formula. Examples: NaCl, H_2SO_4, CH_2, Fe, HO (hydrogen peroxide).

endothermic—describes a reaction during which heat must be furnished to the reacting mixture to maintain its temperature at the initial value; ΔH for the reaction is a positive quantity.

end-point—the point during a reaction, usually in the course of a titration, at which a chemical indicator changes color. Example: The end-point in a titration using phenolphthalein indicator occurs at a *pH* of 9.

energy—a property of a system which is related to its capacity to cause change; can be altered only by exchanging heat or work with the surroundings; given the symbol *E*.

enthalpy—a property of a system which reflects its capacity to exchange heat Q with its surroundings; given the symbol H; defined so that $\Delta H = Q$ for changes in the system that occur at constant pressure.

enthalpy of formation—ΔH_f; heat flow for the reaction in which a pure substance is formed at constant pressure from elementary substances.

entropy—a property of a system related to its degree of organization; highly ordered systems have low entropy; given the symbol *S*.

equilibrium—a state of dynamic balance, where rates of forward and reverse reactions are equal, so system does not change with time. Example: At 100°C, liquid water is in equilibrium with its vapor when the vapor is at a pressure of 1 atm.

equilibrium constant—a number which imposes a condition on reactant and product concentrations in an equilibrium system; formulated according to Law of Chemical Equilibrium; given the symbol K_c. Example: For the reaction $PCl_3(g) + Cl_2(g) \rightleftarrows PCl_5(g)$, $K_c = 20$ at 240°C; therefore, at equilibrium at 240°C, $\frac{[PCl_5]}{[PCl_3][Cl_2]} = 20$.

equivalence point—the point during a reaction between A and B, usually during a titration, when an amount of B has been added that is required to react exactly with the amount of A present. Example: The equivalence point in the reaction $H^+(aq) + OH^-(aq) \rightarrow H_2O(l)$ occurs when the number of moles of OH^- ion added to an acid solution equals the number of moles of H^+ ion in the solution.

equivalent—the amount of a species which, in an electrolysis, will react with or produce one mole of electrons; in acid-base reactions, the amount that will react with a mole of H^+ or OH^- ions; sometimes called a gram equivalent mass. Examples: In the reaction $Cu^{2+} + 2\,e^- \rightarrow Cu(s)$, the gram equivalent mass of copper is one half the gram atomic mass; for the reaction $H_2SO_3(aq) + 2\,OH^-(aq) \rightarrow SO_3^{2-}(aq) + 2\,H_2O$, an equivalent of H_2SO_3 weighs 41 g, one half the *GMM* of H_2SO_3.

ester—the product of the reaction between an alcohol and an acid. Example: When methanol, CH_3OH, reacts with acetic acid, CH_3COOH, the ester called methyl acetate, CH_3—O—CO—CH_3, is formed.

ether—an organic compound containing an oxygen atom connected to two alkyl groups. Examples: CH_3OCH_3, $C_2H_5OC_2H_5$.

exclusion principle—the rule stating that in an atom no two electrons can have the same set of four quantum numbers.

exothermic—describes a reaction during which heat must be removed from the reacting mixture to maintain its temperature at the initial value; ΔH for the reaction is a negative quantity.

F

face-centered cubic—a type of crystal structure in which the unit cell is a cube with identical atoms at each corner and at the center of each face.

Faraday constant—the charge in coulombs carried by one mole of electrons, 96 500 C/mol e^-; alternatively, 96 500 J/(V · mol e^-).

fat—an ester made from glycerol and a long-chain carboxylic acid; found in seeds and in fatty tissue of animals.

fatty acid—a long-chain carboxylic acid. Example: Stearic acid, $CH_3(CH_2)_{16}COOH$.

filtration—a process for separating a solid-liquid mixture by passing it through a barrier with fine pores, such as filter paper.

First Law of Thermodynamics—the statement that the change in energy, ΔE, of a system equals the heat flow, Q, into the system from the surroundings minus the work, W, done by the system on the surroundings ($\Delta E = Q - W$).

first order—a term describing a reaction whose rate depends on reactant concentration raised to the first power. Example: Since the rate of the reaction $2\ N_2O_5(g) \rightarrow 2\ N_2O_4(g) + O_2(g)$ is given by the equation: rate = k(conc. N_2O_5), the reaction is first order.

fixation of nitrogen—any process which converts $N_2(g)$ into a nitrogen-containing compound. Example: The fixation of nitrogen by the Haber process occurs via the reaction of N_2 with H_2 to make ammonia, NH_3.

formula—the expression used to describe the relative numbers of the atoms of the different elements present in a substance; molecular formula is used with substances having molecules; empirical formula is used with nonmolecular substances.

fractional distillation—a procedure used to separate components with different boiling points from a solution; based on passing vapors from a boiling solution up a column along which the temperature gradually decreases; higher boiling components condense on column and return to solution, lowest boiling component goes out of top of column, where it is condensed and collected.

free energy—a property of a system which reflects its capacity to do useful work, given the symbol G; ΔG for a reaction at constant temperature and pressure is equal to minus the amount of useful work the reaction can produce; spontaneous reactions are those for which ΔG is negative.

free radical—a species having an unpaired electron. Examples: The H atom, the NO molecule, and the CH_3 group are all free radicals.

freezing point—the temperature at which a solid and liquid phase can coexist at equilibrium; applies to both pure liquids and solutions. Example: The freezing point of a solution containing one mole NaCl in one cubic decimetre of water is −3.37°C; at that point pure ice and the solution are in equilibrium.

fusion—the melting of a solid to a liquid; also refers to reaction between small atomic nuclei to form a larger one. Example: The fusion reaction $2\ {}^2_1H \rightarrow {}^4_2He$ would produce a large amount of energy.

G

G—see *free energy*.

gas constant—R; the constant which appears in the Ideal Gas Law equation, $PV = nRT$; depends on units of P, V, and T; equals 8.31 kPa · dm³/(mol · K) in the units listed.

Gas Law—see *Ideal Gas Law*.

geometric isomer—a species having the same kind and number of atoms as another species, but in which the geometrical structure is different. Example: There are two geometric isomers with the molecular formula $PtCl_2(NH_3)_2$:

$$Cl-\underset{\underset{NH_3}{|}}{\overset{\overset{NH_3}{|}}{Pt}}-Cl \quad \text{and} \quad Cl-\underset{\underset{NH_3}{|}}{\overset{\overset{Cl}{|}}{Pt}}-NH_3$$

(structures are both planar).

gram—1/1000 of the base unit of mass in the SI, the kilogram.

gram atomic mass—the mass in grams of an element equal numerically to its atomic mass; the mass in grams of 6.022×10^{23} atoms of an element. Example: The gram atomic mass of copper is 63.54 g, since the atomic mass is 63.54.

gram equivalent mass—see *equivalent*.

gram molecular mass—the mass in grams of a molecular substance equal numerically to its molecular mass; the mass of a mole of a molecular substance. Example: Since the molecular mass of N_2 is 28, the gram molecular mass of nitrogen gas is 28 g.

gravimetric—involving mass measurement. Example: In gravimetric analysis, calculations are based on the masses of the sample and some pure compounds obtained from it.

ground state—the lowest allowed energy state of an atom, ion, or molecule.

H

H—see *enthalpy*.

half-cell—half of a voltaic or electrolysis cell, at which either oxidation or reduction occurs. Example: The half-cell reaction at the anode is one of oxidation.

half-life—the time required for a reaction to use up half of the initial reactant.

halogen—an elementary substance in Group 7A. Examples: F_2, Cl_2, Br_2.
hard water—water containing Ca^{2+} or Mg^{2+}.
heat—that form of energy which flows between two samples of matter because of their difference in temperature.
heat capacity—the amount of heat required to raise the temperature of a sample by 1°C.
heat flow—the amount of heat, Q, passing into or out of a system; Q is positive if flow is into system, negative if out of system.
heat of formation—see *enthalpy of formation.*
Henry's Law—a relation stating that the solubility of a gas in a liquid is proportional to the pressure of the gas above the liquid; $[A] = kP_A$.
Hess's Law—a relation stating that the heat flow in a reaction which is the sum of two other reactions is equal to the sum of the two heat flows in those reactions.
heterogeneous—having nonuniform composition.
homogeneous—having uniform composition.
Hund's rule—a relation stating that, ordinarily, electrons will not pair up in an orbital until all orbitals of equal energy contain one electron.
hybrid atomic orbital—an orbital made from a mixture of s, p, d, or f orbitals. Example: An sp^2 hybrid orbital is derived from an s and two p orbitals.
hydrate—a substance containing bound water. Example: $BaCl_2 \cdot 2\ H_2O$ is a common hydrate.
hydrocarbon—a substance containing only hydrogen and carbon atoms.
hydrogen bonds—attractive forces between molecules, arising from interaction between a hydrogen atom in one molecule and a strongly electronegative atom (N, O, F) in a neighboring molecule. Example: Hydrogen bonding in water is due to interaction between the H atoms and O atoms on different H_2O molecules.
hydrolysis—a reaction in which a water molecule is split as a result of interaction with another species. Example: The hydrolysis of CN^- involves the reaction $CN^-(aq) + H_2O(l) \rightleftarrows HCN(aq) + OH^-(aq)$.
hydroxide ion—OH^-.

I

Ideal Gas Law—states the relationship between pressure, volume, temperature, and amount for any gas at moderate pressures; $PV = nRT$.
ideal solution—a solution which obeys Raoult's Law; $P_i = P_i^0 X_i$ for both components i for all values of X_i.
indicator, acid-base—a chemical substance which changes color with pH change; usually color change occurs over about two pH units.
inert complex—a complex ion which exchanges ligands very slowly.
inert gas—noble gas.
infrared—describes light having a wavelength greater than about 700 nm.
intermolecular—between molecules.
ion—a charged species.
ion pair—a species made up of a cation and anion held together by strong electrostatic forces. Examples: In solutions of magnesium sulfate one would find appreciable amounts of $(Mg^{2+}SO_4^{2-})$ ion pairs.
ionic compound—a substance in which component species are cations and anions. Examples: Some common ionic compounds are NaCl, CaO, and NH_4NO_3.
ionic radius—the radius of an ion as based on the assumption that ions in a crystal are in contact with nearest neighbors.
ionization constant—a general term for dissociation constant of a weak acid or base; see *acid dissociation constant.*
isoelectronic (with)—having the same number of electrons as.
isomer—a species having same number and kind of atoms as another species, but having different properties; structural, geometric, optical, and stereoisomers may occur. Example: Dimethyl ether, $CH_3—O—CH_3$ is a structural isomer of ethyl alcohol, $CH_3—CH_2—OH$.
isotope—an atom having same number of nuclear protons as another, but with a different number of neutrons. Example: Ordinary oxygen has three isotopes, all with eight protons in the nucleus, but with eight, nine, and ten neutrons, respectively.

J

joule—basic unit of energy in SI; equal to kinetic energy of a two-kilogram mass moving at a speed of one metre per second.

K

K_a, K_b, K_c, K_{sp}, K_w—see *acid dissociation constant, base dissociation constant, equilibrium constant, solubility product constant,* and *water dissociation constant,* respectively.
Kelvin temperature scale—an absolute temperature scale based on definition that the volume of a gas at constant (low) pressure is directly proportional to temperature and that 100 degrees separate the freezing and normal boiling points of water.
ketone—an organic compound containing a nonterminal carbonyl, C=O, group. Example: The simplest ketone is acetone, $CH_3—CO—CH_3$.
kilo—a prefix on metric units indicating multiple of 1000. Example: One kilojoule equals 1000 joules.
kinetic—associated with motion. Example: The kinetic energy of a particle of mass m moving at speed v is equal to $\frac{1}{2}mv^2$.
kinetics—the study of rates of chemical reactions.

L

labile complex—a complex ion which rapidly reaches equilibrium with ligands in surrounding solution.
lanthanides—elements with atomic numbers 57–71; series in which 4f sublevel is being filled. Example: Europium, atomic number 63, is one of the lanthanides.
Law of Chemical Equilibrium—a relation stating that in a reaction mixture at equilibrium there is a condition, given by the equilibrium constant, relating the concentrations of reactants and products; for

the reaction $aA(g) + bB(g) \rightleftarrows cC(g) + dD(g)$, $K_c = \frac{[C]^c[D]^d}{[A]^a[B]^b}$.

Law of Combining Volumes—a relation stating that relative volumes of gases in a chemical reaction are in the ratio of small integers (all gases at same T and P); also called Gay-Lussac's Law. Example: In the reaction $2\ H_2(g) + O_2(g) \rightarrow 2\ H_2O(g)$, 2 volumes H_2 react with 1 volume O_2 to produce 2 volumes H_2O.

Law of Conservation of Mass—a relation stating that in a chemical reaction the mass of the products equals the mass of the reactants.

Law of Constant Composition—a relation stating that the relative masses of the elements in a given chemical substance are fixed.

Law of Dulong and Petit—a relation stating that the heat capacity of one gram atomic mass of any metal is about 25 J/°C.

Law of Mass Action—same as Law of Chemical Equilibrium.

Law of Multiple Proportions—a relation stating that when two elements A and B form two compounds, the relative amounts of B which combine with a fixed amount of A are in a ratio of small integers. Example: In water and hydrogen peroxide, both of which contain hydrogen and oxygen, there are 8 and 16 g of oxygen, respectively, for each gram of hydrogen.

Le Châtelier's Principle—a relation stating that when a system at equilibrium is disturbed, it will respond in such a way as to counteract the change.

Lewis acid—a species which can accept a pair of electrons. Example: In the reaction $BF_3 + NH_3 \rightarrow BF_3NH_3$, the BF_3 accepts a pair of electrons from NH_3 and so behaves as a Lewis acid.

Lewis base—a species which can donate a pair of electrons. Example: NH_3, in the above reaction.

Lewis structure—electronic structure of a molecule or ion in which each atom has a share in eight electrons and so follows the octet rule.

ligand—a species bonded to the central atom in a complex ion.

lime—calcium hydroxide, $Ca(OH)_2$; also sometimes called slaked lime.

limestone—calcium carbonate, $CaCO_3$.

litre—one cubic decimetre, or 1000 cm^3.

logarithm of a number—the exponent to which another number, usually 10, must be raised to give the number. Examples: The logarithm of 100 to the base of 10 is 2, since $10^2 = 100$; the logarithm of 3.00 is 0.477, since $10^{0.477} = 3.00$.

M

macromolecular—having a structure in which all the atoms in a crystal are linked together by chemical bonds. Example: Since all the atoms in a silicon crystal are bonded chemically into a unit, silicon is macromolecular.

malleable—capable of being shaped, as by pounding with a hammer.

manometer—a device for measuring gas pressure.

mass—a property reflecting the amount of matter in a sample.

mass number—an integer equal to the sum of the number of protons and neutrons in an atomic nucleus. Example: The mass number of a $^{37}_{17}Cl$ isotope is 37; the nucleus of that isotope contains 17 protons and 20 neutrons.

matter—a general term for any kind of material; the stuff of which pure substances are made.

Maxwellian distribution—a relation describing the way in which molecular speeds, or energies, are shared among the molecules in a gas.

mechanism—a sequence of steps that occurs during the course of a chemical reaction.

melting point—same as freezing point.

metal—a substance having characteristic luster, malleability, high electrical conductivity; readily loses electrons to form positive ions.

metre—base unit of length in the SI.

milli—prefix on a metric unit indicating multiple of 1×10^{-3}. Example: One millimetre equals one one-thousandth of a metre, 0.001 m.

miscible (with)—soluble in.

molality—a concentration unit, defined equal to the number of moles of solute divided by number of kilograms of solvent. Example: A solution made by dissolving 0.10 mol KNO_3 in 200 g water would be 0.50 molal in KNO_3 (0.50 m KNO_3).

mole—a convenient chemical mass unit; defined as containing 6.022×10^{23} molecules, atoms, or other units; mass of the mole is equal to the gram formula mass of the substance. Examples: One mole of NH_3 contains 6.022×10^{23} molecules and weighs just about 17 g; one mole Cu contains 6.022×10^{23} atoms and weighs 63.54 g; one mole KNO_3 weighs 101.1 g.

mole fraction—a concentration unit, defined equal to number of moles of component divided by total number of moles in solution. Example: In a solution in which there is 1 mol benzene, 2 mol CCl_4, and 7 mol acetone, the mole fraction of the acetone is 0.7.

mole per cubic decimetre—a concentration unit (abbreviation M). Example: 6 M HCl contains six moles of HCl per cubic decimetre (litre) of solution.

molecular formula—an expression stating the number and kind of each atom present in the molecule of a substance. Example: Since the molecular formula of hexane is C_6H_{14}, there are six C atoms and 14 H atoms in a hexane molecule.

molecular orbital—an orbital involved in the chemical bond between two atoms, and taken to be a linear combination of the orbitals on the two bonded atoms.

molecular mass—a number equal to the sum of the atomic masses of all of the atoms in a molecule; tells the mass of the molecule relative to a ^{12}C atom, taken to have a mass of 12. Example: The molecular mass of C_2H_6 is about 30 $[(2 \times 12) + (6 \times 1)]$, so the molecule is about 2.5 times as heavy as a ^{12}C atom, and the same mass as an NO molecule, whose molecular mass is also 30 (i.e., $14 + 16 = 30$).

molecule—an aggregate of atoms, which is the characteristic component particle in all gases, many pure liquids, and some solids; often contains only a few atoms; has relatively little physical interaction with other molecules. Examples: In nitrogen gas, liquid benzene, and solid glucose, one finds N_2 molecules, C_6H_6 molecules, and $C_6H_{12}O_6$ molecules.

N

natural logarithm—a logarithm based on the number *e*, 2.718 281 8 . . .; if $\log_e X = Y$, then $e^Y = X$, $\log_{10} X = \frac{\log_e X}{2.303}$; base *e* comes from the calculus, where certain derivatives and integrals are most easily expressed in terms of *e*.

Nernst equation—an equation relating the voltage of a cell to its standard potential and the concentrations of reactants and products.

net ionic equation—a chemical equation for a reaction, in which only those species which actually react are listed. Example: When 1 *M* HCl and 1 *M* NaOH solutions are mixed, the net ionic equation for the reaction is $H^+(aq) + OH^-(aq) \rightarrow H_2O$; the Cl^- and Na^+ ions in the solution do not react and so are not in the equation.

neutralization—a reaction of an acid and a base to produce a neutral (*pH* 7) solution.

neutron—one of the particles in an atomic nucleus; mass = 1, charge = 0.

noble gas structure—ns^2np^6 outer electron structure in an atom or ion; a particularly stable structure attained by atoms obeying the octet rule; sometimes called an inert gas structure.

nonelectrolyte—a substance that does not exist as ions in water solution. Example: Since ethyl alcohol does not ionize when dissolved in water, it is a nonelectrolyte.

nonmetal—one of the elements in the upper right-hand corner of the Periodic Table that does not show metallic properties. Example: Nitrogen gas, N_2, is a nonmetal.

nonpolar bond—a chemical bond in which there are no positive and negative ends; found in homonuclear diatomic molecules such as H_2 or O_2.

normal boiling point—boiling point at 1 atm pressure (101.3 kPa).

normality—concentration unit, defined as the number of equivalents of solute per cubic decimetre of solution; see *equivalent*.

nucleus—the small, dense, positively charged region at the center of an atom.

O

octahedral—having the symmetry of a regular octahedron, a solid with six vertices and eight faces, each of which is an equilateral triangle.

octet rule—the principle that bonded atoms tend to have a share in eight outermost electrons.

olefin—a hydrocarbon containing a carbon-carbon double bond. Same as alkene (p. A.41).

optical activity—the ability to rotate the plane of a beam of transmitted polarized light; a property possessed by substances having an asymmetric carbon atom.

orbital—an electron cloud with an energy state characterized by given values of **n**, **ℓ**, and **m_ℓ** quantum numbers; has a capacity for two electrons having paired spins; often associated with a particular region in the atom. Example: In an atom the electrons in the $2p_x$ orbital are in a dumbbell-shaped cloud concentrated along the x axis.

orbital diagram—a sketch showing electron populations of atomic orbitals, including electron spins.

order of reaction—an exponent to which concentration of a reactant needs to be raised to give observed dependence of reaction rate on concentration. Example: If, for the reaction A → products, rate = $k(\text{conc. A})^2$, the reaction is second order.

organic—used to characterize any compound containing carbon, hydrogen, and possibly other elements. Example: Propionic acid, CH_3CH_2COOH, is an organic compound; SiO_2 is not.

osmotic pressure—the excess pressure which must be applied to a solution to prevent the pure solvent from diffusing into the solution through a semipermeable membrane.

oxidation—a half-reaction involving a loss of electrons or, more generally, an increase in oxidation number. Example: If, at an electrode, the reaction is $Ag(s) \rightarrow Ag^+(aq) + e^-$, silver is undergoing oxidation, since its oxidation number is increasing from 0 to +1.

oxidation number—a number which can be assigned to an atom in a molecule or ion which reflects, qualitatively, its state of oxidation; the number is determined by applying a set of rules. Examples: In the NO_3^- ion the oxidation numbers of the N and O atoms are +5 and −2, respectively.

oxidizing agent—a species which accepts electrons from another. Example: In the reaction $Cl_2(aq) + 2\,Br^-(aq) \rightarrow 2\,Cl^-(aq) + Br_2(aq)$, Cl_2 serves as an oxidizing agent.

oxyacid—an acid containing oxygen. Examples: HNO_3, H_2SO_4, and HClO are oxyacids.

oxyanion—an anion containing oxygen. Examples: NO_3^-, SO_4^{2-}, and ClO^- are oxyanions.

ozone—an allotropic form of oxygen, in which the molecule is O_3.

P

paired electrons—two electrons in the same orbital with spins equal to +1/2 and −1/2; an electron pair.

paraffin—a hydrocarbon in which all carbon-carbon bonds are single; an alkane.

paramagnetic—having magnetic properties caused by unpaired electrons. Examples: The NO molecule, the H atom, and the CH_3 radical are paramagnetic.

partial pressure of A—that part of the total pressure in a gaseous mixture which can be attributed to component A. The partial pressure of A is equal to the pressure A would exert in the container if A were there by itself. Example: In a mixture of 3 mol N_2 and 1 mol O_2 at a total pressure of 2 atm, the partial pressure of N_2 is 1.5 atm and that of O_2 is 0.5 atm.

parts per million—for gases, number of molecules solute per million molecules of gas; for liquids and solids, number of grams solute per million grams of sample. Example: If a gas sample contains 6 ppm CO, then in one mole of gas there would be 6×10^{-6} mol CO.

Pauli exclusion principle—see *exclusion principle.*

peptide linkage—the $-\underset{\displaystyle |\atop \displaystyle H}{N}-\underset{\displaystyle \|\atop \displaystyle O}{C}-$ group in proteins; also called peptide bond.

per cent A—parts A in 100 parts of a sample; usually in mass per cent but may be in mole per cent or volume per cent. Example: A 5% NaOH solution contains 5 g NaOH in 100 g solution.

periodic—occurring in cycles.

Periodic Table—an arrangement of the elements into rows and columns in which those elements with similar properties occur in the same column.

pH—alternate way to express H^+ ion concentration; $pH = -\log_{10}[H^+]$.

phase diagram—for one-component systems, a graph of pressure vs. temperature, showing conditions under which the pure substance will exist as a liquid, a solid, or a gas and also the conditions under which two-phase and three-phase equilibria can exist.

photosynthesis—the process by which sunlight makes possible the synthesis of organic compounds from CO_2 and H_2O.

pi (π) bond—a bond in which electrons are concentrated in orbitals which are located off the internuclear axis; one bond in a double bond is a pi bond, and there are two pi (π) bonds in a triple bond.

pi orbital—a molecular orbital in which electron density is concentrated in lobes which do not lie on the internuclear axis.

polar bond—a chemical bond which has positive and negative ends; characteristic of all bonds between nonidentical atoms. Example: In CCl_4 the C—Cl bonds are polar, since the Cl atom tends to attract electrons more than the C atom does.

pollutant—a contaminant, or foreign species, present in a sample; usually has a deleterious effect on quality of sample as far as living things are concerned.

polymer—a molecule made up from many units which are linked together chemically. Example: In the polymer called polyethylene, many $H_2C{=}CH_2$ units become linked together by chemical reaction to form chains which have the structure

```
 H  H  H  H  H  H
 |  |  |  |  |  |
—C—C—C—C—C—C—
 |  |  |  |  |  |
 H  H  H  H  H  H
```

polypeptide—a condensation polymer in which the monomer units are α-amino acids; in the polymer the amino acid residues are linked by peptide bonds; another name for a protein.

positron—a particle identical with an electron except that its charge is opposite in sign.

precipitate—a solid which forms when two solutions are mixed.

pressure—force per unit area; expressed in kilopascals or atmospheres.

principal quantum number—quantum number **n**; the most important quantum number, since it has greatest effect on energy of electron; cited first in the set of four quantum numbers associated with an electron.

product—a substance formed as a result of a chemical reaction. Example: In the reaction $Ag^+(aq) + Cl^-(aq) \rightarrow AgCl(s)$, AgCl is the product.

property—a characteristic of a sample of matter that is fixed by its state. Example: The density and energy of a mole of H_2 at 100°C and 1 atm are properties of that sample of hydrogen gas.

protein—a polypeptide.

proton—the nucleus of a hydrogen atom, the H^+ ion; a component of atomic nuclei, with mass = 1, charge = +1.

Q

Q—see *heat flow.*

qualitative analysis—the determination of the nature of the species present in a sample. Example: By qualitative analysis she found that the solution contained Cu^{2+}, Sn^{4+}, and Cl^- ions.

quantitative analysis—the determination of how much of a given component is present in a sample. Example: The students discovered, by quantitative analysis, that the ore contained 42.45% iron by mass.

quantum number—a number used in the description of the energy levels available to atoms and molecules; an electron in an atom or ion will have four quantum numbers to describe its state.

quicklime—calcium oxide, CaO.

R

R—see *gas constant.*

radical—see *free radical.*

radioactivity—the ability possessed by some natural and synthetic isotopes to undergo reactions involving nuclear transformations to other isotopes.

rare earth—the name sometimes given to members of the lanthanide series.

rate constant—the proportionality constant in the rate equation for a reaction. Example: If the rate equation is rate = $k(\text{conc. A})^n$, then k is the rate constant.

rate of a reaction—the magnitude of the change in concentration of a reactant or product divided by the time required for the change to occur (with both quantities relatively small). Example: For the reaction $A \rightarrow B$, rate $= \frac{\Delta(\text{conc. B})}{\Delta t} = \frac{-\Delta(\text{conc. A})}{\Delta t}$.

reactant—the starting material in a chemical reaction. Example: In the reaction $H_2(g) + \frac{1}{2} O_2(g) \rightarrow H_2O(l)$, H_2 and O_2 are both reactants.

reaction—a chemical change in which new substances are formed. Example: When aluminum burns in air, the chemical reaction that occurs is described by the equation $2\ Al(s) + \frac{3}{2}\ O_2(g) \rightarrow Al_2O_3(s)$.

redox reaction—a reaction involving oxidation and reduction.

reducing agent—a species which furnishes electrons to another. Example: in the reaction $Zn(s) + 2\ H^+(aq) \rightarrow Zn^{2+}(aq) + H_2(g)$, the Zn(s), metallic zinc, is the reducing agent.

reduction—a change in state in which a species gains electrons or, more generally, decreases in oxidation number. Example: In the reaction $Zn(s) + 2\ H^+(aq) \rightarrow Zn^{2+}(aq) + H_2(g)$, the H^+ ions (oxid. no. = +1) are reduced to H_2 (oxid. no. = 0).

relative humidity—$100 \times P/P_0$, where P = pressure of water vapor in air, P_0 = equilibrium vapor pressure of water at same T. Can also be thought of as a measure of the amount of water in the air divided by the amount that sample of air could hold. Example: When the relative humidity is 75%, there is $\frac{3}{4}$ as much water in the air as the air can hold at that temperature.

resonance—used to rationalize properties of octet rule species for which one Lewis structure is inadequate; resonance structure is taken to be an average of two or more Lewis structures which differ only in positions of electrons; species are said to exhibit resonance. Example: The resonance structures of SO_2 are

```
    S          and         S
   / \\                   // \
  O    O                 O    O
```

S

S—see *entropy*.

salt—a solid ionic compound made up from a cation other than H^+ and an anion other than OH^- or O^{2-} Examples: NaCl, $CuSO_4$, NH_4NO_3.

saturated hydrocarbon—an alkane, a paraffin; a hydrocarbon in which all carbon-carbon bonds are single.

second order reaction—a reaction whose rate depends on second power of reactant concentration; may be sum of exponents of two reactant concentrations. Examples: The two expressions rate = $k(\text{conc. A})^2$ and rate = $k(\text{conc. A}) \times (\text{conc. B})$ are both associated with second order reactions.

semiconductor—a substance used in transistors and thermistors whose electrical conductivity depends on presence of tiny amounts of impurities such as As or B in a very pure crystal of an element such as silicon or germanium; conductance increases dramatically as temperature goes up.

semipermeable membrane—a film which will allow passage of solvent molecules such as H_2O but will not pass solute molecules such as proteins or, in some cases, ions.

shielding—a term used to describe effect of inner electrons in decreasing the attraction of an atomic nucleus on outermost electrons.

sigma (σ) bond—a chemical bond in which electron density on internuclear axis is high, which is the case with all single bonds; double bonds contain one sigma and one pi bond, triple bonds contain one sigma and two pi bonds.

sigma (σ) orbitals—molecular orbitals associated with sigma bonds between atoms.

significant figures—meaningful digits in a measured quantity; number of digits in a number when expressed in exponential notation. Example: 1.035×10^3 has four significant figures (exponential doesn't count).

soap—sodium salt of a fatty acid.

soda ash—sodium carbonate; also sometimes called soda.

solubility—the amount of a solute that will dissolve in a given amount of solvent; may be stated in various ways, with moles solute per cubic decimetre of solution being the most common.

solubility product constant—K_{sp}; equilibrium constant for the solution reaction of a relatively insoluble ionic compound. Example: For the reaction $Ca(OH)_2(s) \rightleftarrows Ca^{2+}(aq) + 2\ OH^-(aq)$, $K_{sp} = [Ca^{2+}][OH^-]^2$.

solution—a liquid, gas or solid phase containing two or more components dispersed uniformly throughout the phase.

solvent—a substance, usually a liquid, in which another substance, called the solute, is dissolved.

species—a general term referring to a molecule, ion, or atom.

specific heat—the amount of heat required to raise the temperature of one gram of a substance by one degree Celsius.

spectrum—a pattern of characteristic wavelengths associated with excitation of an atom, molecule, or ion; also used as name of a pattern having a similar appearance obtained as a result of chromatographic or mass spectroscopic experiments.

spontaneous reaction—a reaction which can occur by itself, without input of work from outside; $\Delta G < 0$ for spontaneous reactions at T and P.

stable—will not change spontaneously. Nature of change should be specified. Example: Water is stable at 25°C with respect to decomposition to hydrogen and oxygen.

standard half-cell voltage—voltage associated with a half-reaction when all species are in their standard states. E^0_{ox} is the standard voltage associated with an oxidation half-reaction; E^0_{red} is the standard voltage corresponding to a reduction half-reaction.

standard potential—numerical value of E^0_{red}. Values of E^0_{ox} can be found by changing the sign of the standard potential. Example: For the reduction $Cu^{2+}(aq) + 2\ e^- \rightarrow Cu(s)$, the standard potential is +0.34 V. Consequently $E^0_{red} = +0.34$ V. The standard half-cell voltage, E^0_{ox}, for the corresponding oxidation, $Cu(s) \rightarrow Cu^{2+}(aq) + 2\ e^-$, is −0.34 V.

standard voltage—E^0; voltage of a cell in which all species are in their standard states (solids and liquids are pure and solutes are at unit activity, often taken to be 1 M, gases are at 1 atm).

state—condition of a system when its properties are fixed. Example: A mole of H_2O at 25°C and 1 atm is in a definite state in that all of its properties have values which are fixed.

stereoisomers—isomers which differ in the way bonded

groups are arranged around asymmetric carbon atoms. Example: There are 16 stereoisomers having the same bonding as glucose in its linear form.

stoichiometric—having to do with masses (grams, moles) of reactants and products in a chemical equation.

strong—as applied to acids, bases, and electrolytes, indicates complete dissociation into ions when in water solution. Example: HCl is a strong acid, since it exists as H^+ and Cl^- ions in aqueous solution.

sublimation—a change in state from solid to gas. Example: Iodine slowly sublimes in an open container; the sublimation is endothermic.

substrate—the reactant in an enzyme-catalyzed reaction.

sugar—a carbohydrate containing an aldehyde or ketone group and an OH and H group on all non-carbonyl group carbon atoms; sometimes, loosely, sucrose.

supercool or superheat—to cool or heat a sample beyond the point where a phase change should occur, but under conditions where the change does not occur. Example: Extremely pure water may easily be supercooled to $-10°C$ without freezing.

supersaturated—containing more solute than equilibrium conditions would allow; unstable to addition of solute crystal. Example: It is easy to make a supersaturated solution of sodium acetate by cooling a hot concentrated solution of the salt carefully to 25°C.

surface tension—work which must be done to increase area of surface by unit amount.

suspension—a dispersion of a solid or liquid in a liquid in which the former does not dissolve, but will remain suspended for appreciable periods of time.

system—the sample of matter under consideration.

T

temperature—a property of matter reflecting the amount of energy of motion of its component particles; measured value based on one of several possible scales.

thermal—having to do with heat.

thermodynamics—the study of heat, work, and the related properties of mechanical and chemical systems.

titration—a process in which a solution is added to another solution with which it reacts under conditions such that the volume of added solution can be accurately measured.

trans—as related to geometric isomers, referring to structures in which two identical groups are as far apart as possible, as opposed to *cis,* where they are as close as possible. Example:

```
      NH3
      |
  Cl—Pt—Cl
      |
      NH3
```

would be the structure of the *trans* isomer of the square planar $Pt(NH_3)_2Cl_2$ molecule.

transition metal—any one of the metals in the B groups in the fourth, fifth, and sixth periods in the Periodic Table. Examples: Fe, Zr, and W would all be classified as transition metal atoms.

translational energy—energy of motion through space. Example: A falling raindrop has translational energy.

triple point—that temperature and pressure at which the solid, liquid, and vapor of a pure substance can coexist in equilibrium.

Trouton's rule—a relation stating that, for any liquid, the molar heat of vaporization at the normal boiling point, divided by the boiling temperature in kelvins, is equal to about 88 J/(mol · K).

U

ultraviolet radiation—light having a wavelength less than about 400 nm but greater than about 10 nm.

unit cell—the smallest unit of a crystal which, if repeated indefinitely, could generate the whole crystal.

unsaturated—referring to solutions, being able to dissolve more solute; referring to organic compounds, containing double or triple carbon-carbon bonds.

useful work—the work produced during a change in state in excess of that required to accomplish the change. Example: In the reaction $H_2O(l) \rightarrow H_2O(v)$, since the volume of the vapor is larger than that of the liquid, work of expansion will be required to push back the surrounding air to make room for the water vapor; the useful work will be that work done during the change which is in excess of the work of expansion.

V

valence electrons—usually applied to A-group elements, where it refers to those electrons in the outermost shell. Example: In the carbon atom, with electron configuration $1s^22s^22p^2$, there are four valence electrons, those in the 2s and 2p orbitals.

valence-bond model—the theory that atoms tend to become bonded by pairing, and sharing, their outer, or valence, electrons; also referred to as the atomic orbital model.

van der Waals forces—a general name sometimes given to intermolecular forces.

vapor—a condensable gas.

vapor pressure—the pressure exerted by a vapor when it is in equilibrium with the liquid from which it is derived. Example: If liquid water is admitted to an evacuated container at 60°C, the pressure in the container when the liquid and vapor reach equilibrium becomes 19.92 kPa; therefore, the vapor pressure of water at 60°C is 19.92 kPa.

visible light—light having wavelengths between 400 and 700 nm.

voltage—electric potential; a measure of tendency of a cell or other device to force electrons through an external circuit.

voltaic cell—a device in which a spontaneous chemical reaction is used to produce electrical work.

W

washing soda—sodium carbonate, Na_2CO_3.

water dissociation constant—$K_w = [H^+][OH^-] = 1 \times 10^{-14}$ at 25°C.

water softening—the removal of ions, particularly Ca^{2+} and Mg^{2+}, from water.

wave function—the solution to Schrödinger wave equation, the square of whose magnitude at any point is proportional to probability of finding at that point the particle concerned; name arises from fact that solutions look roughly like waves when plotted on a graph.

wavelength—a characteristic property of light, similar to its color, and equal to the length of a full wave; often expressed in nanometres; can be measured with a spectroscope.

weak—as applied to acids and bases, being partially ionized in water solution. Example: Acetic acid, $HC_2H_3O_2$, is a weak acid because in water solution it is only slightly ionized to H^+ and $C_2H_3O_2^-$ ions.

work—one of the effects which may be associated with an energy change; during a change a system may do work on its surroundings, equivalent to the raising of a mass; work may be electrical, mechanical, or due to expansion or compression, and may be done by, or on, the system.

X

x—the unknown quantity.

x-rays—light rays having a wavelength from 0.01 to 1 nm.

Y

yield—the amount of product obtained from a reaction.

Z

zero order reaction—a reaction whose rate is independent of reactant concentration.

INDEX

Page numbers in *italics* refer to illustrations; those followed by t refer to tables.

C

D

E

F

I

O

S

T

U

V